Epigenetic Principles of Evolution

Epigenetic Principles of Evolution

Nelson R. Cabej
Department of Biology
University of Tirana
Tirana, Albania

ELSEVIER
AMSTERDAM • BOSTON • HEIDELBERG • LONDON • NEW YORK • OXFORD
PARIS • SAN DIEGO • SAN FRANCISCO • SINGAPORE • SYDNEY • TOKYO

Elsevier
32 Jamestown Road, London NW1 7BY
225 Wyman Street, Waltham, MA 02451, USA

First edition 2012

British Library Cataloguing-in-Publication Data
A catalogue record for this book is available from the British Library

Library of Congress Cataloging-in-Publication Data
A catalog record for this book is available from the Library of Congress

ISBN: 978-0-12-415831-3

For information on all Elsevier publications
visit our website at elsevierdirect.com

This book has been manufactured using Print On Demand technology. Each copy is produced to order and is limited to black ink. The online version of this book will show color figures where appropriate.

To my son, Redon

Contents

Introduction: The Explanatory Conundrum and the New Perspective in Modern Biology

Science commits suicide when it adopts a creed.

Huxley (1885)

The evolution of life on Earth is a fact that by consensus is accepted in modern biology. However, the biological community is polarized in regard to the mechanisms of biological evolution. On the one hand the followers of the prevalent view that evolution of the living world is determined by changes in genes and gene frequencies, and on the other, followers of the new, less popular, but already significant and rapidly crystallizing concept, holding that nongenetic, epigenetic factors and mechanisms also play an important role in the evolution of life.

The mainstream paradigm posits that evolution is the result of changes in allele frequencies with gene mutations, genetic recombinations, and gene drift being the source of genetic variability necessary for the action of natural selection. At its extreme form, the theory sees the life and evolution of living forms as a way of propagation and perfection of genes, while the carriers of genes, the organisms themselves, serve as machines or vehicles for survival and reproduction of genes.

Scientific candor requires us to plainly admit that despite the tremendous amount of effort, biology has never succeeded in showing in clear-cut terms how a change in gene(s) during evolution led to an evolutionary change in metazoan morphology or behavior.

In recent decades, a number of new alternative ideas, perspectives, hypotheses, and theories have been propounded (by Eldredge, Gould, West-Eberhard, Piggliucci, Schlichting, Jablonka, Newman, Müller, Cabej, and others) to explain the biological evolution or particular aspects of it. Rather than an unexpected or accidental phenomenon, the development of these hypotheses and theories is the logical result and manifestation of the failure of the existing paradigm to cope with an ever-increasing body of empirical evidence.

At the turn of the twentieth century, evolutionary theory was steered into an explanatory crisis. A general idea of the gravity and proportions of the crisis of the neo-Darwinian explanans of evolution may even be obtained through a glimpse of the following incomplete list of crucial biological phenomena that it cannot account for. (All of them will be considered in some detail in respective chapters of this work.)

1. Sudden evolutionary events, particularly the burst of body plans and emergence of almost 100 phyla within an evolutionarily narrow window of 5–15 million years during the Cambrian (~450 million years ago), whereas almost no new phyla have appeared in the

following 430–445 million years. This, unambiguously, contradicts the basic neo-Darwinian tenet of the gradualism of evolutionary changes.

2. Rapid speciation and morphological changes observed in nature, such as formation of new cichlid species in East African Lakes in a matter of several to hundreds of thousands of years, evolution of the *Drosophila* species in the Hawaii Islands, etc., including the rapid evolution and huge phenotypical (morphological, physiological, behavioral, and cognitive) divergence of humans and chimpanzees during the evolutionarily brief period of several million years.
3. Reproductive isolation and sympatric speciation in the absence of geographic isolation, involving no gene mutations, gene drift, or genetic recombination.
4. Frequent appearance of evolutionary reversions (including atavisms), return of lost ancestral traits even after many millions of years, clearly refuting the neo-Darwinian principle of the irreversibility of evolution.
5. Sudden and systematic appearance in the offspring of inherited changes in morphology, physiology, life history, and behavior (transgenerational developmental plasticity under the influence of environmental stimuli experienced by parents), involving no genetic factors or mechanisms.
6. Developmental polymorphisms, i.e., systematic production of different morphs in the offspring of the same brood and the same genotype, sometimes in adaptive responses to specific environmental stimuli.
7. Rapid and reproducible *experimental reversion* of the lost ancestral morphology, physiology, behavior, and life histories of *Drosophila melanogaster*, within as little as 5–20 generations.
8. Experience-independent generation of the quadrillions of bits of the epigenetic information (exceeding millions of times the amount of the genetic information contained in the human genome) for establishment of specific neuronal connections during the embryonic development of the central nervous system (CNS) in higher vertebrates.

With all the above in view, the reader is asked to judge whether these counterinstances represent "exceptions to the rule," or "exceptions that prove the rule," or whether the paradigm, at its present status, is "ruled by exceptions."

Not only does the neo-Darwinian paradigm fail to account for the above and many other phenomena, but empirical evidence is at odds with some of the basic neo-Darwinian tenets, as will be argued in respective chapters and sections of this work. The failure is the unavoidable result of the doctrinaire attitude, regardless of the challenging developments in various fields of biology, especially the developmental biology and paleontology, and dismissal of the ever-increasing body of evidence on the epigenetic factors involved in the individual development and heredity.

The Neo-Darwinian Paradigm and Darwin's Heritage: Darwin in Procrustes' Bed

Darwinian theory was a major landmark and a turning point in the history of the scientific understanding and study of the world. Due to its unprecedented success from the beginning, about one and half centuries ago, it was conveniently used as a shield for protecting and fending off criticisms by the most diverse biological and even social theories and doctrines from the amorphous Lysenkoist doctrine of inheritance of acquired characters, on the extreme left, to Herbert Spencer's social Darwinism, on the right. Theoretical constructions in the twentieth century could not afford to opt out of using Darwin's imposing scientific reputation even when particular Darwinian ideas contradicted some of their basic tenets.

In this respect, the neo-Darwinian theory is not an exception, despite the incontestable contributions it made in the study of evolution. The attempt to integrate Mendelian genetics into Darwinian doctrine was scientifically highly costly; for the sake of that that integration, they had to cede (part with) some of Darwin's basic ideas that were incompatible with the principles of the new science of genetics, such as the role of changes in the environment and the use/disuse of organs in evolution, evolutionary reversion to ancestral traits, role of behavior and learning in modifying instincts, and sympatric speciation. Although emphasizing their loyalty to Darwinian theory, neo-Darwinians in fact dwarfed Darwin's heritage and excluded from their theory some essential Darwinian tenets. The marriage of genetics with Darwin's theory was made at the latter's expense.

Neo-Darwinians excluded from the theory any nonselectionist mechanism of evolutionary change. According to the neo-Darwinian paradigm, the living organism, the evolving entity, is subject to *random* changes in its genes, while the external environment via natural selection brings order to the random variability and produces innovation. Living organisms, in a figure of speech, produce random variability, only "letters" and "words." It is the environment that via the natural selection uses these letters and words in writing the text of the evolution.

The neo-Darwinian paradigm has already paid a high price for stubbornly holding fast to the strange idea that the creative force of the evolution of living systems is to be found outside the living systems themselves. The cult of chance as the ultimate source of variability and evolution via natural selection was proclaimed to be continuation and advancement of Darwinian legacy. Is this true?

In fact, from the first publication of the *Origin*, and repeatedly in the later editions, Darwin pointed out in a visible part of the Introduction:

> *I am convinced that Natural Selection has been the main but not exclusive means of modification.*
>
> *Darwin (1859)*

Later, Darwin made clear his belief that evolution may occur even without natural selection being involved:

> *It should not, however, be overlooked that certain rather strongly marked variations, which no one would rank as mere individual differences, frequently recur owing to a similar organisation being similarly acted on—of which fact numerous instances could be given with our domestic productions. In such cases, if the varying individual did not actually transmit to its offspring its newly acquired character, it would undoubtedly transmit to them, as long as the existing conditions remained the same,*

> *a still stronger tendency to vary in the same manner. There can also be little doubt that the tendency to vary in the same manner has often been so strong that all the individuals of the same species have been similarly modified without the aid of any form of selection.*
>
> *Darwin (1872)*

Contrary to the neo-Darwinian idea that the loss and vestigialization of organs results from the accumulation of mutations under the action of natural selection, Darwin denied any role of natural selection (and implicitly of the genetic variability) in the process of evolutionary vestigialization and ultimately in the loss of organs:

> *Rudimentary organs, from being useless, are not regulated by natural selection.*
>
> *Darwin (1872, p. 131)*

The neo-Darwinian temptation to dismiss from the Darwinian doctrine all but the principle of natural selection and indefinite changes is not unprecedented. It is as old as the Darwinian theory itself. The genius had to deal with the same misrepresentation problem during his lifetime:

> *But as my conclusions have lately been much misrepresented, and it has been stated that I attribute the modification of species exclusively to natural selection, I may be permitted to remark that in the first edition of this work, and subsequently, I placed in a most conspicuous position—namely, at the close of the Introduction—the following words: "I am convinced that natural selection has been the main but not the exclusive means of modification."*
>
> *Darwin (1872, p. 421)*

Not only did Darwin not exclude the role of nonselectionist factors in evolution but he even elaborated on some of these factors. The neo-Darwinian theory neglects, if it considers at all, the role of use and disuse of organs in evolution of the living world. To the contrary, Darwin pointed out:

> *Disuse, aided sometimes by natural selection, will often tend to reduce an organ, when it has become useless by changed habits or under changed conditions of life.*
>
> *Darwin (1859, p. 479)*

The animal structure may change by mechanisms different from random changes in genes, in the modern biological meaning. Darwin believed that disuse may lead to reduction and vestigialization of organs or parts, biological phenomena that are extremely common throughout nature (Darwin, 1859, p. 450) and which result from the use and disuse:

> *There can be little doubt that use in our domestic animals strengthens and enlarges certain parts, and disuse diminishes them; and that such modifications are inherited.*
>
> *Darwin (1859, p. 134)*

and further:

> *I believe that disuse has been the main agency; that it has led in successive generations to the gradual reduction of various organs, until they have become rudimentary,—as in the case of the eyes of animals inhabiting dark caverns, and of the wings of birds inhabiting oceanic islands, which have seldom been forced to take flight, and have ultimately lost the power of flying.*
>
> *Darwin (1859, p. 454)*

Based on the idea that spontaneous random changes in the hereditary material, i.e., gene mutations, are the only source of inherited variability, neo-Darwinians excluded the possibility of environmental agents acting as stimuli inducing inherited changes in the structure and function of organisms. To the contrary, Darwin accepted, and modern biology has substantiated, the idea that environmental stimuli can induce inherited adaptive changes in living organisms:

> *Changed conditions generally induce mere fluctuating variability, but sometimes they cause direct and definite effects.*
>
> *Darwin (1872, p. 131)*

Translated into modern biological parlance, these "direct and definite effects" are nonrandom, nongenetic, and therefore epigenetic changes. Indeed, firmly established observational facts on transgenerational developmental plasticity have shown that Darwin was right, not the neo-Darwinians (see Chapter 11).

In line with their idea on gene mutations as the exclusive source of inherited variability, neo-Darwinians believe that evolution is irreversible. They are still denying the possibility of evolutionary reversion of lost ancestral structures, as formalized in Dollo's law. To the contrary, Darwin believed that evolutionary reversions are possible, and he hypothesized about the underlying cause of the phenomenon:

> *When a character which has been lost in a breed, reappears after a great number of generations, the most probable hypothesis is, not that the offspring suddenly takes after an ancestor some hundred generations distant, but that in each successive generation there has been a tendency to reproduce the character in question, which at last, under unknown favourable conditions, gains an ascendancy.*
>
> *Darwin (1859, p. 161)*

Then he iterates that

> *there is a tendency in the young of each successive generation to produce the long-lost character, and that this tendency, from unknown causes, sometimes prevails.*
>
> *Darwin (1859, p. 166)*

More-than-adequate evidence on evolutionary reversions shows that Darwin was right and that neo-Darwinians are not (see Chapter 15, Evolution by Reverting to Ancestral Characters).

Exclusion of all the above Darwinian factors from the neo-Darwinian theory of evolution was a timid criticism of Darwin's concept on organic evolution. It is somewhat ironic that by proclaiming Darwinism, the neo-Darwinian approach to the problem of evolution is more Wallaceian than Darwinian. Wallace, the coauthor of the theory of natural selection, has also criticized Darwin's acceptance of nonselectionist factors as important players in the process of living world:

> *Now Mr. Darwin has himself admitted that there are these unknown causes at work, and that "natural selection is the most important but not the exclusive means of modification." There may be some question as to the term "most important," if, as is not improbable, the most radical differences in animals and their most important organs could not have been produced by it alone.*
>
> *Wallace (1880)*

Exclusion of Darwin's nonselectionist heritage (e.g., use and disuse of organs, direct effects of changed conditions of living, loss of organs as a result of disuse, and evolutionary reversions) from neo-Darwinian explanans was a necessity rather than an accident. Any attempt to apply the new knowledge on genes as material carriers of heredity to nonselective factors of Darwinian evolution is confronted with an insuperable difficulty: changes in genes, gene mutations are random events, while the nonselective Darwinian factors implied directed change in evolution. The arbitrary exclusion of all these Darwinian factors was found to be a convenient, but highly costly, "solution" to the problem of introducing genes into the theory of evolution, for adapting the Darwinian theory to the needs of the "evolutionary synthesis."

Now, more than a century after Darwin, his assumptions on the direct influence of the environment on heredity, evolutionary reversion of ancestral phenotypes, and the loss of organs or traits that were excluded from the neo-Darwinian explanans of evolution, are unambiguously validated by empirical evidence, respectively, in the fields of transgenerational developmental plasticity, paleontological record, speciation, and experimental evidence on evolutionary changes, all to be discussed in separate chapters and sections of this work.

The Neo-Darwinian Paradigm and Modern Biology

At the foundation of the neo-Darwinian paradigm is the idea that evolution results from changes in gene frequencies in natural populations, and the source of these changes are basically three genetic phenomena: gene mutations, gene drift, and genetic recombination. Natural selection creates novelty by acting on the genetic variability.

For over half a century, a steadily growing body of evidence contradicting the neo-Darwinian view has been regarded only as a plethora of puzzles and anomalies, although attempts to explain or reconcile these phenomena with the neo-Darwinian paradigm have generally failed. Rather than trying to face and resolve these "puzzles," dealing objectively with the disproving "counterinstances," neo-Darwinians applied the ostrich approach and ignored contravening facts. Remember:

> *Einstein saw as counterinstances what Lorentz, Fitzgerald, and others had seen as puzzles in the articulation of Newton's and Maxwell's theories.*
>
> *Kuhn (1996)*

Genes and Phenotypic Characters in Metazoans

In our time, the central role of genes in the evolution of unicellulars is proven. So is the role of genes in the process of biochemical evolution of multicellulars that, to a great extent, results from gene mutations accumulating via natural selection during long evolutionary periods of time.

An enormous amount of energy and intellectual resources have been, and continue to be, devoted to finding and identifying the gene(s) responsible for various morphological characters in metazoans. In thousands of experiments, it has been demonstrated that spontaneous and induced mutations lead to anomalies in morphology, physiology, and behavior. We know of numerous genes in various species of invertebrates and vertebrates that are clearly related to specific characters. No one can deny that mutations, deletions, or generally nonfunctioning and malfunctioning genes lead to specific abnormal morphologies. Such experimental facts clearly show that functionally normal genes are necessary for the development of particular normal morphologies. But not more than that. In other words, it is correct to say that in metazoans the normal gene is a *necessary condition* for the development of a specific normal character, but it would be logically erroneous to grant the "necessary conditions" attributes of the *cause* when these conditions are *not sufficient* for the development of normal morphology. Many biologists believe that demonstration that a gene can induce ectopic expression of a phenotype proves that it is sufficient for the development of the phenotype (Baker et al., 2001). But even this "sufficiency test" is methodologically flawed and misleading. For ectopic expression of the phenotype in experiments does not exclude participation of other local nongenetic factors in the ectopic development of the phenotype; the gene in this case as well is not the cause or "all what is needed" for the development of that phenotype.

Development of phenotypic characters in metazoans is a function of signal cascades and gene regulatory networks (GRNs) with each network comprising from several to hundreds of genes. The presence of each of these genes is a necessary condition for the normal development of the specific character, but none of these genes is *the cause* of that character. Moreover, even these several to hundreds of genes, in their entirety, do not rise to the level of a cause. Contemporary biological knowledge says that GRNs generally represent downstream entities of signal cascades that are activated by neurohormonal signals, and the primary source of the information for starting these signal cascades is a chemical output released as a result of the processing of electrical signals into which all external and internal stimuli are converted in neural circuits. Signal cascades represent causal chains in which the preceding event is the cause of the next event. One cannot reasonably talk or think about *the cause* of a phenotypic result determined by such signal cascades. But we can instead distinguish between the last link in the causal chain, which is the *proximate cause* of the phenotypic result and the first (or initiating) link, which is the *ultimate cause*.

Any malfunction of an upstream element of the cascade could lead to defects of the character determined by the GRN. The higher the position of the affected element in the cascade, the greater the effect of the change could be. So the first conclusion to be drawn is that no single gene is responsible for the formation of the affected character. To single out a particular gene as the cause of the phenotypic character in such cases is to attribute to the part the functions of the whole.

The fact that the signal cascades that activate GRNs start with electrical/chemical outputs of the processing of internal/external signals in neural circuits suggests that the activation/inactivation of GRNs is determined by the computational activity of neural circuits, which ultimately determine whether, where, and when to turn on/off GRNs (see Processing of External/Internal Stimuli in the CNS Generates Information for Adaptive Phenotypic Responses, in Chapter 2). Thus, in metazoan development, genes are tools which need epigenetic instructions on whether, where, and when they should be expressed.

Gene Mutations Versus Darwinian Variability

The neo-Darwinian identification of spontaneous mutations with Darwin's gradual individual variations or differences caused by the principle of divergence is questionable. Despite the lack of knowledge on the nature of these variations at the time, intuitively Darwin denied the possibility that these variations were random and spontaneous occurrences. According to Darwin, modification of species

> *has been effected chiefly through the natural selection of numerous successive, slight, favourable variations; aided in an important manner by the inherited effects of the use and disuse of parts; and in an unimportant manner, that is in relation to adaptive structures, whether past or present, by the direct action of external conditions, and by variations which seem to us in our ignorance to arise spontaneously (my emphasis - N.C.) It appears that I formerly underrated the frequency and value of these latter forms of variation, as leading to permanent modifications of structure independently of natural selection.*
>
> *Darwin (1872, p. 421)*

Note that Darwin believed that the apparent spontaneous inherited variations (presumably not related to external stimuli) are not genuine random events but products of our ignorance. He felt that even if random changes would occur, they could not lead to the degree of variations observed between species:

> *Mere chance, as we may call it, might cause one variety to differ in some character from its parents, and the offspring of this variety again to differ from its parent in the very same character and in a greater degree; but this alone would never account for so habitual and large an amount of difference as that between varieties of the same species and species of the same genus.*
>
> *Darwin (1859, p. 111)*

Translated into the modern biological language, Darwin believed that what we call "spontaneous variability" is neither spontaneous nor random because random events, such as gene mutations, "would never account for so habitual and large a degree of difference as that between the species of the same genus." This unequivocally contradicts the neo-Darwinian tenet on gene mutations as the source of evolutionary change and speciation and the belief of the semantic equivalence of gene mutations to Darwinian variations.

The neo-Darwinian synthesis has been extremely successful in explaining the evolution of unicellular life and the tremendous biochemical evolution in both unicellulars and multicellulars, but it has failed to bring any understanding of the evolution of multicellulars at the supracellular level, especially the evolution of animal morphology, the most visible aspect of metazoan evolution. One century of research in the fields of evolutionary biology and induced experimental mutagenesis did not provide any proof that a particular mutation can lead to an *adaptive* morphological change in metazoans.

What is the reason for the discrepancy between the success and the failure of the theory of dealing with two aspects of a single process, the evolution of the living world?

Exaggeration of the importance of the knowledge of the time, and extension of that knowledge beyond the scientifically justified limits, have always been flaws of human judgment. Genes became a biological "jack of all trades" and were used for explaining anything related to the living world, from protein biosynthesis to the nature of human behavior and intelligence and even the evolution and functioning of human societies. Extrapolation of the role of genes as carriers of information for protein biosynthesis for explaining a totally different process, the spatial arrangement of cells of various types from which the metazoan morphology arises, represents a rare case of resorting to *Deus ex machina* for resolving problems in science.

Almost half a century ago, at the zenith of the progress in the fields of molecular genetics, after the discovery of the material basis of the gene and genetic code, Chargaff predicted (and warned against) the rise of a mechanomorphic view of living nature and a molecular mythology (Chargaff, 1963). The science of biology, under the influence of the *Zeitgeist*, for half a century would focus on the study of genes as separate and almost independent entities, underestimating the dynamics of their interactions with other cellular and extracellular components and ignoring the wider epigenetic context that determines even the expression of genes in metazoans.

The Neo-Darwinian Explanation of Evolution Put to the Test of Praxis

What is the status of the neo-Darwinian theory now, about 80 years after its formulation, on strictly scientific terms? Does the neo-Darwinian paradigm, in all its slightly varying forms, provide a verified and verifiable genetic mechanism of evolution?

From the neo-Darwinian view, evolution is a statistical process of the accumulation of favorable genetic variability under the action of natural selection, and the sources of that variability are gene mutations, gene drift, and genetic recombination. Gradual accumulation of changes in allele frequencies in populations that are geographically isolated, over time, leads to reproductive (postzygotic and less frequently prezygotic) isolation of the geographically isolated populations and formation of species and higher taxa. While it is theoretically argued and empirically demonstrated that changes in allele frequencies do occur, *in no single case* has it been possible to show that one or a number of changes in genes have led to an adaptive change in a morphological trait in metazoans.

Validating Neo-Darwinian Predictions

The neo-Darwinian paradigm would predict that the *tempo of evolution would be steady with no periods of evolutionary stasis*. This prediction has been refuted by the large body of paleontological evidence. In the early 1970s, Eldredge and Gould (1972) presented paleontological evidence demonstrating that evolution is characterized by short periods of rapid change and longer periods of morphological stability, and formulated the theory of punctuated equilibria. Schindewolf also believed that

> *evolutionary development is episodic—it proceeds in phases, or in quantum leaps; it exhibits an unmistakable periodicity. The unfolding of lineages is divided into evolutionary periods or cycles .... At the onset of a cycle, there is a brief period of abrupt development of forms .... We call this first phase the origin of types or typogenesis. This is followed by a second phase, one of type constancy, or typostasis.*
>
> *Schindewolf (1993)*

The neo-Darwinian paradigm would predict that, with mutation rates comparable in various organisms and with other factors equal, *the tempo of evolution will be faster in organisms that have higher rates of reproduction*. This prediction is almost universally refuted: invertebrates that have shorter life cycles have slower evolutionary rates than vertebrates, and the class of mammals, which has longer generation times, has evolved faster than any class of vertebrates that have shorter life cycles.

The neo-Darwinian prediction that the same genotype under the same environmental conditions will produce the same phenotype is no longer valid. Vast evidence shows that in species, populations, or individuals of the same/identical genotype reared under the same conditions may exhibit very different phenotypes, as is observed in numerous cases of developmental polymorphisms, predator-induced defenses, and phase transition in locusts. Confronted with this and two essential facts that

1. Gene mutations are very rare and deleterious; according to one of the leading neo-Darwinists of the twentieth century, a vast majority of the mutants observed in any organism are detrimental to its welfare (Dobzhansky, 1971); and
2. Gene mutations that are evolutionarily useful are even more rare to account for the sudden morphological changes observed in nature. (Dobzhansky believed that finding a useful mutation in nature is as difficult as finding "a needle in a haystack"; Dobzhansky, 1971, p. 138).

By the middle of the twentieth century, many evolutionists came to the conclusion that genetic recombinations, not gene mutations, are the main source of genetic variability on which natural selection acts. The idea is that since individuals of a species are in possession of different alleles, genetic recombination may lead to reshuffling of alleles and thus create new genetic variability. It was assumed that natural selection determines the evolution of the living world by acting on this genetic variability (and the related phenotypic variability) provided by genetic recombinations. But, theoretically, it has been argued that even if recombination would produce new favorable genetic variability and phenotypic variability (with the latter not being demonstrated), the genetic variability would hardly be maintained in further generations because of the extremely low probability of the recombinant to mate with another individual of the same recombinant genotype:

> *There is no guarantee that such an exceptional individual will engage in genetic recombination only with individuals having a similar adaptive genotype, it is inevitable that this exceptionally favorable genotype will eventually be destroyed by recombination during reproduction.*
>
> *Mayr (1964)*

Furthermore, there is no evidence that evolution of the parthenogenetic species, in which genetic recombination is excluded, is slower than that of sexually reproducing organisms, suggesting that genetic recombination may be neither involved in, nor necessary for, the evolution of metazoans.

If the evolution of metazoans would depend on, or be related to, the evolution of genes, it will be expected that a correlation will exist between the number of genes and the position of organisms in the evolutionary ladder. (This is what was predicted before the gene sequencing era.) While a clear correlation is observed to exist between the number of genes and the structural and functional complexity of unicellulars, no such correlation has been observed in metazoans; the number of genes in the human genome is comparable to that of the simple worm, *Caenorhabditis elegans*, but smaller than that of sponges, the simplest known metazoans. Even an evolutionist like Maynard Smith (1986) would express his surprise that the biochemical difference between a bony fish, such as the carp, and a jawless fish, such as the lamprey, is about the same as that between the human and the lamprey (Wesson, 1991).

According to the neo-Darwinian view on the existence of a genetic program and on the genotype–phenotype relationship, it would be predicted that evolution of similar morphologies (parallel evolution, convergent evolution, and homology) would be related to the evolution and presence of similar genes. This prediction has been rejected by observations on individual development. Long before, de Beer (1971)concluded that "Homologous structures need not be controlled by identical genes," and that "the inheritance of homologous structures from a common ancestor … cannot be ascribed to identity of genes."

The reverse has also been observed. Complex GRNs for the development of parts or organs of animals are discovered, which are surprisingly conserved among phyla, subphyla, and lower taxa, and still with the same genes, with the same genetic information, different organisms and even organisms of the same species can produce very different morphologies (in the case of metamorphosis, polyphenisms, and developmental plasticity).

All of the above facts strongly suggest that a causal relationship between changes in genes and evolutionary events in metazoans might not exist, hence the almost century-old hypothesis on changes in genes as causes of evolution is still waiting to be validated.

The Modern Synthesis and the Rise of the New Trend

The neo-Darwinian paradigm correctly predicts that a favorable inherited change will be positively selected, but this is the second and easy part of the question on the mechanics of evolution. In the beginning, there is the change; selection trails it in the process of evolution. The crucial problem to resolve in modern evolutionary biology, it is the generation of evolutionary change, identification of the source of new information for producing the heritable adaptive change, not the process of natural selection, the universally accepted and intelligible Darwinian concept.

Gene mutations, random changes in the nitrogen base sequences of nucleic acids, normally produce defects rather than adaptive changes (although very rarely they can) in animal phenotype. There is evidence that the genetic information determines the position of amino acids in peptide chains, but there is no indication that it might determine the spatial order of cells, the building blocks of metazoan structure. It may be argued, however, that the lack of evidence is not evidence but failure to find that evidence in more than half a century of research is not promising. The burden of proof, the obligation to prove that genes may code for the spatial arrangement of cells, is upon the supporters of that idea. Affirmative statements must be substantiated.

For half a century after the discovery of the chemical nature of the gene, biologists attempted to fill gaps in our understanding of the processes of individual development in metazoans by applying the concept of "genetic program," without showing how it works, but vaguely positing that the genome is programmed to regulate individual development. The idea of the genetic program has never been formalized into a hypothesis that would elaborate on where in the genome instructions for the spatiotemporal patterns of gene activation/inactivation are, and how these patterns might translate into spatial order of cells of different types from which the animal morphology emerges.

To the contrary, we know that the entire early development is regulated not by the genome or genetic information (otherwise, not gametes alone, but all the somatic cells that are in possession of the same genome would be able to start individual development and produce embryos) but by the epigenetic information deposited in gametes in the form of mRNAs, proteins, hormones, neurotransmitters, and nutrients.

For the sake of argument, however, let us take it for granted that a developmental program is somehow set up in the genome. But this assumption would raise the formidable sphinxoid riddle: How could the DNA of a few billion bits of information code for an edifice, whose erection requires an amount of information that is millions of times greater?

One century of studies on mutations has not provided a single verified example of a gene mutation that led to an adaptive morphological change in metazoans. On the contrary, examples are described of evolutionary changes having suddenly occurred in whole populations without changes in allele frequencies (e.g., numerous cases of transgenerational predator-induced defenses in invertebrates, sympatric speciation in sibling species involving no changes in allele frequencies, evolutionary reversions, atavisms, and experimental induction of reversion of ancestral traits in as little as five generations in *Drosophila*).

Having failed to prove any *causal link* between changes in genes and the evolution of animal phenotype, modern biology has instead succeeded in demonstrating that changes in developmental pathways, involving no relevant changes in DNA or genes, produce evolutionary changes.

Population genetics posits that evolution results from changes in the allele frequencies of populations due to gene mutations, gene drift, and gene recombination under action of natural selection. On this theoretical basis, a whole mathematical edifice for explaining evolution by changes in allele frequencies has been erected. However, the basic tenet of the neo-Darwinian paradigm that changes in genes are responsible for morphological evolution, on which this edifice rises, is not substantiated, hence the empirical foundation of the structure is questionable at best.

We should always bear in mind that evolutionary biology is a science of realized potentialities, a science of the actuality rather than theoretical possibilities. While these methods can determine whether, under certain conditions, an event can happen, the conditions are sometimes selected at discretion, so as to allow the model to function in theory without first determining whether these conditions exist in nature. Even in such cases, the theoretical probability of an event to occur is not a proof that the event has ever occurred. The theoretical probability of an evolutionary event calculated in mathematical models is one thing, and the factual occurrence, another. Additionally, the resolving power of mathematical methods is not unlimited. The empiry of experimentation and observation is the ultimate benchmark for testing theoretical models. Science, biology included, has an Antean dimension: its tremendous cognitive power vanishes as soon as it loses contact with the empiry of observation. The gold standard for verification of hypotheses and theoretical constructions is their congruity with the available empirical evidence.

With these limitations in mind, Norbert Wiener cautioned mathematicians themselves:

> *One of the chief duties of a mathematician in acting as an advisor to scientists is to discourage them from expecting too much of mathematicians.*
>
> *O'Connor and Robertson (2003)*

The theoretical biology cannot be in step with the advances of experimental biology as long as it is not taking full advantage of the ever-increasing evidence on the epigenetic inheritance accumulated in the second half of the twentieth century and especially during the last two to three decades. That evidence shows that, although changes in *genetic* information, in genes, and in gene frequencies do unavoidably occur, they are not necessarily related to the evolution of animal morphology, behavior, and life history. In West-Eberhard's aphorism, genes are followers, not leaders, in evolution (West-Eberhard, 2003).

Why has the neo-Darwinian paradigm been so unresponsive to the multitude of empirically established counterinstances? As with most phenomena in human enterprise, there is more than one single factor, but probably the most fundamental reason for ignoring the disproving evidence is a Kuhnian one:

> *To reject one paradigm without simultaneously substituting another is to reject science itself.*
>
> *Kuhn (1996)*

Recognition of the Role of Epigenetic Factors in Evolution

In early 1940s, Conrad Waddington coined the term "epigenetics" (from the Greek epi- *upon*, *on*, *over* but also *after*) for describing heritable changes in the expression of genes and the mechanisms that translate genotypes into phenotypes in the process of individual development. Due to slow progress in the study of the nature of inherited changes in expression of genes, Waddington's contribution to epigenetics is only mentioned as an episode in the history of experimental evolution and epigenetics.

By the early 1960s, it was demonstrated that even in unicellulars, where the genetic basis of the cell division and inheritance was definitely demonstrated, epigenetic information of some kind was responsible for certain phenotypic traits. By grafting pieces of *DNA-free cortex* from one individual of the ciliate protozoa *Paramecium aurelia* to another, Sonneborn and Beisson succeeded in obtaining protozoans with reverse polarity of fibers and with a different cortical organization, which was transmitted to the offspring for more than 700 generations (Sonneborn, 1964; Beisson and Sonneborn, 1965).

> *This is perhaps enough to show the extreme stability and determinism of a merely structural intracellular rearrangement in the absence of differences in genes or gene action.*
>
> *Beisson and Sonneborn (1965)*

By early 1980s, biologists discovered another epigenetic mechanism of heredity, which is known as "gene imprinting," a phenomenon of the parental determination of expression of alleles in the zygote and embryo. Presently, the concept of epigenetics is basically reduced to DNA methylation, gene imprinting, and chromatin remodeling.

As it is used in this work, epigenetics additionally encompasses a series of nongenetic *processes*, ranging from nongenetic mechanisms of gene expression to deposition of cytoplasmic factors in gametes, mechanisms of developmental plasticity, transgenerational plasticity, and evolutionary change (including the speciation process), all of which involve no changes in genes or genetic information in general. It comprises *generation of epigenetic information* (parental cytoplasmic factors) *and its role* in gametogenesis, gene imprinting, as well as the crucial role of parental cytoplasmic factors in activating expression of zygotic genes, determining the early individual development; it deals with the role of the postphylotypic nervous system in activation of nonhousekeeping genes and GRNs, cell differentiation, organogenesis, and morphogenesis in the process of individual development as well as the maintenance of homeostasis. To put it more explicitly, epigenetic processes direct (control and regulate) individual development and reproduction, which is another way of saying that epigenetic processes determine the inheritance at the supracellular level that is manifested in the visible metazoan phenotype (morphology, physiology, behavior, and life history). This statement implies the existence of an epigenetic system of heredity, predicted by Maynard Smith, and to be discussed in the first part of this work.

Although more-than-adequate evidence shows that epigenetic information is involved in the evolution of metazoans, even in the most recent standard works on evolution (for instance, Barton et al., 2007), epigenetic inheritance (for instance, parental cytoplasmic factors and changes in the expression pattern of genes or GRNs) is reduced to the phenomenon of gene imprinting, and even this is done in contexts that are not related to the generation of evolutionary change and novelties.

As a result of the neglect of the experimental evidence on the role of epigenetic factors in inheritance and evolution, the meaning of the adjective "genetic" expanded to such an extent that it is used synonymously with "hereditary" and "inherited." Many biologists still continue to take it for granted that any Mendelian ratios of the inheritance of characters is related to the presence of genes or genetic factors alone. No rationale has been presented for the "double standard" assumption that epigenetic factors, which are also transmitted to the offspring via gametes, could not be inherited in a Mendelian mode.

Toward a New Paradigm of Evolution

On the Source of Information for Metazoan Structure

Transition from unicellular life to multicellularity required the solution of a difficult problem, unique to multicellular organisms. In clear distinction from unicellulars, whose *building blocks are proteins*, the products of their activity and macromolecular structures, multicellular organisms represent complex material structures with *cells as building blocks*. Erection of multicellular structures needs something that unicellulars do not: huge amounts of information for the strictly determined spatial arrangement of a myriad of cells of different types and a mechanism for transmitting that information to the offspring.

The genetic information, encoded in the form of specific sequences of nitrogen bases in nucleic acids, determines the spatial order of amino acids in peptide chains, but there is no indication, let alone proof, that it can also determine the specific spatial arrangement of billions or trillions of cells of various types in the animal body. Even if, for the sake of argument, one were to assume that genetic information could determine the spatial order of cells in metazoans, the amount of information contained not only in genes but in the whole metazoan genome, including the "junk" DNA, quantitatively represents only a negligible fraction of the information necessary for molding a metazoan structure. A human brain alone has one trillion nerve cells (Kandel, 2000). Before birth, that is experience-independently, each neuron establishes an average of 10,000 *specific* connections with specific neurons, implying that information for establishing these connections alone is of the order of quadrillions of bits, millions of times greater than the total amount of information contained in the genomic DNA. An even greater amount of information would be needed to erect the whole human body in the process of individual development. Qualitative unsuitability and quantitative negligibility of genetic information excludes the possibility that it may be responsible to mold the metazoan structure.

For an evolutionarily long time, this informational constraint was an insuperable barrier for the evolution of multicellularity. The evolution of an appropriate type of information capable of coding multicellular structures was a *sine qua non* for the evolution of multicellularity. Hence, from an informational point of view, the evolution of multicellularity required a radical informational revolution.

The Control System and the Epigenetic System of Heredity in Metazoans

One of the greatest enigmas of modern biology has been how an organism, whose structure at the molecular, cellular, and supracellular levels is continually disintegrating, succeeds in maintaining that complex structure and function during its lifetime. The fact that metazoans succeed in doing so unambiguously proves that they

- Continually monitor the state of the system,
- Figure out structural losses, based on the presence of information about the normal structure,
- Figure out restitutive necessities, and
- Start signal cascades and activate GRN for replacing the lost structures at the right time and place.

The above functions performed by metazoans are typical functions of control systems, in principle similar to the control systems used in engineering.

No metazoan would exist as such in absence of a control system. The presence of a control system is one of the fundamental features of living, as opposed to anorganic, systems. That control system makes possible the development and maintenance of the metazoan organism, a thermodynamically improbable structure.

In unicellulars, the genetic system of heredity, that is, the genome and the epigenetic mechanisms of gene expression, represent at the same time the system control that maintains the cell structure at a steady state. By analogy, the question arises: Is it possible that in multicellulars as well, the control system acts as a system of heredity? Indeed, the maintenance of eroding metazoan structure implies possession by the control system of information on the normal structure. There is no visible reason why that information could not be used in the process of metazoan reproduction.

The metazoan control system, as we know it in extant animals, evolved some 550 million years ago during the Cambrian explosion with the differentiation of the neuron and the evolution of the CNS. In metazoans, this is an integrated control system (ICS) (Cabej, 2004) with the CNS as its controller. During reproduction, the ICS serves as the epigenetic system of heredity, which controls individual development. In the first part of this work (Chapters 1 through 6), I present extensive evidence on the elements of the system and their function in the process of biological reproduction.

The epigenetic system of heredity in metazoans implies the parallel presence of the genetic system of heredity at the cellular level. Although both systems are indispensable and represent fundamental aspects of metazoan biology, the evolution of the epigenetic system came as a solution to the problem of the organization of cells in the supracellular metazoan structures, parts, and organs, and the morphology in general. In the mutual relationship and interaction between the epigenetic and genetic systems, the latter is subordinate to the epigenetic system of heredity.

In a succinct description, the ICS, which also serves as an epigenetic system of heredity in the process of reproduction, is a hierarchical system that in vertebrates looks as follows:

Genes interact and influence the function of each other via their products at a *genetic level of control* on two (transcriptional and translational) sublevels. Most of the nonhousekeeping genes in metazoans are activated/inactivated by extracellular signals (hormones, growth factors, secreted proteins, neurotransmitters/neuromodulators, etc.), which act by binding specific membrane receptors or, in the case of small molecule hormones, by binding their nuclear receptors. This is the control at the *hormonal level* (including growth factors, neuropeptides, neurotransmitters). Most of the hormones secreted by the peripheral endocrine glands are under the control of specific pituitary hormones representing the *second level of hormonal control.* Almost all of the pituitary hormones are induced by the hypothalamic hormones, representing the *third level of neurohormonal control.* Still higher is another, *fourth cerebral level of control*, sending neural signals for the synthesis of most of hypothalamic neurohormones.

As pointed out earlier, in the process of metazoan reproduction, the ICS functions as the epigenetic system of heredity. The individual development from the egg/zygote to adulthood is a bigenerational process in which early development from the unicellular stage to the phylotypic stage takes place based on the epigenetic information provided parentally to the gametes. In the second, the postphylotypic stage, the embryo is in possession of an operational CNS, which generates and provides information for the rest of the individual development, organogenesis and morphogenesis.

Adequate evidence allows us to inductively conclude that after the phylotypic stage, electrical signals resulting from the processing of external/internal stimuli in neural circuits in the CNS (neural net in lower invertebrates) determine the activation of specific signal cascades leading to the development of specific morphological traits. The epigenetic information for metazoan morphology is computationally generated in the CNS by processing the input of internal/external stimuli.

In principle, the processing of internal/external stimuli in neural circuits transforms the stimuli into inducers of gene expression by establishing causal relationships between stimuli and genes that otherwise would not be causally related.

Role of the Nervous System in the Epigenetic Inheritance

Development of the metazoan structure requires considerable investment of matter, free energy, and information. With matter and free energy taken from the environment in the form of food, where does the information for the individual development and restitution of the disintegrating metazoan's *supracellular structure* come from? This is one of the fundamental questions to be addressed in this work.

One great enigma that puzzled biologists of the twentieth century was the source of the information necessary for the prenatal, i.e., experience-independent, establishment of trillions or quadrillions of specific connections among neurons. Now, a number of biologists believe that this information is generated in the nervous system, based on its self-organizing properties and "best guess" (Katz and Shatz, 1996).

The possibility of an involvement of the CNS in the development of animal morphology, at least tacitly, has been rejected, if not considered taboo. But now such a view is hardly defensible. Vast empirical evidence presented in this work shows that the development of various organs during embryogenesis is induced from signals and signal cascades originating in the CNS. The CNS also determines and regulates the development of postnatal development, including the onset of secondary sexual traits.

Despite the fact that no investigations specially aimed at discovering the possible role of the nervous system in individual development have ever been conducted (and "you hardly could find what you are not looking for"), coincidental evidence on the control of the CNS in the postphylotypic development is not simply adequate but surprisingly comprehensive.

As an intellectual remnant of the "one gene–one trait" concept of classical genetics, many biologists are still speaking of particular genes being responsible for particular traits. The tremendous progress made in identifying the complex GRNs has recently shifted attention from genes to the study of GRNs as determinants of phenotypic traits. Their study is considered to be sufficient for understanding the mechanism of the development of phenotypic traits. This concept brushes off the source of information for activation of GRNs and the spatiotemporal restriction of the activation.

The evolution of an information-generating contrivance and of a mechanism that would transmit the huge amount of epigenetic information from parents to the offspring has been a prerequisite of metazoan life. This has been an extremely difficult problem to resolve as it may be inferred from the fact that it took more than three-fourths of the time of the existence of life on Earth to evolve, occurring only about half a billion years ago, during the Cambrian explosion, with the emergence of the nerve cells and their organization into nervous systems.

The fact that only gametes are uniquely capable of developing into adult organisms, while somatic cells of the same genotype are not, unambiguously suggests that some form of epigenetic information is transmitted from parents to the offspring via gametes. But gametes as well are not capable of carrying all the information necessary for developing the metazoan structure. Hence, the evolutionary solution to the informational problem of individual development in metazoans was one of compromise; parent(s) would provide gamete(s) with a limited amount of epigenetic information in the form of parental cytoplasmic factors required for early development up to the phylotypic stage. At the phylotypic stage, the embryonic CNS is operational and takes over the postphylotypic development until adulthood, by generating the epigenetic information necessary for the later stages of individual development.

Ample empirical evidence shows that the inductive signals for the development of tissues and organs during individual development originate in the CNS. During the adult life, as well, signal cascades for the maintenance of animal morphology and homeostasis come from the CNS, via neuroendocrine cascades, often with essential participation of the local innervation. In this work, I will present representative examples of the CNS control of the development and evolution of the metazoans. Could all of this extensive evidence be irrelevant or accidental? Take a look at numerous signal cascades regulating patterns of expression of genes and GRNs. Generally, they are under hormonal control. One might admit that it may be a coincidence that the secretion of hormones by terminal endocrine glands is regulated by the secretion of pituitary hormones. I would admit that it may be another coincidence that the secretion of pituitary hormones is induced by the secretion of the hypothalamic releasing neurohormones. After all, signals have to come from somewhere, and the hypothalamus happens to be a part of the brain. For the sake of argument, I would admit that the secretion of hypothalamic neurohormones in response to the input from other brain regions, and not by any other part of the body, might be a mere coincidence. But could anyone believe that the fact that all of the fully known signal cascades ultimately start in the brain, and not in any other part of the body, represents nothing more than a coincidence? Why must the stroke of luck always fall on the CNS rather than on the other parts of the body? I believe that this "coincidence obsessed" approach defies common sense. For, by the same token, one might proclaim that the fact that genetic information flows from DNA down to RNA and proteins, and not the other way around, might also be a coincidence. Hence, DNA may not be the source of genetic information!

It would contradict any statistic and common sense to believe that mere coincidences could determine:

- The cerebral origin of morphogenetic information in numerous cases of developmental plasticity, including predator-induced defenses;
- The cerebral determination of several described cases of transgenerational developmental plasticity;
- The cerebral origin of signal cascades for numerous described cases of metamorphosis in both invertebrates and vertebrates;
- The central nervous origin of signals inducing development of organs during the individual development in numerous experimentally verified cases;
- The neurocognitive determination of reproductive isolation in numerous described cases of incipient species that manifests itself in changed mating preferences without changes in genes.

When considered in their entirety, all these examples of neurally controlled cases of developmental, circumevolutionary, and evolutionary phenomena logically lead to the conclusion that the morphogenetic information, the epigenetic information necessary for the development of metazoan structures, originates in the brain.

The Epigenetic Mode of Evolution

As a general trend, metazoans have evolved from simple to more complex organisms. They evolved increasingly complex structures because they must accomplish more complex functions in the "struggle for life." Complex structures arise to perform complex functions, and the function is the *raison d'être* of the structure.

Evolutionary changes result from specific changes in developmental pathways, which do not arise randomly, but specific epigenetic information is required for such changes to take place. As I have shown in *Neural Control of Development* (2004), this information is generated in neural circuits. Adaptation to new conditions of life requires modifications of functions as well as modifications of morphology and behavior.

Conditions of living can change so suddenly and drastically that entire populations or even species may go extinct before having a chance to evolve adaptive modifications in their morphology. However, in the course of evolution, metazoans evolved and perfected a complex system of stress response for dealing with the antihomeostatic effects of the adversely changing environment. The stress response is expressed in full and complex form in higher vertebrates, although its evolution can be traced back to invertebrates. It is function of the brain and of the neuroendocrine system characterized by the increased activity and the hypothalamic–pituitary–adrenal (HPA) axis.

Besides changes in the activity of the HPA axis, intended to restore the impaired homeostasis, the stress response is characterized by a series of quick adaptive behavioral changes. Metazoans have a unique ability to adapt instantly to the changed environment by appropriately changing their behavior, e.g., by fleeing the hostile environment and predators or by using alternative food sources. Behavioral changes in metazoans are related to learning, including perception, conditioning, and memory. In the process of learning and behavioral change, animals use existing neural circuits and fixed action patterns. The behavioral adaptation by learning may be a quick remedy, but is not a long-term solution to the problems related with the adversely changed conditions of living. However, theoretically at least, it may enable metazoans to "buy time" until inherited physiological and morphological adaptations can evolve. There is also a possibility that the behavioral adaptations, acquired by learning, may evolve into innate behaviors.

Evolution of heritable adaptations, which may arise over time, imply the generation and investment of new information for evolving morphological, physiological, and life history novelties. We know that the information for changed behavior is a product of neurocognitive processes taking place in the CNS. The question now arises: Where is the new information for adaptive evolutionary changes in morphology generated?

Although we cannot hope to find direct evidence on the source of information for inherited changes that occurred in the past, we can use other related empirical evidence for drawing relevant scientific inferences.

First, the fact that metazoan morphology is determined by developmental pathways during ontogeny suggests that adaptive evolutionary changes in morphology will also be determined during ontogeny. And the fact that during ontogeny many developmental pathways for animal structures are activated by signals that ultimately start in the CNS (see Chapter 5) suggests that the information necessary for the evolution of these structures may also originate in the CNS.

Second, developmental pathways that determine dramatic morphological transformations in metamorphosizing organisms (see Sections Neural Control of Metamorphosis in Invertebrates and Neural Control of Metamorphosis in Vertebrates in Chapter 5) are, ultimately, activated by CNS signals.

Third, many discrete changes in metazoan morphology that are observed in numerous cases of phenotypic plasticity are induced by signal cascades starting in the CNS (see Chapter 10).

Fourth, the biological phenomenon of transgenerational plasticity, that is the appearance in the offspring of inherited morphological (usually adaptive) modifications in response to environmental stimuli that have affected their parent(s). In a number of described cases it has been demonstrated that the information for these inherited changes in the offspring morphology and behavior comes in the form of brain signals (see Chapter 11).

"Inherited changes in morphology" during the transgenerational developmental plasticity are determined by brain signals! But evolutionary changes in morphology are nothing more than "inherited changes in morphology." This fact, logically, suggests that the mechanism of the induction of the evolutionary change may be a neural mechanism. It would be against the logic of evolution to believe that metazoans would have evolved two different mechanisms for attaining the same result (the inherited change in morphology). Evolution is inherently parsimonious.

The new information necessary for inducing specific inherited changes in cases of transgenerational developmental plasticity is generated in the CNS and results from the processing of internal/external stimuli in neural circuits. Mechanisms of the generation of the new information necessary for changes in developmental pathways represent a black box. We know "what" these neural circuits do, but we do not know "how." In the initial link of the causal chain, we see the stimuli being encoded in sensory neurons in the form of electrical spike trains at the entrance of the black box and, at the other end, we see electrical and chemical outputs that activate specific signal cascades in the offspring.

Although in metazoans, the evolution of genes is decoupled from the morphological and behavioral evolution, the epigenetic mechanism of evolution makes use of favorable changes in genes in the course of evolution and of the genetic mechanism of heredity with the genome as its central machinery. In their mutual interaction, the genetic mechanism of inheritance is subordinate to the epigenetic mechanism.

Adaptive evolutionary changes in the phenotype (behavior, morphology, physiology, and life history) in metazoans start with complex neurobiological processes of reception, informational conversion, integration, and processing of external/internal stimuli in the neural system, the CNS or neural net, and investment in gametes of new epigenetic information for inducing specific changes in the developmental pathways in the process of individual development of the offspring. It is encouraging that modern biology, slowly but steadily, is coming to recognize that

> *irrespective of their different nature, most animal adaptation mechanisms share the involvement of the central nervous system and often include endocrine activity.*
>
> *Kolk et al. (2002)*

On the Principles and the Method

The development of biological science, as well as science in general, is a continuous, but not a linear, process. Ideas, hypotheses, and theories of the past are not infallible. Change and continuity in science are two sides of the same coin: the continuity of scientific progress also implies discontinuities with the previous ideas and hypotheses, many of which expire in the long and laborious path to the truth. The same goes for the authorities. The only immutable and enduring bedrock in biology, the ultimate and incontrovertible authority, is the scientific fact as acquired by observation and experiment, not its interpretation or inferences, which are liable to the flaws of human judgment.

My attempt to reconstruct and visualize the developmental mechanisms of evolutionary changes in metazoans is chiefly based on empirical evidence. Biological facts have been the starting point of the inquiry that logically and unavoidably led me to the theory I present in this work, and empirical evidence is the foundation on which my theory arises. I have not built on previous theoretical structures that are still waiting to be validated.

Any phenotypic change, before being fixed as an evolutionary change, goes through two stages. First, it is generated, and then it must undergo a process of selection, in which useful changes, those that improve the fitness of the organism to the environment, or at least do not reduce it, will be conserved and propagated to the progeny. This process, in which the evolutionary "chaff is separated from the grain," is natural selection. Modern evolutionary biology deals predominantly with the second stage, with the process of selection of the change under particular environmental conditions. With a tolerable risk of exaggeration, I would say that by focusing on the process of selection, it has paid little attention to the crucial event in the process: the mechanism of generation of the change that is to be selected.

Given that Darwin and generations of biologists after him have exhaustively discussed and elaborated on the role of natural selection in determining the spread and elimination of evolutionary changes, my focus in this work will be on the mechanisms of the generation of the evolutionary innovations rather than their selection. To build on the metaphor on natural selection as editor, in this work I focus on the process of "writing" rather than on the "editing" of the evolutionary changes.

Evolutionary changes in morphology, resulting from changes in developmental pathways, must be produced sometime during the ontogeny, in the process of the individual development. Thus, the evolutionary change in morphology is the result of a specific change in a developmental pathway. This is the reason why key evolutionary events, the appearance of evolutionary novelties, in Part IV of this work (Chapters 13 through 16) will be considered at a developmental level, and to the extent that the molecular mechanisms, cascades, and gene regulatory mechanisms are known, signal pathways and sources of information for evolved structures, functions, and behaviors are examined.

In modern treatises on evolution, the observational evidence is in the function of validating the standard set of established hypotheses and theories, i.e., the theory is given preponderance, and the function of empirical evidence is to illustrate the theory.

As the intention of this book is not to illustrate an established theory but rather to present the available evidence on which I have erected my theory, the empirical evidence gets the lion's share in this book. Presenting a theory that needs to be substantiated, in this work I naturally focus on the evidence substantiating my theory, but *evidence* that seems to contradict my theory is not ignored.

In cases when I have felt that the interpretation of results of experiments or observations may be contentious (and this has not rarely been the case), accounts and conclusions drawn from leading investigators are quoted in appropriate contexts. This has imposed a relative abundance of quotations in this book.

In a number of cases, competing hypotheses are not discussed at length or even considered at all. In any case, this is related to one or more of the following reasons:

1. Failure of the hypotheses to find experimental and observational support for a long time since first presented,
2. Insufficient relevance to the discussed topic, or
3. Author's unawareness of the hypotheses.

Given that the neo-Darwinian theory still represents the most widely accepted explanans of organic evolution, description of almost all the evolutionary and developmental phenomena is followed by a comparative presentation of the neo-Darwinian view and my epigenetic explanation, in order to enable the reader to assess the relative explanatory power of both the genetic and epigenetic approaches to the mechanism of metazoan evolution.

In order to avoid any misinterpretation, whenever possible, the neo-Darwinian explanation is presented as rationalized by neo-Darwinian authors. However, if in some cases the reader will find my neo-Darwinian interpretation to be misleading or incorrect, the reason will be anything but the intention of creating a neo-Darwinian straw man.

The material included in the work for supporting the theory comprises evidence from widely different fields of biological research, which may sometimes make the reading difficult. To help the reader in overcoming these difficulties, figures are extensively used as a part of the explanatory apparatus. Whenever it has been possible, figures are reproduced without modifications.

I am aware that the literature used might not be in every case the most representative and that studies more important than the ones cited in this work often may be missing. Sometimes, abstracts have been used instead of full papers which have been out of my reach.

References

Baker, B.S., Taylor, B.J. and Hall, J.C. (2001). Are complex behaviors specified by dedicated regulatory genes? Reasoning from *Drosophila*. Cell *105*, 13–24.

Barton, N.H., Briggs, D.E.G., Eisen, J.A., Goldstein, D.B. and Patel, N.H. (2007). *Evolution*. Cold Spring Harbor Laboratory Press, Cold Spring Harbor, NY.

Beisson, J. and Sonneborn, T.M. (1965). Cytoplasmic inheritance of the organization of the cell cortex in *Paramecium aurelia*. Proc. Natl. Acad. Sci. U.S.A. *53*, 275–282.

Cabej, N.R. (2004). *Neural Control of Development*. Albanet, Dumont, NJ, pp. 35–47.

Chargaff, E. (1963). *Essays on Nucleic Acids*. Elsevier, London, pp. 175–176.

Darwin, C.R. (1859). *The Origin of Species by Means of Natural Selection or The Preservation of Favored Races in the Struggle for Life*, first edition. John Murray, London (first issue), p. 6.

Darwin, C.R. (1872). *The Origin of Species by Means of Natural Selection*. John Murray, London, p. 131.

Darwin, C.R. (1872). *The Origin of Species by Means of Natural Selection or The Preservation of Favored Races in the Struggle for Life*, sixth edition. John Murray, London (with additions and corrections), p. 72.

de Beer, G. (1971). *Homology, An Unsolved Problem*. Oxford University Press, London, pp. 15–16.

Dobzhansky, T. (1971). *Mankind Evolving*. Yale University Press, New Haven, CT, pp. 133–137, 138.

Eldredge, N. and Gould, S.J. (1972). Punctuated equilibria: an alternative to phyletic gradualism. In: *Models in Paleobiology*, T.J.M. Schopf (ed.). Freeman, Cooper & Co, San Francisco, CA, pp. 82–115.

Huxley, T.H. (1885). Quoted in Spencer, H. (1886). The Factors of Organic Evolution. Nineteenth Century *19*, 570–589.

Kandel, E.R. (2000). The Molecular Biology of Memory Storage: A Dialog Between Genes and Synapses (Nobel Lecture). Bioscience Reports *24*, 475–522.

Katz, L.C. and Shatz, C.J. (1996). Synaptic activity and the construction of cortical circuits. Science *274*, 1133–1138.

Kolk, S.M., Kramer, B.M., Cornelisse, L.N., Scheenen, W.J., Jenks, B.G. and Roubos, E.W. (2002). Multiple control and dynamic response of the *Xenopus* melanotrope cell. Comp. Biochem. Physiol. B Biochem. Mol. Biol. *132*, 257–268.

Kuhn, T.S. (1996). *The Structure of Scientific Revolutions*, third edition. The University of Chicago Press, Chicago, IL, p. 79.

Maynard Smith, J. (1986). *The Problem of Biology*. Oxford University Press, New York, NY, pp. 35–36.

Mayr, E. (1964). The evolution of living systems. Proc. Natl. Acad. Sci. U.S.A. *51*, 934–941.

O'Connor, J.J. and Robertson, E.F. (2003). MacTutor History of Mathematics. Available online: http://www-history.mcs.st-andrews.ac.uk/Biographies/Wiener-Norbert.html.

Schindewolf, O.H. (1993). *Basic Questions in Paleontology*. Chicago University Press, Chicago, IL, p. 193.

Sonneborn, T.M. (1964). The differentiation of cells. Proc. Natl. Acad. Sci. U.S.A. *51*, 915–929.

Wallace, A.R. (1880). The Origin of Species and Genera S322: 1880, nineteenth century. Available online: http://www.wku.edu/~smithch/wallace/S322.htm.

Wesson, R. (1991). *Beyond Natural Selection*. The MIT Press, Cambridge, MA, pp. 201–202.

West-Eberhard, M.J. (2003). *Developmental Plasticity and Evolution*. Oxford University Press, Oxford, p. 29.

Part I

Epigenetic Basis of Metazoan Heredity

1 Control Systems and Determination of Phenotypic Traits in Metazoans

The fixity of the internal environment is in short the condition of mental activity.
J. Barcroft (1934)

The Thermodynamic Enigma of Living Systems

The ability of living systems to maintain their structure, in apparent defiance of the second law of thermodynamics, has long amazed and perplexed scientists. What makes living systems so extraordinary and unique?

In *What Is Life?* Schrödinger (1944) considered the issue and popularized the idea of entropy in living systems. According to the second law, all processes occurring in material systems, living systems included, lead to increase of the entropy or the loss of order in the system. Schrödinger attempted to resolve the enigma of the existence and maintenance of the thermodynamically improbable structure of living organisms by assuming that they "suck" order or negative entropy (negentropy) from the environment. In other words, they use an external source of order (food for animals and sunlight and carbon dioxide for plants) to compensate for the order they inescapably lose, thus avoiding structural degradation.

With the benefit of hindsight, we know that his pioneering idea is incongruous with some basic physiological knowledge: the first step an organism undertakes after taking the food is to break the ingested organic compounds down to simpler, low-molecular organic compounds (e.g., amino acids, monosaccharides, fatty acids, glycerol, nitrogen bases, and pentoses). In doing so, it destroys the original order, i.e., increases the entropy of ingested nutrients for creating its own species-specific order by synthesizing *de novo* proteins, carbohydrates, lipids, and other compounds resulting from the activity of its proteins.

Biological systems do not maintain their unavoidably eroding structure by absorbing external "order." Most nutrients contain no *utilizable* order of the specific type that metazoans need for building their structures. What the organism utilizes from the ingested food for generating its species-specific molecular and cytological building blocks are the *raw material* and the *free energy* it contains. The organism is a generator of its own order rather consumer of external negentropy. The acquisition of this ability represents a crucial step in transition from anorganic systems to living organisms.

Epigenetic Principles of Evolution. DOI: 10.1016/B978-0-12-415831-3.00001-X

On the Meaning of Biological Information

Genetic Versus Epigenetic Information

In anorganic systems, the concept of order is often used as an equivalent of the negative entropy (negentropy). Both terms were used to measure the degree of atomic order in the system. Later, with the advent of the science of communication systems, a controversial triple relationship among entropy, order, and information arose.

The concept of information is an equivocal one that is intuitively used in widely different meanings in different fields of science.

From a conventional thermodynamical view, information is a measure of structural order in a system: the greater the entropy of the system, the lower the order and the amount of information in the system will be, and the reverse. A freezing temperature, for example, transforms water into ice, implying an increase of the order of the system resulting from restriction of movements of the water molecules within the system. The order in physical systems, thus, is temperature dependent and in the zero absolute (−273.15°C), when the system is devoid of almost all the forms of energy, and movement of molecules comes to an "end," the system will have its maximal amount of order and information.

In physical systems, thus, the order of the system is a function of the temperature. A rise in temperature increases the degree of freedom of the components of the system, thus decreasing the order or, what is equivalent, increasing negentropy and information content of the system, and vice versa.

In communication theory, the concept of information implies its transmission from the sender to the receiver, and entropy measures the fidelity of transmission of the information to the receiver. Due to intrinsic causes, noise or errors occur in the process of transmission through communication channels and in the process of decoding of the message by the receiver.

Even within the field of the study of biological systems, the concept of information is used in different meanings. To a large extent, the confusion on the meaning of information in biological systems derives from the fusion of two distinct concepts of information, the thermodynamic and the communication systems concepts.

As it will be used in this work, biological information is an entity that, when communicated to a system or to parts of it, can produce new and specific structural order. Biological information in metazoans consists of genetic information and epigenetic information. Both of these forms of the biological information functionally transcend or go beyond themselves, displaying some kind of intentionality, or purpose (a teleonomic purpose *sensu* Mayr (1961)).

Essential differences, however, exist as far as the function, origins, and nature of the genetic and epigenetic information is concerned. Genetic information is encoded in sequences of nitrogen bases in genes, which code for specific proteins. Unlike anorganic systems, there is no strict equivalence between the order (negentropy) and the information in the DNA molecule (gene). For instance, a randomly occurring mutation in a nitrogen base does not change the amount of order in the molecule, but it may cause a loss or gain in the biological information of the gene. Thus, a mutational event may not produce *loss of order* in the gene *per se* but may cause a *loss of information* that manifests itself in the loss of function of the product of the gene. In other words, the product of the mutated gene will lose the "meaning" and cannot serve its purpose that is the building of a functioning protein molecule.

Genetic information thus has a new attribute, unknown in anorganic systems: it has a "meaning"; a mutation in the gene may not change the order of the gene but may make the gene lose or gain information. This special quality of information in living systems is related to its biological utility.

Epigenetic information is operational at two different levels. At the cell level, it is a product of the epigenetic structures (centrosome/centrioles and cytoskeleton) responsible for the formation and spatial arrangement of cell organelles, including chromosomes, and for the mechanics of cell reproduction and division. At the metazoan organismic level, the epigenetic information is computationally generated in neural circuits. Its "purpose" is to activate a particular signal cascade to achieve a particular phenotypic change at a supracellular or organismic level by inducing

- Expression of a nonhousekeeping gene,
- Cell proliferation,
- Cell differentiation,
- Molding or restoring morphological traits,
- Maintaining physiological states,
- Determining behaviors, and
- Determining life history traits.

While the new genetic information is acquired via randomly occurring changes in the sequence of nitrogen bases, the new epigenetic information is an adaptation-oriented product of computational activity of neural circuits. It manifests itself in the form of a chemical outcome resulting from the processing of the internal/external stimuli in neural circuits. The release of the chemical product by the neural circuit serves as a command for activating a particular signal cascade that leads to an adaptive phenotypic result, aimed at maintaining or restoring the structural and functional order or the equilibrium between the organism and its environment. (See Chapter 2 for further details.)

The genetic and epigenetic forms of information produce order at different levels of structural organization, at the molecular and at the supracellular level, respectively.

Novel Properties of the Biological Information

Essential differences also exist between the biological (genetic and epigenetic) concepts of information, on the one hand, and the thermodynamic and communications theory concept of information, on the other. As presently used in respective fields, these concepts are polysemic, if not homonymous.

At variance with anorganic systems, the order in living systems is not temperature dependent; within the physiological range, the temperature has no influence on the amount of the order and information in the living system.

Unlike physical systems, living systems can create and maintain different patterns of order (spatial arrangements of cells of specific types) in different parts of the body by differential investment of information.

By using free energy of the food, they are also capable of transmitting order and information outside the system in the process of biological reproduction.

The biological concept of information is related to, but distinct from, the communications concept of information. The order in communication systems depends on the fidelity of the transmission of information from the sender and its decoding by the receiver. In an ideal case, the information transmitted by the sender to the receiver would be identical to the information the receiver gets. At variance with the symmetry of transmission of information in communication systems, transmission of information in biological systems is asymmetrical, in the sense that the information transmitted from the sender (the source of information) produces an order that corresponds to, but is different from, the order generated by the target organ (receiver of the information). This asymmetry and symbolism of the transmission of the genetic and epigenetic information are at the base of what is called the "meaningfulness of biological information."

While the information in the communication systems is decoded by the receiver, the epigenetic information is decoded by the signal cascade it chooses to activate, i.e., it reaches the receiver in a decoded form as an "instruction" or "command" telling target cells what to do.

At the organismic level, epigenetic mechanisms essentially involve activation of specific signal cascades in response to internal/external stimuli in the central nervous system (CNS) (to be discussed in detail in Chapter 2). Activation of a particular signal cascade by a neural circuit represents the solution to the problem that a specific input on the state of the system or an environmental agent poses to the CNS. The solution, which implies a (teleonomic) purpose, results from the stepwise processing of the input resulting in a chemical output (release of a neurotransmitter, neuromodulator, or a neuropeptide) for activating a specific signal cascade that produces a specific phenotypic outcome. This epigenetic mechanism of control of the phenotype evolved more than half a billion years ago, during the Cambrian explosion, with the differentiation of the neuron and the advent of the nervous system.

The Control System in Metazoans and the von Neumann's Machine

As already mentioned, while continually losing molecular and cellular components, metazoans succeed in maintaining a steady state, a flowing equilibrium (Fließgleichgewicht *sensu* von Bertalanffy), continually restoring the lost structural order at all the levels, from the molecular to the cellular and supracellular levels. Easy as it is to say that living systems manage to circumvent the second law, the continual replacement of inevitably degrading components in metazoans, the most complex structures known in the universe, is a formidable task. That replacement implies that metazoans

1. Continually monitor the state of the system,
2. Are in possession of information on the normal state of the system,

3. Are capable of comparing the actual state with the "normal" state, and on this basis,
4. Quantify the lost/degraded molecular and cytological components,
5. Determine the kinds and amounts of molecular and cytological components to be produced, and
6. Via signal cascades, send to the right places, at the right time, instructions for producing and replacing the lost components.

The above functions are basic functions of control systems, similar to the human-engineered control systems. This strongly suggests that a control system may be operational in metazoans.

Due to the multiplicity of the levels of controls and the hierarchical nature, the metazoan control system may be characterized as an integrated control system (ICS).

The ability of the ICS to continually monitor the status of the system in metazoans implies, and is based on, the omnipresence of the neural tissue throughout the animal body down to the level of individual cells.

The integration and processing of the interoceptive input on the state of the system, as well as decision-making for restoring the normal state, are functions of the *controller* of the ICS, which is the CNS (Figure 1.12). By comparing the information on the actual state with the "normal" state, the CNS identifies deviations from the norm and sends instructions for activating signal cascades for restoring the normal state.

Evolution of the control system was a *sine qua non* for the emergence of life on earth and for maintaining the structure and function of living systems. *The control system is a novel and essential feature of living systems*, unknown in the anorganic world. All of the rest of basic properties of the multicellular life depend on the presence and function of the control system. No growth, development, reproduction, metabolism, or evolutionary change would be possible in the absence of a control system.

In unicellulars, the control system is represented by the genome and the epigenetic apparatus of chromosome duplication and gene expression, which is also responsible for the spatial arrangement of cell organelles and cell morphology (see also Sections Epigenetic Information in Unicellulars and Unicellular Precursors of the Metazoan Epigenetic Informational Structure in Chapter 12).

As argued before (Cabej, 2005), the genetic information is qualitatively inappropriate and quantitatively negligible for determining the spatial order of cells of various types in the multicellular metazoan structure. Transition to multicellularity required an additional type of information for determining the spatial arrangement of the myriad of cells of different types in multicellular structures. Evolution of mechanisms for generating that new form of information necessary for determining spatial arrangement of the astronomical number of cells of hundreds of different types of cells in multicellulars has been a long, complex, and very difficult task as witnessed by the fact that the evolution of these information-generating structures took more than three-fourths of the time of the existence of life on Earth. It was a basic requirement for the evolution of the ICS, which would take over the control of all basic functions in metazoans, from the maintenance of the structure and function to their reproduction.

The evolution of the ICS, this uniquely biological antientropic contrivance, which figures out when, where, how much, and what kind of information to invest in the system, first for progressively producing the adult structure in the process of individual development and later, during the adult life, for restituting the unavoidably degrading normal structure and function, marks the most critical moment for the evolution of metazoans.

Recognition of the control systems in living organisms in general and the ICS in metazoans, resolves the thermodynamical paradox of the evolution and survival of metazoans as highly improbable structures.

The solution to the problem of determining the spatial arrangement of billions or trillions of cells of the most different types for molding animal morphology came sometime around the Cambrian explosion by the evolution of a type of biological replicator reminiscent of the von Neumann machine.

In the history of the kingdom Animalia, this began with the differentiation of the nervous cell and the nervous system (see also Section The Diffuse Control System and the Neural Controller: The Eumetazoan's Eureka in Chapter 12). After the evolution of the genetic code with the emergence of life on earth, more than 3 billion years ago, the differentiation of the neuron and the nervous system represents the greatest informational revolution.

Metazoans do not produce copies of themselves as unicellulars do. They cannot do so, due to the formidable problems related to production of copies of multicellular structures. Instead of producing copies of themselves, metazoans produce specialized cells, gametes, which can develop into adult metazoan organisms by themselves or by uniting with their sexual counterpart. Parents provide gametes with the epigenetic information necessary for developing up to an early embryonic stage, the phylotypic stage, when only one organ system, the nervous system, is operational. The CNS at this early embryonic stage, when the maternal epigenetic information provided with gametes is exhausted, can already stepwise computate the epigenetic information necessary for the development of the adult metazoan supracellular structures.

The evolution of metazoans, and multicellular organisms in general, was a complex task that transcends even the most courageous dreams of human genius. von Neumann dreamed of a machine (replicator) that would be able to build copies of its own when provided with the necessary parts. The machine then would install and switch on the operating program in the completed daughter machine, which could then enter a self-repeating cycle of production of self-replicating von Neumann machines. Even this overambitious dream of building a self-replicating machine looks quite unsophisticated when compared with what a metazoan organism actually accomplishes during its development and lifetime.

In the case of the metazoan "living machine," the parent(s) build(s) neither the machine nor the machine in miniature that, by growing in size, would become a complete operating and replicating "machine" of its kind; nor does the parent insert operating program in it. It rather provides epigenetic information for building a Bauplan at the phylotypic stage.

What is beyond the reach of even the wildest human imagination is that at the phylotypic stage, the embryo is in possession of an awesome information-generating machine, the CNS, which starts functioning and generating information for building its own species-specific metazoan structure and its own operating program, long before the construction of "living machine" is completed.

Now, let us try to substantiate the functioning of the metazoan ICS with some empirical evidence on the central neural control of the development and maintenance of the animal phenotype (physiology behavior and morphology).

I anticipate the neo-Darwinian counterargument that even the central control of animal phenotype is ultimately determined by genes. At this juncture, I would simply state that it is inaccurate, and in the next chapter, I elaborate extensively on the non-genetic, computational, epigenetic nature of that control.

Central Control of Animal Physiology

Physiological knowledge accumulated during the last two centuries shows that the vital functions of all the organs and organ systems in higher invertebrates and vertebrates are controlled and regulated by the CNS. An initial theoretical breakthrough in understanding that control represents the development of the concept of *la fixeté du milieu intérieur* by the French physiologist Claude Bernard (1813–1878) in the second half of the nineteenth century, now generally known as *homeostasis*, in a term coined by Walter Cannon in the 1930s. According to this concept, living organisms can keep constant, within certain limits, their interior environment, i.e., the chemical composition of body fluids (blood, lymph, and intercellular fluid), temperature, and other variables. The maintenance of homeostasis is a function of the CNS.

Experimental studies on the mechanisms that enable metazoans to maintain numerous components of their internal environment within "physiological" limits offer some of the best examples of the function of the ICS, with the CNS as its controller.

The term "homeostasis" here will be used in a broader sense to comprise not only the maintenance of the steady state of the body fluids but also the maintenance of the normal structure and function of the animal organism in its entirety.

Modern physiology provides ample evidence on the CNS control of vital functions of all the organs and organ systems in animals, including heart work, blood circulation and pressure, respiration, digestion, and endocrine activity. To illustrate, let us only consider some well-known examples of CNS control in homeostasis, including control of expression of nonhousekeeping genes in cells throughout the animal body.

Neural Control of Water Content

The drop of water content in metazoans is sensed as a rise above a threshold of the osmotic pressure (rise in the concentration of particles) of body fluids. This is the function of osmoreceptor neurons in the circumventricular organ *organum vasculosum* of the *lamina terminalis* and *subfornical nucleus*, which can detect changes in osmolality and "transmute" these changes into electrical signals (Verney, 1947). The animal brain also receives afferent information from peripheral osmoreceptors along the alimentary tract and blood vessels. The information on osmolality is transmitted to many parts of the brain, where it is integrated with information on the blood volume, Na^+ concentration in body fluids, and blood pressure for optimizing the osmoregulatory response to the overall body homeostasis (Bourque, 2008). In response to increased osmolality, respectively via the *median preoptic nucleus* and directly, these osmoreceptors send signals to the "osmometers," magnocellular neurosecretory cells (MNCs), in the hypothalamic paraventricular nucleus (PVN) and supraoptic nucleus.

> *VP-releasing MNCs in the SON and PVN are thus the "command" neurons that regulate diuresis.*
>
> *Bourque (2008)*

Hypothalamic magnocellular neurons interpret the information on the increased osmolality as "thirst," to which they respond by releasing vasopressin (antidiuretic neurohormone) that is released into the body fluids via the pituitary. The hormone acts as antidiuretic by binding its receptors on the membranes of the cells of the distal kidney tubules. Elevated levels of vasopressin stimulate the reabsorption of water in kidney tubules and production of highly concentrated urine, leading to lower osmotic pressure. When the water content increases and the osmolality drops, this is again sensed by osmoreceptor neurons, and in a negative feedback loop, MNCs inhibit vasopressin secretion, leading to less-concentrated urine and increased urination (Figure 1.1).

Neural Control of Glucose and Insulin Biosynthesis

A hypothalamic mechanism of regulation of glucose level, a "glucostat" involving the endocrine pancreas, liver, and adrenal gland was identified almost three decades ago (Benzo, 1983). The hypothalamus plays a critical role in hypoglycemia-induced responses of adrenomedullary, sympathoneuronal, and other systems (Pacak and Palkovits, 2001). Experimental lesions of the hypothalamic PVN result in hypoglycemia. From PVN starts a descending pathway to the intermediolateral cell column and the dorsal vagal nucleus of the vagus innervating ventromedial hypothalamus (VMH) stimulates pancreatic secretion. Electrical stimulation of the hypothalamus inhibits insulin secretion (Li et al., 2003) neurons with pancreatic projections (Figure 1.2).

Glucose sensors for detecting hypoglycemia have been identified in the forebrain and the brain stem and especially in the VMH, which responds to the hypoglycemic state by activating hormonal mechanisms, including secretion of glucagon and food intake. The lateral and posterior parts of the hypothalamus release the neuropeptide orexin (Miyasaka et al., 2002), which stimulates food intake and via the vagal efferent nerve stimulates exocrine hormonal secretion in the pancreas (Wu et al., 2004). Ablation of the lateral hypothalamus inhibits the vagal pancreatic nerve firing and pancreatic secretion.

The nervous system also exerts a direct control on the insulin secretion via the sympathetic and parasympathetic pathways (Figure 1.3). Insulin-producing cells of the Langerhans islets in pancreas are innervated by vagal cholinergic nerves and by sympathetic nerves. The sympathetic activity is mediated by norepinephrine (NE),

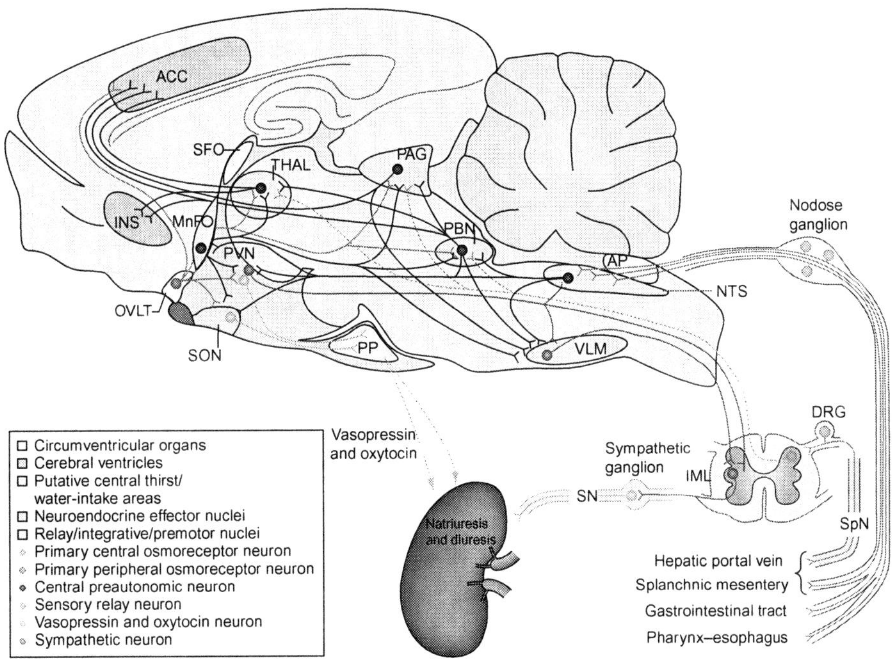

Figure 1.1 Osmoregulatory circuits in the mammalian brain and the periphery. Sagittal illustration of the rat brain, in which the relative positions of relevant structures and nuclei have been compressed into a single plane. Only structures that have been directly implicated in the osmotic control of osmoregulatory responses are shown. Neurons and pathways are color-coded to distinguish osmosensory, integrative, and effector areas. Although visceral sensory pathways that relay information from dorsal root ganglion neurons are known to ascend through the spinal cord, specific evidence that peripheral osmosensory information ascends through this route is only partial, and this tract is therefore illustrated as a dashed line. *Abbreviations*: ACC, anterior cingulate cortex; AP, area postrema; DRG, dorsal root ganglion; IML, intermediolateral nucleus; INS, insula; MnPO, median preoptic nucleus; NTS, nucleus tractus solitarius; OVLT, organum vasculosum lamina terminalis; PAG, periaqueductal gray; PBN, parabrachial nucleus; PP, posterior pituitary; PVN, paraventricular nucleus; SFO, subfornical organ; SN, sympathetic nerve; SON, supraoptic nucleus; SpN, splanchnic nerve; THAL, thalamus; VLM, ventrolateral medulla. (For interpretation of the references to color in this figure legend, the reader is referred to the web version of this book.)
Source: From Bourque (2008).

via the adrenergic receptor, α_{2a}AR, and the parasympathetic activity—by acetylcholine (Ach) via its receptor, M3R. The parasympathetic system is activated in response to the elevation of glucose level, after feeding, leading to increased production and secretion of insulin by β-cells, whereas a drop in the glucose level triggers activation of the sympathetic nervous system and resulting decline in the insulin synthesis and secretion (Norman and Litwack, 1997; Mitrani et al., 2006).

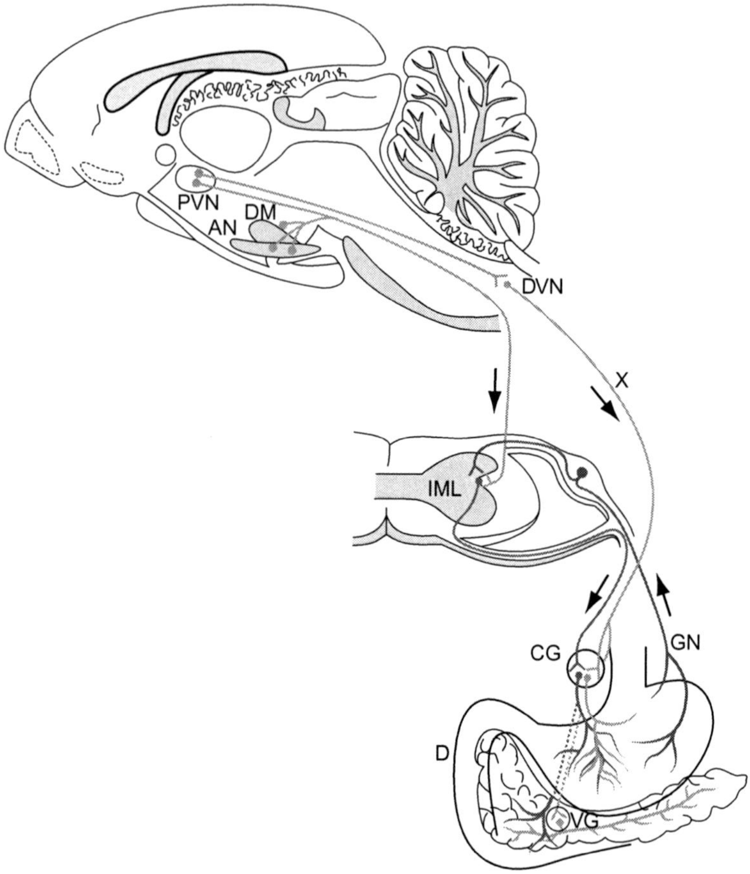

Figure 1.2 Proposed neuronal pathways involved in the response to experimental hypoglycemia by a single insulin injection. Hypothalamic projections to autonomic centers, and autonomic innervation of the pancreas. *Abbreviations*: AN, arcuate nucleus; CG, celiac ganglion; D, duodenum; DM, dorsomedial nucleus; DVN, dorsal motor nucleus of the vagus; GN, sensory gastric fibers; IML, intermediolateral cell column; PVN, paraventricular nucleus; VG, vagal ganglionic cells; X, vagus nerve.
Source: From Pacak and Palkovits (2001).

The above evidence clearly demonstrates that the CNS directly controls the secretion of hormones insulin and glucagon in pancreas.

Neural Control of Thermoregulation

Regulation of body temperature in vertebrates is the function of a central mechanism, and the main thermoregulatory organ is again a part of the brain, the hypothalamus, particularly the preoptic area (POA), where the sensory input on the brain temperature and core temperature are integrated (Boulant, 2000). Other parts of the

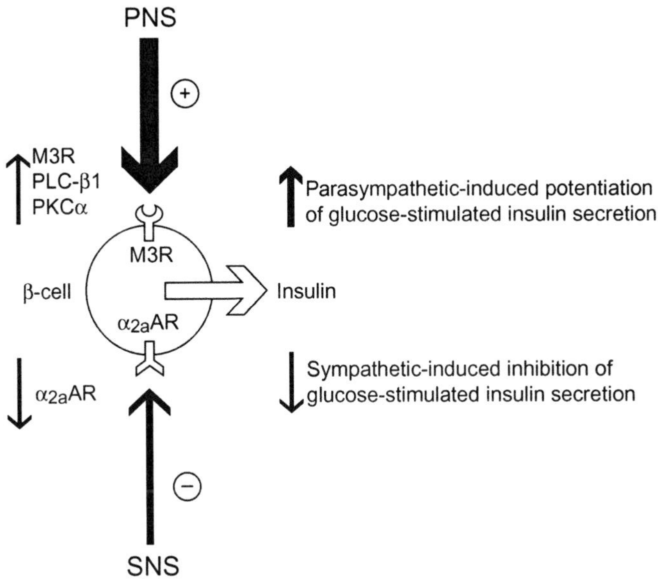

Figure 1.3 Summary of study findings related to β-cell insulin secretion pathways. Increased parasympathetic activity, through increased sensitivity to cholinergic stimulation and increased levels of M3R (muscarinic type 3 receptor), PLC-β1 (inositide-specific phospholipase Cβ1), and PKCα (an isozyme of the PKC family) mRNA, results in enhanced insulin secretion in response to glucose by β-cells from 12-day-old hemicastrated rats. In addition, reduced sensitivity to sympathetic stimulation results in further increases in insulin secretion, in association with decreased α_{2a}AR mRNA levels.
Source: From Mitrani et al. (2006).

CNS, such as the brain stem and spinal cord, are also involved in thermoregulation. There are three functionally different groups of neurons in the hypothalamic POA (Figure 1.4): ~20% warm-sensitive neurons (W), which increase the firing rate (FR) in response to higher temperature and decrease it when the brain temperature drops. Cold-sensitive neurons (C) that represent ~10% of POA neurons are normally inhibited from the warm-sensitive neurons but are activated when the FR of warm-sensitive neurons decreases, during cooling. The remaining 70% of neurons of the POA are temperature insensitive (I) (Boulant, 2000).

High temperature as well as experimental warming of POA decreases thermogenesis via inhibition of thyroid gland secretion and increases heat loss via neural cholinergic stimulation of sweating, vasodilatation, as well as panting and behavioral abandonment of the warm environment. By contrast, lower temperatures and experimental cooling of POA stimulate increased thermogenesis by inducing secretion of thyroid hormones, by stimulating heat production by neural (adrenergic) induction of constriction of skin blood vessels, as well as by neural stimulation of shivering and movement toward warmer places.

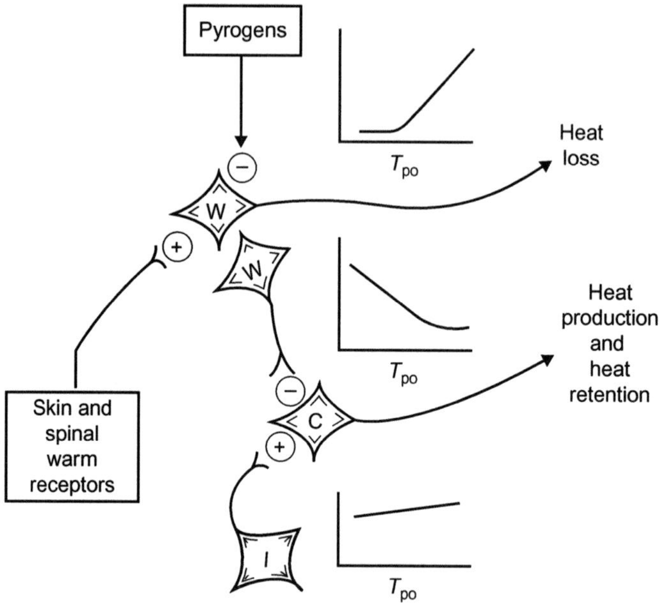

Figure 1.4 FR activity of three types of preoptic neurons. Warm-sensitive neurons (W) increase their FRs during increases in preoptic temperature (T_{po}). Warm-sensitive neurons also synaptically inhibit cold-sensitive neurons (C), which increase their FRs during decreases in T_{po}. It has been postulated that some cold-sensitive neurons play a partial role in heat-production and heat-retention responses, which also increase during decreases in T_{po}. Temperature-insensitive neurons (I) show little change in their FRs during changes in T_{po}, and these neurons may provide tonic synaptic input that is compared with synaptic input from warm-sensitive neurons. Warm-sensitive neurons are also affected by pyrogens and afferent synaptic input from skin and spinal thermoreceptors. *Symbols*: +, excitation; −, inhibition. *Source*: From Boulant (2000).

Soon after the discovery of thermosensory neurons in the hypothalamus, in 1963, a set point for body temperature was discovered in this brain gland (Hammel et al., 1963). How metazoans establish set points is poorly understood. However, Hammel et al. discovery of the existence of set points for normal temperature in the brain of vertebrates provides some important clues for understanding the mechanisms of the change of temperature set points.

The set point temperature, T_{set}, is determined by a reference signal that is the constant FR of the temperature-insensitive neurons (I) (Figure 1.5) while the effector neurons receiving antagonistic excitatory and inhibitory signals from W and I function as an "error-comparator" for generating their adaptive neural output. The cold-sensitive neuron (C) is not cold-sensitive *per se*, but it is the inhibitory input from the warm-sensitive neuron (W) that provides it with cold-sensitivity. The thermosensitive neurons of POA also receive information on the body temperature from the spinal cord, via the brain stem. The set point temperature, T_{set} for effector neurons

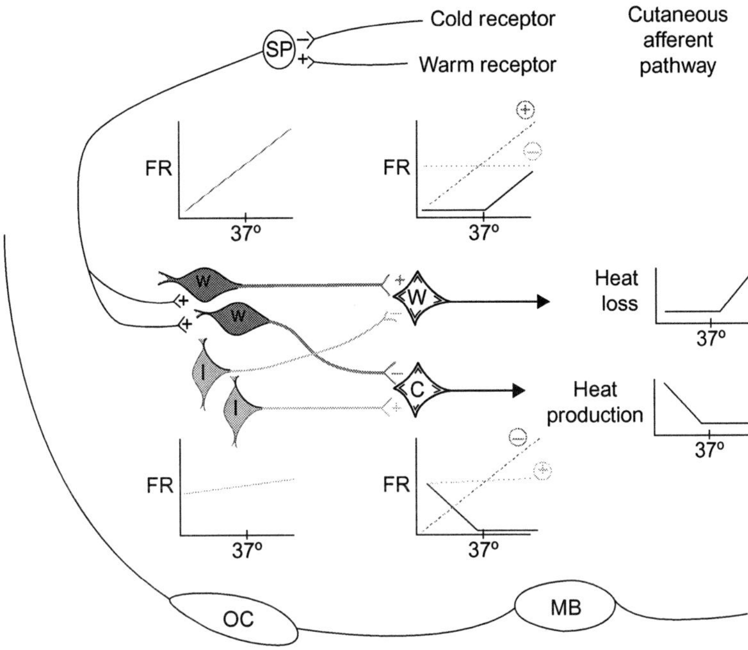

Figure 1.5 Modification of a neuronal model proposed by H.T. Hammel (1965) to explain set-point temperature regulation by a synaptic network of hypothalamic neurons. W, warm-sensitive neuron; I, temperature-insensitive neuron; w, heat loss effector neuron having synaptically derived warm sensitivity; C, heat production effector neuron having synaptically derived cold-sensitivity; SP, dorsal horn spinal neuron; OC, optic chiasm; MB, mammilary body. Graphs show the FRs of each neuron, as well as thermoregulatory responses (*right*) during changes in hypothalamic temperature. Dotted lines indicate the frequency of excitatory (+) and inhibitory (−) synaptic inputs.
Source: From Boulant (2006).

(w and c), is the temperature at which the excitatory and inhibitory inputs are balanced, and any change in the FR of W and I neurons will cause a shift in the T_{set}; when the FR increases, the T_{set} will be higher, and the contrary will happen when the FR decreases. Several internal and external stimuli that increase the FR of W and C neurons are known to lead to corresponding elevations of the T_{set}.

Temperature set points, thus, are not permanent but "adjustable" and can adaptively change during the individual development. For example, rabbits acclimated to lower temperatures show lower temperature set points than the controls (Tzschentke and Nichelmann, 1997). Lower-than-optimal temperatures of incubation induce increased hypothalamic thermosensitivity and stimulate increase in the number of warm-sensitive neurons. The opposite effect is observed when bird (*Gallus domesticus*) eggs are incubated at higher-than-optimal temperatures (Tzschentke and Basta, 2002). The incubation temperature has a considerable influence on the hypothalamic thermosensitivity: during ontogeny, it is observed that birds incubated at higher than the optimal temperature (37.5°C), during the 10 first days posthatching, have higher preferred temperatures than birds incubated at lower temperatures. In birds, the epigenetic temperature adaptation, as a result of changes in incubation temperature at the end of the incubation period, is associated with changes in the thermosensitivity of the relevant hypothalamic neurons and the peripheral thermoregulatory mechanisms (Tzschentke, 2007).

The hypothalamic thermosensitivity, i.e., the hypothalamic set points for body temperature, may change during the evolution of species and even during the adult life of an individual. Under stress conditions, and depending on the degree of the stress, metazoans may modify the neural set points in order to adapt the physiology to the stressful conditions. The normal temperature set points are also changed in response to cyclical variations in hormonal levels. (Warm-sensitive neurons increase their activity in response to estrogens and cool-sensitive neurons—in response to progesterone, leading to lower and higher temperature set points, respectively.) It is possible that the fever may be an adaptive mechanism of the hypothalamus, which, in response to various bacterial endotoxins and other pyrogenic substances, raises the temperature set point (Boulant, 2000), thus inhibiting bacterial division. Boulant concluded that it is in the hypothalamic POA where

> *peripheral temperature information is compared with central temperature information. As a result of this integration, the preoptic region controls the level of output for a set of thermoregulatory responses that are most appropriate for the given internal and environmental temperatures.*
>
> *Boulant (2000)*

There is no evidence that any genes are involved in the determination of set points for the normal or abnormal temperature in warm-blooded animals or in the mechanisms of thermoregulation. Both the comparison of central and peripheral temperatures and the integration and processing of temperature information for producing the neural output that activates the mechanisms of thermoregulation are computational, hence nongenetic, epigenetic processes.

Central Control of Animal Behavior

Animal behavior comprises motor actions of metazoans in response to external and internal stimuli. It may be innate or learned. Innate behaviors are stereotyped motor acts that the animal performs in a determined sequence. Innate behaviors usually represent adaptive responses to external and internal stimuli that trigger their release. Any innate or instinctive behavior is based on the activation of a set of so-called fixed action patterns determined by specific circuits in the nervous system triggered by external stimuli known as *sign stimuli or releasers* (from Konrad Lorenz's concept of *angeborener auslösender Schema*—innate releasing pattern).

Learned behaviors, which only are characteristic of metazoans (with some exceptional cases observed in eukaryotic unicellulars), are modifications of normal behaviors as a result of experience. While learning makes use of species-specific behavioral elements, it is experimentally demonstrated that learning may significantly influence (strengthen or weaken) innate behaviors (Mery and Kawecki, 2004), and this results from modifications of neural circuits/networks for innate behaviors. Imprinting is a classical example of the close relationship between the learned and innate behaviors. At the basis of learning is memory, which is an epigenetic phenomenon.

Neural Control of Locomotion

Vertebrate locomotion results from coordinated movement patterns produced by central pattern generators (CPGs) in the CNS. It is estimated that mammal locomotion involves activation of hundreds of muscles, and, in particular, phases of the movement cycle, which are under control of the spinal cord CPGs (Grillner et al., 1995). Similar CPGs regulate swimming in lampreys by alternatively activating motor neurons in both sides of each of 100 segments, successively but with a phase delay, thus producing an undulatory locomotion. Initiation of locomotion is function of brain stem neurons. They project to the spinal cord, thus activating motor neurons and interneurons (Grillner et al., 1995; Figure 1.6).

Locomotion in metazoans is epigenetically regulated by the activity of specific neural circuits, and there is no evidence that any differences in genes may be responsible for different forms of locomotion in metazoans.

Neural Determination of Monogamy in Prairie Voles

The prairie vole, *Microtus ochrogaster*, is a monogamous species, displaying strong pair bonding. Exposure of female voles to male olfactory cues stimulates estrus and mating-induced pair bonding. The social attachment in voles is triggered by reception of olfactory cues by sensory neurons of the vomeronasal organ. This is proven by the fact that female prairie voles with experimentally lesioned vomeronasal organs, induced to estrus and mating by administration of estrogens, do not exhibit monogamy (Curtis et al., 2001). Sensory neurons transform the olfactory signal into electrical signals that, via vomeronasal nerves, are sequentially transmitted for processing to the accessory olfactory bulb, the medial amygdala, the stria terminalis, and the anterior hypothalamus (Keverne, 1999), with the latter determining the monogamous behavior.

The social attachment is maintained by a distinct pattern of expression of receptors for oxytocin (Ross et al., 2009), vasopressin (Young et al., 1998), and dopamine (Smeltzer et al., 2005), but is modulated by the neuropeptide corticotropin-releasing factor (DeVries et al., 2002) in monogamous and in promiscuous species. The neural circuits related to the interaction between oxytocin and dopamine responsible for monogamous behavior are located in the brain area known as the *nucleus accumbens* (Liu and Wang, 2003; Figure 1.7).

No changes in genes or regulatory regions are known to be involved in determining monogamy or lack of it in prairie voles.

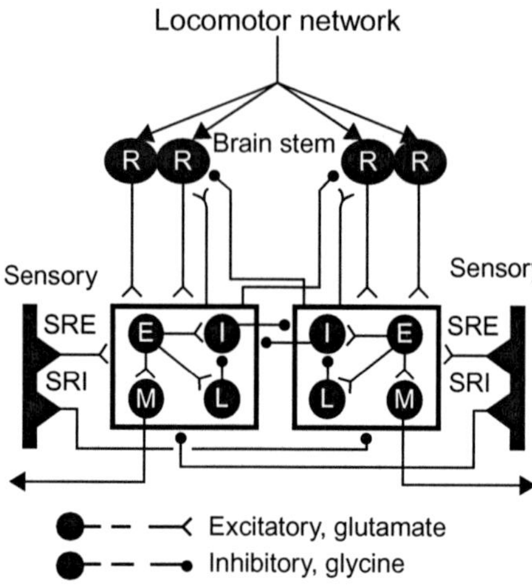

Figure 1.6 The segmental neuronal network that coordinates locomotion. A schematic representation of the brain stem, segmental and sensory components of the neural system that generate burst activity. The reticulospinal glutamatergic brain stem (R) project to the spinal cord and excite all the spinal neurons that are depicted within the black box. The excitatory interneurones (E) excite all types of spinal neurones within the box, i.e., inhibitory glycinergic interneurones (I) that cross the midline and inhibit all neurones within the contralateral box, the lateral interneurone (L), which inhibits the I interneurone, and motoneurones (M), which are cholinergic. The stretch-receptor neurones are of an excitatory (SRE) type that excites neurones within the ipsilateral box and an inhibitory (SRI) type that inhibits all neurones within the contralateral box. Synapses that are shown to terminate on the frame of the box indicate effects that are common to all neurones within the box. The excitatory and inhibitory effects from the spinal neurones (box) back to the phasic neurones are indicated as monosynaptic but might have additional relay interneurons. The tonic neurons receive no feedback from the spinal cord. Note that only one cell of each type is indicated in the scheme, although each cell represents a group of cells.
Source: From Grillner et al. (1995).

Central Control of Metazoan Morphology

Examples presented earlier in this chapter unambiguously show that a number of physiological/homeostatic parameters and behaviors are under central neural control and regulation. Indeed, the neural control of physiological functions and behavior in animals is textbook knowledge. If two aspects of the metazoan phenotype, the animal behavior and animal physiology, are centrally regulated and maintained, it is reasonable to inquire whether the third aspect, the development of metazoan morphology, the most visible feature of metazoan evolution, may also be centrally controlled. My research and my conclusions on this issue I have made known in my previous work (Cabej, 2004). However, the scope of this book requires that I now include the basic evidence and arguments I have already used for substantiating the

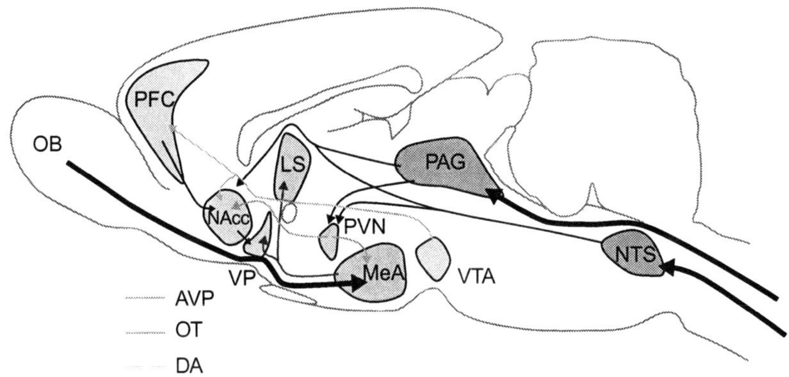

Figure 1.7 Proposed neural circuitry of social bonding in monogamous prairie voles. Somatosensory input from the penis or vagina impinge on the nucleus tractus solitarius (NTS) and midbrain periaqueductal gray (PAG), which then project to the nucleus accumbens (NAcc) and paraventricular nucleus (PVN). Olfactory information is conveyed via the olfactory bulb to the medial amygdala (MeA), where oxytocin facilitates social recognition. The MeA sends vasopressinergic projections to the ventral pallidum (VP) and lateral septum (LS), which is also involved in the formation of social memory. Activation of the ventral tegmental area (VTA) results in dopamine (DA) release within the prefrontal cortex (PFC) and NAcc. The simultaneous activation of the dopaminergic and peptidergic pathways results in the formation of the selective pair bond.
Source: From Young et al. (2005).

idea of the central neural control of animal morphology. I will present here only three examples that illustrate the CNS control of metazoan morphology and body size, but I will present an extensive discussion of the CNS control of metazoan morphology and organogenesis in Chapter 5.

The myriad of coordinated chemical and cell interactions, cell proliferation, cell differentiation, and cell apoptosis taking place in the metazoan body are information-requiring processes rather than random events. As genes are effectors and the focal point of all processes at the molecular level, one basic prediction of the hypothesis of the neural control of morphology is that expression of nonhousekeeping genes, genes that perform organismic functions, other than survival and reproduction of the cell itself, must be under central control.

The science of biology has demonstrated the existence of controls in many different levels, such as the level of genes and signal transduction pathways at the cellular level, as well as the endocrine, neuroendocrine, and neural control at organismic levels. I have shown that all these levels of control form a single hierarchical control system with each downstream level of control being subordinate to the next upstream level of control. Hence, one would expect that by tracing back intracellular transduction pathways and signal cascades that activate them, we might finally reach the source of that information for expression of nonhousekeeping genes.

Control of Expression of Nonhousekeeping Genes

No metazoan cell lives exclusively by itself and for itself. Each of them lives a double life: as an autoregulated cell *per se*, and as a virtuoso that, under the control of extracellular signals, performs one or more specific functions not for itself, but for the organism. This double life of cells has determined two different ways of the control of gene expression in metazoans. On the one hand, as a heritage from their unicellular ancestry, there is a group of *housekeeping genes*, which are necessary for cell's own metabolism, subsistence, and reproduction. These genes, whose minimal number, depending on the metazoan species, is estimated to be in the range of a few thousands, are autoregulated in response to signals on the presence of their products and respective substrates in the cell nucleus or cytoplasm. Such autoregulatory circuits are demonstrated to exist and be operational in both unicellulars and multicellulars.

On the other hand, there is a group of *nonhousekeeping genes*, which are responsible for specific functions of each cell within the general scheme of the division of labor at the organismic level. These specific functions (mainly the synthesis and secretion of specific proteins) of differentiated cells are not related to the needs of the cell itself, but to the demands of the organism. No cell can know what the organism needs at any moment in time. Instructions to the differentiated cells for performing those functions are obligatorily extracellular in origin.

Examples of extracellular signals known to induce expression of nonhousekeeping genes are numerous. The orderly activation of *Hox* genes is known to specify the establishment of anterior–posterior axis during gastrulation in vertebrates. But, adequate empirical evidence demonstrates that the immediate source of their activation is an extracellular signal, the hormone retinoic acid (RA). The hormone is a common immediate regulator of the activity of almost all of the known *Hox* genes, since the earliest stages of the embryonic development and during the postnatal development in metazoans (Conlon, 1995; De Luca and Ross, 1996; Marshall et al., 1996; Clagett-Dame and Plum, 1997; Cupp et al., 1999; Malpel et al., 2000). In many species, RA is provided maternally, but from the stage of early gastrula it is secreted by Hensen's node (Hogan et al., 1992). Other hormones secreted by endocrine glands, as well, are known to act as switches for various *Hox* genes. So, for example, members of the abdominal B (AbdB) *Hox* gene family, which control the morphogenesis of the reproductive tract in vertebrates, have been demonstrated to be themselves under the hormonal control of the estrogen and progesterone (Ma et al., 1998). Administration of diethylstilbestrol in mice, via estrogen receptors (ERs), alters the pattern of expression of several *Hox* genes involved in the patterning of the reproductive tract, leading to developmental anomalies (Block et al., 2000).

No nonhousekeeping gene is master of its destiny; all of them are downstream elements of signal cascades. Far from being "selfish" or "altruistic," a nonhousekeeping gene is subordinate to upstream elements of signal cascades. It is these upstream, epigenetic, signals that determine whether a nonhousekeeping gene will be expressed or not.

Chromatin Remodeling in Control of Gene Expression

The advent of chromatin in unicellular eukaryotes is a crucial event in the evolution of the control of gene expression. Adequate evidence demonstrates that acetylation/deacetylation and methylation of histones by remodeling the chromatin exert an epigenetic control on gene expression (Jenuwein and Allis, 2001; Nakayama et al., 2001; Reik et al., 2001; Rideout et al., 2001; Arney et al., 2002). Acetylases, enzymes that induce acetylation of histones, form complexes with transcription factors, making the binding of the latter to the regulatory regions of the DNA and the expression of respective genes possible. Deacetylase enzymes, on the other hand, by deacetylating histones cause them to wrap the DNA and prevent transcription of genes (Jenuwein and Allis, 2001). Acetylation/deacetylation enzymes represent the link between the transcription factors and genes.

Due to their small molecular mass, some hormones, such as steroid hormones, RA, and thyroid hormones, are essentially involved in the process of chromatin remodeling (Bhattacharyya et al., 1997; Minucci et al., 1997; Collingwood et al., 1999; Sachs and Shi, 2000; Aranda and Pascual, 2001). Mediators of the effects of these hormones on gene expression are their nuclear receptors. In the absence of the nuclear hormone, its specific receptor recruits a histone deacetylase complex, thus compacting the chromatin and preventing gene transcription. In the presence of the hormone, a complex ligand-receptor forms, which can recruit a protein with acetylase activity that looses the chromatin structure, making the gene transcription possible.

It has been demonstrated that a histone acetyltransferase (HAT), which acetylates histones H3 and H4, represents a downstream element of a signal cascade starting with electrical signals in the retina, and via retinohypothalamic tract (glutamate) stimulates a transduction pathway leading to expression of clock genes, thus regulating the circadian physiology.

In order to carry out chromatin remodeling, acetylases and deacetylases first must be activated by upstream signals, i.e., by transcription factors. As regulatory proteins, transcription factors can activate (or inhibit) transcription of specific genes by binding to their regulatory regions (promoters and enhancers). But to do so they first need to be activated. As shown earlier, they may be activated by combining with specific inducers (e.g., nuclear hormones). Other times, their activation results from phosphorylation of their molecules by terminal elements of signal transduction pathways, which link transcription factors with extracellular signals. These pathways consist of chains of intracellular signaling proteins. They are activated by specific extracellular signals. When an extracellular signal (protein hormone, growth factor, or secreted protein) binds to its specific cell membrane receptor, it activates enzymatically the cytoplasmic portion of the receptor. In this active form, the receptor phosphorylates the first protein of the pathway, thus starting a chain of phosphorylation of other proteins downstream the pathway, including the transcription factor. The phosphorylation transforms the transcription factor into an active form that can induce expression of specific gene(s). In some cases, signals from the brain, via signal cascades, are known to be responsible for chromatin remodeling (Figure 1.8). So, for example, signals from the nonhypothalamic

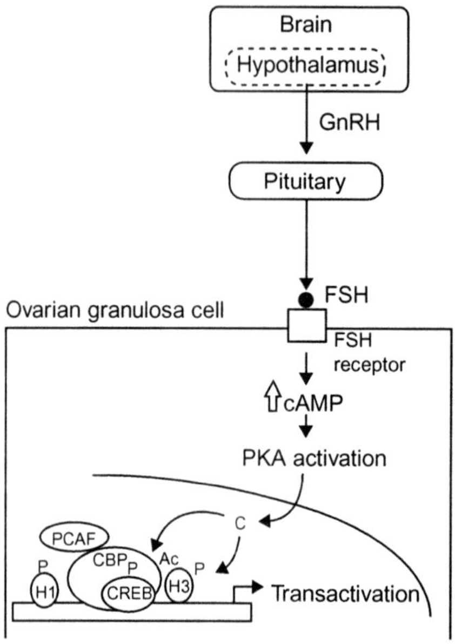

Figure 1.8 Proposed model for FSH signaling to activate histone H3. FSH via the catalytic (C) subunit of PKA catalyzes both histone H3 and CREB phosphorylation as well as histone H1 phosphorylation. Phosphorylated CREB, possibly in conjunction with SF-1 and Sp1, recruits CBP and possibly other HATs such as PCAF, inducing histone H3 acetylation and gene transactivation. *Abbreviations*: CBP, p300/CREB-binding protein; CREB, cAMP-response element-binding protein; FSH, follicle-stimulating hormone; HAT, histone acetyltransferase; PCAF, CBP-associated factor; PKA, protein kinase A; SF-1, steroidogenic factor-1.
Source: Modified from Salvador et al. (2001).

brain induce the hypothalamus to secrete GnRH (gonadotropin-releasing hormone), which stimulates the pituitary to secrete follicle-stimulating hormone (FSH). The latter, in turn, binds its membrane receptor on ovarian granulosa cells and via a signal transduction pathway induces phosphorylation and acetylation of histone H3, leading to a selective gene expression (Salvador et al., 2001).

It is observed that even a pure neural process as novel taste learning elicits histone acetylation and chromatin remodeling (Swank and Sweatt, 2001). Contextual fear conditioning (animals learned to associate a new environment with an aversive stimulus when placed again in that environment in the absence of the aversive stimulus they show a freezing response) is associated with acetylation of the histone H3 in the hippocampus (Levenson et al., 2004).

A number of neuroactive substances, antipsychotics as valproic acid and mood stabilizers, widely used in clinical practice, produce their effects by stimulating/inhibiting synthesis of acetylases/deacetylases and inducing chromatin remodeling and gene expression in neurons. Benzamide MS-275 (Simonini et al., 2006) and valproic acid (Jeong et al., 2003) inhibit deacetylases, dopamine antagonists induce acetylases (Li et al., 2004), and so do pilocarpine and kainic acid (Crosio et al., 2003).

Extracellular Control of Signal Transduction Pathways and Transcription Factors

Extracellular signals for gene expression belong to two main classes: hormones (including growth factors and neuropeptides) and secreted signal proteins. Both of these groups of inducers are conveyors of upstream messages. Protein hormones, having a larger mass, cannot enter the cell and bind specific cell membrane receptors. These receptor molecules are proteins with a transmembrane and a cytoplasmic region. Binding of the protein causes phosphorylation of the cytoplasmic domain of the receptor molecule. This leads to phosphorylation and activation of the first protein and then, sequentially, of the rest of the elements of the transduction pathway. Phosphorylation of the terminal element of the pathway makes it able to enter the nucleus where it activates a specific transcription factor by phosphorylating it. One of the most important signal transduction pathways is the protein kinase C (PKC) pathway (10 proteins of the PKC family are known so far). The pathway is used by various hormones, neurotransmitters, and growth factors.

Having shown the causal link between controls at the levels of genes, chromatin remodeling, transcription factors, and signal transduction pathways at the lower portion of the causal chain leading to activation of nonhousekeeping genes, now let us attempt to reconstruct and visualize the upstream signal cascade. The next upstream step is represented by secreted proteins, growth factors, and protein hormones.

Hormonal Control of Secreted Proteins and Growth Factors

The most important families of secreted proteins and growth factors act by binding their specific receptors on the cell membrane, thus activating intracellular signal transduction pathways and transcription factors. Adequate evidence shows that the synthesis of secreted proteins and growth factors is under control of hormones of the terminal endocrine glands, the pituitary, or neuropeptides.

The Wnt family of secreted glycoproteins is involved in modulating cell proliferation, cell polarity, cell differentiation, and cell migration during the embryonic and postnatal development. Their specific cell surface receptors are frizzled proteins and the intracellular β-catenin, downstream the pathway (Cadigan and Nusse, 1997). Wnt proteins act as mediators of hormone functions. So, for example, formation of the mammary gland during adolescence, puberty, and pregnancy is under hormonal control especially of estrogen and progesterone. But it is *Wnt-4* that mediates the morphogenetic functions (ductal branching and budding) of the progesterone in the development of the mammary gland during puberty and pregnancy. Via its nuclear receptor in the mammary epithelial cells, progesterone stimulates paracrine expression of the *wnt-4* that induces the ductal side branching of the mammary gland (Brisken et al., 2000; Robinson et al., 2000; see later Figure 1.10).

The epidermal growth factor (EGF) family. The mitogenic effect of the estrogen, manifested in the growth and cell differentiation, is mediated by the growth factor, EGF, and administration of EGF alone mimics all of the effects of the estrogen (Nelson et al., 1991). The function of testosterone in differentiation of the reproductive tract is also mediated by EGF, which phosphorylates its own receptor and other proteins, resulting in activation of genes that make the formation of the Wolffian duct and its components, the epididymal duct, and seminal vesicle, possible (Gupta et al., 1991). Estrogen induces the growth response of the breast epithelial cells by activating an inactive EGF receptor signaling pathway (Briand et al., 1999). It is also observed that overexpression of *ER-α* in a cell line inhibits secretion of vascular endothelial growth factor (VEGF) and the growth of the vascular wall (Ali et al., 1999). Estrogens and androgens stimulate transcription of the *vegf* gene and secretion of VEGF protein which promotes formation of new blood vessels (Rouhola et al., 1999). Both EGF and transforming growth factor (TGF) (Nelson et al., 1991) mediate the proliferative and differentiation functions of the estrogen in the reproductive tract of rodents.

The TGF-β superfamily. Secretion of inhibin B, a member of the TGF-β family, by Sertoli cells is stimulated by FSH and inhibited by luteinizing hormone (LH) (Ramaswamy and Plant, 2001).

Expression of three growth factors, TGF-β1, TGF-β2, and TGF-β3, is stimulated by the hormone RA. RA itself is synthesized from vitamin A by enzymes aldehyde dehydrogenase 1 (Aldh 1), retinaldehyde dehydrogenase 2 (Raldh 2), and cytochrome P450 (Cyp 26), but expression of these enzyme genes in the female mouse reproductive system is also under hormonal control of gonadotropin and can be stimulated by administration of chorionic gonadotropin (Vermot et al., 2000).

The fibroblast growth factor (FGF) family. In mammals this family is represented by 22 secreted proteins. FGF2 mediates the cell proliferating effect of the hormone dihydrotestosterone in certain prostate cancer cells (Kassen et al., 2000). Expression of FGF-8 and sonic hedgehog (*SHH*) are necessary for the formation of mid- and upper face and brain in the chick embryo. It is demonstrated that in sites of their expression, the level of the hormone RA is increased as well. Moreover, it is observed that facial and brain anomalies in the embryos that do not express genes for those factors may be rescued by local treatment not only with FGF-8 and SHH proteins but also by RA (Schneider et al., 2001).

Insulin-like growth factor-1 (Igf-1). Growth hormone (GH) expression contributes to the tissue-specific expression of the *Igf-1* gene during development (Shoba et al., 1999). It is also suggested that the neurohormone vasoactive intestinal peptide (VIP) may function as a regulator of *Igf-1* gene expression (Hill et al., 1999; Servoss et al., 2001).

Neural Control of the Endocrine Function

We have already shown that what have been considered to be discrete controls at the gene level via transcription factors, at the epigenetic level of chromatin remodeling (via acetylation/deacetylation of nucleosomal histones), at the level of nuclear hormones and signal transduction pathways, at the level of protein hormones, secreted proteins and growth factors are only levels of a single hierarchical control system.

Identification of the remaining upstream portion of the control system mostly belongs to the history of the twentieth-century biology. Production of hormones by target endocrine glands (e.g., gonads, thyroid gland, parathyroid, adrenal, pancreas, and thymus) in vertebrates is under control of hormones secreted by the pituitary gland, a fact that led to the long-held belief that the pituitary was the master gland, which regulates secretion of hormones by the target endocrine glands.

By the middle of the twentieth century, biologists found that secretion of pituitary hormones is itself under control of a part of the brain, the hypothalamus. With more than 15 "releasing" hormones, and a number of other protein and peptide hormones (e.g., gastrin, angiotensin II, substance P, and enkephalins) secreted by this gland in higher vertebrates, the hypothalamus controls the synthesis of most of the pituitary hormones and appeared to be at the top of the hierarchy of the neuroendocrine system, but it is not. The hypothalamus itself is "instructed" to secrete its hormones by chemical and electrical signals it receives from brain stem centers (Scott et al., 1999), the hippocampus (Lathe, 2001), the limbic system, and aminergic or cholinergic neurons of various brain centers (Norman and Litwack, 1997).

The hypothalamus is not the only producer of hormones (neuropeptides) in the nervous tissue. Numerous neuropeptides are secreted by various neurosecretory cells in many brain centers, a fact that led to the concept of the "endocrine brain." Neurosecretory cells in various organs, such as the gonads, pancreas, peripheral nerves, and gastrointestinal tract, also produce important neuropeptides.

In distinction from the hypothalamic hormones that perform their functions via the pituitary-terminal endocrine glands axis, these neuropeptides act directly on target cells by binding to their membrane receptors. This is the earliest form of the neural control, which evolved in lower eumetazoans and is still involved in their growth, metabolism, reproduction (Strand, 1999), morphogenesis, organogenesis, and regeneration. This control, characteristic for extant primitive metazoans, such as cnidarians and other lower invertebrates, is known as *hormonal system of the first order*, and, as part of the ICS, it is still conserved in the higher vertebrates, including humans. An example of direct control via neuropeptides is the control (independently of insulin) of the basal glucose blood level in rats by the suprachiasmatic nucleus (la Fleur et al., 1999).

At an intermediary and evolutionarily more advanced position is the endocrine system of other invertebrates. In insects, for instance, the prothoracic gland that secretes ecdysteroids is under control of brain neurohormone prothoracicotropic hormone (PTTH) and other neuropeptides, while the secretion of juvenile hormone by corpora allata (CA) is stimulated by neurohormones allatotropins (and is inhibited by allatostatins), by neurogenic signals via *nervi corporis allati I* (NCA-I) (from *corpora cardiaca*) and *nervi corporis allati II* coming from the subesophageal ganglion, and probably by other brain neuropeptides.

Numerous examples of brain signals regulating hormonal secretion in the hypothalamus (sometimes in the pituitary, as well) are known. So, for example, *locus coeruleus* in the brain stem exerts an excitatory influence on the hypothalamic–pituitary–adrenal axis by promoting neuroendocrine mechanisms controlling the physiological mechanisms of the acute stress (Ziegler et al., 1999). Various receptors in the hippocampus bind ligands that reflect the physiological status of the organism (e.g., stress, blood pressure, ion balance, and reproductive activity) (Lathe, 2001), an interoceptive process that enables the brain to constantly monitor the physiological state of body fluids. By processing that input, it generates its output in the form of chemical/electrical signals that are sent to the hypothalamus. In response to that output and afferent information it receives from other parts of the body on the state of physiological parameters, the hypothalamus and other brain centers, activate signal cascades for restoring normal levels of various homeostatic parameters. It is demonstrated that three neurotransmitter systems are directly (not via the hypothalamus) involved in regulation of the normal pulsatile LH release by the pituitary in the female guinea pigs (Gore and Terasawa, 2001). In experiments on ewes, Scott et al. (1999) have shown that the noradrenergic neurons in certain regions of the brain stem possessing ERs sense the level of estrogen, process that input, and project the information to the GnRH neurons in the hypothalamus.

The brain–hypothalamic–pituitary pathway is used for secretion of several pituitary hormones. For example, the GH is synthesized in response to hypothalamic stimulatory- and inhibitory-releasing neurohormones, GH-releasing hormone and GH releasing-inhibiting hormone, also known as somatostatin (SS) (Glavaski-Joksimovic et al., 2002). In turn, GH by binding its receptor, GHR, in the presence of other compounds, phosphorylates other proteins of the pathway, leading to activation genes for various proteins, including the IGF.

The described control of expression of the gene in muscles, liver, and adipose tissue is schematically presented in Figure 1.9.

The diagram presents not a signal cascade but the main neuroendocrine pathways involved in gene expression. It is a systemic response of the organism to external/internal stimuli, and the diagram comprises some basic components and features of the ICS in vertebrates. Note that the direction of the flow of information for adaptive changes is from the CNS all the way down to the target cells and tissues.

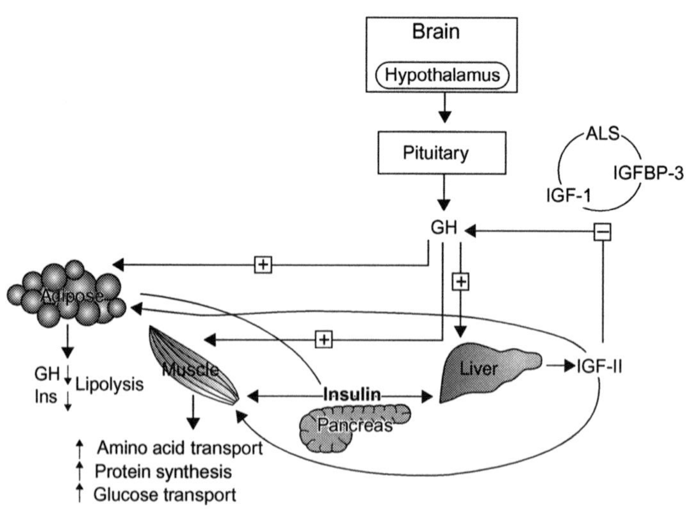

Figure 1.9 Diagram of regulation of insulin, GH, and IGF-1. *Abbreviations*: ALS, acid-labile subunit; GH, growth hormone; IGF-1, insulin-like growth factor-1; IGFBP-3, IGF-binding protein-3.
Source: Slightly extended from Clemmons (2004).

The mechanism of expression of nonhousekeeping genes in metazoan organisms includes not only linear signal cascades but also cross talk between cascades, activation of neural nonhypothalamic pathways, and downstream gene regulatory networks involved in complex interactions between them.

Summarizing the evidence presented so far, it may be said that specific hormones control expression of the growth factor- and secreted protein genes and/or activation of their receptors and, consequently, expression of nonhousekeeping genes. Secretion of the hormones of the target endocrine glands is induced by specific pituitary hormones, which in turn are induced by specific hypothalamic neurohormones. The ultimate upstream signals in this hierarchic system of control are signals from various parts of the brain that determine secretion of the hypothalamic "releasing" hormones.

It may be argued that no generalization on the extracellular control of expression of nonhousekeeping genes can be made because neither all the signal cascades triggering transcription of nonhousekeeping genes have been demonstrated to be centrally regulated, nor are all metazoan species investigated in this respect. I remind the reader that the same anti-inductive judgment could be used against most of the general concepts of modern biology. The common origin of metazoans and the amount of evidence accumulated so far provides sufficient grounds to infer that the *extracellular and central regulation is the general mode of transcription of nonhousekeeping genes in metazoans*. Nature does not make tricks.

Examples of Central Control of Metazoan Morphology

Now, let us consider just three representative examples of the CNS control of the *development* of animal morphology by visualizing the signal cascades (including associating feedback loops and cross talk) and intracellular pathways in the process of the translation of brain signals into animal morphology. An extensive review of the neural control of the development of metazoan organs and morphology is presented in Chapter 5 (*Neural Control of Postphylotypic Development*). Hence, for illustration, I will present here just three examples of the neural control of organogenesis and body size.

Development of the Mammary Gland

The mammary gland is a unique mammal structure that evolved some 200 million years ago to provide the mammalian newborns with milk. Although mammals are born with a mammal anlage, the gland develops predominantly after the puberty and *postpartum* (Hennighausen and Robinson, 1998).

Mammogenesis, development of the mammary gland, is under neurohormonal control (Akers, 2002). The gland develops under a complex control of systemic and local hormonal signals from the pituitary (prolactin) and the ovary (estrogen and progesterone). More than 100 genes are involved in the process of mammary gland

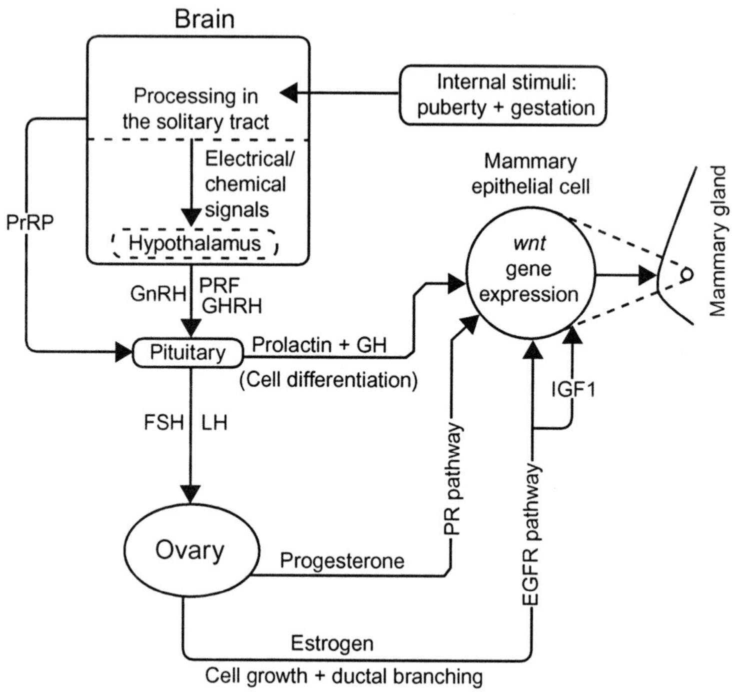

Figure 1.10 Simplified diagram of the development of the mammary gland during puberty and gestation in mammals. All of the developmental pathways converge to expression of *wnt* and other genes in mammary epithelial cells. *Abbreviations*: FSH, follicle-stimulating hormone; GnRH, gonadotropin-releasing hormone; IGF-1, insulin-like growth factor-1; LH, luteinizing hormone; PRF, prolactin-releasing factor, in postpartum animals; PrRP, prolactin-releasing peptide.

development (Hennighausen and Robinson, 2005). Local mediators of these hormones are growth factors such as Wnt (Sassoon, 1999), EGF, TGF-β proteins, and hormones like VIP and hypothalamic thyrotropin-releasing hormone (Gilbert, 1997). The developmental pathway starts in the CNS (Figure 1.10): in response to internal signals related to the activation of the GnRH pulse generator in the hypothalamus, and especially to gestation. Critical for the development of the gland is the release of a neuropeptide, prolactin-releasing peptide by neurons of the caudal part of the solitary tract. The neurohormone, by acting directly on the pituitary, or indirectly, by stimulating secretion by the hypothalamus of the prolactin-releasing factor and prolactin-releasing inhibition factor, induces secretion of pituitary prolactin.

Central Determination of Body Mass

Besides the known hormonal mechanisms of the control of body mass (Pelleymounter et al., 1995; Halaas et al., 1995; Norman and Litwack, 1997, p. 240), recent studies suggest the existence in the vertebrate CNS of a set point for the body mass.

Intraperitoneal implantation of metabolically inert masses in the deer mouse, *Peromyscus maniculatus*, causes an equivalent compensatory loss of body weight. In the third day after the removal of the implant, the animals regained the pre-experimental body weight. Investigators argue that the changes in the body weight suggest the existence of a set point that is sensitive to changes in the perception of mass and that is transduced *via neural* pathways (Adams et al., 2001).

They also believe that numerous mechanoreceptors located within muscles and tendons that have afferent pathways to cerebral cortex could provide the input necessary for changing the set point. Since the set point is located in the brain and is related to the processing of the sensory input by the brain, the growth processes adjusting the body weight to the changed set point must start with brain signals as well (Adams et al., 2001).

Administration of some centrally acting substances, such as sibutramine (an inhibitor of serotonin reuptake), lower the body's weight set point and were for some time used for weight loss in humans. In response to administration of sibutramine, rats increase the sympathetic activity, reset and decrease levels of neuropeptide Y (NPY) and proopiomelanocortin. Since sibutramine is a serotonin reuptake blocker, rats also respond to its administration by elevating the extracellular level of serotonin in the rat hypothalamic arcuate nucleus, thus centrally inducing body weight loss (Levin and Dunn-Meynell, 2000). A similar central effect for inducing body weight loss has experimental administration of another neuroactive substance, the serotonin-releasing agent, fenfluramine (Fantino et al., 1986).

Nijhout (2003) reasons that the developmental control of body size essentially is control of the moment when to stop growing and in insects this is related to the timing of the secretion of ecdysteroid hormones, which depends on the timing of secretion of the brain hormone, PTTH, which, in turn, depends on the brain photoperiodic clock. While insulin-like peptides/insulin growth factors, produced mainly by neurosecretory cells, stimulate the insect growth, the brain controlled secretion of ecdysone and its receptor counteracts that effect in a neurally controlled circuit (Colombani et al., 2005).

For further information on the neural control of the body mass, see Sections Neural Control of the Development of Body Mass (Chapter 5) and Evolution of Body Size in *Manduca sexta* (Chapter 13).

Central Determination of Normal Obesity

In adult vertebrates, fluctuations in the body weight chiefly result from fluctuations of the fat storage. Maintenance of the normal weight implies, among other things, the existence of a mechanism that regulates the food intake. This regulation is very complex but, reduced to its essentials, the regulatory mechanism consists of specific neurons of the hypothalamic arcuate nucleus and neurons in another region of the brain. The arcuate nucleus contains two types of neurons, eating-stimulating and eating-inhibiting neurons. Two different groups of signals are received by those neurons. In response to increases in fat store, the appetite-inhibiting neurons are activated, and the appetite-stimulating neurons in the arcuate nucleus are inhibited. The signals from the appetite-inhibiting neurons will dominate the input in the neurons that control the food intake and energy expenditure. Thus, by decreasing the food intake and energy expenditure, the brain decreases the excess of fat in the body. Besides this long-term mechanism, in vertebrates there is a short-term mechanism based on the release of three kinds of hormones, ghrelin (released by the stomach), cholecystokinin, and the appetite suppressant peptide hormone PYY_{3-36}, released by the colon in an immediate response to food intake. The input of these three hormones is processed in the appetite-stimulating neurons of the arcuate nucleus and via the shown pathway (Figure 1.11) determines food intake (Batterham et al., 2002; Schwartz and Morton, 2002).

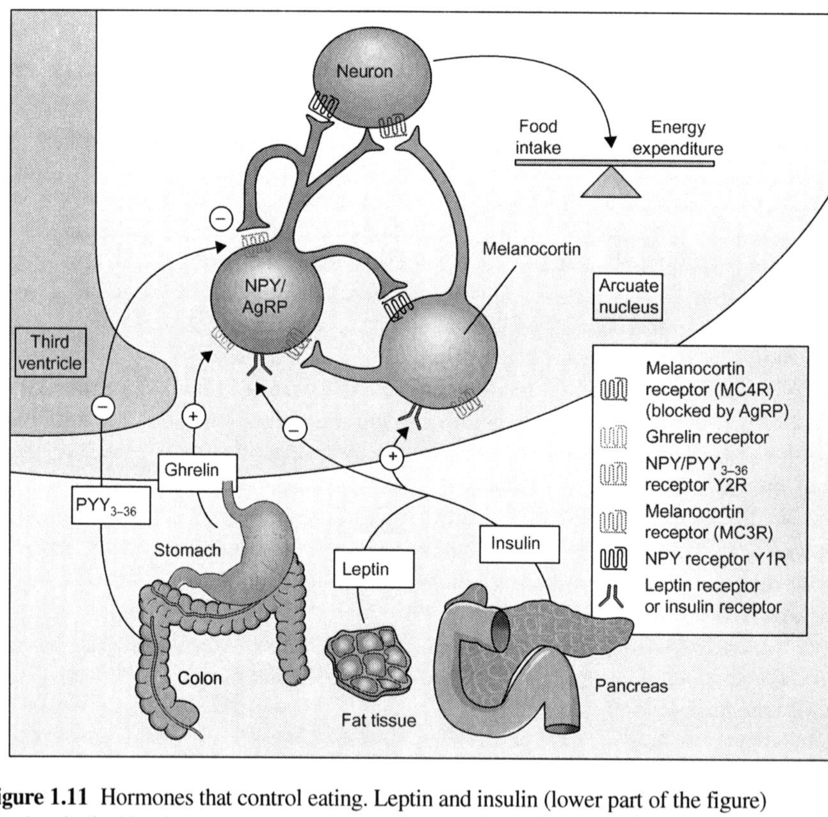

Figure 1.11 Hormones that control eating. Leptin and insulin (lower part of the figure) circulate in the blood at concentrations proportionate to body-fat mass. They decrease appetite by inhibiting neurons (center) that produce the molecules NPY and AgRP (Agouti-related peptide) while stimulating melanocortin-producing neurons in the arcuate nucleus region of the hypothalamus, near the third ventricle of the brain. NPY and AgRP stimulate eating, and melanocortins inhibit eating, via other neurons (top). Activation of NPY/AgRP-expressing neurons inhibits melanocortin-producing neurons. The gastric hormone ghrelin stimulates appetite by activating the NPY/AgRP-expressing neurons. Batterham et al. have now shown that PYY_{3-36}, released from the colon, inhibits these neurons and thereby decreases appetite for up to 12h. PYY_{3-36} works in part through the autoinhibitory NPY receptor Y2R. (For interpretation of the references to color in this figure legend, the reader is referred to the web version of this book.)
Source: From Schwartz and Morton (2002).

In response to PYY_{3-36} hypothalamic circuits modify their synaptic morphology (Batterham et al., 2002) and probably their output. Neurons expressing the receptor for melanocortin-4 (MC4-R) in the arcuate nucleus respond rapidly to signals of satiety that do not require changes in leptin and control organism's metabolic and behavioral responses to the fat diet (Butler et al., 2001).

The Integrated Control System

As discussed earlier, metazoan organisms avoid structural and functional degradation and maintain their biological identity notwithstanding continual loss of their structural order at the molecular and cytological levels. It is impossible to imagine how a structure of unmatched complexity, such as a metazoan organism, would be maintained for a considerable period of time in absence of a control system. Human experience shows that even the incomparably simpler man-made systems cannot maintain their structure and function without a control system in the form of built-in mechanisms or human supervision and operation. A control system for maintaining the structure and function is a *sine qua non* of metazoan life as it is for any other living system. The control system I am delineating for metazoans has the CNS as its controller, but nonneural, epigenetic control systems, still less studied and understood, exist in plants and unicellulars as well.

Adequate evidence, a tiny part of which was presented in this chapter, shows that there is a central neural control, which regulates not only metazoan behaviors and physiological/homeostatic variables but also expression of nonhousekeeping genes. I have also presented three examples of the central control of the development and maintenance of metazoan morphology, size, and morphometry. Based on that evidence, I argued that the maintenance of the eroding metazoan structure implies the existence in metazoans of an ICS, with the CNS as its controller, which compensates for losses and directs the restoration of the normal structure.

Extensive evidence on the neural control of the development of animal phenotype will be presented in Chapters 5 (Neural Control of Postphylotypic Development), 8 (Behavioral Adaptation to Changed Conditions of Living), 10 (Intragenerational Developmental Plasticity), 11 (Transgenerational Developmental Plasticity—An Epitome of Evolutionary Change), 13 (Origins of Evolutionary Novelty), and 19 (Epigenetics of Sympatric Speciation—Speciation as a Mechanism of Evolution).

A conceptual barrier separating what have been considered discrete and independent mechanisms of control at the genetic, intracellular-epigenetic, endocrine, paracrine, autocrine, and neural levels is rapidly crumbling. The emerging picture shows them all to be elements of a single ICS.

The concept of the ICS in metazoans implies a system that controls and regulates their function and structure at the supracellular level. In general terms, the control system of metazoans performs:

1. Monitoring of the status of the system,
2. Comparison of the input on the state of the system with the norm (neurally determined set points),

3. Identification and assessment of deviations from the norm, and
4. Decision-making and transmission to effectors of "instructions" or commands for restoring the "norm" or for adapting the organism to the new conditions of living by activating appropriate signal cascades.

In the cases of phenotypic (behavioral, physiological, morphological, and life history) characters considered in this chapter, the above functions are performed by the nervous system and the endocrine system as its physiological extension.

The Hypothesis of the Epigenetic System of Heredity and its Predictions

The idea on the presence of control systems in metazoans is not conjectural. In this chapter, I have presented evidence demonstrating its role in controlling and regulating the animal phenotype: behavior, physiology, and morphology. This, and the extensive evidence on the essential involvement of the controller (CNS) in individual development (Chapters 3–5), led me to the hypothesis that the ICS, with the CNS as its controller, functions as an epigenetic system of heredity, providing the information for animal morphology in the process of animal reproduction (Cabej, 2004, pp. 35–47).

The ICS is a hierarchical system with the CNS functioning as the *controller* of the system (Figure 1.12). Vast experimental evidence to be presented later in this work shows that signal cascades for the development of many organs during individual development, as well as for developmental plasticity, originate in the CNS, suggesting that the latter is in possession of information on the normal supracellular structure and morphology. This raises a crucial question on the nature and origin of this information. How this nongenetic information is generated in the CNS/nervous system will be subject of a special section in Chapter 2.

The hypothesis is empirically verifiable and falsifiable; it is open to scientific inquiry and experimentation. The basic predictions of the hypothesis are as follows:

1. The expression of nonhousekeeping genes in metazoans is controlled and regulated by signal cascades that ultimately originate in the CNS.
2. Initial signals for starting signal cascades in the CNS result as output of the processing of internal/external stimuli in neural circuits.
3. The reproduction cycles and gametogenesis, production of egg and sperm cells, in metazoans are under control of the ICS, with the CNS as the controller.
4. The transfer of the epigenetic information (parental cytoplasmic factors and gene imprinting) in gametes is regulated by the ICS, which in the process of reproduction serves as the parental epigenetic system of heredity.
5. At the phylotypic stage, when the function of maternal cytoplasmic factors terminates, the embryonic CNS is operational and takes over the control of the embryonic development up to adulthood.
6. Signal cascades determining the postphylotypic development originate in the embryonic CNS.

As part of an attempt to validate the hypothesis, in the following chapters I will present evidence that the ICS, which in the process of metazoan reproduction serves as

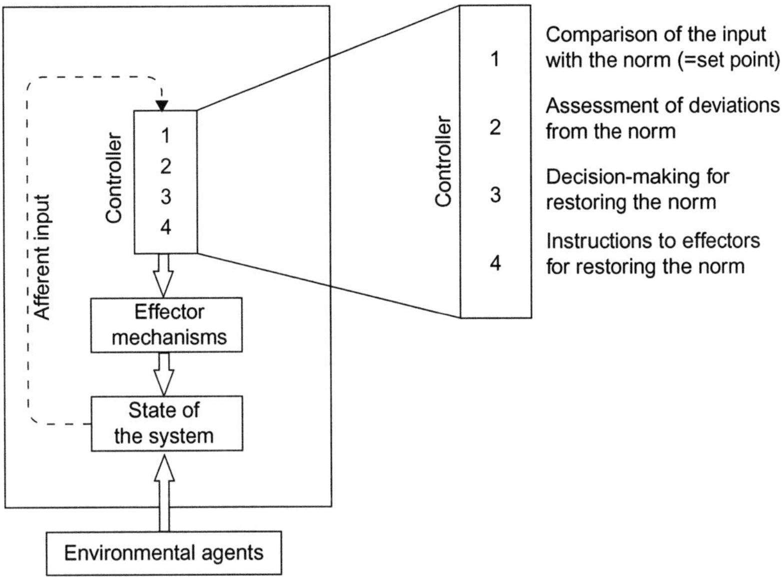

Figure 1.12 Generalized and simplified diagram of the ICS in metazoans with CNS acting as controller of the system. Metazoan structure degrades continually due to intrinsic thermodynamically determined causes as well as a result of adverse influences of the environment. Changes in the structure and function of the organism and environmental changes are monitored by a pervasive network of interoceptors and exteroceptors and communicated to the controller. In the CNS, the input is compared with the neurally determined set points (1). Deviations from the norm are identified (2), and pathways for restoring the norm are determined (3). "Instructions" or commands for restoring the norm (4) are sent to effectors in target tissues and cells through signal cascades. Via the molecular and afferent feedback input, the controller continually receives information on the restored/degraded state of the system.

an epigenetic system of heredity, controls and regulates the basic stages of the process of metazoan reproduction:

- Reproduction cycles and gametogenesis (Chapter 3),
- Early embryonic development, up to the phylotypic stage (Chapter 4), and
- Postphylotypic development (Chapter 5).

References

Adams, C.S., Korytko, A.I. and Blank, J.L. (2001). A novel mechanism of body mass regulation. J. Exp. Biol. *204*, 1729–1734.

Akers, R.M. (2002). *Lactation and the Mammary Gland.* Blackwell Publishing, Ames, IA, p. 129.

Ali, S.H., O'Donnell, A.L., Mohamed, S., Mousa, S. and Dandona, P. (1999). Stable overexpression of estrogen receptor-alpha in ECV 304 cells inhibits proliferation and levels of secreted endothelin-1 and vascular endothelial growth factor. Mol. Cell. Endocrinol. *152*, 1–9.

Aranda, A. and Pascual, A. (2001). Nuclear hormone receptors and gene expression. Physiol. Rev. *81*, 1269–1304.

Arney, K.L., Bao, S., Bannister, A.J., Kouzarides, T. and Surani, M.A. (2002). Histone methylation defines epigenetic asymmetry in the mouse zygote. Int. J. Dev. Biol. *46*, 317–320.

Barcroft, J. (1934). Features in the Architecture of Physiological Function. Cambridge University Press, London 86.

Batterham, R.L., Cowley, M.A., Small, C.J., Herzog, H., Cohen, M.A., Dakin, C.L., et al. (2002). Gut hormone PYY_{3-36} physiologically inhibits food intake. Nature *418*, 650–654.

Benzo, C.A. (1983). The hypothalamus and blood glucose regulation. Life Sci. *30*, 2509–2515.

Bhattacharyya, N., Dey, A., Minucci, S., Zimmer, A., John, S., Hager, G., et al. (1997). Retinoid-induced chromatin structure alterations in the retinoic acid receptor beta 2 promoter. Mol. Cell. Biol. *17*, 6481–6490.

Block, K., Kardana, A., Igarashi, P. and Taylor, H.S. (2000). *In utero* diethylstilbestrol (DES) exposure alters *Hox* gene expression in the developing muellerian system. FASEB J. *14*, 1104–1108.

Boulant, J.A. (2000). Role of the preoptic-anterior hypothalamus in thermoregulation and fever. Clin. Infect. Dis. *31*(Suppl. 5), S157–S161.

Boulant, J.A. (2006). Neuronal basis of Hammel's model for set-point thermoregulation. J. Appl. Physiol. *100*, 1347–1354.

Bourque, C.W. (2008). Central mechanisms of osmosensation and systemic osmoregulation. Nat. Rev. Neurosci. *9*, 519–531.

Briand, P., Lundholt, B.K., Skouv, J. and Lykkesfeldt, A.E. (1999). Growth response of breast epithelial cells to estrogen is influenced by EGF. Mol. Cell. Endocrinol. *153*, 1–9.

Brisken, C., Heineman, A., Chavarria, T., Elenbaas, B., Tan, J. and Dey, S.K., et al. (2000). Essential function of *Wnt-4* in mammary gland development downstream of progesterone signaling. Gene. Dev. *14*, 650–654.

Butler, A.A., Marks, D.L., Fan, W., Kuhn, C.M., Bartolome, M. and Cone, R.D. (2001). Melanocortin-4 receptor is required for acute homeostatic responses to increased dietary fat. Nat. Neurosci. *4*, 605–611.

Cabej, N.R. (2004). *Neural Control of Development*. Albanet, Dumont, NJ, pp. 35–47.

Cabej, N.R. (2005). *Neural Control of Development*. Albanet, Dumont, NJ, pp. 15–17.

Cadigan, K.M. and Nusse, R. (1997). Wnt signaling: a common theme in animal development. Genes Dev. *11*, 3286–3305.

Clagett-Dame, M. and Plum, L.A. (1997). Retinoid regulated gene expression. Crit. Rev. Eukaryot. Gene Expr. *7*, 299–342.

Clemmons, D.R. (2004). The relative roles of growth hormone and IGF-1 in controlling insulin sensitivity. J. Clin. Invest. *113*, 25–27.

Collingwood, T.N., Urnov, F.D. and Wolffe, A.P. (1999). Nuclear receptors: coactivators, corepressors and chromatin remodeling in the control of transcription. J. Mol. Endocrinol. *23*, 255–275.

Colombani, J., Bianchini, L., Layalle, S., Pondeville, E., Dauphin-Villemant, C. and Antoniewski, C., et al. (2005). Antagonistic actions of ecdysone and insulins determine final size in *Drosophila*. Science *310*, 667–670.

Conlon, R.A. (1995). Retinoic acid and pattern formation in vertebrates. Trends Genet. *11*, 314–319.

Crosio, C., Heitz, E., Allis, C.D., Borrelli, E. and Sassone-Corsi, P. (2003). Chromatin remodeling and neuronal response: multiple signaling pathways induce specific histone H3 modifications and early gene expression. J. Cell Sci. *116*, 4905–4914.

Cupp, A.S., Dufour, J.M., Kim, G., Skinner, M.K. and Kim, K.H. (1999). Action of retinoids on embryonic and early postnatal testis development. Endocrinology *140*, 2343–2352.

Curtis, J.T., Liu, Y. and Wang, Z. (2001). Lesions of the vomeronasal organ disrupt mating-induced pair bonding in female prairie voles (*Microtus ochrogaster*). Brain Res. *901*, 167–174.

De Luca, L.M. and Ross, S.A. (1996). Retinoic acid response elements as positive and negative regulator of the expression of the homeobox 1 gene. Nutr. Rev. *54*, 61–63.

DeVries, A.C., Guptaa, T., Cardillo, S., Cho, M. and Carter, C.S. (2002). Corticotropin-releasing factor induces social preferences in male prairie voles. Psychoneuroendocrinology *27*, 705–714.

Fantino, M., Faion, F. and Rolland, Y. (1986). Effect of dexfenfluramine on body weight set-point: study in the rat with hoarding behaviour. Appetite *7*(Suppl. 1), 115–126. (abstract).

Gilbert, S.F. (1997). *Developmental Biology*, fifth edition. Sinauer Associates, Inc. Publishers, Sunderland, MA, p. 765.

Glavaski-Joksimovic, A., Jeftinija, K., Jeremic, A., Anderson, I.I. and Jeftinija, S. (2002). Mechanism of action of the growth hormone secretagogue, L-692, 585, on isolated porcine somatotropes. J. Endocrinol. *175*, 625–636.

Gore, A.C. and Terasawa, E. (2001). Neural circuits regulating pulsatile luteinizing hormone release in the female guinea-pig: opioid, adrenergic and serotonergic interactions. J. Endocrinol. *13*, 239–248.

Grillner, S., Deliagina, T., Ekeberg, O., el Manira, A., Hill, R.H. and Lansner, A., et al. (1995). Neural networks that co-ordinate locomotion and body orientation in lamprey. Trends Neurosci. *18*, 270–279.

Gupta, C., Siegel, S. and Ellis, D. (1991). The role of the EGF in testosterone-induced reproductive tract differentiation. Dev. Biol. *146*, 106–116.

Halaas, J.L., Gajiwala, K.S., Maffei, M., Cohen, S.L., Chait, B.T. and Rabinowitz, D., et al. (1995). Weight-reducing effects of the plasma protein encoded by the obese gene. Science *269*, 543–546.

Hammel, H.T., Jackson, D.C., Stolwijk, J.A.J., Hardy, J.D. and Stroeme, S.B. (1963). Temperature regulation by hypothalamic proportional control with an adjustable set point. J. Appl. Physiol. *18*, 1146–1154.

Hennighausen, L. and Robinson, G.W. (1998). Think globally, act locally: the making of a mouse mammary gland. Genes Dev. *12*, 449–455.

Hennighausen, L. and Robinson, G.W. (2005). Information networks in the mammary gland. Nat. Rev. Mol. Cell Biol. *6*, 715–725.

Hill, J.M., Glazner, G.W., Lee, S.J., Gozes, I., Gressens, P. and Brenneman, D.E. (1999). Vasoactive intestinal peptide regulates embryonic growth through the action of activity-dependent neurotrophic factor. Ann. N.Y. Acad. Sci. *897*, 92–100.

Hogan, B.L.M., Thaller, C. and Eichele, G. (1992). Evidence that Hensen's node is a site of retinoic acid synthesis. Nature *359*, 237–241.

Jenuwein, T. and Allis, C.D. (2001). Translating the histone code. Science *293*, 1074–1080.

Jeong, M.R., Hashimoto, R., Senatorov, V.V., Fujimaki, K., Ren, M. and Lee, M.S., et al. (2003). Valproic acid, a mood stabilizer and anticonvulsant, protects rat cerebral cortical neurons from spontaneous cell death: a role of histone deacetylase inhibition. FEBS Lett. *542*, 74–78.

Kassen, A.E., Sensibar, J.A., Sintich, S.M., Pruden, S.J., Kozlowski, J.M. and Lee, C., et al. (2000). Autocrine effect of DHT on FGF signaling and cell proliferation in LNCaP cells: role of heparin/heparan-degrading enzymes. Prostate *44*, 124–132.

Keverne, E.B. (1999). The vomeronasal organ. Science *286*, 716–720.

la Fleur, S.E., Kalsbeek, A., Wortel, J. and Buijs, R.M. (1999). A suprachiasmatic nucleus generated rhythm in basal glucose concentrations. J. Neuroendocrinol. *11*, 643–652.

Lathe, R. (2001). Hormones and the hippocampus (Review). J. Endocrinol. *169*, 205–231.

Levenson, J.M., O'Riordan, K.J., Brown, K.D., Trinh, M.A., Molfese, D.L. and Sweatt, J.D. (2004). Regulation of histone acetylation during memory formation in the hippocampus. J. Biol. Chem. *279*, 40545–40559.

Levin, B.E. and Dunn-Meynell, A.A. (2000). Sibutramine alters the central mechanisms regulating the defended body weight in diet-induced obese rats. Am. J. Physiol. Regul. Integr. Comp. Physiol. *279*, R2222–R2228.

Li, Y., Wu, X., Zhu, J., Yan, J. and Owyang, C. (2003). Hypothalamic regulation of pancreatic secretion is mediated by central cholinergic pathways in the rat. J. Physiol. *552*, 571–587.

Li, J., Guo, Y., Schroeder, F.A., Youngs, R.M., Schmidt, T.W. and Ferris, C., et al. (2004). Dopamine D2-like antagonists induce chromatin remodeling in striatal neurons through cyclic AMP-protein kinase A and NMDA receptor signaling. J. Neurochem. *90*, 1117–1131.

Liu, Y. and Wang, Z.X. (2003). Nucleus accumbens oxytocin and dopamine interact to regulate pair bond formation in female prairie voles. Neuroscience *121*, 537–544.

Ma, L., Benson, G.V., Lim, H., Dey, S.K. and Maas, R.L. (1998). Abdominal B (AbdB) *Hoxa* genes: regulation in adult uterus by estrogen and progesterone and repression in muellerian duct by the synthetic estrogen diethylstilbestrol (DES). Dev. Biol. *197*, 141–154.

Malpel, S., Mendelsohn, C. and Cardoso, W.V. (2000). Regulation of retinoic acid signaling morphogenesis. Development *127*, 3057–3067.

Marshall, H., Morrison, A., Studer, M., Popperl, H. and Krumlauf, R. (1996). Retinoic acid and *Hox* genes. FASEB J. *10*, 969–978.

Mayr, E. (1961). Cause and effect in Biology. Science *134*, 1501–1506.

Mery, F. and Kawecki, T.J. (2004). The effect of learning on experimental evolution of resource preference in *Drosophila melanogaster*. Evolution *58*, 757–767.

Minucci, S., Horn, V., Bhattacharyya, N., Russanova, V., Ogryzko, V.V. and Gabriele, L., et al. (1997). A histone deacetylase inhibitor potentiates retinoid receptor action in embryonal carcinoma cells. Proc. Natl. Acad. Sci. U.S.A. *94*, 11295–11300.

Mitrani, P., Srinivasan, M., Dodds, C. and Patel, M.S. (2006). Role of the autonomic nervous system in the development of hyperinsulinemia by high-carbohydrate formula feeding to neonatal rats. Am. J. Physiol. Endocrinol. Metab. *292*, E1069–E1078.

Miyasaka, K., Masuda, M., Kanai, S., Sato, N., Kurosawa, M. and Funakoshi, A. (2002). Central Orexin-A stimulates exocrine secretion via the vagus. Pancreas *25*, 400–404.

Nakayama, N., Rice, J.C., Strahl, B.D., Allis, C.D. and Grewal, S.I. (2001). Role of histone H3 lysine 9 methylation in epigenetic control of heterochromatin assembly. Science *292*, 110–113.

Nelson, K.G., Takahashi, T., Bossert, N.L., Walmer, D.K. and McLachlan, J.A. (1991). Epidermal growth factor replaces estrogen in the stimulation of female genital-tract growth and differentiation. Proc. Natl. Acad. Sci. U.S.A. *88*, 21–25.

Nijhout, H.F. (2003). The control of body size in insects. Dev. Biol. *261*, 1–9.

Norman, A.W. and Litwack, G. (1997). *Hormones*, second edition. Academic Press, San Diego, CA , p. 201.

Pacak, K. and Palkovits, M. (2001). Stressor specificity of central neuroendocrine responses: implications for stress-related disorders. Endocr. Rev. *22*, 502–548.

Pelleymounter, M.A., Cullen, M.J., Baker, M.B., Hecht, R., Winters, D. and Boone, T., et al. (1995). Effects of the obese gene product on body weight regulation in ob/ob mice. Science *269*, 540–543.

Ramaswamy, S. and Plant, T.M. (2001). Operation of the follicle-stimulating hormone (FSH)-inhibin B loop in the control of primate spermatogenesis. Mol. Cell. Endocrinol. *180*, 93–101.

Reik, W., Dean, W. and Walter, J. (2001). Epigenetic reprogramming in mammalian development. Science *293*, 1089–1093.

Rideout, W.M., Eggan, K. and Jaenisch, R. (2001). Nuclear cloning and epigenetic reprogramming of the genome. Science *293*, 1093–1098.

Robinson, G.W., Hennighausen, L. and Johnson, P.F. (2000). Side-branching in the mammary gland: the progesterone-Wnt connection. Gene Dev. *14*, 889–894.

Ross, H.E., Freeman, S.M., Spiegel, L.L., Ren, X., Terwilliger, E.F. and Young, L.J. (2009). Variation in oxytocin receptor density in the nucleus accumbens has differential effects on affiliative behaviors in monogamous and polygamous voles. J. Neurosci. *29*, 1312–1318.

Ruohola, J.K., Valve, E.M., Karkkainen, M.J., Joukov, V., Alitalo, K. and Härkönen, P.L. (1999). Vascular endothelial growth factors are differently regulated by steroid hormones and antiestrogens. Mol. Cell. Endocrinol. *149*, 29–40.

Sachs, L.M. and Shi, Y.B. (2000). Targeted chromatin binding and histone acetylation *in vivo* by thyroid hormone receptor during amphibian development. Proc. Natl. Acad. Sci. U.S.A. *97*, 13138–13143.

Salvador, L.M., Park, Y., Cottom, J., Maizels, E.T., Jones, J.C. and Schillace, R.V., et al. (2001). Follicle-stimulating hormone stimulates protein kinase A-mediated histone H3 phosphorylation and acetylation leading to select gene activation in ovarian granulosa cells. J. Biol. Chem. *276*, 40146–40155.

Sassoon, D. (1999). *wnt* genes and endocrine disruption of the female reproductive tract: a genetic approach. Mol. Cell. Endocrinol. *158*, 1–5.

Schneider, R.A., Hu, D., Rubenstein, J.L., Maden, M. and Helms, J.A. (2001). Local retinoid signaling coordinates forebrain and facial morphogenesis by maintaining FGF 8 and SHH. Development *128*, 2755–2767.

Schrödinger, E. (1944). *What Is Life?* Cambridge University Press, Cambridge.

Schwartz, M.W. and Morton, G.J. (2002). Keeping hunger at bay. Nature *418*, 595–597.

Scott, C.J., Rawson, J.A., Pereira, A.M. and Clarke, I.J. (1999). Oestrogen receptors in the brainstem of the female sheep: relationship to noradrenergic cells and cells projecting to the medial preoptic area. J. Neuroendocrinol. *11*, 745–755.

Servoss, S.J., Lee, S.J., Gibney, G., Gozes, I., Brenneman, D.E. and Hill, J.M. (2001). IGF-I as a mediator of VIP/activity-dependent neurotrophic factor-stimulated embryonic growth. Endocrinology *142*, 3348–3353.

Shoba, L., An, M.R., Frank, S.J. and Lowe, W.L. (1999). Developmental regulation of insulin-like growth factor-I and growth hormone receptor gene expression. Mol. Cell. Endocrinol. *152*, 125–136.

Simonini, M.V., Camargo, L.M., Dong, E., Maloku, E., Veldic, M. and Costa, E., et al. (2006). The benzamide MS-275 is a potent, long-lasting brain region-selective inhibitor of histone deacetylases. Neuroscience *103*, 1587–1592.

Smeltzer, M.D., Curtis, J.T., Aragona, B.J. and Wang, Z. (2005). Dopamine, oxytocin, and vasopressin receptor binding in the medial prefrontal cortex of monogamous and promiscuous voles. Neurosci. Lett. *394*, 146–151.

Strand, F.L. (1999). *Neuropeptides*. MIT Press, Cambridge, MA , p. 508.

Swank, M.W. and Sweatt, J.D. (2001). Increased histone acetyltransferase and lysine acetyltransferase activity and biphasic activation of the ERK/RSK cascade in insular cortex during novel taste learning. J. Neurosci *21*, 3383–3391.

Tzschentke, B. (2007). Attainment of thermoregulation as affected by environmental factors. Poult. Sci. *86*, 1025–1036.

Tzschentke, B. and Basta, D. (2002). Early development of neuronal hypothalamic thermosensitivity in birds: influence of epigenetic temperature adaptation. Comp. Biochem. Physiol. A *131*, 825–832.

Tzschentke, B. and Nichelmann, M. (1997). Influence of prenatal and postnatal acclimation on nervous and peripheral thermoregulation. Ann. N.Y. Acad. Sci. *813*, 87–94 (abstract).

Vermot, J., Fraulob, V., Dolle, P. and Niederreither, K. (2000). Expression of enzymes synthesizing aldehyde dehydrogenase 1 and retinaldehyde dehydrogenase 2 and metabolizing (Cyp26) retinoic acid in the mouse reproductive tract. Endocrinology *141*, 3638–3645.

Verney, E.B. (1947). The antidiuretic hormone and the factors which determine its release. Proc. R. Soc. Lond. B Biol. Sci. *135*, 25–106.

Wu, X., Gao, J., Yan, J., Owyang, C. and Li, Y. (2004). Hypothalamus–brain stem circuitry responsible for vagal efferent signaling to the pancreas evoked by hypoglycemia in rat. J. Neurophysiol. *91*, 1734–1747.

Young, L.J., Wang, Z. and Insel, T.R. (1998). Neuroendocrine bases of monogamy. Trends Neurosci. *21*, 71–75.

Young, L.J., Murphy Young, A.Z. and Hammock, E.A.D. (2005). Anatomy and neurochemistry of the pair bond. J. Comp. Neurol. *493*, 51–57.

Ziegler, D.R., Cass, W.A. and Herman, J.P. (1999). Excitatory influence of the locus coeruleus in hypothalamic–pituitary–adrenocortical axis responses to stress. J. Neuroendocrinol. *11*, 361–369.

2 Neural Manipulation of Gene Expression

The ability to create activity patterns may underlie the brain's ability to generate insight and the "trials" of trial-and-error problem solving.

W.J. Freeman (1991)

Predetermined Versus Manipulative Expression of Genes

In Chapter 1, I showed that the activation/inactivation of nonhousekeeping genes, which are responsible for differentiation and coordination of activities of each cell with other cells, is regulated by an integrated control system (ICS), with the central nervous system (CNS) as its controller. Neural signals, which, via the signal cascades, activate/inactivate these genes, ultimately, originate in the CNS. A glimpse at the temporal order of events in the cascade clearly shows that the direction of the information flow is from the CNS to effector genes.

So far, however, we have neither shown nor even speculated on how the CNS can accomplish such a huge and extremely complex task: provision of information for gene activation/inactivation to billions/trillions of cells throughout the animal body. In the case of gene activation under the influence of external stimuli, but internal stimuli as well (changes in the level of estrogens and other hormones), no direct physical contact exists between the gene and the stimulus that triggers its expression. Moreover, even the experimental physical contact of the stimulus with the gene does not induce expression of the gene, implying that no chemical affinity exists between them. This is an uncommon situation: in direct contact with the gene, the stimulus does not induce the expression of the gene, but at a distance it does! However, this is a paradox only at first sight. We know that it is not the stimulus *per se* but a neural product of processing of the stimulus in the nervous system that, by activating a specific signal cascade, induces expression of a specific gene in the cell nucleus. By activating the signal cascade, the nervous system establishes a causal relationship between the stimulus and the stimulus-inaccessible gene that did not previously exist.

This contrived, unorthodox expression of genes is a unique function of the neuron and the CNS, i.e., a new property of metazoans, unknown in unicellulars. It is true that in *eukaryotic* unicellulars as well, a number of external stimuli do not act directly on genes whose expression they trigger, but use existing signal transduction pathways in order to control and regulate expression of target genes. However, in protozoans, the genes that will be expressed are predictable from the nature of the stimulus: if we know the stimulus, we can predict the signal transduction pathway that can be activated and the gene(s) that will be expressed. The action of the

Epigenetic Principles of Evolution. DOI: 10.1016/B978-0-12-415831-3.00002-1

stimulus in protozoans is predetermined by the existence of a natural causal relationship of the stimulus with the transduction pathway inducing expression of the gene.

In clear distinction from this, metazoans can fashion new, otherwise nonexisting, causal relationships between the stimulus and the gene in order to produce adaptive expression of genes. The stimulus *per se* represents no message to any gene, and no gene responds to the contact with the stimulus *per se* for there is no chemical affinity between them. But the stimulus is taken as a message by exteroceptors, proprioceptors, or interoceptors, converted into electrical spike trains (=nongenetic symbols) and, in this form, transmitted to the CNS. By processing these electrical representations of stimuli, neural circuits generate chemical outputs (e.g., neurotransmitter, neuromodulator, neuropeptide), which essentially decipher the genetically meaningless stimulus into a message for expression of a particular gene or a number of genes. The CNS, thus, transforms a genetically inert stimulus into a gene inducer. Both the initial conversion of stimuli into electrical spike trains and their processing in neural circuits are computational nongenetic processes whose mechanisms are only little known.

Neural circuits, functional units of the nervous system, are ensembles of several to millions of neurons connected by specific synaptic contacts. By processing information coming from afferent neurons in the form of electrical signals, neural circuits perceive the "information generated by stimuli arising from both the external and internal environment," which is one of the main functions of the brain and facilitate the transfer of information according to a neural, nongenetic code, which is represented by the spike rate of neurons (Shadlen and Newsome, 1994).

Neural circuits release their output in the form of electrical or chemical signals (neurotransmitters and neuromodulators) that are discharged on neurosecretory cells of the "endocrine brain," the hypothalamus, the pituitary, or, via nerve endings, directly to the target tissues and organs. These neural signals act as "instructions" (=epigenetic information) for selectively activating specific algorithms or signal cascades ultimately leading to selective expression of particular gene(s) out of a variety of genes available for transcription (Figure 2.1).

The ability of the nervous system to respond to the input of external and internal stimuli based on the processing of that input, and to establish naturally nonexistent causal relationships between stimuli and specific genes, enabled metazoans to respond very flexibly to environmental changes and provided them with an additional degree of freedom by overcoming the genetic determinism of gene expression. It enabled metazoans to respond in different ways to the same external or internal stimulus, and represents a goal-oriented, hence adaptive, mechanism of gene expression.

Thus, expression of genes in the CNS is not "stimulus dependent", as commonly described; their expression is *processing dependent, hence nongenetic, epigenetic*. Selection of the gene to be expressed is normally adaptive, i.e., it tends to accommodate the organism to the stimulus-changed environment or even to the situations that the stimulus might presage. The connection between the gene and the stimulus is communicative rather than direct and is computationally determined. As pointed out earlier, the processing in the nervous system establishes a causal relationship between the stimulus and expression of the gene, which otherwise would not exist.

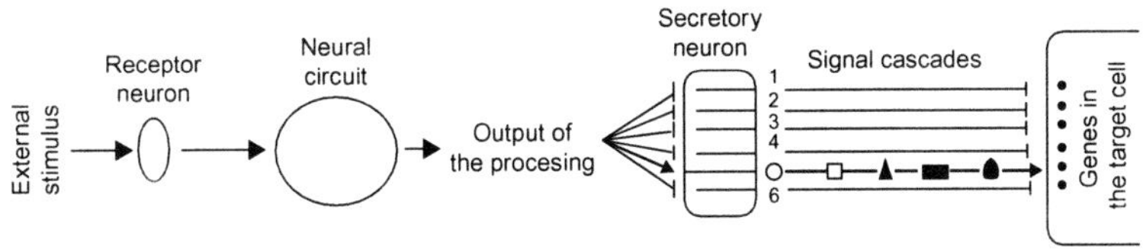

Figure 2.1 Simplified diagram of the fashioning of a relationship between the external stimulus and a particular gene. An external stimulus is received from specific neurons of the sensory organ. Receptor neurons convert the stimulus into a train of electrical spikes and, in this form, transmit it to a specific neural circuit for processing or interpreting. The mechanism of processing is still almost a black box. What we know is the output of the processing, which is an electrical/chemical signal that leads to selective activation of a specific algorithm or adaptive signal cascade out of a number of cascades (1–6) that the secretory neuron can potentially activate.

I call this "nonclassical" expression of genes *manipulative* since the expression of the gene is contrived rather than determined by chemical affinity of the stimulus for the gene; no special thermodynamic and stereochemical relationships exist between the stimulus and the gene that is expressed in response to it. It is not the nature of the external/internal stimulus but the adaptive requirement of the organism that determines expression of the specific gene in the CNS. In response to different stimuli, different organisms may respond by turning on/off the same gene and, in response to the same stimulus, different species may respond by expressing (or not) different genes and activating or inactivating different signal cascades.

By processing external/internal stimuli in neural circuits, metazoans are able to figure both the gene and the signal cascade they have to activate. The function of the processing in the neural circuit is to perceive, to create a gestalt of the stimulus, and interpret it, to give it a "meaning," and find a solution to the problem it may present to the organism by selectively activating a specific signal cascade out of a number of available cascades.

Processing of External/Internal Stimuli in the CNS Generates Information for Adaptive Phenotypic Responses

The fact that many adaptive signal cascades in metazoans begins with external/internal stimuli raises the question of whether the environmental stimuli themselves may contain information for gene expression, despite the Lamarckian implications of recognition of an instructing role of the environment in metazoan development and evolution.

Theoretical contemplations aside, the examples of expression of genes under the influence of external and internal stimuli to be presented in this subsection show that none of them *per se* does or can induce gene expression. The activation of signal cascades by external/internal stimuli is mediated by the CNS; i.e., relevant signal cascades are activated by the output of the processing of the stimuli by the CNS. If external stimuli (e.g., light, darkness, temperature, humidity, crowding, other social stimuli) would provide information for selective activation of signal cascades, the whole costly processing of the stimuli would serve no purpose. Elementary knowledge of the way evolution works suggests that costly neural processing would have not evolved if it did not generate a benefit outweighing the cost.

However, it might be argued that any talk about the CNS control of gene expression in such cases is irrelevant since, ultimately, brain signals that start signal cascades are genetic in origin. Is this really so?

Any dispute on the nature of information for selective activation of signal cascades in the CNS cannot be resolved by theoretical arguments. It can only empirically be determined by a closer examination of activation of signal cascades under the influence of external/internal stimuli in some paradigmatic examples of the so-called stimulus-dependent expression of genes in the brain.

Example 1

Most wild birds, from temperate to arctic zones, are reproductively active (i.e., ovulate and lay eggs) in the spring time. The timing of the reproductive activity in these birds is clearly adaptive, since spring is the best time for rearing the offspring. The beginning of the reproductive activity at that time of the year is mainly related to the prolongation of photoperiod (length of the day) and rise of temperature in the environment. Although the reproductive activity starts with expression of hypothalamic gonadotropin-releasing hormone (GnRH) and activation of the hypothalamic–pituitary–gonadal axis, neither the gene for the hypothalamic GnRH nor the gene for the pituitary gonadotropins or for gonadal hormones can be induced by the direct action of thermal or photic stimuli. In homeothermic animals, such as birds, neither the ovary/testis nor the egg cell/sperm cell (and less so their genes) receive those external stimuli. Nevertheless, birds perfectly respond to these stimuli by a precise spatiotemporal activation/inactivation of the genes along the hypothalamic–pituitary–gonadal axis, inducing expression of thousands of specific genes in gametes as well as neighboring relevant cells (e.g., follicle cells, Sertoli cells), related to the preparation of the genital organs for gametogenesis, mating, and oviposition. Both the temperature and photic stimuli are "unintelligible" to genes, which generally do not respond to the physical contact with these stimuli.

The fact that these genes are activated, despite being inaccessible to the stimuli, would be a paradox if we did not know what goes on "behind the scenes." The temperature is perceived in the hypothalamic preoptic area (POA), in a center containing warm-sensitive neurons and cold-sensitive neurons, whereas the length of the day is sensed in another hypothalamic center, the suprachiasmatic nucleus (SCN). The perception of the temperature and day length in these brain centers comprises reception of the electrical input on the temperature and length of the day and processing of these stimuli in respective neural circuits. As a result of these computational processes, neural circuits generate chemical outputs (= epigenetic information) for starting signal cascades that determine activation of hundreds and thousands of specific genes in the genital tract and the organism in general during the reproductive season.

One of the ways the photoperiod is used for the reproductive timing in vertebrates is by neural inhibition of melatonin biosynthesis and secretion in the pineal gland. This is suggested by the fact that melatonin injections in photoresponsive rats, even during a stimulatory (long) photoperiod, induce reproductive inhibition (Heideman et al., 2001).

The photic stimulus (shortening of the photoperiod) in the pineal gland induces expression of genes for the synthesis of melatonin, although no light, no photons reach those cells. Under the influence of the photic stimulus (day–night cycles) birds switch off/on the genes for melatonin biosynthesis in the pineal gland (Figure 2.2). If the photic stimulus, day–night cycles *per se* would be information for gene expression, it would be expected that dermis cells, which are directly affected by the stimulus (alternation light/darkness), not pineal cells which are not, would synthesize melatonin. As we know, the opposite takes place: dermis cells do not express those genes, while pineal cells, to which the stimulus has no access, do. Moreover, retinal neurons as well, which receive the stimulus and transmit it to the CNS, do not express the genes for melatonin biosynthesis. The fact that the stimulus fails to induce expression of melatonin genes in dermis cells or retinal cells, which are directly affected by the stimulus, and induces expression of these genes in pineal cells to which it has no access, excludes the possibility that the stimulus itself, the external agent, might represent information for expression of the melatonin genes.

The suppression/induction of melatonin results not from a direct action of the photic stimulus on the gland, but from the processing of the photic information in

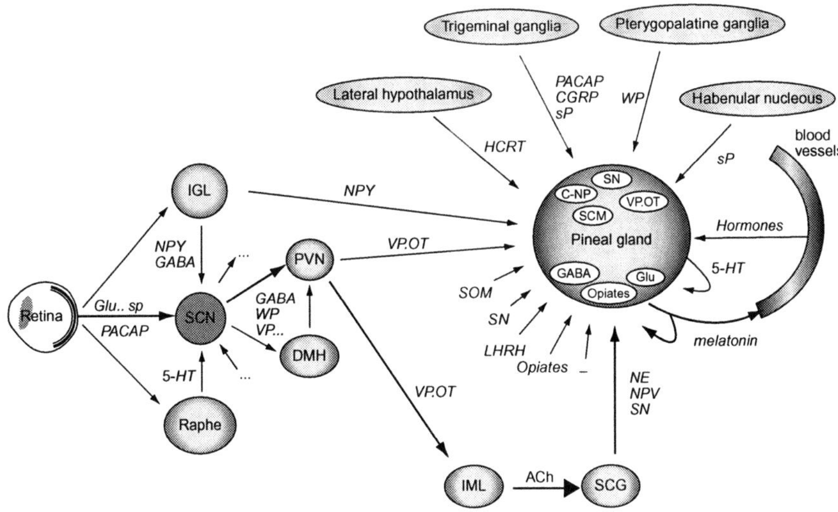

Figure 2.2 Schematic representation of the various neural, endocrine, and paracrine inputs of the mammalian pineal gland. The main neural pathway, which transmits light information to the pineal gland, is shown with thick arrows. In addition, numerous other neural or endocrine inputs are known to reach the pineal gland. Note that there are interspecies differences in the density and origin of the afferent pineal nerve fibers and the nature of the different pineal transmitters.
Abbreviations: CGRP, calcitonin gene-related peptide; DMH, dorsomedial hypothalamic nucleus; HCRT, hypocretin; IGL, thalamic intergeniculate leaflet; IML, intermediolateral nucleus of the upper thoracic spinal cord; LHRH, luteinizing hormone-releasing hormone; NE, noradrenaline; NPY, neuropeptide Y; OT, oxytocin; PACAP, pituitary adenylate cyclase-activating peptide; PVN, paraventricular nucleus; SCG, superior cervical ganglion; SCN, suprachiasmatic nucleus; SN, secretoneurin; SOM, somatostatin; VP, vasopressin; 5-HT - serotonin.
Source: From Simonneaux and Ribelayga (2003).

neural circuits, i.e., not from a genetic, but from a computational epigenetic, process. The photic stimulus is received by retinal photoreceptors, sent to the SCN in the form of electrical signals. By processing the photic stimulus (Larsen et al., 1998), the SCN sends to the paraventricular nucleus a combination of stimulatory and inhibitory circadian signals (Perreau-Lenz et al., 2003), and the latter sends to the pineal signals for starting the synthesis of melatonin (Figure 2.2).

Example 2

In response to seasonal cues and to a *social auditory stimulus*, the conspecific male mating call, female green tree frogs increase the release of the hypothalamic GnRH. The auditory system of the tree frog projects to the ventral hypothalamus (VH) and POA (Figure 2.3).

Investigators have concluded that experimental results

> *provide functional evidence for such a sensory-endocrine circuit by showing that acoustic signals influence GnRH neurons.*
>
> *Burmeister and Wilczynski (2005)*

Obviously auditory signals *per se* cannot induce expression of any gene; it is the informational output of the processing of acoustic signals in the CNS, released in the form of a neurotransmitter/neuromodulator that via its receptors on the hypothalamic GnRH neurons activates a transduction pathway inducing expression of the GnRH

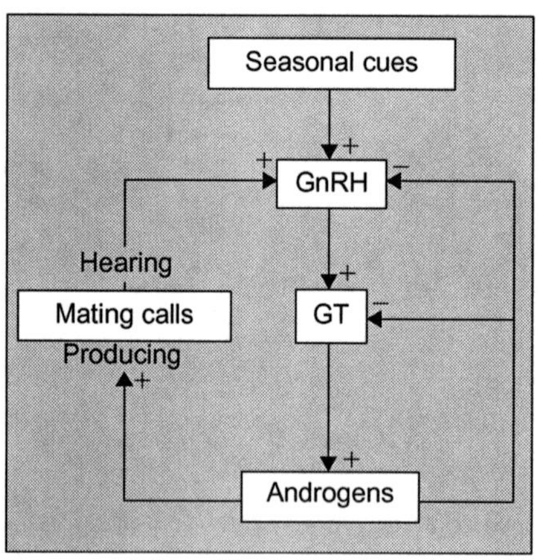

Figure 2.3 A proposed model for regulation of the anuran hypothalamus–pituitary–gonad (HPG) axis during the breeding season. In the model, seasonal cues initiate the annual reactivation of the GnRH neurons, which are further stimulated during the breeding season by hearing mating calls. GnRH neurons regulate androgen levels through their action on gonadotropin (GT) release from the pituitary. Androgens, in turn, exert negative feedback on the HPG axis. Androgens also influence brain regions that increase the production of mating calls, creating a positive feedback loop between the communication system and the endocrine system. *Source*: From Burmeister and Wilczynski (2005).

gene. In turn GnRH, via the pituitary–gonadal axis, starts the signal cascade that stimulate the reproductive activity in female tree frogs.

Clearly, expression of GnRH gene in response to auditory stimuli is not genetically but epigenetically determined by the processing of auditory stimuli (species-specific mating calls).

Example 3

Detection of chilling in the Hawaiian cockroach *Diploptera punctata*, and in other insects, is function of antennal thermoreceptors, which send chilling signals to the *pars intercerebralis* and *pars lateralis* of the insect protocerebrum. Chilling experienced by the antennae has no specific effect on gene expression on the antennae, but it triggers a "long-range" action on another organ that is not directly affected by the chilling; it causes suppression of mitotic divisions in corpora allata, which precedes the increase in the volume of corpora allata and secretion of juvenile hormone by the glands of mated females. Unilateral ablation of antenna does not prevent the contralateral corpus allatum from proceeding normally with the wave of mitotic divisions, while the mitotic divisions in the ipsilateral gland are suppressed. Unilateral disconnection of neurons of the pars intercerebralis of the insect brain from corpora allata suppresses the effect of chilling of antennae on the mitotic division of the contralateral corpus allatum, clearly demonstrating that a neural, not genetic, mechanism determines the suppressing effect of chilling on cell division in corpora allata. The information on the chilling of antennae is transmitted in the form of electric signals, perceived and processed in the *pars intercerebralis* of the insect's brain, which send signals for suppressing the mitotic division in corpora allata (Pszczolkowski and Brown, 2003).

In another experiment, it has been demonstrated that

> *severance of the ventral nerve cord prior to chilling or to application of 20E (hydroecdysone) prevented suppression of CA cell division, indicating that effects of either chilling or 20E application are mediated by the ventral nerve cord.*
>
> *Pszczolkowski and Gelman (2004)*

Example 4

The sound energy of the song of the male canary, *Serinus canaria*, in the form of mechanical waves is transformed into mechanical energy in the tympanum and transmitted through the middle and inner ears before it reaches the auditory receptor neurons, which transform it into electrical spike trains transmitted for processing in the auditory area of the forebrain. The sound waves induce no specific song-related expression of genes along the auditory pathway (tympanum, the outer, middle, and inner ear). The specific song-related expression of the transcriptional regulatory gene *ZENK* in female canaries occurs only in the auditory area of the forebrain after the processing of electrical signals (Jarvis and Nottebohm, 1997; Figure 2.4) but not in the cells that are directly affected by sound waves of singing. Not only hearing of the conspecific song but their own singing in male canaries induces a 60-fold increase in expression of the gene *ZENK* in the high vocal center and other song nuclei. At a histological level, the singing is associated with an increased electrical activity in these same centers (Jarvis

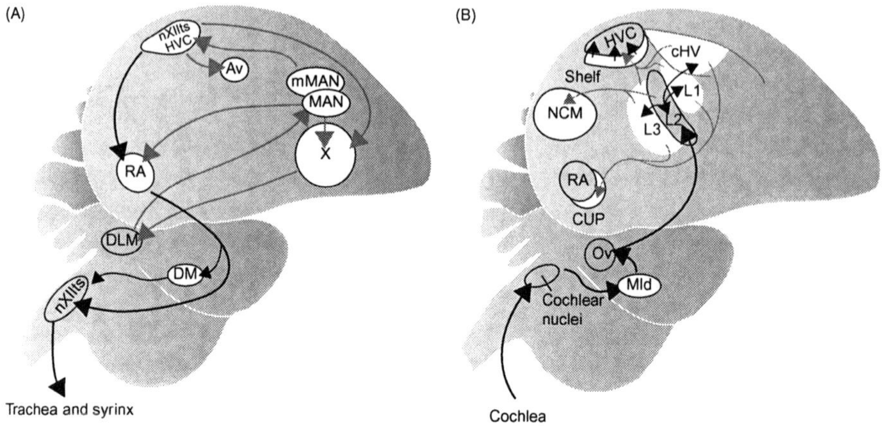

Figure 2.4 Sagittal diagrams of songbird brain. (A) Anatomical relations between song nuclei (white) in which ZENK expression was induced by the act of singing. Black arrows show the direct motor pathway that innervates the vocal organs and produces learned song. Shaded arrows show links between other song system nuclei. (B) Anatomical relations between forebrain auditory relays (white) in which ZENK expression was induced by hearing song. Thick black arrows show the major ascending auditory pathway; thin black and gray arrows show some of the connections between forebrain auditory regions and the channeling of this input into HVC. *Abbreviations*: Av, nucleus avalanche; cHV, the caudal hyperstriatum ventrale; CUP, a neural structure apposed to RA; DLM, medial nucleus of dorsolateral thalamus; DM, dorsomedial nucleus of the intercollicular complex; HVC, high vocal center; MAN, magnocellular nucleus of the anterior neostriatum; Mld, nucleus mesencephalicus lateralis pars dorsalis; nXIIts, tracheosyringeal portion of the hypoglossal (12th) nucleus; NCM, caudomedial neostriatum; Ov, nucleus ovoidalis; RA, the robust nucleus of the archistriatum; X, area X of lobus parolfactorus. *Source*: From Jarvis and Nottebohm (1997).

and Nottebohm, 1997). A similar vocalizing-induced increase in expression of *ZENK* is also observed in a number of hummingbird species (Jarvis et al., 2000).

A recent study has shown that in the caudomedial nidopallium (NCM), song-induced Zenk expression colocalizes with c-fos, c-jun, and Arc expression. Mediators of the song-regulated expression of genes are Ca^{2+} kinases, protein kinase A, and MAP (mitogen-activated protein) kinase as well as MEK1/2 (Velho et al., 2005).

Example 5

Nonshivering hibernating animals use brown adipose tissue (BAT) for producing heat during hibernation. In human newborns, this tissue represents about 25% of the body, and it is their most important source of heat for avoiding postnatal hypothermia. BAT cells (adipocytes) produce a protein, UCP-1, which triggers the mitochondrial thermogenesis. It was known that the proximate inducer of the UCP-1 gene is the release of norepinephrine by postganglionic nerve endings in BAT tissue (Morrison, 2004; Bartness and Song, 2005). Now we know that induction of UCP-1 synthesis is centrally regulated via a long processing of the cold stimulus in the CNS. Cold temperature is received as a stimulus by dermal cold thermoreceptors (sensory nerve endings), which increase the

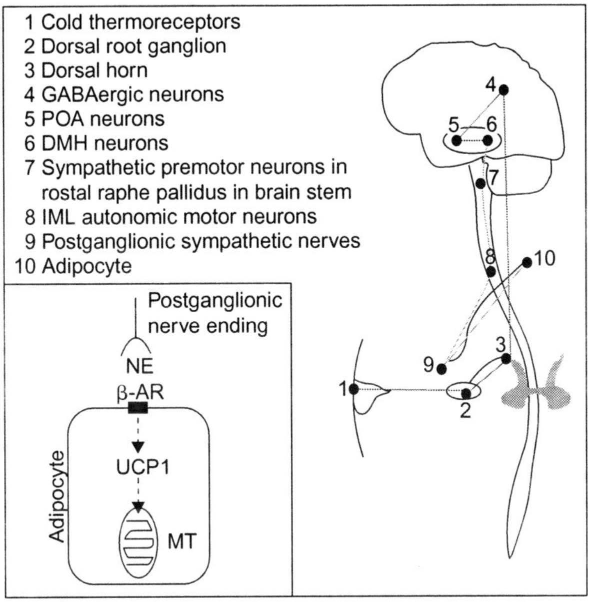

Figure 2.5 The simplified neural circuit of the processing of the signal on cold temperature transmitted by cold thermosensory nerve endings. Numbers in the figure show the sequential steps of processing of the stimulus in the nervous system. The output of the processing is NE released by the postganglionic sympathetic nerves on individual adipocytes. By binding β-AR on the adipocyte membrane, NE triggers a signal transduction pathway that leads to induction of UCP-1, which in turn induces mitochondrial thermogenesis. *Abbreviations*: β-AR, beta-adrenergic receptor; DMH, hypothalamic dorsomedial nucleus; IML, intermediolateral nucleus; MT, mitochondrion; NE, norepinephrine; POA, hypothalamic preoptic area; UCP-1, uncoupling protein gene(s).

firing rate and transmit the stimulus in the form of electric signals to the dorsal root ganglion (DRG). In a simplified form, the neural circuit (Figure 2.5) for processing the stimulus and starting the signal cascade for thermogenesis looks as follows:

DRG → dorsal horn (DH) → GABAergic neurons in a still unidentified region of the brain → neurons of the POA → neurons of the hypothalamic dorsomedial nucleus → sympathetic motor neurons in rostral raphe pallidus nucleus of the brain stem → autonomic motor neurons of the intermediolateral cell column → epinephrine released by postganglionic sympathetic nerves → a signal transduction pathway in brown adipocytes → expression of UCP-1 gene in brown adipocytes → adipocyte mitochondrial thermogenesis.

Thus, since the stimulus itself (cold temperature) is not taken as a signal to start thermogenesis in adipocytes, the task is assigned to the nervous system, which uses a "devious" route to extract the "meaning" of the cold temperature for the organism and find the appropriate adaptive response, i.e., to induce adipocyte thermogenesis. The whole "purpose" of the energetically costly processing of the stimulus is to figure out the appropriate gene necessary to activate mitochondrial thermogenesis.

Example 6

This is an example of an *internal* stimulus that induces expression of a gene without having access to the cells that express the gene in the CNS. In ewes and other mammals, estradiol is critical for the preovulatory luteinizing hormone (LH) surge. In nonneural cells, estradiol forms complexes with its nuclear receptors (estrogen receptor, ER) and, in this form, it binds regulatory regions that activate "estrogenic genes" (*cyclin D1*, *igf-1*, etc.). Nothing of the kind, no classical expression of "estrogenic genes," has been observed so far in the hypothalamus, and there is no evidence whatsoever that estradiol directly regulates GnRH neurons (Smith and Jennes, 2001).

Most of GnRH neurons do not even express nuclear ERs, mediators of the action of estrogen, and this is why even the implantation of estrogen in regions of GnRH neurons does not induce LH surge (Herbison, 1998).

> *It is suggested that the steroid signal is conveyed to the GnRH neurones by a subset of neurones that are regulated by oestradiol.*
>
> *Eyigor et al. (2004)*

What acts on the hypothalamic GnRH neurons is not the estrogen itself but the chemical output of its processing in estrogen-sensitive neurons in other parts of the brain and the hypothalamus. The rise in the estradiol level is received as a stimulus and converted into an electrical signal by non-GnRH neurons, processed in estrogen-sensitive neural circuits inducing release of stimulating neurotransmitters (dopamine by the neural circuit of the ventromedial POA (Anderson et al., 2001; Goubillon et al., 2002) and noradrenaline by a brain stem noradrenergic circuit (Scott et al., 1999)) and inhibitory neurotransmitters (γ-amino butyric acid (GABA) by a circuit in the rostral hypothalamus, and neuropeptide Y (NPY) by two circuits in the brain stem and arcuate nucleus). It is the informational output generated by processing of the estrogen stimulus (a perceptible change in the level of the hormone), rather than estrogen itself, that induces/suppresses expression of the GnRH gene (Herbison, 1998; Smith and Jennes, 2001; Pompolo et al., 2003; Figure 2.6).

Numerous examples of manipulative expression of genes in response to external cues are also described in invertebrates. So, for example, changes in the photoperiod cause the nerves originating in the subpedunculate lobe of the brain of cephalopods to stimulate or inhibit secretion of a gonadotropin by their optic glands, thus inducing or preventing the gonadal enlargement (Strand, 1999).

In all of the examples presented in this section, the expression of relevant genes occurs in neurons and other cells that have no physical contact with the external/internal agent rather than in those that come in direct contact with it. These genes are induced by the chemical output of the processing of the stimuli/agents in respective neural circuits. The so-called stimulus-dependent/activity-dependent expression of nonhousekeeping genes in the CNS is essentially *processing dependent*, *computational*, and hence *epigenetic*.

Although numerous proteins and metabolic products feed back on the CNS, this would question the role of the CNS as the source of that epigenetic information no more than the similar feedback from proteins on genes might cast any doubt upon DNA's attribute as the source of genetic information for protein biosynthesis.

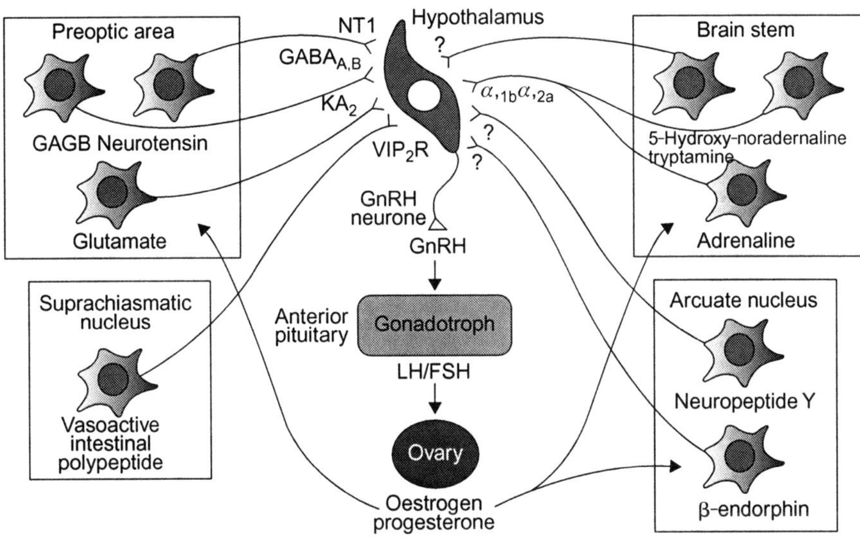

Figure 2.6 Neuronal cell populations within the GnRH network implicated in transmitting estrogen input to GnRH neurons in the rat. This may be direct or indirect on the GnRH neuron and may involve cell body or terminal levels of regulation. Note that the neurochemical identity of estrogen-receptive neurons within the GnRH network is not fully established. Neurons with *black* nuclei express nuclear ERs. An estrogen-receptive neuronal population in the AVPv is hypothesized to project to, and coordinate, neuronal activity within the arcuate nucleus. Note that GnRH neurons do not express ER and respond only to the informational input of neurons of other areas that express ERs (e.g., GABA, NPY, NE). *Abbreviations*: AVPv, anteroventral periventricular nucleus; βEND, β-endorphin neuron; ER, estrogen receptor; GABA, GABA neurons; NE, norepinephrine neuron; NPY, neuropeptide Y neuron.
Source: From Herbison (1998).

Molecular Mechanisms of Manipulative Expression of Genes in the CNS

The CNS expresses many genes (Kornhauser et al., 1990; Morris et al., 1998) that are not expressed in extraneural tissues, and adaptively modifies gene expression (Bito, 1998; West et al., 2001) by modifying the chemical output of neural circuits. For instance, electrical signals and neurotransmitters released by neurons regulate the opening of ion channels (especially by inducing the Ca^{2+} influx), thus stimulating expression of a group of up to 100 immediate early genes (Johnson et al., 1997; Finkelbeiner and Greenberg, 1998; Greenberg and Ziff, 2001), most of them coding for transcription factors. This is followed by induction, after several hours, of 500 to 1,000 late-response genes (Nedivi et al., 1993; Li et al., 2004). By processing circadian stimuli (day/night cycles), the hypothalamic SCN regulates the circadian expression of thousands of genes that control the circadian physiology of metazoans (Yamaguchi et al., 2003). More than 460 processing-dependent genes are induced in activated rat embryonic cortical neurons, with 63 genes found to be nitric oxide dependent. The processing-dependent induction of these genes is mediated by cytosolic calcium as is suggested by the fact that their induction is prevented by blocking voltage-sensitive calcium channels (Li et al., 2004).

Essentially, the unique ability of the neural tissue for manipulative expression of genes is based on the computational capabilities of neurons and neural circuits. This computation enables selection of the appropriate signal cascade that induces a gene whose expression tends to adapt the animal to the stimulus. The processing of the stimulus consists in a series of electrical and chemical transformations in neural circuits, which are necessary for fashioning the novel causal relationship between the stimulus and the gene. In a linguistic metaphor, the processing is an epigenetic restatement of the nature of the stimulus in order to make it intelligible to a particular gene only.

While we do not yet know the exact mechanisms of neural processing and the exact ways the neurons and neural circuits computate, we know what the processing does. As shown in the preceding section, the phenomenological essence of processing is the establishment of a previously nonexisting causal relationship between a stimulus and a particular gene, thus making possible manipulative expression of genes: in response to the same stimulus, and independently of its physical properties, the neural circuit can manipulatively secrete one of a number of chemicals that induce expression of a particular gene via signal cascades.

What are the mechanisms underlying this unique ability of the nervous system?

In response to various stimuli, the nervous system modifies its electrical activity. One of the consequences of electrical activity in the nervous system and of the release and reception of neurotransmitters in postsynaptic neurons is the elevation of the calcium level in the postsynaptic neurons, as a result of the opening of cell membrane Ca^{2+} channels. Transcription of a large number of genes, encoding both transcription factors and molecules that function at synapses, is induced by synaptic activity and subsequent calcium influx (West et al., 2001).

The elevation of calcium level stimulates calcium binding to calmodulin, which by activating calcium–calmodulin-dependent protein kinases (CaMKs) and calcium-sensitive adenylate cyclases, starts signal transduction pathways, resulting in induction of specific genes (Ghosh and Greenberg, 1995). One well-known example is the expression of the brain-derived neurotrophic factor (*BDNF*) gene as a result of electrical activity induced by visual input. The Ca^{2+} influx activates transduction pathways that phosphorylate cyclic adenosine monophosphate (cAMP)-response element binding protein (CREB), a transcription factor, which by binding to a Ca^{2+} response element on the *BDNF* gene makes its transcription possible (Tao et al., 1998; Figure 2.7).

CREB is induced in response to a variety of external (extracellular) stimuli, such as stress, hypoxia, neurotransmitters (in a Ca^{2+}-dependent manner), growth factors, and spontaneous electrical synaptic activity. Each of these factors uses one of several pathways for inducing CREB. In turn, CREB, by binding cAMP-response binding element (CRE), induces expression of more than 100 genes that have CRE, including genes involved in neurotransmission, signal transduction, and metabolism (Lonze and Ginty, 2002).

In response to a patterned visual stimulus, after a short period of darkness, the nervous system displays electrical activity and activates extracellular regulated kinase (ERK), by phosphorylating it in visual cortical neurons in a specific spatiotemporal pattern before returning to the baseline within 40 min. The activation of ERK leads to CRE-mediated gene expression, i.e., expression of genes bearing a CRE promoter (Cancedda et al., 2003).

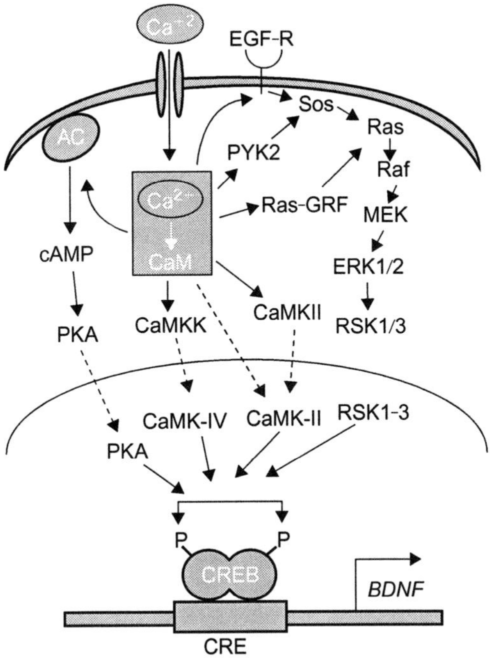

Figure 2.7 Calcium-activated signaling pathways that regulate gene transcription. In neurons, neurotransmitter reception and membrane depolarization lead to the opening of ligand- and voltage-gated calcium channels. Subsequent calcium influx across the plasma membrane drives the activation of a number of signaling molecules, including the calcium-sensitive adenylate cyclase, calcium/calmodulin-activated kinases, and Ras. Each of these molecules activates a cascade of signaling proteins that amplifies the calcium signal and carries it to the nucleus. Dashed lines represent the components of each pathway that are proposed to translocate into the nucleus. Nuclear kinases, including protein kinase A, CaMK-IV, and members of the Rsk family phosphorylate CREB at Ser-133, rendering it competent to mediate transcription of genes such as *BDNF*. *Abbreviations*: *BDNF*, brain-derived neurotrophic factor; CaMK-IV, calcium–calmodulin-dependent protein kinase type IV; CRE, cAMP-response element; CREB, cAMP-response element binding protein; ERK, extracellular signal-regulated kinase; PKA, protein kinase A; PYK, Sos, Raf, MEK, protein kinases; Ras, small G (guanine nucleotide-binding) proteins, acting as secondary messengers; RSK, ribosomal kinase.
Source: From West et al. (2001).

As is shown in the figure, expression of BDNF results not only from any direct action of the visual stimuli on the gene but also from instructions generated by neurons of the visual cortex in response to the visual input, which are communicated to specific neurons for expressing BDNF, via specific channels (signal transduction pathways).

Epigenetic Manipulation of Genetic Information in the CNS

Besides the *manipulative expression of genes* described so far, the CNS exhibits another equally impressive capability, generally known as *gene splicing*, that I would characterize as *manipulation of genetic information*. This is alternative splicing that allows *adaptive* production of different variants of proteins by the same gene through the combination of specific exons into the mature mRNA. Manipulative splicing is a widespread mechanism of generating protein diversity in metazoans. It is used for expression of more than one-third of human genes (Hanke et al., 1999).

From a theoretical point of view, this defies a basic tenet of the classical genetics on the gene as carrier of information for a protein, RNA, or polypeptide: epigenetic mechanisms make possible the use the "raw" genetic information contained in a gene for manipulatively producing a varying number of proteins.

Neural wiring specificity in *Drosophila melanogaster* is thought to be determined by the fact that each neuron can produce a great number of specific protein isoforms. So, for example, in the fruit fly, the Down syndrome cell adhesion molecule (*Dscam*) gene encodes a transmembrane receptor protein necessary for axon guidance. The *Dscam* gene alone can potentially generate 38,016 different protein isoforms (Celotto and Graveley, 2001; Graveley et al., 2004; Kreahling and Graveley, 2004), i.e., almost twice as many transcripts as the total number of genes in the whole human genome.

Certainly, pre-mRNA splicing, excision of introns and assemblage of exons into a translation-ready mRNA, is a *constitutive* process that takes place in cells all over the animal body rather than in the CNS alone. However, in our context, the following should be kept in mind:

First, only eukaryotes, and no prokaryote unicellulars, are capable of mRNA splicing. Second, unlike the *constitutive splicing*, occurring in cells throughout the animal body, *manipulative splicing* is most common in the CNS (Lee and Irizarry, 2003) and it may be the main mechanism for generating the great number of protein isoforms necessary for complex brain functions (Lipscombe, 2005).

In this work, I will use the term *manipulative splicing* instead of the common designation of the phenomenon as *alternative splicing* because it better expresses the process of the adaptive specification of pre-mRNA out of the great number of alternatives. (In the Longman Advanced American Dictionary, the verb *manipulate* is defined as follows: "to work with, or change information, systems, etc., to achieve the result that you want.")

Manipulative pre-mRNA splicing is regulated by extracellular signals and is controlled in a tissue-specific manner (Celotto and Graveley, 2001), implying that it is an epigenetic phenomenon. Although manipulative splicing is observed in other organs (and the possible role of the local innervation in these organs is not investigated), it is the CNS where this form of splicing is extensively used:

> *Alternative splicing in neurons was first discovered in the CA/CGRP gene encoding two peptides that differ in the terminal exon: calcitonin in thyroid cells and CGRP (calcitonin gene-related peptide) gene in specific neurons. Since then, alternative splicing has been found to be a common mechanism used in the generation of isoforms in a variety of functionally important neuronal genes. Although the biological significance of much of the alternative splicing is not always readily evident, in many instances it is thought to be important in the fine tuning of neuronal function.*
>
> *Lisbin et al. (2001)*

Manipulative splicing in the nervous system is regulated by a neuron-specific system of splicing. The electrical activity seems to be at the basis of production of different protein isoforms in the CNS (Mu et al., 2003; Lipscombe, 2005). Calcium elevation resulting from the electrical activity induces specific changes in splicing and translation processes. Among the brain-specific RNA-binding proteins regulating manipulative splicing in metazoans are Nova, Fox1, Fox2, NaPOR, and ELAV. Some typical examples of manipulative splicing leading to formation of multiple protein isoforms involve proteins like neurexins, stress-axis regulated exon (STREX), and Slo (Slowpoke).

Neurexins are surface cell proteins encoded by three genes, but expressed exclusively in the brain (Ushkaryov et al., 1992). In the brain, these genes are subject to intense manipulative splicing for producing specific neurexin isoforms that are involved in synaptogenesis as has been concluded from the fact that their expression changes during synaptic remodeling (Gorecki et al., 1999). Estimates based on experimental work with expression of neurexins in various parts of the brain have shown that three neurexin genes in the brain of the developing *Xenopus laevis* produce between 600 and 3,000 distinct splicing-generated neurexins (Ullrich et al., 1995).

The gene *ania-6* is transcribed into two distinct mRNAs (long *ania-6* mRNA and short *ania-6* mRNA) coding for two different proteins via differential splicing in the process of their transcription. These proteins belong to a new family of cyclins. Sgambato et al. (2003) have shown that production of the long- or short-splice variants of *ania-6* depend on the intracellular pathway: the long variant is produced when CaMK pathway is used, and the short-splice variant results from the activation of the extracellular signal-regulated kinase (ERK) pathway. Which of these two pathways will be used is determined by the neurotransmitters and neurotrophins that are released from the presynaptic neuron, which, in turn, are computationally determined by the output of the processing of the electrical signals: release of glutamate activates CaMK pathway, while BDNF activates ERK pathway.

Experimental administration of neurotransmitters, glutamate or dopamine, induce differential splicing of exons, leading to production of two different proteins by gene *ania-6* (Berke et al., 2001). The role of neurotransmitters in determining the type of protein isoforms that are produced by splicing, in view of the neural epigenetic mechanisms of neurotransmitter synthesis and secretion, suggests that an epigenetic mechanism is also responsible for the manipulative expression of protein isoforms in the nervous system.

Most of the proteins involved in manipulative splicing are neuron specific. So, for example, Nova-1 is a brain-specific splicing factor involved in manipulative splicing of the pre-mRNA for glycine $\alpha 2$ exon and GABAa exon $\gamma 2L$ receptor (Lisbin et al., 2001). *Drosophila melanogaster* ELAV, a product of the gene *elav*, is specifically expressed in all neurons and all the members of the ELAV family, with only one exception, are exclusively expressed in the nervous system. ELAV is implicated in the manipulative splicing of transcripts of genes *nrg* (*neuroglia*), *erect wing*, and *armadillo*, resulting in production of neural-specific protein isoforms.

For a long time biologists wondered (and still do) how the synapses might generate the great molecular diversity necessary for recognizing each of their potential target neurons out of the myriad of neurons in the CNS. Now we know that this may be enabled by the manipulative splicing, which generates numerous protein isoforms (with different binding affinities) from a single gene. So, for example, it is found that three *neurexin* (*nrxn*) genes in mammalian neurons potentially encode over 1,000 isoforms of neuron membrane proteins (Zeng et al., 2006).

Some leading investigators in the field of the pre-mRNA splicing believe that by investing in the direction of generating greater numbers of protein isoforms in neurons, animals have increased chances that each neuron will produce protein isoforms that are different from neighboring neurons, enabling a neuron's dendrites and axons to distinguish between themselves and their neighbors. A more stringent regulation would require a tremendous investment of genetic regulation (Michalowski, 2005).

In view of the tremendous differences in morphology and function of neurons, manipulative splicing is very important for the development and function of the nervous system. As an illustration of this role of manipulative splicing would serve the example of the mechanism of hearing in chicken. Hair cells are mechanosensory cells that communicate with dendrites of efferent neurons by releasing (and receiving) neurotransmitters via calcium channels. Hair cells within cochlea exhibit widely varying properties according to their position along the basilar papilla (Figure 2.8).

Each hair cell is connected to a particular nerve fiber from cochlear ganglion neurons and is more sensitive to a particular sound resonance. The specific frequencies of each hair cell depend partly on the number and properties of calcium-activated potassium (K_{Ca}) channels, which, in turn, are determined by different proteins generated by manipulative splicing at seven sites of the chicken gene *cSlo* (homologue of *Drosophila Slowpoke*).

The splice variants of the Slo protein are numerous and different types of spliced proteins in each hair cell, by controlling the nature of channels, determine the specific sound frequency to which each of those cells responds (Rosenblatt et al., 1997; Black, 1998).

Manipulative splicing of genes in the CNS may also change under the influence of social factors. So, for example, chronic stress is demonstrated to induce changes in the manipulative splicing in male tree shrews exposed to a dominant male. The exposure is accompanied by general symptoms similar to human depression and, at the cellular level, by a decline in the proportion of mRNAs containing the STREX in their adrenals (McCobb et al., 2003). Rats also respond to the transient abrupt decline in the corticosterone level during the stress hyporesponsive period (SHRP) in the early postnatal life by decreasing the proportion of STREX splice variant in Slo K^+ channels in the pituitary (Lai and McCobb, 2006).

Besides stress, various hormones are known to be involved in the manipulative splicing. For example, chromaffin cells, postganglionic secretory neurons in the adrenal gland, respond to adrenal glucocorticoids and to androgens by modulating splicing of *Slo*, thus fine-tuning their excitability and secreting catecholamine (Lai and McCobb, 2002). As for the hormonal regulation of splicing in nonneural tissues (Varayoud, 2005) and organs, let us remember that the hormonal regulation upstream is under neural control via various hypothalamic–pituitary–target gland axes.

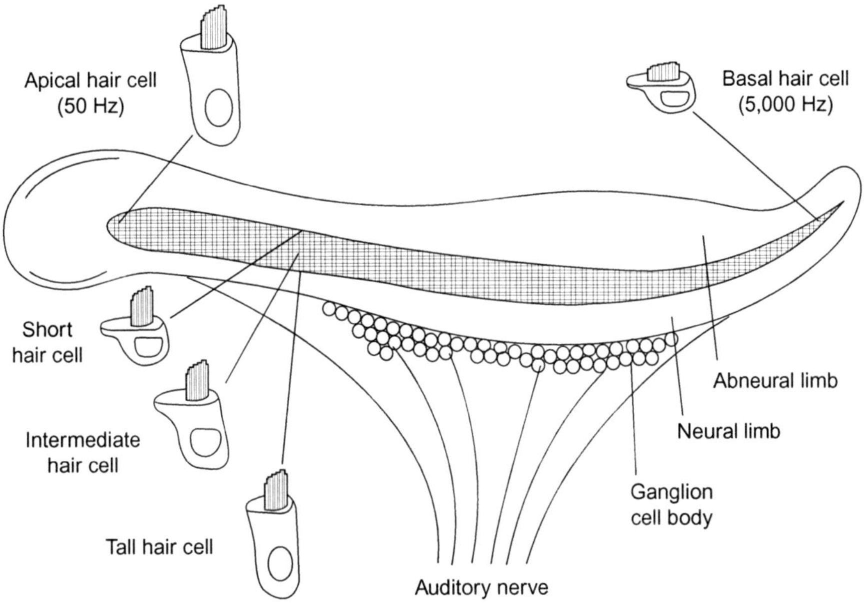

Figure 2.8 Tonotopic and morphological gradients of the chicken's cochlea. Schematic illustration of the chicken's cochlea. In an adult chicken, the cochlea is a 5 mm tubular structure comprising the basilar papilla (stippled), the tegmentum vasculosum (a vascularized secretory tissue), three scalae (fluid-filled compartments), and the cochlear ganglion. The avian homolog of the mammalian organ of Corti, the basilar papilla is an auditory epithelium containing about 10,000 hair cells, and twice as many supporting cells, but no neuronal cell bodies. It rests atop a mechanically tuned basilar membrane that forms the base of the scala media. The hair cells are arranged in a tonotopic gradient along the cochlea, responding to sounds whose frequencies range from 50 to 5,000 Hz; low frequencies are detected at the organ's apical end and high frequencies at its base. These hair cells also manifest two morphological gradients. First, hair cells at the apical end of the cochlea have relatively few, long stereocilia, and those at the base almost 10 times as many shorter stereocilia. Second, there is a gradation in the size and shape of hair-cell somata across the width of the basilar papilla, from tall hair cells overlying the cochlear ganglion to short cells on the abneural edge. The tall and short hair cells are dominantly innervated by afferent and efferent nerve fibers, respectively; they are believed to be analogous to the inner and outer hair cells, respectively, of the mammalian cochlea.
Source: From Rosenblatt et al. (1997).

Closely related to manipulative splicing is A-to-I editing (RNA editing), the process of inosine substitution of particular adenosines, catalyzed by adenosine deaminases acting on RNAs (ADARs), which deaminate adenosine to inosine. Abundant A-to-I editing is unique to primates, which have the most complex brain structure among vertebrates (e.g., in humans A-to-I editing is much more prevalent than in mice; Mattick, 2007).

RNA editing is a general epigenetic mechanism of modification of expression of genetic information in the CNS; by changing the amino acid sequence of proteins, it may lead to changes of their function as it occurs in processes of determination of neuronal identity and maturation. RNA editing is used for modifying ion channels and ligand-gated receptors in the processes of fast neural transmission (Seeburg, 2002).

Selective Elimination of Genetic Information in the CNS

In vertebrates, the CNS evolved another extraordinary ability for controlled loss of whole chromosomes. Thus, differences evolved between the nervous system and other systems or organs not only at the level of gene expression and protein production but also at the level of the number of chromosomes.

The discovery that the number of chromosomes in neurons is not constant has been one of the paradoxical results of the recent studies on the karyotype of neurons in the CNS. Defying a basic genetic tenet that all somatic cells of an organism are genotypically identical, those studies have shown that ~33% of mouse neuroblasts have aneuploid number of chromosomes (most commonly they lack one chromosome) (Rehen et al., 2001), with an average aneuploidy of 1.21–1.45% per chromosome (Yurov et al., 2007). This genetic mosaicism is a normal feature of the human brain (Rehen et al., 2005; Westra et al., 2008) and can be exclusively confined to the brain (Yurov et al., 2007). Besides, it has been demonstrated that when one partner of the chromosome pair 15 is lost, the remaining chromosome of the pair cannot express the gene for the enhanced green fluorescent protein (eGFP), and expression levels of several genes relative to the GFP-expressing controls are permanently altered (Kaushal et al., 2003).

This is an additional mechanism of control and regulation of gene expression, unique for the CNS. It is suggested that these permanent genomic changes may be responsible for physiological and behavioral differences among individual organisms not accounted for by classical genetics (Rehen et al., 2001).

The genetic mosaicism of neurons in the CNS may have contributed to the elaboration of the neural signaling apparatus in metazoans: a network composed of intermixed euploid and aneuploid neurons might produce unique signaling properties distinct from a network composed of purely euploid cells. Another function of aneuploid neurons may be to provide brain circuits with selective advantages, analogous to models of aneuploid tumor cell growth (Kingsbury et al., 2005).

While we have no real knowledge on the mechanism and rules governing the elimination of those chromosomes in the CNS, the fact that the CNS, with one exception, is the only organ system in the body whose cells normally exhibit systematically variable numbers of chromosomes, may suggest that *a neural mechanism is responsible for regulating this variability in the number of chromosomes*. Selective elimination of chromosomes outside the CNS is observed in the process of control of the sex in the offspring. In aphids, for example, elimination of the X chromosome in gametes is used for producing male insects. There is significant evidence that elimination of chromosomes during male meiosis in sciarid flies *Trichosia pubescens* and *Sciara ocellaris* is an epigenetic function performed by microtubules of the cytoplasmic bud (Amabis et al., 1979; Esteban et al., 1997; Fuge, 1997).

Neural Circuits Generate Epigenetic Information

Formation of trillions of prenatal specific, experience-independent, synaptic connections between neurons in the brain of vertebrate embryos and their postnatal experience-dependent refinement is still an enigma. Where does that vast amount of information for establishing these connections in an experience-independent way, before birth, come from?

Let us start by examining the material carriers of that information. There is evidence suggesting that synaptic configuration of neural circuits may code and store information in metazoans. So, for example, in both *in vitro* and *in vivo* experiments a correlation is observed between the number, rearrangement, and shape of synapses and the long-term (day to weeks) memory (Hawkins et al., 2006; Benfenati, 2007; Villareal et al., 2007; Bailey and Kandel, 2008). Synaptic plasticity is a response of circuits to the input of firing rate and spike timing (Sjöström et al., 2001), *but there are reported cases when synapses are established between the pre- and postsynaptic neurons* even before the onset of spontaneous presynaptic activity (Blagburn et al., 1996).

In view of the fact that the chemical output of neural circuits determines the signal cascades that will be activated and ensuing morphophysiological outcomes, it is important to know what determines those outputs. Generally, it was believed that the "neurotransmitter profile," the type of neurotransmitter that a neuron releases, is genetically determined during the early embryogenesis, when the neuron is differentiated. But recent experimental evidence shows that the neurotransmitter phenotype of neurons can epigenetically change, according to the electrical activity of the neuron. Experimental disruption of Ca^{2+} spike activity pattern in the embryonic spinal cord alters neurotransmitter expression in neurons, and suppression of that activity increases the number of glutamatergic and cholinergic neurons. Now it is admitted that, as a rule, changes in the electrical activity can change the type of neurotransmitter the neuron releases (Borodinsky et al., 2004).

Neural circuits respond to various stimuli by reconfiguring their synaptic morphology (Gould et al., 1990; Turrigiano et al., 1994; Turrigiano, 1999; Calizo and Flanagan-Cato, 2000; Widmer et al., 2003). Synaptic changes may act synergistically to produce a 25-fold increase in the spontaneous firing rate that is observed in the layer 4 of the visual cortical circuitry, which may enhance its ability to amplify sensory signals when sensory drive is reduced (Maffei et al., 2004). Neural circuits respond to afferent input on the level of sexual hormones (estrogen and progesterone) by reconfiguring their synaptic morphology. Such is the case with an elevated level of estrogen, to which neurons of the circuit for female sexual behavior in rats respond by increasing the number of dendritic spines in the hypothalamic ventromedial nucleus (VMH) (Frankfurt et al., 1990; Figure 2.9).

Besides the VMH, changes in the number and shape of dendritic spines in response to female sex hormones also occur in specific neuronal populations of the adult female hippocampus. Removal of circulating gonadal steroids by gonadectomy results in a decrease in CA1 pyramidal cell dendritic spine density, but this can be prevented by estradiol treatment (Gould et al., 1990).

It has been possible to experimentally induce reconfiguration of synaptic morphology by applying various substances or by depriving animals of other substances (Michaelson et al., 1996).

Reconfiguration of synaptic morphology may change computational properties of neural circuits. Such is the case, for example, in CHL1-deficient mice, where synaptic reconfigurations lead to serious consequences in the information processing and computational properties of neural circuits (Montag-Sallaz et al., 2003).

There is a song-learning circuit in male zebra finches that connects medial dorsolateral nucleus of the thalamus (DLM) and cortical lateral magnocellular nucleus of the anterior neostriatum (lMAN) (Iyengar and Bottjer, 2002). This circuit underlies song learning from tutors by male birds. It is believed that DLM/lMAN circuit functions by detecting errors in the process of comparison of the male's own motor output with the

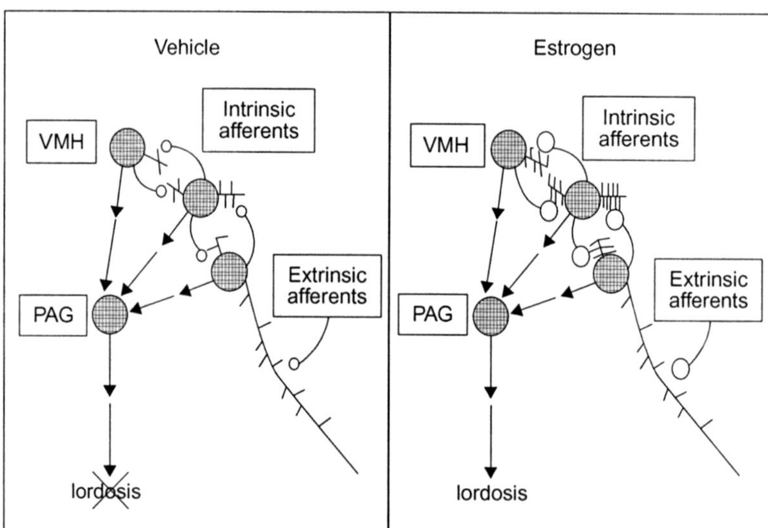

Figure 2.9 Model of possible mechanisms of spine induction on short primary dendrites in the ventrolateral VMH. The effect of estrogen on cells lacking ERs suggests that estrogen acts transynaptically, rather than directly, to induce dendritic spines. The afferents that stimulate spine formation may be either extrinsic or intrinsic to the VMH. Extrinsic afferents are mainly found in the neuropil surrounding the VMH. Long primary dendrites extending to the ventrolateral border of the VMH may be innervated by extrinsic afferents in the neuropil. Intrinsic afferents may innervate short primary dendrites. Estrogen affects local intrinsic afferents to increase spine density on short primary dendrites. *Open circles* indicate excitatory synapses; *arrows* indicate direction of information flow through the circuit. *Abbreviations*: PAG, periaqueductal gray matter in the midbrain; VMH, ventromedial nucleus of the hypothalamus.
Source: From Calizo and Flanagan-Cato (2000).

tutor's song. During the sensitive period of vocal learning, between days 20 and 35, the shell subregion of lMAN and the DLM terminal field experience a threefold increase in the volume that is followed by a sharp decrease between day 35 and adulthood. The dramatic changes in the volume of the above structures related to the song learning reflect dynamic rearrangements of expansion and retraction of axonal structures rather than changes in the number of individual axons. Large-scale experience-dependent pruning of axon branches occur in the course of song learning:

> *The changes in complexity and spatial extent of DLM axon arbors described in this study indicate synaptic remodeling within the DLM → lMAN circuit, which may represent structural correlates of vocal learning. For example, arbor regression may represent a morphological correlate of the auditory tuning of lMAN neurons to the bird's own song and the decreased involvement of lMAN neurons in vocal learning.*
>
> *Iyengar and Bottjer (2002)*

Let us remember that song learning, as a learned behavior, is a phenotypic character that takes place in the specific circuit of the bird's brain. Singing a species-specific song in zebra finches implies the presence of specific information for performing the song. There is only one rational answer that I can think of, to the question: Where that information is stored? The observed correlation between the changes in synaptic morphology and the song learning suggests that the information for performing the song is encoded in the specific synaptic morphology that develops in the brain song circuit of male zebra finches in the process of song learning.

A problem related to the information that may be encoded in the synaptic morphology is the mechanism by which the specific synaptic morphology, which changes easily under the influence of activity, is maintained for determined periods of time. It is believed that neural circuits have their own homeostatic regulatory mechanisms for stabilizing and maintaining their morphology under circumstances of increased or decreased activity (Turrigiano and Nelson, 2004).

The synaptic morphology remains unchanged in the absence of appropriate stimuli, i.e., as long as the perception of the environment remains unchanged. The reconfiguration of synaptic morphology, which is function of large-protein signal-processing machines at the postsynaptic membrane (Kennedy, 2000), may modify the computational properties (von der Malsburg, 1999, 2002) of the neural circuit. It is believed that changes in the structure of the postsynaptic membrane represent the fundamental mechanism for the processing and storage of information for learning and memory in the brain (Luscher et al., 2000). It has been determined that reconfiguration of synaptic morphology of a circuit in the hypothalamic SCN "codes" the information for day length in rats (Schaap et al., 2003).

In experiments, it has been shown that reconfiguration of synaptic connections and the resulting changes in computational properties of circuits induce specific changes in the behavioral or morphological output of the circuits (Getting, 1985). A number of investigators have reported cases of reconfiguration of synaptic morphology in response to changes in the embryonic structure during the individual development (Matsumoto and Arai, 1986; Kent and Levine, 1993) and the transformation of the Bauplan during the metamorphosis in anuran amphibians is accompanied by a "complete reorganization" of neural circuits (Alley, 1990), which also implies reconfiguration of their synaptic morphology.

A correlation exists between the morphological changes taking place during metamorphosis of crickets and the refinement of brain synaptic morphology, via rearrangement of synapses, involving the selective elimination of some synapses and strengthening of others (Lnenicka and Murphey, 1989).

Smoltification is the process of physiological, morphological, and behavioral changes that young salmonids undergo for facilitating transition from freshwater to saltwater during their migration. At a neurobiological level, these phenotypic changes are preceded by changes in synaptic connections in many brain circuits in response to photic cues (day length) (Ebbesson et al., 2003).

Under the influence of mechanical (touching hairs on their hind limbs), olfactory (smell of conspecifics), and social stimuli (overcrowding), the solitary desert locusts, *Schistocerca gregaria*, change their behavior, morphology, color, and morphometry and transform into gregarious locusts within a few to 24 h. The phase change involves activation of 532 genes (Kang et al., 2004). Phase transition in the desert locust is preceded by important changes in the CNS, in the synaptic morphology and in the amount and types of neurotransmitters produced (Rogers et al., 2004) as well as in secretion of neuroparsins, small neuropeptides, by pars intercerebralis of the locust CNS (Claeys et al., 2006). All of the gregarizing stimuli, received via hind legs, eyes, and antennae, converge on the thoracic ganglia for processing in a neural circuit whose output is a neurotransmitter, serotonin. The neurotransmitter is demonstrated to be both necessary and sufficient to induce gregarization (Anstey et al., 2009). With secretion of serotonin starts a complex network of biochemical reactions that determines dramatic phenotypic changes in behavior, morphology, physiology, and life history that characterize the transition from the solitary to the gregarious phase (Clynen et al., 2006).

Sometimes, neural networks of different configurations can produce very similar outputs (Prinz et al., 2004; Hooper, 2004). Such is the case with the pyloric rhythms of the crustacean stomatogastric ganglion: by using the most different combinations of synaptic strengths and neuron properties, investigators succeeded in obtaining the same pyloric rhythm (Prinz et al., 2004). Other times, the same crustacean pyloric network, in response to different modulators, can produce different rhythms.

Here is an interspecific example of the ability of neural circuits to generate different and even opposite outputs by processing the same stimulus. In response to lengthening of the photoperiod, cows respond by increasing their reproductive activity in the spring, while the reproductive season in sheep starts with the shortening of the photoperiod in the summer. Under the apparent differences in response to the stimulus (caused by differences in their gestation periods) lies the identity of responses in respect to the "purpose," *sensu* Mayr, of rearing their offspring in the most favorable period of the year. In both cases, the beginning of the reproductive activity is not stimulus dependent but goal oriented.

What all of the above examples have in common is that, in response to environmental stimuli, neural circuits are reorganized so that by processing these stimuli generate chemical outputs that trigger activation of specific signal cascades or gene regulatory networks (GRNs) determining the phenotypic changes in behavior, physiology, morphology/morphometry, and life history without changes in genes or DNA. Thus, neural circuits, by selecting specific signal cascades and GRNs, create new causal relationships.

The fact that, in principle, neural circuits are capable of relating any external stimulus to any signal cascade, and of determining the timing of the activation of the signal cascade, implies that they generate *de novo* information to develop new phenotypic characters. To say that neural circuits, by modifying their synaptic connections and computational properties, release alternative chemical signals and activate diverse signal cascades, which lead to alternative morphological results, is another way of saying that these circuits generate epigenetic information for animal morphology. Thus, in the course of evolution, the nervous system emerged as the generator of the huge amount of epigenetic information for evolution of metazoan phenotype.

The ability of metazoans to generate epigenetic *information* for the supracellular structure by reconfigurating synaptic morphology will not come as a surprise to us, when we remember that the CNS for a long time is known to determine the behavior, homeostasis, and physiology of animals and can generate and store *knowledge* ("as is the reigning opinion, neural connections are the repository of knowledge in the brain"; von der Malsburg, 2002).

Neural Processing of Stimuli Generates Information for Postphylotypic Development

After learning that genes are quantitatively insufficient and qualitatively inappropriate for determining animal morphology, biologists began grappling with two enigmas that still continue to puzzle them:

1. What is the function of trillions of specific prenatally established connections between neurons?
2. Where and in what form is the huge amount of information (exceeding millions of times the amount of genetic information contained in the metazoan genome) necessary for building the metazoan structure stored?

Formation of primitive neural circuits in the embryo may be a function of parental epigenetic information provided with the gamete(s) as part of the development of the embryonic CNS during the phylotypic stage. At that stage, neurons from various regions of the CNS extend their axons to form connections with specific neurons, thus creating primitive functioning neural circuits. This process takes place in absence of electrical activity (Table 2.1), based on the parental epigenetic information provided via gamete(s) and/or transplacental in placental organisms. The beginning of the spontaneous and sensory-driven electrical activity in response to the developing embryonic structure enables fine-tuning of the imprecise synaptic connections of the primitive neural circuits. Experimental blockade of the endogenous and sensory-driven activity prevents formation of the normal patterns of neuronal connections (Penn and Shatz, 1999; Penn, 2001). The development and refinement of the neural circuits and neuronal connections during the embryonic period of individual development occur in an experience-independent mode. The process of fine-tuning of the neural circuits continues during the postembryonic development and even after birth in an experience-dependent mode.

The fact that the myriad of prenatally (i.e., experience independent) established neuronal connections in an animal's CNS are not random, but strictly determined, clearly indicates that information of some kind is invested for establishing these specific connections and neural circuits. The high informational and energetic cost of that investment implies that the resulting evolutionary and developmental advantages outweigh the cost of the investment.

What could be the functions and advantages of the huge network of quadrillions of synaptic connections in the brain?

Let us remember that this network is established during the embryonic development, implying that it is created in an experience-independent manner and it does not serve the embryo's communication with the environment. What purpose might it serve at that time? In an attempt to answer that question, I put forward the hypothesis that:

The huge and dynamic network of the embryonic neuronal connections established during embryogenesis, before the embryo starts communicating with the external environment, may be necessary for generating the morphological information for the postphylotypic development, organogenesis, and morphology in metazoans.

The overwhelming majority of these connections are determined by the spontaneous electrical activity and computational activity of the CNS during the individual development, in response to the input of internal/external stimuli. This is demonstrated by the fact that, in experimentally verified cases, blockage of afferent input in embryos prevents formation of neural circuits and dendritic spines (Katz and Shatz, 1996; Penn et al., 1998; Segal et al., 2003).

It is interesting to observe that, after the phylotypic stage, when the reserve of the parental epigenetic information (cytoplasmic factors) is consumed, the embryonic development proceeds at an accelerated pace. Rapidly, all the tissues and organs begin to take shape.

Table 2.1 Formation and Refinement of Neural Circuits

Embryonic Stage	Development of Neural Circuit	Processes Driving Formation	Source of Information
Prephylotypic stage	Primitive neural circuits develop	Activity independent and experience independent	Parentally provided epigenetic information
Postphylotypic stage (critical periods)	Refinement of connections and sculpting of new neural circuits	Activity dependent (spontaneous or sensory)	Neurally generated epigenetic information in response to internal sensory input
Postnatal development	Further refinement of neural connections	Activity dependent and experience dependent	Neurally generated epigenetic information in response to external sensory input

At first sight, this explosive development, at a time when the parental epigenetic information is exhausted, is paradoxical. Where might that tremendous amount of information necessary for histogenesis and organogenesis, for erecting the immensely complex metazoan structure after the phylotypic stage, come from? The fact that exactly at this juncture, when parentally provided epigenetic information is exhausted, the embryonic CNS becomes functional (i.e., starts patterned electrical activity) might not be a sheer coincidence.

Speculations aside, is there evidence that the CNS generates information for the postphylotypic development? From the phylotypic stage forward, the CNS becomes functionally active; a continuous input of internal stimuli (changes in the rapidly developing embryonic structure), via afferents, is communicated to the CNS. The embryonic CNS responds to that input by "spontaneous" electrical activity and by establishing specific neural circuits (McAllister, 2000; Peinado, 2000; Zhang and Poo, 2001).

It is not always clear what drives the spontaneous activity in the embryo, but, in the spinal cord at least, spontaneous burstings of ~1 min duration are followed by interepisode pauses of minimal electrical activity. These "episodes" are preceded by motoneuron firing and experimental stimulation of motoneurons also induced Renshaw cells interneurons in the process of starting an episode (Wenner and O'Donovan, 2001). This activity occurs spontaneously without any descending or afferent input with motoneurons playing a critical role in generating spontaneous activity (Hanson and Landmesser, 2003). For example, the precise specification of projections from the retina to the lateral geniculate nucleus starts during embryogenesis at a time when there is no visual input from the environment, and although the retina still cannot respond to photic stimuli, the retinal ganglion neurons show spontaneous activity (Penn and Shatz, 1999).

Spontaneous activity may represent training and learning patterns (Albert et al., 2008) of neurons before the birth, i.e., before the animal can receive external sensory information. This activity is important for the assembly of the nervous system, neuronal proliferation, neurotransmitter phenotype, migration of neurons, and formation of synaptic morphology of neural circuits (Spitzer, 2006). Adequate evidence suggests that spontaneous electrical activity is responsible for sculpting circuits on the basis of the brain's "best guess" (Katz and Shatz, 1996), i.e., computationally, and that activity might represent a "self-organizing property" (Weliky, 1999). In the embryonic retina, for example, this activity "can produce highly stereotyped patterns of connections before the onset of visual experience" (Penn et al., 1998) and "instruct" formation of eye-specific layers (Shatz, 1996). By contrast, in the absence of afferent excitatory input, normal neural circuits are not formed (Katz and Shatz, 1996; Penn et al., 1998), and rat striatal neurons do not develop dendritic spines (Segal et al., 2003). Experimental deafferentiation causes retraction of postsynaptic dendrites, and a particular level and/or pattern of afferent activity is necessary for postsynaptic dendritic morphology to develop.

The fact that in response to the input of afferent stimuli (changes in the developing embryonic structure) neural circuits modify their synaptic morphology suggests that synaptic connections are somehow related to the developing embryonic structure. Indeed, experimental delay of muscle development causes suspension of synaptic branching of respective motoneurons (Fernandes and Keshishian, 1998). Male rats castrated on day 1, implying that they do not receive testis input, have significantly reduced numbers of shaft and spine synapses in the VMH (Matsumoto and Arai, 1986), and removal of ovaries causes a profound decrease in the dendritic spine density of pyramidal cells in the hippocampus. In *D. melanogaster*, where the larva lacks the target muscle (no afferent input from the muscle) the axon of the motoneuron MN5 develops no dendritic connections (Consoulas et al., 2002). Experiments on the moth, *Manduca sexta*, also show that the input of stimuli from the adult leg is involved in shaping the growth of motoneuron dendrites (Kent and Levine, 1993).

All of the above evidence suggests the existence of a causal relationship between the afferent input from the developing embryonic structures to the CNS and the patterns of neuronal connections and synaptic morphology. As pointed out earlier, modification of the synaptic morphology in response to the changes in embryonic structures is a costly process that hardly could have evolved if it did not serve any important purpose, which would be inconsistent with the way organic evolution works.

While adequate evidence has shown that the synaptic morphology changes in nonrandom, strictly determined, ways, is there evidence that the CNS provides information for sequential stages of the development of embryonic structures? At an empirical level, in Chapter 5 (Neural Control of Postphylotypic Development), I will provide substantial evidence that the embryonic CNS, the brain and the spinal cord, is the source of numerous inductions for the development of organs and parts of the embryo and the homotopic transplantation of parts of the embryonic CNS between species leads to transformation of the host morphology into donor-like morphology.

The electrical activity of the CNS, and the synaptic morphology related to it, are necessary for the embryonic development. In experiments, it has been shown that the spontaneous electrical activity arising in response to the input of internal stimuli (changes in the embryonic structure) is necessary for the development of the embryonic structure. So, for example, experimental paralysis of chick embryos prevents normal development of synovial (diarthroidal) joints, of articular structures and articular surfaces, causing these structures to coalesce (Persson, 1983; Pitsillides, 2006). Suppression of the spontaneous activity by paralyzing motoneurons in chick embryos (Hall and Herring, 1990) and duck embryos (Creazzo and Sohal, 1983) results in reduction of the bone- and muscle growth, whereas denervation totally prevents the development of muscles in duck embryos (Sohal and Holt, 1980; Creazzo and Sohal, 1983) and *Drosophila*.

Why does denervation prevent the development of embryonic muscles? If matter, energy, and information are all that are needed for the embryonic muscle and bone development, in the above cases, matter and energy are normally supplied by blood circulation in the local vasculature. What is suppressed by denervation is the normal flow of information necessary for osteo- and myogenesis. Indeed, based on their experimental work, Wenner and O'Donovan (2001) have shown that

> *many developing networks exhibit a transient period of spontaneous activity that is believed to be important developmentally ... in the spinal cord, spontaneous activity has been implicated in the development of limb muscles, bones and joints.*
>
> *Wenner and O'Donovan (2001)*

In its entirety, the above experimental evidence allows us to conclude logically that, with matter and energy supplied by body fluids, the necessity of innervation for developing embryonic structures may indicate that local innervation conveys information for the development of those supracellular structures (muscles, bones, and joints).

Now, let us see how the CNS, by processing the internal stimuli, coming via afferents from the developing embryonic structures, starts its electrical "spontaneous activity" and the information-generating activity.

During embryonic development, each of ~1 trillion mammal cortical neurons establishes specific connections with an average of 10,000 other neurons, leading to formation of the cortical networks. The immature neuronal networks apparently use the inherited information and the input of information from the developing embryonic structure for generating epigenetic information in an experience-independent mode. Their electrical activity not only determines establishment of synaptic connections but is also believed to control, via some prenatal experience-independent learning, several developmental processes of cell differentiation and also embryonic cell migration (Khazipov and Luhmann, 2006). In the early stages of rat development, the primary auditory neurons, before the onset of hearing, respond with "spontaneous" electrical activity to the "spontaneous" release of ATP by the developing cochlea cells and to the release of glutamate by inner hair cells (Tritsch et al., 2007; Blankenship and Feller, 2010). Removal of cochlea (no input from the developing cochlea) prevents the spontaneous firing of the auditory neurons (Jones et al., 2001).

Numerous experimental studies have shown that the morphology of neuronal connections in the lateral geniculate nucleus, whose establishment requires a huge amount of information, is determined predictively in an experience-independent mode, before any visual experience. Interestingly, about 60–80% of the trillions or quadrillions of specific neuronal connections of higher vertebrates are preformed during the embryonic life, i.e., before the embryo starts to interact neurally with the external environment. Whereas that evidence demonstrates experience-independent information-generating properties of the CNS, principles of neural computation enabling the CNS to generate that information remain obscure.

The mechanism of generation of the epigenetic information is not known, but it has been generally related to the "self-organizing property" of the CNS. There is adequate evidence that the formation of neural circuits as functional units of the nervous system during the embryonic development is determined by the CNS in an experience-independent mode primarily in response to internal stimuli. The CNS takes as stimuli not any afferent input but only inputs of stimuli, neurobiologically perceivable changes in the structure and function of the embryo. The *developing* embryo is thus the source of ever-changing flow of afferent input of stimuli to the embryonic CNS, which is used for generation of the epigenetic information for consecutive stages of the development of morphology.

The mechanism of generation of epigenetic information in the CNS will be illustrated later, in the Section Generation of Information for Adaptive Camouflage in *Xenopus*.

The central idea developed in this section is that the establishment of the early neural networks, coinciding with the termination of the activity of parental cytoplasmic factors (epigenetic information) at the phylotypic stage, as part of the early embryonic development, is determined by the regulatory activity of the parental epigenetic information provided with gametes (and also transplacental in placental mammals). The huge amount of the epigenetic information necessary for organogenesis is generated by the embryonic CNS, which is operational from the phylotypic stage forward.

Processing-Dependent Expression of Genes and the Blood–Brain Barrier

The complex and intense communication between neurons in the CNS requires a noise-free medium for reliably transmitting and receiving chemical signals. The processing-dependent expression of genes in the CNS would be of no use if extracellular signals would freely circulate in the CNS and unrestrictedly use signal transduction pathways for "classically" activating neuronal genes in the brain. An evolutionary pressure for utilizing advantages of the mechanism of the processing-dependent, manipulative expression of genes, by minimizing the availability of circulating extracellular inducers in the CNS, led to the evolution of the blood–brain barrier, the Great Wall that separates the CNS from the rest of the body. Essentially, the blood–brain barrier consists of a unique arrangement of endothelial cells in the brain capillaries, which by forming tight junctions among them, make a selective penetration of circulating substances in the brain possible.

The evolution of the blood–brain barrier, which prevents high molecular compounds (above 5,000 daltons, such as protein molecules) from entering the CNS, makes the CNS inaccessible to many signaling molecules what would seriously impair systemic monitoring capability of the CNS as controller of the ICS.

As a solution to the problem of inaccessibility of various protein inducers to the CNS, the latter evolved a long-range monitoring capability. Internal/external signals are generally received by interoceptors and exteroceptors, mainly located outside the CNS, which convert them into a "common currency," i.e., electrical signals, which via afferents are transmitted to specific neural circuits for processing and decision-making. In other words, the CNS succeeded in maintaining its monitoring capability on a whole range of substances that have not access to the CNS by developing a communicative monitoring system via electrical signals.

The evolutionary pressure for making the CNS inaccessible to circulating, genomically active substances was compromisingly resolved. The fact that the CNS itself needs to get both matter and energy, nutrients and oxygen, determined the fact that the barrier would allow numerous, generally nonprotein, small molecules to reach the brain and spinal cord.

Generation of Information for Adaptive Camouflage in *Xenopus*

The clawed frog, *X. laevis*, is a South African aquatic amphibian that cryptically changes its body patterning according to the background in order to become less visible in its location and reduce the probability of being spotted by its predators. In light background, it becomes lighter, and in darker background, darker. The background adaptation is based on the ability of the frog to regulate hormonally the pigment dispersion in skin melanophores according to the visual perception of the background shades.

How does *Xenopus* reformat the information on the background it receives visually into information for changing its body pigmentation, and how does the frog translate it into a corresponding cryptic pigmentation for background adaptation? In a simplified form (Figure 2.10), the pathway comprises reception of visual information on background by retina, its transmission, via optic tract, to the higher brain centers and hypothalamus for integration and secretion of melanophore-stimulating hormone (α-MSH) by the pituitary, which by modifying dispersion of pigments in melanocytes repatterns the skin to match the background.

The hypothalamic SCN consists of ~10,000 neurons. Different neurons in the SCN vary substantially in their free-running circadian periods from 20 to 28 h with an average 24 h period. The variation in the circadian rhythms in individual SCN neurons may provide information for perceiving the day length. However, it is thought that, to some extent, the SCN neurons also show synchronized activity as a function of distance and the connections with other neurons. Many SCN neurons show similar rhythms with neurons that are closer to them. Similar rhythms are also observed in neurons projecting to the same areas (Saeb-Parsy and Dyball, 2003).

Xenopus laevis receives and processes the optic stimuli on its background color/pattern via the retinal–hypothalamic circuit. When in light-colored background, suprachiasmatic melanotrope-inhibiting neurons (SMINs) in the ventrolateral part of the hypothalamic SCN send inhibitory signals to the melanotrope cells in the pituitary, preventing secretion of melatonin, thus determining a lighter body color. In these cases, at the level of the retino-hypothalamic circuit, the clawed frog shows reduced innervation and synaptic contacts in SMINs, which results from retraction and degeneration of synapses that are present in dark background–adapted frogs (Kramer et al., 2002). This modification of the synaptic morphology seems to be causally related, via pituitary secretion of melanotropin, with the adaptive change in the body coloration of *X. laevis*.

It has been shown that the reconfiguration of synaptic morphology (neuronal plasticity) of the retino-hypothalamic circuit, taking place in response to the visual input on the environmental background, is responsible for generation of information for body repatterning in *X. laevis* (Kramer et al., 2002). That reconfiguration is also responsible for the morphological changes in the fish dermis during background adaptation (van Eys and Peters, 1981; Tuinhof et al., 1994; Sugimoto et al., 2000; Sugimoto, 2002). This suggests that the reconfiguration of the synaptic morphology of the retino-hypothalamic circuit generates the information necessary for the background adaptation in *X. laevis*, and this information is generated by processing the visual information in three brain circuits that send inhibitory signals (Tuinhof et al., 1994) and one in the raphe nucleus that sends stimulatory signals (Ubink et al., 1999) to the melanotrope cells of the pituitary.

Melanotrope cells in *X. laevis* are found in the pars intermedia of the pituitary, and they synthesize and secrete the α-MSH as a component of its precursor,

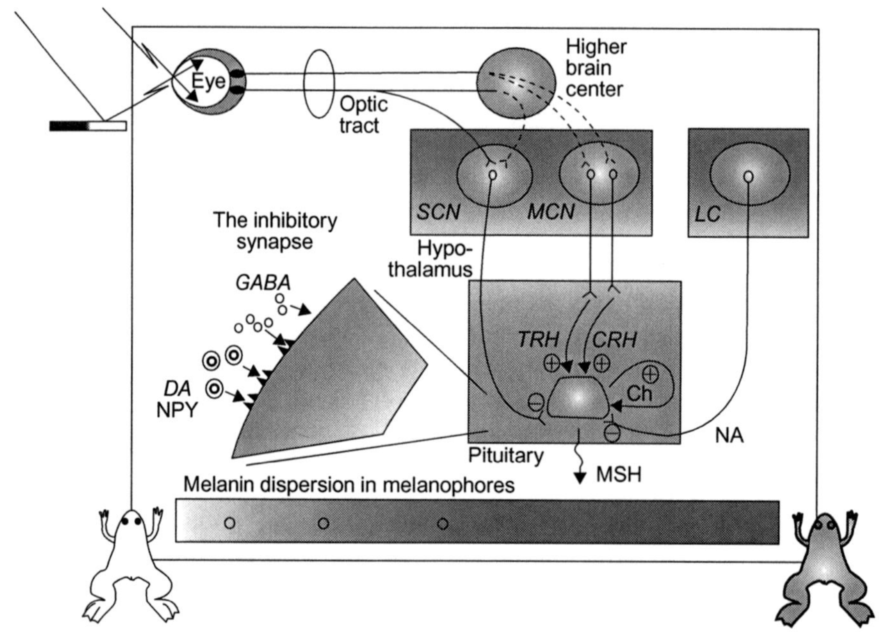

Figure 2.10 Schematic representation of the neuroendocrine reflex regulating pigment dispersion in dermal melanophores of *X. laevis* during the process of background adaptation. Optic input is integrated in the brain, leading to activation or inactivation of neuronal centers in the hypothalamus. Both inhibitory (DA, GABA, NPY) and stimulatory (CRH, TRH) factors are released from the hypothalamic neurons and control α-MSH release from the melanotrope cells. Neurons in the locus coeruleus are assumed to inhibit α-MSH by releasing NA. Finally, acetylcholine (ACh) has an autoexcitatory action on the melanotropes. *Abbreviations*: CRH, corticotropin-releasing hormone; DA, dopamine; GABA, γ-aminobutyric acid; LC, locus coeruleus; MCN, magnocellular nucleus; NA, noradrenaline; NPY, neuropeptide Y (Yl receptor); pn, pars nervosa; pi, pars intermedia; SCN, suprachiasmatic nucleus; TRH, thyrotropin-releasing factor.
Source: From Roubos (1997).

proopiomelanocortin (POMC). The estimated number of melanotrope cells in *Xenopus* is 70,000. A number of neurons in the ventrolateral part of the hypothalamic SCN form synaptic connections with the pituitary melanotrope cells and release their neurotransmitters, dopamine, NPY, and GABA (γ-amino butyric acid). These neurotransmitters have inhibitory effect on the secretion of α-MSH and hence are known *as SMINs*. SMINs respond to reception by the retina of visual input of the light background by releasing GABA, NPY, and dopamine to the synaptic contacts with melanotropes, suppressing α-MSH secretion, thus leading to a lighter body color in *Xenopus*. In contrast, when the background is dark, the SCN responds to the retinal visual input by activating a group of interneurons in the dorsal part of the SCN, which send signals that inhibit secretion of α-MSH-inhibiting neurotransmitters by the SMIN. In fact, the neural control of adaptation of color of *X. laevis* to the background is more complex: to the pars intermedia of the pituitary also project neurons from locus coeruleus, raphe nucleus, and magnocellular nucleus releasing on the melanotrope cells their respective neurohormones/neurotransmitters: noradrenalin, serotonin, and thyrotropin-releasing hormone and corticotropin-releasing hormone, all with stimulating effect on melanotrope cells (Kramer et al., 2002; Figure 2.11).

From a neuroendocrine viewpoint, the synthesis of α-MSH is determined by the release of stimulatory signals (neurohormones/neurotransmitters) on the pituitary melanotrope cells, but the mediators of these neural functions are two converging and interacting second messengers, Ca^{2+} oscillations and cAMP. Ca^{2+} oscillations are involved in the expression of the α-MSH gene, but their pattern is determined by the nature of neurotransmitters on the melanotrope cells.

In summary, it may be said that camouflage in *Xenopus* is epigenetically regulated. It implies a complex computational processing of the visual input in several brain centers and culminates with the release from neurons of these centers of their chemical output via projections on the melanotrope cells of the pars nervosa in the pituitary.

What Do Neural Circuits Do: Sum Up Stimuli or Figure Out Adaptive Responses?

Metazoans, like other living forms, are under the influence of external and internal stimuli and appropriately respond to them. A stimulus is a challenge to which the organism tends to respond adaptively. After being received by the sensory organ, the stimulus in an electrically encoded form is transmitted to a specific neural circuit, which takes it as a problem requiring solution.

The concept of stimulus is a relative one. In the meaning used in this work, a "stimulus" (*stimulus*—Latin for *goad*) is a neurally perceived, significant but not necessarily real, change in a variable, signal, or condition of the external or internal environments to which the CNS responds adaptively or otherwise. The mere presence of a condition, a signal, a variable, or a factor in the environment does not necessarily represent a stimulus. It is the perception by the CNS of a change in those entities that makes a stimulus out of them. A species-specific quantitative threshold or set point exists in the nervous system that determines when an agent will be perceived as a stimulus and will be appropriately responded to. Therefore, what may be stimulus for a species might not be for another.

But the fact that even false perceptions may lead to the same output that the real perception does, unambiguously shows that the product of processing depends not simply on the stimulus, but is a neural product of an intrinsic drive to adapt the organism to the stimulus or to its consequences or presagings.

Stimuli exist only as *perceived changes*. So, for example, the mere presence of estrogen in the blood circulation is not a stimulus for the CNS; only a rise of hormone level above the upper limit of an estrogen set point, as determined in the CNS, will be received as a stimulus, processed in a specific neural circuit, and will lead to expression of the GnRH gene in hypothalamic neurons.

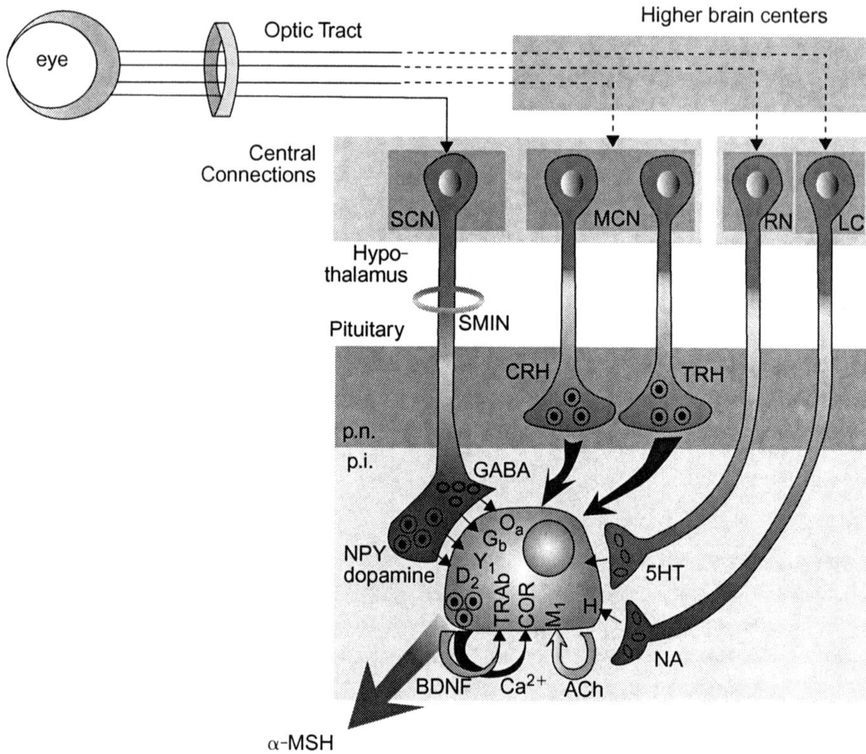

Figure 2.11 Schematic view of the neuroendocrine interface regulating the melanotrope cell in *X. laevis*. The release of MSH and other POMC-derived peptides is regulated by an inhibitory neuron from the suprachiasmatic nucleus (SCN), termed the *suprachiasmatic melanotrope-inhibiting neuron* (SMIN), and by stimulatory neurons from the magnocellular nucleus (MCN), the raphe nucleus (RN), and the locus coeruleus (LC). The SMIN makes synaptic contact with the melanotrope cell and releases the colocalized inhibitory factors neuropeptide Y (NPY), dopamine, and γ-aminobutyric acid (GABA). NPY inhibits via a Y1 receptor, dopamine via a D2 receptor, and GABA via both GABAa (Ga) and GABAb (Gb) receptors. The stimulatory neurons producing thyrotropin-releasing hormone (TRH) and corticotropin-releasing hormone (CRH) terminate in the pars nervosa (p.n.), from where the neuropeptides would diffuse to the pars intermedia (p.i.) to stimulate secretion. The nerve terminals for the serotonin (5HT) and noradrenaline (NA) containing neurons are found within the p.i. The receptor type for serotonin has not been identified; NA stimulates via a β-adrenergic receptor. Acetylcholine (ACh) is a stimulatory autocrine factor working through the muscarinic M1 receptor and brain-derived neurotrophic factor (BDNF), cosequestered with α-MSH, also acts in an autocrine fashion, likely through the TRKb receptor. The melanotrope also expresses the stimulatory G-protein-coupled Ca^{2+}-sensing receptor (CaR), which possibly responds to local extracellular Ca^{2+} increases in the vicinity of exocytosis events.
Source: From Jenks et al. (2003).

By determining set points, the CNS, in fact, sets standards for itself on what may and what may not be taken for stimuli. The thresholds are species-specific, but being neurally determined, they may exhibit intraspecific variability. Individuals of the same species may exhibit different (within-species limits) thresholds or set points, and the same individual may change (reset) thresholds and set points within its lifetime or in response to particular environmental factors.

We take it for granted that a stimulus, after being received by the sensory neurons/organs, must be transformed into trains of electric spikes and processed in neural circuits, sometimes very complex, for obtaining a chemical output that starts a signal cascade that leads to a specific phenotypic change. But if the causal law is relevant in biology, we must ask: Why should the stimuli be processed at all? Why should the stimulus and its electrical representation be led through the intricate mazes of brain circuits? What are the evolutionary logic and pressure behind such an energetically costly processing of external and internal stimuli in brain circuits? Natural selection would not favor evolution of neural processing of environmental stimuli if it would not offer advantages that overweigh the high cost of processing.

An answer to the above questions may be found by looking at the result to which the processing of the stimuli in neural circuits leads. The output of the neural circuit ultimately leads to expression/suppression of a gene or a group of genes, which might not be predicted from the viewpoint of the classical mechanism of gene expression. As argued earlier, there is no physical contact and no direct causal relation between the external/internal stimulus and the gene that will be expressed. Whether the gene will be expressed, or which of genes will be expressed, depends not on the nature of the stimulus but on the result of the processing of the stimulus in the neural circuit(s). In such cases, the phenotypic result is processing dependent rather than stimulus dependent, as is commonly assumed. The observed causal chain of events from the stimulus to the expression of the gene is not predetermined; it is determined by the neural processing, and it cannot occur independently of the nervous system. *Hence, it is the nervous system, rather than the particular stimulus, that determines which of the thousands of genes in the genome will be induced in response to the stimulus.*

The processing of the stimulus in the nervous system bridges the physical gap between the stimulus and the gene, by relating in a contrived causal chain two elements, which otherwise would be causally unrelated.

The function of the processing in the nervous system is to convert the stimulus that *per se* is inert in relation to gene expression, into a chemical representation—a message for inducing its expression. The nervous system transforms the genetically meaningless stimulus in a way that makes it a genetically intelligible message. This is the most profound significance of the neural processing.

What is the processing of the stimulus essentially intended to do?

Let us consider, once again, the case of the shortening of photoperiod. As a stimulus, the change in the length of the day is perceived in the CNS by special hypothalamic circuits. As pointed out earlier, the light or darkness *per se* do not affect expression of any gene in cells that are directly affected (e.g., dermis cells, retinal cells), but it affects expression of the GnRH in the hypothalamus in opposite ways in cows and ewes, which respond to the same stimulus, i.e., the shortening of the day length, in opposing ways by inhibiting and inducing expression of the GnRH gene in hypothalamic GnRH neurons, respectively. Two different results from the same stimulus! This would be paradoxical if the stimulus *per se* would be the cause of expression/suppression of the GnRH gene. As argued in length earlier, the stimulus is not the cause, for were it a cause or reproductive information, it would lead to the same result in both the above species. Far from a paradox, both opposite results are very adaptive as far as the reproduction success of both species is concerned: they are computed to produce offspring at the optimal time of the year for rearing them. These different results show the following:

First, that the external stimulus is not the cause of opposite effects observed in the two species, and it represents no information or instruction for these organisms,
Second, that the information for timing the sexual activity, mating, and parturition is generated in the brain (hypothalamus), and
Third, that the relevant computation in hypothalamic circuits does not consist simply in summation of external stimuli (otherwise, the same result would be observed in both species).

Such diverse responses to the same stimulus show that what induces or inhibits expression of the GnRH gene is not the photoperiod *per se* (day light has no access to those hypothalamic neurons and genes, which are permanently in darkness), but the chemical output resulting from the manipulative, adaptation-oriented processing of the stimulus in specific neural circuits. *Differences in the output result from differences in the processing of the same stimulus in both species, which, ultimately, are determined by differences in computational properties of their neural circuits.*

Now let us put the question more explicitly: Is the computational processing in the nervous system a summation and integration of the input of stimuli, or is it a solution-oriented complex calculation, estimation, or reckoning? If the first were the case, then it would be expected that the result of computation, i.e., the chemical signal released by neural circuits in different species, in response to the same stimulus, would be similar. But, as we all know, the chemical signal(s) released, and the ensuing morphological, physiological, and behavioral characters induced in different metazoan species in response to the same stimulus, often are different or even, in our case, opposite.

What computation of the stimulus in the neural circuit is intended to do is to start an adaptive response to the stimulus by establishing, via a signal cascade, a new, previously/naturally nonexistent relationship between the stimulus and a specific gene or group of genes. The function of the computation in neural circuits is to foreordain a morphological, physiological, behavioral, or life history modification that would adapt the organism to the new or changing element (stimulus) of the external or internal environment.

The end result of the complex processing of the stimulus in the neural circuit is secretion of a chemical that is necessary for starting a signal cascade for expression of a specific gene or group of genes. The processing, and the manipulative expression of genes it induces, represents a solution to the problem of adaptation of the organism to the stimulus.

Manipulative responses to external/internal stimuli evolved from a strong evolutionary pressure for both maintaining the continuously eroding metazoan structure as well as for rapid adaptation to the adversely changing environment. These responses became the predominant mode of expression of nonhousekeeping genes and a crucial element of developmental and adaptational processes in metazoans.

The ability of the CNS to respond manipulatively to various external and internal stimuli is a qualitatively new property of metazoans. By enabling the organism to produce different responses (signal outputs) to the same stimulus, and the reverse, to produce the same response to different stimuli, the CNS created an unprecedented repertory of adaptive capabilities.

Computationally determined manipulative response of metazoans to external/internal stimuli, based on self-organizing properties of the nervous system, stands in sharp contrast with the genetic determinism of responses observed in unicellulars, where the relationship between internal/external stimuli is almost totally determined by thermodynamic–stereochemical properties of the stimuli (in some eukaryotic unicellulars, also signal transduction pathways) and genes. Revelation of this higher, epigenetic, adaptive capability of metazoans is creating a new intellectual situation. Characterizing it von der Malsburg writes:

> *According to an old mode of thinking a Creator, or Mother Nature or a genetic program, are in control of every molecular reaction with exquisite algorithmic foresight. Although this thought pattern of "hetero-organization" under control of a preexisting plan is sub-consciously still governing many a thought, it now is quickly losing its dominance, giving way to models of self-organization on all levels – evolution, ontogenesis, learning and functional organization.*
>
> *von der Malsburg (2002)*

In Chapter 5, adequate evidence is presented to show that by processing internal stimuli from the developing embryo, the embryonic CNS generates the epigenetic information necessary for the sequential stages of individual development.

References

Albert, M.V., Schnabel, A. and Field, D.J. (2008). Innate visual learning through spontaneous activity patterns. PLoS Comput. Biol., *4*(8), p. e1000137.

Alley, K.E. (1990). Retrofitting larval neuromuscular circuits in the metamorphosing frog. J. Neurobiol., *21*, pp. 1092–1107.

Amabis, J.M., Reinach, F.C. and Andrews, N. (1979). Spermatogenesis in *Trichosia pubescens* (Diptera: Sciaridae). J. Cell Sci., *36*, pp. 199–213.

Anderson, G.M., Connors, J.M., Hardy, S.L., Valent, M. and Goodman, R.L. (2001). Oestradiol microimplants in the ventromedial preoptic area inhibit secretion of luteinizing hormone via dopamine neurones in anoestrous ewes. J. Neuroendocrinol. *13*, 1051–1058.

Anstey, M.L., Rogers, S.M., Ott, S.R., Burrows, M. and Simpson, S.J. (2009). Serotonin mediates behavioral gregarization underlying swarm formation in desert locusts. Science, *323*, pp. 627–630.

Bailey, C.H. and Kandel, E.R. (2008). Synaptic remodeling, synaptic growth and the storage of long-term memory in *Aplysia.* Prog. Brain Res., *169*, pp. 179–198.

Bartness, T.J. and Song, C.K. (2005). Innervation of brown adipose tissue and its role in thermogenesis. Can. J. Diabetes, *29*, pp. 420–428.

Benfenati, F. (2007). Synaptic plasticity and the neurobiology of learning and memory. Acta Biomed., *78*(Suppl. 1), pp. 58–66.

Berke, J.D., Sgambato, V., Zhu, P.P., Lavoie, B., Vincent, M., Krause, M., et al. (2001). Dopamine and glutamate induce distinct striatal splice forms of Ania-6, an RNA polymerase II-associated cyclin. Neuron, *32*, pp. 174–176.

Bito, H. (1998). The role of calcium in activity-dependent neuronal gene regulation. Cell Calcium, *23*, pp. 143–150.

Black, D.L. (1998). Splicing in the inner ear: a familiar tune, but what are the instruments? Neuron, *20*, pp. 165–168.

Blagburn, J.M., Sosa, M.A. and Blanco, R.E. (1996). Specificity of identified central synapses in the embryonic cockroach: appropriate connections form before the onset of spontaneous afferent activity. J. Comp. Neurol., *373*, pp. 511–528.

Blankenship, A.G. and Feller, M.B. (2010). Mechanisms underlying spontaneous patterned activity in developing neural circuits. Nat. Rev. Neurosci., *11*, pp. 18–29.

Borodinsky, L.N., Root, C.M., Cronin, J.A., Sann, S.B., Gu, X., and Spitzer, N.C. (2004). Activity-dependent homeostatic specification of transmitter expression in embryonic neurons. Nature, *429*, pp. 523–530.

Burmeister, S.S. and Wilczynski, W. (2005). Social signals regulate gonadotropin-releasing hormone neurons in the green tree frog. Brain Behav. Evol., *65*, pp. 26–32.

Calizo, L.H. and Flanagan-Cato, L.M. (2000). Estrogen selectively regulates spine density within the dendritic arbor of rat ventromedial hypothalamic neurons. J. Neurosci., *20*, pp. 1589–1596.

Cancedda, L., Putignano, E., Impey, S., Maffei, L., Ratto, G.-M. and Pizzorusso, T. (2003). Patterned vision causes CRE-mediated gene expression in the visual cortex through PKA and ERK. J. Neurosci., *23*, pp. 7012–7020.

Celotto, A.M. and Graveley, B.R. (2001). Alternative splicing of the *Drosophila Dscam* pre-mRNA is both temporally and spatially regulated. Genetics, *159*, pp. 599–608.

Claeys, I., Breugelmans, B., Simonet, G., Van Soest, S., Sas, F., and De Loof, A., et al. (2006). Neuroparsin transcripts as molecular markers in the process of desert locust (*Schistocerca gregaria*) phase transition. Biochem. Biophys. Res. Commun., *341*, pp. 599–606.

Clynen, E., Huybrechts, J., Verleyen, P., De Loof, A. and Schoofs, L. (2006). Annotation of novel neuropeptide precursors in the migratory locust based on transcript screening of a public EST database and mass spectrometry. BMC Genomics, *7*, p. 201.

Consoulas, C., Restifo, L.L. and Levine, R.B. (2002). Dendritic remodeling and growth of motoneurons during metamorphosis of *Drosophila melanogaster*. J. Neurosci., *22*, pp. 4906–4917.

Creazzo, T.L. and Sohal, G.S. (1983). Neural control of embryonic acetylcholine receptor and skeletal muscle. Cell Tissue Res., *228*, pp. 1–12.

Ebbesson, L.O.E., Ekström, P., Ebbesson, S.O.E., Stefansson, S.O. and Holmqvist, B. (2003). Neural circuits and their structural and chemical reorganization in the light–brain–pituitary axis during parr–smolt transformation in salmon. Aquaculture, *222*, pp. 59–70.

Esteban, M.R., Campos, M.C., Perondini, A.L. and Goday, C. (1997). Role of microtubules and microtubule-organizing centers on meiotic chromosome elimination in *Sciara ocellaris*. J. Cell Sci., *110*, pp. 721–730.

Eyigor, O., Lin, W. and Jennes, L. (2004). Identification of neurones in the female rat hypothalamus that express oestrogen receptor-alpha and vesicular glutamate transporter-2. J. Neuroendocrinol., *16*, pp. 26–31.

Fernandes, J.J. and Keshishian, H. (1998). Nerve–muscle interactions during flight muscle development in *Drosophila*. Development, *125*, pp. 1769–1779.

Finkelbeiner, S. and Greenberg, M.E. (1998). Ca^{2+} channel-regulated neuronal gene expression. J. Neurobiol., *37*, pp. 171–189.

Frankfurt, M., Gould, E., Woolley, C.S. and McEwen, B.S. (1990). Gonadal steroids modify dendritic spine density in ventromedial hypothalamic neurons: a Golgi study in the adult rat. Neuroendocrinology, *51*, pp. 530–535.

Freeman, W.J. (1991). The physiology of perception. Sci. Amer. *264*, 78–85.

Fuge, H. (1997). Nonrandom chromosome segregation in male meiosis of a sciarid fly: elimination of paternal chromosomes in first division is mediated by non-kinetochore microtubules. Cell Motil. Cytoskeleton, *36*, pp. 84–94. (abstract).

Getting, P.A. (1989). Emerging principles governing the operation of neural networks. Annu. Rev. Neurosci. *12*, 185–204.

Ghosh, A. and Greenberg, M.E. (1995). Calcium signaling in neurons: molecular mechanisms and cellular consequences. Science *268*, 239–247.

Gorecki, D.C., Szklarczyk, A., Lukasiuk, K., Kaczmarek, L. and Simons, J.P. (1999). Differential seizure-induced and developmental changes of neurexin expression. Mol. Cell. Neurosci., *13*, pp. 218–227.

Goubillon, M.L., Caraty, A. and Herbison, A.E. (2002). Evidence in favour of a direct input from the ventromedial nucleus to gonadotropin-releasing hormone neurones in the ewe: an anterograde tracing study. J. Neuroendocrinol. *14*, 95–100.

Gould, E., Woolley, C.S., Frankfurt, M. and McEwen, B.S. (1990). Gonadal steroids regulate dendritic spine density in hippocampal pyramidal cells in adulthood. J. Neurosci. *10*, 1286–1291.

Graveley, B.R., Kaur, A., Gunning, D., Zipursky, S.L., Rowen, L., and Clemens, J.C. (2004). The organization and evolution of the Dipteran and Hymenopteran *Down syndrome cell adhesion molecule* (*Dscam*) genes. RNA, *10*, pp. 1499–1506.

Greenberg, M.E. and Ziff, E.B. (2001). Signal transduction in the postsynaptic neuron-activity-dependent regulation of gene expression. In: *Synapses*, V.M. Cowan, T.C. Sudhof and C.F. Stevens (eds.). The Johns Hopkins University Press, Baltimore, MD, pp. 358–391.

Hall, B.K. and Herring, S.W. (1990). Paralysis and growth of the musculoskeletal system in the embryonic chick. J. Morphol., *206*, pp. 45–56.

Hanke, J., Brett, D., Zastrow, I., Aydin, A., Delbrück, S., Lehmann, G., et al. (1999). Alternative splicing of human genes: more the rule than the exception? Trends Genet. *15*, 389–390.

Hanson, M.G. and Landmesser, L.T. (2003). Characterization of the circuits that generate spontaneous episodes of activity in the early embryonic mouse spinal cord. J. Neurosci., *23*, pp. 587–600.

Hawkins, R.D., Kandel, E.R. and Bailey, C.H. (2006). Molecular mechanisms of memory storage in *Aplysia*. Biol. Bull., *210*, pp. 174–191.

Heideman, P.D., Bierl, C.K. and Sylvester, C.J. (2001). Photoresponsive Fischer 344 rats are reproductively inhibited by melatonin and differ in 2-[^{125}I] iodomelatonin binding from nonphotoresponsive Sprague–Dawley rats. J. Neuroendocrinol., *13*, pp. 223–232.

Herbison, A.E. (1998). Multimodal influence of estrogen upon gonadotropin-releasing hormone neurons. Endocr. Rev., *19*, pp. 302–330.

Hooper, S.L. (2004). Multiple routes to similar network output. Nat. Neurosci. *7*, 1287–1288.

Iyengar, S. and Bottjer, S.W. (2002). Development of individual axon arbors in a thalamocortical circuit necessary for song learning in zebra finches. J. Neurosci., *22*, pp. 901–911.

Jarvis, E.D. and Nottebohm, F. (1997). Motor-driven gene expression. Proc. Natl. Acad. Sci. U.S.A., *94*, pp. 4097–4102.

Jarvis, E.D., Ribeiro, S., da Silva, M.L., Ventura, D., Vielliard, J. and Mello, C.V. (2000). Behaviourally driven gene expression reveals song nuclei in hummingbird brain. Nature, *406*, pp. 628–632.

Jenks, B.G., Roubos, E.W. and Scheenen, W.J.J.M. (2003). Ca^{2+} oscillations in melanotropes of *Xenopus laevis*: their generation, propagation, and function. Gen. Comp. Endocrinol., *131*, pp. 210–219.

Johnson, C.M., Hill, C.S., Chawla, S., Treisman, R. and Bading, H. (1997). Calcium controls gene expression via three distinct pathways that can function independently of the Ras/mitogen-activated protein kinases (ERKs) signaling cascades. J. Neurosci., *17*, pp. 6189–6202.

Jones, T.A., Jones, S.M. and Paggett, K.C. (2001). Primordial rhythmic bursting in embryonic cochlear ganglion cells. J. Neurosci., *21*, pp. 8129–8135.

Kang, L., Chen, X., Zhou, Y., Liu, B., Zheng, W. and Li, R., et al. (2004). The analysis of large-scale gene expression correlated to the phase changes of the migratory locust. Proc. Natl. Acad. Sci. U.S.A., *101*, pp. 17611–17615.

Katz, L.C. and Shatz, C.J. (1996). Synaptic activity and the construction of cortical circuits. Science, *274*, pp. 1133–1138.

Kaushal, D. Contos, J.J.A., Treuner, K., Yang, A.H., Kingsbury, M.A. and Rehen, S.K. (2003). Alteration of gene expression by chromosome loss in the postnatal mouse brain. J. Neurosci., *23*, pp. 5599–5606.

Kennedy, M.B. (2000). Signal-processing machines at the postsynaptic density. Science, *290*, pp. 750–754.

Kent, K.S. and Levine, R.B. (1993). Dendritic organization of an identified neuron during metamorphosis of the moth *Manduca sexta*; the influence of interactions with the periphery. J. Neurobiol., *24*, pp. 1–22.

Khazipov, R. and Luhmann, H.J. (2006). Early patterns of electrical activity in the developing cerebral cortex of humans and rodents. Trends Neurosci., *29*, pp. 414–418.

Kingsbury, M.A., Friedman, B., McConnell, M.J., Rehen, S.K., Yang, A.H., Kaushal, D., et al. (2005). Aneuploid neurons are functionally active and integrated into brain circuitry. Proc. Natl. Acad. Sci. U.S.A., *102*, pp. 6143–6147.

Kornhauser, J.M., Nelson, D.E., Mayo, K.E. and Takahashi, J.S. (1990). Photic and circadian regulation of c-fos gene expression in the hamster suprachiasmatic nucleus. Neuron, *5*, pp. 127–134.

Kramer, B.M., Song, J.Y., Westphal, N.J., Jenks, B.G. and Roubos, E.W., (2002). Regulation of neurons in the suprachiasmatic nucleus of *Xenopus laevis*. Comp. Biochem. Physiol. B Biochem. Mol. Biol., *132*, pp. 269–274.

Kreahling, J. and Graveley, B.R. (2004). The origins and implications of alternative splicing. Trends Genet., *20*, pp. 1–4.

Lai, G.-J. and McCobb, D.P. (2002). Opposing actions of adrenal androgens and glucocorticoids on alternative splicing of *Slo* potassium channels in bovine chromaffin cells. Proc. Natl. Acad. Sci. U.S.A., *99*, pp. 7722–7727.

Lai, G.-J. and McCobb, D.P. (2006). Regulation of alternative splicing of Slo K^+ channels in adrenal and pituitary during the stress hyporesponsive period of rat development. Endocrinology, *147*, pp. 3961–3967.

Larsen, P.J., Enquist, L.W. and Card, J.P. (1998). Characterization of the multisynaptic neuronal control of the rat pineal gland using viral transneuronal tracing. Eur. J. Neurosci., *10*, pp. 128–145.

Lee, C.J. and Irizarry, K. (2003). Alternative splicing in the nervous system: an emerging source of diversity and regulation. Biol. Psychiatry, *54*, pp. 771–776.

Li, H., Gu, X., Dawson, V.L. and Dawson, T.M. (2004). Identification of calcium- and nitric oxide-regulated genes by different analyses of library expression (DAzLE). Proc. Natl. Acad. Sci. U.S.A., *101*, pp. 647–652.

Lipscombe, D. (2005). Neuronal proteins custom designed by alternative splicing. Curr. Opin. Neurobiol., *15*, pp. 358–363.

Lisbin, M.J., Qiu, J. and White, K. (2001). The neuron-specific RNA-binding protein ELAV regulates *neuroglian* alternative splicing in neurons and binds directly to its pre-mRNA. Genes Dev., *15*, pp. 2546–2561.

Lnenicka, G.A. and Murphey, R.K. (1989). The refinement of invertebrate synapses during development. J. Neurobiol., *20*, pp. 339–355. (Longman Advanced American Dictionary. Pearson Education Limited, Barcelona, 2000.)

Lonze, B.E. and Ginty, D.D. (2002). Function and regulation of CREB family transcription factors in the nervous system. Neuron, *35*, pp. 605–623.

Luscher, C., Nicoll, R.A., Malenka, R.C. and Muller, D. (2000). Synaptic plasticity and dynamic modulation of the postsynaptic membrane. Nat. Neurosci., *3*, pp. 545–550.

Maffei, A., Nelson, S.B. and Turrigiano, G.G. (2004). Selective reconfiguration of layer 4 visual cortical circuitry by visual deprivation. Nat. Neurosci., *7*, pp. 1353–1359.

Matsumoto, A. and Arai, Y. (1986). Male–female difference in synaptic organization of the ventromedial nucleus of the hypothalamus in rat. Neuroendocrinology, *42*, pp. 232–236.

Mattick, J.S. (2007). A new paradigm for developmental biology. J. Exp. Biol., *210*, pp. 1526–1547.

McAllister, A.K. (2000). Cellular and molecular mechanisms of dendrite growth. Cereb. Cortex, *10*, pp. 963–973.

McCobb, D.P., Hara, Y., Lai, G.-J., Mahmoud, S.F. and Flügge, G. (2003). Subordination stress alters alternative splicing of the *Slo* gene in tree shrew adrenals. Horm. Behav., *43*, pp. 180–186.

Michaelson, M.D., Bieri, P.L., Mehler, M.F., Xu, H., Arezzo, J.C., Pollard, J.W., et al. (1996). CSF-1 deficiency in mice results in abnormal brain development. Development, *122*, pp. 2661–2672.

Michalowski, J. (2005). Alternative splicing. HHMI Bull., *18*, pp. 22–27.

Montag-Sallaz, M., Baarke, A. and Montag, D. (2003). Aberrant neuronal connectivity in CHL1-deficient mice is associated with altered information processing-related immediate early gene expression. J. Neurobiol., *57*, pp. 67–80.

Morris, M.E., Viswanathan, N., Kuhlman, S., Davis, F.C. and Weitz, C.J. (1998). A screen for genes induced in the suprachiasmatic nucleus by light. Science *279*, 1544–1547.

Morrison, S.F. (2004). Central pathways controlling brown adipose tissue thermogenesis. News Physiol. Sci., *19*, pp. 67–74.

Mu, Y., Otsuka, T., Horton, A.C., Scott, D.B. and Ehlers, M.D. (2003). Activity-dependent mRNA splicing controls ER export and synaptic delivery of NMDA receptors. Neuron, *40*, pp. 581–594.

Nedivi, E., Hevroni, D., Naot, D., Israeli, D. and Citri, Y. (1993). Candidate plasticity-related genes revealed by differential cDNA cloning. Nature, *363*, pp. 718–722.

Peinado, A. (2000). Traveling slow waves of neural activity: a novel form of network activity in developing neocortex. J. Neurosci., *20*, pp. 1–6.

Penn, A.A. (2001). Early brain wiring: activity-dependent processes. Schizophr. Bull., *27*, pp. 337–347.

Penn, A.A., Riquelme, P.A., Feller, M.B. and Shatz, C.J. (1998). Competition in retinogeniculate patterning driven by spontaneous activity. Science *279*, 2108–2112.

Penn, A.A. and Shatz, C.J. (1999). Brain waves and brain wiring: the role of endogenous and sensory-driven neural activity in development. Pediatr. Res., *45*, pp. 447–458.

Perreau-Lenz, S., et al. (2003). Suprachiasmatic control of melatonin synthesis in rats: inhibitory and stimulatory mechanisms. Eur. J. Neurosci., *17*, pp. 221–228.

Persson, M. (1983). The role of movements in the development of sutural and diarthroidal joints tested by long-term paralysis in chick embryos. J. Anat., *137*, pp. 591–599.

Pitsillides, A.A. (2006). Early effects of embryonic movement: "a shot out of the dark". J. Anat., *208*, pp. 417–431.

Pompolo, S., Pereira, A., Kaneko, T. and Clarke, I.J. (2003). Seasonal changes in the inputs to gonadotropin-releasing hormone neurones in the ewe brain: an assessment by conventional fluorescence and confocal microscopy. J. Neuroendocrinol., *15*, pp. 538–545.

Prinz, A.A., Bucher, D. and Marder, E. (2004). Similar network activity from disparate circuit parameters. Nat. Neurosci., *7*, pp. 1345–1352.

Pszczolkowski, M.A. and Brown, J.J. (2003). Chilling affects allatal cell proliferation via antennae and protocerebral neurons in the cockroach *Diploptera punctata*. J. Insect Physiol., *49*, pp. 123–129.

Pszczolkowski, M.A. and Gelman, D.B. (2004). Chilling stress effects on corpus allatum proliferation in the Hawaiian cockroach, *Diploptera punctata*: a role for ecdysteroids. J. Insect Physiol., *50*, pp. 203–208.

Rehen, S.K. McConnell, M.J., Kaushal, D., Kingsbury, M.A., Yang, A.H. and Chun, J. (2001). Chromosomal variation in neurons of the developing and adult mammalian nervous system. Proc. Natl. Acad. Sci. U.S.A., *98*, pp. 13361–13366.

Rehen, S.K., Yung, Y-C., McCreight, M.P., Kaushal, D., Yang, A.H., Almeida, B.S.V., et al. (2005). Constitutional aneuploidy in the normal human brain. J. Neurosci., *25*, pp. 2176–2180.

Rogers, S.M., Matheson, T., Sasaki, K., Kendrick, K., Simpson, S.J. and Burrows, M., et al. (2004). Substantial changes in central nervous system neurotransmitters and neuromodulators accompany phase change in the locust. J. Exp. Biol., *207*, pp. 3603–3617.

Rosenblatt, K.P., Sun, Z.-P., Heller, S. and Hudspeth, A.J. (1997). Distribution of Ca^{2+}-activated K^+ channel isoforms along the tonotopic gradient of the chicken's cochlea. Neuron, *19*, pp. 1061–1075.

Roubos, E.W. (1997). Background adaptation by *Xenopus laevis*: a model for studying neuronal information processing in the pituitary pars intermedia. Comp. Biochem. Physiol. A Physiol., *118*, pp. 533–550.

Saeb-Parsy, K. and Dyball, R.E. (2003). Defined cell groups in the rat suprachiasmatic nucleus have different day/night rhythms of single-unit activity *in vivo*. J. Biol. Rhythms, *18*, pp. 26–42.

Schaap, J., Albus, H., VanderLeest, H.T., Eilers, P.H., Detari, L. and Meijer, J.H., et al. (2003). Heterogeneity of rhythmic suprachiasmatic nucleus neurons: implications for circadian waveform and photoperiodic encoding. Proc. Natl. Acad. Sci. U.S.A., *100*, pp. 15994–9.

Scott, C.J., Rawson, J.A., Pereira, A.M. and Clarke, I.J. (1999). Oestrogen receptors in the brainstem of the female sheep: relationship to noradrenergic cells and cells projecting to the medial preoptic area. J. Neuroendocrinol., *11*, 745–755.

Seeburg, P.H. (2002). A-to-I editing: new and old sites, functions and speculations. Neuron, *35*, pp. 17–20.

Segal, M., Greenberger, V. and Korkotian, E. (2003). Formation of dendritic spines in cultured striatal neurons depends on excitatory afferent activity. Eur. J. Neurosci., *17*, pp. 2573–2585.

Shadlen, M.N. and Newsome, W.T. (1994). Noise, neural code and cortical organization. Curr. Opin. Neurobiol., *4*, pp. 569–579.

Sgambato, V., Minassian, R., Nairn, A.C. and Hyman, S.E. (2003). Regulation of *ania-6* splice variants by distinct signaling pathways in striatal neurons. J. Neurochem., *86*, pp. 153–164.

Shatz, C. (1996). Emergence of order in visual system development. Proc. Natl. Acad. Sci. U.S.A., *93*, pp. 602–608.

Simonneaux, V. and Ribelayga, C. (2003). Generation of the melatonin endocrine message in mammals: a review of the complex regulation of melatonin synthesis by norepinephrine, peptides, and other pineal transmitters. Pharmacol. Rev., *55*, pp. 325–395.

Sjöström, P.J., Turrigiano, G.G. and Nelson, S.B. (2001). Rate, timing, and cooperativity jointly determine cortical synaptic plasticity. Neuron, *32*, pp. 1149–1164.

Smith, M.J. and Jennes, L. (2001). Neural signals that regulate GnRH neurones directly during oestrous cycle. Reproduction, *122*, pp. 1–10.

Sohal, G.H. and Holt, R.K. (1980). Role of innervation on the embryonic development. Cell Tissue Res., *210*, pp. 383–393.

Spitzer, N.C. (2006). Electrical activity in early neuronal development. Nature, *444*, pp. 707–712.

Strand, F.L. (1999). *Neuropeptides* MIT Press, Cambridge, MA , p. 521.

Sugimoto, M. (2002). Morphological color changes in fish: regulation of pigment cell density and morphology. Microsc. Res. Tech., *58*, pp. 496–503.

Sugimoto, M., Uchida, N. and Hatayama, M. (2000). Apoptosis in skin pigment cells of medaka, *Oryzias latipes* (Teleostei), during long-term chromatic adaptation: the role of sympathetic innervation. Cell Tissue Res., *301*, pp. 205–216. (abstract).

Tao, X., Finkbeiner, S., Arnold, D.B., Shaywitz, A.J. and Greenberg, M.E. (1998). Ca^{2+} influx regulates BDNF transcription by a CREB family transcription factor-dependent mechanism. Neuron, *20*, pp. 709–726.

Tritsch, N.X., Yi, E., Gale, J.E., Glowatzki, E. and Bergles, D.E. (2007). The origin of spontaneous activity in the developing auditory system. Nature, *450*, pp. 50–56.

Tuinhof, R., Artero, C., Fasolo, A., Franzoni, M.F., Ten Donkelaar, H.J. and Wismans, P.G., et al. (1994). Involvement of retinohypothalamic input, suprachiasmatic nucleus, magnocellular nucleus and *locus coeruleus* in control of melanotrope cells of *Xenopus laevis*: a retrograde and anterograde tracing study. Neuroscience, *61*, pp. 411–420.

Turrigiano, G.G. (1999). Homeostatic plasticity in neuronal networks: the more things change, the more they stay the same. Trends Neurosci., *22*, pp. 221–227.

Turrigiano, G., Abbott, L.F. and Marder, E. (1994). Activity-dependent changes in the intrinsic properties of cultured embryos. Science, *264*, pp. 974–977.

Turrigiano, G.G. and Nelson, S.B. (2004). Homeostatic plasticity in the developing nervous system. Nat. Rev. Neurosci., *5*, pp. 97–107.

Ubink, R., Buzzi, M., Cruijsen, P.M., Tuinhof, R., Verhofstad, A.A., Jenks, B.G., et al. (1999). Serotonergic innervation of the pituitary pars intermedia of *Xenopus laevis*. J. Neuroendocrinol., *11*, pp. 211–219.

Ullrich, B., Ushkaryov, Y.A. and Südhof, T.C. (1995). Cartography of neurexins: more than 1000 isoforms generated by alternative splicing and expressed in distinct subsets of neurons. Neuron, *14*, pp. 497–507.

Ushkaryov, Y.A., Petrenko, A.G., Geppert, M. and Südhof, T.C. (1992). Neurexins: synaptic cell surface proteins related to the (–Latrotoxin receptor and laminin. Science, *257*, pp. 50–56.

van Eys, G.J. and Peters, P.T. (1981). Evidence for a direct role of alpha-MSH in morphological background adaptation of the skin in *Sarotherodon mossambicus*. Cell Tissue Res., *217*, pp. 361–372.

Varayoud, J. (2005). The estrogen receptor alpha sigma3 mRNA splicing variant is differentially regulated by estrogen and progesterone in the rat uterus. J. Endocrinol., *186*, pp. 51–60.

Velho, T.A.F., Pinaud, R., Rodrigues, P.V. and Mello, C.V. (2005). Co-induction of activity-dependent genes in songbirds. Eur. J. Neurosci., *22*, pp. 1667–1678.

Villareal, G., Li, Q., Cai, D. and Glanzman, D.L. (2007). The role of rapid, local postsynaptic protein synthesis in learning-related synaptic facilitation in *Aplysia*. Curr. Biol., *17*, pp. 2073–2080.

von der Malsburg, C. (1999). The what and why of binding: the modeler's perspective. Neuron, *24*, p. 95–104.

von der Malsburg, C. (2002). Self-organization in the brain. In: *The Handbook of Brain Theory and Neural Networks*, M. Arbib (ed.)., second edition. MIT Press, Cambridge, MA, p. 5.

Weliky, M. (1999). Recording and manipulating the *in vivo* correlational structure of neuronal activity during visual cortical development. J. Neurobiol., *41*, pp. 25–32.

Wenner, P. and O'Donovan, M.J. (2001). Mechanisms that initiate spontaneous network activity in the developing chick spinal cord. J. Neurophysiol., *86*, pp. 1481–1498.

West, A.E., Chen, W.G., Dalva, M.B., Dolmetsch, R.E., Kornhauser, J.M., Shaywitz, A.J., et al. (2001). Calcium regulation of neuronal gene expression. Proc. Natl. Acad. Sci. U.S.A., *98*, pp. 11024–31.

Westra, J.W., Peterson, S.E., Yung, Y.C., Mutoh, T., Barral, S. and Chun, J. (2008). Aneuploid mosaicism in the developing and adult cerebellar cortex. J. Comp. Neurol., *507*, pp. 1944–1951.

Widmer, H., Ludwig, M., Bancel, F., Leng, G. and Dayanithi, G. (2003). Neurosteroid regulation of oxytocin and vasopressin release from the rat supraoptic nucleus. J. Physiol., *548.1*, pp. 233–244.

Yamaguchi, S., Isejima, H., Matsuo, T., Okura, R., Yagita, K., Kobayashi, M., et al. (2003). Synchronisation of cellular clocks in the suprachiasmatic nucleus. Science, *302*, pp. 1409–1412.

Yurov, Y.B., Iourov, I.Y., Vorsanova, S.G., Liehr, T., Kolotii, A.D., Kutsev, S.I., et al. (2007). Aneuploidy and confined chromosomal mosaicism in the developing human brain. PLoS One, *2*(6), p. e558.

Zeng, Z., Sharpe, C.R., Simons, J.P. and Gorecki, D.C. (2006). The expression and alternative splicing of alpha-neurexins during *Xenopus* development. Int. J. Dev. Biol., *50*, pp. 39–46.

Zhang, L.I. and Poo, M. (2001). Electrical activity and development of neural circuits. Nat. Neurosci. Suppl., *4*, pp. 1207–1214.

3 Epigenetic Control of Reproduction

She can do this because she is sensitive to diverse cues from the environment (photoperiod, temperature, relative humidity, substrate odor, food, proximity and state of mates, etc.), is able to integrate this information in her nervous system, and via her endocrine system can use it to develop and deposit her eggs at the right time and place.

B.S. Heming (2003)

Epigenetic Control of Reproductive Physiology and Behavior

One crucial moment of biological reproduction in metazoans is the production of haploid gametes or diploid eggs (in parthenogenetic organisms), which are uniquely able to develop into a new organism.

The zygotic genome *per se*, as well as the genome of any metazoan cell, is incapable of inducing the development of a multicellular organism. Increasing evidence points to the indispensable role of the epigenetic parental (maternal and paternal) information provided in the form of cytoplasmic factors (messenger RNAs, proteins, hormones, neurotransmitters, nutrients, etc.) in the zygote for early embryonic development.

Neuroendocrine Regulation of the Gonadal Function in Insects

Extensive studies on insects show that the insect brain determines "whether," "when," and "how much of" gonadal hormones will be synthesized and secreted. The control of the development of gonads in insects essentially implies processes of cell differentiation resulting from differential patterns of gene expression in different cells of the ovary and the reproductive organs in general. These processes are regulated by juvenile hormone (JH), ecdysteroids (E), and neuropeptide gonadotropins (De Loof et al., 2001; Figure 3.1).

In turn, secretion of JH is cerebrally regulated by various types of neuropeptides that reach corpora allata (CA), not only via the hemolymph but directly as well, via the nervi corporis allati I (NCA I) innervating CA.

In other insects, the subesophageal ganglion (SOG), via NCA II, carries out a similar regulatory function on the JH secretion by CA (Kou and Chen, 2000). In *Manduca sexta*, dopamine released by nerve endings in the CA stimulates JH secretion *in vitro* for the first 2 days of the last larval prepupal period (Granger et al., 1996). Besides its well-known role in insect metamorphosis, in insects, JH acts as a regulator of the gonadotropic cycle. From this standpoint, JH itself is a gonadotropic hormone. Its secretion is regulated by two types of neuropeptides secreted by neurosecretory cells in the insect brain, allatostatins and allatotropins. These neurohormones, released via the nerve endings in CA cause, respectively, inhibition and stimulation of JH synthesis

Epigenetic Principles of Evolution. DOI: 10.1016/B978-0-12-415831-3.00003-3

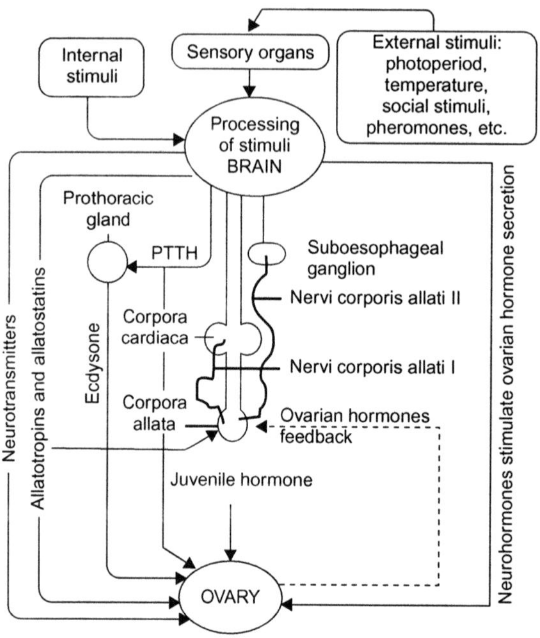

Figure 3.1 Diagrammatic representation of the neural control of ovarian function in insects. *Abbreviation*: PTTH, prothoracicotropic neurohormone. *Source*: From Cabej (2004, p. 72).

(Stay et al., 1996). In the brain of the cricket *Gryllus bimaculatus*, five members of the family of allatostatins are identified, and 14 other neuropeptides have been deduced from cDNA, all of them capable of inhibiting the JH biosynthesis *in vitro* (Witek and Hoffmann, 2001; Gäde and Hoffmann, 2005).

Besides the brain neuropeptides, allatotropins and allatostatins, in the cerebral control of the function of CA, production of JH, ovarian development, ovulation, and oviposition are essentially involved universal neurotransmitters, such as octopamine, a biogenic amine that inhibits the CA function in the cricket *Gryllus bimaculatus* and stimulates oviposition in the common house cricket, *Acheta domesticus* (Linnaeus, 1758; Adamo, 1999) and dopamine that is necessary for the ovarian development and CA activation (Bloch et al., 2000).

In eusocial insects, such as some ants, reproduction is the exclusive function of queens, which produce a pheromone that prevents reproduction of the rest of genetically female ants in the colony. As long as these ants remain in the colony, their JH level is lower than that is necessary for the maturation of oocytes. This inhibitory role of the pheromone may be mediated by biogenic amines such as dopamine; the pheromone is perceived in the insect brain, which responds to it by lowering the level of the dopamine and, consequently, the reproductive activity (Boulay et al., 2001).

Production and secretion of sex pheromones by the sex pheromone gland in some moths is under control of the neuropeptide pheromone biosynthesis-activating neurohormone (PBAN) (Gäde and Hoffmann, 2005), secreted by secretory neurons brain–subesophageal ganglion (Br–SOG). The neurohormone performs its pheromonotropic function by regulating production of pheromone-synthesis enzymes. The Br–SOG stimulates secretion of PBAN in response to external cues, such as photoperiod.

Since the neurotransmitter octopamine is demonstrated to induce pheromone production (Christensen et al., 1992), it is possible that its direct target cells are PBAN-synthesizing neurons. Pheromone secretion is inhibited after mating as a result of the neural signal arising from the sperm storage in spermatheca (Delisle et al., 2000).

In honey bees, the queen excretes from its mandibular gland a pheromone, consisting of a blend of chemicals, which prevents the development of ovaries in workers and inhibits development of new queens in the colony. It is demonstrated that the pheromone performs its effects on the behavior (reduced locomotor activity and nursing) and physiology of worker bees by acting on central neurons, altering their cellular properties and changing the output of neural circuits in the brain (Beggs et al., 2007). The key component of the pheromone is homovanillyl alcohol (HVA) or 4-hydroxy-3-methoxyphenylethanol, which interferes with the brain dopamine system that is responsible for the development of the ovary (Beggs et al., 2007; Harano et al., 2008).

Mating stimulates sexual maturation in *Drosophila* females. It has been shown that a male sex peptide (SP), produced by the male accessory gland, and introduced with seminal fluid during mating, induces reduction of the female receptivity and increased oviposition. The peptide induces secretion of JH by CA (Moshitzky et al., 1996), alongside other brain signals, allatotropins. Besides the control of gonadal function via production of JH by CA, cerebral allatostatins may act directly on the oocytes, as it is suggested by the fact that they (allatostatins A1 and B1) are found in the cortical cytoplasm of the oocytes (Witek and Hoffmann, 2001).

Recently, it has been found that the female irreceptivity to courting males, after the introduction of the male SP with the seminal fluid, is determined by the insect central nervous system (CNS). From the female reproductive tract, via the hemolymph, SP circulates throughout the insect body, including the insect brain, by crossing the blood–brain barrier. However, only the insect CNS is receptive to it because it is only the brain and the ventral nerve cord (VNC) that expresses its specific receptor, SPR. Male-derived SP binds its specific receptor, SPR, in the fru-expressing neurons of the SOG and of the VNC, which determine the switch to the postmating behavior of unresponsiveness to male courting (Yapici et al., 2008).

Two neuroactive substances produced in the brain of the primitively eusocial wasp (*Polistes chinensis*), dopamine and serotonin, are found in its ovary and are involved, the first, in the ovarian development and appearance of egg-laying behavior, and the second, probably, in regulation of reproductive states of the insect (Sasaki et al., 2007).

External cues, such as social conditions and/or specifically male sexual stimuli, are necessary for the reproductive development and ovarian maturation in the dampwood termite *Zootermopsis angusticollis* (Brent and Traniello, 2001). These stimuli, via sensory pathways, reach the CNS, where they are processed in neural circuits, whose chemical output starts signal cascades determining reproductive changes in the genitalia.

In the fire ant, *Solenopsis invicta* Buren, the queen releases a pheromone that prevents reproduction of virgin females as long as they remain in the parental nest (Fletcher and Blum, 1981). The pheromone is thought to decrease JH production in

CA by decreasing the dopamine level in the brain (Robinson and Vargo, 1997), or via the dopaminergic innervation of CA (Granger et al., 1996). A brain neuropeptide hormone, *Lymnatria testis* ecdysiotropin, secreted by the neurosecretory cells of the brain, and by some ganglia in *Lymnatria dispar*, is found to stimulate ecdysteroid synthesis in testis but not in the prothoracic gland. The testis ecdysteroid then induces the synthesis of a growth hormone (GH) that stimulates the development of the male genital tract (Loeb et al., 2001).

In all the above cases, the environmental cues trigger signal cascades that induce reproductive activity and oogenesis in insects. All these cascades start with the release of brain signals, neurotransmitters or neuropeptides.

Local Neural Control of the Reproductive Function in Insects

In insects, the CNS is involved in the reproductive behavior and in the reproductive activity, not only via neuroendocrine pathways but also directly via the local gonadal innervation.

For almost a century, since it was first described, no specific function was attributed to the nerve innervating the ovotestis (OT) in pulmonate snails, but recent studies on the slug *Helix aspersa* have shown that the OT branch of the intestinal nerve consists of as many as 3,025 axons, suggesting that approximately 8% of the total number of neurons of the CNS are involved in the innervation of the slug OT.

From an evolutionary viewpoint, such a costly investment could not have evolved if it did not offer an outweighing advantage. Indeed, recently it has been demonstrated that the innervation has important functions for receiving and sending information from the OT to the brain, and vice versa. The innervation is necessary for maturation of oocytes and ovulation. It has both sensory and motor functions. The fine terminals of the OT nerve branch have stretch sensors in the acinar walls, which monitor the degree of stretching as a result of the concurrent growth of more than 100 oogonia. That information on the stretch, via afferents, is sent to the brain, where it is assessed and, when it exceeds a threshold, efferent brain signals are sent back for increasing peristaltic contractions in the hermaphroditic duct and the beat of the cilia in the internal lining of the duct, thus facilitating the movement down the duct and oviposition of 58–108 egg cells (Antkowiak and Chase, 2003; Chase et al., 2004; Geoffroy et al., 2005).

Neuroendocrine Regulation of Reproductive Function in Vertebrates

In vertebrates as well, reproductive functions are cerebrally controlled, mainly via the hypothalamic–pituitary–gonadal axis. In cyclical ovulators, the reproductive activity is stimulated by environmental factors such as temperature and moisture, but above all by the seasonal changes of the day length (in most cases, its prolongation): the onset of reproductive activity starts with displays of mating behavior, which is determined by activation of specific neural circuits, in response to visual and auditory stimuli. In birds, for example, as a result of the activation of the hypothalamus–pituitary–testis axis, gonads are stimulated to synthesize testosterone, and high levels of this hormone play a crucial role for the onset of mating behavior in males. This leads to increased conversion of testosterone into estrogens in the birds' brains, thus contributing to the full display of the male mating behavior, in which several neural circuits are involved. Neurons of the hypothalamic preoptic area (POA) and the periaqueductal gray represent the most important centers of mating behavior. They send projections to the brain stem and spinal cord, thus determining the motor aspects of the male mating behavior (Ball and Balthazart, 2004).

The input of the environmental stimuli is integrated and processed in particular hypothalamic neurons, where information about the photoperiod and temperature trend (increasing or decreasing) is generated. This information is relayed to the hypothalamic gonadotropin-releasing hormone (GnRH) neurons, which then respond by stimulating (or inhibiting) GnRH secretion. The sensory stimuli are selectively relayed to the hypothalamus via "the medial cortex that determines what kind of sensory information will reach the hypothalamic centres" (Hoogland and Martinez-Garcia, 1999).

One great manifestation of the role of the nervous system in reproduction of vertebrates is the strong sexual dimorphism of those parts of the brain that are involved in the regulation of their reproduction. Sex differences exist in the structure and function of the human brain, and these are believed to provide the basis for the differences in reproduction (Swaab et al., 2001). These differences are very evident in the SDN-POA (sexually dimorphic nucleus of the POA) as part of the neural circuitry underlying masculine sexual behavior and reproductive functions. Hypothalamic suprachiasmatic nucleus (SCN) as well is involved in the control of reproductive and metabolic phenomena and is the principal component of the central clock mechanism. It also changes the morphology and volume (in humans, it is twice as large in the autumn as in the summer) under the influence of photoperiodic cycles (Hofman and Swaab, 1992).

In response to external cues and internal stimuli, various neuropeptides and neurotransmitters, including opioid peptides (Gearing and Terasawa, 1991) and mono- and indoleamines (Johnson and Crowley, 1986), are released by specific neurons to control the GnRH pulse generator in the hypothalamus (Gore and Terasawa, 2001), which regulates reproductive cycles.

GnRH genes have probably evolved through duplication of an ancestral gene, since all vertebrates have at least two of the three GnRH genes (GnRH-1, GnRH-2, and GnRH-3) known thus far in vertebrates. GnRH-1 is the most important of them. It is responsible for the onset of puberty in mammals and is especially responsive to olfactory signals transduced by the olfactory system. GnRH-1 neurons communicate with most diverse neurons in various regions of the brain, and the GnRH-1 system is considered to be the master regulator of reproduction in vertebrates. GnRH-2 is the best conserved of the GnRHs; it is found throughout the brain, a fact that is consistent with its role as neuromodulator. GnRH-3 may also have neuromodulatory function, but it is less known (Hofmann, 2006; Figure 3.2).

Ovulation-inducible animals, as well, display a remarkable neurogenic control of oogenesis and ovulation. In female rabbits, the sensory input of the vaginal stimulation activates a neural circuit that stimulates GnRH neurons, whose secretion

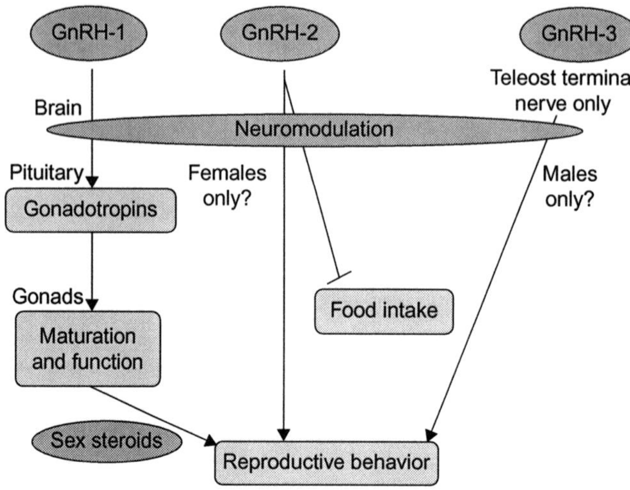

Figure 3.2 The three GnRH subtypes influence reproductive behavior through hormonal and neuromodulatory pathways. GnRH-1, which mainly controls gonadal maturation through gonadotropin release from the pituitary, probably also has neuromodulatory functions throughout the brain. Both GnRH-2 and GnRH-3 influence reproductive and, probably, other behaviors through neuromodulatory actions in the CNS.
Source: From Hofmann (2006).

(GnRH) induces the pituitary luteinizing hormone (LH) surge determining the ovulation (Strand, 1998) as well as the morphological, physiological, and behavioral changes related to ovulation.

In cyclic ovulators, including humans, the regular ovulatory cycles are under strict control of the neuroendocrine system, which is influenced by environmental factors (e.g., seasonal factors, emotional factors, nutritional factors, and stress) as well as internal factors (the levels of steroid hormones and other products in the body fluids).

In humans, for example, oogonia stop their development at the diplotene stage of the prophase of the first meiosis, to resume meiosis years later at the onset of puberty. The resumption of meiosis at puberty is triggered by progesterone—its secretion results from a cascade of signals starting from various opioid, serotonergic, and glutamatergic circuits in the brain (Villa-Diaz and Barrell, 1999; Johnson and Crowley, 1986) along the hypothalamic–pituitary–ovarian (HPO) axis:

$$\text{GnRH} \rightarrow \text{FSH} + \text{LH} \rightarrow \text{estradiol}$$

Then the estradiol (E2) secreted by the ovarian granulosa cells stimulates theca cells to synthesize androgens, which diffuse into granulosa cells to be converted there into progesterone.

In female opossums, a male pheromone is identified, in response to which the CNS stimulates the development of ovarian follicle and body growth, and prematurely induces estrus (Harder and Jackson, 2003).

Estrogen is known to activate the GnRH neurons not directly but via its receptor in the hypothalamic ventromedial nucleus (VMN) neurons, which then provide the stimulating transsynaptic input to the hypothalamic GnRH neurons. Anterograde tracing studies by injecting biotinylated dextran amine (BDA) in the VMN have shown that VMN fibers are associated with GnRH neuron cell bodies, suggesting the existence of a transsynaptic pathway from estrogen receptor-expressing VMN neurons to GnRH-producing neurons in the hypothalamus. Dense networks of BDA-labeled fibers were found in the diagonal band of Broca, POA, and anterior hypothalamic areas (Goubillon et al., 2002).

Suppression of the GnRH synthesis in the hypothalamus by E2 is also processing dependent; E2 binds to specific membrane receptors of the neurons of the ventromedial POA (VMPOA). By processing the E2 stimulus (rise in the E2 level), these neurons provide the transsynaptic input to stimulate a system involving dopamine neurons to inhibit the GnRH/LH pulsatility in anestrous ewes (Anderson et al., 2001). Naloxone, an opiate antagonist, stimulates LH secretion by reducing endogenous opiate neurotransmission (opiates stimulate dopamine release) or "by enhancing hypothalamic catecholamine turnover" (Adler and Crowley, 1984) and is an important modulator of the reproductive function in ewes (Fuentes et al., 2001).

Other observations suggest that in female rats, E2 binds to the neuron receptors of the dorsal raphe nucleus. These serotonergic neurons determine the effects of E2 via their projections to the medial POA neurons (Lu et al., 2001). Noradrenaline and dopamine systems, via the hypothalamus, negatively regulate secretion of prolactin by the pituitary in rams (Lincoln and Clarke, 2002).

In response to the rise in the level of progesterone, sensed indirectly by the hypothalamic GnRH neurons (these neurons have no progesterone receptors), this gland inhibits secretion of the GnRH. It is the dopaminergic neurons that sense the level of progesterone and other sensory input, process them, and mediate the inhibitory effects of progesterone on the GnRH pulse generator and, consequently, on the secretion of follicle-stimulating hormone (FSH) and LH by the pituitary. Moreover, secretion of these hormones at both the hypothalamic and pituitary levels is under brain "nitrinergic influence" (Dixit and Parvizi, 2001).

Among different cues known to stimulate the reproductive activity in vertebrates are visual and olfactory cues. Exposure of anestrous ewes to rams or their odor (fleece) activates the HPO axis, leading to ovulation. The olfactory pathway that activates the HPO axis includes a number of brain structures, the main and accessory olfactory bulb, anterior olfactory nucleus, cortical and basal amygdala, dentate gyrus, and hypothalamic GnRH neurons (Gelez and Fabre-Nys, 2006).

The preovulatory secretion of GnRH in cycling ewes regulates their receptivity by prolonging the initial receptivity-triggering effect of 17-β E2 and the estrous behavior (Caraty et al., 2002). In bulls, neutralization of the function of the neurohormone GnRH causes silencing of testicular function and testicular atrophy (D'Occhio et al., 2001).

Secretion of GnRH stimulates the pituitary to synthesize FSH and LH. But upstream brain signals from serotonergic and glutamatergic circuits are the ultimate stimulants of the pituitary LH via the GnRH pulse generator in the hypothalamus (Villa-Diaz and Barrell, 1999; Ford and Ebling, 2000).

The synthesis and secretion of GnRH in the hypothalamus is upregulated by adrenergic agonists and inhibited by opioid peptides. Its synthesis by GnRH neurons is induced by the input from norepinephrine (NE) circuits (Gearing and Terasawa, 1991). Besides, the hypothalamus responds to secretion of E2 in the ovary by suppressing the synthesis of GnRH. The function of the hypothalamic GnRH neurons is regulated by the informational input coming from specific brain centers:

> *The complexity of neural inputs to LHRH (GnRH—N.R.C.) neurones enables the organism to integrate information about its environment and to relate this to reproduction. For example, information regarding stress, nutritional status, time of the day, season or the presence of a conspecific, needs to be coordinated to increase the likelihood of reproductive success. The information conveyed by these neuronal inputs is integrated at the level of the LHRH neurons, and ultimately determines the animal's reproductive state and response.*
>
> *Gore and Terasawa (2001)*

As far as other internal stimuli are concerned, it is suggested that some of them, those that reflect the physiological status of the organism, are monitored by the hippocampus, which

> *operates, in parallel with amygdala, to modulate body physiology in response to cognitive stimuli.*
>
> *Lathe (2001)*

The fact that the gene for the melatonin receptor is expressed in mice granulosa (but not in theca) cells shows that melatonin secreted by the pineal gland is involved and has a pivotal role in the folliculogenesis (Lee et al., 2001). However, further upstream it is experimentally determined that melatonin secretion in the pineal gland is triggered by signals from a neural circuit involving neurons of the superior cervical ganglion, intermediolateral (IML) nucleus of the upper thoracic spinal cord, parvicellular subdivisions of the hypothalamic paraventricular nucleus (PVN), and the hypothalamic SCN (Larsen et al., 1998).

Together, the above data unambiguously suggest that the processing of the exteroceptive and interoceptive input of external and internal stimuli in specific neural circuits results in chemical outputs that induce the hypothalamic neurosecretory cells to express reproduction-related genes that otherwise would not be expressed. All the functional and morphological transformations related to reproductive activity taking place in the ovary of vertebrates are related to signal cascades starting in the CNS in response to external and internal stimuli.

Local Neural Control of the Reproductive Function in Vertebrates

Vertebrate ovary is densely innervated. In mammals, the innervation of the ovary starts during prenatal life, and catecholaminergic fibers are associated with primordial ovarian cells before the ovary becomes sensitive to gonadotropins (FSH and LH; Malamed et al., 1992). During the reproductive life, that innervation plays important roles in the regulation of ovarian physiology. The follicle growth, as a result of the action of the pituitary FSH, starts when NE and vasoactive intestinal peptide (VIP) released by ovarian nerve fibers induce synthesis of FSH receptor (Mayerhofer et al., 1997) in a process of binary control by the neuroendocrine axis and ovarian nerve fibers.

The ovary is innervated by the ovarian plexus, superior ovarian nerve (SON), vagus, etc., which, depending on species, may have stimulatory or inhibitory effects on follicular development (Morales et al., 1998; Trujillo and Riboni, 2002). Adrenergic nerve terminals establish close contacts (50–100 nm) with cells in the theca externa (Sporrong et al., 1985).

Splanchnic innervation regulates blood flow and expression of LH receptors in testicles, thus enabling/preventing induction of testosterone synthesis by the pituitary LH. The existence of this immediate neural pathway (in addition to the hypothalamus–pituitary axis) that "originates in the brain and travels through the spinal cord, and that plays a crucial role in Leydig cell function" is demonstrated by viral retrograde tracer studies, which have shown that viral staining from testes affects several parts of the CNS (hypothalamus, brain stem, and spinal cord; Lee et al., 2002). Neuron-like elements of catecholaminergic nature are also identified in testes (Mayerhofer et al., 1997).

Since the mid-1990s, it has been reported the existence in the ovary of primates, and in that of at least one rat strain, of isolated or clustered neurons interconnected in a neuronal network (Dees et al., 1995; D'Albora et al., 2000). In rats, these neurons are observed immediately after birth in the ovarian hilum and medulla and appear in the cortex only during juvenile period. Intrinsic ovarian neurons in humans are observed as early as the 24th week of gestation and are believed to be involved in the regulation of ovarian functions (Anesetti et al., 2001).

Epigenetic Control of Gametogenesis

Neural Control of Oogenesis in Insects and Gastropod Molluscs

Signal cascades for oogenesis in invertebrates originate in the CNS. In the female fire ant, *Solenopsis invicta*, for example, electrical activation of the dopamine system, resulting from the processing in the brain of an external stimulus (queen pheromone), controls oogenesis and oviposition (Boulay et al., 2001). The uptake of a blood meal by the predatory bug of the Reduviidae family, *Panstrongylus megistus*, acts as a stimulus to which certain median neurosecretory cells of the *pars intercerebralis* (cells A) respond by releasing a neurohormone that stimulates proliferation of oogonia and prefollicular cells in the fifth instar until day six, and a neuron of another type (cells A′), which responds by stimulating secretion of ecdysone by prothoracic glands, thus determining the differentiation of ovarioles between days 16 and 24 (Heming, 2003).

Oogenesis in *Drosophila* is proximately regulated by the ecdysteroid hormone and JH (Carney and Bender, 2000) but both hormones, ecdysone and JH, are cerebrally regulated by brain signals, prothoracicotropic hormone (PTTH; Zitnan et al., 1993) and allatotropins/allatostatins, respectively. There are indications that upstream these brain signals are neurotransmitters/neuromodulators released as a result of the processing of internal/external stimuli in neural circuits. So, for example, a queen pheromone in colonies of *Solenopsis invicta* ants prevents the reproductive activity in virgin females via a neural pathway (olfactory signals are transmitted from the antennal lobe for processing in the brain). Separation of virgin ant females from the colony (this prevents the influence of the queen pheromone) for 15 days induces more than 80% increase in the dopamine level in their brain, stimulating wing shedding and reproductive activity, including ovarian development and egg laying (Boulay et al., 2001).

In *Drosophila melanogaster*, the development of the oocyte arrests just before vitellogenesis but vitellogenin endocytosis begins only after resumption of the oocyte development as a result of hormonal and environmental cues (Schonbaum et al., 2000). The environmental cues received via the sensory pathways are processed in the CNS, and what reaches the ovary and the oocyte is not the cue itself but its neural representation, i.e., a hormonal signal (ecdysone + JH) secreted under the ultimate control of the insect CNS.

As mentioned earlier, the oogenesis in *Drosophila* is regulated by ecdysone and JH (Buszczak et al., 1999). The ecdysone early response genes, E75, E74, and BR-C, are expressed in stages, according to the cerebrally regulated ecdysone levels during oogenesis, and the stage-specific expression of these genes regulates the egg chamber development and determines specification of the dorsal follicle cell fates (Buszczak et al., 1999).

In molluscs, reproduction is seasonally regulated in response to environmental stimuli, with the photoperiod being the most important cue. The neuronal structures for processing these reproductive stimuli in molluscs are diffusely located in the brain and the neuroendocrine circuitry is highly complex (Wayne, 2001).

The CNS in molluscs comprises varied numbers of ganglia, each of them composed of thousands of neurons. When continually kept under laboratory short-day conditions, males and females of the slug *Limax maximus* remain sexually immature, but soon after they are transferred in long-day conditions they develop the mature reproductive tract and produce mature gametes. Transplantation of the ganglia of long day–stimulated slugs into the immature short day–inhibited slugs induced the development of the mature reproductive tract in recipients, whereas transplantation of short-day ganglia did not induce the development of the reproductive tract (Wayne, 2001). Under the influence of stimulating long days, the slug brain secretes a maturation-inducing hormone (MH), which induces the synthesis of one or more hormones by the gonads (Figures 3.3 and 3.4). Implantation of the brains of long day–stimulated slugs into the brains of short day–inhibited slugs, even under short-day conditions, stimulates development of gonads and accessory sex organs in the latter (McCrone and Sokolove, 1986).

Biologists have known for some time that in many invertebrates, hormones, as mediators of neural signals, regulate the ovulation and the number of eggs to be deposited, but recently it has been demonstrated that in the snail, *Helix aspersa*, a branch of the intestinal nerve, rather than any hormones, is immediately responsible for that regulation (Antkowiak and Chase, 2003; Figure 3.5). By sensing and integrating internal and external stimuli (photoperiod, etc.), the CNS sends to the OT (ovotestis) signals for egg laying. Brain signals via the intestinal nerve are necessary and sufficient to release sperm from the seminal vesicle (Geoffroy et al., 2005).

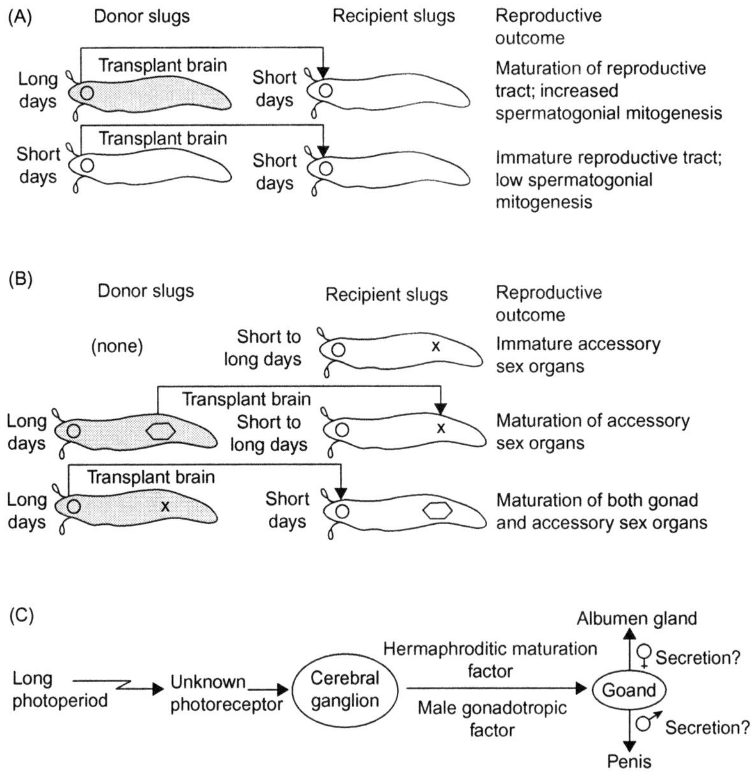

Figure 3.3 (A) Effects of photoperiod and brain transplants on the reproductive system of *Limax maximus*. Brains transplanted from donor slugs maintained on stimulatory long days stimulated the reproductive system in recipient slugs maintained on inhibitory short days. Brains from short-day animals had no effect on maturation of the reproductive system. These findings indicate that the brain mediates the effects of photoperiod on maturation of the reproductive system. (B) Effects of brain and gonad transplants on the reproductive system of *Limax*. Transfer of gonadectomized slugs from inhibitory short days to stimulatory long days had no effect on maturation of the accessory sex organs, indicating that secretions from the gonad are important for mediating the effect of photoperiod on this aspect of the maturation process. Transplanting gonads from slugs maintained on long days to gonadectomized animals stimulated maturation of the accessory sex organs. Likewise, transplanting brains from long-day slugs that were gonadectomized to gonad-intact animals kept on inhibitory short days stimulated maturation of the gonad and accessory sex organs in the recipient animals. These experiments suggest that photoperiod activates neurons in the brain that, in turn, activate the gonad, thereby stimulating maturation of the accessory sex organs. (C) A model for photoperiodic stimulation of the reproductive system of *Limax*. Long photoperiods stimulate unknown photoreceptors, activating neurons in the cerebral ganglion to secrete a hermaphroditic maturation factor and male gonadotropic factor. These secretions from the cerebral ganglion stimulate the gonad to release additional hormones that stimulate maturation of the female and male accessory sex organs.
Source: From Wayne (2001).

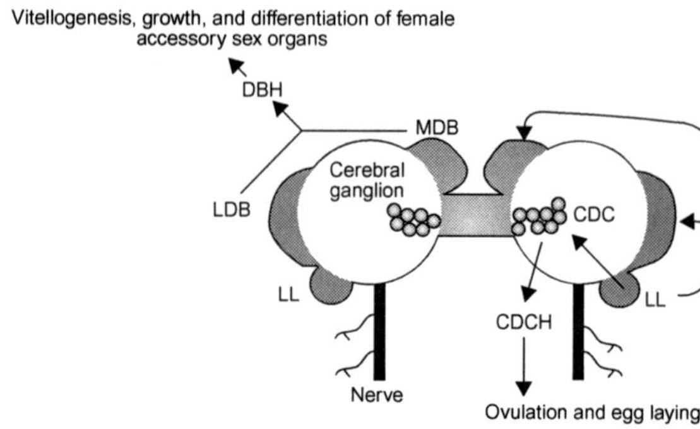

Figure 3.4 Organization of the female reproductive endocrine system of *Lymnaea stagnalis*. Neurons in the lateral lobes activate the endocrine dorsal bodies and caudodorsal cells. Dorsal body hormone, secreted from the lateral and medial dorsal bodies, stimulates vitellogenesis, and growth and differentiation of the female accessory sex organs. Caudodorsal cell hormone, secreted by the caudodorsal cells, stimulates ovulation and egg-laying behaviors. *Abbreviations*: CDC, caudodorsal cells; CDCH, caudodorsal cell hormone; LDB, lateral dorsal body; LL, lateral lobe; MDB, medial dorsal body.
Source: From Wayne (2001).

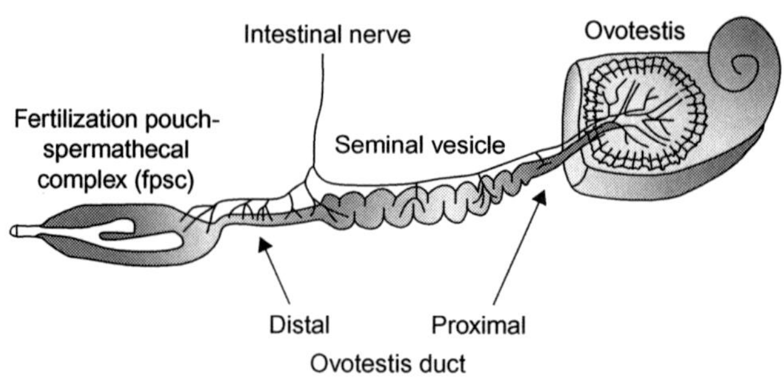

Figure 3.5 The hermaphroditic components of the reproductive system in *Helix aspersa*. The OT duct, together with the organs it connects, is innervated by branches of the intestinal nerve. The approximate location and relative density of each innervation field is indicated.
Source: From Geoffroy et al. (2005).

This brief review of the production of egg cells in insects and gastropod molluscs shows that it results from a complex process involving numerous systems and pathways, all of which point to the controlling role of the CNS in the process. Taking as an example the production of the egg cell by a female insect, Heming described the essence of the oogenesis as follows:

> *She can do this because she is sensitive to diverse cues from the environment (photoperiod, temperature, relative humidity, substrate odor, food, proximity and state of mates, etc.), is able to integrate this information in her nervous system, and via her endocrine system can use it to develop and deposit her eggs at the right time and place.*
>
> *Heming (2003)*

Neural Control of Oogenesis in Vertebrates

Neuroendocrine Control via the Brain–Hypothalamic–Pituitary Axis

Oogenesis and ovulation, in both ovulation-inducible vertebrates such as rabbits (Strand, 1999) and cyclic ovulators such as humans (Johnson and Crowley, 1986; Villa-Diaz and Barrell, 1999), are under the CNS control (Adler and Crowley, 1984; Kalra et al., 1987; Gore and Terasawa, 2001) with the HPO axis (hypothalamic GnRH → pituitary FSH → ovarian estrogen and progesterone) as the main neuroendocrine axis.

In fish, the neurohormonal control of oocyte growth and maturation is mediated, respectively, by estradiol-17-β and -17-α, 20 β-DHP (17-α, 20-β-dihydroxy-4-pregnen-3-one) that are synthesized in granulosa cells by steroid precursors provided by theca cells. The latter is the naturally MH (maturation inducing hormone) in the medaka fish. A variety of other neuromodulatory factors are involved in steroidogenesis in fish ovarian follicles (Fukada et al., 1994; Nagahama et al., 1995).

In sheep, it has been demonstrated that granulosa cells, in response to FSH stimulation, start production of inhibin A (Campbell and Baird, 2001; Drummond et al., 2000). Inhibin production seems to be downregulated by follicle synthesis of estrogen (Drummond et al., 2000), but it has endocrine effects on the pituitary FSH. In humans, paracrine insulin-like growth factors (IGFs) are involved in the synthesis of inhibin A and B, but the main regulators of their production are gonadotropin hormones, FSH and LH, via protein kinase A signal transduction pathway (Vänttinen et al., 2000). As it is well known, gonadotropins themselves are under the control of CNS signals.

The synthesis of progesterone by the granulosa cells is stimulated by various hormones and growth factors, with gonadotropins and IGF-1 as the most important (Khamsi and Roberge, 2001). The immediate effect of neurohormones on progesterone secretion, as well as the fact that gonadotropins and IGF-1 are under neuroendocrine control (hypothalamic GHRH→pituitary GH→IGF-1), shows that progesterone secretion is a downstream element of a signal cascade that starts with an electrical/chemical output of the neural circuits in the CNS. Experiments on rat granulosa cells have also demonstrated the stimulating effect of adrenergic agents on progesterone secretion (Selvaraj et al., 2000). They are also found to be involved in the regulation of oocyte maturation by inhibiting secretion of E2 and the FSH-stimulated progesterone (Salmassi et al., 2000).

Local Control by Ovarian Innervation

In vertebrates, the innervation of the ovary by SON, vagus, and the ovarian plexus stimulatorily regulates the effects of human chorion gonadotropin (hCG) and pregnant mare serum gonadotropin in the ovary (Morales et al., 1998). Unilateral and bilateral vagotomy in rats delays the onset of puberty and influences the levels of E2 and progesterone, but it increases the number of shed ova when it is performed at 28 days of age (Morales-Ledesma et al., 2004).

In regard to the role of neurotransmitters in the function of luteinized granulosa cells *in vitro*, it is observed that activation of the muscarinic receptor by the Ach agonist carbachol stimulates progesterone secretion by granulosa cells, apparently via steroid acute regulatory protein, whose synthesis increases 2–10 times (Fritz et al., 2001). The neurotransmitter Ach, on the other hand, potentiates the maturation effect of the progesterone on *Xenopus* oocytes (Dascal et al., 1984), suggesting that the CNS regulates progesterone secretion and the reproductive physiology in humans not only via the HPO axis but also via an evolutionarily conserved mechanism of direct neural signaling. This is corroborated by the fact that *Xenopus* oocytes are in possession of "native" receptors for Ach (Parker et al., 1987) and maternal serotonin is present in oocytes (Veselá et al., 2003). Serotonin and melatonin, besides their actions via the brain–HPO axis, influence the oocyte maturation and reproductive function by acting directly on the ovarian cells (Sirotkin and Schaeffer, 1997).

The effects of these neurotransmitters and of their agonists and antagonists may be different in different species. So, for example, serotonin inhibits oocyte maturation stimulated by progesterone in amphibians *Xenopus laevis* and *Bufo viridis* (Nikitina et al., 1993; Nikitina and Buznikov, 1996), but it stimulates oocyte maturation and germinal vesicle breakdown (GVBD) in nemertean worms (Stricker and Smythe, 2000), in the starfish (Buznikov, 1993) and in the estuarine teleost fish, *Fundulus heteroclitus heteroclitus* (Cerda et al., 1997).

All the available data on the presence and activation of oocyte receptors for various neurotransmitters, their binding of the respective neurotransmitters, and activities of neurotransmitters as stimulators/inhibitors of the oocyte maturation and GVBD, strongly suggest the presence and functioning of an ovarian neurotransmitter system as a mediator of the central nervous control of the reproductive physiology.

Neural Control of Spermato-/Spermiogenesis

For at least two decades, it has been acknowledged that spermatogenesis "is ultimately controlled by neurons in the CNS" (Sharp and Gow, 1983) and that the CNS exerts that control mainly via the GnRH pulse generator and the hypothalamic–pituitary–testicular axis (Vander et al., 2001). Environmental and somatic stimuli are integrated, processed, and a signal is sent to the hypothalamus triggering production of GnRH, which stimulates secretion of gonadotropins by the pituitary (Figure 3.6). These pituitary hormones are essential for the function of gonads and processes of spermatogenesis and spermiogenesis. Spermiogenesis is a long process of transformation of spermatids into sperm cells that in some birds comprises 12 distinct morphological steps, including: formation of an acrosome, an acroneme, the loss of almost all (97%) of cytoplasm, and replacement of nucleohistones with nucleoprotamine (Kirby and Froman, 2000). Besides the GnRH pulse generator and the hypothalamic–pituitary–testicular axis, a direct hypothalamic–testicular (pituitary-independent) neural pathway controlling the processes of spermato- and spermiogenesis has been found to be operational in mammals.

It has long been known that the synthesis of testosterone by Leydig cells is cerebrally regulated by the brain–hypothalamus–pituitary axis with the pituitary LH as proximate cause, but recently it has been suggested the existence of a LH-independent

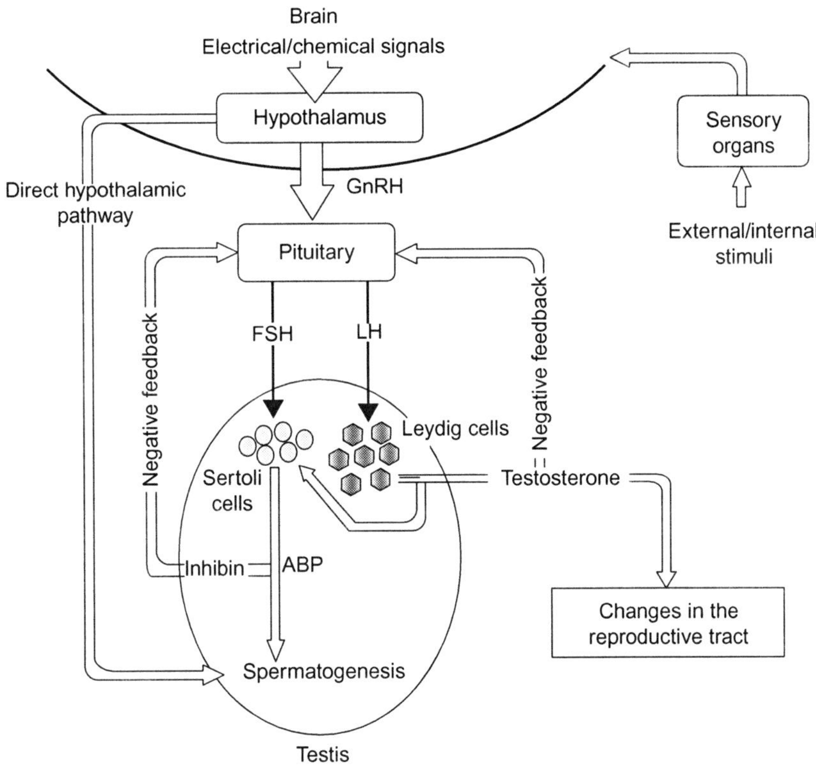

Figure 3.6 The CNS control of the reproductive function in men. By processing specific external and internal stimuli, the nonhypothalamic brain sends to the hypothalamus electrical/chemical signals that stimulate GnRH secretion. In the pituitary, GnRH induces secretion of FSH and LH, which act, respectively, on the Sertoli cells and Leydig cells in testes. The latter secrete testosterone, which induces Sertoli cells to produce inhibin and androgen binding protein (ABP), which binds testosterone and in this form stimulates spermatogenesis. Note also a direct brain regulation of the function of testes via the hypothalamus. *Abbreviations*: FSH, follicle-stimulating hormone; GnRH, gonadotropin-releasing hormone; LH, luteinizing hormone.
Source: Modified from Vander et al. (2001).

neural "brain–testicular circuit." It is also demonstrated that splanchnic innervation regulates testicular LH receptors and blood flow. (Lee et al., 2002)

Moon cycles are used as visual cues for synchronization of the release of mature gametes by several rabbitfish (Rahman et al., 2000a,b, 2003).These fish use different stages of the lunar periodicity for spawning. So, for example, the golden rabbitfish, *Siganus guttatus*, uses as a cue the first quarter moon, whereas *Siganus argenteus* uses the last quarter moon.

An interesting correlation has also been observed to exist between the lunar periodicity and spermatogenesis, as well as hormonal influences regulating it, in the forktail rabbitfish, *S. argenteus*. The lunar cue does not directly act on the gonadal hormonal system but in the CNS:

> *The effects of lunar factors in the rabbitfish would occur in higher parts of the hypothalamus–pituitary–gonadal axis.*
>
> *Rahman et al. (2003)*

The lunar cues are integrated and processed in the hypothalamus, which, by stimulating the release of pituitary gonadotropin, induces DHP (dihydroxy-4-pregnen-3-one) spawning around the last quarter moon (Rahman et al., 2003). The function of the hypothalamic region controlling these processes in male rats, the PVN (paraventricular nucleus), is under control of brain catecholamines (Selvage et al., 2004), also suggesting that the signal cascade inducing spermatogenesis originates in higher regions of the brain.

Neural Control of Deposition of Parental Epigenetic Information in Gametes

What makes the zygote (the egg cell in parthenogenetic metazoans) uniquely capable of developing into an adult metazoan is not the genetic information, its genes or DNA, which are the same in all the cells of the body, but the presence in the egg and sperm cytoplasm of epigenetic information, in the form of thousands of orderly arranged types of parental factors (e.g., mRNAs, proteins, hormones, secreted proteins, and neurotransmitters) as well as imprinting of specific genes in the gametes. It is this epigenetic information, not the genetic information, that endows gametes with the monopoly of biological reproduction.

The idea on the presence in the egg cell of maternal factors determining phenotypic characters, such as muscles in ascidians, is a century old, but the nature of these factors (cytoplasmic mRNAs) was identified in 1968, when Kalthof and Sander reported that embryos from midge eggs, whose RNA was inactivated by ultraviolet (UV) radiation, lacked head and thorax (Kalthof and Sander, 1968), although their genome was functional.

In 1982, G. Freeman and J.W. Lundelius succeeded in obtaining right-coiling embryos by injecting *cytoplasm* from right-coiling snail eggs into the eggs of left-coiling mothers, demonstrating that a maternal cytoplasmic factor in the egg cytoplasm was responsible for right/left coiling in the snail. In an equally illuminating experiment, putative somatic cells were converted into germ cells by simply transplanting germ cytoplasm into them (Ilmensee and Mahowald, 1974).

In the process of gametogenesis, in all metazoans, thousands of different maternal substances are synthesized and orderly deposited or transferred from follicle cells and nurse cells into the oocyte. However, the ordered placement and transfer of maternal cytoplasmic factors requires information that is different from the genetic information. What is the nature and origin of that epigenetic information?

In 2001, I put forward the hypothesis that the maternal CNS determines deposition of maternal cytoplasmic factors in the egg cell (Cabej, 2001). As presented at the time, it was rather a prediction of my wider view on the epigenetic component of evolution. Ever since then, adequate evidence has accumulated for substantiating it. I will briefly review that evidence here.

Neural Control of Deposition of Maternal Factors in Insect Oocytes

The deposition of the epigenetic information in the egg cells of insects occurs via

- Oocytic receptor-mediated endocytosis,
- Squeezing of the cytoplasmic content of nurse cells into the oocyte, and
- Transfer of maternal factors via communicating channels from the nurse and follicle cells into the oocyte.

Neural Control of Deposition in the Egg Cell of Maternal Factors via Receptor-Mediated Endocytosis

The most important form of receptor-mediated endocytosis is the clathrin-mediated endocytosis. Clathrin is a polypeptide molecule located in particular regions of the cell membrane and represents the main component of the coated pits.

One well-known example of the CNS control of the uptake and deposition of maternal factors in the oocyte is vitellogenin (Vg), the precursor of vitellin (Vt), containing the basic proteins of egg yolk in insects. In the ixodid tick, *Amblyomma hebraeum* (Koch, 1844), vitellin consists of seven major and several minor polypeptides. The synthesis of vitellogenin in this tick is under proximate control of the hormone ecdysteroid, which, in turn, is under control of the brain neurohormone, PTTH (Friesen and Kaufman, 2002). Besides ecdysone, JH (its synthesis in CA is also induced by a number of brain neurohormones, allatotropins) is demonstrated to induce vitellogenesis in the fat body and ovary of *Drosophila* (Handler and Postlethwait, 1977) and of the red cotton stainer, *Dysdercus koenigii* (Fabricius; Venugopal and Kumar, 2000).

Experiments on the fall army worm, *Spodoptera frugiperda* (J.E. Smith), suggest that ecdysone is crucial for inducing vitellogenesis in insects and JH is necessary for the uptake of vitellogenin by the oocyte (Sorge et al., 2000). Ecdysone regulates the uptake of vitellogenin and other maternal cytoplasmic factors from the hemolymph by the oocyte. As for the mode of action of the ecdysone, it is demonstrated that in the female mosquito, *Aedes aegypti*, *vg* (vitellogenin) gene, in its regulatory region, has a functional ecdysteroid responsive element (VgEcRE1) to which the complex of the ecdysteroid with its heterodimeric receptor binds and induces transcription of the *vg* gene and vitellogenin synthesis (Martin et al., 2001).

At early stages of oogenesis, vitellogenin is transferred to the oocyte from surrounding follicle cells via gap junctions (Waksmonski and Woodruff, 2002; see later on the neural control of formation of gap junctions). Later on, oocytes take up circulating vitellogenins from the hemolymph. The oocyte and nurse cells sequester circulating vitellogenins in a process of endocytosis, which may be constitutive or induced (Brown and Greene, 1991).

During the receptor-mediated endocytosis, that takes place in late oogenesis, the Vg receptor, a low density lipoprotein receptor (LDLR) is essential for vitellogenin uptake. LDLRs represent a group of receptors closely related to cell surface receptors that regulate the uptake of extracellular ligands (Herz and Bock, 2002).

The synthesis of vitellogenin in *Xenopus* oocytes requires estrogen (McKenzie et al., 1990), and recently it is demonstrated that the estrogen upregulates LDLR and ezrin, a cytoskeletal linker protein with which LDLR forms a functional complex that enables endocytosis (Smith et al., 2004). Thus, ultimately, the synthesis of LDLR in *Xenopus* is regulated by the HPO axis, via the estrogen.

Vitellogenin proteins bind to their receptors on the oocyte surface in an endocytic complex of clathrin and α-adaptin vesicles and, in this form, are engulfed into the oocyte cytoplasm and Golgi lysosomes, where they are processed into yolk proteins.

The oocyte needs JH stimulation for starting vitellogenin uptake. It has been demonstrated that a brain signal, a "cephalic event" (Handler and Postlethwait, 1977), via ecdysone, controls the uptake of vitellogenin (Richard et al., 2000) as well as other maternal cytoplasmic factors (Chapman, 1998) of the hemolymph by the egg cell.

It has been shown that in the mosquito *Aedes aegypti*, ablation of CA inhibits formation of the endocytic complex (Venugopal and Kumar, 2000), and "the formation of the endocytic complex is controlled by juvenile hormone from the corpora allata" (Raikhel and Lea, 1985), whose synthesis and secretion, as shown earlier, is regulated by brain allatotropins.

Neural Control of the Squeezing of Nurse Cell Content into the Oocyte

The larger part of maternal cytoplasmic factors in the insect oocyte comes from nurse cells, which, stimulated by signals that ultimately originate in the CNS, dump their cytoplasmic content into the oocyte. Maternal cytoplasmic factors include transcripts (mRNAs), non-housekeeping proteins, growth factors, hormones, neurotransmitters, and nutrients. As such, their expression is induced by extracellular signals, which, as we have shown in Chapter 1, are downstream elements of signal cascades starting in the CNS. Ecdysteroids and JH play crucial roles in expression of genes in nurse cells during embryogenesis and it is well known that these hormones are mediators of the action of a number of neurohormones synthesized in the insect brain.

Carney and Bender (2000) reported that *Drosophila* nurse cells and follicle cells express the ecdysone receptors (EcR-A and EcR-B) throughout oogenesis, suggesting that maternal ecdysone is active in these cells. When the activity of ecdysone in these cells is experimentally disrupted (e.g., in EcR mutant females), development of abnormal egg chambers and defects of oogenesis, including loss of vitellogenic egg stages, occurs (Carney and Bender, 2000).

The transport of maternal mRNAs from nurse cells into the oocyte begins in early stages of the oogenesis in the form of the so-called slow transport. A common microtubule complex extends from the posterior part of the oocyte to nurse cells. The complex serves as a scaffold on which maternal mRNAs are transported (Theurkauf et al., 1992) from nurse cells into the oocyte. After the "slow transport," toward the end of *Drosophila* oogenesis (stages 10–12), begins the "fast transport" when the cytoplasmic mass of the nurse cell is forced into the oocyte (Figure 3.7).

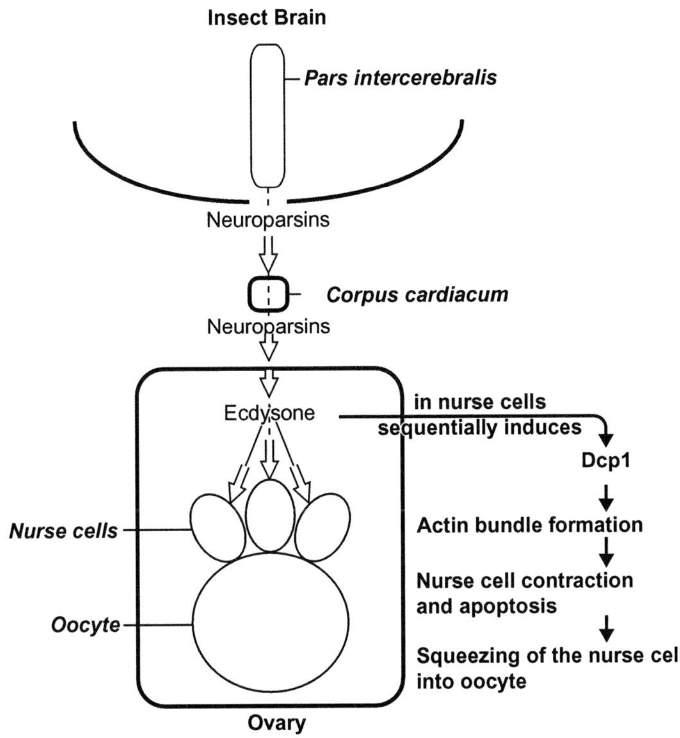

Figure 3.7 Simplified diagram of the neural regulation of the squeezing of nurse cell cytoplasm in the insect oocyte.

The apoptosis of insect nurse cells, far from a suicide constitutive event, is centrally regulated by ecdysone (Soller et al., 1999), whose synthesis, needless to say, is cerebrally regulated by the neurohormone PTTH. In vertebrates as well, apoptosis is controlled by signal cascades originating in the CNS (see Section Neural Control of Apoptosis in Chapter 5).

At this juncture, the actin cytoskeleton of the nurse cell, an essential component of ring canals, forms actin bundles, extending from the oocyte cell membrane to the nuclear membrane of the nurse cell, making the later permeable and thus causing release of nuclear compounds and Ca^{2+} in the cytoplasm. This causes nurse cells to contract and squeeze their cytoplasmic content into the oocyte as part of nurse cell apoptosis, or as a concurrent process (Cavaliere et al., 1998).

It is the ecdysone receptor–ultraspiracle (EcR–Usp) heterodimer that, by binding the promoter region of the *caspase Dronc*, regulates its expression and the beginning of the apoptotic process (Cakouros et al., 2004). Ecdysone, via caspase Dcp-1 (caspase Dronc in the histolytic processes during metamorphosis—Dorstyn et al., 1999; Figure 3.8) controls the crucial process of formation of actin bundles and the squeezing of the cytoplasmic content of nurse cells (Buszczak and Cooley, 2000), thus leading to accumulation of thousands of types of maternal cytoplasmic factors into the oocyte.

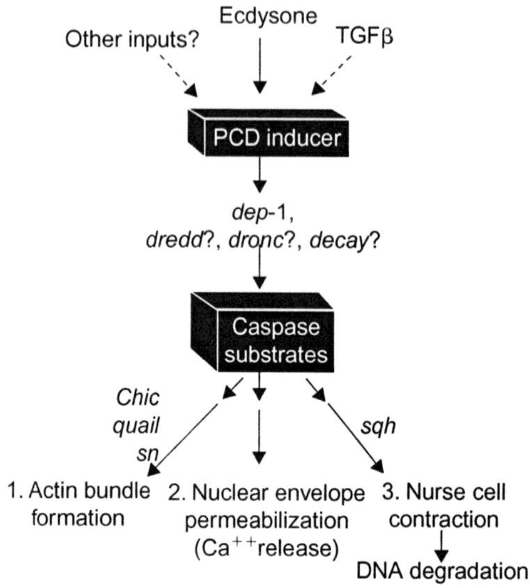

Figure 3.8 A model for apoptosis driving rapid cytoplasm transport. Available evidence suggests that several signaling pathways may regulate the expression or activity of an unidentified inducer of apoptosis. This inducer activates DCP-1 and possibly other caspases, including DREDD, DRONC, and DECAY. Activated caspases cleave unknown downstream substrates, which, in turn, trigger parallel genetic pathways leading to rapid cytoplasm transport. *Source*: From Buszczak and Cooley (2000).

It is demonstrated that the brain-regulated hormone ecdysone regulates Ras and three other small GTPases, which control signal transduction pathways, linking membrane receptors to the assembly and disassembly of actin cytoskeleton. But these inducers themselves are activated by growth factors, such as epidermal growth factor (EGF; Mochizuki et al., 2001), that are under neurohormonal control.

Unlike insects, in *Hydra vulgaris,* it is not the nurse cell that squeezes its content into the oocyte; instead, it is the oocyte that actively phagocytoses the apoptotized nurse cells (Miller et al., 2000).

There is adequate evidence demonstrating that in vertebrates, phagocytosis is neurally regulated. Neuropeptides released by sensory nerves, somatostatin, substance P, and calcitonin gene-related peptide inhibit migration and phagocytic activity of macrophages, whereas neuropeptides of the autonomic nervous system, VIP, and neuropeptide Y stimulate phagocytic activity and migration of macrophages toward infection sites (Ahmed et al., 1998, 2001). Release of the neuropeptide pituitary adenylate cyclase activating polypeptide in lymphoid tissues by autonomic nerves prevents the phagocytic activity of macrophages (Ganea and Delgado, 2003). Neuropeptides somatostatin, substance P, and the pituitary adrenocorticotropic hormone (ACTH) potentiate the tumoricidal effect of macrophages (Peck, 1987), but ACTH and NE suppress their interferon-stimulated tumoricidal effect in mice and the autonomic neuropeptide VIP increases the effect of NE. (Koff and Dunegan, 1985).

Microtubules Are Involved in Transportation and Placement of Maternal Factors in the Oocyte

Follicle cells communicate with the oocyte via gap junctions, and nurse cells mainly via ring canals, which are wider than gap junctions and allow passage of macromolecules, and even whole organelles, into the oocyte.

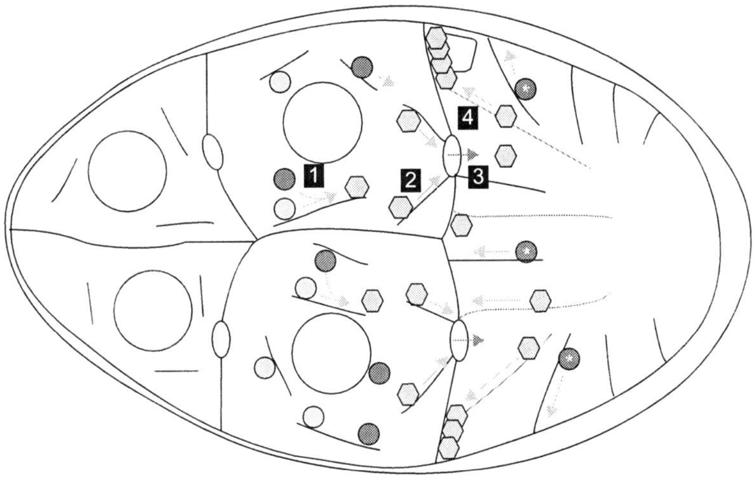

Figure 3.9 A revised model for microtubule-dependent *bcd* mRNA localization. Within the nurse cell cytoplasm, *bcd* RNA forms particles (black circles) through a microtubule and Exu-independent process. Microtubules then mediate the association of Exu protein (white) and at least one additional factor to form mature transport complexes (step 1, gray). These particles are transported to the ring canals along microtubules originating near the cell–cell junctions (step 2), and pass through these junctions by a microtubule-independent process. Mature RNA particles entering the oocyte encounter a dense network of microtubules that are organized by most of the oocyte cortex (step 3) but are preferentially transported along a subset of microtubules (black) originating at the anterior cortex (step 4). By contrast, *bcd* mRNA particles that have not passed through the nurse cell cytoplasm (black, asterisk) are transported on all oocyte microtubules to produce a nonpolar cortical distribution.
Source: From Cha et al. (2001).

One of the earliest maternal factors transported from nurse cells to the oocyte in *Drosophila* is *bicoid* mRNA. It is expressed during early oogenesis (stages 4 and 5) and, in the form of the *bicoid* mRNA–Exu complex (Cha et al., 2001), is transported from nurse cells (Cha et al., 2001; Figure 3.9) to the anterior end of the oocyte via the microtubule skeleton by dynein (Schnorrer et al., 2000).

Staufen, a maternal protein, is required for both *bicoid* mRNA localization to the anterior end of the oocyte and *oskar* mRNA to the posterior end (Micklem et al., 2000) by kinesin I (Brendza et al., 2000). In *Xenopus*, Staufen and kinesin, as well as vegetally localized mRNAs, Vg-1 and VegT, form a ribonucleoprotein complex, which is responsible for the transport of *oskar* mRNA to the posterior pole (Yoon and Mowry, 2004). The localization of the *bicoid* mRNA and *oskar* mRNA at two opposite ends of the oocyte at this stage determines early polarization of the oocyte and the establishment of the first A–P body axis of the future embryo. These maternal transcripts, as well as others, are transported via the oocyte cytoskeleton, i.e., their localization in the oocyte is microtubule dependent (Figure 3.9), as it is suggested by the fact that drugs that depolymerize oocytic microtubules perturb all aspects of the *bicoid* mRNA localization (Pokrywka and Stephenson, 1991; Cha et al., 2001).

The mechanism of the transport of *bcd* mRNA to the egg cell involves both microtubules and the actin cytoskeleton. Actin seems to anchor microtubules of the anterior cortex that localize *bcd* mRNA there but *bcd* mRNA molecules there seem to be nucleated by a local microtubule organizing center (MTOC). Additional *bcd* mRNA is transported to the oocyte during nurse cell dumping. Dynein transports *bcd* mRNA on microtubules (Weil et al., 2006, 2008), and *bcd* mRNA granules are uniformly dispersed as fine particles in the anterior cortex (Weil et al., 2005).

We do not know the particular mechanism of transcription of the *bicoid* in insects, but we know that, in experiments in other animals, its expression is hormonally stimulated by E2 (Burz et al., 1998). We also know that *Drosophila bicoid* (as well as *goosecoid* and *orthodenticle*) belongs to the same family of homeobox genes with vertebrate genes pituitary homeobox 1 (*Ptx1*), *Ptx2*, and *Ptx3* (Drouin et al., 1998). The vertebrate gene *Ptx1*, like *Drosophila bicoid* gene, is a transcription factor that activates gene transcription upon binding a sequence related to the *Drosophila bicoid* target sites (Lamonerie et al., 1996).

Another important maternal factor that is transported from *Drosophila* nurse cells to the oocyte during early oogenesis is *gurken* mRNA, which codes for a TGF-α-like protein. It is also transported to the posterior end of the oocyte via the microtubule skeleton. As a growth factor, its expression must be under control of extracellular signals (see Section Hormonal Control of Secreted Proteins and Growth Factors in Chapter 1). Indeed, expression (both transcription and translation) of *gurken* in nurse cells requires another growth factor, Vasa (Tomancak et al., 1998). In turn, expression of *vasa* mRNA in the ovary of the marine fish, *Sparus aurata*, is stimulated by the pituitary GH, by its combination with hypothalamic GnRH, as well as by E2 (Cardinali et al., 2003). As we know, in vertebrates, all these steps, upstream the cascade, lead to the CNS.

Neurohormonal Regulation of Length of Microtubules

Microtubules are dynamic structures assembled from tubulin proteins. Microtubules grow or shrink in response to the presence of specific molecules. The ability to regulate the length of microtubules and the morphology of the cytoskeleton makes the movement and specific localization of maternal factors in the oocyte possible.

In insects, the neuropeptide PTTH stimulates β-tubulin synthesis. In view of the role of tubulin composition in the structure of microtubules (Panda et al., 1994), controlled secretion of tubulin may be involved in controlled modification of the cytoskeleton and the differential localization in the oocyte of maternal factors moving on the microtubule scaffold.

It is known that Ca^{2+} induces changes in the structure and the length of microtubules (Mäthger et al., 2003), an important component of the cytoskeleton. The neurotransmitter Ach increases the level of Ca^{2+} in cells, thus changing the length and structure of microtubules. It is noteworthy that catecholaminergic innervation is present in the ovary and testes and that gonads, the ovary and testicles, have an intrinsic catecholaminergic local network of neuron-like cells (Mayerhofer et al., 1997).

A number of hormones are known to regulate (polymerize/depolymerize) the length of microtubules. Human GH induces polymerization of all tubulin isoforms

(Goh et al., 1998), and so do prolactin (Ravindra and Grosvenor, 1992), insulin, and EGF, etc.

It is experimentally demonstrated that the structure and function of microtubules may be modified by modifying the relative amounts of different tubulin isotypes (Panda et al., 1994), and it is observed that two α-tubulin isotypes have opposing effects on microtubule dynamics *in vitro* (Bode et al., 2003).

A number of proteins are involved in the process of regulation of microtubule structure and dynamics. The most intensively investigated group of these proteins is that of microtubule-associated proteins (MAPs), which bind microtubules and stabilize (polymerize) the cytoskeleton (e.g., MAPs, cadherin) or destabilize (depolymerize) it (stathmin (Op18, XKCM1), a kinesin-related protein (Walczak et al., 1996), etc.).

In *Xenopus* egg extracts, for instance, the microtubule dynamics is regulated "by a balance between a stabilizing factor, XMAP215, and a destabilizing factor, XKCM1" (Tournebize et al., 2000). The presence of XMAP310 decreases the microtubule shrinkage twofold (Anderssen and Karsenti, 1997).

There is direct evidence that MAPs themselves are regulated by extracellular signals. Tumor necrosis factor (Vancompernolle et al., 2000) and c-AMP-linked agonists (Gradin et al., 1998), by hyperphosphorylating stathmin, turn off its microtubule-destabilizing (depolymerizing) action. It is demonstrated that maturation-promoting factor p34cdc2/cyclin B complex exerts its control on starfish oocyte microtubule networks via MAPs (Ookata et al., 1993). *Xenopus* XMAP215 also associates with p34cdc2 kinase and directs it to the microtubule cytoskeleton, thus acting as a regulator of the cytoskeleton structure (Charrasse et al., 2000). But MAPs can also perform the opposite function, i.e., depolymerization (shortening) of microtubules when they are phosphorylated, by permitting the action of destabilizing proteins (Anderssen, 1998).

By regulating the structure and the length (polymerization/depolymerization) of microtubules, MAPs can control the localization of specific maternal factors in specific regions of the oocyte. A number of neural factors, neurohormones and neurotransmitters, are known to influence the structure and binding of MAPs to microtubules and thus modify the microtubule network. So, for example, the hypothalamic neuropeptide GnRH (Drouva et al., 1998), neuropeptide VIP, as well as monamine neurotransmitters, dopamine and noradrenaline, via their receptors, phosphorylate stathmin (Chneiweiss et al., 1992), thus facilitating its binding to microtubules and inducing their shortening (depolymerization; Rubin and Atweh, 2004; Figure 3.10).

The neurosteroid, dehydroepiandrosterone, binds to MAP2 in the brain (Laurine et al., 2003). Pregnenolone is another neurosteroid that binds to MAP2 and stimulates polymerization and elongation of microtubules in neurons (Murakami et al., 2000; Fontaine-Lenoir et al., 2006). Various hormones, as well, are known to modify the structure and length of microtubules. Testosterone inhibits depolymerization of tubulin molecules by directly targeting them, while E2 only modifies the polymerization of microtubules (Kipp and Ramirez, 2003); progesterone withdrawal causes tubulin polymerization or elongation (Loizzi, 1985).

Human pituitary GH transiently increases polymerization of microtubules, thus modifying the microtubule network (Goh et al., 1998). The synthetic estrogen diethylstilbestrol also has a demonstrated microtubule-depolymerizing effect (Can and Semiz, 2000).

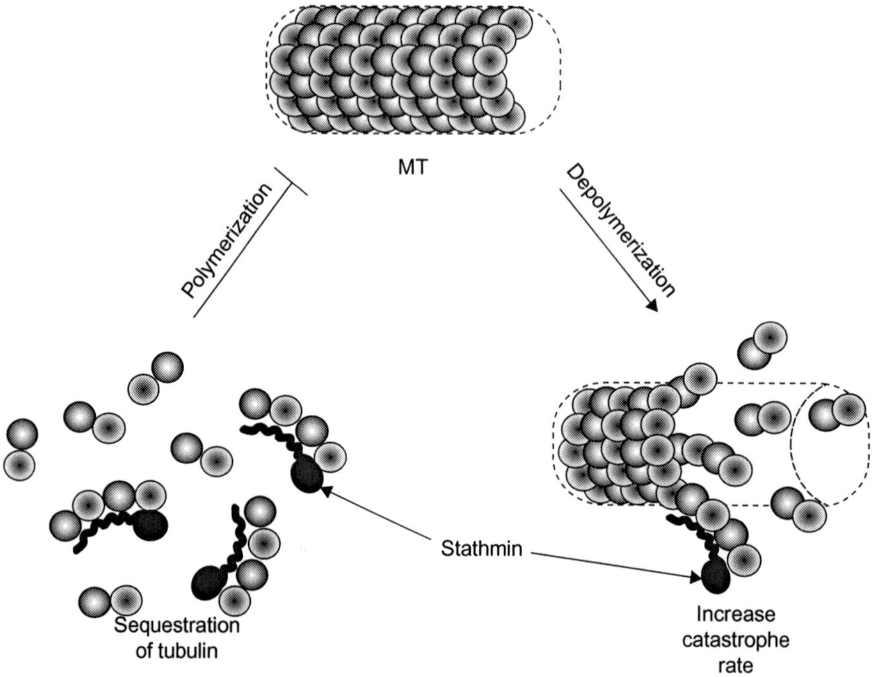

Figure 3.10 Model for the role of stathmin in the regulation of microtubule dynamics. Microtubules (MT) continuously switch between phases of polymerization and depolymerization. Stathmin can sequester unpolymerized tubulin by binding two α/β-tubulin heterodimers (represented by light- and dark-shaded circles), thus diminishing the pool of tubulin heterodimers available for polymerization. Stathmin can also bind to the end of polymerized microtubules and increase the rate of catastrophe by inducing a conformational change that promotes microtubule depolymerization.
Source: From Rubin and Atweh (2004).

Both estrogen and testosterone act on tubulin molecules; the estrogen by binding to two sites on tubulin molecule and progesterone to only one site of lower affinity. Recent evidence shows that estrogen and testosterone exhibit a high binding affinity to two of the MAP family members (MAP2 and Tau), which are modulators of microtubule dynamics. Estrogen activates MAPK (mitogen-activated protein kinase) via its receptors ER-α and -β and progesterone as well activates the MAP kinase pathway via its receptor, PR (Edwards, 2005). G-protein subunits bind tubulin dimers on the one hand, and couple cell surface receptors on the other. Many hormones and neurotransmitters use these proteins to influence, among other things, the polymerization of tubulin molecules via direct transfer of GTP.

There is also evidence that the CNS, besides the above signal cascades along the hypothalamic–pituitary axis, exerts a direct neural control on the structure of microtubule networks. So, for example, the suckling, an innate behavior, in lactating rats leads to simultaneous dramatic increases of the amount of microtubuli in the anterior pituitary (Ravindra and Grosvenor, 1986), as well as to increased tubulin-dependent GTPase activity (Ravindra and Grosvenor, 1992).

To summarize, there is evidence that a number of hormones and neurotransmitters are involved in structural changes of microtubules. The fact that the transport of maternal factors is accomplished via microtubules suggests that a mechanism that regulates the dynamic structure and length of these cytoskeleton structures may convey the necessary information for the placement of maternal factors in strictly determined sites of the oocyte cytoplasm. The evidence presented in this subsection on the chemicals inducing changes in the structure and length of microtubules shows that a number of neuropeptides, neurotransmitters, and hormones are essentially involved in the process. Not only involvement of neurohormones and neurotransmitters but of hormones of the target endocrine glands (which are secreted in response to neural signals cascades via the hypothalamic–pituitary axis) as well suggests that the nervous system plays an important role in the dynamics of microtubules in the oocyte and, consequently, in the process of the ordered placement of maternal cytoplasmic factors in the oocyte.

In the above context, it is plausible to assume that the nervous system may have a similar function in determining the length of microtubules of the oocyte cytoskeleton and, consequently, being involved in the regulation of the transport of maternal factors to their destinations in the oocyte cytoplasm.

The Nervous System Manipulates the Length and Orientation of Microtubules

Is there evidence of neurally adaptive regulation of the microtubule patterns in metazoans? The answer to this question may come from a brief examination of examples of adaptive body coloration and patterning of several fish and cephalopod species. The mechanism of the adaptive coloration in fish evolved more than 400 million years ago (Fujii, 2000).

A number of fish species can change their body color and pattern, in seconds and "at will," to match the perceived color of their background, and some of them can even mimic the body colors and patterns of heterospecific fishes in order to avoid detection within their crowds. This is the case with several fishes, such as the neon tetra, *Paracheirodon innesi*, blue damselfish, *Chrysiptera cyanea*, the paradise whiptail, *Pentapodus paradiseus*, and several cephalopods.

The cells responsible for the adaptive change in body color in these fishes are iridophores, pigmentless light-reflecting cells of the skin that contain piles of thin transparent guanine platelets, which form multilayer reflectors. Any change in the length (increase or decrease) of microtubules between plates will produce a change in the color reflected by iridophores. The distance between guanine plates is determined by the length of microtubules that connect plates with each other (Kasukawa et al., 1987; Mäthger et al., 2003). It was found that neurotransmitters like noradrenaline shift the light reflected by iridophores toward longer wavelength, while Ach and noradrenaline antagonists—such as adenosine—toward the shorter wavelength (Kasukawa et al., 1987; Mäthger et al., 2003). Mediators of changes in the distance between platelets are neurotransmitter receptors. The shifts in the light wavelengths reflected by guanine platelets are determined by corresponding changes (increase and decrease) in the length of microtubules and, consequently, of the distance between reflecting platelets (Kasukawa et al., 1987). Thus, the change in the length of interplate microtubules changes the wavelength and the color of light reflected by the skin. This "Venetian blind model" of adaptive color change, as a result of changed spacings between guanine plates, has recently been quantitatively verified (Yoshioka et al., 2011). This instantaneous change of body color is not instinctive in the classical meaning of the word, i.e., it does not result from any "fixed action pattern"; far from a standard indivisible response, it implies practically an infinite number of responses, i.e., an ability to change its body color in a multitude of variants so that it matches the perceived background or body color and patterns of other animals.

How do animals translate the perceived color and pattern into their own body color and pattern? How do they verify whether the induced change matches the perception? While the neurobiological mechanism of the processing of these stimuli and their "translation" in the brain is still a black box, we know the neural pathway that leads to this form of astonishing adaptive pigmentless coloration.

The visual input from the retina is transmitted via the optic nerve to the CNS, where its processing results in an adaptive neural response that determines the adaptive coloration of the fish. The CNS selects the adaptive response with the sympathetic nervous system being the mediator of the adaptive changes of skin color (Goda and Fujii, 1998). Severance of the sympathetic nerves causes elimination of the chromatic effects of chromatophores in general, leading to darkening of the skin (which also results from depressed activity of the sympathetic activity) that is interpreted as an effect of the parasympathetic system on chromatophores (Fujii, 2000).

The adaptive changes in the distance between plates are controlled and regulated by the local sympathetic innervation which releases locally noradrenaline. Another neurotransmitter, Ach, is responsible for the control of iridophores' activity in the squid *Alloteuthis subulata* (Hanlon et al., 1990; Mäthger et al., 2004).

Other Modes of Neural Control of Deposition of Parental Epigenetic Information in Vertebrate Eggs

Neural Regulation by Modifying Maternal Hormone Level

Maternal plasma level of testosterone in the canary (*Serinus canaria*) progressively increases in the yolk of the consecutively laid eggs: the later-laid eggs have higher testosterone concentration in the yolk than first-laid eggs. Higher concentrations of maternal yolk testosterone (also experimental injection of testosterone) not only increase the viability of the offspring but also increase the mass of the hatching muscle (*musculus complexus*), making thus possible an earlier hatching and effectiveness of begging for parentally delivered food (Lipar and Ketterson, 2000).

Increased yolk testosterone in successively laid eggs is an adaptive maternal investment that compensates for the disadvantaged later-hatched nestlings and represents

> *... a mechanism which communicates environmental conditions from the mother to the offspring, and this mechanism serves to optimize reproduction and/or modify offspring traits.*
>
> *Schwabl (1996)*

The amount of maternal testosterone and androstenedione (a precursor of testosterone) also increases with the order of laying in the eggs of American kestrel (*Falco sparverius*) what has been suggested to be a form of "parental favoritism" (Sockman and Schwabl, 2000) because higher concentrations of testosterone in yolk enhance the survival rate and fitness of the offspring (Pilz et al., 2004).

A similar variation in relation to the laying sequence of eggs is also observed for estrogens: the concentration of E2 in the eggs decreases with the laying sequence (Williams et al., 2005).

In many birds, the yolk concentration of testosterone is regulated by the pituitary prolactin (Sockman et al., 2001). In turn, prolactin synthesis is cerebrally stimulated by a hypothalamic-releasing hormone according to the following signal cascade:

Signals generated by the processing of internal and external stimuli in the birds' CNS → prolactin-releasing peptide secreted by neurons of the nucleus of the solitary tract → hypothalamic secretion of prolactin-releasing factor (PRF)/prolactin-releasing inhibiting factor → pituitary prolactin.

In turkeys, prolactin synthesis is also induced by the hypothalamic neurohormone, VIP , which is the avian PRF (Kang et al., 2004).

A threefold increase in the prolactin level occurs in domestic ducks during egg laying, which doubles and then remains steady during the incubation period, but falls dramatically after hatching. The increase in the level of prolactin is neurally determined and is related to the tactile stimulation of the bird by her own eggs as it is proven by the fact that anesthesia of incubation patches sharply decreases the hormone level (Hall and Goldsmith, 1983). In viviparous animals, such as mammals, cerebrally regulated maternal signals (hormones) pass through placenta and essentially influence the course of early development (Das et al., 1994).

In a number of studies, it has been determined that in birds, sex steroid hormones involved in the embryonic development are not produced by the embryo only, but many of them are of maternal origin (Schwabl, 1993).

In response to various stress conditions (hunger, diseases, presence of predators, etc.), the CNS, via the hypothalamus–pituitary–adrenal (HPA) axis induces a rise in the level of corticosterone in blood (corticotropin-releasing hormone → corticotropin → corticosterone). This is the case with the Japanese quail (*Coturnix coturnix japonica*) where the rise in the blood corticosterone level leads to a proportional increase of the hormone content in the egg yolk. This is considered to be an adaptive response for maximizing the reproductive success under the unfavorable conditions, by producing offspring that grow more slowly and show higher activity of the HPA axis (Hayward and Wingfield, 2004).

Thus, by regulating the temporal patterns of concentration of testosterone in the blood, birds are capable of differentially depositing testosterone in eggs of the same clutch according to the order in which eggs are laid. This differential deposition of maternal factors is adaptive and is neurally regulated, via the HPO axis.

Neural Regulation by Differential Uptake from the Blood

Differential uptake of maternal factors from the blood is a function of a specialized selective mechanism, which, as explained in the section of the receptor-mediated endocytosis, is under neural control.

Leghorn chickens (*Gallus gallus domesticus*) determine the amount of testosterone they deposit in eggs according to their social rank. Dominant females have higher maternal capacities (e.g., body mass, egg mass) than subdominant ones. As the reproductive success of males is more variable than that of females, dominant mothers allocate more testosterone to male-producing eggs, while subdominant chickens invest more in female eggs (Muller et al., 2002).

Female zebra finches mated with attractive males deposit maternal hormones, testosterone and 5α-dihydrotestosterone, in increasing amounts in successively laid eggs, leading to the production of offspring of increased survivorship or fitness (Gil et al., 1999).

Females of the Chinese quail *Coturnix chinensis* paired with males that have more sexual ornaments lay larger eggs than those mated with males with fewer sexual ornaments (Uller et al., 2005).

Female canaries (*Serinus canaria*), when exposed to attractive song repertoires, deposit greater amounts of testosterone than when exposed to unattractive repertoires (Gil et al., 2004; Tanvez et al., 2004). In the interpretation of the investigators

> *this implies that song repertoires convey important information about the reproductive value of a given male.*
>
> *Gil et al. (2004)*

There is no other imaginable way for song repertoires of birds to convey to the offspring information of any type, besides via the neural pathways starting in the CNS.

The collared flycatcher, *Ficedula albicollis*, provides more testosterone to eggs when mates with younger males (Michl et al., 2005) and female mallards (*Anas platyrhynchos*) produce larger eggs after copulating with preferred males, and smaller eggs with less-preferred males (Cunningham and Russell, 2000).

These results have been conventionally explained with "good-gene" hypotheses, which, by taking for granted that more attractive males are genetically superior, predicts that their offspring as well would be beneficiary of the good paternal heritage. But this prediction has been rejected by the analysis of the experimental results, which led investigators to the conclusion that their experiments show no effect of the father on the condition of the offspring (Cunningham and Russell, 2000).

Similar phenomena are also observed in fish. Female Banggai cardinal fish produce heavier eggs, and presumably better offspring, when paired with bigger preferred males (Kolm, 2001).

Females of the zebra finch, *Taeniopygia guttata*, invest differentially in the egg mass, proportionally to the attractiveness of the male mate. Moreover, the offspring traits that were measured in experiments varied in accordance with that differential investment:

> *The father's attractiveness, via maternal effects, influences several aspects of offspring development.*
>
> *Gilbert et al. (2006)*

The amount of the maternal testosterone in hatchling tree lizards (*Urosaurus ornatus*) determines the proportion of males that develop as orange- or blue-throated morphs, and "these naturally occurring phenotypes, differ in territoriality, aggressiveness, and body size" (Ketterson and Nolan, 1999).

In the bonnethead shark, *Sphyrna tiburo*, the egg yolk concentrations of estrogen and progesterone are higher, and testosterone is lower, than the level of these hormones in the blood (Manire et al., 2004). Differences between the blood levels and concentration of estrogens in the egg yolk have also been observed in other studies (Williams et al., 2004). For example, it is reported that in the green anole (*Anolis carolinensis*), a lizard with genotypic sex determination, yolk testosterone (T) level is higher in male-producing eggs than in female-producing ones (Lovern and Wade, 2003a,b).

No explanation has been presented on how the mother figures out and accomplishes this biased allocation of the hormones in male- and female-producing eggs. Performing this differential allocation clearly requires epigenetic information, on "What?", "Where?", and "How much?" of sex hormones to provide to each egg, as well as instructions for activating appropriate executing mechanisms. The fact that differential uptake of gonadal hormones in the oocyte in examples presented in this subsection is related to, and triggered by, the visual and acoustic perception and assessment of characters (e.g., size, attractiveness, social rank, and song repertoire) of mates clearly indicates that the epigenetic information for activating respective signal cascades/gene regulatory networks of the differential uptake originates in the CNS.

Regulation Through Sex-Biased Egg Laying

Another way of differential deposition of maternal factors in eggs has been recently observed in finches. While the mother modulates the level of steroids in blood, it simultaneously exhibits a sex-biased egg-laying order, thus leading to their differential sex-specific allocation in eggs.

A very young, 36-year-old population of house finches (*Carpodacus mexicanus*) in Montana, at the northern limit of species' range, exhibits maternally determined sex-specific ovulation sequence: about 90% of the first-laid eggs are females, and about 80% of the second-laid eggs are males. This differential ovulation according to the sex was not observed in house finches of the same species in the southern extreme of the range, in Arizona. Differences between the two northern and southern populations of house finches were observed also in the temporally differential allocation of yolk in Montana but not Arizona (Badyaev et al., 2006). Let us remember that ovulation in vertebrates is under strict neurohormonal control.

Regulation via Intercellular Communication Channels

Maternal cytoplasmic factors with low molecular weight may be transported from follicle cells to the oocyte via gap junctions. Empirical evidence shows that formation of gap junctions, far from a random process, is an orderly information-requiring process of communication between cells.

Gap junctions may be required for mediating hormonal responses. The channel of the gap junctions consists of two connexons formed of six connexin molecules.

Several hormones and growth factors are known to regulate connexin-43 and/or gap junctions: parathyroid hormone in osteoblasts (Schiller et al., 1992, 1997), growth factors, TGF-β (Chiba et al., 1994), FGF5, and FGF9 in cultured astroglial cells (Reuss et al., 2000), and platelet-derived growth factor in mesangial cells (Yao et al., 2000).

Khan-Dawood et al. (1998) have demonstrated the neuroendocrine control of the formation of gap junctions in the case of baboon *corpus luteum*. The two main cell types secreting progesterone in human and baboon *corpora lutea* form several types of junctions, including gap junctions, which allow communication between cells only when connexins are phosphorylated. The pituitary LH and hypothalamic neurohormone GnRH agonists stimulate phosphorylation or dephosphorylation of connexin-43, depending on the time of their administration. A number of endocrine factors, such as the LH and hCG, as well as paracrine/autocrine hormones, such as oxytocin, E2, and progesterone, control the expression and phosphorylation of connexin-43 and, consequently, the formation of gap junctions between the two cell types, LH/hCG responsive cells and LH/hCG nonresponsive cells in the primate *corpus luteum* (Khan-Dawood et al., 1998). The hCG also regulates gap junction channels of Leydig cells in rats by redistributing connexin-43 to the periphery of those cells and stimulates steroidogenesis in them (You et al., 2000).

In view of the cerebral control of the secretion of these hormones via the HPO axis, the evidence on hormonal regulation of the formation of gap junctions implies neural regulation of those channels of intercellular communication through which follicle cells and nurse cells in vertebrates provide the oocyte with maternal cytoplasmic factors to the oocyte.

Role of the Ovarian Innervation in the Ovarian Function

From the view of the role of the CNS in the physiology of gonads, a recent study on the sensory innervation of the OT in the snail *Helix aspersa* (Figure 3.5) is very illuminating. The brain of the snail consists of ~40,000 neurons, but, as already mentioned, the OT branch of the intestinal nerve alone, which innervates the OT, contains 3,025 axon profiles representing almost 8% of the total number of the brain neurons. This by far exceeds that expected from the size of the nerve (Antkowiak and Chase, 2003).

To sense the stretch resulting from the growth of oocytes is a task that would only need a few neurons, rather than thousands of them. There is a clear threshold of 87 mature oocytes accumulated in the gonad, which marks a ~10-fold increase in afferent electrical spikes, and a correspondingly higher efferent spike rate, and the beginning of oviposition. The clear correlation between the afferent activity of the OT innervation and the efferent response from the CNS indicates that this innervation is of the sensory type (stretch sensors for monitoring the oocyte growth), but other evidence suggests that it might also have motor functions (Antkowiak and Chase, 2003) for oviposition in this snail species.

The endocrine function of the ovary has generally been considered to be under control of the CNS via the hypothalamic–pituitary axis, where the secretion of androgens and estrogens is regulated by the pituitary FSH, LH, prolactin, and corticotropin. However, it appears that, important as it is for the reproductive function, the neurohormonal control would be insufficient for the regulation of the complex process of production and deposition of maternal factors in the oocyte. A closer, immediate monitoring of the process of deposition of parental factors seems to be a *sine qua non*. Indeed, in vertebrates, innervation by the SON, ovarian plexus, and vagus nerve form a dense network of nerve fibers coming into close contact with follicle cells of the theca externa and theca interna surrounding the Graaf follicle. It is acknowledged that the ovarian innervation plays important role in the physiology of this organ (Aguado, 2002). Besides, a local network of intrinsic neurons and innervation has been discovered in the ovary (D'Albora et al., 2002).

Viral transneuronal tracing studies have shown that extrinsic ovarian innervation and the network of intrinsic ovarian neurons, besides the spinal cord (e.g., IML cell column, dorsal horn) are connected, via multisynaptic neuronal pathways, with most different regions of the brain (various structures of the brain stem, pons, mesencephalon, telencephalon, diencephalon, especially hypothalamic structures; Gerendai et al., 1998, 2002). These multisynaptic connections are necessary for continually monitoring the physiological status of the ovary, Graaf follicle, and oocyte and probably for the efferent CNS output for controlling and coordinating the activity of various cell types for the development of the oocyte.

Secretion of ovarian hormones is under parallel local control of a second mechanism from the CNS, via vagus and SON. Involved in this mechanism are nervous fibers descending from the hypothalamus to the spinal cord, sympathetic and vagal preganglionic neurons, as well as sympathetic neurons in para- and prevertebral ganglia (De Bortoli et al., 1998), which influence the peripheral neurons of the autonomic system. Electrical stimulation of a number of brain centers (ventromedial hypothalamus and medial basal prechiasmatic area), even in absence of FSH, LH, and ACTH (hypophysectomized and adrenalectomized animals) stimulates significant increase of plasma E2 and progesterone in the contralateral ovarian venous blood. Ovarian denervation blocks estrogen secretion induced by stimulating the medial basal prechiasmatic area. The surprising differences observed in the secretion of ovarian hormones and functions in left and right ovaries are also explained with differences observed in the neurotransmitters released by innervation of the left and right ovaries, which modulate effects of pituitary gonadotropins (Cruz et al., 2006).

Granulosa cells in the ovary (Mayerhofer et al., 2003, 2006) and epithelial oviducal cells in pregnant pigs (Steffl et al., 2006) are endocrine nonneuronal cells secreting the neurotransmitter Ach, but it is again a central neural mechanism that, via the hypothalamic–pituitary axis, with the FSH as proximate inducer, stimulates secretion of this "cyto-transmitter" and ensuing growth of the ovarian follicle (Mayerhofer et al., 2006).

Paternal Cytoplasmic Factors in the Sperm Cell

In the process of transformation of spermatids into spermatozoa, most of the cytoplasm is lost, and this was considered to exclude the possibility that spermatozoa could contribute to the zygote epigenetic information in the form of paternal mRNAs and other factors. Since the mid-1990s, however, evidence has been accumulating on the presence of paternal mRNAs in the relatively small volume of sperm cell cytoplasm (Kramer and Krawetz, 1997). In a recent study, it was shown that the human sperm cell contains more than 2,500 paternal mRNA types, some of which represent important messages for the early development. From that large repertoire of paternal (spermatozoal) mRNAs, 10 were found to lack in the egg cell (Ostermeier et al., 2002).

The sperm cell is in possession of a developed endocytic apparatus for taking up paternal mRNAs, proteins, etc. from its maternal environment (Qiao et al., 2004). Among the mRNAs that spermatozoa contribute to the zygote are c-Myc mRNA (a transcription factor), β-actin mRNA, estrogen receptor mRNA, as well as mRNAs for N-cadherin, integrins, aromatase, nitric oxide synthase, etc. (Miller et al., 2005). Recently, evidence has been presented of "a surprising degree of paternal control" (Wagner et al., 2004) of the early embryonic development.

It is noteworthy that thin peptidergic nerve fibers in the testes of the fish *Oreochromis niloticus* reach steroid-producing Leydig cells and lobules containing cysts of spermatogenic cells (Nakamura and Nagahama, 1995).This fact suggests the existence of a binary neural control (via the hypothalamus–pituitary axis and via the local innervation of testicles) of spermiogenesis in this species.

Neural Control of Gene Imprinting

An imprinted gene is one that is expressed by only one of the two zygotic alleles. This results from the fact that one of the parents provides the zygote with a methylated, untranscribable allele. Depending on the species, the number of imprinted genes is believed to vary from 30 to 200. A typical example of imprinted genes in several animal species, including humans, for instance, is the IGF-2, expressed by the maternal allele only and its receptor, IGF-2r, that is expressed by the paternal allele only.

Methylation of the DNA is function of specific DNA methyltransferases. Methyltransferases are themselves regulated by extracellular signals, glucocorticoid hormones and synthetic glucocorticoids, such as dexamethazone (Biswas et al., 1999). Glucocorticoids now are considered to be indispensable for the expression of methyltransferases (Laborie et al., 2003). In rats, for example, glucocorticoid hormones specifically induce phenylethanolamine *N*-methyltransferase, and this specificity is related to the presence of two glucocorticoid response elements in the promoter region of the gene for the enzyme (Tai et al., 2002). Needless to say, the synthesis of glucocorticoids is the terminal step of a signal cascade originating in the CNS, along the HPA axis.

Another inducer of the gene for methyltransferase is a transcription factor, the early growth response protein-1 (Egr-1; Ebert et al., 1994). In turn, transcription of the Egr-1 is cerebrally regulated by hypothalamic GnRH (Duan et al., 2002) and by glucocorticoids (Ebert et al., 1994). In granulosa cells of the rodent ovary, Egr-1 is induced by the pituitary FSH (Russell et al., 2003). Hypoxic stress also induces synthesis of Egr-1 and the methyltransferase (Wong et al., 1992). Induction of the methyltransferase genes by the neural crest cell factor, AP-2, is also made possible by a glucocorticoid-dependent mechanism (Ebert et al., 1998).

A further confirmation of the crucial role of the CNS in regulation of the methyltransferase synthesis, and consequently of gene imprinting, via the HPA axis, comes from the observation that hypophysectomy causes a drastic loss of methyltransferase activity (Burke et al., 1983; Pegg and Wiest, 1983).

Epigenetic modification of DNA and nuclear histones in general enables "differential access to the underlying genetic information" and provides an alternative inheritance system (Richards, 2006).

Summarizing the above evidence: the synthesis of methyltransferases that are responsible for gene imprinting in vertebrates is under the neural control via the hypothalamic–pituitary–adrenal/gonadal axes.

Despite the impressive progress in understanding the mechanisms of gene imprinting, the key process of how the organism chooses the sites of DNA methylation/demethylation and histone acetylation/deacetylation is not clear. This is an information-requiring process, and the only way of obtaining empirical evidence on the possible source of that information is to examine proximate causes of these changes and trace back the possible causal chain. Preliminary evidence shows that these changes in epigenetic structures (methylated DNA and acetylated/deacetylated histones) of metazoans represent effects of causal chains starting in the nervous system.

Let us illustrate this with only two clear examples of DNA demethylation and histone acetylation. The electrical activity arising in response to stress in hippocampal neurons (or photic stimuli in neurons of the SCN in hypothalamus) induces expression of the stress response gene Gadd45b (growth arrest and DNA damage–inducible protein-45-β), a neural activity-induced early gene, which by demethylating specific promoters in DNA, induces expression of genes for neurogenesis (Ma, 2009; Wu and Sun, 2009). Thus, neural activity, via Gadd45b, causes DNA demethylation in hippocampal and SCN neurons (Ma, 2009). The causal chain from stressful stimuli to DNA demethylation and epigenetic modification of genes looks as follows:

Stressful stimulus → Processing of the stimulus in the stress neurocircuitry → induction of Gadd45b → Gadd45b synthesis → DNA demethylation → Expression of neurogenesis genes.

The pituitary FSH, by binding its membrane receptor in ovarian granulosa cells, activates a signal cascade that leads to phosphorylation and acetylation of histone H3 and thus to transcription of FSH-responsive genes necessary for granulosa cell differentiation (Salvador et al., 2001). The causal chain of events from the CNS to acetylation of H3 and expression of FSH-responsive genes in ovarian granulosa cells is diagrammatically presented in Figure 3.11.

Due to the neural origin of the information for changes in epigenetic structures (methylated DNA and acetylated histones), I have avoided a full description of these changes in this work.

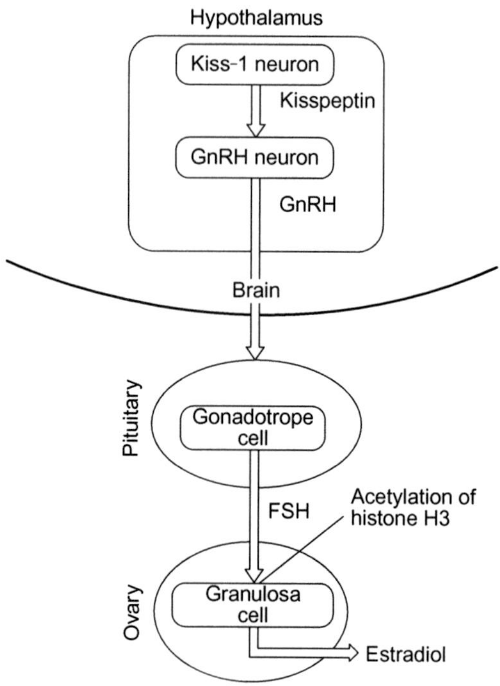

Figure 3.11 Neural control of the signal cascade that leads to acetylation of histone H3 and, consequently, induction of FSH-responsive genes and the synthesis of estradiol by granulosa cells surrounding the oocyte. Note that the epigenetic information necessary for H3 acetylation flows from hypothalamic neurons, via the FSH-secreting pituitary cells, to granulosa cells, where it induces FSH-responsive genes.

References

Adamo, S.A. (1999). Evidence for adaptive changes in egg laying crickets exposed to bacteria and parasites. Anim. Behav. *57*, 117–124.

Adler, B.A. and Crowley, W.R. (1984). Modulation of luteinizing hormone release and catecholamine activity by opiates in the female rat. Neuroendocrinology *38*, 248–253.

Aguado, L.I. (2002). Role of the central and peripheral nervous system in the ovarian function. Microsc. Res. Tech. *59*, 462–473.

Ahmed, A.A., Wahbi, A., Nordlind, K., Kharazmi, A., Sundqvist, K.G. and Mutt, V., et al. (1998). *In vitro Leishmania major* promastigote-induced macrophage migration is modulated by sensory and autonomic neuropeptides. Scand. J. Immunol. *48*, 79–85.

Ahmed, A.A., Wahbi, A.H. and Nordlin, K. (2001). Neuropeptides modulate a murine monocyte/macrophage cell line capacity for phagocytosis and killing of *Leishmania major* parasites. Immunopharmacol. Immunotoxicol. *23*, 397–409. (abstract).

Anderson, G.M., Connors, J.M., Hardy, S.L., Valent, M. and Goodman, R.L. (2001). Oestradiol microimplants in the ventromedial preoptic area inhibit secretion of luteinizing hormone via dopamine neurones in anoestrous ewes. J. Neuroendocrinol. *13*, 1051–1058.

Anderssen, S.S. (1998). *Xenopus* interphase and mitotic microtubule-associated proteins differentially suppress microtubule dynamics *in vitro*. Cell Motil. Cytoskeleton *41*, 202–213.

Anderssen, S.S. and Karsenti, E. (1997). XMAP310: a *Xenopus* rescue factor localized to the mitotic spindle. J. Cell Biol. *139*, 875–983.

Anesetti, G., Lombide, P., D'Albora, H. and Ojeda, S.R. (2001). Intrinsic neurons in the ovary. Cell Tissue Res. *306*, 231–237.

Antkowiak, T. and Chase, R. (2003). Sensory innervation of the ovotestis in the snail *Helix aspersa*. J. Exp. Biol. *206*, 3913–3921.

Badyaev, A.V., Acevedo Seaman, D., Navara, K.J., Hill, G.E. and Mendonça, M.T. (2006). Evolution of sex-biased maternal effects in birds: III. Adjustment of ovulation order can enable sex-specific allocation of hormones, carotenoids, and vitamins. J. Evol. Biol. *19*, 1044–1057.

Ball, G.F. and Balthazart, J. (2004). Hormonal regulation of brain circuits mediating male sexual behavior in birds. Physiol. Behav. *83*, 329–346.

Beggs, K.T., Glendining, K.A., Marechal, N.M., Vergoz, V., Nakamura, I., Slessor, K.N., et al. (2007). Queen pheromone modulates brain dopamine function in worker honey bees. Proc. Natl. Acad. Sci. U.S.A. *104*, 2460–2464.

Biswas, T., Ramana, C.V., Srinivasan, G., Boldogh, I., Hazra, T.K., Chen, Z., et al. (1999). Activation of O^6-methylguanine-DNA methyltransferase gene by glucocorticoid hormone. Oncogene *18*, 525–532.

Bloch, G., Simon, T., Robinson, G.E. and Hefetz, A. (2000). Brain biogenic amines and the reproduction dominance in bumble bees (*Bombus terrestris*). J. Comp. Physiol. *A186*, 261–268.

Bode, C.J., Gupta, M.L., Suprenant, K.A. and Himes, R.H. (2003). The two alpha-tubulin isotypes in budding yeast have opposing effects in microtubule dynamics. EMBO Rep. *4*, 94–99.

Boulay, R., Hooper-Bui, L.M. and Woodring, J. (2001). Oviposition and oogenesis in virgin fire ant females *Solenopsis invicta* are associated with a high level of dopamine in the brain. Physiol. Entomol. *26*, 294–299.

Brendza, R.P., Serbus, L.R., Duffy, L.R. and Saxton, W.M. (2000). A function for kinesin I in the posterior transport of *oskar* mRNA and Staufen protein. Science *289*, 2120–2122.

Brent, C.S. and Traniello, J.F.A. (2001). Influence of sex-specific stimuli on ovarian maturation in primary and secondary reproductives of the dampwood termite *Zootermopsis angusticollis*. Physiol. Entomol. *26*, 239–247.

Brown, V.I. and Greene, M.I. (1991). Molecular and cellular mechanisms of receptor-mediated endocytosis. DNA Cell Biol. *10*, 399–409.

Burke, W.J., Davis, J.W., and Joh, T.H. (1983). The role of compartmentalization of epinephrine in the regulation of phenylethanolamine N-methyltransferase synthesis in rat adrenal medulla. Endocrinology *113*, 1102–1110.

Burz, D.S., Rivera-Pomar, R., Jaeckle, H. and Hanes, S.D. (1998). Cooperative DNA-binding by Bicoid provides a mechanism for threshold-dependent gene activation in the *Drosophila* embryo. EMBO J. *17*, 5998–6009.

Buszczak, M. and Cooley, L. (2000). Eggs to die for: cell death during *Drosophila* oogenesis. Cell Death Differ. *7*, 1071–1074.

Buszczak, M., Freeman, M.R., Carlson, J.R., Bender, M., Cooley, L. and Segraves, W.A. (1999). Ecdysone response genes govern egg chamber development during mid-oogenesis in *Drosophila*. Development *126*, 4581–4589.

Buznikov, G.A. (1993). The control of oocyte maturation in the starfish and amphibians by serotonin and its antagonists. Int. J. Dev. Biol. *37*, 363–364.

Cabej, N. (2001). *Quantic Evolution*. Albanet, Dumont, NJ, pp. 89–96.

Cabej, N.R. (2004). *Neural Control of Development*. Albanet, Dumont, NJ, p. 72.

Cakouros, D., Daish, T.J. and Kumar, S. (2004). Ecdysone receptor directly binds the promoter of the *Drosophila* caspase dronc, regulating its expression. J. Cell Biol. *165*, 631–640.

Campbell, B.K. and Baird, D.T. (2001). Inhibin A is a FSH-responsive marker of granulosa cells differentiation which has both autocrine and paracrine actions in sheep. J. Endocrinol. *169*, 333–345.

Can, A. and Semiz, O. (2000). Diethylstilbestrol (DES)-induced cell cycle delay and meiotic spindle disruption in mouse oocytes during *in-vitro* maturation. Mol. Hum. Reprod. *6*, 154–162.

Caraty, A., Delaleu, B., Chesneau, D. and Fabre-Nys, C. (2002). Sequential role of E2 and GnRH for the expression of estrous behavior in ewes. Endocrinology *143*, 139–145.

Cardinali, M., Gioacchini, G., Candiani, S., Pestarino, M., Yoshizaki, G., Carnevali, O., et al. (2003). Hormonal regulation of *Vasa*-like mRNA expression in the ovary of the marine teleost, *Sparus aurata*. Biol. Reprod. *70*, 737–743.

Carney, G.E. and Bender, M. (2000). The *Drosophila* ecdysone receptor (AcR) is required maternally for normal oogenesis. Genetics *154*, 1203–1211.

Cavaliere, V., Taddei, C. and Gargiulo, G. (1998). Apoptosis of nurse cells at the late stage of oogenesis of *Drosophila melanogaster*. Dev. Genes Evol. *208*, 106–112.

Cerda, J.P., Petrino, T.R., Greenberg, M.J. and Wallace, R.A. (1997). Pharmacology of the serotonergic inhibition of steroid-induced reinitiation of oocyte meiosis in the teleost *Fundulus heteroclitus*. Mol. Reprod. Dev. *48*, 282–291.

Cha, B.J., Kopetsch, B.S. and Theurkauf, W.E. (2001). *In vivo* analysis of *Drosophila bicoid* mRNA localization reveals a novel microtubule-dependent axis specification pathway. Cell *106*, 35–46.

Chapman, R.F. (1998). *The Insects—Structure and Function*, fourth edition. Cambridge University Press, Cambridge, p. 306.

Charrasse, S., Lorca, T., Doree, M. and Larroque, C. (2000). The *Xenopus* XMAP215 and its human TOG proteins interact with cyclin B1 to target p34cdc2 to microtubules during mitosis. Exp. Cell Res. *254*, 249–256.

Chase, R., Antkowiak, T., Geoffroy, E. and Weatherill, D. (2004). Why the ovotestis of *Helix aspersa* is innervated. Acta Biol. Hung. *55*, 239–249.

Chiba, H., Sawada, N., Oyamada, M., Kojima, T., Iba, K., Ishii, S., et al. (1994). Hormonal regulation of connexin43 expression and gap junctional communication in human osteoblastic cells. Cell Struct. Funct. *19*, 173–177.

Chneiweiss, H., Cordier, J. and Sobel, A. (1992). Stathmin phosphorylation is regulated in striatal neurons by vasoactive intestinal peptide and monoamines via multiple intracellular pathways. J. Neurochem. *58*, 282–289.

Christensen, T.A., Lehman, H.K., Teal, P.E.A., Itagaki, H., Tumlinson, J.H. and Hildebrand, J.G. (1992). Diel changes in the presence and physiological actions of octopamine in the female sex-pheromone glands of heliothine moths. Insect Biochem. Mol. Physiol. *22*, 841–849.

Cruz, M.E., Flores, A., Palafox, M.T., Meléndez, G., Rodríguez, J.O., Chavira, R., et al. (2006). The role of the muscarinic system in regulating estradiol secretion varies during the estrous cycle: the hemiovariectomized rat model. Reprod. Biol. Endocrin. *4*, p. 43. http://www.rbej.com/content/4/1/43.

Cunningham, E.J. and Russell, A.F. (2000). Egg investment is influenced by male attractiveness in the mallard. Nature *404*, 74–77.

D'Albora, H., Lombide, P. and Ojeda, S.R. (2000). Intrinsic neurons in the rat ovary: an immunohistochemical study. Cell Tissue Res. *300*, 47–56.

D'Albora, H., Anesetti, G., Lombide, P., Dees, W.L. and Ojeda, S.R. (2002). Intrinsic neurons in the mammalian ovary. Microsc. Res. Tech. *59*, 484–489.

Das, S.K., Tsukamura, H., Paria, B.C., Andrews, G.K. and Dey, S.K. (1994). Differential expression of epidermal growth factor receptor (EGF-R) gene and regulation of EGF-R bioactivity by progesterone and estrogen in the adult mouse uterus. Endocrinology *134*, 971–981.

Dascal, N., Yekuel, R. and Oron, Y. (1984). Acetylcholine promotes progesterone-induced maturation of *Xenopus* oocytes. J. Exp. Zool. *230*, 131–135.

De Bortoli, M.A., Garraza, M.H. and Aguado, L.I. (1998). Adrenergic intracerebroventricular stimulation affects progesterone concentration in the ovarian vein of the rat: participation of the superior ovarian nerve. J. Endocrinol. *159*, 61–68.

Dees, W.L., Hiney, J.K., Schultea, T.D., Mayerhofer, A., Danilchik, M., Dissen, G.A., et al. (1995). The primate ovary contains a population of catecholaminergic neuron-like cells expressing nerve growth factor receptors. Endocrinology *136*, 5760–5768.

Delisle, J., Picimbon, J.-F. and Simard, J. (2000). Regulation of pheromone inhibition in mated females of *Christoneura fumiferana* and *C. rosaceana*. J. Insect Physiol. *46*, 913–921.

De Loof, A., Baggerman, G., Breuer, M., Claeys, I., Cerstiaens, A., Clynen, E., et al. (2001). Gonadotropins in insects: an overview. Arch. Insect Biochem. *47*, 129–138.

Dixit, V.D. and Parvizi, N. (2001). Nitric oxide and the control of reproduction. Anim. Reprod. Sci. *65*, 1–16.

D'Occhio, M.J., Aspden, W.J. and Trigg, T.E. (2001). Sustained testicular atrophy in bulls actively immunized against GnRH: potential to control carcase characteristics. Anim. Reprod. Sci. *66*, 47–58.

Dorstyn, L., Colussi, P.A., Quinn, L.M., Richardson, H. and Kumar, S. (1999). DRONC, an ecdysone-inducible *Drosophila* caspase. Proc. Natl. Acad. Sci. U.S.A. *96*, 4307–4312.

Drouin, J., Lanctôt, C. and Tremblay, J.J. (1998). La famille Ptx des facteurs de transcription à homéodomaine. Médecine/Sciences *14*, 335–339.

Drouva, S.V., Poulin, B., Manceau, V. and Sobel, A. (1998). Luteinizing hormone-releasing hormone-signal transduction and stathmin phosphorylation in the gonadotrope αT3-1 cell line. Endocrinology *139*, 2235–2239.

Drummond, A.E., Dyson, M., Thean, E., Groome, N.P., Robertson, D.M. and Findlay, J.K. (2000). Temporal and hormonal regulation of inhibin protein and subunit mRNA expression by post-natal and immature rat ovaries. J. Endocrinol. *166*, 339–354.

Duan, W.R., Ito, M., Park, Y., Maizels, E.T., Hunzicker-Dunn, M. and Jameson, J.L. (2002). GnRH regulates early growth response protein 1 transcription through multiple promoter elements. Mol. Endocrinol. *16*, 221–233.

Ebert, S.N., Balt, S.L., Hunter, J.P., Gashler, A., Sukhatme, V. and Wong, D.L. (1994). Egr-1 activation of rat adrenal phenylethanolamine *N*-methyltransferase gene. J. Biol. Chem. *269*, 20885–20898.

Ebert, S.N., Ficklin, M.B., Her, S., Siddall, B.J., Bell, R.A., Ganguly, K., et al. (1998). Glucocorticoid-dependent action of neural crest factor AP-2: stimulation of phenylethanolamine *N*-methyltransferase gene expression. J. Neurochem. *70*, 2286–2295.

Edwards, D.P. (2005). Regulation of signal transduction pathways by estrogen and progesterone. Ann. Rev. Physiol. *67*, 335–376.

Fletcher, D. and Blum, M. (1981). Pheromonal control of dealation and oogenesis in virgin queen ants. Science *212*, 73–75.

Fontaine-Lenoir, V., Chambraud, B., Fellous, A., David, S., Duchossoy, Y., Baulieu, E.E., et al. (2006). Microtubule-associated protein 2 (MAP2) is a neurosteroid receptor. Proc. Natl. Acad. Sci. U.S.A. *103*, 4711–4716.

Ford, H. and Ebling, F.J.P. (2000). Glutamatergic regulation of gonadotropin-releasing hormone mRNA levels during development in the mouse. J. Neuroendocrinol. *12*, 1027–1033.

Friesen, K.J. and Kaufman, W.R. (2002). Quantification of vitellogenesis and its control by 20-hydroxyecdysone in the ixodod tick, *Amblyomma hebraeum*. J. Insect Physiol. *48*, 773–782.

Fritz, S., Grunert, R., Stocco, D.M., Hales, D.B. and Mayerhofer, A. (2001). StAR protein is increased by muscarinic receptor activation in human luteinized granulosa cells. Mol. Cell. Endocrinol. *171*, 49–51.

Fuentes, V.O., Sanchez, V., Rosiles, R. and Fuentes, P.I. (2001). The effect of low doses of naloxone on the preovulatory surge of LH and on the onset and duration of oestrus in ewe. Anim. Reprod. Sci. *65*, 225–230.

Fujii, R. (2000). The regulation of motile activity in fish chromatophores. Pigment Cell Res. *13*, 300–319.

Fukada, S., Sakai, N., Adachi, S. and Nagahama, Y. (1994). Steroidogenesis in the ovarian follicle of medaka (*Oryzias latipes*, a daily spawner) during oocyte maturation. Dev. Growth Differ. *36*, 81–88.

Gäde, G. and Hoffmann, K.-H. (2005). Neuropeptides regulating development and reproduction in insects. Physiol. Entomol. *30*, 103–121.

Ganea, D and Delgado, M. (2003). The neuropeptides VIP/PACAP and T cells: inhibitors or activators?. Curr. Pharm. Des. *9*, 997–1004.

Gearing, M. and Terasawa, E. (1991). The alpha-1-adrenergic neuronal system is involved in the pulsatile release of luteinizing hormone-releasing hormone in the ovariectomized female *Rhesus* monkey. Neuroendocrinology *53*, 373–381.

Gelez, H. and Fabre-Nys, C. (2006). Neural pathways involved in the endocrine response of anestrous ewes to the male or its odor. Neuroscience *140*, 791–800.

Geoffroy, E., Hutcheson, R. and Chase, R. (2005). Nervous control of ovulation and ejaculation in *Helix aspersa*. J. Molluscan Stud. *71*, 393–399.

Gerendai, I., Toth, I.E., Boldogkoi, Z., Medveczky, I. and Halasz, B. (1998). Neuronal labeling in the rat brain and spinal cord from the ovary using viral transneuronal tracing technique. Neuroendocrinology *68*, 244–256.

Gerendai, I., Kocsis, K. and Halasz, B. (2002). Supraspinal connections of the ovary: structural and functional aspects. Microsc. Res. Tech. *59*, 474–483.

Gil, D., Graves, J., Hazon, N. and Wells, A. (1999). Male attractiveness and differential testosterone investment in zebra finch eggs. Science *286*, 126–128.

Gil, D., Leboucher, G., Lacroix, A., Cue, R. and Kreutzer, M. (2004). Female canaries produce eggs with greater amounts of testosterone when exposed to preferred male song. Horm. Behav. *45*, 64–70.

Gilbert, L., Williamson, K.A., Hazon, N. and Graves, J.A. (2006). Maternal effects due to male attractiveness affect offspring development in the zebra finch. Proc. Biol. Sci. *273*, 1765–1771.

Goda, M. and Fujii, R. (1998). The blue coloration of the common surgeonfish, *Paracanthurus hepatus*—II. Color revelation and color changes. Zool. Sci. *15*, 323–333.

Goh, E.L.K., Pircher, T.J. and Lobie, P.E. (1998). Growth hormone promotion of tubulin polymerization stabilizes the microtubule network and protects against colchicine-induced apoptosis. Endocrinology *139*, 4364–4372.

Gore, A.C. and Terasawa, E. (2001). Neural circuits regulating pulsatile luteinizing hormone release in the female guinea-pig: opioid, adrenergic and serotonergic interactions. J. Neuroendocrinol. *13*, 239–248.

Goubillon, M.-L., Caraty, A. and Herbison, A.E. (2002). Evidence in favour of direct input from the ventromedial nucleus to gonadotropin-releasing hormone neurones in the ewe: an anterograde tracing study. J. Neuroendocrinol. *14*, 95–100.

Gradin, H.M., Larsson, N., Marklund, U. and Gullberg, M. (1998). Regulation of microtubule dynamics by extracellular signals: cAMP-dependent protein kinase switches off the activity of oncoprotein 18 in intact cells. J. Cell Biol. *140*, 131–141.

Granger, N.A., Sturgis, S.L., Ebersohl, R., Geng, C. and Sparks, T.C. (1996). Dopaminergic control of corpora allata activity in the larval tobacco hornworm, *Manduca sexta*. Arch. Insect Biochem. *32*, 449–466.

Hall, M.R. and Goldsmith, A.R. (1983). Factors affecting prolactin secretion during breeding and incubation in the domestic duck (*Anas platyrhynchos*). Gen. Comp. Endocrinol. *49*, 270–276.

Handler, A.M. and Postlethwait, J.H. (1977). Endocrine control of vitellogenesis in *Drosophila melanogaster*: effects of the brain and corpus allatum. J. Exp. Zool. *202*, 389–402.

Harano, K-I., Sasaki, M., Nagao, T. and Sasaki, K. (2008). Dopamine influences locomotor activity in honeybee queens: implications for a behavioural change after mating. Physiological Entomology *33*, 395–399.

Harder, J.D. and Jackson, L.M. (2003). Male pheromone stimulates ovarian follicular development and body growth in juvenile female opossums. Reprod. Biol. Endocrinol. *1*, p. 21.

Hayward, L.S. and Wingfield, J.C. (2004). Maternal corticosterone is transferred to avian yolk and may alter growth and adult phenotype. Gen. Comp. Endocrinol. *135*, 365–371.

Heming, B.S. (2003). *Insect Development and Evolution*. Comstock Publishing Association. A division of Cornell University Press, Ithaca, NY, p. 58.

Herz, J. and Bock, H.H. (2002). Lipoprotein receptors in the nervous system. Annu. Rev. Biochem. *71*, 405–434.

Hofman, M.A. and Swaab, D.F. (1992). The human hypothalamus: comparative morphometry and photoperiodic influences. Prog. Brain Res. *93*, 133–147.

Hofmann, H.A. (2006). Gonadotropin-releasing hormone signaling in behavioral plasticity. Curr. Opin. Neurobiol. *16*, 343–350.

Hoogland, P.V. and Martinez-Garcia, F. (1999). Species-specific differences in the cortico-hypothalamic connections of lizards. Eur. J. Morphol. *37*, 85–88.

Ilmensee, K. and Mahowald, A.P. (1974). Transplantation of posterior polar plasm in *Drosophila*. Induction of germ cells at the anterior pole of the egg. Proc. Natl. Acad. Sci. U.S.A. *71*, 1016–1020.

Johnson, M.D. and Crowley, W.R. (1986). Role of central serotonin systems in the stimulatory effects of ovarian hormones and naloxone on luteinizing hormone release in female rats. Endocrinology *118*, 1180–1186.

Kalra, P.S., Crowley, W.R. and Kalra, S.P. (1987). Differential *in vitro* stimulation by naloxone and K^+ of luteinizing hormone and catecholamine release from the hypothalami of intact and castrated rats. Endocrinology *120*, 178–185.

Kalthoff, K., Sander, K. (1968). Der Entwicklungsgang der Missbildung "Doppelabdomen" im partiell UV-bestrahlten Ei von *Smittia parthenogenetica* (Diptera, Chironomidae). Wilhelm Roux' Archiv *161*, 129–146.

Kang, S.W., Gazzillo, L.C., You, S., Wong, E.A. and El Halawani, M.E. (2004). Turkey prolactin gene regulation by VIP through 35-bp *cis*-acting element in the proximal promoter. Gen. Comp. Endocrinol. *138*, 157–165.

Kasukawa, H., Oshima, N. and Fujii, R. (1987). Mechanism of light-reflection in blue damselfish motile iridophore. Zool. Sci. *4*, 243–257.

Ketterson, E. and Nolan, V., Jr. (1999). Adaptation, exaptation and constraint: a hormonal perspective. Am. Nat. *154*(Suppl.), 4–25.

Khamsi, F. and Roberge, S. (2001). Granulosa cells of *cumulus oophorus* are different from mural granulosa cells in their response to gonadotrophins and IGF-1. J. Endocrinol. *170*, 565–573.

Khan-Dawood, F.S., Yang, J. and Dawood, M.Y. (1998). Hormonal regulation of connexin-43 in baboon corpora lutea. J. Endocrinol. *157*, 405–414.

Kipp, J.L. and Ramirez, V.D. (2003). Estradiol and testosterone have opposite effects on microtubule polymerization. Neuroendocrinology *77*, 258–272.

Kirby, J.D. and Froman, D.P. (2000). Reproduction in male birds. In: *Sturkie's Avian Physiology*, G. Causey Whittow (ed.). fifth edition. Academic Press, p. 608.

Koff, W.C. and Dunegan, M.A. (1985). Modulation of macrophage-mediated tumoricidal activity by neuropeptides and neurohormones. J. Immunol. *135*, 350–354.

Kolm, N. (2001). Females produce larger eggs for large males in a paternal mouthbrooding fish. Proc. Biol. Sci. *268*, 2229–2234.

Kou, R. and Chen, S.J. (2000). Allatotropic and nervous control of corpora allata in the adult male loreyi leafworm, *Mythimna loreyi* (Lepidoptera: Noctuidae). Physiol. Entomol *25*, 273–278.

Kramer, J.A. and Krawetz, S.A. (1997). RNA in spermatozoa: implications for the alternative haploid genome. Mol. Hum. Reprod. *3*, 473–478.

Laborie, C., Van Camp, G., Bemet, F., Montel, V. and Dupouy, J.P. (2003). Metyrapone-induced glucocorticoid depletion modulates tyrosine hydroxylase and phenylethanolamine *N*-methyltransferase gene expression in the rat adrenal gland by a noncholinergic transsynaptic activation. J. Neuroendocrinol. *15*, 15–23.

Lamonerie T., Tremblay, J.J., Lanctôt, C., Therrien, M., Gauthier, Y. and Drouin, J., (1996). *Ptx1*, a bicoid-related homeobox transcription factor involved in transcription of the pro-opiomelanocortin gene. Gene. Dev. *10*, 1284–1295.

Larsen, P.J., Enquist, L.W. and Card, J.P. (1998). Characterization of the multisynaptic neuronal control of the rat pineal gland using viral transneuronal tracing. Eur. J. Neurosci. *10*, 128–145.

Lathe, R. (2001). Hormones and the hippocampus (Review). J. Endocrinol. *169*, 205–231.

Laurine, E., Lafitte, D., Grégoire, C., Sérée, E., Loret, E., Douillard, S., et al. (2003). Specific binding of dehydroepiandrosterone to the N terminus of the microtubule-associated protein MAP2. J. Biol. Chem. *278*, 29979–29986.

Lee, C.J., Do, B.R., Lee, Y.H., Park, J.H., Kim, S.J., Kim, J.K., et al. (2001). Ovarian expression of melatonin (1a) receptor mRNA during mouse development. Mol. Reprod. Dev. *59*, 126–132.

Lee, S., Miselis, R. and Rivier, C. (2002). Anatomical and functional evidence for a neural hypothalamic-testicular pathway that is independent of the pituitary. Endocrinology *143*, 4447–4454.

Lincoln, G.A. and Clarke, I.J. (2002). Noradrenaline and dopamine regulation of prolactin secretion in sheep: role in prolactin homeostasis but not photoperiodism. J. Neuroendocrinol. *14*, 36–44.

Lipar, J.L. and Ketterson, E.D. (2000). Maternally derived yolk testosterone enhances the development of the hatching muscle in the red-winged blackbird *Agelaius phoeniceus*. Proc. Biol. Sci. *267*, 2005–2010.

Loeb, M.J., De Loof, A., Gelman, D.B., Hakim, R.S., Jaffe, H., Kochansky, J.P., et al. (2001). Testis ecdysiotropin, and insect gonadotropin that induces synthesis of ecdysteroid. Arch. Insect Biochem. *47*, 181–188.

Loizzi, R.F. (1985). Progesterone withdrawal stimulates mammary gland tubulin polymerization in pregnant rats. Endocrinology *116*, 2543–2547.

Lovern, M.B. and Wade, J. (2003). Yolk testostertone varies with sex in eggs of the lizard, *Anolis carolinensis*. J. Exp. Zool. *295*, 206–210.

Lovern, M.B. and Wade, J. (2003). Maternal plasma and egg yolk testosterone concentrations during embryonic development in green anoles (*Anolis carolinensis*). Gen. Comp. Endocrinol. *124*, 226–235.

Lu, H., Ozawa, H., Nishi, M., Ito, T. and Kawata, M. (2001). Serotonergic neurones in the dorsal raphe nucleus that project into the medial preoptic area contain oestrogen receptor beta. J. Neuroendocrinol. *13*, 839–845.

Ma, D.K. (2009). Neuronal activity–induced Gadd45b promotes epigenetic DNA demethylation and adult neurogenesis. Science *323*, 1074–1077.

Malamed, S., Gibney, J.A. and Ojeda, S.R. (1992). Ovarian innervation develops before initiation of folliculogenesis in the rat. Cell Tissue Res. *270*, 87–93.

Manire, C.A., Rasmussen, L.E., Gelsleichter, J. and Hess, D.L. (2004). Maternal serum and yolk hormone concentrations in the placental viviparous bonnethead shark, *Sphyrna tiburo*. Gen. Comp. Endocrinol. *136*, 241–247.

Martin, D., Wang, S.F. and Raikhel, A.S. (2001). The vitellogenin gene of the mosquito *Aedes aegypti* is a direct target of ecdysteroid receptor. Mol. Cell. Endocrinol. *173*, 75–86.

Mäthger L.M., Collins T.F.T. and Lima, P.A. (2004). The role of muscarinic receptors and intracellular Ca^{2+} in the spectral reflectivity changes of squid iridophores. J. Exp. Biol. *207*, 1759–1769.

Mäthger, L.M., Land, M.F., Siebeck, U.E. and Marshall, N.J. (2003). Rapid colour changes in multilayer reflecting stripes in the paradise whiptail, *Pentapodus paradiseus*. J. Exp. Biol. *206*, 3607–3613.

Mayerhofer, A., Dissen, G.A., Costa, M.F. and Ojeda, S.R. (1997). A role for neurotransmitters in early follicular development: induction of functional follicle-stimulating hormone receptors in newly formed follicles of the rat ovary. Endocrinology *138*, 3320–3329.

Mayerhofer, A., Dimitrijevic, N. and Kunz, L. (2003). The expression and biological role of the non-neuronal cholinergic system in the ovary. Life Sci. *72*, 2039–2045.

Mayerhofer, A., Kunz, L., Krieger, A., Proskocil, B., Spindel, E., Amsterdam, A., et al. (2006). FSH regulates acetylcholine production by ovarian granulosa cells. Reprod. Biol. Endocrinol. *4*, p. 37. http://www.rbej.com/content/4/1/37.

McCrone, E.J. and Sokolove, P.G. (1986). Photoperiodic activation of brains in castrates and the role of the gonad in reproductive maturation of *Limax maximus*. J. Comp. Physiol. *A158*, 151–158.

McKenzie, E.A., Cridland, N.A. and Knowland, J. (1990). Activation of chromosomal vitellogenin genes in *Xenopus* oocytes by pure estrogen receptor and independent activation of albumin genes. Mol. Cell. Biol. *10*, 6674–6682.

Michl, G., Torok, J., Peczely, P., Garamszegi, L.Z. and Schwabl, H. (2005). Female collared flycatchers adjust yolk testosterone to male age, not to attractiveness. Behav. Ecol. *16*, 383–388.

Micklem, D.R., Adams, J., Gruenert, S. and Johnston, D.S. (2000). Distinct roles of two conserved Staufen domains in *oskar* mRNA localization and translation. EMBO J. *19*, 1366–1377.

Miller, D., Ostermeier, G.C. and Krawetz, S.A. (2005). The controversy, potential and roles of spermatozoal RNA. Trends Mol. Med. *11*, 156–163.

Miller, M.A., Technau, U., Smith, K.M. and Steele, R.E. (2000). Oocyte development in hydra involves selection from competent precursor cells. Dev. Biol. *224*, 326–338.

Mochizuki, N., Yamashita, S., Kurokawa, K., Ohba, Y., Nagai, T. and Miyawaki, A. (2001). Spatio-temporal images of growth factor-induced activation of Ras and Rap1. Nature *411*, 1065–1068.

Morales, L., Chavez, R., Ayala, M.E. and Dominguez, R. (1998). Effects of unilateral or bilateral superior avian nerve section in peripubertal rats on the ovulatory response of gonadotrophin administration. J. Endocrinol. *158*, 213–219.

Morales-Ledesma, L., Betanzos-Garcia, R. and Dominguez-Casala, R. (2004). Unilateral vagotomy performed on prepubertal rats at puberty onset of female rat deregulates ovarian function. Arch. Med. Res. *35*, 279–283.

Moshitzky, P., Fleischmann, I., Chaimov, N., Saudan, P., Klauser, S., Kubli, E., et al. (1996). Sex peptide activates juvenile hormone biosynthesis in the *Drosophila melanogaster* corpus allatum. Arch. Insect Biochem. *32*, 363–374.

Muller, W., Eising, C.M., Dijkstra, C. and Groothuis, T.G. (2002). Sex differences in yolk hormones depend on maternal social status in Leghorn chickens (*Gallus gallus domesticus*).. Proc. Biol. Sci. *269*, 2249–2255.

Murakami, K., Fellous, A., Baulieu, E.-E. and Robel, P. (2000). Pregnenolone, a neurosteroid binds to microtubule-associated protein 2 and stimulates microtubule assembly. Proc. Natl. Acad. Sci. U.S.A. *97*, 3579–3584.

Nagahama, Y., Yokishuni, M., Yamashita, M., Tokumoto, T. and Katsu, Y. (1995). Regulation of oocyte growth and maturation in fish. Curr. Top. Dev. Biol. *30*, 103–145.

Nakamura , M. and Nagahama, Y. (1995). Innervation of testes in the tilapia. Aquaculture *135*, 41–49.

Nikitina, L.A. and Buznikov, G.A. (1996). Serotonin inhibits phorbol ester-induced oocyte maturation in the green toad (*Bufo viridis*). Ontogenez. *27*, 122–125.

Nikitina, L.A., Trubnikova, O.B. and Buznikov, G.A. (1993). The action of neurotransmitters and their antagonists on oocyte maturation. Ontogenez *24*, 29–38.

Ookata, K., Hisanaga, S., Okumura, E. and Kishimoto, T. (1993). Association of p34cdc2/ cyclin B complex with microtubules in starfish oocytes. J. Cell Sci. *105*, 873–881.

Ostermeier, G.C., Dix, D.J., Miller, D., Khatri, P. and Krawetz, S.A. (2002). Spermatozoal RNA profiles of normal fertile men. Lancet *360*, 772–777.

Panda, D., Miller, H.P., Banerjee, A., Luduena, R.F. and Wilson, L. (1994). Microtubule dynamics *in vitro* are regulated by the tubulin isotype composition. Proc. Natl. Acad. Sci. U.S.A. *91*, 11358–11362.

Parker, I., Sumikawa, K. and Miledi, R. (1987). Activation of a common effector system by different brain neurotransmitter receptors in *Xenopus* oocytes. Proc. Biol. Sci. *231*, 37–45.

Peck, R. (1987). Neuropeptides modulating macrophage function. Ann. N.Y. Acad. Sci. *496*, 264–270.

Pegg, A.E. and Wiest, L. (1983). Regulation of O6-methylguanine-DNA methyltransferase levels in rat liver and kidney. Cancer Res. *43*, 972–975.

Pilz, K.M., Quiroga, M., Schwabl, H. and Adkins-Regan, E. (2004). European starling chicks benefit from higher yolk testosterone levels during a drought year. Horm. Behav. *46*, 17–192.

Pokrywka, N.J. and Stephenson, E.C. (1991). Microtubules mediate the localization of *bicoid* RNA during *Drosophila* oogenesis. Development *113*, 55–66.

Qiao, Y., Yang J-X., Zhang, X-G., Liu, Y., Zhang, J-C., Zong, S-D., et al. (2004). Characterization of rtSH3p13 gene encoding a development protein involved in vesicular traffic in spermiogenesis. Cell Res. *14*, 197–207.

Rahman, M.S., Takemura, A. and Takano, K. (2000a). Correlation between plasma steroid hormones and vitellogenin profiles and lunar periodicity in the female golden rabbitfish, *Siganus guttatus* (Bloch). Comp. Biochem. Physiol. *127B*, 113–122.

Rahman, M.S., Takemura, A. and Takano, K. (2000b). Lunar synchronization of testicular development and plasma steroid hormone profiles in the golden rabbitfish. J. Fish Biol. *57*, 1065–1074.

Rahman, M.S., Morita, M., Takemura, A. and Yakano, K. (2003). Hormonal changes in relation to lunar periodicity in the testis of the forktail rabbitfish, *Siganus argenteus*. Gen. Comp. Endocrinol. *131*, 302–309.

Raikhel, A.S. and Lea, A.O. (1985). Hormone-mediated formation of the endocytic complex in mosquito oocytes. Gen. Comp. Endocrinol. *57*, 422–33.

Ravindra, R. and Grosvenor, C.E. (1986). Microtubules are mobilized in the lactating rat anterior pituitary gland during suckling. Endocrinology *118*, 1194–1199.

Ravindra, R. and Grosvenor, C.E. (1992). Influence of suckling on tubulin-dependent GTPase activity in the anterior pituitary lobe of the lactating rat. Endocrinology *130*, 263–268.

Reuss, B., Hertel, M., Werner, S. and Unsicker, K. (2000). Fibroblast growth factors-5 and -9 distinctly regulate expression and function of the gap junction protein connexin43 in cultured astroglial cells from different brain regions. Glia *30*, 231–241.

Richard, D.S. Jones, J.M., Gilbert, M., Barbarito, M., Brannigan, D. and Fischer, S., (2000). Hormonal regulation of yolk protein uptake by vitellogenic oocytes in diapausing and mutant female *Drosophila melanogaster. International Conference in Honour of Professor David Saunders*, Edinburgh, March 20–24, 2000 (abstract).

Richards, E.J. (2006). Inherited epigenetic variation—revisiting soft inheritance. Nat. Rev. Genet. *7*, 395–401.

Robinson, G.E. and Vargo, E.L. (1997). Juvenile hormone in adult eusocial hymenoptera: gonadotropin and behavioral pacemaker. Arch. Insect Biochem. *35*, 559–583.

Rubin, C.I. and Atweh, G.F. (2004). The role of stathmin in the regulation of the cell cycle. J. Cell. Biochem. *93*, 242–250.

Russell, D.L., Doyle, K.M.H., Gonzales-Robayna, I., Pipaon, C. and Richards, J.S. (2003). Egr-1 induction in rat granulosa cells by follicle-stimulating hormone and luteinizing hormone: combinatorial regulation by transcription factors cyclic adenosine 3′,5′-monophosphate regulatory element binding protein, serum response factor, sp1, and early growth response factor-1. Mol. Endocrinol. *17*, 520–533.

Salmassi, A., Lü, S., Hedderich, J., Oettinghaus, C., Jonat, W. and Mettler, L. (2000). Interaction of interleukin-6 on human granulosa cell steroid secretion. J. Endocrinol. *170*, 471–478.

Salvador, L.M., Park, Y., Cottom, J., Maizels, E.T., Jones, J.C. and Schillace, R.V., et al. (2001). Follicle-stimulating hormone stimulates protein kinase A-mediated histone H3 phosphorylation and acetylation leading to select gene activation in ovarian granulosa cells. J. Biol. Chem. *276*, 40146–40155.

Sasaki, K., Yamasaki, K. and Nagao, T. (2007). Neuro-endocrine correlates of ovarian development and egg-laying behaviors in the primitively eusocial wasp (*Polistes chinensis*). J. Insect Physiol. *53*, 940–949.

Schiller, P.C., Mehta, P.P., Roos, B.A. and Howard, G.A. (1992). Hormonal regulation of intercellular communication. Mol. Endocrinol. *6*, 1433–1440.

Schiller, P.C., Roos, B.A. and Howard, G.A. (1997). Parathyroid hormone up-regulation of connexin 43 gene expression in osteoblasts depends on cell phenotype. J. Bone Miner. Res. *12*, 2005–2013. (abstract).

Schnorrer, F., Bohmann, K. and Nusslein-Wohlhard, C. (2000). The molecular motor dynein is involved in targeting *swallow* and *bicoid* mRNA to the anterior pole of *Drosophila* oocytes. Nat. Cell Biol. 2, 185–190.

Schonbaum, C.P., Perrino, J.J. and Mahowald, A.P. (2000). Regulation of the vitellogenin receptor during *Drosophila melanogaster* oogenesis. Mol. Biol. Cell. *11*, 511–521.

Schwabl, H. (1993). Yolk is a source of maternal testosterone for developing birds. Proc. Natl. Acad. Sci. U.S.A. *90*, 11446–11450.

Schwabl, H. (1996). Environment modifies the testosterone levels of a female bird and its eggs. J. Exp. Zool. *276*, 157–163.

Selvage, D.J., Lee, S., Parsons, L.H., Seo, D.O. and Rivier, C.L. (2004). A hypothalamic-testicular neural pathway is influenced by brain catecholamines, but not testicular blood flow. Endocrinology *145*, 1750–1759.

Selvaraj, N., Dantes, A. and Amsterdam, A. (2000). Establishment and characterization of steroidogenic granulosa cells expressing beta 2-adrenergic receptor: regulation of adrenoxin and steroidogenic acute regulatory protein by adrenergic agents. Mol. Cell. Endocrinol. *168*, 53–63.

Sharp, P.J. and Gow, C.B. (1983). Neuroendocrine control of reproduction in the cockerel. Poult. Sci. *62*, 1671–1677.

Sirotkin, A.V. and Schaeffer, H.-J. (1997). Direct regulation of mammalian reproductive organs by serotonin and melatonin. J. Endocrinol. *154*, 1–5.

Smith, P.M., Cowan, A. and White, B.A. (2004). The low-density lipoprotein receptor is regulated by estrogen and forms a functional complex with the estrogen-regulated protein ezrin in pituitary GH3 somatolactotropes. Endocrinology *145*, 3075–3083.

Sockman, K.W. and Schwabl, H. (2000). Yolk androgens reduce offspring survival. Proc. Biol. Sci. *267*, 1451–1456.

Sockman, K.W., Schwabl, H. and Sharp, P.J. (2001). Regulation of yolk-androgen concentrations by plasma prolactin in the American kestrel. Horm. Behav. *40*, 462–471.

Soller, M., Bownes, M. and Kubli, E. (1999). Control of oocyte maturation in sexually mature *Drosophila* females. Dev. Biol. *208*, 337–351.

Sorge, D., Nauen, R., Range, S. and Hoffmann, K.H. (2000). Regulation of vitellogenesis in the fall armyworm, *Spodoptera frugiperda* (Lepidoptera: Noctuidae). J. Insect Physiol. *46*, 969–976.

Sporrong, B., Kannisto, P., Owman, C., Sjoberg, N.O. and Walles, B. (1985). Histochemistry and ultrastructure of adrenergic and acetylcholinesterase-containing nerves supplying follicles and endocrine cells in the guinea-pig ovary. Cell Tissue Res. *240*, 505–511.

Stay, B., Fairbairns, S. and Yu, C.G. (1996). Role of allatostatins in the regulation of juvenile hormone synthesis. Arch. Insect Biochem. *32*, 287–297.

Steffl, M., Schweiger, M., Wessler, I., Kunz, L., Mayerhofer, A. and Amselgruber, W.M. (2006). Non-neuronal acetylcholine and choline acetyltransferase in oviductal epithelial cells of cyclic and pregnant pigs. Anat. Embryol. (Berl). *211*, 685–690.

Strand, F.L (1999). *Neuropeptides*. MIT Press, Cambridge, MA, pp. 206–207.

Stricker, S.A. and Smythe, T.L. (2000). Multiple triggers of oocyte maturation in nemertean worms: the roles of calcium and serotonin. J. Exp. Zool. *287*, 243–261.

Swaab, D.F., Chung, W.C., Kruijver, F.P., Hofman, M.A. and Ishunina, T.A. (2001). Structural and functional sex differences in the human hypothalamus. Horm. Behav. *40*, 93–98.

Tai, T.C., Claycomb, R., Her, S., Bloom, A.K. and Wong, D.L. (2002). Glucocorticoid responsiveness of the rat phenylethanolamine *N*-methyltransferase gene. Mol. Pharmacol *61*, 1385–1392.

Tanvez, A., Beguin, N., Chastel, O., Lacroix, A. and Leboucher, G. (2004). Sexually attractive phrases increase yolk androgens deposition in canaries (*Serinus canaria*). Gen. Comp. Endocrinol. *138*, 113–120.

Theurkauf, W.E., Smiley, S., Wong, M.L. and Alberts, B.M. (1992). Reorganization of the cytoskeleton during *Drosophila* oogenesis: implications for axis specification and intercellular transport. Development *115*, 923–936.

Tomancak, P., Guichet, A., Zavorsky, P. and Ephrussi, A. (1998). Oocyte polarity depends on regulation of *gurken* by Vasa. Development *125*, 1723–1733.
Tournebize, R., Popov, A., Kinoshita, K., Ashford, A.J., Rybina, S. and Pozniakovsky, A., et al. (2000). Control of microtubule dynamics by the antagonistic activities of XMAP215 and XKCM1 in *Xenopus* egg extracts. Nat. Cell Biol. *2*, 13–19.
Trujillo, A. and Riboni, I. (2002). Effects of functional peripheral sympathetic denervation induced by guanethidine on follicular development and ovulation of the adult female guinea pig. Gen. Comp. Endocrinol. *127*, 273–278.
Uller, T., Eklöf, J. and Andersson, S. (2005). Female egg investment in relation to male sexual traits and the potential for transgenerational effects in sexual selection. Behav. Ecol. Sociobiol. *57*, 584–590.
Vancompernolle, K., Boonefaes, T., Mann, M., Fiers, W. and Grooten, J. (2000). Tumor necrosis factor-induced microtubule stabilization mediated by hyperphosphorylated oncoprotein 18 promotes cell death. J. Biol. Chem. *275*, 33876–33882.
Vander, A., Sherman, J. and Luciano, D. (2001). *Human Physiology*, eighth edition. McGraw Hill, Boston, MA, pp. 658–659.
Vänttinen, T., Liu, J., Liu, J., Hydén-Granskog, C., Parviainen, M. and Penttilä, I., et al. (2000). Regulation of immunoreactive inhibin A and B secretion in cultured human granulosa-luteal cells by gonadotropins, activin A, and insulin-like growth factor type-1 receptor. J. Endocrinol. *167*, 289–294.
Venugopal, K.J. and Kumar, D. (2000). Role of juvenile hormone in the synthesis and sequestration of vitellogenins in the red cotton stainer, *Dysdercus koenigi* (Heteroptera: Pyrrhocoridae). Comp. Biochem. Phys. *C127*, 153–163.
Veselá, J., Rehák, P., Mihalik, J., Czikková, S., Pokorný, J. and Koppel, J. (2003). Expression of serotonin receptors in mouse oocytes and preimplantation embryos. Physiol. Res. *52*, 223–228.
Villa-Diaz, L.G. and Barrell, G.K. (1999). Involvement of serotoninergic pathways in the control of luteinizing hormone secretion in red deer hinds. Reprod. Fertil. Dev. *11*, 95–103.
Wagner, D.S., Dosch, R., Mintzer, K.A., Wiemel, A.P. and Mullins, M.C. (2004). Maternal control of development at the midblastula transition and beyond: mutants from the zebrafish II. Dev. Cell *6*, 781–790.
Waksmonski, S.L. and Woodruff, I. (2002). For uptake of yolk precursors, epithelial cell-oocyte gap junctional communication is required by insects representing six different orders. J. Insect Physiol. *48*, 667–675.
Walczak, C.E., Mitchison, T.J. and Desai, A. (1996). XKCM1: a *Xenopus* kinesin-related protein that regulates microtubule dynamics during mitotic spindle assembly. Cell *84*, 37–47.
Wayne, N.L. (2001). Regulation of seasonal reproduction in molluscs. J. Biol. Rhythm. *16*, 391–402.
Weil, T.T., Forrest, K.M. and Gavis, E.R. (2006). Localization of bicoid mRNA in late oocytes is maintained by continual active transport. Dev. Cell *11*, 251–262.
Weil, T.T., Parton, R., Davis, I. and Gavis, E.R. (2008). Changes in bicoid mRNA anchoring highlight conserved mechanisms during the oocyte-to-embryo transition. Curr. Biol. *18*, 1055–1061.
Williams, T.D., Ames, C.E., Kiparissis, Y. and Wynne-Edwards, K.E. (2005). Laying-sequence-specific variation in yolk oestrogen levels, and relationship to plasma oestrogen in female zebra finches (*Taeniopygia guttata*). Proc. Biol. Sci. *272*, 173–177.
Williams, T.D., Kitaysky, A.S. and Vezina, F. (2004). Individual variation in plasma estradiol-17beta and androgen levels during egg formation in the European starling *Sturnus*

vulgaris: implications for regulation of yolk steroids. Gen. Comp. Endocrinol. *136*, 346–352.

Witek, G.K. and Hoffmann, H. (2001). Immunological evidence for FGLamide- and w^2w^9-allatostatins in the ovary of *Gryllus bimaculatus* (Ensifera, Gryllidae). Physiol. Entomol. *26*, 49–57.

Wong, D.L., Lesage, A., Siddall, B. and Funder, J.W. (1992). Glucocorticoid regulation of phenylethanolamine *N*-methyltransferase *in vivo*. FASEB J. *6*, 3310–3315.

Wu, H. and Sun, Y.E. (2009). Reversing DNA methylation: new insights from neuronal activity–induced Gadd45b in adult neurogenesis. Sci. Signal. *2*, p. 64.

Yao, J., Morioka, T. and Oite, T. (2000). DGF regulates gap junction communication and connexin43 phosphorylation by PI 3-kinase in mesangial cells. Kidney Int. *57*, 1915–1926.

Yapici, N., Kim, Y.-J., Ribeiro, C. and Dickson, B.J. (2008). A receptor that mediates the post-mating switch in *Drosophila* reproductive behaviour. Nature *451*, 33–37.

Yoon, Y.J. and Mowry, K.L. (2004). *Xenopus* Staufen is a component of a ribonucleoprotein complex containing *Vg1* RNA and kinesin. Development *131*, 3035–3045.

Yoshioka, S., Matsuhana, B., Tanaka, S., Inouye, Y., Oshima, N. and Kinoshita, S. (2011). Mechanism of variable structural colour in the neon tetra: quantitative evaluation of the venetian blind model. J. R. Soc. Interface *8*, 56–66.

You, S., Li, W. and Lin, T. (2000). Expression and regulation of connexin43 in rat leydig cells. J. Endocrinol. *166*, 447–453.

Zitnan, D., Sehnal, F. and Bryant, P.J. (1993). Neurons producing specific neuropeptides in the central nervous system of normal and pupariation-delayed *Drosophila*. Dev. Biol. *156*, 117–135.

4 Epigenetic Control of Early Development

Unlike unicellulars, metazoans do not produce copies of themselves. They simply provide gametes with information for building the Bauplan and the CNS at the phylotypic stage, when the embryo becomes competent of generating the information for individual development up to adulthood.

Early embryonic development in metazoans is regulated by epigenetic information that is parentally provided to gametes (egg and sperm cells) in the form of parental cytoplasmic factors. Initially by their own activities, and later on by determining the temporal order and the spatial patterning of expression of zygotic genes, these factors determine cell divisions, establishment of body axes, formation of the embryonic layers, and gastrulation, culminating at the phylotypic stage with formation of the operational central nervous system (CNS) and the resulting integrated control system (ICS). Thus, the epigenetic program put in the zygote is an interim developmental program for early embryonic development, until the phylotypic stage, when the embryonic CNS is operational and takes over the individual development. It bridges the physical gap between the parental and embryonic CNSs.

The absence of that epigenetic information is the reason why somatic cells cannot develop into an organism of their kinds in the way an egg (in parthenogenetic organisms) or a zygote does, even though all of them are in possession of the complete species-specific genetic information.

Earlier I have argued and presented supporting evidence that the synthesis of maternal cytoplasmic factors and their placement in the egg cell by the follicle and/or nurse cells, via the active uptake by the oocyte from the intercellular environment, takes place under the control of the ICS. Now I will attempt to prove that the epigenetic information (parental cytoplasmic factors) determines the early individual development up to the phylotypic stage.

Epigenetic Control of Formation of Primordial Germ Cells and Formation of Zygote

Among the first differentiated cells that form in embryos are primordial germ cells (PGCs). Their fate as gamete precursors is epigenetically determined by maternal factor(s) (Weidinger et al., 2003). In most species studied so far, a maternal factor, *vasa mRNA*, deposited in the vegetal pole of the egg is the most important factor for differentiation of gametes (PGCs) during the early development. In *Drosophila*, it is involved in formation of the pole plasm at the posterior tip of the oocyte (Tomancak

Epigenetic Principles of Evolution. DOI: 10.1016/B978-0-12-415831-3.00004-5

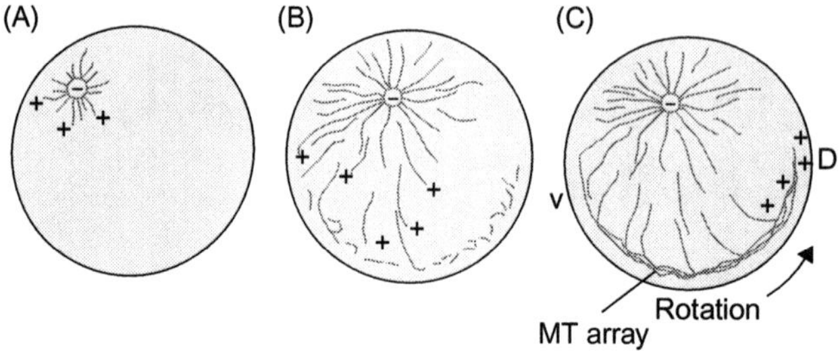

Figure 4.1 Formation of microtubules and dorsalization after fertilization of the egg cell. Microtubules (MTs) of the vegetal array arise from several sources. (A) The sperm centriole introduces polarity by acting as a minus-end MT-organizing center (−). The resulting radial array of MTs is called the *sperm aster*. (B) MTs from the sperm aster grow toward the periphery of the egg, as do additional MTs from unknown sources in the core cytoplasm. In addition, short disorganized MT polymers arise in the vegetal shear zone. (C) During rotation, MTs from deep in the cytoplasm bend into the vegetal shear zone and align with peripheral MTs to form the parallel array, with the plus-ends (+) of the rowing MTs pointing toward the future dorsal (D) side of the embryo. *Abbreviation*: V, ventral.
Source: From Weaver and Kimelman (2004).

et al., 1998), where the information for germ cell differentiation and specification of the abdomen is stored (Vanzo and Ephrussi, 2002).

vasa mRNA is found in the germ line cells of zebrafish (Yoon et al., 1997; Knaut et al., 2000), and vasa proteins are detected in chicken (Tsunekawa et al., 2000) and human (Castrillon et al., 2000) germ cell lineage. As shown earlier, expression of *vasa mRNA* in the ovary of the marine fish, *Sparus aurata*, is hormonally stimulated by the pituitary growth hormone (GH) and by estradiol (Cardinali et al., 2003), both synthesized and secreted under neural control.

The entry of the sperm cell into animal hemisphere of the oocyte detaches the cortex from the egg cytoplasm, thus enabling the cortex to rotate by about 30° in a process that contributes to the elongation and organization of microtubules into parallel bundles with their plus (+) ends toward the opposite side of the sperm entry. This determines the movement of maternal cytoplasmic factors and formation of the dorsal organizer there (Weaver and Kimelman, 2004; Figure 4.1).

Epigenetic Control of Migration of PGCs

As pointed out earlier, the fate of PGCs is determined by deposition of the maternal *vasa mRNA* in the vegetal pole and its asymmetric allocation among the early blastula cells. In order to fully differentiate into germ cells, PGCs must migrate to the genital ridge, the anlage of the future gonads.

The directed migration of PGCs from the vegetal pole to their target sites, traveling distances exceeding thousand times their diameter, is made possible by the presence on these cells of specific membrane receptors, chemokine receptors. PGCs find their way to the target site by binding their receptor, the specific ligand, chemokine stromal cell-derived factor-1 (SDF-1). This almost universal ligand of cell migration is a downstream element of signal cascades originating in the CNS. At the level of hormonal control, recently it has been demonstrated that 17-β-estradiol, via its nuclear receptor, induces expression of the gene coding for SDF-1 (Coser et al., 2003; Hall and Korach, 2003), and the latter is a direct target of that hormone (Hall and Korach, 2003). A downregulator of the SDF-1 is transforming growth factor-β1 (TGF-β1), which inhibits transcription of the *sdf-1* gene in the bone marrow stromal cells, thus affecting their migration and adhesion ability (Wright et al., 2003).

Chick and mouse PGCs use a different way of reaching their target sites; they use the blood flow as a transport vehicle and leave the blood vessels at specific sites where they sense the presence of the chemoattractant SDF-1α and, passing through the blood vessels' walls, they follow the chemokine to find their way to the genital ridge (Stebler et al., 2004).

By experimentally inhibiting the normal pattern of SDF-1 expression and by supplying SDF-1 to sites where it normally is absent, changes in the direction of migration are induced so that PGCs reach ectopic sites of the SDF-1 sources (Doitsidou et al., 2002; Stebler et al., 2004). On arrival in the genital ridge, the PGC population proliferates under the influence of fibroblast growth factors (FGFs) (Kawase et al., 2004) and TGF-β1, but these processes are regulated by a balanced proliferation/programmed cell death or apoptosis (Tres et al., 2004).

Epigenetic Control of Early Development in Insects

In *Drosophila*, like in insects in general, the early embryonic development is epigenetically determined by parental cytoplasmic factors. Cell division during the cleavage is regulated by maternal cyclins and the maternal String proteins, whose transcripts are present in the zygote and are among the earliest mRNAs to be translated.

Because of the uniform distribution of these factors, the early cycles of nuclear divisions are uniformly executed but no cell divisions occur. The resulting naked nuclei move to the outer edge of the embryo, thus leading to formation of a syncytial blastoderm. This lasts until the 14th cell cycle, when normal cell divisions begin to occur and the syncytial blastoderm is transformed into a cellularized blastoderm. This change is related to the fact that, at this point in time, the maternal reserve of the String protein is exhausted, and differential expression of the gene for the String protein in various parts of the embryo begins.

> *After the 17th or 18th cycle, cells in the epidermis and mesoderm stop dividing, and differentiate. This cessation of proliferation is caused by the exhaustion of maternal cyclin E, originally laid down in the egg, which is required for progression through the cell cycle.*
>
> *(Wolpert et al., 1998)*

In insects, the anterior–posterior axis is established by the combined action of three systems of maternal cytoplasmic factors placed in the egg by nurse cells: the *bicoid mRNA* in the anterior part of the egg cell, *nanos mRNA* and *caudal mRNA* in the posterior, and the *hunchback mRNA* initiate a transcription sequence of genes *gap-pair-rule-engrailed homeotic*. In *Drosophila*, these genes are expressed in phase with the gene for the neurotransmitter serotonin and serotonin receptor in characteristic stripes pattern (Colas et al., 1995).

Maternal cytoplasmic factors also induce the synthesis of neurotransmitters, which are among the first products of zygotic gene activity in the early stages of embryonic development (Shmukler et al., 1999). A serotonin receptor gene and the serotonin gene are intensely expressed at 3 h (cell blastoderm stage) of *Drosophila* embryogenesis. Their expression is organized in a seven-stripe pattern of the blastoderm stage and in phase with the pair-rule gene *fushi-tarazu* (Colas et al., 1995). Application of serotonin antagonists causes a transient regression of the first cleavage furrow in sea urchin embryos, suggesting a morphogenic role of the neurotransmitter in the early embryonic development (Shmukler et al., 1999).

Maternal signals determine a peak of expression of the neurotransmitter serotonin and its receptors in *Drosophila*, which is coincident with the onset of the germ band extension (corresponding to the phylotypic stage). How important that peak of serotonin is for the normal gastrulation is shown by the fact that mutant *Drosophila* embryos that fail to reach that peak do not develop proper germ band extension and die with a cuticular organization that is characteristic of embryos that do not express the serotonin receptor (Colas et al., 1999).

Early during the phylotypic stage, the incipient CNS is involved in patterning of neighboring embryonic mesoderm and underlying ectoderm and in determining their cell fates. Among the inductive effects of the CNS midline cells is the formation of somatic muscles from mesodermal progenitors (Lüer et al., 1997). Cells of the ventral midline region of the *Drosophila* CNS, and probably of other regions of the CNS, play a central inductive role in patterning the mesoderm as well as the underlying ectoderm (Lüer et al., 1997).

Drosophila CNS midline cells secrete Spitz, which induces the synthesis of the epidermal growth factor (EGF) receptor in the neighboring ventral ectoderm, thus determining different fates for cells of the ventral ectoderm (Golembo et al., 1996; Zhou et al., 1997).

Epigenetic Control of Early Development in Vertebrates

The dorsal side of the embryo is established opposite of where the sperm cell enters the egg cell. *β-Catenin*, a maternal transcription factor, and other maternal cytoplasmic factors such as mRNAs for Vg1, Xwnt11, Noggin, and Activin, are accumulated in the dorsal side of the developing embryo, as a result of the cortical rotation. The first divisions of the zygote are stimulated by the synthesis of the cyclin protein Cdc6 by the maternal *Cdc6 mRNA* that is stored in the egg cell during the maturation of the oocyte shortly after the germinal vesicle breakdown. The maternal *Cdc6 mRNA* makes possible the replication of zygotic chromosomes by activating the minichromosome maintenance helicase complex (Lemaitre et al., 2002).

On each cell division, cyclins are destroyed and new cyclins are synthesized from the reserve of maternal *cyclin mRNAs*. They combine with a cyclin-dependent kinase to form the maturation promoting factor (MPF), a phosphoprotein that is responsible for cell division since the early stages of the embryonic development. When treated with MPF complex, the cleaved cells arrested in S phase resume their mitotic cycle (Gilbert, 1997). In this context, one must remember that MPF is an active form of the pre-MPF, which, in turn, is induced by progesterone (Murray and Kirshner, 1989).

By the 14th cycle of cell divisions, the reserve of maternal *cyclin mRNA* and the String, a protein phosphatase, are exhausted. The zygotically coded String protein (cdc25 phosphatase) is rhythmically translated to phosphorylate the pre-MPF and transform it into active MPF, just before the mitotic cell division. The zygotic *string* gene will be differentially expressed, only in the cells that have inherited transcription factors of the *gap*, *pair ruled*, and other early-patterning genes, which are expressed in phase with the gene for the neurotransmitter serotonin and serotonin receptor in characteristic stripes pattern. This, ultimately, explains the fact that some parts of the embryo (those that express the *string*) will grow faster than others.

Expression of various types of retinoic acid (RA) receptors in the inner cell mass and trophoectoderm of blastocysts suggests that maternal RA is likely to directly regulate gene expression during preimplantation development (Mohan et al., 2001) of bovine embryos. In experiments on murine embryos during the latter stages of organogenesis, it is demonstrated that excess RA causes phagocytosis of the migrating neural crest cells, leading to craniofacial malformations (Yasuda et al., 1989).

In some species, RA is present as maternal factor in the egg cytoplasm, and during the early gastrulation in mammals the main site of RA synthesis may be the mesoderm adjacent to the node and primitive streak (Niederreither et al., 1997). Later in the mouse embryogenesis (E11), RA is produced in an endocrine way in the incipient adrenal gland. RA penetrates cell membrane and binds to its specific nuclear receptors with which it forms complexes that act as transcription factors. RA may cooperate with growth factors to provide positional information (Cho and De Robertis, 1990; Langston et al., 1997). It has also been suggested that TGF-β may be a mediator of the stimulating effect of RA on cell differentiation (Danielpour, 1996).

Regulatory functions of RA are chiefly related to its ability to regulate expression of *Hox* genes, which have retinoic acid response elements (RARE) in their enhancers (Langston et al., 1997). RA is a common immediate regulator of the activity of almost all of the known homeotic genes, since the earliest stages of the embryonic development and during the postnatal development in metazoans (Conlon, 1995; De Luca and Ross, 1996; Marshall et al., 1996; Clagett-Dame and Plum, 1997; Cupp et al., 1999; Malpel et al., 2000). By regulating expression of the *Hox* genes, RA is crucially involved in establishing the anterior–posterior axis during gastrulation in vertebrates.

Administration of RA in mouse embryos causes anterior *Hox* genes to be expressed more posteriorly, and posterior *Hox* genes to be expressed in more anterior segments.

Vitamin A (RA precursor) deficiency causes early death of quail embryos but administration of retinoids to those bird embryos up to the five-somite stage (not later) rescues embryonic development, suggesting the existence of a narrow developmental window in early development during which presence of retinoids is necessary for the embryonic development (Kostetskii et al., 1998; Knezevic and Mackem, 2001). Based on the facts that the hormone RA synergistically induces expression of *Hox* genes in the blastoderm cell culture and that exogenous RA mimics the effects of hypoblast rotation on primitive streak extension, Knezevic and Mackem (2001) suggest that RA (maternal and/or embryonic—N.C.) plays some role in the development of the pregastrula embryo.

Epigenetic Control of Early Development in Mammals

Like other metazoan groups, the early embryonic development in mammals is under control of parental epigenetic information (parental cytoplasmic factors, imprinted genes, and probably other factors).

The close contact of the developing embryo with the mother and the maternal dependence of the embryonic development in these animals led to some specific features of the maternal control of early development in this group. One of the most visible features of the maternal control in mammals is the comparably early termination of the role of maternal cytoplasmic factors provided with the egg cell in mammalian embryos (Gilbert, 2000).

In mice, the embryonic genome is active from the two-cell stage (Piko and Clegg, 1982) and in rabbits transcription of zygotic genes starts from the one-cell stage (Brunet-Simon et al., 2001), whereas in the clawed frog, *Xenopus laevis*, expression of the zygotic genes first starts in the ~5,000-cell embryo. In other mammals, however, the transition to zygotic gene expression takes place after several cell cycles (Henrion et al., 2000).

The physical continuity of the mother and the developing embryo makes it possible for mammals to depend less on the deposition of maternal factors and rely more on direct maternal control, via the neuroendocrine system, a "real time" control and regulation of the embryonic development. Maternal hormones, growth factors, and other secreted proteins that reach the embryo transplacentally are essentially involved in mammal embryogenesis from its beginning.

Implantation involves interactions between the endometrium and blastocyst. Endometrial secretions are considered to be regulators of implantation and placentation. Among the early-secreted maternal proteins that are involved in the implantation of the blastocyst, during the "implantation window" in mammals, are growth factors, cytokines and *Hox* genes.

Numerous growth factors are specifically expressed in the maternal reproductive tract (Hardy and Spanos, 2002). In preparation for implantation of the blastocyst, at the site of blastocyst attachment to the endometrium, the latter expresses 22 genes for growth factors (Paria et al., 2001). EGF induces expression of its receptor, EGFR, in the mouse eight-cell stage blastocyst (Kim et al., 1999). EGFR is expressed in oviducal and endometrial membranes of pregnant pigs during the preimplantation period. This, and the fact that its receptor, EGFR, is also present in the zygote, suggest that maternal

EGF acts on the blastocyst at this early stage. Expression of EGF in the pig oviduct is stimulated by estradiol (Wollenhaupt et al., 1999), a downstream element of a signal cascade that starts with an epigenetic brain signal that is communicated to the ovary via the hypothalamic–pituitary axis. The same is true for the expression of the *egf-R* gene that is induced by the pituitary GH, as well as for other members of the EGF family expressed in preimplantation uterine tissues, such as heparin-binding EGF-like growth factor and amphiregulin (Giudice, 1999), β-cellulin and epiregulin (Das et al., 1997), galectin (Choe et al., 1997), cytokines such as leukemia inhibitory factor (LIF), macrophage colony-stimulating factor, interleukin-1 (IL-1), hepatocyte growth factor, and insulin-like growth factors (IGFs) (Giudice, 1999; Kauma, 2000).

Ultimately, it is the maternal CNS that, via the hormones of the target endocrine glands, regulates all of the above factors. So, for example, progesterone alone regulates several specific factors such as TGF-β, IL-1, IGF binding protein-1, tissue inhibitors of metalloproteinases, and fibronectin (Johansson et al., 1989). Estradiol and progesterone regulate synthesis of the transmembrane receptor for all the IL-6 type cytokines, which are necessary for blastocyst implantation (Classen-Linke et al., 2004).

Pituitary prolactin, estrogen, and progesterone increase expression of estrogen receptors (ERs), but this does not explain why this expression is spatially restricted mainly to the antimesometrial site of the uterus, where the *deciduas capsularis* forms in rats (Tessier et al., 2000). It is possible that, as it is observed in many other cases, the limited site-specific expression of ER may be a function of the neural arm of the binary neural control of gene expression, performed by local innervation. Under hormonal influence, during the implantation in sheep, changes occur in the expression pattern of endometrial connexins (Gabriel et al., 2004). Proteins of the endometrial extracellular matrix are necessary for binding trophoblast integrins and for adhesion of trophoblast to the uterus (Rout et al., 2004).

Maternal progesterone, via its receptor (PR), induces expression of a number of genes, playing a crucial role in the process of implantation in mammals. However, during the periimplantation period, endometrial epithelial cells lose their PR, what leads to reduced synthesis of MUC1 (mucin 1, a transmembrane peptide), which would prevent adhesion of trophoblast cells to the uterus wall. At the same time, progesterone induces endometrial epithelium to secrete adhesion-promoting proteins such as galectin and osteopontin (Spencer et al., 2004). Note that the "primary effectors," estrogen and progesterone, are under ultimate CNS control via the hypothalamus–pituitary–ovarian axis and by the local ovarian innervation, while endocannabinoids are signaling molecules mainly released by nerves that act on G-protein-coupled receptors.

A number of *Hox* genes encode transcription factors that are essential for the uterus receptivity and implantation. Such a function has been demonstrated for *Hox10* (Cermik et al., 2001; Daftary and Taylor, 2000) and *Hox11* (Daftary and Taylor, 2001). In the preimplantation endometrium and myometrium, *Hox10* is downregulated by maternal progesterone (Cermik et al., 2001). In mice, the administration of diethylstilbestrol, a synthetic nonsteroidal estrogen, via its ERs, alters the pattern of expression of several *Hox* genes involved in the patterning of the reproductive tract, thus leading to developmental anomalies (Block et al., 2000). Maternal hormones insulin and IGF-1 are present in secretions of oviduct and uterus and bind to morula cells that express their respective receptors, thus stimulating the blastocyst formation (Herrler et al., 1998).

Expression in the mesometrium and uterine epithelium of maternal activin and its receptor in pigs shows that "both embryonic and uterine activin are involved in intrauterine processes, such as attachment and early embryonic development" (van de Pavert et al., 2001). Let us remember that the synthesis of activin itself may be neurally regulated via the hypothalamus–pituitary axis by gonadotropins, follicle-stimulating hormone (FSH) and luteinizing hormone (LH) (Demura et al., 1993).

FGF-4 (Leunda-Casi et al., 2001) and nitric oxide (Sengoku et al., 2001) are also involved in the regulation of mouse trophoblast development. TGF-β1 is downregulated by neuroactive opioids and their specific receptors (Chatzaki et al., 2000). Under the control of ovarian hormones, a group of angiogenic factors (vascular endothelial growth factor, FGF-2) and their specific receptors, as well as EGFR, are expressed in and around the human endometrial blood vessels, a fact that is considered to be related to their contribution to the "regulation of angiogenesis" (Möller et al., 2001). The neurohormonally regulated pattern of expression of genes of the EGF family during the menstrual cycle and early pregnancy may be involved in monkey embryo implantation (Yue et al., 2000).

The deciduation or transformation of the mammal endometrium into a receptive state for implantation of the blastocyst requires expression of a number of cytokines. LIF is hormonally induced by estradiol and cytokines (Sawai et al., 1997) and IL-11, as well as by genes for protein D9K, as it may be concluded from the fact that null mutant mice for those factors fail to implant embryos in the endometrium (Salamonsen et al., 2001). The sequence of expression LIF, IL-11, and prolactin in the murine endometrium "suggests a synchronized role for each in the differentiation of endometrium" (Dimitriadis et al., 2000). It has also been reported that optimal levels of the neurotransmitter nitric oxide are crucial for endometrial function and embryo implantation (Ota et al., 1999).

Hormones estradiol and estriol are found to stimulate connexin-43 synthesis in myometrium and increase the formation of gap junctions between myometrial cells in pregnant women (Di et al., 2001). In mice, the transplacentally provided maternal estradiol upregulates expression of the zygotic w*nt11* gene during the morula–blastula transitional period as it is demonstrated by the fact that lack of estrogen (in ovarioectomized mice) prevents expression of *Wnt11* (Mohamed et al., 2004; Chen et al., 2009; Figure 4.2).

Placenta is believed to be regulated by imprinted genes in a process in which paternally expressed genes promote and maternally expressed genes restrain its growth (Coan et al., 2004). It has been shown that decrease of LH and prolactin during the second half of pregnancy in mares is centrally regulated by deactivation of the opioidergic system (Aurich et al., 2001).

Brain neurohormones are also expressed in the placenta. So, for example, the neuropeptide galanin is expressed there throughout the gestation period (Kleine et al., 2001), where it is believed to stimulate the gonadotropin-releasing hormone (GnRH) synthesis, which may be an important "regulator of placental function." GnRH is expressed in both placenta and cultured trophoblasts, and it seems to inhibit expression of its

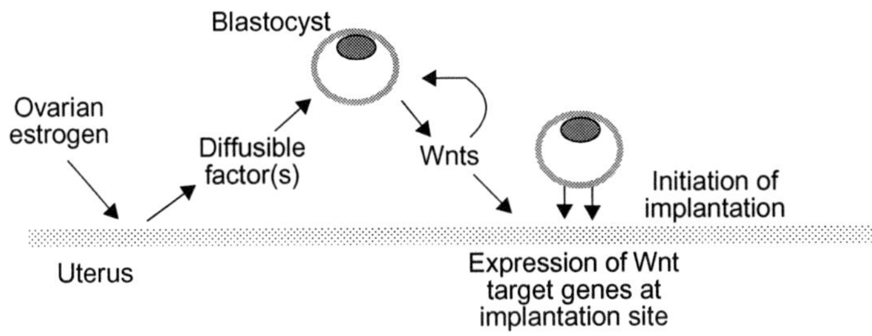

Figure 4.2 Uterine-embryonic communication at implantation. Estradiol is proposed to act on uterine cells to induce synthesis of a secreted molecule that acts on the blastocyst to trigger expression of estradiol-regulated genes encoding factors such as Wnt11. These may induce expression of target genes in the uterus or in the blastocyst that facilitate implantation. *Source*: From Mohamed et al. (2004).

own receptor, GnRH-R, autocrinely secreted in the placenta (Wolfahrt et al., 2001). However, GnRH stimulates the synthesis of human chorionic gonadotropin in the first trimester trophoblast (Islami et al., 2001).

In mice embryos, the neurotransmitter serotonin starts to be produced by E11 (embryonic day 11). Before this stage, it is the maternal serotonin that circulates in the embryo and, due to the incomplete formation of the blood–brain barrier, the maternal serotonin enters the brain and participates in the development of particular embryonic brain structures (Côté et al., 2007).

Maternal thyroid hormones (T3 and T4) are detected in the human fetus since the first gestation trimester, i.e., before the fetus starts producing thyroid hormones. At this stage, maternal T3 is detected only in the fetal brain. Even after activation of the fetal thyroid gland in the second trimester, maternal thyroid hormones are still detected in the human fetal brain. There is evidence that the maternal hypothalamic neurohormone thyroid-releasing hormone transplacentally regulates the function of the fetal thyroid gland before the maturation of the fetal hypopthalamic–pituitary–thyroid axis (Chan and Kilby, 2000). Maternal thyroid hormone (T4) crosses the placenta and enters the embryonic brain, where it regulates neuronal proliferation, neuronal migration, process outgrowth, and myelin formation (Porterfield and Hendry, 1998). It may also induce several genes, including *RC3/neurogranin* gene, in the brain cortex of rat fetus (Dowling and Zoeller, 2000). The facts that growth hormone receptor (*GHR*) *mRNA* is expressed in preimplantation bovine embryos and that supplementation of the GH to those embryos *in vitro* favorably affects their cleavage, blastocyst formation, and hatchability (Izadyar et al., 2000) suggest that the maternal pituitary GH, which is regulated by secretion of the neurohormone growth hormone-releasing hormone (GHRH), plays an essential role in regulation of the preimplantation development of bovine embryos.

Embryo transfer experiments have shown a strong maternal influence on the skeleton development, especially of the embryonic craniofacial skeleton and teeth (Nonaka et al., 1993; Sasaki et al., 1995). Under experimental conditions, various stressors, via the neuroendocrine system, induce hormonal imbalances in pregnant laboratory mammals, causing nonadaptive behavioral and morphological changes in their offspring (Hopper and Hart, 1985).

The representative facts presented herein on the regulation of the activity of fertilized egg and early embryonic development in mammals show that, besides the control of the embryonic development via parental cytoplasmic factors deposited in gametes, this group of animals transplacentally exerts a maternal epigenetic control on the early development of the embryo.

Epigenetic Control of Formation of Embryonic Germ Layers

Maternal Control of Endoderm Formation

It is generally admitted that formation of endoderm is induced by Nodal signaling (Schier, 2003; Poulain et al., 2006). But Nodal subfamily is not where the signal cascade and regulatory network for endoderm formation begins: three main groups of maternal factors are involved in formation of endoderm: *VegT*, *β-catenin*, and *Otx* (Grapin-Botton and Constan, 2004). Maternal *vegT mRNA* is concentrated in the vegetal pole of the *Xenopus* egg. This transcript codes for *VegT*, which is a transcription factor (Zhang and King, 1996) that induces expression of numerous endodermal genes (*Bix1*, *Bix2*, *Bix3*, *Bix4*, *Mix1*, *Mix2*, *Mixer*, *Xsox17α*, *Gata4*, *Gata5*, *Gata6*, including the anterior endodermal genes, *Xhex* and *cerberus*) in the vicinity, leading to the formation of endoderm (Xanthos et al., 2001; Figure 4.3).

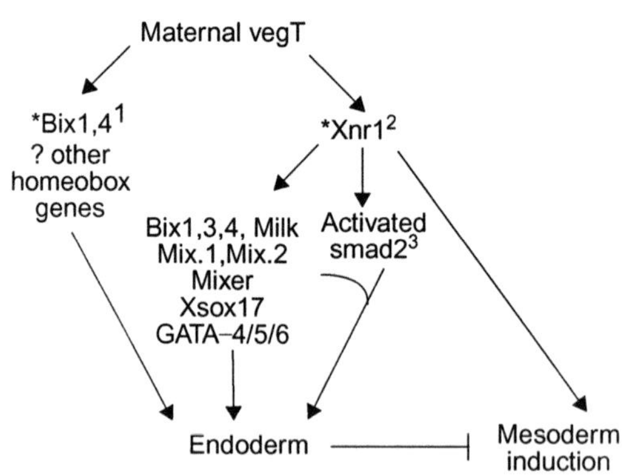

Figure 4.3 Maternal VegT is responsible for initiating mesoderm induction as well as endoderm formation through zygotic TGF-βs. These TGF-βs activate the expression of endodermal transcription factors, including the homeobox genes, GATA factors, and *Xsox17*. Zygotic endodermal transcription factors such as Mixer and GATA5 can act alone and/or in cooperation with Smad2 to activate endodermal gene expression. Transcription factors such as Xsox17 may require a partner such as Smad2 to activate gene expression.
Source: From Xanthos et al. (2001).

Another maternal factor involved in endoderm induction in zebrafish and amphioxus is eomesodermin (Grapin-Botton and Constan, 2004). A self-limiting mechanism enables *VegT* to inhibit induction of endodermal genes, thus determining the endoderm boundary in *Xenopus* (Clements and Woodland, 2003).

The above evidence proves that it is the epigenetic information deposited in strictly determined sites of the egg cell in the form of maternal *VegT*, *β-catenin*, and *Otx* mRNAs, rather than the genetic information (respective zygotic genes are not involved in the process of development of endoderm), that determines formation of endoderm.

Maternal Control of Mesoderm Formation

Maternal cytoplasmic factors (transcripts for growth factors of the TGF-β superfamily of the secreted proteins, Vg1 and activin) of the dorsal region are responsible for mesoderm induction (Kofron et al., 1999). The activin type II receptor is expressed in all blastula cells and a number of maternal TGF-β growth factors, including Vg1, may bind it. FGF and bone morphogenetic protein-4 (BMP-4) are considered to be mesoderm-inducing secreted proteins present in the Nieuwkoop center (Stennard et al., 1996; Zhang and King, 1996).

Maternal *vegt* transcripts are translated into *VegT*, which induces formation of mesoderm by activating expression of zygotic growth factors of the TGF-β superfamily (Nodal subfamily), *Xnr1*, *Xnr2*, *Xnr4*, and *derriere*. Elimination or lack of maternal *VegT* prevents formation of mesoderm (Clements et al., 1999; Kofron et al., 1999; Kimelman and Bjornson, 2004). Ectopic expression of *VegT* converts prospective ectoderm into ventral mesoderm (Zhang and King, 1996), and elimination of *vegT* mRNA prevents formation of mesoderm (Zhang et al., 1998).

Maternal β-catenin is also involved in expression of *Xnr* genes and formation of the organizer, which induces mesoderm formation along the dorsoventral axis (Kimelman and Bjornson, 2004).

Epigenetic Control of Neural Induction and Formation of the CNS

Formation of the CNS is part of the early development, which, as shown earlier, is epigenetically determined by maternal cytoplasmic factors deposited in gametes under control of the parental nervous system(s).

> *The development of the nervous system is a largely epigenetic phenomenon in which events induce subsequent events in what is usually a highly orderly sequence.*
>
> *(Tierney, 1996)*

Earlier, the neural induction was believed to take place during gastrulation. Now we know that neural induction and specification of the neural cells starts much earlier, during the blastula stage (Wilson and Edlund, 2001) if not earlier:

> *The cascade of inductive interactions leading to the formation of the central nervous system starts in the uncleaved egg.*
>
> *(Müller, 1996)*

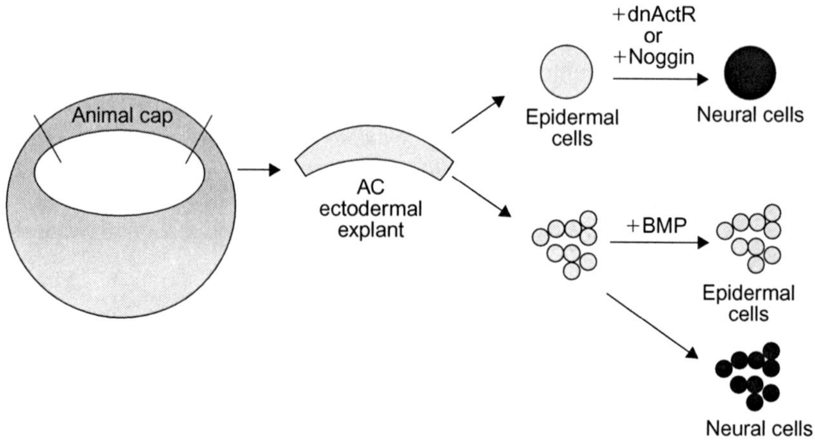

Figure 4.4 The default model of neural induction in *Xenopus*. An ectodermal explant dissected from the animal cap of a blastula stage embryo forms epidermis (light blue). If bone morphogenetic protein (BMP) signaling is inhibited in those cells by expressing a dominant-negative Activin receptor (dnActR) or the extracellular antagonist Noggin, then neural tissue forms (dark blue). Dissociation of the explant into single cells without the addition of exogenous factors also leads to the formation of neural tissues. However, if exogenous BMP is added to the dispersed cells, they adopt an epidermal fate (light blue). (For interpretation of the references to color in this figure legend, the reader is referred to the web version of this book.)
Source: From Rogers et al. (2009).

The role of the CNS in the metazoan individual development is of paramount importance, but here I will only briefly outline the epigenetic control of the neural induction by maternal factors.

The development of the CNS starts before any other organ or system of organs in metazoans. β-Catenin, a maternal transcription factor, accumulates in the dorsal part of the egg and falls to the vegetal cells of the *Xenopus* embryo (Schneider et al., 1996). Ectodermal explants from the animal cap in the absence of BMP differentiate into neural cells whereas in the presence of BMP are transformed into epidermal cells (Rogers et al., 2009; Figure 4.4). As early as the 32-cell blastula, these cells represent the Nieuwkoop center where maternal FGF and BMP-4 are also present (Stennard et al., 1996). There, the maternal β-catenin forms a complex with the Tcf3, thus enabling the transcription of several genes (Yang et al., 2002). The product of one of those genes, Siamois, then induces expression of *goosecoid*, producing a protein that triggers expression of a number of genes in the cells dorsal to Nieuwkoop center, thus determining the formation of the Spemann's organizer (Schneider et al., 1996; Laurent et al., 1997).

The Spemann–Mangold organizer is formed in the blastula stage of the amphibian embryo by cells that have responded to two maternal agents: a general meso-endoderm inducer (involving the TGF-β signaling pathway) and a dorsal modifier (probably involving the Wnt signaling pathway). (Harland and Gerhart, 1997). Formation of the neural plate results from expression of maternal *Fgf* mRNA which blocks expression of maternal *Bmp* mRNA in the medial epiblast cells (prospective neural plate) but not in lateral epiblast cells. In lateral epiblast cells, maternal *Wnt* mRNAs (*Wnt3A* mRNA and *Wnt8C* mRNA?) are translated, and by attenuating the response of those cells to *Fgf* mRNA, allow them to translate maternal *Bmp* mRNA, thus determining their epidermal fate (Wilson et al., 2001; Stern, 2005).

At early blastula stage a blastula Chordin- and Noggin-expressing (BCNE) group of cells in the dorsal animal cap represents both the prospective neuroectoderm and Spemann organizer. Expression of the BMP antagonists, Chd and Nog, in these cells is determined by accumulation of maternal β-catenin. The BCNE center contains much of the presumptive anterior CNS, and is required for brain formation in *Xenopus* embryos (Kuroda et al., 2004).

Neural induction ends by the end of the gastrula stage in different vertebrate classes (Stern, 2004). In *Xenopus laevis*, for example, it ends before the beginning of large-scale expression of zygotic genes (after 12 cell divisions, i.e., at 4,096-cell embryonic stage), implying that neural induction in this amphibian is determined by epigenetic information provided parentally to the zygote in the form of cytoplasmic factors. In humans, neurulation starts with the formation of a flat sheet of about 125,000 cells in the dorsal side of the embryo, known as *neural plate* "from which all the neurons and glial cells derive" (Dowling, 2004). Soon, the neural plate rises on both sides to form the neural tube/CNS, along almost the whole length of the embryo leading to the formation of the neural tube/CNS along almost the whole length of the embryo in the process of the primary neurulation.

Even this glimpse of the processes of neural induction and development of the CNS reveals some thought-provoking facts and coincidences.

First, formation of the CNS coincides with the exhaustion of the parental epigenetic information (parental cytoplasmic factors) at the phylotypic stage, right before the beginning of organogenesis, when the demand of the developing embryo for morphogenetic information is greater than ever.

Second, the nervous system is the first organ system (with neurons being the first differentiated cells) to develop, although conventional wisdom says that systems of blood circulation and excretory system would be necessary to develop before all else. The overearly embryonic development of the CNS suggests that it might serve something other than the communication with the external world, during early embryogenesis when that communication is next to absent.

Third, initially, the CNS is excessively large (in some cases, representing almost a quarter of the overall embryonic mass), a fact that in view of the insignificant low level of communication of the embryo with the external environment suggests some other important role.

Fourth, formation of the neural ectoderm marks the establishment of the primary embryonic axis as supreme source of inductive radiations in metazoans. The incipient CNS immediately engenders a network of inductions that give rise to the different cells, tissues, and organs of embryos and adults (Hall, 1998a).

> *Its further structures arise in relation to this central axis. This is especially evident in the development of paired elements, such as the somites, that presage the vertebrae, and paired organ rudiments, such as left and right limb buds and the primordia of the gonads, kidney, lung and heart.*
>
> *(Hall, 1998b)*

If all the above coincidences are not determined by chance, as seems to be the case, they may point to a central role of the nervous system in the early individual development.

References

Aurich, C., Aurich, J.E. and Parvizi, N. (2001). Opioidergic inhibition of luteinising hormone and prolactin release changes during pregnancy in pony mares. J. Endocrinol. *169*, 511–518.

Block, K., Kardana, A., Igarashi, P. and Taylor, H.S. (2000). *In utero* diethylstilbestrol (DES) exposure alters *Hox* gene expression in the developing muellerian system. FASEB J. *14*, 1104–1108.

Brunet-Simon, A., Henrion, G., Renard, J.P., and Duranthon, V. (2001). Onset of zygotic transcription and maternal transcript legacy in the rat embryo. Mol. Reprod. Dev. *58*, 127–136.

Cardinali, M., Gioacchini, G., Candiani, S., Pestarino, M., Yoshizaki, G., and Carnevali, O. (2004). Hormonal regulation of *vasa-like* mRNA expression in the ovary of the marine teleost, *Sparus aurata*. Biol. Reprod. *70*, 737–743.

Castrillon, D.H., Quade, B.J., Wang, T.Y., Quigley, C. and Crum, C.P. (2000). The human VASA gene is specifically expressed in the germ cell lineage. Proc. Natl. Acad. Sci. U.S.A. *97*, 9585–9590.

Cermik, D., Karaca, M. and Taylor, H.S. (2001). HOXA10 expression is repressed by progesterone in the myometrium: differential tissue-specific regulation of HOX gene expression in the reproductive tract. J. Clin. Endocrinol. Metab. *86*, 3387–3392.

Chan, S. and Kilby, M.D. (2000). Thyroid hormone and central nervous system development. J. Endocrinol. *165*, 1–8.

Chatzaki, E., Margioris, A.N., Makrigiannakis, A., Castanas, E., Georgoulias, V. and Gravanis, A., (2000). Kappa opioids and TGFbeta1 interact in human endometrial cells. Mol. Hum. Reprod. *6*, 602–609.

Chen, Q., Zhang, Y., Lui, J., Wang, Q., Wang, S., Cao, Y., et al. (2009). Embryo–uterine cross-talk during implantation: the role of Wnt signaling. Mol. Hum. Reprod. *15*, 215–221.

Cho, K.W. and De Robertis, E.M. (1990). Differential activation of *Xenopus* homeobox genes by mesoderm-inducing growth factors and retinoic acid. Gene. Dev. *4*, 1910–1916.

Choe, Y.S., Shim, C., Choi, D., Lee, C-S., Lee, K-K. and Kim, K. (1997). Expression of galectin-1 mRNA in the mouse uterus is under the control of ovarian steroids during blastocyst implantation. Mol. Reprod. Dev. *48*, 261–266.

Clagett-Dame, M. and Plum, L.A. (1997). Retinoid regulated gene expression. Crit. Rev. Eukaryot. Gene Expr. *7*, 299–342.

Classen-Linke, I., Mueller-Newen, G., Heinrich, P.C., Beier, H.M. and von Rango, U. (2004). The cytokine receptor gp 130 and its soluble form are under hormonal control in human endometrium and deciduas. Mol. Hum. Reprod. *10*, 495–504.

Clements, D. and Woodland, H.R. (2003). VegT induces endoderm by a self-limiting mechanism and by changing the competence of cells to respond to TGF-beta signals. Dev. Biol. *258*, 454–463.

Clements, D., Friday, R.V. and Woodland, H.R. (1999). Mode of action of VegT in mesoderm and endoderm formation. Development *126*, 4903–4911.

Coan, P.M., Burton, G.J. and Ferguson-Smith, A.C. (2004). Imprinted genes in the placenta—a review. Placenta *26*, S10–S20.

Colas, J.F., Launay, J.M., Kellermann, O., Rosay, P. and Maroteaux, L. (1995). *Drosophila* 5-HT2 serotonin receptor: coexpression with fushi-tarazu during segmentation. Proc. Natl. Acad. Sci. U.S.A. *92*, 5441–5445.

Colas, J.F., Launay, J.M. and Maroteaux, L. (1999). Maternal and zygotic control of serotonin biosynthesis are both necessary for Drosophila germband extension. Mech. Dev. *87*, 67–76.

Conlon, R.A. (1995). Retinoic acid and pattern formation in vertebrates. Trends Genet. *11*, 314–319.

Coser, K.R., et al. (2003). Global analysis of ligand sensitivity of estrogen inducible and suppressible genes in MCF7/BUS breast cancer cells by DNA microarray. Proc. Natl. Acad. Sci. U.S.A. *100*, 13994–9.

Côté, F., Fligny, C., Bayard, E., Launay, J.M., Gershon, M.D., Mallet, J., et al. (2007). Maternal serotonin is crucial for murine embryonic development. Proc. Natl. Acad. Sci. U.S.A. *104*, 329–334.

Cupp, A.S., Dufour, J.M., Kim, G., Skinner, M.K. and Kim, K.H. (1999). Action of retinoids on embryonic and early postnatal testis development. Endocrinology *140*, 2343–2352.

Daftary, G.S. and Taylor, H.S. (2000). Implantation in the human: the role of *Hox* genes. Semin. Reprod. Med. *18*, 311–320. (abstract).

Daftary, G.S. and Taylor, H.S. (2001). Molecular markers of implantation: clinical implications. Curr. Opin. Obstet. Gynecol. *13*, 269–274.

Danielpour, D. (1996). Induction of transforming growth factor-beta autocrine activity by all-trans-retinoic acid and 1 alpha, 25-dihydroxy vitamin D3 in NRP-152 rat prostatic epithelial cells. J. Cell Physiol. *166*, 231–239.

Das, S.K., Das, N., Wang, J., Lim, H., Schryver, B., Plowman, G.D., et al. (1997). Expression of betacellulin and epiregulin genes in the mouse uterus temporally by the blastocyst solely at the site of its apposition is coincident with the "window" of implantation. Dev. Biol. *190*, 178–190.

De Luca, L.M. and Ross, S.A. (1996). Retinoic acid response elements as positive and negative regulator of the expression of the homeobox 1 gene. Nutr. Rev. *54*, 61–63.

Demura, R., Suzuki, T., Tajima, S., Mitsuhashi, S., Odagiri, E., Demura, H. (1993). Activin and inhibin secretion by cultured porcine granulosa cells is stimulated by FSH and LH. Endocr. J. *40*, 447–451.

Di, W.L., Lachelin, G.C., McGarrigle, H.H., Thomas, N.S. and Becker, D.L. (2001). Oestriol and oestradiol increase cell-to-cell communication and connexin43 protein expression in human myometrium. Mol. Hum. Reprod., 671–679.

Dimitriadis, E., Salamonsen, L.A. and Robb, L. (2000). Expression of interleukin-11 during the human menstrual cycle: coincidence with stromal cell decidualization and relationship to leukemia inhibitory factor. Mol. Hum. Reprod. *6*, 907–914.

Doitsidou, M., Reichman-Fried, M., Stebler, J., Köprunner, M., Dörries, J., Meyer, D., et al. (2002). Guidance of primordial cell migration by the chemokine SDF-1. Cell *111*, 647–659.

Dowling, A.L. and Zoeller, R.T. (2000). Thyroid hormone of maternal origin regulates expression of RC3/neurogranin mRNA in the fetal rat brain. Mol. Brain Res. *82*, 126–132.

Dowling, J.E. (2004). *The Great Brain Debate—Nature or Nurture?*. Joseph Henry Press, Washington, DC, p. 8.

Gabriel, S., Winterhager, E., Pfarrer, C., Traub, O. and Leiser, R. (2004). Modulation of connexin expression in sheep endometrium in response to pregnancy. Placenta *25*, 287–296.

Gilbert, S.F. (1997). *Developmental Biology*, fifth edition. Sinauer Associates, Inc. Publishers, Sunderland, MA, p. 197.

Gilbert, S.F. (2000). *Developmental Biology*, sixth edition. Sinauer Associates, Inc. Publishers, Sunderland, MA, p. 223.

Giudice, L.C. (1999). Genes associated with embryonic attachment and implantation and the role of progesterone. J. Reprod. Med *44*(2 Suppl.), 165–171.

Golembo, M., Raz, E. and Shilo, B.Z. (1996). The *Drosophila* embryonic midline is the site of Spitz processing and induces activation of the EGF receptor in the ventral ectoderm. Development *122*, 3363–3370.

Grapin-Botton, A. and Constan, D. (2004). Endoderm development. In: *Gastrulation: From Cells to Embryo*, C.D. Stern (ed.), Cold Spring Harbor Laboratory Press, Cold Spring Harbor, NY, pp. 433–448.

Hall, B.K. (1998a). Evolutionary developmental biology, second edition. Chapman & Hall, London, p. 134.

Hall, B.K. (1998b). Ibid. p. 163.

Hall, J.M. and Korach, K.S. (2003). Stromal cell-derived factor 1, a novel target of estrogen receptor action, mediates the mitogenic effects of estradiol in ovarian and breast cancer cells. Mol. Endocrinol. *17*, 792–803.

Hardy, K. and Spanos, S. (2002). Growth factor expression and function in the human and mouse preimplantation embryo. J. Endocrinol. *172*, 221–236.

Harland, R. and Gerhart, J. (1997). Formation and function of Spemann's organizer. Annu. Rev. Cell Dev. Biol. *13*, 611–667.

Henrion, G., Renard, J.P., Chesné, P., Oudin, J.F., Maniey, D., Brunet, A., et al. (2000). Differential regulation of the translation and the stability of two maternal transcripts in preimplantation rabbit embryos. Mol. Reprod. Dev. *56*, 12–25.

Herrler, A., Krusche, C.A. and Henning, M.B. (1998). Insulin and insulin-like growth factor-I promote rabbit blastocyst development and prevent apoptosis. Biol. Reprod. *59*, 1302–1310.

Hopper, A.F. and Hart, N.H. (1985). *Foundations of Animal Development*. Oxford University Press, New York, NY, p. 286.

Islami, D., Chardonnens, D., Campana, A. and Bischof, P. (2001). Comparison of the effects of GnRH-I and GnRH-II on HCG synthesis and secretion by first trimester trophoblast. Mol. Hum. Reprod. *7*, 3–9.

Izadyar, F., Van Tol, H.T., Hage, W.G. and Bevers, M.M. (2000). Preimplantation bovine embryos express mRNA of growth hormone receptor and respond to growth hormone addition during *in vitro* development. Mol. Reprod. Dev. *57*, 247–255.

Johansson, S., Husman, B., Norstedt, G. and Andersson, G. (1989). Growth hormone regulates the rodent hepatic epidermal growth factor at a pretranslational level. J. Mol. Endocrinol. *3*, 113–120.

Kauma, S.V. (2000). Cytokines in implantation. J. Reprod. Fertil. *55*(Suppl.), 31–42.

Kawase, E., Hashimoto, K. and Pedersen, R.A. (2004). Autocrine and paracrine mechanisms regulating primordial germ cell proliferation. Mol. Reprod. Dev. *68*, 5–16.

Kim, C.H., Chae, H.D., Cheon, Y.P., Kang, B.M., Chang, Y.S. and Mok, J.E. (1999). The effect of epidermal growth factor on the preimplantation development, implantation and its receptor expression in mouse embryos. J. Obstet. Gynaecol. Res. *25*, 87–93.

Kimelman, D. and Bjornson, C. (2004). Vertebrate mesoderm induction. In: *Gastrulation: From Cells to Embryo*, C.D. Stern (ed.), Cold Spring Harbor Laboratory Press, Cold Spring Harbor, NY, pp. 363–372.

Kleine, B., Wolfahrt, S., Lotsch, M., Gantner, T. and Rossmanith, W.G. (2001). Expression of galanin in human placenta. Mol. Hum. Reprod. *7*, 379–385.

Knaut, H., Pelegri, F., Bohmann, K., Schwarz, H. and Nuesslein-Volhard, C. (2000). Zebrafish *vasa* RNA but not its protein is a component of the germ plasm and segregates asymmetrically before germline specification. J. Cell Biol. *149*, 875–888.

Knezevic, V. and Mackem, S. (2001). Activation of epiblast gene expression by the hypoblast layer in the prestreak chick embryo. Genesis *30*, 264–273.

Kofron, M., Demel, T., Xanthos, J., Lohr, J., Sun, B., Sive, H., et al. (1999). Mesoderm induction in *Xenopus* is a zygotic event regulated by maternal VegT via TGFbeta growth factors. Development *126*, 5759–5770.

Kostetskii, I., Yuan, S.Y., Kostetskaia, E., Linask, K.K., Blanchet, S., Seleiro, E., et al. (1998). Initial retinoid requirement for early avian development coincides with retinoid receptor coexpression in the precardiac fields and induction of normal cardiovascular development. Dev. Dyn. *213*, 188–198.

Kuroda, H., Wesely, O. and De Robertis, E.M. (2004). Neural induction in *Xenopus*: requirement for ectodermal and endomesodermal signals via Chordin, Noggin, beta-Catenin, and Cerberus. PloS Biol. *2*, e92.doi.

Langston, A.W., Thompson, J.R. and Gudas, L.J. (1997). Retinoic acid-responsive enhancers located 3′ of the *Hox A* and *Hox B* homeobox gene clusters. J. Biol. Chem. *272*, 2167–2175.

Laurent, M.N., Blitz, I.L., Hashimoto, C., Rothbacher, U. and Cho, K.W. (1997). The *Xenopus* homeobox gene *twin* mediates Wnt induction of *goosecoid* in establishment of Spemann's organizer. Development *124*, 4905–4919.

Lemaitre, J.-M., Bocquet, S. and Mechall, M. (2002). Competence to replicate in the unfertilized egg is conferred by Cdc6 during meiotic maturation. Nature *419*, 718–722.

Leunda-Casi, A., de Hertogh, R. and Pampfer, S. (2001). Control of trophectoderm differentiation by inner cell mass-derived growth factor-4 in mouse blastocysts and corrective effect of FGF-4 on high glucose-induced trophoblast disruption. Mol. Reprod. Dev. *60*, 38–46.

Lüer, K., Urban, J., Klämbt, C. and Technau, G.M. (1997). Induction of identified mesodermal cells by CNS midline progenitors in *Drosophila*. Development *124*, 2681–2690.

Malpel, S., Mendelsohn, C. and Cardoso, W.V. (2000). Regulation of retinoic acid signaling morphogenesis. Development *127*, 3057–3067.

Marshall, H., Morrison, A., Studer, M., Popperl, H. and Krumlauf, R. (1996). Retinoic acid and Hox genes. FASEB J. *10*, 969–978.

Möller, B., Rasmussen, C., Lindblom, B. and Olovsson, M. (2001). Expression of angiogenic growth factors VEGF, FGF-2, EGF and their receptors in normal human endometrium during the menstrual cycle. Mol. Hum. Reprod. *7*, 65–72.

Mohamed, O.A., Dufort, D. and Clarke, H.J. (2004). Expression and estradiol regulation of Wnt genes in the mouse blastocyst identify a candidate pathway for embryo-maternal signaling at implantation. Biol. Reprod. *71*, 417–424.

Mohan, M., Malayer, J.R., Geisert, R.D. and Morgan, G.L. (2001). Expression of retinol binding protein messenger RNA and retinoic acid receptors in preattachment bovine embryos. Mol. Reprod. Dev. *60*, 289–296.

Müller, W.A. (1996). *Developmental Biology*. Springer, New York, NY, p. 90.

Murray, A.W. and Kirshner, M.W. (1989). Cyclin synthesis drives the early embryonic cell cycle. Nature *339*, 275–280.

Niederreither, K., McCaffery, P., Dräger, U.C., Chambon, P., Dollé, P. (1997). Restricted expression and retinoic acid-induced downregulation of the retinaldehyde dehydrogenase type-2 (RALDH-2) gene during mouse development. Mech. Dev. *62*, 67–78.

Nonaka, K., Sasaki, Y., Yanagita, K., Matsumoto, T., Watanabe, Y. and Nakata, M. (1993). Intrauterine effect of dam on parental development of craniofacial complex of mouse embryo. J. Craniofac. Genet. Dev. Biol. *13*, 206–212. (abstract).

Ota, H., Igarashi, S., Oyama, N., Suzuki, Y. and Tanaka, T. (1999). Optimal levels of nitric oxide are crucial for implantation in mice. Reprod. Fertil. Dev. *11*, 183–188.

Paria, B.C., Ma, W., Tan, J., Raja, S., Das, S.K., Dey, S.K., et al. (2001). Cellular and molecular responses of the uterus to embryo implantation can be elicited by locally applied growth factors. Proc. Natl. Acad. Sci. U.S.A. *98*, 1047–1052.

Piko, I. and Clegg, K.B. (1982). Quantitative changes in total RNA, total poly(A), and ribosomes in early mouse embryos. Dev. Biol. *89*, 362–378.

Porterfield, S.P. and Hendry, L.B. (1998). Impact of PCBs (polychlorinated biphenyls) on thyroid hormone directed brain development. Toxicol. Ind. Health *14*, 103–120.

Poulain, M., Fürthauer, M., Thisse, B., Thisse, C. and Lepage, T. (2006). Nodal, FGF and BMP signaling. Development *133*, 2189–2200.

Rogers, C., Moody, S.A. and Casey, E. (2009). Neural induction and factors that stabilize a neural fate. Birth Defects Res. C Embryo Today *87*, 249–262.

Rout, U.K., Wang, J., Paria, B.C. and Armant, D.R. (2004). Alpha5beta1, alphaVbeta3 and the platelet-associated integrin alphaIIbbeta3 coordinately regulate adhesion and migration of differentiating mouse trophoblast cells. Dev. Biol. *268*, 135–151.

Salamonsen, L.A., Nie, G., Dimitriadis, E., Robb, L., Findlay, J.K. (2001). Genes involved in implantation. Reprod. Fertil. Dev. *13*, 41–49.

Sasaki, Y., Nomaka, K. and Nakata, M. (1995). The strain effect of dam on intrauterine incisal growth in mouse fetuses. J. Craniofac. Genet. Dev. Biol. *15*, 140–145. (abstract).

Sawai, K., Matsuzaki, N., Okada, T., Shimoya, K., Koyama, M. and Azuma, C., et al. (1997). Human decidual cell biosynthesis of leukemia inhibitory factor: regulation by decidual cytokines and steroid hormones. Biol. Reprod. *56*, 1274–1280.

Schier, A.F. (2003). Nodal signaling in vertebrate development. Annu. Rev. Cell Dev. Biol. *19*, 589–621.

Schneider, S., Steinbeisser, H., Warga, R.M. and Hausen, P. (1996). Beta-catenin translocation into nuclei demarcates the dorsalizing centers in frog and fish embryos. Mech. Dev. *57*, 191–198.

Sengoku, K., Takuma, N., Horikawa, M., Tsuchiya, K., Komori, H., Sharifa, D., et al. (2001). Requirement of nitric oxide for murine oocyte maturation, embryo development, and trophoblast outgrowth *in vitro*. Mol. Reprod. Dev. *58*, 262–268.

Shmukler, Y.B., Buznikov, G.A. and Whitaker, M.J. (1999). Action of serotonin antagonists on cytoplasmic calcium levels in early embryos of sea urchin *Lytechinus pictus*. Int. J. Dev. Biol. *43*, 179–182.

Spencer, T.E., Johnson, G.A., Bazer, F.W. and Burghardt, R.C. (2004). Implantation mechanisms: insights from the sheep. Reproduction *128*, 657–668.

Stebler, J., Spieler, D., Slanchev, K., Molyneaux, K.A., Richter, U., Cojocaru, V., et al. (2004). Primordial germ cell migration in the chick and mouse embryo: the role of the chemokine SDF-1/CHCL12. Dev. Biol. *272*, 351–361.

Stennard, F., Carnac, G. and Gurdon, J.B. (1996). The *Xenopus* T-box gene, antipodean, encodes a vegetally localised maternal mRNA and can trigger mesoderm formation. Development *122*, 4179–4188.

Stern, C.D. (2004). Neural induction. In: *Gastrulation*, C.D. Stern (ed.), Cold Spring Harbor Laboratory Press, Cold Spring Harbor, NY, p. 427.

Stern, C.D. (2005). Neural induction: old problem, new findings, yet more questions. Development *132*, 2007–2021.

Tessier, C., Deb, S., Prigent-Tessier, A., Ferguson-Gottschall, S., Gibori, G.B., Shiu, R.P., et al. (2000). Estrogen receptors α and β in rat decidua cells: cell-specific expression and differential regulation by steroid hormones and prolactin. Endocrinology *141*, 3842–3851.

Tierney, A.J. (1996). Evolutionary implications of neural circuit structure and function. Behav. Processes *35*, 173–182.

Tomancak, P., Guichet, A., Zavorsky, P. and Ephrussi, A. (1998). Oocyte polarity depends on regulation of *gurken* by Vasa. Development *125*, 1723–1733.

Tres, L.L., Rosselot, C. and Kierszenbaum, A.L. (2004). Primordial germ cells: what does it take to be alive? Mol. Reprod. Dev. *68*, 1–4.

Tsunekawa, N., Naito, M., Sakai, Y., Nishida, T. and Noce, T. (2000). Isolation of chicken vasa homolog gene and tracing the origin of primordial germ cells. Development *127*, 2741–2750.

van de Pavert, S.A., Boerjan, M.L., Stroband, H.W., Taverne, M.A., van den Hurk, R. (2001). Uterine-embryonic interaction in pig: activin, follistatin, and activin receptor II in uterus and embryo during early gestation. Mol. Reprod. Dev. *59*, 390–399.

Vanzo, N.F. and Ephrussi, A. (2002). Oskar anchoring restricts pole plasm formation to the posterior of the *Drosophila* oocyte. Development *129*, 3705–3714.

Weaver, C. and Kimelman, D. (2004). Move it or lose it: axis specification in *Xenopus*. Development *131*, 3491–3499.

Weidinger, G., Stebler, J., Slanchev, K., Dumstrei, K., Wise, C., Lovell-Badge, R., et al. (2003). Dead end, a novel vertebrate germ plasm component, is required for zebrafish primordial germ cell migration and survival. Curr. Biol. *13*, 1429–1434.

Wilson, S.I. and Edlund, T. (2001). Neural induction: toward a unifying mechanism. Nat. Neurosci. *4*, 1161–1168.

Wilson, S.I., Rydström, A., Trimborn, T., Willert, K., Nusse, R., Jessell, T.M., et al. (2001). The status of Wnt signalling regulates neural and epidermal fates in the chick embryo. Nature *411*, 325–330.

Wolfahrt, S., Kleine, B., Jarry, H., Rossmanith, W.G. (2001). Endogenous regulation of the GnRH receptor by GnRH in the human placenta. Mol. Hum. Reprod. *7*, 89–95.

Wollenhaupt, K., Einspanier, R., Gabler, C., Schneider, F., Kanitz, W. and Brüssow, K.P. (1999). Identification of the EGF/EGF-R system in the oviduct and endometrium of pigs in early stages of pregnancy and early conceptus. Exp. Clin. Endocrinol. Diabetes *107*, 530–538.

Wolpert, L., Beddington, R., Lawrence, P. and Jessell, T.M. (1998). *Principles of Development*. Current Biology Ltd. and Oxford University Press, Oxford, p. 259.

Wright, N., de Lera, T.L., García-Moruja, C., Lillo, R., García-Sánchez, F., Caruz, A., et al. (2003). Transforming growth factor-beta 1 down-regulates expression of chemokine stromal cell-derived factor-1: functional consequences in cell migration and adhesion. Blood *102*, 1978–1984.

Xanthos, J.B., Kofron, M. and Heasman, J. (2001). Maternal VegT is the initiator of a molecular network specifying endoderm in *Xenopus laevis*. Development *128*, 167–180.

Yang, J., Tan, C., Darken, R.S., Wilson, P.A. and Klein, P.S. (2002). Beta-catenin/Tcf-regulated transcription prior to midblastula transition. Development *129*, 5743–5752.

Yasuda, Y., Konishi, H., Matsuo, T., Kihara, T. and Tanimura, T. (1989). Aberrant differentiation of neuroepithelial cells in developing mouse brains subsequent to retinoic acid exposure *in utero*. Am. J. Anat. *186*, 271–284.

Yoon, C., Kawamaki, K. and Hopkins, N. (1997). Zebrafish vasa homologue RNA is localized to the cleavage planes of 2- and 4-cell stage embryos and is expressed in the primordial germ cells. Development *124*, 3157–3165.

Yue, Z.P., Yang, Z.M., Li, S.J., Wang, H.B. and Harper, M.J. (2000). Epidermal growth factor family in rhesus monkey uterus during the menstrual cycle and early pregnancy. Mol. Reprod. Dev. *55*, 164–174.

Zhang, J. and King, M.I. (1996). *Xenopus* VegT RNA is localized to the vegetal cortex during oogenesis and encodes a novel T-box transcription factor involved in mesoderm patterning. Development *122*, 4119–4129.

Zhang, J., Houston, D.W., King, M.L., Payne, C., Wylie, C. and Heasman, J. (1998). The role of maternal VegT in establishing the primary germ cell-layers in *Xenopus* embryos. Cell *94*, 515–524.

Zhou, L., Xiao, H. and Nambu, J.R. (1997). CNS midline to mesoderm signaling in *Drosophila*. Mech. Dev. *67*, 59–68.

5 Neural Control of Postphylotypic Development

Perhaps the most interesting thing about having a hormonal regulation of development is that development comes under the control of the central nervous system.

Nijhout (2003)

Postphylotypic development is defined here in the broadest sense to comprise all the morphological, physiological, and behavioral transformations that characterize individual development from the phylotypic stage to adulthood.

Early stages of development up to the formation of gastrula and neurula may considerably vary among species and other taxa of a phylum. But, after that point, all of them converge to a morphologically common phylotypic stage (extended germ band stage in insects and tail bud stage in vertebrates).

Before the phylotypic stage, only a few inductive events (establishment of body axes, formation of mesoderm, gastrulation, and induction of the nervous system) take place based on global interactions, and all of them are epigenetically determined by parental epigenetic information (cytoplasmic factors). At that juncture, when that epigenetic information is exhausted, multiple inductions take place all over the embryo, leading to formation of organs, organ systems, and other metazoan structures.

What is the origin of the vast amount of information necessary for the postphylotypic development when the embryo runs out of the parentally provided cytoplasmic factors at the phylotypic stage? Temporally, this coincides with the development of the operational central nervous system (CNS) and the related integrated control system (ICS) at the phylotypic stage.

Neuroblasts in the neural tube are the first fully differentiated cells to appear in *Drosophila* embryos, and the nervous system is the first organ system to develop and function in metazoan embryos. Temporally, it precedes or coincides with the beginning of embryonic inductions and organogenesis. Formation of somites, limb buds, gonads, kidneys, lungs, and heart takes place only after, and in relation to, the formation of the CNS (Hall, 1998).

Before presenting the representative evidence on the role of the CNS as the originator of inductive signals for metazoan organogenesis and morphogenesis, I will briefly review the postphylotypic development in invertebrates and vertebrates, focusing on the phenomenon of metamorphosis, a widespread mode of individual development in invertebrates and amphibians that represents an impressive manifestation of the control of the postphylotypic development by the CNS as controller of the ICS. As a form of individual development, metamorphosis is more accessible to observation, hence suitable for investigating the mechanisms of individual development.

Epigenetic Principles of Evolution. DOI: 10.1016/B978-0-12-415831-3.00005-7

It offers a simpler picture of the flow of information from the CNS for the postphylotypic development in metazoans.

Neural Control of Metamorphosis

All types of metamorphosis in invertebrates and vertebrates, however different they might appear, have in common apoptosis, programmed death of cells, as a special morphogenetic mechanism for sculpting their structure by getting rid of parts and organs of previous stages and for developing the adult morphology. Hence, a glimpse of the mechanisms of control of apoptotic processes is necessary for understanding the nature of metamorphosis.

Apoptosis in Invertebrates

Apoptosis, as a form of programmed cell death, was described first by C. Vogt (Seipp et al., 2001) in 1842. It is essential for morphogenesis and organogenesis in all metazoans, invertebrates and vertebrates alike. While metazoan cells might have retained the constitutive mechanism of apoptosis, in metazoans apoptosis is primarily a regulated phenomenon, indispensable during the embryonic development and all the later stages of life.

The proximate agents of the apoptotic cell death are a number of proteases (five proteases in *Drosophila*), known under the common name of *caspases*. The process starts with activation of "initiator" caspases (caspases 2, 8, and 9), followed by activation of effector caspases (caspases 3, 6, and 7), which act proteolytically on cell proteins and enzymes responsible for cell metabolism and reproduction. Initiator caspases cause the permeabilization of mitochondria, which simply amplify the caspase activity but have no regulatory function in apoptosis (Lee and Baehrecke, 2001).

In *Drosophila*, products of three genes, *rpr* (*reaper*), *grim* (genes associated with retinoid–IFN (interferon)-induced mortality), and *hid* (*head involution defective*), are known to act as signals for activating the downstream mechanism of apoptosis, and two other proteins (dIAP1 and dIAP2) act as inhibitors of apoptosis (Lassus et al., 2002). The expression of these preapoptotic genes in *Drosophila* always follows the ecdysone pulses, mediated by specific nuclear receptors, ecdysone receptors (EcRs). These pulses induce a few early-response genes, such as *Broad-Complex* (*BR-C*), *E74*, and *E75*, whose products (transcription factors), in turn, activate more than 100 late-response genes (Jiang et al., 1997). During the *Drosophila* metamorphosis, ecdysone pulses

> *direct the destruction of obsolete larval tissues and their replacement by tissues and structures that form the adult fly ... via the precise stage- and tissue-specific regulation of key death effector genes.*
>
> *Draizen et al. (1999)*

While ecdysone controls apoptosis by regulating expression of preapoptotic genes (Draizen et al., 1999; Namba et al., 1997), it is a well-established fact that the ecdysone pulses and ecdysone synthesis in insects are under strict control of the CNS.

The programmed cell death indicates that somehow the embryo identifies the excess cells and corrects the state by apoptotically eliminating them. Furthermore, metazoans seem to control not only the number of their cells, but by neural mechanisms they also control the size of their body (see later on the control of body size) by making the necessary adjustments in the number or the size of their cells. (Cells of pentaploid salamanders have five times the volume of the cells of haploid species, but both of them have approximately the same size because the pentaploid species have five times fewer cells in their body.)

What estimates the number of cells and perceives excess cells that must be eliminated? What sends the apoptotic message to the organs and tissues that will undergo apoptosis? It is clear that it is not individual cells but a supracellular entity capable of monitoring the status of the embryonic organism as a whole.

Neural Control of Metamorphosis in Cnidarians

After fertilization, eggs of *Hydractinia echinata* start the cleavage. The embryo has only two germ layers. Within 2–3 days from fertilization, a ciliated spindle-shaped planula develops (Walther et al., 1996), which stops cell proliferation and differentiation but is competent of starting metamorphosis (Plickert et al., 1988). The ~1 mm planula, consisting of only ~10,000 cells, settles before starting metamorphosis.

The environmental cue inducing the larva to enter metamorphosis and transform into a primary polyp is a chemical signal released by a bacterium, but metamorphosis can be induced by a variety of other chemical agents. The bacterial signal is received by neurosecretory cells in the anterior part of the planula. The external cue triggers secretion of an internal signal, which is transmitted all the way to the posterior end and "synchronizes the events of metamorphosis" (Schwoerer-Böhning et al., 1990). *Hydractinia echinata* starts cell proliferation by the ninth hour after induction of metamorphosis, starting from the middle gastric region (Plickert et al., 1988).

In 1981, a "morphogenetic peptide" was isolated from *Hydra attenuata* and *Anthopleura elegantissima* that was characterized as "head-inducing morphogen" (Schaller and Bodenmuller, 1981). This is a neuropeptide synthesized by, and stored in, the secretory granules of the cnidarian neurons. Another isolated peptide is the neuropeptide metamorphosin A (MMA), which induces metamorphosis of the treated intact larvae.

> *"MMA" is among the most potent peptides known to have a function in controlling animal development.*
>
> *Leitz et al. (1994)*

The subsequent discovery of the role of GLWamide neuropeptides in metamorphosis of *H. echinata* led to the notion of the existence in these lower invertebrates of a neuropeptide signal system controlling metamorphosis (Schmich et al., 1998). In *H. echinata*, sensory neurons secrete two antagonistically acting neuropeptides: GLWamides (stimulators of metamorphosis) and RFamides (inhibitors of metamorphosis) in adaptive responses to environmental conditions (Katsukura et al., 2003a,b).

Hydra neurons secrete another neuropeptide, the "head activator," which is a signal molecule (growth hormone, GH) (Schaller, 1976a). The neuropeptide determines the transformation of the interstitial cells into nerve cells (Schaller, 1976b). The head activator binds to a receptor protein, head-activator binding protein (HAB), which is present both as a secreted protein and as a membrane-anchored receptor (Hampe et al., 1999).

Based on the experimental observation that serotonin stimulates metamorphosis in the planulae of the cnidarian *Eudendrium racemosum*, which normally does not metamorphose, as well as in other evidence, it is concluded that serotonin is involved in the perception of the metamorphosis-triggering environmental cue, through serotonin containing sensory neurons in hydroid planulae of different species (Zega et al., 2007).

Thus, all of the signals for metamorphosis in Hydra come from the neural net, and all of them are neuropeptides secreted by secretory neurons of the neural net in adaptive responses to environmental cues.

Neural Control of Metamorphosis in Molluscs

A metamorphosis-inducing role is demonstrated for several neurotransmitters in molluscs.

The free-swimming veliger (a stage in mollusc metamorphosis) larvae have a functioning nervous system with the rudiments of almost all the adult ganglia and an apical sensory organ (ASO) (Lacalli, 1994), containing, among other cells, serotonergic secretory neurons. These neurons respond to environmental cues by releasing neurotransmitters, such as serotonin and/or catecholamines (epinephrine, norepinephrine (NE), and dopamine (DA)), γ-amino butyric acid (GABA), all of them with proven metamorphosis-inducing action in molluscs.

Molluscan larvae exposed to these neurotransmitters are induced to metamorphose. So, for example, application of serotonin induces metamorphosis in 80–100% of the competent larvae of coenogastropod *Ilyanassa obsoleta* (Leise et al., 2001). Induction of metamorphosis by high levels of NE and DA in metamorphically competent *Phestilla* larvae (Pires et al., 1997) and by L-3,4-dihydroxyphenylalanine in *Crepidula fornicata* (Pires et al., 2000), as well as inhibition of metamorphosis after depletion of these neurotransmitters, indicates that neurotransmitters participate in the control of gastropod development (Pires et al., 1997).

In summary, it may be said that metamorphosis in molluscs is under direct control of neurotransmitters released by secretory neurons.

Neural Control of Metamorphosis in Insects

Two major hormones, ecdysone (Ec) and juvenile hormone (JH), control the complex processes of metamorphosis in insects. Extensive studies carried out especially during the past two to three decades show that hormonal response in insects is under strict cerebral control.

Ecdysone is a steroid hormone secreted by prothoracic gland that, in its active form, stimulates metamorphosis and regulates molting in insects. In the tobacco hornworm,

Manduca sexta, these hormones control metamorphosis, especially processes of muscle degeneration and programmed neuron death (Weeks and Truman 1986). In *Drosophila*, ecdysone

> *induces the histolysis of nearly all of the larval tissues and differentiation and morphogenesis of the structures composing the adult fly.*
>
> *Baehrecke (1996)*

These functions are mediated by its heteromeric receptor, EcR, which is implicated in the reorganization of imaginal and larval tissues at the onset of metamorphosis (Li and Bender, 2000). Let us remember that secretion of ecdysone by the prothoracic gland is regulated by a brain neuropeptide, the prothoracicotropic hormone (PTTH). Altered neuropeptide synthesis leads to low levels of ecdysteroids, and delay or blockage of metamorphosis in *Drosophila* mutants (Zitnan et al., 1993). In the butterfly *Precis coenia*, the neuropeptide bombyxin acts together with ecdysone to stimulate growth of the wing imaginal discs.

> *The level of bombyxin in the hemolymph is modulated by the brain in response to variation in nutrition and is part of the mechanism that coordinates the growth of internal organs with overall somatic growth.*
>
> *Nijhout and Grunert (2002)*

The process of cuticle shedding during metamorphosis requires strict temporal coordination, transformation, and loosening of the cuticle and a stereotyped sequence of pre-ecdysis and ecdysis behaviors. The specific ecdysis behavior is regulated by the hormone ecdysis-triggering hormone (ETH) released by the endocrine Inka cells, under stimulation of the neurohormone eclosion hormone (EH) released by EH neurons (Figure 6.1). The absence of ETH in *Drosophila* null mutants for the respective hormone leads to "incomplete ecdysis and 98% mortality at the transition from the first to second larval instar." When those mutants are treated with a synthetic ETH, they start metamorphosis (Yoonseong et al., 2002). ETH and crustacean cardioactive peptide (CCAP) secreted by CCAP neurons elicit the first two motor behaviors, the pre-ecdysis and ecdysis behaviors (Gammie and Truman, 1997). In response to ETH, specific neurons in the brain secrete the EH which, in turn, increases secretion of the neuropeptide CCAP by CCAP neurons.

The sequential performance of the two behaviors arises from one modulator activating the first behavior and also initiating the release of the second modulator. The second modulator then turns off the first behavior while activating the second (Gammie and Truman, 1997).

The endocrine Inka cells of the insect epitracheal glands, at the end of each developmental stage, respectively secrete the pre-ecdysis- and ecdysis-triggering hormones (PETH and ETH), to which each abdominal ganglion responds by starting, within a few minutes, pre-ecdysis II and I, respectively.

> *The initiation of preecdysis and the transition to ecdysis are regulated by stimulatory and inhibitory factors released within the central nervous system after the initial actions of PETH and ETH.*
>
> *Zitnan and Adams (2000)*

These hormones determine the sequential stereotyped behaviors of cuticle shedding, but ETH secretion itself is controlled by the eclosion neurohormone, EH (Kingan et al., 2001), secreted by two pairs of ventromedial (VM) brain neurons. EH released by the neurosecretory VM cells in the brain stimulates the release of ETH by Inka cells, which, via a positive-feedback loop, stimulates the neurosecretory cells to secrete more EH. This causes activation of peptidergic neurons that release CCAP (Ewer et al., 1997) (Figure 5.1).

The idea that secretion of ETH by Inka cells is under control of the EH has been challenged recently. Although EH stimulates secretion of ETH by Inka cells, these cells secrete ETH even in the absence of EH (Clark et al., 2004). However, the central neural control of the initiation of the ETH secretion may be exerted in an alternative way. Indeed, it is observed that insects cannot initiate ecdysis behavior if their brain is disconnected, even when EH is injected. Novicki and Weeks (1996) assume:

> *Neural input from the brain to the abdominal ganglia is necessary to initiate ecdysis.*
> *Novicki and Weeks (1996)*

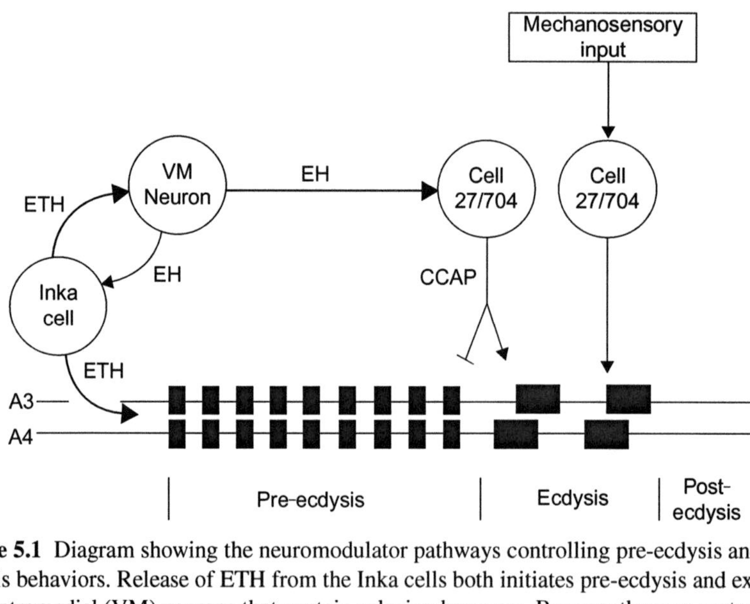

Figure 5.1 Diagram showing the neuromodulator pathways controlling pre-ecdysis and ecdysis behaviors. Release of ETH from the Inka cells both initiates pre-ecdysis and excites the ventromedial (VM) neurons that contain eclosion hormone. Because they are part of a positive-feedback loop, the Inka cells and VM neurons release almost all of their peptide stores. EH release within the CNS triggers cGMP upregulation in the Cell 27/704 group, causing the central and peripheral (not shown) release of CCAP. Centrally released CCAP both activates the ecdysis motor program and terminates pre-ecdysis. Sensory input (possibly from bristle hairs deformed by the pressure of the old cuticle) may maintain excitation of the Cell 27/704 group to insure CCAP release and the continuation of ecdysis until the cuticle is shed. Removal of the cuticle eliminates the sensory input, resulting in the cessation of CCAP release and of ecdysis behavior.
Source: From Gammie and Truman (1997).

In the tobacco hawkmoth, *Manduca sexta*, wing expansion and cuticle tanning take place in the posteclosion period. Both of these processes are neurally determined by the secretion of the neurohormone bursicon in B_{AG} neurons (Luan et al., 2006; Dai et al., 2007).

Neural Control of Metamorphosis in Vertebrates

Neural Control of Metamorphosis in Amphibians

Specific neural circuits in the tadpole's CNS process the input of internal stimuli, such as growth beyond a threshold and environmental cues, such as photoperiod, rise of the water temperature, and lowering of the water level (Denver, 1997a). By processing the input of those stimuli, neurons in the nonhypothalamic brain

> *release catecholamines at their synaptic contacts with the cell bodies and dendrites of TRH (thyrotropin-releasing hormone—N.C.)-containing neurons.*
>
> *Strand (1999)*

in the hypothalamus. In response to these neural signals, hypothalamic neurons secrete thyrotropin-releasing hormone (TRH), which stimulates the tadpole pituitary to secrete the thyroid-stimulating hormone (TSH), which in turn regulates physiological levels of the thyroid hormones (THs), triiodthyronine (T3) and thyroxine (T4), secreted by the thyroid gland.

TH is responsible for initiating all of the physiological manifestations of metamorphosis (for example, morphogenesis, cell death, body restructuring) (Tata, 1998).

> *TH receptors mediate both early and late developmental programs of metamorphosis as diverse as growth in the brain, limb buds, nose and Meckel's cartilage, remodeling of the intestine, and death and resorption of the gills and tail.*
>
> *Tata (1998)*

The functional versatility of the TH during amphibian morphogenesis derives from the crucial role its receptor (TR) plays in gene expression by inducing chromatin remodeling via histone acetylation/deacetylation (Li, 1999; Sachs and Shi, 2000).

Mediators of the action of TH in metamorphosis are matrix metalloproteinases (MMPs) (Damjanovski et al., 2000) that induce remodeling of the extracellular matrix via integrins, as well as their membrane receptors, which in turn send signals for expression of specific genes. These signals make possible the apoptotic remodeling of the intestine during *Xenopus laevis* metamorphosis (Shi et al., 1998). The signal cascade that starts in the tadpole brain looks as follows:

Catecholamines in the nonhypothalamic brain → hypothalamic TRH → pituitary TSH → TH → MMPs → integrins → signals for gene expression → apoptotic remodeling of the *X. laevis* intestine.

T3 is the most active form of THs, which, by binding its receptor, forms the active complex T3/TR that can recruit one of the chromatin-remodeling enzymes. This enables the receptor to specifically bind its DNA response element in the target gene in the form of monomers, homodimers, or even heterodimers with the retinoic acid (RA). This is known as the *primary gene expression response*. Then, during the *secondary gene expression response*, the products of gene expression act on a downstream set of genes. The differential expression of genes induced by this signal cascade during metamorphosis enables the tadpole to specify differential developmental responses, from cell proliferation and cell differentiation to the programmed cell death.

Transcription factors, cellular enzymes, a cytoskeletal element, and secreted signaling molecules of four general classes of genes are activated by the T3 in *Xenopus* tadpoles. It is estimated that TH alone upregulates ~35 genes and downregulates ~10 genes involved in the tail degeneration program of *X. laevis* (Denver, 1997b).

Postphylotypic Development in Vertebrates

The Embryonic CNS Controls the Postphylotypic Development in Vertebrates—Empirical Evidence

Termination of the function of parental cytoplasmic factors at the phylotypic stage coincides with the beginning of organogenesis, when the embryo enters the phase of accelerated development requiring investment of ever-increasing amounts of epigenetic information. Where does the enormous information for erecting the extraordinarily complex metazoan structure come from?

As noted earlier, the most important event coincident with the termination of the regulatory function of the parental cytoplasmic factors, which might be relevant to the above question, is probably the development of the operational CNS. However, to prove a CNS role in controlling the postphylotypic development is far from an easy task. Out of many questions arising in such an attempt, the most dismaying is the following: How could the CNS know *what to do* and *how to do it* during the sequential stages of individual development?

My approach to this issue will be empirical, to trace the source of information for organogenesis by identifying the origin of inductive signals rather than trying to argue theoretically the possibility that the CNS may be in possession of, or can generate, the epigenetic information necessary for individual development.

In Chapter 1, I presented evidence on the role of the CNS in maintaining and restoring homeostatic/physiological parameters and described the CNS control of a few physiological, behavioral, and morphological traits; I have also argued and substantiated the idea that an ICS, with the CNS as its controller, is operational in metazoans. In this chapter, I will expose a representative portion of available evidence on the CNS origin of signals for the development of tissues and organs in the course of the postphylotypic development. Despite the fact that the possible role of the CNS in embryonic development has not been an authentic object of biological research, and almost no experiments designed to investigate that possibility have ever been carried out, surprisingly, the evidence for substantiating that role seems to be adequate.

Somitogenesis

Formation of somites, metameric structures of the vertebrate embryo, represents a fundamental aspect of the vertebrate body segmentation. Somites arise from the presomitic mesoderm (PSM) and develop sequentially in rostral–caudal order. The development of somites is one of the earliest events in the postphylotypic development of vertebrates.

In zebrafish, soon after the formation of the neural tube, a localized expression of the genes for the hypothalamic GH-releasing hormone (GHRH) and pituitary adenylate cyclase-activating polypeptide (PACAP) occurs in the neural tube and the cerebellum. It has been proposed that the temporal and spatial expression of those genes suggests that:

> *these hormones may modulate patterning during development.*
>
> *Krueckl et al. (2002)*

In *X. laevis*, segmentation begins in the PSM with rhythmic expression of the gene for Mesp-like bHlH protein, also known as *thylacine1*, in a pattern that determines both segment boundaries and polarity (Moreno and Kintner, 2004). Expression of myogenic *bHlH* (basic helix–loop–helix) genes is induced by signaling proteins of the Wnt family, originating in the dorsal regions of the neural tube, and sonic hedgehog (Shh) secreted from both the neural tube floor plate and the notochord (Fan and Tessier-Lavigne, 1994) (Figure 5.2).

A combination of the above factors is capable of inducing myogenesis in somites *in vitro* (Munsterberg et al., 1995). Besides thylacine1, components of the Notch signaling pathway (expression of Notch is closely related to the morphogenetic processes in the CNS) are expressed in a rhythmic pattern (Kim et al., 2000), thus determining the identity of the incipient PSM segments. The inducer of the expression of the segmentation bHlH gene, thylacine1, is the hormone RA (Moreno and Kintner, 2004). This fact adds another link to the causal chain (originating in the CNS) of segmentation signaling in the PSM.

Expression of *β-catenin mRNA* in somites is regulated by positive and negative signals (BMP4, Shh, and Wnt/Wnt-3) from the neural tube, notochord, and lateral plate mesoderm (Schmidt et al., 2000). Within somites themselves, myogenesis, formation of muscles, is also induced by neural tube signals, probably directly by Wnt-1 (Stern et al., 1995) or via growth factors, basic fibroblast growth factor (bFGF), transforming growth factor-β1 (TGF-β1), and dorsalin (ds/1) (Stern et al., 1997). Neural tube signals, such as Wnt-1, may also induce expression of Noggin, a BMP antagonist, which may induce somite myogenesis by allowing expression of MyoD and Myf5 in somite cells (Reshef et al., 1998). Expression of all of the investigated MyoDs (Xu et al., 2000), including bHlH proteins, esr9 and esr10, in *Xenopus* somites (Li et al., 2003), occurs rhythmically and is rhythmically activated by the Notch signaling pathway (Li et al., 2003).

Together, these facts suggest that a segmentation clock of the neural tube/notochord axis regulates rhythmic expression of PSM segments. Adequate evidence shows that the segmentation clock consists of the Wnt/β-catenin signaling (Aulehla

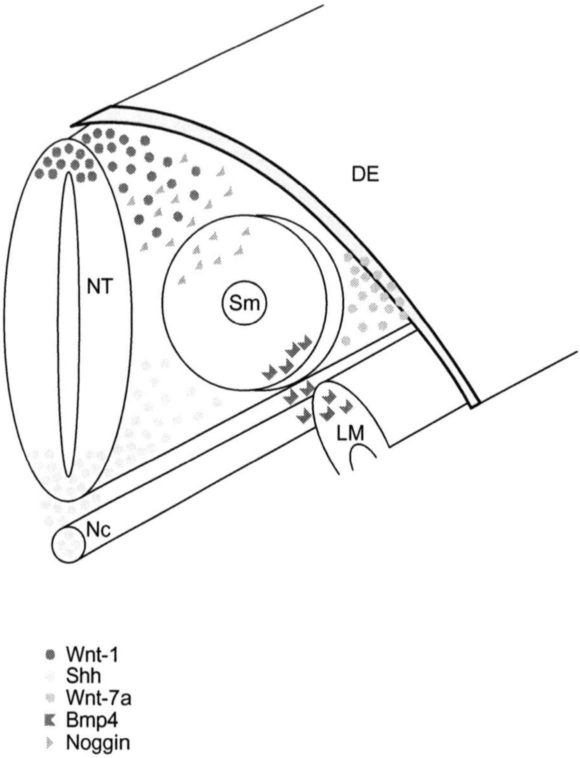

Figure 5.2 A simplified scheme of signaling molecules in newly formed epithelial somite. Shh (gray dots), produced by notochord (Nc) and floor plate, acts on the ventral domain of newly formed epithelial somites, inducing sclerotome, and also on the dorsomedial domain, inducing medial dermomyotome. Wnt-1 (black dots), produced by dorsal neural tube (NT), acts (with Shh) on the dorsomedial domain of newly formed somites (Sm), where *Myf5* expression is observed soon after epaxial progenitors are specified. Wnt-7a (semilunar dots), produced by dorsal ectoderm (DE), acts on the dorsolateral domain, where hypaxial progenitors are specified. BMP4 (black polygons), produced by lateral mesoderm (LM), prevents *MyoD* activation and early differentiation in the lateral domain of somites. Its action is counteracted by direct binding of Noggin (gray triangles) produced by dorsal neural tube. *Source*: From Cossu and Borello (1999).

et al., 2003) and Axin2 protein, a negative regulator of the Wnt cascade (Aulehla and Herrmann, 2004).

Development of the Dermomyotome and Sclerotome

Ablation of the neural tube leads to the cell death and degeneration of the medial dermomyotome, indicating that it is responsible for the development of dermomyotome. The neural tube is also responsible for the formation of dermis (Figure 5.3).

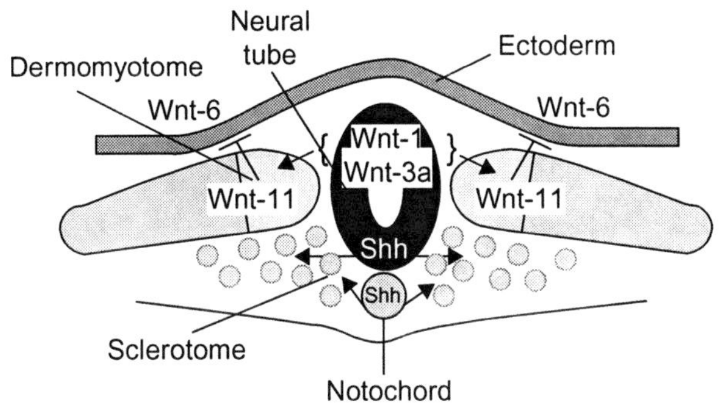

Figure 5.3 Simplified diagram of the neural signals for the development of dermis (Wnt-1 and Wnt-3a) from the central dermomyotome cells and of the sclerotome Shh from ventral neural tube and notochord.

Wnt-1 signals from the dorsal neural tube induce expression of Wnt-11 in the central dermomyotome. Expression of Wnt-11, in turn, induces deepithelialization of cells of the central region of the dermomyotome and downregulates expression of Wnt-6 in the neighboring ectoderm, thus allowing emigration of dermomyotome cells (precursors of dermis) (Olivera-Martinez et al., 2001; Linker et al., 2005; Geetha-Loganathan et al., 2006).

Neural Control of Myogenesis

Control of Myogenesis in Insects

During metamorphosis, insects develop pools of myoblasts in muscle anlagen, which are closely associated with local nerves. Local innervation is essential for the development of abdominal muscles of insects in general. In *Drosophila*, for instance, the patterning of segmental muscles is determined not by myoblasts involved in muscle development but by the motor and sensory neurons innervating these muscles (Lawrence and Johnston, 1986). It is observed that abdominal myoblasts arrive at the target sites by their association with, and migration along, the nerves (Currie and Bate, 1991).

Local denervation causes a decrease in the number of muscle nuclei, suggesting that innervation is necessary for regulating myoblast division and survival. However, denervated muscles also undergo differentiation and establish basic muscle patterns, a fact that may be explained by a central regulation via the neurohormonal pathways.

There is at least one case when the role of the innervation in the development of muscle is of the all-or-none type. In *Drosophila*, male flies have an additional male-specific muscle (msm), or the muscle of Lawrence, in the dorsal part of the abdominal segment 5 (A5), which results from aggregation of three to five adjacent muscles. In distinction from abdominal muscles, the muscle of Lawrence does not form at all in the absence of innervation (Currie and Bate, 1995).

In denervated legs of *Manduca sexta*, myoblasts cannot proliferate and move to the proper locations for developing leg muscles, what, in a considerable number of cases, leads to the development of legs lacking muscles (Consoulas and Levine, 1997). In the same insect, denervation of respecified muscles during metamorphosis leads to dedifferentiation of their cells, loss of muscle fibers, and, at times, muscle death (Bayline et al., 1998). Denervation of the anlage of the longitudinal flight muscle in this insect prevents the development of that muscle, due to the failure of myoblasts to accumulate in the muscle anlage (Bayline et al., 2001).

It is also noteworthy that null mutations of the *single-minded* gene in *Drosophila* are associated with abnormal development of the ventral oblique muscles above the CNS. Investigators found that this defect

> *is not due to the absence of single-minded expression in muscle precursor cells and likely results from an influence of the central nervous system on ventral muscle development.*
>
> *Lewis and Crews (1994)*

In the tobacco hawkmoth, *Manduca sexta*, an LIM (limit)-only protein (DALP) induces elimination of excess myoblasts of the intersegmental muscles (ISMs) at the end of metamorphosis. Again, the causal chain, the signal cascade responsible for the death of ISMs starts in the insect's CNS, as is suggested by the fact that the death of ISMs is triggered by a drop in the level of the molting hormone, 20-hydroxyecdysone (Hu et al., 1999), which, as is well known, is under strict cerebral control.

Maturation of oviposition properties of the longitudinal muscle in female locusts implies acquisition of the ability by the muscle to tolerate large extensions of more than 8 mm. Experimental inactivation of corpora allata (CA) inhibits maturation of oviposition properties of the longitudinal muscle. This suggests that JH, whose synthesis in CA is triggered by brain signals, is necessary for maturation of oviposition properties of that muscle. Indeed, administration of JH in female locusts with inactivated CA reverses the situation by enabling maturation of oviposition properties in the longitudinal muscle (Rose et al., 2001).

The indirect flight muscles (IFMs) of *Drosophila* consist of two muscle groups, the dorsal longitudinal muscles (DLMs) and dorsoventral muscles (DVMs). The development of IFMs takes place in two stages: the first stage of generation of the pool of myoblasts, which is nerve independent, and the second nerve-dependent stage, when motoneurons determine the size of the myoblast pool by establishing a critical threshold of the myoblast pool. However, differences are observed in the effect of innervation between DVMs and DLMs (Figure 5.4).

Denervation of the DVMs leads to a myoblast pool that falls below the threshold so that neither fusion takes place nor mature muscle fibers develop (Fernandes and Keshishian, 2005).

Unilateral denervation of these muscles leads to two major effects. First, there is a significant decline in the proliferation rate of myoblasts and smaller muscle size in denervated hemisegments. Second, it prevents myoblast patterning, leading to failure to form muscle anlagen and muscles.

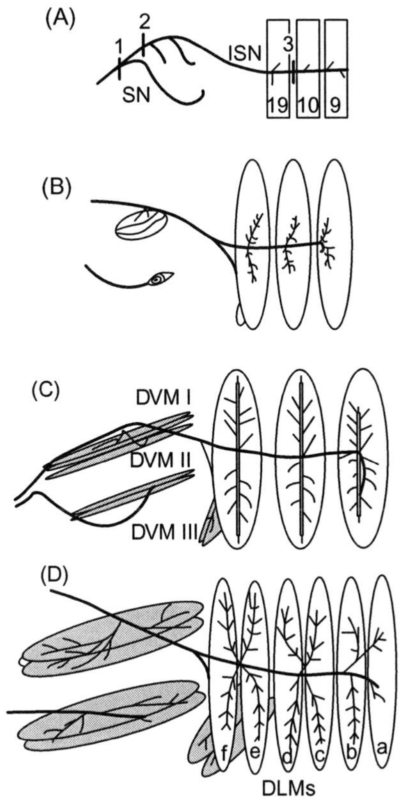

Figure 5.4 Neuromuscular development of the indirect flight muscles (IFMs) during the first 24 h of pupal development. The IFMs consist of the dorsal longitudinal muscles (DLMs) and the dorsoventral muscles (DVMs, I, II, III). (A) At 8 h after pupa formation (APF). Three persistent larval muscles (9, 10, 19′) give rise to the DLMs. The larval nerves (intersegmental nerve, ISN, innervating dorsal targets and the segmental nerve, SN, innervating ventral and lateral targets) have retracted their larval neuromuscular junctions. Numbers 1, 2, and 3 indicate regions of nerve cuts that resulted in complete, partial, and transient denervation, respectively. (B) At 12 h APF. The larval muscles flatten and elongate, and adult-specific nerve outgrowth is seen. In the region of the DVMs, smaller outgrowths are noticeable. (C) At 16 h APF. The larval muscles split as myoblasts fuse with them to begin formation of the six DLM fibers. Simultaneously, the nerve also undergoes reorganization. Higher-order nerve branches arise at this time. This is the earliest time that DVM fibers can be seen. (D) At 24 h APF. The adult neuromuscular pattern is formed. The DLMs (a–f) and DVM III are innervated by the posterior dorsal mesothoracic nerve (PDMN), which arises from the restructuring of the ISN. DVMs I and II are innervated by the anterior dorsal mesothoracic nerve (ADMN) and the mesothoracic accessory nerve, respectively.
Source: From Fernandes and Keshishian (1998).

Local innervation is directly and crucially involved in the proliferation and distribution of myoblasts and myogenesis in *Drosophila* (Currie and Bate, 1991).

In *Drosophila*, denervation causes reduction of myoblast proliferation and alteration of segregation of myoblasts in DVMs (Fernandes and Keshishian, 2005). Neurectomy of the larval leg nerve before metamorphosis in the moth *Manduca sexta* prevents proliferation and normal migration and accumulation of myoblasts in respective regions, so that in 26% of cases no muscles develop in adult legs (Consoulas and Levine, 1997).

During metamorphosis, motoneurons not only withdraw larval synapses and respecify adult nerve branches and synaptic morphology, but they also respecify adult dendritic arbors, neuronal connections, and circuitries in the brain (Kent and Levine, 1993). Denervation of IFMs (dorsolateral muscle, which forms on larval muscle scaffold and DVM, which forms *de novo*) has shown that innervation is necessary for maintaining the size of myoblast population (Fernandes and Keshishian, 1998). The breakdown of the abdominal ISMs in the saturniid silkmoths takes place in the presence of hormone ecdysone and in the absence of JH, but investigators have concluded that

> *the actual breakdown is triggered by a neural mechanism. The latter consists of a sudden curtailing or cessation of the outflow of impulses in the motor nerves which innervate the abdominal muscles .... By chronic electrical stimulation of the nerves, the breakdown of the muscles can be opposed or prevented.*
>
> *Lockshin and Williams (1965)*

In *Drosophila*, the IFMs, DLMs and DVMs, develop in different ways: DLMs develop from myoblasts of existing larval muscle fibers, while DVMs develop *de novo*. Motoneurons regulate muscle formation not by providing any essential survival factor or by stimulating myoblast migration but by regulating imaginal pioneer cells, which serve as myoblast fusion targets and are believed to prefigure the DVM muscle fibers and myoblast proliferation, which is correlated with the expansion of motoneuronal terminal arbors on the muscle fiber surface. Denervation inhibits myoblast proliferation (Fernandes and Keshishian, 2005; Figure 5.5).

> *Based on experimental evidence, it is concluded that the motoneuron influences both the number of cells available for fusion, as well as potentially regulates the fusion events themselves. This in our view is an elegant mechanism for controlling muscle fiber differentiation during myogenesis, and may have evolved as a way to ensure that muscle primordia develop into muscles that meet the diverse demands placed on them by the nervous system.*
>
> *Fernandes and Keshishian (2005)*

Development of secondary myotubes (elongated multinucleate structures, arising from the fusion of several myoblasts) during rat embryogenesis always occurs in contact with, never at a distance from, the primary myotube innervation zone. Additionally, at least some, if not all, secondary myotubes make direct contacts with a nerve terminal (Duxson et al., 1989). Based on the fact that formation of primary and secondary myotubes *in vivo* occurs exclusively at the zones of innervation, it has

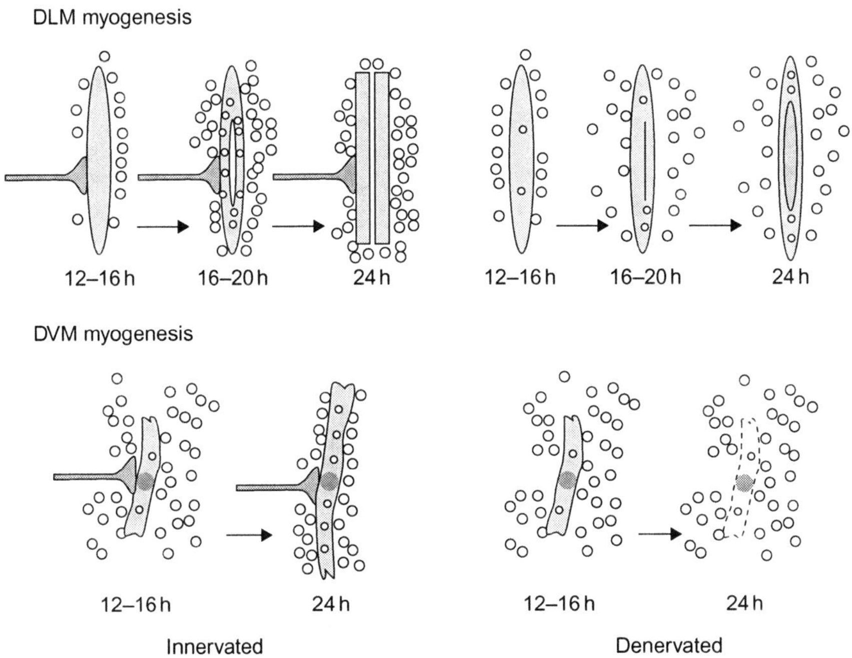

Figure 5.5 The roles of innervation during indirect flight muscle myogenesis. The events associated with the DLM and DVM formation are compared. For both muscle types myoblast proliferation occurs in a nerve-independent fashion prior to the onset of fusion events (0–12 h APF) to generate the initial pool of myoblasts. DLM myogenesis: the persistent larval fibers are responsible for segregating the myoblasts into fibers, which eventually split lengthwise. Nerve-dependent proliferation of myoblasts replenishes the myoblast pool, so that the appropriate fiber size is attained. When denervated, segregation of myoblasts still occurs normally. However, there is a delay in fusion and a reduction in subsequent proliferation of myoblasts, accounting for reduced muscle sizes. DVM myogenesis: the innervating motoneuron, along with the Duf-positive founder cells, is responsible for segregation of myoblasts into the distinct DVM primordia. Nerve-dependent proliferation replenishes the myoblast pool. When denervated, the myoblasts are incapable of segregating into discrete primordia, and proliferation is also reduced. In addition, Duf-positive cells are no longer reliably evident by 24 h APF. *Abbreviations*: APF, after puparium formation; DLMs, dorsolateral muscles; Duf, a type of "imaginal pioneer" cells; DVMs, dorsoventral muscles. *Source*: From Fernandes and Keshishian (2005).

been concluded that nerve terminals regulate the fusion of myoblasts and formation of myotubes (Duxson and Sheard, 1995). In later stages of development, innervation suppresses expression of myogenin, a muscle-specific transcription factor, in muscles of the fetal hind limb. But if denervation of the muscle is performed, myogenin expression resumes (Buonanno et al., 1993). Development of muscle-stretch receptors also requires, as a necessary condition, the presence of sensory innervation (Soukup and Zelena, 1985).

Muscle development often requires both hormonal and neural regulation, the binary neural control system. *In vitro* experiments show that both ecdysone and neurons separately stimulate proliferation of myoblasts (Luedeman and Levine, 1996).

One extremely curious and compelling case of the CNS control and regulation of the muscle development is the development of the dorsal external oblique 1 (DEO1) muscle during metamorphosis in the tobacco hawkmoth, *Manduca sexta* (Hegstrom et al., 1998). During metamorphosis, larval muscles are radically remodeled; out of the five muscle fibers, which the larval DEO1 muscle consists of, only one serves as anlage for the development of the adult muscle, while the rest of them degenerate and are eliminated. The remaining fiber starts myoblast proliferation to grow into the adult muscle.

Investigators now have an answer to the question: What let that specific muscle fiber to survive, although it was under the same hormonal regulation as the eliminated fibers? That muscle fiber is the only one that, during metamorphosis, increases expression of a specific isoform of EcR (EcR-B1), which by binding to ecdysone stimulates myoblast proliferation. But this statement is just a description of an experimental fact, not an explanation, hence it leads us to the next question: What selects this particular muscle fiber to express that specific ecdysone receptor, while four remaining, equally vibrant fibers do not express it?

Experimental evidence has shown that the selection is made by the local innervation. The terminal arbor of the motoneuron innervating the DEO1 muscle during metamorphosis recedes from four of the muscle fibers and remains in contact only with the surviving fiber, on which the axon terminal arbor releases a diffusible agent of still unknown nature (Hegstrom et al., 1998). The conclusion that expression of the EcR (EcR-B1) is determined by the motoneuron is also corroborated by the fact that experimental denervation prevents both EcR-B1 expression and myoblast proliferation. In the absence of innervation, the four muscle fibers, just before ecdysis, express only EcR-A, which is known to be related to apoptosis, determining thus their fate, i.e., elimination of those fibers.

That the innervation is essential for the fiber to respond to ecdysteroids is also shown by the fact that denervation entirely eliminates the upregulation of EcR-B1 if performed before the upregulation occurred. If performed after the upregulation had commenced, denervation dramatically reduces EcR-B1 expression and eliminates it by 48 h later (Hegstrom et al., 1998).

The development of the DEO1 muscle in *Manduca sexta*, thus, is a function of a binary neural control mechanism; on the one hand, it is mediated via the cerebral PTTH–ecdysone axis and, on the other, it is controlled by local innervation.

Control of Myogenesis in Vertebrates

The essential relationship between the nervous system and myogenesis has also been observed in vertebrates. The differentiation of myoblasts, which starts the development of skeletal muscles within somites, depends on the expression of *MyoD* genes (Alves et al., 2003) coding for transcription factors that induce expression of genes for muscle-specific proteins (MSPs) (Te and Reggiani, 2002; Alves et al., 2003) and Myf5.

These MSPs are also expressed according to an identical timetable in each somite (Xu et al., 2000). Again, signals for expression of MyoD genes (among them, *Hedgehog* and *Wnt*) originate in the neural tube/notochord adjacent to somites (Te and Reggiani, 2002) as is also concluded from the fact that, when separated from the axial structure, somites do not express MyoD genes (Alves et al., 2003). Experimental misexpression of Myf5 and MyoD in the chick neural tube results in ectopic skeletal muscle development (Delfini and Duprez, 2004) and in *X. laevis* maternal MyoD (acting downstream the Pax3; Arnold and Winter, 1998) is present in the egg, although the gene for MyoD is the earliest muscle-specific gene to be expressed during gastrulation (Hopwood et al., 1989).

Studies on the development of muscles in the paraxial mesoderm (on both sides of the neural tube) and somites in chicks have shown that signals from the adjacent dorsal neural tube, and, to a limited extent, from the ventral neural tube (VENT), are basic inducers of myogenesis in these embryonic structures (Figure 5.6). Ablation of the neural tube prevents formation of muscles in these structures (Stern et al., 1995). Innervation may also have suppressive effect on gene expression in muscles. This is the case, for example, in rats where denervation of particular muscles, within 1 week, increases 150- to 200-fold expression of muscle transcription factors, such as MyoD, Myf5, and myogenin (Voytik et al., 1993).

The nervous system plays an integral role in determining gene expression (Ravel-Chapuis et al., 2007), cellular activity, and the size of skeletal muscles (Hyatt et al., 2006) in vertebrates.

Genes of the myogenic regulatory family (MRF) are transcription factors that regulate expression of muscle-specific genes, and their expression is very sensitive to denervation and spinal cord isolation (Hyatt et al., 2006). Denervation increases expression of MRF in skeletal muscles. Transcription of MRF and expression of muscle-specific genes in electrocytes of electric fishes, similarly, depend on the neural input from electromotoneurons innervating them (Kim et al., 2008).

Myogenic progenitor cells migrate from the dermomyotome to limb buds (Bajard et al., 2006). While the prospective myotome adjacent to the neural tube is innervated only after the formation of the myotome, the innervation of limbs/limb buds by the ventral ramus of the spinal nerve takes place before the myogenic cells have differentiated (Figure 5.7). Electron microscopy studies on amphibians and mammals have shown that nerve bundles and arborization of axons occurred in the proximal third of the limb bud before condensation and differentiation of mesenchymal cells (Cameron and McCredie, 1982). It is observed that the small number of differentiated myocytes that appear initially occur "strikingly" in close proximity to the growth cones of the nerve (Deries et al., 2008).

Before being innervated, the limb bud is colonized by a population of neural crest cells; a second population later invades the limb bud following the path of nerves for extensor and flexor limb muscles. The first population gives rise to melanocytes and Merkel cells and from the second Schwann cells and terminal glial cells differentiate (Grim and Christ, 1993).

Innervation is necessary for the fusion of muscle fibers and their transdifferentiation into electrocytes (denervation prevents these processes) in the weak electric fish,

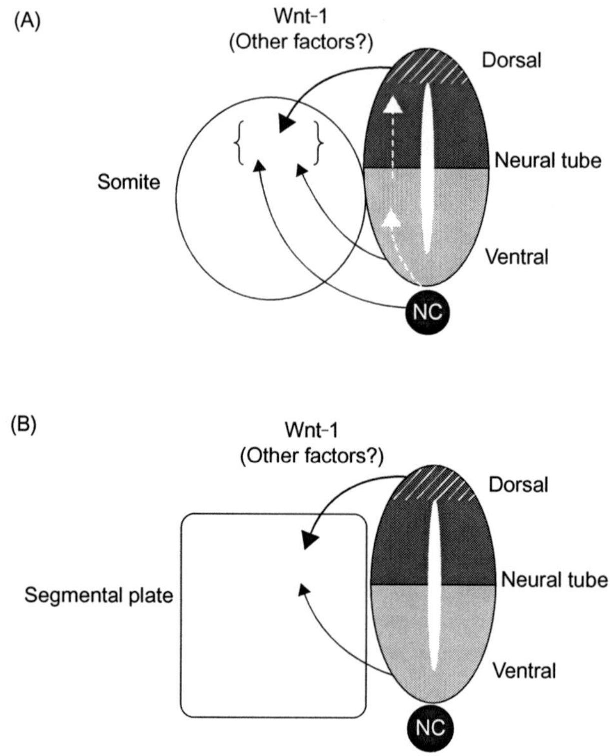

Figure 5.6 Proposed models of axial structure-dependent paraxial mesoderm myogenesis. For somites, (A) the major signal (thick, solid arrow) comes from the dorsal neural tube and may be mediated in part by *Wnt-1* or other *Wnts* in this region of the neural tube, such as *Wnt-3a*. The expression pattern of *Wnt-1* and *Wnt-3a* as determined by Hollyday et al. (1995) is indicated by white hatching. Weaker myogenic signals (thin, solid arrows) emanate from the ventral neural tube and notochord and act directly on somites. These ventral signals may interact in some way (bracket) with the dorsal signal, causing a synergistic increase in the number of myosin heavy chain (MHC)-positive cells induced. Alternatively, the synergy could be due to intra-axial signaling. For example, the ventral neural tube/notochord might signal the dorsal neural tube to boost the dorsal signal (white broken arrows). For segmental plate tissue (B), the dorsal signal remains the major signal (thick arrow) and could be mediated in part by *Wnt-1* or other *Wnts*. The ventral neural tube may emit a weak positive myogenic signal (thin arrow), but our study shows no evidence for synergy between dorsal and ventral signals for segmental plate myogenesis. *Abbreviation*: NC, notochord.
Source: From Stern et al. (1995).

Sternopygus macrurus, and interruption of the neural input leads to the dedifferentiation of electrocytes back into muscle fibers (Unguez and Zakon, 2002).

The CNS controls and regulates the development of target muscles not only directly, via peripheral nerves, but also indirectly, by long-range action, via the neurohormonal signal cascades. Often, both modes of control are operational in the development of the

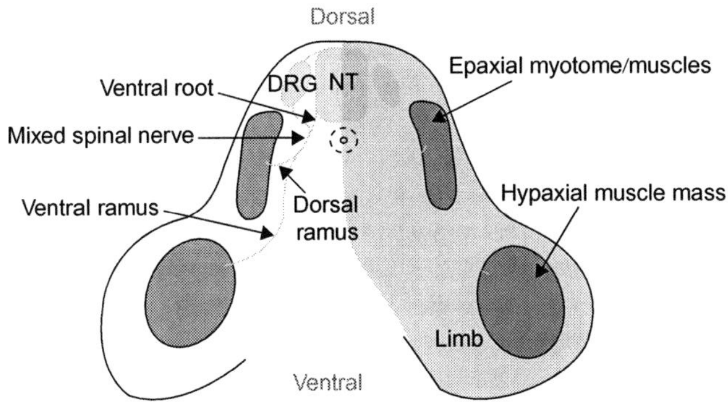

Figure 5.7 Relationship between developing spinal nerves and muscle masses as seen in a transverse section of an embryo at limb level. The ventral (motor) and dorsal (sensory) spinal roots join to form the mixed spinal nerves; these then divide into the dorsal ramus and the ventral ramus, which innervate, respectively, the epaxial muscles and the hypaxial muscles. The epaxial muscles comprise the epaxial component of the myotomes in the embryonic period; these later transform into the deep back muscles. The hypaxial muscles comprise the hypaxial component of the myotome embryonically, but at later times, two distinct hypaxial compartments arise from the myotome. The first includes muscles of the ventral axis such as intercostals and body wall muscles; the second is the limb muscles that form from an evolutionarily "new" migratory myogenic population.
Source: From Deries et al. (2008).

same muscles. This is the case, for example, in the development of the laryngeal muscle in *X. laevis*. In juvenile animals, the laryngeal muscle is similar, female-like, in both sexes. After metamorphosis, as a result of androgen secretion, male individuals develop the mlm. Denervation of the muscle, however, causes its atrophy in male amphibians, whereas androgen administration causes its hypertrophy (Tobias et al., 1993).

It is noteworthy that the laryngeal nerve, under normal conditions, may be involved in the effects produced by the androgen, since both laryngeal muscle and motoneurons, after metamorphosis, express androgen receptors. Experimental evidence has shown that

> *the nerve is required for maintenance of existing fibers and denervation results in cell death.*
>
> *Tobias et al. (1993)*

In chicks, motor neurons, upon entering the chick limbs, release the RA-synthesizing retinaldehyde dehydrogenase-2 (RALDH-2) enzyme, inducing their muscle cell differentiation and muscle formation (Berggren et al., 2001). The development of muscle fiber types in chick embryos coincides with the penetration of nerves in these muscles. Rightly, this has been interpreted as indicative of the role of the local nerves in the differentiation of muscle fiber types. However, the fact that the embryonic denervation performed before the nerves enter the muscles impairs the muscle growth indicates that the local innervation is necessary for the proper growth and survival of muscles, but not for the initial muscle differentiation (Butler et al., 1982).

Skeletal muscle fibers are of two subtypes, slow and fast muscles, depending on ultrastructural morphology, contractile physiology, and susceptibility to fatigue. These differences are related to expression of different types of proteins and enzymes in each myocyte subtype. The specific programs of gene expression for each subtype are determined by the activity of motor nerves innervating each subtype of skeletal muscle fibers. Firing patterns of motoneurons innervating slow muscles sustain Ca^{2+} at levels that activate the calcineurin-nuclear factor of activated T cells (NFAT) pathway leading to dephosphorylation and nuclear localization of NFAT proteins (Chin et al., 1998).

Mechanical stress also may change the patterns of gene expression in vertebrates, and it has been observed that it can activate synthesis of insulin-like growth factor-1 (IGF-1), known for its growth-promoting activity in muscle fibers (Figure 5.8). However, it is not known whether the observed changes in the size and properties of muscle fibers under conditions of mechanical stress are direct results of that stress on muscles or related to changes in other tissues or organs and signals of neural origin. Ideas have been expressed that both pathways are operational and overlap in metazoans (Tidball, 2005).

Left–Right Asymmetry

There is a general consensus that in vertebrates the left–right (LR) asymmetry is determined by asymmetric expression of the *nodal* gene. This gene downstream induces expression of *pitx*, but it is also known that *nodal* expression is induced by the transcription factor, Vg1, a member of the TGF-β family, which are induced by hormones (estrogen and gonadotropins) of signal cascades that ultimately originate in the CNS (see Section Endocrine Control of Secreted Proteins and Growth Factors in Chapter 1).

In chicken, along with the asymmetric expression of Shh in Hensen's node, the *Lefty-1* gene is involved in determining the LR asymmetry, and its expression is induced by a hormone, the RA (Tsukui et al., 1999). It is known that during early development the neural tube is the main source of RA; it produces and releases more RA than any other structure (the RA level in the neural tube is 29 times higher than in the heart) (Maden et al., 1998).

Recent evidence from studies on frogs and chicks shows that the LR asymmetry is determined as early as the four-cell stage, before the initiation of the expression of zygotic genes, i.e., epigenetically, by a maternal factor, the neurotransmitter serotonin (Fukumoto et al., 2005).

LR asymmetry comprises not only asymmetric position of organs and parts in the body but also the asymmetric looping of many organs and parts, which is the most general marker of LR asymmetry in metazoans (Adam et al., 2003). Organ looping in insects is determined by the level of JH secreted by corpus allatum (or the ring gland) in response to brain signals (allatostatins/allatotropins). By investigating the looping

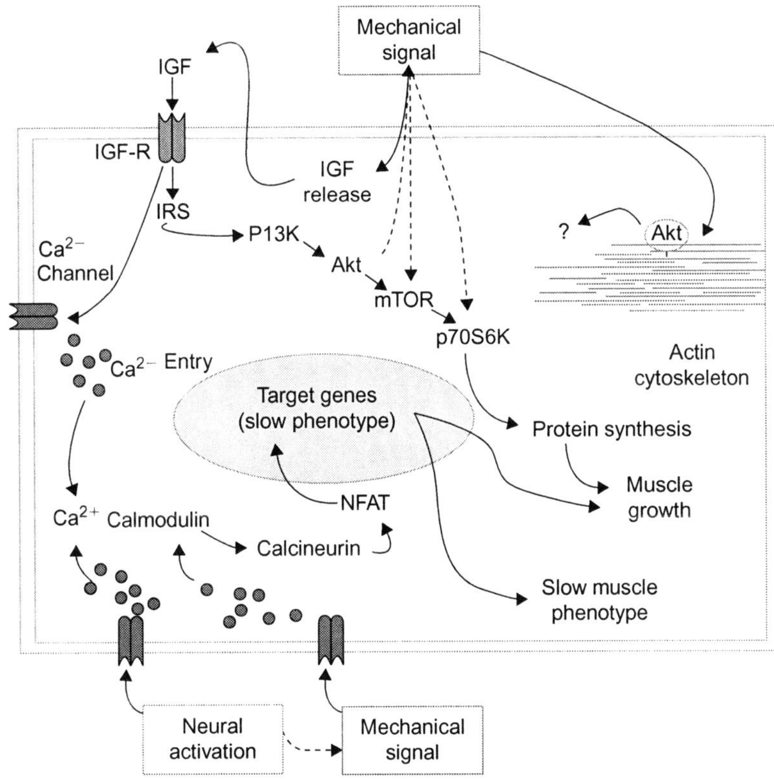

Figure 5.8 Potential mechanisms through which mechanical signals and insulin-like growth factor (IGF)-1-derived signals may be mediated through overlapping pathways. Solid arrows indicate cause and effects or interactions that are likely to be direct. Dotted arrows indicate interactions for which there may be unknown intermediates. *Abbreviations*: IRS, insulin-response substrate; NFAT, nuclear factor of activated T cells; mTOR, mammalian target of rapamycin; PI3K, phosphatidyl-inositol-3 kinase; ?, unknown downstream events that result from Akt association with the actin cytoskeleton.
Source: From Tidball (2005).

asymmetry of genitalia in *Drosophila*, Adam et al. (2003) found that a mutation in the gene *Fas2* (fasciciclin 2) in *Drosophila* males leads to abnormal levels of JH during the pupal stage and rotation defects of their genitalia and spermiduct (Figure 5.9). They concluded that a link exists between organ looping and the retinoic-like JH (Adam et al., 2003). They also presented evidence on the similarities between the invertebrate and vertebrate asymmetries in organ looping, and the fact that a terpenoid, JH, determines organ looping in invertebrates may suggest that the vertebrate terpenoid, RA, may also be involved in organ-looping asymmetry in vertebrates.

Asymmetry of the first pair of chelipeds is characteristic of many crustaceans. Snapping shrimps, *Alpheus heterochelis*, have a pair of asymmetric chelae (claws)—the big one, the snapper, and the smaller one, the pincer—with different morphology

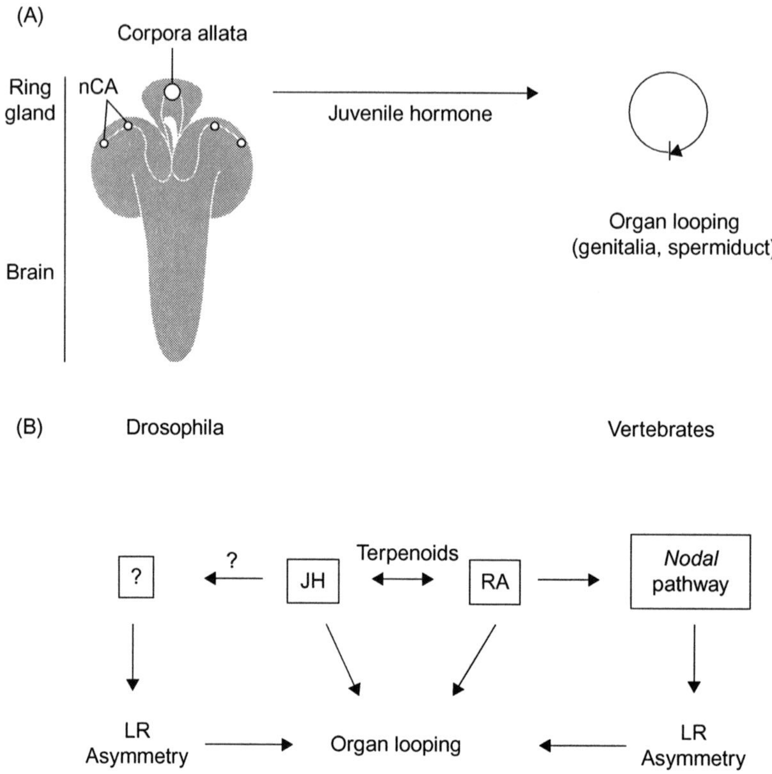

Figure 5.9 Model for Fas2 and JH action in *Drosophila* organ looping. (A) Fas2 is specifically expressed and required in neurosecretory cells innervating the corpora allata, to control JH titers. JH is released into the circulating hemolymph and reaches the genital disc to control its rotation nonautonomously. (B) Parallels between the vertebrate and the *Drosophila* pathways controlling asymmetric organ looping. In vertebrates, two conserved pathways have been implicated in LR asymmetry, the retinoic acid (RA) and the nodal pathways. RA plays a dual role as it is also involved in organ looping. In *Drosophila*, mutants exist that can lead to a reversion of genitalia rotation, suggesting that they are involved in LR asymmetry (collectively represented by a question mark). The analysis of $Fas2^{spin}$ shows a role for JH, an RA analog, in the control of organ looping and suggests an evolutionary conserved function of terpenoids in the control of asymmetric organ looping.
Source: From Adam et al. (2003).

and functions (Figure 5.10). The first serves for defensive responses, and the second is used for burrowing and feeding. Both chelae start as identical structures during the early juvenile stages, and only by the sixth juvenile stage do they start differentiating into snapper and pincer. What determines which of the chelae will become the snapper, and which will become the pincer? Genes and their products, humoral factors (such as hormones), growth factors, or other secreted proteins in the body fluids are excluded from playing any essential role. It was observed that removal of one of the claws before their

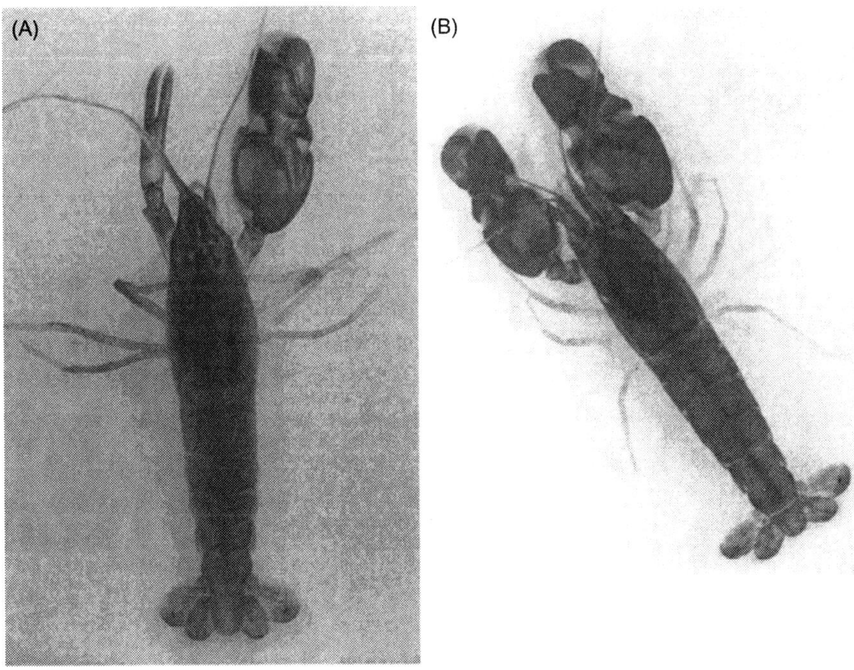

Figure 5.10 *Alpheus heterochelis*, the common snapping shrimp of the east coast. (A) Dimorphism in the claws of the first pair of pereiopods. (B) Symmetrical snappers develop in the shrimp as a result of severance of snapper limb nerves.
Source: From Mellon (1999).

differentiation (until the fourth juvenile stage) induces the remaining claw to develop into a snapper and, in adult shrimps, removal of the snapper induces transformation of the pincer into a snapper, while a new pincer develops on the removed snapper's site. In other experiments, it is observed that cutting the relevant nerve in the snapper claw induces transformation of the pincer into a snapper (Read and Govind, 1997).

Severing the snapper limb nerves, strikingly, resulted in development of symmetrical snappers (Mellon, 1999). Based on the experimental evidence, it has been proposed that

> *neural influences from the transforming pincer-to-snapper claw restrict regeneration of the contralateral claw to a pincer type thereby ensuring bilateral asymmetry in adult shrimps.*
>
> *Young et al. (1994)*

It is concluded that the inhibition by the intact snapper claw is centrally determined (Read and Govind, 1997) and

> *the snapper-based inhibition of the pincer claw is neural in origin.*
>
> *Read and Govind (1997)*

The observation that the intact snapper claw inhibits the development of the contralateral claw to a snapper has been explained with the more intense innervation of the snapper (the snapper has over 13,000 axons, whereas the pincer has only 10,000) (Read et al., 1991).

Neural Control of the Development of the Neuroendocrine System

In humans, the hypothalamus begins to develop by the 6th gestational week and in mice by the embryonic day 13. The hypothalamic anlage is encircled by neurons and fibers containing GABA and its synthetizing enzyme glutamic acid decarboxylase67 (GAD67). It is suggested that GABA from the nearby brain region regulates neuronal movement and is directly involved in the development, cellular organization, and establishment of neural circuits in the ventromedial hypothalamic nucleus (VMH) (Tobet et al., 1999; Dellovade et al., 2001).

The pituitary develops from Rathke's pouch, ectodermal dilatation of the oral portion of the alimentary tract. The pituitary anlage is induced by signals (BMP4 and FGF8) from the adjacent diencephalon (Takuma et al., 1998; Treier et al., 2001) or infundibulum (the extension of the third ventricle of the brain to the pituitary, from which the neural pituitary arises) and juxtapituitary mesenchyme (Ericson et al., 1998). Later, another cerebral signal, FGF8, from the diencephalon (consisting mainly of thalamus and hypothalamus), combined with a BMP2 signal from a ventral pituitary organizing center, induces the second step in the embryonic development of the pituitary. Brain secretion of FGF8 activates homeobox genes Lhx3 and Lhx4 (Sheng et al., 1997), with the product of the first gene controlling the pituitary commitment of anlage cells and transformation of the pouch rudiment into a definitive pouch. In the third stage of the pituitary development, suspension of BMP2 signal from the diencephalon allows differentiation of various pituitary hormone–secreting cells (Treier et al., 1998). The ventral diencephalon also provides growth factors of the Fgf family, especially Fgf3, which are involved in specification of the progenitor cells of the adenohypophysis and formation of the adenohypophyseal anlage (Herzog et al., 2004).

The ultimobranchial body, whose fusion with the thyroid diverticulum leads to formation of the thyroid gland, develops from mesenchymal neural crest cells of the fourth pharyngeal pouch (Manley and Capecchi, 1995).

The rapid growth of the fetal adrenal gland in humans is under indirect control of the pituitary adrenocorticotropic hormone (ACTH). From the 40th embryonic day, this hormone stimulates local secretion of IGF-11 and bFGF. As it is well known, the secretion of the pituitary ACTH is under cerebral control of the hypothalamic corticotropin-releasing hormone (CRH), initially secreted by the placenta (under maternal neurohormonal control). Adrenal medulla cells, as well, are derived from migrating neural crest cells.

The development of the embryonic pancreas in mice begins with differentiation of pancreatic cells from the dorsal gut endoderm. At the same time in the region, neural crest cells arrive, which initially express *Sox10* and its target *Phox2b*. At the E12.5, these cells cease expressing *Phox2b* as a result of the inhibitory action of expression of Nkx2.2 by neighboring endodermally derived pancreatic cells. At this stage, the

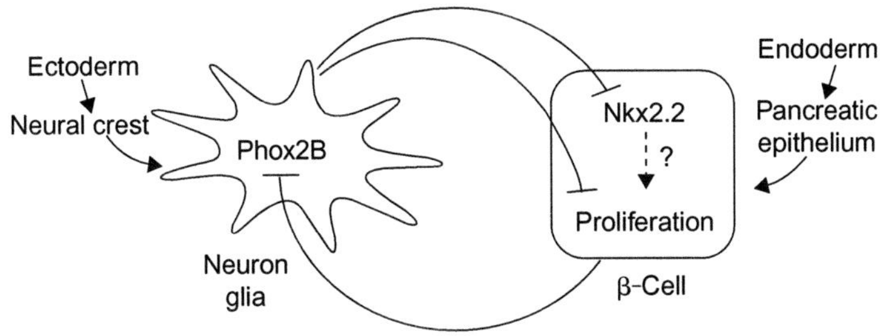

Figure 5.11 Model for the noncell-autonomous interactions between Phox2b and Nkx2.2 in the developing pancreas.
Source: From Nekrep et al. (2008).

neural crest–derived cells differentiate into neurons and glia cells. The silencing of *Phox2b* expression by pancreatic cells is believed to occur via a noncell-autonomous signaling pathway. In turn, neural crest–derived cells inhibit expression of Nkx2.2 by pancreatic cells (Figure 5.11).

The CNS is involved in secretion of insulin by β-cells, and it is suggested that the CNS "may play an essential role ... in regulating the size of the beta cell population" (Nekrep et al., 2008).

Development of the thymus is a typical example of cerebral control of organogenesis. It is experimentally demonstrated that thymus development, and even its aging and atrophy, depends on the balance in the serum levels of two hypothalamic releasing neurohormones, GHRH and GH-releasing inhibitory hormone (Hirokawa et al., 1998, 2001).

Neural Control of the Development of Sensory Organs

Sensory organs in vertebrates mainly develop from cranial placodes, thickenings of the embryonic head ectoderm, which in turn develop from an earlier structure, the preplacodal ectoderm (PPE). The PPE develops next to the neural tube and is characterized by neurogenic potential and expression of specific genes before the formation of the neural plate. It forms within the cranial neural plate border region and may be induced by the same maternal factors that determine the formation of the neural plate (Baker and Bronner-Fraser, 2001), but other interactions between the neural plate and epidermis may also contribute to the differentiation of placodes and to their later subdivisions.

Let us illustrate the neural tube/CNS control of the development of the sensory organs with the development of the otic vesicle and inner ear in mammals.

Inner Ear

It was believed that the otic placode is induced by Fgfs and Wnts of the chordamesoderm and hindbrain, but recent studies show that it is Fgf8 and Fgf3 signals, released from the hindbrain rhombomere 4, that determine the early induction and maintenance of the otic placode and inner ear patterning in the zebrafish (Leger and Brand, 2002). Hindbrain signals are necessary and sufficient for establishing the dorsoventral axis of the inner ear. In this context, it is noteworthy that a specific change in the dorsoventral axis of the inner ear can be induced by reversing the dorsoventral axis of the hindbrain (Figure 5.12).

Shh signals from the neural floor plate and notochord are required for ventral inner ear structures and differentiation of auditory cells (Riccomagno et al., 2002; Bok et al., 2005).

The establishment of the axial polarity and compartmentalization of the otic vesicle as well is believed to be a function of the hindbrain. The otic cup develops adjacent to two hindbrain compartments, rhombomeres 5 and 6 (r5 and r6), whose boundary is aligned with the middle of the otic cup. r5 and r6 may send different signals to anterior and posterior otocyst, respectively. They may send signals exclusively to the dorsal half of the vesicle, thus determining formation of its separate anteromedial and posteromedial compartments. Lateral identity is determined later by the lateral-most part of the invaginating cup and could be influenced by the distance from the hindbrain (Brigande et al., 2000).

Maternal and Neural Control of Heart Development

After the formation of the neural tube and the CNS, in many vertebrate species, the heart is the next organ to develop. The neural tube sends signals (Wnt-3 and Wnt-8) that inhibit the induction of cardiogenesis and promote blood cell differentiation of mesoderm along the whole of its length, except for the region where the heart normally develops. Ectopic heart is experimentally induced by simply activating Wnt antagonists (Schneider and Mercola, 2001; Tzahor and Lassar, 2001; Marvin et al., 2001; Foley and Mercola, 2005). Wnt antagonists from the endoderm permit the development of the heart in a restricted region of mesoderm, where they overwhelm the neural tube Wnt signaling, leading to expression of Nkx2.5 and synergistic action of BMP (Zaffran and Frasch, 2002; Figure 5.13). In invertebrates such as *Drosophila*, by contrast, Wnts are necessary for the development of the heart.

The neurotransmitter serotonin, via its receptor 5-HT^{2B}, regulates differentiation and proliferation of the developing and adult heart (Nebigil et al., 2000). Suppression of 5-HT^{2B} synthesis causes developmental anomalies and midgestational death of the embryo. The neurotrophic factor, brain-derived neurotrophic factor (BDNF), besides its differentiative actions on neurons expressing the Tk receptor tyrosine kinase, has an angiogenic effect on endothelial cells and is necessary for maintaining the stability of the intraventricular walls (Donovan et al., 2000).

The hormone RA signaling determines the proportion of cells that will differentiate into myocardial progenitor cells from a pool of multipotential cells in a way that the population size of cardiomyocytes is inversely related to the RA level (Keegan et al., 2005). The very low level of RA in the heart suggests that the source of the hormone may be the RA-rich neural tube/spinal cord.

Apoptosis, the programmed cell death, also has a special role in heart morphogenesis. It is observed to occur in heart of mouse (myocardium and myocardial

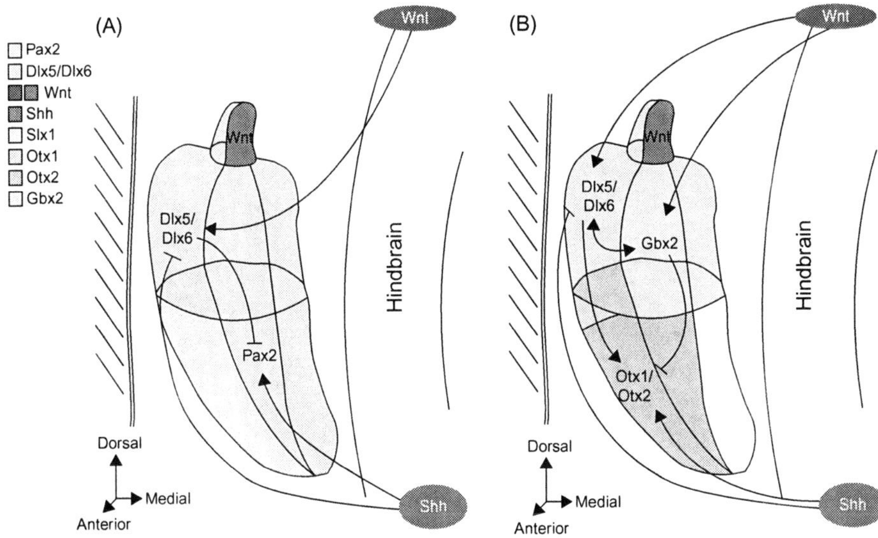

Figure 5.12 Representative expression patterns of genes controlling cochlear and vestibular specifications. (A) Shh functions to maintain *Pax2* and restrict *Dlx5/Dlx6* in the medial wall of the otic vesicle in order to specify cochlear fate. *Dlx5/Dlx6* specify the medial to dorsal-most cells of the otic epithelium that give rise to the endolymphatic duct and vestibular apparatus. (B) Secretion of Shh from the notochord specifies the ventral-most cells of the otic epithelium that express *Otx1/Otx2* and possibly *Pax2* which contribute to cochlear morphogenesis and outgrowth. In addition, *Dlx5/Dlx6*-dependent vestibular specifications and morphogenesis are dependent upon the activation of *Gbx2* and *Bmp4* function (not shown), and partial activation/ expression of *Otx1*. *Dlx5/Dlx6* also functions to restrict *Pax2* expression to the medial wall of the otic vesicle epithelium. Thus, *Dlx5/Dlx6* and *Shh* may functionally antagonize each other, through repression, to generate compartments of activities that specify the vestibular and cochlear cell fates.
Source: From Chatterjee et al. (2010).

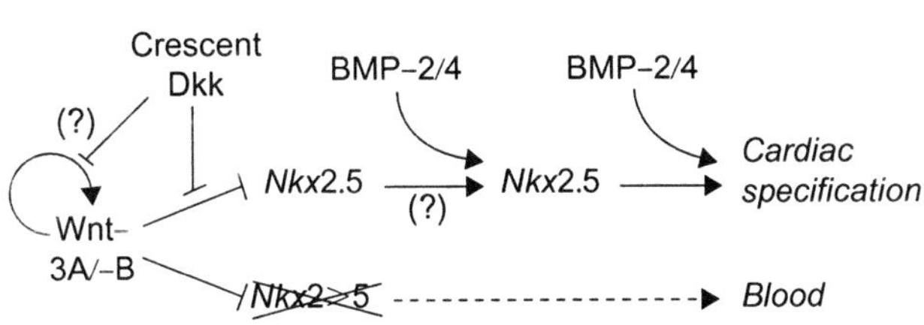

Figure 5.13 Pathways of cardiac induction in vertebrates.
Source: From Zaffran and Frasch (2002).

epithelium) on days 11 through 16 of the embryonic development (Abdelwahid et al., 1999). Apoptosis is essential for the developmental remodeling and shortening of the complex embryonic outflow tract, i.e., for the transformation of the initial tubular structure between the single primitive ventricle and the aortic sac into a permanent outflow tract connecting the ventricles with the arterial trunks in 4- to 8-day chick embryos (Watanabe et al., 1998). (see Neural Control of Apoptosis later in this chapter.)

Besides the neural tube inductions, neurotransmitters, and apoptosis, an essential role in the development of the cardiovascular system is played by neural crest cells, which migrate there to participate in the formation of the heart. The region of the neural crest that is involved in the process is known as *cardiac neural crest*. Cardiac neural crest cells originate in the cardiac crest along the neural tube between the r6, r7, and r8, corresponding to somites S1, S2, and S3. They migrate from the neural tube/CNS, via the caudal pharynx, to the heart region to differentiate into various mesenchymal cell types that initiate the formation of the outflow tract and the smooth muscle of the aortic arches (Phillips et al., 1987; Kirby et al., 1993; Creazzo et al., 1998; Hutson and Kirby, 2007) (Figure 5.14). Due to the important contributions of neural crest cells in the development of the vertebrate heart, the latter is regarded as a new heart, part of a new cardiovascular system composed of modular units, only some of which existed in ancestral invertebrates (Fishman and Chien, 1997). Neural crest cells also surround the pharyngeal arch arteries and "populate aortico-pulmonary septum

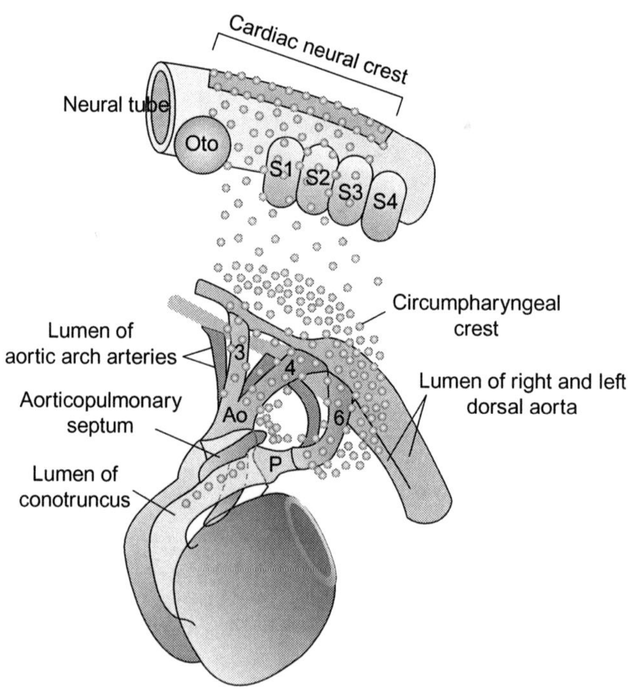

Figure 5.14 Migration and distribution of cardiac neural crest cells from their origin to the caudal pharynx and from there into the outflow tract. *Abbreviations*: Ao, aorta; Oto, otic placode; S1, S2, etc.; somites 1, 2, etc; P, pulmonary artery.
Source: From Hutson and Kirby (2007).

and conotruncal cushions prior to and during overt septation of the outflow tract and surround the thymus and thyroid as these organs form" (Jiang et al., 2000).

Neural crest cells mutants for the type I receptor ALK2 in mouse embryos show impaired migration leading to morphological defects in the heart outflow tract and aortic arch (Kaartinen et al., 2004). Ablation of the cardiac neural crest leads to anomalies of the outflow and inflow tracts of the heart and the aortic arch arteries (Miyagawa-Tomita et al., 1991; Figure 5.15).

Vasculogenesis and Angiogenesis

The processes of vasculogenesis (formation of a primary network of capillaries or the vascular plexus) and the differentiation of endothelial cells from their precursors begin approximately at the time that the embryonic heart starts to develop. A group of five vascular endothelial growth factor (VEGF) types, all of them downstream elements of signal cascades under ultimate control of the CNS, have essential roles in vasculogenesis and hematopoiesis (Carmeliet et al., 1996; Liang et al., 2001) in stimulating proliferation of vascular endothelial cells (Ash and Overbeek, 2000), and migration of endothelial cells (Esser et al., 1998). Their action is mediated by receptor tyrosine kinases Flk1 (VEGF-R2) and Flt1 (VEGF-R1) (Millauer et al., 1993), with the latter being essential for the formation of embryonic capillaries (Fong et al.,

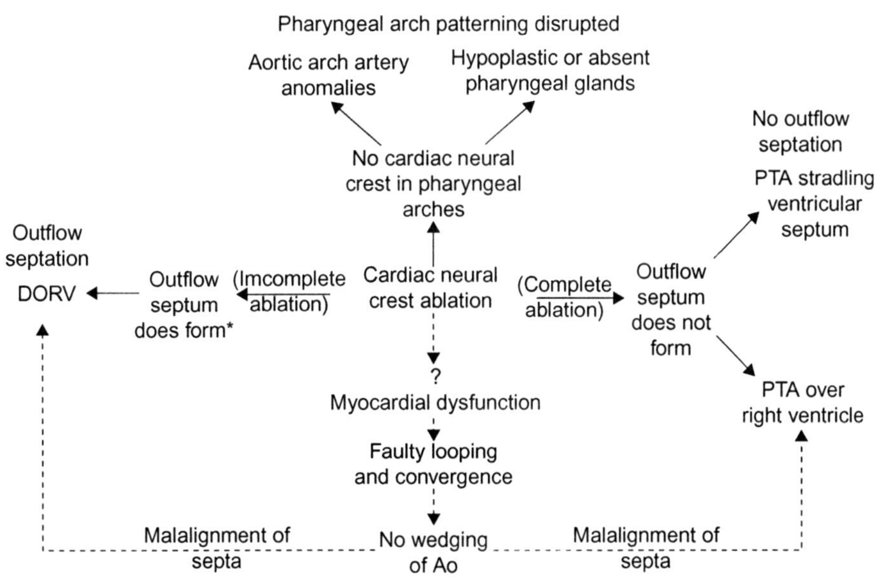

Figure 5.15 Flow chart to illustrate the functional and structural consequences of cardiac neural crest ablation. *Abbreviations*: Ao, aorta; DORV, double-outlet right ventricle; PTA, persistent truncus arteriosis.
Source: From Kirby and Waldo (1995).

1995). Another secreted protein, FGF2, also known as *basic FGF*, has been demonstrated to be a potent stimulator of the embryonic angiogenesis when applied to the quail chorioallantoic membrane at embryonic day 7 (Parsons-Wingerter et al., 2000). FGF2 is also secreted by chick chorioallantoic membrane and has a rate-limiting role in the vascularization of the chorioallantoic membrane (Ribatti et al., 1995). A spatially differential supply of VEGFA determines vascular branching pattern (Ruhrberg et al., 2002).

Evidence about the immediate regulation of angiogenesis by the nervous system has been presented recently. The neurotransmitter DA inhibits the vasculogenic and angiogenic actions of the cytokine vascular permeability factor/VEGF by inducing the endocytosis of the VEGF receptor 2 (Basu et al., 2001). However, probably the most compelling fact so far on the direct involvement of the CNS in the embryonic angiogenesis was reported by Mukoyama et al. (2002). They observed that in mutant mice lacking sensory nerves in limbs, large-diameter vessels do not branch normally into intermediate vessels, but directly into small-diameter vessels. Moreover, they observed that during embryogenesis, the sensory neurons themselves secrete VEGF, which determines the cell differentiation and patterning of arteries in their vicinity (Mukoyama et al., 2002), thus explaining the old anatomic observation on the general association of arteries with peripheral nerves. In the embryonic limb skin, the nerve is the principal source of VEGF. The nerve-derived VEGF164 induces in vessels' endothelium the synthesis of neuropilin-1 (NRP1), an artery-specific coreceptor of VEGF164. This fact may also explain why the arteriogenic effects of sensory nerves are restricted to vessels in close proximity to the nerves (Mukoyama et al., 2005; Figure 5.16).

Based on the experimental evidence, investigators concluded that

> *peripheral nerves provide a template that determines the organotypic pattern of blood vessel branching and arterial differentiation in the skin, via local secretion of VEGF.*
>
> *Mukoyama et al. (2002)*

It is demonstrated that in coculture with PSM (paraxial mesoderm), the neural tube can induce formation of a perineural vascular plexus (PNVP). Based on the demonstrated roles of the neural tube in vascular patterning, the neural tube is considered to be the midline signaling center for vascular patterning in higher vertebrates (Hogan et al., 2004). During embryogenesis, the brain and spinal cord release signals (with VEGF being a central player) for differentiation and migration of somitic angioblasts in their direction and form around them the so-called PNVP (Figure 5.17), as a necessary source of oxygen and nutrients for the development of the CNS in vertebrates. Another example of the neural control of vasculogenesis is the formation of the retinal vasculature. The lower portion of the signal cascade regulating the development of the retinal vascular pattern includes retinal neurons, which, by secreting platelet-derived growth factor A, stimulate proliferation of astrocytes. Astrocytes, in turn, by secreting VEGF, determine blood vessel formation in retina (West et al., 2005). Note that the signal cascade for the development of the retinal vasculature starts with retinal ganglion neurons.

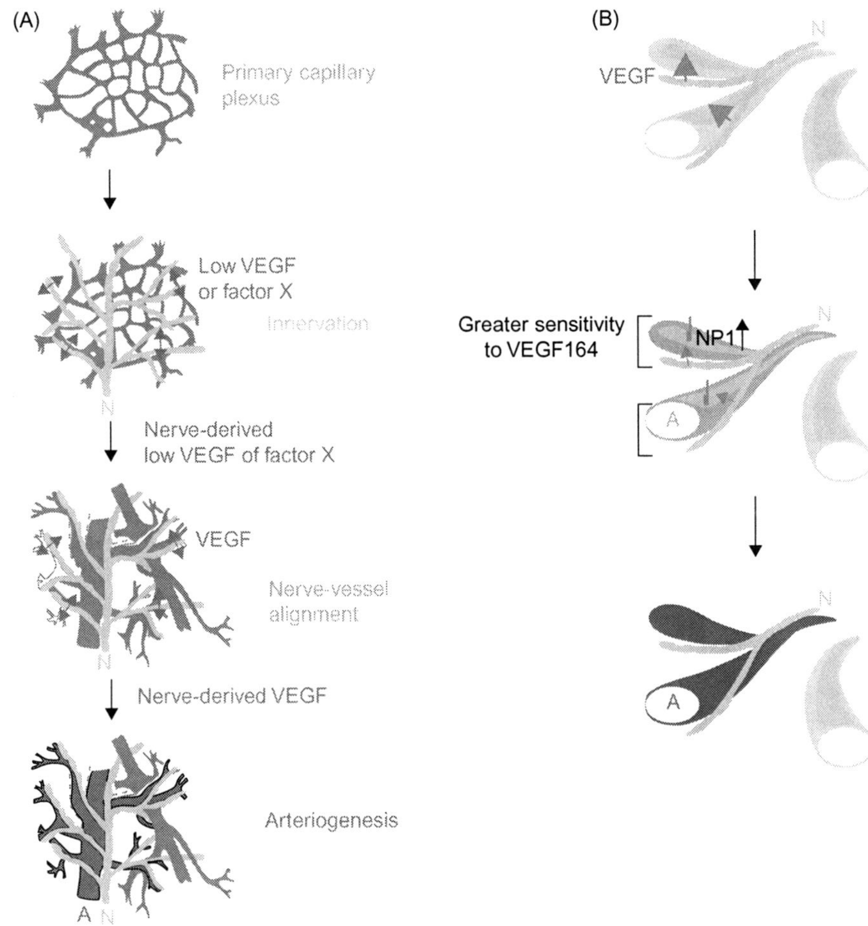

Figure 5.16 Schematic models for nerve-mediated arterial differentiation and vascular branching in the limb skin. (A) Proposed sequence of events in vascularization of limb. A low concentration of VEGFA, or a distinct nerve-derived signal ("factor X"), promotes nerve-vessel alignment, followed by VEGFA/NP-1-dependent arteriogenesis in nerve-aligned vessels. (B) VEGF promotes arteriogenesis via an NRP1-mediated positive-feedback loop. All vessels are initially equivalent. Nerve-derived VEGFA promotes arterial differentiation, and NRP1 amplifies the VEGFA effect due to increased sensitivity to VEGF164 in vessels in close proximity to nerves (N). *Abbreviations*: A, artery; NPR-1, neuropilin-1, an artery-specific coreceptor for VEGF that is induced by VEGF.
Source: From Mukoyama et al. (2005).

Development of the Gastrointestinal Tract

The specification of different parts of this tract is related to the spatiotemporal pattern of expression of various *Hox* genes along the embryonic anterior–posterior (A–P) axis. But the ordered expression of *Hox* genes during the embryonic

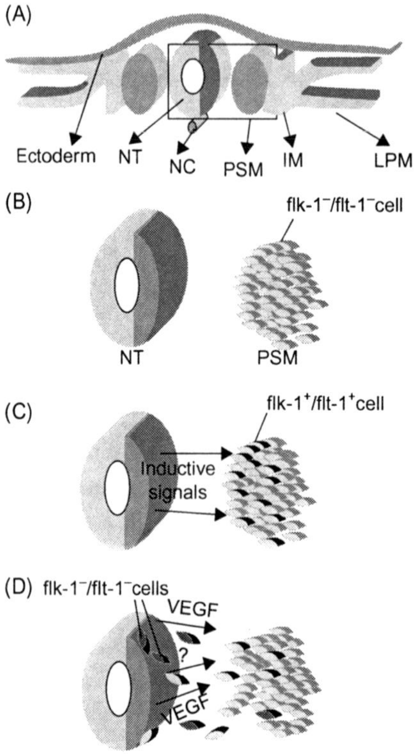

Figure 5.17 Neural tube patterning of blood vessels—a model. (A) Diagram showing cross-sectional view of a mouse embryo at 8.5 dpc. The boxed area is enlarged in (B–D). (B) Initially, the PSM does not contain committed vascular precursor cells. (C) A VEGF-independent signal from the neural tube induces FLK1 expression in a subset of PSM cells. (D) FLK1-expressing angioblasts are now competent to respond to the neural tube–derived VEGFA containing signal. The angioblasts now express additional vascular markers, such as FLT1, and they migrate and assemble the PNVP. Other unidentified signals emanating from the neural tube may also contribute to PNVP patterning *in vivo*. *Abbreviations*: dpc, day postcoitum; IM, intermediate mesoderm; LPM, lateral plate mesoderm; NC, notochord; NT, neural tube; PNVP, perineural vascular plexus; PSM, presomitic mesoderm.
Source: From Hogan et al. (2004).

development in vertebrates is regulated by the hormone RA. In turn, RA is synthesized by enzymes RALDH-1, -2, and -3 (Niederreither et al., 2002a,b). RA signaling plays a crucial role for the establishment of the A–P axis, as well as for the development of numerous organs, including the nervous system, lungs, digestive tract, kidneys, and eyes. RA synthesis is chiefly a function of RALDH enzymes. Starting with the early embryonic development, the synthesis of RALDH-2 is closely related to the neural tissue and motor neurons, which extend their axons to the periphery,

> *indicating a potential role of retinoic acid in nerve influences on peripheral differentiation.*
>
> *Berggren et al. (1999)*

The neural tube, which is adjacent to the gastrointestinal endoderm, may be an important source of RA for the development of the gastrointestinal tract. In chick embryos, the neural tube/spinal cord in its length up to the hindbrain has the highest levels of RA, which gradually decrease with the increased distance from the neural tube (somites → lateral plate) (Maden et al., 1998) and in RA-deficient mice lack of RA in endoderm causes agenesis of the dorsal pancreas (Martin and Luo, 2005).

The neural tube is also crucially involved in the development of the gastrointestinal tract in another direct way. Migratory cells from the neural tube reach the region of the prospective gastrointestinal tract and participate in the development of the organs of the tract. Very early during hepatogenesis, VENT cells originating in the ventral regions of the neural tube appear in liver trabeculae, where they express hepatocyte markers and display hepatocyte morphology. At stage E5, VENT cells reach the stomach and duodenum, where they contribute to formation of their mucosa and, almost at the same time (E6), VENT cells from the VENT reach other regions of the gastrointestinal tract (Dickinson et al., 2004). The above evidence shows that development of the gastrointestinal tract depends on the patterns of RA synthesis as well as on the cellular and inductive contributions of the migratory neural tube cells.

Pneumogenesis

It is becoming clear that the angiogenesis and vasculogenesis of the pulmonary circulation and capillary network are closely linked with each other and may be necessary for lung epithelial morphogenesis. Like epithelial morphogenesis, pulmonary vascularization depends on a fine balance between positive and negative factors. Angiogenic and vasculogenic factors include VEGF, which signals through cognate flk and flt receptors, while endothelial monocyte-activating polypeptide-II acts as anti-angiogenic factor (Warburton, 2000).

In *Drosophila*, the peripheral nervous system and the neighboring CNS sends Dalmatian signals to trachea and the salivary duct, which are necessary for the normal development of these organs. It is concluded that

> *a properly formed nervous system is required for normal tracheal and salivary duct development. Neurons may modify the local environment or may provide cues for proper tracheal and salivary duct formation through signaling molecules.*
>
> *Kerman and Andrews (2006)*

In humans, RA "is indispensable for the formation of primary lung buds and the oesophago-tracheal septum" (Mollard et al., 2000), and the human embryo may extract RA from the maternal circulation (Morris-Kay and Sokolova, 1996). By the 20th gestational week, the embryonic bronchial epithelium differentiates into alveolar epithelium (type-II pneumocytes). At that point in time, besides the Wnt-7b signaling, other hormonally regulated signals, such as FGFs, TGFs, and RA, are also involved in the development of the tracheobronchial tree and alveoli (Chytil, 1996; Cardoso, 2001). In both humans and *Drosophila*, during the development of the respiratory system, an FGF pathway is "reiteratively used" to pattern successive rounds of branching (Metzger and Krasdnow, 1999).

Other hormonal influences on lung development are also known. The hormonally induced epidermal growth factor (EGF) has a stimulating effect on the lung development, while the hormone dihydrotestosterone is an inhibitor of pneumogenesis. As shown in Chapter 1, the above growth factors are elements of signal cascades epigenetically starting in the CNS.

The lung anlage grows out of the foregut endoderm into adjacent mesenchyme as an initial stage in the process of the formation of the bronchial tree. Cells of the primitive lung epithelium are undifferentiated pluripotent cells. The first differentiated cells to appear in the anlage are pulmonary neuroendocrine cells (PNECs) (Linnoila, 2006; Pan et al., 2006; Cutz et al., 2008). In human embryos, their differentiation takes place by the 8th week of gestation. Their differentiation is characterized by expression of the same genes that characterize differentiation of neurons in *Drosophila* (Linnoila, 2006). PNECs appear as solitary cells distributed throughout the airways mucosa of mammalian lungs or in small clusters known as neuroepithelial bodies (NEBs) that appear later in lung development by the 10th gestational week and are located mainly in branching points of bronchioles. Based on the recent evidence that PNECs are the first cell type to differentiate and produce substances with growth factor and mitogen properties (for example, serotonin, bombesin, calcitonin gene-related peptide (CGRP)), and release neurotransmitters and neuropeptides in response to intrauterine hypoxia and mechanical strain, it is suggested that they may play a critical role in cell proliferation and differentiation and lung morphogenesis in general (Pan et al., 2006; Cutz et al., 2008).

Both NEBs and PNECs use NADPH (nicotinamide adenine dinucleotide phosphate) oxidase as a molecular oxygen sensor and respond to hypoxia by releasing the neurotransmitter serotonin, CGRP, and others. Recognition of NEBs as sensory airways receptors required demonstration of the presence in them of sensory innervation, which now is indisputably established to be vagal myelinated innervation (Figure 5.18). NEBs are also innervated by spinal afferents (Adriansen et al., 2006). Changes in oxygen (for example, hypoxia) are converted in action potentials and afferently transmitted for processing in the brain stem, while the efferent input provides NEBs information for secreting serotonin, CGRP, and others. In response to mechanical strain as a result of secreted fluids, diaphragm contraction, or fetal breathing movements, PNECs and NEBs release serotonin, which stimulates proliferation of fetal lung cells and affects lung maturation by inducing production of extracellular matrix and differentiation of alveolar cells (Pan et al., 2006) and has potential effects on lung development and resorption of the lung fluid at the time of birth.

Production of the surfactant (a mixture of proteins and phospholipids that is essential for survival at the time of birth) by alveolar cells in lungs results from the following cerebrally induced signal cascade:

Hypothalamic CRH → pituitary corticotropin → adrenal cortisol → the fibroblast pneumocyte factor (secreted by lung fibroblasts) → surfactant (secreted by type-II pneumocytes) (Figure 5.19).

Nephrogenesis

Since early 1950s, Grobstein (1956) observed that explants of the mouse embryonic spinal cord, separated from the metanephrogenic mesenchyme by a cell-non-penetrable filter membrane, induced formation of tubules in the mesenchyme. Later, it was found that the nephrogenic action of the spinal cord was related to secretion by the spinal cord of Wnt-1 protein, which, at that stage of embryonic development, is detected exclusively in the CNS (Herzlinger et al., 1994). Other studies have shown that another secreted protein, Wnt-4, was the most important inducer of

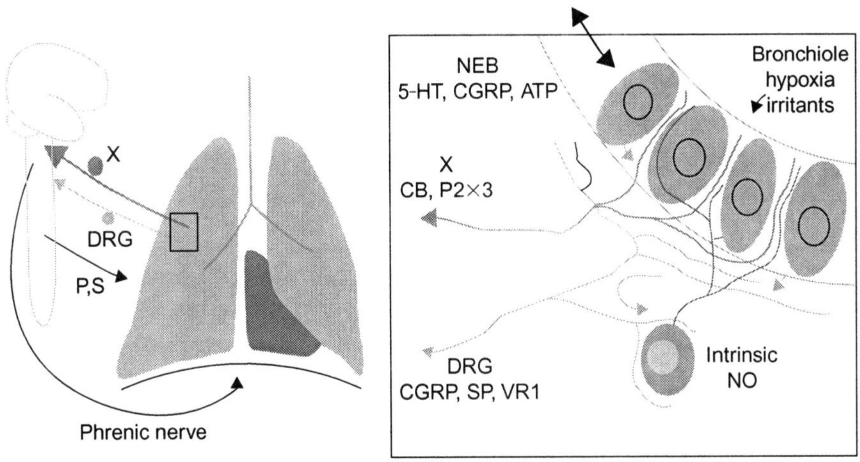

Figure 5.18 Complex functions require complex structures. Schematic representation of dual afferent innervation of the lung and a proposed local NE modulatory arrangement. Left panel: Dual afferent innervation of the lung. Vagal (X, red) and dorsal root ganglionic (DRG, green) primary afferents innervate intrapulmonary airways and project to brain stem and spinal cord (blue). Various reflex responses are relayed through parasympathetic (P), sympathetic (S), and phrenic nerve pathways regulating respiration and cardiopulmonary homeostasis. Right panel: Close-up view of an NEB. Thick caliber vagal afferents (red) distribute terminal branches between cells, while DRG afferent fibers (green) form a subepithelial plexus. Axons originating from nitrergic intrinsic neurons (dark blue) also ramify between NEB cells. Stimuli from the bronchiolar lumen may trigger release of, e.g., ATP from NEB cells, thus exciting vagal afferents through P2X3 receptors. Likewise, subepithelial DRG afferents may also be stimulated and could modulate activity of intrinsic neurons via an axon reflex (curved green arrow). Intrinsic neurons, in turn, may regulate sensitivity of NEB cells to luminal stimuli by release of NO from their intraepithelial terminals. Black double-headed arrow indicates mechanical stimulation of NEB during distension of the bronchiole, which also may trigger mediator release from NEB cells. (For interpretation of the references to color in this figure legend, the reader is referred to the web version of this book.)
Source: From Neuhuber (2003).

nephrogenesis (Vainio and Uusitalo, 2000; Vainio, 2003). In experiments with metanephric mesenchyme of null mutants for Wnt-4, the mesenchyme degenerated, but when brought in contact with the spinal cord of the same embryo, it continued to develop normally, indicating that other Wnt proteins (Wnt-1, Wnt-3a, Wnt-7a, and Wnt-7b) secreted by the spinal cord may replace Wnt-4 in mesenchyme tubulogenesis (Kispert et al., 1998; Figure 5.20). Cultures of the metanephric ridge in presence of the spinal cord are induced to differentiate into renal epithelium (Karp et al., 1994). *In vitro* experiments have shown that when nephrogenic mesenchyme is separated from various embryonic tissues by a membrane filter, of all the tissues tested for induction of nephrogenesis,

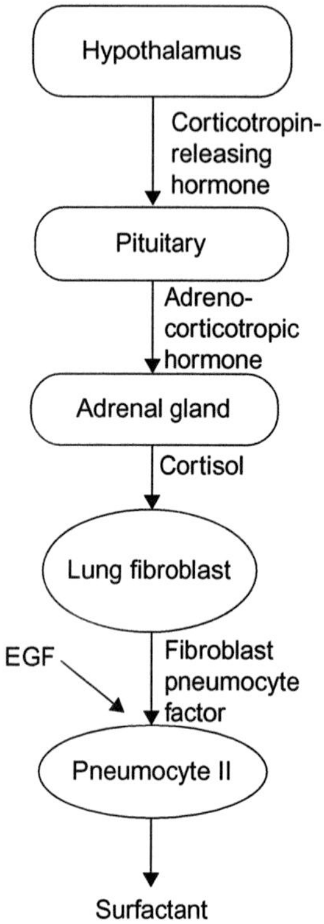

Figure 5.19 The signal cascade for production of surfactant starts in the brain. In the secretion of the surfactant are also involved EGF, via its receptor (EGF-R) (Dammann and Nielsen, 1998), TGFβ-R, dihydrotestosterone (DHT) (Damman et al., 2000), as well as thyroid hormones (DiFiore and Wilson, 1994), paracrine hormones parathyroid hormone-related protein (PTHrP), and leptin (Torday et al., 1998; Torday and Rehan, 2002).

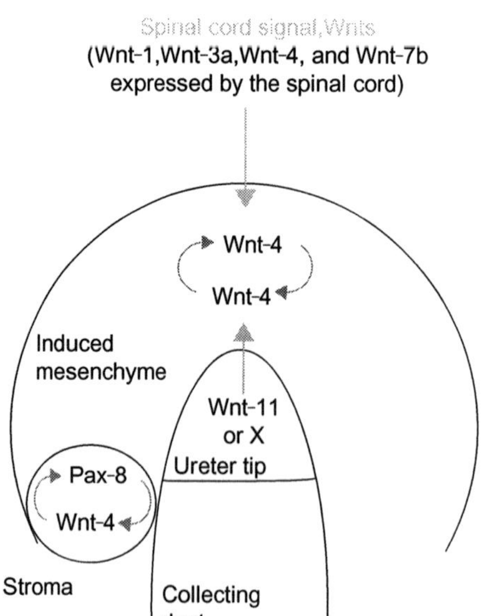

Figure 5.20 A model showing how Wnt-4 operates as "a second" tubule inductive signal. A ureter-derived signal, in the form of Wnt-11 or X leads to induction of expression of Wnt-4, which autoregulates itself and triggers tubule morphogenesis. Wnt-4 signaling involves Pax-8. The spinal cord acts as an inducer tissue as it expresses a panel of Wnts including Wnt-4, which triggers the autoregulated Wnt-4 gene expression to induce tubules.
Source: From Vainio et al. (1999).

only embryonic spinal cord and brain were effective, whereas the ureter bud did not induce .... These studies suggest that embryonic neurons are the most effective inducers of nephrogenic mesenchyme in vitro.

Sariola et al. (1989)

The fact that neurons appear early in the mesonephric mesenchyme and may be present in the ureter bud may explain the results of Grobstein's experiments (1956) that suggested a role for the ureter bud in kidney differentiation (Figure 5.21):

It is demonstrated that neurons may have been dissected from the mesenchyme with the ureter bud.

Sariola et al. (1988)

Additionally, TGF-β2, FGF2 (their secretion is hormonally regulated), and leukemia inhibitory factor (LIF) are known to induce renal tubulogenesis via a common Wnt mechanism (Perantoni et al., 1995; Plisov et al., 2001). But FGF2 can induce epithelial conversion of metanephrogenic mesenchyme during tubulogenesis only if combined with a factor contained in pituitary extracts (Perantoni et al., 1995) and, needless to say, pituitary hormones, being induced by respective hypothalamic

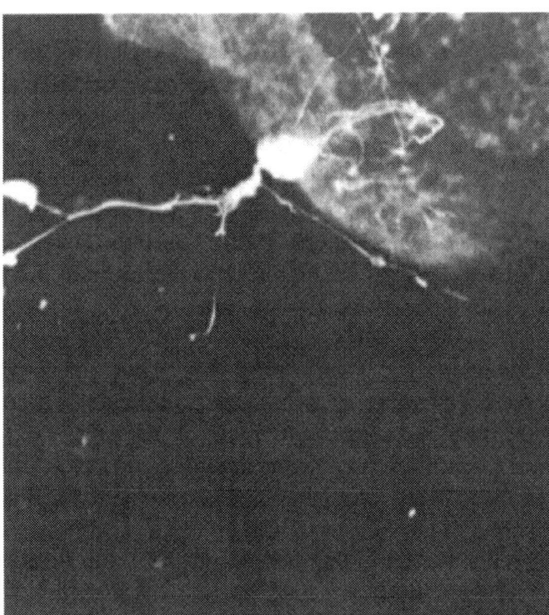

Figure 5.21 Primary culture of an isolated ureter bud cultivated for 2 days on a glass cover slip. The microsurgically dissected ureter bud consists of an epithelial bud surrounded by a sheet of loose mesenchymal cells. Neurofilament staining reveals neurones in the mesenchyme around to the isolated ureter bud.
Source: From Sariola et al. (1989).

releasing hormones, are under CNS control. Wnt-4, hormonally regulated by prolactin, is also involved in the initial stage of tubulogenesis (Wolpert et al., 1998), whereas TGF-β1 prevents formation of ureteric buds (Ritvos et al., 1995).

The development of kidney in mammals involves a close and intricate interaction between the Wolffian duct and metanephrogenic mesenchymal cells. The latter are induced by growth/differentiation factor-11 to release glial cell-line–derived neurotrophic factor (GDNF), which promotes the development of the primary ureteric buds from segments of the Wolffian duct and attracts ureteric bud branches toward the source of GDNF (Sainio et al., 1997). But GDNF itself is known to be induced by an RA signal.

In *X. laevis*, the development of the early embryonic kidney, protonephros, can be induced by treatment of ectoderm with RA and activin A (Uochi and Asashima, 1996). Even the undifferentiated animal pole ectoderm of this amphibian *in vitro* is induced to form protonephros when treated with RA and activin A (Osafune et al., 2002).

Hormones such as pituitary GH and growth factors such as IGF-1 and IGF-2 induce kidney growth during normal renal development (Feld and Hirschberg, 1996). Let us remember that secretion of GH by the pituitary is cerebrally stimulated by the hypothalamic neurohormone GHRH, while GH stimulates synthesis of IGFs in the liver.

Neural Control of the Primary Sex Determination

All human embryos are morphologically indistinguishable until about the 7th week, when they begin to follow divergent patterns of male and female development. Fourteen years ago, in a seemingly *tour de force*, biologists discovered that the process of sexual differentiation starts by the 6th gestational week with the activation of *sry* gene, a testis determining factor, localized in the chromosome Y. Accordingly, expression of this gene stimulates differentiation of Sertoli cells and testis (Koopman et al., 1990; Jeske et al., 1995; Mittwoch, 2000). However, several other genes have been discovered ever since, which, similar to *sry*, are expressed in the genital anlage (genital tubercle): genes responsible for RA receptors, *Fgf8, cyp26* (Ogino et al., 2001) as well as *Shh*, which regulates expression of *Ptch1*, *Bmp4*, *Hoxd13*, and *Fgf10* (Haraguchi et al., 2001).

To cope with these and other experimental results, a more complex explanation was offered: the development of testis requires that SRY, directly or indirectly, induces the gene for the transcription factor, steroidogenic factor-1 (Sf1). The signaling network would also comprise cooperation between the Sf1 and Sox9 proteins for upregulating production of another gene, *AMH* (Gilbert, 2000). Anyway, it was generally acknowledged that the *sry* in the Y chromosome is "the primary sex determining gene in mammals" (Wolpert et al., 1998, pp. 372–374).

Contrary to conventional wisdom, it is now reported that sexually dimorphic gene expression in the brain precedes differentiation of gonads. Out of 12,000 genes expressed in the embryonic brains of mice, investigators have identified 50 candidate genes and confirmed at least seven (Dewing et al., 2003) that are differentially expressed in brains of male and female mice "before any gonadal hormone influence" and before the start of the *sry* gene expression.

Sex-specific expression of genes in the embryonic chick brain starts at embryonic day 4, that is earlier, than in any other embryonic structure, including the genital ridge, where *sex-specific* expression of genes start only at embryonic day 5 (Scholz et al., 2006). This suggests that, notwithstanding the widely held opinion that secretion of sex hormones in the embryonic gonads determines the sex-dependent expression of genes in the brain, it is the sexuality of gonads that may be determined by brain signals.

There is additional strong evidence that the *sry* gene, the chromosome Y, and the genital anlage may not be responsible for determining embryo's maleness. In recent experiments with Japanese quails, it has been demonstrated that transplantation of the female forebrain primordium into male embryos, before gonadal differentiation, suffices to prevent differentiation of male sex organs (Gahr, 2003) in genetically male embryos. It seems that sexual differentiation, in birds and mammals, is cerebrally, rather than hormonally, determined.

Early in embryogenesis, mesonephric cells migrate to the male gonad anlage to differentiate into gonadal somatic cells. It is believed that this migration is induced by MMPs, which are regulated by the Sry–TIMP-3 (tissue inhibitors of metalloproteinase-3) cascade (Nishino et al., 2002), but expression of TIMP-3 mRNA is induced by the hypothalamic neurohormone gonadotropin-releasing hormone (GnRH) (Bakke et al., 2002). Mesonephric endothelial cells migrate to form the coelomic blood vessel, and steroidogenic adrenal precursor cells migrate to convert into steroidogenic testicular cells. The migration of cells to the gonadal anlage is regulated by the Wnt (Jeays-Ward et al., 2003), and the neural tissue is the exclusive provider of the Wnt-1 at that stage of development (Herzlinger et al., 1994).

There are numerous studies in other vertebrates showing that sex may be determined by environmental stimuli, such as the temperature of the environment where eggs are laid, rather than genetically. This is the case in many reptiles: the embryonic brain perceives the temperature, processes it, and converts it into a signal in the brain, which induces expression of aromatase (Willingham et al., 2000; Milnes et al., 2002), which in turn converts androgens into estrogen, determining the female sex of the embryo even when the embryo is genetically male.

Larvae of the genotypically females of *Pleurodeles waltlii*, reared at 32°C before formation of the genital ridge (stages 43–51), convert into male newts (Dournon and Houillon, 1985).

In the sea turtle, *Lepidochelys olivacea*, it is the diencephalon, not the gonads, that starts the sexual differentiation by responding to the female-promoting temperature with elevated estradiol levels, and it is the diencephalon that senses temperature during sex determination (Salame-Mendez et al., 1998). Still undifferentiated gonads of the sea turtle *L. olivacea* are pervaded by acetylcholinesterase-positive nerve fibers from spinal cord during the thermosensitive period of sex determination (stages 20–27), and it is believed that

> *the spinal cord and the innervation derivating from it could play a role in driving or modulating the process of temperature-dependent gonadal sex determination and/or differentiation.*
>
> *Gutierrez-Ospina et al. (1999)*

In many insects, transplantation of male brains alone into female abdomens induces presumptive ovaries to develop apical tissue and male secondary sexual characters (Gorbman and Davey, 1991).

It has also been reported that brain aromatase has no effect on gonadal differentiation in amphibians (for review, see Pieau and Dorizzi, 2004), but this does not raise doubts about the epigenetic nature of the temperature-dependent sex determination.

Neural Control of Sex Conversion

Sex conversion is another illustration of the epigenetic determinism of sex in vertebrates. It is common in a number of teleost fish. It may be protogynous (from female to male) or protandrous (from male to female) and ranges from permanent protogynous/protandrous change of sex more than once in either direction (Grober and Sunobe, 1996) to extremely quick and reversible gender change in a matter of minutes (Bass and Grober, 2001). The phenomenon is associated with radical changes in the direction of the opposite sex in the morphology, morphometry, and behavior of the sex converts.

Sex conversion is induced by specific changes in the social environment, but only recently has it been possible to discover the chain of events that link the social environment with the sex change. At least two different cascades of events have been identified so far. On the one hand, in response to social cues, fish of the African sex-changing cichlid species, *Astatotilapia* (*Haplochromis*) *burtoni*, modify electrical properties and the size of the hypothalamic GnRH neurons, as well as their reproductive activity (Greenwood and Fernald, 2004). On the other, sex conversions (female → male and male → female) in teleost fish depend on the increase and decrease in the number of GnRH-secreting neurons. The increase in the number of GnRH neurons in teleost fish is part of the masculinizing mechanism, which is responsible for mediating the socially determined changes in the morphology of gonads (Elofsson et al., 1997a).

Studies on sex conversion in protogynous (*Labrus berggylta*) and protandrous (*Amphiprion melanopus*) fish have shown that GnRH neurons in the cerebral regions of the preoptic area (POA) are responsible for the socially mediated sex conversion (Elofsson et al., 1997b; Bass and Grober, 2001).

Masculinization of protogynous fish of the species *Lythrypnus dalli* (Gobiidae) is associated with increased size of particular secretory neurons in the forebrain (Reavis and Grober, 1999). Based on the fact that pituitary luteinizing hormone (LH) levels are correlated with conversion from female to male, it is suggested that LH in the black porgy, *Acanthopagrus schlegeli*, may also have a masculinizing effect (Lee et al., 2004). Now, it is generally acknowledged that sex conversion in fish is determined by a central nervous mechanism:

> *Such changes in the POA GnRH cell population phenotype may reflect a proximate central mechanism in the induction of the dramatic gonadal and behavioural transformations that are associated with sex change in hermaphroditic fish.*
>
> *Elofsson et al. (1997b)*

Sometimes, social stimuli trigger a response from the arginine/vasotocin neuropeptide system in the POA region, which is involved in the cascades of events determining sex change in teleost fish (Bass and Grober, 2001; Semsar and Godwin, 2003). Other cerebral factors may be involved in sex conversions in teleost fish, as demonstrated by the fact that intraperitoneal administration of the neuropeptide Y (NPY) in protogynous bluehead wrase, *Thalassoma bifasciatum*, induces degeneration of ovaries and their conversion into males (Kramer and Imbriano, 1997). The stoplight parrotfish, *Sparisoma viride*, may, spontaneously or experimentally, by androsterone injections, change sex during its lifetime. Pollutants such as the herbicide atrazine are reported to have feminizating and/or hermaphroditizing effect on frogs, in both nature and experiments (Hayes et al., 2002a,b).

Generalizing the extensive evidence on the role of social interactions and related behavior in sex conversion in teleost fish, experts in the field have concluded that

> *the initiation of the sex reversal is often controlled by social, behavioural factors, and since the only way behaviour can affect the gonads is through the brain, there must be central neuronal mechanisms underlying the gonadal change (my italics—N.C.).*
>
> *Elofsson et al. (1997a)*

Reproductive activity of males of the African cichlid fish, *Astatotilapia burtoni*, depends on the individual's status. Subordinate fish display submissive (but not courtship) behavior and have dull body coloration, while dominant males display aggressive and courtship behavior and bright coloration and defend their spawning territory. Dominant males also have larger testes, larger GnRH1 neurons in the hypothalamus, higher levels of GnRH1 and LH, and higher expression of the GnRH1 receptor in the pituitary. However, when the social environment allows it, they ascend to the dominant social status by activating the brain–pituitary–gonad axis and displaying the dominant phenotype in behavior, body coloration, and hormonal parameters, in testes cytology and spermiogenesis, thus becoming reproductively competent (Maruska and Fernald, 2011; Figure 5.22).

Osteogenesis

Bone formation takes place in two distinct ways: by endochondral osteogenesis (*osteogenesis cartilaginea*) and intramembraneous osteogenesis (*osteogenesis membranacea*).

During endochondral osteogenesis, mesenchymal cells accumulate to form condensations, cartilaginous anlagen, on the site of the future bone. The pituitary GH, which is induced by the hypothalamic neurohormone GHRH, stimulates proliferation of prechondrocytes by shortening 50% of their cell cycle in the epiphyseal growth plates (between the epiphysis and diaphysis of long bones). GH binds its receptor in the cells of the growth plate (Wolpert et al., 1998, p. 425) and induces local synthesis of IGF-1 in these cells, but the fact that administration of systemic recombinant IGF-1 stimulates bone growth in hypophysectomized (hence, unable to

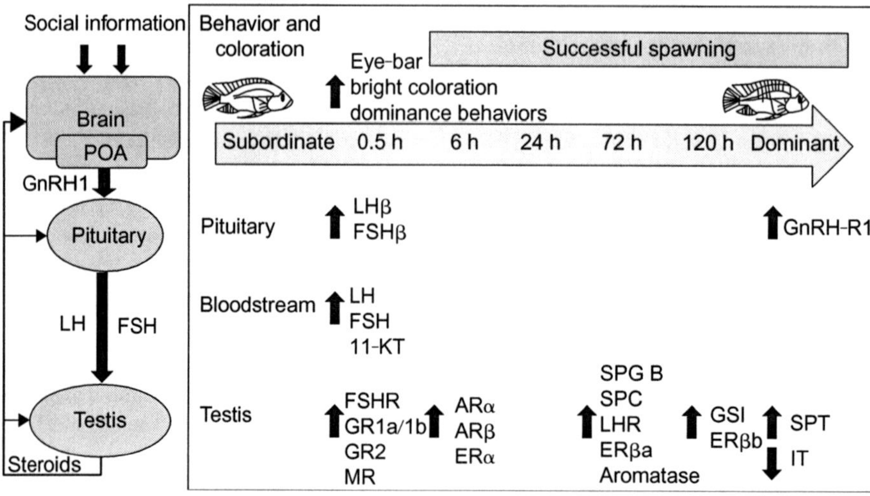

Figure 5.22 Temporal summary of physiological changes along the reproductive axis during social ascent in male *Astatotilapia burtoni*. *Arrows* indicate the time point at which the first significant increase (*up arrows*) or decrease (*down arrows*) from stable subordinate male values was observed. Note that any further significant differences after this initial change in each measure are not shown. Schematic simplified diagram of the brain–pituitary–gonad (BPG) axis is shown at left. *Abbreviations*: 11-KT, 11-ketotestosterone; IT, interstitial tissue; SPC, spermatocytes; SPG B, type B spermatogonia; SPT, spermatids.
Source: From Maruska and Fernald (2011).

produce GH) rats suggests that IGF-1 has itself endocrine function in osteogenesis (Ohlsson et al., 1998). Thus, the pituitary GH stimulates longitudinal bone growth by directly inducing division of prechondrocytes in the epiphyseal growth plates and by inducing local and hepatic synthesis and secretion of IGF-1 (Ohlsson et al., 1998; Figure 5.23).

Osteoblasts (Gr. *osteon*—bone and *blastos*—germ, sprout) are the main cytological players in osteogenesis. They produce the extracellular matrix of the bone to which calcium salts are deposited. About 90% of the proteins of the cellular matrix they secrete are represented by type I collagen. Collagen fibers serve as crystallization germs for mineral components that are delivered in the form of matrix vessels containing calcium as well as proteins, phospholipids, pyrophosphatase, and alkaline phosphatase (Beck, 2003).

The neuropeptide CGRP, which is released by nerve fibers in contact with osteoclasts, promotes bone growth by decreasing formation of osteoclasts (Burt-Pichat et al., 2005). Remember, while IGF-1 inhibits GH synthesis, it is the hypothalamic neurohormone GHRH that stimulates GH synthesis and secretion.

Adrenal glucocorticoids decrease bone formation by downregulating the type I collagen, osteocalcin, and IGF-1 (Delany et al., 1994). These adrenal hormones cause bone loss by stimulating osteoclastogenesis via expression of RANK ligand

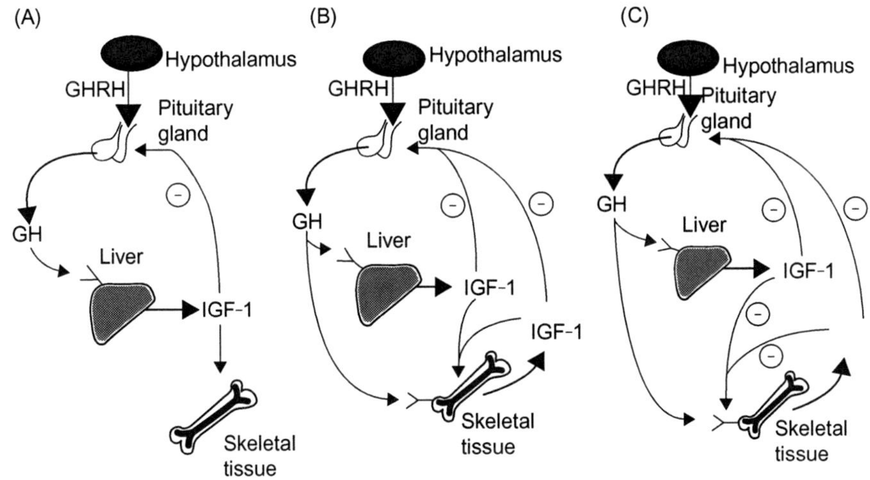

Figure 5.23 Schematic representation of regulation of skeletal tissue by GH and IGF-1. Y depicts GHRs. (A) Classic negative feedback mechanism. In accordance with the somatomedin hypothesis, circulating IGF-1 derived from the liver in response to GH stimulates skeletal tissue growth and feeds back centrally to inhibit pituitary GH secretion. (B) Modified classic negative feedback mechanism. In addition to the endocrine effects of IGF-1 derived from the liver, GH can directly stimulate skeletal tissue growth through local production of IGF-1. Hepatic and extrahepatic sources of IGF-1 contribute to feedback inhibition of GH release. (C) Proposed peripheral negative feedback loop. IGF-1 produced by skeletal tissue in response to GH feeds back to inhibit the local action of GH by reducing GHR availability. *Abbreviations*: GH, growth hormone; GHR, growth hormone receptor; IGF-1, insulin-like growth factor-1.
Source: Slightly expanded from Ohlsson et al. (1998).

and by decreasing expression of osteoprotegerin (Canalis, 2003). They also cause bone loss by decreasing division and differentiation of osteoblastic cells and by inducing apoptosis of mature osteoblasts (Canalis and Delany, 2002; Weinstein et al., 2002) by stimulating osteoclastogenesis (Canalis and Delany, 2002) and by extending the life span of osteoclasts (Weinstein et al., 2002).

Recent evidence shows that the traditional view on the osteoporosis as a result of hyperfunction of the thyroid gland and its THs needs to be revised. It is reported that the pituitary TSH acts directly (not via THs) on osteoblasts and osteoclasts via its receptor (TSHR), which is expressed in both of these cell types. TSH inhibits both differentiation of osteoblasts and formation of osteoclasts, thus leading to osteoporosis (Abe et al., 2003).

The hormonal regulation of osteogenesis considered so far involves four main regulatory axes: the hypothalamic–pituitary–gonadal, hypothalamic–pituitary–adrenal, hypothalamic–pituitary–thyroid, and hypothalamic–pituitary (in the case of the direct action of the pituitary TSH) axes. Signal cascades along all the three axes have their origin in the CNS (hypothalamus). Biologists marvel at the accuracy of the symmetrical development of long bones:

> *In view of the complexity of the growth plate, it is remarkable that human bones in the limbs on either side of the body can grow for some 15 years independently of each other, and yet eventually match to an accuracy of about 0.2%.*
>
> *Wolpert et al. (1998, p. 424)*

All of the evidence presented above on the control of the bone development by signal cascades that ultimately originate in the CNS points in the direction of this information-receiving, -processing, and -generating organ. To the same direction points the maintenance of the continually eroding bone structure during the adulthood and also pathological processes leading to the loss of its normal structure.

Regulation of Bone Homeostasis

Bone homeostasis implies bone formation and bone resorption, carried out, respectively, by osteoblasts and osteoclasts. Two main components, a neural and a hormonal, are involved in the maintenance of the normal adult bone mass and structure. Needless to reiterate, hormones generally represent downstream elements of signal cascades originating in the CNS.

Studies on the relationship between the hormone leptin and bone formation in mammals led investigators to the conclusion that leptin performs its antiosteogenic action via the CNS (Ducy et al., 2000). Until recently, it was not known how this central CNS message was delivered to the bones. Now, the experimental evidence shows that the regulation of bone mass is not hormonal, because leptin does not act on bone osteoblasts. When leptin was intracerebrally infused in one of a pair of cross-circulated homozygous leptinless (*ob/ob*) mice, significant decrease of bone mass was observed only in the cerebrally infused mouse. However, in absence of Arβ2 receptor, intracerebrally administered leptin fails to produce its bone resorptive action (Takeda et al., 2002; Elefteriou et al., 2005).

When the level of leptin is high, it binds its receptor ObRb in serotonergic neurons of the brain stem stimulating these neurons to secrete serotonin which attenuates activation of noradrenergic neurons in locus coeruleus (see Figure 6.8 in Chapter 6). Serotonin binds its receptor Htr2c on neurons of the VMHs, which, via medulla oblongata, decrease sympathetic activity, thus leading to bone growth. When the level of leptin is low, the electrical activity in serotonergic neurons of the brain stem decreases, and serotonin secretion is inhibited. This increases the noradrenergic tone and results in bone loss. It was found that brain-derived serotonin (BDS) represents only 5% of the total serotonin pool of the body but, despite that fact, the role of BDS in bone remodeling was greater than that of gut-derived serotonin (Yadav et al., 2009).

Hypothalamic regulation of bone formation/remodeling also involves the central hypothalamic molecular clocks and peripheral clocks in osteoblasts, but the latter act downstream the sympathetic tone and inhibit osteoblast proliferation (Takeda and Karsenty, 2008). Central stimulation of the sympathetic innervation in bones increases

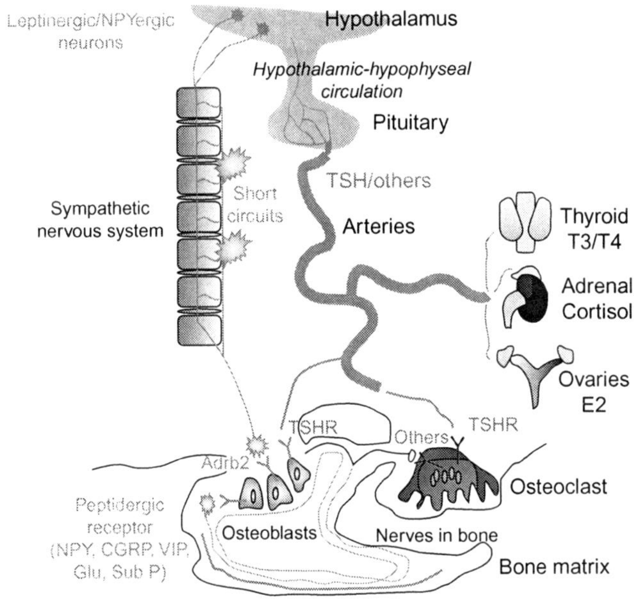

Figure 5.24 Frequency (FM)- and amplitude (AM)-modulated arms for hypothalamic surveillance of skeletal homeostasis. Short circuits refer to nonleptinergic, non-NPYergic release of adrenergic neurotransmitters, as well as epinephrine release from the adrenals. In addition, neuropeptide-containing nerves richly innervate bone tissue and skeletal blood vessels; however, their physiological significance in skeletal homeostasis is unclear. *Abbreviations*: *Adrb2*, β2-adrenergic receptor; CGRP, calcitonin gene-related peptide; E2, estradiol; FSH, follicle-stimulating hormone; Glu, glutamate; LH, luteinizing hormone; NPY, neuropeptide Y; Sub P, substance P; TSH, thyroid-stimulating hormone; TSHR, TSH receptor; T3 and T4, thyroid hormones; VIP, vasoactive intestinal peptide.
Source: From Zaidi (2005).

the release of noradrenaline on osteoblasts, where it binds its receptor β2-AR and activates a signal transduction pathway involving circadian genes *Per* and *Cry*, which, by suppressing expression of cyclins, inhibit mitotic division and proliferation of osteoblasts leading to loss of bone mass.

The central neural control is complemented with a network of interacting signal cascades involving the endocrine glands as can be seen in Figure 5.24.

The maintenance of the normal bone structure and mass is related to, and depends on, the load they bear and the normal structure of the tissues they support. The effect of load bearing on the structure and mass of bones may be direct (i.e., it derives from strains it produces on the bones) or indirect, resulting from signal cascades that start in the CNS in response to the input of information of load changes. Mechanical loading (for example, body weight, exercise) has an important effect on the development and maintenance of the normal state of skeleton bones. Changes may affect cell metabolism and even expression of genes. Periodical deformation of the bone leads to the flow of extracellular fluid in the periosteal space, shear stress on osteocytes, and expression of *prostaglandin G/H synthase-2 (Cox-2) mRNA*. The latter stimulates proliferation of osteoblasts and osteoclasts (Hakeda et al., 2000). The sympathetic nervous system (β-adrenergic pathway) is crucial in mediating homeostatic effects of mechanical loading in bones (Levasseur et al., 2003).

Immediate Neural Regulation of Bone Homeostasis

For a long time, it was believed that bones represent a poorly innervated tissue. Now we know that bones are pervaded by an intense network of sensory and autonomic nerve fibers, not only in periosteum, but in the cortical bone and bone marrow as well (Lerner, 2002). Nerve fibers come into direct contact with osteoclasts (Hara-Irie et al., 1996) and release there a number of neuropeptides/neurotransmitters (for example, vasoactive intestinal peptide (VIP), PACAP, substance P (SP), CGRP, BDNF, serotonin, glutamate, catecholamine) with effects that are

> *similar to those induced by osteotropic hormones and cytokines and suggest that the nervous system may be part of the endo- and paracrine control of bone metabolism.*
>
> *Lerner (2002)*

Neuroactive substances released by local nerves on osteoblasts and osteoclasts perform their osteoregulatory functions by binding specific neuropeptide receptors, such as VIP-1R and VIP-2R (Lundberg et al., 2001), the CGRP receptor (Lerner, 2002), and adrenergic receptors (Togari, 2002), respectively, to vasoactive intestinal peptide (VIP); CGRP; adrenalin; and other neuropeptides, neurotransmitters, and growth factors secreted by adjacent nerve fibers. It has been demonstrated that nerve fibers in rat bones secrete VIP, NPY, tyrosine hydroxylase (Bjurholm et al., 1988), SP (Hill and Elde, 1991), and CGRP (Hill and Elde, 1991; Irie et al., 2002).

Besides the direct mechanical effect on bone structure and cells, mechanical loading influences bone homeostasis in an indirect way.

Changes in bone loading are sources of mechanosensation. After reception by the sensory neurons, mechanical stimuli arising from changes in bone loading, in the form of electrical signals, are transmitted to specific centers in the CNS, where the sensory information is integrated and processed. The CNS determines activation of signal cascades for processes of bone remodeling "from normal remodeling to fracture healing and nonunion."

> *Bone nerves/neuropeptides may explain why various inputs/outputs are transformed in a meaningful way to altered mass and quality of bone.*
>
> *Konttinen et al. (1996)*

Given both direct and indirect (via neural pathways) effects of the mechanical load on bones, it is not easy to determine the relative role in bone changes resulting from the direct action of loading/unloading on the bones from that of signal cascades that start with processing of loading/unloading stimuli in neural circuits in the CNS. The simplest test for assessing their relative role would be to see how neurectomy of sensory nerves innervating bones affects the symptoms of unloading.

Bone formation and mineral apposition in endosteum increases in loaded (but not in unloaded) control rats (Cheline et al., 2002). Experimental loss of gravitational loading by hindlimb elevation in adult rats reduces bone formation and is characterized by an expected rise in the calcium blood level and a drop in the vitamin D level (Dehority et al., 1999).

The fact that the sensory information of bone loading/unloading from sensory neurons is transmitted for processing in specific brain centers implies that experimental neurectomy or accidental disruption of the transmission of that information to the CNS would cause sensation of unloading. Furthermore, it might be predicted that neurectomy and/or neural injury leads to bone mass losses similar to those observed in the cases of experimental unloading. Indeed, experimental evidence has shown that this is the case. Sciatic neurectomy causes substantial decrease in bone mass in adults (Zeng et al., 1996) and inhibits bone formation during postnatal development (Zeng et al., 1996; Edoff et al., 1997). Spinal injury, causing paraplegia, leads to an increase in the number of osteoclasts, which are specialized in bone resorption, thus causing loss of bone mass (Demulder et al., 1998).

Autonomic nerve fibers are normally present in bone epiphysis and periosteum (Bjurholm et al., 1988), where they release various neuropeptides. *In vitro* experiments have shown that the neuropeptide VIP is a regulator of bone matrix proteins; hence, it is also involved in bone formation. By binding to its receptors, VIP stimulates osteoclasts to resorb bone tissue (Lundberg et al., 2000). Neuropeptide VIP also decreases transcription of the gene for interleukin-6, which, as well as its receptor, enhances the action of osteogenic protein-1, which, in turn, stimulates bone formation (Yeh et al., 2002). And the fact that VIP stimulates alkaline phosphatase activity suggests that it is also involved in regulation of anabolic processes and calcium accumulation in bones (Lundberg et al., 1999).

The neuropeptide CGRP is identified in the sensory nerve fibers of bones. It has the opposite effect of VIP, i.e., it inhibits bone resorption (Lerner, 2002). CGRP binds to specific membrane receptors of osteoblasts and osteoclasts. Stimulation of autonomic sympathetic nerve fibers induces release of osteoclast differentiation factor and osteoclastogenesis inhibitory factor by osteoblasts (Togari, 2002), thus stimulating bone resorption.

Neural Control of the Development of the Mammary Gland

Formation of this gland is under a complex control of systemic neurohormonal signals via hypothalamus–pituitary–ovarian axis (prolactin) and the ovary (estrogen and progesterone). The signals are executed by local mediators, Wnt (Sassoon, 1999), EGF, and TGF-β proteins, neuropeptide VIP, hypothalamic hormone TRH (Gilbert, 1997), and others.

Thus, the cascade of events for the mammary gland development starts in the CNS: in response to internal signals, related to the activation of the hypothalamic GnRH pulse generator, and especially to gestation, neurons of the caudal part of the solitary tract release a neuropeptide, prolactin-releasing peptide, which stimulates the hypothalamus to secrete prolactin-releasing factor/prolactin-releasing inhibition factor. The hypothalamus, via these hormones and by acting directly on the pituitary,

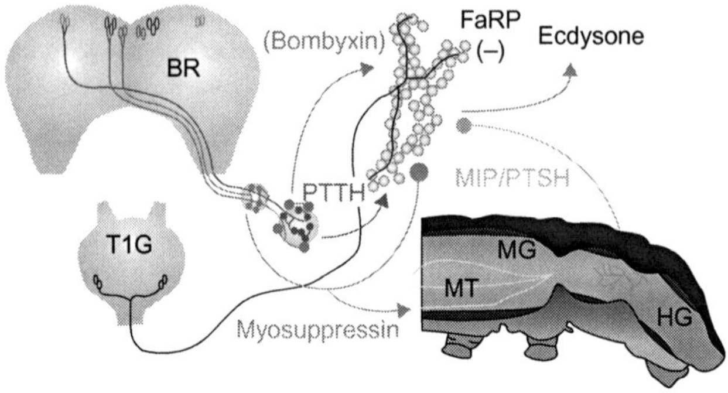

Figure 5.25 Control of ecdysone secretion by the prothoracic glands. The prothoracicotropic hormone (PTTH) from the brain (BR) provides the primary drive for steroid secretion, but the steroid profile is sculpted by the inhibitory effects of circulating myosuppressin from the brain and myoinhibitory peptide–prothoracicostatic peptide (MIP/PTSP) from (secretory neurons—N.C.) of the hindgut (HG). FMRFamide-related peptides (FaRPs) from regulatory neurons in the first thoracic ganglion (T1G) also directly suppress ecdysone secretion. Bombyxin may have an indirect effect on steroidogenesis through stimulation of gland growth. Midgut (MG) and Malpighian tubules (MT) are other targets of the myosuppressin.
Source: From Truman (2006).

induces the latter to secrete prolactin, which has an indispensable role in the mammary gland development. Hormonal signals from the ovary are also involved in the development of the gland (see also Section Development of the Mammary Gland in Chapter 1, and Figure 1.10).

Neural Control of the Development of Body Mass

The fact that metazoans can maintain their body size within relatively narrow limits clearly suggests that they can, among other things, assess their size. It is argued, for insects, that the mechanism that regulates organ or body size "cannot reside at the cellular level" but "resides at a higher level" (Nijhout, 2003).

Where is the center of body size control located in the animal body?

Since insulin signaling is essential for cell growth, and many insulin-like peptides in insects are neurohormones, the brain neurosecretory system may be involved in the mechanism of body mass control (Nijhout, 2003).

In order to determine when "to grow or stop growing," the control system must be able to monitor the body mass continually. Under normal conditions, during insect metamorphosis this function is performed by the CNS. Upon reaching the species-specific body mass, as perceived in the insect CNS, based on afferent information via interoceptors, the CNS, on the one hand, starts secretion of allatostatins, which suppress secretion of the JH by CA, and, on the other, its brain secretes the neuropeptide PTTH (Figure 5.25), which, under the influence of the daily recurring photoperiodic stimulus, induces secretion by the prothoracic gland of ecdysone, which in turn prevents insect's growth.

Figure 5.26 Overall size decrease in adult *Drosophila* flies as a result of ablation of IPCs in the protocerebrum. Left, IPC-ablated fly; right, wild type (wt). *Abbreviation*: IPC, insulin-producing neurons in the protocerebrum.
Source: From Rulifson et al. (2002).

Ecdysone and insulin-like peptides act antagonistically: while ecdysone, above the basal level, acts as inhibitor of cell growth and proliferation, the latter stimulate cell growth and proliferation (Colombani et al., 2005a).

Drosophila expresses seven insulin-like peptides, and the main source of insulin-like peptides in *Drosophila* and other insect larvae is a cluster of insulin-producing secretory neurons in the pars intercerebralis of the protocerebrum (Rulifson et al., 2002; Krieger et al., 2004). These neurons extend processes and release insulin-like peptides at the lateral protocerebrum, subesophageal ganglion, corpora cardiaca and the heart, from where, via hemolymph, they reach all of the regions of the body (Rulifson et al., 2002).

In the brain of *Drosophila*, seven median secretory neurons (m-NSCs) express three insulin-like genes, but expression of only two of these genes depends on the food availability. Ablation of the insulin-producing cells of the *pars intermedialis* of the protocerebrum prevents the growth of the fly with symptoms of starvation phenotype (Ikeya et al., 2002). It also leads to reduction of the wing size to 61% of the norm, while wing cell number and size are reduced to 72% and 85%, respectively (Rulifson et al., 2002) (Figure 5.26). The decrease in the size of the flies also results from decrease in both the number and the size of cells.

The insulin-signaling pathway in *D. melanogaster* plays a central role in regulating growth, development, reproduction, and aging in response to nutrition and metabolism (Krieger et al., 2004).

Insulin-like peptides act by binding their membrane receptors, thus activating the kinase PI3K and protein kinase B transduction pathway. By experimentally increasing levels of PI3K in the prothoracic gland, investigators have been able to increase the speed of cell proliferation and growth (without changing the period of the larval growth), thus producing adult *Drosophila* that were 17% bigger than their parents. Similar results of producing offspring of larger body size were obtained by silencing EcR in insulin-producing neurons of the brain (Colombani et al., 2005b).

The body weight set point beyond which the yellow-spotted longicorn beetle, *Psacothea hilaris*, can enter diapause is 600 mg (Munyiri et al., 2004). In the dung beetle, *Onthophagus taurus*, only males that reach a particular body mass set point develop long horns (Emlen and Nijhout, 2001). A similar cerebrally established set point for body mass seems to exist in other insects: when they reach a weight of 5 g, the JH titer drops to allow metamorphosis to take place,

> *suggesting that proprioceptive information, depending perhaps on stretch, is integrated by the brain causing the CA (corpora allata—N.C.) to switch off.*
>
> *Gorbman and Davey (1991), p. 721*

What essentially takes place when insects of the genus *Rhodnius* stop growing at the onset of metamorphosis has been summarized as follows:

> *A blood meal causes the abdomen to expand, which stimulates stretch receptor nerves in the abdomen, which in turn stimulate specialized cells in the brain to secrete prothoracicotropic hormone, which is released into the blood and stimulates the prothoracic gland to secrete the molting hormone, ecdysone, which stimulates molting.*
>
> *West-Eberhard (2003)*

Another paradigmatic example of the brain control of growth in invertebrates is observed in the freshwater snail *Lymnaea stagnalis*. In 1976, Geraerts reported that cauterization of the neurosecretory light green cells (LGCs) in the brains of juvenile snails of this species results in marked retardation of body growth that may be rescued by implanting cerebral ganglia containing LGCs. Studies on the function of LGCs have shown that these paired groups of 150–200 giant neurons (Smit et al., 1998) express four member types of molluscan insulin-like peptides (MIPs), which then are deposited in axon terminals and released in the periphery of the paired median lip nerves, which serve as their neurohemal area (Figure 5.27). Brain lateral lobes are connected with LGCs and seem to have an inhibitory effect on the function of LGCs, as is indicated by the fact that cauterization of lateral lobes results in giant growth of the snail (Geraerts, 1976). MIPs may act both as neurohormones via hemolymph and in a paracrine way within the CNS (van Heumen and Roubos, 1990).

Neurohormonal mechanisms for the control of the body mass also exist in vertebrates (Halaas et al., 1995; Pelleymounter et al., 1995; Norman and Litwack, 1997), where the CNS with the hypothalamic–pituitary–liver axis functions as the main regulatory axis (Figure 5.28).

These mechanisms are under the CNS control, and recent evidence suggests the existence in the CNS of higher vertebrates of a body mass set point. In experiments with deer mouse (*Peromyscus maniculatus*), intraperitoneal implantation of metabolically inert masses causes a compensatory and equivalent loss of body weight. In

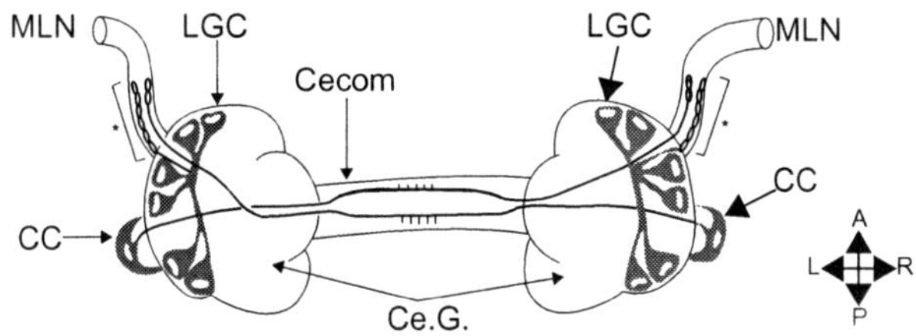

Figure 5.27 Schematic representation of the light green cells (LGCs) and canopy cells (CCs) of the cerebral ganglia of *Lymnaea stagnalis*. Dorsal view of a pair of cerebral ganglia. The LGCs release their materials from the regions indicated by asterisks, which are the axonal plates. The CCs release MIPs at two different sites, i.e., the axon terminals in the cerebral commissure and the contralateral MLNs. *Abbreviations*: CC, canopy cell; CeCom, cerebral commissure; Ce.G. cerebral ganglion; LGC, light green cell; MLN, median lip nerve.
Source: From Hatakeyama et al. (2000).

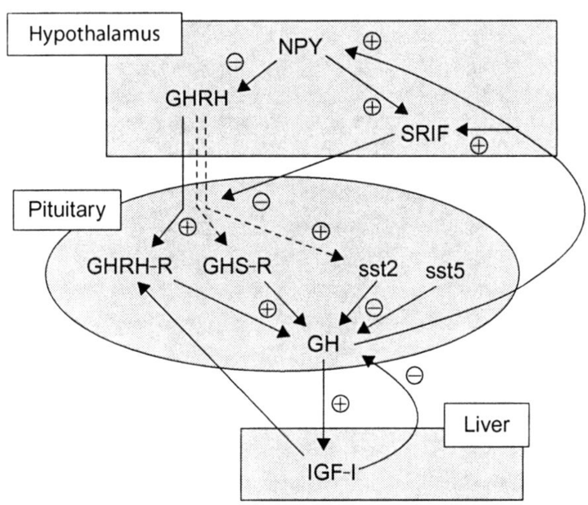

Figure 5.28 Schematic representing the putative interactions of the various components of GH axis. Interactions representing direct effects are shown by the *solid lines*, whereas the *dashed lines* show interactions that may be indirect. +, −, stimulatory, or inhibitory actions on hormone synthesis and/or release, respectively. *Abbreviations*: GH, growth hormone; GHRH, growth hormone-releasing hormone; GHRH-R, growth hormone-releasing hormone receptor; GHS-R, secretagogue receptor; IGF-1, insulin-like growth factor-1; NPY, neuropeptide Y; SRIF, somatostatin; sst2 and sst5, SRIF receptor subtypes.
Source: From Peng et al. (2001).

the 12th day after removal of the implant, the animals regained the pre-experimental body weight. Changes in the body weight suggest the existence of a set point that is sensitive to changes in the perception of mass and transduced via neural pathways (Adams et al., 2001).

It is suggested that numerous mechanoreceptors located within muscles and tendons that have afferent pathways to cerebral cortex provide to the CNS the input on the body mass (Adams et al., 2001). Since the processing of the sensory input is performed by the brain, the processes of growth and/or apoptosis for adjusting the body weight, according to the changed perception of the body mass, must also start with brain signals. Indeed, experimental evidence has localized, in the hypothalamus, a center that controls food intake and body weight via $GABA_B$ receptors (Baeckberg et al., 2003).

Neural Control of the Life Span

The IGF (insulin-like)-signaling pathway is conserved across metazoan taxa. Besides its well-established functions in the control and regulation of body size, this pathway seems to have another important function in determining life span in the animal kingdom. In *D. melanogaster*, ablation of the m-NSCs (medial neurosecretory cells) of the pars intercerebralis, which secrete *Drosophila* insulin-like peptides-2, -3, and -5 (dilp2, dilp3, and dilp5), increased the life span by 10.5% in males and 18.5% and 33.5% in virgin and mated females, respectively. The increase in life span was accompanied by an increase of stress resistance and reduced fecundity (Broughton et al., 2005).

Thirty-seven genes encoding insulin-like peptides are produced in the nervous system of the nematode worm, *Cenorhabditis elegans* (Pierce et al., 2001). It is believed that this family of genes evolved some 600 million years ago, but some of them evolved as a result of recent gene duplications (Nelson and Padgett, 2003). In *C. elegans*, DAF-2, which is similar to vertebrate IGF receptors (Foxo), is the only insulin/IGF receptor, and via Akt-1,2 and Sgk-1 it phosphorylates DAF-16, leading to nuclear exclusion of DAF-16 and physiological aging. This is the normal "reproductive mode" of *C. elegans* life history. Suppression of insulin-like peptides and IGFs, or their receptor DAF-2, switches the worm to "waiting mode" by increasing life span and delaying age-related functional decline (Tatar et al., 2003; Panowski and Dillin, 2009; Figure 5.29).

Based on the above mechanism, in response to environmental conditions, *C. elegans* may adopt one of two alternative modes of life history: under favorable conditions, it adopts the reproductive mode (pro-aging mode), whereas under unfavorable stressful conditions, the waiting mode, which is an antiaging mode (for example, diapause, hibernation).

In vertebrates, the body growth is determined mainly by the hypothalamus–pituitary axis (GHRH → GH → IGFI). It is observed that the reduced signaling along that axis, i.e., lower levels of relevant hormones, resulting from spontaneous or induced mutations, leads to longer life span. Hypopituitary mice, as well as GH-resistant mutants, often live for 3–4 years, while average mice life span is 2 years (Bartke, 2005).

Neural Control of the Onset of Puberty

The onset of puberty, which is associated with substantial changes in the morphology and physiology of the reproductive system and the development of the

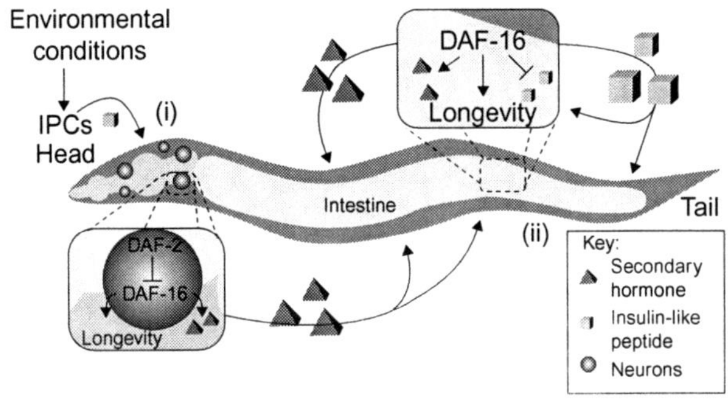

Figure 5.29 A model of organismal aging by the insulin/IGF-1 signaling pathway in the worm. (i) In response to environmental cues, insulin-producing cells (IPCs) produce insulin-like peptides, such as INS-1, INS-7, and DAF-28, to regulate DAF-2 and DAF-16 in the neurons. Active DAF-16 promotes longevity and releases secondary hormone signals that are sent throughout the entire organism. (ii) In response to decreased insulin signaling, DAF-16 in the intestine localizes to the nucleus, where it promotes longevity through target genes, such as sod-3, releases secondary hormone signals to regulate aging of the whole organism, and directly influences its own activity (both in the intestine and in other tissues) by downregulation of ins-7.
Source: From Panowski and Dillin (2009).

secondary sexual traits, results from the activation of the GnRH pulse generator, a network of 3,000–4,000 GnRH-secreting neurons in the hypothalamus of most studied animals. Although the GnRH pulse generator, in primates, is operational at birth, the onset of puberty in this group is delayed for years until a later stage of the individual development.

The delay of puberty in primates is related to the activation, soon after birth, of a central neurobiological brake or mechanism, upstream the GnRH pulse generator, which holds in check the release of the GnRH (not the transcription of the *GnRH* gene) (El Majdubi et al., 2000) and probably the presence of inhibitory GABA (Plant and Barker-Gibb, 2004), whose secretion increases during the prepubertal and, especially, the juvenile period (Plant and Shahab, 2002). In young rats, 23–29 days after birth, activation of the GABAergic system inhibits GnRH synthesis and the hypothalamic–pituitary–ovarian axis, thus preventing the onset of puberty (Feleder et al., 1999). The decrease of the NPY secretion at the onset of the puberty withdraws the neurobiological brake, thus allowing activation of the hypothalamic GnRH pulse generator.

The above central neurobiological mechanism might not be the only mechanism for timing the sexual maturation and reproductive activity in metazoans.

A recent study in the European eel (*Anguilla anguilla*) shows that in this teleost fish, the function of the central neurobiological brake is performed by the neurotransmitter DA. Experimental removal of that brake induces secretion of GnRH. DA inhibition of the GnRH-stimulated LH secretion is an ancient evolutionary component of the neuroendocrine regulation of the reproductive function, which appears

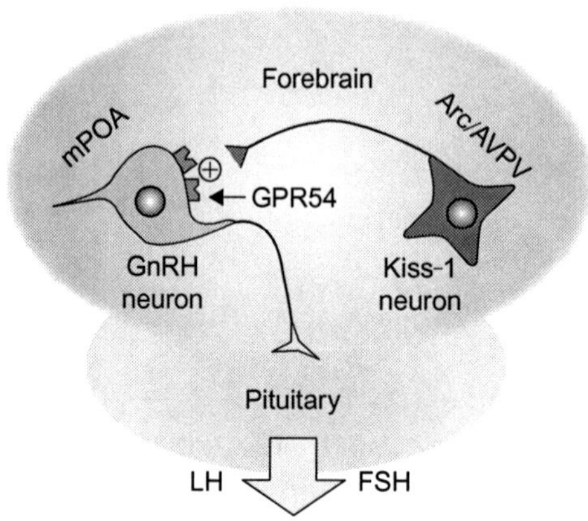

Figure 5.30 Proposed interactions between kisspeptin-secreting neurons and GnRH neurons. In this model, KiSS1 mRNA-expressing neurons, from the arcuate nucleus (Arc) and the anteroventral periventricular nucleus (AVPV), make synaptic contact with GnRH neurons within the preoptic area (POA). Upon activation of the kisspeptin receptor GPR54, GnRH neurons are stimulated to release GnRH into the portal circulation, which in turn stimulates the release of gonadotropins, luteinizing hormone (LH), and follicle-stimulating hormone (FSH) from the pituitary.
Source: From Smith et al. (2006).

to be a powerful counterpart to the well-known stimulatory role of GnRH (Vidal et al., 2004).

Recently, a novel neurohormone, kisspeptin, has been identified in specialized neurons of the hypothalamic arcuate nucleus and the anteroventral periventricular nucleus (AVPV) (Figure 5.30). Kisspeptin, via its receptor GPR54, controls the activation/inactivation of GnRH neurons in response to the level of estrogens in the blood. It is the most potent excitatory stimulus of GnRH neurons. The fact that, at the onset of puberty, GnRH neurons show increased sensitivity and electrophysiologic response to kisspeptin suggests that the neurohormone may have a role in the onset of puberty (Smith et al., 2006).

Administration of kisspeptins in prepubertal mice shows limited effect on the GnRH neurons, but in adult mice it exhibits very strong stimulating effect, and central injection of lower doses stimulates secretion of GnRH and, consequently, LH (Han et al., 2005). This differential action of the neurohormone on GnRH neurons is related to an increase of the sensitivity of GnRH neurons to kisspeptin at the onset of puberty. Experiments in guinea pigs suggest that the sympathetic nerves innervating the ovary are also involved in the appearance of puberty by regulating the follicular growth as well as oocyte maturation and atresia (Riboni et al., 1998).

In mice and rats, Kiss1 neurons in the AVPV (anteroventral periventricular nucleus) are sexually differentiated since early prenatal development, with males insufficiently expressing Kiss1 in the AVPV. Kisspeptin is critical for the sexual differentiation of the CNS and the peripheral nervous system, as well as for a number of sex-specific behaviors (Kauffman, 2009).

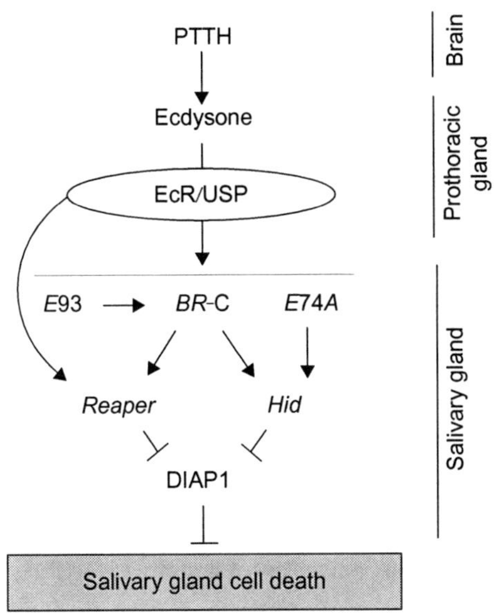

Figure 5.31 A model for the regulation of salivary gland cell death in *Drosophila*. In response to internal/external stimuli, during metamorphosis the brain secretes prothoracicotropic hormone (PTTH), which stimulates the prothoracic gland to secrete ecdysone. Ecdysone functions through the heterodimeric ecdysone receptor (EcR)/ultraspiracle (USP) receptor to directly induce primary response transcription factors, *E93*, *BR-C*, and *E74A*, which are required for proper *rpr* and *hid* induction. Ecdysone can also regulate *rpr* directly through an ecdysone response element in its promoter. To overcome the inhibitory effect of DIAP1, *reaper* and *hid* act together in a combinatorial manner, triggering salivary gland cell death. *Source*: Slightly extended from Yin and Thummel (2004).

Neural Control of Apoptosis

The programmed cell death, apoptosis, is an evolutionarily old mechanism of elimination of excess cells with indispensable functions in sculpting animal organs and general morphology during embryogenesis, postembryonic development, and adult life.

Central regulation of apoptosis in invertebrates may be illustrated by the programmed cell death in *Drosophila* salivary gland during metamorphosis. Ecdysone is the key hormone in apoptosis of the salivary gland. The cascade of the salivary gland apoptosis may be succinctly described as follows: In response to internal and external stimuli, the CNS secretes the neurohormone PTTH, which induces the prothoracic gland to secrete ecdysone. The latter, via its receptor EcR, in a dimer form with the ultraspiracle, induces expression of the primary response genes BR-C, E74A, and E93 (Lee and Baehrecke, 2001; Yin and Thummel, 2004) (Figure 5.31). Products of these genes induce apoptotic genes *reaper* and *hid*, whose products overcome the antiapoptotic effects of diap1 (Yin and Thummel, 2004).

Before the formation of the functioning embryonic CNS, apoptotic phenomena are determined by maternal factors. For example, in *Xenopus*, a maternal cell death program is activated at a maternally determined checkpoint as early as the midblastula stage (Sible et al., 1997; Hensey and Gautier, 1999) if the expression of zygotic genes does not occur or is delayed. The program compensates for the lack of cell cycle checkpoints in the pre-midblastula transition embryos (Hensey and Gautier, 1997). Apoptosis is believed to serve also as a mechanism of elimination of metabolically severely damaged cells in *Xenopus* (Shiokawa et al., 2000).

Among the most visible transformations taking place during metamorphosis in *Manduca sexta* is the degeneration of larval muscles and development of new adult muscles. This process is correlated with drastic regression of motor neurons innervating these muscles. The surviving muscle fibers later start respecification in a process that leads to formation of adult musculature. The fact that denervation of these adult muscles leads to their regression, and sometimes to the death of muscle fibers (Bayline et al., 1998), suggests that innervation is necessary for cell differentiation and proliferation in the process of muscle metamorphosis.

Many insects are capable of shedding their damaged hindlegs. The loss of normal proprioceptive input from the leg, after hindleg autotomy, in the grasshopper *Barytettix humphreysii* leads to induction of apoptosis of fibers of the thoracic muscles. Apoptosis can also be neurally induced in parts of insect's body by severing axons of proprioceptive organs in these parts (Arbas and Weidner, 1991; Personius and Chapman, 2002). (For further evidence of the control of apoptosis in insects, see earlier in the Section Apoptosis in Invertebrates.)

When the medaka fish, *Oryzias latipes*, is kept on a white background, its body color becomes lighter, because of a decrease in the number of melanophores as a result of the programmed cell death. However, when medaka fish, on the same white background, are chemically denervated, they exhibit no apoptosis. The absence of melanophore apoptosis in denervated fish suggests:

> *Sympathetic innervation has an important role in the regulation of apoptosis in melanophores.*
>
> *Sugimoto et al. (2000) and Uchida-Oka and Sugimoto (2001)*

Another example of central neural regulation of apoptosis is that of glucocorticosteroids (products of the target glands of the hypothalamus–pituitary–adrenal axis), which, by binding their cell surface receptors, induce signal transduction pathways that activate mitochondrial caspases leading to cell apoptosis (Distelhorst, 2002). The apoptotic property of glucocorticosteroids to kill lymphocytes is used in modern medicine for immunosuppressive purposes.

During amphibian metamorphosis, the TH (T3) regulates tail regression in tadpoles by activating expression of the *Bax* gene, which stimulates apoptosis of tail myocytes (Sachs et al., 1997). The TH upregulates ~35 genes and downregulates ~10 genes involved in the tail degeneration program of *X. laevis* (Denver, 1997a). Amphibian metamorphosis is under control of, and regulated by, TH, which itself is regulated by the pituitary TSH, which, in turn, is regulated by the hypothalamic TRH.

Downstream, the metamorphic action of the TH is mediated by MMPs (Damjanovski et al., 2000), enzymes that induce the remodeling of the extracellular

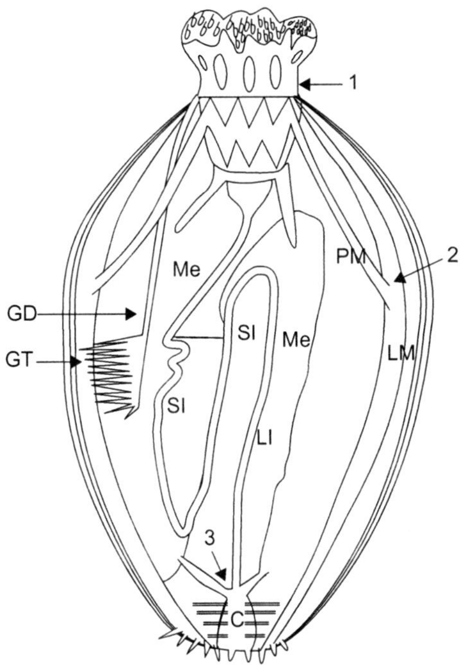

Figure 5.32 Diagrammatic view of the autotomy structures of *Eupentacta quinquesemita*. 1, introvert; 2, P–L tendon (linking the pharyngeal retractor muscle to the longitudinal body wall muscle); 3, intestine–cloacal junction. *Abbreviations*: C, cloaca; GD, gonoduct, GT, gonad tubules, LI, large intestine; LM, longitudinal body wall muscle; Me, mesentery: PM, pharyngeal retractor muscle; SI, small intestine.
Source: From Byrne (2001).

matrix via their membrane receptors, integrins, which in turn send to cells specific signals for gene expression. These signals affect apoptosis in different systems, including the apoptotic remodeling of the intestine during *X. laevis* metamorphosis and postlactation involution of the mouse mammary gland (Shi et al., 1998).

The pituitary hormone, prolactin, induces apoptosis in the penultimate stage of the spermatogonium in newts by activating an unidentified caspase (Yazawa et al., 2000). It is also reported that progesterone is essentially involved in inducing apoptosis in human endometrial epithelial cells during implantation of the embryo in endometrium (Li et al., 2001). Another hormone, 17-β estradiol, prevents apoptosis of cardiomyocytes (Pelzer et al., 2000).

Apoptotic elimination of ovarian germ cells, and degeneration of ovarian follicles in rats, involves caspase-3 and caspase-7, but expression of the respective genes is regulated by the pituitary hormones, LH and follicle-stimulating hormone (FSH), which also are under cerebral control of the hypothalamic GnRH (Yacobi et al., 2004).

Evisceration. Some invertebrates can suddenly get rid of whole organs as it occurs in some bizarre cases of autotomy known as *evisceration*. In the typical case of the sea cucumber, *Eupentacta quinquesemita* (Figure 5.32), this occurs in response to predation threats by sea stars or for discarding waste or parasites accumulated in the digestive tract.

The process starts with a "catastrophic softening" of mutable connective tissues (MCTs)

> *that undergo rapid muscle-like, nerve-mediated changes in their mechanical properties.*
>
> *Byrne (2001)*

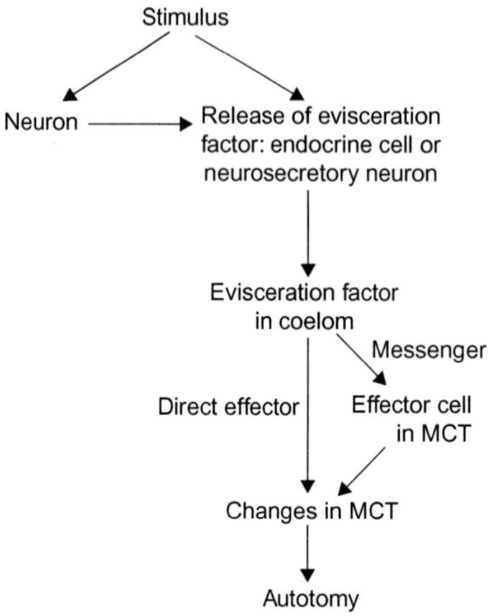

Figure 5.33 Hypothetical sequence of events leading to autotomy in *Eupentacta quinquesemita*. *Abbreviation*: MCT, mutable connective tissue.
Source: From Byrne (2001).

MCTs are rich in neurosecretory-like processes containing large dense vesicles (LDVs), which, because of the association with nervous and muscle tissues, are thought to be of neural origin. The neurally regulated activity of the muscles of the sea cucumber is essential in the process of evisceration. External (predator) or internal (waste or parasite accumulation in the digestive tract) evisceration stimuli may induce activity in upstream neurons or may act directly on evisceration-factor-producing cells. There is strong evidence for the involvement of active factors distributed through the coelom that have a direct transmitter-like or neurosecretory-like mode of operation (Byrne, 2001; Figure 5.33).

Neural Control of Regeneration

In a broad meaning of the term, metazoan regeneration is a special case of the ability of metazoans to maintain their structure. Since biological regeneration implies use of epigenetic information for rebuilding lost structures, it is relevant to the understanding of the neural mechanisms of individual development.

Local Control of Regeneration by Nerve Fibers

Local innervation is essential for the process of regeneration of the amputated limb in amphibians. Soon after amputation, the stump is covered by a thin layer of epidermis, under which the apical ectodermal cap (AEC) forms. All different types of cells in the vicinity of the stump, under the cap, including migrating fibroblasts, dedifferentiate, i.e., revert to a pluripotent state, similar to the state of cells during the early development. They form the regeneration blastema (a cluster of undifferentiated cells, which forms at the place of injury).

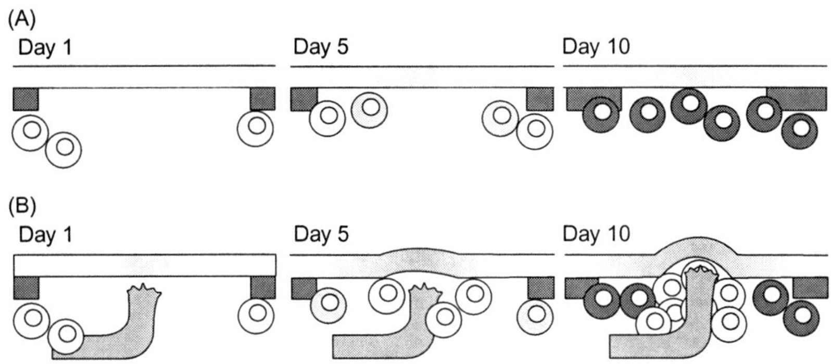

Figure 5.34 Model for the regulation of dedifferentiation and blastema formation by nerve signaling. (A) Wound healing without blastema formation. In the absence of nerve signals, the wound epithelium (yellow) and fibroblasts coordinately reform the dermal matrix (type I collagen, blue dots), leading eventually to regeneration of the dermis (solid blue regions). Fibroblast-derived cells in the wound express *AmTwist* at low levels 5 days after wounding (pink) and at higher levels after 10 days (red). (B) During blastema formation, signals from the nerve (purple) induce dedifferentiation of the wound epithelium to form the regeneration epithelium (RE)/apical epithelial cap (AEC) (green) while coordinately inhibiting expression of *AmTwist* in the underlying blastema cells. Cells in the periphery of the wound and the base of the forming blastema are outside the influence of the nerve signals and express *AmTwist* (pink to red) and undergo dermis regeneration as in (A). (For interpretation of the references to color in this figure legend, the reader is referred to the web version of this book.)
Source: From Satoh et al. (2008).

In lower invertebrates, undifferentiated cells are present in many locations, and they migrate and form regeneration blastema in any damaged part of the body, but in higher vertebrates it seems that the main, and often the only, source of undifferentiated cells is a process of dedifferentiation of differentiated cells in the stump.

We do not know how the formation of the blastula is induced, but we know that it will not develop in the absence of local innervation, suggesting that the nervous system is essentially involved in the process of dedifferentiation and, later, in the redifferentiation of the blastema cells for producing the regenerate (Figure 5.34).

Innervation is necessary for the initial development of blastema and, according to Hall,

> *dedifferentiation of differentiated cells such as chondrocytes, osteoblasts, and fibroblasts at the stump is under the control of the factors released by the nerves in the epithelial cap.*
>
> *Hall (1998, p. 339)*

Blastema cells then start to redifferentiate and serve as a source of progenitor cells of the regenerate. After denervation, postamputation formation of the regeneration blastema does not occur in early *Xenopus* larvae, but it occurs in late larvae. The reason for the difference is the fact that *fgf2 mRNA*, which is released by local nerves

(Mullen et al., 1996) in early larvae, is present in high levels, while in late larvae it is low (Cannata et al., 2001).

Experimental limb amputation in *X. laevis* froglets is followed by formation of blastema, characterized by a series of nerve-dependent and nerve-independent events. Nerve supply is necessary for expression of regeneration markers Fgf8, Fgf10, and Msx1 in the blastema, whereas expression of two other markers, Tbx5 and Prx1, may occur in the absence of nerves. The presence of nerves is necessary for cell survival, cell proliferation, and the growth of the blastema. In the absence of nerve supply, redifferentiation of cells and tissue regeneration may take place, but limb regeneration does not (Suzuki et al., 2005).

Denervation of the tail simultaneously with its amputation in the weakly electric fish, *Sternopygus macrurus*, prevents regeneration to a large extent, and this has been interpreted as another indication that blastema formation is nerve-dependent (Unguez and Zakon, 2002).

> *Denervation of limbs which have already formed blastemas does not stop future morphogenetic progress.*
>
> *Goss (1969)*

This may suggest that, by the time the blastema is developed, it has received some key epigenetic information necessary for blastema, and subsequent regeneration may be performed by neuroendocrine factors.

One of the most important genes for limb regeneration in amphibians is the group of homeobox gene *Dlx*, a homolog of the *Drosophila distal-less* (Dll). The expression of *Dlx-3*, which is essential for regeneration of limb in axolotl, is demonstrated to depend on the presence of innervation in the nerve-dependent phase (up to the late bud stage) (Mullen et al., 1996). The distribution of nerve fibers is in the same proximodistal gradient that Dlx-3 is expressed in AEC (a single layer of cells that develops on the amputation site), suggesting that the expression of Dlx-3 is upregulated by nerves penetrating the AEC (Mullen et al., 1996). One hour after denervation, the gene is downregulated, and its expression cannot be detected 12–24 h after denervation. Experimental evidence has shown that the presence of nerves is necessary not only for expression of the gene *Dlx-3* but also for expression of FGF, as is inferred from the fact that application of FGF in the denervated AEC enables axolotl to regenerate the denervated limb. Investigators believe that

> *FGF-2, or other members of the FGF family, is the neurotrophic factor required for regeneration, or is a mimic of the endogenous neurotrophic factor operating in the limb ... nerves could release FGF into the blastema or, in the case of sensory nerves, into the epidermis, and receptors are appropriately located to respond.*
>
> *Mullen et al. (1996)*

That the release of FGF is correlated with the nerve fiber regeneration has also been demonstrated in *in vitro* experiments with cocultures of spinal cord and blastema where interaction between the regenerating nerve fiber and blastema induced

Figure 5.35 Axolotl with two left limbs. The additional left limb induced experimentally by deviation of the brachial nerve to the lateral wound and grafting of a posterior skin graft.
Source: From Satoh et al. (2009).

the release of FGF1, and inhibition of nerve growth caused a drop in FGF level (Zenjari et al., 1997).

Studies on the European newt, *Triton*, show that there is a region around the limb where deviation of nerves to wound sites induces formation of supernumerary limbs (Figure 5.35). Deviation of the sciatic nerve into the area close to hind leg, at a dorsal location, and then posteriorly, induces formation of an additional leg, dorsal crest, and tail, respectively.

Examples have been presented, however, which seem to reject the idea that the nervous tissue is necessary for inducing lens differentiation and development. This seems to be the case for species known to form lenses after eye-cup is removed. But again, it could be argued that the necessary lens-inducing information might have been provided to the iris before the removal of the eye-cup. Moreover, it is demonstrated that the adjacent brain is the best substitute for the eye-cup (Goss, 1969, p. 80).

Many invertebrates respond to limb injury by autotomy, by reflexively casting off the limb from predetermined points of the limb. This is the case, for example, with the crustacean, *Uca pugilator*. Regeneration of limbs following autotomy in this crustacean occurs in two phases. The first phase immediately follows the loss of the limb and is called *basal growth*, which can take place at any time during the molt cycle. The second phase is called *proecdysial growth* and occurs only as the animal is preparing for ecdysis (Figure 5.36).

Strong analogies between the regeneration of limbs and the limb embryonic development are related to the fact that similar molecular mechanisms are used in both cases, and these mechanisms are conserved in tetrapods (Figure 5.37).

The most visible difference between the embryonic development and regeneration of limbs is the fact that the limb bud in tetrapod embryos starts developing before it is innervated, while for regeneration the presence of nerves in the AEC is indispensable. It has been argued that, in the developing limb, a permissive epidermis expressing both *FGF* and *Dlx* genes exists from the outset, while the regenerating limb must rely on the presence of local innervation to provide these essential factors (Gardiner et al., 1998). However, it should be kept in mind that the absence of the local innervation in early

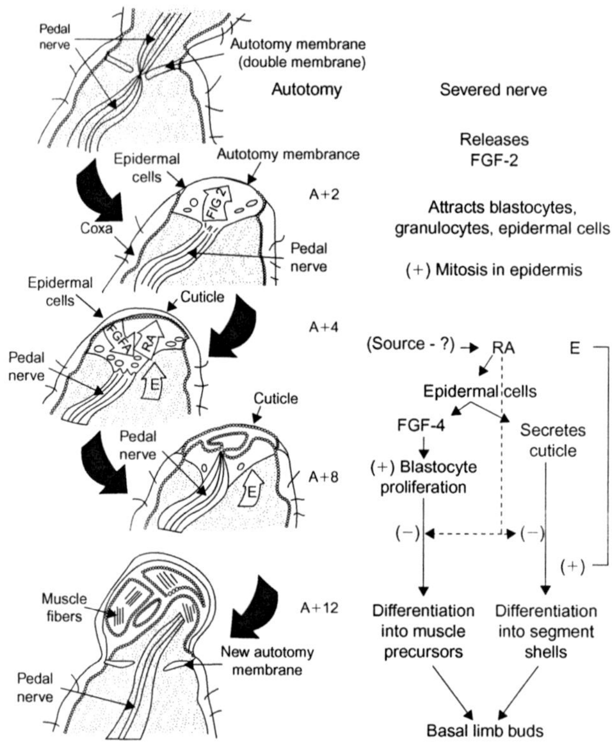

Figure 5.36 Drawing of a limb and a limb bud with possible mechanisms of control suggested. Top figure is a longitudinal representation of a limb that includes the preformed plane of autotomy (with double autotomy membrane) prior to autotomy. Two days after autotomy (A + 2), epidermal cells are migrating from sides of coxa under distal edge of autotomy membrane, and fibroblast growth factor-2 (FGF2) is seen in basin formed by proximal edge of autotomy membrane, severed pedal nerve, and distal portion of autotomy membrane. Four days after autotomy (A + 4), a thin cuticle is secreted by the epidermal cells, and FGF4 is seen in epidermal cells. Endogenous retinoic acid (RA) is also present in the blastemas at this time. By 8 days after autotomy (A + 8), infoldings from the earlier cuticle are seen. These are believed to be the first indication of leg differentiation into segments. Pedal nerve regeneration is evident at that time. At 12 days after autotomy (A + 12), developing muscle fibers are seen in the individual segments of the regenerating limb. A new autotomy membrane begins to form, and the regenerating pedal nerve can be seen moving into the newly formed limb segments. *Abbreviation*: E, ecdysis.
Source: From Hopkins (2001).

stages of limb development does not exclude the possibility of the neurohormonal regulation of limb development by the CNS. In annelids as well, the amputated segments sometimes regenerate in the absence of the nerve cord. Here is the explanation by Goss:

> *Probably the infrequent persistence of regeneration in the absence of the ventral nerve cord may be attributed to residual nerve fibers inadvertently left intact, or to*

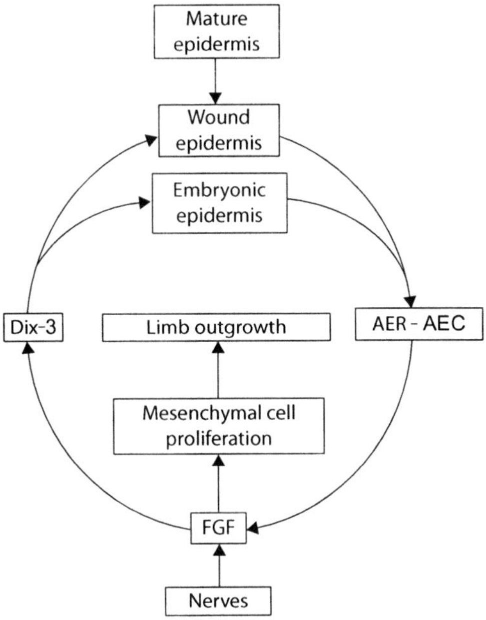

Figure 5.37 Model of the relationship between nerves, *Dlx-3* expression, FGF, and the specialized apical epidermis of developing chick and mouse limb buds (AER), developing axolotl limb buds (AEC) and regenerating axolotl limbs (AEC). The interaction between these factors controls limb outgrowth during both limb development and regeneration.
Source: From Gardiner et al. (1998).

axon regeneration across several segments to reach the wound site. It does not take very many fibers to promote regeneration, although suboptimal innervation results in the production of imperfect regenerates, which is what is formed after removal or deflection of the nerve cord.

Goss (1969, p. 152)

To the same conclusion would probably lead the fact that initial results of denervation of amputated amphibian limbs depend on the time when denervation is performed; when denervation is carried out at the time of amputation, it has no effect on the wound closure, but in amphibians denervated 5 days before amputation, tissue loss and regression are observed (Mescher and Munaim, 1984). The only relevant difference between the two cases is that in the predenervated (denervation performed 5 days before amputation) amphibians, no input on amputation is received by the CNS; hence, no normal neurohormonal response by the brain is to be expected (see below on the hormonal regulation of regeneration by the CNS). The observed loss of tissue implies cell death, a programmed cell death that seems to be part of the general mechanism of regeneration for elimination of regenerationally unusable cells, especially during the early regeneration. This is suggested by recent studies on regeneration of amputated limbs in salamanders (Mescher et al., 2000) and bodies in planarians (Platyhelminthes) (Hwang et al., 2004).

Finally, it is relevant to point out that evidence exists for at least two blastemagenic inducers released by nerves in the stump, glial growth factor (GGF) and transferrin. GGF is normally present in the extracts of newt nervous system and of regeneration blastema. Its level drops after denervation. When added to blastema cells *in vitro*, GGF from newt brain dramatically increases their proliferation (Brockes and Kintner,

1986). Transferrin is another secreted protein released by nerves that stimulates regeneration. By facilitating the supply of blastema cells with iron, it reduces to half the blastula cell cycle time (Mescher and Munaim, 1984).

Neurohormonal Control of Regeneration by the Nervous System

By the middle of the twentieth century, it became clear that, besides its immediate influence on the process of regeneration, the nervous system might also exert significant humoral effects. At that time, it was observed that, after removal of the eyes (photoreceptors), the flatworm *Polycelis nigra* regenerates new eyes (the worm has about 70 photoreceptors) within a week. But worms whose brains were removed failed to regenerate eyes. Experimentally, it was also demonstrated that the regeneration of eyes depended not on nerve connections with the brain, and the brain did not supply any cells for regeneration of eyes. This and the fact that the regeneration of eyes takes place normally when brains from other flatworm species, or even necrotic brains of the same species, are implanted, suggested that a hormonal substance released by the worm brain stimulates the eye regeneration. Planarian head lysate (but not tail lysates, for example) stimulates regeneration in brainless planaria, again suggesting that a hormonal substance released by the brain is responsible for eye regeneration in these organisms. Similar regeneration-stimulating effects of brain extracts are also observed in amphibians: when newt forelimb regenerates are denervated, DNA synthesis in blastema drops dramatically, but its synthesis reaches almost normal rates when brain extract is infused (Jabaily and Singer, 1977; Reddien and Alvarado, 2004).

Another suggestive observation was made on oligochaetes, which cannot regenerate their tails during the diapause, i.e., when they are sexually inactive. It was discovered that a neurohormone with gonadotropic effects is secreted by neurons of the supraesophageal ganglion, which inhibits both diapause and regeneration. The removal of the brain, by preventing secretion of the neurohormone, makes possible a condition of diapause and the regeneration of the tail. The inverse situation is observed in some polychaetes where the removal of the brain prevents tail regeneration. Again, it was demonstrated that a hormonal substance secreted by neurosecretory cells in the ganglia of these worms is necessary for the regeneration of the tail. There is evidence that both asexual reproduction and regeneration in these worms use the same cellular mechanisms (Martinez et al., 2005).

Regeneration of the polychaete worm *Nereis* depends on the presence of the brain, and it has been demonstrated that the brain exerts its influence on regeneration hormonally (the same is true for oligochaetes) (Gorbman and Davey, 1991, p. 739).

A strict polarity of regeneration is observed in annelids. In an amputated annelid, the posterior end of the anterior piece would regenerate posterior segments, while the anterior part of the caudal remnant would regenerate anterior segments and head. It is demonstrated that even the same region may have the capacity to grow both head or tail, and what actually regenerates in each particular case depends on the direction of its subjacent nerve cord. A head regenerates where the forward end of the caudal fragment of nerve cord comes into contact with the wounded body wall and a posteriorly directed end of an anterior piece of nerve cord induces tail regeneration.

It is argued that when the concentration of the molting hormone, ecdysone (produced in response to neural signals by the prothoracic gland in insects and by the Y-organ in crustaceans), in intermolding periods is higher, it makes impossible the process of regeneration. In the beginning of the molting cycle the production of ecdysone is minimal, and this favors the process of regeneration. After the metamorphosis is completed, in insects the prothoracic gland degenerates, and the complete absence of ecdysone in the mature insects makes regeneration impossible. It is demonstrated that the bilateral removal of the Y-organ in the crab, *Sesarma haematocheir*, inhibits the premolt growth of the regenerating limb (Suzuki, 1985).

Two protein molecules, extracted from the leech CNS, systematically change their distribution at distinct times during regeneration in leeches (Lüthi, 1994). When the concentration of the molting hormone, ecdysone, in intermolting periods in leeches is higher, it makes the regeneration impossible.

Finally, it is necessary to address the problem of the "aneurogenic limb." It is known that limbs of embryos start developing before they are innervated, and embryologists have been able to produce amphibians with noninnervated limb. To their surprise, such "aneurogenic" limbs are fully capacitated to regenerate if amputated. This great puzzle was solved in 1963, when Steen and Thornton experimentally demonstrated in *Ambystoma* larvae that it was the skin that made the regeneration of the "aneurogenic" limb possible. Further experiments have shown that the apical cap of cells that forms on wound epidermis on the end of the stump is intensely innervated. It has been suggested that nerves influence the epidermis, which, in turn, promotes blastema formation immediately beneath it. It is known that no regeneration takes place in the absence of epidermis. While a number of experimental facts suggest an instructive role of the nervous system in regeneration, the need for epidermis is thought to be *permissive* rather than *instructive*: when taken from diverse locations such as abdomen, flank, or tail, the epidermis always permits limb regeneration in frogs (Goss, 1969, p. 32). What should not be forgotten in the case of the regeneration of "aneurogenic limbs" is the fact that the neural control is performed not only directly, i.e., via local innervation, but neurohormonal mechanisms under the CNS control are also operational.

References

Abdelwahid, E., Pelliniemi, L.J., Niinikoski, H., Simell, O., Tuominen, J., Rahkonen, O., et al. (1999). Apoptosis in the pattern formation of the ventricular wall during mouse heart organogenesis. Anat. Rec., *256*, 208–17.

Abe, E., Marians, R.C., Yu, W., Wu, X.B., Ando, T. and Li, Y., et al. (2003). TSH is a negative regulator of skeletal remodeling. Cell, *115*, 151–62.

Adam, G., Perrimon, N. and Noselli, S. (2003). The retinoic-like juvenile hormone controls the looping of left-right asymmetric organs in *Drosophila. Development*, *130*, 2397–406.

Adams, C.S., Korytko, A.I. and Blank, J.L. (2001). A novel mechanism of body mass regulation. J. Exp. Biol., *204*, 1729–34.

Adriaensen, D., Brouns, I., Pintelon, I., De Proost, I. and Timmermans, J.-P. (2006). Evidence for a role of neuroepithelial bodies as complex airway sensors: comparison with smooth muscle-associated airway receptors. J. Appl. Physiol., *101*, 960–70.

Alves, H.J., Alvares, L.E., Gabriel, J.E. and Coutinho, L.L. (2003). Influence of MyoD and cellular proliferation in chicken embryos. Braz. J. Med. Biol. Res., *36*, 191–97.

Arbas, E.A. and Weidner, M.H. (1991). Transneuronal induction of muscle atrophy in grasshoppers. J. Neurobiol., *22*, 536–46.

Arnold, H.H. and Winter, B. (1998). Muscle differentiation: more complexity to the network of myogenic regulators. Curr. Opin. Genet. Dev., *8*, 539–44.

Ash, J.D. and Overbeek, P.A. (2000). Lens-specific VEGF-A expression induces angioblast migration and proliferation and stimulates angiogenic remodeling. Dev. Biol., *223*, 383–98.

Aulehla, A. and Herrmann, B.G. (2004). Segmentation in vertebrates: clock and gradient finally joined. Gene. Dev., *18*, 2060–67. Available online: http://genesdev.cshlp.org/content/18/17/2060.long - fn-3.

Aulehla, A., Wehrle, C., Brand-Saberi, B., Kemler, R., Gossler, A., Kanzler, B., et al. (2003). Wnt3a plays a major role in the segmentation clock controlling somitogenesis. Dev. Cell, *4*, 395–406.

Baeckberg, M., Collin, M., Ovesjoe, M.L. and Meister, B. (2003). Chemical coding of GABAB receptor-immunoreactive neurones in hypothalamic regions regulating body weight. J. Neuroendocrinol., *15*, 1–14.

Baehrecke, E.H. (1996). Ecdysone signaling cascade and regulation of *Drosophila* metamorphosis. Arch. Insect Biochem., *33*, 231–44.

Bajard, L., Relaix, F., Lagha, M., Rocancourt, D., Daubas, P. and Buckingham, M.E. (2006). A novel genetic hierarchy functions during hypaxial myogenesis: Pax3 directly activates *Myf5* in muscle progenitor cells in the limb. Gene Dev., *20*, 2450–64.

Baker, C.V. and Bronner-Fraser, M. (2001). Vertebrate cranial placodes I. Embryonic induction. Dev. Biol., *232*, 1–61.

Bakke, L.J., Dow, M.P., Cassar, C.A., Peters, M.W., Pursley, J.R. and Smith, G.W. (2002). Effect of the preovulatory gonadotropin surge on matrix metalloproteinase (MMP)-14, MMP-2, and tissue inhibitor of metalloproteinases-2 expression within bovine periovulatory follicular and luteal tissue. Biol. Reprod., *66*, 1627–34.

Bartke, A. (2005). Minireview: role of the growth hormone/insulin-like growth factor system in mammalian aging. Endocrinology, *146*, 3718–23.

Bass, H. and Grober, M.S. (2001) Social and neural modulation of sexual plasticity in teleost fish. Brain Behav. Evol., *57*, 293–300.

Basu, S., Nagy, J.A., Pal, S., Vasile, E., Eckelhoefer, I.A., Bliss, V.S., et al. (2001). The neurotransmitter dopamine inhibits angiogenesis induced by vascular permeability factor/vascular endothelial growth factor. Nat. Med., *7*, 569–74.

Bayline, R.J., Khoo, A.B. and Booker, R. (1998). Innervation regulates the metamorphic fates of larval abdominal muscles in the moth, *Manduca sexta*. Dev. Genes Evol., *208*, 369–81.

Bayline, R.J., Duch, C. and Levine, R.B. (2001). Nerve–muscle interactions regulate motor terminal growth and myoblast distribution during muscle development. Dev. Biol., *231*, 348–63.

Beck, T. (2003). The bone—a metabolic active fabric. Available online: www.pharmazeutische-zeitung. Accessed on 10.4.2004.

Berggren, K., McCaffery, P., Drager, U. and Forehand, C.J. (1999). Differential distribution of retinoic acid synthesis in the chicken embryo as determined by immunolocalization of the retinoic acid synthetic enzyme, RALDH-2. Dev. Biol., *210*, 288–304.

Berggren, K., Ezerman, E.B., McCaffery, P. and Forehand, C.J. (2001). Expression and regulation of the retinoic acid synthetic enzyme RALDH-2 in the embryonic chicken wing. Dev. Dyn., *222*, 1–16.

Bjurholm, A., Kreicbergs, A., Terenius, L., Goldstein, M. and Schultzberg, M. (1988). Neuropeptide Y-, tyrosine hydroxylase- and vasoactive intestinal polypeptide-immunoreactive nerves in bone and surrounding tissues. J. Auton. Nerv. Syst. *25*, 119–25.

Bok, J., Bronner-Fraser, M. and Wu, D.K. (2005). Role of the hindbrain in dorsoventral but not anteroposterior axial specification of the inner ear. Development, *132*, 2115–24.

Brigande, J.V., Kiernan, A.E., Gao, X., Iten, L.E. and Fekete, D.M. (2000). Molecular genetics of pattern formation in the inner ear: do compartment boundaries play a role? Proc. Natl. Acad. Sci. U.S.A., *97*, 11700–06.

Brockes, J.P. and Kintner, C.R. (1986). Glial growth factor and nerve-dependent proliferation in the regeneration blastema of Urodele amphibians. Cell, *45*, 301–06.

Broughton, S.J., Piper, M.D.W., Ikeya, T., Bass, T.M., Jacobson, J., Driege, Y., et al. (2005). Longer life span, altered metabolism, and stress resistance in *Drosophila* from ablation of cells making insulin-like ligands. Proc. Natl. Acad. Sci. U.S.A., *102*, 3105–10.

Buonanno, A., Edmondson, D.G. and Hayes, W.P. (1993). Upstream sequences of the myogenin gene convey responsiveness of skeletal muscle denervation in transgenic mice. Nucleic Acid Res., *21*, 5684–93.

Burt-Pichat, B., Lafage-Proust, M.H., Duboeuf, F., Laroche, N., Itzstein, C., Vico, L., et al. (2005). Dramatic decrease of innervation density in bone after ovariectomy. Endocrinology, *146*, 503–10.

Butler, J., Cosmos, E. and Brierley, J. (1982). Differentiation of muscle fiber types in aneurogenic brachial muscle of the chick embryo. J. Exp. Zool., *224*, 65–80.

Byrne, M. (2001). The morphology of autotomy structures in the sea cucumber *Eupentacta quinquesemita* before and during evisceration. J. Exp. Biol., *204*, 849–63.

Cameron, J. and McCredie, J. (1982). Innervation of the undifferentiated limb bud in rabbit embryo. J. Anat., *134*, 795–808.

Canalis, E. (2003). Mechanisms of glucocorticoid-induced osteoporosis. Curr. Opin. Rheumatol., *15*, 454–57.

Canalis, E. and Delany, A.M. (2002). Mechanisms of glucocorticoid action in bone. Ann. N.Y. Acad. Sci., *966*, 73–81.

Cannata, S.M., Bagni, C., Bernardini, S., Christen, B. and Filoni, S. (2001). Nerve-independence of limb regeneration in larval *Xenopus laevis* is correlated to the level of fgf-2 mRNA expression in limb tissues. Dev. Biol., *231*, 436–46.

Cardoso, W.V. (2001). Molecular regulation of lung development. Annu. Rev. Physiol., *63*, 471–94.

Carmeliet, P., Ferreira, V., Breier, G., Pollefeyt, S., Kieckens, L., Gertsenstein, M., et al. (1996). Abnormal blood vessel development and lethality in embryos lacking a single VEGF allele. Nature, *380*, 435–39.

Chatterjee, S., Kraus, P. and Lufkin, T. (2010). A symphony of inner ear developmental control genes. BMC Genet., *11*, 68.

Cheline, A.J., Reddi, A.H. and Martin, R.B. (2002). Bone morphogenetic protein-7 selectively enhances mechanically induced bone formation. Bone, *31*, 57–574.

Chin, E.R., Olson, E.N., Richardson, J.A., Yang, Q., Humphries, C., Shelton, J.M., et al. (1998). A calcineurin-dependent transcriptional pathway controls skeletal muscle fiber type. Genes Dev., *12*, 2499–509.

Chytil, F. (1996). Retinoids in lung development. FASEB J., *10*, 986–92.

Clark, A.C., del Campo, M.L. and Ewer, J. (2004). Neuroendocrine control of larval ecdysis behavior in *Drosophila*: complex regulation by partially redundant neuropeptides. J. Neurosci., *24*, 4283–92.

Colombani, J., Bianchini, L., Layalle, S., Pondeville, E., Dauphin-Villemant, C., Antoniewski, C., et al. (2005). Antagonistic actions of ecdysone and insulins determine final size in *Drosophila*. Science, *310*, 667–70.

Colombani, J., Raisin, S., Pantalacci, S., Radimerski, T., Montagne, J. and Léopold, P. (2005). A nutrient sensor mechanism controls *Drosophila* growth. Cell, *114*, 739–49.

Consoulas, C. and Levine, R.B. (1997). Accumulation and proliferation of adult leg muscle precursors in *Manduca* are dependent on innervation. J. Neurobiol., *32*, 531–53.

Cossu, G. and Borello, U. (1999). Wnt signaling and the activation of myogenesis in mammals. EMBO J., *18*, 6867–72.

Creazzo, T.L., Godt, R.E., Leatherbury, L., Conway, S.J. and Kirby, M.L. (1998). Role of cardiac neural crest cells in cardiovascular development. Annu. Rev. Physiol., *60*, 267–86.

Currie, D.A. and Bate, M. (1991). The development of adult abdominal muscles in *Drosophila*: myoblasts express *twist* and are associated with nerves. Development, *113*, 91–102.

Currie, D.A. and Bate, M. (1995). Innervation is essential for the development and differentiation of a sex-specific adult muscle in *Drosophila melanogaster*. Development, *121*, 2549–57.

Cutz, E., Yeger, H., Pan, J. and Ito, T. (2008). Pulmonary neuroendocrine cell system in health and disease. Curr. Resp. Med. Rev., *4*, 174–86.

Dai, L., Dewey, E.M., Zitnan, D., Luo, C.W., Honegger, H.W. and Adams, M.E. (2007). Identification, developmental expression, and functions of bursicon in the tobacco hawkmoth, *Manduca sexta*. J. Comp. Neurol., *506*, 759–74. (abstract).

Damjanovski, S., Amano, T., Li, Q., Ueda, S., Shi, Y.B. and Ishizuya-Oka, A. (2000). Role of ECM remodeling in thyroid hormone-dependent apoptosis during anuran metamorphosis. Ann. N.Y. Acad. Sci., *926*, 180–91.

Dammann, C.E. and Nielsen, H.C. (1998). Regulation of the epidermal growth factor receptor in fetal rat lung fibroblasts during late gestation. Endocrinology, *139*, 1671–77.

Dammann, C.E.L., Ramadurai, S.M., McCants, D.D., Pham, L.D. and Nielsen, H.C. (2000). Androgen regulation of signaling pathways in late fetal mouse lung development. Endocrinology, *141*, 2923–29.

Dehority, W., Halloran, B.P., Bikle, D.D., Curren, T., Kostenuik, P.J., Wronski, T.J., et al. (1999). Bone and hormonal changes induced by skeletal unloading in the mature male rat. Am. J. Physiol., *276*, E62–69.

Delany, A.M., Dong, Y. and Canalis, E. (1994). Mechanisms of glucocorticoid action in bone cells. J. Cell Biochem., *56*, 295–302.

Delfini, M.-C. and Duprez, D. (2004) Ectopic Myf5 or MyoD prevents the neuronal differentiation program in addition to inducing skeletal muscle differentiation, in the chick neural tube. Development, *131*, 713–23.

Dellovade, T.L., Davis, A.M., Ferguson, C., Sieghart, W., Homanics, G.E. and Tobet, S.A. (2001). GABA influences the development of the ventromedial nucleus of the hypothalamus. J. Neurobiol., *49*, 264–76.

Demulder, A., Guns, M., Ismail, A., Wilmet, E., Fondu, P. and Bergmann, P. (1998). Increased osteoblast-like cells formation in long-term bone marrow cultures from patients with a spinal cord injury. Calcif. Tissue Int., *63*, 396–400. (abstract).

Denver, R.J. (1997). Environmental stress as a developmental cue: corticotropin-releasing hormone is a proximate mediator of adaptive phenotypic plasticity in amphibian metamorphosis. Horm. Behav., *31*, 169–79.

Denver, R.J. (1997). Thyroid hormone-dependent gene expression program for *Xenopus* neural development. J. Biol. Chem., *272*, 8179–88.

Deries, M., Collins, J.J.P. and Duxson, M.J. (2008). The mammalian myotome: a muscle with no innervation. Evol. Dev., *10*, 746–55.

Dewing, P., Shi, T., Horvath, S. and Vilain, E. (2003). Sexually dimorphic gene expression in mouse brain precedes gonadal differentiation. Mol. Brain Res., *118*, 89–90.

Dickinson, D.P., Machnicki, M., Ali, M.M., Zhanying, Z. and Sohal, G.S. (2004). Ventrally emigrating neural tube (VENT) cells: a second neural tube-derived cell population. J. Anat., *205*, 79–98.

DiFiore, J.W. and Wilson, J.M. (1994). Lung development. *Semin. Pediatr. Surg.*, *3*, 221–32.

Distelhorst, C.W. (2002). Recent insights into the mechanism of glucocorticosteroid-induced apoptosis. Cell Death Differ., *9*, 6–190.

Donovan, M.J., Lin, M.I., Wiegn, P., Ringstedt, T., Kraemer, R., Hahn, R., et al. (2000). Brain derived neurotrophic factor is an endothelial cell survival factor required for intramyocardial vessel stabilization. Development, *127*, 4531–40.

Dournon, C. and Houillon, C. (1985). Thermosensibilité de la différenciation sexuelle chez l'amphibien Urodèle, *Pleurodeles waltlii* Michah. Reprod. Nutr. Dev., *25*, 671–88.

Draizen, T.A., Ewer, J. and Robinow, S. (1999). Genetic and hormonal regulation of the death of peptidergic neurons in the *Drosophila* central nervous system. J. Neurobiol., *38*, 455–65.

Ducy, P., Amling, M., Takeda, S., Priemel, M., Schilling, A.F., Beil, F.T., et al. (2000). Leptin inhibits bone formation through a hypothalamic relay: a central control of bone mass. Cell, *100*, 197–207.

Duxson, M.J. and Sheard, P.W. (1995). Formation of new myotubes occurs exclusively at the multiple innervation zones of an embryonic large muscle. Dev. Dyn., *204*, 391–405.

Duxson, M.J., Usson, Y. and Harris, A.J. (1989). The origin of secondary myotubes in mammalian skeletal muscles: ultrastructural studies. Development, *107*, 743–50.

Edoff, K., Hellman, J., Persliden, J. and Hildebrand, C. (1997). The developmental skeletal growth in the rat foot is reduced after denervation. Anat. Embryol., *195*, 531–38.

Elefteriou, F., Ahn, J.D., Takeda, S., Starbuck, M., Yang, X., Liu, X., et al. (2005). Leptin regulation of bone resorption by the sympathetic nervous system and CART. Nature, *434*, 514–20.

El Majdubi, M., Sahu, A., Ramaswamy, S. and Plant, T.M. (2000). Neuropeptide Y: a hypothalamic brake restraining the onset of puberty in primates. Proc. Natl. Acad. Sci. U.S.A., *97*, 6179–84.

Elofsson, U., Winberg, S., Francis, R.C. and Nilsson, G.E. (1997a). Neuronal correlates of socially induced sex change in teleost fish. In: *Experimental Biology Online*. The Society for Experimental Biology Annual Meeting, University of Kent at Canterbury, 7–11 April 1997, A8.20 (abstracts).

Elofsson, U., Winberg, S. and Francis, R.C. (1997b). Sex differences in number of preoptic GnRH-immunoreactive neurones in a protandrously hermaphroditic fish, the anemone fish *Amphiprion melanopus*. J. Comp. Physiol. A, *181*, 484–92.

Emlen, D.J. and Nijhout, H.F. (2001). Hormonal control of male horn length dimorphism in *Onthophagus taurus* (Coleoptera: Scarabaeidea): a second critical period of sensitivity to juvenile hormone. J. Insect Physiol., *47*, 1045–54.

Ericson, J., Norlin, S., Jessell, T.M. and Edlund, T. (1998). Integrated FGF and BMP signaling controls the progression of progenitor cell differentiation and the emergence of pattern in the embryonic anterior pituitary. Development, *125*, 1005–15.

Esser, S., Lampugnani, M.G., Corada, M., Dejana, E. and Risau, W. (1998). Vascular endothelial growth factor induces VE-cadherin tyrosine phosphorylation in endothelial cells. J. Cell Sci., *111*, 1853–65.

Ewer, J., Gammie, S.C. and Truman, J.W. (1997). Control of insect ecdysis by a positive feedback system: roles of eclosion hormone and ecdysis-triggering hormone. J. Exp. Biol., *200*, 869–81.

Fan, C.M. and Tessier-Lavigne, M. (1994). Patterning of mammalian somites by surface ectoderm and notochord: evidence for sclerotome induction by a hedgehog homolog. Cell, *79*, 1175–86.

Feld, S. and Hirschberg, R. (1996). Growth hormone, the insulin-like growth factor system, and the kidney. Endocr. Rev., *17*, 423–80.

Feleder, C., Ginzburg, M., Wutke, W., Moguilevsky, J.A. and Arias, P. (1999). GABAergic activation inhibits the hypothalamic–pituitary–ovaric axis and sexual development in the immature female rat. Associated changes in hypothalamic glutamatergic and taurinergic systems. Dev. Brain Res., *116*, 151–57.

Fernandes, J.J. and Keshishian, H. (1998). Nerve-muscle interactions during flight muscle development in *Drosophila*. Development, *125*, 1769–79.

Fernandes, J.J. and Keshishian, H. (2005). Motoneurons regulate myoblast proliferation and patterning in *Drosophila*. Dev. Biol., *277*, 493–505.

Fishman, M.C. and Chien, K.R. (1997). Fashioning the vertebrate heart: earliest embryonic decisions. Development, *124*, 2099–117.

Foley, A.C. and Mercola, M. (2005). Heart induction by Wnt antagonists depends on the homeodomain transcription factor Hex. Genes Dev., *19*, 387–96.

Fong, G.H., Rossant, J., Gertsenstein, M. and Breitman, M.L. (1995). Role of the Flt-1 receptor tyrosine kinase in regulating the assembly of vascular endothelium. Nature, *376*, 66–70.

Fukumoto, T., Kema, I.P. and Levin, M. (2005). Serotonin signaling is a very early step in patterning of the left–right axis in chick and frog embryos. Curr. Biol., *15*, 794–803.

Gahr, M. (2003). Male Japanese quails with female brains do not show male sexual behavior. Proc. Natl. Acad. Sci. U.S.A., *100*, 7959–64.

Gammie, S.C. and Truman, J.W. (1997). Neuropeptide hierarchies and the activation of sequential motor behaviors in the hawkmoth, *Manduca sexta*. J. Neurosci., *17*, 4389–97.

Gardiner, D.M., Torok, M.A., Mullen, L.M. and Bryant, S.V. (1998). Evolution of vertebrate limbs: robust morphology and flexible development. Am. Zool., *38*, 659–71.

Geetha-Loganathan, P., Nimmagadda, S., Huang, R., Christ, B. and Scaal, M. (2006). Regulation of ectodermal Wnt6 expression by the neural tube is transduced by dermomyotomal Wnt11: a mechanism of dermomyotomal lip sustainment. Development, *133*, 2897–904.

Geraerts, W.P.M. (1976). Control of growth by the neurosecretory hormone of the light green cells in the freshwater snail *Lymnaea stagnalis*. Gen. Comp. Endocrinol., *29*, 61–71.

Gilbert, S.F. (1997). *Developmental Biology*, fifth edition. Sinauer Associates, Inc. Publishers, Sunderland, MA, p. 765.

Gilbert, S.F. (2000). *Developmental Biology*. Sinauer Associates, Inc. Publishers, Sunderland MA, pp. 294–296.

Gorbman, A. and Davey, K. (1991). Endocrines. In: *Neural and Integrative Animal Physiology*, fourth edition. C.L. Prosser (ed.) Wiley-Liss, pp. 693–754, p. 733.

Goss, R.J. (1969). *Principles of Regeneration*. Academic Press, London, p. 156.

Greenwood, A.K. and Fernald, R.D. (2004). Social regulation of the electrical properties of gonadotropin-releasing hormone neurons in a cichlid fish (*Astatotilapia burtoni*). Biol. Reprod., *71*, 909–18.

Grim, M. and Christ, B. (1993). Neural crest cell migration into the limb bud of avian embryos. Prog. Clin. Biol. Res., *383A*, 391–402.

Grober, M.S. and Sunobe, T. (1996). Serial adult sex change involves rapid and reversible changes in forebrain neurochemistry. Neuroreport, *7*, 2845–949.

Grobstein, C. (1956). Trans-filter induction of tubules in mouse metanephrogenic mesenchyme. Exp. Cell Res., *10*, 424–40.

Gutiérrez-Ospina, G., Jiménez-Trejo, F.J., Favila, R., Moreno-Mendoza, N.A., Granados Rojas, L. and Barrios, F.A., et al. (1999). Acetylcholinesterase-positive innervation is present at undifferentiated stages of the sea turtle *Lepidochelys olivacea* embryo gonads: implications for temperature-dependent sex determination. J. Comp. Neurol., *410*, 90–8.

Hakeda, Y., Arakawa, T., Ogasawara, A. and Kumegawa, M. (2000). Recent progress in studies on osteocytes–osteocytes and mechanical stress. Kaibogaku Zasshi, *75*, 451–56. (abstract).

Halaas, J.L., Gajiwala, K.S., Maffei, M., Cohen, S.L., Chait, B.T., Rabinowitz, D., et al. (1995). Weight-reducing effects of the plasma protein encoded by the obese gene. *Science*, *269*, 543–46.

Hall, B.K. (1998). *Evolutionary Developmental Biology*, second edition. Chapman & Hall, London, p. 163.

Hampe, W., Urny, J., Franke, I., Hoffmeister-Ullerich, S.A., Herrmann, D., Petersen, C.M., et al. (1999). A head-activator binding protein is present in hydra in a soluble and a membrane-anchored form. Development, *126*, 4077–86.

Han, S.K., Gottsch, M.L., Lee, K.J., Popa, S.M., Smith, J.T., Jakawich, S.K., et al. (2005). Activation of gonadotropin-releasing hormone neurons by kisspeptin as a neuroendocrine switch for the onset of puberty. J. Neurosci., *25*, 11349–56.

Hara-Irie, F., Amizuka, N. and Ozawa, H. (1996). Immunohistochemical and ultrastructural localization of CGRP-positive nerve fibers at the epiphyseal trabecules facing the growth plate of rat femurs. Bone, *18*, 29–39.

Haraguchi, R, Mo, R., Hui, C., Motoyama, J., Makino, S., Shiroishi, T., et al. (2001). Unique function of sonic hedgehog signaling during external genitalia development. Development, *128*, 4241–50.

Hatakeyama, D., Ito, I., Kojima, S., Fujito, Y. and Ito, E. (2000). Complement receptor 3-like immunoreactivity in the light green cells and the canopy cells of the pond snail, *Lymnaea stagnalis*. Brain Res., *865*, 102–06.

Hayes, T., Haston, K., Tsui, M., Hoang, A., Haeffele, C. and Vonk, A. (2002). Herbicides: feminization of male frogs in the wild. Nature, *419*, 895–96.

Hayes, T.B., Collins, A., Lee, M., Mendoza, M., Noriega, N., Stuart, A.A., et al. (2002). Hermaphroditic, demasculinized frogs after exposure to the herbicide atrazine at low ecologically relevant doses. Proc. Natl. Acad. Sci. U.S.A., *99*, 5476–80.

Hegstrom, C.D., Riddiford, L.M. and Truman, J.W. (1998). Steroid and neuronal regulation of ecdysone receptor expression during metamorphosis of muscle in the moth, *Manduca sexta*. J. Neurosci., *18*, 1786–94.

Hensey, C. and Gautier, J. (1997). A developmental timer that regulates apoptosis at the onset of gastrulation. Mech. Dev., *69*, 183–95.

Hensey, C. and Gautier, J. (1999). Developmental regulation of induced and programmed cell death in *Xenopus* embryos. Ann. N.Y. Acad. Sci., *887*, 105–19.

Herzlinger, D., Qiao, J., Cohen, D., Ramakrishna, N. and Brown, A.M.C. (1994). Induction of kidney epithelial morphogenesis by cells expressing *Wnt-1*. Dev. Biol., *166*, 815–18.

Herzog, W., Sonntag, C., von der Hardt, S., Roehl, H.H., Varga, Z.M. and Hammerschmidt, M. (2004). Fgf3 signalling from the ventral diencephalon is required for early specification and subsequent survival of the zebrafish adenohypophysis. Development, *131*, 3681–92.

Hill, E.L. and Elde, R. (1991). Distribution of CGRP-, VIP-, D beta H-, SP-, and NPY immunoreactive nerves in the periosteum of the rat. Cell Tissue Res., *264*, 469–80.

Hirokawa, K., Utsuyama, M. and Kobayashi, S. (1998). Hypothalamic control of development and aging of thymus. Mech. Ageing Dev., *100*, 177–85.

Hirokawa, K., Utsuyama, M. and Kobayashi, S. (2001). Hypothalamic control of thymic function. Cell. Mol. Biol., *47*, 97–102.

Hogan, K.A., Ambler, C.A., Chapman, D.L. and Bautch, V.L. (2004). The neural tube patterns vessels developmentally using the VEGF signaling pathway. Development, *131*, 1503–13.

Hollyday, M., McMahon, J.A. and McMahon, A.P. (1995). *Wnt* expression patterns in the chick embryo nervous system. Mech. Dev. *52*, 9–25.

Hopkins, P.M. (2001). Limb regeneration in the fiddler crab, *Uca pugilator*: hormonal and growth factor control. Am. Zool., *41*, 389–98.

Hopwood, N.D., Pluck, A. and Gurdon, J.B. (1989). MyoD expression in the forming somites is an early response to mesoderm induction in *Xenopus* embryos. EMBO J., *8*, 3409–17.

Hu, Y., Cascone, P.J., Cheng, L., Sun, D., Nambu, J.R. and Schwartz, L.M. (1999). Lepidopteran DALP, and its mammalian ortholog HIC-5, function as negative regulators of muscle differentiation. Proc. Natl. Acad. Sci. U.S.A., *96*,10218–23.

Hutson, M.R. and Kirby, M.L. (2007). Model systems for the study of heart development and disease cardiac neural crest and conotruncal malformations. Semin. Cell Dev. Biol., *18*, 101–10.

Hwang, J.S., Kobayashi, C., Agata, K., Ikeo, K. and Gojobori, T. (2004). Detection of apoptosis during planarian regeneration by the expression of apoptosis-related genes and TUNEL assay. Gene, *333*, 15–25.

Hyatt, J.-P.K., Roy, R.R., Baldwin, K.M., Wernig, A. and Edgerton, V.R. (2006). Activity-unrelated neural control of myogenic factors in a slow muscle. Muscle Nerve, *33*, 49–60.

Ikeya, T., Galic, M., Belawat, P., Nairz, K. and Hafen, E. (2002). Nutrient-dependent expression of insulin-like peptides from neuroendocrine cells in the CNS contributes to growth regulation in *Drosophila*. Curr. Biol., *12*, 1293–300.

Irie, K., Hara-Irie, F., Ozawa, H. and Yajima, T. (2002). Calcitonin gene-related peptide (CGRP)-containing nerve fibers in bone tissue and their involvement in bone remodeling. Microsc. Res. Tech., *58*, 85–90. (abstract).

Jabaily, J.A. and Singer, M. (1977). Neurotrophic stimulation of DNA synthesis in the regenerating forelimb of the newt, *Triturus*. J. Exp. Zool., *199*, 251–56.

Jeays-Ward, K., Hoyle, C., Brennan, J., Dandonneau, M., Alldus, G., Capel, B., et al. (2003). Endothelial and steroidogenic cell migration are regulated by WNT4 in the developing mammalian gonad. Development, *130*, 3663–70.

Jeske, Y.V., Bowles, J., Greenfield, A. and Koopman, P. (1995). Expression of a linear *Sry* transcript in the mouse genital ridge. Nat. Genet., *10*, 480–82.

Jiang, C., Baehrecke, E.H. and Thummel, C.S. (1997). Steroid-regulated programmed cell death during *Drosophila* metamorphosis. Development, *124*, 4673–83.

Jiang, X., Rowitch, D.H., Soriano, P., McMahon, A.P. and Sucov, H.M. (2000). Fate of the mammalian cardiac neural crest. Development, *127*, 1607–16.

Kaartinen, V., Dudas, M., Nagy, A., Sridurongrit, S., Lu, M.M. and Epstein, J.A. (2004). Cardiac outflow tract defects in mice lacking ALK2 in neural crest cells. Development, *131*, 3481–90.

Karp, S.L., Ortiz-Arduan, A., Li, S. and Neilson, E.G. (1994). Epithelial differentiation of metanephric mesenchymal cells after stimulation with hepatocyte growth factor or embryonic spinal cord. Proc. Natl. Acad. Sci. U.S.A., *91*, 5286–90.

Katsukura, Y., Ando, H., David, C.N., Grimmelikhuijzen, C.J.P. and Sugiyama, T. (2003). Control of planula migration by LWamide and RFamide neuropeptides in *Hydractinia echinata*. J. Exp. Biol., *207*, 1803–10.

Katsukura, Y., David, C.N., Grimmelikhuijzen, C.J. and Sugiyama, T. (2003). Inhibition of metamorphosis by RFamide neuropeptides in planula larvae of *Hydractinia echinata*. Dev. Genes Evol., *213*, 579–86.

Kauffman, A.S. (2009). Sexual differentiation and the *Kiss1* system: hormonal and developmental considerations. Peptides, *30*, 83–93.

Keegan, B.R., Feldman, J.L., Begemann, G., Ingham, P.W. and Yelon, D. (2005). Retinoic acid signaling restricts the cardiac progenitor pool. Science, *307*, 247–49.

Kent, K.S. and Levine, R.B. (1993). Dendritic organization of an identified neuron during metamorphosis of the moth *Manduca sexta*; the influence of interactions with the periphery. J. Neurobiol., *24*, 1–22.

Kerman, B.K. and Andrew, D.J. (2006). Analysis of Dalmatian suggests a role for the nervous system in *Drosophila* embryonic trachea and salivary duct development. 47th Drosophila Research Conference Abstracts, p. 275.

Kim, S.H., Jen, W.C., DeRobertis, E.M. and Kintner, C. (2000). The protocadherin PAPC establishes segmental boundaries during somitogenesis in *Xenopus* embryos. Curr. Biol., *10*, 821–30.

Kim, J.A., Laney, C., Curry, J. and Unguez, G.A. (2008). Expression of myogenic regulatory factors in the muscle-derived electric organ of *Sternopygus macrurus*. J. Exp. Biol., *211*, 2172–84.

Kingan, T.G., Cardullo, R.A. and Adams, M.E. (2001). Signal transduction in eclosion hormone-induced secretion of ecdysis-triggering hormone. J. Biol. Chem., *276*, 25136–25142.

Kirby, M.L. and Waldo, K.L. (1995). Neural crest and cardiovascular patterning. Circ. Res., *77*, 211–15.

Kirby, M.L., Kumiski, D.H., Myers, T., Cerjan, C. and Mishima, N. (1993). Backtransplantation of chick cardiac neural crest cells cultured in LIF rescues heart development. Dev. Dyn., *198*, 296–311.

Kispert, A., Vainio, S. and McMahon, A.P. (1998). Wnt-4 is a mesenchymal signal for epithelial transformation of metanephric mesenchyme in the developing kidney. Development, *125*, 4225–34.

Konttinen, Y., Imai, S. and Suda, A. (1996). Neuropeptides and the puzzle of bone remodeling. State of the art. Acta Orthop. Scand., *67*, 632–39.

Koopman, P., Muensterberg, A., Capel, B., Vivian, N. and Lovell-Badge, R. (1990). Expression of a candidate sex-determining gene during mouse testis differentiation. Nature, *348*, 450–52.

Kramer, C.R. and Imbriano, M.A. (1997). Neuropeptide Y (NPY) induces gonad reversal in the protogynous bluehead wrase, *Thalassoma bifasciatum* (Teleostei: Labridae). J. Exp. Zool., *279*, 133–44.

Krieger, M.J.B., Jahan, N., Riehle, M.A., Cao, C. and Brown, M.R. (2004). Molecular characterization of insulin-like peptide genes and their expression in the African malaria mosquito, *Anopheles gambiae*. Insect Mol. Biol., *13*, 305–15.

Krueckl, S.L., Fradinger, E.A. and Sherwood, N.M. (2002). Developmental changes in the expression of growth hormone-releasing hormone and pituitary adenylate cyclase-activating polypeptide in zebrafish. J. Comp. Neurol., *455*, 396–405.

Lacalli, T.C. (1994). Apical organs, epithelial domains, and the origin of chordate nervous system. Am. Zool., *34*, 533–41.

Lassus, P., Opitz-Araya, X. and Lazebnik, Y. (2002). Requirement for caspase-2 in stress-induced apoptosis before mitochondrial permeabilization. Science, *297*, 1352–54.

Lawrence, P.A. and Johnston, P. (1986). The muscle pattern of a segment of *Drosophila* may be determined by neurons and not by contributing myoblasts. Cell, *45*, 505–13.

Lee, C. and Baehrecke, E.H. (2001). Steroid regulation of autophagic programmed cell death during development. Development, *128*, 1443–55.

Lee, Y.H., Wu, G.-C., Du, J.-L. and Chang, C.-F. (2004). Estradiol-17beta induced a reversible sex change in the fingerlings of protandrous black porgy, *Acanthopagrus schlegeli* Blecker: The possible roles of luteinizing hormone in sex change. Biol. Reprod., *71*, 1270–78.

Leger, S. and Brand, M. (2002). Fgf8 and Fgf3 are required for zebrafish ear placode induction, maintenance and inner ear patterning. Mech. Dev., *119*, 91–108.

Leise, E.M., Thavaradhara, K., Durham, N.R. and Turner, B.E. (2001). Serotonin and nitric oxide regulate metamorphosis in the marine snail *Ilyanassa obsoleta*. Am. Zool., *41*, 258–67.

Leitz, T., Morand, K. and Mann, M. (1994). A novel peptide controlling development of the lower metazoan *Hydractinia echinata* (Coelenterata, Hydrozoa). Dev. Biol., *163*, 440–46.

Lerner, U.H. (2002). Neuropeptidergic regulation of bone resorption and bone formation. J. Musculoskelet. Neuronal Interact., *2*, 440–47.

Levasseur, R., et al. (2003). Sympathetic nervous system as transmitter of mechanical loading in bone. Joint Bone Spine, *70*, 515–19.

Lewis, J.O. and Crews, S.T. (1994). Genetic analysis of the *Drosophila single-minded* gene reveals a central nervous system influence on muscle development. Mech. Dev., *48*, 81–91.

Li, Q. (1999). Modification of chromatin structure by the thyroid hormone receptor. Trends Endocrinol. Metab., *10*, 157–64.

Li, T. and Bender, M. (2000). A conditional rescue system reveals essential functions for the ecdysone receptor (EcR) gene during molting and metamorphosis in *Drosophila*. Development, *127*, 2897–905.

Li, H.Y., Chang, S.P., Yuan, C.C., Chao, H.T., Ng, H.T. and Sung, Y.J. (2001). Nitric oxide induces extensive apoptosis in endometrial epithelial cells in the presence of progesterone: involvement of mitogen-activated protein kinase pathways. Mol. Hum. Reprod., *7*, 755–63.

Li, Y., Fenger, U., Niehrs, C. and Pollet, N. (2003). Cyclic expression of esr9 gene in *Xenopus* presomitic mesoderm. Differentiation, *71*, 83–9.

Liang, D., Chang, J.R., Chin, A.J., Smith, A., Kelly, C., Weinberg, E.S., et al. (2001). The role of vascular endothelial growth factor (VEGF) in vasculogenesis, angiogenesis, and hematopoiesis in zebrafish development. Mech. Dev., *108*, 29–43.

Linker, C., Lesbros, C., Gros, J., Burrus, L.W., Rawls, A. and Marcelle, C. (2005). β-Catenin-dependent Wnt signalling controls the epithelial organisation of somites through the activation of paraxis. Development, *132*, 3895–905.

Linnoila, R.I. (2006). Functional facets of the pulmonary neuroendocrine system. Lab. Invest., *86*, 425–44.

Lockshin, R.A. and Williams, C.M. (1965). Programmed cell death—III. Neural control of the breakdown of the intersegmental muscles of silkmoths. J. Insect Physiol., *11*, 601–10.

Luan, H., Lemon, W.C., Peabody, N.C., Pohl, J.B., Zelensky, P.K., Wang, D., et al. (2006). Functional dissection of a neuronal network required for cuticle tanning and wing expansion in *Drosophila*. J. Neurosci., *26*, 573–84.

Luedeman, R. and Levine, R.B. (1996). Neurons and ecdysteroids promote the proliferation of myogenic cells cultured from the developing adult legs of *Manduca sexta*. Dev. Biol., *173*, 51–68.

Lundberg, P., Boström, I., Mukohyama, H., Bjurholm, A., Smans, K. and Lerner, U.H. (1999). Neuro-hormonal control of bone metabolism: vasoactive intestinal peptide stimulates ALP activity and mRNA expression in mouse calvarial osteoblasts as well as calcium accumulation in mineralized bone nodules. Regul. Pept., *85*, 47–58.

Lundberg, P., Lie, A., Bjurholm, A., Lehenkari, P.P., Horton, M.A., Lerner, U.H., et al. (2000). Vasoactive intestinal peptide regulates osteoclast activity via specific binding sites on both osteoclasts and osteoblasts. Bone, *27*, 803–10.

Lundberg, P., Lundgren, I., Mukohyama, H., Lehenkari, P.P., Horton, M.A. and Lerner, U.H. (2001). Vasoactive intestinal peptide (VIP)/pituitary adenylate cyclase-activating peptide receptor subtypes in mouse calvarial osteoblasts: presence of VIP-2 receptors and differentiation-induced expression of VIP-1 receptors. Endocrinology, *142*, 339–47.

Lüthi, T.E. (1994). Molecules that become redistributed during regeneration of the leech central nervous system. J. Exp. Biol. *186*, 43–54.

Maden, M., Sonneveld, E., van der Saag, P.T. and Gale, E. (1998). The distribution of endogenous retinoic acid in the chick embryo: implications for developmental mechanisms. Development, *125*, 4133–44.

Manley, N.R. and Capecchi, M.R. (1995). The role of Hox-3 in mouse thymus and thyroid development. Development, *121*, 1989–2003.

Martin, T. and Luo, Z.-X. (2005). Homoplasy in the mammalian ear. Science, *307*, 861–62.

Martinez, V.G., Menger, G.J., III and Zoran, M.J. (2005). Regeneration and asexual reproduction share common molecular changes: upregulation of a neural glycoepitope during morphallaxis in *Lumbriculus*. Mech. Dev., *122*, 721–32.

Maruska, K.P. and Fernald, R.D. (2011). Plasticity of the reproductive axis caused by social status change in an African cichlid fish: II. testicular gene expression and spermatogenesis. Endocrinology, *152*, 291–302.

Marvin, M.J., di Rocco, G., Gardiner, A., Bush, S.M. and Lassar, A.B. (2001). Inhibition of Wnt activity induces heart formation from posterior mesoderm. Genes Dev., *15*, 316–27.

Mescher, A.L. and Munaim, S.I. (1984). "Trophic" effect of transferring on amphibian limb regeneration blastemas. J. Exp. Zool., *230*, 485–90.

Mescher, A.L., White, G.W. and Brokaw, J.J. (2000). Apoptosis in regenerating and denervated, nonregenerating urodele forelimbs. Wound Repair Regen., *8*, 110–16.

Metzger, R.J. and Krasdnow, M.A. (1999). Genetic control of branching morphogenesis. Science, *284*, 1635–39.

Mellon, D., Jr. (1999). Muscle restructuring in crustaceans: myofiber death, transfiguration and rebirth. Am. Zool., *39*, 527–40.

Millauer, B., et al. (1993). High affinity VEGF binding and developmental expression suggest Flk-1 as a major regulator of vasculogenesis and angiogenesis. Cell, *72*, 835–36.

Milnes, M.R., Roberts, R.N. and Guillette, L.J. (2002). Effects of incubation and estrogen exposure on aromatase activity in the brain and gonads of embryonic alligators. Environ. Health Perspect., *110*(Suppl. 3), 393–96.

Mittwoch, U. (2000). Genetics of sex determination: exceptions that prove the rule. Mol. Genet. Metab., *71*, 405–10.

Miyagawa-Tomita, S., Waldo, K., Tomita, H. and Kirby, M.L. (1991). Temporospatial study of the migration and distribution of cardiac neural crest in quail–chick chimeras. Am. J. Anat., *192*, 79–88.

Mollard, R., Ghyselinck, N.B., Wendling, O., Chambon, P. and Mark, M. (2000). Stage-dependent responses of the developing lung to retinoic acid signaling. Int. J. Dev. Biol., *44*, 457–62.

Moreno, T.A. and Kintner, C. (2004). Regulation of segmental patterning by retinoic acid signaling during *Xenopus* somitogenesis. Dev. Cell., *6*, 205–18.

Morris-Kay, G.M. and Sokolova, N. (1996). Embryonic development and pattern formation. FASEB J., *10*, 961–68.

Mukoyama, Y.S., Shin, D., Britsch, S., Taniguchi, M. and Anderson, D.J. (2002). Sensory nerves determine the pattern of arterial differentiation and blood vessel branching in the skin. Cell, *109*, 693–705.

Mukoyama, Y., Gerber, H.-P., Ferrara, N., Gu, C. and Anderson, D.J. (2005). Peripheral nerve-derived VEGF promotes arterial differentiation via neuropilin 1-mediated positive feedback. Development, *132*, 941–52.

Mullen, L.M., Bryant, S.V., Torok, M.A., Blumberg, B. and Gardiner, D.M. (1996). Nerve dependency of regeneration: the role of distal-less and FGF signaling in amphibian limb regeneration. Development, *122*, 3487–97.

Munsterberg, A.E., Kitajewski, J., Bumcrot, D.A., McMahon, A.P. and Lassar, A.B. (1995). Combinatorial signaling by Sonic hedgehog and Wnt family members induces myogenic bHlH gene expression in the somites. Gene Dev., *9*, 2911–22.

Munyiri, F.N., Shintani, Y. and Ishikawa, Y. (2004). Evidence for the presence of a threshold weight for entering diapause in the yellow-spotted longicorn beetle, *Psacothea hilaris*. J. Insect Physiol., *50*, 295–301.

Namba, R., Pazdera, T.M., Cerrone, R.L. and Minden, J.S. (1997). *Drosophila* embryonic pattern repair: how embryos respond to bicoid dosage alteration. Development, *124*, 1393–403.

Nebigil, C.G., Choi, D., Dierich, A., Hickel, P., Le Meur, M., Messaddeq, N., et al. (2000). Serotonin 32B receptor is required for heart development. Proc. Natl. Acad. Sci. U.S.A., *97*, 9508–13.

Nekrep, N., Wang, J., Miyatsuka, T. and German, M.S. (2008). Signals from the neural crest regulate beta-cell mass in the pancreas. Development, *135*, 2151–60.

Nelson, D.W. and Padgett, R.W. (2003). Insulin worms its way into the spotlight. Genes Dev., *17*, 813–18.

Neuhuber, W.L. (2003). Lung sensors–complex functions require complex structures. Am. J. Respir. Cell Mol. Biol., *28*, 265–66.

Niederreither, K., Vermot, J., Fraulob, V., Chambon, P. and Dolle, P. (2002). Retinaldehyde dehydrogenase 2 (RALDH2)-independent patterns of retinoic acid synthesis in the mouse embryo. Proc. Natl. Acad. Sci. U.S.A., *99*, 16111–16.

Niederreither, K., Vermot, J., Schuhbaur, B., Chambon, P. and Dollé., P. (2002). Embryonic synthesis of retinoic acid is required for forelimb growth and anteroposterior patterning in the mouse. Development, *129*, 3563–74.

Nijhout, H.F. (2003). Development and evolution of adaptive polyphenisms. Evol. Dev., *5*, 9–18.

Nijhout, H.F. (2003). The control of body size in insects. Dev. Biol., *261*, 1–9.

Nijhout, H.F. and Grunert, L.W. (2002). Bombyxin is a growth factor for wing imaginal disks in *Lepidoptera*. Proc. Natl. Acad. Sci. U.S.A., *99*, 15446–50.

Nishino, K., Yamanouchi, K., Naito, K. and Tojo, H. (2002). Matrix metalloproteinases regulate mesonephric cell migration in developing XY gonads, which correlates with the inhibition of tissue inhibitor of metalloproteinase-3 by Sry. Dev. Growth Differ., *44*, 35–43.

Norman, A.W. and Litwack, G. (1997), *Hormones*, second edition. Academic Press, p. 240.

Novicki, A. and Weeks, J.C. (1996). The initiation of pre-ecdysis and ecdysis behaviors in larvae *Manduca sexta*. J. Exp. Biol., *199*, 1757–69.

Ogino, Y., Suzuki, K., Haraguchi, R., Satoh, Y., Dolle, P. and Yamada, G. (2001). External genitalia formation. Ann. N.Y. Acad. Sci., *948*, 13–31.
Ohlsson, C., Bengtsson, B.-A., Isaksson, O.G.P., Andreassen, T.T. and Slootweg, M.C. (1998). Growth hormone and bone. Endocr. Rev., *19*, 55–79.
Olivera-Martinez, I., Thélu, J., Teillet, M.-A. and Dhouailly, D. (2001). Dorsal dermis development depends on a signal from the dorsal neural tube, which can be substituted by Wnt-1. Mech. Dev., *100*, 233–44.
Osafune, K., Nishinakomura, R., Komazaki, S. and Asashima, M. (2002). *In vitro* induction of the pronephric duct in *Xenopus* explants. Dev. Growth Differ., *44*, 161–68.
Ogino, Y., Suzuki, K., Haraguchi, R., Satoh, Y., Dolle, P. and Yamada, G. (2006). Mechanical stretch-induced serotonin release from pulmonary neuroendocrine cells: implications for lung development. Am. J. Physiol. Lung Mol. Physiol., *290*, L185–93.
Panowski, S.H. and Dillin, A. (2009). Signals of youth: endocrine regulation of aging in *Caenorhabditis elegans*. Trends Endocrinol. Metab., *20*, 259–64.
Parsons-Wingerter, P., Elliott, K.E., Clark, J.I. and Farr, A.G. (2000). Fibroblast growth factor-2 selectivity stimulates angiogenesis of small vessels in arterial tree. Arterioscler. Thromb. Vasc. Biol., *20*, 1250–56.
Pelleymounter, M.A., et al. (1995). Effects of the obese gene product on body weight regulation in ob/ob mice. Science, *269*, 540–43.
Pelzer, T., et al. (2000). 17-Beta-estradiol prevents programmed cell death in cardiac myocytes. Biochem. Biophys. Res. Commun., *268*, 192–200.
Peng, X.-D., et al. (2001). The growth hormone (GH)-axis of GH receptor/binding protein gene-disrupted and metallothionein-human GH-releasing hormone transgenic mice: hypothalamic neuropeptide and pituitary receptor expression in the absence and presence of GH feedback. Endocrinology, *142*, 1117–23.
Perantoni, A.O., Dove, L.F. and Karavanova, I. (1995). Basic fibroblast growth factor can mediate the early inductive events in renal development. Proc. Natl. Acad. Sci. U.S.A., *92*, 4696–700.
Personius, K.E. and Chapman, R.F. (2002). Control of muscle degeneration following autotomy of a hindleg in the grasshopper, *Barytettix humphreysii*. J. Insect Physiol., *48*, 91–102.
Phillips, M.T., Kirby, M.L. and Forbes, G. (1987). Analysis of cranial neural crest distribution in the developing heart using quail–chick chimeras. Circ. Res., *60*, 27–30.
Pieau, C. and Dorizzi, M. (2004). Oestrogens and temperature-dependent sex determination in reptiles: all is in the gonads. J. Endocrinol., *181*, 367–77.
Pierce, S.B., Costa, M., Wisotzkey, R., Devadhar, S., Homburger, S.A., Buchman, A.R., et al. (2001). Regulation of DAF-2 receptor signaling by human insulin and ins-1, a member of the unusually large and diverse *C. elegans* insulin gene family. Genes Dev., *15*, 672–86.
Pires, A., Coon, S.L. and Hadfield, M.G. (1997). Catecholamines and dihydroxyphenylalanine in metamorphosing larvae of the nudibranch *Phestilla sibogae* Bergh (Gastropod: Opistobranchia). J. Comp. Physiol. A, *181*, 187–94.
Pires, A., Guibault, T.R., Mitten, J.V. and Skiendzielewski, J.A. (2000). Catecholamines in larvae and juveniles of the prosobranch gastropod, *Crepidula fornicate*. Comp. Biochem. Phys. C, *127*, 37–47.
Plant, T.M. and Barker-Gibb, M.L. (2004). Neurobiological mechanisms of puberty in higher primates. Hum. Reprod. Update, *10*, 67–77.
Plant, T.M. and Shahab, M. (2002). Neuroendocrine mechanisms that delay and initiate puberty in higher primates. Physiol. Behav., *77*, 717–22.
Plickert, G., Kroiher, M. and Munck, A. (1988). Cell proliferation and early differentiation during embryonic development and metamorphosis of *Hydractinia echinata*. Development, *103*, 795–803.

Plisov, S.Y., Yoshino, K., Dove, L.F., Higinbotham, K.G., Rubin, J.S. and Perantoni, A.O. (2001). TGFbeta2, LIF and FGF2 cooperate to induce nephrogenesis. Development, *128*, 1045–57.

Ravel-Chapuis, A., Vandromme, M., Thomas, J.-L. and Schaeffer, L. (2007). Postsynaptic chromatin is under neural control at the neuromuscular junction. EMBO J., *26*, 1117–28.

Read, A.T. and Govind, C.K. (1997). Claw transformation and regeneration in adult snapping shrimp: test of the inhibition hypothesis for maintaining bilateral asymmetry. Biol. Bull., *193*, 401–09.

Read, A.T., McTeague, J.A. and Govind, C.K. (1991). Morphology and behavior of an unusually flexible thoracic limb in the snapping shrimp, *Alpheus heterochelis*. Biol. Bull., *181*, 158–68.

Reavis, R.H. and Grober, M.S. (1999). An integrative approach to sex change: social, behavioural and neurochemical changes in *Lythrypnus dalli* (Pisces). Acta Ethol., *2*, 51–60.

Reddien, P.W. and Alvarado, A.S. (2004). Fundamentals of planarian regeneration. Ann. Rev. Cell Dev. Biol., *20*, 725–57.

Reshef, R., Maroto, M. and Lassar, A.B. (1998). Regulation of dorsal cell fates: BMPs and Noggin control the timing and pattern of myogenic regulator expression. Genes Dev., *12*, 290–303.

Ribatti, D., Urbinati, C., Nico, B., Rusnati, M., Roncali, L. and Presta, M. (1995). Endogenous basic fibroblast growth factor is implicated in the vascularization of the chick embryo chorioallantoic membrane. Dev. Biol., *170*, 39–49.

Riboni, L., Escamilla, C., Chavira, R. and Dominguez, R. (1998). Effects of peripheral sympathetic denervation induced by guanethidine administration on the mechanisms regulating puberty in the female guinea pig. J. Endocrinol., *156*, 91–8.

Riccomagno, M.M., Martinu, L., Mulheisen, M., Wu, D.K. and Epstein, D.J. (2002). Specification of the mammalian cochlea is dependent on sonic hedgehog. Genes Dev., *16*, 2365–78.

Ritvos, O., Tuuria, T., Erämaaa, M., Sainiob, K., Hildéna, K., Saxén, L., et al. (1995). Activin disrupts epithelial branching morphogenesis in developing glandular organs of the mouse. Mech. Dev., *50*, 229–45.

Rose, U., Ferber, M. and Hustert, R. (2001). Maturation of muscle properties and its hormonal control in an insect. J. Exp. Biol., *204*, 3531–45.

Ruhrberg, C., Gerhardt, H., Golding, M., Watson, R., Ioannidou, S., Fujisawa, H, et al. (2002). Spatially restricted patterning cues provided by heparin-binding VEGF-A control blood vessel branching morphogenesis. Genes Dev., *16*, 2684–98.

Rulifson, E.J., Kim, S.K. and Nusse, R. (2002). Ablation of insulin-producing neurons in flies: growth and diabetic phenotypes. Science, *296*, 1118–20.

Sachs, L.M. and Shi, Y.B. (2000). Targeted chromatin binding and histone acetylation *in vivo* by thyroid hormone receptor during amphibian development. Proc. Natl. Acad. Sci. U.S.A., *97*, 13138–43.

Sachs, L.M., Abdallah, B., Hassan, A., Levi, G., De Luze, A., Reed, J.C., et al. (1997). Apoptosis in *Xenopus* tadpole tail muscles involves Bax-dependent pathways. FASEB J., *11*, 801–08.

Sainio, K., Suvanto, P., Davies, J., Wartiovaara, J., Wartiovaara, K., Saarma, M., et al. (1997). Glia cell-line-derived neurotrophic factor is required for bud initiation from ureteric epithelium. Development, *124*, 4077–87.

Salame-Mendez, A., Herrera-Munoz, J., Moreno-Mendoza, N. and Merchant-Larios, H. (1998). Response of diencephalons but not the gonad to female-promoting temperature

with elevated estradiol levels in the sea turtle *Lepidochelys olivacea*. J. Exp. Zool., *280*, 304–13.
Sariola, H., Holm, K. and Heinke-Fahle, S. (1988). Early innervation of the metanephric kidney. Development, *104*, 589–99.
Sariola, H., Ekblom, P. and Henke-Fahle, S. (1989). Embryonic neurons as *in vitro* inducers of differentiation of nephrogenic mesenchyme. Dev. Biol., *132*, 271–81.
Sassoon, D. (1999). *wnt* genes and endocrine disruption of the female reproductive tract: a genetic approach. Mol. Cell. Endocrinol., *158*, 1–5.
Satoh, A., Bryant, S.V. and Gardiner, D.M. (2008). Regulation of dermal fibroblast dedifferentiation and redifferentiation during wound healing and limb regeneration in the Axolotl. Dev. Growth Differ., *50*, 743–54.
Satoh, A., James, M.A. and Gardiner, D.M. (2009). Axolotl limb regeneration. J. Bone Joint Surg. Am., *91*, 90–8.
Schaller, H.C. (1976). Action of the head activator as growth hormone in Hydra. Cell Differ., *5*, 1–11.
Schaller, H.C. (1976). Action of the head activator on the determination of interstitial cells in Hydra. Cell Differ., *5*, 13–20.
Schaller, H.C. and Bodenmuller, H. (1981). Isolation and amino acid sequence of a morphogenetic peptide from hydra. Dev. Biol., *78*, 7000–04.
Schmich, J., Trepel, S. and Leitz, T. (1998). The role of GLWamides in metamorphosis of *Hydractinia echinata*. Dev. Genes Evol., *208*, 267–73.
Schmidt, M., Tanaka, M. and Munsterberg, A. (2000). Expression of beta-catenin in the developing chick myotome is regulated by myogenic signals. Development, *127*, 4105–13.
Schneider, V.A. and Mercola, M. (2001). Wnt antagonism initiates cardiogenesis in *Xenopus laevis*. Genes Dev., *15*, 304–15.
Scholz, B., et al. (2006). Sex-dependent gene expression in early brain development of chicken embryos. BMC Neurosci., *7*, 12.
Schwoerer-Böhning, B., Kroiher, M. and Müller, W.A. (1990). Signal transmission and covert prepattern in the metamorphosis of *Hydractinia echinata*. Roux's Arch. Dev. Biol. *198*, 245–251.
Seipp, S., Schmich, J. and Leitz, T. (2001). Apoptosis: a death-inducing mechanism tightly linked with morphogenesis in *Hydractinia echinata* (Cnidaria, Hydrozoa). Development, *128*, 4891–98.
Semsar, K. and Godwin, J. (2003). Social influences on the arginine vasotocin system are independent of gonads in a sex-changing fish. J. Neurosci., *23*, 4386–93.
Sheng, H.Z., Moriyama, K., Yamashita, T., Li, H., Potter, S.S., Mahon, K.A., et al. (1997). Multistep control of pituitary organogenesis. Science, *278*, 1809–12.
Shi, Y.B., Damjanovski, S., Amano, T. and Ishizuya-Oka, A. (1998). Regulation of apoptosis during development: input from extracellular matrix. Int. J. Mol. Med., *2*, 273–82.
Shiokawa, K., Kai, M., Higo, T., Kaito, C., Yokoska, J., Yasuhiko, Y., et al. (2000). Maternal program of apoptosis activate shortly after midblastula transition by overexpression of *S*-adenosylmethionine decarboxylase in *Xenopus* early embryos. Comp. Biochem. Phys. B, *126*, 149–55.
Sible, J.C., Anderson, J.A., Levellyn, A.L. and Maller, J.L. (1997). Zygotic transcription is required to block maternal program of apoptosis in *Xenopus* embryos. Dev. Biol., *189*, 335–46.
Smit, A.B., van Kesteren, R.E., Li, K.W., Van Minnen, J., Spijker, S. and Van Heerikhuizen, H., et al. (1998). Towards understanding the role of insulin in the brain: lessons

from insulin-related signaling systems in the invertebrate brain. Prog. Neurobiol., *54*, 35–54.

Smith, J.T., Clifton, D.K. and Steiner, R.A. (2006). Regulation of the neuroendocrine reproductive axis by kisspeptin-GPR54 signaling. Reproduction, *131*, 623–30.

Soukup, T. and Zelena, J. (1985). The differentiation of muscle stretch receptors in the rat after neonatal de-efferentiation. Physiol. Bohemoslov., *34*(Suppl.), 153–56. (abstract).

Stern, H.M., Brown, A.M.C. and Hauschka, S.D. (1995). Myogenesis in paraxial mesoderm: preferential induction by dorsal neural tube and by cells expressing Wnt-1. Development, *121*, 3675–86.

Stern, H.M., Lin-Jones, J. and Hauschka, S.D. (1997). Synergistic interactions between bFGF and a TGF-beta family member may mediate myogenic signals from the neural tube. Development, *124*, 3511–23.

Strand, F.L. (1999) *Neuropeptides*. MIT Press, p. 183.

Sugimoto, M., Uchida, N. and Hatayama, M. (2000). Apoptosis in skin pigment cells of medaka, *Oryzias latipes* (Teleostei), during long-term chromatic adaptation: the role of sympathetic innervation. Cell Tissue Res., *301*, 205–16. (abstract).

Suzuki, S. (1985). Effect of Y-organ removal on limb regeneration and molting in the terrestrial crab, *Sesarma haematocheir*. Gen. Comp. Endocrinol. *58*, 202–210.

Suzuki, M., Satoh, A., Ide, H. and Tamura, K. (2005). Nerve-dependent and -independent events in blastema formation during *Xenopus* froglet limb regeneration. Dev. Biol., *286*, 361–75.

Takeda, S. and Karsenty, G. (2008). Molecular bases of the sympathetic regulation of bone mass. Bone, *42*, 837–40.

Takeda, S., Elefteriou, F., Levasseur, R., Liu, X., Zhao, L., Parker, K.L., et al. (2002). Leptin regulates bone formation via the sympathetic nervous system. Cell, *111*, 305–17.

Takuma, N., Sheng, H.Z., Furuta, Y., Ward, J.M., Sharma, K., Hogan, B.L., et al. (1998). Formation of Rathke's pouch requires dual induction from the diencephalon. Development, *125*, 4835–40.

Tata, J.R. (1998). Amphibian metamorphosis as a model for studying the developmental actions of thyroid hormone. Cell Res., *8*, 259–72.

Tatar, M., Bartke, A. and Antebi, A. (2003). The endocrine regulation of aging by insulin-like signals. Science, *299*, 1346–51.

Te, K.G. and Reggiani, C. (2002). Skeletal muscle fibre type specification during embryonic development. J. Muscle Res. Cell Motil., *23*, 65–9.

Tidball, J.G. (2005). Mechanical signal transduction in skeletal muscle growth and adaptation. J. Appl. Physiol., *98*, 1900–08.

Tobet, S.A., Henderson, R.G., Whiting, P.J. and Sieghart, W. (1999). Special relationship of gamma-aminobutyric acid to the ventromedial nucleus of the hypothalamus during embryonic development. J. Comp. Neurol., *405*, 88–98.

Tobias, M.L., Marin, M.L. and Kelley, D.B. (1993). The roles of sex, innervation, and androgen in laryngeal muscle of *Xenopus laevis*. J. Neurosci., *13*, 324–33.

Togari, A. (2002). Adrenergic regulation of bone metabolism: possible involvement of sympathetic innervation of osteoblastic and osteoclastic cells. Microsc. Res. Techniq. *58*, 77–84. (abstract).

Torday, J.S. and Rehan, V.K. (2002). Stretch-stimulated surfactant synthesis is coordinated by the paracrine actions of PTHrP and leptin. Am. J. Physiol. Lung Cell Mol. Physiol., *283*, 130–35.

Torday, J.S., Sanchez-Esteban, J. and Rubin, L.P. (1998). Paracrine mediator of mechanotransduction in lung development. Am. J. Med. Sci., *316*, 205–08.

Treier, M., Gleiberman, A.S., O'Connell, S.M., Szeto, D.P., McMahon, J.A., McMahon, A.P. and Rosenfeld, M.G. (1998). Multistep signaling requirements for pituitary organogenesis *in vivo*. Genes Dev., *12*, 1691–704.

Treier, M., O'Connell, S., Gleiberman, A., Price, J., Szeto, D.P., Burgess, R., et al. (2001). Hedgehog signaling is required for pituitary gland development. Development, *128*, 377–86.

Truman, J.W. (2006). Steroid hormone secretion in insects comes of age. Proc. Natl. Acad. Sci. U.S.A., *103*, 8909–10.

Tohru Tsukui, T., Capdevila, J., Tamura, K., Ruiz-Lozano, P., Rodriguez-Esteban, C., Yonei-Tamura, S., et al. (1999). Multiple left–right asymmetry defects in *Shh -/-* mutant mice unveil a convergence of the Shh and retinoic acid pathways in the control of *Lefty-1*. Proc. Natl. Acad. Sci. U.S.A., *96*, 11376–81.

Tzahor, E. and Lassar, A.B. (2001). Wnt signals from the neural tube block ectopic cardiogenesis. Genes Dev., *15*, 255–60.

Uchida-Oka, N. and Sugimoto, M. (2001). Norepinephrine induces apoptosis in skin melanophores by attenuating cAMP-PKA signals via alpha2-adrenoceptors in the medaka, *Oryzias latipes*. Pigment Cell Res., *14*, 356–61.

Unguez, G.A. and Zakon, H.H. (2002). Skeletal muscle transformation into electric organ in *S. macrurus* depends on innervation. J. Neurobiol., *53*, 391–402.

Uochi, T. and Asashima, M. (1996). Sequential gene expression during pronephric tubule formation *in vitro* in *Xenopus* ectoderm. Dev. Growth Differ., *38*, 625–34.

Vainio, S.J. (2003). Nephrogenesis regulated by Wnt signaling. J. Nephrol., *16*, 279–85.

Vainio, S.J. and Uusitalo, M.S. (2000). A road to kidney tubules via the Wnt pathway. Pediatr. Nephrol., *15*, 151–56.

Vainio, S.J., Itäranta, P.V., Peräsaari, J.P. and Uusitalo, M.S. (1999). Wnts as kidney tubule inducing factors. Int. J. Dev. Biol., *43*, 419–23.

van Heumen, W.R. and Roubos, E.W. (1990). Ultrastructural evidence for synthesis, storage and release of insulin-related peptides in the central nervous system of *Lymnaea stagnalis*. Neuroscience, *39*, 493–500.

Vidal, B., Pasqualini, C., Le Belle, N., Holland, M.C., Sbaihi, M., Vernier, P., et al. (2004). Dopamine inhibits luteinizing hormone synthesis and release in the juvenile European eel: a neuroendocrine lock for the onset of puberty. Biol. Reprod., *71*, 1491–500.

Voytik, S.L., Przyborski, M., Badylak, S.F. and Konieczny, S.F. (1993). Differential expression of muscle regulatory factor genes in normal and denervated adult rat hindlimb muscles. Dev. Dyn., *198*, 214–24.

Walther, M., Ulrich, R., Kroiher, M. and Berking, S. (1996). Metamorphosis and pattern formation in *Hydractinia echinata*, a colonial hydroid. Int. J. Dev. Biol., *40*, 313–22.

Warburton, D. (2000). The molecular basis of lung morphogenesis. Mech. Dev., *92*, 55–8.

Watanabe, M., Choudhry, A., Berlan, M., Singal, A., Siwik, E., Mohr, S., et al. (1998). Developmental remodeling and shortening of the cardiac outflow tract involves myocyte programmed cell death. Development *125*, 3809–20.

Weeks, J.C. and Truman, J.W. (1986). Steroid control of neuron and muscle development during the metamorphosis of an insect. J. Neurobiol., *17*, 249–67.

Weinstein, R.S., Chen, J., Powers, C.C., Stewart, S.A., Landes, R.D., Bellido, T., et al. (2002). Promotion of osteoclast survival and antagonism of biphosphonate-induced apoptosis by glucocorticoids. J. Clin. Invest., *109*, 1041–48.

West, H., Richardson, W.D. and Fruttiger, M. (2005). Stabilization of the retinal vascular network by reciprocal feedback between blood vessels and astrocytes. Development, *132*, 1855–62.

West-Eberhard, M.J. (2003) *Developmental Plasticity and Evolution.* Oxford University Press, Oxford, p. 462.

Willingham, E., Baldwin, R., Skipper, J.K. and Crews, D. (2000). Aromatase activity during embryogenesis in the brain and adrenal–kidney–gonad of the red-eared slider turtle, a species with temperature-dependent sex determination. Gen. Comp. Endocrinol., *119*, 202–07.

Wolpert, L., Beddington, R., Lawrence, P. and Jessell, T.M. (1998) *Principles of Development.* Current Biology Ltd. and Oxford University Press, Oxford, pp. 335–6.

Xu, Y., He, J., Wang, X., Lim, T.M. and Gong, Z. (2000). Asynchronous activation of 10 muscle-specific protein (MSP) genes during zebrafish somitogenesis. Dev. Dyn., *219*, 201–15.

Yacobi, K., Wojtowicz, A., Tsafriri, A. and Gross, A. (2004). Gonadotropins enhance caspase-3 and -7 activity and apoptosis in the theca-interstitial cells of rat preovulatory follicles in culture. Endocrinology, *145*, 1943–51.

Yadav, V.K, Oury, F., Suda, N., Liu, Z., Gao, X., Confavreux, C., et al. (2009). A serotonin-dependent mechanism explains the leptin regulation of bone mass, appetite, and energy expenditure. Cell, *138*, 976–89.

Yazawa, T., Yamamoto, T. and Abé, S.-I. (2000). Prolactin induces apoptosis in the penultimate spermatogonial stage of the testes in Japanese red-bellied newt (*Cynops pyrrhogaster*). Endocrinology, *41*, 2027–32.

Yeh, L.C., Zavala, M.C. and Lee, J.C. (2002). Osteogenic protein-1 and interleukin-6 with its soluble receptor synergistically stimulate rat osteoblastic cell differentiation. J. Cell Physiol., *190*, 322–31.

Yin, V.P. and Thummel, C.S. (2004). A balance between the *diap1* death inhibitor and *reaper* and *hid* death inducers controls steroid-triggered cell death in *Drosophila.* Proc. Natl. Acad. Sci. U.S.A., *101*, 8022–27.

Yoonseong, P., Filippov, V., Gill, S.S. and Adams, M.E. (2002). Deletion of the ecdysis-triggering gene leads to lethal ecdysis deficiency. Development, *129*, 493–503.

Young, R.E., Pearce, J. and Govind, C.K. (1994). Establishment and maintenance of claw bilateral asymmetry in snapping shrimps. J. Exp. Zool., *269*, 319–26.

Zaffran, S. and Frasch, M. (2002). Early signals in cardiac development. Circ. Res., *91*, 457–69.

Zaidi, M. (2005). Neural surveillance of skeletal homeostasis. Cell Metab., *1*, 219–21.

Zega, G., Pennati, R., Fanzago, A. and de Bernardi, F. (2007). Serotonin involvement in the metamorphosis of the hydroid *Eudendrium racemosum.* Int. J. Dev. Biol., *51*, 307–13.

Zeng, Q.Q., Jee, W.S., Bigornia, A.E., King, J.G., Jr., D'Souza, S.M., Li, X.J., et al. (1996). Time responses of cancellous and cortical bones to sciatic neurectomy in growing female rats. Bone, *19*, 13–21.

Zenjari, C., Boilly, B., Hondermarck, H. and Boilly-Marer, Y. (1997). Nerve–blastema interactions induce fibroblast growth factor-1 release during limb regeneration in *Pleurodeles waltl.* Dev. Growth Differ., *39*, 15–22.

Zitnan, D. and Adams, M.E. (2000). Excitatory and inhibitory roles of central ganglia in initiation of the insect ecdysis behavioural sequence. J. Exp. Biol., *203*, 1329–40.

Zitnan, D., Sehnal, F. and Bryant, P.J. (1993). Neurons producing specific neuropeptides in the central nervous system of normal and pupariation-delayed *Drosophila.* Dev. Biol., *156*, 117–35.

6 The Epigenetic System of Inheritance—An Outline

The metaphor of the "genetic tool kit" is an unfinished one. Tools are only effectual when operated by a user.

In Chapter 1, I argued and presented substantiating evidence that an integrated control system (ICS) maintains the normal structure and function in metazoans. On that basis, I put forward the hypothesis that in the process of animal reproduction, the ICS, with the CNS as its controller, functions as an epigenetic system of inheritance, providing the information for animal morphology.

I conditioned the verification of the hypothesis on the epigenetic system of heredity with validation of the following testable predictions of the hypothesis:

1. The expression of nonhousekeeping genes in metazoans is controlled and regulated by signal cascades that ultimately originate in the CNS.
2. Initial signals for starting signal cascades in the CNS result as output of the processing of internal/external stimuli in neural circuits.
3. The reproduction cycles and gametogenesis, production of egg and sperm cells, in metazoans are under control of the ICS, with the CNS as the controller.
4. The transfer of the epigenetic information (parental cytoplasmic factors and gene imprinting) in gametes is regulated by the ICS, which, in the process of reproduction, serves as the parental epigenetic system of inheritance.
5. At the phylotypic stage, when the function of maternal cytoplasmic factors terminates, the embryonic CNS is operational and takes over the control of the embryonic development up to adulthood.
6. Signal cascades determining the postphylotypic development originate in the embryonic CNS.
7. The evidence for validating predictions 1, 2, 3, 4, 5, and 6 is presented in Chapters 1, 2, 3, 4, and 5, respectively.

I believe that the empirical evidence and the related theoretical inferences presented have validated the hypothesis and warrant the following narrative of the epigenetic system of inheritance and the epigenetic mechanism of the development of morphological traits in the process of metazoan reproduction.

The Epigenetic System of Inheritance

Metazoan structure, at all the levels of organization (molecular, subcellular, cellular, supracellular, and organismic), is unavoidably eroding and disintegrating. The fact that despite the continual erosion of their structure, metazoans can maintain that structure in a steady state suggests that they compensate for the lost and

Epigenetic Principles of Evolution. DOI: 10.1016/B978-0-12-415831-3.00006-9

disintegrated structures. But maintenance at a steady state of an extremely complex and continuously eroding structure requires as a *sine qua non* an appropriate control system. We have identified that control system as an ICS, with the CNS as its controller. It enables metazoans to continually compensate for the (physiologically, apoptotically, or accidentally) dying cells and degraded compounds as well as replace the damaged structures.

In vertebrates, the ICS is a hierarchical system consisting of the CNS, at the vertex of the hierarchy, the peripheral nervous system and the neuroendocrine system. Neurons of the peripheral nervous system carry information to (sensory neurons) the CNS and from the latter to organs (motor neurons).

The function of the neuroendocrine branch is based on the neural regulation of secretion of hormones of the target endocrine glands (e.g., thyroid, parathyroid, thymus, pancreas, adrenal, ovaries, and testes) and their downstream inducers (growth factors and secreted proteins), as well as on the endocrine brain, hypothalamus, pituitary, and clusters of secretory cells in organs such as the liver, kidney, and intestines.

The peripheral branch performs its functions by selectively releasing inducers (e.g., growth factors, secreted proteins, and neurotransmitters) via its innumerable nerve endings, in strictly determined sites of the animal body. The targeted release of inducers at specific sites by the peripheral branch plays a crucial role in regulation of the function of the neuroendocrine branch; it determines the specificity of the action of hormones in organs and tissues of the animal body, by preventing their action in all but their target sites. In many of the known cases, this is done by controlling, i.e., switching on or off, expression of specific receptors that mediate the function of hormones and other inducers.

Based on the pervasive presence of the nervous system all over the metazoan body, the CNS monitors the actual state of the structure and function of the animal, by continuously receiving a huge input of data on the state of the system, down to the level of individual cells. Its enormous computational capability allows the CNS to compare data on the state of the system with the normal state as determined by set points (information for the normal structure) it establishes. That comparison allows the CNS to detect deviations from the norm, to make decisions, and to send correcting signals to the aberrant structures. Via specific signal cascades, its signals are translated into messages for restoring the normal state.

The CNS, as controller of the ICS, determines the normal structure, i.e., the patterns of spatial organization of the enormously large numbers of cells of tens to hundreds of different types. By definition, a control system that provides information for the normal structure along with the capability to communicate that information to the offspring, is a mechanism of inheritance. Indeed, this is what takes place in the process of metazoan reproduction: *the ICS functions as an epigenetic system of inheritance.* As was shown in Chapters 3 and 4, the parental CNS controls the reproductive activity and related morphophysiological changes taking place there in the process of reproduction, as well as production of gametes and the early individual development.

In vertebrates, the input of various favorable environmental stimuli (e.g., lengthening of the photoperiod, warmer temperature, and social factors) and internal stimuli

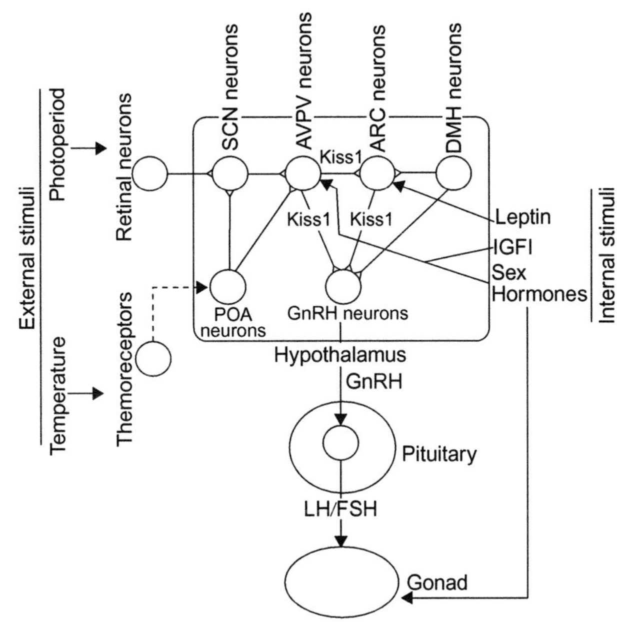

Figure 6.1 Simplified model of neural regulation of reproductive activity in vertebrates. The hypothalamus responds to the input of the information on the internal (nutrition status, hormonal balance, age, body weight) and external (day length, environmental temperature, and social cues) conditions by activating or not the hypothalamus–pituitary–gonadal axis. The activation starts with secretion of the neuropeptide kisspeptin by AVPV and ARC, both of which release kisspeptin on GnRH neurons. Binding of kisspeptin to its receptor GPR54 induces GnRH neurons to secrete and release GnRH in the pituitary, stimulating the latter to secrete gonadotropins FSH and LH, which via blood circulation induce sexual activity and secretion of sex hormones by gonads. *Abbreviations*: ARC, arcuate nucleus; AVPV, anteroventral periventricular nucleus; DMH, hypothalamic dorsomedial nucleus; FSH, follicle-stimulating hormone; GnRH, gonadotropin-releasing hormone; IGFI, insulin growth factor I; LH, luteinizing hormone; SCN, suprachiasmatic nucleus; POA preoptic area.

(e.g., nutritional state, changes in the hormonal balance, age, and body weight) is continually processed in the neural circuits in the CNS and particularly in hypothalamic nuclei (Figure 6.1). In vertebrates, the output of the processing of the input of the internal and external stimuli in specific neural circuits in the form of electrical/chemical signals stimulates specific hypothalamic neurons to secrete the neuropeptide kisspeptin (Dungan et al., 2006; Popa et al., 2008; Smith et al., 2006). It is demonstrated that poor nutrition status (e.g., experimental fasting) causes decline of secretion of kisspeptin and its receptor GPR54 (Luque et al., 2007). By binding its specific receptor GPR54, kisspeptin stimulates hypothalamic gonadotropin-releasing hormone (GnRH) neurons to secrete GnRH, which stimulates the pituitary to secrete luteinizing hormone/follicle-stimulating hormone (LH/FSH), which, in turn, induces the activity of the reproductive organs. These neurohormonal messages, along with the reproductive messages sent by the gonadal innervation, induce specific morphophysiological and behavioral changes that prepare these organs for starting the reproductive activity, including gametogenesis, production of ova and sperm cells.

The timing of the sexual reproductive cycle is crucial for the reproductive success in cyclic ovulatory species. In moderate climate regions, the reproductive activity must be timed so that, with the gestation period factored in, the animal produces offspring at the time of year when food availability, climate, and other conditions are optimal for raising their young. This clearly implies that these species have an internal mechanism, or central clock, for timing the reproductive phenomena. Indeed, not only vertebrates but even lower invertebrate species, such as insects, can ascertain the time of the year, based on the processing of data on the photoperiod or temperature, for example, in their brains and, accordingly, regulate their reproductive cycles and life histories.

In vertebrates, the central clock in the hypothalamic suprachiasmatic nucleus (SCN) determines the circadian changes in the expression of thousands of genes fashioning the circadian physiology, including reproductive physiology (Yamaguchi et al., 2003). In most of the cyclic ovulatory vertebrates, the hypothalamus, in response to signals from other parts of the brain, also determines the annual or seasonally increased activity of the GnRH pulse generator, which is responsible for activating the pituitary–gonadal axes and the ensuing morphophysiological and behavioral changes related to their reproduction, including the production of gametes.

Gametogenesis and Provision of Maternal Epigenetic Information to Oocytes

Formation of egg cells and sperm cells, from lower invertebrates to higher vertebrates, including humans, is under strict neural control and regulation. Signals released by specific neural circuits, as output of the processing of internal (e.g., age, health, and physiological and psychological state) and external (e.g., environmental temperature, photoperiod, and social factors) stimuli, start signal cascades, which induce the morphological, physiological, and behavioral changes necessary for regulating sexual behavior, the function and structure of the reproductive tract and gametogenesis, as well as ovulation and oviposition.

Obviously, the control of the structure and function of the reproductive system and gametogenesis requires that the neural signals and/or neural tissue have access to the reproductive organs. There are two main pathways through which the CNS might provide instructions to the reproductive tract and direct the orderly deposition into gametes of the epigenetic information (cytoplasmic factors) that control and regulate the early individual development up to the phylotypic stage. One is the neuroendocrine pathway, regulated mainly, but not entirely, by the hypothalamic–pituitary axis, and the other is the direct neural pathway of control of the function of reproductive organs and gametogenesis (Figure 6.2). Both of these pathways are considered in Chapter 3 of this work, and I will briefly present here only essentials of the direct neural control.

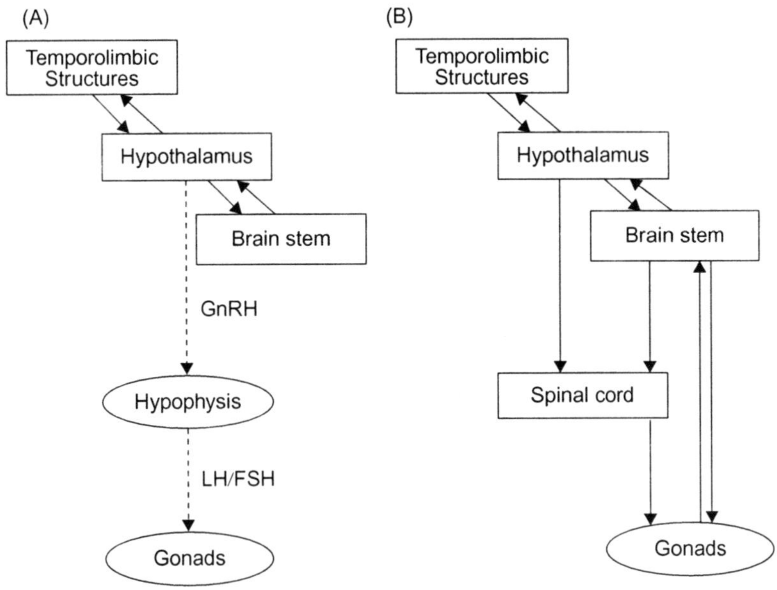

Figure 6.2 Simplified schematic drawing illustrating the neuroendocrine (A) and the direct neural (B) control of the gonads.
Source: From Gerendai (2004).

Crucial in processes of gametogenesis, in both dioecious and parthenogenetic species, is provision of egg cells and sperm cells with the appropriate epigenetic information in the form of thousands of cytoplasmic factors in the form of long-lived *mRNAs*, hormones, secreted proteins, neurotransmitters, *tRNAs*, and nutrients arranged in specific spatial patterns (Figure 6.3) that enables these cells, and only these cells among more than 200 types of cells in higher vertebrates, to develop into adult metazoan organisms. As mentioned earlier, the unique ability of gametes to develop into adult individuals depends on the presence of the gametic cytoplasm of epigenetic information in the form of parental factors rather than genetic information (which is the same in all cells of the body) in strictly determined spatial patterns. The investment of this form of epigenetic information is absolutely necessary for reproduction in both dioecious and parthenogenetic organisms. However, the mechanism of the ordered deposition of parental factors in gametes remains one of the great enigmas of modern biology. In the following text, I will argue that the developmentally indispensable process of ordered deposition of epigenetic information in the oocyte is neurally regulated.

The direct neural control of reproductive function and gametogenesis is function of the extrinsic and intrinsic innervation of the ovary and Graaf follicle. The extrinsic innervation consists mainly of the sympathetic and sensory nerves originating in the CNS from neurons of particular brain nuclei (hypothalamus, amygdala, locus coeruleus, and insular cortex) and spinal cord (intermediolateral column). Most of

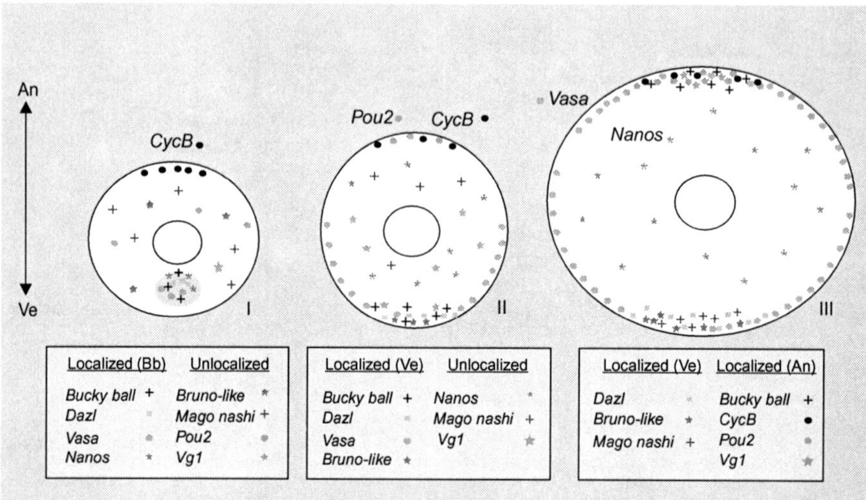

Figure 6.3 mRNA localization during oocyte development. During stage I of oogenesis *buc*, *nanos*, *vasa*, and *dazl* transcripts localize to the Balbiani body (Bb, pink), while cyclinB begins to be localized to the animal pole. By stage II of oogenesis, the Bb has disassembled, leaving *buc*, *vasa*, and *dazl* mRNAs at the vegetal cortex, whereas *nanos* becomes unlocalized, and *pou2* becomes localized to the animal pole. Note that *vasa* has a broad vegetal cortical domain at stage II. By stage III, *bruno-like* and *mago nashi* become vegetally localized (late pathway); now *buc* and Vg1 are localized to the animal pole, and *vasa* is localized radially at the cortex. Animal (An) pole is to top, and vegetal (Ve) to bottom. (For interpretation of the references to color in this figure legend, the reader is referred to the web version of this book.) *Source*: From Abrams and Mullins (2009).

these sympathetic nerves are adrenergic in nature (Madekurozwa, 2008). The sympathetic nerves form the ovarian plexus, which represents the main neural communication pathway between the CNS and the ovary (Gerendai et al., 2002). Similar nuclei are found to control the function of testis and spermiogenesis in male vertebrates (Gerendai, 2004). Extrinsic innervation penetrates deep into the ovary up to the individual granulosa cells neighboring the oocyte The experimental fact that lesioning of the left side of the anterior hypothalamus (Cruz et al., 1990) and dorsal raphe nucleus (Ayala et al., 1994) produces specific effects in the left ovary, and the lesions in the right side produce such effects in the right ovary alone clearly show a functional asymmetry in the brain centers of the direct neural control of ovaries. At the same time, this fact indicates that each brain side provides different types of information to the ovary.

It took a century to confirm (in 1995) an early observation of Winterhalter on the existence of an intrinsic innervation by neurons located within the ovary (Dees et al., 1995, 2006). Intrinsic neurons of the ovary are mainly catecholaminergic (Anesetti et al., 2001). The fact that these neurons secrete catecholamines (adrenalin or noradrenaline) suggests that they are involved in the process of steroidogenesis, and the fact that human granulosa cells express four of five known dopamine receptors

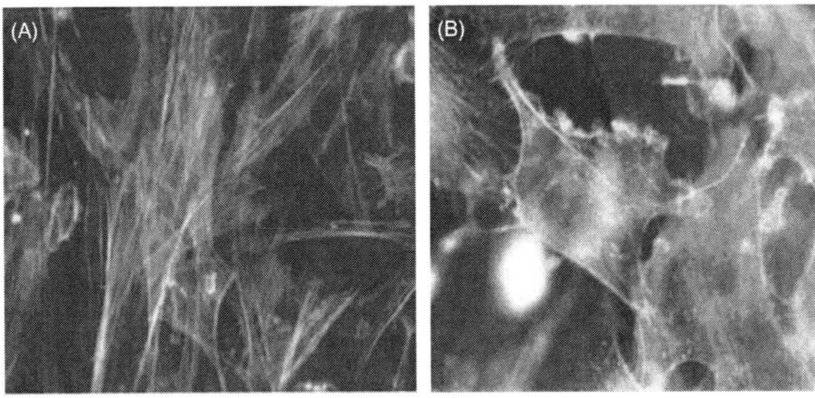

Figure 6.4 Serotonin (B) induces modification of actin cytoskeleton of pulmonary artery smooth muscle cells.
Source: From Day et al. (2006).

(Rey-Ares et al., 2007) and adrenergic receptors (Čikoš et al., 2005) suggests that the neurotransmitters themselves may have access to the oocyte. Serotonin is found in the follicular fluid of humans and rats (Clausell and Soliman, 1978; Amenta et al., 1992; Bódis et al., 1993) and the intrinsic and extrinsic innervation may be responsible for its secretion and fluctuations with the ovulatory cycle in the follicular fluid that surrounds the oocyte (Bódis et al., 1992).

Certainly, the molecular mechanism of the specific spatial arrangement of thousands of various types of cytoplasmic factors in egg and sperm cells is an amazing event, and it requires use of appropriate information. Genetic information, based on the triplet code, clearly does not come into account as a possible explanans. However, from the perspective of the present-day epigenetic knowledge, it would be an overstatement to call the mechanism an enigma.

The deposition of parental factors into gametes implies their spatially and temporally ordered transport. These factors are transported to specific sites along the oocyte skeleton with microtubules and actin filaments as molecular rails. The factors are driven by molecular motors such as dynein, kinesin, and myosin. Where these factors will be deposited in the oocyte depends on the length and direction of the "rails" (microtubules or actin filaments) on which they move. One simple way to control the distance of transport from one point to another in the oocyte is by modifying or increasing or decreasing the length of the rails.

There is evidence that some neurotransmitters can influence the configuration of the cytoskeletal rails, microtubules, and actin filaments. For example, serotonin modifies actin cytoskeleton in pulmonary artery smooth muscle cells (Day et al., 2006; (Figure 6.4) and human intestinal epithelial cells (Gill et al., 2007; 2008), but it also has effects on the microtubule skeleton. Changes in the microtubule structure and dynamics induced by serotonin are observed to occur in neurons as a result of chronic unpredicted mild stress, which are reversed by administration of fluoxetine, a serotonin reuptake inhibitor (Yang et al., 2009).

There is adequate empirical evidence showing that the nervous system does adaptively change the length of microtubules in the pigment cells, chromatophores, in order to instantly create, "at will," a practically infinite number of patterns of pigment organelles within chromatophores in order to produce body colorations and patterns to match the visual perception of their background, of individuals of other species, or to produce other socially impressive images of their own.

As indicated by the name, chromatophores are skin pigment cells that are responsible for body colors in numerous invertebrates and vertebrates. They are neural crest-derived cells that leave the neural crest early during the individual development. Chromatophores owe their color to the presence in their cytoplasm of pigment organelles. According to their color, chromatophores may be black (melanophores), yellow (xanthophores), blue (cyanophores), red (erythrophores), white (leukophores), or iridescent (iridophores).

Chromatophores can modify their color by dispersing pigment granules throughout the cytoplasm, which makes them more intensely colored, or by aggregating them around the cell nucleus, thus making the cell paler. The first process implies movement of pigment organelles toward the periphery, involving first their transport along microtubule plus ends, then their switching onto actin filaments (Figure 6.5).

But how can the nervous system regulate the details of the spatial arrangement of maternal factors in the oocyte? There are studies on the transport of various maternal factors to particular sites in the oocyte and on the motors that make their transport possible, but, unfortunately, there are no studies designed specially for identifying the mechanism of control of this transport. Nevertheless, we are already in possession of evidence on the neural control and regulation of the length of the cytoskeleton rails (both microtubules and actin filaments) in somatic cells.

The example of the modification of body pigmentation presented in Chapter 2 (Section Generation of Information for Adaptive Camouflage in *Xenopus*) illustrates the neural control of the transport of pigment particles along the cytoskeleton (microtubule and actin filaments) in both directions (toward the periphery and the cell nucleus). As already described, background adaptation of *Xenopus* depends totally on neural mechanisms and neuroendocrine cascades that control the dispersion/aggregation of pigment organelles in skin melanophores. We have seen that the proximate cause of the transport of melanosomes (pigment organelles) in melanophores is the neurally regulated release of the hormone α-melanophore-stimulating hormone (α-MSH). The hormone performs its function by influencing the cytoskeleton of melanophores.

The hormonal regulation of dispersion/aggregation of pigment particles in melanophores, however, is a slow mechanism of camouflage that cannot explain cases of instantaneous adaptive changes in body coloration and patterning observed, in many cases, in a matter of seconds, especially in cephalopods and fishes. Such rapid changes indicate that a neural mechanism acting directly on chromatophores must be involved. Indeed, studies of iridescence and iridophores in fish and cephalopods have shown that this is a neural mechanism.

Iridophores owe their adaptive coloration to the presence in their cytoplasm of stacks of guanine platelets. The light reflected by the platelet stacks is commonly

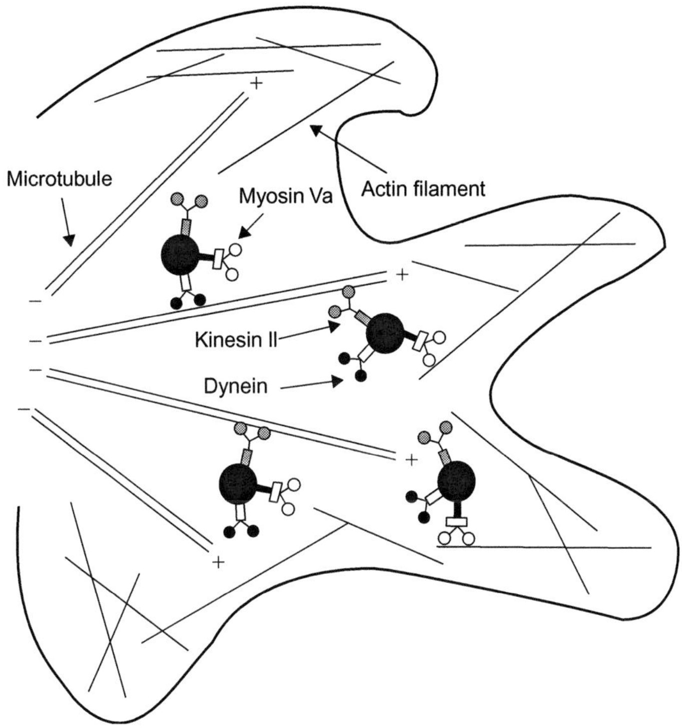

Figure 6.5 Distribution of microtubules and actin filaments in a *Xenopus laevis* melanophore. Retrograde transport of melanosomes (aggregation) is driven by the molecular motor dynein along microtubules and anterograde movements (dispersion) by kinesin II and myosin Va along both microtubules and actin filaments. The actin filaments trap melanosomes at the cell periphery, assuring an even distribution of pigment throughout the cell.
Source: From Eriksson (2009).

colored. Any change in the distance between guanine platelets or of the angle of reflection produces a change in the reflected color. There are fishes and cephalopods that can change the body color they reflect, "at will" and instantancously, by appropriately changing the distance between guanine platelets in iridophores. The animals achieve this by regulating (increasing or decreasing) the length of microtubules that keep platelets apart. This ability enables them not only to match the background color or pattern but also to mimic the image of individuals of other species. The details of the neural processing of the visual perception of the background or the body patterning/coloration of individuals of other species are not known, but the output of that processing of the visual input is delivered to individual iridophores in the form of detailed instructions (patterns of neurotransmitters release) by the sympathetic innervation (Kasukawa et al., 1986; Mäthger et al., 2004; for further evidence see in Chapter 10, Section Camouflage—Adaptive coloration, Cryptic Coloration, Crypsis).

The above evidence on the direct neural regulation (increase or decrease) of the length the microtubules and the ordered transport of pigment organelles on cytoskeleton rails makes it plausible that local innervation similarly might also regulate the transport of cytoplasmic factors in specific sites of the oocyte by lengthening, or shortening of microtubule rails. Biological evolution is too parsimonious to neglect the use in oocytes of a precious mechanism of strict localization of macromolecules that it uses in other types of cells.

Biphasic CNS Control of Reproduction

Sound experimental evidence shows beyond doubt that the epigenetic information provided to gametes in the form of parental cytoplasmic factors and a considerable number of imprinted genes controls and regulates the early development in vertebrates, from the first cleavage to the phylotypic stage. Initially, the parental cytoplasmic factors act on their own (maternal *mRNAs* after being translated into proteins), but later they perform their functions mainly by regulating expression of zygotic genes.

> *In vertebrates, maternal gene products direct fertilization, egg activation, the first cell division(s), and the initiation of zygotic transcription.*
>
> *Dosch et al. (2004)*

The function of the parental epigenetic information in the individual development terminates with the exhaustion of the reserve of the maternal cytoplasmic factors at the phylotypic stage, when an embryonic CNS is already operational. After the phylotypic stage, it is the embryonic CNS that provides the epigenetic information for all of the developmental processes, organogenesis, and histogenesis, in the form of inductive signals that activate developmental signal cascades.

Thus, the biological reproduction in metazoans results from a biphasic, transgenerational relay process, in which epigenetic system(s) of heredity of the parent(s) and the embryo are successively and complementarily involved (Figure 6.6).

The Parental CNS-Controlled Phase of Reproduction: Formation of Bauplan and the CNS

Besides the direct participation of the products of their translation in the establishment of the embryonic axes and formation of germ layers, the epigenetic information (maternal mRNAs and other products) in parthenogenetic eggs and in gametes of dioecious metazoan species represents indispensable instructions on when, where, and how to activate zygotic genes in the process of the early development. It is the epigenetic information, which is parentally provided in the processes of gametogenesis, that endows the egg cell and the sperm cell with the monopoly of procreation.

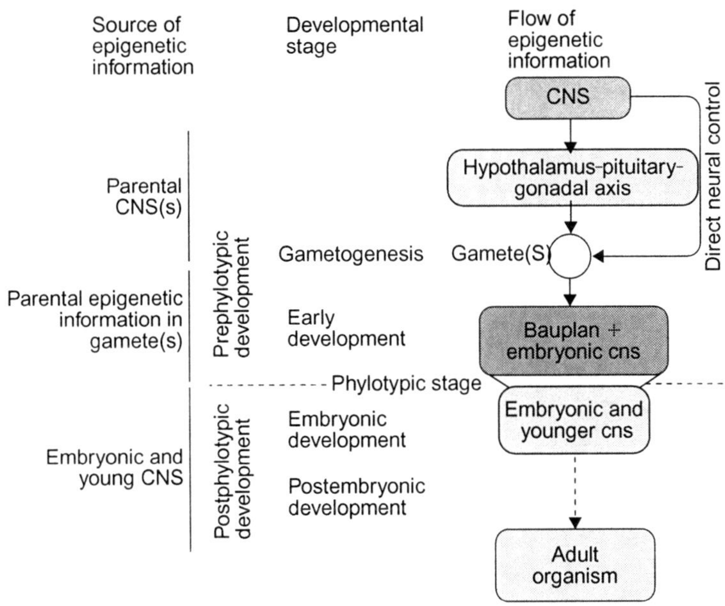

Figure 6.6 Diagrammatic representation of the biphasic mode of reproduction in oviparous metazoans (in viviparous organisms, maternal influence on the embryonic development continues after the phylotypic stage, during the whole intrauterine life). Note that the parental system only provides epigenetic information for gametogenesis and the early development. Epigenetic information for the postphylotypic development is generated from the embryonic and young CNS.
Source: Modified from Cabej (2004).

The epigenetic information for early development is provided to gametes in two basic forms:

1. In the form of thousands of parental cytoplasmic factors (e.g., gene transcripts, proteins, various protein and nonprotein inducers, neurotransmitters, and nutrients) deposited in the egg in strict spatial arrangements.
2. In the form of imprinted genes, i.e., genes that are expressed from one allele (maternal or paternal) only.

The idea of the CNS control of the placement of parental cytoplasmic factors and gene imprinting in both types of gametes is one of the basic pillars of the epigenetic theory of heredity presented in this work. Hence, its substantiation has been a major objective of the research for verifying it.

Depending on the species, the maternal/paternal CNS regulates the deposition of cytoplasmic factors in gametes in a number of experimentally identified ways:

1. Via communicating channels, gap junctions, and ring canals, with the surrounding follicle cells. In insects and other groups, most of the maternal factors in the oocyte come from nurse cells, which, in the process of their programmed death, by the end of oogenesis, force their content into the oocyte (see Chapter 3, Section Neural Control of Deposition of

Maternal Factors in Insect Oocytes). In contrast, in lower invertebrates like Hydra, it is the oocyte that phagocytizes the apoptosized nurse cells (Miller et al., 2000).

2. By neuroendocrine regulation of the level of maternal cytoplasmic factors in body fluids (blood, hemolymph, intercellular fluid). The oocyte actively takes up parental factors from body fluids largely via the receptor-mediated endocytosis (see Chapter 3, Section Neural Regulation by Modifying Maternal Hormone Level).
3. By differential uptake of maternal hormones and other substances from the blood, which, in a number of experimentally observed cases, determines differences in the size of eggs, body size, vitality, morphology, and behavior of the offspring (see Chapter 3, Section Neural Regulation by Differential Uptake from the Blood).
4. In viviparous animals, under maternal neural control, placenta regulates passage of maternal hormones, growth factors, and secreted proteins as well as their respective receptors. Controlled transplacental passage of maternal factors to the embryo takes place throughout the gestational period (see Chapter 4, Section Epigenetic Control of Early Development in Mammals).

After fertilization, the zygote starts cleavage divisions. Depending on species, parental cytoplasmic factors control first cleavage divisions until the midblastula stage. In all metazoan embryos, even after the activation of zygotic genes, parental cytoplasmic factors continue to function as inducers of the early development for varying periods, until the phylotypic stage, when their reserve is exhausted. They induce transcription of zygotic genes and interact with zygotic gene products. Recent empirical evidence not only corroborates the considerable maternal control, but also has demonstrated a significant paternal contribution after the activation of the zygotic genome (Wagner et al., 2004).

In different species, the control of the early development by parental cytoplasmic factors continues for different periods, and for different numbers of cell divisions, until the midblastula transition (MBT) stage, when expression of zygotic genes starts. In zebrafish, this occurs at the 10th cycle of cell divisions (~1,024 cells) but in oviparous vertebrates, such as *Xenopus laevis*, it persists up to the 12th cycle of cell divisions (>4,000 cell stage) (Wagner et al., 2004). Transcription of zygotic genes starts at the MBT stage, but 47 maternal cytoplasmic factors are still found to be active in the embryonic development of zebrafish at or after the MBT stage.

In placental species, the "silent" period of zygotic genes is shorter, one to two cleavage divisions, but the maternal control, to some extent, continues during the whole embryonic intrauterine development: cerebrally induced maternal factors reach the embryo transplacentally during the whole intrauterine period.

Maternal factors present in the cytoplasm of the egg cell include maternal transcripts (mRNAs) of most important genes, such as cyclins, which determine the first cleavage divisions, hence are among the earliest maternal transcripts to be translated. Other maternal factors induce expression of genes for the synthesis of neurotransmitters, and this takes place very early, at the blastoderm stage.

Three systems of maternal transcripts (*bicoid* mRNA, *nanos* mRNA, *caudal* mRNA, and *hunchback* mRNA) determine the establishment of the anterior–posterior axis in insects. In insects, most of these transcripts that come into the egg cell from nurse cells initiate the well-known transcription sequence of zygotic genes (*gap-pair ruled-engrailed-homeotic*).

In vertebrates, the first divisions of the zygote are stimulated by translation of the maternal *Cdc6* mRNA, which makes possible the replication of zygotic chromosomes by activating the minichromosome maintenance (MCM) helicase complex (Lemaitre et al., 2002).

Where the sperm cell enters the egg cell, it induces the cortical reaction, characterized by accumulation of a number of maternal cytoplasmic transcripts (*beta-catenin-*, *Wnt-*, *Vg1-*, *Xwnt11-*, *Noggin-*, and *Activin* mRNAs) determining, thus, formation of the dorsal side of the future embryo. Signals that induce mesoderm formation are also maternal cytoplasmic factors of the dorsal region, probably fibroblast growth factor 2 (FGF2) and bone morphogenetic protein 4 (BMP4). To this group also belong maternal mRNAs for growth factors of the TGF-beta superfamily of secreted proteins, Vg1 and activin. The activin type II receptor is expressed in all the blastula cells, and a number of maternal TGF-beta growth factors, including Vg-1, may bind it. Maternal retinoic acid (RA) is present in bovine oocytes during all stages of their development. Via its receptors, RA regulates expression of *Hox* genes, which have RA response elements (RARE) in their enhancers and are involved in establishing the anterior–posterior axis during early gastrulation in vertebrates.

Maternal cytoplasmic factors also play an essential role in the early development of mammals, but, in distinction from other groups of animals, the placental mode of reproduction enables them to exert an additional, "real-time" maternal control on the embryonic development. In preparation for implantation of the blastocyst, at the site of blastocyst attachment to the endometrium, the latter expresses 22 genes for growth factors (Paria et al., 2001), whose induction is under the ultimate maternal CNS control. In mice, epidermal growth factor (EGF) induces expression of its receptor, EGFR, as early as the eight-cell stage blastocyst (Kim et al., 1999). EGF is also expressed in oviducal and endometrial membranes of pregnant pigs during the preimplantation period. This, and the fact that EGFR is also present in the zygote, suggests that the maternal EGF is active in the blastocyst at this early stage. A maternal neurotransmitter, serotonin, plays an important role in mouse neurogenesis and in the formation of serotonin circuitries in the CNS, where it may act as a "differentiation signal" (Lauder et al., 1981).

The early embryonic development ends at the phylotypic stage, at a time when the parental epigenetic information in the form of cytoplasmic factors is exhausted. Temporally, the consumption of maternal cytoplasmic factors coincides with the emergence of the functioning CNS at the phylotypic stage. At that point in time, the CNS is capable of rising to occasion to generate and provide epigenetic information for molding the extraordinarily complex species-specific structures and organs that develop after the phylotypic stage.

The Embryonic CNS-Controlled Phase: The Postphylotypic Development

The embryonic operational CNS, with established circuits, is capable of neural computation. By computationally determining the release of electrical/chemical signals for activating specific signal cascades, it starts a long series of inductive events, all over the embryonic structure, stimulating the development of all the tissues, organs, and other parts, the whole metazoan morphology. After the phylotypic stage, the lion's share of the inductions that determine the development of metazoan morphology goes to the embryonic CNS (see Chapter 5 for relevant evidence).

These crucial functions of the embryonic CNS as inducer of the development of all the organs and organ systems in metazoans may account for its puzzling over early emergence during the individual development. These functions may be the reason why the CNS (whose main function is considered to be communication with the external environment) is consistently the first organ system to develop in metazoans, although the blood circulation system and excretory system, related with nutritive and excretory functions, from the conventional physiological view, would be required to develop before any other system. Its morphogenetic functions in the postphylotypic development might also account for the disproportionately large size of the embryonic CNS in early development.

The phylotypic stage is a crucial moment in the individual development, when embryos of different species, no matter how different their initial development might have been, converge toward a recognizable common Bauplan, displaying the basic morphological features of the phylum. An informational crisis unfolds. The maternal epigenetic information is exhausted at the moment when the postphylotypic development requires incomparably greater amount of information for morphogenetic processes (development of organs and other metazoan structures) than the early development did. Temporally, the informational crisis coincides with the emergence of the functioning CNS at the phylotypic stage, which appropriately responds to the input of data from the developing embryonic structure, with "spontaneous" electric activity, necessary for setting up neural circuits, for modifying their synaptic morphology (Peinado, 2000; Zhang and Poo, 2001) and sequentially generating the information necessary for successive stages of the development.

The enormous amount of information necessary for establishing these circuits and specific neuronal connections is not inherited parentally via gametes. The parental epigenetic information in the form of cytoplasmic factors controls the development of the initial structure of the CNS (and probably the assembly of the initial neural circuits) at the phylotypic stage, but the epigenetic information for the development of postphylotypic structures is not inherited. What the embryo parentally inherits is not an epigenetic program but an antientropic contrivance, i.e., the CNS, which is capable of computationally generating the epigenetic information necessary for the postphylotypic development based on its interaction with the developing embryonic structure. Essential in this interaction is the processing of the afferent input (spontaneous electrical activity) from the developing embryonic structure according to the brain's "best guess" (Katz and Shatz, 1996) based on the "self-organizing properties" (Weliky, 1999) of the nervous system.

That the afferent input is necessary for formation of neural circuits in the CNS is indicated by the experimental evidence that in the absence of the afferent input on the actual embryonic structure, normal neural circuits are not formed (Penn et al., 1998), and rat striatal neurons do not develop dendritic spines (Segal et al., 2003). Empirically, it is demonstrated that disruption of that input from reaching the CNS (via denervation or otherwise) may fully or partially prevent the embryonic development of corresponding embryonic structures.

At the time of birth, the vertebrate embryo has established an estimated 70–80% of trillions of nonrandom, specific neuronal connections that it will have as an adult organism.

The operational embryonic CNS at the phylotypic stage is developmentally self-reliant and takes over the postphylotypic development. By processing the input of internal and external stimuli in neural circuits, it generates its output in the form of chemical signals, which activate signal cascades, or via the nerve endings, is communicated to various parts of the developing embryo for inducing expression of specific genes and cell differentiation. To substantiate this, in Chapter 5 I presented adequate evidence that such neural signals induce the development of tissues (e.g., muscle, bone, and adipose), organs (e.g., heart, endocrine glands, gastrointestinal tract, liver, lungs, kidneys, and gonads), and related developmental phenomena (e.g., directed cell migration, programmed cell death, and left-right asymmetry).

Additionally, in vertebrates, the CNS expanded its organogenetic functions by evolving a do-it-yourself mode of operation. The novel structures performing that function are the neural crest (and likely ventrally emigrating neural tube cells), which form on (and from) the embryonic neural tube/CNS. Neural crest cells leave the CNS early during embryogenesis and migrate to precisely determined sites throughout the animal body to contribute to the development of most of the vertebrate tissues, organs, and structures. Before leaving the neural tube/CNS and starting their migration, neural crest cells are provided with epigenetic information for finding their way to specific sites of migration and for transforming themselves, as well as cells in the target sites, into cell types characteristic of the organs to which they give rise (see Chapter 16).

In the process of postphylotypic development, the embryonic CNS in vertebrates extensively uses neurohormonal mechanisms along the hypothalamic–pituitary–target glands. The CNS also performs inductive/suppressive functions in targeted regions of the body, via inducers released by local nerve endings of the peripheral nervous system in order to restrict the action of hormones in specific organs and parts of the body.

The Binary Neural Control of Gene Expression

The fact that hormones, growth factors, neurotransmitters, and other inducers, while circulating throughout the animal body, display inductive activity only in particular regions of the animal body, while the rest of the regions of the body remain insensitive to their action, suggests that metazoans have evolved a mechanism for targeted action of these inducers in the animal body.

All of these inducers perform their functions by binding to specific membrane or nuclear receptors in a process that triggers activation of signal transduction pathways or nuclear factors, with expression of specific genes as end result. Hence, the theoretical possibility exists that animals could spatially restrict the action of hormones in target sites by simply expressing specific receptors for hormone receptors in these sites alone, while preventing their expression in the rest of the animal body. *Adequate experimental evidence shows this to be the case*: sensitization of the target tissues to hormones results from expression of specific hormone receptors in the target cells and tissues, but not in other parts of the body.

The temporal correlation of expression of hormones and their specific receptors in particular regions of the body is not self-explanatory: the fact that only target cells express the respective hormone receptor suggests that the information necessary for secretion of the hormone and expression of its specific receptor in target cells alone may ultimately come from the same source. Besides, in numerous cases, it is observed that the spatially restricted expression of genes in particular regions of the body is determined by the fact that inducers are released not globally but by the local innervation. This led to the concept of the binary, global (Figure 6.7) and local

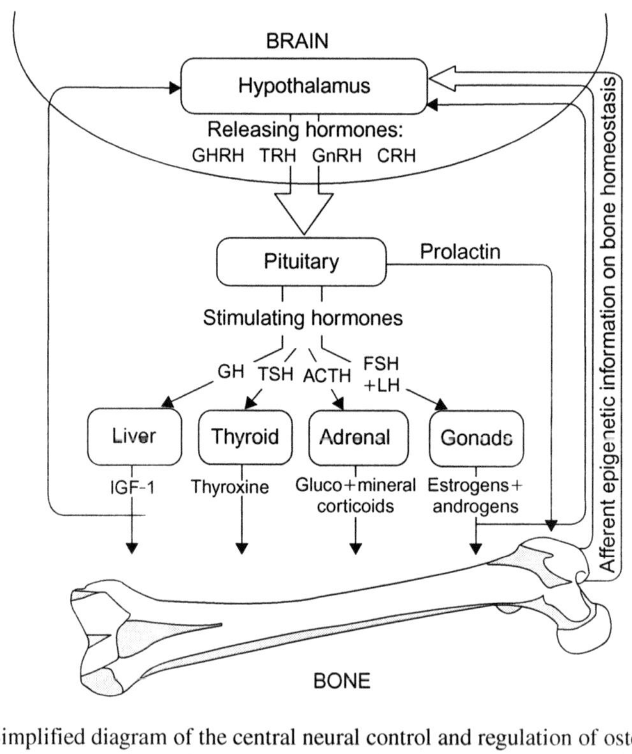

Figure 6.7 Simplified diagram of the central neural control and regulation of osteogenesis in vertebrates via the hypothalamic–pituitary–terminal endocrine glands (thyroid, gonads, adrenals) axes and via the hypothalamic–pituitary axis (prolactin). Note that the orders for activation/inactivation of the five signal cascades or central regulatory axes for bone homeostasis ultimately originate in the brain. *Abbreviations*: ACTH, adrenocorticotropic hormone; CRH, corticotropin-releasing hormone; FSH, follicle-stimulating hormone; GH, growth hormone; GHRH, growth hormone-releasing hormone; GnRH, gonadotropin-releasing hormone; IGFI, insulin-like growth factor-1; LH, luteinizing hormone (lutropin); TRH, thyrotropin-releasing hormone; TSH, thyroid-stimulating hormone (thyrotropin).

(Figure 6.8) neural control of gene expression illustrated in the example of osteogenesis in vertebrates.

It is known that the synthesis and secretion of a key insect hormone, the ecdysteroid, is regulated by a neurohormone, prothoracicotropic hormone (PTTH) produced in the insect's brain. But there is also evidence that the targeted expression

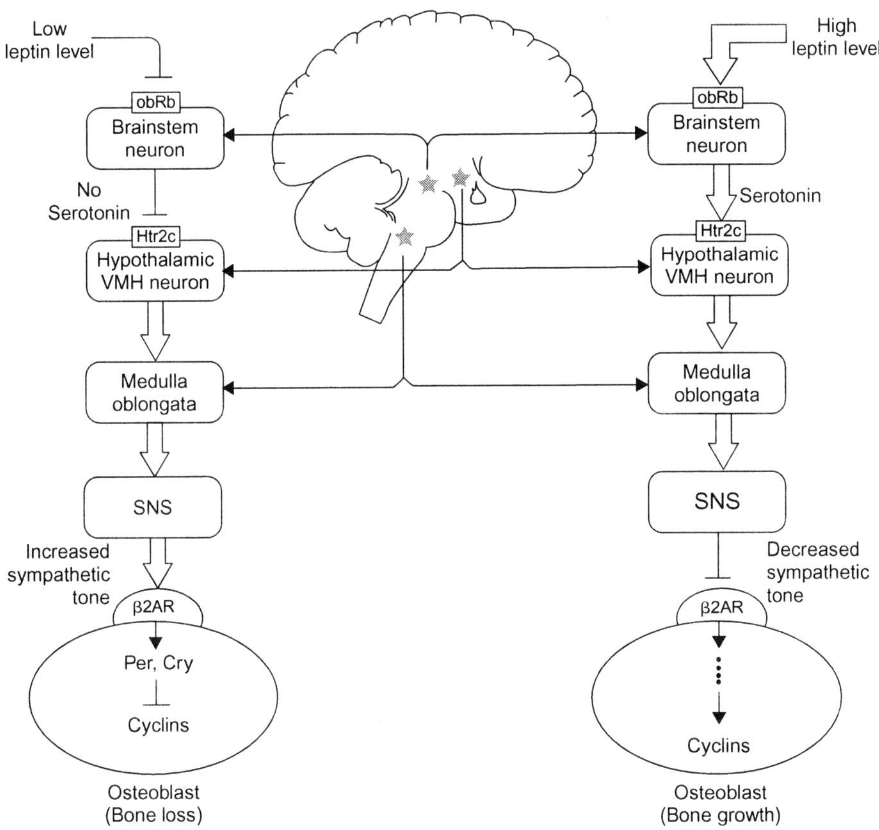

Figure 6.8 Diagram of the mechanism of local sympathetic regulation of bone mass. (Left) When the level of leptin is low (nonobese individuals), it does not bind its receptor ObRb in brain stem neurons, which consequently secrete no serotonin, thus leading to strengthening of the sympathetic tone (increased noradrenalin secretion). Higher sympathetic tone of the SNS leads to inhibition of osteoblast proliferation and increase of osteoclast proliferation, resulting in bone resorption and bone loss. (Right) When the level of leptin in body fluids is high (obesity), it binds its receptors ObRb of serotonergic neurons of the brain stem, stimulating their electrical activity and secretion of serotonin and its binding to Htr2c receptor in specific hypothalamic neurons. This leads to decreased activity of the sympathetic system, i.e., decreased secretion of noradrenaline, thus stimulating osteoblast proliferation, with bone growth as a result. *Abbreviations*: β2-AR, β2-adrenergic receptor; Htr2c, serotonin receptor; ObRb, leptin receptor; *Per* and *Cry*; period and cryptochrome genes; SNS, sympathetic nervous system; VMH, ventromedial hypothalamus.

and secretion of the receptor for the ecdysteroid hormone, EcR, in dorsal external oblique1 (DEO1) muscle in the moth, *Manduca sexta*, is regulated by corresponding motoneurons (Hegstrom et al., 1998). This is a case of binary neural regulation of gene expression based, on the one hand, on the neurally controlled expression and secretion of hormones and, on the other, by the release (or not) of signals for expression of corresponding receptors by local nerve endings.

The mechanism of binary neural control of gene expression is demonstrated in well-established examples of myogenesis (Lawrence and Johnston, 1986; Currie and Bate, 1995; Bayline et al., 1998), osteogenesis (Goss, 1969; Zeng et al., 1996; Edoff et al., 1997; Demulder et al., 1998), regeneration (Goss, 1969; Hall, 1998), puberty in mammals (Riboni et al., 1998), oogenesis in vertebrates (Morales et al., 1998), expression of receptors for pituitary LH in testicles (Lee et al., 2002), regulation of progesterone synthesis and reproductive physiology in humans (Fritz et al., 2001), secretion of estrogen and progesterone under influence of nervous fibers descending from the hypothalamus to the spinal cord, as well as sympathetic and vagal preganglionic neurons (De Bortoli et al., 1998), secretion of the juvenile hormone by corpora allata in insects (Stay et al., 1996; Kou and Chen, 2000), ovulation and determination of the number of deposited eggs in insects (Antkowiak and Chase, 2003), regulation of ecdysone synthesis by the neurohormone PTTH, and by direct neural control (Chapman, 1998), the development of the laryngeal muscle in male *Xenopus* (Tobias et al., 1993).

Metazoans seem to have resolved, early in the course of their evolution, the problem of the selective restriction of hormone action to particular parts of the animal body at particular times by evolving a *binary, humoral and adjacent, control of gene expression.* In this control system, the action of centrally induced signal cascades is permitted by the action of local nerve endings, which induce the expression of specific hormone receptors only in target regions of the body.

Currently, it is impossible to assert firmly how widespread (or general) the mechanism is in Animalia, but from an evolutionary viewpoint it is plausible that such a valuable and, highly likely indispensable, mechanism of the spatial restriction of hormones and other circulating inducers invented and conserved in so many species would be wasted and not put into general use by evolution.

The adjacent control of gene expression is carried out by the peripheral nervous system, which via nerve endings establishes a ubiquitous presence all over the animal body, often down to the cell level. Nerve endings in specific regions of the body, by selectively inducing expression of specific hormone receptors, are responsible for limiting the action of circulating hormones within the regions of prospective organs.

Solid empirical evidence demonstrates that, from the postphylotypic stage forward, the embryonic CNS is systematically involved in the formation of all other organs and organ systems, not only in its vicinity but throughout the animal body. However, the evidence on the control of organogenesis by the embryonic CNS deriving from experimental studies in the developmental biology and related fields raises a new and more difficult question: How does the embryonic CNS generate the epigenetic information necessary for erecting the metazoan supracellular structure?

Whereas an answer to this question is beyond the scope of this work, a modest attempt was made in Chapter 2 to address this fundamental question descriptively, based on the current knowledge on the generation and storage of information in the CNS. We have shown that this information is processing dependent by origin, while the precise computational mechanisms determining the release of signals (information) that start developmental pathways for organogenesis remain to be determined.

Novel Features of the Epigenetic System of Inheritance

Genetic (Watson–Crick) inheritance determined by the genome, as exemplified in the reproduction of unicellular organisms, is also operative in the reproduction of individual cells in Metazoa and Metaphyta (Plantae). In the case of unicellular organisms, the full amount of genetic (in some well-known cases, also nongenetic) information for the two daughter cells is produced by the mother cell and provided to them in the process of cell division.

Although proteins are necessary components of metazoans, their structural units or building blocks are not proteins but cells, billions and trillions, arranged in strict spatial patterns requiring huge amounts of nongenetic information. As argued in Chapter 1, the genetic information is qualitatively unsuitable for determining the relative position of cells in metazoans, and quantitatively negligible compared to the amount of information necessary for building their multicellular structure.

The lack of the information for arranging the myriad of cells of different types in specific spatial patterns from which the metazoan morphology arises represented a tremendous barrier for transition from unicellular life to metazoan multicellularity. This barrier stimulated a pressure for evolving a new type of information and an incomparably larger source of information of a new type. The solution came with evolution of a specialized structure capable of computating and generating the information for the multicellular metazoan structure as well as for transmitting it to the offspring. Crucial for the solution was the differentiation of the neuron and the nervous system.

This was a key evolutionary innovation that would enable parent(s) to *communicate* to the offspring only a part of the necessary information, i.e., information for the early embryonic development until the phylotypic stage, including the formation of the operational nervous system. At that juncture, the embryonic and young nervous system takes over the further, postphylotypic development until adulthood.

Thus, from an informational view, a great distinction exists between the Watson–Crick and epigenetic systems of heredity. In unicellulars, the mother cell produces two copies of itself, physically passing on to each daughter cell complete genetic information, the genome and the extragenomic genetic and epigenetic factors. Metazoans, in distinction, do not provide the offspring with the full amount of epigenetic information necessary for the individual development. Their contribution to the offspring is by far more modest: metazoans only communicate to the zygote (egg cell in parthenogenetic organisms) epigenetic information for advancing to a stage of development when the embryo becomes informationally self-reliable, a stage that coincides with the formation of the Bauplan at the phylotypic stage, including the development of the incipient nervous system. In order to emphasize the distinction from the *replicative heredity*, i.e., the fact that in metazoans parents do not physically transmit to the offspring the full amount information for erecting their structure, I will tentatively designate this *communicative heredity*.

This mechanism of the communicative heredity is reminiscent of the von Neumann machine that, when provided with the necessary parts, could reproduce itself by appropriately assembling these parts and then providing them with a tape containing operating instructions. But being incomparably more sophisticated than the von Neumann machine, all the "zoomachines," including us human beings, are different in some essential ways:

1. In clear distinction from the von Neumann machine, "zoomachines" can reproduce themselves by using "parts" (cells, tissues, or organs) they produce themselves rather than supplied parts.
2. While the von Neumann machine is assembled from the "parental von Neumann machine," the "zoomachine" is self-assembled.
3. While the von Neumann machine operates on instructions provided by the parental machine, the "zoomachine" computationally generates itself all the instructions (epigenetic information) for its function.
4. In distinction from the von Neumann machine, which would be operative only after being completely assembled, the "zoomachine" is operative before being completely assembled, from the phylotypic stage forward.

The information that controls the postphylotypic development is epigenetic in origin. That information is generated in the embryonic CNS in the form of electrical/chemical outputs resulting from the processing in neural circuits of enormous input of the stimuli from the developing embryonic structure and its environment. The computational activity of the CNS is essential for the epigenetic system of heredity.

As an evolutionary innovation, the epigenetic system of heredity in metazoans represents a groundbreaking event in the evolution of life on Earth, whose importance in the evolution of life cannot be overestimated.

Essential in the evolution of the bigenerational epigenetic system of heredity in metazoans was the acquisition of the ability to produce self-developing unicellular structures, gametes/zygotes. As mentioned earlier, metazoans do not physically pass on the whole amount of the epigenetic information, but via gametes they communicate to the offspring part of that information in the form of parental cytoplasmic factors (e.g., mRNAs, proteins, hormones, growth factors, neurotransmitters, nutrients) and imprinted genes. Under the influence of internal and external stimuli, by their own action and by activating specific zygotic genes, the parental epigenetic information deposited in the zygote controls and regulates the early development of metazoans up to formation of the Bauplan of the phylum and the species-specific operational CNS at the phylotypic stage. From that point forward, the developing embryonic CNS is capable of determining cell differentiation, cell proliferation, and the related processes of organogenesis and individual development in general. The postphylotypic metazoan embryo is informationally a self-reliable structure that self-organizes into an adult metazoan structure.

It is the parentally determined incipient CNS that, while developing, stepwise generates the huge amount of information necessary for the postphylotypic development, for regulating the differentiation and proliferation of cells as well as for assembling billions or trillions of differentiated cells in that highly specific spatial arrangement that, in our visual perception, effectuates the animal morphology.

The epigenetic system of heredity is neither an extension nor a development of the Watson–Crick system. It is essentially product of an informational revolution and represents a qualitatively new property of metazoan life. The Watson–Crick system and the epigenetic system of heredity are evolutionarily as discontinuous as the genetic code and computational generation of epigenetic information are.

The epigenetic system of heredity is different from the Watson–Crick system of heredity with respect to the source, nature, and function of information it provides.

In contrast with the genetic code, which is believed to have been the result of a "frozen accident," the epigenetic information generated in neural circuits is adaptation oriented and highly flexible. While a particular DNA segment codes for a strictly determined amino acid sequence, in the CNS, the same operational neural unit, the same circuit, depending on the stimulus and on its computational properties, may adaptively induce a number of different phenotypic results.

The computational processing of external stimuli and processing-dependent adaptive generation of the epigenetic information in neural circuits added a novel dimension to the relationship between the living systems and their environment. It enabled metazoans to virtually relate any external/internal stimulus to any gene, as it is manifested in numerous facts of expression of the same gene in response to different stimuli and the same stimulus inducing expression of different genes in the CNS of different species.

In the Watson–Crick system, in the process of the mitotic division, parents pass on to the daughter cells not only the total amount of genetic information in the form of DNA, the genome, but also the matter and energy necessary for their independent life. Both daughter cells are qualitatively finished cells. The daughter cells may grow, but they do not go through a developmental process; they are born "developed," in that from the beginning they are morphologically and functionally full-fledged individual organisms.

The physical continuity of the Watson–Crick system from the parent to the offspring is in contrast with the physical discontinuity of the parental and embryonic CNS in metazoans; metazoans do not transmit to the offspring the full amount of information necessary for developing species-specific structure. What metazoans instead pass on to the offspring via gametes is epigenetic information in the form of parental cytoplasmic factors and imprinted genes that regulate the formation of the zygote, the cleavage, and the early embryonic development. The parentally provided epigenetic information is consumed during the early development and has no role in the postphylotypic development, histogenesis, organogenesis, or development of species-specific adult morphology.

The huge amount of information for the postphylotypic development of metazoan supracellular structure is generated in the embryonic CNS. It is the CNS that in the process of the postphylotypic development, in response to the input of information on the developing embryonic structure throughout the animal body, and the external stimuli, computationally, stagewise, generates the information necessary for the individual development until adulthood. In this meaning, the notion that in the process of their reproduction, metazoans, like unicellulars, "produce copies of themselves" is inaccurate and misleading. In distinction from unicellulars, metazoans produce self-organizing, information-generating structures rather than "copies of themselves."

Essential differences between the Watson–Crick and epigenetic systems of heredity exist not only as far as the nature and the mode of transmission of information are concerned, but also in relation to the mechanisms of generation of new information. In the Watson–Crick system, new information emerges by random changes occurring in the structure of the gene (DNA in general), as a result of thermodynamic factors. If mutations in the DNA structure happen to increase the fitness of the carrier in its environment, which is a very rare occurrence, they will be favored, saved, and propagated in the offspring as new information; otherwise, they will be discarded in the process of elimination of their carriers under the action of natural selection.

In distinction, the new epigenetic information results from changes in the computational properties of circuits. These changes are not random, but adaptive, resulting from the adaptation-oriented processing of the input of external/internal stimuli in neural circuits. But, similarly to the Watson–Crick system, whether the new epigenetic information will be sustained, propagated, or lost in future generations depends on natural selection.

During the template replicative Watson–Crick heredity, the number of offspring a unicellular can produce per mitotic generation is strictly limited; it can produce only one copy (two daughter cells) in its lifetime. In communicative heredity, the possibility exists for the metazoan organism to produce more than one (depending on species, from one to thousands) offspring during its lifetime.

As mentioned earlier, despite the essential differences between the two systems of heredity, the evolution of the epigenetic system of heredity does not mark the end of the Watson–Crick system of heredity in metazoans. The Watson–Crick system is still necessary and responsible for the heredity and reproduction at the cellular and molecular levels, for the biochemical evolution of metazoans in general. Hence, it is conserved in metazoans as a system that is complementary to the epigenetic system of heredity.

As for the relationship between the two systems in metazoans is concerned, the epigenetic system of heredity comprises, and is inseparable from, the genetic Watson–Crick system of heredity. Both the genetic and epigenetic systems of heredity are present and operational at the cellular and supracellular levels, respectively. They are functionally complementary, and their mutual relationship has found a vivid expression in Medawars' aphorism: "Genetics proposes, epigenetics disposes." This is unambiguously manifested in the fact that, in metazoans, all of the signal cascades leading to expression or suppression of nonhousekeeping genes, and signals for cell reproduction and differentiation *ultimately* start with generation of neurally derived electrical/chemical signals (epigenetic information). Thus, while complementary to the epigenetic system, the Watson–Crick system of heredity is subordinate to it.

The study of the epigenetic system of heredity and the epigenetic information, an almost *terra incognita* so far, could provide new insights into the nature of life and enable us to understand many of the unexplained biological phenomena, including the mechanisms of their evolutionary adaptation.

References

Abrams, E.W. and Mullins, M.C. (2009). Early zebrafish development: it's in the maternal genes. Curr. Opin. Genet. Dev., *19*, 396–403.

Amenta, A.U., Vega, F., Ricci, J.A. and Collier, W.L. (1992). Localization of hydroxytryptamine-like immunoreactive cells and nerve fibers in the rat female reproductive system. Anat. Rec., *223*, 478–84.

Anesetti, G., Lombide, P., D'Albora, H. and Ojeda, S.R. (2001). Intrinsic neurons in the human ovary. Cell Tissue Res., *306*, 231–37.

Antkowiak, T. and Chase, R. (2003). Sensory innervation of the ovotestis in the snail. *Helix aspersa*. J. Exp. Biol., *206*, 3913–21.

Ayala, M.E., Rosas, P. and Domınguez, R. (1994). Different effects of unilateral and bilateral lesions of the dorsal raphe nucleus on puberty and first ovulation. Brain Res. Bull., *40*, 149–55.

Bayline, R.J., Khoo, A.B. and Booker, R. (1998). Innervation regulates the metamorphic fates of larval abdominal muscles in the moth. *Manduca secta*. Dev. Genes Evol., *208*, 369–81.

Bódis, J., Török, A., Tinberger, H.R., Volker, H., Hamori, M. and Cledon, P. (1992). Influence of serotonin on progesterone and estradiol secretion by cultured human granulosa cells. Fertil. Steril., *57*, 1008–11.

Bódis, J., Török, A., Tinneberg, H.-R., Hanf, V., Papenfuß, F. and Schwarz, H. (1993). Serotonin induces progesterone release from human granulosa cells in a superfused granulosa cell system. Arch. Gynecol. Obstet., *253*, 59–64.

Bódis, J., Hartman, G., Török, A., Bognar, Z., Tinnberg, H-J., Cledon, P., et al. (1993). Relationship between the monoamine and gonadotropin content in follicular fluid of preovulatory Graafian follicles after superovulation treatment. Exp. Clin. Endocrinol., *101*, 178–82.

Cabej, N.R. (2004). *Neural Control of Development*. Albanet, Dumont, NJ, p. 167.

Chapman, R.F. (1998) *The Insects—Structure and Function*, fourth edition. Cambridge University Press, Cambridge, p. 578.

Cikos, Š., Veselá, J., Ilková, G., Rehák, P., Czikková, S. and Koppel, J. (2005). Expression of beta adrenergic receptors in mouse oocytes and preimplantation embryos. Mol. Reprod. Dev., *71*, 145–53.

Clausell, D.E. and Soliman, K.F.A. (1978). Ovarian serotonin content in relation to ovulation. Experientia, *34*, 410–11.

Cruz, M.E., Moran, J.L., Jaramillo, L.P. and Domınguez, R. (1990). Differential effects of a unilateral hypothalamic lesion on ovulation and compensatory ovarian hypertrophy in hemiovariectomized adult rats. J. Endocrinol., *124*, 37–41.

Currie, D.A. and Bate, M. (1995). Innervation is essential for the development and differentiation of a sex-specific adult muscle in *Drosophila melanogaster*. Development, *121*, 2549–57.

Day, R.M., Agyeman, A.S., Segel, M.J., Chévere, R.D., Angelosanto, J.M. and Suzuki, Y.J., et al. (2006). Serotonin induces pulmonary artery smooth muscle cell migration. Biochem. Pharmacol. *71*, 386–97.

De Bortoli, M.A., Garraza, M.H. and Aguado, L.I. (1998). Adrenergic intracerebroventricular stimulation affects progesterone concentration in the ovarian vein of the rat: participation of the superior ovarian nerve. J. Endocrinol., *159*, 61–8.

Dees, W.L., Hiney, J.K., Schultea, T.D., Mayerhofer, A., Danilchik, M., Dissen, G.A. and Ojeda, S.R. (1995). The primate ovary contains a population of catecholaminergic neuron-like cells expressing nerve growth factor receptors. Endocrinology, *136*, 5760–68.

Dees, W.L., Hiney, J.K., McArthur, N.H., Johnson, G.A., Dissen, G.A. and Ojeda, S.R. (2006). Origin and ontogeny of mammalian ovarian neurons. Endocrinology, *147*, 3789–96.

Demulder, A., Guns, M., Ismail, A., Wilmet, E., Fondu, P. and Bergmann, P. (1998). Increased osteoblast-like cells formation in long-term bone marrow cultures from patients with a spinal cord injury. Calcified Tissue Int., *63*, 396–400. (abstract).

Dosch, R., Wagner, D.S., Mintzer, K.A., Runke, G., Wiemelt, A.P. and Mullins, M.C. (2004). Maternal control of vertebrate development before the midblastula transition mutants from the zebrafish. Dev. Cell, *6*, 771–80.

Dungan, H.M., Clifton, D.K. and Steiner, R.A. (2006). Kisspeptin neurons as central processors in the regulation of gonadotropin-releasing hormone secretion. Endocrinology, *147*, 1154–58.

Edoff, K., Hellman, J., Persliden, J. and Hildebrand, C. (1997). The developmental skeletal growth in the rat foot is reduced after denervation. Anat. Embryol., *195*, 531–38.

Eriksson, T.L. (2009). Organelle movement in melanophores—effects of *Panax ginseng*, ginsenosides and quercetin. Linköping University, Thesis No. 103, Linköping, Sweden.

Fritz, S., Grunert, R., Stocco, D.M., Hales, D.B. and Mayerhofer, A. (2001). StAR protein is increased by muscarinic receptor activation in human luteinized granulosa cells. Mol. Cell. Endocrinol., *171*, 49–51.

Gerendai, I. (2004). New Data on the Neural Control of Gonadal Functions. Supraspinal Innervation of the Gonads. Timisoara Med. J., *54*, 218–24.

Gerendai, I., Kocsis, K. and Halász, B. (2002). Supraspinal connections of the ovary: structural and functional aspects. Microsc. Res. Techniq., *59*, 474–83.

Gill, R., Turner, J., Saksena, S., Alrefai, W., Annaba, F., Pant, N., et al. (2007). Serotonin modifies actin cytoskeleton in human intestinal epithelial cells via involvement of PKCα. FASEB J. 21, 617.17.

Gill, R.K., Shen, L., Turner, J.R., Saksena, S., Alrefai, W.A., Pant, N., et al. (2008). Serotonin modifies cytoskeleton and brush-border membrane architecture in human intestinal epithelial cells. Am. J. Physiol. Gastr. L., *295*, G700–08.

Goss, R.J. (1969) *Principles of Regeneration*. Academic Press, New York and London, , p. 198.

Hall, B.K. (1998) *Evolutionary Developmental Biology*, second edition. Chapman & Hall, London, p. 339.

Hegstrom, C.D., Riddiford, L.M. and Truman, J.W. (1998). Steroid and neuronal regulation of ecdysone receptor expression during metamorphosis of muscle in the moth, *Manduca sexta*. J. Neurosci., *18*, 1786–94.

Kasukawa, H., Oshima, N. and Fujii, R. (1986). Control of chromatophore movements in dermal chromatic units of blue damselfish—II. The motile iridophore. Com. Biochem. Phys. C., *83*, 1–7.

Katz, L.C. and Shatz, C.J. (1996). Synaptic activity and the construction of cortical circuits. Science, *274*, 1133–38.

Kim, C.H., Chae, H.D., Cheon, Y.P., Kang, B.M., Chang, Y.S. and Mok, J.E. (1999). The effect of epidermal growth factor on the preimplantation development, implantation and its receptor expression in mouse embryos. J. Obstet. Gynaecol. Res., *25*, 87–93.

Kou, R. and Chen, S.J. (2000). Allatotropic and nervous control of corpora allata in the adult male loreyi leafworm, *Mythimna loreyi* (Lepidoptera: Noctuidae). Physiol. Entomol., *25*, 273–78.

Lauder, J.M., Sze, P.Y. and Krebs, H. (1981). Maternal influences on tryptophan hydroxylase activity in embryonic rat brain. Dev. Neurosci., *4*, 291–95.

Lawrence, P.A. and Johnston, P. (1986). The muscle pattern of a segment of *Drosophila* may be determined by neurons and not by contributing myoblasts. Cell, *45*, 505–13.

Lee, S., Miselis, R. and Rivier, C. (2002). Anatomical and functional evidence for a neural hypothalamic–testicular pathway that is independent of the pituitary. Endocrinology, *143*, 4447–54.

Lemaitre, J.-M., Bocquet, S. and Mechall, M. (2002). Competence to replicate in the unfertilized egg is conferred by Cdc6 during meiotic maturation. Nature, *419*, 718–22.

Luque, R.M., Kineman, R.D. and Tena Sempere, M. (2007). Regulation of hypothalamic expression of Kiss-1 and GPR54 genes by metabolic factors: analyses using mouse models and a cell line. Endocrinology, *148*, 4601–11.

Madekurozwa, M.-C. (2008). An immunohistochemical study of ovarian innervation in the emu (*Dromaius novaehollandiae*). Onderstepoort J. Vet. Res., *75*, 59–65.

Mäthger, L.M., Collins, T.F.T and Lima, P.A. (2004). The role of muscarinic receptors and intracellular Ca^{2+} in the spectral reflectivity changes of squid iridophores. J. Exp. Biol., *207*, 1759–69.

Miller, M.A., Technau, U., Smith, K.M. and Steele, R.E. (2000). Oocyte development in Hydra involves selection from competent precursor cells. Dev. Biol., *224*, 326–38.

Morales, L., Chávez, R., Ayala, M.E. and Domínguez, R. (1998). Effects of unilateral or bilateral superior ovarian nerve section in prepubertal rats on the ovulatory response to gonadotrophin administration. J. Endocrinol. *158*, 213–19.

Nijhout, H.F. (1999). Polyphenic development in insects. Bioscience, *49*, 181–92.

Noh, S.J. and Han, J.K. (1998). Inhibition of the adenylcyclase and activation of the phosphatidylinositol pathway in oocytes through expression of serotonin receptors does not induce oocyte maturation. J. Exp. Zool., *280*, 45–56.

Paria, B.C., Ma, W., Tan, J., Raja, S., Das, S.K., Dey, S.K., et al. (2001). Cellular and molecular responses of the uterus to embryo implantation can be elicited by locally applied growth factors. Proc. Natl. Acad. Sci. U.S.A., *98*, 1047–52.

Peinado, A. (2000). Traveling slow waves of neural activity: a novel form of network activity in developing neocortex. J. Neurosci., *20*, 1–6.

Penn, A.A., Riquelme, P.A., Feller, M.B. and Shatz, C.J. (1998). Competition in retinogeniculate pattern driven by spontaneous activity. Science, *279*, 2108–12.

Popa, S.M., Clifton, D.K. and Steiner, R.A. (2008). The role of kisspeptins and GPR54 in the neuroendocrine regulation of reproduction. Annu. Rev. Physiol., *70*, 213–38.

Rey-Ares, V., Lazarov, N., Berg, D., Berg, U., Kunz, L. and Mayerhofer, A. (2007). Dopamine receptor repertoire of human granulosa cells. Reprod. Biol. Endocrin., *5*, 40.

Riboni, L., Escamilla, C., Chavira, R. and Domínguez, R. (1998). Effects of peripheral sympathetic denervation induced by guanethidine administration on the mechanisms regulating puberty in the female guinea pig. J. Endocrinol. *156*, 91–8.

Segal, M., Greenberger, V. and Korkotian, E. (2003). Formation of dendritis spines in cultured striatal neurons depends on excitatory afferent activity. Eur. J. Neurosci., *17*, 2573–85.

Smith, J.T., Popa, S.M., Clifton, D.K., Hoffman, G.E. and Steiner, R.A. (2006). *Kiss1* neurons in the forebrain as central processors for generating the preovulatory luteinizing hormone surge. J. Neurosci., *26*, 6687–94.

Stay, B., Fairbairns, S. and Yu, C.G. (1996). Role of allatostatins in the regulation of juvenile hormone synthesis. Arch. Insect Biochem., *32*, 287–97.

Tobias, M.L., Marin, M.L. and Kelley, D.B. (1993). The roles of sex, innervation, and androgen in laryngeal muscle of *Xenopus laevis*. J. Neurosci., *13*, 324–33.

Veselá, J., Rehák, P., Mihalik, J., Czikková, S., Pokorný, J. and Koppel, J. (2003). Expression of serotonin receptors in mouse oocytes and preimplantation embryos. Physiol. Res., *52*, 223–28.

Wagner, D.S., Dosch, R., Mintzer, K.A., Wiemel, A.P. and Mullins, M.C. (2004). Maternal control of development at the midblastula transition and beyond: mutants from the zebrafish II. Dev. Cell, *6*, 781–90.

Weliky, M. (1999). Recording and manipulating the *in vivo* correlational structure of neuronal activity during visualcortical development. J. Neurobiol., *41*, 25–32.

Yamaguchi, S., Isejima, H., Matsuo, T., Okura, R., Yagita, K. and Kobayashi, M., et al. (2003). Synchronization of cellular clocks in the suprachiasmatic nucleus. Science *302*, 1408–12.

Yang, C., Wang, G., Wang, H., Liu, Z. and Wang, X. (2009). Cytoskeletal alterations in rat hippocampus following chronic unpredictable mild stress and re-exposure to acute and chronic unpredictable mild stress. Behav. Brain Res., *205*, 518–24.

Zeng, Q.Q., Jee, W.S., Bigornia, A.E., King, J.G., Jr., et al., D'Souza, S.M., Li, X.J., et al. (1996). Time responses of cancellous and cortical bones to sciatic neurectomy in growing female rats. Bone, *19*, 13–21.

Zhang, L.I. and Poo, M. (2001). Electrical activity and development of neural circuits. Nat. Neurosci. Suppl., *4*, 1207–14.

Part II

Neural-Developmental Premises of Evolutionary Adaptation

7 Evolution and Stress Responses to Changes in Environment

But environments do change, sometimes a lot, sometimes a little, daily, seasonally, yearly, and over longer, geological periods. And these changes are important. The heart of the evolutionary process lies within local ecosystems, for it is here that ecological and evolutionary processes intersect and here that the winners and losers in the game of life are recorded.

Eldredge (1998)

Evolution of Animal Kingdom Is Related to Changes in Environment

Paleontological evidence shows that all the major diversifications in metazoans, during the last ~600 million years, including the Cambrian explosion, coincide with major shifts in the climatic and geological conditions that significantly affected their conditions of living.

The Cambrian explosion, as estimated by fossil record, began some 544 million years ago with the sudden appearance of almost 100 phyla, including all the 37 extant phyla (Erwin et al., 1997). Many paleontologists believe that eumetazoans evolved even earlier, but, being soft-bodied, those animals left no fossils.

The Ordovician radiation, which took place by the end of the Cambrian period, ~500 million years ago, and by some estimates is considered to be larger than the Cambrian explosion, is characterized by a rapid appearance of numerous families (including crinoids, articulate brachiopods, cephalopods, and corals), genera, and species that dominated marine ecosystems for 250 million years (Droser et al., 1996). The extraordinary biological innovation of Ordovician coincided with two drastic changes in the marine and terrestrial environments: a drop in the sea level, and the intense orogenic activity of the earth that led to formation of many mountain ranges, which is suggested by the fact that the number of the marine taxa that diversified at the period was larger in the regions in the proximity of orogenic activity (Miller and Mao, 1995). The large-scale climatic and ecological changes determined by those events might have led to the global diversification of Ordovician radiation (Droser et al., 1996).

Devonian was characterized by rapid diversification of fish and hence is also known as the *Age of Fish*. During this period, ~380–360 million years ago, began the colonization of land by plants, invertebrates, such as scorpions, and vertebrate tetrapods (archaic amphibians, such as *Ichthyostega*, and later reptiles). It is believed that during this period, the biggest mass extinction of metazoans occurred.

Epigenetic Principles of Evolution. DOI: 10.1016/B978-0-12-415831-3.00007-0

A paleontologically recorded retreat of the sea, with its ecological and climatic consequences in the ocean and on the land, might have been the common cause for both the recorded mass extinction and the colonization of land during Devonian.

Next, *Carboniferous* period, with its hyperoxic atmosphere, is believed to have favored evolution of tetrapod locomotion and evolution of flight in both vertebrates and insects (Dudley, 1998).

During the hypoxic Permian period, ~250 million years ago, the extinction of the last Cambrian fauna occurred, and Paleozoic fauna rapidly declined. It is estimated that 96% of all species, and 50% of families, became extinct because of the formation of Pangea II (the bringing together of the continents as a result of plate tectonics). This marked the beginning of the Mesozoic Era, immediately followed by the modern fauna.

At the beginning of the Mesozoic, ~200 million years ago, marine life was dominated by molluscs. Reptiles, including dinosaurs, sustained by a thriving plant life, were the dominant terrestrial vertebrate class. Extinction of dinosaurs and the rise of mammals (which, up to that point in time, were represented by nocturnal insectivorous animals) by the end of Cretaceous, some 65 million years ago, has generally been related to various groups of geological factors, such as intensification of volcanic activity, the fall of a huge meteor, and a drop in the sea level, which drastically changed the climate and conditions of living on earth (Wilf et al., 2003).

No convincing reason, however, is given as to why those factors that led to extinction of dinosaurs did favored the mammals at the same time.

The role of environmental changes in taxonomic diversification of metazoans may also be illustrated by the evolution of eutherian mammals, which have been documented for the Late Cretaceous, 85–90 million (Archibald et al., 2001), to Early Cretaceous, 125 million years ago (Ji et al., 2002). Global warming during the Late Cretaceous (99–65 million years ago) is considered to have stimulated the rapid diversification and the dispersal of salamanders (Vieites et al., 2007).

Earliest placentals are believed to have evolved ~86–90 million years ago (Archibald, 2003) and earliest placental mammals ~65 million years ago (Easteal, 1999). According to Meng and McKenna (1998), major evolutionary changes in placental mammals (eutherians) also took place during short periods coinciding with global climatic changes. First, during the warming (the mean annual temperature of 30°C) humid period of Palaeocene/Eocene, ~55 million years ago, which was characterized by dense forest areas. Then, during the Eocene/Oligocene boundary, 33.5 million years ago, a cooler and more arid period came that was characterized by disappearance of perissodactyl-dominant fauna, which was "replaced by rodent/lagomorph-dominant faunas of the Oligocene" (Meng and McKenna, 1998). A 4°C drop of the winter temperature during that period is considered to have been the cause of one of the largest extinctions of marine invertebrates during the whole Cenozoic (Ivany et al., 2000).

Radical climatic changes during the Miocene, ~20 million years ago, transformed the Sahara region from a tropical to an arid environment. This transformation is believed, for example, to have been the cause of the splitting of a single elephant species into two: the only elephant shrew species in the north of Sahara and the south of Sahara–residing species (*Petrodromus tetradactylus*), now belonging to a different genus (Douady et al., 2003).

Continued contraction of the wooded fruit-rich habitat to the low latitudes of Africa and Southeast Asia by late Miocene, ~10 million years ago, which coincided with a decrease in the diversity of great apes to these parts of the world and the dynamics of these habitats, is considered to have been an environmental factor to which these species responded by increasing their sociality and cognitive capabilities (Potts, 2004).

A study on the influence of ecological factors on evolution rates in 20 grouse taxa has shown that habitat changes have induced higher rates of evolution of body proportions (Drovetski et al., 2006). Repeated glaciation processes also have had considerable influence in acceleration of evolutionary rates in metazoans:

> *Deep-sea organisms, such as benthic ostracods, exhibit fluctuations in diversity over time scales of 10,000 to 100,000 years that correlate with glaciation, diversity decreasing as glaciers advance, and recovering during interglacial periods.*
>
> *Cronin and Raymo (1997)*

Possible climatic influences on morphology, speciation, and extinction have been extensively discussed by Vrba (1983, 1996) and Vrba et al. (1996).

The paleontological evidence, at a geological scale, strongly suggests that a causal relationship between the drastic changes in the environment and diversification of animal taxa exists. The extraordinarily high tempo of formation of new taxa, *taxogenesis*, during the periods of drastic environmental changes and *taxostasis*, during the intervening periods, suggests that diversification and morphological evolution have not always been gradual (Eldredge and Gould, 1972).

Neuroendocrine System

Integration and coordination of actions of all the cells in multicellular organisms require maintenance of a stable "physiological" intercellular medium in which cells must carry out their functions. The maintenance of this medium, as a part of body homeostasis, is function of the integrated control system (ICS). The term *homeostasis* will be used here in a broader sense, so as to include not only the maintenance of the physiological state of body fluids but the maintenance of the structural integrity of the organism as well.

The function of the ICS requires continuous input of data in the central nervous system (CNS) on the state of the system, whose structure is continually and unavoidably eroding. The input is compared with the normal state (such as neurally determined set points or thresholds), and on that basis decisions are made, and instructions are sent for restoring the normal state. The nervous system uses specific signal cascades as communication channels for sending instructions to the target cells or organs.

Generally, the adaptation to drastic changes in environment begins with behavioral changes, which usually are accompanied by, or related to, a state of environmental stress with all the characteristic neuroendocrine and behavioral correlates. This stress condition itself is an adaptive response of the organism to cope with adversely changing conditions of living and to reestablish the disturbed homeostasis.

The maintenance of homeostasis in metazoans is function of the neuroendocrine system, which evolved at the dawn of the metazoan life as a result of the evolution of the nerve cell, the neural net, and the nervous system. Primitive metazoans had a nervous net but not a separate endocrine system, which evolved later as extension of the nervous system. Even the most primitive nervous net, as we know it in cnidarians, is capable of monitoring changes in the external and internal environments, processing the input, and communicating its output throughout the body in the form of chemical signals, neurosecretions, for inducing adaptive changes in behavior, physiology, and morphology.

It is a widely held opinion that the first nerve cells to evolve might have been neurosecretory by function. The nerve cell, nerve net, and the CNS evolved in response to the evolutionary pressures for maintaining homeostasis of evolving metazoan systems of ever-increasing structural and functional complexity, for dealing with impredictabilities of the changing environment and for avoiding their harmful effects. It is believed that a first step in differentiation of the neuron has been the differentiation of specialized receptor cells for monitoring changes in the external environment and communicating these changes to the rest of cells via chemical mediators they release (Gorbman and Davey, 1991).

Most of the hormonal systems in lower invertebrates are first-order systems, i.e., neurosecretory systems, with neurons being the only source of hormones for regulating the growth, metabolism, water balance, and reproduction.

In chordates and vertebrates, the hypothalamus took over most of the functions of communication with the rest of the body and regulation of homeostasis as well as adaptive responses to the changed environment. A process of centralization of the nervous system and formation of brain began very early in the evolution of metazoans. Parts of the brain would later specialize in performing adaptive-regulatory functions, and an endocrine system would later evolve as an extension of the nervous system.

The neuroendocrine system in vertebrates regulates all of the basic biological functions of growth, metabolism, and reproduction that were regulated by the neurosecretory system of the first order in lower invertebrates. Additionally, chiefly via the autonomous nervous system, it regulates the visceral activities (Strand, 1999, p. 188).

By the mid-1930s, Ernst and Bertha Scharrer introduced the concept of an integrated neuroendocrine system, in which neurons secreted their hormones into circulation (Strand, 1999). The neural and nonneural tissues secrete relatively a large number of neuropeptides necessary for the integration of the functions of all the systems of organs in metazoans.

By the middle of the 20th century it was demonstrated that a small neurosecretory part of brain (weighing only 4–5 g in humans), the hypothalamus, was specialized for the communication between the CNS and the endocrine system. It is the most ancient part of the brain shared by species from the lower vertebrates to mammals.

The endocrine system consists of ductless glands that release their hormones into the bloodstream. The classic endocrine glands in higher vertebrates are the pituitary, the thyroid, the parathyroid, the pancreas, the thymus, and the gonads. Other organs in which clusters of secretory cells produce hormones are the liver (insulin-like growth factor 1), the kidney (renin, erythropoietin, and vitamin D3), the pineal gland (melatonin), and the endocrine cells of the gut. The total number of hormones in humans exceeds 130 (Norman and Litwack, 1997).

For a long time, the pituitary was considered to be the "master" gland, but now we know that this gland itself is subordinate to the hypothalamus, which controls the synthesis of pituitary hormones via specific releasing and release-inhibiting hormones and neurotransmitters. (The synthesis of some pituitary hormones is under the direct or indirect neuronal control.) (Norman and Litwack, 1997, p. 134). Each of the hypothalamic-releasing hormones controls the synthesis of a specific pituitary hormone, which, in turn, stimulates secretion of a specific hormone by the target endocrine glands.

The hypothalamus itself secretes its "releasing" hormones from nerve endings of its peptidergic neurons in response to electrical/chemical signals it receives from aminergic or cholinergic neurons in various brain centers (Norman and Litwack, 1997, p. 99), often through the limbic system. The median eminence (ME) alone secretes more than 40 neuropeptides and other chemical messengers.

The hypothalamus monitors and regulates the body temperature, sodium chloride, glucose levels, and chemistry of body fluids in general. It controls most of the involuntary activities in the animal body, including innate behaviors. Through the pituitary gland, it determines the whole hormonal activity of the target endocrine glands, which is crucial for performing reproductive, developmental, behavioral, and other physiological functions.

As mentioned earlier, besides the hormonal control, the hypothalamus exerts a direct control on the pituitary, via a special anatomical structure enabling the close interaction between the nervous and endocrine systems, represented by nerve endings of hypothalamic neurosecretory neurons that project into *pars nervosa* of the pituitary. This enables the hypothalamus to discharge its neurohormones directly into the portal vessels of the pituitary. The hypothalamus–pituitary–target endocrine glands axes represent the core of the neuroendocrine system, which plays a crucial role in maintaining homeostasis in vertebrates.

The close relationship between the nervous and endocrine systems makes possible the well-known influence of external and internal environments on the activity of the endocrine system; the hypothalamus is a coordinating center that integrates various inputs to ensure a well-organized, coherent, and appropriate set of autonomic and endocrine responses.

Changes in the external environment are perceived in the brain through the animal senses, and the hypothalamus is connected to the external world through the forebrain (Kandel and Schwartz, 1985). Based on the input of information it receives from other brain regions, the hypothalamus, by releasing specific hormones, via the pituitary, adaptively modifies the activity of the target endocrine glands, and it also serves as the site where other parts of the CNS interact with the autonomic system (Delcomyn, 1998, p. 86). Along with the frontal and parietal eyes (associated with the pineal gland), the hypothalamus appears as low in chordates as amphioxus, where its neurosecretory cells control basic physiological functions and reproduction (Allman, 1999).

The hypothalamus regulates secretion of many neurohormones in response to the temporal cyclicalities of the external environment. The two more important cycles are the daily (circadian) rhythm of light and darkness and the annual cycle of seasonal changes in temperature, length of the day, rainfall, salinity of the water, and other variables. Light, temperature, and other parameters of the external environment are received by sense organs, and the information is transmitted to the hypothalamus over neural pathways. Hypothalamic neurosecretions regulate the production of hormones of the anterior lobe of the pituitary, which promote reproductive activities such as gametogenesis and different forms of behavior (migration, territory defense, mating behavior, nest building, and care of eggs and young).

The hypothalamus–pituitary–adrenal (HPA) axis is a major player in the vertebrate response to stress.

Stress and Stressors in Vertebrates

Sudden changes in environment are often so rapid that organisms do not have the necessary time for the gradual changes to accumulate for the adaptation and survival of the animal in the new environment. Such environmental changes can invalidate various phenotypic adaptations that species have evolved in their course of phylogeny, and individuals may suddenly find themselves in life-threatening situations. This is a time when the "struggle for life" may take its fiercest form. Each individual organism, population, and species, sometimes the whole ecotype, can face a survival crisis—often being doomed to extinction. At an organismic physiological level, this situation is characterized by a general stress condition, and by effort, not always successful, to maintain homeostasis.

Stress is defined as "a threat, real or implied, to the psychological or physiological integrity of an individual" (McEwen, 1999). According to Goldstein (1990):

> *Stress is a condition where expectations—whether genetically programmed, established by prior learning, or deduced from circumstances—do not match the current or anticipated perceptions of the internal or external environment, and this discrepancy between what is observed or sensed and what is expected or programmed elicits patterned, compensatory responses... During stress, many body systems—including the sympathoadrenal, parasympathetic, and hormonal homeostatic systems—are activated or inhibited in primitively specific patterns regulated by physiological, biochemical, and psychological homeostats.*

From an evolutionary point of view, stress is a centrally determined adaptive response to the changed conditions in the environment that threaten species-specific homeostasis.

The concept of stress incorporates both an environmental agent (the stressor) adversely influencing the living system (stress condition) and a response of the affected animal for overcoming injurious effects of the stressor (stress response). The effects of the stressor depend not only on the nature of the stressor but also on the nature of the stressed organism, in the meaning that what might be a stressor for one species might not be for another.

According to the duration of their action, two general types of stressors are distinguished:

1. Long-term stressors
2. Short-term stressors.

Long-term stressors comprise drastic changes in the environment, such as transformation of a terrestrial habitat into an aquatic one (or *vice versa*), the introduction of a specific predator into the habitat, severe changes in the climate and temperature, and overcrowding and resulting food shortage.

Short-term stressors, by acting for short periods of time, are irrelevant from an evolutionary point of view.

Currently, many biologists consider stress to be a complex biological process, an organismic response fundamentally determined by neural (central and peripheral) mechanisms with the limbic–hypothalamus–pituitary–adrenal (LHPA) axis as the major player. However, different stressors might be, there is a common pattern of stress condition in higher vertebrate species, characterized by a neuroendocrine response (activation of the HPA axis) for restoring the disturbed homeostasis and by adaptation of the behavior to the changed environment.

Stressful situations activate two basic interacting and overlapping stress–response circuits: the central neural circuit and the neuroendocrine circuit. The central neural circuit receives stress signals, which converge to the amygdala as the central stress processor. Activation of this circuit leads to the so-called emotional or cognitive stress. The neuroendocrine circuit consists of the LHPA axis and leads to the best-known "physiological" stress (Avishai-Eliner et al., 2002).

Neuroendocrine Response to Stress

Stress condition is a systemic condition arising when the intensity of the action of the stressor exceeds an intrinsically determined threshold. The threshold beyond which a noxious agent would act as a stressor varies not only among species, but individuals of the same species may vary in their sensibility to stressors. An adaptive response to the stress condition, i.e. to the adverse effects of environmental factors, is triggered by the perception that the threshold is reached. The perception and evaluation of the stressor or its effects take place in the nervous system.

Both vertebrate and invertebrate organisms have evolved special systems for coping with such adverse environmental influences, *neuroendocrine stress-controlling mechanisms*, representing an essential part of the ICS, with the CNS as its controller.

The stress response in higher vertebrates is essentially a neurally controlled response, and this has also found its expression in the modern definition of stress as a response to real or *perceived* changes in the organism (Greenberg et al., 2002). It is the CNS that must detect the disturbed state by comparing it to "anticipated perceptions of the internal or external environment" (Goldstein, 1990) at a systemic level and trigger an appropriate neurohormonal response for restoring the normal state of the internal environment. The input of the stressful sensory and somatosensory stimuli is transmitted to the CNS via spinal and cranial sensory nerves.

Signals from the limbic system induce the synthesis and secretion of the hypothalamic neurohormone corticotropin-releasing hormone (CRH) by a group of neurons in the hypothalamus (Figure 7.1). The hormone is a mediator of behavioral and physiological responses to stress.

> *CRH is an important coordinator of the endocrine, neuroendocrine, autonomic, and behavioral responses to stress ... and the innervation of the PVN (hypothalamic paraventricular nucleus—N.C.) reflects the extensive afferent input necessary for such integrated action. Through neural input from visceral, somatic, and special sensory systems, the CRH neuron in the PVN becomes the center of an information highway bringing data about a variety of stresses. Information from blood-borne molecules also is received by those cells.*
>
> *Strand (1999, p. 188)*

In response to stress condition, the activity of hypothalamic CRH-producing neurons in mammals is increased. In desert tadpoles, under conditions of dessication,

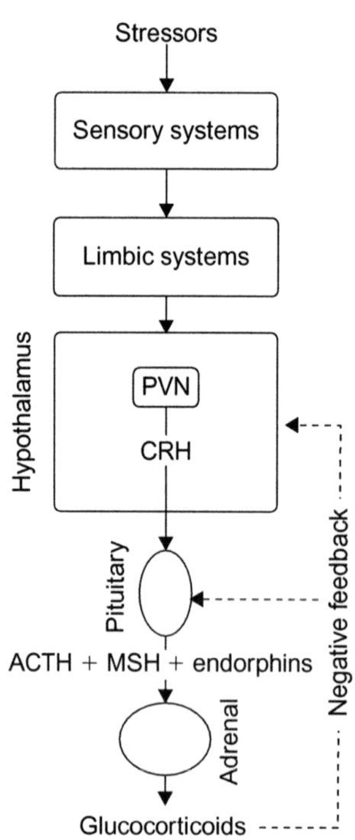

Figure 7.1 Simplified mechanism of the neuroendocrine response to stress condition.

there could be a complex interaction among food intake (which is suppressed during metamorphosis because the gut at that time is undergoing deep structural changes), behavior, and morphogenesis that is coordinated by CRH-producing neurons (Denver, 1997).

Because of the central role CRH plays in the stress phenomena, it is called the *stress neurohormone*. It acts as a mediator for many behavioral and physiological responses to stress.

CRH secreted in the hypothalamus, through the portal system, reaches the anterior pituitary, where it induces secretion of adrenocorticotropic hormone (ACTH), melanocyte-stimulating hormone (MSH), and endorphins (opioid peptides). In the adrenal cortex, ACTH, through a complex conversion of lipoproteins, leads to the synthesis and secretion of cortisol into blood. High levels of glucocorticoids and dehydroepiandrosterone have negative feedback on both the hypothalamus and pituitary to inhibit the production of CRH and ACTH and, consequently, the stress response (Strand, 1999, p. 143). CRH inhibits appetite and, via somatostatin, inhibits growth hormone (GH), thyrotropin-releasing hormone (TRH), and thyroid-stimulating hormone (TSH) secretion (Tsigos and Chrousos, 2002) as well as secretion of the gonadotropin-releasing hormone (GnRH), thus inhibiting the HPA axis and secretion of gonadal hormones with resulting inhibition of reproductive activity.

Essential for acute stress responses is the release of catecholamines by the sympathetic and central nervous systems (Greenberg et al., 2002). Beta-endorphin, secreted by the pituitary during the environmental stress, helps the organism to be less sensitive to the traumatic pain, while stress-released ACTH and MSH "increase attention and motivation, also improve neuromuscular performance, all important components of the successful response to stress" (Strand, 1999, p. 144). Recent evidence shows that electrical field stimulation of the peptidergic innervation of the anterior pituitary alone can suppress the ACTH secretion (Gao et al., 1999). The oral administration of antalarmin (an antagonist of the CRH type I receptor) significantly diminishes the increase of CRH in cerebrospinal fluid and inhibits the repertory of behaviors associated with anxiety and fear (body tremors, grimaces, teeth gnashing, urination, and defecation) during the psychological stress in primates. At the same time, it increases exploratory and sexual behaviors that are normally suppressed during stress (Habib et al., 2000).

Stress condition induces rapid changes in the behavior of animals, and the behavioral changes are an integral component of the stress response (Orchinik, 1998). The activation of the HPA axis and the fluctuations in the levels of hormones during stress could elicit various adaptive neurophysiological responses to sensory (visual, tactile, auditory, olfactory, somatosensory) stimuli, which may be a mechanism of altering brain function during exposure to a stressor.

Administration of corticosterone inhibits the courtship behavior of male amphibians of the roughskin newt, *Taricha granulosa* (Moore and Miller, 1984; Orchinik et al., 1991; Orchinik, 1998). Corticosterone treatment of Gambel's white-crowned sparrow, *Zonotrichia leucophrys gambelii*, within 15 min, causes a rapid transient increase in perch-hopping behavior of this bird (Breuner et al., 1998).

Stress-controlling mechanisms evolved early in the evolution of vertebrates (Crespi and Denver, 2005), and there is evidence that CRH-like and cortisol-like

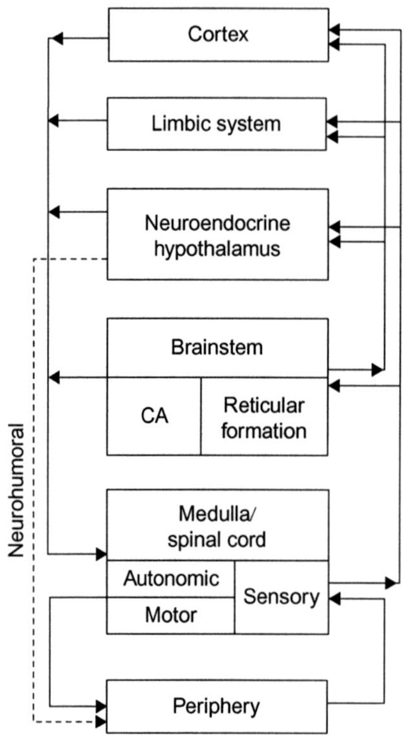

Figure 7.2 Neuronal circuits in the organization of stress responses. *Horizontal thick and thin lines* indicate "short circuit": autonomic (sympathoadrenal and/or parasympathetic) and defense (withdrawal) spinal reflexes in response to stressful stimuli. *Thin lines* represent "long circuit": ascending (afferent) and descending (efferent) neuronal loops between the spinal cord/medulla and "higher" brain centers. The *dashed line* indicates neurohumoral hypothalamo-pituitary outflow. Abbreviation CA, Brainstem catecholaminergic neurons. *Source*: From Pacak and Palkovits (2001).

substances are present in the hemocytes of molluscs (Ottaviani et al., 1998). In *Xenopus laevis*, CRH is secreted throughout the brain–spinal cord, and recent evidence shows that expression of CRH in the brain arose early in vertebrate evolution (Yao et al., 2004).

At the core of the stress-response mechanism is the classical stress circuit: the LHPA axis (Loez et al., 1999) or locus ceruleus/norepinephrine (LC/NE) neurons as well as "their peripheral effectors, the pituitary-adrenal axis, and the limbs of the autonomic system" (Tsigos and Chrousos, 2002), with the hypothalamic CRH neurons.

The final result of the activation of the HPA axis is the release of glucocorticoids into blood, to which the hypothalamus responds by suppressing synthesis and secretion of CRH (Miller and O'Callaghan, 2002).

According to the neural mechanisms involved, stress responses may be determined by short or long neuronal circuits, respectively, also known as *spinal stress responses*, based on spinal reflexes, and supraspinal stress responses, which involve brain centers, the hypothalamus, the limbic system, and the cerebral cortex (Figure 7.2).

Stress condition and stress responses also influence the relations between the mother and the embryo in placentals. Humans respond to prenatal stress condition by secreting the neurohormone CRH and glucocorticoids, thus activating their respective receptors in the hippocampus (Figure 7.3) (Avishai-Eliner et al., 2001, 2002).

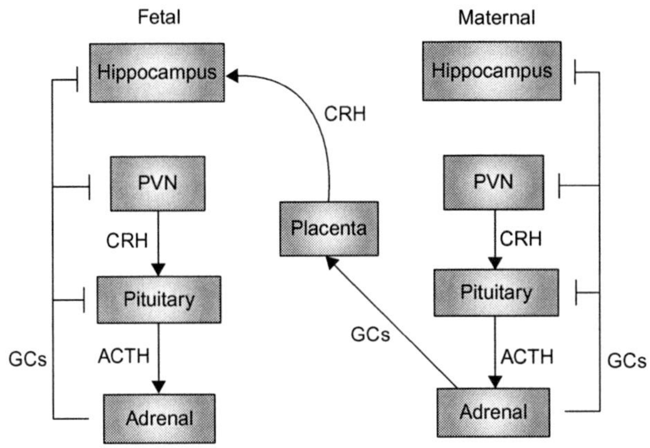

Figure 7.3 In the prenatal human, CRH derived from maternal placenta could influence the fetal hippocampus. Sustained stress during pregnancy activates the maternal neuroendocrine stress axis, resulting in increased production and release of placental CRH into the bloodstream. In contrast to hypothalamic CRH production, which is suppressed by stress-induced glucocorticoids (GCs), CRH-gene expression in placenta is enhanced by GCs, so that maternal stress leads to progressively higher fetal plasma CRH levels. This maternal-origin CRH reaches the fetal brain (black curved arrow), influencing fetal learning and/or memory functions, presumably by activating hippocampal CRH receptors. Arrows indicate facilitatory pathways but do not imply monosynaptic connections. Blunt-ended lines denote inhibitory feedback loops.
Source: From Avishai-Eliner et al. (2002).

Environmental Stress as a Cue for Adaptive Developmental Plasticity

In response to stressful stimuli, vertebrates sometimes activate mechanisms of morphological adaptation. Desert amphibians, for instance, have evolved adaptive mechanisms that allow them to adapt to particular unpredictable changes in environment. When ponds dry unusually early, tadpoles can accelerate metamorphosis and thus escape death by reaching earlier a stage in their metamorphosis when they can survive on dry land.

Robert J. Denver has reproduced the phenomenon in laboratory and has elucidated the underlying physiological mechanism. Tadpoles exposed to earlier habitat dessication, being under strong environmental stress, exhibited elevated hypothalamic CRH content and acceleration of metamorphosis. When the level of water in aquarium was lowered (environmental stress), tadpoles went through metamorphosis in 26 days after hatching, while tadpoles kept in high water needed 36 days. Acceleration of metamorphosis occurred also when CRH-like peptides were injected into spadefoot toad tadpoles. These injections led to a rise in the levels of thyroxine, triiodthyronine, and corticosterone that are considered to be crucial for the hormonal regulation of metamorphosis. Injection, on the other hand, of CRH antagonists, such as anti-CRH serum, minimized the acceleration of metamorphosis caused by dessication of the environment (Denver, 1997). These experiments are very important for demonstrating the existence of a relationship between the hormonal stress pathway and adaptive changes in amphibian morphology during metamorphosis. According to Denver, in all vertebrates, "CRH is exquisitely sensitive to stress in environment," and "the lowering of the level of the water in the environment makes the hypothalamus to produce more CRH, which stimulates pituitary to produce hormones that stimulate thyroid and adrenal glands whose products help organisms to cope with the stress, in this case by losing their tail and beginning the growth of their limbs." Like the tadpoles, the human fetus produces CRH when things are becoming unfavorable and this may stimulate birth (Mlot, 1997).

> *The neuroendocrine stress axis represents a phylogenetically ancient signaling system that allows the fetus or larva to match its rate of development to the prevailing environmental conditions. In diverse vertebrate species, including humans, CRF (corticotropin-releasing factor—N.C.) and corticosteroids act both centrally and peripherally to alter the rate of development in response to unfavorable environmental conditions. This neuroendocrine response generates life history transitions necessary for immediate survival.*
>
> *Crespi and Denver (2005)*

In fact, depending on the environmental conditions and species, the animal response to stress condition always consists of adaptive changes in behavior, physiology, morphology, and life history.

Environmental Stress Induces Evolutionary Changes Without Changes in Genes

Environmental stress, heat shock, and also chemical stress (Liu et al., 1995), osmotic stress (Fiorenza et al., 2004), and neuronal/metabolic stress (Guzhova et al., 2001) sometimes induce expression of a group of heat shock factors (HSFs), such as Hsp70 and Hsp90. HSFs are believed to represent a buffering system necessary for normal development by acting as chaperones contributing to normal folding of other proteins. Hsp90, for example, is known to interact with some relatively unstable developmental switches, such as cyclin-dependent kinases and steroid hormone receptors. Hence, malfunction of the system may lead to phenotypic changes. Stimulation of expression of heat shock proteins (HSPs) increases resistance to heat stress (Mitchell-Olds and Knight, 2002), but suppression of their synthesis or pharmacological blockade of their activity leads to appearance of phenotypic changes.

In 1998, Rutherford and Lindquist observed that in *Drosophila*, pharmacological inhibition of Hsp90 produced phenotypic changes similar to those observed in Hsp90 mutants in nature. They suggested that both the insufficient availability of the Hsp90 and its impaired function can induce appearance of phenotypic changes. They also expressed the idea that "Hsp90 may link developmental programs to environmental contingency" (Rutherford and Lindquist, 1998), implying that the HSF may play a role in the evolution of *Drosophila* under natural conditions:

> *Chaperones also provide a plausible molecular mechanism for regulating the capacity of populations and lineages for evolutionary adaptation.*
>
> *Rutherford (2003)*

Echoing the above observations, it was reported that reduction of the activity of Hsp90 in *Drosophila melanogaster* induces an epigenetic change in chromatin, leading to ectopic expression of *wingless* in eye imaginal disks and abnormal eye development, which are inherited in the following generations (Sollars et al., 2002; Rutherford and Henikoff, 2003). No changes in genes were involved in the evolutionary change. They believe that the chromatin remodeling in this case is related to an interaction with the TrxG protein, which is then "fixed" epigenetically, and speculate that

> *A combination of both epigenetic and genetic mechanisms is probably required to explain rapid changes in body plans that are observed in the fossil record.*
>
> *Sollars et al. (2002)*

The phenomenon of induction of evolutionary changes in the phenotype without changes in genes has also been described in plants (Quetsch et al., 2002).

Stress-induced epigenetic remodeling of the chromatine is not mediated by HSPs alone. A relationship is found to exist between the shear stress (SS) and gene expression in endothelial cells, and recently evidence is obtained indicating that the human umbilical vein endothelial cells (HUVECs) respond to SS by modulating their chromatin, making specific genes accessible to transcription factors. Mediators of the SS-dependent chromatin modulation via modification of histone H3 and H4 by acetylases and deacetylases in endothelial cells are mechanosensors (Illi et al., 2003). One should be reminded that mechanosensory cells and neurons innervating them have similar embryonic origin (placodal ectoderm) (Satoh and Fekete, 2005). They have ion channels in their cell membrane, which allow them to detect mechanical stimuli and convert them into electrical signals that are sent to the brain (Gillespie and Walker, 2001; Dumont and Gillespie, 2003).

In rats and mice, it is demonstrated that stressful circumstances, such as forced swimming, neurally induce histone H3 phosphorylation, which correlates with expression of acetylases and specific genes in mature dentate gyrus neurons and may be related to the behavioral adaptation to the stressful situation (Bilang-Bleuel et al., 2005).

In response to various stressors, animals can modify the synaptic morphology and, consequently, the properties of respective neural circuits. This may lead to changes in stress set points. So, for example, a prior high-temperature stress in a muscle of *Locusta migratoria* modifies the future synaptic physiology and morphology of the muscle by increasing the upper temperature limit at which it can function normally:

> *The prior history of an organism's environment can alter how neural circuitry operates in the long term.*
>
> *Barclay and Robertson (2001)*

In the CNS, glial cells release Hsp70, thus affecting neuronal function and probably enhancing neuronal stress tolerance (Guzhova et al., 2001). Evidence has also been presented suggesting that an inactive form of the maternal Hsp70 mRNA is present in unstressed *Xenopus* oocytes, and it may play a role in expression of the HSPs during early stages of oogenesis (Gordon et al., 1997).

Maltreatment of rat puppies results in methylation of the *bdnf* gene, thus reducing its expression across their life span, and this can be rescued with administration of inhibitors of DNA methylation. More importantly, this leads to a maternally inherited behavior in the offspring: the offspring, like their mothers, maltreat their puppies as well and show the same hypermethylation of the *bdnf* gene (Roth et al., 2009). Similarly, maternal high LG (licking and grooming) in rats during the first postnatal week is being processed in the puppy's brain cortex and via a described pathway (see later in Chapter 8) leads to demethylation and expression of the glucocorticoid receptor (GR) gene, thus determining inherited high LG behavior toward their own offspring (Meaney and Szyf, 2005).

References

Allman, J.M. (1999) *Evolving Brains*. Scientific American Library, New York, p. 7.

Archibald, J.D. (2003). Timing and biogeography of the eutherian radiation: fossils and molecules compared. Mol. Phylogenet. Evol., *28*, 350–359.

Archibald, J.D., Averianov, A.O. and Ekdale, E.G. (2001). Late Cretaceous relatives of rabbits, rodents, and other extant eutherian mammals. Nature, *414*, 62–65.

Avishai-Eliner, S., Brunson, K.L., Sandman, C.A. and Baram, T.Z. (2002). Stressed-out, or in (utero)? Trends Neurosci., *25*, 518–524.

Avishai-Eliner, S., Eghbal-Ahmadi, M., Tabachnik, E., Brunson, K.L. and Baram, T.Z. (2001). Down-regulation of hypothalamic corticotropin-releasing hormone messenger ribonucleic acid (mRNA) precedes early-life experience-induced changes in hippocampal glucocorticoid receptor mRNA. Endocrinology, *142*, 89–97.

Barclay, J.W. and Robertson, R.M. (2001). Enhancement of short-term synaptic plasticity by prior environmental stress. J. Neurophysiol., *85*, 1332–1335.

Bilang-Bleuel, A., Ulbricht, S., Chandramohan, Y., De Carli, S., Droste, S.K. and Reul, J.M. (2005). Psychological stress increases histone H3 phosphorylation in adult gyrus granule neurons: involvement in a glucocorticoid receptor-dependent behavioral response. Eur. J. Neurosci., *22*, 1691–1700.

Breuner, C.W., Greenberg, A.L. and Wingfield, J.C. (1998). Noninvasive corticosterone treatment rapidly increases activity in Gambel's white-crowned sparrows (Zonotrichia leucophrys gambelii*)*. Gen. Comp. Endocr., *111*, 386–394.

Crespi, E.J. and Denver, R.J. (2005). Ancient origins of human developmental plasticity. Am. J. Hum. Biol., *17*, 44–54.

Cronin, T.M. and Raymo, M.E. (1997). Orbital forcing of deep sea benthic species diversity. Nature, *385*, 624–627.

Delcomyn, F. (1998). Foundations of Neurobiology. W.H. Freeman and Co., New York, p. 86.

Denver, R.J. (1997). Environmental stress as a developmental cue: corticotropin-releasing hormone is a proximate mediator of adaptive phenotypic plasticity in amphibian metamorphosis. Horm. Behav., *31*, 169–179.

Douady, C.J., Catzeflis, F., Raman, J., Springer, M.S. and Stanhope, M.J. (2003). The Sahara as a vicariant agent, and the role of Miocene climatic events, in the diversification of the mammalian order *Macroscelidea* (elephant shrews). Proc. Natl. Acad. Sci. U.S.A., *100*, 8325–8330.

Droser, M.L., Fortey, R.A. and Li, X. (1996). The Ordovician radiation. Am. Sci., *84*, 122–131.

Drovetski, S.V., Rohwer, S. and Mode, N.A. (2006). Role of sexual and natural selection in evolution of body size and shape: a phylogenetic study of morphological radiation in grouse. J. Evol. Biol., *19*, 1083–1091.

Dudley, R. (1998). Atmospheric oxygen, giant Paleozoic insects and evolution of aerial locomotor performance. J. Exp. Biol., *201*, 1043–1050.

Dumont, R.A. and Gillespie, P.G. (2003). Ion channels: hearing aid. Nature, *424*, 28–29.

Easteal, S. (1999). Molecular evidence for the early divergence of placental mammals. Bioessays, *21*, 1052–58.

Eldredge, N. (1998). Evolution and environment. Nat. Hist., *107*, 54–55.

Eldredge, N. and Gould, S.J. (1972). Punctuated equilibria: an alternative to phyletic gradualism. In *Models in Paleobiology* (Schopf, In T.J.M. Freeman, Cooper & Co, San Francisco, CA, pp. 82–115.

Erwin, D., Valentine, I. and Jablonsky, D. (1997). The origin of animal body plan. Am. Sci., *85*, 126–137.

Fiorenza, M.T., Bevilacqua, A., Canterini, S., Torcia, S., Pontecorvi, M. and Mangia, F. (2004). Early transcriptional activation of the *hsp70.1* gene by osmotic stress in one-cell embryos of the mouse. Biol. Reprod., *70*, 1606–1613.

Gao, L.Z., Zhang, W.H. and Ju, G. (1999). Suppression of adrenocorticotropic hormone release by stimulation of the nerve fibres in the anterior pituitary. J. Neuroendocrinol., *12*, 753–757.

Gillespie, P.G. and Walker, R.G. (2001). Molecular basis of mechanosensory transduction. Nature, *413*, 194–202.

Goldstein, D.S. (1990). Neurotransmitters and stress. Biofeedback Self-Reg., *3*, 243–271. (abstract).

Gorbman, A. and Davey, K. (1991). Endocrines (fourth edition). In C.L. (1991) Prosser (Ed.),. Neural and Integrative Animal Physiology (693–754, pp. 744). Wiley-Liss, New York, NY, pp. 744

Gordon, S., Bharadwaj, S., Hnatov, A., Ali, A. and Ovsenek, N. (1997). Distinct stress-inducible and developmentally regulated heat shock transcription factors in *Xenopus* oocytes. Dev. Biol., *181*, 47–63.

Greenberg, N., Carr, J.A. and Summers, C.H. (2002). Causes and consequences of stress. Integr. Comp. Biol., *42*, 508–516.

Guzhova, I., Kislyakova, K., Moskaliova, O., Fridlanskaya, I., Tytell, M., Cheetham, M., et al. (2001). *In vitro* studies show that Hsp70 can be released by glia and that exogenous Hsp70 can enhance neuronal stress tolerance. Brain Res., *914*, 66–73.

Habib, K.E., Weld, K.P., Rice, K.C., Pushkas, J., Champoux, M., Listwak, S., et al. (2000). Oral administration of a corticotropin-releasing hormone receptor antagonist significantly attenuates behavioral, neuroendocrine, and autonomic responses to stress in primates. Proc. Natl. Acad. Sci. U.S.A., *97*, 6079–6084.

Illi, B., Nanni, S., Scopece, A., Farsetti, A., Biglioli, P., Capogrossi, M.C., et al. (2003). Shear stress-mediated chromatin remodeling provides molecular basis for flow-dependent regulation of gene expression. Circ. Res., *93*, 155–161.

Ivany, L.C., Patterson, W.P. and Lohmann, K.C. (2000). Cooler winters as a possible cause of mass extinctions at the Eocene/Oligocene boundary. Nature, *407*, 887–890.

Ji, Q., Luo, Z.-X., Yuan, C.-X., Wible, J.R., Zhang, J.-P. and Georgi, J.A. (2002). The earliest known eutherian mammal. Nature, *416*, 816–822.

Kandel, E.R. and Schwartz, J.H. (1985). *Principles of Neural Science*. Elsevier Sci. Publ. Co., Amsterdam, pp. 623–624.

Liu, R.Y., Corry, P.M. and Lee, Y.J. (1995). Potential involvement of a constitutive heat shock element binding factor in the regulation of chemical stress-induced *hsp70* gene expression. Mol. Cell. Biochem., *144*, 27–34.

Loez, J.F., Akil, H. and Watson, S.J. (1999). Neural circuits mediating stress. Biol. Psychiatr., *46*, 1461–1471.

McEwen, B.S. (1999). StressIn *The MIT Encyclopedia of the Cognitive Sciences* (Wilson, R. A., and Keil, F. eds). MIT Press, Cambridge, MA, pp. 1 (according to Greenberg et al., 2002).

Meaney, M.J. and Szyf, M. (2005). Maternal care as a model for experience-dependent chromatin plasticity? Trends Neurosci., *28*, 456–463.

Meng, J. and McKenna, M.C. (1998). Faunal turnovers of Palaeogene mammals from the Mongolian Plateau. Nature, *394*, 364–367.

Miller, A.I. and Mao, S. (1995). Association of orogenic activity with the Ordovician radiation of marine life. Geology, *23*, 305–308.

Miller, D.B. and O'Callaghan, J.P. (2002). Neuroendocrine aspects of the response to stress. Metabolism, *51*, 5–10.

Mitchell-Olds, T. and Knight, C.A. (2002). Chaperones as buffering agents? Science, *296*, 2348–2349.

Mlot, C. (1997). Mammal birth signal cues tadpole to morph. Science News Online, May10, 1.

Moore, F.L. and Miller, L.J. (1984). Stress-induced inhibition of sexual behavior: corticosterone inhibits courtship behaviors of a male amphibian (*Taricha granulosa*). Horm. Behav., *18*, 400–410.

Norman, A.W. and Litwack, G. (1997). *Hormones*, second edition. Academic Press, New York, NY, pp. 525–530.

Orchinik, M. (1998). Glucocorticoids, stress, and behavior: shifting the timeframe. Horm. Behav., *34*, 320–327.

Orchinik, M., Murray, T.F. and Moore, F.L. (1991). A corticoid receptor in neuronal membranes. Science, *252*, 1848–1851.

Ottaviani, E., Franchini, A. and Franceschi, C. (1998). Presence of immunoreactive corticotropin-releasing hormone and cortical molecules in invertebrate gametocytes and lower and higher vertebrate thymus. Histochem. J., *30*, 61–67.

Pacak, K. and Palkovits, M. (2001). Stressor specificity of central neuroendocrine responses: implications for stress-related disorders. Endocr. Rev., *22*, 502–548.

Potts, R. (2004). Paleoenvironmental basis of cognitive evolution in great apes. Am. J. Primatol., *62*, 209–228.

Quetsch, C., Sangstar, T.A. and Lindquist, S. (2002). Hsp90 as a capacitor of phenotypic variation. Nature, *417*, 618–624.

Roth, T.L, Lubin, F.D., Funk, A.J. and Sweatt, J.D. (2009). Lasting epigenetic influence of early-life adversity on the *bdnf* gene. Biol. Psychiatr., *65*, 760–769.

Rutherford, S.L. (2003). Between genotype and phenotype: protein chaperones and evolvability. Nat. Rev. Genet., *4*, 263–274.

Rutherford, S.L. and Henikoff, S. (2003). Quantitative epigenetics. Nat. Genet., *33*, 6–8.

Rutherford, S.L. and Lindquist, S. (1998). Hsp90 as a capacitor for morphological evolution. Nature, *396*, 336–342.

Satoh, T. and Fekete, D.M. (2005). Clonal analysis of the relationships between mechanosensory cells and the neurons that innervate them in the chicken ear. Development, *132*, 1687–1697.

Sollars, V., Lu, X., Xiao, L., Wang, X., Garfinkel, M.D. and Ruden, D.M. (2002). Evidence for an epigenetic mechanism by which Hsp90 acts as a capacitor for morphological evolution. Nat. Genet., *33*, 70–74.

Strand, F.L. (1999). *Neuropeptides*. MIT Press, Cambridge, MA, p. 4.

Tsigos, C. and Chrousos, G.P. (2002). Hypothalamic–pituitary–adrenal axis, neuroendocrine factors and stress. J. Psychosom. Res., *53*, 865–871.

Vieites, D.R., Min, M.-S. and Wake, D.B. (2007). Rapid diversification and dispersal during periods of global warming by plethodontid salamanders. Proc. Natl. Acad. Sci. U.S.A., *104*, 19903–19907.

Vrba, E.S. (1983). Macroevolutionary trends: new perspectives on the role of adaptation and incidental effect. Science, *221*, 387–389.

Vrba, E.S. (1996). In *Beyond Neo-Darwinian. An Introduction to the New Evolutionary Paradigm* (Ho, M.-W., and Saunders, P.T. eds). Academic Press, London, pp. 115–42.

Vrba, E.S., Denton, G.H., Partridge, T.C. and Burckle, L.H. (1996). *Paleoclimate and Evolution, with Emphasis on Human*. Yale University Press, New Haven, CT, 1996.

Wilf, P., Johnson, K.R. and Huber, B.T. (2003). Correlated terrestrial and marine evidence for global climate changes before mass extinction at the Cretaceous–Paleogene boundary. Proc. Natl. Acad. Sci. U.S.A., *100*, 599–604.

Yao, M., Westphal, N.J. and Denver, R.J. (2004). Distribution and acute stressor-induced activation of corticotrophin-releasing hormone neurones in the central nervous dystem of *Xenopus laevis*. J. Neuroendocrinol., *16*, 880–893.

8 Behavioral Adaptation to Changed Conditions of Living

Either chance and selection can explain everything or else behavior is the motor of evolution.

Piaget (1976)

Adaptation to Changed Conditions of Living

Under conditions of environmental stress, metazoans may display two types of behavioral responses: *immediate* or *delayed* response.

Generally, immediate responses are neuroendocrine responses aiming at neutralizing or avoiding noxious factors of the changed environment. Examples of immediate adaptive behaviors are avoiding reflexes, relocation, and migration.

Delayed responses may lead to changes in the morphology, physiology, and life history of animals.

Delayed responses may be *intragenerational*, i.e., taking place within the lifetime of the affected individual, which are normally aimed at restoring the disturbed homeostasis, repairing damaged structure (regeneration), or adaptively changing the morphology of the affected or challenged organs or parts. Such adaptive changes, as a rule, are not inherited in the progeny. Sometimes, delayed responses may be *transgenerational*, i.e., they may arise first in the offspring of the affected or challenged individuals and are inherited in successive generations.

Any drastically changed environment, by disturbing the homeostasis, represents a challenge to the central nervous system (CNS), which is in control of the homeostasis. External stimuli transmitted in the form of electrical spike trains present a problem, to which the CNS may respond in two main ways mentioned above.

Due to the fact that from all the major phenotypic features, animal behavior is, beyond compare, the most plastic, the adaptive change of behavior is the first phenotypic change to take place under conditions of adversely changed environment. Adaptive changes in morphology, physiology, and life history generally may come later.

Neural Basis of Animal Behavior

Types of animal behavior vary over a wide range, from strictly innate behaviors to those that, although innate, can be modified by learning, to behaviors that are exclusively learned. Innate behaviors, with instincts as their extreme form, appear in fully functional form for the first time since they were performed. Instinctive behaviors of

Epigenetic Principles of Evolution. DOI: 10.1016/B978-0-12-415831-3.00008-2

an animal are just as stereotyped and characteristic for its species as its morphology is hence one might expect to find a similar logic underlying the hereditary mechanisms that specify behavior and morphology.

Innate behaviors are automatic stereotyped actions of the organism in response to external releasing stimuli, which trigger *innate releasing mechanisms* (Konrad Lorenz's *angeborenes auslösendes Schema*—innate releasing schema) to produce motor patterns generally known as *fixed action patterns* (FAPs). The pathway from reception of the releasing stimulus, via the CNS, to motor neurons represents the neural circuit responsible for FAP. Any FAP is based on the presence and activation of a specific neural circuit.

In his time, Darwin (1859) observed:

> *As in repeating a well-known song, so in instincts, one action follows another by a sort of rhythm; if a person be interrupted in a song, or in repeating anything by rote, he is generally forced to go back to recover the habitual train of thought: so P. Huber found it was with a caterpillar, which makes a very complicated hammock; for if he took a caterpillar which had completed its hammock up to, say, the sixth stage of construction, and put it into a hammock completed up only to the third stage, the caterpillar simply re-performed the fourth, fifth, and sixth stages of construction. If, however, a caterpillar were taken out of a hammock made up, for instance, to the third stage, and were put into one finished up to the sixth stage, so that much of its work was already done for it, far from deriving any benefit from this, it was much embarrassed, and, in order to complete its hammock, seemed forced to start from the third stage, where it had left off, and thus tried to complete the already finished work.*

The nervous system is organized in neural circuits, which represent functional units rather than anatomic structures. It is generally believed that

> *Neural circuits are the basis of all behavior, from simple reflex withdrawal away from a noxious to a complex mating dance.*
>
> *Delcomyn (1998)*

Various complex behaviors may result from the interaction of various functionally (not necessarily anatomically) linked circuits. Many neurons and their gene products (e.g., neuropeptides, neurotransmitters) are involved in performing these behaviors, and a neuron may be involved in more than one behavior, i.e., in more than one circuit.

Konrad Lorenz observed that, upon seeing an egg outside its nest, a goose tries to roll it back to the nest. The release of this FAP is triggered by a "sign stimulus," which may be represented not only by the goose egg but also by other objects, even those remotely resembling it.

FAPs are hardwired in the brain, as has been experimentally demonstrated by Balaban (1997). By transplanting various parts of the neural tube of the Japanese quail (*Coturnix coturnix*) into domestic chicken (*Gallus gallus domesticus*) embryos, he succeeded in producing chimerae chicken exhibiting quail crowing and head movements, both of the subcomponents of the innate behavior originating from two different regions of their brain (Balaban, 1997). Using similar brain transplant techniques, investigators have been able to transfer an inborn perceptual auditory preference between the same above species (Long et al., 2001).

Innate behaviors may heritably change without changes in genes. Century-long attempts to find genes responsible for particular behaviors have failed. In the meantime, successes in understanding the nature of learning and memory show that animal behavior may be determined by special processing and organizing properties of the nervous system.

Many innate behavior patterns are active immediately after birth. So, for example, after giving birth to its pups, the mother rabbit, *Oryctolagus cuniculus*, releases from its nipples a pheromone, and the pups are born with a fully established neural circuit for identifying that chemical cue and respond by searching behavior for grasping the nipples. That the maternal CNS is involved in the cue release is proven by the fact that injection of prolactin stimulates maximum release of the chemical cue (Moncomble et al., 2005).

In the marine mollusc *Aplysia*, egg laying consists of a number of FAPs: extrusion of a long string of eggs from the reproductive duct, taking of the egg string in the mouth, stereotypic head movements intended at pulling the egg string from the duct, coiling it into a mass glued together by secretions from its mouth, affixing the entire mass of eggs on a solid substrate with a strong head movement.

It was discovered that a neuropeptide composed of 36 amino acids, the egg-laying hormone (ELH), secreted by certain neurons in the nervous system of the mollusc was responsible for performing this complex egg-laying behavior. Later, it turned out that ELH was only a part of a precursor molecule composed of almost 300 amino acids from which other neuropeptides are synthesized, which serve as neural signals controlling other FAPs of egg-laying behavior (Purves et al., 1992).

The overwhelming majority of behaviors studied in vertebrates are related to the function of neural circuitries in the hypothalamus and the hypothalamic–pituitary–target endocrine gland axes:

> *Specialized neuroendocrine circuits for innate behaviors thus seem to process sensory information relevant to ethological contexts and influence sensory perception and processing; integration by these circuits of multiple pathways of information relevant to different behaviors determines the behavioral state of the animal.*
>
> *Manoli et al. (2006)*

The suckling reflex, involving the hypothalamic–pituitary axis, is a typical example of an innate behavior in mammals. Stimulation of nipples by young mammals initiates sensory impulses that reach the CNS and via the hypothalamus end in the PN (*pars nervosa*) of the pituitary-stimulating secretion of oxytocin in the blood within a few seconds. Oxytocin causes contraction of cells in the mammary gland, squeezing milk down to the nipple in less than 1 min (Gorbman and Davey, 1991).

One of the most widespread innate behaviors is the annual migration and repatriation in a large number of animals, invertebrates (e.g., insects, crabs), as well as vertebrates (e.g., fish, reptiles, birds, and mammals). Many of the animals take the round trip journey, although a visible reason or evolutionary advantage for undertaking the trip cannot always be identified. Indeed, often it is not easy to imagine an evolutionary pressure that would reasonably be responsible for the evolution of this migration instinct. Following Darwin, most biologists believe that innate behaviors have evolved from learned behaviors. Migration of birds and other vertebrates and invertebrates, the complex spider web-building behavior, or social behavior in ants, bees, and other animals, are complex innate behaviors, which could reasonably be explained as resulting from learned antecedents:

> *Learning ... is one of the standard, off-the shelf programming tricks available to evolution—and despite the usual dichotomy, this kind of learning is the epitome of the instinct.*
>
> *Gould (1982, p. 274).*

Birds are known to have an innate ability to recognize their conspecific song and to respond more strongly to the song of conspecifics than to the songs of other birds, even when not all phrase types are present in the song and when the song is played in reverse. Identification of specific neurons responding differently to conspecific and heterospecific songs suggests that circuits for song recognition in these birds are established experience independently, i.e., during the embryonic life, probably perfected by learning during the postnatal life (Whaling et al., 1997). Evidence also has been presented showing that an innate song recognition and preference at a subspecific level also exists (Nelson, 2000).

Behavior patterning is commonly determined in the CNS according to the sensory input from the animal's periphery and the environment, but in the rhythmic behaviors (swimming, flight, and chewing), their patterning is totally of central origin and unmodified by the sensory input. The animal can swim, fly, or chew in the complete absence of sensory feedback, as it is demonstrated in deafferented animals (Delcomyn and Prosser, 1991). Their neural circuits are relatively hardwired (Gillette, 1991). This has led to the concept of the central pattern generation (CPG) as basis of FAPs.

Progress has also been made in identifying interneurons and motor neurons in the circuits for the basic locomotory movements (swimming, crawling, shortening, and bending) in the leech (Fan et al., 2005).

Among the best-known examples of CPGs for FAPs is that of locomotion in molluscs. The CPG for swim escape of the marine mollusc, *Tritonia diomedea*, is activated as soon as the slug comes in contact with the predatory seastar *Pycnopodia helianthoides*, and it can also be activated by stimulating any group of peripheral nerves (Frost et al., 2001; Figure 8.1). However, artificial injection of depolarizing current pulses into the interneuron C2 (a crucial member of the swim escape CPG) cannot stimulate the swim escape CPG, because it does not mimic the interneuron's own inherent spike frequency adaptation (SFA). In order for the swim escape to occur, it is necessary to change the SFA. This property of the circuit changes, the firing rate of C2 is regulated, and swim escape behavior occurs when serotonergic interneurons, dorsal swim interneuron, intrinsic to the circuit, are stimulated (Katz and Frost, 1997).

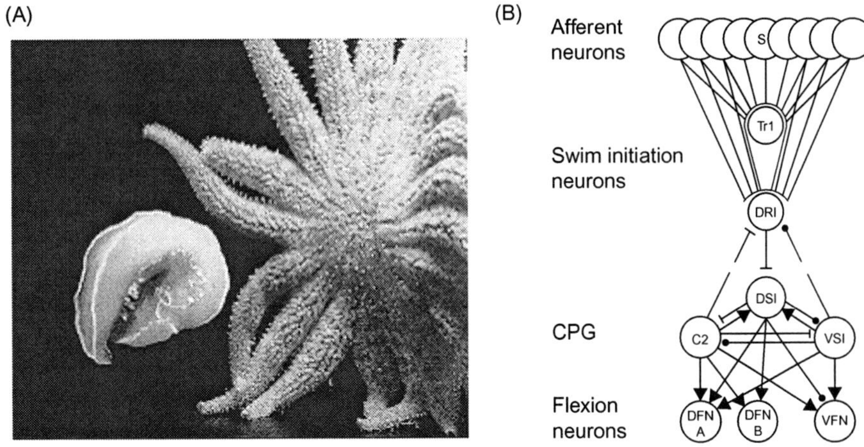

Figure 8.1 The *Tritonia* escape swim and its underlying circuit. (A) Upon contact with a suitably aversive stimulus, such as the tube feet of the seastar *Pycnopodia helianthoides*, *Tritonia* respond with an escape swim consisting of a series of alternating ventral and dorsal whole-body flexions. The photograph shows an animal at a moment of maximum dorsal flexion. (B) The known swim circuit. Solid lines represent direct, monosynaptic connections, broken lines represent indirect connections, or connections not yet confirmed to be monosynaptic. Synaptic symbols: lines, excitatory; black circles, inhibitory; lines and circles, multiple component monosynaptic connections. "VSI" represents both VSI-A and VSI-B; the exact connectivity shown is for VSI-B only. The known number of neurons of each type on each side of the brain are S-cells, ~80; Tr1, 1; DRI, 1; DSI, 3; C2, 1; VSI, 2; FNs, ~55. *Abbreviations*: CPG, central pattern generator; C2, cerebral cell 2; DRI, dorsal ramp interneuron; DSI, dorsal swim interneuron; S, (sensory) afferent neurons; TR1, pre-CPG trigger type 1 interneuron; VSI, ventral swim interneuron.
Source: From Frost et al. (2001).

However, not all neural pattern generators are central. So, for example, the pattern generator of feeding behavior in gastropods is distributed between the buccal and cerebral ganglia and is, therefore, modified by sensory input. Besides, the pattern generation of the buccal ganglion can sustain more than one type of rhythmic pattern (Croll et al., 1985a; Suesswein and Byrne, 1988).

Neural circuits often interact and may interfere with each other's activity. Sometimes activation of a circuit may automatically inactivate the circuit for another behavior. Such is the case of the dominant escape swimming behavior in the marine predator sea slug, *Pleurobranchaea*. Escape swimming is an avoidance behavior in the predatory sea slug *Pleurobranchaea californica*. Stimulation of its swim escape circuit leads to automatic inhibition of the feeding circuit.

In response to environmental stimuli, i.e., to particular types of sensory information, animals secrete neurohormones and other hormones that act as gain-setting devices, biasing the animal behavior toward particular stereotypical responses (e.g., male/female behavior, fight or flee, explore, search for food). By acting both in the nervous system and on effector organs, hormonal substances can modify the input processing, and the output of specific subsets of neurons to enhance the probability of specific outcomes (Kravitz, 1988).

Intragenerational changes in animal behavior result from reversible neuromodulations rather than any changes in genes. Similar behaviors may be produced by convergent, rather than homologous, circuits. Changes in the configuration of the neural network may change the behavioral output of circuitries; neuromodulation, the state of the neural network, thus provides behavior with flexibility, and small changes in neural pathways, involving no changes in genes, may result in dramatic changes in behavior (Nishikawa, 2002).

We have some solid evidence for the molecular mechanism of innate behaviors. So, for example, it is proven that the female sexual behavior in rodents is controlled by specific neuronal networks. These neuronal networks, themselves, are under the control of nuclear receptors in the hypothalamus. In response to changes in levels of estrogen and progesterone, the receptors are activated and modulate the expression of genes involved in the respective sexual behaviors. It is demonstrated that nuclear receptors may also be activated by neurotransmitters such as dopamine, which mimics the effect of progesterone in facilitating sexual behavior in female rats (Mani, 1994).

Neural circuits responsible for innate behaviors in animals are established prenatally. For example, the circuit for bending behavior in the leech, *Hirudo medicinalis*, which consists of 45 neurons (17 interneurons, 24 stimulatory and inhibitory motor neurons, and 4 mechanosensory neurons), is established during the embryonic development. The connections between the neurons of the circuit are precisely established from the beginning, i.e. experience independent (Marin-Burgin et al., 2006).

The modern concept of neural circuits implies a high degree of flexibility of the neural networks, which may result from the ability of the receiving neurons to interpret the stimulus differentially. External stimuli reach the CNS via the relatively fast-acting and rapidly reversible signaling molecules (e.g., insect locomotor patterns and molluscan feeding movements). In other cases, the neuromodulatory network can be reconfigured by slow-acting and slowly reversible neuromodulatory signals such as peptides and hormones.

Most rhythmic behaviors, such as walking, swimming, chewing, or scratching, are patterned by CPGs. A single CPG may produce several types of rhythmic behavior, depending on the sensory and descending input. It is believed, for example, that lampreys use the same basic CPG to produce different locomotory behaviors, such as swimming, crawling, and burrowing (Ayers et al., 1983).

When a specific sensory input is lost in the course of evolution, the respective neural circuit may be conserved and used for another purpose, as is the case of blind cavefish; within ~10,000 years since they lost their eyes, the tectum has lost the function of visual processing and is used for processing somatosensory input (Voneida and Fish, 1984).

Although most of investigators believe that CPGs are discrete functional entities, evidence also exists that specific CPGs may be formed by a pool of individual neurons of diverse origin (Meyrand and Simmers, 1991). Contrary to the early concept of neural circuits as static networks of interconnected neurons, they are seen now as dynamic information-processing networks interacting with other networks that may be functionally reconfigured under the influence of external stimuli.

> *Many types of change are intrinsic to the machinery of synaptic terminals ... This type of variability may allow much use-dependent alteration in the functioning of sets of neurons. When one superimposes on these intrinsic elements of flexibility the possibility of hormonal modification of transmitter release or of target responsiveness, then the capacity of even small sets of neurons to produce a wide range of processing and output becomes great. A particularly elegant example of this comes from recent work on the stomatogastric ganglion in the lobster. The small group of neurons linked together in a well-defined circuit gives completely different physiological outputs in response to the application of different hormones.*
>
> *Kravitz (1988)*

Neuromodulation for generating different outputs is accomplished in response to external stimuli, but evidence suggests that modulation may also intrinsically occur, without external stimuli:

> *There are two sources of neuromodulation for neuronal circuits: extrinsic inputs and intrinsic components of the circuits themselves. Extrinsic neuromodulation is known to be pervasive in nervous systems, but intrinsic neuromodulation is less recognized, despite the fact that it has now been demonstrated in sensory and neuromuscular circuits and in central pattern generators. By its nature, intrinsic neuromodulation produces local changes in neuronal computation, whereas extrinsic neuromodulation can cause global changes, often affecting many circuits simultaneously.*
>
> *Katz and Frost (1996)*

Among complex behaviors regulated by neural circuits are those related to the electrogenic and electroceptive phenomena in fish, some amphibians, and the monotreme mammal, platypus (*Ornithorhynchus anatinus*). Some fish, which are typical in this respect, have electrical organs, consisting of electrocytes (which may be modified nerve or muscle cells). They generate electrical fields by producing electric organ discharge (EOD). They also have subcutaneous electroreceptor cells on the body surface for receiving electric discharges from other fish. These fish use the capability for emitting weak electrical discharges for communicating with conspecifics as well as for navigating and electrolocating prey in the dark. Some fish are specialized in emitting strong electrical discharges for stunning prey.

In order to avoid errors of electrolocation in cases when another fish generates an electrical field at a frequency close to its own, the fish can also change the frequency of their discharge. This ability is known as *jamming avoidance response* (JAR). Fish use complex algorithms to figure out whether they must increase or decrease the frequency of their electric discharges. This assessment is done in other areas of the brain, which then send signals to the medullar pacemaker to modulate the frequency of EOD.

Apteronotus and *Eigenmannia* are two closely related groups of gymnotiform fish. These fish use EODs for communication and for localizing and identifying objects.

Both of them have similar electric organs that produce EODs at a regular, resting frequency that is under neural control of the medullar pacemaker nucleus (Pn) (Figure 8.2).

Both *Apteronotus* and *Eigenmannia* can regulate their EOD in order to avoid jamming electric signals emitted by conspecifics, based on a CPG. However, on detecting a jamming signal of slightly higher frequency than theirs, both species respond in different ways. When the jamming signal is of higher frequency, *Eigenmannia* activates a torus subnucleus nE↑ *nucleus electrosensorius*, which sends signals for raising the frequency to the pacemaker nucleus (Pn), and when detects a signal of slightly lower frequency it activates another torus subnucleus, nE↓, which lowers the frequency. The behavior of *Apteronotus*, instead, is asymmetrical; fish of this group can respond to jamming signals by only raising the frequency of the electrical organ discharge via activation of the normally inhibited sublemniscal nucleus (Heiligenberg et al., 1996; Metzner, 1999).

The input from electroreceptors is processed in a complex brain circuit that, in a simplified form, is represented in Figure 8.2. It is a branched circuit that comprises the hindbrain electrosensory lateral line (ELL)→torus semicircularis→diencephalon's nucleus electrosensorius (NE)→sublemniscal prepacemaker nucleus (SPPn) (and parallelly, prepacemaker nucleus, PPn)→pacemaker→electromotor neurons→electrocytes in the electrical organ.

The whelk, *Fusitriton oregonensis*, displays another curious innate behavior. Mating pairs form seasonally and are maintained for up to 4 months. Then a parent attaches its clutch of eggs to a rock and patrols them against predators. Potential predators attack with twisting movements of the whelk's shell to dislodge them; failing that, the whelk may directionally squirt an aversive acid secretion (Gillette, 1991).

Electric organs and respective JAR behavior in fish have evolved repeatedly and independently, i.e., in many families whose ancestors lacked it (Rose, 2004).

Complex innate behaviors expose the failure of the gradualist explanation of evolution of phenotypes: such behaviors usually offer their evolutionary advantages when they are fully developed, and it is impossible to imagine why and how a number of gradual steps, each of them not advantageous in itself, could be accumulated over time. Ethologists, for example, calculate that a cocoon-spinning spider must perform 6,400 individual movements to make its egg cocoon by spinning a base plate, building up walls, laying its eggs inside, and spinning a lid to close the cocoon. This is done by an innate, closed program not modifiable by experience (Purves et al., 1992).

Neural Basis of Learned Behavior

Owing to the flexibility of neurobiological mechanisms, the earliest response of animals to changes in environment is behavioral. The new behavior is a learned behavior intended to avoid or neutralize harmful effects of the changed environment and represents not an off-the-shelf solution of the type of the innate behavior.

Despite the striking differences from innate behaviors, learned behaviors as well are products of the activity of neural circuits. For patterning and executing learned

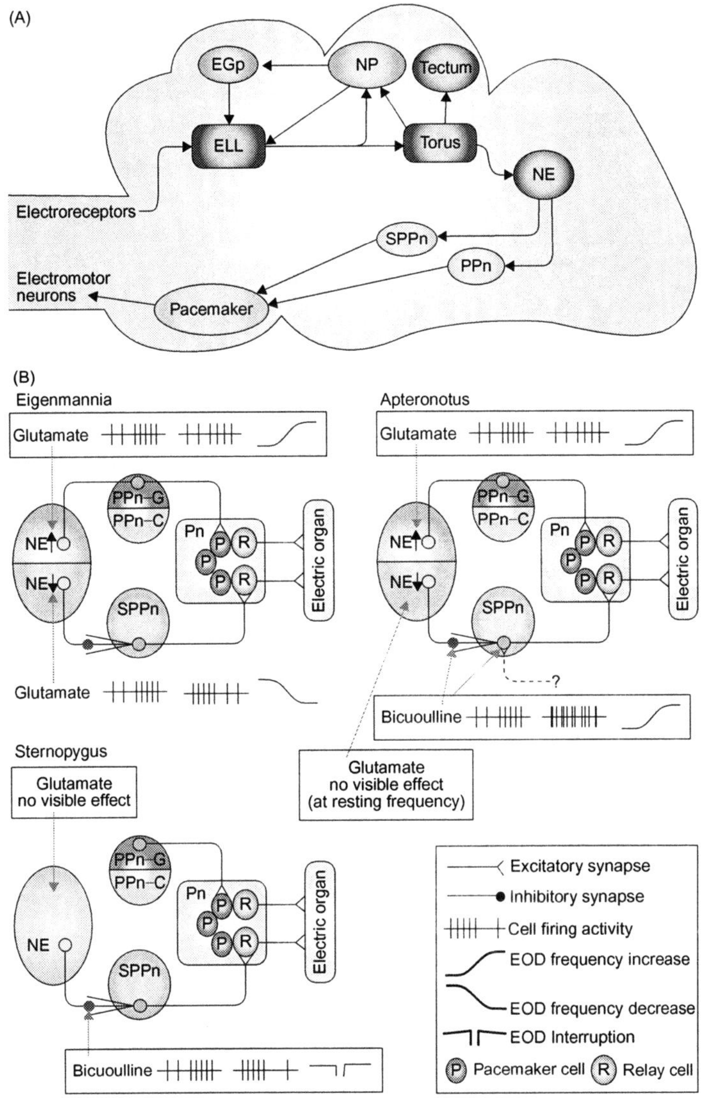

Figure 8.2 Electrosensory circuits. (A) Electrosensory pathway in *Eigenmannia*. *Abbreviations*: ELL, electrosensory lateral line lobe; EGp, *eminentia granularis pars posterior*; NP, *nucleus praeeminentialis*; NE, *nucleus electrosensorius*; SPPn, sublemniscal prepacemaker nucleus; PPn, prepacemaker nucleus. (B) Premotor–motor circuits for controlling EOD modulations in three genera of gymnotiform fish. NE↑ and NE↓ regions of the NE, when stimulated by iontophoresis of glutamate in *Eigenmannia*, cause slow increases or decreases in the EOD frequency. When active, PPn-G (gradual) and PPn-C (chirp) (subdivisions of the PPn) elicit slow, smooth increases in EOD frequency or abrupt cessations of EODs, respectively. Cells in the NE↓ area make inhibitory contacts with the dendrites of SPPn neurons. Blocking inhibition at the SPPn (with bicuculline) causes EOD frequency increases in *Apteronotus* or large EOD frequency increases that culminate in interruptions in *Sternopygus*. Prepacemaker neurons project to the pacemaker nucleus (Pn), where they synapse with either pacemaker (P) or relay (R) cells.
Source: From Rose (2004).

behaviors, animals use existing species-specific FAPs by modulating the respective neural circuits. Learned behavior might not be perfectly adaptive under the changed conditions of environment, for acquisition of learned behavior is not an "all-or-none" process. Although learned behaviors may be initially imperfect, they may be beneficial because the survival in a drastically changed environment does not necessarily require a perfect behavioral adaptation from the beginning; once the changed behavior makes the survival possible, perfection of the behavior and adaptive morphophysiological changes may follow.

Acquisition of new behaviors is facilitated by the fact that

First, changes in the environment often are represented by contrasting conditions of living, for example aquatic/terrestrial habitat, moist/dry terrain, cold/warm weather, short/long photoperiod, abundance/scarcity of food, herbivorous/carnivorous diet, or presence/absence of predators, which are not as numerous as their intermediate states would be, and

Second, often the same FAPs can serve more than a single purpose:

> *Learning involves a single set of basic processes which are shared throughout the animal kingdom from molluscs to man.*
>
> *Gould (1982, p. 261)*

In this sense, the learning and innate behaviors share a common neural basis. As pointed out earlier, lampreys can modulate the same central motor program of a basic undulatory pattern to perform different forms of locomotion such as swimming, burrowing, and crawling. The same CPG is conserved across metazoans, and it would, in principle, allow them to switch from an aquatic to a terrestrial form of locomotion in cases when the habitat changes from aquatic to a telluric one. Further, when an animal changes its behavior, it can do this by switching to new patterns of connections between the same neurons of the circuit as it occurs with neurons that control the lobster's eating and digestion.

Third, there is experimental evidence showing that neural elements and connections for performing a behavior may be conserved after a species has lost the behavior. This may greatly facilitate the "learning" of the new behavior: activating an ancestral circuit may be all that is required for performing the new behavior.

So, both innate and learned behaviors are based on *essentially* similar neurobiological mechanisms. We know that even an innate behavior can be modified as a result of feedback action, i.e., of its disadvantageous consequences, and to the contrary, behaviors learned at early stages of life are very little, if at all, modifiable (the phenomenon of behavioral imprinting).

Any learned behavior is generally based on the existence and use of innate motor programs or motor program preadaptations. These programs may exist as "out of service" preadaptations still available for reactivation. Over time, by repeated practicing for varying periods of time, the learned behavior may be stereotyped and performed automatically, similarly to an innate behavior, or to its elementary unit, FAP. If it is indeed the case, we might reasonably assume that the learned behavior is based on formation of a specific neural circuitry. If the learned behavior happened to be a lost ancestral innate behavior, the relevant circuit may be identical or similar to the neural circuitry used by ancestors for performing the innate behavior.

The fact that the learned behavior is based on specific motor programs on which the innate behaviors are based suggests that their neural circuitry, as a system of nerve cells that cooperate in generating and controlling that behavior, may be similar to the circuitry for the innate behavior. Says Gould (1982, p. 177):

> *The emerging picture of motor behavior, then, whether it be the wholly innate performance of spiders, the goal-directed-learning of infants, or the plastic learning we see in piano playing, is that all are routines stored, either from the outset or ultimately, as discrete neural programs.*

and

> *Beethoven's continued mastery throughout his encroaching deafness is a triumphant instance of how hardwired a motor program can become. (Gould, 1982, p. 176)*

Simple behaviors, based on FAPs, can be learned, and there is no alternative left on how a "complex behavior" could be learned by animals, but by integrating the respective FAPs, i.e., their neural circuits, in interacting neural networks.

The phenomenon of imprinting also demonstrates the common physiological basis and the continuity of innate and learned behaviors. In the Konrad Lorenz's example, Graylag geese are born with an instinct for adopting as mother something that is bigger than themselves, that is closer to them, and that provides food, rather than a mother-resembling figure. By fulfilling all these criteria on his own, the biologist got the newly hatched geese to follow him as their own mother. He succeeded in diverting the natural course of the innate behavior of the young birds because they inherited an innate behavior to look for their mother, neglecting any sensory cues for recognizing her.

This change in behavior is epigenetic by nature, i.e., it involves no change in genes. Such epigenetic changes in both learned and innate behaviors may be induced not only during postnatal life but also during the embryonic and intrauterine development in placentals and oviparous animals. Maternal glucocorticoids from stressed mothers may transplacentally affect the postnatal phenotype (behavior, morphology, and life history) of the offspring. Experimental administration of corticosterone in eggs of the ovoviviparous lizard *Lacerta vivipara* leads to altered antipredator behavior (less-risky behavior indicative of increased fear or anxiety) in the corticosterone-manipulated offspring. Investigators believe that early hormone exposure as a result of maternal stress could be an important evolutionary factor generating epigenetic variation in natural populations (Uller and Olsson, 2006).

Experimental increase of the testosterone level in female dark-eyed juncos (a small American sparrow), besides its known consequences in male offspring, increases the responsiveness of the hypothalamic–pituitary–adrenal (HPA) axis and intrasexual aggression but weakens the cell-mediated immune response (Zysling et al., 2006). Injection of the neuropeptide corazonin in locusts *L. migratoria* induces their transition into the gregarious phase, which implies gregarious behavior (including flying), changes in the morphology, and dark pigmentation. In this context, it is important to remember that production of corazonin in the locust's nervous system is the result of a manipulative expression of the gene through activation of a circuit of the gene for the neurohormone.

Often, learned behaviors are associated with functional and structural changes in nerve cells and in the CNS. In canaries, for example, singing is related to a part of their forebrain. Three forebrain nuclei are involved in normal singing, *nucleus hyperstriatum ventralis, pars centralis* (HVc), and *nucleus robustus archistriatalis* (RA). Two nuclei (HVc and RA) normally are, respectively, 99% and 76% larger in the spring (when birds learn new songs) than in fall (Prosser, 1991). Male canaries are the only sex that learns to sing, and their *nucleus hyperstriatum ventralis* is larger than that of females. Moreover, males that have a longer repertoire of songs have that part three times larger than others (Alcock, 1989).

In zebra finches, the circuits responsible for vocal learning and singing in the cortical nucleus lMAN undergo striking changes in morphology, in size, and axonal arbors during juvenile male development (Iyengar et al., 1999). Although neural circuits are conserved in the course of metazoan evolution, even small changes in neuronal circuits in various species can produce new species-specific behaviors (Katz and Harris-Warrick, 1999).

Animal Behavior Is Not Determined by Genes

Although genes and their products, as well as other metabolic products, are involved in expression of specific behaviors, the information necessary for expression of a behavior is not in genes, i.e., in the sequence of nitrogen bases in DNA segments. That information is generated by activation of specific neural circuits/networks and related FAPs in response to external/internal stimuli. Even proponents of the view that genes are involved in the expression of behaviors are now compelled to admit that

> *The issue is not whether genes control all behaviors; they do not, but rather whether some/many/most behaviors in the former category might be controlled by genes.*
>
> *Baker et al. (2001)*

However, while genes are recruited for, and involved in, expression of any behavior, no gene has been identified to be responsible for expression of any behavior. The conventional view that the fact that mutations in specific genes prevent animals from performing behaviors proves that genes are responsible for the behavior is inaccurate. That negative evidence proves only that products of these genes are necessary conditions for performing respective behaviors, but not necessarily their cause. For genes to be considered the cause of a behavior it is not sufficient to prove that they are necessary for expression of that behavior. (Many other nongenetic factors are necessary for performing even simple behaviors.) In order to be considered "the cause," they must also be sufficient to induce the behavior, which they are not.

Theoretical arguments are certainly less important than the fact that no positive evidence has ever been presented to show that one or a number of genes is sufficient for performing a behavior. While such evidence is lacking, a determining role of genes in animal behavior could be demonstrated by showing that changes in behaviors are causally related to corresponding changes in genes. Again, no such evidence has ever been presented.

Fireflies rely on precise control of flash timing for finding mates. In the beetles of the genus *Photuris*, flashing serves an additional purpose. Females of firefly *Photuris versicolor* are highly specialized predators that can mimic flash signals given by females of other species. Using this false signal, the predatory females can lure in unsuspecting males of *Photinus tanytoxus* (Trimmer et al., 2001) and then eat them. The timing of the flash signal is computationally determined in the nervous system. *Photuris* females can change the flash patterning *behavior* into flash patterns of another genus instantaneously, without changing their genes, by processing in neural circuits allospecific flashing patterns.

Despite the considerable evolution of genes, particular neurons and even entire circuits in metazoans are conserved across the phylogeny, suggesting that evolution of neural circuits is not related to evolution of genes. So, for example, in crustaceans, not only the neuronal circuitry of the stomatogastric ganglion (STG) but even the synaptic connections of the circuitry have been conserved to a remarkable extent during 350 million years of evolution of the group, despite the evolution of genes (Katz and Harris-Warrick, 1999).

Three sympatric morphs of the electric fish flock *Brienomyrus*, which have evolved recently, are genetically and morphologically indistinguishable but have different EOD rates and EOD waveform patterns although they occupy the same niche, implying that they are under the same evolutionary selective pressure and still show this behavioral polymorphism (Arnegard et al., 2005). The fact that these morphs are genetically identical and occupy the same niche unambiguously suggests that their different EOD behaviors do not depend on genes, but on neural, computational, epigenetic mechanisms.

Prairie voles *Microtus ochrogaster* are monogamous. They display a social bond that manifests itself in the tendency of male and female individuals to have only one mate, stay together, share the same nest, continue pair bonding beyond the reproductive period, and show biparental care for pups. Other closely related vole species, such as the montane vole, *Microtus montanus*, and meadow vole, *Microtus pennsylvanicus*, are not monogamous. No evidence has ever been presented, and no assumptions have been made, that the difference between monogamous and polygamous species is related to relevant differences in genes. The only difference demonstrated to exist between them is that several dopamine circuits, which are active in the prairie voles, are not active in the montane and meadow voles. Monogamous behavior in voles is impaired by administration of neuroactive agents such as haloperidol, an antagonist of the dopamine receptor, and the monogamous behavior can be experimentally induced by administration of apomorphine, a dopamine agonist (Curtis et al., 2006).

In some fish, like the ocellated wrasse, *Symphodus ocellatus*, males display four distinct reproductive behaviors, which may be fixed, irreversible, or reversible, depending on "the dynamics of the state of the individual or environment" (Henson and Warner, 1997), a fact that clearly excludes genes as possible determinants of different behaviors during the lifetime of the fish.

Recently, however, it has been claimed that, finally, an innate behavior has been identified that is determined by the *fru* gene. It is about the *Drosophila* sexual behavior, which male flies perform without previous experience. Male mutants for the *fru* gene display defective sexual behavior and do not develop a male-specific muscle known as the *muscle of Lawrence* (MOL).

The *fru* gene is only expressed in neurons. Some 500 out of the total of 10^5 neurons, i.e., about 5% of the *Drosophila* CNS neurons, express *fru* and these neurons are found in the regions believed to be responsible for the male courtship behavior in the *Drosophila* brain (Ryner et al., 1997). Besides the CNS neurons, *fru* is also expressed in specific neurons of the olfactory, gustatory, and auditory organs, belonging to the male sexual behavior circuitry (Stockinger et al., 2005).

The fact that *fru* is expressed in male flies, but not in females, led some biologists to the conclusion that the product of *fru* specifies sexual differences in the CNS that determine male sexual behavior (but they fail to demonstrate it and show how):

> *Male sexual behavior is the result of sensory information entering the CNS via largely sex-nonspecific sensory systems, which is processed by sex-specific (fru-specified) circuitry in higher order neuropils.*
>
> *Baker et al. (2001)*

The authors do not explain why they believe that it is the expression of the gene *fru* that specifies the sex-specific circuitry and not vice versa, i.e., that male sexuality of the brain (which in many experimentally demonstrated cases appears before the primary sexual traits) enables the latter to express the male-specific isoforms of the *fru*, as normally occurs with manipulative expression of genes in the CNS (see Chapter 2). However, they correctly point out:

> *Any behavior requires the functioning of a multicellular circuit beginning with input to the nervous system, propagation and interpretation of that input in the CNS, and output via neurons that directs a response via neuromuscular, or neuroendocrine systems, or both. Impairment of any part of such a circuit is likely to cause decrements in the behavior it subserves.*
>
> *Baker et al. (2001)*

Thus, investigators explicitly state that impairment of any part of the male sexual behavior circuitry (not any change in the *fru* gene) "is likely to cause decrements in the behavior it subserves." They make it clear that the *fru* gene is involved in the expression of the male sexual behavior to the extent that it is involved in the function of the "multicellular circuit." Hence, the *fru* gene is, a necessary, but not sufficient, condition for effectuating the male sexual behavior in *Drosophila*.

Many other products are necessary for the onset of the male sexual behavior, but all of them are not more than necessary conditions, among a number of other conditions, for expression of the male sexual behavior. To classify something as a cause of an effect, you must show that it is both necessary and sufficient for inducing the effect. The *fru* gene is not the cause of the behavior, but the circuit, the processing of olfactory stimuli in the circuit, is.

Both males and females have the same gene *fru*, and alternative splicing of the *fru* locus produces a number of transcripts, most of them being common for both sexes. fru has at least 4 promoters (P1-P4) but only transcripts initiated from the most distal promoter (P1) are spliced differently in males and females. (Demir and Dickson, 2005). These transcripts are potential transcription factors, but as Dulac pointed out,

> *Most of the transcription factors affected in behavioral mutants have broad distributions and general functional properties that seem incompatible with a role as genetic determinants of specific behaviors...a specific subset of olfactory neurons, presumably detecting pheromones, are essential triggers of the male courtship behavior, and that they are functionally connected to other FRU^M-positive neurons in the brain.*
>
> *Dulac (2005)*

The circuit for male sexual behavior also exists in female *Drosophila* flies. Since both males and females possess the same genes and the same circuit, but only males display courtship behavior, it is concluded that the courtship behavior in males is related not to any specific male genes but to modulation of the respective circuit in males as a result of the different splicing of *fru*. Indeed, experimental "male splicing" of *fru* in females causes the latter to behave like males by courting other females. In contrast, males undergoing experimental "female splicing" of *fru* cannot display male courtship behavior (Demir and Dickson, 2005). The fact that both males and females have the same *fru* locus suggests that the underlying cause of the male courtship behavior is not any genes but an epigenetic mechanism that makes possible the manipulative male-specific *fru* splicing, which is function of a neuron-specific system in the nervous system, based on electrical activity and the related changes in Ca^{2+} levels of relevant neurons (see Chapter 2, Section Epigenetic Manipulation of Genetic Information in the CNS).

The fact that the mating behavior is nongenetically, epigenetically, determined is also corroborated by ample evidence that individuals of sibling insect species with identical genetic information, including regulatory sequences, display different mating signalings, mating biases, and behaviors. These phenotypic differences arise without changes in genes, from minute differences in specific neural circuits, by changing the number of neurons or the patterns of synaptic connections (Katz and Harris-Warrick, 1999), and it has been shown that single circuits can generate the same behavior, and that different circuits "can combine to produce a single, new behavior" (Tierney, 1995).

All of the evidence on the expression of innate behaviors in animals shows that expression of specific behaviors is a function of neural circuits, of computational properties and activities of those circuits, which can change, and quite commonly change, without changes in genes. Extensive evidence on evolution of new innate behaviors (sexual behaviors) in insects without changes in genes will be presented in Chapter 19.

Learned Behaviors Evolve into Innate Behaviors

Like innate behaviors, learned behaviors result from the activity of specific neural circuits that often arise in appropriate responses to changes in the environment. So, for example, some nonvolatile male mice pheromones are innately attractive to female mice, but volatiles from male-soiled bedding are not. However, repeated exposure of female mice to male-soiled bedding "learns" them to confer pheromonal properties to the male urine-borne volatiles. Learning, thus, might confer "pheromonal" properties to every odorant, as a result of the ability of the CNS "to relate virtually any stimulus to any adaptive result," and this suggests that innate and acquired olfactory attractiveness relies on similar neural mechanisms (Moncho-Bogani et al., 2002).

Both innate and learned behaviors are products of the activity of neural circuits. Although differences between these two types of animal behavior do exist, they do not amount to a "Chinese Wall" between them. For a number of biological phenomena (imprinting, modification of song circuits, and circuits of other behaviors), it has been shown that both of these forms of behavior are based on essentially similar neurobiological mechanisms. Like evolution of innate behaviors, the learning of new behaviors is associated with structural and functional changes in neurons and neural circuitries.

The fact that similar neural mechanisms determine both learned and innate behaviors raises the question of whether learned behaviors may evolve into innate behaviors. Darwin believed that the learned behaviors could be inherited and could lead to evolution of innate behaviors:

> *If we suppose any habitual action to become inherited—and it can be shown that this does sometimes happen—then the resemblance between what originally was a habit and an instinct becomes so close as not to be distinguished.*
>
> *Darwin (1859, p. 209)*

Even more explicitly, he stated:

> *Some intelligent actions, after being performed during several generations, become converted into instincts and are inherited.*
>
> *Darwin (1874)*

Moreover, Darwin believed that learning of new behaviors may even modify innate behaviors:

> *Habit no doubt sometimes comes into play in modifying instincts; but it certainly is not indispensable, as we see, in the case of neuter insects, which leave no progeny to inherit the effects of long-continued habit.*
>
> *Darwin (1859, p. 474)*

Currently, as well, it is generally believed that innate behaviors have evolved from learned behaviors. However, contrary to Darwin's view that cerebral organization is involved in evolution of instincts, the prevailing neo-Darwinian view is that, in order for a learned behavior to be transmitted to the offspring and become an innate behavior, spontaneously arising gene mutations must occur. It is quite ironic that on Darwin's behalf, neo-Darwinians reject Darwin's idea that variations of innate behaviors are not accidental and that randomness and spontaneity of the variations in instincts is a product of "our ignorance." More than adequate experimental evidence proves that no changes in genes are necessary for evolution of new innate behaviors, and later in this work additional evidence will be presented in support of this idea (see Chapter 19).

Being a response to changed conditions of living, a new behavior is often associated with a stress condition. It is possible that neuroendocrine mechanisms of stress facilitate the transmission of the *learned behavior* to the offspring as an *innate behavior.* Reliable experimental evidence in support of that hypothesis is modest, but now it is almost consensually admitted that any instinct or innate behavior has evolved from a learned behavior at some point in the past. Hence, it is worth reviewing the modest evidence on transformation of learned behaviors into innate behaviors.

Example 1

Although evolution of behavior of unicellulars is beyond the scope of this work, an experimental example of transmission of a learned behavior into an innate behavior in a unicellular, involving no change in genes, bears some relevance to the study of the possibility of transformation of learned behavior into an innate behavior. For, if one believes in the origin of multicellulars from unicellulars, then it is beyond imagination, and incompatible with the principles of organic evolution, to think that evolution would have lost or not used such a highly adaptive property (transmission of learned behavior to offspring) in organisms standing higher on the evolutionary ladder.

In 1971, S.R. Bergstrom reported his experiments on learning and transmission of learned behavior in a unicellular. Those experiments were carried out under strictly controlled conditions in *Tetrahymena*, a freshwater ciliated Protozoa, 0.2 mm long. Normally, the microorganism does not react to the light, but it avoids electric impulses. In order to learn *Tetrahymena* to avoid the light, both light flashes and electric current impulses were simultaneously applied. After a training period of time, the microorganism learned to associate the light with electric impulses, so that it was able to display avoiding behavior to the light even when the light was applied alone, not combined, with an electric impulse. The most important observation was that after cell division, daughter cells of the animals which learned to avoid the light displayed the same avoiding behavior when exposed to light alone (Bergstroem, 1970).

Bergstroem's experiments on *Tetrahymena* demonstrated that the unicellular organism transmitted to the first-generation offspring (two daughter cells), a new character, without any change in genes.

Nevertheless, one must be wary of premature generalizations. Indeed, there is a strong argument against such a generalization: in a unicellular whose reproduction is based on mitotic division, there are no great barriers for the transformation of a learned behavior into an innate behavior; the acquired epigenetic structure (it cannot be a gene mutation, since no particular gene mutation would occur so systematically in whole populations) that makes this transformation possible would be easily divided between both daughter cells. Unlike this, in a sexually reproducing multicellular animal, a similar transformation of a learned behavior into an innate behavior would require that the germ cell(s) somehow inherit the epigenetic information for that behavior rather than the epigenetic structure *per se*.

Fortunately, examples of unambiguous sudden evolution of instincts in metazoan organisms, however scarce, are also known.

Example 2

When the cane toad *Bufo marinus* was first introduced to Australia, it was so toxic to the Australian black snake, *Pseudechis porphyriacus*, that ingestion of even a small toad was lethal to the snake. Now, about 60 years (~23 generations) after introduction of the toad on the continent, the snake has evolved an *innate* avoidance behavior toward the toxic cane toad (Phillips and Shine, 2006).

> *The shift in prey preference indicates either a congenital disposition to avoid toads or an evolved ability to learn from a single noxious encounter.*
>
> *Phillips and Shine (2006)*

Example 3

The solitary sedentary form of the locust *Schistocerca gregaria* does not practice flying. However, when forced to live in sites crowded with conspecifics, the locust changes its behavior from solitary to gregarious, preferring to live in crowd and fly over distances with other locusts of its species. This behavioral transformation is related to perception by the locust, under conditions of crowding, of the aggregation pheromone, which binds specific proteins on dendrites of the olfactory receptor neurons. Perception of aggregation pheromone in the brain induces mutual attraction of locusts, while cuticular hydrocarbons perceived via antennal olfactory neurons stimulate insects' activity and group formation. The behavioral change is correlated with secretion by the brain of a factor (*agoratropic factor*) whose injection in locusts also induces transition from the solitary to the gregarious behavior. Transition to the larval gregarious behavior occurs within 0.5–4 h, but the reverse transition to the solitary behavior is slower (Applebaum and Heifetz, 1999).

Phase transition is followed by specific changes in several morphological, morphometrical, and physiological characters, and, what is even more surprising, *the new, learned behavior of living in group and flying is maternally transmitted to the offspring*, together with all the morphophysiological characters acquired by the mother. Although the acquisition of this innate behavior may be reversed to the original if the locust would be exposed to respective conditions, it clearly shows that a parental *learned behavior* is passed on to the offspring in an *innate behavior* within a single generation, without changes in genes.

Example 4

The predatory snake, *Natrix maura*, was first introduced to the Spanish Mediterranean island of Mallorca by Romans about 2,000 years ago as part of an ancient Roman fertility ritual. Ever since, the native Mallorcan midwife toad, *Alytes muletensis*, inhabiting natural ponds, has evolved an adaptive *innate* behavior of suppressing its movement upon visually detecting the presence of the predator snake. Moreover, it has been shown experimentally that the toad displays the same adaptive behavior even when it does not see the predator but only perceives the chemical cues the snake releases in the water, at the same time that it does not respond to chemical cues of other midwife toad-eating snakes of the mainland Spain (Griffiths et al., 1998).

It may be assumed that introduction by ancient Romans of the predator snake in Mallorca added a threat and an environmental stress to the population of the native *A. muletensis*, which first learned that by suppressing its movements it becomes less visible to the predator. Later it might have learned to relate the presence of the predator snake to the chemicals it releases in the environment. (That it has *learned* to recognize chemical cues of the predator is proven by the fact that the toad does not react to chemical cues of other nonpredator snakes.) Now, both initially learned behaviors have evolved into innate behaviors: Mallorcan midwife toads are born with the instinct of "freezing" in the presence of the predator. There is no indication or evidence that any changes in genes might have been involved in evolution of this innate behavior.

There is no reason to believe that the above examples (as well as some other examples from insects not presented here) of transformation of learned behaviors into innate behaviors may be "exception to the rule"; for from an evolutionary point of view, i.e., from the point of view of the advantages it offers, once evolved in unicellulars and multicellulars, the ability to transform a learned behavior into an innate behavior would have been conserved in multicellulars.

Example 5

Licking and grooming (LG) and arched back nursing (ABN) of pups is an innate behavior in rats. Offspring of high LG–ABN dams tend to display the same high LG–ABN toward their offspring, and the offspring of the low LG–ABN display low LG–ABN. However, the offspring of the high LG–ABN mothers display low LG–ABN toward their offspring when reared by low LG–ABN mothers in the first postnatal week, and the offspring of low LG–ABN mothers switch to high LG–ABN if reared by high LG–ABN dams. Thus, the offspring "learn" from the adopted mother and transmit to the offspring as an innate behavior the LG–ABN behaviour, which is different from that of their biological mother.

Essentials of the mechanism of transformation of the learned behavior in an innate behavior are known. Processing of the tactile stimuli of the maternal LG stimulates serotonin secretion by particular hippocampal neurons, which induces there the synthesis of the protein NGFI-A, which by demethylating the glucocorticoid receptor (GR) gene stimulates its acetylation and increased expression of GR (Meaney and Szyf, 2005). Binding of adrenal glucocorticoids to GR in hippocampus and other brain centers inhibits expression of corticotropin-releasing factor (CRF) (Zhang and Meaney, 2010; Caldji et al., 2011) (Figure 8.3) and sequentially glucocorticoid secretion by adrenal glands. Lower levels of corticosteroids moderate pups' response to acute stress and determine high LG–ABN behavior. The reverse occurs with pups reared by low LG–ABN mothers.

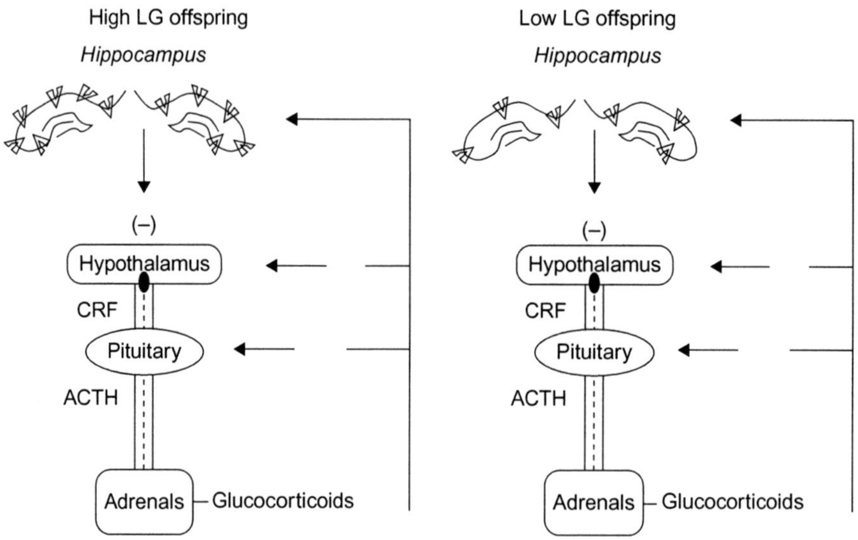

Figure 8.3 A schema outlining the function of the HPA axis, the nexus of which are the CRF neurons of the paraventricular nucleus of the hypothalamus. CRF is released into the portal system of the anterior pituitary stimulating the synthesis and release of adrenocorticotrophin (ACTH), which then stimulates adrenal glucocorticoid release. Glucocorticoids act on GRs in multiple brain regions, including the hippocampus, to inhibit the synthesis and release of CRF (i.e., glucocorticoid negative feedback). The adult offspring of high LG mothers, by comparison to those of low LG dams, show (1) increased GR expression, (2) enhanced negative-feedback sensitivity to glucocorticoids, (3) reduced CRF expression in the hypothalamus, and (4) more modest pituitary–adrenal responses to stress (references provided in text).
Source: From Zhang and Meaney (2010).

Behavioral Atavisms—Activation of Ancestral Behavioral Circuitries

Essentially, evolution of behavior in metazoans is evolution of the structure and connectivity of neural circuits. We have shown that the same behavior may be determined by distinct neural circuits, and the same circuit may be modulated to produce different elementary behaviors (FAPs).

All innate behaviors (e.g., avoiding behavior, mating behavior, courting behavior) result from activation of specific neural circuits. Under changed conditions of living, the animal may be forced to change a particular behavior and later, in the course of its phylogeny, the organ(s) performing the behavior.

Let us consider a hypothetical gradual transformation of a terrestrial habitat into an aquatic habitat. Under such conditions, walking will gradually become impossible, and terrestrial vertebrates will be forced to swim. As for the possibility of activation of a swimming motor pattern, recall that studies in invertebrates have shown that the same CPG might serve several different locomotory behaviors, such as swimming, crawling, and burrowing, implying that, from an evolutionary view, the reverse transition of burrowing and crawling vertebrates into an aquatic swimming-requiring habitat would not be impossible. Swimming circuits are still functional in most terrestrial vertebrates as suggested by the fact that most of them can swim when forced to do so. The stressfully changed conditions in the new aquatic environment may only stimulate their modification for better swimming.

When some 380 million years ago an adventurous fish decided (or was forced) to explore the land, it found locomotion in the terrestrial environment very difficult, but its fins, to some extent, might have supported its crawling movements. In experiments (Ayers et al., 1983), the isolated lamprey spinal cord bathed in D-glutamic acid (an amino acid that also serves as a neurotransmitter) generated a motor pattern that has been assumed to represent the central motor program underlying swimming, but their analysis shows that undulations produced by exposure of the spinal cord to D-glutamate solution are different from those observed during normal behavior, and the investigators believe that this central motor program might represent

> *a fundamental undulatory pattern that is modulated by different descending systems to produce the complete undulatory behavioral repertoire.*
>
> *Ayers et al. (1983)*

Thus, behaviors mediated by front-to-rear lateral undulations, including swimming, burrowing, and crawling movements of ancestral lampreys, may be regulated by a single motor pattern. The same central motor program might have been used by land-exploring fish to switch from swimming behavior to crawling behavior and holding up their body. It must have been a very painful journey, but nevertheless extremely beneficial and rewarding. The innate half-crawling locomotion of the first colonizer of land might have started as a learned behavior, based on the swimming central motor program.

Limbs of terrestrial vertebrates, as we see them currently, have lost the hydrodynamic features of appendages of aquatic vertebrates, such as interdigital membranes that are still used by aquatic birds and frogs for paddling in water. Nevertheless, there is evidence that having lost these ancestral webbed feet, many mammals have conserved not only the swimming circuit and swimming behavior but also the developmental pathways for developing webbed feet. So, for example, embryos of all the vertebrate terrestrial classes, including us humans, still develop webbed feet (with interdigital membranes), which are later apoptotically eliminated at various stages of embryogenesis.

Let us also remember that the evolution of webbed feet, while adapting them for swimming, would not prevent animals from walking, as can be seen in modern ducks and other webbed-feet birds. Terrestrial mammals that had to adapt to aquatic conditions of life stepwise returned to ancestral webbed feet by simply preventing apoptosis of interdigital membranes they develop as embryos. In all likelihood, this is what is actually happening in mammals such as otters and sea lions that seem to be in a transition stage of adaptation to aquatic habitat.

Reversal of lost ancestral behaviors is not only a theoretical possibility. Konrad Lorenz obtained a hybrid duck by crossing two races, Chiloe teals and Bahama pintails. To his surprise, he observed that elements of the courtship display of the hybrid duck resembled those of neither of two parental species, but a primitive precursor of both parental species. According to Gould, the ancestral behavior had not been replaced, but simply repressed (Gould, 1982, pp. 302–303).

Another experimental example of reversion of an ancestral behavioral response: when female guppies of the species *Poecilia reticulata* visually detect the presence of their cichlid predator, *Cichlasoma biocellatum*, in the environment, they reverse to the ancestral preference for duller instead of bright-colored males (Gong and Gibson, 1996).

In the course of their phylogeny, animals have shifted to different habitats. In the process of evolving new behaviors and structures, animals lose previous behaviors that are no longer adaptive to the new habitat, but they conserve the circuits for the "lost" behaviors. Species that happen to return to ancestral or quasi-ancestral habitats might activate the conserved ancestral circuit and reverse to the lost ancestral innate behavior.

This is not a purely speculative idea. There is evidence that, while losing particular behaviors and even the structures performing these behaviors, metazoans retain the structural basis of the "lost" behaviors. The flightless grasshopper, *Barytettix psolus*, and *Schistocerca* locusts are in possession of two similar large interneurons, the descending contralateral movement detector (DCMD) and the tritocerebral commissure giant (TCG). These interneurons, which are homologous in both species, enable locusts to fly, while the grasshopper cannot. It has been observed that differences exist only in the connections made by a single first-order axonal branch of the DCMD interneuron with the flight motoneurons; while in locusts, the DCMD sends branches to the dorsolateral neuropile and forms synapses with flight motor neurons, such branches form only in 52% of cases in the grasshopper, and all of the connections have abnormal projections in comparison with locusts:

> *The differences in DCMD projection suggest that a discrete set of output connections may have been modified in Barytettix by the alteration of a single first-order axonal branch.*
>
> *Arbas (1983)*

The grasshopper, *Barytettix psolus*, lost the ability to fly as a result of the loss of hind wings, reduction of immobile forewings to a vestigial state, and the loss of the indirect flight muscle, the metathoracic dorsal longitudinal muscle, which develops in the nymphal stage but is lost in adult grasshoppers. However, the motor neurons for flight muscles and wings are retained, although their target muscles are lost (Arbas, 1983; Arbas and Tolbert, 1986). The conservation of motor neurons that are directly related to the lost structures (muscles and wings) and functions (flight) suggests that, despite the changed connections, the flight circuitry in nonflying grasshoppers is preserved.

Here is another impressive illustration of the conservation of neural circuits in the course of the evolution of metazoans. Innate fear of snakes is common not only among humans but among other primates as well. Eleven species of primates exhibit fear-related responses (e.g., avoidance, alarm calls, mobbing) in virtually all instances in which they were observed confronting large snakes (Oehman and Mineka, 2003). Limbic structures, related to the snake-fear neural circuits, emerged during the evolutionary transition from reptiles to mammals (first mammals were small insectivorous tetrapods) but before the evolution of neocortex, as can be concluded by the irrationality of the snake fear observed under natural and experimental conditions and the immodifiability of the snake aversiveness. Although the snake-fear circuit evolved long before the evolution of man, during transition from reptilians to mammals, the circuit still exists in most mammals but has been later modified in mammals that preyed on reptiles, including snakes (Oehman and Mineka, 2003). For more than 200 million years, the "snake-fear" circuit is still functional in the rest of mammals!

The fact that the structure of the nervous system in metazoans and even specific neural circuits are conserved to a great extent represents a crucial premise of the evolutionary ability of metazoans to revert to ancestral behaviors and may be an important asset of the metazoan evolutionary-adaptive strategy.

If we should believe, as most biologists do, that the swim bladder in teleost fish evolved from the lung of ancestral lung fish, the breathing CPG must have been modified when they substituted the swim bladder for the lung. The reevolution of lungs in tetrapods required reinvention of the lost circuitry for breathing for replacing the "buoying" motor pattern used for controlling the water depth-graded swimming by inflating and/or deflating the swim bladder in teleost fish. In the Urtetrapod, this might have required just reactivation of an ancestral silenced "breathing" circuitry for terrestrial life. Recall that even the same circuit in metazoans (e.g., *Aplysia* circuitry) may be modified to generate more than one motor pattern.

Developmental and Evolutionary Relationship Between Behavior and Morphology

Each phenotypic structure is related to one or a number of innate behaviors: a fin to swimming, a wing to flying, a digestive tract to eating, a lung to breathing, secondary sexual traits—the sexual behavior of males and females, and many other examples. For, ultimately, the structure evolves not for its own sake but essentially for the sake of the function, of the behavior it must perform. As Evans put it:

> *One cannot intelligently discuss behavior and structure separately. Behavior is what an animal does with its structure; structure is what an animal uses to behave.*
>
> *Evans (1966)*

Behavior is the ultimate cause of the structure. In the first part of this work, I have presented empirical evidence that the CNS and neural circuits are essentially involved in the development of morphological, physiological, and behavioral characters. The question now arises whether neural circuits for behavioral and morphophysiological characters may be developmentally and evolutionarily related.

Adaptation of metazoans to sudden environmental changes and the ensuing environmental stress begins with adaptive changes in behavior, which precede all other forms of the phenotypic (morphological, physiological, and life history) adaptation.

In view of the fact that, ultimately, the information for both performing behaviors and developing animal morphology comes from the nervous system, it is to be expected that, in the course of evolution, causal relationships might have been established between behavioral circuits and circuits involved in the development of morphological traits. Is there evidence of a causal or noncausal relationship between the evolution of behavior and the evolution of metazoan morphology?

If evolutionary change is transmission to the offspring of a character that the parents have not inherited but have acquired during their lifetime, or that appears for the first time in the offspring, modern biology offers a considerable number of demonstrated and demonstrable cases of evolutionary change, which now are the object of an interesting field of the study, the transgenerational developmental plasticity. Based on the fact that such cases meet the basic criterion of the evolutionary change, i.e., transmission of the new character to the offspring, identification of the mechanisms of their emergence may provide important clues to understanding mechanisms of metazoan evolution.

Lex parsimoniae tells us that there is no reason to suspect that the mechanisms of transgenerational developmental plasticity might be different from mechanisms of long-term evolutionary changes. Evolution is too economical to waste resources for the luxury of evolving two different mechanisms for a single end, i.e., generation of inherited changes.

It is generally believed that evolution of behavior precedes evolution of the structure for performing the behavior. In an overused aphorism, "Behavior evolves first."

From an evolutionary point of view, a causal relationship between evolution of behavior and morphology that might arise from performing the new behavior would clearly be advantageous.

Evidence of a close relationship between evolution of behavior and animal morphology and physiology has been presented earlier by a number of authors. In experiments on functional mechanisms of predator-induced changes in morphology and behavior of *Hyla versicolor* tadpoles, Van Buskirk and McCollum (2000) have observed that changes in behavior, on the one hand, and the color and relative length and depth of tadpole body and tail, on the other, vary as an integrated unit, and they conclude that behavior, color, and morphology are highly correlated in naturally occurring tadpoles. Fuchs et al. also have described the existence of a relationship between the behavior and morphological and physiological changes and have pointed out the role of behavior in inducing physiological changes in the case of phase transition in locusts:

> *Locusts are capable of extreme behavioral plasticity; in response to changes in population density, they dramatically alter their behavior. These changes in behavior facilitate the appearance of various morphological and physiological changes, cumulatively termed density-dependent phase characteristics... the behavioral changes are, on the one hand, a response to specific environmental changes, and on the other, stimulant-catalysts of various other environmentally induced physiological changes.*
>
> *Fuchs et al. (2003)*

Theoretically, it might be argued that the experimentally confirmed correlation between changes in behavior and morphology is inherently determined by the fact that morphologies in general are means for performing specific functions and behaviors, such as feeding; preying; hiding; or aquatic, air, or terrestrial locomotion (swimming, flying, or walking).

As shown, I cannot claim to be the first to have presented evidence on the existence of a close relationship between behavior and animal morphology and physiology. What I claim here, instead, is that adequate empirical evidence exists for validating my hypothesis on the existence of a causal relationship in the evolution of metazoan behavior and morphology.

It was pointed out earlier (and will be discussed in some detail in Chapters 10 and 11) that, in response to specific stimuli, locusts of the species *Schistocerca gregaria* (Forskål) switch between two behaviorally and morphologically distinct forms in a phenomenon known as *phase transition*. The solitary form, which lives isolated, away from other conspecifics, when put under crowding conditions or under influence of pheromones or tactile stimuli, switches to the gregarious form, which not only displays several new behavioral traits (e.g., tendency to swarm and fly with locust crowds, to feed on a toxic alkaloid-containing plant that they used to avoid before (Despland and Simpson, 2005) but also exhibits several changes in morphology, morphometry, and body coloration. Behavioral changes may appear within one to several hours, and they precede the morphological, morphometric, and color changes during phase transition.

All the phase change-inducing factors act via the insect CNS. (Crowding in this insect is a stressor that also acts via the CNS, as is concluded by the experimental evidence that antennectomized locusts do not change phase under conditions of crowding (Applebaum and Heifetz, 1999).) Moreover, the stressed locusts transmit the acquired traits to the offspring. The full-scale phase transformation takes several generations and occurs probably only in nature (Pener et al., 1997).

At a neuroendocrine level, this transformation is related to an elevated level of juvenile hormone (JH) under stimulation of neurohormones allatotropins and nerves innervating the corpora allata as well as cerebral secretion of [His^7]-corazonin, also known as *dark-color-inducing neurohormone* (DCIN) (Grach et al., 2003). Intense changes are also observed in the levels of numerous neurotransmitters in the locust brain (Rogers et al., 2004).

All of the changes during transition to the gregarious phase are triggered by sensory stimuli (visual, olfactory, and tactile), which are perceived in the insect brain where the information for activating the signal cascade for changes in behavior and morphology is generated by processing the afferent neural input from sensory neurons. Aggregation pheromones received by the olfactory neurons are converted into electrical spike trains, in which form they are transmitted for processing, first to the frontal antennal lobe and then to the mushroom body, and further to the lateral protocerebrum (Anton and Hansson, 1996). Tactile stimuli (touch on the outer side of the upper portion of a hind leg, for example) from mechanosensory trichoid sensillae on the hind limb, via metathoracic nerve 5, are also transmitted to the CNS (Rogers et al., 2003). At a neurobiological level the phase transition this is determined by "a substantial but transient (<24 h) increase in the amount of serotonin" in the CNS (thoracic ganglia) (Anstey et al., 2009; Figure 8.4).

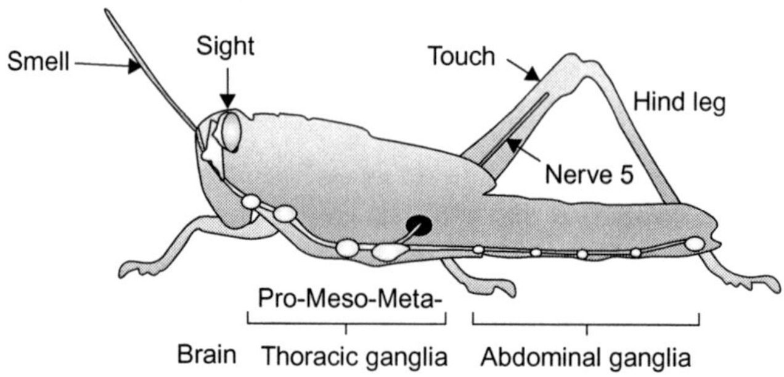

Figure 8.4 The CNS of a locust consists of the brain, which receives olfactory and visual information, and a chain of segmental ganglia linked by paired connectives. The three thoracic ganglia receive direct mechanosensory and proprioceptive inputs from the legs.
Source: From Anstey et al. (2009).

Transmission of the gregariousness to the offspring is correlated with deposition of a five- to tenfold greater amounts of ecdysteroids in the eggs of the gregarious locusts than in the eggs of the solitarious locusts (Tawfik et al., 1999; Tawfik and Sehnal, 2003; Hägele et al., 2004).

The facts that the circuitry for gregarious behavior and circuitries for gregarious morphology in locusts are activated by the same stimulus, and that changes in behavior are always followed by changes in morphology and body coloration, suggest that at some level of the brain function or structure, behavioral circuits are related to circuitries that, via signal cascades, determine the development of gregarious morphologies.

The temporal correlation between changes in behavior and changes in morphology observed during phase transition in locusts is not unique. Examples of such correlations abound in the field of developmental plasticity. In most cases of predator-induced developmental plasticity, changes in morphology are also preceded by, or accompanied with, changes in behavior. For example, larvae of the pipevine swallowtail butterfly, *Battus philenor*, show a phenotypic plasticity in the Southwest of the United States: in California, they are predominantly black, while in western Texas and Arizona, predominantly red. Recently, investigators have observed that California butterfly larvae, in an adaptive response to the higher summer temperature, exhibit a double (behavioral and morphological) phenotypic plasticity. In order to avoid the excessive summer heat, they switch to a new climbing behavior by climbing higher on nonhost plants and change their body color from black to red (Figure 8.5). These changes are adaptive, for both color change and climbing allow the larvae to escape the higher temperatures. The critical temperature for the onset of the polyphenism lies between 30°C and 36°C, and the polyphenism is reversible. Both red color and climbing behavior are components of a thermoregulatory strategy intended

> *to avoid internal temperatures above the thermal maximum temperature for growth and development in B. philenor or to maintain body temperatures in the optimum range for facilitating maximum growth rate... The maintenance of maximum growth rate may be critical for insect larvae susceptible to larval predators or parasites.*
>
> *Nice and Fordyce (2006)*

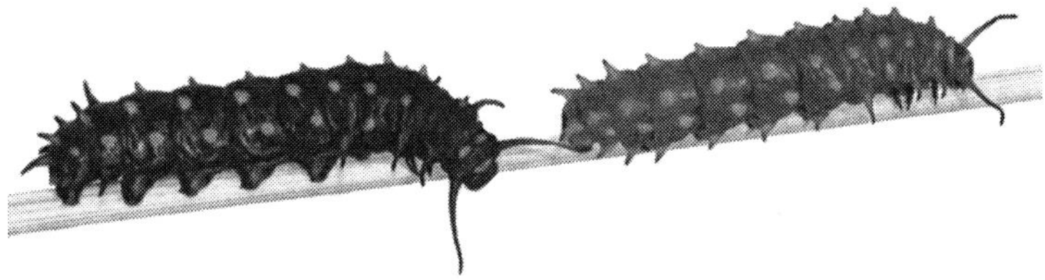

Figure 8.5 The black (left) and red larval phenotypes of *B. philenor* observed in two half-siblings from Texas. The black larva was reared at 30°C, and the red larva was reared at 36°C. (For interpretation of the references to color in this figure legend, the reader is referred to the web version of this book.)
Source: From Nice and Fordyce (2006).

The systematic correlation of the change in climbing behavior with the change in body color suggests the existence of a causal relationship between them.

The neotropical tadpole, *Rana palmipes*, in response to the presence of its predator water bug, or even of predator cues alone, changes its behavior by strongly reducing its activity, darkening its body color, and increasing the size of its muscles and tail (McIntyre et al., 2004).

In response to the presence of its predators, the freshwater snail, *Helisoma trivolis*, simultaneously changes its behavior (preference for a particular habitat and the timing of the onset of the reproductive behavior) and morphology (the form of the shell) (Hoverman et al., 2005).

Acyrthosiphon pisum (Harris, 1776) is a pea aphid that in the presence of predators emits a volatile alarm pheromone, which, when perceived in the brains of females, induces the latter not only to shift to walking behavior and drop off the plants but also to increase the proportion of winged morphs in the offspring (Dixon and Agarwala, 1999; Kunert and Weiser, 2003).

North American frogs of the genus *Scaphiopus* are omnivorous amphibians that, as tadpoles, inhabit ephemeral ponds and flooded areas, which only exist for short periods of time, often before the tadpoles could develop into adult terrestrial individuals. These species exhibit an adaptive strategy, a developmental plasticity that enables a proportion of tadpoles to develop an alternative carnivorous behavior and mouth morphology. According to Pfennig (1990), the proportion of tadpoles that develop carnivorous behavior and mouth morphology depends on the amount of shrimp they eat, and shrimps are more abundant in ephemeral ponds (Pfennig, 1990). The carnivorous tadpole morphology is similar to mouth adaptations of *Hoplobatrachus* tadpoles. Tadpoles of both groups have longer intestines than those of other carnivorous species (Grosjean et al., 2004). All these facts support the hypothesis that carnivorous behavior and mouth morphology in anuran tadpoles evolved in correlation, as an adaptation to the temporal unpredictability of desiccation of the pond. Tadpoles of desert amphibians live in temporary ponds that contain water for unpredictable periods of time. In the years of low precipitation, the pond dries up earlier than usual. This causes a habitat stress to which the tadpoles of that and some other species respond by changing their behavior and speeding up their metamorphosis to transform into adult amphibians, able to live on dry land.

I have already mentioned the example of the Mallorcan midwife toad, *Alytes muletensis*, which, in response to the presence of its viperine predator, and even upon detecting a chemical released by the predator, induces rapid changes in its behavior and later changes in morphology, which make the toad less vulnerable to the snake.

Konrad Lorenz has shown that birds that make nodding movements while courting eventually develop highly colored feathers or crests, which draw attention to these movements—not the reverse (Taylor, 1983, p. 218).

It is believed that a tendency to reptate (wriggle), instead of walking on their reduced limbs, which is observed in some lizards, is an indication that this "wriggling" behavior is causally related to the reduction of their limbs and may predict their future evolutionary loss (Taylor, 1983, p. 219).

The systematic correlation of specific behavioral changes with specific changes in morphology in all of the above cases strongly suggests the existence of a causal relationship between the evolution of the behavior and the structure(s) performing it.

What seems to have in common in all the examples of correlated change in behavior and morphology is the fact that these changes are stimulated by drastic changes in the environment, which trigger a stress response, a response that, as has been shown, is neurally determined. The immediate change in behavior often is itself an integral part of the stress response. As will be later shown, the stress response leads to developmental instability that is an important permissive factor for ensuing morphophysiological changes.

References

Alcock, J. (1989). *Animal Behavior*, fourth edition. Sinauer Associates Inc. Publishers, Sunderland, MA, p. 46.

Anstey, M.L., Rogers, S.M., Ott, S.R., Burrows, M. and Simpson, S.J. (2009). Serotonin mediates behavioral gregarization underlying swarm formation in desert locusts. Science *323*, 627–630.

Anton, S. and Hansson, B.S. (1996). Antennal lobe interneurons in the desert locust *Schistocerca gregaria* (Forskål): processing of aggregation pheromones in adult males and females. J. Comp. Neurol. *370*, 85–96.

Applebaum, S.W. and Heifetz, Y. (1999). Density-dependent physiological phase in insects. Annu. Rev. Entomol. *44*, 317–341.

Arbas, E.A. (1983). Neural correlates of flight loss in a Mexican grasshopper, *Barytettix psolus*. II. DCMD and TCG interneurons. J. Comp. Neurol. *216*, 381–389.

Arbas, E.A. and Tolbert, L.P. (1986). Presynaptic terminals persist following degeneration of "flight" muscle during development of a flightless grasshopper. J. Neurobiol. *17*, 627–636.

Arnegard, M.E., Bogdanowicz, S.M. and Hopkins, C.D. (2005). Multiple cases of striking genetic similarity between alternative electric fish signal morphs in sympatry. Evolution *59*, 324–343.

Ayers, J., Carpenter, G.A., Currie, S. and Kinch, J. (1983). Which behavior does the lamprey central motor program mediate? Science *221*, 1312–1314.

Baker, B.S., Taylor, B.J. and Hall, J.C. (2001). Are complex behaviors specified by dedicated regulatory genes? Reasoning from *Drosophila*. Cell *105*, 13–24.

Balaban, E. (1997). Changes in multiple brain regions underlie species differences in a complex, congenital behavior. Proc. Natl. Acad. Sci. U.S.A. *94*, 2001–2006.

Bergstroem, S.R. (1970). Lernen bei Einzellern. Bild der Wissenschaft *7*, 687–692.

Caldji, C., Hellstrom, I.C., Zhang, T.-Y., Diorio, J. and Meaney, M.J. (2010). Environmental regulation of the neural epigenome. FEBS Lett. (2011 March 18) [Epub ahead of print].

Croll, R.P., Davis, W.J. and Kovac, M.P. (1985). Neural mechanisms of motor program switching in the mollusc *Pleurobranchaea*. I. Central motor programs underlying ingestion, egestion, and the "neutral" rhythm(s). J. Neurosci. *5*, 48–55.

Curtis, J.T., Liu, Y., Aragona, B.J. and Wang, Z. (2006). Dopamine and monogamy. Brain Res. *1126*, 76–90.

Darwin, C.R. (1859). *The Origin of Species by Means of Natural Selection or The Preservation of Favored Races in the Struggle for Life*, first edition. John Murray, London (1st issue), pp. 208, 209, 474.

Darwin, C.R. (1874). *The Descent of Man and Selection in Relation to Sex* 1909, second edition, revised and augmented. D. Appleton and Co., New York, NY, p. 629.

Delcomyn, F. (1998). *Foundations of Neurobiology*. W.H. Freeman and Co., New York, NY, p. 602.

Delcomyn, F. and Prosser, C.L. (1991). Central nervous system, section A. In: *Neural and Integrative Animal Physiology*, C.L. Prosser (ed.), fourth edition. Wiley-Liss, New York, pp. 547–554.

Demir, E. and Dickson, B.J. (2005). *Fruitless* splicing specifies male courtship behavior in *Drosophila*. Cell *121*, 785–794.

Despland, E. and Simpson, S.J. (2005). Surviving the change to warning colouration: density-dependent polyphenism suggests a route for the evolution of aposematism. Chemoecology *15*, 69–75.

Dixon, A.F.G. and Agarwala, B.K. (1999). Ladybird-induced life-history changes in aphids. Proc. Roy. Soc. B Biol. Sci. *266*, 1549–1553.

Dulac, C. (2005). Sex and the single splice. Cell *121*, 664–666.

Evans, H.E. (1966). *The Comparative Ethology and Evolution of the Sand wasps*. Harvard University Press, Cambridge, p. 2, Quoted by Wcislo, W.T. (1989). Behavioral environments and evolutionary change. Annu. Rev. Ecol. Syst. *20*, 137–169.

Fan, R.-J., Marin-Burgin, A., French, K.A. and Friesen, W.O. (2005). A dye mixture (Neurobiotin and Alexa 488) reveals extensive dye-coupling among neurons in leeches; physiology confirms the connections. J. Comp. Physiol. A *191*, 1157–1171.

Frost, W.N., Hoppe, T.A., Wang, J. and Tian, L.-M. (2001). Swim initiation neurons in *Tritonia diomedea*. Am. Zool. *41*, 952–961.

Fuchs, E., Kutsch, W. and Ayali, A. (2003). Neural correlates to flight-related density-dependent phase characteristics in locusts. J. Neurobiol. *57*, 152–162.

Gillette, R. (1991). Central nervous system, section D. In: *Neural and Integrative Animal Physiology*, C.L. Prosser (ed.)., fourth edition. Wiley-Liss, New York, pp. 574–611.

Gong, A. and Gibson, R.M. (1996). Reversal of a female preference after visual exposure to a predator in the guppy, *Poecilia reticulate*. Anim. Behav. *52*, 1007–1015.

Gorbman, A. and Davey, K. (1991). Endocrines. In: *Neural and Integrative Animal Physiology*, C.L. Prosser (ed.), fourth edition. Wiley-Liss, New York, pp. 693–754.

Gould, J.L. (1982). *Ethology—The Mechanisms and Evolution of Behavior*. W.W. Norton & Co., pp. 274, 261, 177, 176, 302–303.

Grach, C., Wang, Y., Barda, Y., Gilon, C. and Pener, M.P. (2003). *In vivo* structure–activity studies on the dark-color-inducing neurohormone of locusts. J. Peptide Res. *62*, 135–142.

Griffiths, R.A., Schley, L., Sharp, P.E., Dennis, J.L. and Roman, A. (1998). Behavioural responses of Mallorcan midwife toad tadpoles to natural and unnatural snake predators. Anim. Behav. *55*, 207–214.

Grosjean, S., Cences, M. and Dubois, A. (2004). Evolutionary significance of oral morphology in the carnivorous tadpoles of tiger frogs, genus *Hoplobatrachus* (Ranidae). Biol. J. Linn. Soc. *81*, 171–181.

Hägele, B.F., Wang, F.H., Sehnal, F. and Simpson, S.J. (2004). Effects of crowding, isolation, and transfer from isolation to crowding on total ecdysteroid content of eggs in *Schistocerca gregaria*. J. Insect Physiol. *50*, 621–628.

Heiligenberg, W., Metzner, W., Wong, C.J.H. and Keller, C.H. (1996). Motor control of the jamming avoidance response of *Apteronotus leptorhynchus*: evolutionary changes of a behavior and its neuronal substrates. J. Comp. Physiol. A *179*, 653–674.

Henson, S.A. and Warner, R.R. (1997). Male and female alternative reproductive behaviors in fishes. Annu. Rev. Ecol. Syst. *28*, 571–592.

Hoverman, J.T., Auld, J.R. and Relyea, R.A. (2005). Putting prey back together again: integrating predator-induced behavior, morphology, and life history. Oecologia *144*, 481–491.

Iyengar, S., Viswanathan, S.S. and Bottjer, S.W. (1999). Development of topography within song control circuitry of zebra finches during sensitive period for song learning. J. Neurosci. *19*, 6037–6057.

Katz, P.S. and Frost, W.N. (1996). Intrinsic neuromodulation: altering neuronal circuits from within. Trends Neurosci. *19*, 54–61.

Katz, P.S. and Frost, W.N. (1997). Removal of spike frequency adaptation via neuromodulation intrinsic to the *Tritonia* escape swim central pattern generator. J. Neurosci. *17*, 7703–7713.

Katz, P.S. and Harris-Warrick, R.M. (1999). The evolution of neuronal circuits underlying species-specific behavior. Curr. Opini. Neurobiol. *9*, 628–633.

Kravitz, E.A. (1988). Hormonal control of behavior. Science *241*, 1775–1781.

Kunert, G. and Weiser, W.W. (2003). The interplay between density- and trait-mediated effects in predatory-prey interactions: a case study in aphid wing polymorphism. Oecologia *135*, 304–312.

Long, K.D., Kennedy, G. and Balaban, E. (2001). Transferring an inborn auditory perceptual predisposition with interspecies brain transplants. Proc. Natl. Acad. Sci. U.S.A. *98*, 5862–5867.

Mani, S.K. (1994). Convergent pathways for steroid hormone- and neurotransmitter-induced rat sexual behavior. Science *265*, 1246–1249.

Marin-Burgin, A., Eisenhart, F.J., Kristan, W.B., Jr. and French, K.A. (2006). Embryonic electrical connections appear to prefigure a behavioral circuit in the leech CNS. J. Comp. Physiol. A *192*, 123–133.

Manoli, D.S., Meissner, G.W. and Baker, B.S. (2006). Blueprints for behavior: genetic specification of neural circuitry for innate behaviors. Trends Neurosci. *29*, 444–451.

McIntyre, P.B., Baldwin, S. and Flecker, A.S. (2004). Effects of behavioral and morphological plasticity. Oecologia *141*, 130–138.

Meaney, M.J. and Szyf, M. (2005). Maternal care as a model for experience-dependent chromatin plasticity? Trends Neurosci. *28*, 456–463.

Metzner, W. (1999). Neural circuitry for communication and jamming avoidance in gymnotiform electric fish. J. Exp. Biol. *202*, 1365–1375.

Meyrand, P. and Simmers, J. (1991). Construction of a pattern-generating circuit with neurons of different networks. Nature *351*, 60–63.

Moncho-Bogani, J., Lanuza, E., Hernández, A., Novejarque, A. and Martínez-García, F. (2002). Attractive properties of sexual pheromones in mice: innate or learned? Physiol. Behav. *77*, 167–176.

Moncomble, A.-S., Coureaud, G., Quennedey, B., Langlois, D., Perrier, G. and Schaal, B. (2005). The mammary pheromone of the rabbit: from where does it come? Anim. Behav. *69*, 29–38.

Nelson, D.A. (2000). A preference for own-subspecies' song guides vocal learning in a song bird. Proc. Natl. Acad. Sci. U.S.A. *97*, 13348–13353.

Nice, C.C. and Fordyce, J.A. (2006). How caterpillars avoid overheating: behavioral and phenotypic plasticity of pipevine swallowtail larvae. Oecologia *146*, 541–548.

Nishikawa, K.C. (2002). Evolutionary convergence in nervous systems: insights from comparative phylogenetic studies. Brain Behav. Evolut. *59*, 240–249.

Oehman, and Mineka, S. (2003). The malicious serpent: snakes as a prototypical stimulus for an evolved module of fear. Curr. Dir. Psychol. Sci. *12*, 5–9.

Pener, M.P., Ayali, A. and Golenser, E. (1997). Adipokinetic hormone and flight fuel related characteristics of density-dependent locust phase polymorphism: a review. Comp. Biochem. Phys. B *117*, 513–524.

Pfennig, D. (1990). The adaptive significance of an environmentally-cued developmental switch in an anuran tadpole. Oecologia *85*, 101–107.

Phillips, B.L. and Shine, R. (2006). An invasive species induces rapid adaptive change in a native predator: cane toads and black snakes in Australia. Proc. Roy. Soc B Biol. Sci. *273*, 1545–1550.

Piaget, J. (1976). *Biology and Knowledge*. The University of Chicago Press, Chicago, IL, p. 87, Quoted by Vidal, F., Buscaglia, M. and Vonèche, J.J. (1983). Darwinism and decvelopmental psychology. J. Hist. Behav. Sci. *19*, 81–94.

Prosser, C.L. (1991). Central nervous system. Section H. In: *Neural and Integrative Animal Physiology*, C.L. Prosser (ed.), fourth edition. Wiley-Liss, New York, pp. 678–691.

Purves, W.K., Orians, G.H. and Heller, H.C. (1992). *Life: The Science of Biology*, third edition. Sinauer Associates, Inc/W.H. Freeman and Company, Sunderland, MA, p. 984.

Rogers, S.M., Matheson, T., Despland, E., Dodgson, T., Burrows, M. and Simpson, S.J. (2003). Mechanosensory-induced behavioural gregarization in the desert locust *Schistocerca gregaria*. J. Exp. Biol. *206*, 3991–4002.

Rogers, S.M., Matheson, T., Sasaki, K., Kendrick, K., Simpson, S.J. and Burrows, M. (2004). Substantial changes in central nervous system neurotransmitters and neuromodulators accompany phase change in the locust. J. Exp. Biol. *207*, 3603–3617.

Rose, G.J. (2004). Insights into neural mechanisms and evolution of behaviour from electric fish. Nat. Rev. Neurosci. *5*, 943–951.

Ryner, L.C., Goodwin, S.F., Castrillon, D.H., Anand, A., Villella, A. and Baker, B.S., et al. (1997). Control of male sexual behavior and sexual orientation in *Drosophila* by the *fruitless* gene. Cell *87*, 1079–1089.

Stockinger, P., Kvitsiani, D., Rotkopf, S., Tirián, L. and Dickson, B.J. (2005). Neural circuitry that Governs *Drosophila* male courtship behavior. Cell *121*, 795–807.

Suesswein, A.J. and Byrne, J.H. (1988). Identification and characterization of neurons initializing patterned neural activity in the buccal ganglia in *Aplysia*. J. Neurosci. *5*, 48–55.

Tawfik, A.I., Vedrova, A. and Sehnal, F. (1999). Ecdysteroids during ovarian development and embryogenesis in solitary and gregarious *Schistocerca gregaria*. Arch. Insect Biochem. *41*, 134–143.

Taylor, G.R. (1983). *The Great Evolution Mystery*. Harper & Row Publishers, New York, NY, and London, pp. 218, 219

Tierney, A.J. (1995). Evolutionary implications of neural circuit structure and function. Behav. Process. *35*, 173–182.

Trimmer, B.A., Aprille, R., Dudzinski, D.M., Lagace, C.J., Lewis, S.M. and Michel, T., et al. (2001). Nitric oxide and the control of firefly flashing. Science *292*, 2486–2488.

Uller, T. and Olsson, M. (2006). Direct exposure to corticosterone during embryonic development influences behaviour in an ovoviviparous lizard. Ethology *112*, 390–397.

Van Buskirk, J. and McCollum, S.A. (2000). Functional mechanisms of an inducible defence in tadpoles: morphology and behaviour influence mortality risk from predation. J. Evolution. Biol. *13*, 336–347.

Voneida, T.J. and Fish, S.E. (1984). Central nervous system changes related to the reduction of visual input in a naturally blind fish (*Astyanax hubbsi*). Am. Zool. *24*, 775–782.

Wcislo, W.T. (1989). Behavioral environments and evolutionary change. Annu. Rev. Ecol. Syst. *20*, 137–169.

Whaling, C.S., Solis, M.M., Doupe, A.J., Soha, J.A. and Marler, P. (1997). Acoustic and neural bases for innate recognition of song. Proc. Natl. Acad. Sci. U.S.A. *94*, 12694–12698.

Zhang, T.-Y. and Meaney, M.J. (2010). Epigenetics and the environmental regulation of the genome and its function. Annu. Rev. Psychol. *61*, 439–466.

Zysling, D.A., Greives, T.J., Breuner, C.W., Casto, J.M., Demas, G.E. and Ketterson, E.D. (2006). Behavioral and physiological responses to experimentally elevated testosterone in female dark-eyed juncos (*Junco hyemalis carolinensis*). Horm. Behav. *50*, 200–207.

9 Ontogeny: The Workshop of Evolutionary Change

Ontogenies evolve, not genes or adults.

McKinney and Gittleman (1995)

Sexual reproduction is by far the predominant form of reproduction in metazoans. Parthenogenesis is a less-frequent phenomenon, sometimes alternated with sexual reproduction. Asexual reproduction by budding is restricted to a limited number of primitive invertebrates like hydra.

In distinction from unicellulars which, via binary fission, produce "copies" of themselves (two daughter cells), metazoans, due to constraints imposed by the complexity of their structure, cannot produce copies of themselves and instead produce unicellular structures, egg and sperm cells, capable of independently developing into adult organisms of their kind.

While unicellulars rely for their reproduction on a template, Watson–Crick, form of heredity (the parental organism replicates itself), metazoan heredity is of a communicative type. Metazoans do not build the structure of the young but only communicate, via the gametes, epigenetic information that enables the zygote (the egg cell in parthenogenetic organisms) to develop up to the phylotypic stage, when the embryo is capable of generating itself the epigenetic information for the rest of individual development.

The development of the zygote (the egg cell in parthenogenetic organisms) in metazoans is enabled by deposition of the epigenetic information necessary for the early development from the egg/zygote to the phylotypic stage. This information is provided to the gamete(s) parentally in the form of cytoplasmic factors, which control and regulate the blastula, gastrula, and neurula stages initially by their own activity and later by regulating expression of zygotic genes. In placentals, maternal factors continue to influence ontogenetic processes for the whole period of intrauterine development.

The process of individual development from a single cell, an egg cell or a zygote, to an adult organism is known as *ontogeny*. From an evolutionary theoretical point of view, it represents the first laboratory for antenatally testing new developmental solutions and evolutionary changes, before they are put through the sieve of natural selection under natural conditions.

Recapitulationist followers of Ernst Haeckel (1834–1919) held that in embryonic stages of all animals we can distinguish features of their *adult* ancestors. Accordingly, the study of ontogeny would give biologists information for reconstructing animal phylogenies. Now, most biologists, following Karl Ernst von Baer (1792–1876), believe

Epigenetic Principles of Evolution. DOI: 10.1016/B978-0-12-415831-3.00009-4

that only embryonic, not adult, ancestral features appear in the embryonic stages of animals.

Haeckel's biogenetic law has been continually criticized, especially its central tenet that ontogeny recapitulates phylogeny. However, Mayr has warned us against a "water-bathist" approach in dealing with the law:

> *The invalidity of (Haeckel's) law has been demonstrated so often, and so conclusively, that it is easy to fall into the opposite extreme and ignore the fact that many organisms that are highly dissimilar as adults go through similar larval stages.*
>
> *Mayr (1963)*

And the fact that evolutionary relationships between species and higher taxa often are reflected in similarities of early stages of their embryonic development is the reason why

> *Haeckelian concepts have survived, usually misattributed to von Baer, in the disciplines using a comparative approach to the study of morphological pattern, such as comparative anatomy, paleontology, and classical comparative embryology. These are areas where the neodarwinian "populational" approach has not influenced much the classical methodology. For example, Haeckelian concepts underlie most ontogenetic arguments utilized in the determination of homologies in comparative morphology.*
>
> *Alberch and Blanco (1996)*

Despite the theoretical value, the almost 200-year dispute between supporters of these opposing views has contributed little to the deeper causal question: Why must ontogeny occur at all?

One of the basic tenets of the neo-Darwinian paradigm is that the zygote possesses the whole amount of genetic information necessary for the development of the adult organism. If one were to follow this still-prevailing view, the conservation of the phylotypic stage and repetition of some embryonic stages in species standing higher on the tree of evolution would evolutionarily make no sense. Logically, the question would arise: If the genetic information or the so-called genetic program for developing the species-specific structure is present in gamete(s) from the beginning, why has the embryo to follow that circuitous and very costly path often involving development of structures that are apoptotically eliminated later in the individual's development? This is not the way evolution works. Evolution would have done away with the costly building of ancestral embryonic structures that must be repeatedly replaced or transformed.

The cost of developing embryonic structures that later are apoptotically lost is so high that where they are unnecessary or "vestigial" biological phenomena, strong evolutionary pressures would have arisen for not building them at all. When this is considered in the context of the universal occurrence of that ontogenetic "replay" of embryonic features in Animalia, it would be quite logical to believe that the "recapitulation" has functions that outweigh its excessive biological cost.

Ontogeny and the Phylotypic Stage—Why Do All Developmental Pathways Converge to the Common Bauplan?

The development of vertebrate embryos starts with the zygote going through a few intermediate stages (blastula, gastrula, neurula) before reaching the phylotypic stage, when embryos of different species, which heretofore may display distinct morphologies, converge to the common Bauplan of the phylum known as pharyngula, identified by William Ballard as essentially "an early post-neurula 'larva' with paired pharyngeal slits and initiation of the basic vertebrate organ systems" (Hall, 1999, p. 228). Pharyngula possesses the dorsal nerve cord, notochord, pharyngeal arches, somites, and tail. A similar stage of embryonic development, when all the species converge to a common Bauplan, is also observed in most invertebrates. In arthropods, it is known as *segmented germ band*, which also appears after gastrulation.

The phylotypic stage is a watershed in the individual development from the point of view of the source of information. At this point in time, the parental epigenetic information, i.e., parental mRNAs, proteins, hormones, neurotransmitters, nutrients, and other parental chemicals provided via gametes in the form of cytoplasmic factors, is totally consumed or no longer functional. This moment of "informational crisis" coincides with the beginning of the accelerated process of development characterized by numerous inductions of neural origin, leading to complex processes of cell differentiation, histogenesis and organogenesis, which require vast and ever-increasing amounts of information. Exactly at this moment, at the phylotypic stage (whether it is a vertebrate pharyngula, an insect segmented germ band, or an annelid "segmented germ band"), the embryo has succeeded in developing a functioning central nervous system (CNS). It takes over the individual development up to adulthood.

In defining the vertebrate phylotypic stage, Ballard points out:

> *The pharyngula exhibits the basic anatomical pattern of all vertebrates in its simplest form: a set of similar organs, similarly arranged with respect to a bilaterally symmetrical body axis, possessing chiefly the characters that are common to all the vertebrate classes ... One sees in them [the pharyngulas of vertebrates] epidermis but no scales, hair or feathers; kidney tubules and longitudinal kidney ducts are there, but no metanephros; all the little hearts have the same four chambers and there is at least a transient cloaca; there are no middle ears, no gills on the pharynx segments, no tongue, penis, uterus, etc. Basically just vertebrate anatomy, unobscured by the vast array of characters that appear later in development to distinguish the various classes, orders, and families.*
>
> *Ballard (1981)*

Recently, the conventional concept of the phylotypic stage has been criticized by a number of authors. Based on a review of the pertinent evidence, Richardson (1995) believed that it is inappropriate even to talk about a phylotypic stage, which reflects Haeckel's inaccurate views on ontogeny as recapitulation of phylogeny. What he proposes instead is a concept of "a phylotypic 'period'" because this avoids the idea of a narrow timepoint implied by the word "stage." However, from a semantic aspect, his critique is hardly justified, for commonly in English dictionaries the word *stage* defines not a point in time but a period of time. The word *stage* is explained as "one of the distinguishable periods of growth and development of a plant or animal" (Merriam Webster). A number of other authors agree with Richardson but I find convincing Hall's argument that:

> *Ballard's phylotypic stage is essentially a visceral animal—head, branchial arches, bilateral symmetrical organs. His criteria were presence of somites, a neural tube: "Basically just vertebrate anatomy, unobscured by the vast array of characters that appear later in development." Richardson et al. focus on features such as size, numbers of somites, presence or absence of fin or limb buds, types of neurulation, and time of appearance of structures. Their rejection of the phylotypic stage therefore is not based on criteria used by Ballard to define the stage. The features they emphasize are, in large part, embryonic, larval, or life history adaptations. Arriving at conserved stages by different mechanisms or with different numbers of repeated units is not a reason for negating the existence of the stages. Rather, temporal variability in appearance of conserved characters directs us to seek the phylogenetic suite of characters that typifies a taxon: to ask why those characters so often appear at a conserved phylotypic stage: and to understand which developmental and evolutionary processes regulate the temporal and spatial appearance of those characters.*
>
> *Hall (1997)*

von Baer's laws say that the development of vertebrate embryos proceeds from general features, which are shared by all of them, to more specific features. In other words, the laws predict that the earliest embryonic stages must be uniform. However, numerous exceptions are known when, starting from different initial states, embryos of different taxa of the same phylum converge to the morphologically common type, the Bauplan of the phylum. So, for example the hylid frogs of the genus *Gastrotheca* display a pattern of gastrulation from an embryonic disc and an egg size that are typical of birds, not of amphibians. But this affects neither the *Bauplan* nor the adult frog morphology. Prephylotypic stage modifications of developmental patterns that do not affect the adult morphology are also observed in taxa which were subject to modification or even to elimination of their larval structure. So, for example a species developing via a complex feeding larva and its congener, which develops directly, have different embryonic cell lineages and divergent patterns of early development, but converge on the adult sea urchin body plan (Raff, 1998).

The morphoconvergent phase is followed by a morphodivergent phase after the phylotypic stage with a clear tendency toward divergent types of species-specific morphologies.

The phylotypic stage is characterized by inductions initiated by the neural tube/CNS. These inductions lead to global interactions with the target tissues that are observed during the phylotypic stage, which narrow the field of the possibilities of evolutionary changes. Constraints imposed on the evolutionary changes during this stage are best illustrated by the conservation of the structure of organs (e.g., the number of digits and bones in amniotes and of cervical vertebrae in mammals) that are determined during the phylotypic stage. Not only is the phylotypic stage itself a conserved stage, but the organs that develop during that stage also are refractory to evolutionary change (Galis et al., 2001).

Why do metazoan embryos converge to a common Bauplan before they start developing class-, order-, and species-specific features? As of yet, modern biology has no firm or convincing answer to this question.

In attempting to understand the developmental significance of the phylotypic stage, one should start by examining the spatiotemporal pattern of events associated with the stage. The formation of the Bauplan of the phylum at that stage coincides with two crucial events:

First, termination of the function of the parental epigenetic information, i.e., the parental cytoplasmic factors and
Second, the development of the embryonic CNS, which at this stage is operational and triggers a series of inductive events.

The systematic correlation of these three events (the Bauplan, the exhaustion of parental epigenetic information, and the formation of the operative embryonic CNS) in the early vertebrate ontogeny suggests that a causal relationship between them might exist.

We may reasonably relate the two last events to each other, for both represent sequential and complementary sources of epigenetic information necessary, respectively, for pre- and postphylotypic development. The fact that the exhaustion of parentally provided epigenetic information coincides with the formation of the operational CNS is unlikely to be a mere coincidence, for vast evidence shows that the CNS immediately starts a series of inductions exactly when the reserve of epigenetic information is exhausted.

But may we relate the appearance of the Bauplan to the function of the embryonic CNS? The coincidence of the formation of the Bauplan with formation of the operational CNS at the phylotypic stage would suggest that this rudimentary phylotypic outline of the future organism may be necessary for the CNS as a preparatory groundwork and as a point of reference for the future direction of the individual development.

My hypothesis is that the Bauplan serves as developmental outline of the phylum that the embryonic CNS requires for fashioning initial patterns of synaptic connections and neural circuitries as its functional information-generating units. Additionally, the Bauplan may provide the CNS with a general sense of direction for the postphylotypic development.

The common phylotypic structure (Bauplan of the phylum), on the one hand, and the specificity of the incipient operational CNS at that stage, on the other, determine the divergent species-specific paths of postphylotypic development. By continually interacting with the developing embryonic structure, the CNS is able, in a self-sustainable mode, to generate stage-specific epigenetic information for the sequential stages of the postphylotypic development up to adulthood.

In view of the fact that the CNS directs the postphylotypic development by sequentially "recapitulating" ancestral embryonic structures, one is tempted to imagine that "recapitulation" may be the simplest or the only possible way the parentally "designed" phylotypic CNS is "programmed" to "memorize" ancestral structures.

It is also possible that the CNS, while determining the sequential development of quasi-ancestral structures, uses small differences in the Bauplan and later stages of each species as cues for gradually developing species-specific structures.

Additionally, the conservation of ancestral developmental programs for these structures is neither a biological luxury nor designed to help us reconstruct phylogenies. It may be a potential adaptive mechanism worth its high biological cost; metazoans may retrieve these ancestral programs in a "rainy day," when the environment changes in direction of ancestral or quasi-ancestral conditions. Vast evidence on evolutionary reversions (see Chapter 15, Evolution by Reverting to Ancestral Characters) may represent a validation of the hypothesis.

Many changes in morphology during the evolution of Animalia are related to drastic changes in the environment, such as transitions from one type of environment to another (e.g., terrestrial–aquatic; aquatic surface to bottom dwelling; seashore to open sea; low to high altitudes above the sea level; forest to prairies; and cold, moderate, or warm climates).

It is not a rare occurrence that species in the course of evolution suddenly find themselves in very different, or even contrasting, habitats that may happen to be similar to those inhabited by their evolutionary ancestors. Under such drastically changed conditions, species survival would be in danger if individual organisms could not rapidly (in evolutionary "instants") adapt their behavior, morphology, and physiology to the changed conditions of life. Under such circumstances, there is no time for gradual, long-term evolutionary adaptation in nature. It is one of those ecological crises that resolve uncompromisingly: change "instantly" or go extinct. At such junctures, the ontogenetic retrieval and replay of ancestral developmental programs during the individual development may offer off-the-shelf solutions for rapid phenotypic adaptation.

It is impossible to prove now that this is what has occurred in the cases of evolutionary reversions, but we have adequate corroborating evidence that metazoans, in response to stressful changes in environment, can switch to alternative and ancestral developmental pathways and life histories (see Chapters 11 and 15 on transgenerational plasticity and evolutionary reversions, respectively).

Ontogeny May Change Without Changes in Genes

The species-specific ontogeny is not immutable. There is a long, and still growing, list of cases when organisms, in adaptive responses to significant changes in environment, or even in the absence of changes in environment, produce offspring of more than one morphotype, omit certain embryonic stages, and even undergo the so-called reverse ontogeny, which certainly requires new structural information. These discrete and often drastic transgenerational changes in morphology, physiology, behavior, and life history do not involve changes in genes.

Vast evidence on the lability of the ontogenic processes is accumulated in the rapidly expanding field of developmental plasticity (see Chapter 10, Intragenerational Developmental Plasticity). All of the predatory-induced defenses in the offspring of animals that have perceived the presence of the predator in the environment imply significant changes in developmental pathways or even activation of new developmental pathways in the process of the ontogeny. All of these cases of inherited changes in ontogeny involve absolutely no changes in genes.

Production by insects of a number (two to four) of distinct morphs within the same clutch, as well as changes in the proportion of morphs in the offspring, in response to environmental stimuli, demonstrate the astounding plasticity of the ontogenetic processes in metazoans. Cases of facultative paedomorphosis, when salamanders of several species can switch between the metamorphic (legged terrestrial) and pedomorphic (gilled aquatic) life histories; cases of the development of two (carnivorous or herbivorous) behaviorally and morphologically distinct types of tadpoles in frogs of some species in response to specific environmental cues, among others, all are impressive illustrations of the high lability and adaptability of the ontogenetic processes in metazoans. Many such examples will be described and discussed in Chapter 10 of this work. Here I will only present an extraordinary example of the plasticity of ontogenetic processes, the reversion of the ontogeny.

The phenomenon of the "reverse ontogeny" (*Rückbildung*) was first described in Hydrozoa by H.C. Müller (1913). The life cycle of hydrozoans comprises three stages: the planula (a motile gastrula stage), the polyp (the true larval stage), and the adult medusa stage (Piraino et al., 2004). In many species of this group, such as *Hydractinia (Podocoryna) carnea*, *Eleutheria dichotoma*, and a number of *Cladonema* spp., gonozooids, early medusae buds (but not late medusae buds), when they are artificially detached from reproductive polyps, regress into the larval polyp stage. In other hydrozoans as well, deviations may occur from the normal course of development: medusae to stolon and polyps (Piraino et al., 2004). A schema of reverse and regular development in particular cnidarians is presented in Figure 9.1.

In addition to the artificial detachment of medusae buds, various environmental stimuli or conditions lead to similar processes of "reverse ontogeny." Strong water movements, for example cause fragmentation of gonozooids bearing medusae buds, of which the early ones transform into polyps. Retrogression of medusae buds into polyps has been possible to induce experimentally in *Sarsia tubulosa* (M. Sars, 1835) by sudden shifts to low temperatures of 6–8°C.

It was thought that the ability of hydrozoans to shift to the reverse ontogeny is limited to the early medusae buds, and that a constraint on ontogeny reversion is established after the onset of the sexual reproduction. Studies on the hydrozoan *Turritopsis nutricula* McCrady, 1859 (Anthomedusae, Clavidae), however, have shown that, under certain conditions, this species has no temporal constraints for entering the reverse ontogeny.

The onset of sexual reproduction in this species does not represent a point of no return for entering the process of "reverse ontogeny." Several stressors, which lead to "sublethal stress" are known to induce "reverse ontogeny." Among the environmental stressors known to trigger that developmental reversion are starvation, mechanical stress, temperature changes, water salinity, exposure to caesium (an inducer of metamorphosis in this species), and even an intrinsic physiological stressor such as senescence (starting with the sexual maturation of gonads) (Piraino et al., 2004).

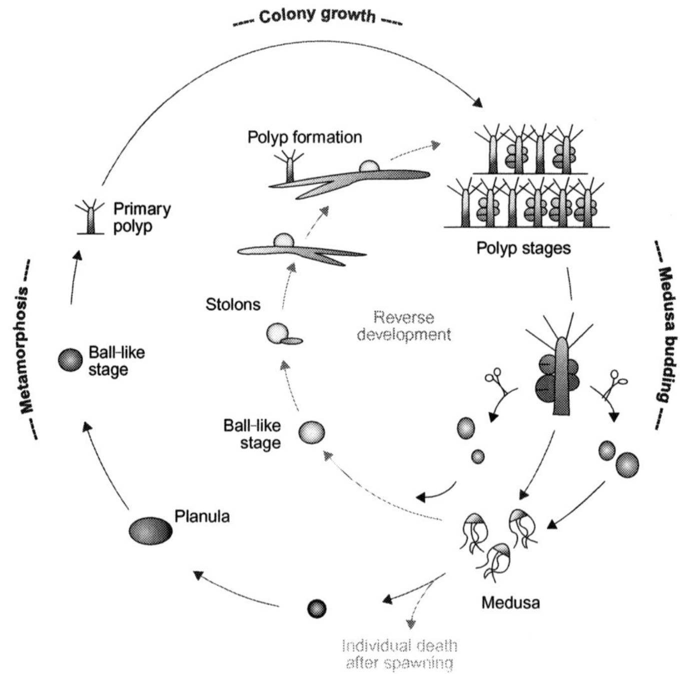

Figure 9.1 Life cycle of *Turritopsis dohrnii* and *Hydractinia carnea*. During the normal life cycle, polyp colonies bud gonochoric medusae, which are eventually liberated into the free water column. After a period of planktonic life, mature medusae release their gametes and, following fertilization, a planula larva is formed in a couple of days. Usually, medusae die after the release of the gametes. The mature planula larva undergoes metamorphosis upon an external bacterial signal and forms the primary polyp, the starting unit to form a new polyp colony. *Turritopsis dohrnii* shows an extraordinary variation to this regular life cycle, i.e., reverse development: the medusae are able to transform back into the polyp stage by asexual processes. *Hydractinia carnea* has a reduced potential to undergo this process, limited to young medusa buds.
Source: From Schmich et al. (2007).

In the interpretation of the investigators, the reversion to the polyp stage may be "a tool to increase jellyfish population growth rate during the favorable season" (De Vito et al., 2006).

Reverse ontogeny from the medusa stage to the juvenile polyp stage implies transdifferentiation, i.e., transformation of the medusa cells into polyp cells (Piraino et al., 2004). Reversion of medusae to the polyp stage in *Turritopsis nutricula* takes place in stages: healthy medusa, unhealthy medusa, four-leaf clover, and cyst. The first and fourth stages are characterized by programmed cell death, and the second and third stages are characterized by degenerative processes (Carlà et al., 2003).

A number of chemicals, and especially environmental stress, may affect the ontogeny and produce inherited change in the offspring, especially via the main neuroendocrine hypothalamic–pituitary axis (see Transgenerational Developmental Plasticity in Experiments, in Chapter 11).

The high plasticity of ontogenic processes demonstrated in examples presented in this section, and in examples to be presented later in this work, point to the adaptability of the ontogeny to the changing environment and its potential to induce evolutionary change without changes in genes.

It is within the phenomenon of ontogeny, regarded here as a workshop of evolutionary change, where the metazoan potential for multiple adaptations of the species to sudden adverse changes of the environment unfolds. Each individual organism is endowed with a repertory of potential ontogenetic solutions for adapting to ancestral and quasi-ancestral environments.

Sudden changes in ontogeny challenge one of the basic tenets of the neo-Darwinian doctrine that inherited phenotypic changes result from changes in genes or genetic information in general. Accordingly, it would be predicted that, in order for adaptive evolutionary changes to occur, changes must take place in the genetic information, i.e., *favorable* mutations must be accumulated over generations. This neo-Darwinian prediction is contradicted by a considerable body of evidence demonstrating that such sudden evolutionary changes have occurred and are even reproducibly induced in the laboratory (see on experimental evolution of characters in Chapter 13, Origins of Evolutionary Novelty).

Neo-Darwinian synthesis, for a long time, and at its own expense, neglected studies in the field of developmental biology, although any gene mutation and any change in the genetic information first must be tested in the process of individual development:

> *By viewing evolution as a branching tree of adults or genes, theorists have omitted what selection really acts upon: ontogeny. Ontogenies evolve, not genes or adults. Mutated genes are passed on only to the extent that they promote survival of ontogenies; adulthood is only a fraction of ontogeny.*
>
> *McKinney and Gittleman (1995)*

Generation of Epigenetic Information for Organogenesis After the Phylotypic Stage

The pervasive presence of the nervous tissue throughout the organism, often down to the level of individual cells, and the continuous input of the information from animal structures to the central and peripheral nervous system, endows the CNS with the ability to monitor the status of the structure and functions of the embryo in "real time." As shown earlier, the CNS compares the input of information on the actual structure with the information on the normal structure it possesses, and makes decisions for restoring the continually disintegrating structures and normal chemistry of the body fluids (the neuroendocrinely regulated homeostasis) (see Chapter 1).

On the one hand, the CNS responds to the input from the developing embryonic structures by modifying its neural circuits and synaptic morphology (when the afferent input to the CNS is prevented, this does not take place or occurs abnormally) and, on the other, in response to the changing embryonic structure, it sends epigenetic information for regulating successive stages of the development by activating specific signal cascades. The epigenetic information is provided to the developing structures via both the local innervation and/or signal cascades, starting with chemical signals in the CNS. And it is experimentally demonstrated that prevention of these brain signals from reaching embryonic structures impairs or even prevents their development.

Changes induced by brain signals in the developing embryonic structures via afferents are sent for processing in neural circuits. In response, neural circuits modify their synaptic morphology, which in turn modifies their processing properties and the outcome of the processing, i.e., start new signal cascades for regulating consecutive stages of the embryonic development.

Adequate evidence of the CNS as the source of information for individual development is presented in Chapter 2 of this work (especially Section Neural Processing of Stimuli Generates Information for Postphylotypic Development). Here I will only present some experimental evidence demonstrating that just before the phylotypic stage the CNS is operative, i.e., the neural circuits and neurons show "spontaneous" electrical activity and respond to the developing embryonic structure by modifying their synaptic morphology and by inducing development of various organs and parts.

The embryonic CNS responds to the input of internal stimuli (changes in the developing embryonic structure) by spontaneous electrical activity, which determines the wiring and the establishment of neural circuits in the brain (Peinado, 2000; Zhang and Poo, 2001). In retina, for example this activity produces highly stereotyped patterns of connections before the onset of visual experience (Penn et al., 1998) and experience independently instructs formation of eye-specific layers (Shatz, 1996). By contrast, in the absence of afferent input, normal neural circuits are not formed (Penn et al., 1998), and rat striatal neurons do not develop dendritic spines (Segal et al., 2003).

Prenatally, the embryonic CNS of higher vertebrates establishes trillions or quadrillions of *specific* connections between specific neurons. Establishment of this huge amount of specific neuronal connections in the CNS requires information, which is neither inherited from parents via gametes nor acquired by the sensory experience. That information is computationally, experience independently, generated in the embryonic CNS, by converting internal stimuli into trains of electrical spikes and processing them in neural circuits. The conversion of internal stimuli into electrical signals and generation of epigenetic information for the establishment of specific synaptic connections by processing electrical signals led biologists to the idea that spontaneous electrical activity might represent a "self-organizing property" (Weliky, 1999) of the nervous system. Indeed, formation of specific neuronal connections is prevented, or is drastically reduced, when the embryonic CNS is deprived of the input of afferent stimuli from the developing organs. So, for example the experimental delay of muscle development causes suspension of synaptic branching of respective motor neurons (Fernandez and Keshishian, 1998), and male rats castrated on day 1 (no afferent input from testicles) have significantly reduced numbers of shaft and spine synapses in the ventrolateral part of the hypothalamic ventromedial nucleus (Matsumoto and Arai, 1986). Ovariectomy (no afferent input from the ovary) causes a profound decrease in the dendritic spine density of pyramidal cells in the hippocampus. In *Drosophila melanogaster*, where the larva lacks the target muscle (no input of stimuli from the muscle), the axon of the motoneuron MN5 develops no dendritic connections (Consoulas et al., 2002). In experiments on the moth *Manduca sexta*, it is also observed that this insect shapes the growth of its motoneuron dendrites in response to afferent stimuli coming from the leg (Kent and Levine, 1993).

Such facts indicate that a clear relationship does exist between the embryonic structure and the synaptic morphology, that information for patterning neuronal connections and neural circuits is computationally generated by processing the afferent input from developing embryonic structures, according to the "brain's best guess" (Katz and Shatz, 1996).

The above evidence shows that the epigenetic information necessary for the establishment of the myriad of specific neuronal connections in the embryonic CNS is an experience-independent function of the spontaneous activity of the CNS. Designing and constructing that great complex of synaptic morphology requires investment of an enormous amount of epigenetic information, and the high cost of generation of that information suggests that the advantages it offers outweigh the cost of its erection. Otherwise, evolution, which is so parsimonious, would not have favored its appearance. Hence, the question arises: Does that synaptic structure serve any purpose during the embryonic–fetal development, long before the organism starts the independent life and communication with the external environment?

This causal relationship between the afferent input from the developing embryonic structure and the synaptic morphology may, *inter alia*, suggest that synaptic morphology codifies the actual embryonic structure. The idea that synaptic morphology may store epigenetic information is not new: patterns of synaptic connections are demonstrated to be at the basis of the long-term memory, which implies formation of new synaptic connections and induction of new protein synthesis and of short-term memory (strengthening of preexisting synaptic connections) (Milner et al., 1998). Could the synaptic morphology also represent epigenetic information for the developing embryonic structure?

If so, then the formation of that costly and immense structure of neuronal connections in the embryonic nervous system might have a meaning: it may be necessary for figuring consecutive stages of embryonic development in a neuro-computational process of self-organization.

Now I will attempt to substantiate the hypothesis that signal cascades starting with epigenetic signals in the CNS and local innervation convey instructions for the development of embryonic tissues and organs.

There is significant evidence that the spontaneous activity, which arises in response to the input of internal stimuli (changes in the developing embryonic structure), is necessary for the further development of the embryo. For example, suppression of the spontaneous activity (and the accompanying change in synaptic morphology—N.C.) by paralysis in chick embryos (Hall, 1975), by paralysis of motor neurons in chick embryos (Hall and Herring, 1990) and duck embryos (Creazzo and Sohal, 1983), results in the reduction of the bone- and muscle growth and abnormalities in the development of bones (Hosseini and Hogg, 1991). Inhibition of skeletal muscle contractions *in ovo* in chicks by a neuromuscular blocking agent prevents the normal development of synovial joint cavities leading to formation of stiff and fused joints (Hall, 1975; Persson, 1983). Development of the leg musculature is severely retarded in paralyzed embryos, and denervation experiments lead to total prevention of the development of muscles in duck embryos (Sohal and Holt, 1980; Creazzo and Sohal, 1983) and *M. sexta* (Consoulas and Levine, 1997). Based on experimental observations, it has been concluded that spontaneous electrical activity in the spinal cord is involved in the development of limb muscles, bones, and joints (Wenner and O'Donovan, 2001). All of this evidence points to an instructive role of the efferent output of the CNS (generated by processing the afferent input) in molding embryonic morphology.

Dramatic changes in neurogenesis, death of neurons, and reduction of the number of neurons as well as radical changes in the synaptic morphology of sensory neurons (in the sensory system and the CNS) and motor neurons are closely related to changes in the structure of muscles during metamorphosis in *M. sexta* (Consoulas et al., 2000).

The theoretical implication of the above evidence is that, with matter and energy supplied by the blood, the necessity of the local innervation for developing embryonic structures is an indication that it provides epigenetic information for the embryonic development.

Indeed, adequate evidence on the neural origin of signals for postphylotypic development presented in this section and in Chapter 2 shows that the CNS is the ultimate source of the epigenetic information necessary for the postphylotypic development. The fact that no other organ or organ system has been shown to be systematically involved in the control of the development of other organs, as the CNS is, cannot be an accidental occurrence.

Based on the present knowledge on

1. The role of the embryonic CNS in induction of organogenesis and morphogenesis in general,
2. Negative effects of experimental prevention of afferent information from reaching the CNS on the structure of neural circuits, and
3. Negative effects of the blockage of efferent information transmitted from the CNS to the target tissues on the development of animal structure,

we can reconstruct a hypothetical generalized scheme of the interaction of the nervous system with the developing embryo.

The full message that the parental CNS(s) (via the epigenetic information put in gamete(s)) conveys to the embryo is embodied in the Bauplan of the phylum, including the embryonic CNS, that develops just before the phylotypic stage. The common Bauplan implies no identicalness of the structure of all embryos at the phylotypic stage. Behind the general design, there is a multitude of differences in the details of the structure of the Bauplan, including the CNS. By processing the afferent input from the developing embryo, the embryonic CNS reconfigurates the synaptic morphology, thus modifying computational properties of neural circuits. Modification of computational properties implies modification of outputs of the circuits, which represent new epigenetic information necessary for the subsequent stage of development. The epigenetic information is communicated to embryonic structures in the form of inductive signals by the CNS and/or directly by local innervation. The new input on the developing embryonic structure is again processed for generating and sending information for the consecutive stages of embryonic development in a self-sustaining cycle of interaction between the CNS and the developing embryonic structure. Modification of the neural circuits in response to sequentially changed embryonic structures, as well as formation of new circuits, generates the epigenetic information for starting signal cascades that bring about stage-specific transformations in the morphology of the embryo.

Stress-Induced Developmental Instability and Evolution

In the course of evolution, metazoans evolved not only mechanisms of adaptation to the changed conditions in the environment but also mechanisms for preserving evolutionary achievements, which are manifested in the developmental stability of their structure and function, in production of offspring that express all the basic parental morphological, physiological, behavioral, and life history characters. These mechanisms serve as developmental anchorages for preserving and reinforcing evolutionary achievements and for preventing evolution in certain directions, often known as *developmental constraints*.

In the absence of a better criterion, developmental stability of metazoans is commonly assessed indirectly by evaluating “fluctuating asymmetry” (FA), which measures random deviations (developmental noise) from the perfect asymmetry of bilateral traits. Although measurement of FA is far from an ideal means of assessing the developmental stability, for more than half a century the disturbed developmental stability has been the object of considerable interest.

Under normal, relatively stable conditions, the ontogeny is conservative, and developmental pathways are generally stable, resistant to change or “noise,” as it is seen in the precise implementation of the extremely complex developmental programs. Ontogeny and, especially, its “embryonic stages” are so deeply rooted that, under normal circumstances, their course is invariable and precisely determined. As a rule, organs or parts of the animal body that are patterned during the phylotypic stage are refractory to evolutionary changes. This is the case, for example with the number of digits and limb bones in amniotes and probably with cervical vertebrae in mammals. The evolutionary constraints imposed by the phylotypic stage might have been the cause of the frequent appearance of modularity in evolution of vertebrates (Galis et al., 2001).

However, drastic changes in environment, by disturbing the homeostasis and inducing stress response, can affect the developmental stability. The unpredictability of environments to which metazoans are exposed during their evolution has given rise to an evolutionary pressure for evolving mechanisms that would also provide some degree of flexibility to the normal development and allow for changes to occur in the individuals experiencing stressful conditions of the environment or even in their offspring. Under conditions of environmental stress, metazoans can switch to mechanisms that induce discrete adaptive developmental changes (developmental plasticity) in their own phenotype (intragenerational changes) or may first induce such changes in the offspring (transgenerational, inherited changes). Cases are also known when metazoans reactivate dormant ancestral developmental pathways and revert to lost ancestral phenotypes (see Chapter 15, Evolution by Reverting to Ancestral Characters).

One important factor that contributed to reduction of constraints on evolutionary change (and related higher rates of evolution) in vertebrates has been evolution of the neural crest. The class of vertebrates, as it is well known, owes much of its *unprecedented* morphological evolution to neural crest cells, and some of the structures produced through specific migration and differentiation of these cells are among the most diversified (e.g., jaws, craniofacial structures, heart, the peripheral nervous system) in vertebrates.

There is no evidence of any role of gene mutations in evolution of the neural crest and even it is impossible, by any stretch of the imagination, to relate the evolution of neural crest cells to the accumulation of mutational changes under the action of natural selection. Evolution of the neural crest in vertebrates is related to recruitment of genes that existed in invertebrates rather than invention of new or changed genes. And there is evidence suggesting that gene recruitment in metazoans is under neural control (Cabej, 2011).

Neural crest cells have a common origin with nerve cells, and before leaving the neural tube they somehow are provided with epigenetic information on where to go, and into what to transform themselves and other cells in the migration sites. Evolution of these cells is related to the evolutionary pressure for more efficacious vertebrate structures necessary for their new and increasingly predatory lifestyle. The evolution of the neural crest and use of neural crest cells for developing most of the evolutionary novelties in vertebrates indicates that preexisting neural mechanisms alone (centrally controlled signal cascades and the activity of the local innervation) would not enable formation of complex vertebrate structures, such as cranium and the increased general complexity observed in most vertebrate organs.

Both observational evidence from nature and experimental evidence show that any change in the conditions of living that disturbs the normal metazoan physiology, i.e., that leads to environmental stress of any form, interferes with the homeostasis and may disturb developmental stability of the embryonic/fetal and even postnatal development in metazoans. The environmental stress creates a developmentally less-stable state, prone to changes in the epigenetic programming, as exemplified in numerous cases of stress-induced changes during developmental plasticity that will be extensively considered in the next chapter (Intragenerational Developmental Plasticity).

Evolutionary paleontological evidence presented in Chapter 7 (Metazoan Response to Changes in Environment), on the one hand, and vast evidence from the study of developmental plasticity to be presented in the next chapter, on the other, show that a close relationship exists between changes in the metazoan phenotype and the stressful changes or cues presaging stressful changes in the environment. In response to such environmental changes and stimuli, animals might not only change or modify developmental pathways for particular traits but also perform radical changes by switching, for example from the indirect metamorphic development to direct development and the reverse.

Vast evidence shows that environmental (including nutritional and energetic) stress leads to higher FA and developmental instability in fish (Leary and Allendorf, 1989; Swaddle and Witter, 1994; Vøllestad and Hindar, 1997).

Axolotls (*Ambystoma mexicanum*) skip metamorphosis in nature, but axolotls kept under conditions of captivity stress in a Paris museum produced normal metamorphosed *Ambystoma* adults (Gould, 1977). In some neotenic salamander species, metamorphosis can be induced even after the reproductive maturity of the brachiate adult by various natural and experimental stresses, such as starvation and drought.

An interesting case of developmental instability induced by a stressful event has been reported by Badyaev. The sudden closure of garbage dumps in Yellowstone National Park caused a high mortality rate in the local bear population as well as increased asymmetric growth in the canine, but not the premolar, teeth of bears (Badyaev, 1998). The relationship between the FA and environmental stress is so strong that the measurement of the FA is considered a good indicator of the environmental stress in natural populations (Leary and Allendorf, 1989).

In relation to the point in time in ontogeny when FA occurs, experiments on an aquatic holometabolous insect, *Hydropsyche exocellata* (Dufour, 1841), have shown that FA occurs early in the larval development and gradually decreases to finally cease from the last larval instar (Piscart et al., 2005). However, there is some evidence showing that the asymmetry may remain constant in later stages of development (Collin, 1997; Servia et al., 2002).

Observations on the ontogeny of FA of primary feathers of European starling, *Sturnus vulgaris* (Linnaeus), led Swaddle and Witter to the conclusion that of six hypotheses proposed for explaining the FA, the hypothesis of the compensational growth provides the best explanation of the FA of primary feathers (Swaddle and Witter, 1994). According to that hypothesis, asymmetries that arise during early development later decrease as a result of a compensational growth that is regulated by a feedback mechanism.

Developmental Constraints and Adaptive Evolutionary Innovations

A serious objection could be raised against the idea that reversion to ancestral phenotypes (morphology, behavior, physiology, and life history) is related to a return to ancestral or quasi-ancestral environments. It may be argued that if ancestral epigenetic programs are conserved in ontogeny, it would be expected that reversions to ancestral phenotypes would occur more frequently than are actually observed in nature.

In order to survive and succeed in an unpredictable environment, metazoans must maintain a balance between their ability to change and their ability to transmit their characters accurately to the offspring. This gave rise to an evolutionary pressure for evolving mechanisms for both resisting changes and providing the necessary epigenetic variability during the individual development. Evolutionary pressure must have gravitated toward the stability of epigenetic programs for avoiding accidents in developmental programs, which might have detrimental consequences under generally little or insignificantly changed environments. This would be the case when individual organisms might experience short-term, ephemeral, quasi-ancestral changes in the environment.

As a result of the two opposing evolutionary pressures, developmental constraints in metazoans rose not as devices for absolute prevention of changes in developmental pathways, but as mechanisms for sustaining a certain level of probability for such events to occur. From this view, developmental constraints represent statistical phenomena.

As for their origin, developmental constraints may be

- *constitutive constraints* resulting from increased complexity and integration of structures and functions in metazoans, or
- *functional constraints* determined by evolutionary inhibition of particular pathways in response to the pressure for higher stability in particular pathways.

When the first is the case, the integration of functions of various structures, organs, and parts and their functional interdependency related to the cell differentiation, specialization, and division of labor, may make the morphological change extremely difficult or impossible. For example, in experiments with several shrew species, a strong resistance to changes has been observed in intensely integrated structures, while changes in less-integrated structures are more frequent. This fact led to the prediction that related species would be more similar in highly integrated structures and exhibit more changes in less-integrated structures. This prediction has been validated by observations on the experimentally induced changes in shrew mandibles under the influence of the environmental stress (Badyaev and Foresman, 2000, 2004).

Functional constraints, imposed by the inhibition of developmental processes, evolve in response to evolutionary pressures for higher stability in developmental pathways. Kirschner and Gerhart (1998) have observed that functional constraints are imposed chiefly by inhibitions, but these inhibitions are sometimes relieved by other inhibitions. So, for example blockade during gastrula stage of bone morphogenetic proteins, inhibitors of neurogenesis, is sufficient to initiate neuralization of the ectoderm in *Xenopus laevis*. Sometimes, constraints can also be weakened by what they call "weak linkage" (a process that is minimally dependent on other processes or components). Such "weak linkages" are characteristic for the regulatory pathways that evolved during the evolution of Metazoa, such as signal transduction and neural relays (Kirschner and Gerhart, 1998).

Generally, developmental stability is a relative stability, which means that limits exist beyond which actions of environmental or internal factors may weaken or break it.

With the organism under the constant influence of these two opposing tendencies, the evolutionary problem "To switch or not to switch" to new or ancestral epigenetic programs has apparently found a compromising solution: the switching to a new available epigenetic program appears to be the "last resort" in the survival effort under stressful conditions of drastically changed environment.

Switching to ancestral epigenetic programs seems to be under excessive epigenetic constraints. This is the reason why, even when a mechanism for reversion to an ancestral structure, behavior, or life history is available, reacquisition of ancestral traits normally requires a varying number of generations, if ever, to occur. One typical example of the delayed evolution of a character in response to environmental stimuli is the delayed transition from amictic to mictic (sexual) reproduction in some rotifer species (e.g., *Brachionus calyciflori*s, *Brachionus angularis*, *Epiphanes senta*). In response to a social stimulus such as crowding, these rotifer species produce mictic female offspring, whose eggs develop into males or, when fertilized, into diapausing eggs. However, in the first generation after being exposed to crowding, the proportion of mictic (sexually reproducing) females produced by amictic females is very low until the fifth generation, after which that proportion increases dramatically (Schröder and Gilbert, 2004). This seems to be the case even in some examples of transgenerational plasticity, such as the adaptation of the aphid, *Dysaphis anthrisci*, to a hostile plant, *Chaerophylum maculatum*, on which initially the death rate of the aphid population was almost 100%, but after rearing on this plant for four to eight generations of asexual reproduction, the aphid heritably changed its morphology (proboscis and body size) and used the ex-hostile plant as its natural host (Shaposhnikov, 1965). The case is considered to be one of the most unambiguous examples of experimental speciation.

Evolutionary reversions in metazoan populations are not "all-or-none" but statistical phenomena only affecting certain individuals of the population per generation. In experiments on *D. melanogaster*, populations that have evolved in a number of biochemical and life history characters under stressful conditions within a few hundred generations (several decades) in laboratory, it has been observed that the return to the ancestral (wild-type) environment does not lead to an immediate reversion of ancestral traits and does not affect all of the individuals of the population. Reversion to ancestral, wild-type traits may take from 5 to 50 generations to occur, and only certain proportions of the population revert to the ancestral characters (Teotonio and Rose, 2000). The degree of reversion to ancestral characters for different traits also varied from complete to incomplete, and sometimes no reversion occurred at all.

The process of experimental evolutionary reversion of the lost ancestral traits in *Drosophila* followed four different patterns:

1. Full convergence to the ancestral character in little more than 20 generations (e.g., in the case of the female developmental time),
2. Initial rapid reversion with partial convergence to the ancestral trait followed by arrest of reversion (e.g., starvation resistance),
3. Changes that did not converge to the ancestral traits within 50 generations, and
4. Lack of significant changes after 50 generations (fecundity under high density) (Teotonio and Rose, 2000, 2001; Teotonio et al., 2002).

Due to the statistical nature of the reversion of ancestral characters, under natural conditions, evolution in the reverse takes place as a process of natural selection.

The above empirical evidence reveals some general features of evolutionary adaptations to changed conditions of living:

- The evolutionary change usually does not appear in F1 of the organisms that experienced the environmental change/stress but may first appear in latter generations,
- The evolutionary change appears not in the population as a whole but only in limited portions of the population,
- The evolutionary change might not be complete from the beginning but may unfold stepwise,
- There is always a number of individuals in the population that do not show the evolutionary change, and
- There may be a proportion of individuals that may change in directions other than the expected one.

All of the features of the appearance of the evolutionary change observed in the above experimental examples seem to epitomize evolution of metazoans in nature under the action of natural selection.

References

Alberch, P. and Blanco, M.J. (1996). Evolutionary patterns in ontogenetic transformation: from laws to regularities. Int. J. Dev. Biol. *40*, 845–858.

Badyaev, A. (1998). Environmental stress and developmental stability in dentition of the Yellowstone grizzly bears. Behav. Ecol. *9*, 339–344.

Badyaev, A.V. and Foresman, K.R. (2000). Extreme environmental change and evolution: stress-induced morphological variation is strongly concordant with patterns of evolutionary divergence in shrew mandibles. Proc. Biol. Sci. *267*, 371–377.

Badyaev, A.V. and Foresman, K.R. (2004). Evolution of morphological integration. I. Functional units channel stress-induced variation in shrew mandibles. Am. Nat. *163*, 868–879.

Ballard, W.W. (1981). Morphogenetic movements and fate maps of vertebrates. Am. Zool. *21*, 391–392. Quoted by Hall, B.K. (1997). Phylotypic stage or phantom: is there a highly conserved embryonic stage in vertebrates? Trends Ecol. Evol. *12*, 461–463.

Cabej, N.R. (2011). Neural control of gene recruitment in metazoans. Dev. Dyn. *240*, 1–8.

Carlà, E.C., Pagliara, P., Piraino, S., Boero, F. and Dini, L. (2003). Morphological and ultrastructural analysis of *Turritopsis nutricula* during life cycle reversal. Tissue Cell *35*, 213–222.

Collin, R. (1997). Ontogeny of subtle asymmetries in individual larvae of the sand dollar *Dendraster excentricus*. Evolution *51*, 999–1005.

Consoulas, C. and Levine, R.B. (1997). Accumulation and proliferation of adult leg muscle precursors in *Manduca* are dependent on innervation. J. Neurobiol. *32*, 531–553.

Consoulas, C., Duch, C., Bayline, R.J. and Levine, R.B. (2000). Behavioral transformations during metamorphosis: remodeling of neural and motor systems. Brain Res. Bull. *53*, 571–583.

Consoulas, C., Restifo, L.L. and Levine, R.B. (2002). Dendritic remodeling and growth of motoneurons during metamorphosis of *Drosophila melanogaster*. J. Neurosci. *22*, 4906–4917.

Creazzo, T.L. and Sohal, G.S. (1983). Neural control of embryonic acetylcholine receptor and skeletal muscle. Cell Tissue Res. *228*, 1–12.

De Vito, D., Piraino, S., Schmich, J., Bouillon, J. and Boero, F. (2006). Evidence of reverse development in Leptomedusae (Cnidaria, Hydrozoa): the case of *Laodicea undulata* (Forbes and Goodsir 1851). Mar. Biol. *149*, 339–346.

Fernandes, J.J. and Keshishian, H. (1998). Nerve–muscle interactions during flight muscle development in *Drosophila*. Development *125*, 1769–1779.

Galis, F., van Alphen, J.J.M. and Metz, J.A.J. (2001). Why five fingers? Evolutionary constraints on digit numbers. Trends Ecol. Evol. *16*, 637–646.

Gould, S.J. (1977). *Ontogeny and Phylogeny*. Belknap Press, Cambridge, MA . According to West-Eberhard, M.J. (2003). *Developmental Plasticity and Evolution*. Oxford University Press, New York, NY, p. 511.

Hall, B.K. (1975). A simple, single-injection method for inducing long-term paralysis in embryonic chicks, and preliminary observations on growth of the tibia. Anat. Rec. *181*, 767–777.

Hall, B.K. (1997). Phylotypic stage or phantom: is there a highly conserved embryonic stage in vertebrates?. Trends Ecol. Evol. *12*, 461–463.

Hall, B.K. (1999). *Evolutionary Developmental Biology*, second edition. Chapman & Hall, London , p. 135.

Hall, B.K. and Herring, S.W. (1990). Paralysis and growth of the musculoskeletal system in the embryonic chick. J. Morphol. *206*, 45–56.

Hosseini, A. and Hogg, D.A. (1991). The effects of paralysis on skeletal development in the chick embryo. I. General effects. J. Anat. *177*, 159–168.

Katz, L.C. and Shatz, C.J. (1996). Synaptic activity and the construction of cortical circuits. Science *274*, 1133–1138.

Kent, K.S. and Levine, R.B. (1993). Dendritic organization of an identified neuron during metamorphosis of the moth *Manduca sexta*; the influence of interactions with the periphery. J. Neurobiol. *24*, 1–22.

Kirschner, M. and Gerhart, J. (1998). Evolvability. Proc. Natl. Acad. Sci. U.S.A. *95*, 8420–8427.

Leary, R.F. and Allendorf, F.W. (1989). Fluctuating asymmetry as an indicator of stress: implications for conservation biology. Trends Ecol. Evol. *4*, 214–217.

Matsumoto, A. and Arai, Y. (1986). Male–female difference in synaptic organization of the ventromedial nucleus of the hypothalamus in rat. Neuroendocrinology *42*, 232–236.

Mayr, E. (1963). *Animal Species and Evolution*. Harvard University Press, Cambridge, MA , p. 606.

McKinney, M.L. and Gittleman, J.L. (1995). Ontogeny and phylogeny: tinkering with covariation in life history, morphology and behavior. In: *Evolutionary Change and Heterochrony*, K.J. McNamara (ed.). Wiley, Chichester, pp. 21–47.

Merriam Webster Dictionary. Available online: http://www.merriam-webster.com/dictionary/stage.

Milner, B., Squire, L.R. and Kandel, E.R. (1998). Cognitive neuroscience and the study of memory. Review. Neuron *20*, 445–468.

Peinado, A. (2000). Traveling slow waves of neural activity: a novel form of network activity in developing neocortex. J. Neurosci. *20*, 1–6.

Penn, A.A., Riquelme, P.A., Feller, M.B. and Shatz, C.J. (1998). Competition in retinogeniculate pattern driven by spontaneous activity. Science *279*, 2108–2112.

Persson, M. (1983). The role of movements in the development of sutural and diarthroidal joints tested by long-term paralysis in chick embryos. J. Anat. *137*, 591–599.

Piraino, S., De Vito, D., Schmich, J., Bouillon, J. and Boero, F. (2004). Reverse development in Cnidaria. Can. J. Zool. *82*, 1748–1754.

Piscart, C., Moreteau, J.-C. and Beisel, J.-N. (2005). Decrease of fluctuating asymmetry among larval instars in an aquatic, holometabolous insect. C. R. Biol. *328*, 912–917.

Raff, R.A. (1998). Larval homologies and radical evolutionary changes in early development. In: *Symposium on Homology, held at the Novartis Foundation*, G.R. Bock and G. Cardew (eds.). London, 21–23 July 1998.

Richardson, M.K. (1995). Heterochrony and the phylotypic period. Dev. Biol. *172*, 412–421.

Schmich, J., Kraus, Y., De Vito, D., Graziussi, D., Boero, F. and Piraino, S. (2007). Induction of reverse development in two marine hydrozoans. Int. J. Dev. Biol. *51*(1), 45–56.

Schröder, T. and Gilbert, J.J. (2004). Transgenerational plasticity for sexual reproduction and diapause in the life cycle of monogonont rotifers: intraclonal, intraspecific and interspecific variation in response to crowding. Funct. Ecol. *18*, 458–466.

Segal, M., Greenberger, V. and Korkotian, E. (2003). Formation of dendritis spines in cultured striatal neurons depends on excitatory afferent activity. Eur. J. Neurosci. *17*, 2573–2585.

Servia, M.J., Cobo, F. and Gonzalez, M.A. (2002). Ontogeny of individual asymmetries in several traits of larval *Chironomus riparius* Meigen, 1804 (Diptera, Chironomidae). Can. J. Zool. *80*, 1470–1479.

Shaposhnikov, G.Kh. (1965). Morphological divergence and convergence in an experiment with aphids (Homoptera, Aphidinae). Entomol. Rev. *44*, 1–12. According to Görür, G. (2000). The role of phenotypic plasticity in host race formation and sympatric speciation in phytophagous insects, particularly in aphids. Turk. J. Zool. *24*, 63–68.

Shatz, C. (1996). Emergence of order in visual system development. Proc. Natl. Acad. Sci .U.S.A. *93*, 602–608.

Sohal, G.H. and Holt, R.K. (1980). Role of innervation on the embryonic development. Cell Tissue Res. *210*, 383–393.

Swaddle, J.P. and Witter, M.S. (1994). Food, feathers and fluctuating asymmetries. Proc. Biol. Sci. *255*, 147–152.

Teotónio, H. and Rose, M.R. (2000). Variation in the reversibility of evolution. Nature *408*, 463–466.

Teotónio, H. and Rose, M.R. (2001). Perspective: reverse evolution. Evolution *55*, 653–660.

Teotónio, H., Matosi, M. and Rose, M.R. (2002). Reverse evolution of fitness in *Drosophila melanogaster*. J. Evol. Biol. *15*, 608–617.

Weliky, M. (1999). Recording and manipulating the *in vivo* correlational structure of neuronal activity during visual cortical development. J. Neurobiol. *41*, 25–32.

Wenner, P. and O'Donovan, M.J. (2001). Mechanisms that initiate spontaneous network activity in the developing chick spinal cord. J. Neurophysiol. *86*, 1481–1498.

Vøllestad, L.A. and Hindar, K. (1997). Developmental stability and environmental stress in *Salmo salar* (Atlantic salmon). Heredity *70*, 215–222.

Zhang, L.I. and Poo, M. (2001). Electrical activity and development of neural circuits. Nat. Neurosci. Suppl. *4*, 1207–1214.

10 Intragenerational Developmental Plasticity

The central nervous system can integrate information about the animal's internal and external environment and use this information to regulate the secretion of hormones. In this way, development can become responsive to a wide variety of environmental signals, without the need to have developmental processes themselves be sensitive to the environment.

Nijhout

Developmental Plasticity: Beyond the Reaction Norm

Metazoans are dynamic systems that can change during the lifetime, both *in response* to external or internal stimuli and *as a result* of external influences. This is the *phenotypic plasticity* in the broadest sense of the term (Figure 10.1). Phenotypic plasticity may occur in the form of the *norm of reaction* (Woltereck, 1909), consisting of a continuum of incrementally varying phenotypes involving no qualitative changes, or in the form of *developmental plasticity*, which implies appearance of *discrete* alternative phenotypes.

Both the norm of reaction and developmental plasticity involve no changes in genes or genetic information in general. However, developmental plasticity and reaction norm are two qualitatively different phenomena as far as the nature of the change they bring about and the underlying mechanisms are concerned.

First, the norm of reaction, as the term itself indicates, is a *reaction* to changes in the environment, whereas developmental plasticity does not always/necessarily depend on environmental stimuli (e.g., developmental polymorphisms). In cases when the developmental plasticity is related to, or depends on, environmental stimuli, it is an adaptive *response* rather than a reaction, implying that it is determined not by the nature of the stimulus but by the adaptive needs of the organism, which switches to a new developmental pathway for achieving a discrete phenotypic change.

Second, developmental plasticity is related to switches in developmental pathways and mechanisms, while the reaction norm is not.

Third, and as mentioned earlier, the norm of reaction implies existence of a continuum of phenotypes displaying only *quantitative differences* between them, whereas developmental plasticity usually implies *qualitative changes* of the phenotype.

Some authors, however, believe that differences between polyphenisms (herein to be considered under the general term of *developmental plasticity*) and the norm of reaction are only quantitative differences (Nijhout, 2003). However, the fact that the appearance of polyphenisms depends on switches in developmental pathways, while the norm of reaction implies no such switches, is an essential difference and

Epigenetic Principles of Evolution. DOI: 10.1016/B978-0-12-415831-3.00010-0

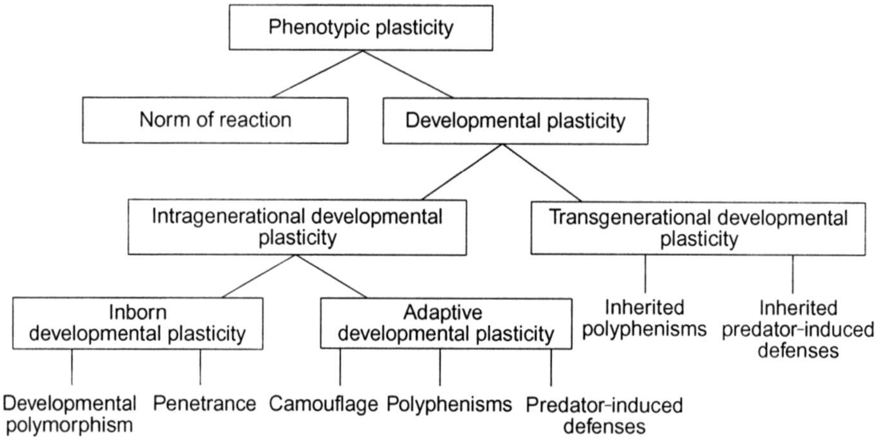

Figure 10.1 Main forms of phenotypic plasticity in metazoans under natural conditions.

the reason of differences in the phenotypic results of these two forms of phenotypic plasticity: the norm of reaction generally produces incremental *quantitative* changes in expression of the same character, whereas polyphenisms, under natural conditions, imply development of new or discrete changes, emergence of *qualitatively* different characters without intermediate forms that characterize the norm of reaction. The discontinuous changes observed in cases of developmental plasticity often are of the type "all-or-none" as it occurs in numerous cases of predator-induced defenses, developmental polymorphisms (animals give birth to offspring of two or more distinct morphs), or the development of different castes of insects with conspicuously different morphology, physiology, and behavior in social insects.

It is true that in a *limited* number of cases, when the environmental stimulus responsible for the norm of reaction is experimentally escalated beyond an upper limit, discrete changes of the type of developmental plasticity are produced. But this does not prove that there are no differences between the mechanisms of the norm of reaction and developmental plasticity. To the contrary, it proves that the norm of reaction and developmental plasticity result from clearly different mechanisms, with the later only resulting from activation of a new developmental pathway.

Most of the cases of developmental plasticity (polyphenisms) are related to attainment of a threshold in the sensitivity of the organism to external agents or conditions. But what determines a threshold or a set point? Metazoan organisms are under constant action of numerous external agents, but not all of them are perceived as stimuli. Whether an external agent will be perceived as a stimulus to which the organism responds by a developmental plasticity or not that is neurally determined. An external agent acts as a stimulus only when it is perceived as such in the central nervous system (CNS), i.e., when it attains a certain level of intensity, which represents a threshold or the set point beyond which the organism responds by a specific developmental plasticity. The threshold or the set point is determined by the specific neural circuit that processes the action of that agent or the stimulus in the CNS. When the agent is perceived as a stimulus (the intensity of action is above a neurally determined set point), the CNS activates a developmental pathway that leads to the sudden (not gradual or incremental) development of the new, discrete phenotypic trait, which is characteristic of the developmental plasticity rather than of the norm of reaction. By processing the stimulus in a specific neural circuit, the organism generates an output that switches on a specific developmental pathway leading to the development of different discrete phenotype.

Essential to keep in mind when the norm of reaction and the developmental plasticity are compared is the fact that the developmental plasticity, as a rule and as opposed to the norm of reaction, excludes intermediates between the two alternative forms.

Both the norm of reaction and developmental plasticity are biological concepts intended to categorize the ways an organism reacts to external stimuli: by incremental or by discrete changes in the phenotype. Concepts are created for organizing our information on the material and immaterial world and phenomena. In the case of the norm of reaction and developmental plasticity, we have two clearly defined, related but not overlapping, concepts. Hence, no change in our concepts and definitions of the norm of reaction and developmental plasticity is necessary. To blur the distinction is possible but hardly useful.

In the majority of described cases of developmental plasticity, the discrete phenotypic changes are not passed on to the offspring. Hence, they represent *intragenerational developmental plasticity* (e.g., polyphenisms, seasonal polyphenisms, predator-induced defenses, camouflage, changes in life history, discrete changes in the body mass). In cases when discrete phenotypic changes are transmitted to the offspring (or appear first in the offspring) of individuals that have experienced specific environmental stimuli, we speak of *transgenerational developmental plasticity*.

Intragenerational developmental plasticity may be *inborn* (appearance of more than one morph in the same brood), which represents *developmental polymorphism* (also known as *genetic polymorphism*) or may be acquired during the lifetime, what is known as *adaptive developmental plasticity* (e.g., camouflage, seasonal polyphenisms, predator-induced defenses, changes in the life history, changes in the age of maturity).

In developmental plasticity, I also include the so-called genetic polymorphisms. This is certainly controversial; hence, I must present the rationale for the inclusion. What have been considered to be genetic polymorphisms, the presence of two or more distinct morphs in the offspring of the same brood, by definition is not genetically determined, for all of the individuals of the brood are of the same genotype.

As a validation of the hypothesis that inborn polymorphisms are genetically determined is usually presented the appearance of Mendelian ratios but, as argued by West-Eberhard (2003), such ratios also appear in unambiguous cases of environmentally determined polyphenisms, such as caste determination in insects, suggesting that these ratios may be artifacts of constant experimental conditions.

Appearance of Mendelian ratios for specific phenotypic characters in the offspring is not necessarily an indication of existence of specific genes determining the characters. There is no scientific justification for deducing the presence of one or more genes based solely on the appearance of these ratios. In the modern era of genome sequencing, it is scientifically very risky and substandard to infer existence of genes for particular phenotypic characters solely based on the experimental ratios of the appearance of phenotypic characters in the offspring.

Mendelian ratios are determined by the fact that genetic factors, genes are transmitted to the offspring via the parental gametes, but now we know that in the same way are transmitted other hereditary nongenetic, epigenetic factors, such as thousands of parental cytoplasmic factors, whose presence can lead to similar Mendelian ratios.

Let us consider only two examples of the fallacy of the genetic approach of inferring presence of genes based solely on the appearance of the Mendelian ratios. Based on the appearance of the 3:1 ratio of workers to queens in the bee *Melipona marginata*, and 7:1 in other *Melipona* species, in 1950 Kerr concluded that these ratios were the results of differences in genotypes of larvae that produce workers from those that produce queens (Kerr, 1950).

Later, after the recognition of the role of the juvenile hormone (JH) in caste determining in *Melipona*, the same investigator put forward the hypothesis that in *Melipona quadrifasciata* there were two sets of sex genes: one set that acts in the embryo and determines ovary or testis, and another that acts in the prepupal stage determining the transformation of the imaginal discs and tegument in adult female or adult male structures. Next Kerr et al. (1975) recognized JH as an integral part of the gene control for caste determination in *Melipona*.

Then, in order to reconcile the new data on the JH control of caste determination in *Melipona*, they had to postulate that JH activates some unidentified "feminizing genes, inducing differentiation of female larvae into queens" (Bonetti et al., 1995). It has also been postulated that JH secretion is regulated by hypothetical genes in two loci. However, more than half a century of studies to demonstrate the existence of specific genes that in heterozygous state could regulate secretion of these hormones (Kerr, 1950) have failed.

Later studies have shown that caste determination in *Melipona* is determined by the amount of food received (van Veen, 2000) rather than any differences in genotypes of gyne-producing and worker-producing eggs. (Individuals of all the castes are in possession of the same set of genes.) Changes in the diet exert their influence on caste determination via the CNS, which regulates secretion of hormones JH and ecdysteroids. Caste determination in *Melipona* is a function of switches in endocrine developmental programs (Pinto et al., 2002), and JH and ecdysteroid switches are under control of neural epigenetic mechanisms (see Chapter 2 on the mechanisms of gene expression in the CNS) rather than under control of any hypothetic "feminizing" or "masculinizing" genes.

Now we know that another example of the unfortunate inference of the presence of genes from the observed Mendelian ratios of phenotypes in the offspring comes from studies on the cave-dwelling fish. Based on results of crosses between the blind hypogean and eyed epigean forms of *Astyanax fasciatus mexicanus*, in 1973 Sadoglu came to the conclusion that the loss of eyes in cavefish was caused by mutations in genes responsible for eye development (Wilkens, 1971) and that the number of degenerative mutations determines the degree of reduction or the loss of eyes (Wilkens, 1971). Now we know that no loss or mutations in genes involved in the loss of eyes has occurred in the blind hypogean form of *A. fasciatus mexicanus* (Jeffery, 2005).

Developmental Plasticity and Possible Evolutionary Implications

Here I will briefly discuss the possible involvement in the evolutionary process of the intragenerational developmental plasticity alone. The transgenerational developmental plasticity virtually represents a special form of evolutionary change, and therefore it will be considered as a separate issue in the next chapter.

Discussing the role of developmental plasticity in evolution, West-Eberhard (2003, p. 147) has argued that new phenotypic traits, arising in response to the changed environment or as a result of mutational events, may become evolutionarily relevant via processes of phenotypic and genetic accommodation. Phenotypic accommodation implies adjustments that are necessary for integrating the novel phenotypic trait to the general phenotype, in order "to reduce the amount of functional disruption occasioned by a developmental novelty," whereas genetic accommodation implies the presence of variation in alleles whose frequency will increase under new selection regime determined by the phenotypic novelty. Other authors also believe that novel phenotypic traits arise by environmental rather than genetic factors, and they become evolutionarily relevant via the process of genetic assimilation (Pigliucci and Murren, 2003; Pigliucci et al., 2006). Both hypotheses hold that genetic variation is necessary for phenotypic/developmental plasticity to be evolutionarily relevant. The difference is that the first hypothesis implies that the genetic variation is present in natural populations, and selection enables the evolutionary fixation of the phenotypic novelty, whereas the latter holds that phenotypic plasticity increases the chances of survival under changed conditions, and mutations that may occur thereafter (genetic assimilation) enable the evolutionary fixation of the novelty.

Theoretically, it might be argued that metazoans are capable of producing with the same genes not only small evolutionary changes but even different Bauplan (recall amphibian metamorphosis: during its lifetime, i.e., with the same genes, a frog sculptures such radically different Bauplan as that of the class of fish (as a tadpole) and amphibians (as an adult organism)), and flies, during their lifetime, sequentially develop both worm and insect Bauplan. The fact that in many cases the plasticity is not adaptive, indicates that accumulation of gradual mutational changes under the action of natural selection is not necessary for the evolution of the developmental plasticity. One should always bear in mind that developmental plasticity involves no changes in genes or DNA and hence is essentially a nongenetic, epigenetic phenomenon.

Cases of developmental plasticity described so far, generally, show no signs of "functional disruption" that would require any mutations in genes (Pigliucci and Murren, 2003).

But the fact that the developmental plasticity commonly is adaptive to the changed environment in response to which it arises suggests that it arises not spontaneously or randomly. The plastic response is somehow computed in the meaning that a relation is established between two unrelated elements, the environmental stimulus and the appearance of developmental plasticity. The establishment of this relationship implies as a *sine qua non* use of new information. The origin and nature of that information is essential for understanding the origin and nature of the epigenetic plasticity, the mechanism by which animals translate environmental stimuli into information for the phenotypic change.

All of the discrete morphological changes to be presented in this chapter start with, and essentially involve, reception and processing in neural circuits of visual, olfactory, tactile, or interoceptive input, perception or some other sensing by animals of environmental stimuli, including their predators. Since perception, or sensing, in any case takes place in the CNS, it is logical to assume that the brain is the place where the causal chain leading from the stimulus to the changed morphology or life history starts (see Chapter 2, Neural Manipulation of Gene Expression).

Adaptive Intragenerational Developmental Plasticity

Camouflage (Adaptive Coloration, Cryptic Coloration, and Crypsis)

Camouflage is generally defined as morphological adaptation "intended" to make an individual less visible in its background. It serves mostly preys to hide from their predators, but predators, too, may extract some hunting profits by being less visible. The chain of events from the visual perception of the environmental background to generation of a body pattern resembling the background is often very complex but already known in some details.

Let us consider some paradigmatic examples of adaptive coloration in animals and use current knowledge for understanding its mechanism.

> *Under strong disturbance or provocation, the cuttlefish Sepia officinalis undergoes a series of colour-changes as conspicuous and complete as they are rapid. The first of the successive patterns is one in which two large black spots appear on the dorsal surface of the mantle. Then a rapid and complete paling of the rest of the animal follows, while the black spots themselves become still more saturated and intense ... At the same time a black crescent forms beneath each eye, the pupil dilates and the edges of the fins become strongly outlined in black—the rest of the body remaining white.*
>
> *Cott (1966)*

The body pattern of a cephalopod, its gross appearance, is determined by skin chromatophores (from Greek χρωμα (hroma)—color and φορος (foros)—bear, carry). The change of color or patterning of the body in cephalopods occurs instantaneously and is accomplished by millions of chromatophores, multicellular organs consisting of a central pigment-containing cell attached to a set of 6–20 radially

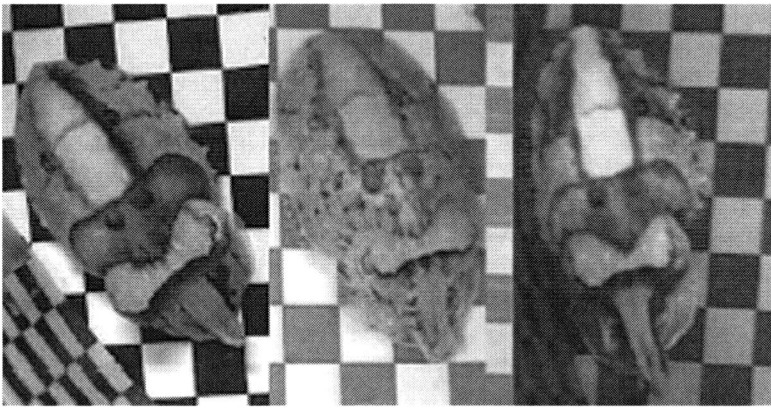

Figure 10.2 Images of cuttlefish showing disruptive body patterning on black and white, green and white, and green and black checkerboards.
Source: Selected from Mäthger et al. (2006).

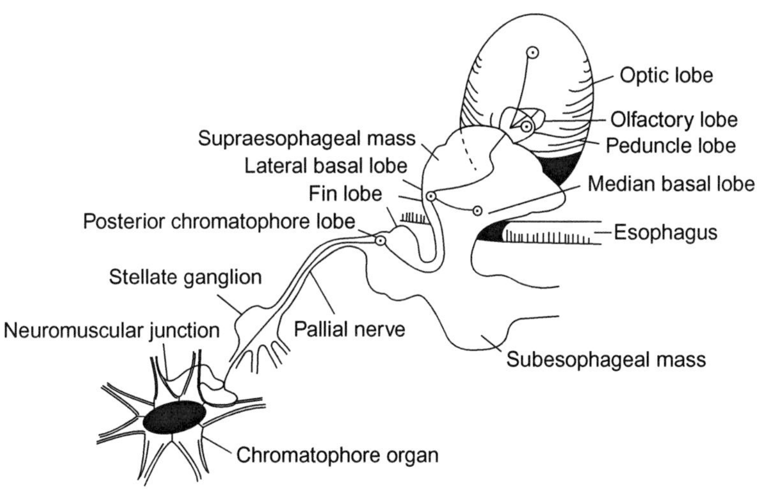

Figure 10.3 Diagrammatic representation of lobes in the cephalopod CNS that are thought to control chromatophore patterning. Only one side of the brain is represented, and the anterior chromatophore lobe is omitted for the sake of clarity. The main pathway is considered to be: optic lobes to lateral-basal lobes to chromatophore lobes to chromatophores.
Source: From Dubas et al. (1986).

arrayed chromatophore muscles. These muscles are innervated directly by the CNS, by motor neurons whose cell bodies in the European cuttlefish, *Sepia officinalis*, are located in the posterior subesophageal mass of the brain (Dubas et al., 1986; Gaston and Tublitz, 2006) (Figures 10.2 and 10.3).

Neural signals from the brain determine patterns of muscle contraction and, consequently, the dispersion of pigment granules within the chromatophore, so that in their totality the skin chromatophores produce a pattern that matches the background, achieves general resemblance to the substrate, or breaks up the body's outline. Neural control of chromatophores enables a cephalopod to change its appearance almost instantaneously, which is a key feature in some escape behaviors and during agonistic signaling. This rapid body pattern polyphenism is adaptive because it may hinder search-image formation by predators.

By releasing neuropeptides of the FMRFamide-related peptides family and glutamate in the chromatophore neuromuscular junction, these motor neurons regulate contraction of muscles and the color patterning according to the perception of the background that forms in the cuttlefish's brain (Loi et al., 1996; Loi and Tublitz, 2000). Thus, *S. officinalis* first draws the camouflaging image of its body in the brain. The whole pathway of body patterning is neurally determined (Messenger, 2001).

Chromatophore lobes in *Octopus vulgaris* contain over half a million neurons. The fact that those animals change their body pattern to match their background within a very short time of about a second in cuttlefish (Hanlon and Messenger, 1988), and 2–8 s in fish, seems to exclude involvement of any inducer molecules (hormones or secreted proteins) as a possible regulator of body patterning:

> *A neural rather than humoral, mechanism must be involved.*
>
> *Ramachandran et al. (1996)*

Now we know that specific small molecule and peptide neurotransmitters released at the cephalopod neuromuscular junction induce chromatophore functions. The first neurotransmitter to be recognized to have a role in the functioning of chromatophores has been the glutamate. Recently, it has been shown that endogenous neuropeptides of the FMRFamide family are involved in the body patterning of *S. officinalis* (Hanlon and Messenger, 1988; Loi et al., 1996).

And since this form of camouflage results from disruptive coloration, i.e., from a combination of a varied number of chromatic units, it has been proposed that a physiological unit in the brain must be responsible for each chromatic-morphological unit in the skin (Packard, 1982). The size, contrast, and number of white squares in a black background are cues the cuttlefish (order Sepiidae) use to switch from uniformly stippled skin patterns (general resemblance) to disruptive skin patterns (Chuan-Chin and Hanlon, 2001).

The pathway of patterning signals from the brain physiological units to the skin chromatophores in cuttlefish, within the broader schema of information flow along the body patterning circuitry, has been schematized as follows:

> *visual input → eyes → brain optical lobes → brain lateral basal lobes → brain chromatophore lobes → skin.*
>
> *Chuan-Chin and Hanlon (2001)*

Intraspecific signaling and communication is another function of chromatophores that is well documented in several inshore species, and interspecific signaling, using ancient, highly conserved patterns, is also widespread. Neurally controlled chromatophores lend themselves supremely well to communication, allowing rapid, finely graded, and bilateral signaling.

Many crustaceans can also rapidly modify their color so that it matches the background. Blood-borne factors involved in their phenotypic adaptive coloration belong to two main groups. There is a group of neuropeptides from the X organ–sinus gland complex, which acts directly on chromatophores, causing either dispersion or concentration of the pigment granules. The second group consists partly of neurotransmitters acting within the nervous system by triggering the release of chromatophorotropic neuropeptides. The rest of them, primarily amines, are released in hemolymph from pericardial organs (Gorbman and Davey, 1991). Two main neurohormones involved in adaptive coloration of crustaceans are light-adapting distal retinal pigment hormone, with light-adapting function, and its antagonist neurohormone, red pigment-concentrating hormone (RPCH), with the latter being responsible for adaptive coloration of all pigment cells (melanophores, erythrophores, leukophores, and xanthophores) (Josefsson, 1983).

It is demonstrated that electrical stimulation of eyestalks in Crustacea results in the release of peptides with activity on specific types of chromatophores (Gorbman and Davey, 1991, p. 738). The neuropeptide, RPCH, also controls the color changes in shrimps (Strand, 1999) and modulates the swimmeret activity rhythms in the crayfish. Neurohormones regulate color changes and movement of pigments during light-and-dark adaptation of eyes in other arthropods (Pearse et al., 1987).

The CNS also determines skin color changes in many fish and amphibians. Fish scales and amphibian skin also contain specialized pigment cells, *melanophores*. The neurohormonal stimulation of these cells, under conditions of stress/alertness, stimulates melanophores to transport (along microtubules) membrane-enclosed pigment granules toward the periphery (producing darkened cell color) or center (producing pale cell color) (Ramachandran et al., 1996). In response to artificial changes in the background, fish (Ramachandran et al., 1996), like cuttlefish (Chuan-Chin and Hanlon, 2001), repattern their body to match the background if the backgrounds display "appreciable differences" (Marshall and Messenger, 1996). This clearly implies visual input and perception of the background, and since that perception takes place in the brain, the color change and repatterning of the skin are clearly under control of the CNS:

> *The fish must have independent visual control of each set (or subset) of markings, a possibility that requires verification. There may be "feature detectors" in the fish visual centres that are specialized for detecting different spatial frequencies of textures in the environment and these might exert direct control over the corresponding set of marks on the skin surface. Indeed, there might be a map of effector neurons in (say) the tectum, so that focal electrical stimulation might produce selective contrast enhancement of specific spatial frequency components on the skin.*
>
> *Ramachandran et al. (1996)*

The paradise whiptail, *Pentapodus paradiseus*, is a fish inhabiting the coastal waters off Queensland, Australia. The fish has colored stripes on its body and can change its color from blue to red within less than 1 s (Mäthger et al., 2003). The coloration of this and many other fish is determined by iridophores (light-reflecting cells) in the skin. They contain thin guanine plates, which are multilayer reflectors with refractive index higher than spaces separating them. Plates are connected with each other by microtubules. Any change in the distance between plates contained within the reflective cells will produce a change in the reflected color. Changes in the structure or length of microtubules change the distance between guanine plates, and this leads to changes in the color reflected by iridophores. It has been found that the sympathetic nervous system regulates the distance between plates and the body color by regulating the length (polymerization/depolymerization) of microtubules. Under stress conditions, the release of noradrenaline shifts the reflection toward red (longer wavelength), whereas in response to topical application of the neurotransmitter acetylcholine, spectral reflections shift toward shorter wavelengths (Oshima and Fujii, 1987; Mäthger et al., 2004). Fluctuations in intracellular Ca^{2+} also change the structure of microtubules (Mäthger et al., 2003) and, consequently, the distance between plates and the color reflected by iridophores. In experiments, it has been found that the neurotransmitter acetylcholine increases the Ca^{2+} level (Mäthger et al., 2004). Studies on the color change (from blue to green) in the blue damselfish, *Chyrsiptera cyanea*, and in the neon tetra (Nagaishi and Oshima, 1989) have led the investigators to the conclusion that

> *the motile iridophores are solely under the control of the sympathetic adrenergic system, and that the co-transmitter, adenosine, may function to antagonize quickly the true transmitter-induced colored state of the cells.*
>
> *Kasukawa et al. (1986)*

Monocirrhus polyacanthus is a little Nandid fish in the Lower Amazon Valley. It resembles a dead leaf. Populations of this fish, known as *Peche de folha* by the Brazilian natives, consist of individuals of three main color groups (light gray, golden with only a few mottlings of dark brown and brown). All of them are capable of changing within 1 h their color to darker or lighter tint according to the tones of the background (Cott, 1966). At his time, O. v. Frisch marveled: "How the fish brain can command the pigment cells so skillfully challenges comprehension" (Frisch, 1973).

When on a white background, the teleost fish medaka, *Oryzias latipes*, gradually acquires a lighter body color. This is the result of two processes: initially, reduction of the size of skin melanophores and later the programmed cell death, apoptosis of dark pigment cells, melanophores (Sugimoto, 2002). The opposite is observed when medaka fish are kept on a dark background: size reduction and apoptosis of leukophores—white pigment cells. Experimental chemical denervation suppresses apoptosis of melanophores, suggesting that the process of melanophore apoptosis is also regulated by skin sympathetic innervation (Sugimoto et al., 2000).

Some teleost fish also change their body color/patterning as a way of communication in their social interactions. In Arctic char, *Salvelinus alpinus*, subordinate individuals, in the presence of dominant individuals, experience constant stress and activation of the hypothalamus–pituitary–interrenal axis via serotonergic neurons. This is associated with a submissive behavior and darkening of the skin as a result of increased secretion of the pituitary α-melanophore-stimulating hormone (α-MSH) and adrenocorticotropic hormone. By contrast, release of neurotransmitters catecholamine, dopamine, and norepinephrine suppresses secretion of the above pituitary hormones and induces aggressive behavior and lighter body color. The darkening of the skin in subordinate Arctic char fish has been interpreted as being intended to reduce unnecessary fights and energy loss "in an established dominance hierarchy" (Höglund et al., 2002).

The clawed toad, *Xenopus laevis*, also modifies its body color so that it matches various environmental backgrounds. The hormone determining this adaptation is α-MSH, synthesized and secreted by the pituitary under central neural control. Various neurotransmitters are involved in the release of this hormone by the pituitary. Using axonal tract tracing, Tuinhof et al. (1994a,b) have identified details of the neural circuit (from retina to the hypothalamus), whose signals to the pituitary stimulate the release of the hormone and make the adaptive coloration possible (Tuinhof et al., 1994a,b). Neurons of the suprachiasmatic nucleus project to the pituitary α-MSH-producing cells, where they release neurotransmitters regulating α-MSH synthesis (Tuinhof et al., 1994a,b). Another serotonin center in raphe nucleus innervating the pituitary is also involved in the regulation of the secretion of the hormone (Ubink et al., 1999) (see also Section Generation of Information for Adaptive Camouflage in *Xenopus* in Chapter 2).

Larvae of salamanders of the family Ambystomatidae, when kept in total darkness, blanch after about 1 h. When illuminated, they darken in a dark background and brighten when the background is white. Predictably, removal of eyes prevents these camouflage responses in salamanders.

Mimicry is a special form of camouflage whereby a species (the mimic) develops morphologies and behaviors resembling those of other species (the model). The adaptation of the mimic to the model may be profitable to the mimic in the case of Batesian mimicry, or may be profitable for both mimicry partners in cases of Müllerian mimicry. A hypothesis presented by Darwin (1874) posits that Batesian mimics began to evolve at a remote past, when the model and mimic were much more similar. Even if right, the hypothesis would only account for a limited number of known cases of mimicry (Turner, 1977).

Juveniles of the predator coral fish, fang blenny (*Plagiotremus rhinorhynchos*), develop a striking resemblance to their prey, cleaner wrasse (*Labroides dimidiatus*) but only when about 1 m from the prey; when removed from the prey/model, they lose the mimetic coloration and restore the normal color (Moland and Jones, 2004).

Environmental stress (e.g., heat shock or cold shock) can modify the wing patterns in butterflies in such a way that they resemble certain wing pattern mutants, implying that morphological changes attributed to mutations are not always or necessarily related to genes, as it has been for a long time taken for granted.

There are many known cases of coloration and mimicry that do not offer any advantage to the carriers, what casts doubt on the role of the natural selection in the evolution of the mimicry. As Morgan would put it more than a century ago:

> *There is no need to question that in some cases animals may be protected by their resemblance to other animals, but it does not follow, despite the vigorous assertions of some modern Darwinians, that this imitation has been the result of selection.*
>
> *Morgan (1903)*

With neural mechanisms proven to regulate camouflage (cryptic coloration), there is no visible reason for excluding the possibility of the involvement of the CNS in the evolution of mimicries. Any case of camouflage and mimicry implies, as a *sine qua non*, perception of the body color and neural representation of the model in the brain of the camouflant and mimic. While the neural processing of visual stimuli leading to the cascade of events that change the color and pattern to produce cryptic coloration is experimentally determined in numerous cases, we are still ignorant about the mechanisms used by the mimics for transmitting the mimicry to their offspring. It could be argued that the insect, the fish, and the clawed frog are aware of the fact that the mimicry and adaptive coloration contributes to their survival no more than an amoeba knows that its debris-engulfing routine is necessary for its survival. This argument is hardly relevant, since the control of the CNS on adaptive changes of color in both fish and *Xenopus* takes place on an unconscious level, and for the color adaptation to take place it is not necessary for these animals to "know" what they are doing. The unconscious instinctive "knowledge" is all they need to adapt their color or pattern to the background.

The conscious–unconscious dualism is an anthropocentric view rather than an evolutionary principle. No matter at what level of the CNS activity cryptic coloration might be controlled and regulated, it is an *adaptive*, nonrandom phenomenon, based on the innate instinct that contributes to animal's survival.

Commenting on the insect mimicry, more than a century ago, Alfred Tylor wrote:

> *Here, in this common British butterfly, we have the whole problem set before us—vivid colour, the result of intense and long continued effort; grand display, the object of that colour; dusky, indefinite colour, for concealement; and the "instinctive" pose, to make that protective colour profitable. The insect knows all this in some way.*
>
> *Tylor (1886)*

Summarizing the evidence on the mechanisms of adaptive coloration presented in this section, it is noteworthy that no changes in genes are involved and that no hypothetic mechanism has ever been presented to show how changes in genes may induce adaptive coloration. The vast majority of the known cases of adaptive coloration could reasonably be explained based on the experimental evidence on the neural mechanisms of coloration determined for a number of species.

Polyphenisms in Invertebrates

Populations of a marine bryozoan, *Membranipora membranacea*, are polymorphic for inducible spine type and consist of a constitutively spined type (6.2%), which produces spines in the absence of the predator stimulus, an unspined type (13.4%), and an inducible type (80%), which produces spine when it detects the presence of its predator in the environment (Harvell, 1998). This is a complex case that comprises both polyphenism and predator-induced plasticity (in the case of the inducible type) triggered by the perception of the predator or its kairomones in the bryozoan's brain.

Butterflies of the families Nymphalidae, Pieridae, and Papilionidae produce pupae that display different body colors, depending on the color of the pupation site.

The peacock butterfly *Inachis io* (Nymphalidae), the white butterfly, *Pieris brassicae* (Pieridae), and *Papilio polyxenes* (Papilionidae), as pupae, display one of two alternative cryptic colors, green-yellow or brown-black, depending on the prevailing color at the pupation site. In order for the pupae to be able to adopt the body color that better matches the color of its surroundings, they first must perceive the prevailing color of the pupation site. This perception takes place in the insect's brain. In the insect's brain is also synthesized the first signal of the signal cascade leading to production of the cryptic body color (green or brown). That signal is a neuropeptide, pupal melanization-reducing factor (PMRF), which is released from the brain into the hemolymph. PMRF is located throughout the entire CNS, but its release during the prepupal stage is controlled by neural stimulation from the brain. Experimental inhibition of PMRF release by applying ligatures results in pupae with brown color anterior to the ligature and green color posterior (Jones et al., 2007) (Figure 10.4).

Figure 10.4 *Papilio glaucus* pupa that was ligatured as a prepupa 2 h after proleg retraction. Larval skin was removed 2 days after pupation to expose pupal cuticle.
Source: Photo by Robert Stark from Jones et al. (2007).

When pupation of butterflies of the species *I. io* and *P. brassicae* takes place in a light-green background, the pupa synthesizes PMRF, which starts a signal cascade that leads to accumulation of lutein into cuticle and appearance of the green-yellow body color. When pupation takes place on a dark site, PMRF is not produced, lutein is not incorporated into the cuticle, and the pupa appears black. Butterflies of the Papilionidae family, although using the same neurohormone for determining their cryptic body color, seem to rely on a different mechanism:

> *If the plasticity in pupal color in I. io and P. polyxenes is controlled by PMRF, then the neural control of the hormone's release would have to be different in the two species. In I. io (and P. brassicae) the hormone is released when pupation is on or in green vegetation, while in P. polyxenes it is released when pupation is on a brown surface.*
>
> *Starnecker and Hazel (1999)*

Experiments on *Graphium sarpedon nipponum* have shown that this butterfly determines its protective body color during the pupal stage, according to the color of the pupation site: green on white background and reddish brown on black background.

Pupae of *Papilio xuthus*, however, do not use the background color but the smoothness of the surface of leaves and twigs, on which they molt, as cues for determining their body color. This indirect cue also leads to adaptive coloration. On leaves and new twigs, which are smooth and green, they become green, while on dead branches, which have rough surfaces and brownish color, they become grayish brown. For explaining the mechanism of the body colors in these cases, Hiraga (2005) has presented the "tactile signals accumulation model," according to which body coloration is related to accumulation of tactile signals and perception in the pupal brain of the smoothness/roughness of the surface of the pupation site (Figure 10.5).

Many butterflies show a marked pupal color plasticity, which depends on the color of the pupariation background. We know that the pigmentation of the pupae is neurally determined. The visual stimuli of the background color are received and transmitted to the brain for processing in neural circuits resulting in the synthesis and release of a neurohormone called PMRF by secretory neurons in the CNS and various nervous ganglia (Bückman and Maisch, 1987).

When pupariated on a light-colored background, neurons in the brain of the pupae of the peacock butterfly *I. io* of the nymphalid family, and the large white butterfly *P. brassicae* of the pierid family, secrete PMRF, the "browning hormone." The neurohormone leads to incorporation in the cuticle of the lutein, which turns their body color yellow. When pupariated on dark background, they do not produce/secrete PMRF, leading to appearance of a green body color. Injection of extracts of ganglion chains in prepupae produces effects of the pupation site of the donor of the ganglia. A model of the neural control of pupal color in swallowtail butterflies is presented in Figure 10.6.

In contrast, when pupariated on the same light-colored background, butterflies of the Papilionidae family, such as *Pa. xuthus* and *Pa. polyxenes*, release no PMRF and

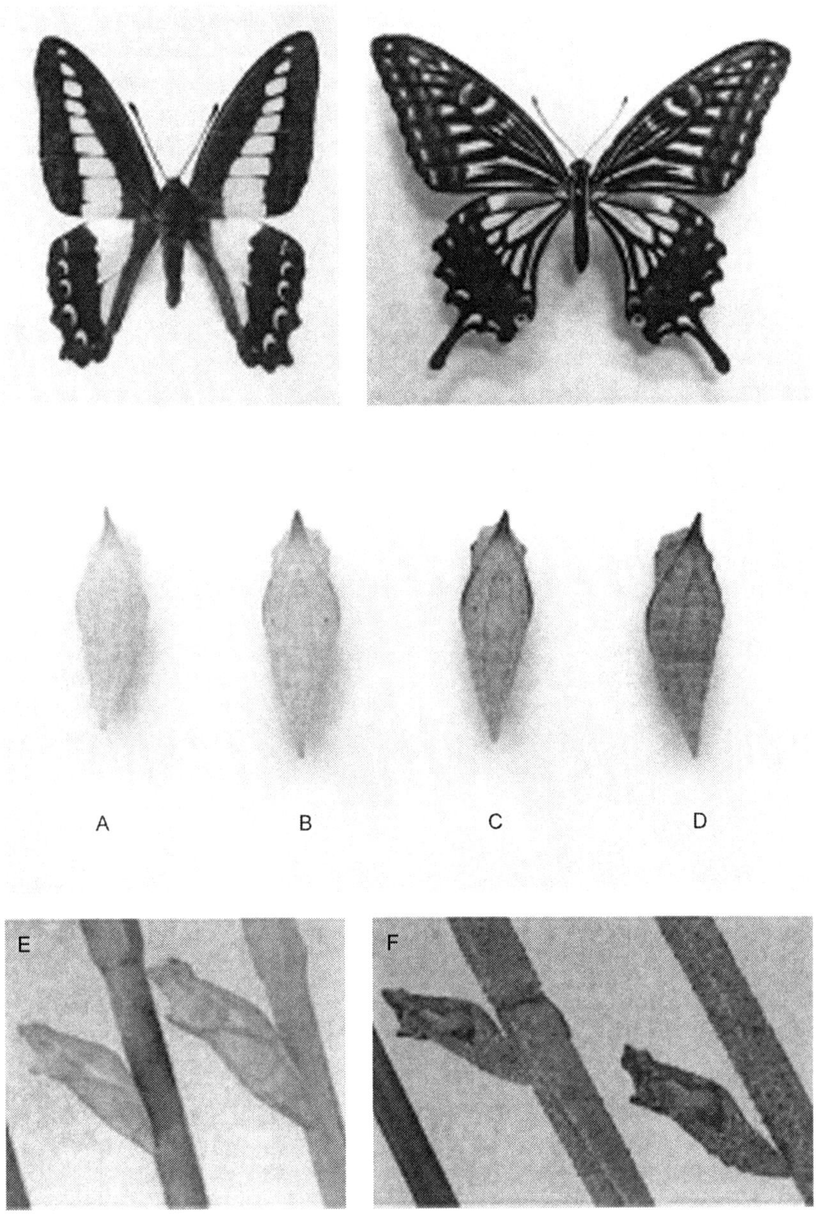

Figure 10.5 Upper panel: adult butterflies of *Graphium sarpedon* (left) and *Papilio xuthus* (right). Middle panel: four pupal color types of *G. sarpedon*. A, bright yellowish green; B, pale green; C, grayish green; D, reddish brown. Lower panel: pupal color types of *Pa. xuthus*. E, green; F, dark brown.
Source: From Hiraga (2005).

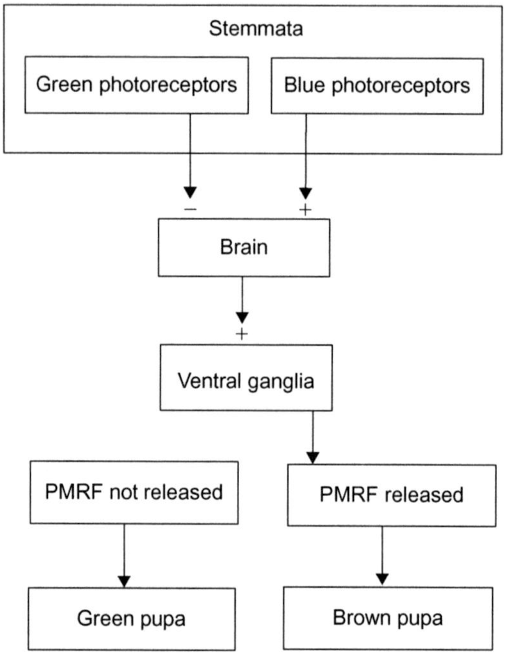

Figure 10.6 Model for the proximate control of pupal color in swallowtail butterflies (family Papilionidae). Plus and minus signs indicate stimulatory and inhibitory effects, respectively. Note that when pupation is in green vegetation, the green photoreceptors will be stimulated, ultimately resulting in the production of a green pupa by inhibiting pupal melanization-reducing factor (PMRF) release.
Source: From Mellencamp et al. (2007).

develop a yellow body color; on dark background, they release PMRF and display a green body color. Thus, secretion of the same neuropeptide, PMRF, determines two contrasting body colorations, yellow coloration in Nymphalidae and Pieridae butterflies, and green coloration in papilionides (Starnecker and Hazel, 1999).

How did the divergence in the body color evolve?

It has been hypothesized that both nymphalides and pierides inherited this plasticity from their common ancestor. But since both these families are more remotely related to papilionides, it is more likely that the contrasting polyphenisms evolved after these families diverged from the common ancestral group. Essential to remember in this context is the fact that initial information for color plasticity is generated by the processing of the visual signals on the pupariation background, which leads to two contrasting results: secretion of PMRF in one species and suppression of PMRF secretion in the other. This suggests that

> *the neural control of the hormone's release would have to be different in the two species.*
>
> *Starnecker and Hazel (1999)*

In some cases, social and olfactory stimuli also stimulate adaptive responses in the form of changes in phenotype and life history. So, for example, exposure of female cockroaches of the species *Nauphoeta cinerea* to conspecific males, or even to male pheromones alone, affects their time of reproduction, increases the number of offspring they produce, and increases the biases of producing parthenogenetic offspring in this facultative parthenogenetic species (Moore and Moore, 2003).

A proportion of eggs of *Drosophila mercatorum* develop parthenogenetically into adult flies. The proportion of unfertilized eggs, which develop parthenogenetically into adult flies, increases with the decrease of the population density (Kramer and Templeton, 2001), suggesting that social or olfactory factors play a role of in determining the parthenogenetic development.

In North American field crickets, *Gryllus firmus* and *Gryllus rubens*, formation of wings is determined by lack of JH, as may be concluded from the fact that in nonpolyphenic winged female crickets, *Gryllus assimilis,* formation of wings is prevented by stimulating JH secretion (Zera et al., 1998), which is under strict control of brain neurohormones (allatotropins and allatostatins). Flightless female offspring with enlarged ovaries have been produced in the nonpolyphenic cricket, *G. assimilis*, by simply applying a JH analog on adult females. Experimental increase of the level of JH in flight-capable crickets not only increases the ovarian size, similarly to flightless morphs, but also leads to the loss of flight muscles by histolysis (Zera and Cisper, 2001).

Populations of the common pond skater (a long-legged insect gliding over water), *Gerris lacustris* (L.) (Heteroptera: Gerridae), in Bavaria, Germany, show remarkable environmentally induced discrete phenotypic differences in life history and morphology. Populations living in the field ponds are bivoltine (i.e., produce two generations annually) with predominantly long-winged individuals, while forest pond populations are univoltine with an increased proportion of short-winged flies (Pfennig and Poethke, 2006).

Seasonal Polyphenism in Insects

Each year, the African butterfly *Bicyclus anynana* produces two different seasonal morphs: a wet-season generation of individuals with large marginal eyespots of several colors and a transverse band on the ventral side of the wings, and individuals with small eyespots of fewer colors and no band on their wings in the dry season (Figure 10.7).

Both patterns are adaptive under respective conditions in environment: the patterning of the dry-season morph is cryptic, resembling the brown leaves on the forest floor, whereas the conspicuous patterning of the mobile wet-season morph might serve as a warning to predatory birds and lizards.

It is observed that earlier secretion and higher levels of ecdysteroids in the young pupae are responsible for the wet-season morph with large eyespots and transverse bands on their wings (Brakefield et al., 1996; Koch et al., 1996) and ecdysteroid treatment, depending on temperature, may enlarge eyespots by locally regulating the synthesis of pigments. Recall that production of ecdysteroids is under the control of the

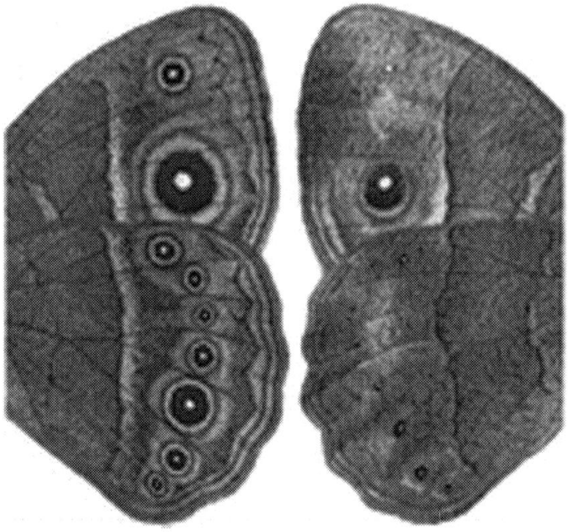

Figure 10.7 *Bicyclus anynana* wet- (left) and dry-season-like (right) phenotypes obtained by rearing larvae at different temperatures. Note that the larger eyespot on the forewing is typically hidden behind the hindwing in resting butterflies (the posture relevant for the antipredatory strategies described). Also, note that wing size (typically larger in dry-season phenotypes) was adjusted to emphasize comparison of color patterns.
Source: From Beldade et al. (2011).

insect CNS: it is stimulated by the prothoracicotropic hormone (PTTH), a brain neurohormone, and inhibited, at least in some insects, by another hormone and by direct neural control (Chapman, 1998).

The tropical butterfly, *B. anynana,* also responds to the predictable seasonal changes in temperature by adaptively changing the egg size, and this change is predominantly nongenetically, maternally, determined (Steigenga et al., 2005). In order to perform the eyespot-inducing function, ecdysteroids must bind their nuclear receptor, ecdysteroid receptor (EcR), which forms a heterodimer with ultraspiracle protein. The final color of the scale cells in *B. anynana* wings coincides with expression of nuclear EcRs, and the patterns of EcR expression in the wet-season and dry-season butterflies are different (Koch et al., 2002, 2003).

The complex patterning of the wing eyespots results from a complex spatial pattern of the activity of the hormone, but ecdysteroid hormones are released in the hemolymph and are uniformly distributed all over the butterfly wing and body. Hence, it is plausible that the patterning and the color of the wings may be determined by the patterns of expression of the EcR in the butterfly wings.

The close correlation between expression of EcRs and colored scale cells in *B. anynana* wings raises the critical question: What determines the expression patterns of EcRs, and ensuing patterning and colors, in *B. anynana*'s wings?

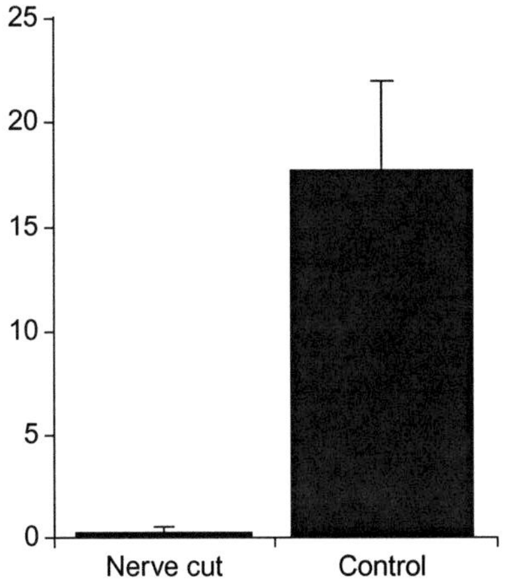

Figure 10.8 Effect of unilateral axotomy on the number of myonuclei expressing EcR-B1 within a 10,000 mm^2 of the central portion of the intact and denervated fiber. Axotomy was performed on diapausing pupae; 72 h later, the pupae were injected with 20E to initiate adult development. *Source*: From Hegstrom et al. (1998).

It is known that ecdysteroids themselves upregulate EcR expression, but given the fact that they circulate with hemolymph, they cannot determine the spatial pattern of expression of EcR in the eyespot foci. One specific inducer of EcR expression in insects is local innervation, as it has been demonstrated in experiments with other insects where denervation prevents upregulation of EcRs by ecdysteroids. This is also corroborated by the fact that denervation (axotomy of the motoneuron) of the dorsal external oblique-1 muscle in *Manduca sexta* almost totally prevents expression of the EcR-B1 (Figure 10.8) and the growth of the muscle. The inducer of EcR-B1expression released from the motoneuron is probably a diffusible factor because the effects on the EcR-B1 expression are observed beyond the contact of the muscle with the arbor (Hegstrom et al., 1998).

Based on their experiments, investigators concluded:

> *Innervation regulates the choice of EcR isoforms expressed in growing muscle.*
>
> *Hegstrom et al. (1998)*

The main external cue activating the neurohormonal mechanism for seasonal diphenism in *B. anynana* is the temperature during larval–pupal stages (temperatures >24°C induce wet-season eyespot morph, and temperatures under 20°C, dry-season, eyespotless morph).

By manipulating temperatures and selecting (for absence of eyespots under low temperature) for up to 20 generations, investigators succeeded in obtaining monophenic forms of *B. anynana*, which produce only one of the phenotypes (with or without eyespots) even when reared under alternative environmental temperature. This is an experimentally induced evolutionary change involving no change in genes.

Seasonal polyphenism is also observed in many other insects. For example, the black swallowtail butterfly, *P. polyxenes* produces larvae of darker color in fall, and larvae of lighter color in summer. The polyphenism seems to be adaptive, since darker larvae, which are produced when the photoperiod is shorter and the environmental temperature lower, have a higher growth rate (Hazel, 2002).

In response to experimental winter-approaching cues and predator cues, the damselfly, *Lestes sponsa*, shifts to the so-called cohort splitting, which is diapausing and overwintering by minimizing the developmental rate (Johansson et al., 2001) until spring. Remember, cues such as photoperiod and temperature are analyzed in the CNS, and the CNS is the site where the external stimuli are related to the changes in color of the larvae.

The European map butterfly, *Araschnia levana* L., exhibits strong seasonal polyphenism. It has two seasonal forms: the short-day spring generation developing from diapausing pupae into red butterflies with black spots, and the long-day summer generation of black color with a vertical white stripe that develops from nondiapausing pupae but also develops larger body and wings and shows greater mobility. The seasonal diphenism of *A. levana* L. is also regulated by timing of ecdysteroid secretion (Koch, 1987), which is under neural photoperiodic control (Fric et al., 2004). Switching from one morph to the other is determined by the length of the photoperiod (Short photoperiods induce formation of the spring morph, and long photoperiods, the summer morph.) (Koch and Bückmann, 1987).

Larvae of a bivoltine race of the silkmoth, *Bombyx mori*, can change from the summer morph into the autumn morph, and the reverse, under action of other neurohormones, such as the brain–subesophageal ganglion (Br–SG) complexes and various neuropeptides, such as summer morph-producing hormone (SMPH) and diapause hormone (DH).

Neo-Darwinian Explanation of the Seasonal Polyphenisms in Insects

The neo-Darwinian explanation of the seasonal polyphenism of *B. anynana* presented here is based on the interpretation of the phenomenon in the case of some *Drosophila* spp. of California by Theodosius Dobzhansky, one of the founders of the neo-Darwinian evolutionary synthesis. These *Drosophila* species in spring produce generations adapted to warm weather, whereas the autumn offspring is adapted to cold weather (Dobzhansky, 1971):

> *The readaptation to warmth need not be any more difficult than the adaptation to cold was; both will depend on the availability of genetic raw materials on which natural selection can act. But the readaptation may occur in two ways.*
>
> *Dobzhansky (1971, p. 170)*

Dobzhansky tried to explain the phenomenon with the presence of genetic variability, i.e., the presence of the necessary adaptive genes in the population's gene pool:

> *If old genes, adaptive to warmth were not completely eliminated from the population during the cold phase, they may now be selected and the gene pool may revert to its old state ... a microevolutionary change occurs every year and is undone as the season changes.*
>
> *Dobzhansky (1971, p. 170)*

Decades after this hypothesis was presented, there is no hint on the existence of "genes adaptive to warmth" in these species. But the strange statement that "a microevolutionary change occurs every year and is undone as the season changes" may be responsible for adaptation of a myriad of *Drosophila* individuals contradicts population genetic knowledge.

A strictly determined "microevolutionary change" that would affect whole populations is an impossible event, hence absurd, from the neo-Darwinian view itself. My imagination is too weak to envisage how any "microevolutionary change" could be related to the seasonal polyphenism of *B. anynana*, but for the sake of argument, let us take it for granted that such an event is possible. This, however, would cost the population an exceptionally high death rate every 6 months. It does not occur, because, if it did, this high death rate would not escape observation.

Epigenetic Explanation of Seasonal Polyphenisms in Insects

In all examples of seasonal polyphenism presented here, a neural component is involved in determining the color and patterning of insect wings. Environmental stimuli related to the appearance of seasonal polyphenism are perceived in the CNS of the insect. For example, the changes in wing patterning between the two seasonal morphs of *B. anynana* are systematic and affect all of the individuals in populations, a fact that excludes any involvement of gene mutations in expression of the seasonal polyphenism. The basic difference (presence or absence of eyespots in the wings) in the East African butterfly is determined, on the one hand, by the timing of expression of the ecdysteroid hormone, which is secreted under strict neural control, via the brain hormone PTTH, and, on the other, by the expression of the EcR exclusively in the wing eyespots, which is regulated by local innervation, at least in the case of expression of this receptor in insect muscles (Hegstrom et al., 1998).

Wing Polyphenism in Insects

The flesh fly *Sarcophaga argyrostoma*, like most flesh flies, is ovoviviparous. In autumn, under short-day conditions, it gives birth to offspring that diapause as pupae, whereas during summer, it produces long-day nondiapausing and direct-developing generations. Using techniques of artificial uterus, investigators found that this photoperiod-related developmental plasticity, although determined during the intrauterine period, is not induced maternally but is determined by the embryonic CNS after it becomes operative during intrauterine life. Investigators conclude that photosensitive period in embryos

> *may begin when the embryonic central nervous system is sufficiently developed and continues through larval development and the period of post-feeding "wandering" to come to an end before or at puparium formation.*
>
> *Kenny et al. (1992)*

Clearly, no specific changes in genes, DNA, or selection are involved in the evolution of the dramatic change of this life history character.

In the aphid *Megoura viciae* Buckton (Homoptera, Aphididae), the photoperiod and temperature determine whether the female will reproduce sexually or parthenogenetically. In response to long (16h) days and temperatures higher than 15°C, the aphid loses its wings and switches from the normal ovipara production of the adult gynopara to production of vivipara. Both temperature and day length are perceived in the CNS. Hence, it is logical to infer that external stimuli (temperature and density) received by sensory organs and processed in the brain enable the latter to respond appropriately by secreting neurohormones that induce (allatotropins) or inhibit (allatostatins) secretion of JH in corpora allata, leading to or inhibiting formation of wings, respectively.

Exposure to long days and 25°C temperature, as well as local application of JH, induces apterization in *Aphis fabae* larvae. The wingless insects switch from ovipara to vivipara production (Hardie, 1981). Apterized insects can be distinguished from juvenilized insects in the fifth instar. Topical application of JH induces both apterization and juvenilization (attainment of sexual maturity at larval stage) of presumptive gynoparae at different times during larval development; JH treatment during the early instars promotes apterization but induces little juvenilization, whereas maximum juvenilization, without apterization, is produced by middle-instar treatment. Thus, the apterizing effects of JH do not depend on its neotenic action. The response profile of JH-induced apterization is similar to that observed with long days and 25°C temperature. It is suggested that such environmental stimuli increase endogenous JH levels in *A. fabae*. Both long-day and JH-apterized insects switch from the normal ovipara production to vivipara production.

Not only is the function of corpora allata and JH secretion by these glands under the control of the CNS, but it is experimentally determined that the photoperiodic mechanism of alternation of sexual and asexual generations in the aphid *M. viciae* Buckton is neurally determined, and its effector is a neurosecretion of the insect protocerebrum (Steel and Lees, 1977). Thus, the seasonal switch to two alternative reproductive modes in this insect is epigenetically regulated by neural mechanisms involving no changes in genes or genetic mechanisms in general (Figure 10.9).

Experimental Polyphenisms in Insects

Daphnia magna females produce only female offspring, but their oocytes, when exposed to aqueous solutions of the crustacean hormone methyl farnesoate (a terepenoid synthesized, under neural control, by the mandibular organ), during the late ovarian development, develop exclusively into males (Olmstead and LeBlanc, 2002, 2003). The reprogramming of the oocytes to produce males by the same genotype that produces females shows that sex in *Daphnia* is determined nongenetically, epigenetically.

Cases of male polyphenisms (production of winged and wingless males) reported in some ant species of the genus *Cardiocondyla*, such as *Cardiocondyla obscurior*, are very interesting, not only because they are observed among individuals of the same sex (males) of the same species, but also because of the different nature of the external stimuli that induce their appearance. Experimental data show that this polyphenism is related to environmental stress rather than any genetic polymorphism (Cremer and Heinze, 2003; Schrempf and Heinze, 2006). The main inducing factor of the caste polyphenism in these ants is a sudden drop of at least 5°C in temperature, but other stressors, such as reduction of the colony size by experimental splitting of the colony and food shortage, also lead to increased production of winged

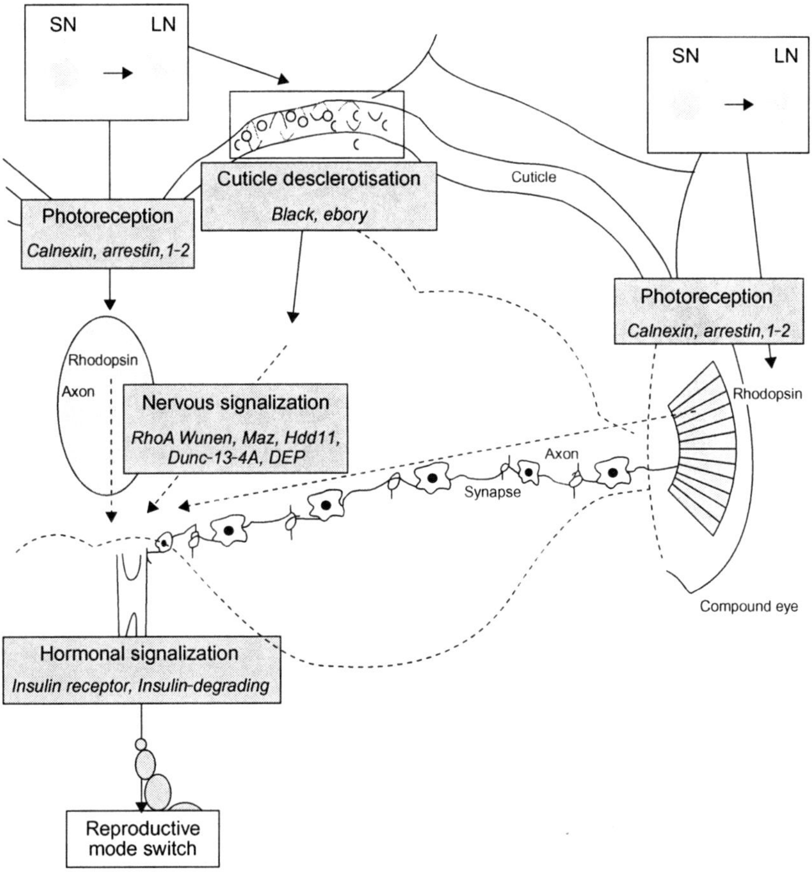

Figure 10.9 Hypothetic model of seasonal photoperiodism transcriptomic regulation in the pea aphid's head. This scheme corresponds to the head of the sexuparae submitted to photoperiod shortening and transmitting the signal to its embryos, the future sexuals. The photoperiod is sensed by still-unknown photoreceptors located either in the brain (left side of the diagram) or in the compound eyes (right side). Perception is followed by a series of nervous signaling through different pathways such as Rho, Wunen, or Dunc. In parallel, modification of the cuticle structure might lead to a higher concentration of dopamine in the brain, which acts as a neurotransmitter. Nervous signalization is relayed by endocrine regulation through the juvenile hormone (JH) signaling: the insulin pathway could be a regulator of the JH signaling pathway. *Abbreviations*: ca, corpora allata; cc, corpora cardiac; LN: long night; SN: short night.
Source: From Le Trionnaire et al. (2009).

males. Depending on the environmental conditions, ant colonies can flexibly allocate resources between two alternate (winged or wingless) male morphs. They invest more in producing exclusively wingless worker males, when conditions are favorable, or, to the contrary, under unfavorable conditions, they invest in producing the expensive dispersal form of winged males. Investigators have proven that this male diphenism is not genetic and not transmitted via the egg cell, but determined during the larval development in an adaptive response to the environmental conditions (Cremer and Heinze, 2003; De Menten et al., 2005). Experimental evidence led them to the conclusion that it is not the male larva itself that determines expression of the winged phenotype, but a change in the workers' social behavior toward larvae, the antennation (touching the larvae with antennae) that "instructs" these larvae to develop into winged male morphs (Schrempf and Heinze, 2006). Obviously, the tactile information that is provided to larvae is transmitted to the sensory organs, and processed in the brain, before larvae determine the developmental pathway they have to switch to for developing into winged or wingless males.

Recent studies on the migratory locust, *Locusta migratoria*, and the silkworm, *B. mori*, have shed additional light on the mechanisms controlling polyphenisms in insects. A larval brain neuropeptide, [His^7]-corazonin (as well as another neuropeptide, [Arg ^{7}]-corazonin, isolated from the larval brain of the silkworm), seems to be responsible for body-color polymorphisms of the migratory locust. Injection of this brain peptide alone in the young adults of the grasshopper *Oedipoda miniata* (Brakefield et al., 1996) and alone or in combination with JH in migratory locust instars (Tanaka, 1995, 2000a,b; Yerushalmi and Pener, 2002) produces a variety of body-color patterns, depending on the time and the dose of injection.

Experimental stimulation of JH secretion, which is induced by brain allatotropins, prevents formation of wings in normally nonpolyphenic winged female crickets (Zera et al., 1998).

A proportion of eggs of the nonparthenogenetic species *D. mercatorum* can parthenogenetically develop into adult flies. The fact that proportion of unfertilized eggs that are able to develop into adult flies increases with the decrease of the population density (Kramer and Templeton, 2001) suggests a role of social factors in determining the parthenogenetic development. Spontaneous transition of whole populations of *D. mercatorum* to parthenogenetic reproduction has been observed in laboratory (Takenaka-Dacanay and Carson, 1991). Remember, social stimuli are perceived and processed in the CNS.

A bivoltine race (Daizo) of the silkmoth, *B. mori*, produces two seasonal moths, autumn and summer morphs, in response to long- and short-day photoperiods, respectively. Transplantation or injection of the extracts of the Br–SG complexes as well as injection of the neurohormone DH from long-day pupae (which normally would develop into autumn morphs) into short-day pupae transforms the latter into autumn morphs. Decerebration of early pupae, like transplantation of Br–SG complexes, also makes short-day males shifting toward developing into autumn morphs. Another neurohormone, SMPH, induces transformation of long-day larvae into summer morphs (Yamanaka et al., 2000).

Polyphenisms in Vertebrates

Cichlid fish are widely known for their exceptionally rapid morphological evolution and speciation in East African lakes, but in the Western hemisphere they exhibit surprising polyphenisms.

Mexican cichlid fish are developmentally extravagant, displaying several phenotypes of teeth and digestive apparatuses in individuals of the same brood, even when raised on soft food in the laboratory (Sage and Selander, 1975). This enables the offspring to take advantage of a variety of edibles: snails, algae, fish, and arthropods. Thus, the parents increase chances that some individuals of their offspring will survive even under unpredictable hostile environmental conditions.

While any imaginable neo-Darwinian explanation fails to account for the phenomenon, an alternative epigenetic explanation would be that generation of different morphs in a single brood can be induced by differential distribution of maternal factors in each individual of the brood, similar to the epigenetic process that is experimentally demonstrated to occur in other cases to be described later (e.g., *Urosaurus ornatus*).

Rearing tadpole larvae of the salamander *Hynobius retardatus* together with heterospecific larvae of the frog *Rana pirica* increases proportion of the broad-headed, cannibal morph, in comparison with the normal salamander morph (Michimae and Wakahara, 2002). Larvae of this salamander develop broad carnivorous head, not only in response to visual stimuli but also in response to experimentally induced mechanical vibrations resembling those of flapping tails of tadpoles, in the water, which helps tadpoles to capture and handle the prey (Michimae et al., 2005). Remember, vibrational stimuli are neurally perceived and processed in the CNS before the developmental pathway for developing carnivorous head is activated.

Some salamanders of the genus *Ambystoma* (Ambystomatidae family), depending on the environment temperature, may undergo indirect metamorphic development, with a larval aquatic stage and a terrestrial reproductively mature stage or remain paedomorphic, i.e., they may reach reproductive maturity while still at the larval stage in water (retaining external gills and tail with fin margin) and avoid the terrestrial stage of life cycle. Other times, salamanders of the above species are polymorphic, i.e., in the same population, individuals of both types (paedomorphic and metamorphic) coexist, even though there seems to be no advantage from the maintenance of this life history polyphenism (Denoel et al., 2005).

It is interesting to point out that, during the last 3,000 years, the ratio of paedomorphic to metamorphic specimens in the population of *A. tigrinum* from Lamar Cave (Yellowstone National Park, WY, USA) has remained unchanged (Bruzgul et al., 2005). Most salamanders develop by metamorphosis, which enables them to use both aquatic and terrestrial habitats. However, there are also known species of salamanders that avoid metamorphosis and remain paedomorphic aquatic species during the whole life cycle (Figure 10.10).

Administration of thyroid hormones (THs) induces metamorphosis in neotenic axolotls. This and the fact that nuclear receptors for the TH are expressed and are functional in this species led investigators to the conclusion that "there is no relationship between TR allelic variation and life cycle modes among other Mexican ambystomatids" (Voss et al., 2000). It is suggested that low levels of THs in blood may be the cause of paedomorphosis in these species (Safi et al., 2004).

Some salamander species express an alternate developmental mode in which they forego metamorphosis and remain in the aquatic habitat throughout their lifetimes.

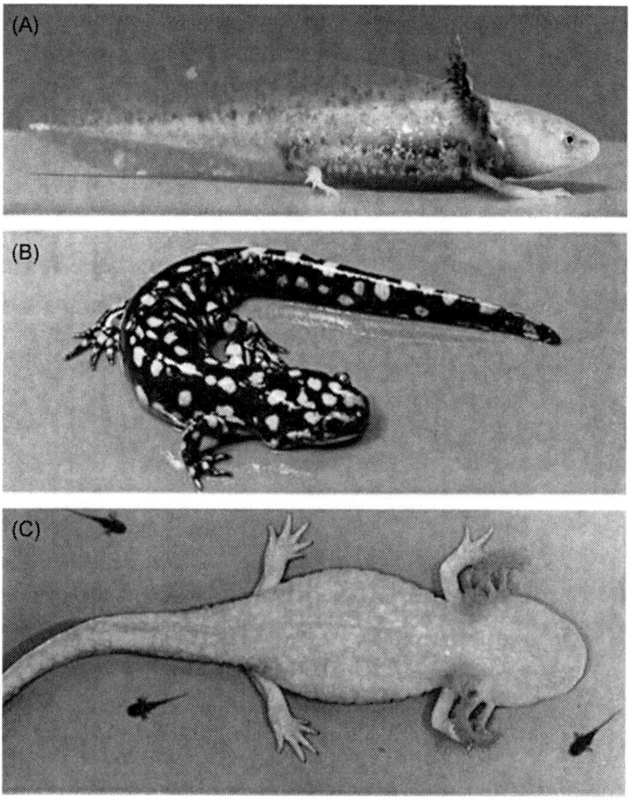

Figure 10.10 Larval and adult stages of *Ambystoma*. (A) Larval *A. mexicanum*. (B) Adult *A. tigrinum* (metamorphic). (C) Adult *A. mexicanum* (paedomorphic).
Source: From Voss and Smith (2005).

Tadpoles of the pinewood tree frog, *Hyla femoralis*, respond with morphological changes to the presence in environment of noncontact cues, such as metabolites (their nature is still unknown) of digestion of conspecific preys released by the predator and alarm pheromones released by the threatened conspecific prey. The morphological changes include deeper and shorter tails, changes in tail fin coloration, and reduced body size (LaFiandra and Babbitt, 2004). Recently, it has been shown that tadpoles of this anuran respond to cues of the larval migratory dragonfly predator *Anax junius* by morphological changes that are graded proportionally to the assessed risk of predation (Richardson, 2006).

Females of the polyandrous lizard, *Uta stansburiana*, display a strange ability to selectively use sperm from large-body sires to produce, within the same clutch, male offspring, and sperm from small-body sires to produce female offspring. It is not known how females make this selection, but they obviously receive visually the information about the body size of sires, and that perception takes place in the CNS (Calsbeek and Sinervo, 2004).

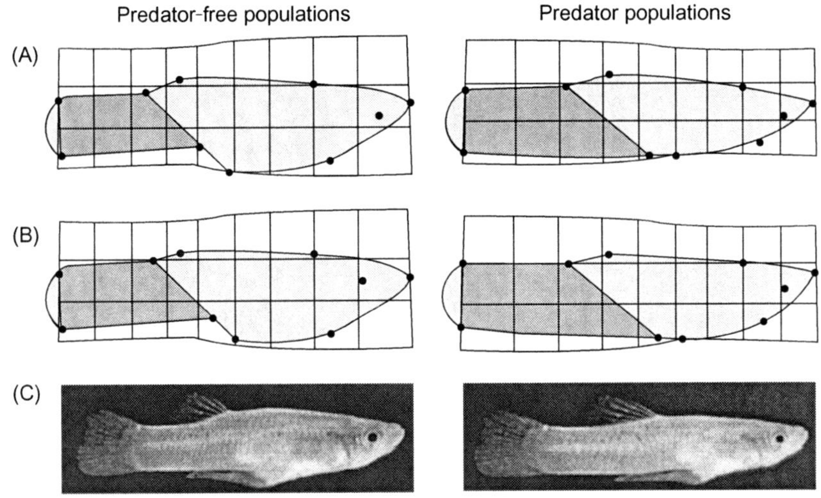

Figure 10.11 Visualization of morphological divergence between predator regimes for *Gambusia affinis* males. Thin-plate spline transformations depict morphological differences in (A) 2001, (B) 2003, and (C) both years combined as described by canonical axes derived from the predator regime effect. Photographs in (C) represent deviations in landmark configurations between predatory environments applied to a single photograph of an individual with an intermediate canonical score.
Source: From Langerhans et al. (2004).

A similar pattern of adapting sex ratio to mate quality has been observed in experiments with the blue tit in Sweden; when mated with males of brighter ultraviolet coloration, blue tit females increase proportion of their male offspring (Griffith et al., 2003).

The tree lizard, *Urosaurus ornatus*, gives birth to offspring that develop as orange- or blue-throated morphs. Both morphs belong to the same genotype. It is demonstrated that it is an epigenetic mechanism, the amount of maternal testosterone in their hatchling, that determines the proportion of the above phenotypes (Ketterson and Nolan, 1999).

Experimental Polyphenisms in Vertebrates

Male offspring of mosquito fish (*Gambusia affinis*) that prey on mosquitoes modify body shape and show better locomotory performance (Figure 10.11). The differences in morphology and swimming performance observed between populations under predation and predation-free populations persist for many years under predation-free laboratory conditions (Langerhans and DeWitt, 2004; Langerhans et al., 2004). Langerhans et al. (2004) succeeded in validating their ecomorphological prediction that fish coexisting with their predator fish "evolve a larger caudal region and a shallower anterior body/head region in order to increase burst-swimming speed region," which is necessary for their antipredator behavior.

The pregnant placentotrophic viviparous scincoid lizard, *Pseudemoia pagenstecheri*, adaptively manipulates the morphology and life history of its offspring according to a number of environmental stimuli it experiences during the pregnancy. In response to the olfactory perception of lizard-eating snakes during pregnancy, it gives birth to offspring of bigger body mass and very long tail, thus reducing the offspring's vulnerability to the predator snake (Shine and Downes, 1999).

The spadefoot toad, *Scaphiopus multiplicatus*, produces offspring of two distinct forms: fast-growing carnivorous and slow-growing omnivorous individuals, at particular ratios. Both morphs have significantly different digestive tract morphology and dietetic requirements. Experimental treatment with thyroxine induces production of a third, quite different morph (Storz, 2004).

Experiments conducted by Denver (1997) shed some light on the signal cascades triggered by environmental stimuli in desert amphibians. As tadpoles, these amphibians live in temporary ponds that contain water for unpredictable periods of time. In the years of low precipitations, the ponds dry up earlier. This causes a habitat stress to which the tadpoles of that and some other species respond by changing their behavior and by speeding up the metamorphosis to transform into adult amphibians, able to live on dry land. Denver attributes this polyphenism to the earlier activation of a cerebrally activated neuroendocrine stress pathway:

> *The lowering of the level of the water in the environment makes the hypothalamus to produce more CRH (corticotropin-releasing hormone—N.C.), which stimulates pituitary to produce hormones that stimulate thyroid and adrenal glands whose products help organism to cope with stress, in this case by losing their tail and beginning the growth of their limbs.*
>
> *Denver (1997)*

Let us remember that the hormone that stimulates the thyroid gland to secrete TH is the pituitary thyroid-stimulating hormone, which, in turn, is upregulated by hypothalamic neurohormone thyrotropin-releasing hormone (TRH). In turn, the hypothalamic TRH neurons

> *receive a dense input from neurons originating in other brain areas, which release catecholamines at their synaptic contacts with the cell bodies and dendrites of TRH-containing neurons.*
>
> *Strand (1999, p. 183)*

In the Australian lizard, *Bassiana dupperreyi*, Scincidae, sex is determined both genetically (the species has heteromorphic sex chromosomes) and by the incubation temperature, but the incubation temperature overrides chromosomal sex determination (Shine et al., 2002).

Administration of corticosterone in pregnant lizards, *Lacerta vivipara*, leads to reduction of juvenile body size and increases survival rates of juvenile male offspring (Meylan and Clobert, 2005).

It has been believed that, in distinction from reptiles and fish, the sex in birds is genotypically determined. Recent experiments with the Australian brush turkey,

Alectura lathami, a megapode with heteromorphic Z and W sex chromosomes, have shown that in birds, too, environmental temperature may considerably change the sex ratio from 1:1, when egg incubation from stage 1 takes place at 34°C, to 3:1 for males at 31°C, and 3:1 for females at 36°C (Goeth and Booth, 2005). Again, the epigenetic determination of the sex in the offspring overrides the genetic sex determination.

Based on the fact that birds deposit glucocorticoids in their eggs, experiments have been conducted to demonstrate the possible roles of glucocorticoids in the offspring phenotype. Administration of corticosterone in the eggs of the yellow-legged gull (*Larus michahellis*) was shown to delay hatching, reduce begging display, and decrease the body weight (Rubolini et al., 2005).

Predator-Induced Defenses

Predator-induced defenses are discrete changes in morphology, physiology, behavior, or life history that invertebrates develop in response to detection in the environment of the natural predators or their chemical cues. It is believed that the developmental plasticity induced by chemical cues in aquatic animals led to evolution of special neural structures for detecting these cues (Wisenden, 2000).

Upon detecting kairomones of its predator, *Daphnia magna* increases production of yolk proteins, shortens the time until the first reproduction, and reduces the size of the brood. These changes lead to production of more yolk proteins than are needed, a phenomenon that is known as *parental optimism* (Stibor, 2002).

In response to filtrates of 10 different species of organisms belonging not only to rotifers but also to remotely related species of other phyla, a rotifer, *Keratella testudo*, produces posterior spines, which make it less vulnerable to its predators (Stemberger and Gilbert, 1987).

The marine colonial bryozoan, *Membranipora membranacea*, produces spines within 2 days of exposure to its waterborne predator extracts, and the type of spines produced varies in accordance with the concentration of the cue (Harvell, 1998).

Snails *Litorina obtusata* in the northern Gulf of Maine have thinner shells than their southern conspecifics. This for a good reason: they do not have to bother carrying a heavier shell in that northern habitat, where their predator, *Carcinus maenus,* is very rare or even absent. But, when the northern snails are transplanted to the southern, predator-rich habitat, their shell becomes thicker, and their body mass larger, and inversely, the southern snails transplanted to the North develop thinner and lighter shells and larger bodies.

The blue mussel, *Mytilus edulis*, when outplanted to the sites of high predation by the crab *Carcinus maenus*, also produces a thicker shell (Leonard et al., 1999). Other marine snails exposed to predatory crabs increase shell thickness at a degree comparable to "historical transitions in thickness previously attributed to selection by invading predator" (Trussell and Smith, 2000; Trussell, 2001).

The freshwater snail, *Helisoma trivolis*, shows a remarkable phenotypic flexibility in response to different predators. It responds to the presence of water bugs with a particular suite of changes in behavior, morphology, and life history (e.g., the extent and type of the habitat, body mass at reproduction, shell width and height, and number of eggs laid), which is distinct from the suite of changes it generates in the presence of crayfish (Hoverman et al., 2005).

On detecting, in nature or under laboratory conditions, the presence of their predator, the brook trout *Salvelinus fontinalis*, the larvae of the mayfly, *Drunella coloradensis*, develop longer caudal filaments, which enhance their survival rate under predation (Dahl and Peckarsky, 2002).

The largemouth bass, *Micropterus salmoides* (Centrarchidae, Perciformes), responds with a behavioral plasticity when it detects the Ostariophysan alarm pheromone released by the finescale dace, *Phoxinus neogaeus* (Cyprinidae); it avoids visiting the space where it smells the pheromone (Brown et al., 2001).

Broenmark and Miner (1992) demonstrated that the crucian carp (*Carassius carassius*) raised in pond sections with piscivorous pike (*Esox lucius*) increased its body depth, which represents a morphological defense against pike. They considered two possible causes of the appearance of the adaptive morphology: (1) selective predation and (2) predator-induced phenotypic modification of body shape. Based on results of their experiments, they concluded against selection as a factor in the experimental evolution of guppies:

> *The small variance in the depth and the absence of overlap between treatments suggested no polymorphism with regard to this trait in the original population; thus, selective pressure on genetically determined morphs does not account for the increase in body depth.*
>
> *Broenmark and Miner (1992)*

The authors, thus, exclude a possible role of natural selection in the rapid evolution of the defensive morphology in the crucian carp. Obviously, one crucial initial step in the evolution of the fish population is perception of the predator threat in the brain of the crucian carp. Although we do not know the developmental pathways and signal cascades linking the perception of the predator to the adaptive change in morphology in this particular case, some relevant experimental evidence will be presented in the next chapter, Transgenerational Developmental Plasticity.

Red-eyed tree frogs, *Agalychnis callidryas*, attach their eggs to vegetation overhanging water. When attacked by egg-eating snakes, *the whole clutch* of eggs of the tree frog hatch immediately and fall into the water below, thus reducing mortality rate from snake predation, although in the water they face aquatic predators. This response is not a general response but a specific response to the mechanical disturbance caused by the snake, as is proven in experiments where touching and pulling of the eggs does not induce hatching (Warkentin, 1995). Embryos seem to distinguish predator-specific vibration patterns and on that basis make their hatching decision (Warkentin, 2005).

In Panama, the social wasps, *Polybia rejecta*, also prey on the eggs of the tree frogs, killing about one-quarter of them. Most undisturbed eggs hatch relatively late in order to survive aquatic predator attacks, but wasp-attacked eggs hatch immediately, and most of hatching embryos escape aquatic predators. In distinction from the case of snake attack, when the whole clutch hatches immediately, in the case of wasp attack, eggs hatch individually, proportionally to the consuming capacity of the predator (Warkentin, 2000).

Tadpoles of the frog *Rana dalmatina* respond to the presence of the fish predator three-spined stickleback, *Gasterosteus aculeatus*, by developing longer tails and more massive tail muscle (for a higher acceleration speed and higher promptness to escape), which helps tadpoles not for swimming (sticklebacks are much faster) but for quickly reaching a refuge, such as mud or plant debris. In response to the presence of dragonfly larvae, they develop a deeper tail fin, which helps them to swim faster (Teplitsky et al., 2005a,b).

On detecting predators, and with no refuge available, two salamander species, *Ambystoma barbouri* and *Ambystoma texanum*, change their body color so that it better matches the background, whereas *A. texanum*, additionally, show a tendency to move toward places that better match their body pattern and color (Garcia and Sih, 2003).

No changes in genes are involved in the development of predator-induced defenses described so far. From this view, the neo-Darwinian idea that "predators are the selecting agent" (Slogget and Weisser, 2002) is not accurate. Natural selection, as it is generally understood, acts by gradually accumulating, over long periods of time, "useful" gene mutations, but there is no evidence that predator-induced defenses evolved gradually or as a result of changes in genes.

Inborn Developmental Plasticity

Developmental Polymorphisms Are Not Genetic Polymorphisms

Presence of different morphs (individuals of distinct discrete morphologies) in the same brood has been considered genetic polymorphism. But the very fact that morphs occur within the same brood, i.e., belong to the same genotype, excludes the possibility that genetic factors may be involved in the phenomenon, making *genetic polymorphism* a misnomer. Hence, and to emphasize the developmental (not genetic) origin of such morphs, the term *developmental polymorphism,* instead of *genetic polymorphism*, will be used here.

The flightless bug, *Pyrrhocoris apterus* (L.), develops nonfunctional wings and produces diphenic morphs for wing length and flight muscles. The macropterous (long-winged) morph develops flight muscles, while the brachypterous (short-winged) morph only develops rudimentary flight muscles because of the arrested growth during development. The flight muscles in macropterous morphs grow until the adulthood, when they are histolyzed, in a process that is determined by a series of neuroendocrine changes. Histolysis of the flight muscles in this insect is prematurely induced by administration of a JH analog (Socha and Šula, 2006). The fact that JH synthesis and secretion are under control of brain neurohormones (allatostatins and allatotropins) suggests that development of morphs, with or without wings and flight muscles, is under ultimate neural control.

In the adult cricket, *Gryllus bimaculatus*, the histolysis of flight muscles is neurally regulated via local innervation. Motor neurons innervating the flight muscle

M112a send signals for timing the histolysis of muscle fibers. Decapitation of insects prevents the muscle histolysis:

> *JH causes breakdown of flight muscles and motor neurons are involved in the onset of degeneration in the presence of JH. Both endocrine and neural factors are important for flight muscle degeneration.*
>
> *Shiga et al. (2002)*

Furthermore, administration of ecdysone retards the degeneration of flight muscles in adult crickets, *Acheta domestica*, but has no effect on denervated flight muscles, indicating that a neural factor is necessary for the action of the hormone in muscle degeneration (Srihari et al., 1975).

A proportion of unfertilized eggs of the mayfly, *Stenonema femora,* develop parthenogenetically, and even some female flies produce mixed broods of sexually and parthenogenetically developing offspring despite the fact that all of them are of the same genotype (Ball, 2001). Cases of spontaneous transformation of whole populations into parthenogenetic populations of *D. mercatorum* have been observed in laboratory (Takenaka-Dacanay and Carson, 1991).

Once a year, some aphids produce a sexual generation. Production of males and females by parthenogenetic XX females necessitates elimination of an X chromosome from the eggs that will produce males; this elimination is again mediated by a brain neuropeptide acting directly on oocytes (Chapman, 1998, p. 401).

One interesting case in the crayfish, *Procambarus clarkii*, may shed additional light on the neural mechanisms underlying the production of more than one morph in polyphenic invertebrates. Male individuals of this small crustacean belong to one of two distinct morphotypes, the reproductive form Is and the nonreproductive form IIs. The morphological differences between these two forms consist of Is having larger chelae (claws) on the ischiopodites of walking legs and also having spines (lacking in IIs) on the ischiopodites. Experimental evidence shows that whether an individual will grow as an Is adult morph or as a nonreproductive IIs morph depends not on any genetic differences but on the level of the hormone methyl farnesoate (with functions similar to the insect JH) secreted by the mandibular organ, which is under strict cerebral control (Laufer et al., 2005).

Dispersal polyphenism (production of various morphs in order to increase the dispersal capability) in the cricket, *G. firmus*, is represented by three morphs: a flight-capable morph with long wings and fully developed flight muscles, a flightless morph with reduced wings and flight muscles, and the third with fully developed wings but incapable of flight because of histolysis of flight muscles. Not any differences in genotype of the individuals of the brood, but neurally regulated levels of JH determine production of different morphs (Zera, 2006). The proximate cause of the wing developmental polymorphism seems to be JH. As mentioned earlier, wing discs in *Precis coenia* cease growing in the presence of JH, and their growth can be experimentally inhibited by JH (Miner et al., 2000). A simple hormonal manipulation, topical administration of JH, leads to formation of wingless females of crickets *G. firmus* and *G. rubens*, which are naturally winged (alates).

Alate females of another insect, the fire ant, *Solenopsis invicta*, are induced to cast their wings when JH is locally applied, and local application of the JH antagonist, precocene, prevents the dealating action of JH (Burns et al., 1999). Precocene also induces development of long wings in 30% of short wing–programmed larvae of the ectoparasite *Melittobia digitata* (Consoli and Vinson, 2004).

Occurring among individuals of the same genotype, the wing developmental polymorphism precludes involvement of changes in genes and any neo-Darwinian explanation of the phenomenon of wing developmental polymorphism in insects.

References

Ball, S.L. (2001). Tychoparthenogenesis and mixed mating in natural populations of the mayfly, *Stenonema femora*. Heredity *87*, 373–380.

Beldade, P., Mateus, A.R.A. and Keller, R.A. (2011). Evolution and molecular mechanisms of adaptive developmental plasticity. Mol. Ecol. *20*, 1347–1363.

Bonetti, A.M., Kerr, W.E. and Matusita, S.H. (1995). Effects of juvenile hormones I, II and III, in single and fractionated dosage in *Melipona* bees. Rev. Bras. Biol. *55*(Suppl. 1), 113–120.

Brakefield, P.M., Gates, J., Keys, D., Kesbeke, F., Wijngaarden, P.J. and Monteiro, A., et al. (1996). Development, plasticity and evolution of eyespot patterns. Nature *384*, 236–242.

Broenmark, C. and Miner, J.G. (1992). Predator-induced phenotypical change in the body morphology in crucian carp. Science *258*, 1348–1350.

Brown, G.E., LeBlanc, V.J. and Porter, L.E. (2001). Ontogenetic changes in the response of largemouth bass (*Micropterus salmoides*, centrarchidae, perciformes) to heterospecific alarm pheromones. Ethology *107*, 401–414.

Bruzgul, J.E., Long, W. and Hadly, E.A. (2005). Temporal response of the tiger salamander (*Ambystoma tigrinum*) to 3,000 years of climatic variation. BMC Ecol. *5*, p. 7.

Bückmann, D. and Maisch, A. (1987). Extraction and partial purification of the pupal melanization reducing factor (PMRF) from *Inachis io* (Lepidoptera). Insect Biochem. *17*, 841–844.

Burns, S.N., Vander Meer, R.K., Teal, P.E.A. and Nation, J.L. (1999). Effects of topical juvenile hormone and precocene treatments on sexually mature *Solenopsis invicta* female alates and the comparison of the rates of dealation of alates of different ages. Proc. XIII International Congress of IUSSI, 1/3/99; p. 89.

Calsbeek, R. and Sinervo, B. (2004). Within-clutch variation in offspring sex determined by differences in sire body size: cryptic mate choice in the wild. J. Evol. Biol. *17*, 464–470.

Chapman, R.F. (1998). *The Insects–Structure and Function*, fourth edition. Cambridge University Press, Cambridge, MA, p. 578.

Chuan-Chin, C. and Hanlon, R.T. (2001). Cuttlefish camouflage: visual perception of size, contrast and number of white squares on artificial checkboard substrata initiates disruptive coloration. J. Exp. Biol. *204*, 2119–2125.

Consoli, F. and Vinson, S.B. (2004). Wing morph development and reproduction of the ectoparasite *Melittobia digitata*: nutritional and hormonal effects. Entomol. Exp. App. *112*, 47–56.

Cott, H.B. (1966). *Adaptive Coloration in Animals*. Methuen and Co. Ltd., pp. 223–224.

Cremer, S. and Heinze, J. (2003). Stress grows wings: environmental induction of winged dispersal males in *Cardiocondyla* ants. Curr. Biol. *13*, 219–223.

Dahl, J. and Peckarsky, B.L. (2002). Induced morphological defenses in the wild: predator effects on a mayfly, *Drunella coloradensis*. Ecology *83*, 1620–1634.

Darwin, C.R. (1874). *The Descent of Man and Selection in Relation to Sex*. John Murray, London, p. 324.

De Menten, L., Cremer, S., Heinze, J. and Aron, S. (2005). Primary sex ratio adjustment by ant queens in response to local mate competition. Anim. Behav. *69*, 1031–1035.

Denoel, M., Joly, P. and Whiteman, H.H. (2005). Evolutionary ecology of facultative paedomorphosis in newts and salamanders. Biol. Rev. *80*, 663–671.

Denver, R.J. (1997). Environmental stress as a developmental cue: corticotropin-releasing hormone is a proximate mediator of adaptive phenotypic plasticity in amphibian metamorphosis. Horm. Behav. *31*, 169–179.

Dobzhansky, T. (1971). *Mankind Evolving*. Yale University Press, New Haven, CT, pp. 170–171.

Dubas, F., Hanlon, R.T., Ferguson, G.P. and Pinsker, H.M. (1986). Localization and stimulation of chromatophore motoneurons in the brain of the squid, *Lollinguculla brevis*. J. Exp. Biol. *121*, 1–25.

Fric, Z., Konvicka, M. and Zrzavy, J. (2004). Red & black or black & white? Phylogeny of the *Araschnia* butterflies (Lepidoptera: Nymphalidae) and evolution of seasonal polyphenism. J. Evol. Biol. *17*, 265–278.

Frisch, O.v. (1973). *Animal Camouflage*. Collins, New York, NY, p. 70. According to Wesson, R. (1991). *Beyond Natural Selection*. MIT Press, Cambridge, MA, p. 68.

Garcia, T.S. and Sih, A. (2003). Color change and color-dependent behavior in response to predation risk in the salamander sister species *Ambystoma barbouri* and *Ambystoma texanum*. Oecologia *137*, 131–139.

Gaston, M.R. and Tublitz, N.J. (2006). Central distribution and three-dimensional arrangement of fin chromatophore motoneurons in the cuttlefish *Sepia officinalis*. Invert. Neurosci. *6*, 81–93.

Goeth, A. and Booth, D.T. (2005). Temperature-dependent sex ratio in a bird. Biol. Lett. *1*, 31–33.

Gorbman, A. and Davey, K. (1991). Endocrines. In: *Neural and Integrative Animal Physiology*, C.L. Prosser (ed.), fourth edition. Wiley-Liss, New York, pp. 693–754.

Griffith, S.C., Omborg, J., Russell, A.F., Andersson, S. and Sheldon, B.C. (2003). Correlations between ultraviolet coloration, overwinter survival and offspring sex ratio in the blue tit. J. Evol. Biol. *16*, 1045–1054.

Hanlon, R.T. and Messenger, J.B. (1988). Adaptive coloration in young cuttlefish (*Sepia officinalis* L.). Phil. Trans. R. Soc. Lond. B *320*, 437–487.

Hardie, J. (1981). Juvenile hormone and photoperiodically controlled polymorphism in *Aphis fabae*. Postnatal effects on presumptive *gynoparae*. J. Insect Physiol. *27*, 347–355.

Harvell, C.D. (1998). Genetic variation and polymorphism in the inducible spines of a bryozoan. Evolution *52*, 80–86.

Hazel, W.N. (2002). The environment and genetic control of seasonal polyphenism in larval color and its adaptive significance in a swallowtail butterfly. Evolution *56*, 342–348.

Hegstrom, C.D., Riddiford, L.M. and Truman, J.W. (1998). Steroid and neuronal regulation of ecdysone receptor expression during metamorphosis of muscle in the moth, *Manduca sexta*. J. Neurosci. *18*, 1786–1794.

Hiraga, S. (2005). Two different sensory mechanisms for the control of pupal protective coloration in butterflies. J. Insect Physiol. *51*, 1033–1040.

Höglund, E., Balm, P.H.M. and Winberg, S. (2002). Behavioural and neuroendocrine effects of environmental background colour and social interaction in Arctic charr (*Salvelinus alpinus*). J. Exp. Biol. *205*, 2535–2543.

Hoverman, J.T., Auld, J.R. and Relyea, R.A. (2005). Putting prey back together again: integrating predator-induced behavior, morphology, and life history. Oecologia *144*, 481–491.

Jeffery, W.R. (2005). Adaptive evolution of eye degeneration in the Mexican blind cavefish. J. Hered. *96*, 185–196.

Johansson, F., Stoks, R., Rowe, L. and De Block, M. (2001). Life history plasticity in a damselfly: effects of combined time and biotic constraints. Ecology *82*, 1857–1869.

Jones, M., Rakes, L., Yochum, M., Dunn, G., Wurster, S., Kinney, K. and Hazel, W. (2007). The proximate control of pupal color in swallowtail butterflies: implications for the evolution of environmentally cued pupal color in butterflies (Lepidoptera: Papilionidae). J. Insect Physiol. *53*, 40–46.

Josefsson, L. (1983). Chemical properties and physiological actions of crustacean chromatophorotropins. Am. Zool. *23*, 507–515.

Kasukawa, H., Oshima, N. and Fujii, R. (1986). Control of chromatophore movements in dermal chromatic units of blue damselfish—II. The motile iridophore. Comp. Biochem. Physiol. C *83*, 1–7.

Kenny, N.A.P., Richard, D.S., Bradley, H.K. and Saunders, D.S. (1992). Photoperiodic sensitivity and diapause induction during ovarian, embryonic and larval development of the flesh fly, *Sarcophaga argyrostoma*. J. Biosci. *17*, 241–251.

Kerr, W.E. (1950). Genetic determination of castes in the genus *Melipona*. Genetics *35*, 143–152.

Kerr, W.E., Akahiras, W. and Camargo, C.A. (1975). Sex determination in bees. IV. Genetic control of juvenile hormone production in *Melipona quadrifasciata* (Apidae). Genetics *81*, 749–756.

Ketterson, E. and Nolan, V., Jr. (1999). Adaptation, exaptation and constraint: a hormonal perspective. Am. Nat. *154*, 4–25.

Koch, P.B. (1987). Hormonal control of seasonal morphs by the timing of ecdysteroid release in *Araschnia levana* L. (Nymphalidae: Lepidoptera). J. Insect Physiol. *33*, 823–829.

Koch, P.B. and Bückmann, D. (1987). Hormonal control of seasonal morphs by the timing of ecdysteroid release in *Araschnia levana* L. (Nymphalidae: Lepidoptera). J. Insect Physiol. *33*, 823–829.

Koch, P.B., Brakefield, P.M. and Kesbeke, F. (1996). Ecdysteroids control eyespot size and wing color pattern in the polyphenic butterfly *Bicyclus anynana* (Lepidoptera: Satyridae). J. Insect Physiol. *42*, 223–230.

Koch, P.B., Merk, R., Brakefield, P.M., French-Constant, R.H. (2002). Ecdysone receptors link pattern and process in butterfly eyespot development. *XV International Ecdysone Workshop*. 30 June to 6 July 2002, Kolymbari Crete, Greece (abstract).

Koch, P.B., Merk, R., Reinhardt, R. and Weber, P. (2003). Localization of ecdysone receptor protein during colour pattern formation in wings of the butterfly *Precis coenia* (Lepidoptera: Nymphalidae) and co-expression with distal-less protein. Dev. Genes Evol. *212*, 571–584.

Kramer, M.G. and Templeton, A.R. (2001). Life-history changes that accompany the transition from sexual to parthenogenesis reproduction in *Drosophila mercatorum*. Evolution *55*, 748–761.

LaFiandra, E.M. and Babbitt, K.J. (2004). Predator-induced phenotypic plasticity in the pinewoods tree frog, *Hyla femoralis*: necessary cues and the cost of development. Oecologia *138*, 350–359.

Langerhans, R.B. and Dewittt, T.J. (2004). Shared and unique features of evolutionary diversification. Am. Nat. *164*, 335–349.

Langerhans, R.B., Layman, C.A., Shokrollahi, A.M. and DeWitt, T.J. (2004). Predator-driven phenotypic diversification in *Gambusia affinis*. Evolution *58*, 2305–2318.

Laufer, H., Demir, N., Pan, X., Stuart, J.D. and Ahl, J.S.B. (2005). Methyl farnesoate controls adult male morphogenesis in the crayfish, *Procambarus clarkii*. J. Insect Physiol. *51*, 379–384.

Leonard, G.H., Bertness, M.D. and Yund, P.O. (1999). Crab predation, waterborne cues, and inducible defenses in the blue mussel *Mytilus edulis*. Ecology *80*, 1–14.

Le Trionnaire, G., Francis, F., Jaubert-Possamai, S., Bonhomme, J., De Pauw, E. and Gauthier, J.-P., et al. (2009). Transcriptomic and proteomic analyses of seasonal photoperiodism in the pea aphid. BMC Genomics *10*, p. 456.

Loi, P., Saunders, R., Young, D. and Tublitz, N. (1996). Peptidergic regulation of chromatophore function in the European cuttlefish *Sepia officinalis*. J. Exp. Biol. *199*, 1177–1187.

Loi, P.K. and Tublitz, N.J. (2000). Roles of glutamate and FMRFamide-related peptides at the chromatophore neuromuscular junction in the cuttlefish, *Sepia officinalis*. J. Comp. Neurol. *420*, 499–511.

Marshall, N.J. and Messenger, J.B. (1996). Colour-blind camouflage. Nature *382*, 408–409.

Mäthger, L.M., Land, M.F., Siebeck, U.E. and Marshall, N.J. (2003). Rapid colour changes in multilayer reflecting stripes in the paradise whiptail, *Pentapodus paradiseus*. J. Exp. Biol. *206*, 3607–3613.

Mäthger, L.M., Collins, T.F.T and Lima, P.A. (2004). The role of muscarinic receptors and intracellular Ca^{2+} in the spectral reflectivity changes of squid iridophores. J. Exp. Biol. *207*, 1759–1769.

Mäthger, L.M., Barbosa, A., Miner, S. and Hanlon, R.T. (2006). Color blindness and contrast perception in cuttlefish (*Sepia officinalis*) determined by a visual sensorimotor assay. Vision Res. *46*, 1746–1753.

Mellencamp, K., Hass, M., Werne, A., Stark, R. and Hazel, W. (2007). Role of larval stemmata in control of pupal color and pupation site preference in swallowtail butterflies *Papilio troilus, Papilio polyxenes, Eurytides marcellus*, and *Papilio glaucus* (Lepidoptera: Papilionidae). Ann. Entomol. Soc. Am. *100*, 53–58.

Messenger, J.B. (2001). Cephalopod chromatophores: neurobiology and natural history. Biol. Rev. Camb. Philos. *76*, 473–528.

Meylan, S. and Clobert, J. (2005). Is cortisone-mediated phenotype development adaptive? Maternal corticosterone treatment enhances survival in male lizards. Horm. Behav. *48*, 44–52.

Michimae, H. and Wakahara, M. (2002). A tadpole-induced polyphenism in the salamander *Hynobius retardatus*. Evolution *56*, 2029–2038.

Michimae, H., Nishimura, K. and Wakahara, M. (2005). Mechanical vibrations from tadpoles' flapping tails transform salamander's carnivorous morphology. Biol. Lett. *21*, 75–77.

Miner, A.L., Rosenberg, A.J. and Nijhout, H.F. (2000). Control of growth and differentiation of the wing imaginal disk of *Precis coenia* (Lepidoptera: Nymphalidae). J. Insect Physiol. *46*, 251–258.

Moland, E. and Jones, G.P. (2004). Experimental confirmation of aggressive mimicry by coral reef fish. Oecologia *140*, 676–683.

Moore, P.J. and Moore, A.J. (2003). Developmental flexibility and the effect of social environment on fertility and fecundity in parthenogenetic reproduction. Evol. Dev. *5*, 163–168.

Morgan, T.H. (1903). *Evolution and Adaptation*. Macmillan, New York, NY, pp. 358–359.

Nagaishi, H. and Oshima, N. (1989). Neural control of motile activity of light-sensitive iridophores in the neon tetra. Pigm. Cell Res. *2*, 485–492.

Nijhout, H.F. (2003). Development and of evolution of adaptive polyphenisms. Evol. Dev. *5*, 9–18.

Olmstead, A.W. and LeBlanc, G.A. (2002). Juvenoid hormone methyl farnesoate is a sex determinant in the crustacean *Daphnia magna*. J. Exp. Zool. *293*, 736–739.

Olmstead, A.W. and LeBlanc, G.A. (2003). Insecticidal juvenile hormone analogs stimulate the production of male offspring in the crustacean *Daphnia magna*. Environ. Health Perspect. *111*, 919–924.

Oshima, N. and Fujii, R. (1987). Motile mechanism of blue damselfish (*Chrysiptera cyanea*) iridophores. Cytoskeleton *8*, 85–90.

Packard, A. (1982). Morphogenesis of chromatophore patterns in cephalopods: are morphological and physiological "units" the same? Malacologia *23*, 193–201.

Pearse, V., Pearse, J., Buchsbaum, M. and Buchsbaum, R. (1987). *Living Invertebrates*. Blackwell Scientific Publications, p. 461.

Pfenning, B. and Poethke, H.J. (2006). Variability in the life history of the water strider *Gerris lacustris* (Heteroptera: Gerridae) across small spatial scales. Ecol. Entomol. *31*, 123–130.

Pigliucci, M. and Murren, C.J. (2003). Genetic assimilation and a possible evolutionary paradox: can macroevolution sometimes be so fast as to pass us by? Evolution *57*, 1455–1464.

Pigliucci, M., Murren, C.J. and Schlichting, C.D. (2006). Phenotypic plasticity and evolution by genetic assimilation. J. Exp. Biol. *209*, 2368–2376.

Pinto, L.Z., Hartfelder, K., Gentile Bitondi, M.M. and Simões, Z.L.P. (2002). Ecdysteroid titers in pupae of highly social bees relate to distinct modes of caste development. J. Insect Physiol. *48*, 783–790.

Ramachandran, V.S., Tyler, C.W., Gregory, R.L., Rogers-Ramachandran, D., Duensing, S. and Pillsbury, C., et al. (1996). Rapid adaptive camouflage in tropical flounders. Nature *379*, 815–818.

Richardson, J.L. (2006). Novel features of an inducible defense system in larval tree frogs (*Hyla chrysoscelis*). Ecology *87*, 780–787.

Rubolini, D., Romano, M., Boncoraglio, G., Ferrari, R.P., Martinelli, R. and Galeotti, P., et al. (2005). Effects of elevated egg corticosterone levels on behavior, growth, and immunity of yellow-legged gull (*Larus michahellis*) chicks. Horm. Behav. *48*, 592–605.

Safi, R., Bertrand, S., Marchand, O., Duffraisse, M., de Luze, A. and Vanacker, J.-M., et al. (2004). The axolotl (*Ambystoma mexicanum*), a neotenic amphibian, expresses functional thyroid hormone receptors. Endocrinology *145*, 760–772.

Sage, R.D. and Selander, R.K. (1975). Trophic radiation through polymorphism in cichlid fishes. Proc. Natl. Acad. Sci. U.S.A. *72*, 4669–4673.

Schrempf, A. and Heinze, J. (2006). Proximate mechanisms of male morph determination in the ant *Cardiocondyla obscurior*. Evol. Dev. *8*, 266–272.

Shiga, S., Yasuyama, K., Okamura, N. and Yamaguchi, T. (2002). Neural- and endocrine control of flight muscle degeneration in the adult cricket, *Gryllus bimaculatus*. J. Insect Physiol. *48*, 15–24.

Shine, R. and Downes, S.J. (1999). Can pregnant lizards adjust their offspring phenotypes to environmental conditions? Oecologia *119*, 1–8.

Shine, R., Elpick, N.J. and Donnellan, S. (2002). Coocurrence of multiple, supposedly incompatible modes of sex determination in a lizard population. Ecol. Lett. *5*, 486–489.

Slogget, J.J. and Weisser, W.W. (2002). Parasitoids induce production of the dispersal morph of the pea aphid, *Acyrthosiphon pisum*. Oikos *98*, 323–333.

Socha, R. and Šula, J. (2006). Flight muscles polymorphism in a flightless bug, *Pyrrhocoris apterus* (L.): developmental pattern, biochemical profile and endocrine control. J. Insect Physiol. *52*, 231–239.

Srihari, T., Gutmann, E. and Novak, V.J.A. (1975). Effect of ecdysterone and juvenoid on the developmental involution of flight muscles in *Acheta domestica*. J. Insect Physiol. *21*, 1–8.

Starnecker, G. and Hazel, W. (1999). Convergent evolution of neuroendocrine control of phenotypic plasticity in pupal colour in butterflies. Proc. Biol. Sci. *266*, 2409–2412.

Steel, C.G.H. and Lees, A.D. (1977). The role of neurosecretion in the photoperiodic control of polymorphism in the aphid *Megoura viciae*. J. Exp. Biol. *67*, 117–135.

Steigenga, M.J., Zwaan, B.J., Brakefield, P.M. and Fischer, K. (2005). The evolutionary genetics of egg size plasticity in a butterfly. J. Evol. Biol. *18*, 281–289.

Stemberger, R.S. and Gilbert, J.J. (1987). Multiple-species induction of morphological defenses in the rotifer *Keratella testudo*. Ecology *68*, 370–378.

Stibor, H. (2002). The role of yolk protein dynamics and predator kairomones for the life history of *Daphnia magna*. Ecology *83*, 362–369.

Storz, B.L. (2004). Reassessment of the environmental mechanisms controlling developmental polyphenism in spadefoot toad tadpoles. Oecologia *141*, 402–410.

Strand, F.L. (1999). *Neuropeptides*. MIT Press, p. 522.

Sugimoto, M. (2002). Morphological color changes in fish: regulation of pigment cell density and morphology. Microsc. Res. Tech. *58*, 496–503.

Sugimoto, M., Uchida, N. and Hatayama, M. (2000). Apoptosis in skin pigment cells of medaka, *Oryzias latipes* (Teleostei), during long-term chromatic adaptation: the role of sympathetic innervation. Cell Tissue Res. *301*, 205–216 (abstract).

Takenaka-Dacanay, J.H. and Carson, H.L. (1991). Sexual behavior in laboratory strains of *Drosophila mercatorum*. Behav. Genet. *21*, 305–316.

Tanaka, S. (1995). A cricket (*Gryllus bimaculatus*) neuropeptide induces dark colour in the locust, *Locusta migratoria*. J. Insect Physiol. *42*, 287–294.

Tanaka, S. (2000). Hormonal control of body-color polymorphism in *Locusta migratoria*: interaction between [His^7]-corazonin and juvenile hormone. J. Insect Physiol. *46*, 1535–1544.

Tanaka, S. (2000). The role of [His^7]-corazonin in the control of body-color polymorphism in the migratory locust *Locusta migratoria* (Orthoptera: Acrididae). J. Insect Physiol. *46*, 1169–1176.

Teplitsky, C., Plenet, S. and Joly, P. (2005). Costs and limits of dosage response to predation risk: to what extent can tadpoles invest in anti-predator morphology? Oecologia *145*, 364–370.

Teplitsky, C., Plenet, S., Lena, J.P., Mermet, N., Malet, E. and Joly, P. (2005). Escape behaviour and ultimate causes of specific induced defences in an anuran tadpole. J. Evol. Biol. *18*, 180–190.

Trussell, G.C. (2001). Phenotypic clines, plasticity, and morphological trade-offs in an intertidal snail. Evolution *54*, 151–166.

Trussell, G.C. and Smith, L.D. (2000). Induced defenses in response to an invading crab predator. Proc. Natl. Acad. Sci. U.S.A. *97*, 2123–2127.

Tuinhof, R., Artero, C., Fasolo, A., Franzoni, M.F., Ten Donkelaar, H.J. and Wismans, P.G., et al. (1994). Involvement of retinohypothalamic input, suprachiasmatic nucleus, magnocellular nucleus and *locus coeruleus* in control of melanotrope cells of *Xenopus laevis*: a retrograde and anterograde tracing study. Neuroscience *61*, 411–420.

Tuinhof, R., Gonzalez, A., Smeets, W.J., Scheenen, W.J. and Roubos, E.W. (1994). Central control of melanotrope cells of *Xenopus laevis*. Eur. J. Morphol. *32*, 307–310.

Turner, J.R.G. (1977). Butterfly mimicry: the genetical evolution of an adaptation. Evol. Biol. *10*, 163–206.

Tylor, A. (1886). *Colouration in Animals and Plants*. Alabaster, Passmore, and Sons, London, p. 29.

Ubink, R., Buzzi, M., Cruijsen, P.M., Tuinhof, R., Verhofstad, A.A. and Jenks, B.G., et al. (1999). Serotonergic innervation of the pituitary pars intermedia of *Xenopus laevis*. J. Neuroendocrinol. *11*, 211–219.

Van Veen, J.H. (2000). Cell provisioning and oviposition in *Melipona beecheii* (Apidae, Meliponinae), with a note on caste determination. Apidologie *31*, 411–419.

Voss, S.R., Shaffer, H.B., Taylor, J., Safi, R. and Laudet, V. (2000). Candidate gene analysis of thyroid hormone receptors in metamorphosing vs. nonmetamorphosing salamanders. Heredity *85*, 107–114.

Voss, S.R. and Smith, J.J. (2005). Evolution of salamander life cycles. Genetics *170*, 275–281.

Warkentin, K.M. (1995). Adaptive plasticity in hatching age: a response to predation risk trade-offs. Proc. Natl. Acad. Sci. U.S.A. *92*, 3507–3510.

Warkentin, K.M. (2000). Wasp predation and wasp-induced hatching of red-eyed treefrog eggs. Anim. Behav. *60*, 503–510.

Warkentin, K.M. (2005). How do embryos assess risk? Vibrational cues in predator-induced hatching of red-eyed treefrogs. Anim. Behav. *70*, 59–71.

West-Eberhard, M.J. (2003). *Developmental Plasticity and Evolution.* Oxford University Press, Oxford, p. 421.

Wilkens, H. (1971). Genetic interpretation of regressive evolutionary processes: studies on hybrid eyes of two Astyanax Cave populations (Characidae, Pisces). Evolution *25*, 530–544.

Wisenden, B.D. (2000). Olfactory assessment of predation risk in the aquatic environment. Phil. Trans. R. Soc. Lond. B *355*, 1205–1208.

Woltereck, R. (1909). Weitere experimentelle Untersuchungen über Artveränderung, speziell über das Wesen quantitativer Artunterschiede bei Daphniden. Verhandlungen der Deutschen Zoologischen Gesellschaft *1909*, 110–173.

Yamanaka, A., Tsurumaki, J. and Katsuhiko, E. (2000). Neuroendocrine regulation of seasonal morph development in a bivoltine race (Daizo) of the silkmoth *Bombyx mori* L. J. Insect Physiol. *46*, 803–808.

Yerushalmi, Y. and Pener, M.P. (2002). Age-dependent response of adults of a homochrome grasshopper, *Oedipoda miniata*, to the dark-colour-inducing neurohormone (DCIN) of locusts. Physiol. Entomol. *27*, 165–169.

Zera, A.J. (2006). Evolutionary genetics of juvenile hormone and ecdysteroid regulation in *Gryllus*: a case study in the microevolution of endocrine regulation. Comp. Biochem. Physiol. A Mol. Integr. Physiol. *144*, 365–379.

Zera, A.J. and Cisper, G. (2001). Genetic and diurnal variation in the juvenile hormone titer in a wing-polymorphic cricket: implications for the evolution of life histories and dispersal. Physiol. Biochem. Zool. *74*, 293–306.

Zera, A.J., Potts, J. and Kobus, K. (1998). The physiology of life-history trade-offs: experimental analysis of a hormonally induced life-history trade-off in *Gryllus assimilis*. Am. Nat. *152*, 7–23.

Part III

Epigenetics of Circumevolutionary Phenomena and the Mechanism of the Evolutionary Change

11 Transgenerational Developmental Plasticity—An Epitome of Evolutionary Change

The epigenetic information for transgenerational phenotypic changes is generated in the CNS.

Phenotypic changes observed in examples of developmental plasticity considered in the previous chapter do not affect the offspring, in the absence of the inducing factors that triggered their appearance in parents. In the classical biological terminology, all of them represent nonheritable "acquired characters." Conventionally, the inability to transmit acquired characters to the offspring has been explained with the fact that no changes in DNA or genes encoding these characters have occurred. Hence, all of them are evolutionarily irrelevant.

However, an ever-increasing number of biologists believe that, in many cases, the developmental plasticity is a component of evolutionary process or a stage in the process of the evolution of morphological innovations and novelties (West-Eberhard, 1989, 2003; Thompson, 1991).

The fact that the evolution of animal morphology, by human life span standards, takes very long periods of time, impedes attempts to directly observe changes in the developmental mechanisms enabling evolution of morphological innovations in metazoans. One empirical way to overcome this difficulty is to look for possible short-term biological phenomena where similar mechanisms may be working. Fortunately, it seems that such a phenomenon exists, and its study may contribute to the understanding of the nature, mechanisms, and origins of the evolutionary change (Cabej, 2004, pp. 192–193). This is the phenomenon of transgenerational developmental plasticity (TDP), where offspring displays and transmits to the next generation new characters that its parent(s) or close predecessors lacked.

Both the evolutionary change and TDP consist in appearance in the offspring, and transmission to the future generations, of phenotypic (behavioral, morphological, physiological, and life history) characters that the parents did not inherit but may (or may not) have acquired during their lifetime. The similarity of results (heritable transmission of new characters to the offspring) may suggest that mechanisms and the source of information for evolutionary change and TDP are similar. For metazoan evolution is inherently too parsimonious to afford for developing two separate mechanisms for achieving the same result.

In view of the fact that any heritable change in a phenotypic character requires investment of new information, and in the case of TDP no changes in genes are involved, the question arises: Where does the new information for transgenerational phenotypic changes come from?

Epigenetic Principles of Evolution. DOI: 10.1016/B978-0-12-415831-3.00011-2

Let us first briefly review the experimental and observational evidence showing that metazoan organisms, in response to specific stressful environmental stimuli, can not only adaptively change their phenotype (behavior, morphology, physiology, and life history) but also transmit phenotypic changes to the offspring.

TDP—Inherited Changes Without Changes in Genes

TDP in Nature

The first report of an experimentally induced transgenerational change in morphology was described in 1909 by Woltereck (1909). By cultivating in Italy, a native Denmark strain of *Daphnia cucullata*, he obtained a *Dauermodifikation* (German—long-lasting modification): within a number of generations, the crustacean changed its morphology and transmitted the morphological change, in the absence of the inducing factor, to the offspring for 40 generations before reverting to its ancestral form.

When *D. cucullata*, olfactorily perceives its predator's kairomones in the environment, it doubles the size of its own (Figure 11.1) and of its offspring's (F_1) helmet (Agrawal et al., 1999) and also doubles its carapace thickness (Laforsch et al., 2004), which make it less vulnerable to the attacks of the predator. This is a general morphological change providing defense against different predators. Under laboratory conditions, on visually perceiving small-scale turbulence, the crustacean also develops extremely large helmets, more elongated than those induced by the predator kairomone (Laforsch and Tollrian, 2004).

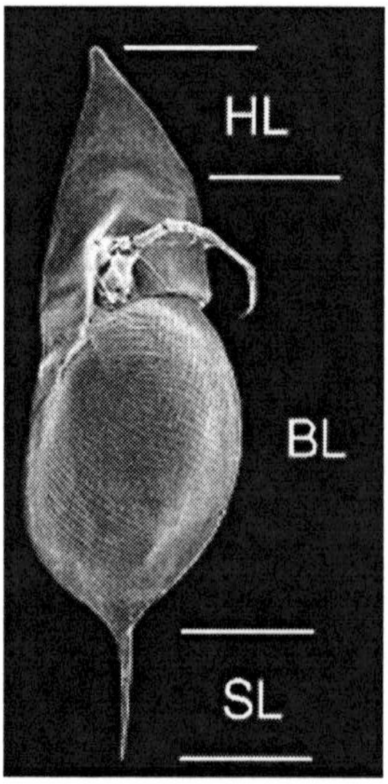

Figure 11.1 Electron micrograph of an adult helmeted *D. cucullata*. *Abbreviations*: BL, body length; HL, helmet length; SL, tail spine length.
Source: From Laforsch and Tollrian (2004).

It is also reported that, in response to stressful external stimuli, such as fish smells and food shortage, *Daphnia* reduces its body as well as that of the offspring (Hanazato et al., 2001).

On detecting the presence of the predator midge (*Chaoborus flavicans*) larvae, *Daphnia pulex* develops a neck spine containing several teeth, delays its reproductive maturity, and increases the body size of the offspring. In the presence of another of its predators, the water bug, *Notonecta glauca*, it produces offspring of smaller body size that reach reproductive maturity at a younger age (Lüning, 1992). In both cases, the inherited change is adaptive and, obviously, involves no changes in genes.

Daphnia magna is an all-female asexually reproducing species but when environmental conditions deteriorate (crowding, depletion of food resources) or presage deterioration (shortening of the photoperiod) (Stelzer, 2008), it gives birth to a sexually reproducing generation (male and female individuals). It is experimentally demonstrated that production of male individuals in this species is a function of increase in the levels of the crustacean juvenile hormone (JH), methyl farnesoate (MF) (Olmstead and LeBlanc, 2007; Figure 11.2). The sexually reproducing generation produces freezing- and desiccation-resistant eggs, which only hatch into male individuals. The transgenerational change is also experimentally induced by administration of insecticidal JH analogs, pyriproxyfen and methoprene (Olmstead and LeBlanc, 2003), and by exposing maternal organisms with maturing oocytes in the ovaries to the crustacean JH, MF (Rider et al., 2005). The nuclear retinoic acid receptor is believed to be the receptor for the hormone MF, i.e., mediator of the sex-determining activity of the hormone.

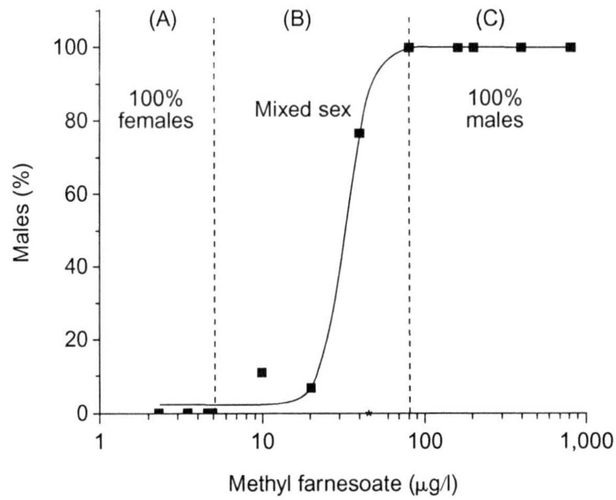

Figure 11.2 Percentage males in broods of offspring produced by maternal daphnids exposed to concentrations of methyl farnesoate. Dashed lines demarcate the methyl farnesoate levels that produced: (A) only female offspring, (B) mixed-sex broods, and (C) only males. The asterisk denotes the methyl farnesoate level at which gynandromorphic offspring were produced. Experiments were performed at 20°C.
Source: From Olmstead and LeBlanc (2007).

The mechanism of the switch from the parthenogenetic reproduction to the sexually reproduction generation is empirically known to a large extent. Unfavorable/stressful environmental stimuli (e.g., crowding, shortening of photoperiod) are processed in the crustacean's central nervous system (CNS). The neural processing results in activation of the cholinergic system (Eads et al., 2008) and inhibition of the synthesis of neurohormones mandibular organ-inhibiting hormone-1 (MO-IH-1) and MO-IH-2 by the secretory neurons of the X-organ–sinus gland complex, leading to secretion of MF by the MO. MF binds its specific receptor during a critical period of oocyte maturation (Rider et al., 2005), thus activating genes for sexual generation.

In response to crowding, amictic female rotifers of the cyclically parthenogenetic species, *Brachionus angularis*, give birth to mictic daughters (sexually reproducing generation), which produce haploid eggs developing into males or, when fertilized, developing into diapausing eggs. Amictic females hatching from diapausing eggs respond very slowly to crowding, to reach the normal sexual reproduction response only after several generations (Stelzer and Snell, 2003; Schröder and Gilbert, 2004; Gilbert and Schröder, 2004). The crowding chemicals, density-dependent kairomones released by *B. angularis*, like other kairomones, act neurally, and may target a maternal tissue, possibly the nervous system, which then stimulates some oocytes to differentiate into mictic females (Gilbert and Schröder, 2004).

Another small rotifer, *Brachionus calycifloris*, when detecting the presence in the environment of its predator, *Asplanchna brightwelli*, produces offspring with an additional pair of spines, which protects them from being eaten by the predator. The character is transmitted for many generations (Gilbert, 1966). The predator itself produces different morphs according to the quantity of prey it perceives in the environment (Gilbert, 1980). It is believed that both the predator-induced defense in the first case and the prey-induced polyphenism in the second are transmitted to the offspring via some maternal cytoplasmic factors, but the factor(s) and the steps of transmission of the information on the presence of a predator from the animal's CNS to its eggs are still unknown. However, we can safely assume that whatever the route might be (visual, olfactory, or via the interoception), the sensing and/or perception of the predator/prey takes place in the CNS, implying that there is a causal chain extending from the CNS to the synthesis and deposition in the egg of the maternal cytoplasmic factor.

The offspring of the cotton aphid, *Aphis gossypii*, consists of four different phenotypes: alatae (winged), normal light green apterae (unwinged), normal dark green apterae (unwinged), and dwarf yellow apterae (unwinged) (Figure 11.3), each with distinct life histories. When on predator-free plants, apterous and alatae aphids produce primarily offspring of apterous and alatae types, respectively, but under increased predation risk (e.g., when exposed to search track of the ladybird beetles, *Hippodamia convergens*), the aphids produce a greater proportion of alatae for several generations (Mondor et al., 2005).

When it is attacked by its predators or when it visually detects their presence, the pea aphid, *Acyrthosiphon pisum* emits a volatile compound, the alarm pheromone, (E)-β-farnesene, which is the only volatile compound emitted by these aphids in the presence of their predators (Kunert et al., 2005). The synthesis of the alarm pheromone in many insects is induced by pheromonotropic neurohormones (Masler

Figure 11.3 Photograph showing a parasitized aphid (a mummy), the green "normal" morph, the yellow "dwarf" morph, and the alatae "winged" morph of the cotton aphid, *Aphis gossypii*. *Source*: From Mondor et al. (2008).

et al., 1994; Rafaeli, 2009). The alarm pheromone is received by olfactory neurons and perceived in the brain of conspecific aphids, inducing an immediate behavioral change, walking or dropping off the plant. In response to the alarm pheromone or to the presence of predators (e.g., ladybirds, hoverfly larvae, lacewing larvae), these aphids also increase the proportion of winged individuals in the offspring (Dixon and Agarwala, 1999; Kunert and Weiser, 2003). This shift in morphology does not protect the prey from the predator, but allows it to avoid the predator by flying to other plants (Weisser et al., 1999).

Recently, investigators have demonstrated that exposure of mother aphids to the alarm pheromone alone also increases the proportion of winged morphs in the offspring (Kunert et al., 2005; Podjasek et al., 2005). *A. pisum* also produces a greater proportion of alatae individuals when exposed to its natural parasitoid enemy, *Aphidius ervi*, but it does not if is parasitized (Sloggett and Weisser, 2002).

There is another impressive example of TDP, that is considered to be a genuine case of speciation by many authors. It comes from experiments conducted by Shaposhnikov with the parthenogenetic aphid, *Dysaphis anthrisci majkopica*, reared on the unsuitable host plant, *Chaerophylum maculatum*, leading to almost 100% death rate. After eight asexual generations, however, a proportion of aphids changed their morphology (their body size and rostrum), became able to survive, and adopted the ex-hostile plant as their host.

In response to deterioration of food quality, crowding, and other adverse factors, female individuals of some parthenogenetic aphids, such as *Sitobion avenae* Fabricius, *A. pisum*, and rosy apple aphid, *Dysaphis devecta*, switch to production of alatae (winged) offspring, but in absence of these factors give birth to apterae (wingless) offspring. The change has adaptive character.

Another aphid, *Aphis fabae*, experiences very high mortality when reared on the garden nasturtium, *Tropaeolum majus*. However, after three generations of rearing on this inhospitable plant, the aphid is able to use that plant as its specific host.

And finally, an example of the TDP determined by mother's behavior. Flying alatae vetch aphid, *Megoura viciae*, that have not flown produce both alatae and apterae offspring, but those that have flown produce almost exclusively apterae (Burns, 1972).

Phase Transition in Locusts

Under natural conditions, the locust species, *Schistocerca gregaria* (Forskål) and *Locusta migratoria*, occur in one of two behaviorally and morphologically alternative and reversible states known as "solitary sedentary" and "gregarious migratory" phases.

Most of the time in the wild, these locusts are in the solitarious phase with a population density of ~3/100 m^2. When solitarious larvae of these locusts experience crowding, they dramatically, within 4 h, change their behavior from the tendency to stay away from, or avoid, other locusts (solitarious phase) to a strong bias of associating with other conspecifics (gregarious phase), reaching a population density of up to 100,000/100 m^2.

These locusts have an extremely strong ability that, in response to a social cue (living alone or in crowd with other conspecifics), as well as to olfactory, visual, tactile, or auditory stimuli, to switch between the above two phases. This phenomenon of phase transition is characterized by a number of behavioral, physiological and morphological changes, including several changes in head size, wing size, and the number of sensilla, as well as morphometric and behavioral characters. All of these new characters are maternally transmitted to the offspring.

Transition from the solitary to the gregarious phase is also associated with a change from the cryptic green to dark coloration. Administration of the brain neurohormone [His^7] corazonin changes to dark the body color of the albino nymphs.

Gregarious locust nymphs also switch to the solitarious phase, with all of the accompanying behavioral and morphological changes in reverse, when they are kept isolated.

A factor from accessory glands of the adult females determines the maternal inheritance of gregarious behavior; by washing eggs of gregarious locust females with a saline solution, a solitarious offspring is produced, but gregarious behavior may be restored by administration of extracts of gregarious females (Hügele et al., 2000). From field observations, it has been concluded that the color of locust hatchlings is correlated with the degree of crowding that the locust mothers have experienced rather than by volatile chemicals in pods (Bouaichi and Simpson, 2003).

One crucial proximate inducer of transformation of the solitary form into the gregarious form is JH. The hormone has a significant effect on the external coloration of locusts (Tawfik et al., 1997; Figure 11.4). In this context, it is important to remember that the JH secretion by corpora allata in insects is under strict cerebral control of stimulating neuropeptides (allatotropins) and inhibiting neuropeptides (allatostatins)

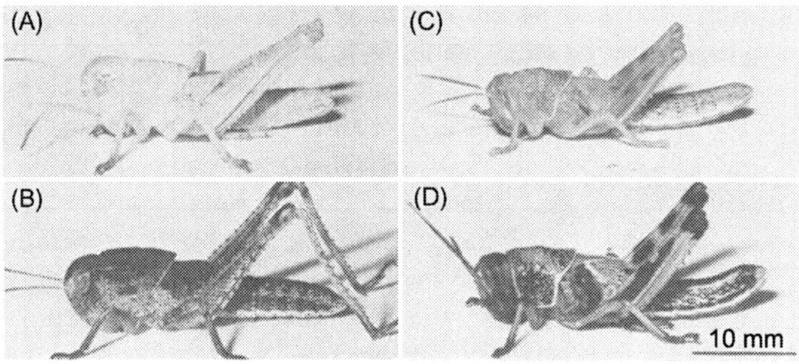

Figure 11.4 Photographs showing the effect of [His^7] corazonin on the body color in *Locusta migratoria* (A and B) and *S. gregaria* (C and D). Locusts were injected with the hormone in 4 μl of oil or with 4 μl of oil alone at day 3 of the third instar and photographed 2 days after the subsequent ecdysis. Hormone was injected into *L. migratoria* and *S. gregaria*, respectively. (A) Oil-injected, crowded albino nymph. (B) Hormone-injected, crowded albino nymph. (C) Oil-injected, isolated nymph. (D) Hormone-injected, isolated nymph.
Source: Selectively from Tawfik et al. (1999).

secreted by secretory neurons in the insect brain (Stay et al., 1996) as well as via nerves innervating corpora allata (*nervi corpori allati I* and *nervi corpori allati II*) (Kou and Chen, 2000).

Besides the above controls via the brain–corpora allata axis (JH secretion) and innervation of corpora allata, the CNS exerts a humoral control on the process of phase transition by secreting in hemolymph the neurohormone dark-color-inducing neurohormone, an 11-amino acid residues long peptide, also known as [His^7]-corazonin (Grach et al., 2003), whose primary structure, to a large degree, is similar to the vertebrate melanophore-stimulating hormone. Administration of [His^7]-corazonin in solitary nymphs of both sexes induces morphometric changes, characteristic of the gregarious form, especially F/C (length of the hind femur/head width) and E/F (length of fore wing/length of the hind femur) ratios (Maeno and Tanaka, 2004). Administration of [His^7]-corazonin in nymphal solitary locusts also induces a gregarious type of reduction in the number of antennal sensillae at the onset of adulthood (Maeno and Tanaka, 2004). Injection of the neuropeptide [His^7]-corazonin, alone or combined with JH, in migratory locust instars (Tanaka, 1996, 2000a,b; Yerushalmi and Pener, 2002) produces various patterns of coloration, depending on the time of administration and the dose. Differences between the two phases are also observed in the number and amount of neuropeptides secreted by *corpora cardiaca* in the hemolymph, suggesting that these neuropeptides may also play a role in the transition to the gregarious phase (Clynen et al., 2002).

Intense changes are also reported on the levels of a number of neurotransmitters in the locust brain (Rogers et al., 2004), but a transient increased secretion of serotonin seems to be responsible for behavioral change (Anstey et al., 2009) (see also figure 8.4).

Summarizing the above evidence, it may be said that a number of neural and endocrine factors, such as the brain neuropeptide [His^7]-corazonin, corpora cardiaca neuropeptides, serotonin, and JH, act as a conveyor of CNS gregarizing messages to target cells and tissues during gregarization in locusts.

At this juncture, for our purpose, it is helpful to switch to the reverse approach, i.e., to follow the chain of events in their natural course, as they unfold under the influence of the external stimuli. External (visual, olfactory, auditory, or tactile) stimuli are received by the sensory neurons (exteroceptors), converted into specific electrical spike trains, and as such transmitted to the respective neural circuits in the CNS, via interneurons. By processing the input of electrical signals, neural circuits generate their chemical output, mostly in the form of neurotransmitters and neuromodulators, which stimulate secretory neurons to release neurohormones ([His^7]-corazonin, allatotropins, etc.).

In the case of the desert locust, *Schistocerca gregaria* (Forskål), the olfactory stimuli (aggregation pheromones) are converted into electrical signals, which are transmitted to interneurons of the frontal antennal lobe for further processing, and then to the mushroom body and the lateral protocerebrum. These interneurons respond with different specificity to different pheromone stimuli (Anton and Hansson, 1996). The CNS response to aggregation pheromones also depends on the presence of JH, as is demonstrated by the fact that in allatectomized or old individuals, which do not secrete JH because their *corpora allata* are atrophied, the neural circuits do not respond to the pheromone (Ignell et al., 2001). Tactile stimuli (e.g., touch on the outer side of the upper portion of a hind leg) from mechanosensory trichoid sensillae on the hind limb, via metathoracic nerve 5, are transmitted to the CNS (Rogers et al., 2003). Tactile stimuli induced in head hair receptors by wind are transmitted to specific interneurons and then to thoracic motor centers, inducing flight behavior in gregarious locusts (Ayali et al., 2004).

Many studies have shown that phase transition in locusts is also associated with differences in the neuronal function. Analysis of responses of the descending contralateral movement detector interneuron have shown that in solitary locusts habituation to approaching objects is five times stronger than in gregarious locusts (Matheson et al., 2004). Changes related to behavioral phase transition are observed especially in the activity of flight-related neurons. So, for example, tritocerebral commissure giant (TCG) interneurons have weaker wind-induced spiking activity in solitary locusts than in gregarious locusts, and the spontaneous activity of the tritocerebral commissure dwarf interneuron in darkness is significantly weaker in solitary than gregarious locusts (Fuchs et al., 2003).

The darker color of gregarious locusts results from deposition of melanin in the cuticle, which is cerebrally regulated by secretion of pheromone biosynthesis-activating neuropeptide (PBAN), also known as "melanization and reddish coloration hormone."

While the olfactory (gregarizing pheromone) and visual (crowding) stimuli can induce gregarious behavior when combined, tactile stimuli *per se*, uncombined with other stimuli, are capable of inducing that behavior even when applied to individual locusts (Rogers et al., 2003). Repeated mechanical stimulation in the hairy femoral region leads to transition of the solitary locusts to the gregarious behavior within a few hours. Electrical stimulation of the metathoracic nerve 5 is very effective in inducing gregarious behavior. However, its branches, the nerves 5A and 5B, are less effective in that respect. Electric stimulation of the 5B nerve branches, 5B1 and 5B2, can induce gregarious behavior when applied on both nerves simultaneously but not separately (Rogers et al., 2003).

The neuronal circuits that integrate mechanosensory gregarizing stimuli should combine exteroceptive signals from the anterior surface of a hind femur with a specific proprioceptive signal that naturally results from the inward displacement of the leg on contact with another locust. These neuronal circuits are the first central elements of the pathway that initiates the rapid and widespread neuronal plasticity that underlies the behavioral phase change (Rogers et al., 2003).

Electric stimulation of the nerve 5B causes gregarization, but stimulation of neither 5B1 nor 5B2 alone leads to gregarization. The nerve 5B1 innervates tactile hairs on the anterior face of the femur and is the provider of the exteroceptive input, whereas the nerve 5B2 innervates hairs on the posterior face and is the main source of the proprioceptive input. Only combination of the exteroceptive and interoceptive input serves as a gregarizing signal (Rogers et al., 2003; Figure 11.5).

Neural tissue (brain and/or corpora cardiaca, subesophageal ganglion (SOG), thoracic ganglion) of insect species belonging to 18 orders, when implanted to albino locusts, induce darkening of their body, as illustrated for the case of the heelwalker, *Hemilobophasma montaguensis* (Tanaka, 2006) in Figure 11.6.

Now, based on the above evidence and the general knowledge on processing of external/internal stimuli in the CNS, we can connect the dots and outline the reconstructed pathway of the development of gregarizing characters in locusts (Figure 11.7; see also Section Reconstructing the Chain of Events in Transgenerational Plasticity, later in this chapter).

Sensory (visual, olfactory, auditory, and tactile) stimuli related with crowding are received by respective locust exteroceptors, converted into electrical signals and via interneurons transmitted to the respective neural circuits in the mushroom body. Further processing in the pars intercerebralis results in the release via axons of specific chemical outputs (neurotransmitters/neuromodulators) in hemolymph or in the corpus cardiacum (the pars intercerebralis–corpora cardiaca axis is considered to be the insect equivalent of the vertebrate hypothalamus–pituitary axis; Clynen et al., 2002). These chemicals stimulate secretion of neurohormones such as [His^7]-corazonin, mainly by protocerebral lateral secretory neurons (Roller et al., 2003) and a number of other neuropeptides, which act directly for developing gregarious characters in parental locusts. The brain neurohormone corazonin is necessary for expression of the green color in solitarious locusts and is involved in the changes in morphometric ratios of F/C (hind femur length/maximum head width) and E/F (elytron length/hind femur length).

Similarly to crowding, experimental injection of corazonin alone induces transition of solitarious locusts to gregarious phase (Tanaka, 2006). Injection of some other neuropeptides secreted in the brain and neural structures, such as corpora cardiaca (as well as implantation of brains from gregarious locusts), at various larval stages of albino locusts, produces all the spectrum of body colors characterizing gregarious locusts. Injection of corpora allata extracts from gregarious locusts into an albino

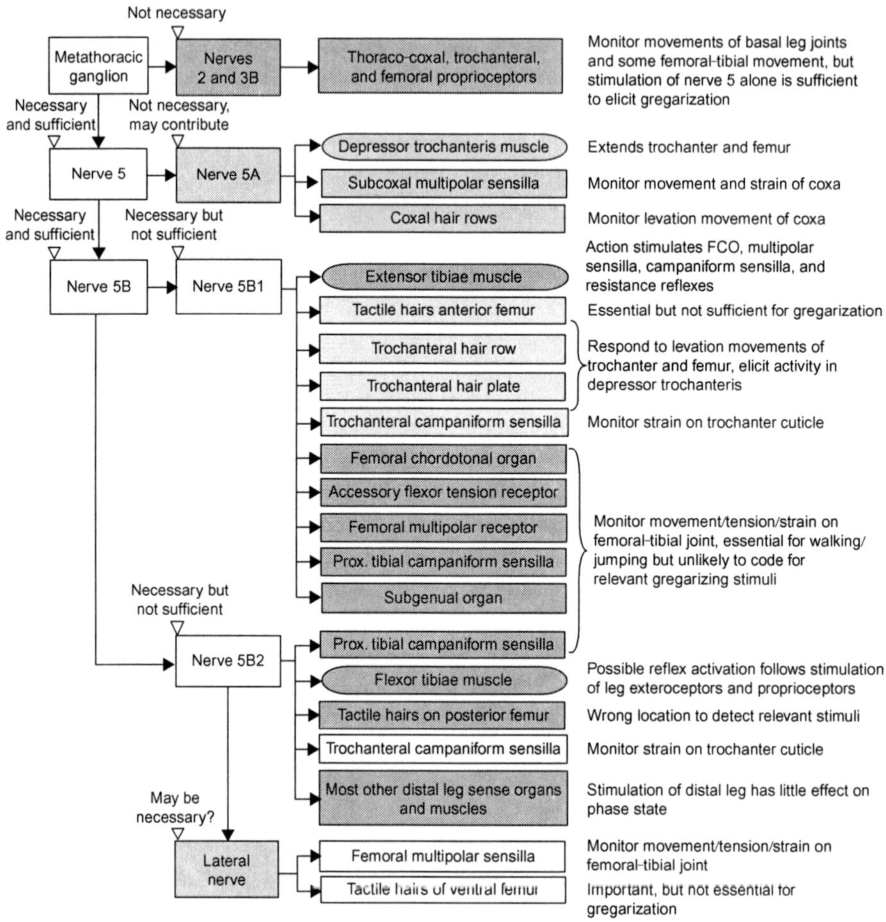

Figure 11.5 Summary of the innervation of the exteroceptors, proprioceptors, and muscles of the hind leg and their possible roles in eliciting behavioral phase change. Nerves or sense organs in white boxes need to be stimulated in order to elicit phase change. Nerves or sense organs in the lightest gray boxes are implicated in signaling appropriate mechanosensory gregarizing stimuli. Nerve 5A and its innervated structures, in mid-gray boxes, may contribute to signaling gregarizing stimuli, but it is not necessary to stimulate this nerve for phase change to occur. Nerves, muscles, and sense organs in the dark gray boxes have no role in mechanosensory-elicited phase change. *Abbreviation*: FCO, femoral chordotonal organ. *Source*: From Rogers et al. (2003).

strain locust that lacks corazonin induces typical morphometric changes but does not produce dark color (Tanaka, 1996, 2000a,b, 2007).

In contrast, activation of the brain–corpora allata JH pathway induces green color, inhibits the pheromone secretion, and generally contributes to a shift to the solitary phase (Tawfik et al., 1997).

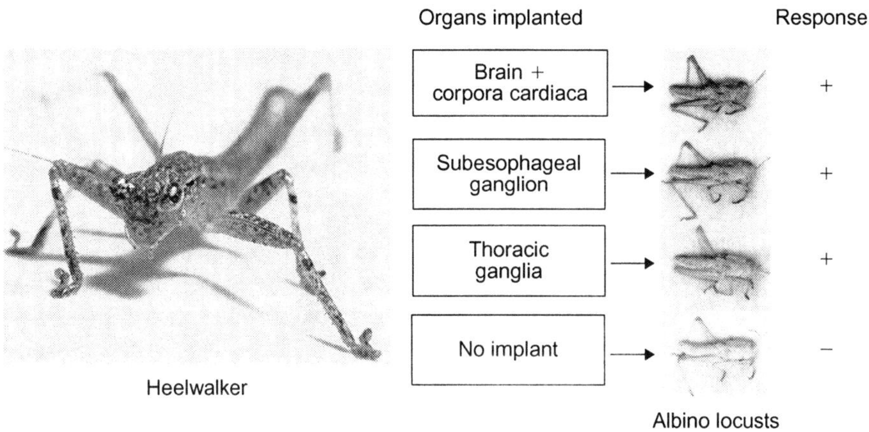

Figure 11.6 Dark-color-inducing activity of various organs from a heelwalker, *Hemilobophasma montaguensis*, when implanted into albino nymphs of *Locusta migratoria*. *Source*: From Tanaka (2006).

As shown in Figure 11.7, the chain of events for phase transition in locusts starts with reception of gregarizing stimuli. The stimuli *per se* do not represent epigenetic information for phase transition. Transformation of these stimuli into gregarizing cues is a function of the locust neural circuits, which by processing these stimuli generate a gregarizing chemical output, in the form of neurotransmitters/neuromodulators. This processing-dependent origin of gregarizing signals in the CNS determines the epigenetic nature of the information for phase transition in locusts.

TDP in Experiments

1. When *D. magna*, a small crustacean organism, is fed with toxic cyanobacteria, during its lifetime, it develops an inducible defense against the bacteria and transmits to the offspring not only that resistance but also a shorter time to maturity and reproduction (thus reducing chances of being eaten by predators). It also increases the number of offspring produced. The inherited defense increases from F_1 to F_2 and is sustained for generations. Here is the explanation of the phenomenon:

   > *Such a pattern is consistent with the presence of maternal effects, in which experienced mothers transferred information about their environment to their offspring to improve their survival. In contrast, offspring of inexperienced mothers matured later, delivered their first clutch approximately 2 days later than the other group, and had fewer offspring.*
   >
   > *LaMontagne and McCauley (2001)*

   Let us remember that the "maternal effects" are anything but changes in genes or genetic information. Hence, the resistance to cyanobacteria is provided to the offspring in the form of epigenetic information.

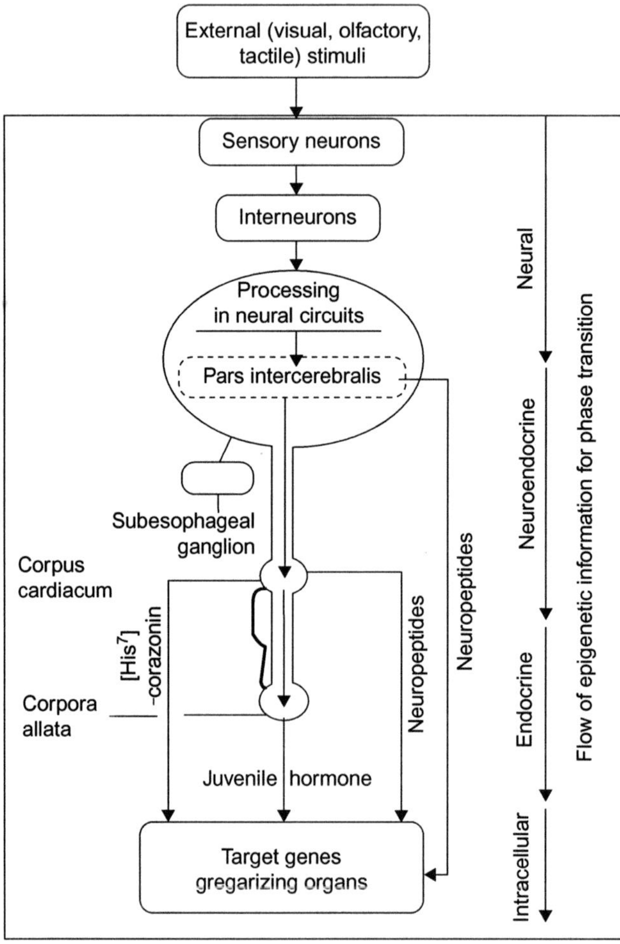

Figure 11.7 A generalized diagram of the chain of events and the flow of epigenetic information for the development of gregarious characters in locusts. Note that the information is computated by the nervous system in brain circuits involving respective sensory neurons and interneurons in response to specific external stimuli.

2. Under normal-to-favorable conditions in the environment, *D. magna* reproduces asexually by producing diploid female offspring. It responds to the environmental stressful conditions (e.g., shortening of the photoperiod, drop in food quality and quantity, crowding) by activating the neuroendocrine cascade crustacean hyperglycemic hormone (CHH)—MF hormone, via the CNS–X-organ/sinus gland complex–ovary axis. Thus, the crustacean transduces unfavorable environmental stimuli into inherited phenotypic changes in the offspring, giving birth to both male and sexually responsive female individuals, leading to sexual reproduction and production of eggs that are different from parental eggs in morphology (contain a protective cover, ephippium), biochemistry (contain substances that prevent them from drying and freezing), and life history (can delay hatching for many years until favorable conditions in the environment return) (Rider et al., 2005).

3. The natural host plant of the rosehip fruitfly, *Rhagoletis alternata*, is the rose, *Rosa canina*. When the fruitfly is experimentally reared on rugosa rose (*Rosa rugosa*), it responds to the new host plant by changing its morphometry (it develops a bigger body mass) and life history (it reduces its developmental time and produces a larger number of eggs) (Leclaire and Brandl, 1994).
4. In 1896 Standfuss demonstrated that heat-shocked Swiss subspecies of scarce swallowtail, *Iphiclides podalirius*, produced offspring that resembled subspecies of warmer parts of Europe (Sicily). By heat-shocking the European form of *Papilio mochon*, investigators succeeded in producing offspring resembling the Syrian form of the butterfly.
5. In experiments on central European subspecies of the butterfly *Aglais urticae*, were obtained offspring resembling the Sardinian subspecies after heat-shocking, and North Scandinavian subspecies after cold-shocking. The same phenomenon of producing characters of subspecies living in warmer (southern) regions or of those living in colder (northern) regions by heat-shock or cold-shock, respectively, was later echoed by Shapiro on brush-footed butterflies *Nymphalis antiopa* (Shapiro, 1976) and on the pupa of *Pieris occidentalis* (Shapiro, 1982).
6. D'Amico et al. (2001) have observed that over a period of 30 years (220 generations) under laboratory condition, the tobacco hawkworm, *Manduca sexta*, increased its body mass by 50% (from 7.8 to 11.1 g) and this coincided with an increase in the critical weight (from 5 to 6 g), the time when corpora allata, under action of brain signals (allatostatins) stops JH synthesis (D'Amico et al., 2001).
7. The offspring (F1) of the marine stickleback fish of the species *Gasterosteus aculeatus*, transferred from marine tide pools to freshwater ponds, when females had ripe ovaries, and males had breeding coloration, showed significant morphological changes manifested in a lower number of armor plates and in general body morphology (Kristjansson, 2005).
8. Developmental stress experienced by female zebra finch (*Taeniopygia guttata*) nestlings reared in crowded broods leads to a maternal transgenerational effect of producing offspring of smaller size, probably as a result of lower testosterone levels in the eggs of stressed zebra finch mothers (Naguib and Gil, 2005).
9. Licking and grooming (LG) of pups during the first week of postnatal life is an inborn behavior in rats and dams may display high or low LG. Generally, the offspring of high-LG and low-LG dams show, respectively, high and low LG toward their offspring. However, when during the first postnatal life females born to low-LG dams, but fostered to high-LG dams, exhibit high LG toward their offspring, and when females born to high-LG dams are reared by low-LG dams, they display the same low-LG behavior to their offspring.
   Two neural epigenetic mechanisms for explaining the inherited change of behavior have been proposed (Meaney and Szyf, 2005; Champagne, 2008).
   According to one of them, this inherited change in behavior of rats is related to "an epigenetic change in the brain of the offspring" (Meaney and Szyf, 2005). The proximate cause of the inherited change is a change in the DNA methylation of the NGFI-A response element of the glucocorticoid receptor (GR) exon-17 promoter in hippocampal neurons of the offspring. Succinctly the pathway from maternal tactile stimuli to transmission of high-LG behavior in the next generation (of female offspring of low-LG behavior dams) may be described as follows. The processing in the rat puppies' brains of the tactile stimuli of the high maternal LG stimulates hippocampal neurons to secrete serotonin receptor, which, by binding its receptor in hippocampal neurons, stimulates NGF1-A secretion.

In turn, NGF1-A by recruiting a histone acetylase looses the chromatin structure, thus exposing the exon-17 promoter to deacetylating enzymes, resulting in increased expression of the GR gene (Meaney and Szyf, 2005). Increased expression of GR activates the hypothalamic–pituitary–adrenal axis, which gives feedback to relevant hippocampal neurons (Figure 11.8). This demethylated state of DNA is epigenetically maintained through adulthood, thus determining the transmission of the maternal high LG to the next generation.

10. Rat embryos exposed, at the time of gonadal sex determination, to the endocrine disruptor, antiandrogenic compound vinclozolin, exhibit pathological conditions (prostate disease, nephritis, breast tumors, hypercholesterolemia) and morphological changes in the development of testes, which were transmitted consistently for at least four generations (F1–F4). Here is the investigators' interpretation of the results of the study:

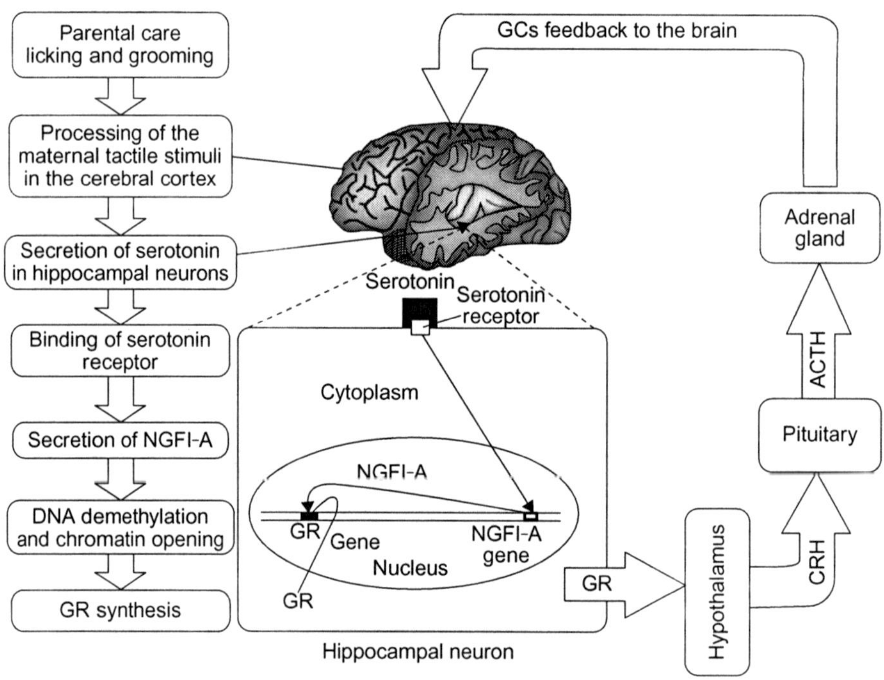

Figure 11.8 Diagrammatic representation of the mechanism of the epigenetic transmission of the maternal high licking and grooming (LG) behavior to the offspring. Tactile stimuli of the high maternal LG are processed in the pup's cerebral cortex. The chemical output of the processing induces secretion of serotonin in hippocampal neurons. By binding its receptor, serotonin induces expression of NGFI-A, which binds its response element of the GR gene exon-17 promoter, thus inducing expression of the GR gene. This leads to the activation of the hypothalamic–pituitary–adrenal axis, which through a signal cascade (GR → CRH → ACTH → GCs) gives feedback to the pup's brain, thus enhancing her GC sensitivity and improving maternal LG behavior toward her offspring. *Abbreviations*: ACTH, adrenocorticotropic hormone; CRH, corticotropin-releasing hormone; GCs, glucocorticoids; GR, glucocorticoid receptor; NGFI-A, nerve growth factor-inducible protein-A.

> *The frequency of the abnormal phenotypes observed ranges from 12–50%. The frequency of a hot spot DNA sequence mutational event has been shown to be approximately 5% at its highest and generally is less than 1%. A genetic DNA sequence mutation also involves segregation with reduced frequency in subsequent generations. Therefore, the high frequency of the disease states and absence of normal Mendelian transmission observed in the current study suggests the transgenerational nature of the phenotype appears to be epigenetic through the germ-line.*
>
> *Anway et al. (2006)*

While the above examples of transgenerational changes in morphology are clearly stress induced, let us remember that stress response is also neurally controlled and regulated (see Chapter 7).

Now, let us attempt to reconstruct the mechanism of TDP in a few of the examples that are known best.

Reconstructing the Chain of Events in Transgenerational Plasticity

The Mechanism of Induction of Sexually Reproducing Generation in D. magna

This transgenerational change results from a neurogenic increase in the level of the crustacean juvenoid hormone, MF (the crustacean equivalent of the insect JHs) produced by the MO, an endocrine gland homologous to the insect corpora allata (Tobe et al., 2004).

External stimuli (e.g., decreasing photoperiod, higher temperature) are received by sensory neurons and processed in neural circuits in the crustacean ganglia/brain. Neurons of the X-organ–sinus gland complex of the eyestalk secrete neuropeptides of the MO-IH family, which negatively regulate secretion of MF (Wainwright et al., 1996) by inhibiting the synthesis of farnesoic acid methyl transferase, which is secreted mainly in the nervous system, X-organ–sinus gland complex of the eyestalk and is necessary for the synthesis of MF by the MO, as is indicated by the fact that removal of the eyestalk causes MF production by the MO to increase (Li et al., 2010). Thus, the epigenetic information for the synthesis of MF flows from the nervous system to the MO.

The sinus gland neurohormone, CHH (crustacean hyperglycaemic hormone), is another negative regulator of the secretion of MF (Wainwright et al., 1996) (its action is similar to the action of allatostatins, neurohormones that suppress secretion of JH in insects). Inhibition of the release of these neurohormones in neurons of the X-organ–sinus gland complex determines pulses of secretion of MF by the MO. MF, in turn, activates a downstream cascade (genes *sex-1*, *dsx*, *csd* (?)), leading to production by female parthenogens of male and female individuals (sexual generation) (Figure 11.9) and production of drought- and freeze-resistant eggs (ephippia).

While the structure of the neural circuit(s) inducing secretion of neurohormones, MO-IHs and CHHs, in *D. magna* is not known, experimental studies on another crustacean, the crayfish, *Procambarus clarkii*, suggest that it is a serotoninergic circuit. Serotonin released by neurons of this circuit bind to its receptors on the membrane of neurons of eyestalk ganglia, thus stimulating secretion of the neuropeptide CHH. Serotonergic neurons are part of the neural network that modulates secretory activity of the X-organ–sinus gland neuroendocrine center (Borst et al., 2002; Olmstead and LeBlanc, 2003).

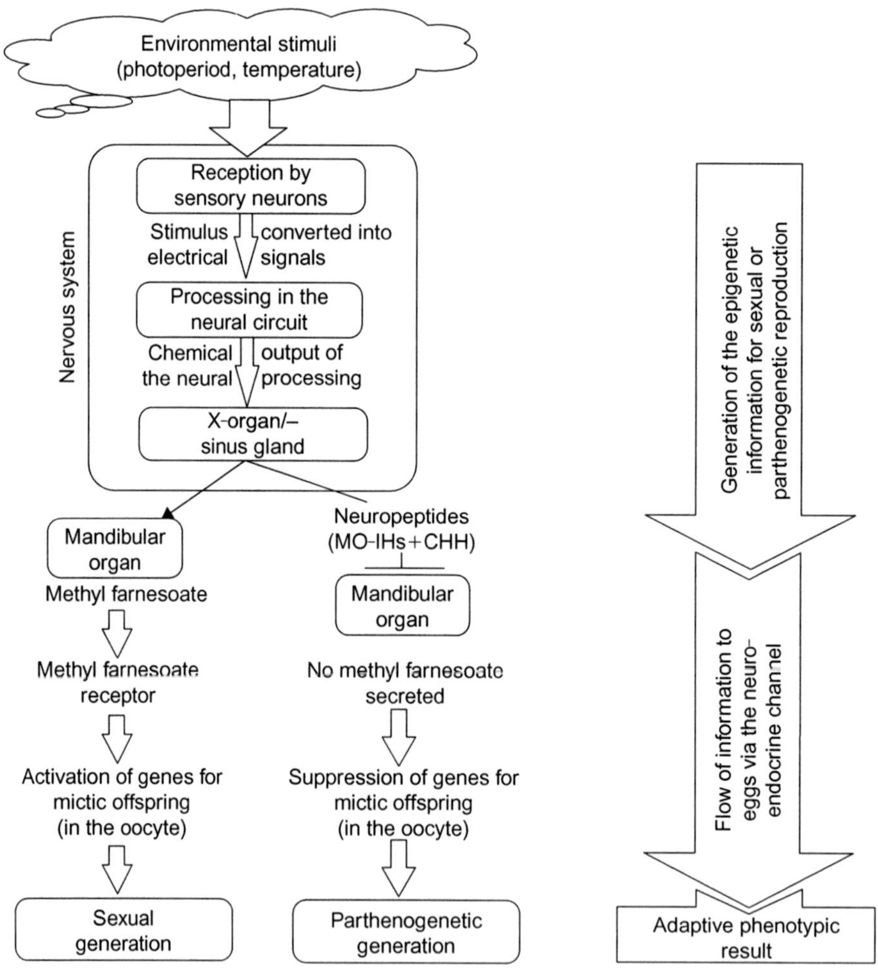

Figure 11.9 Illustration of the recruitment of genes in two different pathways leading to birth of asexual (parthenogenetic) and sexual generations in *Daphnia magna*. The point of divergence between the two pathways starts in the crustacean's brain with the failure of the X-organ–sinus gland to synthesize and secrete neuropeptides MO-IH-1, MO-IH-2, and CHH (left cascade) and recruitment of the methyl farnesoate in the case of sexual generation-producing Daphnias and recruitment of the above genes and prevention of the synthesis of MF in the case of production of the asexual generation.
Source: From Cabej (2011).

Insect Diapause

Insect diapause is a neurally determined arrest of development characterized by extremely low metabolic activity. It is an adaptation arising in response to specific environmental stimuli presaging deterioration of the conditions of living, such as the annual changes in the temperature and photoperiod.

Diapause in the Silkworm

The silkworm, *Bombyx mori*, determines the diapausing/nondiapausing fate of its eggs by synthesizing or not a neuropeptide, the diapause neurohormone (DH), by processing a precursor protein consisting of DH, PBAN, and three other SOG neuropeptides. However, the production of DH is controlled by the dorsal protocerebrum (Shimizu et al., 1997). The neurohormone targets the insect ovaries via the hemolymph (Kitagawa et al., 2005).

A correlation was found to exist between the dopamine levels in the brain–SOG and hemolymph, on the one hand, and the induction and onset of diapause, on the other. Later it was determined that secretion of DH by neurons of the SOG is induced by a dopaminergic circuit (Noguchi and Hayakawa, 2001) and inhibited by a γ-amino butyric acid (GABA)-ergic circuit (Shimizu et al., 1997), in response to the brain photoperiodic clock (Hasegawa and Shimizu, 1987).

Experimental elevation of the dopamine level stimulates switching to diapause of nondiapausing larvae and young pupae of the silkworm (Noguchi and Hayakawa, 2001). Release of dopamine by the brain is correlated with increased firing activity of the SOG labial secretory (Lb) neurons of the diapause-egg producers but not nondiapause-egg producers (Ichikawa, 2003). In response to the onset of illumination, Lb neurons increase the firing rate, a fact that has been explained with the existence in the brain of *B. mori* of extraocular photoperiodic receptor neurons. Ablation of

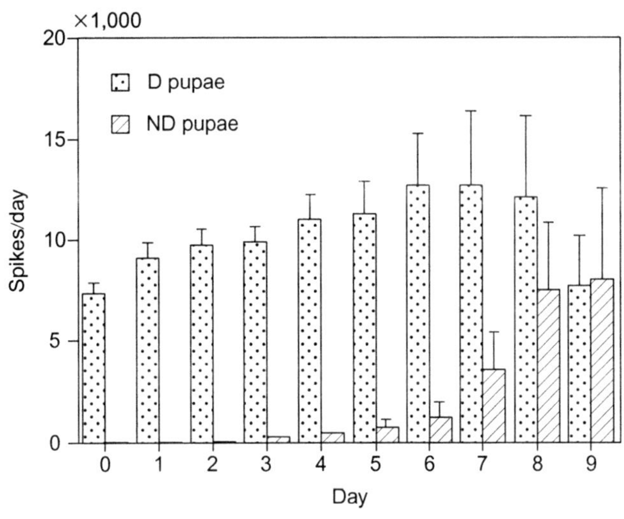

Figure 11.10 Averaged daily firing activities of Lb cells in diapause (D)- and nondiapause (ND)-egg producers during pupal period.
Source: From Ichikawa (2003).

the labial (posterior) neurons of the SOG impairs induction of diapause eggs, and it seems that active Lb neurons are the exclusive source of DH.

A correlation exists between the differences in firing activity of the Lb cells in diapausing and nondiapausing pupae, on the one hand, and the differences in their DH content, on the other. DH content in diapausing (D) pupae begins to decrease just after pupation, unlike nondiapausing (ND) pupae in which it returns to the initial low level as pharate pupae. Electrical activity in Lb (active labial neurons) of the pupae of the diapause type persists during the whole stage, while Lb neurons of the nondiapausing type become electrically active by the last third of pupation (Figure 11.10).

A Neurotransmitter Coding for a Life History Character

Now let us consider an experimentally determined case of a maternal neurotransmitter that starts a brain cascade leading to TDP. This is prevention of diapause in the offspring of the flesh fly (*Sarcophaga bullata*) diapausing mothers reared under short photoperiod (SP). Firing activity patterns in nondiapausing pupae suggest that an inhibitory signal from the brain suppresses almost completely the secretion of DH during the initial stage and higher activity on day 9 may indicate release from that inhibition. GABA is a possible candidate for the brain inhibition signal, as it may be concluded from the experimental prevention of production of nondiapausing eggs in *B. mori*. Thus, a GABAergic circuit may be responsible for cerebral inhibition of DH in the silkworm (Shimizu et al., 1989).

Flesh fly mothers that have experienced pupal diapause give birth to a progeny that does not express pupal diapause even when they experience diapause-inducing SP. Since only mother's photoperiod history matters in determining this transgenerational change in the life history, it has been concluded that the character is determined by a maternal factor deposited in the egg cell (Webb and Denlinger, 1998).

Extracts from brains of mothers that have experienced SP also suppress the appearance of diapause, and recent experiments have shown that a neurotransmitter, GABA, can do the same. Therefore, it is suggested that a GABA-mediated response might control the placement or synthesis in the oocyte of a maternal cytoplasmic factor that determines the nondiapause phenotype. It is believed that

> *the information transfer from mother's brain (the site of photoperiodic reception) to her ovaries occurs sometime after pupariation but before the second day of adult life.*
>
> *Henrich and Denlinger (1982)*

The maternal information is transferred to the ovary long before egg maturation, which is also consistent with the observation of a unique transcript present in the ovary of SP females on the first day of adult life (Denlinger et al., 1995). Later, investigators discovered that the neurotransmitter GABA, when applied to the mothers that would normally diapause, drastically reduces the likelihood of diapause to occur (Webb and Denlinger, 1998).

Diapausing flies can survive the cold weather because they contain cryoprotectants that enable insects and their larvae to overwinter in freezing temperatures despite the high proportions of water in their body. In the case of *S. bullata*, the transition to diapause in response to stressors such as low temperature and desiccation, cryoprotectants, such as glycerol, are synthesized and accumulated in the body and this is cerebrally controlled, as has been shown in experiments with brain-ligated larvae (Yoder et al., 2005).

The nature of the maternal factor suppressing diapause in *S. bullata* is not known yet, but in the silkworm, *B. mori*, a GABAergic circuitry is involved in regulation of DH synthesis and the onset of diapause (Shimizu et al., 1989).

What essentially takes place in the case of the flesh fly, *S. bullata*, is the appearance in the next generation of a new life history character (diapause skipping) without any changes in genes or genotype in general. But this is unheard of and thought provoking. Something different from genes is responsible for the new life history character.

All the data suggest that the information for switching to that new character comes from the nervous system in the form of the neurotransmitter GABA released by the mother's brain. Via hemolymph, it can reach ovaries and determine the synthesis or placement of the diapause-preventing maternal factor(s) in the egg cell.

Thus, whether the flesh fly, *S. bullata*, produces diapausing or nondiapausing eggs depends on the information on the length of the day that is computated in the brain of the mother. Production of diapausing eggs is induced by a dopaminergic circuit and inhibited by a GABAergic circuit (Webb and Denlinger, 1998).

The Mechanism of Transgenerational Phase Transition in Locusts

Crowding intensifies the contact of locusts with the aggregation pheromone secreted by the pheromone gland. The pheromone binds a pheromone-binding protein on the dendritic membrane of antennal olfactory neurons (Klause-de-Pupka et al., 1997).

Most locusts have ~100,000 olfactory receptor neurons (ORNs). Each ORN contains cilia, which have receptor molecules sensitive to pheromones. ORNs respond to the contact of receptor molecules with pheromone molecules by generating electrical spike trains (even a single pheromone molecule can stimulate the receptor neuron to generate nerve impulses; Kaissling, 1996). These trains, via axons, are transmitted to the ipsilateral antennal lobe, which contains neurons of two types: axonless inhibitory GABAergic local neurons (LNs) and excitatory cholinergic projection neurons (PNs), which project and arborize to specific glomeruli. There are about 1,000 glomeruli in the locust macroglomerular complex (MGC), which is the first-order center of the antennal lobe for synaptic processing of olfactory information. The basic neural olfactory structure is similar in all of the insects, and Figure 11.11 illustrates the reception and initial processing of olfactory stimuli in *Drosophila*.

From the MGC, PNs send axons to the brain higher processing centers to the base of ipsilateral mushroom body branching to form contacts in the calyx with PNs

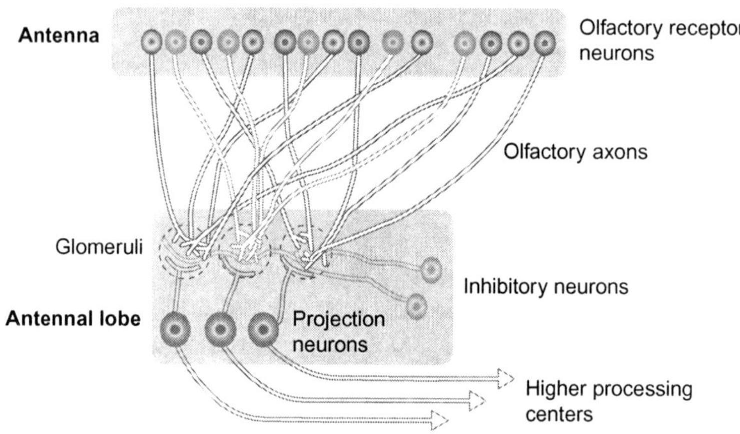

Figure 11.11 Schematic illustration of the glomerular organization of the *Drosophila* antennal lobe. The axons of neurons expressing the same type of odorant receptor converge on glomeruli in the antennal lobe. Blue, green, and red are used to denote olfactory neurons expressing a common odorant receptor. Glomeruli are outlined in purple. In *Drosophila*, each excitatory projection neuron sends its dendrites to a single glomerulus, where it receives extensive input from a single class of olfactory receptor neuron and sends an axon to higher brain centers (purple). The inhibitory (lateral) neurons (orange) in *Drosophila* project to multiple glomeruli. (For interpretation of the references to color in this figure legend, the reader is referred to the web version of this book.)
Source: From Bargmann (2006).

send axons to a higher processing center of the brain, the mushroom body where they branch to form contacts in the calyx with the Kenyon cells (KCs) (Figures 11.12 and 11.13), before projecting to the lateral protocerebral lobes. All KCs (there are ~50,000 KCs in the calyx) send axons ventrally to α and β lobes (Laurent, 1996).

The transmission of the pheromonal stimulus through the olfactory circuit illustrates the complex pathway of the neural processing from ORNs to the calyx neurons and neurons of the inferior lateral protocerebrum lobe. All this convoluted path of the processing of the pheromone stimulus ORNs → LNs → PNs→ neurons of the mushroom calyces → neurons of lateral protocerebrum has a high energy cost. But the fact that this information-processing labyrinth has evolved suggests that the benefit of the processing outweighs the cost. We know what this great benefit is: generation of epigenetic information for starting the cascade(s) that induces the behavioral phase transition, with all the accompanying morphological, morphometrical, and physiological characters to the offspring.

While we, to a satisfactory extent, know the main steps of neural processing that results in the chemical output that determines the locust phase transition, a serious gap exists in our knowledge of the mechanisms of transmission of this neurally generated information to the offspring.

One crucial piece of information for reconstructing the connections between the neural information produced by the locust pheromonal circuit and the adaptive

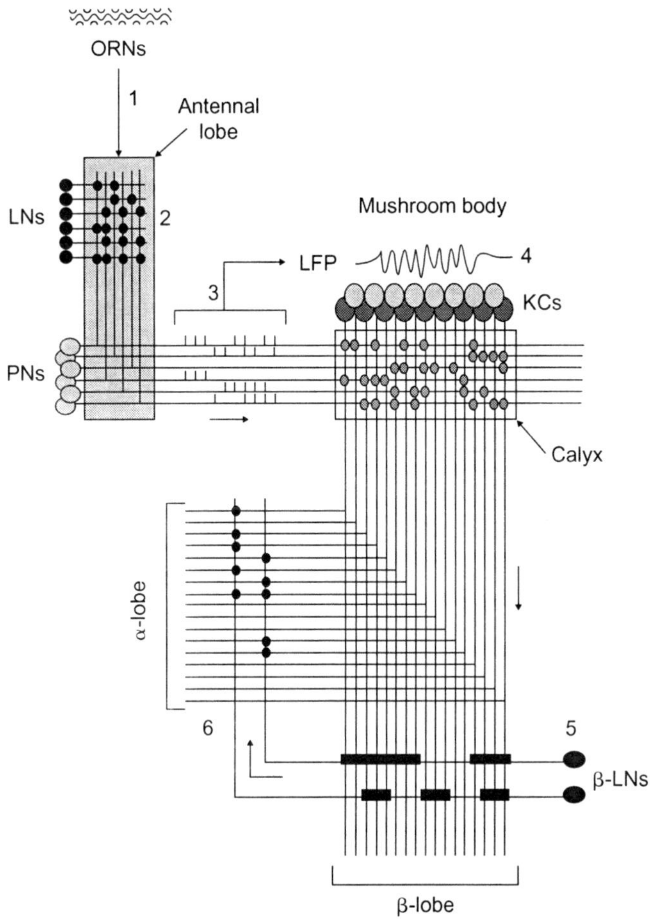

Figure 11.12 Circuit diagram of the olfactory pathway in the locust brain. Odors are transduced by arrays of olfactory receptor neurons (ORNs) in the antenna (1). ORNs activate ensembles of local and projection neurons (LNs and PNs) in the antennal lobe. LNs inhibit PNs (2) through a fast picrotoxin-sensitive GABAergic synapse. This inhibitory synapse is (at least in part) responsible for the oscillatory synchronization of the PNs activated by an odor. PNs responding to a given odor usually display specific and reliable slow temporal activity patterns (3), superimposed on these oscillatory responses. The patterns are neuron and odor specific. The coactivation of PNs during an odor response causes synchronized and rhythmic EPSPs in KCs, the intrinsic neurons of the mushroom bodies. These synchronized EPSPs can be detected as odor-evoked bouts of 20–30 Hz local field potential (LFP) oscillations (4) in the calyx. Arrays of KCs thus activated send action potentials down their axons to the α- and β-lobes. In the β-lobe, the KCs contact β-lobe neurons (5), which send axonal collaterals to the α-lobe (6). *Abbreviations*: EPSP, excitatory postsynaptic current; KC, Kenyon cell.
Source: From Laurent et al. (1998).

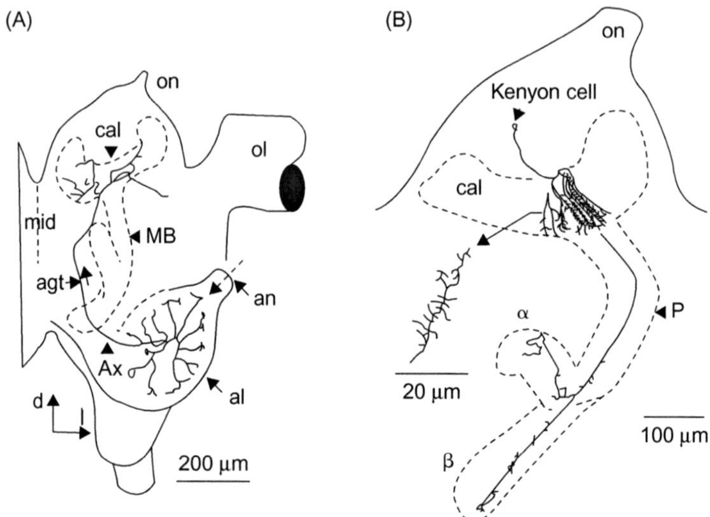

Figure 11.13 Projection of antennal lobe interneurons in antennal lobe and mushroom body and Kenyon cell morphology (camera lucida drawing of neurons stained by iontophoresis of cobalt hexamine). (A) Arborizations of antennal lobe (al) projection neuron, showing tufts of short dendrites in individual glomeruli, radial secondary dendrites, soma, and axon (ax), exiting deutocerebrum via the antennoglomerular tract (agt). The axon sends varicose collaterals in the calyx (cal) of the mushroom body (MB), before projecting to the lateral protocerebral lobe. *Abbreviations*: an, antennal nerve; mid, midline; ol, optic lobes; on, ocellar nerve. (B), Kenyon cell in mushroom body. Note the small soma (7 pm diameter), the primary neurite, the spiny secondary dendrites in the calyx (cal), the axon in the peduncle (p), and the relatively unbranched axonal terminals in the α- and β-lobes.
Source: From Laurent and Naraghi (1994).

phenotypic results it induces is the fact that the only difference found so far between the eggs of the gregarious and solitarious locusts is that the gregarious locusts contain in their eggs amounts of ecdysone that is five times higher than the solitarious eggs, even though the amount of ecdysteroids in the hemolymph of solitary individuals is higher than in gregarious locusts (Tawfik et al., 1999; Tawfik and Sehnal, 2003; Hägele et al., 2004). The fact that ecdysone is increased only in the insect ovary of the gregarious females suggests that the well-known cascade that starts in the insect's brain with the synthesis of prothoracicotropic hormone, which stimulates the prothoracic gland to secrete ecdysone, might not be responsible for the difference in amounts of ecdysone between the eggs of gregarious and solitary locusts.

It also suggests that the high levels of ecdysteroids in the eggs and ovaries of the gregarious females may be determined by local ecdysteroid secretion in their ovaries. Let us try to reconstruct the "missing" links from the processing of the pheromonal stimulus in the locust CNS to the differential deposition of ecdysone in the eggs of gregarious locusts. The structure of the olfactory system "is exceptionally similar across animal phyla—from molluscs to insects, crustaceans and vertebrates,"

> *the functional principles that can be established from studying one system might apply to the others as well, because similar circuit design principles are likely (although by no means certain) to underlie similar functional or computational principles.*
>
> *Laurent (1996)*

This suggests that, cautiously, we may extrapolate the knowledge on neural factors that regulate TDP in other insects to the neural circuits involved in induction of gregarious phase transition.

The phase transition in both directions (toward gregariousness and solitariousness) in locusts is correlated with changes in the level of various neurotransmitters and neuromodulators in the brain and ganglia. Among them, the most impressive are changes in the serotonin level (Anstey et al., 2009) in early stages of gregarization and solitarization. A clear difference is observed in the location of neurochemical changes, mainly in the brain during solitarization and in thoracic ganglia in the case of gregarization. Changes in the synaptic morphology may also be involved:

> *Perhaps phase change may alter the branching patterns and numbers of neurons expressing different neurochemicals.*
>
> *Rogers et al. (2004)*

It is also possible that changes in the synaptic morphology of the behavioral circuits may change the properties of these circuits and lead to the behavioral changes characterizing switches from one phase to the other. So, for example, it has been observed that wind-induced spiking activity in the TCG interneurons that relay input from head hair receptors to thoracic motor centers is weaker in solitary locusts than in gregarious ones (Fuchs et al., 2003).

The experimental evidence on specific changes in neurotransmitters and neuropeptides released from the CNS during phase transition seems to suggest that they may have a role in determining deposition of maternal factors, particularly ecdysone, in the egg cells of convert gregarious locusts.

Biologists have already demonstrated a number of cerebrally regulated mechanisms of deposition of maternal factors in the egg cytoplasm, such as receptor-mediated endocytosis (Handler and Postlethwait, 1977; Raikhel and Lea, 1985; Richard et al., 2001), neurally controlled increase of hormone in each of the successively laid eggs (Lipar and Ketterson, 2000; Sockman et al., 2001; Gil et al., 2004; Hayward and Wingfield, 2004), neurally controlled squeezing of the nurse cell content into the oocyte (Buszczak and Cooley, 2000), etc. The neural regulation of the deposition of maternal factors may be a general mechanism of deposition of those factors in metazoan egg cells (Cabej, 2004, pp. 87–100), but these neural mechanisms do not seem to be related to the differential deposition of ecdysteroids in the locust eggs.

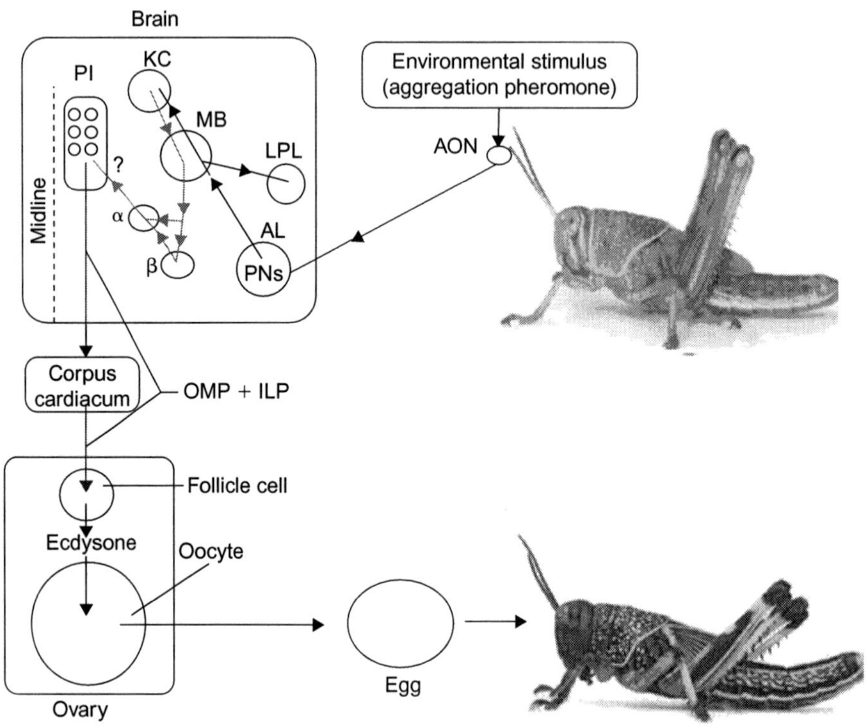

Figure 11.14 Simplified diagram of generation and the flow of information for transgenerational plasticity (phase transition) in *Locusta migratoria*. External stimuli received by antennal olfactory neurons are transformed into electric trains and sequentially transmitted to the antennal lobe, mushroom body, KCs, and pars intercerebralis of the insect brain. Secretory neurons of the pars intercerebralis secrete OMP and ILPs. These inducers, which initially are deposited in the corpus cardiacum, via hemolymph stimulate increased production of ecdysone in the ovary. Increased presence of ecdysone in the egg cells induces production of the gregarious offspring of solitarious locusts. *Abbreviations*: α, α-lobe; β, β-lobe; AON, antennal olfactory neuron; AL, antennal lobe; ILP, insulin-like neuropeptide; KC, Kenyon cell; LPL, lateral protocerebral lobe; MB, mushroom body; OMP, ovarian maturation parsin; PI, pars intercerebralis; PNs, projection neurons.

It is demonstrated that the differential expression of ecdysteroid genes in the insect ovary is controlled by neurohormones synthesized in the pars intercerebralis of the insect brain and released into the hemolymph via corpora cardiaca (Figure 11.14).

The most important in this respect are neuroparsins (Girardie et al., 1989, 1998a,b), such as the brain neurohormone, ovarian maturation parsin (OMP), equivalent of ovary ecdysteroidogenic hormone (OEH) (formerly known as egg development neurosecretory hormone), secreted by locust medial neurosecretory cells in the pars intercerebralis of the insect brain, and deposited in the corpus cardiacum. Its stimulating effect on the synthesis of ecdysteroids in the ovary is also demonstrated in experiments with a number of other insects, such as *Manduca sexta*, and mosquitoes, *Aedes aegypti* and *Aedes taeniorhynchus* (Lea, 1967; Hagedorn et al., 1979; Manière et al., 2000; Brown and Cao, 2001). Brain insulin-like peptides (ILPs) are also essentially involved in the induction of ovarian ecdysteroidogenesis.

Regulation of the ovarian production of ecdysteroids by brain/cephalic factors is also observed in other insects, such as the blowfly, *Phormia regina* (Manière et al., 2000). In *Manduca sexta*, OEH is found throughout the nervous system (Brown and Cao, 2001). In blowfly, *P. regina*, the control of ecdysteroids is a function of two neurohormones, OEH secreted by neurons of the pars intercerebralis and another neurohormone secreted by another, but still unknown, part of the brain (Hetru et al., 1982; Manière, 2000).

Theoretically at least, the high levels of ecdysteroids in the ovary may be determined and maintained by local innervation as part of the binary neural control of gene expression in metazoans, where local innervation may induce expression of ecdysone receptors in the ovary (Cabej, 2004, pp. 35–47) (see Section The Binary Neural Control of Gene Expression in Chapter 6).

Summarizing this brief review on the signal cascades determining transmission of gregarious characters to the offspring (transgenerational phase transition), we have seen that the pheromonal stimulus received by the ORNs in the insect antennae is converted into a train of electrical spikes, which are transmitted to the olfactory circuit. The end result of the elaborate processing of these spike trains in the olfactory circuit is the release of a chemical output, probably a neurotransmitter, on secretory neurons in at least two areas of the insect brain, leading to secretion by the latter of at least two types of neurohormones (OMPs and ILPs) with ecdysteroidogenic action in the ovary. Local innervation in the ovary may be responsible for the observed exclusive ovarian expression of the receptor for ILPs.

Since no changes in DNA or genetic information have been reported to be involved in the phase transition in locusts, the question arises: Where does the new information for transgenerational phenotypic changes come from?

The Origin of the Information for TDP

TDP is an adaptive response to stressful stimuli or to stimuli presaging deterioration of conditions of living. Stressful stimuli are interpreted by the neural circuits as problems requiring solutions. There are two different schools of thought on the nature of problem solving in neural networks. The classical hypothesis posits that neural networks are universal and are remodeled in order to solve the problems they are facing. Another hypothesis holds that not the network remodeling but the flexibility of the inborn networks, their ability to learn and self-organize are the source of the problem-solving capacity, creativity, and ability to solve unpredictable problems that might arise during the life:

> *Universality cannot solve the brain's problem, which is rather characterized by, "given the concrete network that an individual is born with, learn to cope with situations and problems as they arise."*
>
> *von der Malsburg (1999)*

The source and nature of information for erecting the metazoan structure are crucial for understanding evolution of metazoans. Although no changes in genes are involved, TDP is related to specific changes in patterns of gene expressions determined by activation in the offspring of signal cascades that were different, or inactive, or entirely lacked, in their parents. Being not a random but an ordered process, any specific change in the pattern(s) of gene expression during TDP unquestionably requires new information.

Tracing back the temporal sequence of events in the signal cascades is the logical way to identify the ultimate source of the information that determines transgenerational developmental changes and, by extension, evolutionary changes.

Adequate evidence, especially the three examples of reconstructed mechanisms of TDP considered in this section (induction of sexually reproducing generation in *D. magna*, prevention of diapause in the offspring of *S. bullata*, and transgenerational phase transition in locusts) make it clear that the epigenetic information for transgenerational (inherited) developmental plasticity originates in the maternal/paternal CNS. That epigenetic *information is not stored* in the CNS but it *is generated* by processing relevant stimuli in neural circuits.

Looking at Figure 11.14 one wonders: Why must the brain meander the stimulus through before a specific transgenerational signal cascade is activated in response to the stimulus? As explained in Chapter 2, the reason is simple: the environmental stimulus *per se* is causally not related to the specific cascade or the effector gene(s), hence cannot activate them. A chemical secreted as a result of its processing in the brain does. The chemical represents the information for selectively activating a specific cascade. How is that information generated?

The first step in the chain of events for transgenerational plasticity is the external or internal stimulus. All the environmental stimuli involved in cases of transgenerational plasticity considered earlier in this section are received by exteroceptors. These stimuli represent no information: the probability that these stimuli will induce expression of any particular genes, start a signal cascade, or lead to transgenerational plasticity is virtually zero. As data on the environment, the stimuli "tell" nothing to genes, in the meaning that, *per se*, they do not, and cannot, induce any particular gene or activate a particular signal cascade. These stimuli "make no sense" to genes.

In the second step, sensory neurons convert these environmental (e.g., visual, olfactory, tactile, auditory, tactile) stimuli into a universal neural language by encoding them in the form of electrical signals that are transmitted to specific neural circuits in the brain. The conversion generates data suitable for neural processing, but they still represent no information for activating/inactivating any gene or starting a signal cascade. The probability that these electrical signals themselves may induce a specific transgenerational plasticity cascade still remains practically zero.

The third link in the causal chain is the processing of the input of electrical data in neural circuits. This is a computational process, i.e., a process of transformation by a nonlinear dynamic system (nervous system) of the input of electrical data into a chemical output (neurotransmitter/neuromodulator), which is released on particular secretory neurons. In response, the secretory neurons synthesize and/or release a specific neurohormone, which activates the signal cascade that ultimately leads to a change in the epigenetic information (cytoplasmic parental factor) deposited in the gamete(s) or in the epigenetic structure (imprinted gene, centrosome, cytoskeleton) of gametes.

Biologists know what this processing does but they have very little knowledge on how neural circuits computate their final result, the epigenetic information that in the form of a specific chemical starts a specific signal cascade for inducing changes in epigenetic information or epigenetic structures in gametes.

From a biological point of view, the neural processing is a nongenetic, hence epigenetic, process, and so is the information it generates (Cabej, 2004, pp. 35–47). The result of the processing, i.e., the release of neurotransmitter(s)/neuromodulator(s) on specific secretory neurons, represents an "instruction" for selectively activating a signal cascade inducing a specific epigenetic change in the form of parental cytoplasmic factors and changes in epigenetic structures and, probably, other parental factors of unknown nature. The parental epigenetic information is the crucial link in the bigenerational chain of events leading from the environmental stimulus acting on the parent(s) to the heritable phenotypic change appearing in the offspring.

Returning to the problem of the source of epigenetic information for inherited transgenerational changes, as pointed out earlier, external stimuli represent no information that would be somehow related to expression of genes. However, empirical evidence shows that a correlation between specific environmental stimuli and expression of genes does exist. This is anything but paradoxical. Metazoans have evolved the ability to develop novel, extrabiologically nonexisting, relationships between entities that are causally not related, between the external stimuli and expression of specific genes. They manipulate external and internal stimuli by converting them into electrical spike trains, according to still unknown codes, and by processing the input they can adaptively generate a chemical output (neurotransmitter/neuromodulator), which activates particular signal cascades that ultimately lead to changes in epigenetic information or epigenetic structures of gamete(s). As shown in Figure 11.15, the crucial element in the chain of events is the processing-dependent release on specific neurons of the specific chemical output. It is at this juncture that organism's response to the stimulus (the release of a neurohormone that activates the specific signal cascade) is irreversibly determined: the probability of inducing the relevant cascade, thus, increases from 0 to 1, while the possibility of activation of the rest of the possible cascades is excluded. This is new information *sensu* Shannon.

As we have shown (chapter 2, Neural Manipulation of Gene Expression), the nervous system can establish causal relationships between virtually any particular stimulus and any particular gene, by activating a particular signal cascade and, via the binary neural control, restrict the action of the cascade to a specific part of the body. But the exclusion of all the rest of optional cascades for actualizing a specific one implies the generation of epigenetic information for producing a specific phenotypic result.

Thus the processing of the stimulus in the parental neural circuit generates epigenetic information (release of a specific chemical on a specific neuron) that represents an "instruction" to induce a specific signal cascade leading to a specific transgenerational phenotypic change. This is another way of saying that the information necessary for inducing the adaptive phenotypic change in the offspring

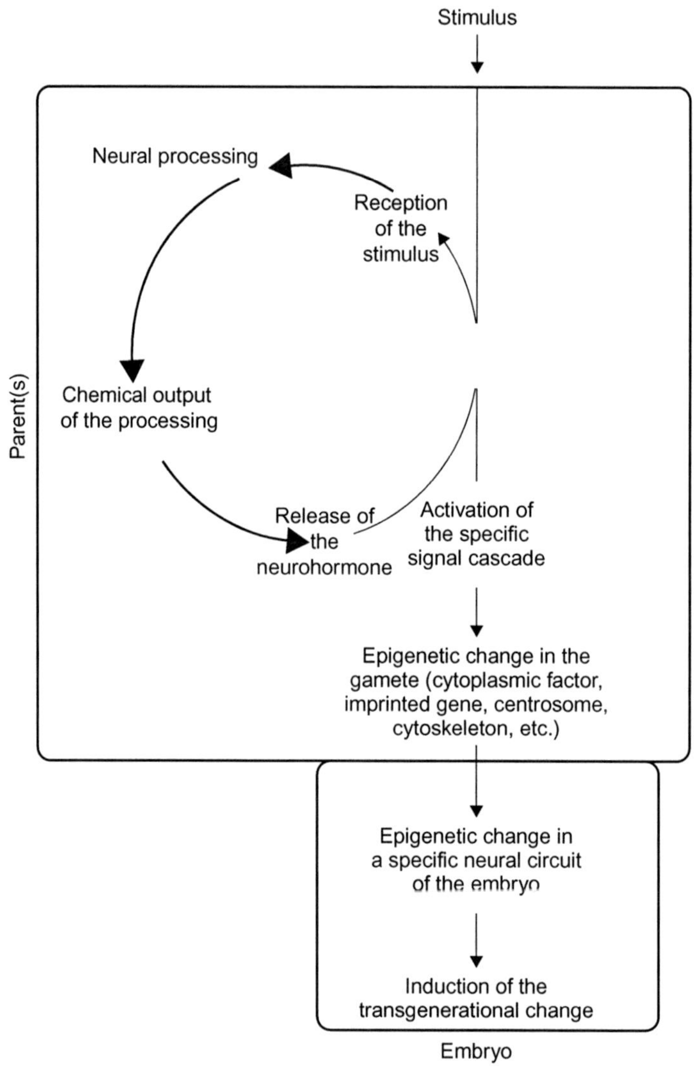

Figure 11.15 Diagramatic representation of the "stimulus detour" mechanism of the induction of transgenerational developmental change. Neural reception of the stimulus and its processing in the neural circuit results in a chemical that induces secretion of a neurohormone, which triggers activation of a specific signal cascade leading to an adaptive epigenetic change in gamete(s). The neural manipulation of the stimulus establishes a previously nonexisting causal relationship between the stimulus and the signal cascade that causes the transgenerational change.

originates in the parental neural circuit. Thus, we have obtained a conjectural answer to the earlier posed question: "Where does the new information for transgenerational phenotypic changes come from?"

The result of the neural processing, as a neural response to the environmental stimulus, does not depend on the nature of the stimulus. For there is nothing in the stimulus that would bias the nervous system toward the actual result, as may be proven by the fact that the same stimulus in different species leads to different and often to opposing results. The processing of the stimulus is a manipulative and goal-oriented solution.

As for how the information for transgenerational plasticity is generated in neural circuits, no answer could be provided at this time of ignorance on the nature of computation taking place in the nervous system. However, the available evidence allows us to reasonably relate the generation of the information for developing new characters in the offspring to the dynamics of structural changes taking place in neural circuits. To reiterate, our hypothetical scenario on the principles of the generation of epigenetic information in the CNS would look as follows:

Environmental stimuli are received by sensory neurons, which convert them into electrical signals, the "common language" into which all stimuli are converted in the nervous system. In response to the electrical signals from sensory neurons, the specific neural circuit reconfigures its synaptic morphology (Gould et al., 1990; Calizo and Flanagan-Cato, 2000; Widmer et al., 2003; Choi et al., 2005) and structure. The reconfiguration results in a modification of computational properties of the neural circuit (von der Malsburg, 1999; Montag-Sallaz et al., 2003), leading to specific changes in the chemical output of the circuit (Getting, 1989), which represent new information or "instructions" for adaptively activating a specific signal cascade leading to a new adaptive phenotypic character(s) in the offspring.

References

Agrawal, A.A., Laforsch, C. and Tollrian, R. (1999). Transgenerational induction of defences in animals and plants. Nature *401*, 60–63.

Anstey, M.L., Rogers, S.M., Ott, S.R., Burrows, M. and Simpson, S.J. (2009). Serotonin mediates behavioral gregarization underlying swarm formation in desert locusts. Science *323*, 627–630.

Anton, S. and Hansson, B.S. (1996). Antennal lobe interneurons in the desert locust *Schistocerca gregaria* (Forskål): processing of aggregation pheromones in adult males and females. J. Comp. Neurol. *370*, 85–96.

Anway, M.D., Leathers, C. and Skinner, M.K. (2006). Endocrine disruptor vinclozolin induced epigenetic transgenerational adult-onset disease. Endocrinology *147*, 5515–5523.

Ayali, A., Fuchs, E. and Kutsch, W. (2004). Neurophysiological studies of flight-related density-dependent phase characteristics in locusts. Acta Biol. Hung. *55*, 137–141.

Bargmann, C.I. (2006). Comparative chemosensation from receptors to ecology. Nature *444*, 295–301.

Borst, D.W., Wainwright, G. and Rees, H.H. (2002). *In vivo* regulation of the mandibular organ in the edible crab, *Cancer pagurus*. Proc. Biol. Sci. *269*, 483–490.

Bouaichi, A. and Simpson, S.J. (2003). Density-dependent accumulation of phase characteristics in a natural population of the desert locust *Schistocerca gregaria*. Physiol. Entomol. *28*, 25–31.

Brown, M.R. and Cao, C. (2001). Distribution of ovary ecdysteroidogenic hormone in the nervous system and gut of mosquitoes. J. Insect Sci. *1*, p. 3.

Burns, M. (1972). Effect of flight on the production of alatae by the vetch aphid, *Megoura viciae*. Entomol. Exp. Appl. *15*, 317–323.

Buszczak, M. and Cooley, L. (2000). Eggs to die for: cell death during *Drosophila* oogenesis. Cell Death Differ. *7*, 1071–1074.

Cabej, N.R. (2004). *Neural Control of Development*. Albanet, Dumont, NJ, pp. 232–249.

Cabej, N.R. (2011). Neural control of gene recruitment in metazoans. Dev. Dyn. *240*, 1–8.

Calizo, L.H. and Flanagan-Cato, L.M. (2000). Estrogen selectively regulates spine density within the dendritic arbor of rat ventromedial hypothalamic neurons. J. Neurosci. *20*, 1589–1596.

Champagne, F.A. (2008). Epigenetic mechanisms and the transgenerational effects of maternal care. Front. Neuroendocrinol. *29*, 386–397.

Choi, S.-Y., Chang, J., Jiang, B., Seol, G.-H., Min, S.-S. and Han, J.-S., et al. (2005). PLC and LTD in visual cortex. J. Neurosci. *25*, 11433–11443.

Clynen, E., Stubbe, D., De Loof, A. and Schoofs, L. (2002). Peptide differential display: a novel approach for phase transition in locusts. Comp. Biochem. Physiol. B Biochem. Mol. Biol *132*, 107–115.

D'Amico, L.J., Davidowitz, G. and Nijhout, H.F. (2001). The developmental and physiological basis of body size evolution in an insect. Proc. Biol. Sci. *268*, 1589–1593.

Denlinger, D.L., Joplin, K.H., Flannagan, R.D., Tammariello, S.P., Zhang, M.-L. and Yocum, G.D., et al. (1995). Diapause-specific gene expression. In: *Molecular Mechanisms of Insect Metamorphosis and Diapause*, A. Suzuki, H. Kataoka and S. Matsumoto (eds.), Industrial Publishing & Consulting, Inc., Tokyo, pp. 289–297.

Dixon, A.F.G. and Agarwala, B.K. (1999). Ladybird-induced life-history changes in aphids. Proc. Biol. Sci. *266*, 1549–1553.

Eads, B.D., Andrews, J. and Colbourne, J.K. (2008). Ecological genomics in *Daphnia*: stress responses and environmental sex determination. Heredity *100*, 184–190.

Forrest, J.M.S. (1970). The effects of maternal and larval experience on morph determination in *Dysaphis devecta*. J. Insect Physiol. *16*, 2281–2292.

Fuchs, E., Kutsch, W. and Ayali, A. (2003). Neural correlates to flight-related density-dependent phase characteristics in locusts. J. Neurobiol. *57*, 152–162.

Getting, P.A. (1989). Emerging principles governing the operation of neural networks. Annu. Rev. Neurosci. *12*, 185–204.

Gil, D., Leboucher, G., Lacroix, A., Cue, R. and Kreutzer, M. (2004). Female canaries produce eggs with greater amounts of testosterone when exposed to preferred male song. Horm. Behav. *45*, 64–70.

Gilbert, J.J. (1966). Rotifer ecology and embryological induction. Science *151*, 1234–1437.

Gilbert, J.J. (1980). Female polymorphism and sexual reproduction in the rotifer *Asplanchna*. Evolution of their relationship and control by dietary tocopherol. Am. Nat. *116*, 409–431.

Gilbert, J.J. and Schröder, T. (2004). Rotifers from diapausing, fertilized eggs: unique features and emergence. Limnol. Oceanogr. *49*, 1341–1354.

Girardie, J., Girardie, A., Huet, J.C. and Pernollet, J.C. (1989). Amino acid sequence of locust neuroparsins. FEBS Lett. *245*, 4–8.

Girardie, J., Geoffre, S., Delbecque, J.P. and Girardie, A. (1998). Arguments for two distinct gonadotropic activities triggered by different domains of the ovary maturating parsin of *Locusta migratoria*. J. Insect Physiol. *44*, 1063–1071.

Girardie, J., Huet, J.C., Atay-Kadiri, Z., Ettaouil, S., Delbecque, J.P. and Fournier, B., et al. (1998). Isolation, sequence determination, physical and physiological characterization of the neuroparsins and ovary maturing parsins of *Schistocerca gregaria*. Insect Biochem. Mol. Biol. *28*, 641–650.

Gould, E., Woolley, C.S., Frankfurt, M. and McEwen, B.S. (1990). Estradiol mediates fluctuation in hippocampal synapse density during estrous cycle in the adult rat. J. Neurosci. *10*, 1286–1291.

Grach, C., Wang, Y., Barda, Y., Gilon, C. and Pener, M.P. (2003). *In vivo* structure-activity studies on the dark-color-inducing neurohormone of locusts. J. Pept. Res. *62*, 135–142.

Hagedorn, H.H., Shapiro, J.P. and Hanaoka, K. (1979). Ovarian ecdysone secretion is controlled by a brain hormone in an adult mosquito. Nature *282*, 92–94.

Hanazato, T., Fueki, K. and Yoshimoto, M. (2001). Fish-induced life-history shifts in the cladocerans *Daphnia* and *Simocephalus*: are they positive or negative responses. J. Plankton Res. *23*, 945–951.

Handler, A.M. and Postlethwait, J.H. (1977). Endocrine control of vitellogenesis in *Drosophila melanogaster*: effects of the brain and *corpus allatum*. J. Exp. Zool. *202*, 389–402.

Hasegawa, K. and Shimizu, I. (1987). *In vivo* and *in vitro* photoperiodic induction of diapause using isolated brain–subesophageal ganglion complexes of the silkworm, *Bombyx mori*. J. Insect Physiol. *33*, 959–966.

Hayward, L.S. and Wingfield, J.C. (2004). Maternal corticosterone is transferred to avian yolk and may alter growth and adult phenotype. Gen. Comp. Endocrinol. *135*, 365–371.

Henrich, V.C. and Denlinger, D.L. (1982). A maternal effect that eliminates pupal diapause in progeny of the flesh fly, *Sarcophaga bullata*. J. Insect Physiol. *28*, 881–884.

Hetru, C.C., Kappler, C., Hoffmann, J.A., Nearn, R., Bang, L. and Horn, D.H. (1982). The biosynthetic pathway of ecdysone: studies with vitellogenic ovaries of *Locusta migratoria* (Orthoptera). Mol. Cell. Endocrinol. *26*, 51–80.

Hügele, B.F., Oag, V., Bouaïchi, A., McCaffery, A.R. and Simpson, S.J. (2000). The role of female accessory glands in maternal inheritance of phase in the desert locust *Schistocerca gregaria*. J. Insect Physiol. *46*, 275–280.

Hügele, B.F., Wang, F.H., Sehnal, F. and Simpson, S.J. (2004). Effects of crowding, isolation, and transfer from isolation to crowding on total ecdysteroid content of eggs in *Schistocerca gregaria*. J. Insect Physiol. *50*, 621–628.

Ichikawa, T. (2003). Firing activities of neurosecretory cells producing diapause hormone and its related peptides in the female silkmoth, *Bombyx mori*. I. Labial cells. Zool. Sci. *20*, 971–978.

Ignell, R., Couillaud, F. and Anton, S. (2001). Juvenile-hormone-mediated plasticity of aggregation behaviour and olfactory processing in adult desert locusts. J. Exp. Biol. *204*(Pt 2), 249–259.

Kitagawa, N., Shiomi, K., Imai, K., Niimi, T., Yaginuma, T. and Yamashita, O. (2005). Establishment of a sandwich ELISA system to detect diapause hormone, and developmental profile of hormone levels in egg and subesophageal ganglion of the silkworm, *Bombyx mori*. Zool. Sci. *22*, 213–221.

Klause-de-Pupka, A., Schmidt, G.H., Brunnemann, U. and Francke, W. (1997). Perception of species-specific volatile substances in the desert locust *Schistocerca gregaria* (Caelifera: Acrididae). Entomol. Gen. *22*, 109–118.

Kou, R. and Chen, S.J. (2000). Allatotropic and nervous control of corpora allata in the adult male loreyi leafworm, *Mythimna loreyi* (Lepidoptera: Noctuidae). Physiol. Entomol. *25*, 273–278.

Kristjansson, B. (2005). Rapid morphological changes in threespine stickleback, *Gasterosteus aculeatus*, in freshwater. Environ. Biol. Fish. *74*, 357–363.

Kunert, G. and Weiser, W.W. (2003). The interplay between density- and trait-mediated effects in predatory–prey interactions: a case study in aphid wing polymorphism. Oecologia *135*, 304–312.

Kunert, G., Otto, S., Roese, U.S.R., Gershenzon, J. and Weisser, W.W. (2005). Alarm pheromone mediates production of winged dispersal morphs in aphids. Ecol. Lett. *8*, 596–603.

Laforsch, C. and Tollrian, R. (2004). Extreme helmet formation in *Daphnia cucullata* induced by small-scale turbulence. J. Plankton Res. *26*, 81–87.

Laforsch, C., Ngwa, W., Grill, W. and Tollrian, R. (2004). An acoustic microscopy technique reveals hidden morphological defenses in *Daphnia*. Proc. Natl. Acad. Sci. U.S.A. *101*, 15911–15914.

LaMontagne, J.M. and McCauley, E. (2001). Maternal effects in *Daphnia*: what mothers are telling their offspring and do they listen? Ecol. Lett. *4*, 64–71.

Laurent, G. (1996). Dynamical representation of odors by oscillating and evolving neural assemblies. Trends Neurosci. *19*, 489–496.

Laurent, G. and Naraghi, M. (1994). Odorant-induced oscillations in the mushroom bodies of the locust. J. Neurosci. *14*, 2993–3004.

Laurent, G., MacLeod, K., Stopfer, M. and Wehr, M. (1998). Spatiotemporal structure of olfactory inputs to the mushroom bodies. Learn. Memory *5*, 124–132.

Lea, A.O. (1967). The medial neurosecretory cells and egg maturation in mosquitoes. J. Insect Physiol. *13*, 419–420.

Leclaire, M. and Brandl, R. (1994). Phenotypic plasticity and nutrition in a phytophagous insect: consequences of colonizing a new host. Oecologia *100*, 379–385.

Li, S., Friesen, J.A., Holford, K.C. and Borst, D.W. (2010). Methyl farnesoate synthesis in the lobster mandibular organ: the roles of HMG-CoA reductase and farnesoic acid-*O*-methyltransferase. Comp. Biochem. Physiol. A Mol. Integr. Physiol. *155*, 49–55.

Lipar, J.L. and Ketterson, E.D. (2000). Maternally derived yolk testosterone enhances the development of the hatching muscle in the red-winged blackbird *Agelaius phoeniceus*. Proc. Biol. Sci. *267*, 2005–2010.

Lüning, J. (1992). Phenotypic plasticity of *Daphnia pulex* in the presence of invertebrate predators; morphological and life history responses. Oecologia *92*, 383–390.

Maeno, K. and Tanaka, S. (2004). Hormonal control of phase-related changes in the number of antennal sensilla in the desert locust, *Schistocerca gregaria*: possible involvement of [His^7]-corazonin. J. Insect Physiol. *50*, 855–865.

Masler, E.P., Raina, A.K., Wagner, R.M. and Kochansky, J.P. (1994). Isolation and identification of a pheromonotropic neuropeptide from the brain–subesophageal ganglion complex of *Lymnatria dispar*—a new member of the PBAN family. Insect Biochem. Mol. Biol. *24*, 829–836.

Masuda-Nakagawa, L.M., Tanaka, N.K. and O'Kane, C.J. (2005). Stereotypic and random patterns of connectivity in the larval mushroom body calyx of *Drosophila*. Proc. Natl. Acad. Sci. U.S.A. *102*, 19027–19032.

Matheson, T., Rogers, S.M. and Krapp, H.G. (2004). Plasticity in the visual system is correlated with a change in lifestyle of solitarious and gregarious locusts. J. Neurophysiol. *91*, 1–12.

Meaney, M.J. and Szyf, M. (2005). Maternal care as a model for experience-dependent chromatin plasticity? Trends Neurosci. *28*, 456–463.

Mondor, E.B., Rosenheim, J.A. and Addicott, J.F. (2005). Predator-induced transgenerational phenotypic plasticity in the cotton aphid. Oecologia *142*, 104–108.

Mondor, E.B., Rosenheim, J.A. and Addicott, J.F. (2008). Mutualist-induced transgenerational polyphenisms in cotton aphid populations. Funct. Ecol. *22*, 157–162.

Montag-Sallaz, M., Baarke, A. and Montag, D. (2003). Aberrant neuronal connectivity in CHL1-deficient mice is associated with altered information processing-related immediate early gene expression. J. Neurobiol. *57*, 67–80.

Naguib, M. and Gil, D. (2005). Transgenerational effects on body size caused by early developmental stress in zebra finches. Biol. Lett. *1*, 95–97.

Noguchi, H. and Hayakawa, Y. (2001). Dopamine is a key factor for the induction of egg diapause of the silkworm, *Bombyx mori*. Eur. J. Biochem. *268*, 774–780.

Olmstead, A.W. and LeBlanc, G.A. (2003). Insecticidal juvenile hormone analogs stimulate the production of male offspring in the crustacean *Daphnia magna*. Environ. Health Perspect. *111*, 919–924.

Olmstead, A.W. and LeBlanc, G.A. (2007). The environmental-endocrine basis of gynandromorphism (intersex) in a crustacean. Int. J. Biol. Sci. *3*, 77–84.

Podjasek, J.O., Bosnjak, L.M., Brooker, D.J. and Mondor, E.B. (2005). Alarm pheromone induces a transgenerational wing polyphenism in the pea aphid, *Acyrthosiphon pisum*. Can. J. Zool. *83*, 1138–1141.

Rafaeli, A. (2009). Pheromone biosynthesis activating neuropeptide (PBAN): regulatory role and mode of action. Gen. Comp. Endocrinol. *162*, 69–78.

Raikhel, A.S. and Lea, A.O. (1985). Hormone-mediated formation of the endocentric complex in mosquito oocytes. Gen. Comp. Endocrinol. *57*, 422–433.

Richard, D.S., Jones, J.M., Barbarito, M.R., Cerula, S., Detweiler, J.P. and Fisher, S.J., et al. (2001). Vitellogenesis in diapausing and mutant *Drosophila melanogaster*: further evidence for the relative roles of ecdysteroids and juvenile hormones. J. Insect Physiol. *47*, 905–913.

Rider, C.V., Gorr, T.A., Olmstead, A.W., Wasilak, B.A. and LeBlanc, G.A. (2005). Stress signaling: coregulation of hemoglobin and male sex determination through a terpenoid signaling pathway in a crustacean. J. Exp. Biol. *208*, 15–23.

Rogers, S.M., Matheson, T., Despland, E., Dodgson, T., Burrows, M. and Simpson, S.J. (2003). Mechanosensory-induced behavioural gregarization in the desert locust *Schistocerca gregaria*. J. Exp. Biol. *206*, 3991–4002.

Rogers, S.M., Matheson, T., Sasaki, K., Kendrick, K., Simpson, S.J. and Burrows, M. (2004). Substantial changes in central nervous system neurotransmitters and neuromodulators accompany phase change in the locust. J. Exp. Biol. *207*, 3603–3617.

Roller, L., Tanaka, Y. and Tanaka, S. (2003). Corazonin and corazonin-like substances in the central nervous system of the pterygote and apterygote insects. Cell Tissue Res. *312*, 393–406.

Schröder, T. and Gilbert, J.J. (2004). Transgenerational plasticity for sexual reproduction and diapause in the life cycle of monogonont rotifers: intraclonal, intraspecific and interspecific variation in response to crowding. Funct. Ecol. *18*, 458–466.

Shapiro, A.M. (1976). Seasonal polyphenism. In: *Evolutionary Biology*, M.K. Hecht and W.C. Steere (eds.), vol. 9. Plenum, New York, NY, pp. 259–333.

Shapiro, A.M. (1982). Redundancy in pierid polyphenisms: pupal chilling induces vernal phenotype in *Pieris occidentalis* (Pieridae). J. Lepidopterists' Soc. *36*, 174–177.

Shimizu, I., Matsui, T. and Hasegawa, K. (1989). Possible involvement of GABAergic neurons in regulation of diapause hormone secretion in the silkworm, *Bombyx mori*. Zool. Sci. *6*, 809–819.

Shimizu, I., Aoki, S. and Ichikawa, T. (1997). Neuroendocrine control of diapause hormone secretion in the silkworm, *Bombyx mori*. J. Insect Physiol. *43*, 1101–1109.

Sloggett, J.J. and Weisser, W.W. (2002). Parasitoids induce production of the dispersal morph of the pea aphid, *Acyrthosiphon pisum*. Oikos *98*, 323–333.

Sockman, K.W., Schwabl, H. and Sharp, P.J. (2001). Regulation of yolk-androgen concentrations by plasma prolactin in the American kestrel. Horm. Behav. *40*, 462–471.

Stay, B., Fairbairns, S. and Yu, C.G. (1996). Role of allatostatins in the regulation of juvenile hormone synthesis. Arch. Insect Biochem. Physiol. *32*, 287–297.

Stelzer, C.P. (2008). Obligate asex in a rotifer and the role of sexual signals. J. Evol. Biol. *21*, 287–293.

Stelzer, C.P. and Snell, T.W. (2003). Induction of sexual reproduction in *Brachionus plicatilis* (Monogononta, Rotifera) by a density-dependent chemical cue. Limnol. Oceanogr. *48*, 939–943.

Tanaka, S. (1996). A cricket (*Gryllus bimaculatus*) neuropeptide induces dark colour in the locust, *Locusta migratoria*. J. Insect Physiol. *42*, 287–294.

Tanaka, S. (2000). Hormonal control of body-color polymorphism in *Locusta migratoria*: interaction between [His^7]-corazonin and juvenile hormone. J. Insect Physiol. *46*, 1535–1544.

Tanaka, S. (2000). The role of [His^7]-corazonin in the control of body-color polymorphism in the migratory locust, *Locusta migratoria* (Orthoptera: Acrididae). J. Insect Physiol. *6*, 1169–1176.

Tanaka, S. (2006). Corazonin and locust phase polyphenism. Appl. Entomol. Zool. *41*, 179–193.

Tanaka, S. (2007). Albino corpus cardiacum extracts induce morphometric gregarization in isolated albino locusts, *Locusta migratoria*, that are deficient in corazonin. Physiol. Entomol. *32*, 95–98.

Tawfik, A.I. and Sehnal, F. (2003). A role for ecdysteroids in the phase polymorphism of the desert locust. Physiol. Entomol. *28*, 19–24.

Tawfik, A.I., Osir, E.O., Hassanali, A. and Ismail, S.H. (1997). Effects of juvenile hormone treatment on phase changes and pheromone production in the desert locust, *Schistocerca gregaria* (Forskål) (Orthoptera: Acrididae). J. Insect Physiol. *43*, 1177–1182.

Tawfik, A.I., Tanaka, S., De Loof, A., Schoofs, L., Baggerman, G. and Waelkens, E., et al. (1999). Identification of the gregarization-associated dark-pigmentotropin in locusts through an albino mutant. Proc. Natl. Acad. Sci. U.S.A. *96*, 7083–7087.

Thompson, J.D. (1991). Phenotypic plasticity as a component of evolutionary change. Trends Ecol. Evol. *6*, 246–249.

Tobe, S.S., Zhang, J.R. and Kwok, R. (2004). Regulation of methyl farnesoate production in crayfish: a possible role for allatostatins. J. Insect Sci. *4*, p. 14.

von der Malsburg, C. (1999). The what and why of binding: the modeler's perspective. Neuron *24*, 95–104.

Wainwright, G., Webster, S.G., Wilkinson, M.C., Chung, J.S. and Rees, H.H. (1996). Structure and significance of mandibular organ-inhibiting hormone in the crab, *Cancer pagurus*. J. Biol. Chem. *271*, 12749–12754.

Webb, M.-L. and Denlinger, D.L. (1998). GABA and picrotoxin alter expression of a maternal effect that influences pupal diapause in the flesh fly, *Sarcophaga bullata*. Physiol. Entomol. *23*, 184–191.

Weisser, W.W., Braendle, C. and Minoretti, N. (1999). Predator-induced morphological shift in the pea aphid. Proc. Biol. Sci. *266*, 1175–1181.

West-Eberhard, M.J. (1989). Phenotypic plasticity and the origins of diversity. Ann. Rev. Ecol. Syst. *20*, 249–278.

West-Eberhard, M.J. (2003). *Developmental Plasticity and Evolution*. Oxford University Press, Oxford, p. 10.

Widmer, H., Ludwig, M., Bancel, F., Leng, G. and Dayanithi, G. (2003). Neurosteroid regulation of oxytocin and vasopressin release from the rat supraoptic nucleus. J. Physiol. *548*(1), 233–244.

Woltereck, R. (1909). Weitere experimentelle untersuchungen über Artveränderung, speziell über das wesen quantitativer artunterschiede bei daphnien. Verhandl. Deut. Zool. Gesell. *19*, 110–173.

Yerushalmi, Y. and Pener, M.P. (2002). Age-dependent response of adults of a homochrome grasshopper, *Oedipoda miniata*, to the dark-colour-inducing neurohormone (DCIN) of locusts. Physiol. Entomol. *27*, 165–169.

Yoder, J.A., Benoit, J.B., Denlinger, D.L. and Rivers, D.B. (2005). Stress-induced accumulation of glycerol in the flesh fly, *Sarcophaga bullata*: evidence indicating anti-desiccant and cryoprotectant functions of this polyol and a role of the brain in coordinating the response. J. Insect Physiol. *52*, 202–214.

12 Evolution of Metazoans and Their Control System

The same genes are used over and over in different contexts and combinations.

West-Eberhard

Principles of Organization of the Multicellular Structure

The complexity arising from aggregation of cells in multicellular structures and coordination of their function unavoidably led to the evolution of hierarchies of organization in metazoans. The structural complexity and hierarchical organization naturally led to functional complexity (Valentine, 2003) to differentiation and specialization of cells based on the division of labor between them.

As argued in Chapter 1, the maintenance of such amazingly complex structures such as metazoan organisms implies the presence in them of a control system that would memorize that structure in the process of their reproduction and maintain it during their lifetime.

One of the biggest challenges for the transition from unicellular life to the multicellular organization is the coordination of the activity of cells that would enable the emergence and maintenance of the multicellular system. Individual cells would be instructed to relinquish their independent existence, show some elementary "altruism," and subject their existence to the need of maintaining the structure and function of the multicellular system. Ultimately, their own life depends on, and is subordinate to, the life of the organism to which they belong.

What is it that essentially brought about this great transition in the evolution of life on earth?

The ability of unicellulars to form colonies, loose aggregates of relatively autonomous cells, seems to have appeared as early as prokaryotes of the type of modern bacteria, with considerable progress shown in the eukaryotic colonies of the *Volvox* type. These aggregations are qualitatively different from what is normally understood to be a metazoan organism; the activity of the cells in the colony is neither coordinated nor subordinate to the "demands of the colony."

Transition from those loose communities of cells (multicellular colonies) to the more complex multicellular structure of parazoans (porifera and parazoans) occurred no later than 600 mya. But even before the emergence of eumetazoans, elements of the coordination at the level of the multicellular system evolved in Porifera. Moreover, nongenetic, epigenetic control evolved early in unicellular eukaryotes (cortical inheritance), as will be shown later in this chapter.

Epigenetic Principles of Evolution. DOI: 10.1016/B978-0-12-415831-3.00012-4

Parazoans (Porifera and Placozoa) display a minimal degree of coordination and cooperation at suborganismic levels and a degree of structural integration at the level of organism, which inspired some biologists to speak of the presence of a still unidentified "neuroid" system in that group. They also display a surprising degree of cellular differentiation, which is comparable to that observed in lower eumetazoans (Cnidaria).

Needless to say, the development of complex structures such as metazoan organisms of this primitive group from unicellular gametes, egg or zygote, implies the presence of a developmental program at the supracellular level, which is necessary not only for the individual development from the unicellular stage to adulthood but also for maintaining the unavoidably eroding adult structure. What both the individual development of the metazoan organism from a unicellular state to adult and the maintenance of that structure, the homeostasis in the broad meaning of the word, have in common is that both imply the use of information for that normal adult structure.

When it comes to the nature and origin of information for erecting the enormously complex metazoan structure, one widely accepted hypothesis is that evolution of genes and genetic information made possible the transition to multicellularity and genes are responsible for both the development of metazoan adult structure starting from a single cell, egg or zygote. According to this hypothesis, it is to be expected that a correlation must exist between the amount of genetic information and genes on the one hand, and the complexity of living organisms on the other. Indeed, this seems to be the case in unicellulars.

Evolution of unicellular life obviously depends on the increase in the number and changes in the structure of their genes: mycoplasmas, a group of the simplest unicellular organisms known so far, contain in their genome several hundred genes, whereas the prokaryote *Escherichia coli* contains more than 4,000 genes. A more complex unicellular eukaryote, the protozoan *Plasmodium falciparum*, has 5,300 genes but that number dramatically increased to 40,000 in an evolutionarily higher unicellular eukaryote, *Paramecium tetraurelia* (Aury et al., 2006). The parallel increase in the number of genes and the degree of functional and structural complexity in the kingdom of unicellular organisms suggests the existence of a causal relationship between genes on the one hand, and the function and morphology in evolution of unicellulars on the other.

However, the unicellular principle of the correlation between the number of genes and degree of structural complexity is not valid for metazoans. The early expectations of geneticists that the number of genes in higher metazoans would be larger than in lower metazoans did not prove to be true. Genome sequencing of various metazoan organisms has shown that no relationship between the number of genes or the size of the genome and the complexity or the position of metazoans in the tree of life exists. A human organism has about 20,000–25,000 genes in its genome (International Human Genome Sequencing Consortium, 2004), i.e., only little more than *Caenorhabditis elegans*, one of the smallest worms, only ~1 mm in length, consisting of fewer than 1,000 somatic cells, but fewer than a sponge with

18,000–30,000 genes and, even more surprisingly, half of the genes of a unicellular eukaryote (Aury et al., 2006).

Metazoans, and multicellular organisms in general, evolved from unicellulars, which clearly lack any "coordinating genes" for they had no cells to coordinate. Evolution does not work prophetically.

If the degree of metazoan structural and functional complexity and the evolutionary progress related to it do not depend on the genetic information or the number of genes, what it may depend on?

In order to behave as, and evolve into, a multicellular organism, a cell aggregation must meet the following minimal requirements:

1. Create a stable internal environment that would sustain the life of all the cells throughout the multicellular system.
2. "Memorize" its own supracellular structure and function and, on this basis, set standards (set points) of the parameters of its structure and functions.
3. Continually monitor the status of its structure and function by receiving a continuous input of information.
4. Process the input of information from the internal and external environment for detecting deviations from the normal structure and function and generate its output in the form of signaling molecules that start signal cascades for maintaining and restoring the unavoidably eroding structure and functions of the organism.
5. Transmit to the germ cells its information on the species-specific structure and function.

The above functions are essentially the functions of a control system.

It seems that the first multicellular systems capable of meeting all the above requirements emerged *ca.* 550–540 mya, during the Cambrian explosion. This seems to have roughly coincided with the differentiation of the nerve cell, the resulting nerve net, and the central nervous system (CNS), the core of the metazoan integrated control system (ICS) (see Chapter 1).

Unicellular Antecedents of Metazoan Life

From an evolutionary point of view, the emergence of metazoans may be considered a natural result of the *ca.* 2.5 billion years evolution of unicellular life and accumulated "evolutionary experience and wisdom". In a specific meaning, Protista was pregnant with multicellular life. Primitive multicellulars used differentiated cells, instead of molecules, as building blocks of their structures, tissues, and organs. Just as unicellulars use different macromolecules in molding their organelles, multicellulars use differentiated cells as building blocks for erecting their structure and must produce these building blocks starting from a single initial cell (an egg cell or zygote). Differentiated metazoan cells adopt different morphologies and specialize to perform different physiological and structural functions, despite the fact that they are genetically identical. Division of labor between differentiated cells in metazoans has its counterpart in a "division of labor" at the molecular level in unicellulars.

Unicellular Premises of Coordination of Cell Activity in Metazoans

If one would accept the prevailing idea that metazoans are monophyletic, then the last common ancestor (LCA) of Porifera and eumetazoans, the Urmetazoa, has been in possession of

> *nearly all developmental gene families used by living animals and that these families evolved before metazoan cladogenesis.*
>
> *Degnan (2006)*

One crucial condition for evolution of the metazoan structure has been establishment *in unicellulars* of a link between the extracellular stimuli with specific signal transduction pathways, which in metazoans are related to the cell differentiation, such as tyrosine kinase pathway, TGF-β, and Wnt pathways. Some biologists believe that these pathways could be activated by morphogen gradients (Degnan et al., 2005), but this seems very unlikely because such gradients could not determine the intricate and convoluted patterns of different cell types in developing metazoan tissues and organs. The presence of signal transduction pathways in the immediate ancestor of the Urmetazoan has been an essential condition for the controlled expression of transcription factors during embryogenesis that is observed as early in the evolution of metazoan life as sponges are.

Although multicellularity evolved about 1 billion years ago, sponges as first metazoans in paleontological record are dated 580 mya, i.e., between 30 and 50 million years before the Cambrian explosion (Müller et al., 2004). Among the metazoan-specific transcription factor families expressed during sponge embryogenesis are members of POU homeobox genes, LIM-HD, Pax, Bar, Prox2, NK-2, T-box, MEF-2, Fox, Sox, and nuclear hormone receptor gene families (Degnan et al., 2005).

Vast evidence shows that many of the genes that are responsible for production of a number of vitally important molecules in unicellulars are conserved in higher metazoans. If metazoans evolved from unicellulars and if the Urmetazoan had "master" control genes, it inherited them from its unicellular ancestor. This is not just a speculation. All *Hox* genes in metazoans, and even in metaphyta, share a *helix–loop–helix* (HLH) and one or more modules containing a 180 bp DNA segment, called *homeobox*. But it is known that regulatory proteins in prokaryotes also contain a similar 60 amino acid (what corresponds to 180 bp in DNA) long HLH that is found in all of their repressor molecules. Similar sequences are contained in yeast protein molecules *MAT (mating type)-a1* and *MAT-a2*. It is noteworthy that switching on/off of each of these genes determines the morphological type of the yeast or mating type or both. In support of the hypothesis that *Hox* genes evolved in protists before the appearance of metazoans and metaphyta also comes the fact that both animals and plants share "master control genes," implying that their unicellular LCA has possessed at least one *Hox* gene.

It is assumed (based on the number of common *Homeobox* genes in arthropods and mammals) that the Urmetazoan had three to five *Hox* genes in a single cluster. Later on, a considerable expansion of these genes took place via gene duplication. In sponges, the number of *Hox* genes surprisingly decreased to one from a calculated five or six in the Urmetazoan. In arthropods, their number is eight, while in mammals the *Hox* gene cluster has been repeatedly duplicated to form four clusters, all slightly different, with a total of 38 genes (Erwin et al., 1997).

In summary, we know that:

1. a unicellular homeobox domain is present in numerous transcription factors in unicellulars,
2. unicellulars are capable of developing different morphological or mating types depending on whether one or the other of two alternative homeodomain genes is expressed,
3. the Urmetazoan had not one or two *Hox* genes but at least five of them, and
4. Metaphyta and Metazoa share common *Hox* genes inherited from their common ancestor, which has been a unicellular.

Hox genes evolved from one (or a few) primitive *Hox* gene(s) that probably appeared in unicellulars *ca.* 1 billion years ago (Lappin et al., 2006). Evolving by gene duplication, in most extant metazoans, *Hox* genes in extant metazoans are found in gene clusters (Figure 12.1).

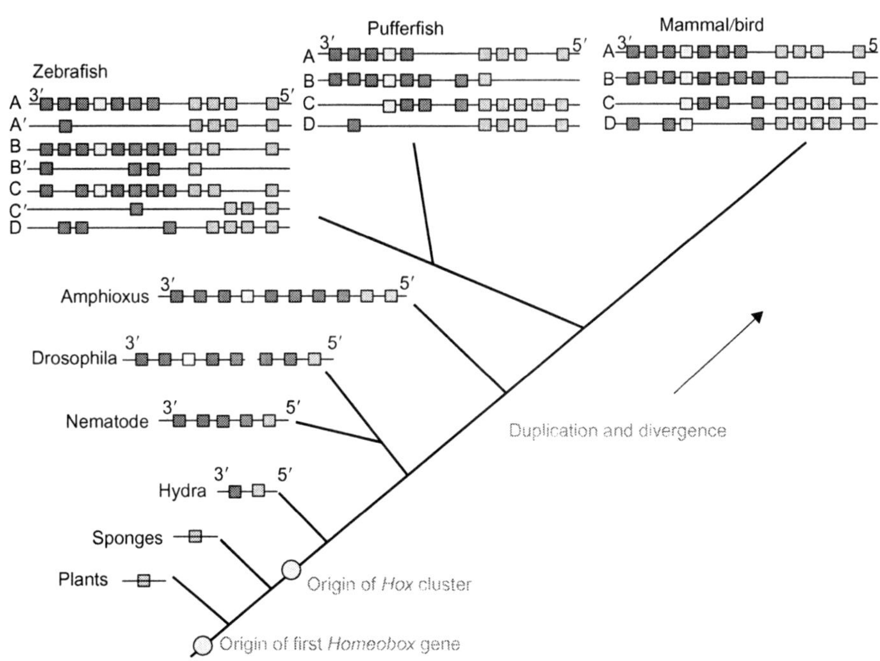

Figure 12.1 A representative dendrogram illustrating the evolution of *Hox* gene clusters from a primordial *Hox* gene in unicellulars.
Source: From Lappin et al. (2006).

Most other gene families, and even gene regulatory networks, are conserved across metazoan taxa with few essential changes. Even species and higher taxa that are only remotely related use similar gene regulatory networks and the same regulatory *Hox* genes for molding widely different morphologies:

> *Thus, strikingly similar gene expression in embryos produces strikingly dissimilar adults. This broad-scale evolutionary dissociation between cause and effect is the Hox Paradox (Wray 2002), so named because the Hox cluster has been emblematic of the phenomenon of conserved developmental genes and expression domains.*
>
> *Wray (2003)*

But paradoxes are expression of wrong views, illusions, and logical incongruities arising from fallacious understanding or erroneous information on the phenomena under study. In dealing with paradoxes, it is always wiser to revise our ideas rather than blame nature for creating them.

Recent studies have added new evidence that increase in the number of genes has not been necessary and is not related to the evolution of metazoans. It came as a great surprise the finding that the number of genes in sponges, the oldest and lower standing phylum of the animal kingdom, a "dead end" of evolution, is about twice the number of genes in higher animal groups, including humans (Müller et al., 2004).

The presence of all the essential gene families related to the development shows that the Urmetazoan had, in an overly used metaphor, the common "genetic tool kit" necessary for cell differentiation and development of metazoan structure. But the tool kit is only a necessary condition, a prerequisite for individual development, neither a sufficient condition nor the cause of development. The "genetic tool kit" metaphor, however, unintentionally, seems to perfectly express the real role of the genetic component in the individual development of metazoans: tools are not more than tools. The tools have no idea or information on what should be done and how to do anything. They can be used for many different purposes by a user that has the information on what needs to be done, and also when and how. But the tool kit is indispensable, too. It may put restrictions on what the user can do with it, and this reminds us of Medawar's aphorism: "Genetics proposes, epigenetics disposes." In a different context, Carroll says:

> *Well, we now understand that animal form is not so much a matter of the genes an animal has, but how they're used during the process of development, of going from a single egg to the complete mature animal, the building of all of its body parts. So these genes are used in every animal, but the choreography is what differs. And it's that different choreography between species that accounts for the endless forms of animals.*
>
> *Carroll (2006)*

To build further on the metaphor, we need to know who is the choreographer, which, with the same actors, produces different shows. Returning from the metaphor to the metazoan cell, we need to know what determines activation of genes, including "master" *Hox* genes; we need to know, not ignore, the origin of the epigenetic information flowing into the cell for using these genetic tools at the right time and at the right place.

Unicellular Precursors of the Metazoan Epigenetic Informational Structure

Metazoans are dynamical multicellular systems capable of maintaining their unavoidably eroding structure and function that, despite the inevitable death, perpetuate their kind by reproducing themselves. The maintenance of an eroding structure and self-reproduction of a system imply possession of a control system.

The metazoan control system implies memorization of the structure of the living system, i.e., heredity (Pattee, 2001; Cabej, 2004) and "memorization" of the species-specific spatial arrangement of billions/trillions of cells, of different types from which complex metazoan structures and functions arise, is a formidable task to accomplish in the process of reproduction.

Being qualitatively inappropriate and quantitatively negligible, genetic information contained in the genomic DNA is excluded as a possible carrier of the tremendous amount of information necessary for building the metazoan structure. Where, then, in the metazoan organism should one look for possible information-storing structures capable of memorizing the metazoan morphology?

As biological information implies a material carrier, the first structure that intuitively comes to mind as a potential carrier or encoder of the huge amount of information invested in metazoan organisms is the nervous system. There is evidence to assume that the specificity of synaptic connections in the nervous system may encode information in the form of the short-term memory (generated by processing of the sensory input) and long-term memory. The number of these synaptic connections is on the order of trillions (in humans and most vertebrates it amounts to quadrillions) of bits, a fact that speaks to the vast information-storing capacity of the metazoan nervous system. We are still far from a full understanding of the underlying chemical and computational mechanisms that enable the generation of the epigenetic information for establishing these highly specific connections mainly before birth that is experience independent. We still wonder how a neuron performs such an incredible function:

> *A single nucleus, with the same DNA, must serve an entire lifetime for the formation and maintenance of tens of thousands of synapses. It seems difficult to imagine a differential distribution of genetic material from a single nucleus to each of these tens of thousands of synapses unless we conjure up a mysterious "demon" who selectively channels this material to each synapse according to a preestablished code! The differential expression of genes cannot alone explain the extreme diversity and specificity of connections between neurons.*
>
> *Changeux (1985)*

How was this astounding cell type differentiated from the original stock of "protozoan" cells the Urmetazoan consisted of?

Let us remember that extremely complex cells evolved before the advent of the multicellularity. In Sonneborn's opinion (1950), a *Paramecium* is "far more complicated in morphology and physiology than any cell of the body of Man" (Sonneborn, 1950) and Gould considered it to be a "promising nerve cell analogue" that seems to have "all the complexity of real nerve cells and more" (Gould, 1982). This implies that protozoans had already built up the potentialities of becoming neurons or any of the major types of metazoan differentiated cells. What they necessitated to accomplish that was a mechanism for controlled actualization of those potentialities, rather than acquisition of additional potentialities.

Impulse Conduction and Sensory–Motor Properties in Protozoans

Based on their ability to transmit throughout the cell electric impulses induced by various environmental stimuli, protozoans move toward or away from various objects by perfectly coordinating movements of their cilia and flagellae. They can distinguish concentrations of chemicals in the environment, sources and intensity of light and to move accordingly (positive and negative chemotaxis and phototaxis), to "recognize" their food and ingest it, etc. They can learn, that is, memorize environmental stimuli, a fact that attests to the presence in these unicellulars of an elementary system for receiving, storing, and using information on their environment. So, for example, when a *Paramecium* collides with an obstacle

> *the cilia that cover its body reverse their beating stroke for a few seconds ... the tactile stimulus has an effect on the electrical charge gradient across the membrane ... this change makes cilia to reverse their beating stroke.*
>
> *Alcock (1989)*

This clearly is an adaptive behavioral response that suggests existence of an integration system in unicellulars.

A protist is both a receptor and an effector cell. On receiving stimuli, it conducts electric impulses to different parts of the cell, particularly to effectors such as cilia and flagellae, and can activate a surprisingly large repertory of neurotransmitters and neuromodulators that are also used in the nervous system of metazoans.

> *Like animal receptors, protozoans often receive external stimuli as signal substances that bind to specific membrane molecules. Binding can cause a specific ion channel to open, allowing ions (often Na^+ and K^+) to flow down their concentration gradient (Na^+ in, K^+ out) ... opening the ion channel depolarizes the membrane ... calcium ions enter the cell ... trigger other changes such as reversal of ciliary beat. Paramecium, for example, has at least nine different ion channels, some of which are localized at the front and others at the rear of the cell. Such localized receptor fields differentiate "head" from tail and are thus analogous to the concentration of receptor organs of many animals. Intracellular chemical signaling (pheromones)*

> *in protozoans, in fact, often involves signal molecules such as serotonin, beta-endorphin, acetylcholine, and cyclic-AMP, which, in animals, function as neurotransmitters and internal messengers.*
>
> *Ruppert and Barnes (1994)*

Perfectly coordinated movements of cilia in protozoans suggest that stimuli are sequentially conducted from one cilium to another (Purves et al., 1992).

In *Paramecium*

> *nerve-like neurofibrils leading to individual cirri coordinate this locomotion. The coordination is lost if the neurofibrils are experimentally cut.*
>
> *Purves et al. (1992, p. 488)*

Generalized protoplasmic conduction is apparently used in amoeboid organisms. In organisms with specialized, permanent locomotor organelles, it is not unreasonable to expect that some kind of permanent conduction structures should evolve. In a few Protozoa, evidence of conducting fibrils appears to be incontestable; in *Euplotes*, severing the fibrils results in a loss of coordination. The fibrillar elements of the infraciliature provide a system that could be, and perhaps is, used for coordination.

All of the above point to the existence of a system of integration, a *neuroid* system in unicellulars such as Protozoa:

> *In their receptiveness to environmental stimuli, protozoans resemble the sensory nerve cells of animals.*
>
> *Ruppert and Barnes (1994)*

From the view discussed here, a nerve cell as those that appear in lower metazoans is essentially a differentiated cell that overexpresses the impulse conduction and storage of information of its molecular ancestor.

Epigenetic Information in Unicellulars

Contrary to conventional wisdom, there is adequate observational evidence that, besides the genetic information, epigenetic information of some kind is participating in transmission of inherited characters in unicellulars. This has been unambiguously demonstrated in experiments with *Stentor*.

A fragment of its cell, containing at least one of the total of 15 or more macronuclear beads, a portion of the cortical ectoplasm with its kinetosomes, and some of the intervening cytoplasm will be able to regenerate into a complete stentor cell (Goss, 1969).

When a piece of cortex is microsurgically removed from a ciliate and then reinserted with its polarity relative to the ends of the cell reversed, it reverses the pattern of cilia that develop in that location, and the reversed pattern persists for many generations, indicating that the new character transmitted to following generations involved cytoplasmic structures rather than any genes.

The hereditary phenomenon known as *cytotaxis* (movement toward or away from other cells) or structural guidance is also attributed to the cytoplasmic organization, excluding gene involvement. It is believed that it is the organization of the cytoskeleton and other components in the cell's cortex that provides the information for the pattern of cilia movement:

> *An enucleated interior piece of a stentor can quickly reorganize a new holdfast on its posterior end .... The contractile vacuole is in the same category, for it too can be regenerated in the absence of the nucleus. Wounds in the ectoplasm can also heal in enucleate pieces, and pigment stripes can reconnect their cut ends.*
>
> *Campbell et al. (1999)*

Potentially, any cortical region of the stentor can form a head, and any head grafted to the anterior part of a decapitated stentor would inhibit the formation of a new head. A polarized anterior control seems to be present in stentor, which prevents the formation in posterior head-forming regions (Goss, 1969, pp. 96–99).

In 1963, Tracy Sonneborn (1905–1981) observed that during conjugation, a *Paramecium* received a piece of its partner's paroral cortex. In subsequent divisions that *Paramecium* formed a clone of double-mouthed *Paramecia*, as a result of what he called *cortical mutation* (Sonneborn, 1964).

In numerous cases of conjugation, pieces of cortex of one *Paramecium* went with the partner, and in other cases pieces of cortex of one of the conjugant partners were experimentally grafted to the other. The grafted cortex was integrated to the host, and the reversed polarity of the donor rows of cortical units was inherited for many generations (Beisson and Sonneborn, 1965).

> *Complete doublets and cells with rows of RP cortical units, can apparently be maintained indefinitely with periodic selection. Even those which revert 100 per cent to normal and do so rapidly (for example, incomplete doublets that revert after 30–40 cell generations) transmit the variant organization to a remarkably large number of progeny cells.*
>
> *Beisson and Sonneborn (1965)*

Based on his extensive experiments on cortical inheritance with *Paramecium*, Sonneborn concluded:

> *For all cortical traits examined, development is hereditarily determined by existing and self-reproducing cortical arrangements: the genes (or DNA) doubtless control synthesis of the molecular building blocks, but not their site of assembly or the position, orientation and number of assemblies.*
>
> *Sonneborn (1970)*

Sonneborn's results on epigenetic cortical determination of the orientation of ciliary rows were later echoed by Ng and Frankel (1977) in experiments on *Tetrahymena*.

Studies on the cortical/cytoplasmic inheritance have shown that no genetic material (genes, DNA, or RNA) is involved in the appearance and inheritance of the new characters:

The available evidence on role of microtubules of the cytoskeleton on impulse conduction, sensory–motor properties, sensibility and adaptive behavior in response to external stimuli led to the hypothesis that the cytoskeleton might play the role of the cell "brain" (Glade, 2008), which looks like an echo of an idea expressed more than 60 years ago by the great neuroscientist Sherington, who, in relation to protozoans, wrote: "of nerve there is no trace, but the cytoskeleton might serve" (Sherrington, 1951). Sherrington also believed that cytoskeleton might process information (Hameroff, 2008).

The "neuroid," sensory and motor, properties of the cortical skeleton and these properties may be related to the computational capability of microtubule arrays, achieving a single final distribution to disparate spatial signals (Gerhart and Kirschner, 1997 p. 160), which is a feature of neurons. Indeed, microtubules are an important factor in the control of the cytoplasmic structure, including the intracellular distribution of endoplasmic reticulum, lysosomes, intermediate filaments as well as most of vesicular transport (Gerhart and Kirschner, 1997, p. 159). They may represent a general mechanism of intracellular communication as well as a mechanism for the transport of chemicals within the cell. The structure of a microtubule array depends not on the properties of its protein units, tubulin molecules, rather than on its location within the cell (Harold 2001), a fact that may point to some computation capabilities of microtubules and cytoskeleton that are responsible for cell shape, intracellular transportation, and processes of cell division.

It has been argued that tubulin dimers, which microtubules consist of, possess at least two different conformations so that the single tubulin dimer can represent the two values of a binary digit. Such binary structures may act as novelty detectors that perceive differences in physical quantities rather than absolute values. Any change in the environment will be "perceived" by the subunit. If the change exceeds a critical value, it may lead to a "quantum jump," which may result in a change of the conformation of the tubulin molecule. The individual tubulin molecules participating in the quantum jumps are integrated to a single information-bearing experience of the microtubule (Pitkanen, 1998).

The above properties seem to underlie the unusual feature of the assembly of microtubules, characterized as "dynamic instability"; they never reach a characteristic length but continually adjust their length to the homeostatic and other fluctuations of the environment. Dynamic instability is also observed when pure tubulin subunits assemble *in vitro*, suggesting that it is an intrinsic property of tubulin molecules (Gerhart and Kirschner, 1997, p. 154).

Unlike metazoans, where microtubules are produced by centrosomes or microtubule organizing centers, in unicellulars they usually grow at particular nucleating points in the cytoplasm and their direction and extent of growth is often controlled so that the final array has a well-defined geometry and polarity.

> *In principle, therefore, any cytoskeletal structure that is passed on from mother cell to daughter cell ... could act as a seed to nucleate the growth of another similar structure. The cytoskeleton would in this case carry epigenetic information.*
>
> *Bray (1992)*

The presence of the epigenetic information in cortical structures of unicellulars can explain the above-mentioned phenomena of nongenomic inheritance in protozoans, such as the inherited reversion of rows of transplanted cilia that beat in opposite direction of the host cilia, and the cortical inheritance in general.

The Origin of Eumetazoan Life

About 540 mya, within the 10–50 million years that the "Cambrian explosion" is believed to have lasted (Fortey, 2001; Cummings, 2006), about 50 phyla (all of the known phyla that readily fossilize), a large number of classes, and about 300 new major body plans evolved. No new phyla have evolved ever since (Cummings, 2006). Such an explosive evolution of phyla is impossible to explain by extrapolation from the known rates of evolution (Carroll, 2000). What gave rise to the extraordinary acceleration of the tempo of evolution during the Cambrian, after 3 billion years long of only limited evolutionary progress, since the emergence of life on earth?

In searching for an answer to that question, it would be helpful to identify essential features that distinguish the Precambrian Porifera, that dead end of evolution, from the Cambrian fauna, which gave rise to the tremendous wealth of forms of eumetazoan life.

From the neo-Darwinian point of view, it would be expected that, ascending the ladder of evolution, from the Urmetazoan on, eumetazoan forms would be characterized by an increase in the number of genes. The new evidence that the number of genes in eumetazoans, including higher vertebrates and humans, is approximately two times smaller than that of sponges, the less evolved of all the known metazoans, unequivocally disproves the long-held view that evolution of metazoans is related somehow with the evolution and increase in the number of genes. No essential changes in genes (both quantitatively and qualitatively) have been involved in the transition from the Urmetazoans to eumetazoans: the Urmetazoa was in possession of all the important gene families we find in modern lower eumetazoans.

The only relevant difference between sponges and eumetazoans is the differentiation of the neuron and evolution of the neural net and the CNS.

Judging by differences in tracks left by Precambrian and Cambrian animals in sediments, a tremendous change occurred in the patterns of their behavior that, in all likelihood, is related to the evolution of the nervous system in the latter. The simple meanders, often *crossing themselves*, which characterized the Neoproterozoic (570–543 mya), are supplanted by nonrandom patterns of movement, spirals, and regular meandering that do not cross each other and burrow with elaborate branches through the sediment (Raff, 1996).

Complex brains and sophisticated sensory systems evolved very rapidly, in a time interval perhaps shorter than 5 million years, during the Early Paleozoic (545–542 mya). It is argued that these brains and sophisticated sensory systems

> *must have been forged via a modular mode of reorganization rather than by invention of a large number of novel elementars (new proteins, new gene control regions, or new cell types).*
>
> *Miklos et al. (1994)*

Indeed,

> *many of the regulatory genes in butterflies, giraffes and squid, for example, are similar, having been inherited from their last common ancestor, the protostome–deuterostome ancestor which lived at least half a billion years ago. Thus the striking changes in body plans have been accompanied by relative modest tinkering with the machinery of early development of that long-extinct precursor.*
>
> *Erwin et al. (1997)*

Genes and even gene regulatory networks are conserved across species for long evolutionary periods. While the size and the composition of metazoan genome experienced relatively modest changes, the nervous system from invertebrates to higher vertebrates evolved and grew rapidly: the nervous system of a rotifer or a nematode has about 300 neurons, that of the octopus—30 million, with whales and elephants in possession of 10 billion to 10 trillion neurons, and humans about 100 billion neurons (Prosser and Greenough, 1991), theoretically capable of performing an incredible amount of operations per second.

Although generally it is thought that all the extant phyla appeared during the Cambrian explosion, there is also evidence that one phylum–level body plan might have evolved afterwards: the inversion of the dorsoventral axis in chordates, which might have been accomplished via a "small step," by switching the position of the mouth from the neural to abneural side (Arendt and Nübler-Jung, 1994). In chordates, the mouth opens—given their dorsoventral inversion—to their former dorsal side, suggesting that the chordate mouth is a new mouth, with the old mouth having been obliterated (Arendt and Nübler-Jung, 1994; Nübler-Jung and Arendt, 1996).

Transition to the multicellular form of life came as the crowning achievement of *ca.* 2.5 billion years of evolutionary achievements. The individual cell in the multicellular organism gave up its original *totipotency*, identity and freedom for the sake of a collective life in a "cell society" of different types of cells. Far from a Virchowian democratic republic of cells, the metazoan organism is hierarchically structured. Differentiated cells express only a fraction of their genetic potential. They only produce a very limited number of specific molecules required for the living system, besides the products of some hundred to a few thousands of genes necessary for cell's own subsistence. The degree of expression of genes ranges in increasing order from an adipose cell or a red blood cell in the lower extreme to a nerve cell in the higher end. Differentiated cells have to sacrifice their initial totipotency and freedom for the sake of the survival and ceded the "totipotency" to the living organism. Thus, free will of multicellular organisms increases at the cost of the loss of the freedom of its cells.

Precambrian Metazoans: An Obscure Control System

It is generally believed that multicellularity evolved *ca.* 1 billion years ago, with the appearance of red algae. The paleontological record suggests that the metazoan kingdom may not have arisen earlier than *ca.* 650 mya (Morris, 1998), with the advent of Ediacaran fauna, including sponges that evolved *ca.* 580 mya (Schütze et al., 1999) (Figure 12.2).

Of all the extant metazoan representatives of the subkingdom Parazoa (placozoans and sponges), placozoans have the simplest Bauplan. They have four types of somatic cells but show no body symmetry, no organs, nor muscle nor nerve cells.

Sponges are in possession of about a dozen of cell types, and some sponge species show an axial symmetry. Sponges may have originated from choanoflagellates. The

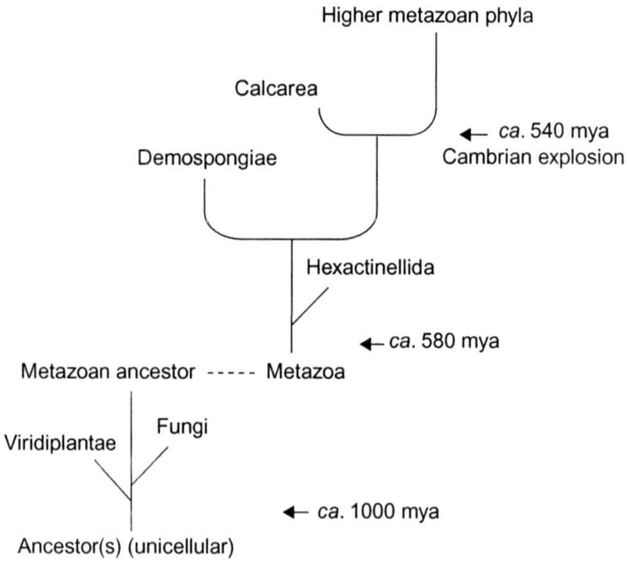

Figure 12.2 Proposed branching order of the three major subkingdoms, Viridiplantae, Fungi, and Metazoa, evolving from ancestral unicellular eukaryotes. Porifera form the basis of the Metazoa. With the classes Hexactinellida, Demospongiae, and Calcarea, Porifera have a common ancestor with higher metazoan phyla. The approximate dates of divergence are indicated.
Source: From Schütze et al. (1999).

differentiation of both types of cells in the primitive sponges may have been facilitated by the preexisting signal transduction systems as it is concluded from the presence in choanoflagellates of a family of tyrosine kinases and components of the tyrosine kinase pathway as well as other signaling protein families (King et al., 2003). However, like paleozoans, sponges also lack nerve and muscle cells (Collins et al., 2005).

It is believed that the transition from the hypothetical Urmetazoan to sponges took place in two stages: first evolved colonial systems with a limited number of differentiated cells, with the appearance of signal transduction pathways and the evolution of communication between cells, and later evolved immune and apoptotic systems, which mark the integration of structure in Porifera (Müller, 2003).

When we speak of a Bauplan, we imply the presence of a control system that directs erection of the multicellular structure. Earlier in this book (Chapter 1), we have shown that the neuroendocrine system represents the basic structure of the ICS that is responsible for both erection and maintenance of metazoan structure. The first metazoan systems, the Urmetazoan and even Parazoa (Porifera and Placozoa), have no nervous system and no differentiated nerve cells, hence they lack a neural control system. But the fact that they evolved as multicellular organisms is clear evidence that they possess a control system, fully capable for developing and maintaining their simple structure. However, the incapability of these primitive metazoans to learn by experience and transmit that experience to the offspring may be the reason why they, unlike eumetazoans, still now, *ca.* 600 million years after their emergence, have made almost no evolutionary progress, representing "living fossils" of the primitive metazoan life.

Although sponges have a greater number of genes than eumetazoans do, and were in possession of all the gene families that enabled eumetazoans to perform the astounding evolution and diversification observed in the animal kingdom, they are stuck in an evolutionary dead end. This fact suggests that the unprecedented rapid evolutionary diversification and progress of eumetazoans proceeded separately from evolution of genes. Hence, if the extraordinarily rapid evolution of eumetazoans has a causal basis, it is clearly nongenetic.

There is reason to believe that the cause of the extraordinarily accelerated rates of evolution during the Cambrian explosion is the emergence of the ICS, which resolved the fundamental informational problem of memorizing the multicellular structure and transmitting instructions for erection of the same structure to the offspring. That system, as explained in Chapter 2, is capable of manipulative expression of genes, thus making it possible for metazoans with the same genes to mold widely varying morphologies. That mechanism is at the basis of what West-Eberhard calls "developmental recombination" during which "the same genes are used over and over in different contexts and combinations" (West-Eberhard, 2005) and which is the cause of the conservation of genes and gene networks over evolutionarily long periods of time.

Sponge cells are autonomous functional units with broad functional capabilities. Each cell must find and use the food themselves, and this is the reason they must stay in contact with the external environment and cannot form tissues or layers of more than two or a few cells thick.

Sponges represent an evolutionarily successful group of 5,000–10,000 species. They possess 6–12 different types of cells, i.e., as many as cnidarians do. If the number of different cell types is one of the main criteria of the structural complexity in metazoans, the question arises: what determined the quite different evolutionary fates and positions of sponges and cnidarians, with the first representing a dead end of evolution and the latter group at the base of the evolutionary tree? The only difference at the level of cell differentiation is that Cnidaria, or a Cnidaria-like extinct group, evolved the nerve cell and neural net, while sponges failed to do so (Mackie, 1990; Ruppert and Barnes, 1994, p. 91). The absence of neurons and neural net may also be the reason why sponges do not show the coordination of activity that is characteristic of eumetazoans.

In sponges, after fertilization, the zygote divides unequally without cell growth, thus the cleavage produces small cells (micromeres) and large cells (macromeres). However, unlike eumetazoans, the process is random with variable patterns of cells within the embryo, with no fixed cleavage pattern, and no cell lineages exist (Degnan et al., 2005). Formation of the "gastrula" results from migration of macromeres into the interior of the embryo and micromeres surrounding them. The mechanism of this migration and later-occurring directed cell migration is not known, but maternal factors may be involved (Degnan et al., 2005).

The absence of a fixed cleavage pattern, that in eumetazoans is determined by parental factors, may explain the fact that sponges, unlike any eumetazoan, show no symmetry (oral–aboral or bilateral) and polarity (anterior–posterior or dorsoventral) at the organismic level (Finnerty, 2003).

As argued earlier (Chapter 1), due to the functional and structural complexity as well as to the unavoidable tendency of the living systems to lose order at all the levels of organization, no multicellular organism can evolve, develop, and exist without a control system at the supracellular and organismic levels. Parazoa are not an exception. A control system for restoring the lost order at the supracellular and organismic levels during their lifetime and for regulating the individual development in the process of biological reproduction is certainly operative in Parazoa. While we have considerable knowledge of the material basis and function of the control system in unicellulars (genome and related structures involved in expression of genetic and epigenetic information) and eumetazoans, the control system in parazoans is not known in any details. A neuroendocrine control of vital phenomena, analogous to that of eumetazoans, is not known to exist in this group.

Despite the absence of the nervous system, in sponges it is observed that a sharp stimulus given at a definite point slowly spreads and evokes responses in nearby cells or in a special group of cells. This so-called *neuroid conduction* inspired some biologists to believe that sponges display a minimal coordination. The idea is that intercellular communication and nonnervous signaling pathways for an elementary coordination of the activity of cells exists even in these "living fossils." It is noteworthy that sponges react by contracting the body and by changing the rhythm of contractions in response to a number of neuroactive substances.

They have an endogenous contraction rhythm that makes possible water exchange, which implies the presence of at least two signaling pathways (Ellwanger and Nickel, 2006). In sponges also neuron-like receptors are identified (Perovic et al., 1999). The presence of a minimal coordination and integration in sponges, in the absence of differentiated neurons and nerve nets, is not surprising, for even their protozoan ancestors evolved mechanisms of signal conductance.

Sponges exhibit some evolutionary premises for performing neuronal conductance of stimuli and differentiation of the neuron. In the hexactinellid sponge, *Rhabdocalyptus dawsoni*, experimentally induced electrical and tactile stimuli are propagated throughout the body, causing the flagellae of the flagellated chamber to stop beating, with the resulting arrest of the feeding current. The proximate cause of the arrest of the flagellae are Ca^{2+} channels and the conducting tissue is the trabecular reticulum, a syncitial structure (a multinucleate protoplasma mass) that pervades the entire body of the sponge, unimpeded by membranous structures. Surprisingly, prenervous/nonnervous tissues conducting electrical impulses are still conserved in lower invertebrates, along the nervous net (Leys et al., 1999).

Very recently, in a number of studies, it has been shown that a specific cell type, epidermal ciliated flask cells, in *Amphimedon* sponges, which, like all sponges, have no neurons (Sakarya et al., 2007), express together the "synaptic" genes, which represent the core of the postsynaptic scaffold of neurons in eumetazoans and are essential for the postsynaptic transmission in neurons (Grant, 2006; Gerrow et al., 2006). It is believed that these epidermal flask cells may have a role as receptors for sensing environmental stimuli and may be evolutionary precursors of the eumetazoan neuron:

> *The few innovations that came with the origination of nervous systems were novel binding partners to an existing scaffold.*
>
> *Sakarya et al. (2007)*

Although sponges have no nervous system or muscles, they are in possession of a complex system of chemical messengers that regulate their contraction behavior and movement in response to various external stimuli, including diurnal rhythms (Ellwanger and Nickel, 2006) and a rudimentary coordination of cell activity, which may be related to a number of evolutionary innovations such as sexual reproduction and embryogenesis (Degnan et al., 2005). It is assumed that the system of chemical messengers in sponges schematically presented in Figure 12.3 represents *the first nonneural control system at the supramolecular level known in metazoans.*

The system is based on intercellular communication via various inducers, including some hormones secreted in eumetazoans (Hartenstein, 2006) and probably other "neuroid" factors, as may be concluded from the presence of the GABA (gamma-aminobutyric acid) receptor in sponges and from the responsiveness of these primitive metazoans to various neurogenic substances. The next step, the great transition in the evolution of metazoan control system, is related to the differentiation of the neuron and evolution of the nerve net.

The origin of cell differentiation observed in sponges, although innovative in some respects, can be traced back to unicellular systems. So, for example, Sonneborn obtained from a single *Paramecium* clone a dozen of different serotypes, each

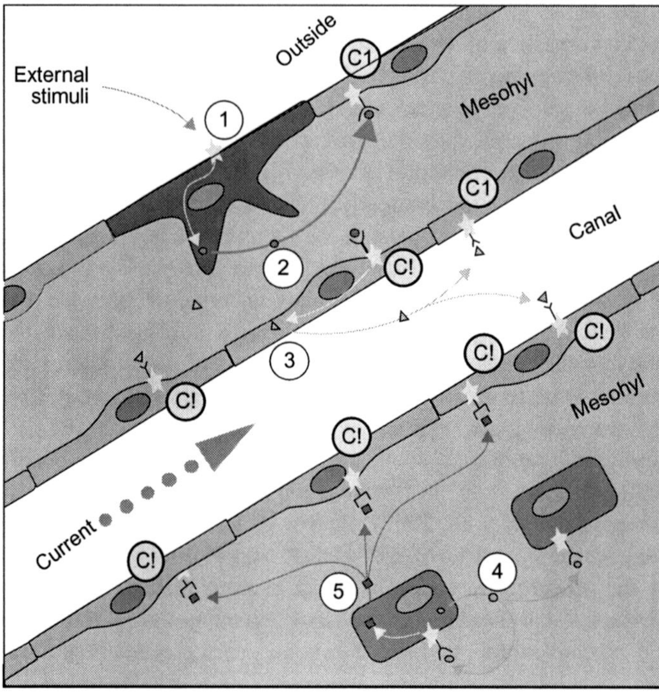

Figure 12.3 Hypothetical signaling pathways in *Tethya wilhelma* involved in coordination of contractions upon external stimuli and endogenous signals. An external stimulus (1) at a putative receptor cell (gray) in the pinacoderm triggers the release of a signal substance (2) which diffuses through the mesohyl (light blue) of the sponge and triggers the contraction (C!) of contractile pinacocytes (blue) via a specific receptor and an intracellular signaling pathway. Eventually, stimulated pinacocytes release a second signal substance (3), which may further diffuse through the mesohyl or be distributed by currents in the canal system. Such a secondary signal would amplify the reaction speed upon external and internal triggering of contraction. The endogenous contraction rhythm may be controlled by numerous trigger cells (red) distributed in the mesohyl. These cells are supposed to release an autocrine–paracrine signal substance (4), which diffuses through the mesohyl to coordinate the release of a signal substance (5), which diffuses to the pinacocytes and triggers contraction and eventually results in a signal amplification as shown in step (3). Signal substances (2) and (5) are likely not identical to allow independent specific coordination of contraction upon endogenous and exogenous stimuli. (For interpretation of the references to color in this figure legend, the reader is referred to the web version of this book.)
Source: From Ellwanger and Nickel (2006).

producing a different antigen (high molecular proteins covering the *Paramecium* membrane, including cilia) and retained this character for several generations. This is essentially similar to the differential expression of genes that leads to cell differentiation in multicellulars.

The Diffuse Control System and the Neural Controller: The Eumetazoan's Eureka

The evolution of the Urmetazoan could not have been a straightforward process. The first extant group evolving from Urmetazoans, Parazoa (sponges and Placozoa) turned out to be an evolutionary "dead end." It took a long time for Urmetazoans to find an open evolutionary pathway, and the success was essentially related to the advent of the neuron and formation of the neural net. As pointed out earlier, transition from Urmetazoans to eumetazoans is not characterized by relevant changes in the genetic information: all of the gene families involved in the individual development of eumetazoans were present in the Urmetazoan, as they are present in sponges. Even the nervous system–related genes have existed in metazoans before the emergence of the nervous system. Bilateral animals share 95% of nervous system–related genes, and "these genes have remained unchanged in the evolutionary lineage from the ancestral CNS in the common ancestor of bilaterians to the highly ordered human brain" (Mineta et al., 2003). More surprisingly, 30% of nervous system–related genes of platyhelminthes have homologues in the plant *Arabidopsis* and yeast, which have no nervous system (Mineta et al., 2003).

It is a widespread opinion that, initially, the neuron differentiated from the stock of pluripotent cells of primitive invertebrates as a secretory cell, which was able to respond to an external or internal stimulus to secrete a diffusible substance that changed the function or the structure of other cells in order to adapt the organism to the changed external or internal environment:

> *The origin of differentiated nervous tissue must have proceeded in a number of steps, of which the first would obviously be the development of a specialized receptor, monitoring changes in the external environment, such as light. In order for the detected changes to influence the organism, the receptor or primitive neuron would have to communicate in some way with the rest of the organism. There is no metazoan taxon from which the neuroendocrine cell is absent .... Chemical communication is an obvious means of establishing the communication, and primitive receptor neuron might be seen as secreting a chemical in response to appropriate stimuli. That scenario defines a minimal neuroendocrine structure, in which a receptor becomes also an independent effector, secreting a molecule that carries a message to all parts of the organism.*
>
> *Gorbman and Davey (1991)*

Differentiation of the neuron, with its potential to communicate with other neurons, led to formation of the pervasive neural net, which could also monitor the status of the internal environment, detect deviations from the norm, and release chemical signals (neurosecretions) for restoring the normal state by stimulating other cells to produce depleted elements in the internal environment, or stimulate proliferation of cell types that are lost, or even induce programmed cell death, when necessary.

Probably, the most important difference between the Precambrian control systems and the neurally determined control systems is learning of new behaviors based on

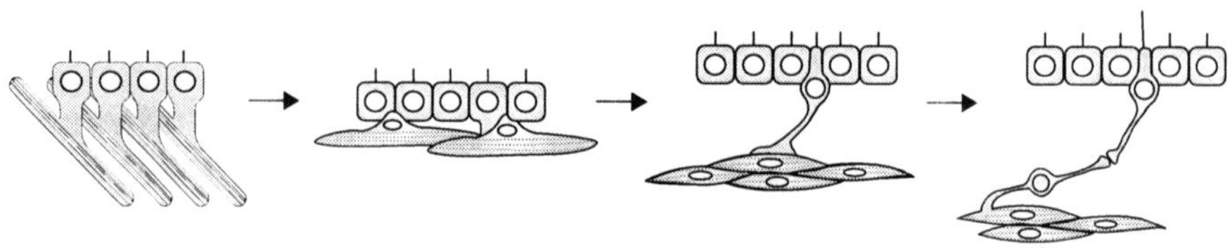

Figure 12.4 The evolution of neuronal circuits by functional segregation. Diversification of the cnidarian myoepithelial cell into the sensory cell, motor neuron, and muscle cell.
Source: From Arendt (2008).

the memorization of the past experience and, under special conditions, transmission of that experience to the offspring. The nervous net made it possible for the metazoan system to coordinate the activity of all of its cells. Accumulation of neurons specifically linked to each other via synaptic connections increased the capacity of the nerve net for processing external and internal stimuli, for producing adaptive responses at the organismic level.

From a functional point of view, the essential difference between sponges, as a dead end of evolution, and cnidarians, is the very weak coordination between parts of the body in the first group, and the perfect coordination of behavior as well as physiology at the organismic and cell levels in the latter. As is well known, these differences are related to the presence in phylum Cnidaria and absence in sponges of the nervous system.

In the course of individual development in modern cnidaria, myoepithelial cells differentiate into sensory cells, motor neurons, and muscle cells (Arendt, 2008; Mackie, 1970). If, as some believe, the sequence of diversification of cell types during development (Figure 12.4) recapitulates, *sensu* Haeckel (Arendt, 2008), the origin of neurons (and other cell types) in the course of evolution, then modern cnidaria may offer a clue to the evolutionary origin of the neuron.

Primitive eumetazoans, from which almost all the great diversity of metazoan life evolved, have been in possession of a diffuse nervous net as we see it in modern cnidarians. This is believed to have coincided with the Cambrian explosion more than 54 mya (Figure 12.5). For the first time in Cnidaria, a network of nerve cells emerges, which is specialized in receiving stimuli from the external environment, transmitting of electric impulses throughout the body and integrating and coordinating of the activities of different types of cells. This system is responsible for maintaining homeostasis in cnidarians. Ultimately, all the behavioral, reproductive, and growth phenomena, including regeneration and metamorphosis, in Cnidaria are under the control of neurohormones released by secretory neurons (Hartenstein, 2006).

The nervous system represents the core of the ICS in eumetazoans (see Chapter 1). It provided the evolution with an alternative source of information, i.e., epigenetic information, capable of meeting the huge informational requirements for erecting extremely complex metazoan structures.

During more than half a billion years of metazoan life, the neuron does not seem to have evolved significantly. Its function is essentially the same from a *Hydra* to *Homo sapiens*.

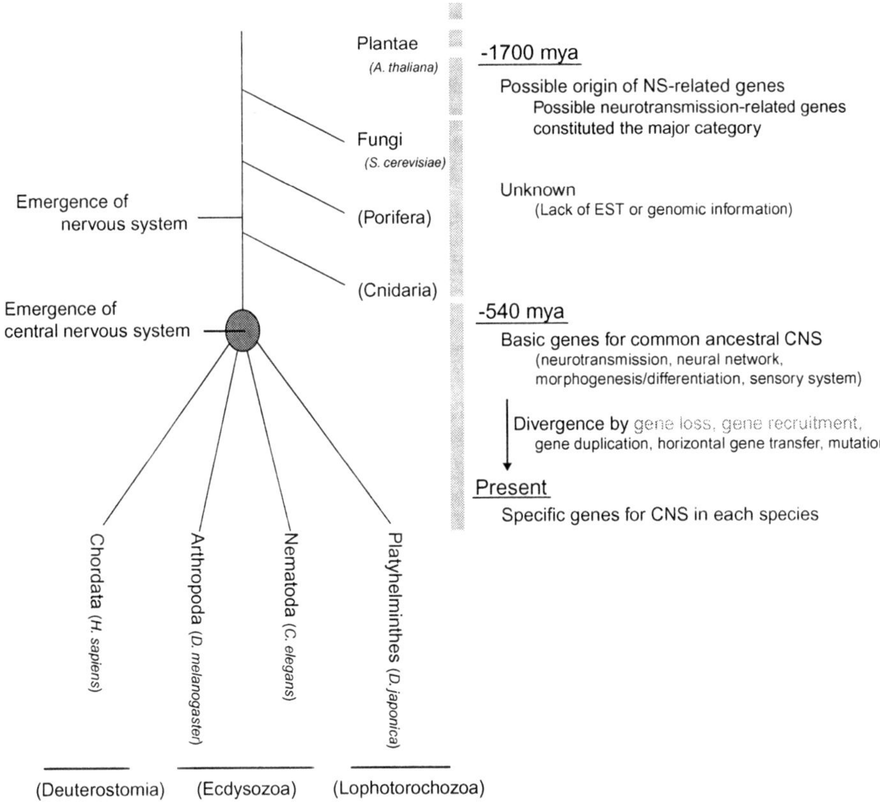

Figure 12.5 A possible scenario of the process of evolution of the CNS. The phyla not analyzed are indicated in parenthesis. Based on our analysis, possible events during evolution of the NS (nervous system)-related genes are noted.
Source: From Mineta et al. (2003).

While genotypically identical to all the cell types of the metazoan organism, a neuron expresses a greater part of the genome than any other cell type. Morphologically, it is a cell with a very branched structure, dendrites on one side, and a long trunk (axon) that is branched only at its tip. Neurons communicate with each other via those branched structures connected in synapses. Generally, the axon transmits information from the presynaptic neuron to the postsynaptic neuron through dendrites of the latter.

Unlike higher organisms, in Cnidaria neurons are multipolar or bipolar nerve cells that can conduct stimuli in either direction. Neurons in *Hydra*, for example, are generalists lacking the functional specialization observed in neurons of higher metazoans. However, cnidarian neurons produce almost all of the neurotransmitters, neurohormones, and nonneural hormones that arthropods and chordates do (Hartenstein,

2006). They have neurites (projections of the neuron in the form of axons or dendrites) but show no morphological differences between axons and dendrites.

In *Hydra*, neurons are multifunctional, i.e., each neuron has sensory, motor, and secretory functions (Figure 12.6). A sensory neuron forms synaptic connections to the muscle cells as a motor neuron and to neurons of the ganglia as interneurons, and at the same time it has a secretory function.

Neurons in *Hydra* form a diffuse nerve net consisting of ~6,700 neurons interspersed among epithelial cells of epidermis and gastrodermis in a specific spatial pattern (Koizumi et al., 2004). The nerve net extends throughout the body. The cnidarian nerve net sends signals *directly* from sensory cells to effectors.

Cnidarians have no specialized nervous organs and form no large concentrations of neurons as ganglia, but the net around the hypostome forms a ganglion-like structure (Schaller et al., 1996), the nerve ring, consisting of a thick nerve bundle, which contains four different subsets of neurons. The nerve ring may be the controller of the control system in cnidarians and is believed to be the evolutionary precursor of the CNS that evolved in bilaterians (Koizumi, 2007).

Some experimental evidence from extant lower metazoans shows that not only undifferentiated cells but differentiated cells as well, when appropriately manipulated, may be transformed into nerve cells: isolated striated muscle cells of jellyfish, for example, are reversibly transdifferentiated into nerve cells by applying a low-molecular weight compound, the *proportion-altering factor* (PAF) just before or after initiation of DNA synthesis (Schmid and Plickert, 1990). Transformation

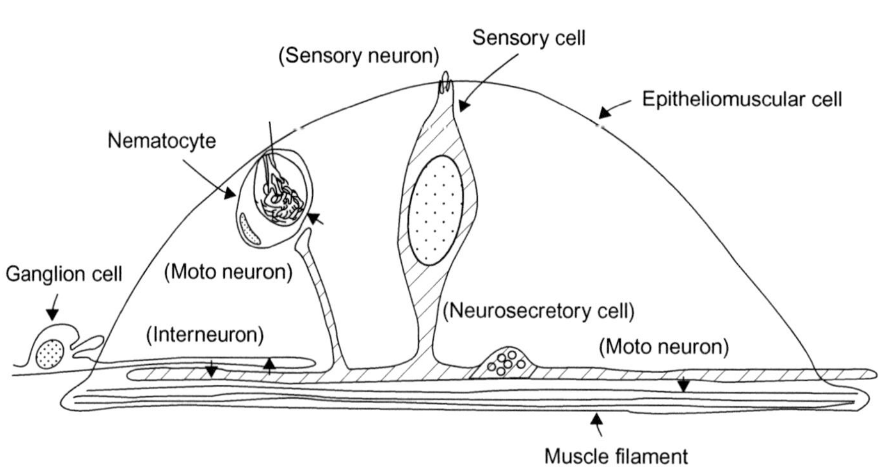

Figure 12.6 Multifunctional nerve cells in *Hydra*. A single sensory cell has synaptic connections to the muscle sheet of an epitheliomuscular cell, a nematocyte, and a ganglion cell. Moreover, it has sensory cilia and neurosecretory granules. The arrows show synapses and their polarities.
Source: From Koizumi (2002).

of other cell types into nerve cells by PAF is demonstrated almost in all stages of the development of *Hydractinia echinata*, in planula and even after metamorphosis (Plickert, 1989).

From an evolutionary viewpoint, the most difficult problem to resolve, in the process of transition to multicellular life, would have been the acquisition by the multicellular system of the ability to memorize its supracellular structure and evolution of mechanisms for transmitting that memory (= information) to the offspring. In the first part of this work, I have shown that this is a function of the control system with the nervous system as its controller. From the beginning of the eumetazoan life, even the simple cnidarian neural net was able to function as a control system, as is proven by the fact that the neural net in species of this group controls and regulates their vital functions, including their behavior, reproduction (individual development, cell differentiation, and growth), cell proliferation, and regeneration. In cnidarians, the proximate regulators of all of these functions are neurohormones released by secretory neurons:

> *The nervous system of hydra serves two functions. It is responsible for the coordination of movement and behaviour, and it controls morphogenesis. For the coordination of movement this evolutionary oldest nervous system uses peptides for neurotransmission. Classical transmitters like acetylcholine, serotonin and catecholamines and their processing enzymes do not seem to be present in nerve cells of hydra, but neuropeptides are abundant both in hydra and in other coelenterates ... two sets of substances, secreted by nerve cells, regulate head- and foot-specific differentiation events and thus pattern formation in hydra, an activator and an inhibitor of head formation, and a second set for foot formation ... The head inhibitor controls where a head structure is induced, primarily by regulating, presynaptically, the release of the head activator and also its own release.*
>
> *Schaller et al. (1996)*

The neuropeptide head activator (HA) at higher concentrations determines differentiation of stem cells to head-specific fates (Schaller et al., 1996).

The nervous net of *Hydra* also regulates the growth and related process of regeneration and budding, and at the same time, it maintains its head-to-foot morphology (Rezgaoui et al., 2006) by regulating patterns of secretion of the neuropeptide HA. As a strong mitogen in the G2-mitosis transition, the neuropeptide regulates cell-cycle progression during embryonic development (Figure 12.7).

Hydra continually produces new neurons from interstitial cells (Bode et al., 1988) for supplying neurons to young buds as well as for replacing neurons that are lost in extremities. Differentiation of neurons from interstitial cells in *Hydra* is under control of the nervous net, whose neurons send signals for transdifferentiation of interstitial pluripotent cells (I stem) into neurons on their way to the head and foot (Figure 12.8). The differentiation signal released by neurons is the neuropeptide Hym355, which overpowers inhibiting action of epithelial peptides. Interstitial cells committed to nerve cell differentiation have information on the site they must migrate (Koizumi, 2002).

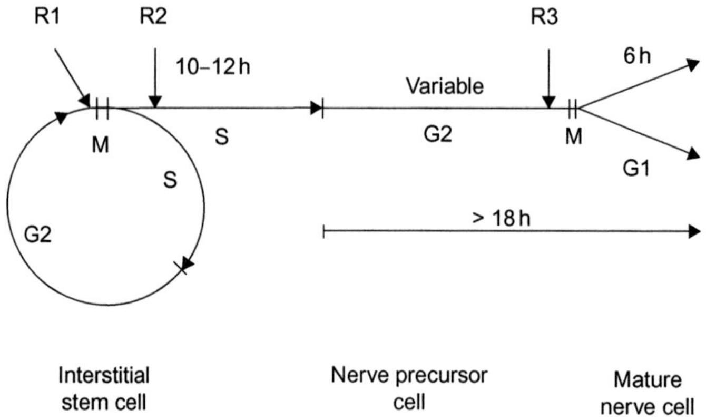

Figure 12.7 Restriction points (R) in the pathway from uncommitted interstitial stem cell to mature nerve cell, at which HA acts as positive signal. *Abbreviations*: G1, S, G2, and M, phases of mitotic cell cycle; HA, head activator.
Source: From Schaller et al. (1996).

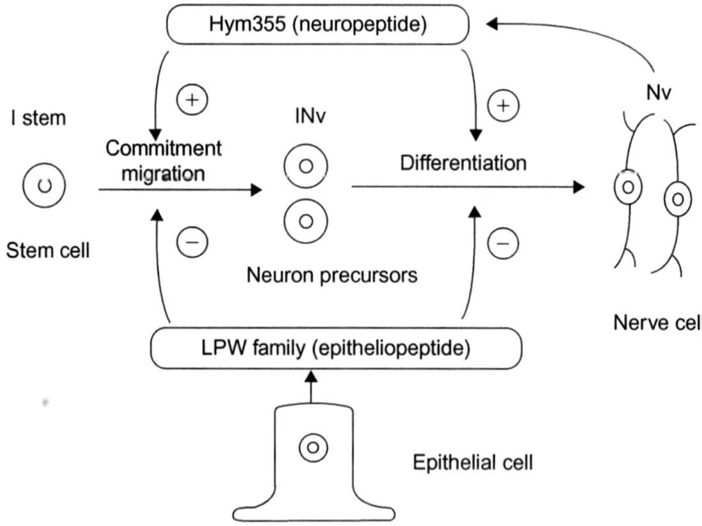

Figure 12.8 Effects of peptide signal molecules on nerve differentiation in *Hydra*. A neuropeptide, neuropeptide Hym355, activates nerve differentiation, while epithelial peptides, the LPW family, inhibit it. *Abbreviations*: I stem, interstitial stem cell; INv, committed nerve precursor cells; Nv, differentiated nerve cells.
Source: From Koizumi (2002).

Many features that are characteristic of higher metazoans, such as precise patterning and axis formation during development, and the presence of a nervous system, were probably "invented" by cnidarians or an equally ranked or related ancestral group.

The epigenetic information for the early stages of development in cnidarians is provided to the egg cell in the form of maternal factors, provided by nurse cells (David et al., 2005) and by ingesting surrounding apoptotized nurse cells (Technau et al., 2003).

Apoptosis in *Hydra* and many other cnidarian species is a widely used process for maintaining cytological homeostasis. It is almost indistinguishable from the programmed cell death in higher metazoans (Böttger and Alexandrova, 2007) and clearly implies presence of epigenetic information on the normal number of cells *Hydra* has to maintain.

There is no evidence on the nature of the control of the number of cells in this organism, but, based on the evidence on the present knowledge on mechanisms of the control of cell number and body mass in higher metazoan organisms (Gorbman and Davey, 1991, pp. 731, 744; Halaas et al., 1995; Pelleymounter et al., 1995; Adams et al., 2001; Baeckberg et al., 2003; Nijhout, 2003), a neural mechanism may be operative in *Hydra* as well.

Recently Nordström et al. (2003) have described an interesting phenomenon: while planula larvae of cnidarians generally have a simple nerve net, they could not discover any neurons in the larvae of the jellyfish (cnidarians of the class Scyphozoa). The extremely simple organization of these larvae, with only five cell types and no neurons, is reminiscent of sponges rather than cnidarians (Nordström et al., 2003).

As already pointed out, the early development in Cnidaria, like other oviparous animals, is determined by the maternal epigenetic information contained in the egg cell in the form of cytoplasmic factors.

The next phase of the individual development, the metamorphosis (Figure 12.9) of the planula into a polyp, is also under epigenetic control, but signals for metamorphosis now come from the embryonic nervous net. Larvae have chemosensitive and mechanosensitive cells. *Hydractinia* planulae mostly settle on hermit crab shells. They have sensory neurons on both extremes of their bodies, which discharge sticky threads of nematocysts to attach to the shells.

If the shell happens to be of a hermit crab, it contains a thin film of bacteria, whose outer membrane chemicals serve as a cue for inducing the planula to metamorphose (several bacterial species are identified to have such an effect) (Müller and Leitz, 2002). Planulae respond specifically only to cues from bacteria of the genus *Alteromonas*.

Upon attaching to the shell, ectodermal neurosensory cells at the anterior of the planula respond to the bacterial cue by secreting the neurotransmitter serotonin (5-HT or 5-hydroxytryptamine), which starts a signal cascade (McCauley, 1997) (Figure 12.10) leading to release by secretory neurons of a neuropeptide. Here is the interpretation of the phenomenon by the investigators:

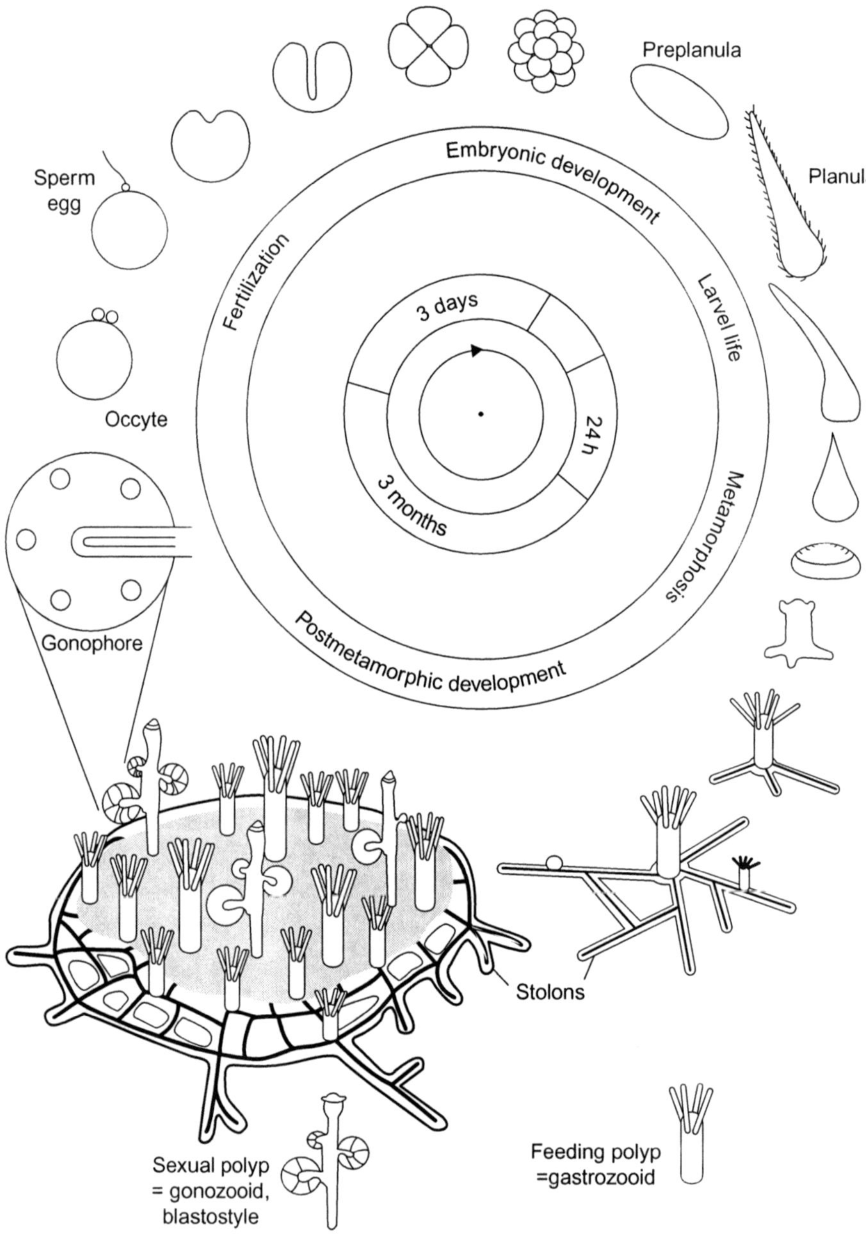

Figure 12.9 Life cycle of the hydroid *Hydractinia echinata* (and *Hydractinia symbiolongicarpus*).
Source: From Müller and Leitz (2002).

If sensory cells perceive a distinct external cue, their function is to inform the other cells in the larval body. They have to emit the following message: it is time to start the program that leads to the decomposition of specific larval characters and the development of adult ones .... Efforts to isolate and identify the molecules that transmit this internal signal in Hydractinia culminated in the identification of a novel class of neuropeptides. These share the carboxy-terminal consensus sequence—Gly-Leu-Trp-NH2 (GLWamide) and cause not only anterior fragments but also posterior fragments of planulae to undergo metamorphosis .... They were neurosensory cells located in a belt near the anterior pole with their axonal fibres extending along the mesogloea to the posterior pole.

Schmich et al. (1998)

These axonal fibers are thought to convey the message to all the body parts. The neuropeptides may be released along the length of the fibers and (or) at their terminals (Müller and Leitz, 2002).

Thus, the transformation of an organless planula into a polyp is the result of a neurally controlled and regulated chain of events that starts with the reception of the specific environmental stimulus and its processing in the planula neural net that functions as a control system.

LWamide peptides are considered to be part of the control system.

Plickert et al. (2003)

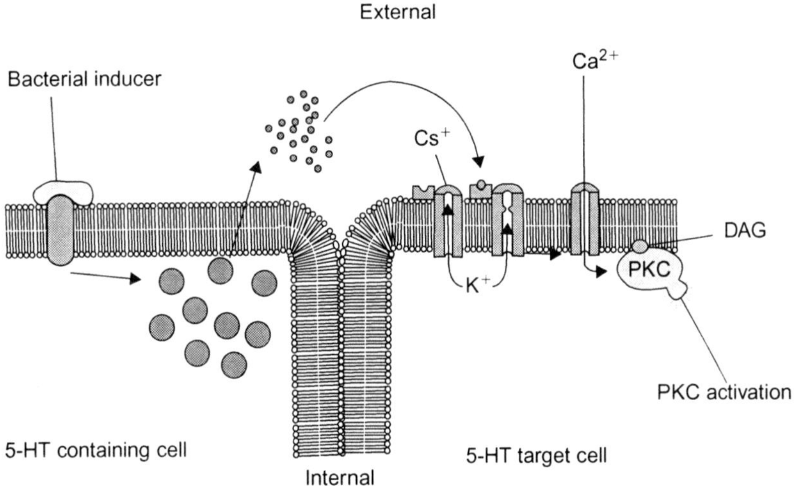

Figure 12.10 Model for action of 5-HT during the initiation of metamorphosis in *Phialidium*. Upon contact with an appropriate external cue (yellow), 5-HT (red) is released from internal stores into the extracellular environment. 5-HT binding at nearby cells closes K^+ leak channels (dark blue), resulting in membrane depolarization. Membrane depolarization triggers Ca^{2+} influx through voltage-gated Ca^{2+} channels (green). Increased Ca^{2+} and membrane-bound diacylglycerol (DAG) activate protein kinase C (light blue) to trigger subsequent metamorphic events.
Source: From McCauley (1997).

It is experimentally demonstrated that metamorphosis of *H. echinata* planulae is prevented by blocking LWamide neuropeptides and can resume by administration of synthetic LWamide neuropeptides (Plickert et al., 2003).

LWamide neuropeptides are secreted by about 40 neurons (out of a total of 450 neurons in the planula) located in the anterior half of the ectodermal cell layer, which function as sensory neurons for monitoring the environment. These sensory neurons send long processes to the posterior part of the planula, thus making it possible that LWamide neuropeptides reach the target larval cells all over the planula body, stimulating them to enter metamorphosis (Katsukura et al., 2003a). A similar number of planula sensory neurons secrete an RFamide neuropeptide, which has an opposite effect, i.e., it inhibits the metamorphosis. Sensory neurons also control and regulate the migration of planulae. The above neuropeptides, in concentrations that are one to two orders of magnitude lower than those affecting metamorphosis, regulate the migration of planulae. Migration of planulae is not continuous. Planulae intermittently experience phases of migration and resting behavior. It has been suggested that migration periodicity may be controlled and regulated by a neural control mechanism in which secretion of LWamide by LWamide neurons stimulates muscle contraction and migration but at the same time stimulates secretion of RFamides, which inhibit migration and determine the resting phase and other phenomena (Katsukura et al., 2003b).

Thus, the cnidarian nervous net represents a neural control system that is responsible for all of the vitally important phenomena, such as behavior, homeostasis, sexual and asexual reproduction, including the growth and cell differentiation, and animal physiology in general. This control system made it possible the evolution of the asymmetry and the strict determinism and patterning of the early development in metazoans as compared to sponges. Most importantly, it provided eumetazoans with the potential for producing and storing the great volume of information necessary for building the metazoan structure and the morphological diversity, based on the ability of the control system to memorize not only its own structure, behavior, and functions, but sometimes also changes in phenotypic traits that the metazoan might develop in response to environmental stimuli. The control system enabled metazoans to escape the evolutionary dead end where Precambrian metazoan life was long trapped.

Centralization of the Neural Control System

Segregation and concentration of neurons of particular types in specific regions of the metazoan body have been universal trends of metazoan evolution. The tendency for a centralized nervous system appeared as low in metazoan evolution as Cnidaria, which, despite the generally diffuse character of the neural net, show simple neuronal concentrations in the form of the nerve ring.

From a physiological point of view, transition from the diffuse neural net of primitive metazoans to centralized nervous system implies transition from processing of the sensory input for generating *local* locomotory and neurosecretory outputs to a higher processing and integration of the sensory input in the CNS for generating systemic responses via neuroendocrine pathways or centrally coordinated motor activities (Arendt et al., 2008). This concentration is the result of the evolutionary pressure for more complex processing of the internal and external stimuli necessary for generating the more complex adaptive behaviors, morphologies, and life histories that characterize evolution of higher metazoans. Concentration of neurons in ganglia and brains is not related to any changes in genes, and more than 95% of 116 nervous system–related genes are shared by organisms of different phyla, such as platyhelminthes (planaria), nematodes, invertebrates (insects), and vertebrates (humans) (Mineta et al., 2003).

An intermediate position in the evolution of the nervous system is occupied by the deuterostome enteropneusts, which display aspects of both central and diffuse organization (Holland, 2003).

One of the lower extant invertebrates that evolved a CNS is the planarian *Dugesia japonica*, a platyhelminth that displays a concentration of neurons in the head region, which has been considered to be a simple brain (Nakazawa et al., 2003). Its CNS consists of a bilobed cephalic ganglion and a ventral nerve cord. The ganglionic neurons exhibit a structural novelty similar to dendritic spines and express *otx* genes, facts that are believed to justify ranking the planarian cephalic ganglion as a brain (Nakazawa et al., 2003).

The planarian brain shows functional regionalization (Figure 12.11), which allows them to perform complex functions, comparable to those observed in higher metazoans (Nakazawa et al., 2003).

While all of the primitive Cnidaria and Ctenophora have a diffuse nervous net, the modern protostomes and deuterostomes evolved the CNS. This fact raises questions about the first metazoan group(s) that evolved the CNS: Did the Urbilaterian, the earliest of bilaterian metazoans, to which both protostomes (ecdysozoa and lophotrochozoa) and deuterostomes (mainly chordates) belong, evolve a CNS? Recent studies on the evolution and ontogeny of the CNS in these groups have shed some light on the issue.

Similarities in the organization of the CNS in all the bilaterian groups have been observed since the nineteenth century, at the dawn of neurobiological studies, but a recent study in Arendt's laboratory, Heidelberg, focused on both similarities and divergences of the CNS in these groups.

Investigation of the CNS of a lophotrochozoan, the polychaete annelid *Platynereis dumerilii*, showed some surprising similarities in the mediolateral patterning and neuron-type distribution of the developing trunk CNS of the worm and the mediolateral architecture of the developing vertebrate neural tube, which "surpasses that documented previously for vertebrate and fly and indicates that a CNS already existed in Urbilateria" (Denes et al., 2007). Their evidence suggests that

> *a large part of the spatial organization of the annelid and vertebrate CNS was already present in their last common ancestor, which implies that Urbilateria had already possessed a CNS.*
>
> *Denes et al. (2007)*

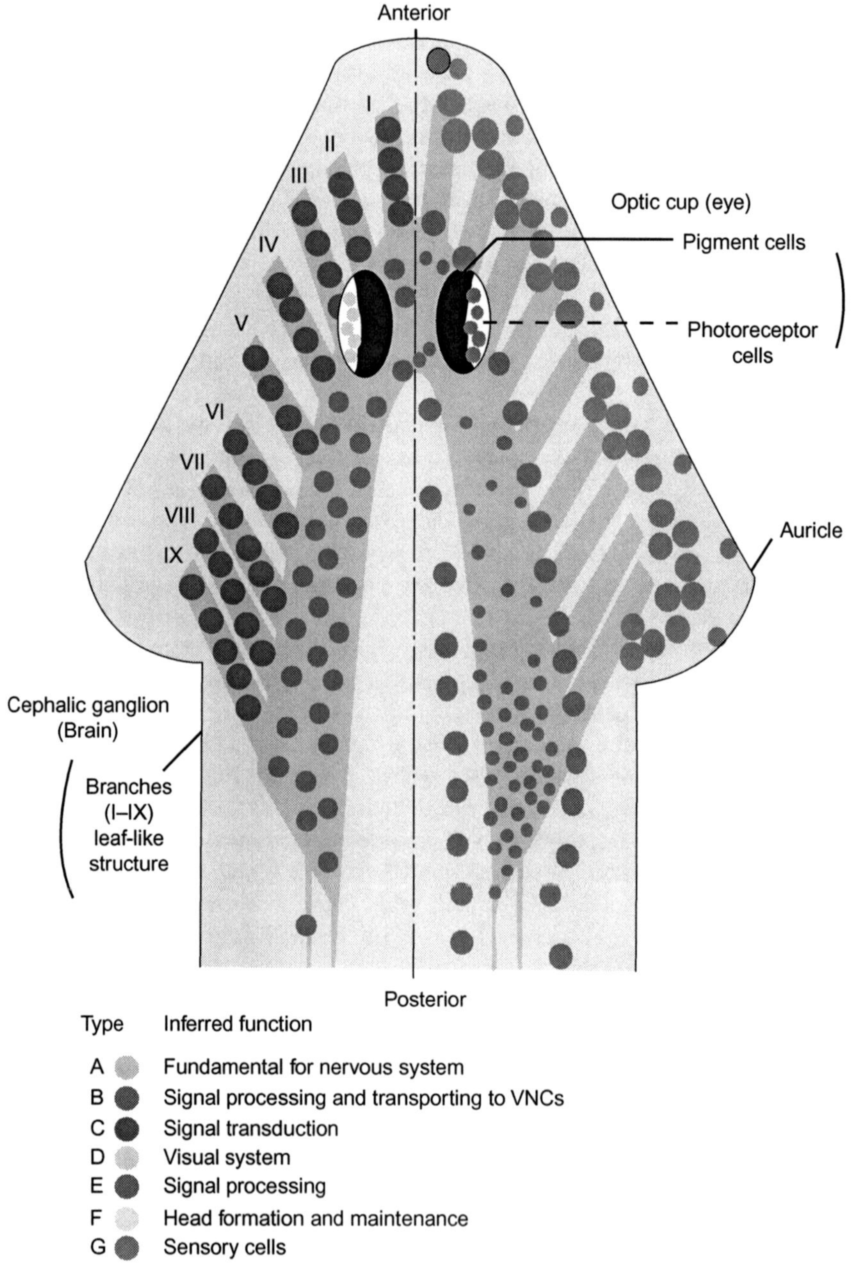

Figure 12.11 Cytoarchitecture map of planarian brain. The represented patterns of each type of gene expression (A to G) are schematically drawn using different shades at a half side. (For interpretation of the references to color in this figure legend, the reader is referred to the web version of this book.)
Source: From Nakazawa et al. (2003).

The fact that ecdysiozoans (flies and nematodes) have lost some transitory neural structures has been explained with an evolutionary pressure for speeding up neurodevelopmental processes (Denes et al., 2007).

Evolution of the Neuroendocrine Control System

Neuroendocrine Control System in Invertebrates

Centralization and functional regionalization of the nervous system in bilaterians might have been stimulated by the evolutionary pressure for coordinating ever-increasing complexity of functions and structure of metazoans, which required more sophisticated neural processing of the sensory input in neural circuits. At the same time, the evolution of more complex metazoan structures required and led to evolution of hierarchies in the control system. One great manifestation of the hierarchical structuration of the neural control system in invertebrates is evolution of the neuroendocrine system, in which the CNS takes control of the three tiers of endocrine glands whose products became mediators and effectors of the neural signals in the control system.

The following excursus on the evolution of the neuroendocrine control system is based on the review of Hartenstein (2006).

The CNS evolved first in bilaterians. The fact that the spatial arrangement of progenitors of the neuroendocrine system is conserved in metazoan groups as different as *Drosophila* and vertebrates suggests that

> *fundamental elements of a primordial "neuroendocrine system" were already present in the Bilaterian ancestor.*
>
> *Katsukura et al. (2003a)*

The chemosensory neuronal structures involved in reproductive and feeding behaviors in invertebrates may be evolutionary precursors of the pituitary and comparable endocrine glands in vertebrates, which were later placed under the control of the hypothalamus, the brain neurosecretory gland.

Specialized neurons in the epidermis and in the gastrodermis of primitive eumetazoans delaminated to form the brain, nerve plexuses, peptidergic endocrine glands, and cell clusters close to the brain (corpora cardiaca in insects and pituitary in vertebrates) gave rise to specialized endocrine glands.

A process of centralization (cephalization) of the nervous systems led to formation of the CNS and evolution of a neuroendocrine system as the central element of the ICS, which in invertebrates included neurally controlled endocrine glands (Figure 12.12D).

All of the reproductive functions (reproductive physiology, gametogenesis, early development and postphylotypic development, including metamorphosis and regeneration) and behavior in invertebrates are under strict control of the neuroendocrine system.

It has been suggested that an intermediate step in the evolution of the vertebrate neuroendocrine control, a precursor of the hypothalamal–pituitary–target endocrine glands

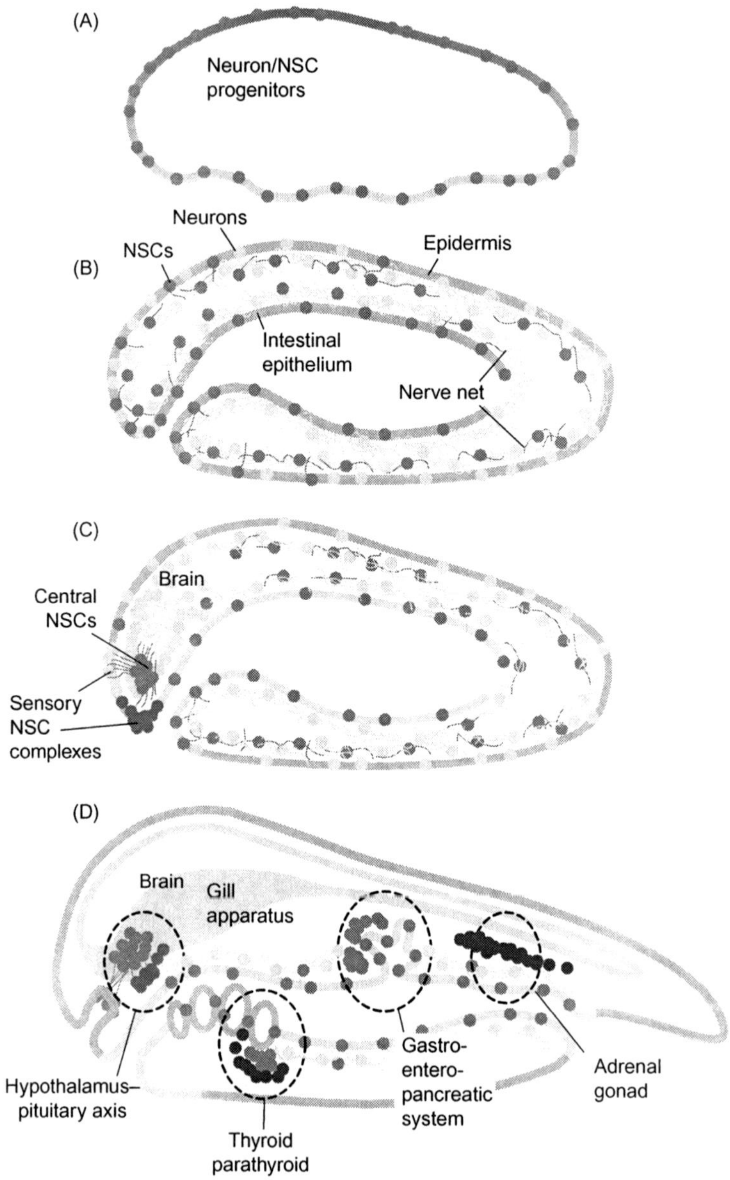

Figure 12.12 Hypothetical stages in the evolution of the endocrine and neuroendocrine systems. (A) Cells releasing endocrine signals are likely to predate the appearance of a nervous system, since they can be found in extant metazoa lacking nerve cells. (B) The first nervous system is thought to have possessed the structure of a basi-epithelial nerve net, similar to the one still found in present-day cnidarians. (C) At this stage, neurons and NSCs (neurosecretory cells—N.R.C.)/endocrine cells most likely had evolved into distinct lineages of sensory cells integrated into the epidermis, the gastrodermis, and the nerve net. A central nervous system integrating multimodal sensory input evolved in bilaterian animals.

is a new brain structure that evolved in the amphioxus. Being in contact with the external environment on the one hand, and having basal blood contact on the other, Hatschek's pit cells are at the same time chemosensory and endocrine cells, secreting both gonadotropins (FSH and LH) and gonadotropin-releasing hormone (GnRH), thus performing olfactory, hypothalamal (GnRH) and pituitary functions (gonadotropins) (Figure 12.13). This fact has been interpreted as an indication of the common evolutionary origin of the olfactory cells, hypothalamal and the pituitary cells. This hypothesis seems to be corroborated by the common developmental origin of these organs (Arendt, 2008).

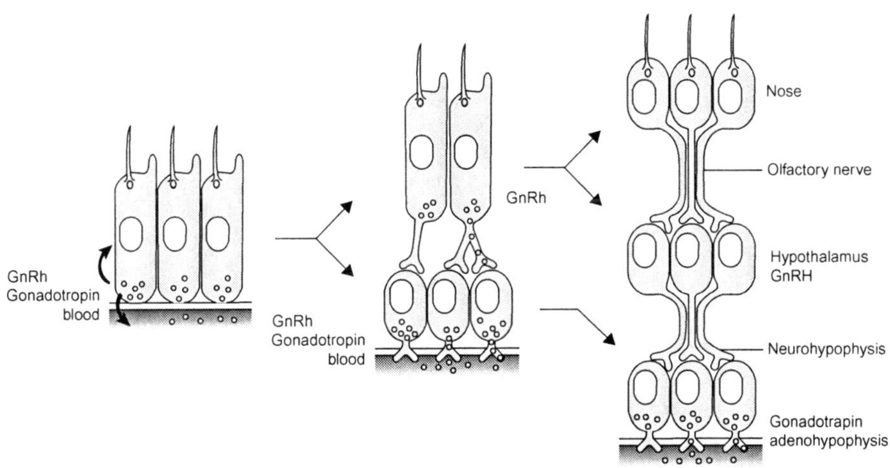

Figure 12.13 Segregation of olfactory, integrative, and neurosecretory functions during the diversification of the olfactory–hypothalamal–pituitary axis. Functionally segregating cell types retain axonal contact, and this contact evolves into a neuronal circuit that interconnects the nose, hypothalamus, and pituitary. The most downstream endocrine cells relocate to the pituitary, where they directly release gonadotropin into the blood. The gonadotropin-releasing hormone (GnRH)-releasing cells that are located more upstream come to lie in the hypothalamus, where they integrate information from other neurons, and the most upstream olfactory sensory neurons form part of the peripheral sensory organ, the nose, which must be in direct contact with the environment.
Source: From Arendt (2008).

◀ (D) Populations of sensory NSCs involved in the regulation of fundamental biological processes, such as feeding and reproduction, may have formed specialized complexes in the brain, pharynx, and gut of early bilaterians. During later stages of evolution of the chordate lineage, NSCs and endocrine cells in general show the tendency of losing their sensory function, delaminating from the surface epithelium (epidermis, pharynx, and intestinal epithelium), and undergoing morphogenetic changes that produced dedicated endocrine glands, such as the pituitary, thyroid/parathyroid, and pancreatic islets. *Abbreviation*: NSCs, neurosecretory cells.
Source: From Hartenstein (2006).

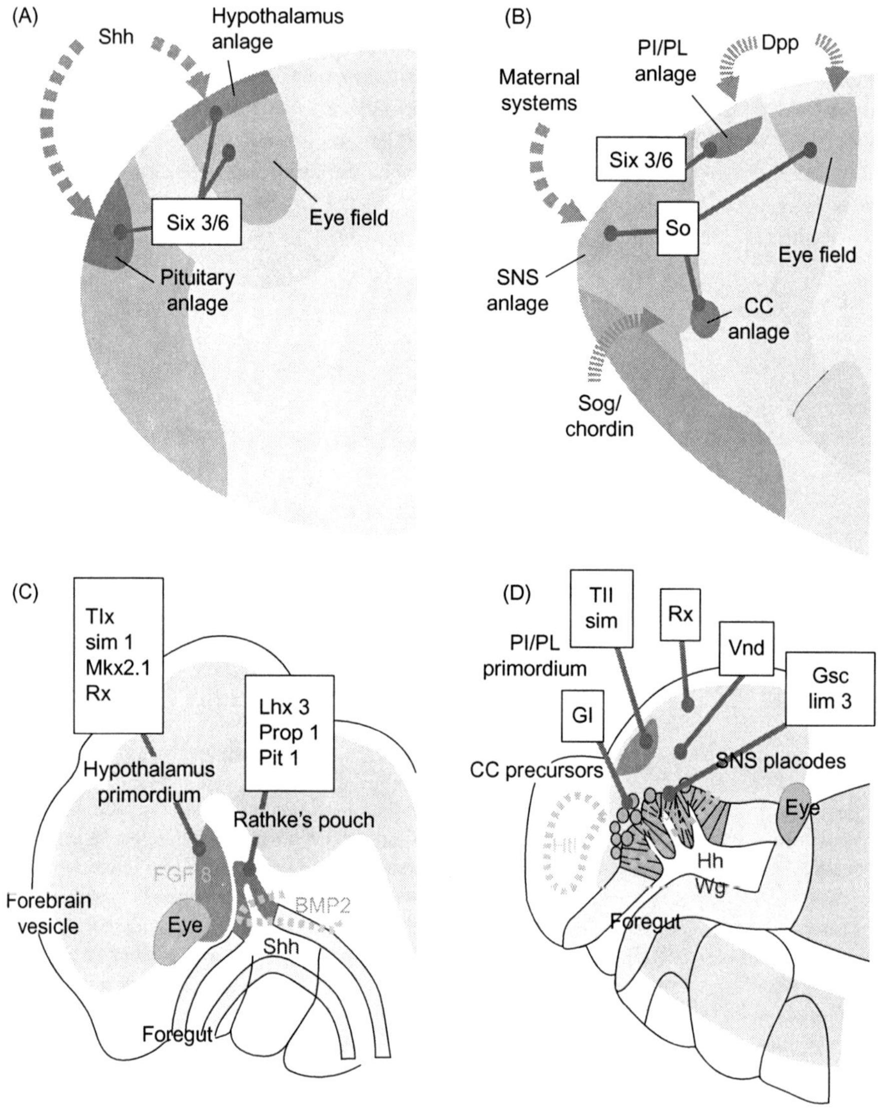

Figure 12.14 Molecular determinants of embryonic neuroendocrine system development in vertebrate (A) and (C) and *Drosophila* (B) and (D). All panels show schematic lateral views of anterior part of embryos (anterior to the left, dorsal up). (A) and (B) Postgastrula stage embryos. The eye field forms in the anterior domain of the neural plate. The anlage of the hypothalamus represents the medial part of the eye field; the anlage of the pituitary (anterior lobe) is located anteriorly adjacent to the hypothalamus. In *Drosophila*, the anlage of the pars intercerebralis/pars lateralis is located near the midline of the head neuroectoderm, anteriorly adjacent to the eye field. The anlagen of the corpora cardiaca and SNS map anterior to the head neuroectoderm. In the later-stage vertebrate embryo (C), hypothalamus and anterior pituitary are seen in their primordial state. The primordium of the anterior pituitary,

Evolution of the Neuroendocrine Control in Vertebrates

In the course of evolution, the neuroendocrine control system in vertebrates became progressively more complex. However, despite the independent evolution for more than half a billion years, astounding similarities are observed between the neuroendocrine system of extant invertebrates and vertebrates. The insect brain–pars intercerebralis/pars lateralis–corpora cardiaca–corpora allata complex is the counterpart of the vertebrate neuroendocrine brain–hypothalamus–pituitary–target glands axis (Veelaert et al., 1998). Like the vertebrate pituitary, the insect's corpora cardiaca consists of a glandular lobe and a neurohemal lobe where nerves projecting from the brain release their neurohormones that are stored or regulate production of other hormones there. Not only the general hierarchy of the neuroendocrine structure is conserved in both groups, but even the patterns of the development of the neuroendocrine structure are surprisingly similar (Figure 12.14). In both invertebrates and vertebrates, the neuroendocrine system arises within the anterior neural plate. In both groups, the pars intercerebralis and the hypothalamus develop behind the corpora cardiaca and the pituitary, respectively. Patterns of gene expression in the process of the development of neuroendocrine systems are also similar to a considerable extent (Veelaert et al., 1998; Hartenstein, 2006).

The general pattern of the organization of the neuroendocrine system in vertebrates is very similar to the one observed in invertebrates and is represented by the hypothalamic–pituitary–target glands system (Figure 12.15), which is highly conserved across vertebrate taxa.

The hypothalamus produces a number of neurohormones, which program and stimulate secretion of the corresponding pituitary hormones. As an evolutionary innovation of vertebrates, evolution of the hypothalamus added another level in the hierarchy of the ICS in vertebrates. In vertebrates a diffuse neuroendocrine level of secretory neurons distributed in various regions of the body also evolved. Peripheral neurons of the autonomic nervous system also participate in the regulation of hormonal secretion of the adrenal medulla, pancreas, and intestinal wall (Hartenstein, 2006). The neuroendocrine system in vertebrates also controls and regulates processes of metabolism, homeostasis, reproduction, and growth and determines the vertebrate neuroendocrine response to stress.

Existence of a collection of such "synthesis islands" in invertebrates led to formulation of the bridging hypothesis according to which preexisting ligand-receptor systems could be bridged to form new networks, and then structural innovations in

◀ called Rathke's pouch, invaginates from the roof of the stomodeum and comes into contact with the primordium of the hypothalamus. In a *Drosophila* embryo of a corresponding stage (D), the primordium of the SNS forms three invaginating placodes in the roof of the foregut. Precursors of the corpora cardiaca are associated with the SNS primordium. The expression pattern of relevant signaling pathways and transcription factors (boxed) involved in neuroendocrine system specification is indicated. *Abbreviations*: PI, pars intercerebralis; PL, pars lateralis; SNS, stomatogastric nervous system.
Source: From Hartenstein (2006).

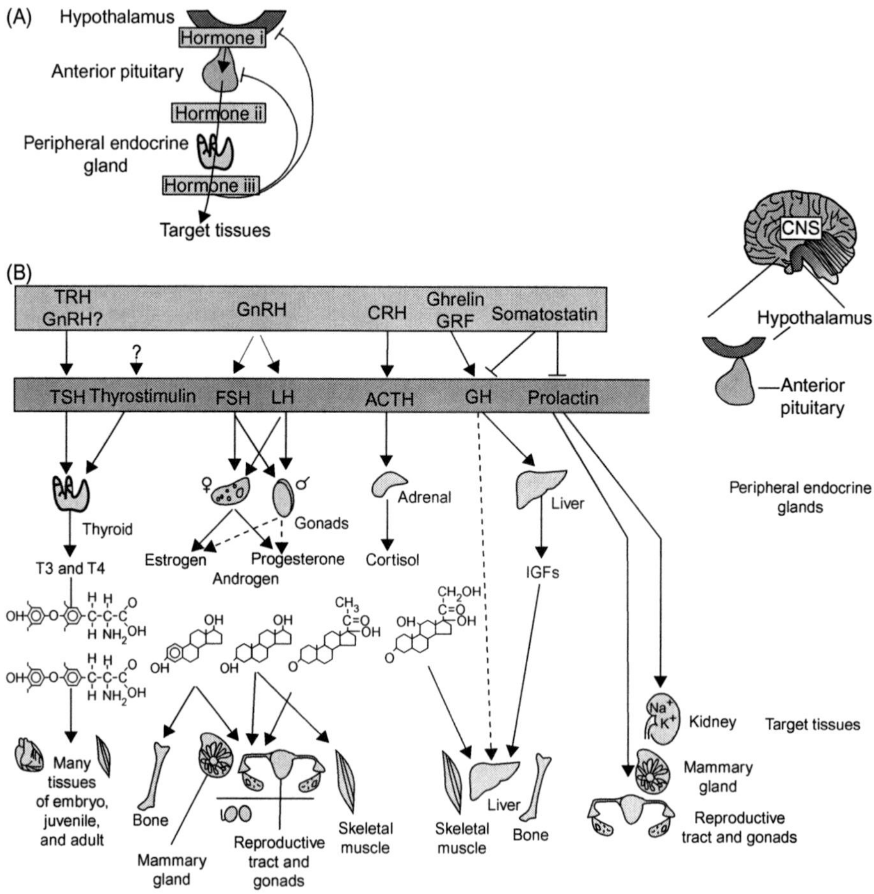

Figure 12.15 (A) Hypothalamic–pituitary–peripheral gland (H–P–PG) system. The vertebrate H–P–PG system generally shows three levels of ligand-receptor systems. Ligands that are produced by the hypothalamus (i) in the central nervous system (CNS) regulate the release of hormones from the nearby pituitary gland (ii). Pituitary hormones regulate the production of other hormones by target endocrine glands (iii). Hormones from the target endocrine glands can, in turn, influence functions of the hypothalamus and pituitary, forming positive or negative feedback loops. (B) Examples of vertebrate H–P–PG system connection to target tissues. The H–P–PG system coordinates physiological functions of diverse tissues in response to internal and external environments. *Abbreviations*: ACTH, adrenocorticotropic hormone; CRH, corticotropin-releasing hormone; FSH, follitropin; GH, growth hormone; GRF, growth hormone-releasing factor; GnRH, gonadotropin-releasing hormone; IGF, insulin-like growth factor; LH, lutropin; TRH, thyrotropin-releasing hormone; TSH, thyrotropin; T3, triiodothyronine; T4, tetraiodothyronine (thyroxine).
Source: From Campbell et al. (2004).

proteins could produce changes in signaling pathways, including formation of the hypothalamal–pituitary–target (peripheral) endocrine glands axis. But the problem is that no one has any idea how that bridging might take place. Neither this nor the model of the sequential capture of preexisting regulatory pathways (Campbell et al., 2004) could satisfactorily explain the evolution of the vertebrate hypothalamal–pituitary–peripheral gland system.

Evolution of the hypothalamus is related to the evolution of the vertebrate brain in general. The central role of the hypothalamus in controlling functions, structures, and behaviors of vertebrates via the hypothalamal–pituitary–target gland axis cannot be considered isolated from the activity of the brain, of which it is an integral part. Ultimately, all of the neuroendocrine signals released by the hypothalamus result from integration of the various inputs of information it receives from many parts of the brain.

By the time the vertebrates split from their ancestral chordates, a series of evolutionary changes in the brain developmental program took place. With vertebrates began a new stage in the evolution of the patterns of development of the CNS, the controller of the ICS: the segmental development of the brain with neuromeres as its basic units. Prosomeres, mesomeres, and rhombomeres are neuromeres from which the forebrain, midbrain, and hindbrain, respectively, develop. Vertebrates have seven or eight rhombomeres, each with specific patterns of gene expression.

The vertebrate neuromeric structure evolved first in agnathans, such as lampreys. This implies that the segmented brain may have first appeared in the common ancestor of agnathans and gnathostomes, i.e., not later than 540 mya (Rosa-Molinar et al., 2005) during the Cambrian explosion.

Another characteristic feature of the evolution of the CNS in vertebrates is evolution of the cerebellum. The midbrain–hindbrain boundary in vertebrates serves as an organizing center of morphogenetic patterning with a characteristic pattern of gene expression that is involved in the anterior–posterior patterning of the midbrain and cerebellum. In lampreys, this organizing center does not induce formation of a cerebellum, probably because lampreys do not express Pax6, whereas the rest of vertebrates do (Murakami et al., 2005). The most recent vertebrate brain structure to evolve is telencephalon, which appeared first in hagfish and lampreys.

References

Adams, C.S., Korytko, A.I. and Blank, J.L. (2001). A novel mechanism of body mass regulation. J. Exp. Biol. *204*, 1729–1734.

Alcock, J. (1989). *Animal Behavior*, fourth edition. Sinauer Associates, Inc. Publishers, Sunderland, MA, p. 7.

Arendt, D. (2008). The evolution of cell types in animals: emerging principles from molecular studies. Nat. Rev. Genet. *9*, 868–882.

Arendt, D. and Nübler-Jung, K. (1994). Inversion of dorsoventral axis? Nature *371*, p. 26.

Arendt, D., Denes, A.S., Jékely, G. and Tessmar-Raible, K. (2008). The evolution of nervous system centralization. Philos. Trans. R. Soc. Lond. B Biol. Sci. *363*, 1523–1528.

Aury, J.-M., Jaillon, O., Duret, L., Noel, B., Jubin, C., Porcel, B.M., et al. (2006). Global trends of whole-genome duplications revealed by the ciliate *Paramecium tetraurelia*. Nature *444*, 171–178.

Baeckberg, M., Collin, M., Ovesjoe, M.L. and Meister, B. (2003). Chemical coding of $GABA_B$ receptor-immunoreactive neurones in hypothalamic regions regulating body weight. J. Neuroendocrinol. *15*, 1–14.

Beisson, J. and Sonneborn, T.M. (1965). Cytoplasmic inheritance of the organization of the cell cortex in *Paramecium aurelia*. Proc. Natl. Acad. Sci. U.S.A. *53*, 275–282.

Bode, H.R., Heimfeld, S., Koizumi, O., Littlefield, C.L. and Yaross, M.S. (1988). Maintenance and regeneration of the nerve net in hydra. Am. Zool. *28*, 1053–1063.

Böttger, A. and Alexandrova, O. (2007). Programmed cell death in *Hydra*. Semin. Cancer Biol. *17*, 134–146.

Bray, D. (1992). *Cell Movements*. Garland Publishers Inc., New York, NY, p. 64.

Cabej, N.R. (2004). *Neural Control of Development*. Albanet, Dumont, NJ, pp. 21–34.

Campbell, N.A., Reece, J.B. and Mitchell, L.G. (1999). *Biology*, fifth edition. p. 529.

Campbell, R.K., Satoh, N. and Degnan, B.M. (2004). Piecing together evolution of the vertebrate endocrine system. Trends Genet. *20*, 359–366.

Carroll, R.L. (2000). Towards a new evolutionary synthesis. Trends Ecol. Evol. *15*, 27–32.

Carroll, S. Interview with Nova conducted on July 14, 2006 by S. Holt, NOVA. Available online: http://www.pbs.org/wgbh/nova/body/genetic-factor.html.

Changeux, J.P. (1985). *Neuronal Man: The Biology of Mind*. Pantheon, New York, NY, pp. 216–217.

Collins, A.G., Cartwright, P., McFadden, C.S. and Schierwater, B. (2005). Phylogenetic context and basal metazoan model systems. Integr. Comp. Biol. *45*, 585–594.

Cummings, F.W. (2006). On the origin of pattern and form in early metazoans. Int. J. Dev. Biol. *50*, 193–208.

David, C.N., Schmidt, N., Schade, M., Pauly, B., Alexandrova, O. and Böttger, A. (2005). *Hydra* and the evolution of apoptosis. Integr. Comp. Biol. *45*, 631–638.

Degnan, B. (2006). Reconstruction the ancestral metazoan genome: structural and functional aspects of the sponge genome. ITEE Seminar: Reconstruction of the ancestral metazoan genome: structural and functional aspects of the sponge genome. Available online: http://www.itee.uq.edu.au/~seminar.

Degnan, B.M., Leys, S.P. and Larroux, C. (2005). Sponge development and antiquity of animal pattern formation. Integr. Comp. Biol. *45*, 335–341.

Denes, A.S., Jékely, G., Steinmetz, P.R., Raible, F., Snyman, H., Prud'homme, B., et al. (2007). Molecular architecture of annelid nerve cord supports common origin of nervous system centralization in bilateria. Cell *129*, 277–288.

Ellwanger, K. and Nickel, M. (2006). Neuroactive substances specifically modulate rhythmic body contractions in the nerveless metazoon *Tethya wilhelma* (Demospongiae, Porifera). Front. Zool. *3*, p. 7.

Erwin, D., Valentine, I. and Jablonsky, D. (1997). The origin of animal body plan. Am. Sci. *85*, 126–137.

Finnerty, J.R. (2003). The origins of axial patterning in the metazoa: how old is bilateral symmetry? Int. J. Dev. Biol. *47*, 523–529.

Fortey, R. (2001). The Cambrian explosion exploded? Science *293*, 438–439.

Gerhart, J. and Kirschner, M. (1997). *Cells, Embryos, and Evolution*. Blackwell Science, Oxford, pp. 154, 159, 160.

Gerrow, K., Romorini, S., Nabi, S.M., Colicos, M.A., Sala, C. and El-Husseini, A. (2006). A preformed complex of postsynaptic proteins is involved in excitatory synapse development. Neuron *49*, 547–562.

Glade, N. (2008). Computing with the cytoskeleton: a problem of scale. Int. J. Unconvent. Comput. *4*, 33–44.

Gorbman, A. and Davey, K. (1991). Endocrines. In: *Neural and Integrative Animal Physiology*, C.L. Prosser (ed.), fourth edition. Wiley-Liss, New York, pp. 693–754.

Goss, R.J. (1969). *Principles of Regeneration*. Academic Press, New York, NY, pp. 24, 96–99.

Gould, J.L. (1982). *Ethology—the Mechanisms and Evolution of Behavior*. W.W. Norton & Co., New York, NY, p. 319.

Grant, S.G.N. (2006). The synapse proteome and phosphoproteome: a new paradigm for synapse biology. Biochem. Soc. Trans. *34*(Part 1), 59–63.

Halaas, J.L., Gajiwala, K.S., Maffei, M., Cohen, S.L., Chait, B.T., Rabinowitz, D., et al. (1995). Weight-reducing effects of the plasma protein encoded by the *obese* gene. Science *269*, 543–546.

Hameroff, S. (2008). That's life! The geometry of π electron resonance clouds. In: *Quantum Aspects of Life*, D. Abbott, P. Davies and A. Pati (eds.), World Scientific, pp. 403–434.

Harold, F.M. (2001). *The Way of the Cell*. Oxford University Press, Oxford - New York, p. 125.

Hartenstein, V. (2006). The neuroendocrine system of invertebrates: a developmental and evolutionary perspective. J. Endocrinol. *190*, 555–570.

Holland, N.D. (2003). Early central nervous system evolution: an era of skin brains? Nat. Rev. Neurosci. *4*, 617–627.

International Human Genome Sequencing Consortium, Finishing the euchromatic sequence of the human genome. Nature *431*, 931–945.

Katsukura, Y., David, C.N., Grimmelikhuijzen, C.J. and Sugiyama, T. (2003). Inhibition of metamorphosis by RFamide neuropeptides in planula larvae of *Hydractinia echinata*. Dev. Genes Evol. *213*, 579–586.

Katsukura, Y., Ando, H., David, C.N., Grimmelikhuijzen, C.J.P. and Sugiyama, T. (2003). Control of planula migration by LWamide and RFamide neuropeptides in *Hydractinia echinata*. J. Exp. Biol. *207*, 1803–1810.

King, N., Hittinger, C.T. and Carroll, S.B. (2003). Evolution of key cell signaling and adhesion protein families predates animal origins. Science *301*, 361–363.

Koizumi, O. (2002). Developmental neurobiology of *Hydra*, a model animal of cnidarians. Can. J. Zool. *80*, 1678–1689.

Koizumi, O. (2007). Nerve ring of the hypostome in *Hydra*: is it an origin of the central nervous system of bilaterian animals? Brain Behav. Evol. *69*, 151–159.

Koizumi, O., Sato, N. and Goto, C. (2004). Chemical anatomy of *Hydra* nervous system using antibodies against *Hydra* neuropeptides: a review. Hydrobiologia *530/531*, 41–47.

Lappin, T.R.J., Grier, D.G., Thompson, A. and Halliday, H.L. (2006). *HOX* GENES: seductive science, mysterious mechanisms. Ulster Med. J. *75*, 23–31.

Leys, S.P., Mackie, G.O. and Meech, R.W. (1999). Impulse conduction in a sponge. J. Exp. Biol. *202*, 1139–1150.

Mackie, G.O. (1970). Neuroid conduction and the evolution of conducting tissues. Q. Rev. Biol. *45*, 319–332.

Mackie, G.O. (1990). The elementary nervous system revisited. Am. Zool. *30*, 907–920.

McCauley, D.W. (1997). Serotonin plays an early role in the metamorphosis of the hydrozoan *Phialidium gregarium*. Dev. Biol. *190*, 229–240.

Miklos, G., Campbell, K.S.W. and Kankel, D.R. (1994). The rapid emergence of bio-electronic novelty, neuronal architectures, and organismal performance. In: *Flexibility and Constraint in Behavioral Systems*, R.J. Greenspan and C.P. Kyriacou (eds.), John Wiley and Sons Ltd, pp. 269–293.

Mineta, K., Nakazawa, M., Cebrià, F., Ikeo, K., Agata, K. and Gojobori, T. (2003). Origin and evolutionary process of the CNS elucidated by comparative genomics analysis of planarian ESTs. Proc. Natl. Acad. Sci. U.S.A. *100*, 7666–7671.

Morris, S.C. (1998). Early metazoan evolution: reconciling paleontology and molecular biology. Am. Zool. *38*, 867–877.

Müller, W.A. and Leitz, T. (2002). Metamorphosis in the Cnidaria. Can. J. Zool. *80*, 1755–1771.

Müller, W.E.G. (2003). The origin of metazoan complexity: porifera as integrated animals. Integr. Comp. Biol. *43*, 3–10.

Müller, W.E.G., Schwertner, H. and Müller, I.M. (2004). Porifera a reference phylum for evolution and bioprospecting: the power of marine genomics. Keio J. Med. *53*, 159–165.

Murakami, Y., Uchida, K., Rijli, F.M. and Kuratani, S. (2005). Evolution of the brain developmental plan: insights from agnathans. Dev. Biol. *280*, 249–259.

Nakazawa, M., Cebrià, F., Mineta, K., Ikeo, K., Agata, K. and Gojobori, T. (2003). Search for the evolutionary origin of a brain: planarian brain characterized by microarray. Mol. Biol. Evol. *20*, 784–791.

Ng, S.F. and Frankel, J. (1977). 180 degrees rotation of ciliary rows and its morphogenetic implications in *Tetrahymena pyriformis*. Proc. Natl. Acad. Sci. U.S.A. *74*, 1115–1119.

Nijhout, H.F. (2003). The control of body size in insects. Dev. Biol. *261*, 1–9.

Nordström, K., Wallén, R., Seymour, J. and Nilsson, D. (2003). A simple visual system without neurons in jellyfish larvae. Proc. Biol. Sci. *270*, 2349–2354.

Nübler-Jung, K. and Arendt, D. (1996). Enteropneusts and chordate evolution. Curr. Biol. *6*, 352–353.

Pattee, H.H. (2001). The physics of symbols: bridging the epistemic cut. Biosystems *60*, 5–21.

Pelleymounter, M.A., Cullen, M.J., Baker, M.B., Hecht, R., Winters, D., Boone, T., et al. (1995). Effects of the obese gene product on body weight regulation in ob/ob mice. Science *269*, 540–543.

Perovic, S., Krasko, A., Prokic, I., Müller, I.M. and Müller, W.E.G. (1999). Origin of neuronal-like receptors in Metazoa: cloning of a metabotropic glutamate/GABA-like receptor from the marine sponge *Geodia cydonium*. Cell Tissue Res. *296*, 395–404.

Pitkanen, M. Oct. 28, 1998. Binary structures as novelty detectors? Biol Outreach Group.

Plickert, G. (1989). Proportion-altering factor (PAF) stimulates nerve cell formation in *Hydractinia echinata*. Cell Differ. Dev. *26*, 19–27.

Plickert, G., Schetter, E., Verhey-Van-Wijk, N., Schlossherr, J., Steinbüchel, M. and Gajewski, M. (2003). The role of α-amidated neuropeptides in hydroid development—LWamides and metamorphosis in *Hydractinia echinata*. Int. J. Dev. Biol. *47*, 439–450.

Prosser, C.L. and Greenough, W.T. (1991). Central nervous system: section G. In: *Neural and Integrative Animal Physiology*, C.L. Prosser (ed.), fourth edition. Wiley-Liss, pp. 639–678.

Purves, W.K., Orians, G.H. and Heller, H.C. (1992). *Life: The Science of Biology*, third edition. Sinauer Associates, Inc.–W.H. Freeman and Company, pp. 487–488.

Raff, R.A. (1996). *The Shape of Life—Genes, Development, and Evolution of Animal Form*. The University of Chicago Press, Chicago, IL, pp. 87–88.

Rezgaoui, M., Süsens, U., Ignatov, A., Gelderblom, M., Glassmeier, G., Franke, I., et al. (2006). The neuropeptide head activator is a high-affinity ligand for the orphan G-protein-coupled receptor GPR37. J. Cell Sci. *119*, 542–549.

Rosa-Molinar, E., Krumlauf, R.K. and Pritz, M.B. (2005). Hindbrain development and evolution: past, present and future. Brain Behav. Evol. *66*, 219–221.

Ruppert, E.E. and Barnes, R.D. (1994). *Invertebrate Zoology*. Saunders College Publishing, pp. 24, 91.

Sakarya, O., Armstrong, K.A., Adamska, M., Adamski, M., Wang, I.F., Tidor, B., et al. (2007). A post-synaptic scaffold at the origin of the animal kingdom. PLoS One *2*(6), p. e506.

Schaller, H.C., Hermans-Borgmeyer, I. and Hoffmeister, S.A. (1996). Neuronal control of development in *Hydra*. Int. J. Dev. Biol. *40*, 339–344.

Schmich, J., Trepel, S. and Leitz, T. (1998). The role of GLWamides in metamorphosis of *Hydractinia echinata*. Dev. Genes Evol. *208*, 267–273.

Schmid, V. and Plickert, G. (1990). The proportion altering factor (PAF) and the *in vitro* trans-differentiation of isolated striated muscle of jellyfish into nerve cells. Differentiation *44*, 95–102.

Schütze, J., Krasko, A., Custodio, M.-R., Efremova, S.M., Müller, I.M. and Müller, W.E.G. (1999). Evolutionary relationships of Metazoa within the eukaryotes based on molecular data from Porifera. Proc. Biol. Sci. *266*, 63–73.

Sherrington, C.S. (1951). *Man on His Nature*, second edition. Cambridge University Press, Quoted by Glade, N. (2008). Computing with the cytoskeleton: a problem of scale. Int. J. Unconvent. Comput. *4*, 33–44.

Sonneborn, T.M. (1950). *Paramecium* in modern biology. Bios *21*, 31–43.

Sonneborn, T.M. (1964). The differentiation of cells. Proc. Natl. Acad. Sci. U.S.A. *51*, 915–929.

Sonneborn, T.M. (1970). Gene action in development. Proc. Biol. Sci. *176*, 347–366.

Technau, U., Miller, M.A., Bridge, D. and Steele, R.E. (2003). Arrested apoptosis of nurse cells during *Hydra* oogenesis and embryogenesis. Dev. Biol. *260*, 191–206.

Valentine, J.W. (2003). Architectures of biological complexity. Integr. Comp. Biol. *43*, 99–103.

Veelaert, D., Schoofs, L. and De Loof, A. (1998). Peptidergic control of the corpus cardiacum–corpora allata complex of locusts. Int. Rev. Cytol. *182*, 249–302.

West-Eberhard, M.J. (2005). Developmental plasticity and the origin of species differences. Proc. Natl. Acad. Sci. U.S.A. *102*(Suppl.), 6543–6549.

Wray, G.A. (2003). Transcriptional regulation and the evolution of development. Int. J. Dev. Biol. *47*, 675–684.

Part IV

Epigenetics of Metazoan Evolution

13 Origins of Evolutionary Novelty

We now understand that animal form is not so much a matter of the genes an animal has, but how they're used during the process of development, of going from a single egg to the complete mature animal, the building of all of its body parts.

Carroll (2006)

The Epigenetic Hypothesis of Evolution and its Predictions

The modern neo-Darwinian theory has not succeeded in its long attempt to account for the mechanism of the emergence of evolutionary change. In *Neural Control of Development* (2004), I put forward an alternative epigenetic theory of heredity of metazoans. Almost all predictions of that theory have been validated and substantiated by adequate evidence presented in the first part of this book (Chapters 1–6).

Substantial research I conducted on relevant observational and experimental evidence of evolutionary change encouraged me to extend and apply the epigenetic theory of heredity in developing an epigenetic hypothesis of metazoan evolution, which essentially posits that evolution of metazoan morphology, behavior, and life history is the result of heritable epigenetic changes in developmental pathways.

These epigenetic changes are essentially different from genetic changes (gene mutations) in an essential aspect. Genetic changes unavoidably and randomly arise from errors occurring in the process of DNA replication. In contrast, epigenetic changes in developmental pathways are anything but random. They arise as solutions to the problems of adaptation of the organism to the changed environment. These solutions result from the computational activity of neural circuits that perform the processing of the external/internal stimuli. This is suggested by three general facts:

First, unlike gene mutations, which are randomly occurring solitary events affecting individuals in population independently of the changes in environment, epigenetically determined changes (intragenerational developmental plasticity, transgenerational developmental plasticity, and evolutionary changes) are responses of the organism to the changed environment that affect many, most, or even all individuals in a population.

Second, epigenetic changes in developmental pathways are the result of epigenetic changes in neural signals (or their spatiotemporal pattern of secretion) that activate developmental pathways.

Third, and in clear distinction from gene mutations that are mostly deleterious, epigenetic changes, as a rule, are adaptive.

In anticipation of the criticism of possible teleological implications of the concept, I will remind the reader that it excludes any teleological agent in the sense that while the change is triggered by neural computation it has nothing to do with the conventional meaning of consciousness, for most of neural computation is not related to consciousness.

Epigenetic Principles of Evolution. DOI: 10.1016/B978-0-12-415831-3.00013-6

The main predictions of the epigenetic theory of evolution developed in this work are:

1. Evolutionary changes in phenotype arise during the individual development as a result of epigenetic changes in developmental pathways occurring without, or independently of, changes in genes.
2. Evolutionary loss of phenotypic traits is epigenetically determined and occurs without loss of genes.
3. Evolution of animal phenotype (morphology, physiology, behavior, and life histories) is reversible.
4. Reversion to ancestral phenotypes is epigenetically determined and is not related to reversion of any lost genes.
5. Sympatric speciation is possible because reproductive isolation can prezygotically evolve by neuro-cognitive mechanisms, without geographic isolation of populations.

The remaining part (Chapters 13–19) of this work is devoted to verification and substantiation of the above predictions of this epigenetic hypothesis of evolution.

The traditional taxonomic classification, determination of nodes in cladograms and cladogenesis in general, whose study has contributed so much to establishing relationships between taxa and elucidating patterns of evolution, is out of the scope of this work. My causal inquiry will overextend at the expense of these descriptive aspects of metazoan evolution. I will focus on the developmental mechanisms of evolutionary change by tracing back the development of these traits as they occur in the course of ontogeny, from the early development to adulthood.

To the extent that the present knowledge allows it, I will consider the evolution of metazoan morphology, behavior, and life histories in concrete examples, focusing on the changes in developmental mechanisms that made these changes possible. There is no alternate way for an evolutionary change in metazoan morphology to arise but through a specific change in a developmental pathway. Hence, what controls changes in developmental pathways in metazoans also controls their evolution.

To enable the reader to assess the explanatory power of the neo-Darwinian and epigenetic paradigms, I will use a comparative approach: the neo-Darwinian explanation of the evolutionary change will be followed by an epigenetic rationalization of the concrete examples of evolutionary change.

The Nature of the Evolutionary Change

Far from an ideal world of a stable, unchanging environment, our planet is a dynamic, ever-changing system. Hence, evolutionary pressure for evolving not only intragenerational phenotypic plasticity but also evolutionary changes, inherited adaptations to change living conditions, would have risen as soon as living systems emerged.

For a long time, biologists have focused on the process of selection of evolutionary changes rather than on their origin and the mechanisms determining their emergence. Natural selection acts on, and even exists for the sake of, evolutionary changes. No selection will occur before the evolutionary change arises. In the beginning, there was evolutionary change. Selection follows it:

> *All novel adaptive phenotypes must originate before they can be molded by selection, and they need not be altered under selection to be adaptive.*
>
> *West-Eberhard (2003)*

The fact that the science of biology focused for such a long time on selection rather than on generation of evolutionary changes that must be selected is anything but surprising. Only in the past few decades have the substantial breakthroughs been made in the molecular mechanisms of individual development, the ontogeny at a deep causal level, where all the evolutionary changes originate.

In their perpetual effort to heritably adapt to the changing environment, metazoans evolved numerous novelties (*sensu* West-Eberhard, "a trait that is new in composition or context of expression relative to established ancestral traits" (West-Eberhard, 2003, p. 198)) and modifications of their parts, traits, and organs. Most of the new evolutionary adaptations arise in the form of modifications of existing structures (Bock and Wahlert, 1998). The idea that most evolutionary novelties are modified versions of older structures (Campbell et al., 1999) still remains almost unchallenged. *De novo* adaptations are relatively rare phenomena in the living world.

By the early 1980s, Raff and Kaufman expressed the revolutionary idea that evolution by DNA mutations "is largely uncoupled from morphological evolution" (Raff and Kaufman, 1983). Indeed, no example has ever been presented of a gene mutation leading to an *adaptive* novelty in metazoans, although gene mutations unavoidably occur and are selected over time. But if there is no genetic information involved in the emergence of evolutionary novelties, then the only remaining alternative is to assume that some form of epigenetic information (in the broad meaning of the word) is responsible for that change.

The discovery of the transgenerational inheritance of methylated DNA by Holliday and Pugh (1975) and Riggs (1975) provided the impetus for a revival of the interest in epigenetic mechanisms as source of evolutionary changes. For obvious reasons, this interest was initially focused on epigenetic changes in DNA and chromatin (Holliday, 1987; Jablonka and Lamb, 1989; Strahl and Allis, 2000; Jaenisch and Bird, 2003, etc.). While this basically chromosomecentric approach to epigenetics made tremendous progress in laboratory research, a holistic approach seems to have been more fruitful in regard to demonstrating the epigenetic origin of changes in developmental pathways that lead to evolutionary changes (West-Eberhard, 1986, 1989, 2003; Matsuda, 1987; Schlichting and Pigliucci, 1998; Cabej, 1999a, b, c, 2001, 2004, 2008, 2011; Newman and Müller, 2000; Pigliucci, 2001; Ginsburg and Jablonka, 2010, etc.).

In the first part of this work, I have presented adequate empirical evidence that the signals for activating developmental pathways determining the normal development of animal tissues, organs, and morphology in general come from the central nervous system (CNS). In Chapter 10, I argued that intragenerational discrete changes in animal morphology, physiology, behavior, and life history also involve no changes in genes or genetic information but result from signal cascades starting in the nervous system in response to specific external/internal stimuli. The epigenetic information, generated by processing these stimuli in neural circuits, flows from the nervous system to the target organs/parts of the organism. In Chapter 11, I have also presented adequate observational and experimental evidence that inherited changes in morphology, i.e., transgenerational developmental plasticity, which, in principle, is indistinguishable from evolutionary changes, also involves no changes in genes. In all the considered cases of transgenerational developmental plasticity, the CNS functions as generator of the *de novo* epigenetic information necessary for inducing inherited phenotypic changes.

Interactions Organism–Environment in Evolution: The Causal Relationship

Unlike anorganic systems, which adequately *react* to external factors, living systems *respond adaptively* to external influences. This is the reason why external influences on biological systems are known as *stimuli* (from Latin *stimulus*—goad). All the cases of stress response, intragenerational and transgenerational developmental plasticity, and evolutionary changes in general are intended to adapt the organism's phenotype to the external change so that it avoids its harmful effects. There is an inherent intentionality in these responses; it is certainly unconscious but "in the best of their interest" of living systems.

The adaptive response to external stimuli is a novel, essential, and unique feature of living systems in general, unknown in anorganic systems. It tends to counteract, avoid, overcome, or compensate for the harmful effects of the action of external agents. The considerable degree of the free will that metazoans, and living systems in general, exhibit in their adaptive reactions to environmental influences is the reason why their responses cannot be fully predicted by the nature or intensity of the external influences.

For at least two centuries, the causal basis of evolution has been a major topic of theoretical biology. While the role of environment in evolution has been recognized from all the parties involved in the debate, disagreements have always arisen, and still arise, on the issue of the relative role of organisms and the environment in metazoan evolution. All the different views on the issue may roughly be grouped in two opposing classes. On the one hand, there is a school of thought holding that the environment is the driving force of the evolutionary change in living systems, and on the other, those believing that it is the living system itself, its inherent properties, that determines the evolutionary change, its nature and extent, with the environmental factors representing conditions that may favor the evolutionary change or not.

The issue is fundamentally related to the problem of causation in biology. The concept of the final cause (*causa finalis* from Latin *causa*—cause and *finis*—purpose) was introduced by Aristotle (384–322 BC) in conformity with his idea of the rule of order and finality in nature. He believed that the aim of biological studies is

> *not merely to discover facts (that things are so), but reveal causes (how and why they are so), and in particular reveal the final causes and the absence of chance in the works of nature.*
>
> *Lloyd (1999)*

Aristotle believed that *causa finalis*, the "end" of the change, could be found within the end-product of the goal-oriented action of the *causa finalis* and that "natural objects have their 'ends' within themselves" (Lloyd, 1999, p. 64). For Aristotle, the purposes and causes of things are in things themselves. Accordingly, the end of the evolutionary change is the organism itself. Here I use the concept of the "end" *sensu* Mayr, implying that, although there is no evidence for teleology, "there is no conflict between the causality and teleonomy" (Mayr, 1961).

Aristotle's concept implies that intrinsic, not external, causes determine the change. Darwin introduced the idea of the end of evolution as a product of natural selection, but even he was skeptical of the idea that it is the environment alone that determines evolutionary changes:

> *Naturalists continually refer to external conditions, such as climate, food, &c., as the only possible cause of variation. In one very limited sense, as we shall hereafter see, this may be true; but it is preposterous to attribute to mere external conditions, the structure, for instance, of the woodpecker, with its feet, tail, beak, and tongue, so admirably adapted to catch insects under the bark of trees.*
>
> *Darwin (1859)*

One important implication of the Aristotelian concept of *causae finales* is that if the study of the developmental mechanism of formation of a new phenotypic trait would lead to a complete knowledge of that mechanism, it would automatically reveal the *causa finalis*. Numerous examples on the developmental plasticity reviewed by West-Eberhard (2003) and others, as well as additional examples to be presented in this work, especially several cases of transgenerational developmental plasticity, seem to prove that *causa finalis* is not an external cause but an information-generating process taking place within the animal organism:

> *The self-organized ontogenesis of brain structures constitutes a natural language, and all evolution had to do is use this language to write the particular text that defines us.*
>
> *von der Malsburg (2002)*

In his attempt to apply Aristotelian principles of proximate and ultimate (final) causes to the neo-Darwinian doctrine, by the 1960s Ernst Mayr came to the conclusion that the causal chains of any evolutionary change start in the distant past in the natural selection, which shaped all the genetic programs, whose change makes evolution of living beings possible.

Identification of *causae finales* in biology makes sense as far as they are helpful to understanding the underlying mechanisms of evolutionary change. Natural selection has always been used, even by Darwin himself (see Introduction of this work), ambiguously, sometimes implying the cause, and sometimes the effect of the evolutionary change. In assessing the problem of the causation and natural selection in evolution, however, it is necessary to avoid the ambiguity.

If natural selection will be conceived as a *result* of the struggle for life, of differential reproductive success, or as a result of the organism–environment interaction in a broad sense of the word, by definition it cannot be a cause, let alone the *causa finalis*. If we, after Mayr, were to identify the *causa finalis* with the principle of natural selection, a principle that is external to the living organism, we would hardly add anything to our modest knowledge on the mechanisms of evolutionary change and to the Darwinian principle of natural selection. Any attempt to reconstruct the causal chain (the numerous untraceable steps of the action of natural selection over generations on the structure and functions of animals) of the evolutionary change is doomed to failure.

However, Mayr implies that the action of natural selection as a causal agent is materialized in the structure and functions of the animal. If this interpretation is correct, if the structure of metazoans embodies causes that acted on metazoans in the past, then developmental mechanisms that determine formation of these structures during the ontogeny may give us crucial clues on the mechanisms of evolutionary changes that occurred in the course of phylogeny. Hence, the study of developmental mechanisms could lead to identification of mechanisms that determined particular evolutionary changes, which, beclouded in the evolutionary past, would otherwise be unidentifiable.

Understood as a means for preserving the "useful" and eliminating the disadvantageous inherited changes, natural selection determines the survival or extinction of a species, but it tells nothing about the deeper cause, about "why" and "how" the internally determined evolutionary change emerges. And the real problem in biology is why and how evolutionary changes are produced rather than how the changes are accumulated or eliminated under the action of natural selection.

Using natural selection as an off-the-shelf answer for explaining any particular evolutionary change adds nothing to our still-modest knowledge on the cause and mechanisms of evolution. Natural selection is not, and cannot be, a substitute for empirical identification of the mechanisms of evolutionary change. Restriction of the scientific inquiry to the action of natural selection would prevent investigation of the underlying causes of the evolutionary changes, of the "raw material" on which natural selection has to act.

It is commonplace in modern biology to speak, sometimes in a Lyssenkoan style, of environmental stimuli (e.g., changes in photoperiod, temperature, humidity, social environment, and intraspecific and interspecific competition) as being in "control" of, or even "regulating," various structures and functions of the organism. This concept implies recognition of the metazoan organism as a passive entity, destined to "obey" "instructions" from external agents. It implies that the environment possesses, and can transmit to metazoan organisms, information on the morphological changes they must accomplish to adapt to current changes in the environment. It is not difficult to show that metazoans are neither "regulated" nor "controlled" by environmental stimuli, by the environment in general. It is easy to prove the opposite, i.e., that metazoan organisms are not "changed by external environment" but rather *respond* to changes in the environment, usually adaptively and antientropically, by changing their behavior, physiology, and morphology at both the developmental and evolutionary levels.

Unlike gene mutations, epigenetic changes are *adaptive responses* to the changed environment, not random changes. Solid evidence from nature and from experiments shows that living organisms can change, sometimes suddenly, developmental pathways and produce adaptive phenotypic changes without changes in genes.

From the neo-Darwinian point of view, *ultimate causes* of evolutionary change have been selective processes that occurred in the past.

> *Where, then, is it legitimate to speak of purpose and purposiveness in nature, and where it is not? To this question we can now give a firm and unambiguous answer. An individual who—to use the language of the computer—has been "programmed" can act purposefully. Historical processes, however, cannot act purposefully. A bird that starts its migration, an insect that selects its host plant, an animal that avoids a predator, a male that displays to a female—they all act purposefully because they have been programmed to do so.*
>
> *Mayr (1961)*

In relation to the cause of bird migration, he adds:

> *The lack of food during the winter and the genetic disposition of the bird, are the ultimate causes. These are causes that have a history and that have been incorporated into the system through many thousands of generations of natural selection. It is evident that the functional biologist would be concerned with analysis of proximate causes, while the evolutionary biologist would be concerned with analysis of ultimate causes.*
>
> *Mayr (1961)*

Migration of birds, like migration of all metazoans, invertebrate and vertebrate species, is an innate behavior. Now, almost half a century since Mayr's publication, we know that, while one or many genes are involved in expression of behaviors, all the attempts to prove that specific gene(s) may be responsible (i.e., both necessary and sufficient) for any behavior have failed (see Animal Behavior Is Not Determined by Genes in Chapter 8). Hence, the neo-Darwinian point of view on accumulation of "useful mutations" in the DNA that leads to new programs and characters is, at best, far from being proven. To the contrary, adequate empirical evidence from the field of developmental plasticity, especially transgenerational plasticity, shows that changes in developmental pathways/programs often occur within one, two, or several generations and affect all the individuals of a population, a fact that excludes gene mutations, existing genetic variability, and the related action of natural selection in the evolution of developmental programs. Adequate empirical evidence also shows that changes in genes and accumulation of "useful gene mutations" have not been necessary for evolution of characters and speciation in metazoans.

In the case of migrating birds, environmental stimuli *per se* provide instructions neither for starting migration nor for the course that the birds must follow during the flight to the migration site. It is the hypothalamic clock that determines the timing of migration based on the environmental cues (which represent only environmental data from which the hypothalamus generates the information for timing the migration) and there are other neural circuits that determine the position of the bird during the flight and the itinerary, based on the processing of other sensory cues (e.g., visual, acoustic, geomagnetic) and turn these cues into instructions (=epigenetic information) for staying or correcting the course of migration. Thus, the *causa finalis* of the bird migration is in the processing activity of migration neural circuits, activation of identified complex neural circuits in the birds' brain (Beason, 2005) rather than any imaginary accumulation of "useful gene mutations" and their selection over generations.

As for the proposition that the *causae finales* may be embodied in the genetic programs that living beings have acquired during their evolution, one should be reminded that all attempts to prove the existence of *genetic programs* have failed. But even if, for the sake of argument, the genome were to embody a "genetic program," then it would be expected that somatic cells, which contain the same genome, the full set of species-specific genes, could develop into an adult organism of their kind. As we all know, somatic cells fail to do so, and only the zygote (and the egg cell in parthenogenetic metazoans), which contains the same set of genes, succeeds in developing into an adult organism. This fact proves two things:

> First, that the zygote and the egg cell of parthenogenetic organisms contains a *developmental program* that enables it to enter the process of individual development.
> Second, that the genome contains no "genetic program" for individual development. Indeed, for more than three decades, we have known that the program determining the development of the egg/zygote during the early stage of embryonic development is an *epigenetic program*, consisting of mRNAs, proteins, hormones, neurotransmitters, nutrients, and other materials arranged in the egg cell/zygote in a strictly determined spatial order. The lack of this epigenetic program in somatic cells, which have all the genetic material the egg/zygote possesses, is why somatic cells fail to enter the individual development and develop into an adult organism. With the benefit of the knowledge accumulated in about half a century since Mayr's publication, we know that all behaviors, such as migration of birds, are determined not by DNA, by any genes, or by any "genetic program," but by information generated by a nongenetic, computational integration and processing of internal/external stimuli in neural circuits in the CNS (see Section Neural Mechanisms of Metazoan Migration in this chapter).

In anticipation of the neo-Darwinian counterargument that the activity of the neural circuits themselves is determined by genes, recall that the properties of neural circuits, and behaviors they determine, can and do change, not only in the course of evolution, but even within the life of an individual, what excludes changes in genes.

By definition, natural selection is the second stage in the process of the evolutionary change: the inherited, evolutionarily relevant change must occur (by changing a developmental pathway or by a gene mutation) before the natural selection can act on it. It exists not for its own sake but for the sake of the evolutionary change. There can be no selection without change: In the beginning, there is the change.

The change in the developmental pathways precedes the action of natural selection and the generation of the evolutionary change is the first event of the causal chain. The ultimate cause in a causal chain is the first link in the chain, which is the emergence of the evolutionary change, not the ensuing selection of the change. Natural selection, although necessary, is always a *post factum* process in relation to the emergence of the evolutionary change. Hence, it is logically inaccurate to believe that "many thousands of generations of natural selection" have been the *causa finalis* of actual phenotypes in metazoans; the inherited change is.

But any evolutionary change at a supracellular level is always a manifestation of a corresponding change in the developmental pathway that takes place during ontogeny. This suggests that the study of the ontogeny of the extant organisms can give us important clues as to the ultimate causes that might have acted long before in the course of the species phylogeny. For, as a rule, the developmental pathway that brought about the evolutionary change is still present and operational, even though often blurred, in the ontogeny of the bearers, as long as they express the evolved character.

The species ontogeny, thus, is a chronicle of the evolutionary history, often written in undeciphered epigenetic codes. It may reasonably be argued that the present ontogeny might not exactly reflect the original change in the developmental pathway. By analogy, identification in the ontogeny of the initial change in the evolution of a structure presents to the biologist the difficulties that a linguist encounters in attempts to reconstruct the ancient root of a modern word. However, comparative developmental studies give sufficient clues to make that identification possible, just as comparative linguistics in etymological studies does.

Obviously, natural selection is not involved in the emergence of the new/modified developmental pathway. What natural selection can do, in exact terms, is to eliminate carriers of the change if less fit under existing conditions of the environment.

Vast empirical evidence (part of which is included in this work) shows that living organisms, in sharp contrast with anorganic systems, *not simply react but adaptively respond* to environmental agents that disturb their homeostasis, by adaptively changing their behavior, physiology, morphology, and life histories. These responses are teleonomic responses (*sensu* Mayr) enabled by the existence in living systems of control systems (see Chapter 1) that actively serve to restore the normal function and structure, their homeostasis, when they are disturbed by adverse actions of environmental agents. Living systems are actively adapting machines rather than passive receivers of external influences.

Adaptive responses are manifestations of the ultimate cause, of an intrinsic drive, which intends to restore the disturbed homeostasis, by counteracting, resisting, or avoiding harmful effects of the changes in environment. The *causa finalis* is realized and embodied in the end-product of evolution, in the adaptively changed animal phenotype.

Crucial in evolving new characters in biological systems is not the matter or free energy, rather than the source of information, "instructions," on how to produce the evolutionary change. Any evolutionary change requires and implies change (acquisition, but sometimes loss) in the informational content of the system. Does the environment possess that information, and can it provide metazoans with specific information to adaptively change their behavior, morphology, or physiology? The answer is clearly no. Metazoans themselves can generate and use the information necessary for erecting and evolving their structure.

The neo-Darwinian distinction between the so-called functional biology (developmental biology) and causal biology (evolutionary biology), with the first answering the question *how*, and the latter the question *why*, is artificial and erroneous. That distinction is based on the wrong premise that *causa finalis* of evolutionary changes is in the environment and natural selection. (See earlier in this section.) The artificial separation of biology into causal and functional has played a negative heuristic role. As West-Eberhard has pointed out:

> *The answer to "why" an organism is, or behaves in a certain way can be answered either in terms of mechanisms (proximate causes) or in terms of selection and evolution (termed "ultimate causes" by Mayr, 1961). This distinction is designed to prevent confusion between levels of explanation in biology. But it was an easy step from this important point to the idea that the mechanisms of development have nothing directly to do with evolution or that they are the focus of a different research approach, one not primarily concerned with evolution and justifiably left aside by those primarily interested in selection and adaptation.*
>
> *West-Eberhard (2003, p. 198)*

There is no other discipline in biology where both the "hows" and "whys" of biological processes unfold so clearly as in developmental biology. The study of individual development could provide answers to many fundamental hows and whys of biological phenomena. There is no other way to understand *how* a structure develops, but by identifying the developmental mechanism that leads to the development of that structure, and there is no other way to understand *why* that structure develops, but by identifying the function it performs.

Behavioral Prelude of Evolutionary Modifications of Animal Morphology

As a rule, *any evolutionary change* triggered by drastic changes in environment, whether it is a modification, a loss of a feature, a reversal to an ancestral feature, or a *de novo* morphology, *begins with, or is preceded by, an adaptive change in the behavior of animals*. This is related to the fact that behavior is the most plastic of all the phenotypic traits in metazoans. It is the environmentally imposed change in behavior and the accompanying stress state that sets the stage for the morphological adaptations. According to the "behavior evolves first" hypothesis, all the major morphological changes in evolution are preceded by changes in the behavior of animals, in response to changes in conditions of living. The idea is not new and has been attractive to a majority of biologists during the past two centuries. Charles Darwin unambiguously pointed out the role of behavioral changes in evolution:

> *Habit also has a decided influence, as in the period of flowering with plants when transported from one climate to another. In animals it has a more marked effect; for instance, I find in the domestic duck that the bones of the wing weigh less and the bones of the leg more, in proportion to the whole skeleton, than do the same bones in the wild-duck; and I presume that this change may be safely attributed to the domestic duck flying much less, and walking more, than its wild parent.*
>
> *Darwin (1859, p. 11)*

To survive in adversely changing environments, animals must adaptively change their behavior. Adaptive behavioral changes represent extemporaneous responses necessary for their survival, but they can also help the species for "buying" the time necessary for evolving corresponding heritable changes in morphology, physiology, and life history.

Normally, under drastically changed conditions of living, animals *learn* to behave differently, adaptively. The animals' learning is aided by the fact that in learning new behaviors, animals quite often use preexisting fixed action patterns (FAPs) and motor patternings. This logically raises a question: Is it possible for an animal to be in possession of such an apparently large number of FAPs and motor patternings it might need under different conditions of an ever-changing environment? The difficulty is obvious but not as insurmountable as it may look at first sight, if one bears in mind that:

First, the survival in a severely changed environment does not necessarily require a perfect behavioral adaptation from the beginning; once the changed behavior makes the survival possible, perfection of the behavior may come later.

Second, radical changes in environment imply contrasting conditions of living, such as aquatic-terrestrial environment, cold–warm weather, low–high degree of illumination, abundance–scarcity of food, presence–absence of predators, or herbivorous–carnivorous diet, which are not as numerous as their intermediate states would be.

Third, there is evidence that the same circuitry for a FAP or motor pattern can often be modified to serve more than a single "purpose," without changes in genes. It is also experimentally demonstrated that the same neuronal circuit may produce different behavior patterns in response to application of different stimuli or hormones.

Fourth, animals can adaptively modify their behavior by switching to new patterns of connections between the same neurons as occurs with neurons that control a lobster's eating and digestion.

Fifth, there is experimental evidence showing that the neural elements and connections for performing an ancestral behavior are conserved even in cases when the species has lost the specific part or organ used to perform the behavior. (See Chapter 8 for an expanded discussion.)

Epigenetic Determination of Modifications of Existing Phenotypes

Any animal structure has evolved for performing some function, which is the ultimate cause of the evolution of the structure. Under changed environmental conditions, metazoans may use an existing organ for performing a new behavior. It is likely that the first fish species that ventured to explore the *terra firma* might have used its fins to support and carry its body weight for locomotion on dry land. There is reason to believe that a switch from the innate swimming behavior to a "learned" shambling, which was facilitated by the existence in the fish brain of a highly plastic neural circuit for different forms of locomotion, including crawling, enabled the early aquatic gambler to survive until it succeeded in remodeling its fins into walking limbs. In all likelihood, the "learned" shambling preceded and facilitated the evolution of tetrapod limbs from swimming fins.

Arguably, new behaviors precede evolutionary modifications, and if the behavioral adaptation is indeed the first step in the process of evolution, it suggests that evolutionary changes start with a neural mechanism. How these neural mechanisms of behavioral adaptation to the changed environment are related to the ensuing specific evolutionary changes in morphology still represents one of the great enigmas of modern biology. Difficult as the understanding of that relationship is, it must be pointed out that examples of neural determination of morphology, and even of evolution of morphology and other phenotypic characters, are not lacking.

As mentioned earlier, it is generally believed that most evolutionary innovations arise by modification of existing structures and probably represent the most widespread mode of evolution of morphology in metazoans. *De novo* structures that are not homologous to any structure in ancestral species are rare events in metazoan evolution. Morphological novelties result from adaptive changes in developmental programs rather than gradual accumulation of useful gene mutations. On a large evolutionary scale, this is monumentally manifested in the "Cambrian explosion": about 540 mya, in the space of only several million years, an incredibly rapid evolution and morphological diversification of the animal world occurred, with the appearance of ~100 phyla, including all of ~30 modern phyla, characterized by distinct major body plans (Baupläne). Abundant paleontological evidence also suggests that new taxa appeared so suddenly that accumulation of extremely rare and randomly occurring and undemonstrated "useful mutations" does not come into account as a possible cause.

Numerous cases of polyphenisms, when individuals of the same genotype, in response to specific environmental stimuli, switch to alternative discrete morphologies, demonstrate the ability of animals *to change their morphology without changing their genotype* in an adaptive response to the changed conditions of life.

Metamorphosis, a widespread phenomenon in the animal kingdom, also demonstrates that no changes in genes are necessary even for radical morphological changes, affecting Baupläne of different classes of invertebrates and vertebrates. For example, premetamorphic anuran amphibians develop a fish Bauplan, and insects a worm Bauplan, before metamorphosing into the Baupläne of their own classes. It is interesting to observe that, during the larval stages, these species along the ancestral morphology also display ancestral behaviors related to that morphology, which suggests that basic neural circuits for performing these ancestral behaviors (fish and worm behaviors, respectively) are also conserved after the loss of ancestral structures.

Although evolutionary modifications of animal morphology are usually triggered by environmental stimuli, the mechanisms that determine modifications, and the source of information necessary for their development, are of intrinsic origin. As argued in Chapter 2, the external stimulus provides no information for the morphological change, as is suggested by the fact that the modification it triggers is not predictable from the nature of the external stimulus: in response to the same environmental stimulus, different organisms respond by developing different, and often opposite, phenotypic modifications.

How is the environmental stimulus related to the evolutionary modification of the morphology that it triggers? As pointed out earlier, there is reason to believe that the change in behavior provides a crucial link in the causal chain of events extending between the environmental stimulus and the evolutionary modification of the morphology it stimulates. This *epigenetic link* among environment, behavior, and morphology is visualized in many observations from nature and experiments. R. Denver, for example, has shown that as tadpoles, desert amphibians live in temporary ponds that contain water for unpredictable periods of time. In the years of low precipitations these ponds dry up earlier. The input of the external stimulus (earlier receding water in the pond) and internal stimuli of growth are integrated and processed in neural circuits in the tadpole's brain (hypothalamus). This leads to a neurally determined stress response (*habitat stress*) in tadpoles and a series of identified changes in the behavior of tadpoles. The epigenetic link between the behavior and morphology in the case of tadpole metamorphosis is provided by the output of the processing of sensory information in the limbic system, which stimulates secretion of the hypothalamic neurohormone corticotropin-releasing hormone (CRH), which is responsible for both the behavioral stress and inducing metamorphosis (Denver, 1997). Secretion of CRH, the principle vertebrate stress neurohormone, by the hypothalamus is the key element for the changed behavior, stress response, and speeding up the morphological transformation of the aquatic tadpole into an adult terrestrial amphibian organism.

Denver demonstrated that the same result of speeding up the transformation of the fish-like tadpole into an adult amphibian is also obtained in laboratory when the water level is experimentally lowered earlier. The lowering of the water level is a "drastic change in the environment," which causes a stress condition and the related changes in the behavior of tadpoles struggling to avoid the threat of drought:

> *The lowering of the level of the water in the environment makes the hypothalamus to produce more CRH, which stimulates pituitary to produce hormones that stimulate thyroid and adrenal glands whose products help organism to cope with the stress, in this case by losing their tail and beginning the growth of their limbs.*
>
> *Denver (1997)*

Evolution of Body Size in *Manduca sexta*

Attainment of species-specific body size in animals roughly coincides with the onset of sexual maturity. Mechanisms of control and regulation of body mass have only recently been thoroughly investigated. Different mechanisms seem to be operative at different steps of the evolutionary ladder. Studies carried out so far show that signal cascades for regulating this basic phenotypic character start in the nervous system.

Among invertebrates, several studies on regulation of body size have been carried out in insects such as *Drosophila* and *M. sexta* (Sphingidae). An experimental case of the evolutionary change in body weight is recorded under laboratory conditions on *M. sexta*. Within ~220 generations (30 years), the insect increased its body mass by 50% (D'Amico et al., 2001). Experimental studies have shown that the adult body size in *M. sexta* depends on a number of factors: the initial size of the last larval instar, the growth rate during that instar, the value of the critical weight, the time required for the clearance of juvenile hormone (JH) during the last instar, and the timing of the photoperiodic gate for secretion of the neuropeptide

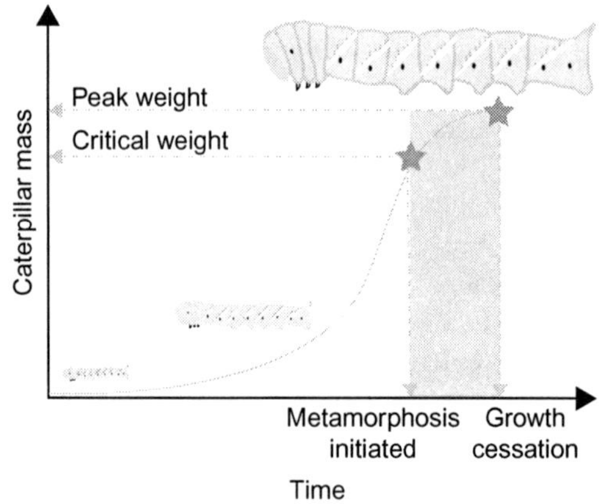

Figure 13.1 Factors that determine body size in *M. sexta*. In *Manduca*, peak larval weight depends on three parameters: the growth rate (the slope of the curve), the weight at which metamorphosis is initiated (the critical weight), and the length of time between attainment of critical weight and the large ecdysone pulse that terminates feeding and growth (shaded yellow). (For interpretation of the references to color in this figure legend, the reader is referred to the web version of this book.)
Source: From Parker and Johnston (2006).

prothoracicotropic hormone (PTTH) (Figure 13.1) by secretory neurons in the insect brain. Evolutionary changes in the three last factors (growth rate, critical weight, and secretion of PTTH) "almost completely account for the evolutionary increase in body size observed" (D'Amico et al., 2001; Davidowitz et al., 2003, 2004).

The difference between the adult body weight and the critical weight results from the fact that cessation of JH synthesis does not automatically lead to cessation of growth. Residual JH and JH mRNA still continue to stimulate growth and to prevent secretion of the neuropeptide PTTH, which stimulates secretion of ecdysone by the prothoracic gland, thus arresting the larval growth (Davidowitz et al., 2003, 2004). Besides, the larva, after achieving competence for PTTH synthesis, must wait for the "photoperiodic gate." The photoperiodic gate is 8 h long, after which one-third of larvae will become competent, while the rest of them have to wait until next photoperiodic gate opens (D'Amico et al., 2001).

Evolutionary changes in the growth rate, critical weight, and PTTH delay time are responsible for 95% of the evolutionary increase in body mass of *M. sexta* (D'Amico et al., 2001) (see also Section The Central Control of Body Mass in Chapter 1).

A closer look at the developmental mechanisms determining growth rate, critical weight, and PTTH delay time may shed some light on the origin and nature of information for the recorded evolutionary change in *M. sexta*.

The Growth Rate

A correlation exists between the patterns of secretion of insulin growth factors (IGFs) (insulin-like peptides, Ilps) and PTTH in the brains of insects and their growth rate during larval stages (Rulifson et al., 2002). The insulin/IGF signaling pathway is the mediator of the function of the CNS in determining the growth rate in *Drosophila melanogaster* (Colombani et al., 2005). Five Ilps genes, homologous to human and mouse insulins, are identified in *Drosophila* (Rulifson et al., 2002), and an insulin supergene family of 37 genes is identified in the nematode worm, *Cenorhabditis elegans* (Leevers, 2001). Although these genes are expressed in other tissues, it seems that their expression in the insulin-producing neurons in the CNS during larval stages is of determining importance for larval growth and development in *Drosophila*. These Ilps are synthesized and secreted by two bilateral clusters of neurons in the pars intercerebralis of the protocerebrum: ablation of these secretory neurons leads to production of flies of normal body proportions but of smaller size, with wing size reduced to 61% and wing cell number reduced to 72% of the normal condition.

Critical Weight

There is a moment during the last instar of *M. sexta* when secretion of JH suddenly stops, and the activity of JH esterase increases so that hemolymph is cleared of JH, thus making secretion of ecdysteroids by the prothoracic gland possible. Secretion of ecdysteroids is regulated by a brain signal, the neuropeptide PTTH, whose synthesis starts during the first photoperiodic gate after the clearance of JH, 2.5 h after onset of the photophase (Davidowitz et al., 2003). With secretion of ecdysteroids, the insect stops feeding and growing. This is also corroborated by the fact that ablation of neurons that secrete PTTH (and, consequently, stimulate ecdysteroid production) results in production of larger flies (McBrayer et al., 2007). Suppression of JH secretion and of JH esterase activity coincides with the time when the insect attains the critical body mass, which is about 55% of the adult body mass. This suggests that a causal relationship exists between the attainment of the critical body size and the cascade of events leading to metamorphosis.

> *The assessment of this critical body size is made in the brain and is based on processing of signals sent by the insect's stretch proprioceptive neurons that receive mechanical stimuli of increasing stretch as a result of increased body size.*
>
> *Gorbman and Davey (1991)*

By integrating and processing these stretch stimuli, neural circuits determine the time for sending signals (neuropeptides of the family of allatostatins) to corpora allata for suppressing the synthesis of JH.

PTTH Delay Time

The growth of the fifth instar larvae continues after the termination of JH secretion until ecdysteroids from the prothoracic gland are released in hemolymph. Although at the time that the larva suppresses JH secretion and it is competent for PTTH synthesis, it does not do so until the brain is stimulated by a photic cue, which is the first photophase after the interruption of JH secretion.

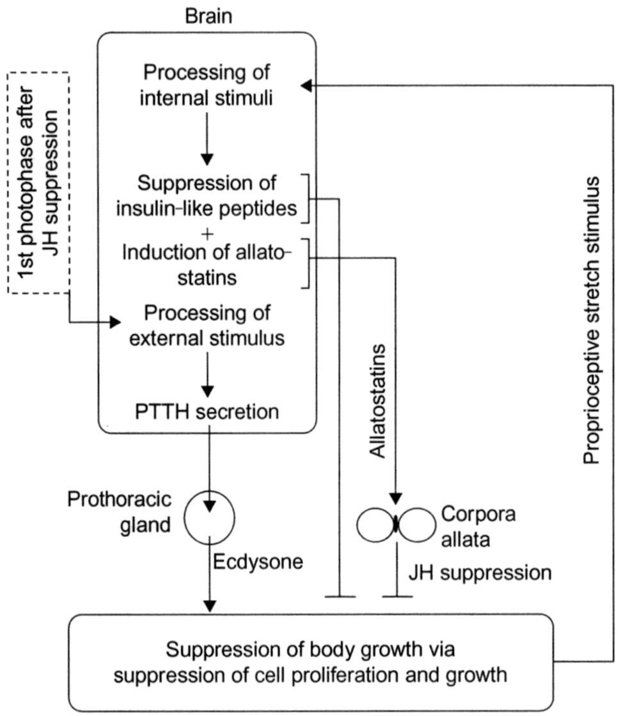

Figure 13.2 Generalized diagrammatic representation of the neural suppression of body growth in insects.

The fact that the brain starts secreting PTTH, not at any time of the day but only during the photoperiodic gate, creates a time span between the suppression of JH secretion and the commencement of PTTH secretion. This interval between the termination of JH secretion (its secretion is also terminated by brain signals) by corpora allata and the beginning of PTTH secretion is known as *PTTH delay time*. During this interval, larvae continue to grow until the PTTH secretion. Evidently, the length of this interval influences the adult body size that the insect attains at the onset of metamorphosis.

A generalized diagram of the mechanisms involved in determination of body size in insects is presented in Figure 13.2.

It is important to know what determines the timing of PTTH release in the insect's brain. As early as 80 years ago, Wigglesworth found that insect proprioceptors may receive and transmit to the brain input on increasing stretch of body growth, and processing of this neural input leads to secretion in the brain of PTTH, which in turn stimulates secretion of ecdysone by the prothoracic gland (Gorbman and Davey, 1991; West-Eberhard, 2003). This suggests that a set point for an upper limit of stretch must exist, beyond which the brain activates the ecdysone cascade and suppresses JH cascade, thus stopping further growth and determining the species-specific adult body size.

Based on the mechanosensory information sent by stretch proprioceptors, the CNS assesses the degree of stretch, which is matched against a neurally determined stretch set point. The existence of this set point is experimentally demonstrated to exist in both vertebrates (Adams et al., 2001) and invertebrates (Munyiri et al., 2004). Evidently, neurally changed stretch set points could lead to corresponding evolutionary changes in an animal's body size.

Summarizing: the evolutionary increase of the body size observed in *M. sexta* laboratory strain within 30 years is determined by epigenetic factors (the growth rate, critical weight, and PTTH delay time), all nongenetic phenomena determined by the computational activity of the insect's CNS. There was no genetic variation for plasticity of critical weight, and no indication exists of mutations having played any role in the evolutionary increase of the body weight of the laboratory strains of the tobacco hawkworm, *M. sexta*. Hence the conclusion:

> *Plasticity of body size must thus be due to plasticity in this underlying endocrine control mechanism.*
>
> *Davidowitz et al. (2004)*

Evolution of Wings in Insects

Although the earliest fossilized insects belong to the Devonian (up to ~410 mya), paleontological evidence shows that wings in insects evolved only ~300 mya. Ancestral insects evolved two pairs of wings. In the course of evolution, various taxa evolved morphological differences between the forewings, which remained fully membranous and flight-capable wings, and hindwings, which were reduced to halteres, morphological adaptations with flight-balancing function.

Two main hypotheses have been proposed for explaining the evolution of wings in insects. One sees the insect wings as *de novo* structures on the body wall, not related to preexisting structures, which is only endorsed by a limited number of biologists. The second one posits that wings in insect evolved by modification of gill-like epipodites of the triple-branched (endopod, exopod, and epipod) limbs of ancestral crustaceans (Wigglesworth, 1973; Averof and Cohen, 1997; Jokusch and Ober, 2004) (Figure 13.3).

The support for the crustacean origin of insect wings has been summarized as follows:

> *Some epipods resemble insect wings more closely than insect legs in the following ways: wg is expressed along the entire margin rather than being restricted to the ventral side, ap and nub are expressed in large domains, and Dll expression is restricted or absent. Significant differences in these expression patterns include ap expression throughout the epipod, whereas in wings it is confined to the dorsal compartment, and wg expression along the dorsoventral margin of Drosophila wings but the anteroposterior margin of branchiopod epipods.*
>
> *Jokusch and Ober (2004)*

Here we will only consider the second, the so-called wings-from-legs hypothesis on the origin of insect wings from ancestral crustacean branched limbs, which has

Figure 13.3 A lobster (*Homarus americanus*) uropod.
Source: From Fox (2006).

found wider support from developmental and phylogenetic studies but has been questioned recently (Niwa et al., 2010). But first, let us recapitulate the present knowledge on the ontogeny of insect wings.

Neural Control of Developmental Pathways and Gene Regulatory Networks of Insect Wings

If insect wings evolved by modification of ancestral crustacean limbs, then it is to be expected that during the ontogeny the same key regulatory genes (*en* (*engrailed*), *pdm* (nubbin), ap (*apterous*), and *Dll* (*Distalless*)) will be used in both groups. In *Drosophila*, the Apterous protein determines separation of dorsal and ventral compartments of the wing. This function of the Apterous is made possible by the activity of the Notch receptor, which, in turn, is activated by Fringe (Milan and Cohen, 2003). Anterior–posterior (A–P) patterning is similar in legs of crustaceans and wings of insects (Averof and Cohen, 1997).

Development of wings in insects involves activation of specific gene regulatory networks (GRNs), which have been conserved across holometabolous insects for 300 million years (Abouheif and Wray, 2002). Quite commonly, GRNs are considered separately from the systems to which they belong (Davidson et al., 2003) and from the upstream channels of communications through which the epigenetic information for activation of these GRNs flows. GRNs represent downstream networks of signal cascades that start with neural signals, which result from the computational processing of internal/external stimuli in the CNS. GRNs *per se*, as commonly studied and described in modern biological literature, are not self-regulated systems. Attributing to the part the properties of the whole (the signal cascade), in the case of GRNs, may be conceptually misleading.

In the case of the development of wings in insects, it is known that a neurohormonal control and regulation is superimposed onto the described wing GRN. The growth of wing disks in insect larvae depends on the level of nutrition, which is sensed not by wing disks themselves but by the insect's CNS. Three insulin-like neurohormones are secreted by seven secretory neurons in the central region of the *Drosophila*'s brain (Ikeya et al., 2002). These neurohormones act as growth factors for regulating the growth and development of wing imaginal discs. (Experimental ablation of the secretory neurons prevents the growth of wing imaginal discs.)

In experiments on *Precis coenia*, it is demonstrated that wing disks may be grown in tissue culture, in presence of the hormone ecdysone and hemolymph. The active principle of the hemolymph is a neurohormone, bombyxin, which also is an insulin-like neuropeptide produced by secretory neurons in the brain. By assessing the nutritional state of the body, the CNS determines the timing of secretion of the neurohormone bombyxin and the hormone ecdysone. (Secretion of the latter is also cerebrally regulated by the neurohormone PTTH.)

What is the evidence that the nutritional status of the organism is sensed by the CNS? Decades ago, it was observed that in some locusts brain neurosecretions double within 10 min to 1 h after the start of feeding, while their transport to corpora cardiaca (CC) doubled (Highnam and Mordue, 1974) or tripled (Friedel and Loughton, 1980). Administration of glucose in *Bombyx mori* also stimulates secretion of insulin-like neuropeptides by CC (Masumura et al., 2000). Neurons that secrete insulin-like neurohormones receive input on the nutritive status of the insect organism based on the changes in the level of glucose in hemolymph, which represents a reliable indicator of the nutrition status and food availability. Based on this assessment, secretory neurons determine whether, when, and how much of the insulin-like neurohormones to secrete (Masumura et al., 2000; Britton et al., 2002).

The CNS, thus, assesses the level of nutrition and, via bombyxin and ecdysone, at the right time, activates the wing development GRN, thus regulating the growth of wing imaginal disks according to the nutritional status.

> *It appears that the level of bombyxin in the hemolymph is modulated by the brain in response to variation in nutrition and is part of the mechanism that coordinates the growth of internal organs with overall somatic growth.*
>
> *Nijhout and Grunert (2002)*

The most widely held hypothesis on wing determination in insects is that the presence of JH above certain levels inhibits wing development (Zera and Denno, 1997). Indeed, topical application of JH III or methoprene (a JH analog) at various developmental stages of the cricket *Gryllus rubens* switches the insect from long- to short-winged morphology. Experiments conducted on the aphid *Aphis fabae* and the brown planthopper, *Nilaparvata lugens*, also demonstrate the role of the neurally controlled JH as inhibitor of wing development (Zera and Denno, 1997). The fact that the same brachypterizing effect is obtained under the influence of a social factor, the transfer from individual to group rearing (Zera and Tiebel, 1988), also corroborates the crucial involvement of a neural mechanism in the development of insect wings.

The Hox protein, Ubx (ultrabithorax), is necessary for specification of the third thoracic segment. It suppresses development of wings and promotes development of halteres by suppressing expression of *sal* (Galant et al., 2002). Evolutionary changes in the target genes of *Ubx* in *Drosophila* and the butterfly *P. coenia* (portions of the expression patterns of genes *DSRF* (*Drosophila serum response factor*), *AS-C*, and *wg*, which are repressed in *Drosophila* halteres, are expressed in the butterfly

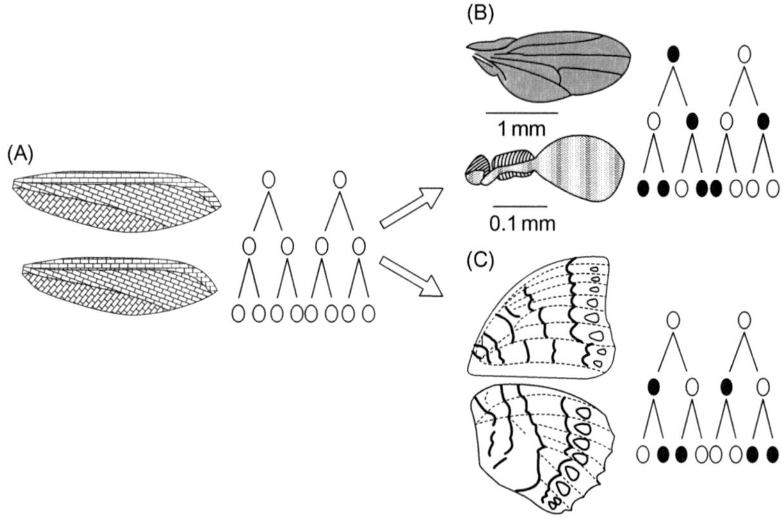

Figure 13.4 The evolution of insect hindwing patterns and the divergence of Ubx-regulated target gene sets. A schematized view of the course of the evolution of the dipteran (B) and lepidopteran (C) lineages from a common four-winged ancestor (A), which had similar forewings and hindwings. On the left of each panel are drawings of wing pairs, and on the right are schematics representing genetic regulatory hierarchies for wing development. In this scenario, Ubx, although expressed in the ancestral hindwing (A), did not yet regulate genes in the wing-patterning hierarchy to differentiate hindwing from forewing morphology. Subsequently, many genes (represented by black ovals) fell under the control of Ubx, and these sets of Ubx-regulated genes differed between the (B) dipteran (e.g., *wg*, *AS-C*, *SRF*) and (C) lepidopteran (e.g., *Dll*, scale morphology genes), and presumably other, insect lineages. *Source*: From Weatherbee et al. (1999).

hindwings) led to the different hindwing morphologies that they evolved ~200 mya when diverged from their common ancestor (Weatherbee et al., 1999; Figure 13.4).

Development of wings in insects takes place in the absence of expression of *Hox* genes. Their expression is prevented in the first thoracic segment (T1) by activation of *Scr* (*Sex combs reduced*). Wings can develop on the second thoracic segment (T2), for the only *Hox* gene expressed there has no effect on wing morphology. Expression of the Hox gene *Ubx* in the third thoracic segment (T3) prevents wing formation, while promoting formation of halteres. Expression of Abdominal A (abdA) in abdominal segments (A) prevents formation of wings in abdominal segments (Figure 13.5).

Ubx represses the expression of the *Vg* quadrant enhancer and two downstream targets of *Vg*, *salr* and *DSRF*, in the haltere, thus suppressing, wing formation and leading to formation of reduced "wings," halteres. Experimental loss of function of the *Ubx* leads to transformation of halteres into wings (Weatherbee et al., 1998). By increasing the level of one of its receptors, the Thickvein (Crickmore and Mann, 2007; Makhijani et al., 2007), Ubx determines the haltere size and shape.

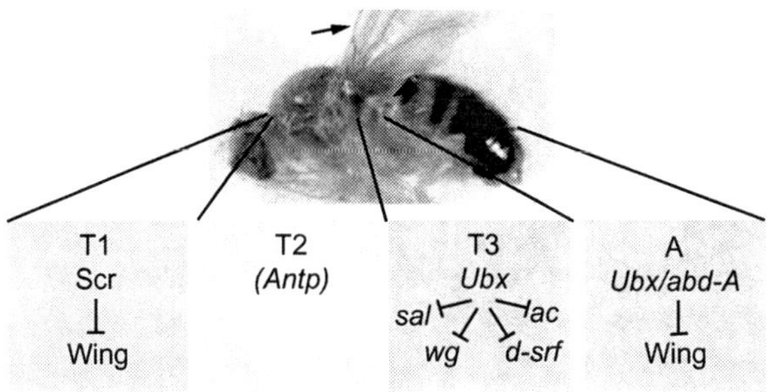

Figure 13.5 Function of *Hox* genes in fore- and hindwing differentiation in insects. A model for fore- and hindwing differentiation in *Drosophila*. Wing (arrow) and haltere (arrowhead) are indicated. *Abbreviations*: d-srf, *Drosophila* serum response factor; wg, wingless. *Source*: From Tomoyasu et al. (2005).

Thus, whether an insect or one of its segments develops wings or not depends on the patterns of expression of existing genes rather than evolution of changes in genes or new genes. What, then, controls the spatial and temporal patterns of expression of wing genes and development of wings in insects? Two lines of evidence show that formation of wings in *Drosophila* is under neural control via ecdysone pathway.

First, at the onset of metamorphosis, ecdysone binds its nuclear receptor ecdysone receptor (EcR), which forms a heterodimer with another nuclear receptor, ultraspiracle (Usp), and in this form it regulates expression of early response genes (six of them known so far, among which *BR-C* (*Broad-Complex*) and *E74*), whose products are transcription factors for inducing late response or effector genes, which determine specific effects of ecdysone on genes in various tissues, including wing disks. Under action of this ecdysone-triggered cascade, the wing imaginal disc evaginates or unfolds to form the wing. Four hundred sixty-eight genes (Li and White, 2003), or 3.4% of the total of 13,600 genes of the *Drosophila* genome (Adams, 2000), are expressed in wing disks.

This set of genes was "remarkably distinct" from the sets of genes induced by ecdysone in other organs and tissues, and 289 genes (~2.1%), among which the *hedgehog*, *Notch*, *EGF*, *dpp*, *vestigial*, and *wingless*, were specifically expressed in the wing disks (Li and White, 2003). Expression of integrins in the wing disks is also induced by EcR, thus making possible cell adhesion (D'Avino and Thummel, 2000). This means that almost all the genes of the second tier, which induce genes involved in wing formation, are induced by ecdysone. Needless to say, the synthesis and secretion of ecdysone is regulated by the neurohormone PTTH released by secretory neurons in the insect's brain.

Formation of the flat two-layered wings from one-layered embryonic discs implies folding of the basal ventral and dorsal epithelia, adhesion of both basal surfaces into a two-layered wing structure, and expansion of the wing via flattening of the existing wing epithelial cells. All these processes are closely related to a rise in the ecdysone titer during pupal development.

Second, besides the direct control by the ecdysone pathway described above, by binding its receptor EcR, ecdysone acts downstream as suppressor of *Ubx* expression (Monier et al., 2005), thus stimulating the development of imaginal wing disks. It is redundant to say that secretion of ecdysone by the prothoracic gland is under brain control, via the neurohormone PTTH.

Wing development in insects is negatively regulated by JH. In the buckeye butterfly, *P. coenia*, a decline in JH titer during the first few days of the last larval instar determines pupal commitment of the wing imaginal disc. Wing disks in *P. coenia* cease growing in the presence of JH, and their growth can be inhibited by application of JH or its analogs (Kremen and Nijhout, 1989; Miner et al., 2000). A simple hormonal manipulation, topical administration of JH, leads to formation of wingless females in crickets *Gryllus firmus* and *G. rubens*, which are naturally winged. So, the development of wing disks is determined upstream the Ubx and is negatively controlled by JH, whose expression or suppression is determined by brain signals, allatotropins, and allatostatins, respectively.

Ilps are also involved in the process of wing disk growth in *Drosophila*. Seven Ilp genes are discovered in *Drosophila*, and the silkworm *B. mori* has at least 38 Ilp genes. The primary source of Ilps are brain medial neurosecretory cells (MNCs), which project their axons to CC, where they release the neurohormone (Wu and Brown, 2006). Bombyxin, a DILP neurohormone, stimulates wing disk growth in *P. coenia* (Nijhout and Grunert, 2002).

A number of environmental stimuli, such as photoperiod, crowding, and temperature, also exert their influence on wing formation via specific changes in endocrine pathways (Consoli and Bradleigh Vinson, 2004). Remember, there is no other way that these external stimuli can modify endocrine pathways but through reception and processing in the CNS, which, via neurohormones, PTTH, other neuropeptides, and allatostatins/allatotropins, control secretion of wing-related hormones ecdysone and JH, respectively.

Neo-Darwinian Explanation

The neo-Darwinian paradigm would predict that evolution of insect wings from crustacean branched limbs has been a gradual process that has been made possible by accumulation of favorable mutations at least in the genes involved in the patterning of wings, under the action of natural selection.

No evidence has ever been provided, and no hypothesis has been presented to show how a mutation in a particular gene or regulatory sequence might cause evolution of crustacean branched appendages into insect wings. All the key genes and GRNs responsible for wing development are shared by insects and their arthropod ancestors (crustaceans) and even vertebrates, but only specific groups of insects develop wings. Within the class of insects, there are no differences in relevant genes, or DNA in general, between the winged and wingless insects. The production of winged and unwinged offspring (morphs) by the same individuals, and even in the same brood, implying the same genotype, makes any imaginable neo-Darwinian explanation unfeasible.

Epigenetic Explanation

From the epigenetic view, it would be predicted that the changes responsible for evolution of wings in insects must occur in the CNS signals that trigger activation of wing signal cascades and GRNs.

Let us remember a few important facts about the mechanism of development of wings in insects:

- Secretion of the ecdysone activates the specific wing GRN.
- The drop of the JH level in the last larval instar of the butterfly *P. coenia* induces formation of the wing imaginal disc.
- Administration of JH and its analogs prevents formation of wings and activation of the wing GRN in insects that normally develop wings (*G. firmus* and *G. rubens*).
- Expression of both hormones, the ecdysone and JH, is under strict control of the insect CNS.
- Expression of ecdysone is stimulated and regulated by the brain neurohormone PTTH and other neuropeptides secreted in insect's brain, which start the signal cascade that activates the GRN for wing development in insects (Figure 13.6).
- JH is also cerebrally controlled by neurohormones allatostatins and allatotropins.

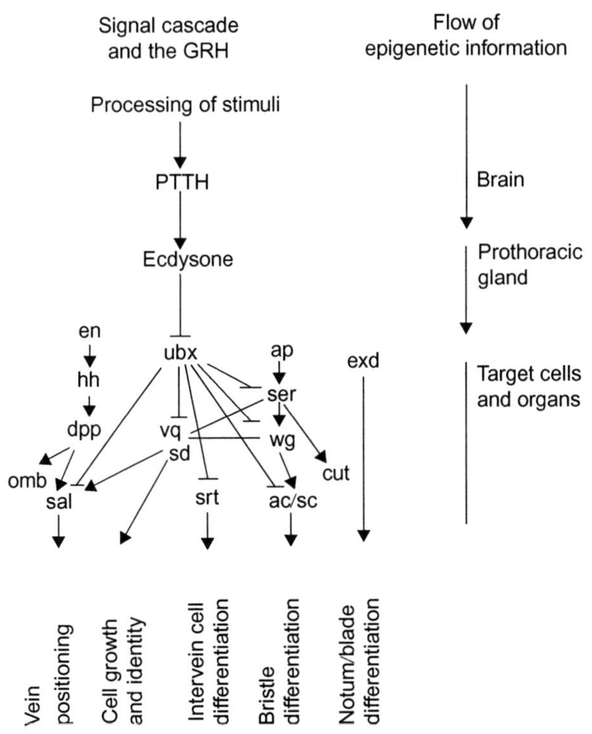

Figure 13.6 Flow of epigenetic information along ecdysone pathway for cell differentiation and growth in the process of the wing development in insects.
Source: Modified from Abouheif and Wray (2002).

The fact that the JH, the ecdysone, and the wing genes remained functionally intact across the insect taxa indicates that the change that led to the evolution of wings in insects has taken place upstream these hormones (Figure 13.6), i.e., in the spatiotemporal patterns of expression and secretion in the insect CNS of the mentioned neurohormones that control secretion of ecdysone, JH, and Ilps.

Given the distribution of ecdysone via the hemolymph all over the insect body, evolutionary changes in insect wing and haltere morphology might have required differential expression of EcR in wings, which in at least some cases is demonstrated to be determined by the local innervation (Hegstrom et al., 1998).

Evolution of Caste Developmental Polymorphisms in Insects

One of the impressive examples of polyphenisms (developmental polymorphisms) is observed in social insects, where castes consist of individuals of distinct morphology and behavior, with the latter acting as biological glue to hold the colony together as a functional unit. The postembryonic production of individuals of morphologically and behaviorally different types from eggs of the same genotype is still not fully understood.

It is thought that wing polyphenism in ants evolved only once, ~125 mya, and wing-patterning network of wingless worker castes is evolutionarily labile although the GRN has been largely conserved across holometabolous insects for the past 325 million years (Abouheif and Wray, 2002).

Wing polyphenism of any castes challenges the basic neo-Darwinian tenet on the determining role of the genotype in the development of phenotypic traits. Individuals belonging to different castes and expressing different behavioral, morphological, and life history characters have the same set of genes! The cause of the caste polymorphism is an epigenetically determined differential expression of genes in individuals of the same genotype. All the wing-patterning genes that are interrupted in workers must still retain the ability to specify and pattern the wings in queens and males, as well as the legs or CNS in all castes (Abouheif and Wray, 2002).

Before dealing with the evolution of the caste wing polymorphism in insects let's first schematize the mechanism of the normal development of wings in insects (Figures 13.7 and 13.8).

Typically, in *D. melanogaster*, the development of the insect wing takes place in three stages: during the first embryonic stage, activation of a specific GRN leads to the establishment of wing disk precursors (Figure 13.8). Then, by the end of the embryonic stage, another GRN is activated, which leads to formation of imaginal wing disks. Finally, during the late larval stage, activation of a more complex GRN determines formation of wings with all their structures. Activation of the three GRNs is under neural control via hormones ecdysone and JH.

Ants of the genus *Pheidole* have four castes: the queen, major workers, minor workers, and soldiers. Whether wings develop or not in each of the castes (queens, workers, and soldiers) of the ant *Pheidole morrisi*, it depends on the presence of the environmental stimuli during the development (Abouheif and Wray, 2002).

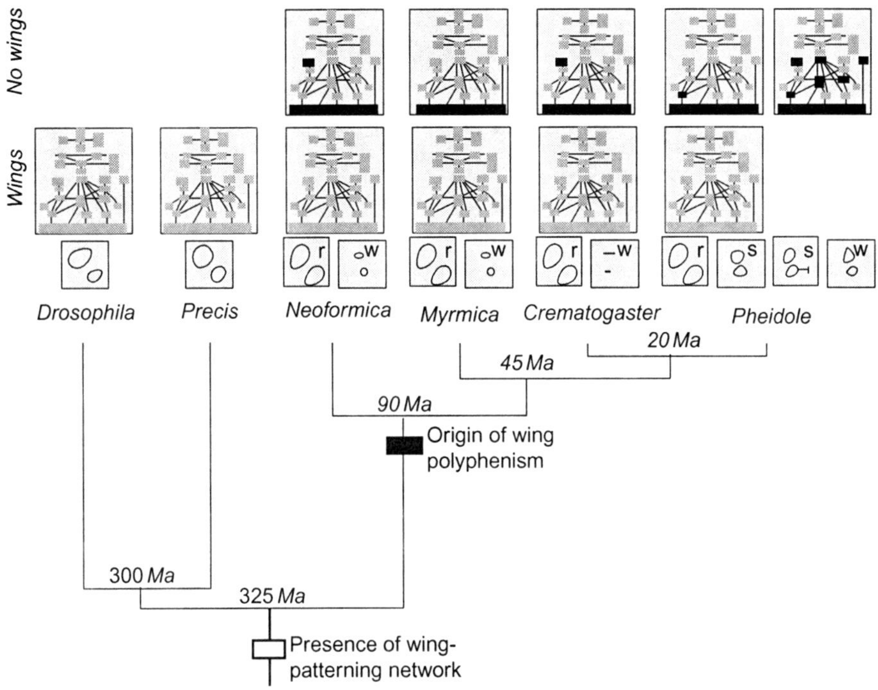

Figure 13.7 Evolutionary history of ant wing polyphenism. Wing polyphenism evolved just once, approximately 125 mya (black bar). The wing-patterning network has been largely conserved across holometabolous insects for 325 million years (empty box). Diagrams of reproductive and worker disc morphology within each species are shown in the first row above the phylogeny (for *P. morrisi*, arrowheads refer to either the soldier forewing or hindwing). Network diagrams for winged reproductive castes are shown in the second row above the phylogeny; network diagrams in wingless sterile castes are shown in the third row. Within each panel, conserved gene expression is indicated in darker gray, and interrupted expression is indicated in black; genes not examined are shown in lighter gray. Note the dissociation among phylogenetic history, rudimentary disc morphology, and points of interruption. Points of interruption have evolved over relatively short time scales, particularly in contrast to the long-term conservation of the wing-patterning network among holometabolous insect orders. *Source*: From Abouheif and Wray (2002).

Experiments with winged and wingless castes of *P. morrisi* have shown that the developmental pathway for wings consists of three switch points; the first one that determines development of queens and workers depends on the level of maternal JH during oogenesis. The second depends on the external stimuli (photoperiod and temperature), to which the embryo responds by generating brain signals (allatotropins/allatostatins) that regulate production of JH. Pulses of JH determine production of (winged) queens, whereas lack of JH pulse determines formation of worker and soldier larvae. Then, and again in response to environmental stimuli, on a specific diet, in the second point of interruption of the gene network, the JH pulse determines

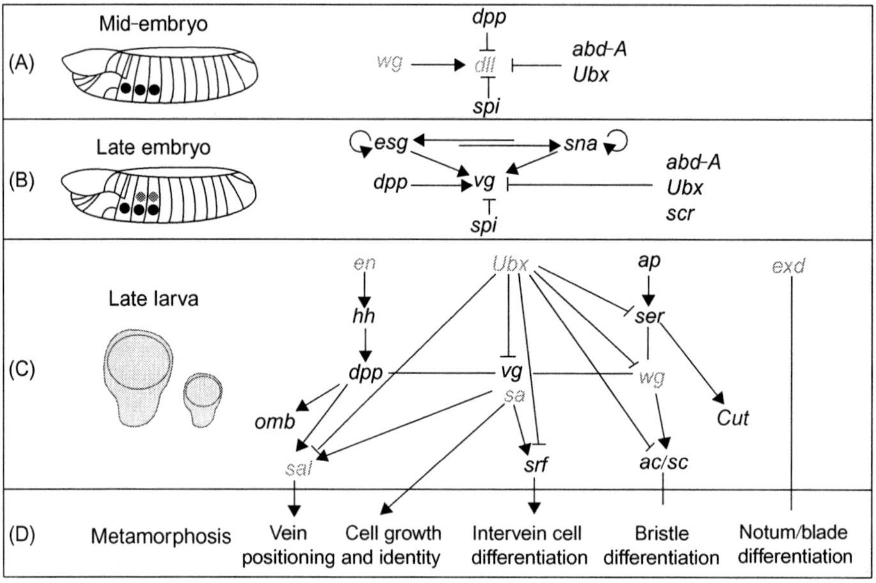

Figure 13.8 The wing-patterning network in *D. melanogaster*. During embryogenesis (A), interacting signaling molecules and transcription factors establish a cluster of about 20 ectodermal cells as precursors of both the leg and wing imaginal discs (dark gray). (B) A second set of interacting gene products then divides these cells into separate clusters that give rise to three pairs of leg (dark) and two pairs of wing (light gray) imaginal discs. During the last larval instar (C), the wing precursor cells proliferate into full-sized imaginal discs. A third set of interacting gene products then patterns these discs, imparts a wing-specific identity, and activates downstream target genes that pattern detailed structures, such as veins and bristles (D). Genes examined are shown in gray. Dashed lines indicate regulatory interactions specific to the hindwing disc, arrowheads indicate activation, and bars indicate repression.
Source: From Abouheif and Wray (2002).

formation of soldiers from worker larvae, and the lack of JH pulse determines formation of worker larvae (Abouheif and Wray, 2002). No genetic factor has been shown to determine any of the three switch points. Switch points are related to the activity of hormones (ecdysone and JH), but brain neurohormones are responsible for their expression in aphids, locusts, and some butterflies (Nijhout, 1999).

Pheidole megacephala has a winged queen caste, two wingless worker castes (major and minor workers), and one wingless soldier caste. The final instar larvae of presumptive queens and of major workers develop normal wing disks, but only 71% of minor larvae develop barely detectable wing disks. Later, during the prepupal stage, only queens' larvae develop intercellular structures, while the wing disks of major workers start degenerating as a result of programmed cell death (PCD). Experimental evidence from *Pheidole* species (*P. megacephala* and *Pheidole bicarinata*) (Figure 13.9) suggests that a neurally determined early pulse of the JH level induces formation of incipient mesothoracic wing disks in both queen and worker

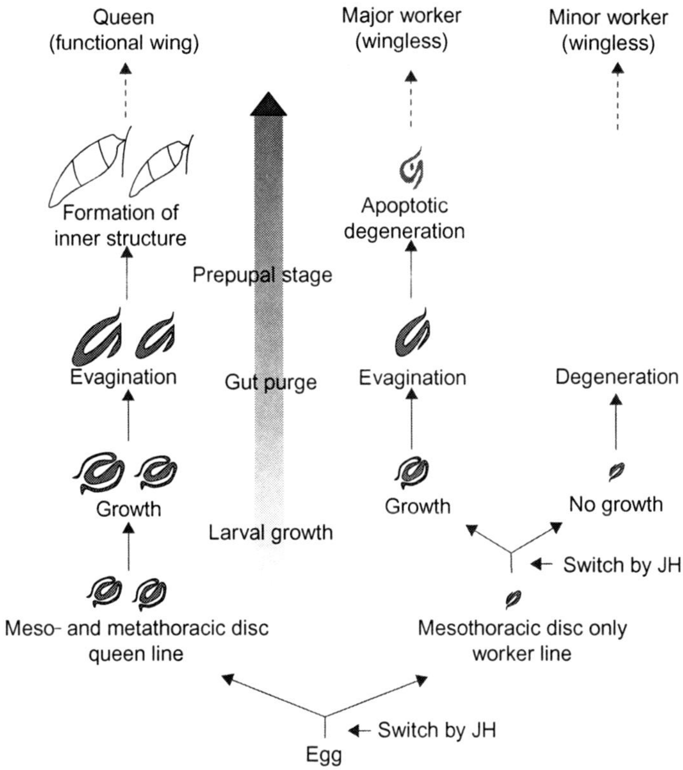

Figure 13.9 A diagram of differential wing formation/degeneration in *P. megacephala*, showing that homologous organs (wings) follow completely different developmental fates according to caste.
Source: From Sameshima et al. (2004).

lines, and a second pulse is responsible for the growth in major workers and the absence of growth in minor workers (Sameshima et al., 2004; Wheeler and Nijhout, 1983). All embryos develop wing disks, which later degenerate during the prepupal stage in all but the presumptive queen, by evagination in the major workers and by PCD in minor workers.

In some cases, the behavior of the colony members has a great role in determining the female individual that becomes queen. For example, at the onset of the prepupal stage in females of the Japanese ponerine ants of various *Diacamma* species, forewing buds of larvae develop into a pair of glandular *gemmae*, secreting pheromones, while the hindwing buds undergo PCD (Gotoh et al., 2005; Miura, 2005). Workers of the colony then clip off or mutilate gemmae from all but one female, which will develop into the sole reproduction-capable queen in the colony (Miura, 2005) in a strange behaviorally determined process. What occurs in the wingless castes of major and minor workers is a drop in the level of JH in the latter and a PCD in major workers.

The high lability observed in evolution of different points of interruption of wing development is in strong contrast with the evolutionarily long periods of conservation of GRNs in holometabolous insects.

Application of JH and juvenile hormone analogues (JHA) in these insects also induces development of soldiers from the prospective winged nymphs, leading even to production of intercaste morphs with combined soldier and winged traits (Figure 13.10).

All the available evidence suggests that the winged/wingless diphenism in *P. megacephala* is determined by pulses of JH and ecdysteroids. However, it is well known that the synthesis and secretion of JH is under neural control via

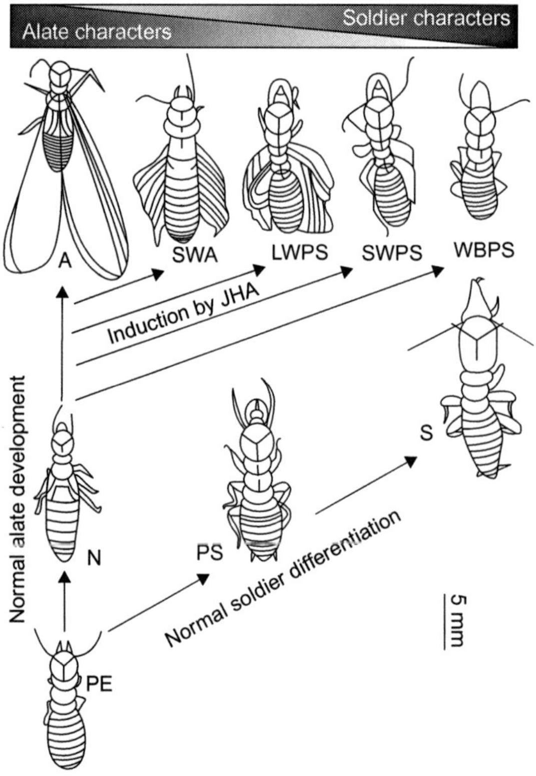

Figure 13.10 JHA induces soldier characters, even from the alate developmental line. Normal developmental pathways are indicated by solid arrows, while developments induced by JHA are indicated by serrated arrows. Normally nymphs (N) develop into alates (A), while soldiers (S) are derived from pseudergates (PE) via presoldiers (PS). Application of JHA to nymphs induced intercastes between alates and presoldiers: shrunk-winged alates (SWA); long-winged presoldiers (LWPS); short-winged presoldiers (SWPS); and wing-budded presoldiers (WBPS). The morphology of intercastes seems to be determined by the developmental stage of nymphs when JHA is applied. The morphological characteristics of alates and soldiers have opposite responses to JHA application.
Source: From Miura (2005).

neurohormones, allatotropin/allatostatins as well as other neuropeptides; an Ilp secreted in the brain of *Drosophila* (Wheeler et al., 2006) and three types of brain Ilps in the mosquito *Culex pipiens* (Sim and Denlinger, 2009) are necessary for JH production by corpora allata.

The GRN for PCD in insects also is hormonally determined by ecdysone. The synthesis and secretion of ecdysone, which regulates the PCD in insects, is also neurohormonally regulated by the neuropeptide PTTH, which via its dimer receptor EcR/Usp induces expression of the apoptosis caspase Dronc and other apoptotic downstream genes (Cakouros et al., 2004).

Another impressive example of the neural/behavioral control of wing development in insects is that of dealation (wing shedding) by virgin queens in colonies of the fire ant, *Solenopsis invicta*. It has long been observed that sexually mature virgin queens cannot begin reproductive activity (start oogenesis and dealate) as long as they remain in the same colony with the mother queen. When the mother queen is removed from the colony, in a number of virgin queens the ovaries are enlarged, numerous oocytes are produced, wings are shed, and at least one of the virgin queens starts laying eggs. In the meantime, workers eliminate the rest of virgin females, again behaviorally determining the maintenance of the caste structure of the colony. It has been determined that both oogenesis and dealation (wing shedding) in virgin queens are determined by the fact that with the removal of the mother queen is removed the source of a pheromone she releases for preventing virgin females to develop into queens (Fletcher and Blum, 1981).

The fact that dealation is induced by a pheromone unambiguously indicates that the signal cascade starts in the insect's CNS where the pheromonal stimulus is perceived and processed. Indeed, there is experimental evidence showing that both the decrease of the neurotransmitter dopamine level in the brain (Robinson and Vargo, 1997) and the dopaminergic innervation of corpora allata (Granger et al., 1996) cause a decrease of the synthesis of JH, thus leading to dealation and oogenesis (Figure 13.11).

In examples presented above, JH switching, PCD (Sameshima et al., 2004), and secretion of pheromones that are responsible for wing polyphenisms are neurally determined. The only confirmed known signals for switching JH production are brain signals, neurohormones allatotropins and allatostatins. The "degeneration process," that is the PCD, in insects as well, is determined by a signal cascade that starts in the CNS with the release of the neurohormone PTTH and other neuropeptides (PTTH + other neuropeptides → ecdysone → preapoptotic genes) (Draizen et al., 1999; Namba et al., 1997) (see also Section Apoptosis in Invertebrates in Chapter 5).

One clear neural pathway for avoiding loss of wings in mature females is observed in colonies of the fire ant, *S. invicta*. The queen releases a pheromone that inhibits dealation and production of eggs by females. The pheromone is received by antennal olfactory neurons and perceived in the insect's brain, where it starts a still-unknown cascade that (presumably via brain hormones, allatostatins) leads to suppression of JH in corpora allata, drop of the JH titer in the hemolymph, and inhibits dealation.

Another cue for preventing dealation is related to the social environment of females: in presence of workers, dealation of alates (winged) is faster. Removal

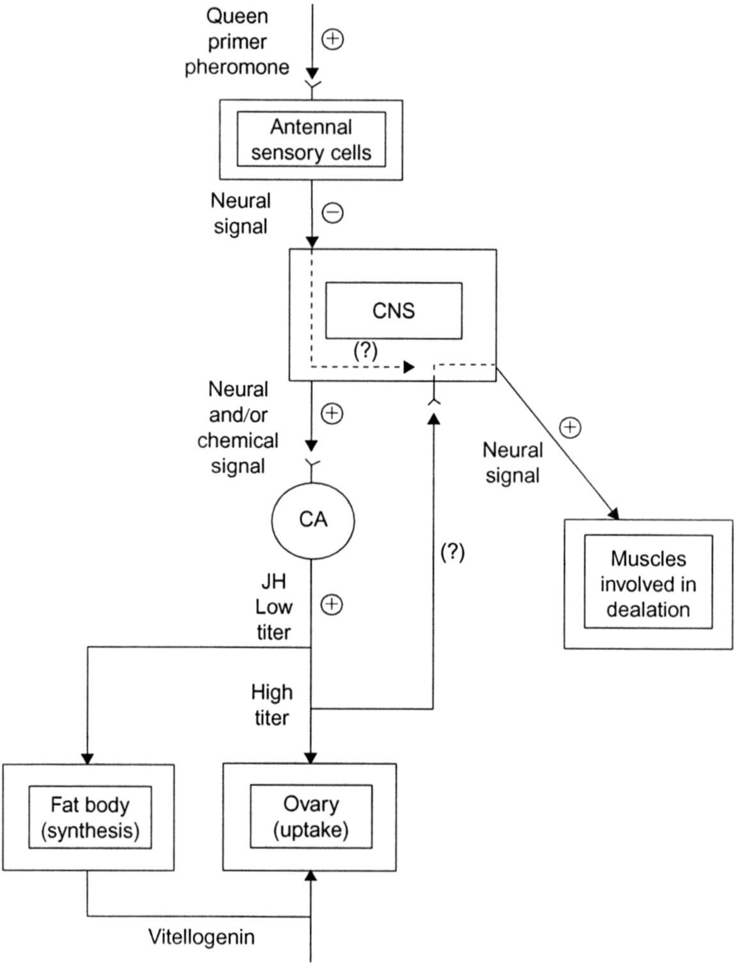

Figure 13.11 Proposed general model for the mode of action of the primer pheromone of queen fire ants that inhibits dealation and ovary development in virgin queens. The pheromone triggers antennal receptors, which send inhibitory signals to the median neurosecretory cells in the brain. Largely inhibited, the median neurosecretory cells only weakly stimulate the corpora allata to synthesize JH, maintaining low titers of this hormone. At low levels, JH stimulates vitellogenin synthesis in the fat body. In the absence of the pheromone, the disinhibited neurosecretory cells send a stronger chemical and/or neural signal that triggers the corpora allata to produce larger quantities of JH. At higher titers, JH stimulates vitellogenin uptake by the ovaries and dealation. The latter process possibly involves an effect of JH on the nervous system. Dealation may result from a JH-independent pathway in the nervous system in lieu of, or in addition to, the JH-mediated pathway. These two possible pathways for control of dealation are flagged with question marks.
Source: From Vargo (1998).

of the queen as well as antennectomy disinhibit dealation of insects (Burns et al., 2005). This triggers drastic neuroendocrine changes, which are reflected in morphological changes in ovaries and reproductive behavior of alates. In the presence of males, an increase in ecdysteroid level in hemolymph is believed to act via a neurohormonal pathway (involving brain and CC and release of allatostatins by the latter) to inhibit JH production in corpora allata (Brent et al., 2005). Under normal conditions, pea aphids, *Acyrthosiphon pisum*, produce winged and wingless offspring. However, under conditions of predator stress, caused by the presence of ladybirds, lacewing larvae, hoverfly larvae, and others, these aphids release a pheromone, the sesquiterpene (E)-β-farnesene (EBF).

The pheromone is received by antennal receptor neurons and is processed in a specific neural circuit in the brain of conspecific aphids (females with amputated antennae produce only few winged offspring) (Weisser et al., 1999; Kunert and Weisser, 2005). On receiving and perceiving the pheromone, females respond adaptively by activating a developmental pathway that leads to production of a larger proportion of winged offspring, which by flying could escape the predator and colonize other predator-free plants. Some plants also release a similar sesquiterpene, but pea aphids can compare the proportion of the sesquiterpenes released by the plant and their conspecifics (Kunert et al., 2005) in their brain circuits and activate signal cascades for production of more winged morphs on assessing that more sesquiterpene comes from conspecifics rather than from plants.

Another example of developmental lability, with significant implications for the evolution of wings in insects, is the marked wing dimorphism observed in the lygaeid bug, *Dimorphopterus japonicus*. This insect can selectively and adaptively regulate proportion of short-winged (brachypterous) and long-winged (macropterous) individuals. In response to social stimuli, such as conspecific crowding, high temperature, and long photoperiod, during the nymphal stage, this bug increases the proportion of macropters in the offspring, or even produces only macropters, depending on the intensity of the perceived stimuli. This is an adaptive response for escaping the deteriorating habitat (Sasaki et al., 2002, 2003).

All the above factors that trigger/suppress wing development in insects are neurally acting/sensed, and a simplified mechanism of various stimuli (external and internal) in the development of insect wings is diagrammatically presented in Figure 13.12.

Neo-Darwinian Explanation of the Caste-Specific Wing Developmental Polymorphism in **P. megacephala**

As discussed earlier, not only castes of these insects share the key wing-patterning genes, but the wing-patterning network has been conserved across holometabolous insects. No changes in genes or in genetic information are involved in production of winged and unwinged individuals. Development of phenotypically different castes in colonies consisting of individuals of the same genotype often under the same conditions of living (e.g., soldiers and workers) cannot be accounted for from a neo-Darwinian view for the phenomenon contradicts one of the basic theoretical tenets of the neo-Darwinian paradigm on genes as determinants of phenotypic characters.

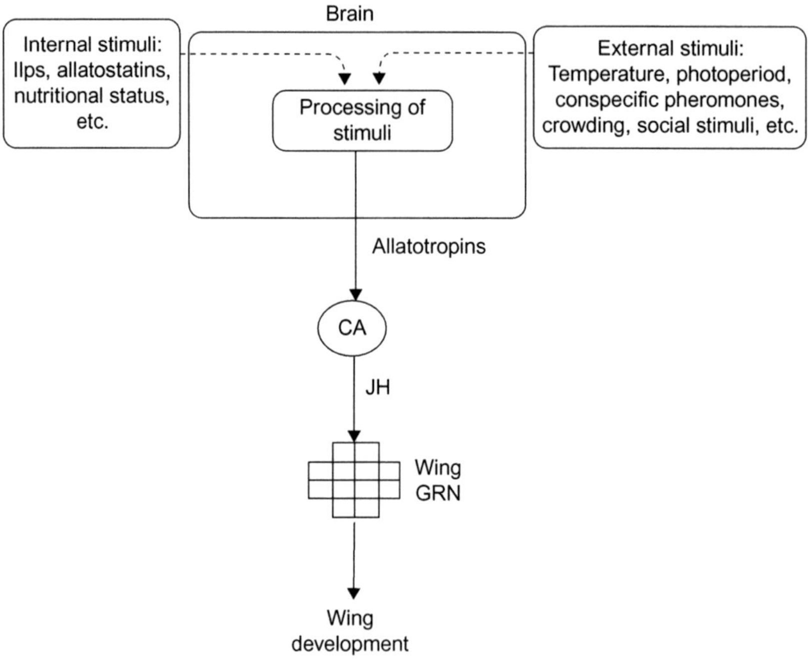

Figure 13.12 Generalized and simplified diagram of the wing development in response to internal and external stimuli in insects.

Epigenetic Explanation of Caste-Specific Wing Developmental Polymorphism in P. megacephala

Signal cascades responsible for the development of wing disks and wings ultimately originate in the brains of insect larvae. However, this fact does not tell us anything about why some of the larvae (queen and major workers) develop full wing disks, and some develop barely detectable wing disks. In *P. megacephala*, the initial differentiation of the queen line from the worker lines is determined by a first JH pulse during the early embryonic development, and the differentiation of the major workers from minor workers is determined by a second JH pulse taking place during larval development in the minor worker line. The wing disk that develops in presumptive major workers is later eliminated by apoptosis. Both JH secretion and the PCD are neurally determined by signal cascades that start in the insect brain, as a result of the processing of external and internal stimuli.

Crucial for understanding the nature and evolution of these morphological adaptations is the mechanism by which an animal, in response to these external biotic and abiotic (physical, chemical, and social) stimuli, generates "instructions" or the epigenetic information for regulating switches in JH secretion. With the nervous system being the site where external stimuli are integrated and processed, and where the signal cascades for caste polymorphisms start, it is logical to conclude that the caste

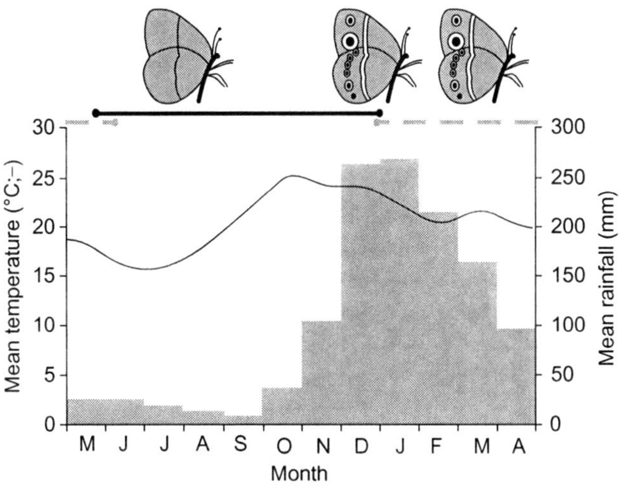

Figure 13.13 A wet–dry seasonal cycle. A dry season is followed by a wet season (brown and green, respectively). Two generations of the wet season form with conspicuous eyespots occur in each rainy season. Larvae of both of these cohorts develop at high average temperatures. The second generation of the wet season form lays eggs before the grass food plants die out, and the larvae develop at progressively declining temperatures. This cohort produces the generation of the dry season form without eyespots that persists through the period of low rainfall.
Source: From Brakefield et al. (2007).

wing polymorphism in *P. megacephala* is determined by an epigenetic mechanism related to the processing of external and internal stimuli in the insect's CNS.

Evolution of the Seasonal Diphenism in the Butterfly *Bicyclus anynana*

The African butterfly *B. anynana* responds adaptively to the semestral cycles (spring–fall) of color changes in its natural background. Its wings change from a spotted wing pattern, which serves as warning against its predators in the colorful lighter spring background, to nonspotted pattern in fall, which makes it less visible in the season's brownish background of fallen leaves (Figure 13.13).

Some biologists believe that the evolution of eyespot patterns in butterflies "requires only single or very few changes in regulatory genes" (Brakefield et al., 1996), but as of yet no relevant changes in genes have been identified. Above all, the change in one or very few regulatory genes is logically excluded by definition: the distinct seasonal morphs are produced by the same genotype.

Environmental stimuli (both visual and thermosensory stimuli) that trigger this developmental plasticity do not, and cannot, act directly on genes in the cells of wing eyespots. Arguably they are received by sensory, retinal neurons and transmitted for processing and integration in the butterfly brain, leading to secretion of neurotransmitters/neuromodulators that start signal cascades determining the seasonal diphenism.

Recent studies demonstrate the determining role of hormones in switching African satyrine *B. anynana* to alternative developmental pathways and the *adaptive* (hence intrinsic) nature of the response to the changing environment. These experiments seem to reconstruct the first links of the causal chain extending from environmental stimuli to the development of the *B. anynana* seasonal camouflage, which, like other insect polyphenisms, is based on hormone-induced switches in developmental pathways (Nijhout, 1996).

While the temperature, which is related to the seasonal weather, is the main trigger of the diphenism, mediators of the switch from one morph to the other are neurally controlled ecdysteroid hormones (Oostra et al., 2011). The switch is determined maternally, i.e., the mother provides eggs with the alternative wing pattern information (Steigenga et al., 2005). The seasonal diphenism appears in the final stage of *B. anynana*'s development, and this might suggest that it is an evolutionarily new (not ancestral) feature. Phylogenetically earlier characters, such as the number of eyespots per wing, which vary among butterfly species, appear in the earlier stages of butterfly development (Brakefield et al., 1996).

Ecdysteroids bind their nuclear receptor, EcR (in the form of the heterodimer EcR-USP), and it is observed that the color of scale cells and wing eyespots coincides with patterns of EcR expression (Koch et al., 2002, 2003). There is evidence that expression of EcR in insects is determined by motoneurons (Hegstrom et al., 1998).

The complex patterning of the wing eyespots results from a complex spatial pattern of the activity of the hormone, but ecdysteroid hormones are released in the hemolymph and are uniformly distributed all over the butterfly wing and body. Hence, it is plausible that the patterning and color of the wings may be determined by the patterns of expression of the ecdysone receptor (EcR) in the butterfly wings.

Neo-Darwinian Explanation

Any gradualist mechanism of the accumulation of favorable hereditary changes under the action of natural selection or genetic drift would immediately be rejected as an explanation of the seasonal diphenism. Indeed, no attempt has been made to explain the phenomenon from a neo-Darwinian view.

The widely held idea that the seasonal polyphenism of *B. anynana* is under control of environment (Brakefield et al., 1996) seems to be too vague and misleading: living systems are under control of their own control system rather than under any unidentified and unidentifiable environmental control. Environment definitely influences the behavior of living systems, but environmentally triggered changes in these systems represent intrinsically determined *responses* for adapting the system to the environment, rather than automatic products of the environment.

Epigenetic Explanation

There can be no doubt that the seasonal polyphenic adaptation did not arise instantly at the moment this butterfly diverged as a separate species from its ancestral stock.

First, the butterfly might have evolved a phenotypic plasticity for wing patterning, where individuals of the same population, or even of the same brood (implying the same genotype), might express one of alternative phenotypes, as often occurs in cases of developmental polymorphisms (see Developmental Polymorphism, Chapter 10), which abound throughout the living nature.

Suppose a situation of developmental polymorphism where populations of the incipient species of *B. anynana* were composed of individuals of two morphs, each of them with one of two alternate wing phenotypes, spotted and nonspotted, in certain proportions. Given that no genetic changes are involved in the appearance of two different morphs, it may reasonably be inferred that an epigenetic factor determines activation of the spring and fall season polyphenism.

What determines the appearance of each of the seasonal morphs in *B. anynana*? To put the question in another way: *How* and *where* are the visual and thermosensory environmental cues related to the activation of alternative developmental pathways for spotted and nonspotted wing patterning?

We can answer this question with a high degree of certainty. Wing patterning in seasonal morphs of *B. anynana* is neurally determined during the individual development according to the principle of the binary neural control (see on the subject in Chapter 6): on the one hand, the CNS control of the wing patterning is mediated by the neurally controlled secretion of ecdysone and, on the other, by expression of the EcR by the local neural innervation.

We know that both ecdysone and its receptor in both the wet and dry season morphs are unchanged and functional. What is essentially different in these morphs is the pattern of expression of the ecdysone receptor in the wing, and this pattern is experimentally demonstrated to be epigenetically determined by the local innervation (Hegstrom et al., 1998) (for further discussion see Section Seasonal Polyphenism in Insects in Chapter 10). But if the development of the eyespot in *B. anynana* wings is epigenetically determined by the nervous system during individual development, its evolution also would have been epigenetically determined by an inherited change in the neural regulation of expression of the gene for the ecdysone receptor. For there is no other way that evolutionary changes in morphology could take place but by changes in developmental pathways.

Evolution of Horns in Beetles

There are thousands of beetle species in which a proportion of male individuals develop horns (Moczek and Nagy, 2005) as cuticle extensions. Some representatives of the dung beetles of the genus *Onthophagus*, consisting of more than 2,000 known species, have been the object of extensive studies on the development of horns in male beetles. Males use horns, which represent more than 10% of their body mass, as weapons in combat with other males to get access to resources accumulated by females in tunnels in the soil beneath dung (Emlen, 2000). Horned males block the tunnel entrance, thus preventing other males from having access to the buried female and the dung deposited at the bottom of the tunnel.

Onthophagus taurus, *Onthophagus nigriventris*, and other species of the *Onthophagus* genus display two distinct morphs: individuals that develop horns after attaining a threshold in body mass, and hornless individuals. Horn Anlagen develop early during the prepupal stage. Large male pupae develop a long pronotal horn (developing on the dorsal exoskeletal plate of the first thoracic segment), whereas small male pupae only form a pronotal outgrowth. Such an outgrowth also develops in female pupae, but it disappears later.

The fact that at least one gene, *Dll* (*distalless*), is expressed in all horn Anlagen, as well as in all arthropod appendages, has been interpreted as indication that the proximal–distal (P–D) development of horns and arthropod limb development may depend on the same ontogenic mechanism (Moczek and Nagy, 2005). Investigators speak about "some upstream activators and downstream targets of Dll" but do not identify the upstream activators. However, studies on the development of insect wings have shown that *Dll* expression is regulated by the Hox gene *Ubx*, which in turn is regulated by the hormone ecdysone: ecdysone binds its receptor (EcR), and in this form they bind the Ubx, which downregulates expression of Dll. Needless to say, the synthesis and secretion of ecdysone is cerebrally regulated by the neurohormone PTTH.

Investigation of the facultative nature of the horn diphenism in male dung beetles has shown that whether male beetles develop horns or not depends on reaching a threshold body size; only males that reach that threshold become horned beetles (Figure 13.14).

A slight increase in the levels of ecdysteroids and a drop in JH titers occur in females and hornless males. This is consistent with the role of ecdysone as a downregulator of the Dll. In contrast, large pupae, which develop horns, show an increase in the hemolymph level of JH at the end of the third larval stage. When methoprene, a JH analog, is topically applied, 80% of the hornless males also develop horns (Emlen and Nijhout, 1999). During a short period of about 30 h, high levels of JH induce proliferation of the epidermal cells and formation of horns in males of the dung beetles, *O. taurus* (Emlen, 2000).

The correlation between horn development and higher level of JH is corroborated by experiments in which local application of the JH analog, methoprene, induces formation of horns even in smaller male pupae of *O. taurus*, which normally do not develop horns (Emlen and Nijhout, 1999).

From an evolutionary view, it is important to point out that horn development in dung beetles is characterized by expression of transcription factors *Dll* (distalless) and *al* (aristaless) qualitatively similarly to the case of the development of insect appendages. Male morphs in both *O. taurus* and *O. nigriventris* express *Dll* in the developing horn or horn rudiment, respectively, and differences in horn development among major and minor morphs in these species are not due to the presence or absence of Dll protein in the horn region (Moczek and Nagy, 2005).

Not only horn-patterning genes but also genes for the relevant hormones (ecdysteroids and JH) are unchanged in dung beetle species that evolved distinct morphologies, and no differences in genes exist between the horned males and hornless

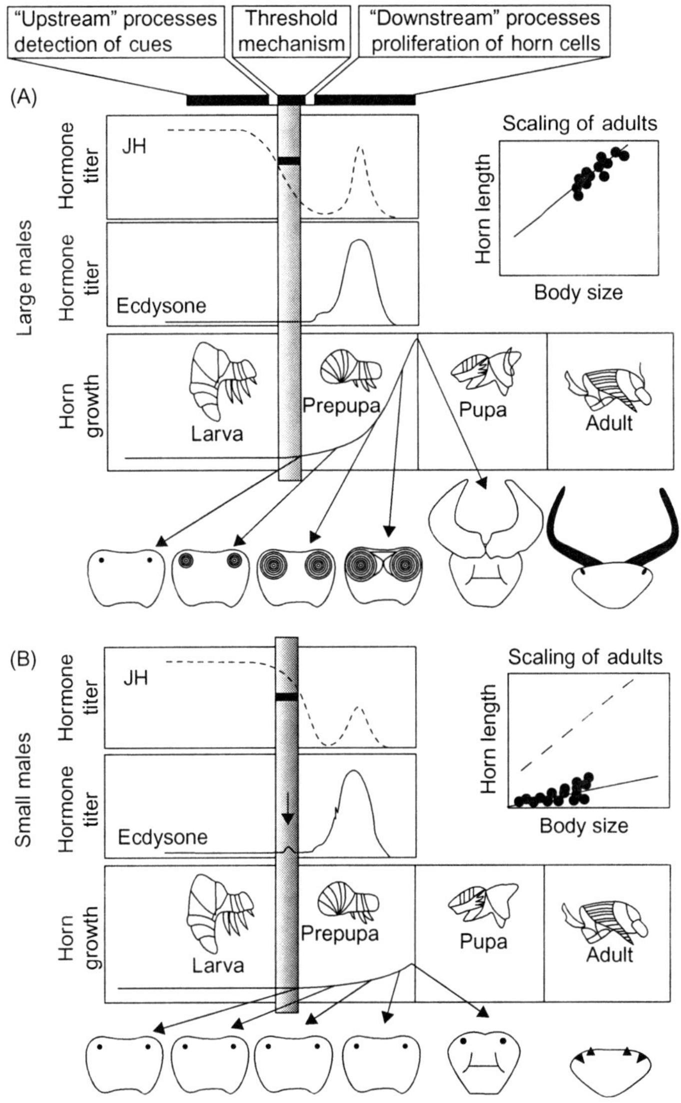

Figure 13.14 Endocrine regulation of male and sexual dimorphism in the beetle *O. taurus*. By the middle of the third larval instar, large and small males differ in circulating levels of JH: large males have lower concentrations than smaller males. JH levels are assessed during a brief sensitive period immediately before the cessation of feeding (*vertical gray bar*), and relatively large males have JH concentrations below the critical threshold (*black horizontal line*) at that time. (A) Cells in the developing horns of these individuals undergo a brief pulse of rapid proliferation during the prepupal period, and these larvae mature into adult males with fully developed horns (*insert*). (B) Small male larvae have JH concentrations above the threshold during the sensitive period, and these animals experience a brief pulse of a second hormone, ecdysone (*arrow* in B). Ecdysone is known to initiate cascades of gene expression, and this tactic-specific pulse appears to affect the fate of horn cells such that they subsequently undergo only minimal proliferation. Small males mature into adults with only rudimentary horns (*inset*). *Source*: From Emlen et al. (2005).

females and small males. Thus, the horn in the dung beetle *O. taurus* is epigenetically determined by shifts in the body size threshold:

> *One major avenue of evolutionary change in this group has involved shifts in the threshold body size regulating horn growth.*
>
> *Emlen (2000)*

Hence, an understanding of the mechanism of the evolution of horn size and morphology requires knowledge of

1. How these insects might change the body size threshold, and
2. How they can modify expression of JH receptors in epidermal cells.

Nijhout believes that "suppression of horns in small males required the evolution of a size-sensing mechanism in the final larval instar" (Nijhout, 2003), and it is believed that the stretch stimulus, after the third larval stage, is received by proprioceptive neurons in the soft regions of the cuticle and transmitted for processing in the insect brain (Gorbman and Davey, 1991), where the set point for body size is believed to be in insects. The location of the set points in the insect brain is also suggested by the fact that all the morphological and physiological changes that follow the reach of the growth threshold start with signal cascades from the brain. Other body mass set points are identified in the brain of *Peromyscus maniculatus* (Adams et al., 2001) and other mammals (Baeckberg et al., 2003), the thermoregulation (e.g., Hammel et al., 1963; Boulant, 2000), and a number of components of body fluids (see Section Central Control of Animal Physiology in Chapter 1).

> *Our current hypothesis is that size-sensing mechanism during the larval stage regulates the program of juvenile hormone level during the prepupal stage. Juvenile hormone normally declines just before the prepupal stage and is absent at the time that ecdysone-stimulated cell division occurs in the horns. It is possible that in small males the decline of juvenile hormone is delayed, so that it is still above threshold and is able to inhibit ecdysone-stimulated horn growth.*
>
> *Nijhout (2003)*

A comparative study on the North American and Western Australian populations of dung beetle, *O. taurus*, has shown that these populations have undergone a very rapid evolution of the horn polyphenism. Individuals of this circum-Mediterranean species were introduced to the above regions by the late 1960s. Within as little as ~40 years, these races have diverged dramatically in the body size threshold for the development of horns in nature, while maintaining the original threshold under laboratory conditions for many generations. West Australian populations now switch to the horned morph at a much larger body size, and North American populations, at a much smaller body than their Mediterranean ancestors (Moczek and Nijhout, 2003). Differences between the Western Australian and North American races are of the magnitude of differences between species. These "race" differences are of a magnitude comparable to differences of the original form with another dung beetle species, *Onthophagus illyricus*. This suggests that North American and West Australian populations have entered separate paths of evolving into two new species in the genus *Onthophagus* (Moczek and Nijhout, 2003) by only changing the body size threshold via a size-sensing mechanism, which, in all likelihood, is a neural mechanism.

Experimental development of horns in dung beetles by application of JH at sensitive stages of the larval development suggests that the body size threshold may be correlated with the increased secretion of JH:

> *Changes in sensitivity to juvenile hormone, or changes in the timing of sensitivity to juvenile hormone relative to other developmental events, could modify the range of larval body weights that would fall above or below the threshold required for horn induction ... The divergence in body size thresholds between exotic and native* O. taurus *populations may have been mediated by evolutionary modifications in the degree and timing of sensitivity to juvenile hormone.*
>
> *Moczek and Nijhout (2003)*

Now, let us bear in mind that the only "size-sensing mechanisms" we are aware of (i.e., that are described so far in metazoans) are neural mechanisms (see Sections The Central Control of Body Mass in Chapter 1 and Neural Control of the Development of Body Mass in Chapter 5).

Emlen et al. (2005) envisage the basic mechanism of horn development in beetles as well as the evolutionarily important switching (horned/hornless) as a hierarchical mechanism consisting of a

1. sensory apparatus, which determines secretion of a hormone (JH) and
2. temporally restricted expression of the receptor for that hormone, which makes possible the downstream expression of secondary hormones and transcription factors.

The fact that during the development at least the first two steps and the fourth step reside/are determined in insect's brain suggests that the evolutionary switching from horned to hornless beetles and the reverse involved a corresponding change in the neural circuits that determine the body size threshold and regulate JH secretion. Indeed, there is no other way for an evolutionary change that involves no changes in genes to occur but via appropriate changes in development.

The neural determination of the "size-sensing mechanisms" could explain the exceptional evolutionary lability of horns in beetles: horn sexual dimorphism in beetles has been gained seven times, lost thirteen times, and regained once, whereas male horn dimorphism has been gained eight times and lost twelve times (Emlen et al., 2005).

Neo-Darwinian Explanation

Neo-Darwinian theory would predict that in order for horns, as a new character, to evolve in dung beetles, one or an undefined number of "useful" mutations need to occur in DNA or genes, whose products play key roles in the development of horns, such as the patterning genes *Distalless* and *aristaless* or the genes for ecdysteroids and JH. There is neither evidence nor a hint that such changes have occurred. On the contrary, all the above genes and their protein products are functionally well conserved. Hence, the basic neo-Darwinian prediction on the evolution of horns in dung beetles is patently refuted.

Epigenetic Explanation

From the epigenetic view, it would be predicted:

1. No changes in genes relevant to development of horns are necessary for evolution of head horns in dung beetles.
2. Evolution of horns in dung beetles is a result of epigenetically determined changes in the patterns of expression of genes that are essentially involved (e.g., ecdysone, JH, Dll) in the development of horns.
3. Evolution of horn sexual dimorphism and horn intrasexual polyphenism in dung beetles of the genus *Onthophagus* involved an epigenetic change in the body size threshold.
4. Epigenetic information necessary for inducing signal cascades determining these specific changes in expression patterns of genes involved in horn development originates in the beetles' CNS.

The first prediction is validated by the experimental evidence that the products of genes (e.g., ecdysone, JH, Dll) involved in the development of horns are conserved and have not changed their function.

The second prediction is also validated by the experimental evidence presented in this section. That evidence shows a clear distinction in the patterns of expression of genes (ecdysteroids and JH and Dll) involved in the process of proliferation of epidermal cells during formation of horns in large-size beetles, compared to small-size male beetles, although all of them share a common genetic background.

The third prediction is also substantiated by the evidence already presented in this section. The regulatory, i.e., epigenetic, changes are the only changes that have been scientifically demonstrated to occur systematically in the process of horn development in dung beetles. The most relevant difference in determining the horned/hornless alternative is the body size, as perceived in the insect brain. This body size threshold of the same circum-Mediterranean species *O. taurus*, in response to different environmental conditions, in North America and Australia, was respectively lowered and elevated, thus leading to formation of two incipient species (Moczek and Nijhout, 2003) within an evolutionary instant of less than 40 years.

The fourth prediction is that the signal cascade for horn development originates in the CNS. The signal cascades for production of ecdysteroids and JH, which are the only known signal cascades that activate genes involved in the development of dung beetle horns, start in the brain after the processing in neural circuits of the stretch input coming from sensory cells (proprioceptors).

Evolution of Appendages and Tetrapod Limbs

The astonishing similarity of GRNs involved in patterning appendages in groups that are phylogenetically so far apart as insects and vertebrates suggests that these network have been present in their common ancestor and have been essentially conserved in both groups for patterning their appendages (Shubin et al., 1997). Changes that have occurred in relevant appendage genes in both insects and vertebrates have not affected their function. What should surprise us, in view of the unavoidable change of genes over the evolutionary time, is the amazing conservation of the function of genes involved in appendage patterning in both groups for hundreds of millions of years. Not only the relevant appendage-specific genes but also the GRNs involved in appendage development have been conserved to an incredibly high degree.

All the above is especially true for the evolution of vertebrate limbs, which also involved no changes in genes or new genes but was based on the use of an evolutionarily preexisting genetic toolkit. The study of the developmental mechanisms of limb development seems to be a rational approach to the problem of the evolution of the tetrapod limbs from fish fins.

Fish and tetrapods share Hox gene modules involved in the development of pectoral fin/forelimb, and the most conspicuous change associated with evolution of tetrapod limb from fish pectoral fin is the decoupling of limb motoneurons from the hindbrain (Ma et al., 2010; Figure 13.15).

Evolution of tetrapod limbs from fish fins has been a critical event in the history of life on earth that led to tetrapod domination of terrestrial life. This transition occurred within the space of less than 40 million years, from the Middle Devonian, ~380 mya (tetrapodomorph osteoichthyans) to ~360 mya (aquatic tetrapods such as *Acanthostega*) until the appearance of the fully terrestrial tetrapod ~340 mya (Long and Gordon, 2004). The transition was characterized by functional and morphological transformations related not only to terrestrial locomotion (e.g., evolution of tetrapod limbs, shoulder girdle, digits, and modification of the pelvis). It was also associated with the aerial respiration, implying the evolution of lungs (Daeschler et al., 2006) and adaptation to aerial hearing, which implies radical morphological and functional diversification of ears as specialized hearing organs. Ears adapted to both underwater and aerial hearing appeared for the first time in *Ichthyostega* (Clack et al., 2003). Appearance of *Ichthyostega* and *Acanthostega* marks a clear transition from fish to aquatic tetrapods with digits (Figure 13.16).

The early stages of fin and limb development show a common pattern of gene expression, and only later differences in that pattern determine formation of tetrapod limbs (Hinchliffe, 2002). The homology of tetrapod limbs with fish fins declines in a P–D direction, i.e., it is almost complete in stylopod to absent in digits (Laurin et al., 2000).

The limb bud in tetrapod embryos represents a conserved developmental stage, in the sense that all tetrapod classes exhibit the same interactive developmental processes (Hinchliffe, 2002). It contains the apical ectodermal ridge (AER) and the underlying zone of polarizing activity (ZPA) of mesodermal-somatic origin at the distal tip of the limb bud (Figure 13.17). The development of the AER on the ectoderm of the presumed limb bud occurs in response to mesodermal fibroblast growth factor (FGF) signals, by stages 16 and 17. In turn, AER signals (FGF-8) induce formation of the underlying ZPA, which is responsible for determining the A–P axis of the limb.

Inhibition of retinoic acid (RA) synthesis with sulphiram before limb bud induction leads to the prevention of limb bud development and the suppression of Shh, suggesting that FGF-8 alone, in the absence of RA, is not sufficient for the expression of Shh (Stratford et al., 1996). Administration of retinoid receptor antagonists "not only abolishes *Hoxb-8* expression, but also prevents the establishment of

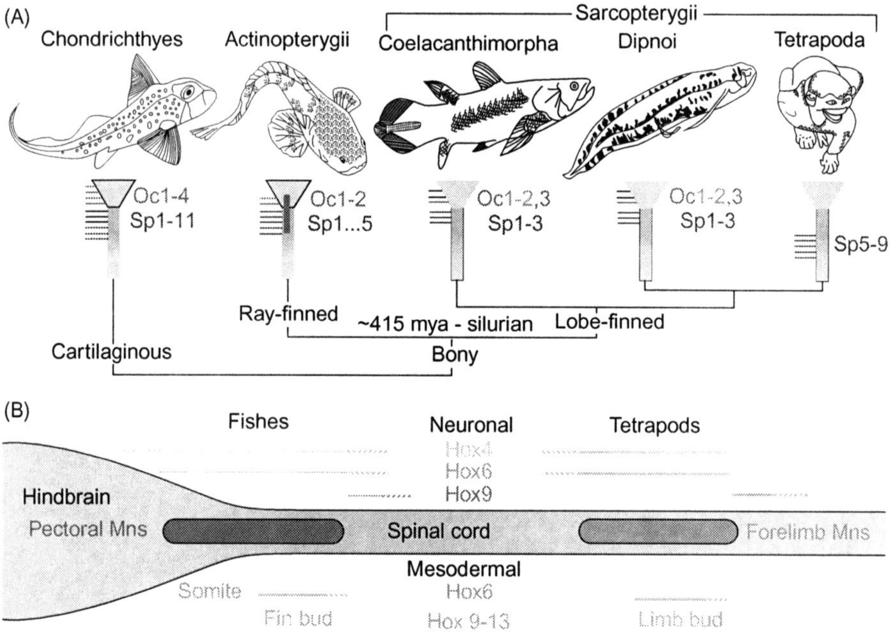

Figure 13.15 Evolution of pectoral innervation. (A) Cladogram of living jawed vertebrates, with vignettes showing innervation patterns of pectoral appendages. Occipital (pectoral, red; hypobranchial, gray) and spinal (blue) nerves are illustrated schematically. (B) Summary of key Hox genes expressed in neuronal (top) and mesodermal (bottom) compartments along the anterior–posterior axis in fish and tetrapods. *Abbreviations*: Oct, occipital nerve; Sp, spinal nerve. (For interpretation of the references to color in this figure legend, the reader is referred to the web version of this book.)
Source: From Ma et al. (2010).

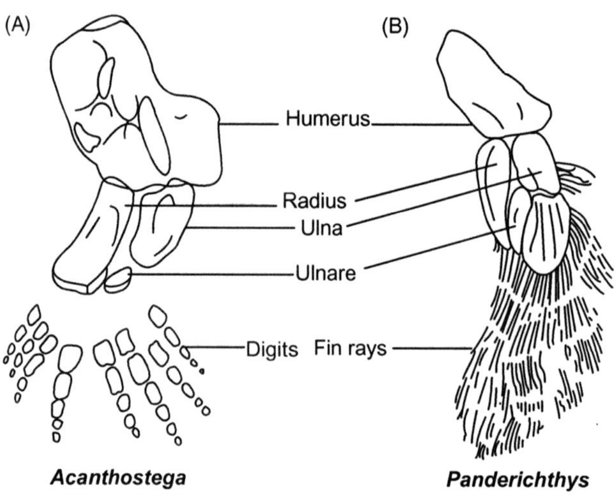

Figure 13.16 Comparison between forelimb structures of (A) a primitive tetrapod (*Acanthostega*) and (B) an advanced sarcopterygian fish (*Panderichthys*).
Source: From Long and Gordon (2004).

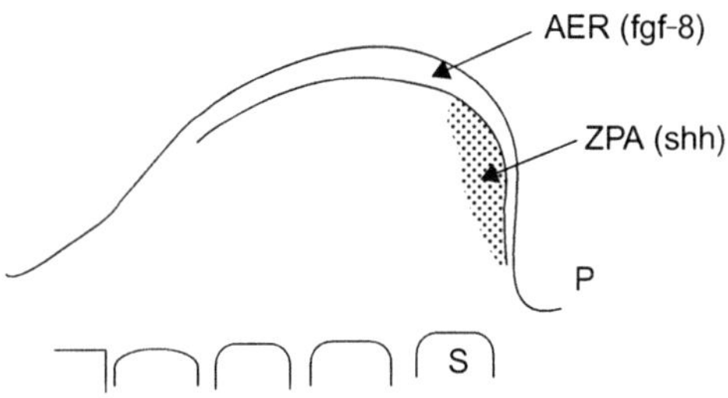

Figure 13.17 Schematic diagram of tetrapod limb bud. (AER and ZPA as sites of chick and mouse *Fgf-8* and *Shh* gene activity are indicated.) *Abbreviations*: AER, apical ectodermal ridge; P, posterior; S somites; ZPA, zone of polarizing activity.
Source: From Hinchliffe (2002).

a ZPA" (Lu et al., 1997). The forelimb bud secretes retinaldehyde dehydrogenase type 2 (RALDH-2) and RA. Secretion of RA by local motor neurons (Berggren et al., 2001) is the triggering signal for both of these crucial events in limb development.

First, RA stimulates the synthesis of Hgf (hepatocyte growth factor/scatter factor), which, along with the Wnt signals from the neural tube, stimulates in somites (lateral dermomyotome) differentiation of epithelial cells into mesenchymal myoblastic cells that migrate to the limb bud. Second, RA stimulates formation of the ZPA, which controls the establishment of the limb A–P axis (Stratford et al., 1999).

Wnt signals from the neural tube and the dorsal ectoderm, via β-catenin, activate the FGF-dependent mechanism that controls limb initiation and formation of AER and the dorsal ectoderm in the chick embryo (Kawakami et al., 2001; Niswander, 2002; Francis-West et al., 2003).

Removal of the AER in tetrapods prevents formation of distal structures and even leads to reduction of cell proliferation and cell death in distal mesenchyme within 4h. This may be responsible for the distal limb truncation. These consequences of the removal of the AER are rescued by local application of FGF (Dudley et al., 2002), one of the developmentally most important products of the AER.

A complex interaction between the ZPA and AER and their central molecular players (SHH, FGFs, and gremlin) determines the establishment of the limb A–P axis and growth of the limb and the extent of the growth (Figure 13.18).

Empirical evidence points to a causal relationship between the change in expression patterns of *Hox* genes and evolutionary transition from fish fins to tetrapod limbs and digits (Sordino et al., 1995; Wagner and Chiu, 2001). We know that *Hox* genes are regulated by RA, but there is no direct evidence yet on the mechanism and time when in the evolution of chordates the RA signaling cascade was coopted for regulating *Hox* genes (Marlétaz et al., 2006). However, indirect evidence indicates that a neural mechanism may be responsible for gene recruitment in metazoans (Cabej, 2011).

Limb buds in tetrapods seem to be initiated by RA signals from lateral plate mesoderm (LPM) (Capdevila and Belmonte, 2001; Mic et al., 2004) and from intermediary mesoderm (IM), as is suggested by the fact that placement of a barrier between the IM and LPM inhibits limb bud formation (Capdevila and Belmonte, 2001). RA signals induce secretion by limb bud mesenchymal cells of Hgf, which in turn

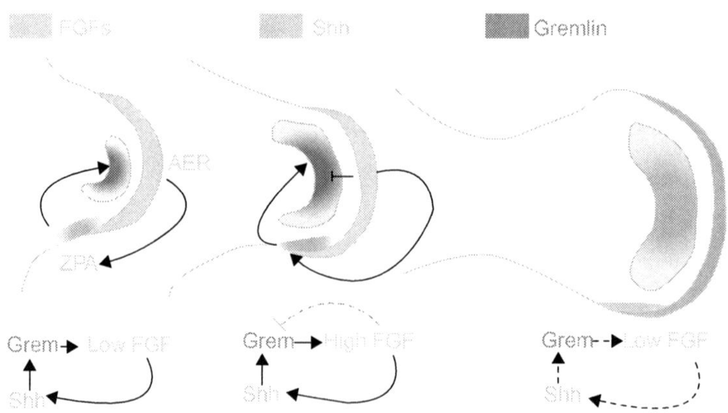

Figure 13.18 Positive- and negative-feedback loops control limb outgrowth and cessation of growth. This model explains how the two loops are used first to promote and then to terminate signals. Arrows indicate activation; the T-shaped red line indicates inhibition. Dashed lines represent diminished regulation. During promotion of outgrowth, the positive regulatory loop increases all signals; Shh produced in the ZPA (green zone) promotes FGF expression in the AER (blue zone). This effect of Shh is mediated through its ability to induce Gremlin. Gremlin, in turn, antagonizes bone morphogenetic proteins (BMPs), which inhibit FGF action (not shown). The overall effect is thus to promote FGF expression in the AER. The trigger to cessation of growth occurs when AER-FGFs are produced at a high-enough level that they repress expression of Gremlin (represented by a T-shaped line in the figure). By this stage of development, the limb mesenchyme has grown, which leads to a larger domain of Gremlin expression (red zone). Once Gremlin expression declines, cessation of limb growth occurs (last panel in figure). Gremlin is no longer present to repress BMP signals, and thus BMPs can repress FGF expression. Loss of FGF leads to an inability of FGF to maintain Shh expression. Thus, growth along all limb axes ceases. *Abbreviations*: AER, apical ectodermal ridge; FGFs, fibroblast growth factors; Grem, Gremlin; Shh, Sonic hedgehog; ZPA, zone of polarizing activity. (For interpretation of the references to color in this figure legend, the reader is referred to the web version of this book.)
Source: From Lyons and Ezaki (2009).

stimulates migration of myoblasts from somites to the limb bud (Mic and Duester, 2003). Migratory myoblasts in the fin and limb bud express identical genes (Haines and Currie, 2001). RA induces *Bmp4*, which in turn regulates *Hgf*, but other signals seem to be involved to fully determine expression domains of *Bmp4* and *Hgf* during the development of the limb bud (Mic and Duester, 2003).

It is demonstrated that RA synthesized in the limb bud (Stratford et al., 1996) stimulates formation of the ZPA and, by inducing expression of *Meis1* and *Meis2* genes (Mercader et al., 2000; Figure 13.19), as well as *Shh* (Riddle et al., 1993), contributes to P–D and A–P patterning of the limb. RA also regulates expression of *Fgf-4* gene (Stratford et al., 1996). It is possible that, besides the direct control on expression of *Hox* genes, RA also regulates their expression indirectly, by inducing expression of the *Cdx1* gene (Houle et al., 2003).

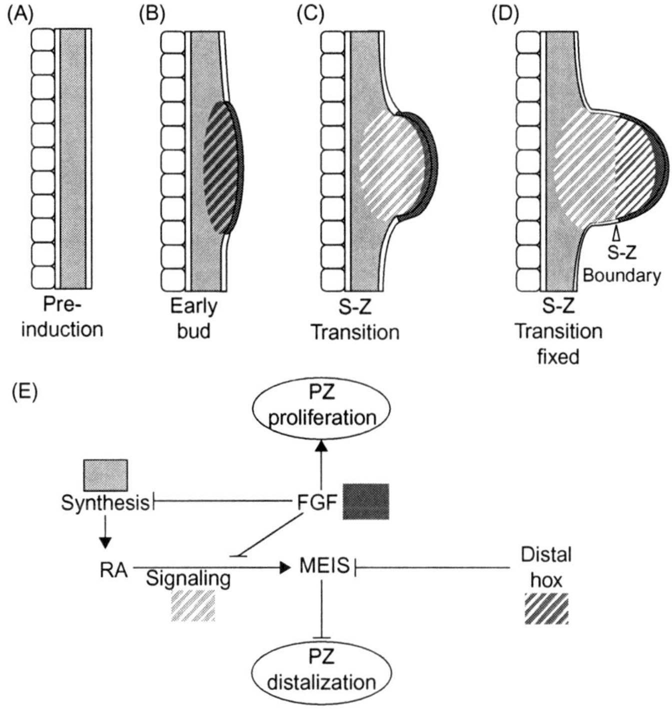

Figure 13.19 A new model integrating signals for limb growth and P–D specification. (A–D) Different stages of chick limb development. Arrowhead in D indicates the stylopod-zeugopod boundary (S-Z boundary). (E) A schematic representation of the interactions taking place between RA- and FGF-signaling pathways during limb development. Arrows indicate activation, and bars indicate repression.
Source: From Mercader et al. (2000).

Development of limb muscles starts in the dorsal part of the somites (dermomyotome) between the neural tube and mesoderm (Figure 13.20), but the development of somites themselves results from the interaction of dorsalizing (Wnt1, Wnt3a, and Wnt4) secreted by the dorsal half of the neural tube, and ventralizing signals consisting of Sonic hedgehog (Shh), Noggin, and FGFs, secreted from the floor plate of the neural tube and the notochord (Christ and Brand-Saberi, 2002).

The neural tube-notochord complex releases signals (among them, Hedgehog and Wnt) for MyoD expression, which codes transcription factors that induce expression of genes for muscle-specific proteins (MSPs) and Myf5, thus leading to differentiation of myoblasts and development of skeletal muscles within somites (Kronnie and Reggiani, 2002; Alves et al., 2003). The MSPs are also expressed according to an identical timetable in each somite (Xu et al., 2000). When separated from the neural tube, somites do not express *MyoD* genes (Alves et al., 2003). Limb muscle cell precursors form in the lateral dermomyotome and, after delaminating, migrate in the direction of the limb where, under the action of Shh and Wnt, signals from the neural

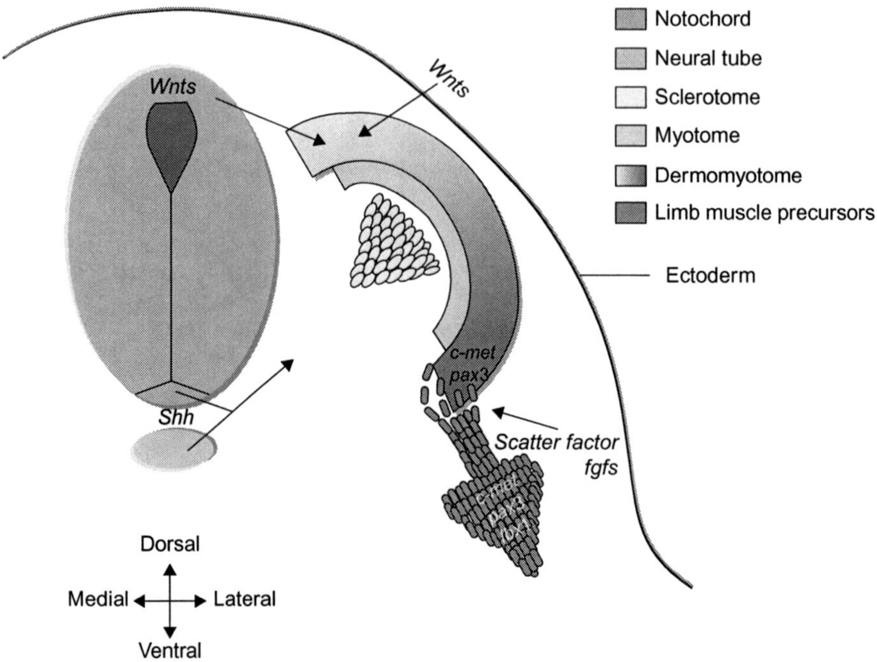

Figure 13.20 Factors that control myogenic induction in the somite. The somite is initially specified into two regions: the sclerotome and dermomyotome. The sclerotome forms the vertebrae and ribs, while the dermomyotome gives rise to the myogenic precursors and the dermis. The dorsomedial edge of the dermomyotome forms the epaxial muscles. This region involutes to give rise to the myotome, consisting of committed myogenic cells (i.e., expressing *Myf5*), which will form the back musculature. The dorsolateral region of the dermomyotome gives rise to the hypaxial muscles. At the limb level, the premyogenic limb precursors delaminate and migrate distally into the developing limb bud (large arrow). The molecular signals that control these events are well characterized. Simplistically, Shh, produced by the notochord and floor plate of the neural tube induces the sclerotome. Wnt proteins are expressed in the dorsal neural tube and the dorsal ectoderm and, together with Shh signaling, promote myogenesis. The limb premyogenic cells are induced to delaminate by scatter factor and FGFs. These factors are also thought to control migration. The premyogenic cells express *Pax3*, *Lbx1*, and the scatter factor receptor c-met and are not committed to myogenic differentiation due to repressive signals from the lateral plate mesoderm (Bmp4). Pax3 and c-met are needed for delamination, while Lbx1 is required for migration.
Source: From Francis-West et al. (2003).

tube and notochord differentiate into muscle cells, proliferate, and form limb muscles (Figure 13.21).

From Fish Fins to Tetrapod Limbs

The most important event for the evolution of the tetrapod limb is formation of broad domains of prespecified fate (Richardson et al., 2004; Figure 13.22) rather than any change in genetic information, of which no hint exists at all.

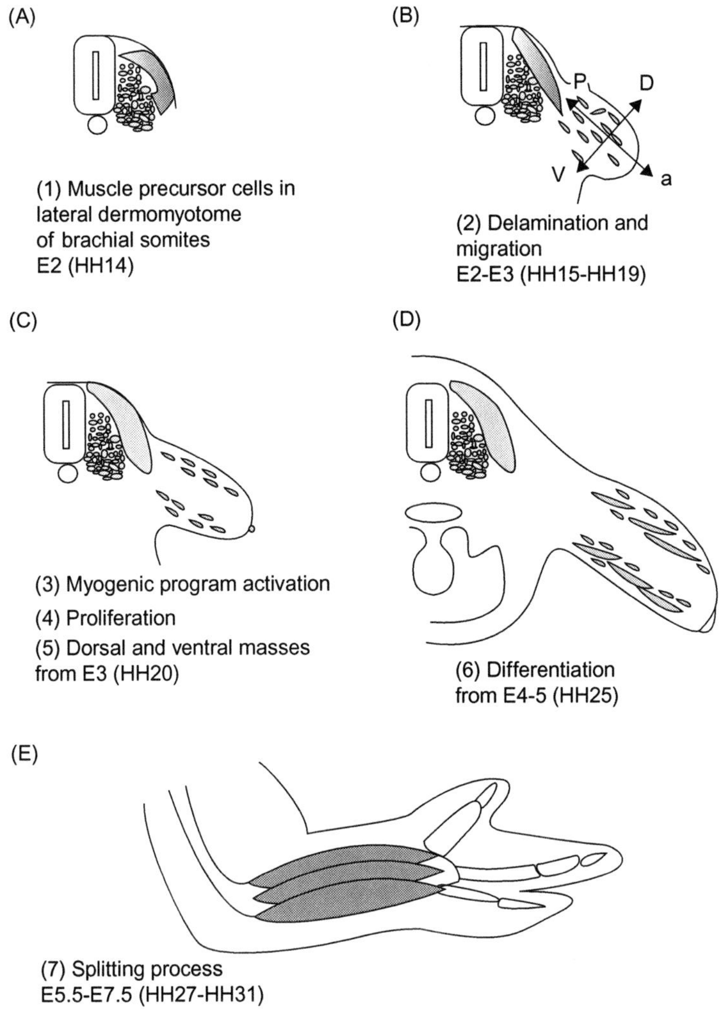

Figure 13.21 Events defining muscle formation in the embryonic chick wing. Somitic muscle cells (A) migrate into the wing somatopleure between HH (Hamilton and Hamburger, 1951) stages E2 and E3 (B) and then segregate into dorsal and ventral muscle masses (C, D) that undergo cleavages from E5.5 to give the final muscle pattern at E7.5 (E). The myogenic program is initiated from E3, concomitant with a proliferation step. The first signs of terminal differentiation are detectable from E4.5, from which time onward the proliferation and terminal differentiation proceed in parallel during embryonic development.
Source: From Duprez (2002).

Although it is impossible to prove with certainty that tetrapod limb bud is homologous to the fish fin (there are no fingers, but only fin rays in fish fins), we know that a similar "genetic toolbox" is used in both cases and even in "analogous" structures such as insect limbs. Positions of pectoral and pelvic fin buds also are similar to those of tetrapod fore- and hindlimbs.

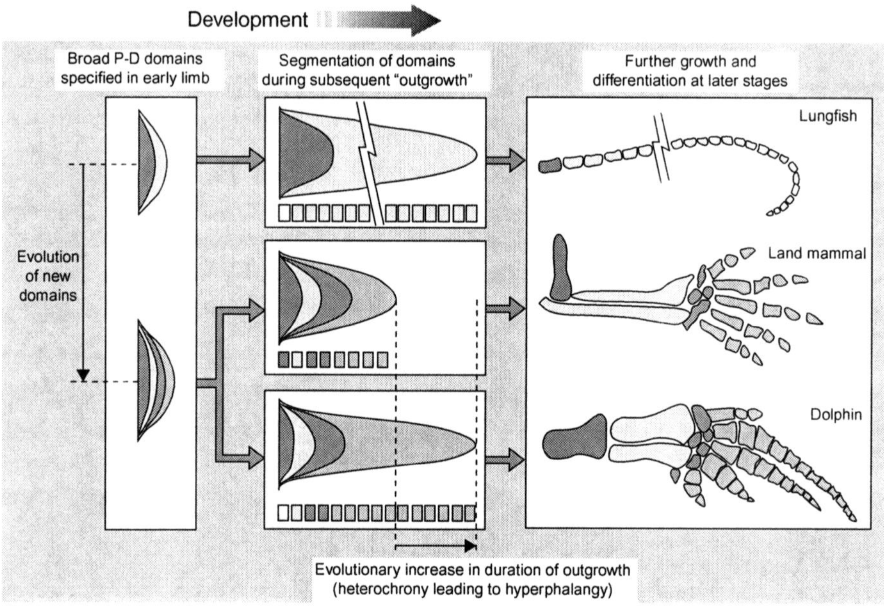

Figure 13.22 Formation and evolution of limb pattern. It is possible that an early stage of limb development involves the specification of broad domains of positional identity (represented in the far left-hand column by differently shaded vertical stripes). It is assumed that tetrapods have evolved a greater number of early domains than lungfish. It is assumed that in the dolphin, outgrowth is prolonged to give more elements with the same distal character. One possible outcome in the formed limb is shown schematically on the far right: Limb regions, corresponding to the early domains, are subdivided into skeletal elements.
Source: From Richardson et al. (2004).

Observations on a limited number of species (zebrafish (*Danio rerio*), mouse, and chicken) have shown that the same *Hox* genes are expressed in fin buds and limb buds (Hinchliffe, 2002). During the development of fins in fish, and limbs in tetrapods, only the early stage (AER-like stage) is similar. In clear distinction from tetrapods, the AER-like ridge in fish persists longer to form the fin fold, which fills with mesenchymal cells of neural crest origin, a unique fish feature. *HoxA* and *HoxD* genes are expressed in the mesenchyme of both fish and tetrapods, but in tetrapods expression of *HoxD* at a later stage shifts from the posterior part of the limb bud to the distal part, a fact that has been related to specification of digits (the autopod). The patterning of hindlimbs in mice is determined by three groups of paralogous genes: *Hox10* for the stylopod (femur), *Hox11* genes for the zeugopod, and the paralogous group of *Hox13* genes for the most distal part of the limb, the autopod (Wellik and Capecchi, 2003).

The study of mice mutant for *Hoxd-13* suggests that the new expression pattern of *Hox* genes in the limb bud led to the evolution of tetrapod limb from fish fins and particularly to the evolution of digits as a *de novo* vertebrate structure (Sordino et al., 1995).

In a series of experiments, it was conclusively shown that *HoxA-13* regulates formation of digits in the autopod by patterning expression of *Bmp2* and *Bmp7* in the interdigital and distal joint tissues, where these BMPs direct interdigital PCD, digit development, and chondrogenesis (Knosp et al., 2004). A clear correlation is also observed to exist between the spatiotemporal shift in the expression of *Hox* genes and morphological transformations in the fossil record (Shubin et al., 1997).

Based on the above facts, it is argued that it was not evolution of new regulatory genes but a change in expression pattern of existing genes (i.e., an epigenetic mechanism) that made the evolution of the tetrapod limbs and digits possible (Daeschler and Shubin, 1995; Long and Gordon, 2004).

Evolution of Digits

There is no consensus on the time when digits evolved in tetrapods, but paleontological evidence shows that the rule of observed pentadactyly is a latter development in the evolution of tetrapod limbs that probably was preceded by a stage of polydactyly.

In a recent model, it is the Shh that determines patterning of digits in two phases. Initially (the long-range phase), Shh primes mesenchyme cells to become competent for digit formation. Later (short-range phase), it regulates expression of *Bmps*, which, in a dose-dependent way, determine the pattern of digit formation (Drossopoulou et al., 2000), while the number of the digits is related to the length of the apical ridge.

But what controls expression of Shh for patterning digits?

The *Shh* gene expression in the limb mesenchyme is activated by RA and FGF-4, but, once induced, *Shh* expression is maintained by FGF-4 alone. This fact led to the hypothesis that retinoids are most likely part of a cascade of signals providing positional information in the limb, which may also include signals from *Hox* genes, *Shh*, and FGF (Ross et al., 2000).

A crucial role in the development and evolution of free digits is played by apoptotic processes, the PCD, which ultimately sculpts the species-specific digit morphology. In amniota (reptiles, birds, and mammals), apoptosis affects mainly interdigital mesodermal cells, but this process is limited to the distal parts of interdigital tissue, thus leading to formation of webbed digits, as is the case with aquatic birds (Zuarte-Luis and Hurle, 2002). Apoptosis of interdigital tissues in amniota is mediated by a signal cascade with caspases as downstream elements and with a few growth factors/hormones playing important roles in the process (Figure 13.29). RA is an upstream element in this cascade in the interdigital tissues. Recall, the synthesis of RA is a function of retinaldehyde dehydrogenase enzymes, whose synthesis, during limb development, is closely related to the neural tissue (Maden et al., 1998; Berggren et al., 2001).

Role of the Nervous System in Limb Development

The early belief that *Hox* genes are master control genes regulating the function of other genes is now replaced by the idea that the function of *Hox* genes themselves is regulated by extracellular signals, hormones. Hence, identification of upstream signals that regulate patterns of expression of *Hox* genes could contribute to our understanding of the causes and mechanisms of the evolution of the tetrapod limb. Experimental evidence in the past two decades has shown that RA is a specific inducer of *Hox* genes (Conlon and Rossant, 1992).

RA is necessary for specification of mesenchymal cells and their interaction with AER. Besides its role in initiating formation of the limb bud and the axial patterning, RA plays a role in the specification of muscle and bone tissues and, depending on its concentration, induces several different cell fates (Berggren et al., 2001).

In all its functions in limb bud and limb development, RA only acts as mediator of the activity of the local innervation (Berggren et al., 1999). Following the pattern of the synthesis of RA during stages 17 through 30 of the chick wing, Berggren et al. observed that there are two distinct phases in the synthesis of RA: a strong presence in the limb bud of RA during stages 17 and 18 that is followed by its absence during stages 19 through 21, and again high levels of RA in the presumptive brachial plexus region after the stages 22 through 30 (Berggren et al., 2001).

The role of the innervation as a source of RA in the developing limb bud and limb development is also inferred by the observation that denervation leads to defects that are similar to those produced by the absence of RA activity in limb development (Berggren et al., 2001). Indeed, studies on limb development in the frog *Rana pipiens* have shown that temporary denervation of limbs causes decreases in muscle and bone tissues and general reduction of the limb size in comparison with the controls (Dietz, 1987). Sciatic denervation performed in *R. pipiens* at stages 14 through 20 led to reduction of bone and foot size (Dietz, 1989).

RALDH-2, the most important enzyme for RA synthesis, is especially abundantly secreted by the cervical and lumbar regions of the spinal cord, corresponding to the forelimbs and hindlimbs, but it is also secreted by the axons of the motor spinal nerves extending toward the limbs (Berggren et al., 2001):

> *Within the embryonic CNS, the limb motor neurons in the spinal cord represent major sites of RALDH2 expression.*
>
> *Ross et al. (2000)*

Empirical evidence shows that RA plays a key role as an upstream inducer of signal cascades that triggers and drives the limb development on tetrapods. The mediators of the RA limb-inducing activity are Hox genes, which, in turn, induce secretion of Shh FGFs and formation of the ZPA and AER (Zakany and Duboule, 2007; Gillis et al., 2009).

As already mentioned, the neural tube and spinal cord are the main producers of RA in embryos. At E13 (embryonic day 13) in mice, the neural tube shows two (brachial and lumbar) RA maxima or "hot spots," which created haloes of RA in the surrounding tissues. These coincide with the sites where limb buds develop (Figures 13.23 and 13.24). RA levels between the maxima are several hundred times lower (McCaffery and Dräger, 1994).

Three other observations are relevant to understanding the mechanism of limb development. First, temporally RA hot spots coincide with the appearance of the limb buds (McCaffery and Dräger, 1994); second, RA levels decline with increased

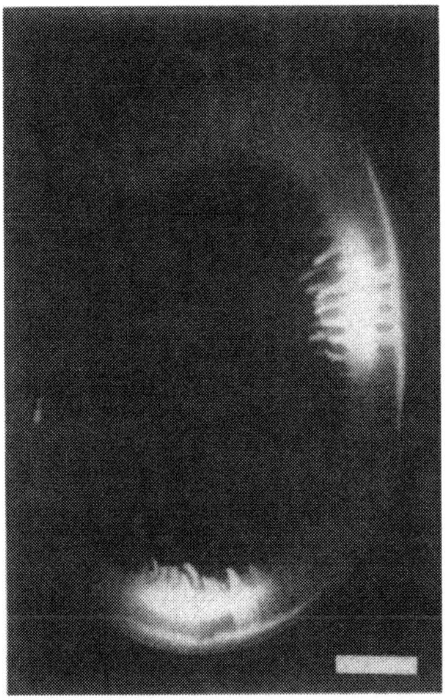

Figure 13.23 Fluorescent view of E12.8 spinal cord labeled with DiAsplO from the limbs showing the brachial and lumbar RA hot spots.
Source: McCaffery and Dräger (1994).

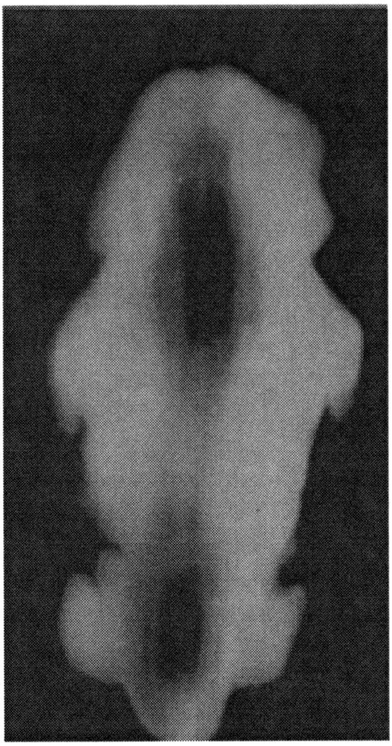

Figure 13.24 Localization of RA in the cervical and lumbar regions of the spinal cord, corresponding to fore- and hindlimbs, in E12.5 mouse embryo.
Source: From Colbert et al. (1993).

distance from the spinal cord (Maden et al., 1998); third, the RA maxima colocalize with the origin of nerves that innervate limbs (McCaffery and Dräger, 1994).

To understand whether these spatial and temporal coincidences of spinal cord RA maxima with limb development may reflect a causal relationship, it is logical to

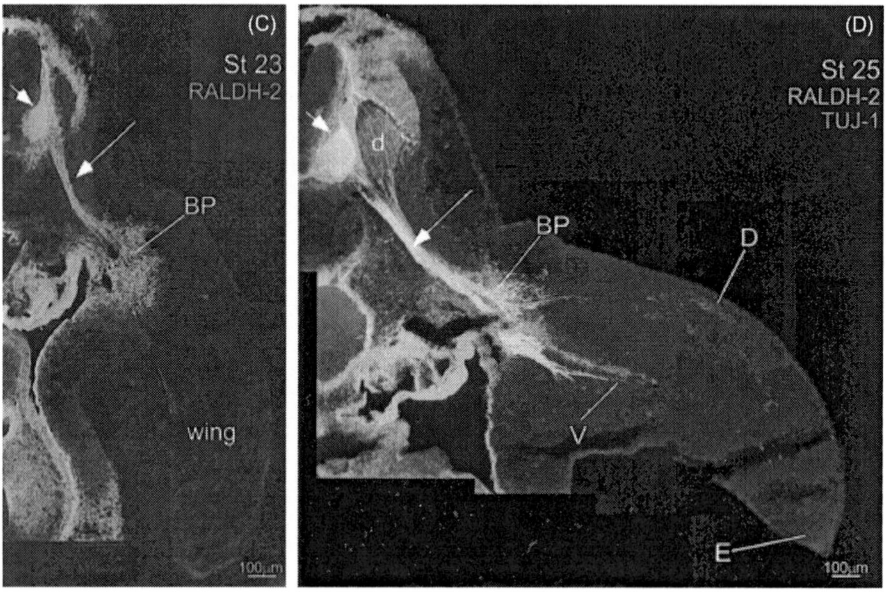

Figure 13.25 Normal series showing retinaldehyde dehydrogenase type 2 immunoreactivity (RALDH-2-IR) (green) in the developing wing. (C) Stage 23 wing section, RALDH-2-IR is present in the brachial plexus but not in the rest of the wing. Long arrow, axons; short arrow, mn. (D) Stage 25 wing section, RALDH-2-IR is present surrounding the axon tips as they extend into the wing. *Abbreviations*: D, RALDH-2-IR surrounding the dorsal nerve branch; V, RALDH-2-IR surrounding the ventral nerve branch; E, RALDH-2-IR at the distal end of the wing bud. Yellow color in motor axons (long arrow) results from colocalization of RALDH-2-IR and TUJ-1 (red). Sensory neurons in the dorsal root ganglion (d) label only with TUJ-1. Short arrow, mn; BP, brachial plexus; TUJ-1, an antibody against class III β-tubulin that serves as neuron-specific marker. (For interpretation of the references to color in this figure legend, the reader is referred to the web version of this book.)
Source: From Berggren et al. (2001).

examine whether the limb innervation from the spinal cord is involved in the embryonic development of the limb. The most interesting study in this respect is made by Berggren et al. (2001) on the development of forelimb (chick wings). RA is the regulator of expression of *Hox* genes and their downstream genes in tetrapod limbs, and that study has shown that, initially, the limb bud expresses no RA, while the motor neurons (but not the sensory neurons (Berggren et al., 1999)) of the adjacent brachial plexus express it (Figure 13.25).

Production of RA by the mesenchyme of the developing limb bud coincides with the penetration of motor axons in the chick wing, and even RA production in the limb mesenchyme is at least partly under control of the motor axons and vasculature (Berggren et al., 2001). Experiments of denervation of chick wings have also demonstrated that limb innervation is necessary for RA production by limb mesenchyme (Figure 13.26) and for limb development in general.

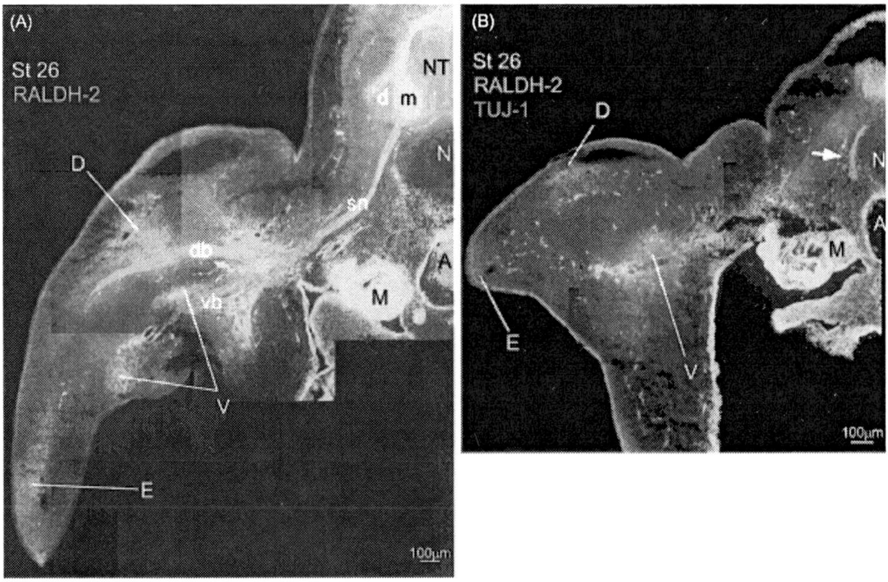

Figure 13.26 Decrease in wing mesenchymal retinaldehyde dehydrogenase type 2 immunoreactivity (RALDH-2-IR) seen after motor denervation. (A) Stage (St) 26 control wing section. Dorsal (D), ventral (V), and distal (E) mesenchymal RALDH-2-IR are seen in relation to the dorsal (db) and ventral (vb) nerve branches. *Abbreviations*: NT, neural tube; m, motor neurons; d, dorsal root ganglia; sn, spinal nerve; A, aorta; M, mesonephros. (B) Decrease in the area of RALDH-2-IR after motor denervation in a stage 26 embryo that had been denervated at stage 16. Sensory nerves stained with TUJ-1 are present in axial regions (TUJ-1 stain, short arrow) but very sparse in the wing.
Source: From Berggren et al. (2001).

It is interesting also to know that

> *The hot spots appear with formation of the limbs and persist until limb innervation is about complete ... the hot spots are a likely factor in the formation of the limb zones.*
>
> *McCaffery and Dräger (1994)*

The role of the spinal cord in limb development is also suggested by experiments in which segments of the spinal cord and central vasculature added in culture stimulate normal wing outgrowth (Tanaka et al., 1996).

According to Berggren et al. (2001), in chick embryos,

> *it seems likely that innervation of the limbs is necessary for both muscle and bone development .... RA is a factor that could mediate this neuronal influence. Between stages 23 and 30, localized regions of RALDH-2 expression develop in the mesenchyme along the vasculature and nerve branches; this RALDH-2 is partially under the control of the blood vessels and motor axons as they enter the wing. This later*

> *regional expression of RALDH-2 provides localized sources of RA in the limb during the period of mesenchymal specification. As RA is known to be involved in many aspects of cellular differentiation, we propose that the RA synthesized locally by RALDH-2 in the wing is involved in the specification of mesenchymal tissues of the limb, including cartilage, skeletal muscle, and vascular smooth muscle.*

The evolutionary and developmental implications of the role of the local innervation in cell differentiation and patterning of limbs in tetrapods cannot be overestimated:

> *If some of this RA-mediated differentiation is under the control of innervation and vasculature, there are broad implications for the role of motor nerves and blood vessels in the development of many other systems of the body.*
>
> *Berggren et al. (2001)*

The development of muscle fiber types in chicks coincides with the penetration of nerves in these muscles. Upon entering the developing chick limbs, motor neurons release RALDH-2 (RA-synthesizing enzyme), inducing muscle cell differentiation and muscle formation there (Berggren et al., 2001).

The role of the nervous system in the development of limbs in tetrapods is not limited to the above, and the evidence presented so far would be incomplete if one were to neglect the neurohormonal activity of the CNS during the latter stages of limb development. Signal cascades starting from the CNS are involved in the formation and specification of limb tissues. So, for example, a cascade starting with secretion of growth hormone-releasing hormone (GHRH), via the pituitary growth hormone (GH), stimulates proliferation of prechondrocytes in the epiphyseal growth plates of developing bones (Wolpert et al., 1998).

Neo-Darwinian Explanation

Transition from fish fins to tetrapod limbs is characterized by tremendous changes and diversity in the morphology of vertebrate limbs. That transition was based on the use of preexisting genes rather than any changes in genetic information or DNA in general. All the essential genes involved in limb development are functionally conserved from lower vertebrates to higher mammals, including humans. This makes impossible any gene-centric explanation.

Epigenetic Explanation

From the epigenetic view, it would be predicted that evolution of limbs in tetrapods:

- Required no new genes or changes in genes involved in the formation of limb tissues and structure,
- Involved changes in the spatiotemporal expression of the preexisting functionally conserved genes,
- May involve participation of neural crest cells,
- Involved neurohormonal regulation of the development of limb tissues,
- Involved participation of local innervation in the limb development.

The evidence presented in this section validates all the above predictions. No changes in genes whatsoever but

> *Changes in the temporal expression, spatial expression, and levels of expression of key developmental regulators such as Bmp and Fgf appear to be important in driving the evolution of vertebrate limbs.*
>
> *Weatherbee et al. (2006)*

Crucial to the evolution of the tetrapod limb, like evolution of any new phenotypic character, is the source of the new information necessary for the development of that character during the ontogeny. With changes in the function of genes or evolution of new genes being excluded as potential sources of information for the evolution of tetrapod limbs from fish fins, the experimental evidence presented in this section shows that all the stages of the limb development are related to epigenetic factors related to the activity of the embryonic and postembryonic CNS.

Evolution of Wings in Bats

Bats appeared during Eocene (56–40 mya) in an extraordinary mammalian Big Bang. They are grouped in 18 families comprising 20% of extant mammal species. The group is characterized by the powered flight, which is unique among mammals, and echolocation abilities, both of which may be functionally and evolutionarily linked (Simmons, 2005). The length of wing bones relative to the body size in bats has remained unchanged during ~50 million years, as is inferred from the bat fossil records (Figure 13.27).

The fact that these species have evolved their present wing skeletal morphology suddenly (in a few million years; Sears et al., 2006) and early in their evolution stimulated some investigators to search for mechanisms that could have facilitated or made the rapid evolution of bat wings possible. In tracking the embryonic development of the bat *Carollia perspicillata*, they observed that all the embryonic parts of the bat forelimbs (humerus, ulna, radius, phalanges) are of the same length as those of mice (Figure 13.28), until the segmentation, when an accelerated rate of proliferation and differentiation of chondrocytes causes the characteristic elongation of forelimb digits in bats. When cultured in the presence of bone morphogenetic protein 2 (BMP2), both bat and mouse embryos grow longer, and in presence of the BMP2 antagonist, Noggin, they grow shorter. The presence of BMP2 stimulates increased proliferation and differentiation of chondrocytes, and the development of longer wing digits in bats is related to higher levels of BMP2 in the flying mammal. The increased expression of BMP2, but not any mutation affecting the function of the *Bmp* gene (which did not occur), seems to have been the critical event for the evolution of wings in bats. This epigenetic (=regulatory) change in the forelimb developmental pathway might have been sufficient for the rapid evolution of these exaggerated mammal structures (Sears et al., 2006).

Bats retain interdigital membranes in the forelimbs, while eliminating them in the hindlimbs. The mechanism of this differential development in the forelimbs and hindlimbs is evidently not related to any mutational changes in genes: the same

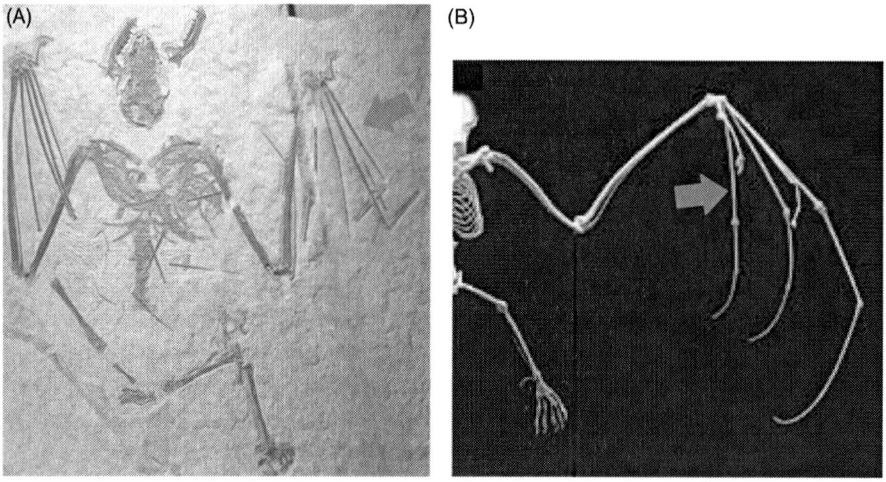

Figure 13.27 The relative length of bat forelimb digits has not changed in 50 million years. (A) *Icaronycteris index* (American Museum of Natural History specimen no. 125000), which is a 50 million-year-old bat fossil. (B) Extant adult bat skeleton. The metacarpals (gray arrows) of the first fossil bats are already elongated and closely resemble those of modern bats. This observation is confirmed by morphometric analysis of bat forelimb skeletal elements.
Source: From Sears et al. (2006).

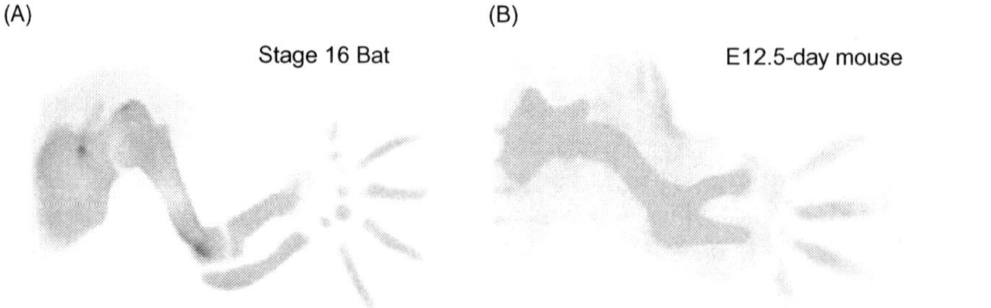

Figure 13.28 Comparable stages of development of cartilaginous forelimbs in bat and mouse.
Source: From Sears et al. (2006).

set of genes is present in all cells throughout the bat organism, including fore- and hindlimbs. While retaining the ancestral mechanism of elimination of interdigital membranes in the hindlimbs, bats evolved an epigenetic mechanism of retaining embryonic interdigital membranes consisting of induction of a BMP antagonist, gremlin (a secreted protein of the group of BMP antagonists), and FGF in the forelimbs. In ducks, the antiapoptotic gremlin expresses strongly in the proximal side of the limb, whereas in mice it expresses still less distally (Weatherbee et al., 2006; Figure 13.29).

The fact that the relevant change in BMP expression pattern occurred in the forelimbs alone, but not in the hindlimbs (despite the fact that all the genes are the same

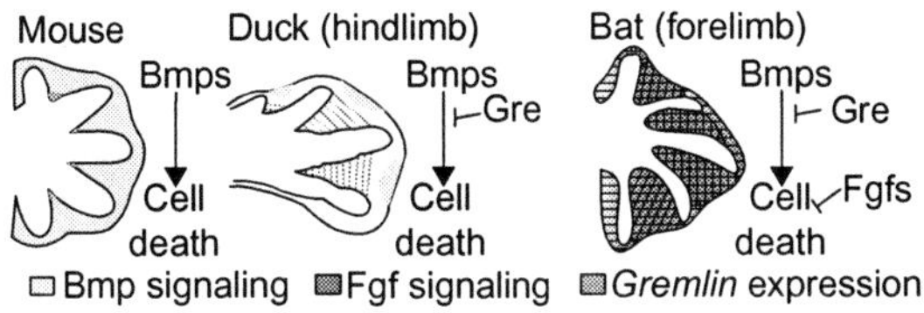

Figure 13.29 Schematic of the differences in gene expression in free-toed mouse limbs and webbed duck and bat limbs. Mouse forelimbs show proximally restricted *Gremlin* expression (red) and high levels of Bmp signaling (yellow) throughout the interdigit, which results in extensive cell death of interdigit tissue and free digits. Duck hindlimbs have strong proximal expression of *Gremlin*, which blocks Bmp-induced gene expression and apoptosis. Bat forelimbs exhibit Bmp signaling, but cell death is blocked, likely because of the widespread expression of *Gremlin* and the unique domain of Fgf-8 signaling (blue) in forelimb interdigit regions. (For interpretation of the references to color in this figure legend, the reader is referred to the web version of this book.)
Source: From Weatherbee et al. (2006).

all over the animal body), clearly points to a manipulative BMP expression in forelimbs, but the only known mechanism of manipulative expression of genes is neurally determined expression (see Section Molecular Mechanisms of Manipulative Expression of Genes in the CNS in Chapter 2).

Neo-Darwinian Explanation

The neo-Darwinian paradigm would predict that

> Evolution of bat wings is a gradual process of adaptation of the mammalian phenotype via accumulation of useful mutations that led to incremental changes in the size of forelimbs and in interdigital membranes.
> Paleontological evidence, however, shows no intermediate forms that would suggest gradual accumulation of phenotypic changes leading to evolution of wings in bats. The explosive evolution of bat wings in the space of a few million years and the astounding conservation of the wing structure ever since (Sears et al., 2006) clearly speak against any gradualistic explanation.
> Besides, there is no evidence of new genes, changes in function of existing genes or regulatory sequences involved in the evolution of bat wings. The same functionally unaffected genes for limb development are used for the development of widely different structures, such as the mouse feet, duck webbed feet, and bat wings.

Epigenetic Explanation

Epigenetic paradigm would predict that

1. *Changes in expression* rather than mutations in key genes in forelimb development would have been necessary for evolution of bat wings from mammalian forelimbs.
2. The source of information necessary for changes in expression of genes for bat wing development originates in the nervous system.

The evidence presented in this section validates the first prediction, but we have no reliable evidence for substantiating the second prediction. However, there is evidence that increased proliferation and differentiation of chondrocytes during embryogenesis, as a result of increased expression of Bmp2, is responsible for elongation of digits in bats (Nishihara et al., 2003; Sears et al., 2006). The fact that RA regulates expression of Bmp2 in chondrocytes (Cohen et al., 2006), in view of the dominant role of the neural tube–derived RA and later RA secreted by nerves innervating limbs in the development of tetrapod limbs, suggests that the CNS may have played an essential role in the evolution of wings in bats.

Evolution of Blood Circulatory System

Blood circulation system in metazoans evolved from open circulation systems with a peristaltic heart in invertebrates into closed systems in vertebrates characterized by a multichambered heart that, via outflow and inflow tracts, pumps blood into a closed system of blood vessels. Surprisingly (not from the present-day view), the developmental pathways used during the development of widely different blood circulation systems of invertebrates and vertebrates are very similar (Yoshida et al., 2010). This led to the idea of a common genetic toolkit that is used to develop the most varied hearts and circulatory systems in metazoans. Key signaling molecules in the development of the cardiovascular system, such as VEGF receptor (VEGFR), FGFR (Cripps and Olson, 2002; Yoshida et al., 2010), and Nkx2.5/Csx (Elliott et al., 2006; Figure 13.30), are shared between cephalopods and vertebrates. This idea is further corroborated by the fact that there are also invertebrates that have evolved closed and complex hearts and circulatory systems. Such is the case with cephalopods, for example.

Based on the above experimental evidence, it is proposed that "the conserved molecular developmental program for cardiovascular systems were recruited independently to the closed circularity systems of cephalopods and vertebrates" (Yoshida et al., 2010).

When this evidence is considered in the context of the fact that VEGFR is induced by VEGF (Zentilin et al., 2010), that the endogenous neurotransmitter dopamine, by binding its receptor D2 in endothelial cells, regulates angiogenetic action of VEGF (Sarkar et al., 2004), that neural tube is the midline signaling center for vascular patterning in higher vertebrates (Hogan et al., 2004), that the brain and spinal cord determine the formation of the perineural vascular plexus around themselves by releasing VEGF and other signals (Hogan et al., 2004), that peripheral nerves by secreting VEGF determine skin blood vessel patterns (Mukoyama et al., 2005), one cannot escape the idea that the nervous system is essentially involved in the development and evolution of the blood circulatory system.

Evolution of Air-Breathing and Surfactant System in Vertebrates

Lungs develop as a pair of evaginations from endodermal buds of the foregut (Stenmark and Gebb, 2003). Lungs and swim bladder in fish evolved from different structures and

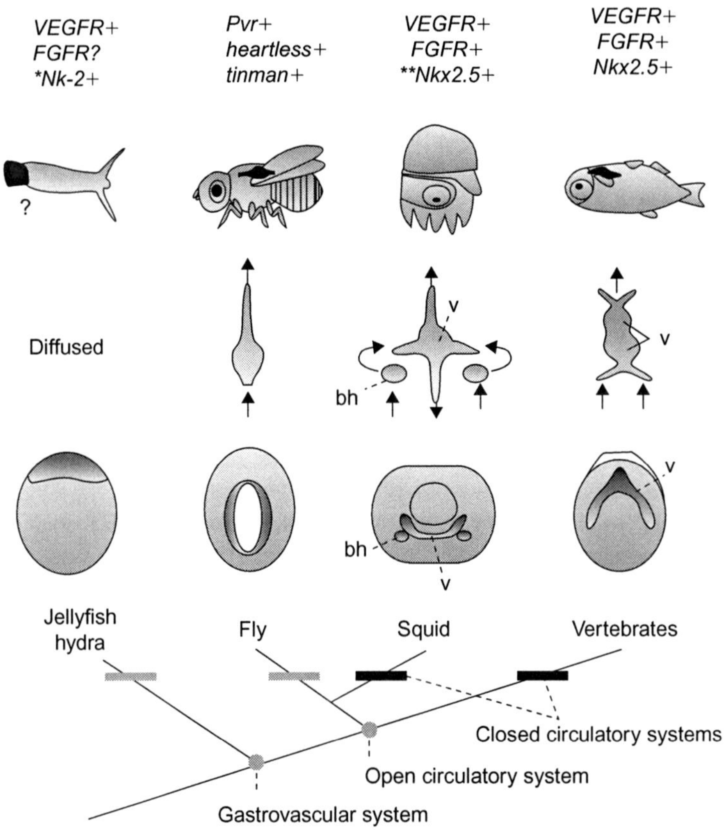

Figure 13.30 Generalized scheme for a conserved molecular signature of the developing hearts in selected animals. A traditional phylogenetic tree and emergence of three types of circulatory systems (bars) are shown below. Situation of the common ancestor is hypothesized as circles. In each adult body plan depicted above, the black area shows the position of hearts or heartlike structure. The embryonic origins of hearts or heartlike tissues are restricted in the embryos as flat anlagen. (Anterior direction is above in figures immediately above the phylogenetic tree.) These anlagen are centralized and finally develop into the diffused gastrovascular system or tube-form hearts in adults. (The main vascular directions of blood flow are shown with arrows in figures immediately below the diagrams of adult body plan.) bh, branchial heart; v, ventricle. Data are sourced for: hydra, jellyfish, Drosophila, vertebrates. In cnidarians, involvement of FGFR in the circulatory formation is not reported. *In jellyfish, VEGFR is expressed in endodermal cells of their gastrovascular system, but coexpression with Nk-2 is not determined. **In squids, Nkx2.5 is also expressed in the developing hearts. *Source*: Elliott et al. (2006).

for different functions, respectively, for air breathing and buoyancy. In modern fish, air breathing evolved independently, at least 38 (and possibly as many as 67) times. Lungs may have appeared first in the armored fish, placoderms (Figure 13.31), but are not homologous with Ostheoichthyan lungs; they differ with respect to both the embryonic origin (placoderm lungs derive from the anterior pharynx), and the function they perform (in placoderms, they serve for buoyancy regulation rather than air breathing).

With extinction of placoderms, primitive cartilaginous fish had no lungs, and fish had to re-evolve lungs. Sarcopterygii and Actinopterygii evolved paired lungs in the ventral part of the body from the posterior pharynx.

One precondition for the evolution of lungs has been evolution of the surfactant, a mixture of phospholipids and proteins. Because of the surface tension from water droplets and the small size, the lung alveoli tend to cave in, which would make breathing impossible. To prevent this from happening, vertebrates secrete the surfactant, which decreases the surface tension, thus avoiding the collapse of the alveoli. Defects in the production of surfactant lead to numerous lung and respiratory pathologies; no lung alveolization takes place in the absence of surfactant and, consequently, embryos cannot survive postnatally. This is why the surfactant excretion in the alveoli and airways is a common characteristic of lungs in all vertebrates.

The surfactant is produced by the pulmonary alveolar type II cells, and experimental evidence has shown that placental leptin is involved in both surfactant production and the increase of the number of alveolar type II cells in lungs (Kirwin et al., 2006). Production of surfactant in lungs is stimulated by parathyroid hormone-related protein (PTHrP), which in turn is induced by stretch. PTHrP receptor is the mediator of the surfactant-producing action of the PTHrP in lungs. Similarly to bones, the stretch affects expression of the hormone PTHrP and, consequently, the

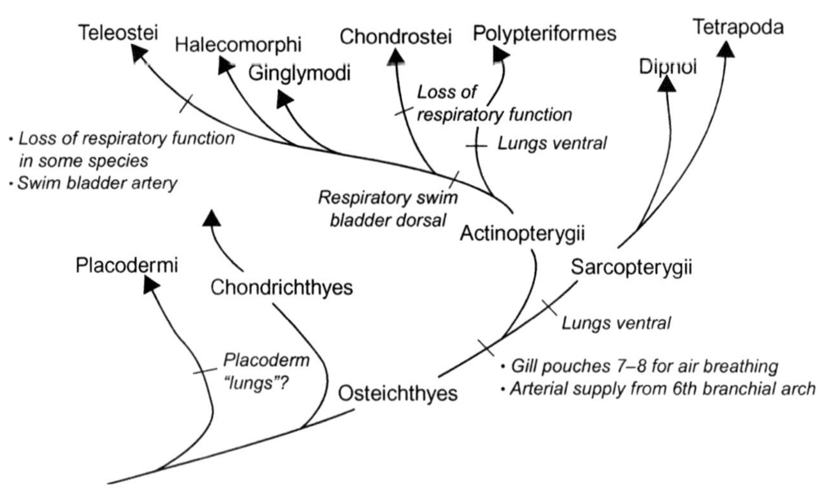

Figure 13.31 Schematic diagram of the evolutionary sequence of air-breathing organs among the fish. The ontogenetic origin (i.e., ventral or dorsal) and the evolution of lungs, swim bladders, and their blood supply and the loss of respiratory function are shown.
Source: From Daniels et al. (2004).

homeostatic control of both bones and lungs (Daniels and Orgeig, 2003). Evolution of the surfactant system is closely related to the evolution of lungs in vertebrates (Torday and Rehan, 2002).

Evolution of air breathing and lungs would have been impossible without evolution of a surfactant system (Daniels et al., 2004). Hence, evolution of surfactant system predated the evolution of lungs. Indeed, surfactant is also present in the swim bladder of fish (Daniels et al., 2004). A look at the evolution of air breathing and lungs in vertebrates shows that the increased metabolic rates observed in the course of vertebrate evolution are accompanied by a tendency for increased surfactant activity, thinning of the alveolar wall, and diminution of alveolar size, which leads to a greater surface area for gas exchange in lungs, all related to an epigenetic increase in the strength of PTHrP signaling (Torday and Rehan, 2002; Figure 13.32).

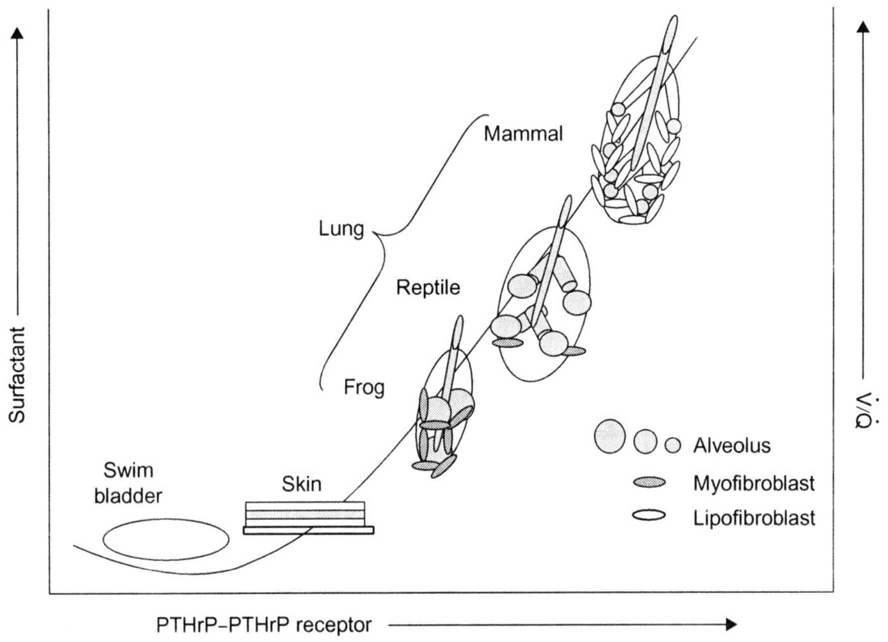

Figure 13.32 Structural evolution of the organ of gas exchange. During phylogeny from fish to mammals, the organ of gas exchange becomes more and more complex, increasing in surface area to accommodate the metabolic demand for oxygen. This is particularly true of the arboreal conducting airways and clustering of alveoli in the mammalian lung. Cellular changes in the interstitium of the lung from amphibians to reptiles and mammals are characterized by a decrease in myofibroblasts and an increase in lipofibroblasts. There is a concomitant decrease in the diameter of the alveoli. We hypothesize that the structural changes are due to the progressive increase in the PTHrP/PTHrP receptor amplification signaling (*x*-axis), which enhances surfactant production and $\dot{V}/\dot{Q}$ matching (*y*-axis). *Abbreviation*: $\dot{V}/\dot{Q}$, the physiological process of ventilation/perfusion matching (increased surfactant secretion and blood flow in alveolar capillaries as a result of alveolar wall distension).
Source: From Torday and Rehan (2002).

Over time, the blood–gas barrier became progressively thinner in land vertebrates from amphibians to reptiles, mammals, and birds (Torday et al., 2009). Despite the lower efficiency and differences in the embryonic origin of the fish surfactant, the surfactant systems of fish and tetrapods are homologous because they "had a single evolutionary origin that predated the evolution of the vertebrates" (Daniels and Orgeig, 2003).

Neural crest cells also migrate to the early budding epithelial tubules and in the developing lungs to develop there the intrinsic lung innervation (Freem et al., 2010). They form a neural tissue in the proximal part of the lung, and nerves follow smooth muscle-covered tubules at the base of the growing lung buds that put "smooth muscle and neural tissue in a prime position to influence growth and development" (Tollet et al., 2001).

Neo-Darwinian Explanation

From the neo-Darwinian view, evolution of the surfactant system and the related evolution of the alveolar structure in vertebrates would depend on occurrence and selection of changes affecting function of PTHrP or PTHrP cascade. The fact that no changes affecting the function of PTHrP or its cascade have occurred during ~500 million years of vertebrate evolution shows that the cause of evolution of the surfactant system is not genetic, and there is no reasonable neo-Darwinian explanation of that evolution.

Epigenetic Explanation

As pointed out earlier, evolution of the surfactant system is not related to evolution of changes in DNA, genes, or allele frequencies, but only to regulatory, epigenetic changes in the expression of PTHrP and to different combinations of the same components (phospholipids and proteins) in the surfactant mixture.

The developmental pathway that determines surfactant production is regulated prenatally by the following brain signal cascade:

hypothalamic CRH → pituitary corticotropin → adrenal cortisol → the fibroblast pneumocyte factor (secreted by lung fibroblasts) → surfactant (secreted by type II pneumocytes) (see also Figure 5.18).

Evolution of Dentition in Vertebrates

About 500 mya, jawed vertebrates evolved exoskeletal body armors. It is believed that teeth of jawed vertebrates evolved by modifications of dentinoid and enameloid odontodes, ridges of tubercles on vertebrate body armor. Dermal bony plates of gnathostomes represent an intermediate stage in the evolution of teeth from odontodes (Hildebrand et al., 1995). Ever since, dentition patterns have diversified widely among vertebrate taxa.

Tooth development results from a coordinated interaction of the oral epithelium and the underlying mesenchyme. Odontogenic mesenchymal cells are neural crest–derived cells of rhombencephalic neural crest origin that migrate to the mandibular process to form the dental papillae (Hildebrand et al., 1995). Before the initiation of tooth formation, when the dental lamina is already formed, the neural crest cells populate the entire region under the oral epithelium (Chai et al., 2000). The odontogenic properties of the underlying mesenchyme depend solely on the neural crest–derived mesenchyme (Luukko, 1998).

Premigratory cranial neural crest (CNC) cells from the mesencephalic and metencephalic regions, as well as the trunk neural crest cells of the neural folds when combined with mandibular arch epithelium, induce tooth formation (Lumsden, 1988). These facts clearly show that neural crest–derived cells of the mandibular mesenchyme are uniquely predisposed or prespecified (Tucker and Lumsden, 2004) for odontogenesis before coming in contact with local cytological-molecular elements in dentition sites. Such facts lead investigators to the conclusion that the "mammalian neural crest has an odontogenic potential" (Lumsden, 1988).

The crucial role of the neural crest cell mesenchyme in odontogenesis must be considered in relation to important experimental evidence that the nervous system is also directly involved in processes of odontogenesis. The neural crest cells involved in the formation of maxillar and mandibular teeth come from the midbrain and hindbrain (rhombomeres r1, r2, r3), start migrating to the maxillar and mandibular regions by the 8-somite stage in mice (Figure 13.33), and are provided with odontogenic information before reaching dentition sites.

In antagonistic (activator/inhibitor) actions of FGFs and BMPs on *Pax9* expression, a shift of balance in favor of FGF expression induces expression of *Pax9* (Neubüser et al., 1997), and *Msx1* (Peters and Balling, 1999) in the mesenchyme,

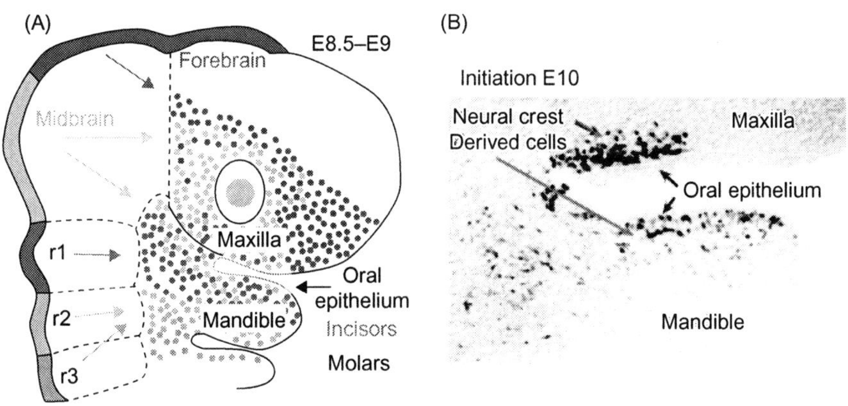

Figure 13.33 (A) Schematic representation of the migration of cranial neural crest cells toward the facial region and the oral cavity. (B) Section showing neural crest–derived cells (red arrows) in contact with the oral epithelium. (For interpretation of the references to color in this figure legend, the reader is referred to the web version of this book.)
Source: Mitsiadis and Graf (2009).

thus determining sites of tooth formation. In latter stages of odontogenesis, it is these mesenchymal signals that induce expression of FGFs and BMPs (Peters and Balling, 1999). Sequential and reciprocal interactions between CNC cells and oral epithelium are required for tooth development (Mitsiadis et al., 2003; Figure 13.34). The proximal boundary of expression of the homeobox gene *Msx1* in the mesenchyme coincides with expression domain of another homeobox gene, *Barx1*. The boundary between the expression domains of these homeobox genes in the mandibular arch corresponds to expression domains of FGF-8 and BMP4 in the overlaying epithelium and determines the two tooth-forming regions of incisors and molars in mouse embryos. Investigators have succeeded in transforming presumptive incisor teeth into molar teeth by implanting FGF-8 beads in the region of the presumptive incisors. Similar results of transforming presumptive incisor tooth germs into molar tooth germs are obtained by implanting beads of Noggin, an inhibitor of BMP (Tucker et al., 1998), thus demonstrating the role of the initial FGF-8 in determining the incisor/molar fate of the teeth. In both cases, investigators stimulated expression of molar-determining gene *Barx1* that was inhibited by BMP. Despite the apparent differences in the number and morphology of teeth in vertebrates, *Bmp* genes involved in odontogenesis are functionally unchanged across the lower and higher vertebrate species (Streelman et al., 2003).

Drastic changes in tooth phenotypes (e.g., loss and size reduction of molars, changes in the form of the crown, lack of enamel and accompanying deformation of incisors, as well as changes in dental formula) have been experimentally induced by downregulating BMP signaling in mice (Plikus et al., 2005). These drastic modifications in the number and morphology of teeth by topical manipulation of the level of BMP, without changes in the *Bmp* and other key genes in tooth patterning, strongly suggest that no changes in genes have been necessary for the evolution of dentition in vertebrates.

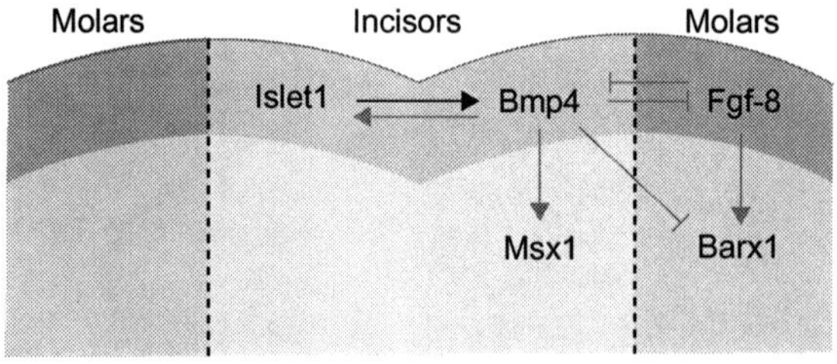

Figure 13.34 A model illustrating the molecular events directing murine dentition patterning. A regulatory loop exists between ISL1 and BMP4 in oral epithelium, where *Isl1* stimulates *Bmp4* expression and *Bmp4* activates *Isl1* expression. Thereafter, *Bmp4* activates *Msx1* expression in mesenchyme, whereas it acts as an antagonist for *Fgf-8* expression from oral epithelium and downregulates *Barx1* expression in mesenchyme. Darker-shaded areas represent the incisor and molar territories, respectively, of the oral epithelium, overlaying the mesenchyme. *Abbreviation*: Isl1 (Islet1), a homeodomain protein of the LIM family.
Source: From Mitsiadis et al. (2003).

Role of the Nervous System in Tooth Development

Two observations attracted the attention of investigators to the possibility that the nervous system is directly involved in vertebrate tooth development. First, that the dental nerve enters the jaw before the beginning of odontogenesis (Hildebrand et al., 1995; Fried et al., 2000) and, second, that axons are detected in the sites where the teeth develop.

When a dental lamina is formed, a plexus of nerve branches is seen in the subepithelial mesenchyme. Shortly thereafter, specific branches to individual tooth primordia are observed. In the bud-stage tooth germs, axon terminals surround the condensed mesenchyme, and in the cap stage, primordia axons grow into the dental follicle (Hildebrand et al., 1995; Figure 13.35). The tooth anlage is innervated by nerves from the trigeminal ganglion, and this innervation is tightly linked to the tooth development (Kettunen et al., 2005). However, conflicting experimental evidence came from studies of odontogenesis in mammals. Experiments on rat embryos have shown that the local innervation is not necessary for the *initiation* of odontogenesis (Luukko, 1997; Lumsden and Buchanan, 1986). According to Luukko (1997), although dental nerves control the rate of dentinogenesis, the main branches emerging from the trigeminal ganglion run parallel to, and beneath, the oral epithelium, i.e., they do not reach the presumptive tooth-bearing area:

> *Trigeminal nerve fibres were not detected in the vicinity of the developing rat tooth germ before the bud stage.*

Hence, his conclusion:

> *Trigeminal nerve fibres do not influence initiation (my emphasis—N.C.) of mammalian tooth development.*
>
> *Luukko (1997)*

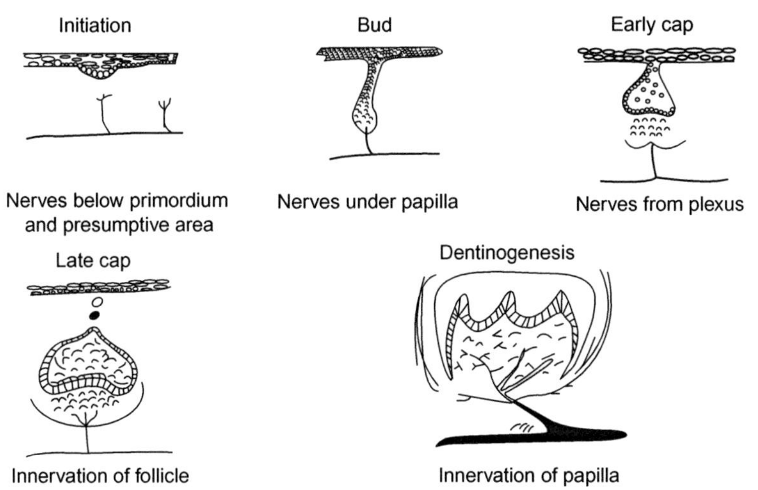

Figure 13.35 A schematic summary of the structural relation between nerves and tooth germs during development.
Source: From Hildebrand et al. (1995).

However, as Tuisku and Hildebrand observed later, these experiments did not justify the conclusion drawn, because the teeth in these experiments were examined after denervation, and investigators did not consider the possibility of the influence of the trigeminal placode, which produces some trigeminal neurons at a very early stage (Hildebrand et al., 1995).

Besides, the murine trigeminal ganglion commences at ED (embryonic day) 8.5–9.5, when the neural crest cells from the rhombencephalic neural crest have formed the tooth-forming mesenchyme in the mandibular process, and by ED10, axons from the trigeminal ganglion enter the mandibular process, and some reach the presumptive incisor and molar regions of the oral epithelium (Hildebrand et al., 1995).

In an attempt to resolve the controversy, investigators undertook a series of experiments on the formation of mandibular tooth germs in the cichlid polyphyodont fish, *Tilapia mariae*. Given the experimental difficulties of an earlier denervation of the mouse embryo, they made use of the fact that in polyphyodont vertebrates, such as cichlid fish, teeth are continuously replaced in cycles of about 100 days. They performed unilateral denervation of the lower jaw of *T. mariae* through neurectomy of the *ramus alveolaris trigemini*. No new (replacement) tooth germs formed in the denervated lower jaw, but "numerous mineralized replacement teeth were present in the innervated jaw cavity on the unoperated side" (Tuisku and Hildebrand, 1994). Prospective dental nerves are present in the fish jaw before the onset of tooth formation, nerve branches are seen just under the odontogenic epithelium, and later specific branches extend to individual tooth primordia (Tuisku and Hildebrand, 1994; Hildebrand et al., 1995). Based on the evidence from their experiments, Tuisku and Hildebrand conclude that

> *de novo formation of tooth primordial in the lower jaw of the cichlid* T. mariae *was arrested following denervation ... the local presence of trigeminal nerve branches in the jaw cavity is necessary for the formation of tooth germs in the lower jaw of the cichlid* T. mariae.
>
> *Tuisku and Hildebrand (1994)*

and that mandibular innervation

> *may have a primary initiating or instructive role in early odontogenesis.*
>
> *Tuisku and Hildebrand (1994)*

Summarizing extensive experimental work done at the time, Tuisku and Hildebrand argue:

> *Histologically, tooth initiation commences as a local thickening of the oral epithelium (dental lamina) and a beginning condensation of the underlying mesenchyme. The subsequent development of tooth primordia runs more or less autonomously. Hence, if nervous influences are involved in the tooth formation, they should occur before or during the emergence of a dental lamina. Indeed, nerve endings have been observed to occur transiently and selectively at loci where epithelial thickenings indicative of tooth development appear some hours later.*
>
> *Tuisku and Hildebrand (1994)*

Based on results of the above experiments for interpreting evolution of the teeth shape in cichlid fish, Streelman et al. believe that

> *Tuisku and Hildebrand (1994) demonstrated that the development of replacement tooth germs in the cichlid lower jaw is dependent on mandibular innervation. It is possible that such neural input directs not only the process of replacement but, in certain cases, the shape of new teeth as well. This might result in rapid changes in dentition given changing feeding behaviors.*
>
> *Streelman et al. (2003)*

Investigators have explained the reason why other studies have overlooked the presence of innervation in incipient tooth germs:

> *First, the initial steps in mammalian and teleost odontogenesis may not be identical. In fact, the odontogenic oral epithelium seems to have partly different roles in mammals and teleosts. In teleosts, the epithelial tooth component does not produce true enamel. The functional surfaces of teleost teeth are covered with a structure different from that of enamel. In actinopterygian teleosts, which comprise* T. mariae, *the dental epithelium and the odontoblasts both contribute extracellular matrix proteins to the enameloid layer. To what extent the oral epithelium participates in initiating odontogenic events in these teleosts is unknown. An instructive or patterning role of local nerves at an early stage of odontogenesis cannot be excluded. Second, some few, tiny trigeminal axons, which were not readily revealed by silver staining of paraffin sections, might possibly have entered the E9 and E10 mouse mandibular arches used for explantation. That this might have been the case is indicated by the absence of stained axons coursing from the host eye to E9 and E10 mandibular arch grafts after 2 weeks in oculo. Third, trigeminal ganglion neurons have a dual origin–the neural crest and the trigeminal epidermal placode. Placode-derived trigeminal neurons develop before crest-derived ones. In the mouse, placode-derived trigeminal ganglion neurons start to differentiate very early in E9. Interestingly, the avian placodal neurons, which develop even if the neural crest is removed, may form peripheral projections before reaching the ganglion proper and before the crest-derived trigeminal sensory neurons emit processes. If the development of the murine trigeminal ganglion is principally similar to that of the chick, as seems to be the case, murine E9 and E10 mandibular arch grafts could possibly have been contacted by placodal trigeminal sensory neurons.*
>
> *Tuisku and Hildebrand (1994)*

Another interesting phenomenon, relevant to the possible involvement of the trigeminal nerve fibers in the regulation of teeth formation in vertebrates, is the fact, that in contrast to the mandibular and maxillar odontogenic areas, where these fibers are abundant, no nerve fibers are detected in the diastema, the space between two teeth, where teeth primordia soon disappear, and no teeth develop (Løes et al., 2002).

Besides the neural crest cells and nerve fibers, neurons localized in the region of the developing teeth are also essentially involved in regulation of mammalian odontogenesis (Luukko et al., 1997), and it is noteworthy that this conclusion is drawn by investigators who previously denied the possibility that local nerves are involved in the initiation of tooth formation. Indeed, in a later study, these investigators have obtained, both *in vivo* and *in vitro*, results suggesting that neurons may also participate in tooth formation in mammals. Now they believe:

> *Neuronal cells are present in the developing rat tooth. Their localization during the bud and cap stages suggests that local neurons may participate in the regulation of mammalian tooth formation. Moreover, the apparent neuronal characteristics of the cells of the dental mesenchyme may reflect the specific ability of neural crest-derived cells to contribute to mammalian tooth formation.*
>
> *Luukko et al. (1997)*

The neural network that is involved in tooth development contains, at its core, a set of over 50 genes (among which at least 12 transcription factors) expressed in the enamel knots–signaling centers of the odontogenic epithelium. These enamel knots secrete BMPs, which act as stimulators, as well as FGFs and Shh, which act as inhibitors (Salazar-Ciudad and Jernvall, 2002) of odontogenesis. A secreted protein, termed *ectadin* (Laurikkala et al., 2003), also acts as inhibitor of BMPs, whose expression pattern regulates formation of cusps (Kassai et al., 2005). The GRN for tooth development seems to have been conserved across vertebrate taxa.

The innervation of odontogenic regions is correlated with specific patterns of expression of odontogenic genes in the region (Figure 13.36). It is possible that the sensory or autonomic nerve terminals in proximity of odontoblasts release specific neurotransmitters/neuroregulators that, by binding to odontoblast receptors, start ontogenetic cascades (Chiego, 1995). Studies on the evolution of molars in two mammal species, mice and voles, have shown that changes in the number and configuration of molars in these species are not related to any mutations in "molar patterning genes" but simply to a regulatory shift in lateral topography in mice and to a greater number of iterations of the established lateral topography in voles. These evolutionarily important changes take place in very early stages of development (Jernvall et al., 2000).

The rapid evolution of East African cichlid fish in lakes Tanganyika, Malawi, and Victoria was accompanied by evolutionary changes in the morphology of dentition consisting of multiple rows of teeth in the mouth and pharynx. *Metriaclima zebra* and *Labeotropheus fuelleborni* are two cichlid species that diverged from their common ancestor ~50,000 to 500,000 years ago. It is observed that interspecific differences in the CNC are correlated with corresponding differences in the feeding morphology (Albertson and Kocher, 2006). Changes in the tooth morphology of the two species (*M. zebra* has bicuspid teeth, and *L. fuelleborni*, tricuspid) are not related to any changes in odontogenic genes. This is corroborated by the observation that both species produce similar first-generation unicuspid teeth, and later in the life *M. zebra* replaces them with bicuspids, and *L. fuelleborni* with tricuspids (Streelman et al., 2003).

Neo-Darwinian Explanation

From the neo-Darwinian point of view, it would be predicted that vertebrate dentition is the result of evolution of new genes and/or mutations in existing ones or their

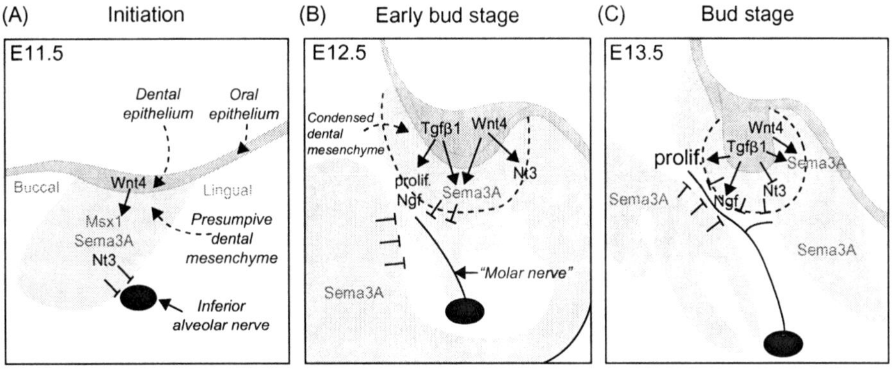

Figure 13.36 Schematic model for coordination of early tooth organogenesis and establishment of nerve supply by epithelial–mesenchymal interactions. The Sema3a exclusion areas regulate timing of tooth innervation and the innervation pattern. Prior to the histological onset of tooth formation (E10.5), the odontogenic oral epithelium, which instructs tooth formation and also possesses information to control tooth-specific nerve supply, induces (mediated by Wnt4) Sema3a in the presumptive dental mesenchyme. During subsequent morphogenesis, epithelial signaling and Wnt4 and Tgfß1 continue to control Sema3a expression domains in the dental mesenchyme target area. Wnt4 and Tgfß1 contribute to the regulation of tooth morphogenesis by maintaining *Msx1* (the effect of Wnt4 on *Msx1* expression at E11.5 is hypothetical) and stimulating dental mesenchymal cell proliferation, respectively. The trigeminal molar nerve located in the mesenchymal axon pathway and tooth target fields are indicated in black.
Source: From Kettunen et al. (2005).

regulatory sequences. This prediction is rejected: genes involved in tooth formation are shared by, and functional in, all tooth-developing vertebrates. Moreover, they are also present in toothless invertebrates and in the whole class of birds that lack teeth.

> *Although teeth are unique to vertebrates, much of their development involves genetic pathways found in invertebrates .... No developmental mechanisms or regulatory molecules have so far been shown to be unique for tooth development.*
>
> *Thesleff and Sharpe (1997)*

Epigenetic Explanation

As mentioned earlier, there are two main histological components involved in tooth formation: dental lamina, the thickening of odontogenic oral epithelium, and the underlying ectomesenchyme of neural crest origin. As far as the neural crest–derived ectomesenchyme is concerned, vast evidence demonstrates that neural crest cells, before delaminating from the neural tube/CNS and starting migration, are provided with *epigenetic* information on not only where to go but also what to do in their migration sites (Trainor et al., 2002; Schneider and Helms, 2003). And the fact that, besides the oral epithelium, tooth, in the presence of neural crest cells, develop in other places, such as pharyngeal slits, demonstrates the "dominant role for the neural crest mesenchyme over epithelia in tooth initiation" (Soukup et al., 2008). Both the "outside-in" hypothesis that propounds the ectodermal origin of dentition from skin odontodes and the "inside-out" hypothesis that relates evolution of dentition to the endodermal epithelium of oro-pharyngeal cavity imply the essential role of neural crest cells in tooth development (Fraser et al., 2010). Indeed, even temporally, the advent of skin odontodes was preceded by evolution of the neural crest cells in vertebrates (Reif, 1982).

This essential role of the neural crest in odontogenesis and in the evolution of dentition must be considered in the context of the experimental observation on the role of local innervation in odontogenesis. Experiments with denervation of the mandibular arch in fish (Tuisku and Hildebrand, 1994) have shown that the local innervation is indispensable for odontogenesis.

The fact that the neural crest and local innervation play such a crucial role in the development of teeth in vertebrates, in the context of the absence of changes in relevant genes and genetic information, strongly suggests that the nervous system is the source of information for the development and evolution of vertebrate teeth.

Sudden Evolution of Morphology in the Threespine Stickleback, *Gasterosteus aculeatus*

When female (with riped ovaries) and male (with breeding colorations) threespine stickleback fish of the species *Gasterosteus aculeatus*, at the beginning of their breeding season, were transferred from marine tide pools to freshwater ponds, they produced offspring of different shape and with fewer armor plates than their parents (Kristjansson, 2005).

In 1987, the Hraunsfjordur, a fjord in northwest Iceland, was dammed to form a freshwater lagoon for cultivating salmon (*Salmo salar*). Ever since, within 12 years, the stickleback (*G. aculeatus*) population of the freshwater lagoon experienced a rapid morphological evolution that led to a remarkable divergence in morphology from its marine ancestral population. The rates of evolution for dorsal spines and for keeled armor plates were comparable only to the exceptionally fast rates of evolution of coloration and decoration of Trinidad guppies, *Poecilia reticulata*, and of the beak in Darwin's finches (Kristjansson, 2005).

In the Queen Charlotte Islands (British Columbia, Canada) as well, populations of the stickleback, *G. aculeatus*, show remarkable differences in morphology. These differences seem to have arisen as adaptive responses to the local habitat and fish predators (Moodie and Reimchen, 1976; Figure 13.37).

Neither changes in gene functions nor selection on the existing genetic variability have been proposed for explaining the exceptionally rapid morphological evolution of this fish species. Hence, epigenetic changes in the regulation of expression of genes is the only alternative explanation of the phenomenon.

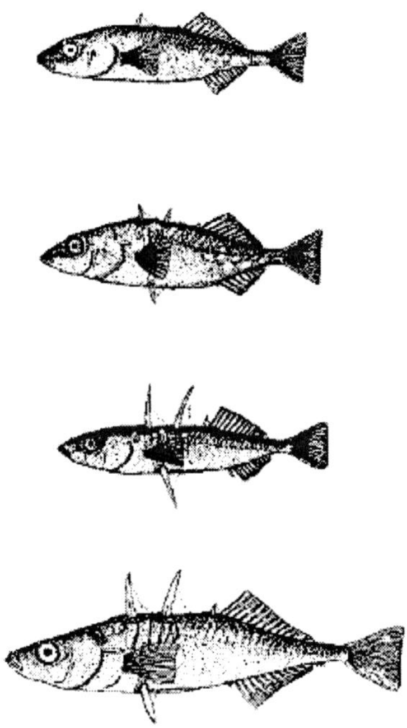

Figure 13.37 Variability among *Gasterosteus aculeatus* populations in the Queen Charlotte Islands. Top to bottom: Boulton Lake, Gold Creek, Yakoun Lake, and Mayer Lake. Typical representatives were drawn to the same scale with the aid of a camera lucida; body proportions were measured and transferred to the drawings. *Source*: From Moodie and Reimchen (1976).

Evolution of the Auditory System

Mechanosensation appeared early in the evolution of metazoans, probably as early as the common diploblastic ancestor of bilateran metazoans. The simplest metazoan mechanosensory structure consists of a ciliated sensory neuron and its associated nonsensory cells (Fritzsch and Beisel, 2004).

Evolution of metazoan ears essentially is evolution of an ancestral mechanosensory system or mechanosensory modules, i.e., hair cells that form epithelia for detecting and transmitting mechanical stimuli via axons directly to the brain (primary sensory cells). At a later stage of evolution, in craniate chordates, the sensory hair cells lose their axons, and these axonless secondary sensory cells are connected to "sensory" neurons, which were used for conducting to the brain the information detected by the secondary sensory cells. This has occurred independently several times during the evolution of metazoans (Fritzsch and Beisel, 2004).

Diploblastic metazoans evolved mechanosensory cells, whereas triploblasts evolved the ability for detecting high-frequency sounds. In the evolution of the auditory system, it seems that the neurosensory specialization preceded the morphological evolution of the system of tubes and associated recesses (Maklad and Fritzsch, 2003).

Crucial for the development and evolution of the mechanosensory cells in metazoans is the product of the fly *atonal* gene, which is conserved across metazoans and even existed in unicellulars (Simionato et al., 2007). In vertebrates it evolved to two atonal orthologs, *Atoh1* and *Atoh5*, but this group also expresses neurogenin genes, which are not expressed in insects and other arthropods. However, the most recent evidence, in fact not at all surprisingly, shows that the absence of neurogenin genes in insects like *Drosophila* is result of the loss of these "vertebrate" neural genes: orthologs of the most important neural bHLH genes known in vertebrates, including the *Olig* and *NeuroD* genes, are found to be active in simpler organisms such as the marine annelid worm, *Platynereis dumerilii* (Simionato et al., 2008). Thus, evolution of the auditory system from lower metazoans to mammals does not seem to be related to (or at least to depend on) evolution of "auditory" genes, and the great morphological evolution of hearing organs is in contrast with the conservation of genes and developmental modules that are used across phyla (Fritzsch and Beisel, 2004).

Evolution of Ears and Ultrasonic Echolocation in Insects

More than 14 families of moths have ears adapted for detecting ultrasonic echolocation calls generated by predatory bats (Waters, 2003). They are located in thoracic or abdominal segments, legs, wings, or mouth parts and have evolved independently at least 19 times (Yager, 1999).

Insect ears consist of a tracheal sac with a tympanum on the front and a tympanal chordotonal organ formed of a group of sensillae, each containing between three and 15 sensory cells. These receptors together provide neural input to the CNS.

The evolutionary precursors of tympanal chordotonal organs are chordotonal proprioceptors. A comparative study with two metamorphic insect species, the eared gipsy moth, *Lymnatria dispar* (Lymnatriidae: Noctoidea) and the earless caterpillar *Malacosoma disstria* (Lasiocampidae: Bombycoidea) has shown that both of them possess a homologous chordotonal organ, which has auditory function in *L. dispar* but lacks such a function in *M. disstria*. Both species have three chordotonal sensory neurons and one nonchordotonal multipolar cell, projecting to the homologous brain sites. Neurectomy of the homologue of the adult auditory nerve (IIIN1b1) in *L. dispar* larvae prevents development of the auditory chordotonal organ (Lewis and Fullard, 1996), suggesting that the auditory nerve (IIIN1b1) might have played a role in the evolution of the auditory chordotonal organ in *L. dispar*. The fact that both species, the eared and the earless ones, at the larval stage possess a homologue chordotonal organ suggests that the putative proprioceptor nonauditory chordotonal organ of *M. disstria* may represent the evolutionary precursor of the auditory chordotonal organ of *L. dispar* and other metamorphic insects. A model of the evolution of the auditory organ from mechanosensory systems in insects is presented in Figure 13.38.

Developmental and comparative studies conducted on eared and earless insects suggest that insect ears represent a later stage in the evolution of the long-range auditory reception system from the mechanosensory, proprioceptive system. The fact that chordotonal and auditory organs are sensitive to respiratory movements suggests that they share a common origin (van Staaden and Römer, 1998).

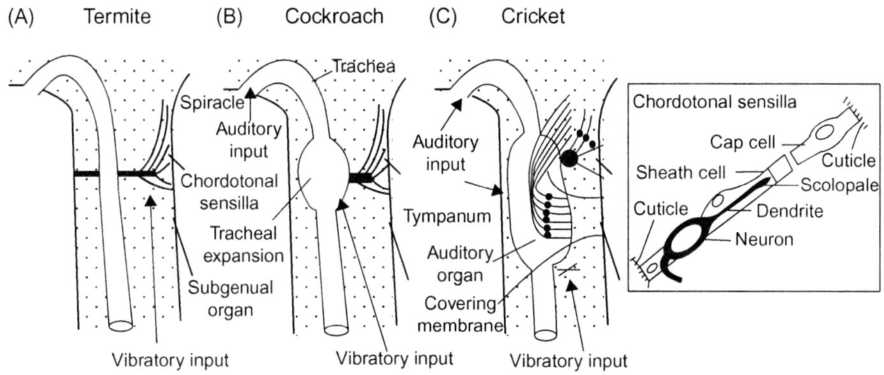

Figure 13.38 Diagram of three major steps (A–C) proposed for the evolutionary transformation of a vibration-sensitive subgenual organ into an auditory organ. (A) The precursor state still found in some extant termites. The subgenual organ, with its inserted chordotonal sensilla, is suspended in hemolymph between opposite walls of the tibia, creating a relatively insensitive detector of vibrations transmitted in the hemolymph. (B) Sensitivity to vibration can be increased by coupling one edge of the subgenual organ to a compressible tracheal expansion, as is found in cockroach tibiae. This pre-adapts the organ to detect auditory stimuli that enter the leg's trachea via spiracles on the body wall, producing a bimodal, combined vibration-and-sound detector. (C) In some crickets, a more specialized auditory organ is elaborated from the distal edge of the subgenual organ. Development of one or more tympana further facilitates sound entry, and enclosure of the organ by a covering membrane may contribute to its vibration isolation. The inset depicts components of a chordotonal sensillium.
Source: From Field and Matheson (1998).

The ability to tune their ears to varying frequencies is an adaptive character that enabled insects to detect hunting insectivorous bats, identify the direction of their flight, and undertake aerial manoeuvres (e.g., change their flight direction or fall out of the sky under the gravity attraction or by powered downward flight) to escape predator bats. These abilities are related to the processing properties of the auditory neural circuits of the insects.

The auditory system, i.e., the neural circuits for recognizing and encoding auditory stimuli, has evolved the capability to detect a wide range of bat echolocation calls (Fullard et al., 2005).

The Dogbane tiger moth, *Cycnia tenera*, a tympanate arctiid species in North America, is able to produce sounds similar to those generated by predatory bats, thus interfering with the bats' echolocation, thus forcing bats to discontinue the chase. Male gypsy moths of *L. dispar*, which fly and are exposed to predation by bats, exhibit high sensitivity to high frequency bat sounds, while females that do not fly, and hence are not threatened by bats, show less sensitivity to bat sounds. Over time, bats also modify the frequencies of echolocation calls (Fenton and Fullard, 1979). This is also corroborated by the sexual dimorphism of acoustic sensitivity in mantises: 63 of 183 mantis genera exhibit sexual dimorphism for acoustic sensitivity.

However, often this is not associated with dimorphism in the structure of the ears (Fong et al., 1995).

During flight, moths themselves produce ultrasounds as a result of hindwing flapping, but they have evolved neural filters for ignoring their own low-level activity ultrasounds, which would interfere with identification of bat ultrasounds (Waters, 2003).

Insects have also evolved mechanisms of modulation of the acoustic signals to adjust those signals to parameters encodable and recognizable in respective neural *circuits*. Female crickets, *Teleogryllus commodus*, have two temporal filters for syllable periods in the CNS, one for chirp and one for trill of the male song. These filters are tuned to the species-specific syllable periods. In distinction, its sibling species, *T. oceanicus*, has a single filter for the chirp part of the male song (Hennig and Weber, 1997). Appearance of a new filter in a sibling species shows the exceptional evolutionary plasticity of the neural mechanisms of hearing in insects.

There is no evidence on a possible involvement of changes in genes, in DNA or existing genetic variability in the process of the evolution and diversification of auditory organs and related behaviors in insects.

Evolution of Ears in Vertebrates

Evolution of the vertebrate ear is characterized by specialization of sensory epithelia within an increasingly complex network of tubes and recesses within which the ancient mechanosensory module is conserved (Figure 13.39). Hair cells are sensory receptors, which can detect motions of atomic dimensions and respond more than 100,000 times per second (Hudspeth, 1997). Having no axons, hair cells in vertebrates are connected to the brain via sensory neurons.

During the development, invagination of the embryonic ectoderm leads to formation of the otocyst from which neurons delaminate that migrate to the presumptive vestibular and auditory ganglion between the hindbrain and the ear. These auditory and vestibular neurons send projections to precisely determined areas of the brain (in the absence of spontaneous activity from hair cells), thus creating central representations of the end organ. As for the origin of the individual auditory nuclei, studies in chickens have shown that neurons of the *nucleus laminaris* in the brainstem originate in the brain, mainly in the rhombomere 5, with neurons of *nucleus magnocellularis* from rhombomeres 5–7 and those of *nucleus angularis* from rhombomeres 4–5 (Maklad and Fritzsch, 2003).

Ear development appears to use the same ear-patterning genes across vertebrates. During the development, the hindbrain secretes FGF signals, which induce formation of the otic placode, a thick area of ectoderm adjacent to the hindbrain, and expression of a number of transcription factors (Pax2, Pax8, Dlx5, Gbx2, Hmx3) in the otic epithelium. The next stage, the development of the dorsal otocyst, which is critical for the development of the vestibulum, is regulated by Wnt signals also emanating from the dorsal hindbrain.

It is believed that other hindbrain signals are involved in formation of the inner ear, and experimental evidence shows that these brain signals restrict Shh expression to ventral and medial regions of the ear epithelium (Bok et al., 2005; Riccomagno

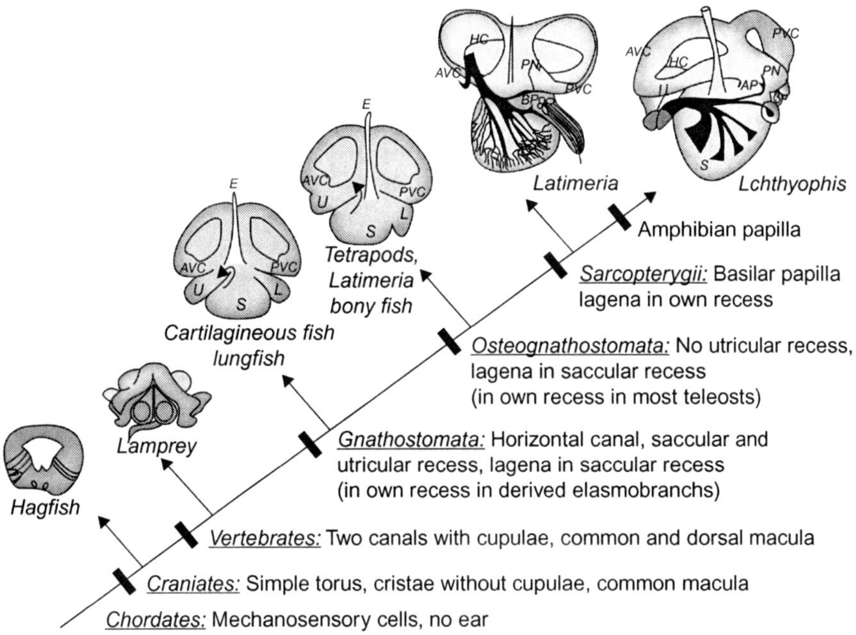

Figure 13.39 The morphological evolution of the craniate ear is shown. It is assumed that the outgroup had mechanosensory cells but no ear. The hagfish ear shows a single torus with only three sensory cell patches, two rings of hair cells forming the anterior and posterior sensory canal crista and the common macula. The sensory cristae of hagfish have no cupula, a unique and likely primitive feature of chordates. Evolution results in multiplication of end organs through developmental segregation, culminating in a total of nine end organs in certain limbless amphibians. In parallel, the ear becomes a labyrinth of as many as three distinct semicircular canals and three distinct recesses harboring the otoconia/otolith bearing saccular, lagenar, and utricular macula. These recesses form two distinct patterns: one pattern is found among chondrychthians and lungfish; the second pattern is found in actinopterygian and sarcopterygian fish. Sarcopterygian fish have evolved a separate organ, the basilar papilla, that exists in most tetrapods and that becomes the mammalian cochlea. *Abbreviations*: AP, amphibian papilla; AVC, anterior vertical crista; BP, basilar papilla; HC, horizontal crista; L, lagena; PN, papilla neglecta; PVC, posterior vertical canal; S, saccula; U, utricle. *Source*: From Fritzsch and Beisel (2004).

et al., 2005; see also Figure 5.11, Chapter 5), thus determining the auditory cell fate in the otic vesicle (Riccomagno et al., 2002). The hindbrain signals provide the positional information necessary for patterns of growth and differentiation leading to formation of vestibulum and cochlea, for sensing, balance, and sound. Also, the FGF cues from the presumptive neurosecretory cristae at the base of each semicircular canal are involved in formation of canals by regulating expression of Bmp2 (Riccomagno et al., 2005).

The development of the ear structure of interconnected tubes and sacks "seems to present a rather clear Haeckelian ontogenetic recapitulation of evolution: evolutionarily late organs, such as the cochlea, also develop last" (Fritzsch and Beisel, 2004).

Neo-Darwinian Explanation

No evidence on mutations in genes or any changes in existing gene frequencies in populations has ever been presented that could explain the differences in the structures of ears, auditory neurons, and neural circuits among species. This makes a neo-Darwinian explanation of the evolution of ears, at this time at least, impossible.

Epigenetic Explanation

Experimental evidence shows that changes in ear morphology, and especially in hearing in vertebrates, are epigenetically determined by changes in involvement and expression of ear-patterning genes, especially by inductive signals emanating from the hindbrain and changes in the patterns of migration of cerebral neural crest cells. The crucial role of the brain signals in the development of ears strongly suggests that the epigenetic information necessary for evolution of ears in invertebrates and vertebrates originates in the CNS. The evolution follows development, not the other way around.

Evolution of Eyes

About one-third of the 33 extant metazoan phyla have light-sensitive organs, and one-third (representing the overwhelming majority of metazoan species) have eyes. Eyes, as vision organs, evolved during the early Cambrian (Nilsson, 2004), about 540–530 mya. Compound eyes and camera eyes have evolved in both invertebrates and vertebrates, but the first group displays greater variation in regard to the number and location of eyes (Figure 13.40).

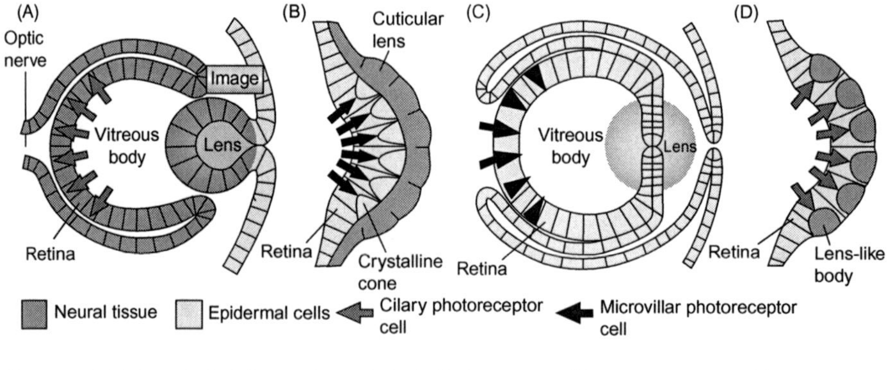

Figure 13.40 Simplified illustrations of the building plans of four types of eye: (A) a vertebrate eye; (B) an arthropod compound eye; (C) a cephalopod lens-eye; (D) a compound eye in polychaete tube-worms and arcoid clams. All are paired cephalic eyes, except those in polychaete tube-worms and arcoid clams, where large numbers of the eyes are spread on the feeding tentacles and along the mantle edge, respectively.
Source: From Nilsson (1996).

Eyes are specialized extensions of the brain that begin developing at late gastrula stages in the form of epithelial Anlagen, the eye field at the anterior end of the body axis (A–P axis) in the prospective forebrain. The eye field splits to form two optic vesicles consisting of neuroectodermal (future retina) and ectodermal (future lens and cornea) cells, which evaginate from the forebrain (Carl et al., 2002; Wittbrodt, 2002). Invagination of the lens placode leads to formation of the lens vesicle, and the optic vesicle forms the bilayered optic cup, which gradually develops into an eye (Fernald, 2004). An important role in formation of the vertebrate eye is played by neural crest-derived mesenchyme cells.

Formation of the neural retina is a crucial moment in the development of eyes in metazoans. Retinogenesis consists in differentiation of retinal progenitor cells (RPCs) into six different neuronal types. As many as ~2,500 genes are involved in the process of the eye development, but expression of Pax6 and a number of other transcription factors (Rx1, Lhx2, Six3, and Six6) in a highly conserved GRN in RPCs is essential for the eye development to take place, as may be inferred from the fact that in absence of Pax6 (in Pax6−/−mutants) no neurons are differentiated and, when *Pax6* gene is switched off, RPCs produce only amacrine cells (Marquardt et al., 2001). The crucial role of Pax6, Six2, and Ath is shared by almost all metazoans, beginning with Urbilateria (Arendt et al., 2002; Arendt and Baptista, 2002).

Differences among species in "nonocculogenic" genes involved in eye development exist, but all of them have those genes and use their products for different purposes. In evolving new eyes, metazoans used the same old genes. Thus, the wide differences in the building plans of different eyes in metazoans result from the fact that different species, in the course of evolution, recruited different genes to the eye developmental GRN. There is evidence suggesting that the CNS may be critically involved in gene recruitment (Cabej, 2011), and this may have been hinted at earlier by Nilsson:

> *when a simple nervous system became more elaborate, and new sensory organs were acquired, the ancient Pax-6 genes gradually changed their role to include new targets which were functionally related to, or physically near, older targets.*
>
> *Nilsson (1996)*

Evolution of Feathers

Feathers are hierarchically branched structures of birds, covering their bodies, insulating them from cold temperatures and water, and providing a necessary means of flight. They are one of the most important defining features of the class of Aves.

Molecular Mechanism of Feather Development

Feather buds develop as a result of complex interactions between the epidermal placode and the underlying mesenchyme. The skin feather macropattern in chickens is determined by the dorsal neural tube, for the dorsal part of the body, and from the ventral neural tube for the ventral side.

Wnt1 signals from the dorsal neural tube induce formation and the maintenance of the dermomyotome (Chang et al., 2004). Around the E3 (embryonic stage 3), dermal precursor cells from the medial somatic dermomyotome, stimulated by Wnt signals from the neural tube, migrate to form the dermal tissue (Olivera-Martinez et al., 2001) (Figure 13.41). Neural tube Wnt1 alone allows both the survival and specification of the medial dermomyotome, giving rise to the feather-inducing dermis (Olivera-Martinez et al., 2001).

Wnt1 from the neural tube also induces Noggin expression in the medial part of the somite, thus preventing the action of Bmp4 secreted by the somatopleural

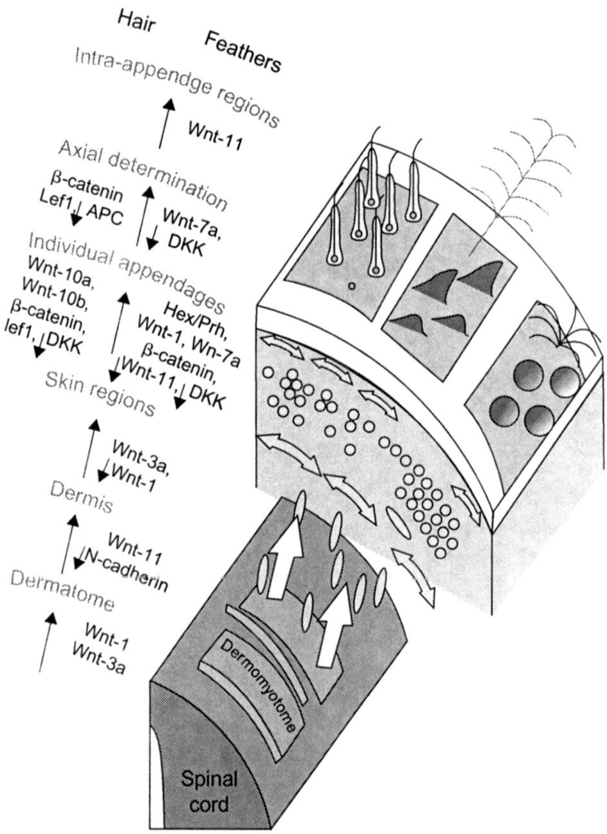

Figure 13.41 Involvement of Wnt signaling in the specification of dermis. Schematic showing the involvement of Wnt signaling involvement in dermis, tract and skin appendage development as determined in mouse hairs (left of arrows) and chicken feathers (right of red arrows indicating developmental progression). Morphogenetic events that take place in the dermomyotome, dermis, feather tracts, and individual feather primordia are shown. Spindles and circles represent dermal cells. Two different tracts with different density and shape of skin appendages are shown. (For interpretation of the references to color in this figure legend, the reader is referred to the web version of this book.)
Source: From Widelitz (2008).

mesoderm and instructing the dermomyotome to form the progenitors of the spinal pteryla (consecutive rows of feather buds) (Fliniaux et al., 2004).

The instructive role of the neural tube in feather development is also shown in experiments, where living pieces of the neural tube or agar implants impregnated with brain extract induce formation of feathers even in the midventral apterium, the unfeathered midventral area (Figure 13.42). While Wnt1 and Wnt3 are secreted by the anterior dorsal neural tube, two signaling factors, Noggin and Shh, which inhibit Bmp signaling, are released by the anterior ventral neural tube and at E13 (embryonic stage 13) are respectively expressed in the intermediate mesoderm and the endoderm.

Noggin and, probably synergistically, Shh (by downregulating Bmp4) determine the ventral skin feather macropattern (Fliniaux et al., 2004). Shh expression may be responsible for the growth (McKinnell et al., 2004) and the shape (Chang et al., 2004) of the feather buds.

As early as 1942, F.R. Lillie, in experiments of transplantation of neural crest melanophores to different bird species, observed that cells of neural crest origin of the donor determined the color and patterns of feathers of the host (Lillie, 1942). Recently, Eames and Schneider performed homotopic transplantation of premigratory neural crest cells from the midbrain and rostral hindbrain between Japanese quails and chicks. Japanese quail CNC mesenchyme transplanted in ducks stimulates the latter to develop quail-like pigmented feathers and developmental timing by advancing the feather morphogenesis by three stages, while the relative timing of expression of genes *bmp4*, *bmp2*, *follistatin*, *bmpr1a*, *shh*, *ptc*, *delta1*, and *notch1*

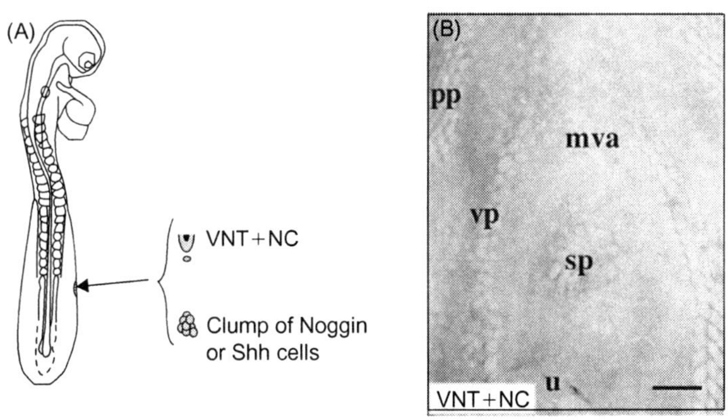

Figure 13.42 Induction and analysis of supplementary pterylae (sp) in the midventral apterium (mva). (A) Microsurgical procedure: fragments of 2-day old ventral neural tube plus chord, or aggregates of cells expressing Shh or Noggin were grafted under the ectoderm of HH13 embryo in the presumptive territory of the midventral apterium, posterior to the level of the 20th somite. (B) Supplementary pteryla obtained 8 days after the graft of (B) ventral neural tube (VNT) plus chord (NC) fragment. *Abbreviations*: mva, midventral apterium; pp, pectoral pteryla; sp, supplementary pteryla; vp, ventral pteryla.
Source: From Fliniaux et al. (2004).

is similar (Eames and Schneider, 2005). These experiments demonstrated that "neural crest cells function as the dominant source of spatial and temporal patterning information via the regulation of genes essential to cranial feather morphogenesis" (Eames and Schneider, 2005).

Successive Stages in Evolution of Feathers

The prevailing opinion now is that feathers initially served to insulate the body and only later evolved into flight structures (Xu et al., 2001; Prum and Brush, 2002). This hypothesis has found considerable support by the fossil evidence showing that filamentous integumental structures evolved in reptiles such as the dinosaur *Sinornithosaurus millenii*, which display two types of branching structures corresponding to two first stages of the embryonic development of feathers (Xu et al., 2001).

Prum and Brush (2002) have proposed a model of evolution of the feather in five stages. In the first stage, the primitive feather grows in the form of a feather papilla surrounded by the feather follicle. In the second stage, the epidermal layer of the follicle (follicle collar) generates barbs. The third stage comprises formation of branched structures, which is followed by differentiation of the distal and proximal barbule plates (stage IV). Additional novelties in the evolution of the feather morphology appear in the stage V. The model is only partly recapitulationist (Prum and Brush, 2002). Two alternative models have been presented in regard to the sequence of events in the evolution of feathers in birds. According to the first, widely accepted model, rachis appeared as a final event in the evolution of feathers (from barbs to barbules and finally to rachis), therefore known as the *barb to rachis model*. According to the second model, rachis represents the first element in the evolution of feathers in birds.

Neo-Darwinian Explanation

The fact that despite the numerous studies on the patterns of gene expression in the dermal mesenchyme, skin placode, and even in the surrounding tissues and in the neural tube, no relation has been found to exist between any changes in genes or DNA and the evolution and development of feathers, currently makes impossible a neo-Darwinian explanation of the evolution of feathers in birds.

Epigenetic Explanation

Evidence of the molecular mechanisms of the feather development presented in the beginning of this section unambiguously shows that all the signals (Wnts, Bmps, Shhs) and signal cascades involved in the process originate in the neural tube. The key genes (*Wnts*, *Bmps*, *Shhs*) involved in feather development are functionally unchanged across Aves. What has changed in the course of evolution of feathers in birds is the expression pattern of these genes, especially expression of Wnt signals from the neural tube and the behavior of the neural crest cells involved in the development of the feather structure as well as in the patterns of distribution of feathers on the bird body. With changes in genes or genetic information excluded as possible agents of the evolution of feathers and plumage in birds, the mechanisms of feather evolution, as would be predicted by the epigenetic view, are related to changes in the activity of the neural tube/CNS and the neural crest cells involved in formation of the dermal mesenchyme.

Evolution of Sexual Dichromatism in Birds

Differential plumage coloration of individuals of two sexes is characteristic of numerous bird species. The evolution of sexual dichromatism in birds often is reduced to selection of sexual traits. However, to say that male-specific characters that are preferred by female birds, over generations, will increase their presence in populations does not add anything to our knowledge of the causal basis of evolution of sexual dichromatism. It is something behind that truism that forms the crux of the problem: How do evolutionary changes in birds' plumage coloration arise, without changes in genes?

The prevailing idea is that sexual dichromatism depends on a combination of hormonal and nonhormonal factors, but in some bird species, plumage coloration is not affected by hormones (Kimball and Ligon, 1999). According to Owens and Short (1995), in species such as the mallard and peacock, where the plumage display is characteristic of males, the male plumage is the default state, whereas the cryptic female plumage results from secretion of estrogens (Owens and Short, 1995). So, e.g., suppression of the secretion of estrogen by ovaries, as a result of salmonellosis, leads to the development of male plumage in female poultry and pheasants.

If one were to believe that sexual dichromatism in birds is hormonally regulated, the fact that in birds no gene mutations affecting the function of hormones have occurred clearly implies that evolution of dichromatism in birds is a nongenetic phenomenon.

Neo-Darwinian Explanation

Based on the fact that in birds and butterflies, where females are heterogametic (ZW), males have more ornate traits than mammals (in fact, male mammals have no ornate traits at all), where females are homogametic (XX), Hastings proposed a model of sexual selection and suggested that sex-linkage genes are responsible for W-linked female preferences and male display traits by indirect selection for good genes (Hastings, 1994). Other authors also argued that male display genes and female preference for male display traits may be sex-linked (Houde, 1992; Reeve and Pfennig, 2003; Kirkpatrick and Hall, 2004).

> *Our comparative analysis revealed an association between sexual dimorphism and genetic system. Hence, taxonomic biases for showy males may stem from differences in sex chromosome systems.*
>
> *Reeve and Pfennig (2003)*

The existence of sex-linked genes for sexual traits is inferred from genetic ratios of crosses, but this might not be a reliable criterion, since, theoretically, similar ratios may also result from epigenetic factors provided in abundance to gametes. Besides, the belief that W-linked genes in birds are involved in evolution of male showy traits seems to be contradicted by the fact that W chromosome in birds is severely truncated (Fridolfsson et al., 1998), contains excessive repetitive DNA (Tone et al., 1982; Saitoh et al., 1991), and is almost devoid of genes (Kirkpatrick and Hall, 2004), hence from the small number of genes in the W chromosome it is not known how many, if any, are sex-linked. Brooks also reported the existence of a correlation between the male guppies' (*P. reticulata*) preference for female ornamentation and the evolution of that ornamentation determined by genes in Y chromosome via mating success (by producing more attractive offspring) (Brooks, 2000), but others, in later experiments with the same guppy species for selection on female preferences and male sexual traits, were not able to find any evidence of such a correlation (Hall et al., 2004).

The theoretical conclusion that female preferences and male signal traits were related to sex-linked genes is supported by some investigators and is rejected by others. Hence, the correlation of sex-linked genes with female preferences and male ornate characters, far from being scientifically validated, remains speculative at best. To add to the confusion,

> *a reporting bias probably exists wherein positive associations between sex chromosome system and sexually selected traits may have appeared in print more often than outcomes in which no such empirical relationship was detected. These factors have complicated efforts to assess any general relationship that might exist between male heterogamety and good-genes processes, or between female heterogamety and Fisherian runaway.*
>
> *Mank et al. (2006)*

Most recently, investigators, including some of the contributors and supporters of the hypothesis on the sex-linked evolution of female and male secondary sexual traits, have come to realize that theoretical predictions of the hypothesis are not supported by the empiry of observation and experiments. Based on Fisher's runaway hypothesis of sexual selection, one would predict that female heterogamety would favor evolution of extreme male traits, but studies on evolution of exaggerated male ornaments in actinopterygiian fish, in which heterogametic systems of sex determination have repeatedly evolved in males and females, have identified no significant correlation between female chromosomes and sexually selected traits in males (Mank et al., 2006).

In direct selection experiments on female mating preferences and male attractiveness on the guppy *P. reticulata*, Hall et al. (2004) observed that, contrary to the expectations from models of sexual selection hypothesis, no significant changes in female preferences and male sexual signals were detected (Hall et al., 2004), suggesting that sexual selection may not be responsible for evolution of female preferences for attractive males and male attractiveness. Even the widely held opinion that male ornate traits in groups where females are heterogametic (butterflies and birds) are more conspicuous than in groups with homogametic females (mammals) is discredited by the empirical evidence:

In conclusion, our phylogenetic analyses suggest that the particular mode of sex determination has had no consistent and discernible impact on the evolution of sexually selected traits in ray-finned fish ... our analysis suggests that female-heterogametic (ZZ–ZW) lineages are not significantly more or less prone to male ornamentation than male-heterogametic (XY–XX) lineages in these fish ....

Mank et al. (2006)

Theoretical arguments against the role of sexual selection in sexual dichromatism have also been presented. One of them derives from the fact that gain and loss of ornamentation often precedes evolution of female preferences, i.e., they evolve independently of sexual selection. Also, sexual dichromatism is often an ancestral state, not a derived state (Badyaev and Hill, 2003).

It is also argued that if mutations in sex chromosomes and their selection were related to the evolution of dichromatism in birds, it would be expected that no bias would be observed in the occurrence of loss and gain of male sexual ornamentation (Badyaev and Hill, 2003). Contrary to this neo-Darwinian prediction, males in numerous bird taxa have lost sexual dichromatism several times more frequently than they have gained it. In an attempt to address and resolve this apparent contradiction, some evolutionists (Mayr, 1942) have resorted to gene drift as a possible cause of the loss of sexual dichromatism in birds, but falling short of substantiating the hypothesis. In addition to the theoretical arguments to be presented in sections of the basic allopatric model and the peripatric model in Chapter 18, that explanation is discredited by the fact that while yellow-green color is lost in the plumage, it is conserved in the bill of many monochromatic birds (Omland, 1997), suggesting that the loss of the plumage color is not related to changes in sex chromosome-linked genes or any other gene. The same is true for other bird groups as well.

Additional evidence shows that very distinct sister oriole species, which have evolved very recently, show no differences in the sex chromosome genes. So, for example, the Baltimore oriole and the black-backed oriole, which diverged from their common ancestor between 5,000 and 150,000 years ago, are not known to hybridize under natural conditions. Although these species, to a considerable extent, have diverged morphologically, genetically they are almost identical, with only one difference in the cytochrome b sequence (that has not affected the function of the molecule), which might have occurred sometime between now and the time of their divergence. Despite the extraordinary genetic similarity, these new species have diverged in 17 plumage characters (Kondo et al., 2004), unambiguously indicating that numerous differences in plumage between these sister species are related to nongenetic processes and involved no changes in sex chromosomes.

In ducks as well, the difference between males and females in plumage and color patterning is not related to differences in DNA or genetic information. One indication of this is the fact that male ducks in dichromatic species, as juveniles, display female plumage and only later, as adults, develop male sexual dichromatism (Omland, 1997). The difference is a result of changes in neurohormonal mechanisms controlling and regulating specific GRNs (Owens and Short, 1995), which will be discussed later in this section.

Another neo-Darwinian explanation for the interspecific variation in the plumage color in birds is that existing variations are selected to avoid or reduce hybridization (the species isolation hypothesis). If this hypothesis were true, it would be expected that sympatric species, which are more prone to hybridization, would display a stronger divergence in plumage coloration than species that have evolved in allopatry. This would be the result of the so-called reproductive character displacement. However, observations in nature have shown that, in the majority of cases, the contrary is true. Studies on birds in Australia, for example, have shown that the divergence in plumage coloration between sympatric species is smaller than between allopatric species (McNaught and Owens, 2002).

Epigenetic Explanation

On the proximal end of the causal chain leading to the development of male sexual traits, in most of the cases studied so far are sex hormones, estrogens, testosterone, and luteinizing hormone (LH) (Kimball and Ligon, 1999). On the other extreme of the causal chain of sexual dichromatism act female preferences. How could one connect these two terminal points of the causal chain of the development of sex-related body coloration in birds?

Different sex steroid hormones are involved in determining sexual dichromatism in different species of birds. The main hormones involved in the appearance of sexual dimorphism in birds are estrogens, androgens, and pituitary LH. In birds of the orders Struthioniformes, Galliformes, and Anseriformes, sexual dichromatism is estrogen-dependent: the bright colors develop in the absence of estrogen, while the estrogen presence determines dull plumage coloration (Figure 13.43). In Charadriiformes, the sexual dimorphism depends on testosterone: the presence of the hormone determines bright-colored plumage, and the reverse, the absence of the hormone, leads to duller plumages. In the fifth order of Passeriformes, the bright color depends on the presence of LH and on unknown nonhormonal factors.

Studies on species of the dabbling ducks of the genus *Anas* have shown that in northern temperate zones, males molt twice per year, alternating between bright and dull plumage. The seasonal change in the coloration of male birds clearly shows that changes in plumage coloration are determined not by changes in genes (these genes are the same during the whole life cycle of the bird) but by changes in the patterns of gene expression, which are epigenetically determined.

Attempts have been made to explain the seasonal dichromatism with aromatization of testosterone into estrogen (Kimball and Ligon, 1999), but the seasonal switches in aromatization as well represent an epigenetic phenomenon rather than a genetic one, for the phenomenon has nothing to do with any changes in DNA. Besides, castrated males and females, in which the absence of circulating estrogen and testosterone is implied, maintain the bright color year-round.

Another widely held hypothesis on the evolution of dichromatism in birds posits that the ancestral state has been bright coloration (dichromatism) in both sexes, with females later losing the brightness and evolving into duller monochromatism.

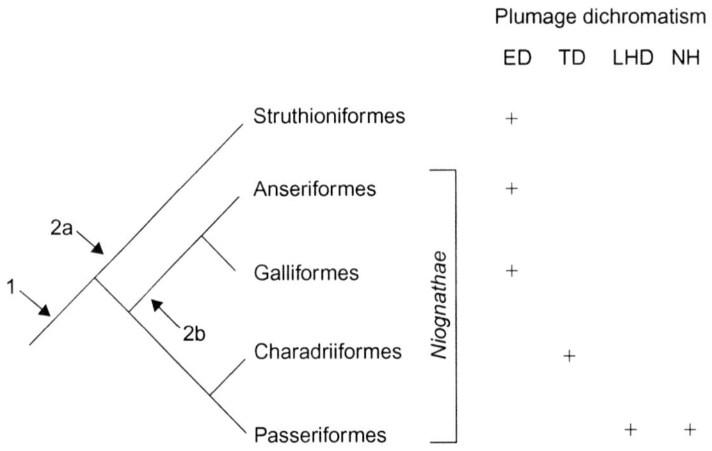

Figure 13.43 Phylogenetic distribution of the proximate mechanisms controlling sexual plumage dichromatism. 1 represents the evolution of estrogen-dependent plumage dichromatism, assuming a single evolutionary event; 2a, 2b represent the evolution of estrogen-dependent plumage dichromatism, assuming it evolved twice. *Abbreviations*: ED, estrogen-dependent; LHD, LH-dependent; NH, nonhormonal plumage dichromatism; TD, testosterone-dependent.
Source: From Kimball and Ligon (1999).

To arrive at estrogen-dependent plumage dichromatism, the pathway requiring the fewest evolutionary steps begins with brighter coloration in both sexes, followed by selection for duller color in one sex. This is in contrast with arguments based on sexual selection that assume that dichromatism is a result of selection for brighter coloration in one sex. If brighter plumage initially developed in the absence of gonadal hormones (as seen in the ostrich, galliforms, and anseriforms), the evolution of estrogen-dependent plumage dichromatism would require selection for duller plumage in one sex, plus a novel mechanism (estrogen-dependent plumage) for the development of this evolutionarily new plumage (Kimball and Ligon, 1999).

Experimental evidence shows that the proximate causes inducing dichromatism and monochromatism in most birds are sex hormones, testosterone, and estrogens. Administration of testosterone in male ducks leads to molting into duller plumage. This transformation of plumage color is explained with the aromatization of the androgens into estrogen (Haase, 1993), a conclusion drawn from the fact that levels of estrogen in male ducks are higher in late spring and early summer, when they molt into duller plumage as a result of changes in the melanin and pheomelanin content (Haase et al., 1995).

Presence or absence of estrogens determines the dull or bright plumage coloration, respectively, but in Charadriiformes it is testosterone that has the determining role. So, for example, a sex-role reversal has occurred in Alascan Wilson's pharalope (*Pharalopus tricolor*). Males of this species have dull plumage, and females have bright plumage coloration, and testosterone has been found to be responsible for the bright female plumage coloration. It is the female of this shorebird that aggressively competes for mates, although the blood level of sex steroids in breeding phalaropes do not differ from other breeding birds. The only differences observed in the Wilson's pharalope breeding females is the aromatase level, which is lower in females (hence insufficient for converting androgens into estrogen) than in males (Schlinger et al., 1989).

In several species of passerine birds, the breeding season plumage is bright, and the nonbreeeding season plumage is henny. Castration of both males and females does not affect the plumage dichromatism, suggesting that gonadal hormones, testosterone and estrogens, are not involved in plumage coloration. Indeed, it is the pituitary LH that determines plumage coloration in these species (Kimball and Ligon, 1999). Let us remember that induction of LH secretion in the pituitary and the level of circulating LH are neurally controlled by hypothalamic neuropeptides.

Another phenomenon with significant bearing on the mechanism of plumage coloration is the occurrence in galliforms, anseriforms, and passerines of gynandromorph birds in which half of the body exhibits plumage of one sex, and the other the plumage of the opposite sex. Obviously, in such cases, the plumage is not determined hormonally, for circulating hormones cannot produce two different plumages in two sides of the same body. Attempts to explain the phenomenon from a genetic point of view (loss or nondisjunction of chromosomes or supernumerary spermatozoa) (Kimball and Ligon, 1999) have failed.

The prevailing idea that *gonadal* steroids only are responsible for the development of sexual morphological traits and sexual dichromatism in birds is no longer tenable. The brain can accomplish all the stages of synthesis of sex steroid hormones (estrogen and testosterone) *de novo* (Keefe, 2002), starting from cholesterol. This seems to be the case not only for male zebra finches, where the circulating estrogens are of cerebral origin and peripheral aromatization of androgens does not significantly contribute to the level of circulating estrogen (Schlinger and Arnold, 1993; Gahr and Hutchison, 1992; Schlinger and Arnold, 1993). Adequate evidence shows that circulating estrogens are synthesized in the brains of male birds and even of male mammals (Carroll et al., 1988).

In the brain of adult songbirds, a neurosteroidogenic pathway is operational, which transforms dihydroepiandrosterone (DHEA) into androstenedione (AE), and finally aromatizes AE into estrogen, via aromatase (CYP19). High brain aromatase activity for conversion of androgens into estrogen has been observed not only in songbird species but also in other nonsongbird species (Silverin et al., 2000).

The role of the brain sex hormones is also corroborated by experiments on castrated males of Gambel's and Scaled quails, which do not change their ornate plumage (Hagelin, 2001), indicating that *brain-derived* testosterone, rather than *gonadal* testosterone, is responsible for the development of male ornate plumage.

A number of neurally active progestogens and androgens are synthesized *de novo* in the brain, and androgens can be converted into estrogens within the brain by the enzyme aromatase. Dull female-like plumage of males in some bird species is a consequence of aromatization of testosterone at the time of the molt (Hagelin, 2001).

It is clear that circulating hormones (testosterone, estrogen, LH) alone cannot regulate patterns of plumage coloration: as circulating hormones, they would tend to produce rather uniform coloration of the plumage all over the bird's body. While the neo-Darwinian paradigm offers no explanation of the spatially restricted coloration patterns in birds, the epigenetic model offers two possible alternative explanations. First, birds may be in possession of regulatory mechanisms for local expression of respective nuclear receptors, mediators of the functions of sex hormones, or second, by neural regulation of local secretion of aromatase. The first alternative is a characteristic of the binary neural control of gene expression in metazoans (Cabej, 2004, pp. 192–193; see also Section The Binary Neural Control of Gene Expression in Chapter 6), where the local innervation switches on/off expression of hormone receptors. The second alternative derives from the recent experimental evidence that the steroidogenic enzyme, aromatase, produced synaptically by neurons in the brain, via axonal innervation can be transported

> *to terminals far from their source. This action combines the relatively long-range characteristic of an endocrine mechanism with the targeted specificity of axonal innervation.*
>
> *Peterson et al. (2005)*

In this regard, according to the epigenetic paradigm would be predicted that:

1. Sexual signaling traits will be more frequently lost than gained, and
2. Transition from dichromatism to monochromatism will occur more frequently than the reverse transition (Badyaev and Hill, 2003).

Both of these predictions are empirically validated.

Evolution of Life Histories

Evolution of Sexual Reproduction and Alternation of Asexual and Sexual Modes of Reproduction

Why living systems evolved sex and sexual reproduction, is still an unresolved problem in biology. Various hypotheses are presented to explain why sex evolved and why natural selection might have favored evolution of sexual reproduction despite its high cost and complexity. One of the most frequent answers to the question of the evolution of sex is that it makes possible genetic recombination, which, as a source of genetic variation, is more important than gene mutations. This view is already rejected. First, because the role of genetic recombination in generating genetic variation is an assumed role that has not been scientifically substantiated, and, second, evolution of parthenogenetic insects and asexual rotifers, which experience no genetic recombination, has proceeded at a pace that is not slower than that of sexually reproducing organisms.

The prevalence of sexual reproduction in the animal world, despite its high cost, related to waste of resources for producing male individuals, the complex processes of meiosis, and the probability of disrupting favorable gene combinations, implies that this mode of evolution offers some advantage, of which we have no clues as of yet.

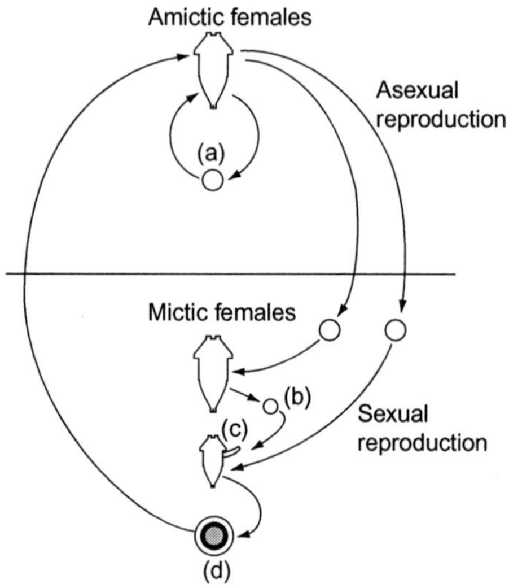

Figure 13.44 Typical heterogonic life cycle of monogonont rotifers. Amictic females produce diploid eggs (A) that develop parthenogenetically into females. Mictic females produce haploid eggs (B) that develop parthenogenetically into males (C) or, if fertilized, develop into thick-shelled diapausing embryos called resting eggs (D).
Source: From Gilbert (2003).

However, recent experiments with rotifers are somewhat illuminating. They clearly suggest that unfavorable conditions and crowding trigger transitions of these invertebrates, from asexually reproducing to sexually reproducing organisms.

The water flea, *Daphnia magna*, is a freshwater crustacean. (In another context, this example is presented in Chapter 11 as an example of transgenerational developmental plasticity.) Under favorable environmental conditions, *D. magna* is an asexually reproducing species that produces female offspring only. When environmental conditions of living deteriorate (crowding, depletion of food resources, shortening of the photoperiod), it produces a sexually reproducing generation (male + female individuals). This generation produces freezing- and desiccation-resistant eggs, which can hatch many years after being released (Figure 13.44).

Amictic females of many rotifer species hatching from diapausing eggs reproduce parthenogenetically (amictic reproduction). However, in response to various environmental cues, such as crowding (*Brachionus*), long photoperiods (*Notommata* and *Trichocerca*), and uptake of tocopherol (*Asplanchna*) amictic females of some species produce male haploid offspring from unfertilized eggs.

In response to crowding, amictic female rotifers of the cyclically parthenogenetic species *Brachionus angularis* give birth to mictic daughters (producing a sexually reproducing generation), which produce haploid eggs developing into males or, when fertilized, developing into diapausing eggs, which hatch into female

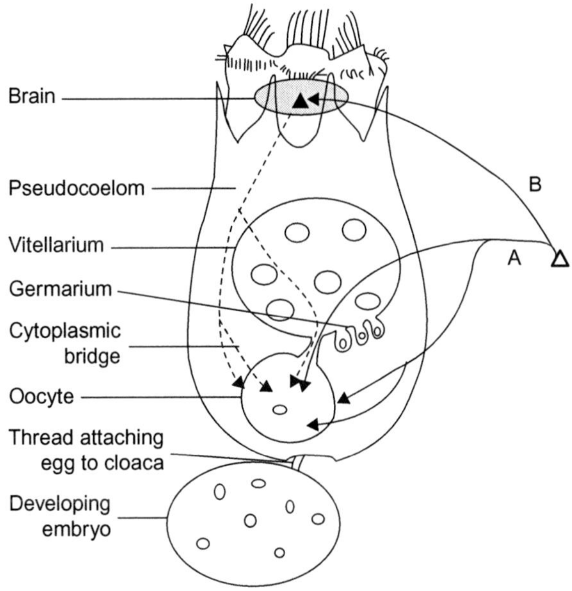

Figure 13.45 Mechanisms by which a crowding stimulus may induce oocytes of *Brachionus* to develop into mictic females. Proposed chemical inducer produced by rotifers is indicated by open triangle at right. Inducer may act directly on growing oocyte within maternal body cavity via various pathways (solid-line arrows A). Alternatively, inducer may act indirectly on maternal physiology, such as by causing the brain (solid-line arrow B) to secrete factor affecting the oocyte (broken-line arrows).
Source: From Gilbert (2003).

individuals of smaller size, incapable of sexual reproduction, for several generations (Stelzer and Snell, 2003; Gilbert and Schröder, 2004). It is determined that an environmental cue triggers the transgenerational transition to sexual reproduction in *Brachionus plicatilis*. It is a chemical cue released by conspecific individuals, which accumulates to attain threshold levels under high density of rotifer population (Stelzer and Snell, 2003). The mixis-inducing signal is a protein (MIP) (Snell et al., 2006), which on binding specific chemoreceptors (probably of corona neurons) is transmitted to the brain for processing and via a still-unknown pathway stimulates some oocytes to differentiate into mictic females (Gilbert and Schröder, 2004) (Figure 13.45).

It is suggested that

> *the crowding stimulus could diffuse into the maternal body cavity and directly affect the oocyte such that the mictic-female phenotype is expressed later on in development .... Alternatively, the chemical could directly affect the physiology of the mother. For example, it could target the nervous system via chemoreceptors, and some secreted factor could then affect the growing oocyte.*
>
> *Gilbert (2003)*

In a number of rotifer species, the increase in the proportion of the mictic offspring appears not in the first generation but several generations after the exposure to the mixis-inducing stimuli. So, e.g., the response of certain clones of *B. calyciflorus* to crowding is very low in females of the first generation after hatching from the diapausing egg, but increases in subsequent generations until it reaches a maximum from the 12th to 18th generation (Gilbert, 2002; Gilbert and Schröder, 2004).

The threshold is determined in the rotifer's brain (see on set point determination in Chapter 2), and the stimulus serves as a forewarning to the rotifer population on depleting food resources and/or increasing predation risk, as a result of the increased density of rotifer population.

The above examples give us some significant clues for understanding the evolution of sexuality and reversion to asexuality in nature. In the case of the water flea, *D. magna*, external stimuli (crowding or scarce food resources) or cues such as shortening or prolongation of the day (forewarning the approach of the winter or drying up of the ponds during the summer, respectively) cause an environmental stress to which daphnids respond adaptively by producing desiccation-resistant eggs and male individuals, thus shifting to the sexual mode of reproduction for surviving in the hostile environment. At an endocrine level, in parthenogenetic females it is observed that the response is characterized by secretion of hormone methyl farnesoate (MF), which determines production of females and males by diploid females (Olmstead and LeBlanc, 2002), thus starting the sexual phase of reproduction in *D. magna*. Experimental administration of the hormone in *Daphnia*, at a critical time of oogenesis, leads to production of eggs developing into male individuals (Rider et al., 2005).

MF is chemically similar to juvenoid hormones of insects and retinoid hormones of vertebrates (Rider et al., 2005), and its action is mediated by a nuclear receptor (Olmstead and LeBlanc, 2002). Downstream, the MF receptor, by binding its response element, induces expression of sex-determining genes (*sex-l*, *dsx*, *csd*). The natural juvenoid hormone, MF, which exists in a few isoforms, is synthesized and secreted by the mandibular organ (Liu and Laufer, 1996) under negative control of two neurohormones, mandibular organ-inhibiting hormone-1 (MO-IH-1) and mandibular organ-inhibiting hormone-2 (MO-IH-2), members of the crustacean hyperglycemic hormone (CHH) family, produced in the X-organ/sinus gland complex in response to deteriorating conditions of living in environment. The X-organ/sinus gland complex is in the eyestalk and consists of a neurosecretory organ containing neurons that secrete various neurohormones and the sinus gland, a neurohemal organ where neurohormones are released (Nagaraju et al., 2005).

Besides MF, JH and several other juvenoid substances (10 substances identified so far) have been demonstrated to induce formation of male offspring when administered to female parthenogenetic water fleas (Tatarazako et al., 2003; Oda et al., 2005).

Neo-Darwinian Explanation

The author is not aware of any attempt to interpret the phenomenon from a neo-Darwinian view. It would be almost impossible, by any stretch of the imagination, to explain from that view such a "major transition," the sudden transformation of a *whole* asexual population into a sexually reproducing population, taking place in a unigenerational "instant," without changes in genes, genetic recombination, or natural or sexual selection.

Epigenetic Explanation

From the epigenetic viewpoint, it could be predicted that transition from asexuality to sexuality in rotifers is related to activation of specific neuroendocrine cascades that start in the CNS in response to mixis-inducing protein (MIP), released by conspecifics or environmental stimuli (photoperiod) presaging deteriorating conditions of living in the environment. It is self-evident that modification of the cascade, i.e., secretion of MF, in a whole population of rotifers requires an epigenetic change, suppression of the release of specific neurotransmitters on MO-IH secretory neurons rather than changes in the downstream "sex-determining genes." A look at the cascade of events leading to transition from the asexual generation (all-female generation) to the sexual reproduction (production of male individuals and females) in *D. magna* (Cabej, 2011; Figure 13.46; see also Figure 11.9, in Chapter 11, Transgenerational Developmental Plasticity) clearly shows that the

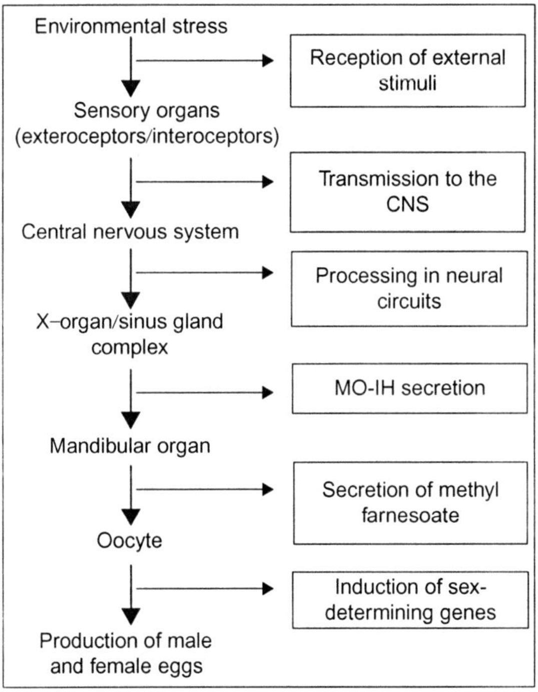

Figure 13.46 Neural mechanism of transition to sexual (male + female) reproduction of the offspring of parthenogenetically reproducing female water fleas (*D. magna*), in response to external stimuli (crowding and cues presaging deteriorating conditions in the environment).

epigenetic information for starting the cascade of events that leads to production of the sexually reproducing generation, is generated by processing of the external stimuli in the brain of these small aquatic animals.

The synthesis and secretion of neuropeptides in the nervous tissue is known to be controlled and regulated by the activity of respective neural circuits that results in the release of neurotransmitters on secretory neurons. Based on this fact, it might be predicted that the effect of the external stimuli (stressors) in inducing secretion of neuropeptides MO-IH-1and MO-IH-2 is mediated by a neurotransmitter released by a neural circuit that processes these stimuli in the crustacean CNS. Indeed, injection of the neurotransmitter dopamine (and serotonin) as well as thermal stress conditions (low temperature) causes increased production of CHH in the crayfish *Procambarus clarkii*.

Recent experimental evidence has unambiguously proven that it is the dopamine that stimulates the synthesis and secretion of CHH by neurons of the X-organ/sinus gland complex: first, by demonstrating that dopamine administration does not produce a rise in the level of CHH in eyestalk-ablated crustacean and, second, that neurotransmitters serotonin (Lorenzon et al., 2005) and dopamine (Zou et al., 2003) stimulate synthesis of CHH in *in vitro* incubated eyestalk ganglia.

The evolutionary pressure for transition to sexual reproduction is related not only to advantages of the sexual selection under unfavorable conditions but also to the fact that *sexuality facilitates reproductive isolation as a prerequisite to speciation and accelerated rates of evolution.*

Now, to summarize, generational alternation of the sexual and asexual modes of reproduction, in invertebrates such as insects and rotifers, is a neurally determined adaptive response to environmental cues or conditions that involves no changes in genes. The nongenetic, developmental mechanism of transition from asexual to sexual reproduction suggests that epigenetic switches alone in developmental pathways might have been necessary for evolution of sexuality in metazoans in general.

Evolution of Paedomorphosis in a Salamander Species

Paedomorphosis is observed in nine out of ten salamander families. The *Eurycea multiplicata* complex, a monophyletic group of salamanders, endemic to the Ozark Plateau and the Ouachita Mountains, south-central North America, comprises a diverse radiation of paedomorphic surface-dwelling (*Eurycea tynerensis*), metamorphic surface-dwelling (*E. multiplicata multiplicata* and *E. m. griseogaster*), and metamorphic subterranean (*Typhlotriton spelaeus*) hemidactyliine plethodontid salamanders. Paedomorphosis in this group evolved three to nine times (Bonett and Chippindale, 2006) as adaptive responses to the local environment.

All three paedomorphic subterranean nominal forms of the Tennessee cave salamander complex (*Gyrinophilus palleucus palleucus*, *G. p. necturoides*, and *G. gulolineatus*) evolved recently in sympatry, without geographical isolation, from the epigean (surface-dwelling) metamorphosizing forms of *G. porphyriticus* (Niemiller et al., 2008; Figure 13.47).

Figure 13.47 The Tennessee cave salamander complex and the spring salamander (bottom): (A) Pale salamander (*G. p. palleucus*), (B) Big Mouth Cave salamander (*G. p. necturoides*), (C) Larval spring salamander (*G. porphyriticus*), and (D) Berry Cave salamander (*G. gulolineatus*). Note the phenotypic differences between the larval epigean form (C) and the paedomorphic subterranean forms (A, B, and D). Larval subterranean salamanders have reduced eyes and broader, more spatulate snouts than larvae of the epigean form.
Source: From Niemiller et al. (2008).

The plethodontid salamander, *E. tynerensis*, an endemic species of the Ozark Plateau, has been considered to be an exclusively paedomorphic species. However, recently it was reported that paedomorphic populations of this species inhabit streams with chert gravel, whereas highly embedded streams (containing chiefly fine sediment) are populated by metamorphic populations (Figure 13.48).

Paedomorphosis, or the ability to switch facultatively to paedomorphosis, probably evolved independently in the ancestor of the *E. tynerensis* group, while the group also maintained the ability to metamorphose. Paedomorphosis enables the exploitation by this species of unique chert gravel–bottom streams that allow continuous access to permanent water, while metamorphosis has permitted the continued colonization of the seasonally ephemeral aquatic habitats throughout the Ozark Plateau (Bonnet and Chippindale, 2006).

Figure 13.48 Paedomorphic (A) and metamorphic (B) forms of *E. tynerensis* and associated substrates.
Source: From Bonett and Chippindale (2006).

The life histories of *E. tynerensis*, considered here as a single salamander species, are clearly correlated with the microhabitat, i.e., with the streambed microstructure; they are paedomorphic in streams with large chert gravel and metamorphic in neighboring streams of sandstone beds.

There is no fully established reproductive isolation of the paedomorphic and metamorphic forms of *E. tynerensis*, but differences in life histories, ecology, and behavior represent reproductive barriers, which, over time, may lead to complete reproductive isolation between them and to ensuing speciation.

This case of two alternative life histories displayed by two sympatric populations of a single species, of the same genotype, whose populations still can interbreed and produce fertile offspring, rejects any neo-Darwinian explanation.

Evolution of Viviparity

Viviparity (live-bearing reproduction) and oviparity (reproduction by oviposited eggs) are two basic modes of sexual reproduction in metazoans. The first implies production of live viable young by the mother, while oviparity implies production of eggs that may proceed with embryogenic development after oviposition.

Viviparity implies matrotrophy with placentotrophy as its most advanced form. Placentotrophy relies on evolution and development of structures that make the nourishment and respiration of the embryo in the reproductive tract possible, and oviparity implies provision to the egg of nutrients in the form of yolk (lecithotrophy) and water necessary for the development until hatching.

Out of ~4,000 cockroach species, only one, *Diploptera punctata*, is known to be viviparous. In this species, embryos are wrapped by a brood sac that provides the embryo with water and also releases nutritive secretions, the "milk" containing proteins of the family of lipocalin. The milk is ingested by the embryo.

Ovoviviparity, where embryogenesis takes place within the mother's body, without special maternal nourishment, is a more common phenomenon in cockroaches. It has been suggested that viviparity in cockroaches evolved from ovoviviparity. Indeed, two ovoviviparous cockroach species, *Byrsotria fumigata* and *Gromphadorhina portentosa* have brood sacks, secretory apparatus with ducts, similar to *D. punctata*. If this has been the ancestral state of *D. punctata*, then it implies that a single nongenetic *behavioral* step, i.e., the evolution of the ability of the embryo to drink, has been necessary for transition of the cockroach ovoviviparous species to viviparity (Williford et al., 2004).

In vertebrates, viviparity is estimated to have originated independently more than 140 times, with 29 of these origins having occurred among fish (Blackburn, 2005) and 98 among reptiles (Blackburn, 1995). Viviparity occurs in every vertebrate class except birds. In invertebrates, it has rarely been described.

Evidence from reptiles lends support to the view of saltational mode of appearance of viviparity, matrotrophy, and placentation (Blackburn, 1992).

In sharks and rays, the ancestral form of parity is oviparity, egg-laying, which is observed in 40% of extant species. Transition from oviparity to viviparity in this group occurred nine to ten times, and maternal input, four to five times. Reversion from viviparity to oviparity has taken place only twice (see Table 13.1).

Placentation in mammals evolved only once, some 100 mya. Among fish placentation was found only in Carcharhiniformes (ground sharks). Investigators have concluded that elasmobranchs (sharks and rays) have a high degree of evolutionary flexibility of reproductive modes. In general, evolution of viviparity in elasmobranchs seems to have been convergent, and evolution of maternal input exhibits a tendency to reverse to lecithotrophic (yolk-only) viviparity (Dulvy and Reynolds, 1997; Figure 13.49).

Fish are mostly oviparous, but some fish species are ovoviviparous, hatching within the female genital tract. In fish of the genus *Poeciliopsis* alone, a complex organ such as the placenta has independently evolved several times, and the estimated time necessary for its evolution is 750,000 years or less. It is interesting to note that species in which the placenta has evolved independently are still interbreeding and produce fertile hybrids, suggesting that the time of evolution of placentas in these species might have been much shorter (Reznick et al., 2007).

Neuroendocrine mechanisms regulating the function of the reproductive tract, which have been considered characteristic of mammals, are believed to have been in place in elasmobranchs for 400 million years, preceding in time, and surpassing in diversity, those known in mammals (Callard and Koob, 1993).

Some 40–80 mya, within the oviparous class of amphibians, a group of marsupial frogs evolved, which now comprises about 60 tree rain forest species belonging to seven genera. These frogs evolved a unique way to develop their eggs within

Table 13.1 Proportion of Live-Bearers, Number of Independent Origins of Live-Bearing and Maternal Input Estimated in Major Vertebrate Groups

Group	Incidence of Live-Bearing (%)	Transition to Live-Bearing	Transition to Maternal Input (Matrotrophy)
Mammals	99	1–2	1
Birds	0	0	0
Reptiles	<15	98	3
Amphibians	<10	5	3
Teleost fishes	2–3	10–13	12
Sharks and rays			
Previous estimates	55	15–18	5
This study	40	9–10	4–5
Total (this study)		123–128	23–24

Source: From Dulvy and Reynolds (1997).
Maternal Input refers to the period between fertilization and birth.

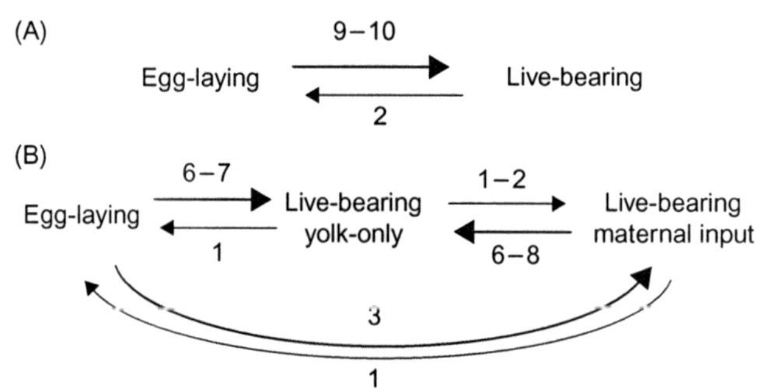

Figure 13.49 (A) Transitions between egg-laying and live-bearing elasmobranchs. (B) Transitions among egg-laying, live-bearing yolk-only (no maternal input), and live-bearing with maternal input. Arrow width is proportional to the number of transitions.
Source: From Dulvy and Reynolds (1997).

a special pouch on the mother's back, where the embryo develops around itself a fluid-filled sac reminiscent of the amniotic sac and fluid of mammal embryos. Furthermore, the pouch lining and the embryo develop intimate contact that allows passage of nutrients from mother to the embryo, essentially similar to the mammal placenta. Remarkable similarities are discovered in the hormonal regulation of embryonic development in both classes (del Pino, 1989).

In 60% of cases, viviparity in squamates (lizards and snakes) is of recent, Pleistocene, origin, as is suggested by the subgeneric level of evolution of viviparity in this group. The prevailing idea that viviparity precedes placentation has not found

Table 13.2 Hypotheses and Predictions About the Evolution of Viviparity in Squamates

Hypotheses	Predictions
Gradual	1. Clades contain species in primitive, intermediate, and advanced evolutionary stages
	2. A continuum exists of developmental stages at partition among living species
Saltational	3. Viviparous and oviparous congeners are similar
	4. Recent origins of viviparity exhibit a bimodal distribution of partition stages
	5. No oviparous for advanced eggs
	6. Facultative viviparity occurs
Punctuated equilibrium	7. A bimodal distribution of partition stages exists, but some species oviposit advanced eggs
	8. Facultative, oviparous egg-retention with intraoviductal development occurs
	9. Viviparous and oviparous congeners are similar

Source: From Blackburn (1995).

empirical support and seems to be discredited by the recently evolved cases of viviparity in lizards (Blackburn, 1995).

Blackburn (1995) has comparatively examined predictions of the three basic hypotheses on the evolution of viviparity in squamates (Table 13.2).

Lacerta vivipara is a viviparous species that evolved very recently, during the ice age, throughout Eurasia, but its populations in the Pyrenees lay eggs. Oviparous and viviparous individuals hybridize in captivity, and the hybrid eggs have half the thickness of the eggs of oviparous females. According to the gradual hypothesis of viviparity, its eggs must be laid at an advanced stage of embryonic development. In fact, they are not, and this validates the prediction number 4 (of the saltational hypothesis).

It is believed that evolution of viviparity is an adaptation to conditions of cold climate, and some empirical evidence from reptiles in support of this hypothesis exists (Shine, 1983; Mathies and Andrews, 1995). However, evidence contradicting the cold-climate hypothesis has also been presented. But although the viviparous species of the North American lizard genus *Sceloporus* (with approximately 68 species, of which 28 are viviparous) generally are found at higher elevations and latitudes, the northernmost species in North America are oviparous (Guillette Jr., 1993).

A widely held gradualistic neo-Darwinian hypothesis holds that thinning of the eggshell precedes the evolution of viviparity (Blackburn, 1998) as an adaptive modification for gradually allowing gas exchange between the increasingly consuming oxygen embryo and the uterus. Studies for testing this hypothesis in lizards have revealed no correlation between the gas permeability of the eggshell and its capacity to support embryonic development.

There are populations of the skink, *Saiphos equalis*, where females produce eggs that hatch within a few days of laying, although their eggs are thick-shelled. Mathies and Andrews believe that these animals can support embryonic development to term within fully shelled eggs in oviducts and that the thinning of the eggshell may be a postviviparity event rather than a prelude to viviparity (Mathies and Andrews, 2000).

Sometimes, transition from oviparity to viviparity may be related to the thinning and elimination of the eggshell. This may have been achieved by decreasing activity of the shell glands, by changing the number of eggs or by shortening the retention of eggs in the uterus, all epigenetic processes involving no changes in genes, genetic information, or genetic mechanisms. So, for example, in clear distinction from amphibians, reptiles have evolved a neural control on prostaglandin-induced uterine contractions, which allowed them to speed up parturition that evidently may lead to thinning and even to absence of the eggshell.

Evolution of viviparity has been considered to be a process of three successive, gradualistic processes: placentotrophy, placentation, and true viviparity. Contrary to that conventional gradualistic model of evolution of viviparity in lizards and snakes, more than 100 clades of these groups have made transition from oviparity to true viviparity (Blackburn, 1996), and recent studies have failed to find intermediate forms between viviparous and oviparous species:

> *Various phenotypic intermediates postulated by the gradualistic model are either scarce or unrepresented among known forms, including those in which viviparity has evolved at specific and subspecific levels ... placentae and a degree of placentotrophy have evolved repeatedly as necessary correlates of viviparity, not as subsequent modifications.*
>
> *Blackburn (1995)*

Transition of squamates (lizards, snakes, and amphisbaenians) to viviparity is associated with changes in the structure and function of the oviduct and uterus, which made possible the viviparity and the establishment of the complex physiological relationship between the mother and embryo (Blackburn, 1998).

In lizards, viviparity evolved in various forms, ranging from lecithotrophic viviparity through viviparity with more complex placentae, to obligate placentotrophy (Stewart and Thompson, 2000; Thompson and Speake, 2006). The last form, although less common, evolved at least five times (Thompson and Speake, 2006). One example of the rapid evolution of the complex trait of viviparity is that of *L. vivipara*, a lizard species that consists of viviparous and oviparous populations/subspecies in various regions of Europe. In Russia and Hungary, they (*Lacerta vivipara pannonica*) reproduce viviparously, whereas neighboring Slovenia and western Europe are populated by the oviparous variant (Surget-Groba et al., 2001).

Both oviparous and viviparous forms of the lizard *L. vivipara* express in their materno-fetal structure cytokines and immunotolerance factors, necessary for protecting the semi-allogenic embryo from mother's immune response, a fact that might have had an important role as a preadaptation for possible transition of the oviparous populations to viviparity (Paulesu, 1997; Paulesu et al., 2005).

In the viviparous reptile, *Chalcides chalcides*, the placenta performs endocrine functions by producing progesterone, which is increased during the late stage of pregnancy for compensating the decline in ovarian progesterone production as a consequence of the regression of corpora lutea (Guarino et al., 2002).

Experimental studies on the viviparous lizard, *Sceloporus jarrovi*, have shown that a complex endocrine maternal–fetal relationship, including a selectively regulated passage of maternal hormones through the placenta, has evolved within an evolutionarily short period of time since the evolution of viviparity in this species. When the level of progesterone of the pregnant female is experimentally elevated 100-fold, the level of the hormone in the fetus is elevated only two-fold, implying that its placenta has a strong buffering function. This function is very important for protecting the fetus from fluctuations at the level of maternal stress hormones during stress conditions, which might interfere with sex-specific development of the fetus, but it also protects the mother from the influence of fetal hormones. During the relatively short period of time since it developed the morphological structure of the placenta, *S. jarrovi*, has evolved a surprisingly complex endocrine mechanism (not new or functionally changed genes) for regulating maternal–fetal interactions (Painter et al., 2002).

A lizard from the lowlands of New Guinea, which is considered to be at an incipient stage of viviparity, develops only a thin eggshell (Guillette, 2005). The scincid lizard, *Lerista bougainvillii*, also is a reproductively bimodal species exhibiting both oviparity and viviparity. The thinning of the eggshell in this species has been considered to be an adaption for transition from oviparity to viviparity (Qualls, 1996).

In Australia, the scincid lizard, *S. equalis*, offers a very interesting example of a species that shows both viviparous and oviparous modes of reproduction. Oviparous and viviparous specimens of the same species were collected in close neighborhood, within 55 km in New South Wales. Populations from the northern highlands (Riamukka) exhibit an intermediate mode of reproduction where females produce offspring that emerge from their birth membranes within 12 h to up to 7 days, which in scincid lizards is considered viviparity. Populations of lizards from the southern coastal area (Sydney), however, produce thick-shelled eggs that have a short incubation period of 1–9 days, a fact that led investigators to the conclusion that this population "is genuinely intermediate between 'oviparity' and 'viviparity,' as these conditions are generally defined in reptiles" (Smith and Shine, 1997).

The further evolution of the intrauterine part of development in metazoans proceeded along two divergent lines, with birds reversing to oviparity, and mammals toward an ever-increasing to full independence of yolk for their embryonic development.

Neo-Darwinian Explanation

The neo-Darwinian interpretation and its predictions in relation to the evolution of viviparity presented herein are based on a review by Blackburn (1995), one of the leading investigators in the field.

For over half of a century, evolution off viviparity and placentation in squamates has been imagined as a three-stage process comprising

> *A gradual increase in the duration of oviductal egg retention, leading to viviparity, a gradual development in viviparous forms of a simple placenta that functions in gas exchange and water uptake, and a progressive reliance on the placenta as a means of supplying inorganic and organic nutrients for development, eventually leading to placentotrophy.*
>
> *Blackburn (1996)*

Transition from oviparity to viviparity is an extremely complex process. According to the neo-Darwinian paradigm, it would be predicted that useful changes in genes must occur to make this transition possible, and accumulation of the genetic changes in populations under the action of natural selection would take long periods of time. Empirical evidence shows that the transition from oviparity to viviparity occurred repeatedly and independently (in about 100 cases in squamates alone) during an evolutionarily short period of about one million years. Moreover, no changes in DNA or genes relevant to evolution of viviparity have been reported, and many genes involved in this transition have been well conserved in taxa that are so distant as insects and humans.

The neo-Darwinian gradualism would also predict that within the extant species, many, if not all, of the intermediate stages of transition from oviparity to viviparity would exist. (The argument of incompleteness of paleontological record in this case is not applicable because, by evolutionary standards, these transitions have occurred very recently.)

> *An important testable prediction that derives from the gradualistic scenario is that extant clades contain species representing primitive\intermediate\and advanced stages in the development of viviparity and placentation .... Available data on squamates do not support this prediction.*
>
> *Blackburn (1995)*

What is observed under natural condition is a wide gap between the viviparity and oviparity rather than a continuum of intermediate states (Figure 13.50).

In view of the recency of the evolution of viviparity in squamates, from the neo-Darwinian standpoint it would be predicted that among extant squamates, species of transitional states between the oviparity and viviparity must exist. However, no evidence to support this prediction has been found: Out of more than 100 clades that have made the transition, many of them in recent times, very few, if any, have shown phenotypic intermediates predicted by the neo-Darwinian model (Blackburn, 1996).

The appearance of interbreeding viviparous and oviparous populations within the same lizard species is unexplainable from the neo-Darwinian view.

Finally, if viviparity evolved because of some selective advantages as a reproductive strategy, it would be incomprehensible what could have favored accumulation of changes and evolution of intermediate forms (between the oviparity and viviparity), which evidently did not offer the advantages of viviparity.

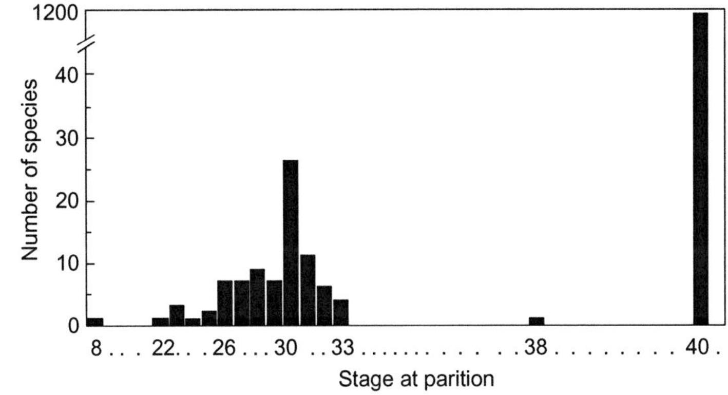

Figure 13.50 Stages of embryonic development at deposition of the reproductive product (egg or neonate) in squamate reptiles. Staging follows the D & H system in which Stage 1 is an unfertilized egg, and Stage 39 represents birth or hatching; thus parition at Stage 39 represents viviparity. For species with a range of reported stages at oviposition, modal values (or, if unavailable, range midpoints) were used. Representation of stages along the horizontal axis approximates the time course of embryonic development.
Source: From Blackburn (1995).

Epigenetic Explanation

The epigenetic view would predict that neither changes in genes or genetic information nor long evolutionary periods of time are necessary for the transition from oviparity to viviparity. This prediction is validated by empirical evidence accumulated so far, which shows that viviparous forms are morphologically similar or almost identical (in the cases of oviparous and viviparous populations of the same species) to their oviparous conspecific/congeners (Blackburn, 1996).

Predictably, phenotypic changes are restricted to the organs and tissues involved in the transition. These changes, in squamates, include:

1. Reduction of egg thickness,
2. A possible increase in oviducal vascularization,
3. Postponement of parition,
4. Suppression of nesting behavior.

None of the above changes requires changes in genes.

The reduction of the eggshell thickness involved (1) "No loss or suppression of the genes for shell membrane deposition" (Blackburn, 1996).

During the individual development and adult life in female vertebrates, vascularization (2) of the oviduct is neurohormonally regulated, and the two other phenotypic changes (3 and 4) necessary for transition to viviparity (postponement of parition and suppression of nesting behavior) are under obvious control of behavioral neural circuits.

The fact that most cases of viviparity in lizards and snakes appeared recently during Pleistocene (1.8 million to 11,500 years ago), and especially the fact that the viviparity in lizard species *L. vivipara* and *Sceloporus aeneus* is estimated to have evolved in the past 11,000–25,000 years, also support the epigenetic-developmental hypothesis.

Evolution of Behavioral Characters in Insects

Most behaviors are motor actions that an animal performs in response to external or internal, real or perceived, stimuli. As a rule, the external/internal stimuli do not directly affect genes related to the behavior, and no gene knows whether in the presence of light, higher temperature, longer day length, social interaction, sensory information, or other conditions it must be transcribed or not. Activation/inactivation of nonhousekeeping genes depends on extracellular "instructions" that reach them at the end of signal cascades that start in the nervous system, where decisions that are made on how to respond adaptively to external/internal stimuli (see Chapter 2).

Evolution of Flight in Insects

The flight-control system, in locusts, for example, involves dozens of neurons in several ganglionic areas. Many of the interneurons involved in producing the flight pattern have their cell somata in the parts of the metathoracic ganglion that represent the embryologically fused abdominal ganglia 1, 2, and 3 (Delcomyn, 1991). It has been proposed that this distribution of neurons suggests that insect wings originated from movable appendages carried by both thoracic and abdominal segments in primitive insects, rather than from paranotal lobes, which were confined to the thorax.

The evidence that sets of homologous flight interneurons occur in abdominal and thoracic ganglia supports theories that insect wings originated from movable appendages, which were serially distributed along the thorax and abdomen and which were under central nervous control (Robertson et al., 1982).

Considerable progress has been made during the last two decades in understanding the mechanism of control of the flight behavior in insects. Besides wings, during flight, insects also use halteres, evolutionary vestiges of hindwings. Halteres oscillate in antiphase to the wing and are subject to different inertial, gravitational, and, especially, Coriolis forces emerging during angular rotation.

There are 17 small steering muscles that are used for controlling the wing stroke, but only one of them, the first basalar muscle (B1), fires an action potential for each wing beat during straight flight, based on the input it receives from afferent nerves of the wing. However, when, as a result of motion of a visual target, the insect must correct the course, it does this very rapidly, in as little as 30ms, by activating the rest of steering muscles. The change in course is sensed by campaniform sensillae at the base of halteres, which are activated, and the input from the haltere nerve becomes stronger so that the motor neuron of the B1 (MNB1) will fire 1–2ms earlier. After the flight course is

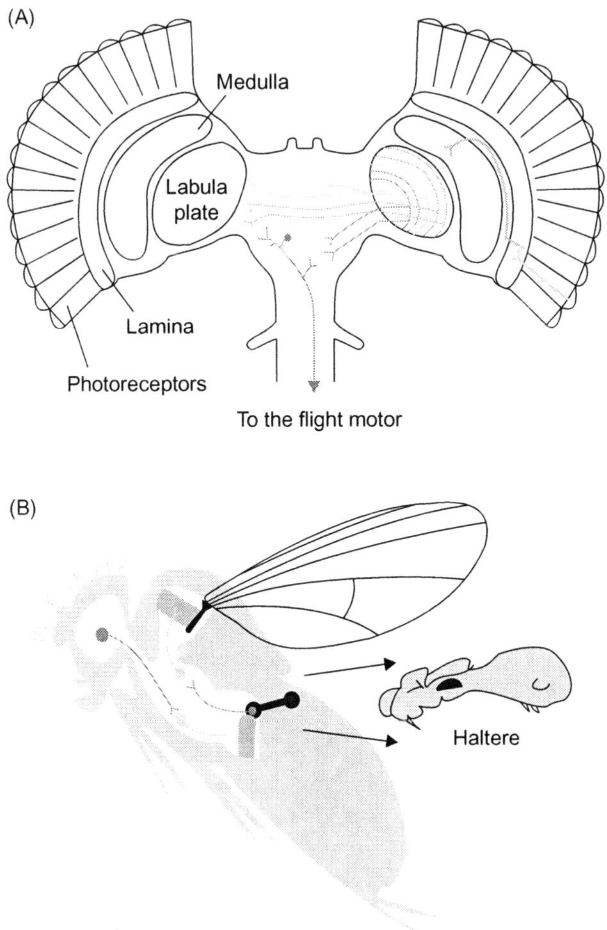

Figure 13.51 A subset of sensorimotor pathways in the fly flight-control system. (A) Early visual pathways project retinotopically from the photoreceptors to the medulla (*brown*), and from the medulla to the lobula complex (*purple*). The last stop for motion processing within the brain, the lobula plate houses at least three major families of motion detecting interneurons. The HS, CH, and VS cells project within the ipsilateral hemisphere (*blue* tones), and H1, H2, and FD cells (see text for definitions) project to the contralateral hemisphere (*green* tones). The terminals of some of these groups visit the sensory dendrites of interneurons projecting to the flight motor (*red*). (B) Descending visual interneurons drive motor neurons innervating steering muscles of the mechanosensory halteres. Haltere sensory afferents, in turn, provide electrotonic input to wing muscle motor neurons, completing a visually gated feedback circuit (sensory neurons indicated in *red*, motor neurons indicated in *yellow*, steering muscles indicated in *brown*). (For interpretation of the references to color in this figure legend, the reader is referred to the web version of this book.)
Source: From Frye and Dickinson (2001).

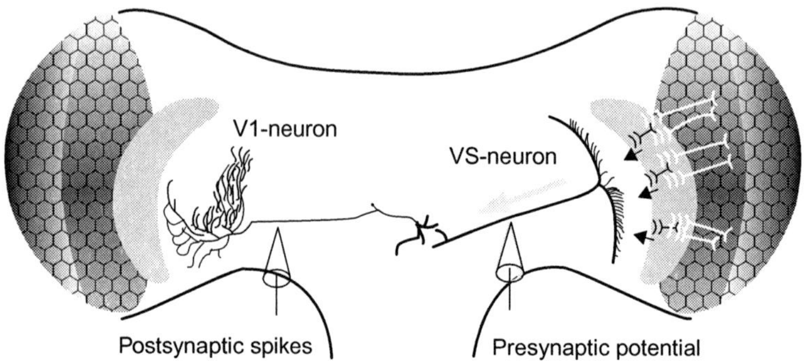

Figure 13.52 Schematic of the neural circuit with one presynaptic VS cell, VS2, and the postsynaptic V1 cell. V1 integrates motion signals from at least four presynaptic VS cells and transforms them into spike activity, which is conveyed to the contralateral brain hemisphere. *Source*: From Kalb et al. (2006).

corrected, the input from wing mechanoreceptors will once again become normal, and the MNB1 returns to the normal firing (Fayyazuddin and Dickinson, 1996).

The insect flight is determined by a motor program, and in this sense it is an inherited behavior. However, while flying insects regulate their flight according to the continual input of sensory information reaching the brain from various sensory systems (Figure 13.51). Visual information during flight is processed in the brain visual centers, and the output resulting from the processing is transmitted to the flight motor for making necessary corrections. Another source of information for the flight motor program is the input from hundreds of mechanoreceptors at the base of halteres.

During the flight, the visual and the mechanosensory systems concurrently send input to the flight-control system. The integration of the visual and mechanosensory input takes place in descending neurons within the brain or in the flight motor neurons within the thorax. There, a particular functional weight is given to the visual and mechanosensory (from halteres) input, with the latter being dominant (Sherman and Dickinson, 2004). The visual input provides feedback for flight stability at lower angular velocities to which halteres are less sensitive (Bender and Dickinson, 2006).

Extraction of relevant information from the above sources of sensory information is a function of synaptic filtering and postsynaptic integration of the sensory input in the insect brain (Kalb et al., 2006; Figure 13.52). Ventral striatum (VS) neurons receive motion sensitivity by many small-field retinotopic motion-sensitive elements that are pooled by their dendritic tree.

A group of four VS motion-sensitive neurons sends motion signals to the postsynaptic V1 neuron, which convert these signals into motion information and transmits it to the contralateral visual system in the form of spike trains (Kalb et al., 2006).

The processing of the sensory information takes place in the flight-control system (Figure 13.53). The sensory (visual and mechanosensory) feedback to the controller is processed in neural circuits for modulating forces to be produced by the wings,

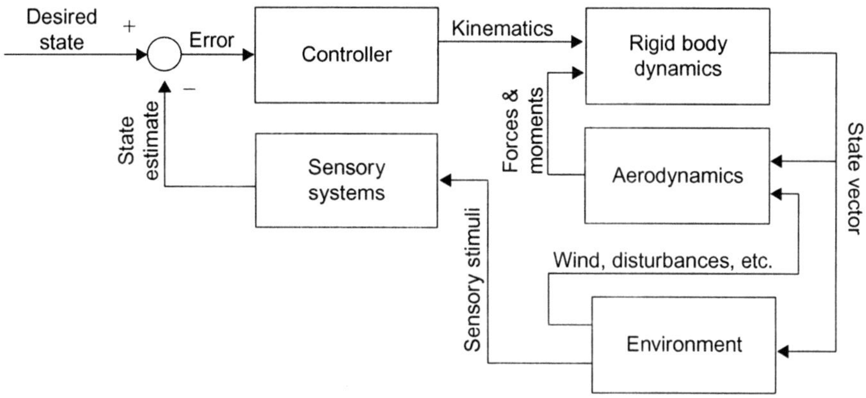

Figure 13.53 Block diagram illustrating the overall architecture of the flight-control system. *Source*: From Dickson et al. (2006).

which is necessary both for the short-term flight stability and for the long-term navigation and guidance (Dickson et al., 2006; Dickson et al., 2008).

There is no evidence whatsoever that evolution and loss of flight behavior in insects is related to any change in genes or DNA. All the available evidence relates the evolution of flight with the computational activity of neural circuits and related flight-control system. The flight-generating function of these circuits is an epigenetic phenomenon related to the properties of the neural circuits, properties that can adaptively change very rapidly, without changes in genes, in the course of evolution.

Evolution of Avoidance Behavior in the Green Tree- and Red-Bellied Snakes

The toxic cane toad, *Bufo marinus*, was introduced in Australia in 1935. This toad is so toxic that if snakes of two native Australian species, the green tree snake, *Dendrelaphis punctulatus*, and the red-bellied black snake, *Pseudechis porphyriacus*, eat even a relatively small toad, they die (Phillips and Shine, 2004). Now, about 60 years, or about 23 generations after colonization of parts of the Australian continent by *Bufo marinus*, the Australian black snake, *P. porphyriacus*, has evolved an innate avoidance behavior toward the toxic cane toad (Phillips and Shine, 2006). From the neo-Darwinian point of view, this is an extremely short period of time and a very small number of generations for an innate avoidance behavior to evolve via mutations in relevant genes.

Experimental procedures have excluded the possibility that the behavior may have been acquired by learning or experience within the lifetime of snakes. Investigators have determined that the behavior is not related to intragenerational learning, because they failed in attempts to learn naïve (not adapted) black snakes from other regions to avoid the toxic prey (Phillips and Shine, 2006).

Besides the adaptive avoidance behavior, within the same incredibly short period of time, in the regions populated by toxic cane toad, the black snake has also evolved adaptive evolutionary morphological and morphometrical characters: a smaller head size and a bigger body size. According to the investigators, the smaller head size prevents them from preying on toxic toads of large body size, which consequently are more toxic, and a bigger body size of the snake implies a lower likelihood of dying as a result of eating a large toxic cane toad (Phillips and Shine, 2004). On the other hand, within the same short period, toads have decreased their body size and toxicity (Phillips and Shine, 2005, 2006).

Neo-Darwinian Explanation

Based on the experimental failure to induce learned avoidance behavior or acquired resistance to the toad toxin, investigators believe that the inherited avoidance behavior the snakes display is a result of selection on the existing variability (Phillips and Shine, 2006). However, their conclusion is based on two flawed premises:

First, they take for granted that snakes cannot learn an avoidance behavior because they will not "survive even a single encounter with a large toad," thus ignoring the empirically demonstrated fact that snakes that happen to eat smaller toads survive in nature, and second, they take into consideration only two alternative possibilities of evolution of the avoidance behavior, natural selection or intragenerational plasticity ("a change acquired within an individual snake's lifetime"), leaving out of consideration transgenerational developmental plasticity.

The neo-Darwinian explanation of evolution of the above behaviors in snakes is extremely difficult, if possible at all, because the probability of evolution of "genes for avoidance behavior" within ~23 generations is negligible although within the realm of possibility. The virtual impossibility aside, there is no report or hypothesis on the existence or evolution of such genes, not only for this special case but for evolution of animal behaviors in general. It has never been claimed that changes in genes, gene drift, genetic recombinations, or changes in allele frequencies (existing genetic variability) may have been involved in this evolutionary behavioral adaptation.

The fact that all snake populations in areas populated by the toxic cane toad have evolved the innate avoidance behavior toward *B. marinus*, in the context of the limited number of generations, suggests that it is unlikely that population genetics knowledge may be of any use in explaining the sudden evolution of the behavior.

Epigenetic Explanation

From a paradigmatic epigenetic viewpoint, evolution of the innate avoidance behavior in the Australian black snake, *P. porphyriacus*, is the result of:

> The transformation of an experience-dependent, learned avoidance behavior (see Section Learned Behaviors Evolve into Innate Behaviors in Chapter 9) into an innate behavior, a hypothesis that is rejected by, and is incompatible with, the neo-Darwinian theory, although embraced by Darwin in his time.

The evolution of the innate avoidance behavior in the Australian black snake resembles cases of transgenerational developmental plasticity, which evolve within one to several generations and needs no changes in genes or genetic information (see Chapter 11, Transgenerational Developmental Plasticity).

The transformation of the experience-dependent, learned avoidance behavior into an innate avoidance behavior in the Australian black snake, *P. porphyriacus*, implies establishment of a relationship between the toxic toad perception in the visual recognition system and the circuit determining the avoidance behavior. Establishment of the relationship required acquisition of the relevant neurobiological (epigenetic) information rather than changes in genes, which did not occur.

Evolution of a New Ovulation Character in House Finches

Within 36 years (36 generations), a whole population of house finches (*Carpodacus mexicanus*) in Montana, at the northern limit of the species' range, evolved a novel ovulation character, a maternally determined sex-specific ovulation sequence: about 90% of the first-laid eggs are females, and about 80% of the second-laid eggs are males (Badyaev et al., 2006). The extremely short period of the time during which the evolution of the trait occurred, and the fact that it took place within the range of species, i.e., under conditions of sympatry and gene flow, excludes the possibility of involvement of gene mutations, gene drift, or genetic recombination.

Such evolutionary changes are related to specific fluctuations in the levels of the pituitary prolactin (PRL) and androgens in the circulation of mothers during oogenesis. Levels of these hormones, which are ultimately regulated by the bird's brain, influence segregation of sex chromosomes during the first meiotic division, thus determining the genetic sex of the egg; oocytes that grow during high androgen levels develop into male eggs, and those that happen to grow during high PRL, into female eggs (Badyaev et al., 2005). Hence, it is a neurally controlled change in the patterns of release of these hormones by mothers, rather than any changes in genes or DNA, that determines the evolutionary change of the sex-biased egg-laying.

Evolution of the Circadian System in Blind Moles

As a result of the subterranean life, about 40 mya, mole rats lost their eyes (Avivi et al., 2004). The blind mole rat, *Spalax ehrenbergi*, lost its normal eye structures and now has rudimentary eyes, covered by skin, that do not respond to visual stimulation (Sanyal et al., 1990). Most of the eye socket is occupied by the harderian gland, a newly evolved photoperiodic organ.

The removal of these rudimentary eyes, however, disturbs photoperiodic responses, suggesting that the reduced eyes may still be operational in the neuroendocrine pathways of photoperiodic behavior and physiology (Sanyal et al., 1990), while the restructuration of the retina into a pineal-like photoreceptive organ seems to have reprogrammed it for a photoperiodicity sensory function, as an adaptation to the underground mode of life (Cernuda-Cernuda et al., 2002; Avivi et al., 2002).

Expression of the gene *sCry1* in this gland is synchronized to its expression in the hypothalamic suprachiasmatic nucleus (SCN) and to the synthesis of melatonin and expression of melatonin receptors.

> *The two synchronous clockworks residing in the SCN and the harderian are connected neurally and via melatonin signaling and may thereby stabilize each other during long periods of zeitgeber (e.g., light) absence.*
>
> *Avivi et al. (2004)*

The retina has conserved the basic retinal structure, but the lateral geniculate nucleus (LGN) in this species receives auditory input (Bronchti et al., 1991; Heil et al., 1991) instead of visual retinal input. Additionally, the LGN gained a somatosensory function (Necker et al., 1992).

It seems that all these evolutionary changes in the structure and function of eyes in the blind mole rat have occurred without any changes in genes, or at least there is no evidence for such changes in genes or allele frequencies to have taken place in these darkness-adapted subterranean animals. This fact, for the time being at least, rules out a neo-Darwinian explanation of the evolution of the circadian system in blind moles.

Neural Mechanisms of Metazoan Migration

This type of orientation is widespread among invertebrates and vertebrates. As early as 1859, von Middendorf hypothesized that migratory birds utilize the earth's magnetic field for orientation during migration, and 30 years ago it was demonstrated that European robins use a magnetic compass for orientation (Ritz et al., 2000).

Empirical evidence suggests the existence of two magnetic senses in birds–one residing in the visual system (Mouritsen and Ritz, 2005; Zapka et al., 2009), and another in the upper beak (Fleissner et al., 2003; Wiltschko and Wiltschko, 2007).

Light-Dependent Magnetic Orientation

Two competing variants of the hypothesis of visual geomagnetic orientation have been presented: one positing that small magnetite particles are utilized by migratory birds for geomagnetic orientation, and the other propounding that geomagnetic orientation is a function of bi-radical reactions of macromolecules excited by light energy. According to the first hypothesis, ferro(i)-magnetite crystals in the oil droplets of the bird retina interact with pigment molecules such as β-carotene and line up parallel to the magnetic field lines, letting light to enter to specific photoreceptors when the bird's head is parallel or antiparallel to the geomagnetic field lines (Muheim et al., 2002). Being very sensitive to changes in the intensity of the geomagnetic field ($<50\,nT$), this system can provide information for the navigational "map," hence it can be used for determining position of the bird during migration (Beason, 2005).

It is the second hypothesis that has found the wider experimental and observational support (Wiltschko et al., 1993; Ritz et al., 2004). It predicts that the magnetic orientation is based on magnetoreception via magnetically sensitive receptors and that the magnetoreceptor should be a pair of photoreceptor molecules that enter a radical-pair reaction upon excitation by high-energy (short-wavelength) light.

Magnetoreception starts with light-dependent processes in retina. The photoreceptor molecules responsible for the magnetic compass in birds are cryptochromes. Behavioral evidence and theoretical considerations have led to the idea that these photoreceptors enable migratory birds to perceive the magnetic field as visual patterns (Mouritsen et al., 2004) or variation in the light intensity or color in their visual field (Muheim et al., 2002).

A dim monochromatic light from the blue-green range of the spectrum allows passerine birds to orient, whereas monochromatic light of longer wavelength (red), beyond 600 nm, causes them to lose orientation (Wiltschko and Wiltschko, 2002). Light in the blue-green range will excite photoreceptors, initiating the radical-pair process in the retina of the migratory bird.

In contrast to the magnetite system, this system shows only low sensitivity to changes in the intensity of the geomagnetic field (Wiltschko et al., 2004), which fits best with its function for determining only direction (with the position of the bird in the "map" determined by the magnetite system). According to Beason (2005), both mechanisms, the ferro(i)-magnetite mechanism and light-dependent excitation of photoreceptors in migratory birds, may be seen as components of a single orientation system in which the first serves position finding (map), and the second direction finding (compass).

The fact that the migratory birds with their right eye covered fail to perform magnetic compass orientation shows that the latter requires night vision. The system of night vision in migratory birds is related to the evolution of the "cluster N" (N—for *night*), an area in the dorsal part of the brain, which is believed to have evolved from the thalamofugal visual pathway, as part of the ancestral visual *Wulst* (German for *bulge*) that transmits information from the retina to the hypothalamus, or as a new adjoining structure (Figure 13.54). For night navigation, migratory birds use sensory input on star constellations.

The thalamofugal pathway in birds transmits information from the retina to the tectum in the midbrain and via thalamus to the entopallium, nidopallium, and ventral mesopallium. The cluster N increases neural firing during the night vision. Evolution of the cluster N with its five subregions is related to an evolutionary pressure for night navigation by using the information about the star constellations and magnetic field of the earth (Mouritsen et al., 2005). A striking difference has been observed recently in the structure and function of the "cluster N" in the brains of migratory and nonmigratory birds. When awake during the night, the migratory songbirds show an increased expression of *ZENK* genes in "cluster N," while the activity of the visual Wulst in nonmigratory birds, which have no "cluster N," is even lower than their low day-time activity.

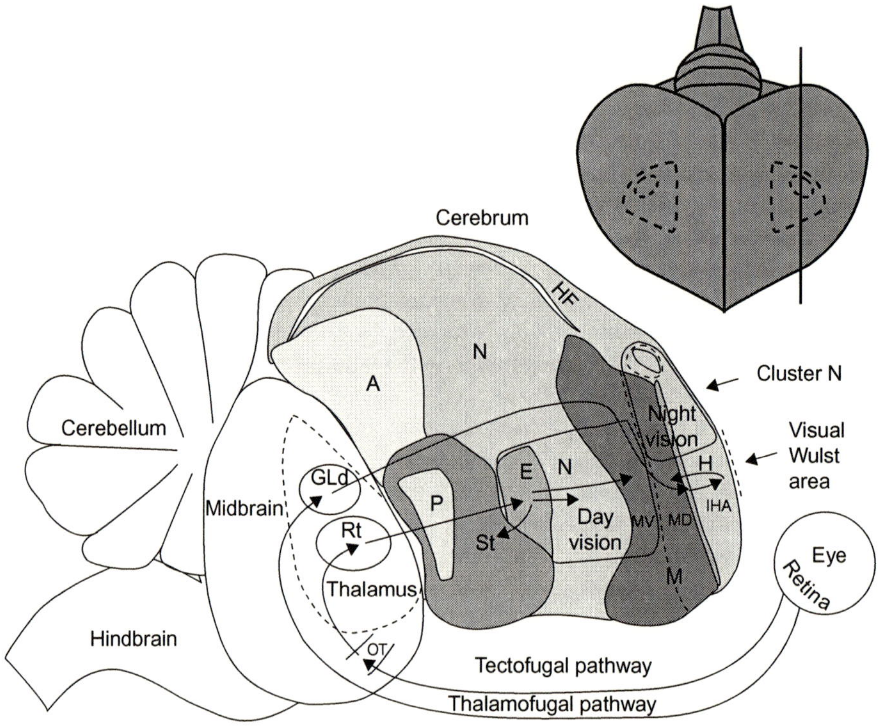

Figure 13.54 Schematic drawing of a brain showing the relative locations of the day- and night-vision (cluster N) activated brain regions in night-migrant songbirds. The thamalofugal and tectofugal visual pathways have been determined in other bird species. Upper right, the extent of cluster N seen from the dorsal surface of the brain. *Abbreviations*: GLd, lateral geniculate nucleus, dorsal part; H, posterior hyperpallium; HA, hyperpallium apicale; IHA, interstitial region of the hyperpallium apicale; M, mesopallium; MD, dorsal mesopallium; MV, mesopallium ventrale; Rt, nucleus rotundus.
Source: From Mouritsen et al. (2005).

Trigeminal System of Geomagnetic Orientation

In the migratory European robin, *Erithacus rubecula*, the ferromagnetic structures of the upper beak, which are the primary magnetic sensor, are innervated by the ophthalmic branch of the trigeminal nerve (V1). It converts changes in the magnetic field into electrical information and transmits them for integrating and processing, on the one hand, to the principal sensory nucleus of the trigeminal nerve (PrV), and almost certainly further to higher brain centers, and on the other, to the spinal trigeminal nucleus (SpV) and further to hindbrain centers (Heyers et al., 2010; Figure 13.55). In response to changes in geomagnetic field, neurons throughout the trigeminal system express ZENK.

A complex dendritic structure that is detected in the beak not only in migratory birds but also in the domestic chicken, containing iron minerals (almost completely Fe III-oxides), may be the primary structure in determining the migration behavior, and

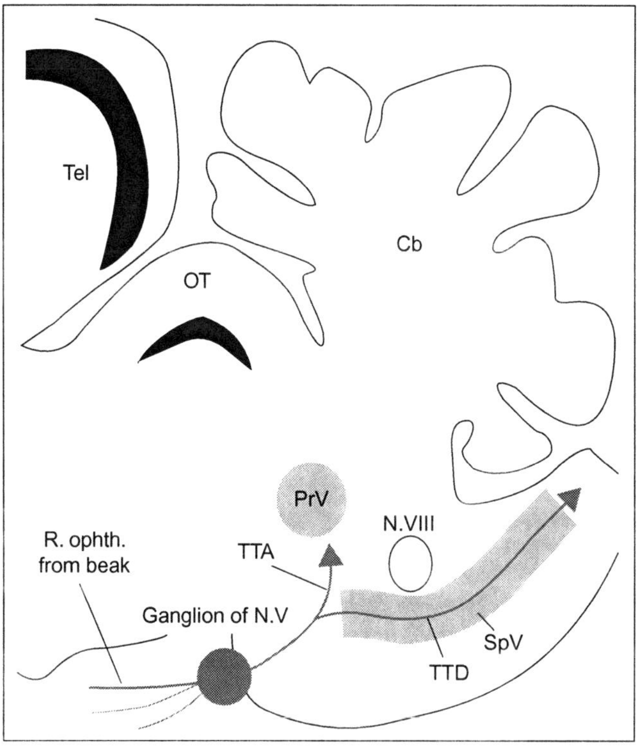

Figure 13.55 Magnetic field changes induce ZENK activation in the trigeminal system. Schematic illustration of the avian trigeminal sensory system; dorsal is up, and anterior is left. Neuronal somata of all three branches of the trigeminal nerve are located in the trigeminal ganglion. Their afferents give rise to an ascending (TTA) and a descending tract (TTD), which terminate in PrV and SpV, respectively. *Abbreviations*: Cb, cerebellum; N.V, trigeminal nerve; OT, optic tectum; PrV, principal sensory nucleus of the trigeminal nerve; SpV, spinal trigeminal nucleus; Tel, telencephalon.
Source: From Heyers et al. (2010).

> *this complex dendritic system in the beak is a common feature of birds, and that it may form an essential sensory basis for the evolution of at least certain types of magnetic field guided behavior.*
>
> *Falkenberg et al. (2010)*

Magnetite crystals are found in the upper beak and the ethmoid region in many bird taxa. A magnetite mechanism of orientation in the geomagnetic field may be operational in the homing pigeon (*Columbia livia*), and a similar location (the trigeminal nerve and the nasal region) of the magnetic sense is identified in teleost fish (Mora et al., 2004).

Evolution of migration and orientation in magnetic field involved no changes in genetic information.

Experimental Induction and Evolution of New Characters

Rapid Evolution of Physiological Characters in Drosophila

Under conditions of demographic and stress selection in laboratory, in a surprisingly small number of generations (fifty (Teotónio and Rose, 2000) to a few hundred (Teotónio et al., 2002)), *D. melanogaster* populations evolved a number of biochemical, physiological, and life history characters, such as desiccation resistance, survival under starvation conditions, and early/late reproduction age.

Selection for desiccation resistance for 200 generations in *D. melanogaster* led to an increase of 35% in wet mass and body size, mainly due to an increase of the water content of the body, which is reflected in increased hemolymph volume. As a result, the survivability of flies under conditions of extreme water stress was enhanced (Folk and Bradley, 2005). At a physiological level, evolutionary changes became manifest in an increase of the carbohydrate content and a decline of the lipid content that have never been observed in natural populations of xeric *Drosophila* species. A considerable increase is also observed in the absolute content of sodium and chloride, critical elements of the insect homeostasis that are under control of brain neurohormones, antidiuretic hormone (ADH) and diuretic hormone (DH), released, respectively, by CC and the median secretory neurons in the insect brain.

The increase in the carbohydrate content is due to increased content of glucose and trehalose, which is oxidized in flies under desiccation stress. It is noteworthy that, in insects, it is the nervous system that determines and regulates the carbohydrate levels via the neuropeptide adipokinetic/hypertrehalosemic family of neurohormones (AKH/HTH) synthesized and secreted by CC (Folk and Bradley, 2005).

Neo-Darwinian Explanation

With no changes in genes, allele frequencies, or genetic recombination involved in the rapid appearance of these evolutionary physiological and morphometric changes, it is needless to say that there is no feasible neo-Darwinian explanation for their evolution.

Epigenetic Explanation

The fact that the above-mentioned suddenly appearing evolutionary changes in *Drosophila* (the resistance to desiccation condition, increase in the water content, as well as in carbohydrate, sodium, and chloride content) are not related to any genetic changes suggests that epigenetic mechanisms may be responsible for evolution of these innovations. This is not a mere inference.

The water content in insects is regulated by Malpighian tubules, whose function is under neural control of the antidiuretic and diuretic neurohormones (Beyenbach, 2005). The synthesis and the level of trehalose and carbohydrates in general are also neurally regulated by the neuropeptides of the adipokinetic/hypertrehalosemic family. So, what really occurred in the experiments was not any changes in genes but an epigenetically determined heritable change in the output of the respective neural circuits, in response to stressful desiccation, that increases the amount of the AKH/HTH neurohormone, which is regulated by another neurohormone, proctolin, secreted by the lateral secretory neurons in the brains of insects (Clark et al., 2005).

What determined the appearance of these innovations in the offspring were neurally determined epigenetic changes in the patterns of expression of the same, functionally unchanged, genes in the insect brain.

Evolution of Life History Characters in *D. melanogaster*

Within three years under laboratory conditions, *Drosophila* strains evolved longer preadult developmental time, increased early fecundity, and decreased late fecundity (Sgrò and Partridge, 2000). Investigators rejected the previous neo-Darwinian explanation that the cause of the evolutionary change could be inbreeding depression or accumulation of deleterious mutations in small populations:

> *Two lines of evidence are against these explanations. First, both types of culture were maintained at high effective population sizes, 2,000–3,000 adults per generation per population in the case of bottles and about 7,000 adults in the case of population cages. Second, several of the traits showed evolutionary change in a direction opposite to that predicted by a hypothesis of inbreeding depression or mutation accumulation. Inbreeding leads to slower preadult development and reduced adult survival—both of which were indeed observed—but also to lowered larval competitive success, lowered body size, and lowered adult fecundity, all of which are the opposite of what was observed during laboratory adaptation. There was therefore no general lowering of fitness by inbreeding or mutation accumulation.*
>
> *Sgrò and Partridge (2000)*

Under these circumstances, there is no alternative but to accept the existence of epigenetic mechanisms; changes in the timing of the developmental processes determining longer preadult developmental time, increased early fecundity, and decreased late fecundity, both under well-known neural control.

References

Abouheif, E. and Wray, G.A. (2002). Evolution of the gene network underlying wing polyphenism in ants. Science *297*, 249–252.

Adams, M.D. (2000). The genome sequence of *Drosophila melanogaster*. Science *287*, 2185–2195.

Adams, C.S., Korytko, A.I. and Blank, J.L. (2001). A novel mechanism of body mass regulation. J. Exp. Biol. *204*, 1729–1734.

Albertson, R.C. and Kocher, T.D. (2006). Genetic and developmental basis of cichlid trophic diversity. Heredity *97*, 211–221.

Alves, H.J., Alvares, L.E., Gabriel, J.E. and Coutinho, L.L. (2003). Influence of MyoD and cellular proliferation in chicken embryos. Braz. J. Med. Biol. Res. *36*, 191–197.

Arendt, D. and Baptista, I. (2002). Evolutionary aspects of eye and nervous system development. In: *Brain Patterning and Eye Development*, J. Wittbrodt (ed.). EMBL Research Reports 2002, pp. 93–96.

Arendt, D., Tessmar, K., de Campos-Baptista, M.-I.M., Dorresteijn, A. and Wittbrodt, J. (2002). Development of pigment-cup eyes in the polychaete *Platynereis dumerilii* and evolutionary conservation of larval eyes in Bilateria. Development *129*, 1143–1154.

Averof, M. and Cohen, S.M. (1997). Evolutionary origin of insect wings from ancestral gills. Nature *385*, 627–630.

Avivi, A., Oster, H., Joel, A., Beiles, A., Albrecht, U. and Nevo, E. (2002). Circadian genes in a blind subterranean mammal II: conservation and uniqueness of the three *period* homologs in the blind subterranean mole rat, *Spalax ehrenbergi* superspecies. Proc. Natl. Acad. Sci. U.S.A. *99*, 11718–11723.

Avivi, A., Oster, H., Joel, A., Beiles, A., Albrecht, U. and Nevo, E. (2004). Circadian genes in a blind subterranean mammal III: molecular cloning and circadian regulation of cryptochrome genes in the blind subterranean mole rat, *Spalax ehrenbergi* superspecies. J. Biol. Rhythms *19*, 22–34.

Badyaev, A.V. and Hill, G.E. (2003). Avian sexual dichromatism in relation to phylogeny and ecology. Annu. Rev. Ecol. Evol. Syst. *34*, 27–49.

Badyaev, A.V., Schwabl, H., Young, R.L., Duckworth, R.A., Navara, K.J. and Parlow, A.F. (2005). Adaptive sex differences in growth of preovulation oocytes in a passerine bird. Proc. Biol. Sci. *272*, 2165–2172.

Badyaev, A.V., Acevedo Seaman, D., Navara, K.J., Hill, G.E. and Mendonca, M.T. (2006). Evolution of sex-biased maternal effects in birds: III. Adjustment of ovulation order can enable sex-specific allocation of hormones, carotenoids, and vitamins. J. Evol. Biol *19*, 1044–1057.

Baeckberg, M., Collin, M., Ovesjoe, M.L. and Meister, B. (2003). Chemical coding of $GABA_B$ receptor-immunoreactive neurones in hypothalamic regions regulating body weight. J. Neuroendocrinol. *15*, 1–14.

Beason, R.C. (2005). Mechanisms of magnetic orientation in birds. Integr. Comp. Biol. *45*, 565–573.

Bender, J.A. and Dickinson, M.H. (2006). A comparison of visual and haltere-mediated feedback in the control of body saccades in *Drosophila melanogaster*. J. Exp. Biol. *209*, 4597–4606.

Berggren, K., McCaffery, P., Dräger, U. and Forehand, C.J. (1999). Differential distribution of retinoic acid synthesis in the chicken embryo as determined by immunolocalization of the retinoic acid synthetic enzyme, RALDH-2. Dev. Biol. *210*, 288–304.

Berggren, K., Ezerman, E.B., McCaffery, P. and Forehand, C.J. (2001). Expression and regulation of the retinoic acid synthetic enzyme RALDH-2 in the embryonic chicken wing. Dev. Dyn. *222*, 1–16.

Blackburn, D.G. (1992). Convergent evolution of viviparity, matrotrophy, and specializations for fetal nutrition in reptiles and other vertebrates. Amer. Zool. *32*, 313–321.

Blackburn, D.G. (1995). Saltationist and punctuated equilibrium models for the evolution of viviparity and placentation. J. Theor. Biol. *174*, 199–216.

Blackburn, D.G. (1998). Structure, function, and evolution of the oviducts of squamate reptiles, with special reference to viviparity and placentation. J. Exp. Zool. *282*, 560–617.

Blackburn, D.G. (2005). Evolutionary origins of viviparity in fishes. In: *Viviparity in Fishes*, H. Grier and M.C. Uribe (eds.). New Life Publications, Homestead, FL, pp. 303–317.

Bock, W. and von Wahlert, G. (1965). Adaptation and the form-function complex. Evolution *19*, 269–299.

Bok, J., Bronner-Fraser, M. and Wu, D.K. (2005). Role of the hindbrain in dorsoventral but not anterioposterior axial specification of the inner ear. Development *132*, 2115–2124.
Bonett, R.M. and Chippindale, P.T. (2006). Streambed microstructure predicts evolution of development and life history mode in the plethodontid salamander *Eurycea tynerensis*. BMC Biology *4*, p. 6.
Boulant, J.A. (2000). Role of the preoptic-anterior hypothalamus in thermoregulation and fever. Clin. Infect. Dis. *31*(Suppl. 5), S157–S161.
Brakefield, P.M., Gates, J., Keys, D., Kesbeke, F., Wijngaarden, P.J., Monteiro, A., et al. (1996). Development, plasticity and evolution of eyespot patterns. Nature *384*, 236–242.
Brakefield, P.M., Pijpe, J. and Zwaan, B.J. (2007). Developmental plasticity and acclimation both contribute to adaptive responses to alternating seasons of plenty and of stress in *Bicyclus* butterflies. J. Biosci. *32*, 465–475.
Brent, C.S., Schalb, C. and Vargo, E.L. (2005). Endocrine changes in maturing primary queens of *Zootermopsis angusticollis*. J. Insect Physiol. *51*, 1200–1209.
Britton, J.S., Lockwood, W.K., Li, L., Cohen, S.M. and Edgar, B.A. (2002). Drosophila's insulin/PI3-kinase pathway coordinates cellular metabolism with nutritional conditions. Dev. Cell *2*, 239–249.
Bronchti, G., Rado, R., Terkel, J. and Wollberg, Z. (1991). Retinal projections in the blind mole rat: a WGA-HRP tracing study of a natural degeneration. Dev. Brain Res. *58*, 159–170.
Brooks, R. (2000). Negative genetic correlation between male sexual attractiveness and survival. Nature *406*, 67–70.
Burns, S.N., Vandermeer, R.K. and Teal, P.E.A. (2005). The effect of age and social environment on dealation in *Solenopsis invicta* (Hymenoptera: Formicidae) female alates. Fla. Entomol. *88*, 452–457.
Cabej, N. (1999a). Is "Central Dogma" incompatible with the inheritance of acquired characters? Sci. Bio. Evol., November 22, 1999. Available online: http://www.aol.com.
Cabej, N. (1999b). Dollo's law and the reversibility of biological design. Sci. Bio. Evol. December 1, 1999. Available online: http://www.aol.com.
Cabej, N. (1999c). Re: Dollo's law and the reversibility of biological design. Sci. Bio. Evol. December 30, 1999. Available online: http://www.aol.com.
Cabej, N. (2001). *Quantic Evolution*. Albanet, Dumont, NJ.
Cabej, N.R. (2004). *Neural Control of Development*. Albanet, Dumont, NJ.
Cabej, N.R. (2008). *Epigenetic Principles of Evolution*. Albanet, Dumont, NJ.
Cabej, N.R. (2011). Neural control of gene recruitment in metazoans. Dev. Dyn. *240*, 1–8.
Cakouros, D., Daish, T.J. and Kumar, S. (2004). Ecdysone receptor directly binds the promoter of the *Drosophila* caspase dronc, regulating its expression. J. Cell Biol. *165*, 631–640.
Callard, I.P. and Koob, T.J. (1993). Endocrine regulation of the elasmobranch reproductive tract. J. Exp. Zool. *266*, 368–377.
Capdevila, J. and Belmonte, J.C.I. (2001). Patterning mechanisms controlling vertebrate limb development. Annu. Rev. Cell Dev. Biol. *17*, 87–132.
Carl, M., Loosli, F. and Wittbrodt, J. (2002). Six3 inactivation reveals its essential role for the formation and patterning of the vertebrate eye. Development *129*, 4057–4063.
Carroll, R.S., Weaver, C.E. and Baum, M.J. (1988). Evidence implicating aromatization of testosterone in the regulation of male ferret sexual behavior. Physiol. Behav. *42*, 457–460.
Carroll, S. (2006). The genetic factor. Nova. Available: http://www.pbs.org/wgbh/nova/body/genetic-factor.html

Cernuda-Cernuda, R., DeGrip, W.J., Cooper, H.M., Nevo, E. and Garcia-Fernandez, J.M. (2002). The retina of *Spalax ehrenbergi*: novel histologic features supportive of a modified photosensory role. Invest. Ophthalmol. Vis. Sci. *43*, 2374–2383.

Chai, Y., Jiang, X., Ito, Y., Bringas, P., Jr., Han, J., Rowitch, D.H., et al. (2000). Fate of the mammalian cranial neural crest during tooth and mandibular morphogenesis. Development *127*, 1671–1679.

Chang, C.-H., Jiang, T.X., Lin, C.M., Burrus, L.W., Chuong, C.M. and Widelitz, R. (2004). Distinct Wnt members regulate the hierarchical morphogenesis of skin regions (spinal tract) and individual feathers. Mech. Dev. *121*, 157–171.

Chiego, D.J. (1995). The early distribution and possible role of nerves during odontogenesis. Int. J. Dev. Biol. *19*, 191–194.

Christ, B. and Brand-Saberi, B. (2002). Limb muscle development. Int. J. Dev. Biol. *46*, 905–914.

Clack, J.A., Ahlberg, P.E., Finney, S.M., Dominguez Alonso, P., Robinson, J. and Ketcham, R.A. (2003). A uniquely specialized ear in a very early tetrapod. Nature *425*, 65–69.

Clark, L., Zhang, J.R., Tobe, S.S. and Lange, A.B. (2005). The distribution and physiological roles of proctolin in the locust, *Locusta migratoria*. In invertebrate neuropeptide abstracts. J. Insect Sci. *5*, p. 22. Available online: insectscience.org/5.22.

Cohen, A.J., Lassová, L., Golden, E.B., Niu, Z. and Adams, S.L. (2006). Retinoids directly activate the collagen X promoter in prehypertrophic chondrocytes through a distal retinoic acid response element. J. Cell. Biochem. *99*, 269–278.

Colbert, M.C., Linney, E. and LaMantia, A. (1993). Local sources of retinoic acid coincide with retinoid-mediated transgene activity during embryonic development. Proc. Natl. Acad. Sci. U.S.A. *90*, 6572–6576.

Colombani, J., Bianchini, L., Layalle, S., Pondeville, E., Dauphin-Villemant, C., Antoniewski, C., et al. (2005). Antagonistic actions of ecdysone and insulins determine final size in *Drosophila*. Science *310*, 667–670.

Conlon, R.A. and Rossant, J. (1992). Exogenous retinoic acid rapidly induces anterior ectopic expression of Hox-2 genes *in vivo*. Development *116*, 357–368.

Consoli, F. and Bradleigh Vinson, S. (2004). Wing morph development and reproduction of the ectoparasite *Melittobia digitata*: nutritional and hormonal effects. Entomol. Exp. Appl *112*, 47–56.

Crickmore, M.A. and Mann, R.S. (2007). Hox control of organ size by regulation of morphogen production and mobility. Science *313*, 63–68.

Cripps, R.M. and Olson, E.N. (2002). Control of cardiac development by an evolutionarily conserved transcriptional network. Dev. Biol. *246*, 14–28.

Daeschler, E.B. and Shubin, N. (1995). Tetrapod origins. Paleobiology *21*, 404–409.

Daeschler, E.B., Shubin, N.H. and Jenkins, F.A., Jr. (2006). A Devonian tetrapod-like fish and the evolution of the tetrapod body plan. Nature *440*, 757–763.

D'Amico, L.J., Davidowitz, G. and Nijhout, H.F. (2001). The developmental and physiological basis of body size evolution in an insect. Proc. Biol. Sci. *268*, 1589–1593.

Daniels, C.B. and Orgeig, S. (2003). Pulmonary surfactant: the key to the evolution of air breathing. News Physiol. Sci. *18*, 151–157.

Daniels, C.B., Orgeig, S., Sullivan, L.C., Ling, N., Bennett, M.B., Schürch, S., et al. (2004). The origin and evolution of the surfactant system in fish: insights into the evolution of lungs and swim bladders. Physiol. Biochem. Zool. *77*, 732–749.

Darwin, C.R. (1859). *The Origin of Species by Means of Natural Selection or The Preservation of Favored Races in the Struggle for Life*, first edition. John Murray, London (first issue), p. 3.

Davidowitz, G., D'Amico, L.J. and Nijhout, H.F. (2003). Critical weight in the development of insect body size. Evol. Dev. *5*, 188–197.

Davidowitz, G., D'Amico, L.J. and Nijhout, H.F. (2004). The effects of environmental variation on a mechanism that controls insect body size. Evol. Ecol. Res. *6*, 49–62.

Davidson, E.H., McClay, D.R. and Hood, L. (2003). Regulatory gene networks and the properties of the developmental process. Proc. Natl. Acad. Sci. U.S.A. *100*, 1475–1480.

D'Avino, P.P. and Thummel, C.S. (2000). The ecdysone regulatory pathway controls wing morphogenesis and integrin expression during *Drosophila* metamorphosis. Dev. Biol. *220*, 211–224.

Delcomyn, F. (1991). Section F. In: *Neural and Integrative Animal Physiology*, C.L. Prosser (ed.), fourth edition. Wiley-Liss, pp. 617–639.

del Pino, E.M. (1989). Marsupial frogs. Sci. Am. *260*, 110–118.

Denver, R.J. (1997). Environmental stress as a developmental cue: corticotropin-releasing hormone is a proximate mediator of adaptive phenotypic plasticity in amphibian metamorphosis. Horm. Behav *31*, 169–179.

Dickson, W.B., Straw, A.D., Poelma, C. and Dickinson, M.H. (2006). An integrative model of insect flight control. *Proceedings of the 44th AIAA Aerospace Sciences Meeting and Exhibit*, pp. 1–19.

Dickson, W.B., Straw, A.D. and Dickinson, M.H. (2008). Integrative model of *Drosophila* flight. AIAA J. *46*, 2150–2164.

Dietz, F.R. (1987). Effect of peripheral nerve on limb development. J. Orthop. Res. *5*, 576–585.

Dietz, F.R. (1989). Effect of denervation on limb growth. J. Orthop. Res. *7*, 292–303.

Draizen, T.A., Ewer, J. and Robinow, S. (1999). Genetic and hormonal regulation of the death of peptidergic neurons in the *Drosophila* central nervous system. J. Neurobiol. *38*, 455–465.

Drossopoulou, G., Lewis, K.E., Sanz-Ezquerro, J.J., Nikbakht, N., McMahon, A.P., Hofmann, C., et al. (2000). A model for anteroposterior patterning of the vertebrate limb based on sequential long- and short-range Shh signaling and Bmp signaling. Development *127*, 1337–1348.

Dudley, A.T., Ros, M.A. and Tabin, C.J. (2002). A re-examination of proximodistal patterning during vertebrate limb development. Nature *418*, 539–544.

Dulvy, N.K. and Reynolds, J.D. (1997). Evolutionary transitions among egg-laying, live-bearing and maternal inputs in sharks and rays. Proc. Biol. Sci. *264*, 1309–1315.

Duprez, D. (2002). Signals regulating muscle formation in the limb during embryonic development. Int. J. Dev. Biol. *46*, 915–925.

Eames, B.F. and Schneider, R.A. (2005). Quail–duck chimeras reveal spatiotemporal plasticity in molecular and histogenic programs of cranial feather development. Development *132*, 1499–1509.

Elliott, D.A., Solloway, M.J., Wise, N., Biben, C., Costa, M.W., Furtado, M.B., et al. (2006). A tyrosine-rich domain within homeodomain transcription factor Nkx2-5 is an essential element in the early cardiac transcriptional regulatory machinery. Development *133*, 1311–1322.

Emlen, D.J. (2000). Integrating development with evolution: a case study with beetle horns. BioScience *50*, 403–413.

Emlen, D.J. and Nijhout, H.F. (1999). Hormonal control of male horn length dimorphism in the dung beetle *Onthophagus taurus* (Coleoptera: Scarabaeidae). J. Insect Physiol. *45*, 45–53.

Emlen, D.J., Hunt, J. and Simmons, L.W. (2005). Evolution of sexual dimorphism in the expression of beetle horns: phylogenetic evidence for modularity, evolutionary lability, and constraint. Am. Nat. *166*(Suppl.), S42–S66.

Falkenberg, G., Fleissner, G., Schuchardt, K., Kuehbacher, M., Thalau, P., Mouritsen, H., et al. (2010). Avian magnetoreception: elaborate iron mineral containing dendrites in the upper beak seem to be a common feature of birds. PLoS One *5*(2), p. e9231.

Fayyazuddin, A. and Dickinson, M.H. (1996). Haltere afferents provide direct, electronic input to a steering motor neuron in the blowfly, *Calliphora*. J. Neurosci. *16*, 5225–5232.

Fenton, M.B. and Fullard, J.H. (1979). The influence of moth hearing on bat echolocation strategies. J. Comp. Physiol. A *132*, 77–86.

Fernald, R.D. (2004). Eyes: variety, development and evolution. Brain Behav. Evol. *64*, 141–147.

Field, L.H. and Matheson, T. (1998). Chordotonal organs of insects. Adv. Insect Physiol. *27*, 1–230.

Fleissner, G., Holtkamp-Rotzler, E., Hanzlik, M., Winklhofer, M., Petersen, N. and Wiltschko, W. (2003). Ultrastructural analysis of a putative magnetoreceptor in the beak of homing pigeons. J. Comp. Neurol. *458*, 350–360.

Flemming, A.F. and Blackburn, D.G. (2003). Evolution of placental specializations in viviparous African and South American lizards. J. Exp. Zool. A Comp. Exp. Biol *299*, 33–47.

Fletcher, D. and Blum, M. (1981). Pheromonal control of dealation and oogenesis in virgin queen ants. Science *212*, 73–75.

Fliniaux, I., Viallet, J.P. and Dhouailly, D. (2004). Signaling dynamics of feather tract formation from the chick somatopleure. Development *131*, 3955–3966.

Folk, D.G. and Bradley, T.J. (2005). Adaptive evolution in the lab: unique phenotypes in fruit flies comprise a fertile field of study. Integr. Comp. Biol. *45*, 492–499.

Fong, D.W., Kane, T.C. and Culver, D.C. (1995). Vestigialization and loss of nonfunctional characters. Annu. Rev. Ecol. Syst. *26*, 249–268.

Fox, R. (2006). Invertebrate Anatomy. Available online: http://webs.lander.edu/rsfox/invertebrates/procambarus.html.

Francis-West, P.H., Antoni, L. and Anakwe, K. (2003). Regulation of myogenic differentiation in the developing limb bud. J. Anat. *202*, 69–81.

Fraser, G.J., Cerny, R., Soukup, V., Bronner-Fraser, M. and Streelman, J.T. (2010). The odontode explosion: the origin of tooth-like structures in vertebrates. Bioessays *32*, 808–817.

Freem, L.J., Escot, S., Tannahill, D., Druckenbrod, N.R., Thapar, N. and Burns, A.J. (2010). The intrinsic innervation of the lung is derived from neural crest cells as shown by optical projection tomography in *Wnt1-Cre; YFP* reporter mice. J. Anat. *217*, 651–664.

Fridolfsson, A.-K., Cheng, H., Copeland, N.G., Jenkins, N.A., Liu, H.C., Raudsepp, T., et al. (1998). Evolution of the avian sex chromosomes from an ancestral pair of autosomes. Proc. Natl. Acad. Sci. U.S.A. *95*, 8147–8152.

Fried, K., Nosrat, C., Lillesaar, C. and Hildebrand, C. (2000). Molecular signaling and pupal nerve development. Crit. Rev. Oral Biol. M. *11*, 318–332. (abstract).

Friedel, T. and Loughton, B.G. (1980). Neurosecretory protein synthesis in relation to starvation and feeding in adult male *Locusta migratoria*. J. Insect Physiol. *26*, 33–37.

Fritzsch, B. and Beisel, K.W. (2004). Keeping sensory cells and evolving neurons to connect them to the brain: molecular conservation and novelties in vertebrate ear development. Brain Behav. Evol. *64*, 182–197.

Frye, M.A. and Dickinson, M.H. (2001). Fly flight a model for the neural control of complex behavior. Neuron *32*, 385–388.

Fullard, J.H., Ratcliffe, J.M. and Guignion, C. (2005). Sensory ecology of predator-prey interactions: responses of the AN2 interneuron in the field cricket, *Teleogryllus oceanicus* to the echolocation calls of sympatric bats. J. Comp. Physiol. A *191*, 605–618.

Gahr, M. and Hutchison, J.B. (1992). Behavioral action of estrogen in the male dove brain: area differences in codistribution of aromatase activity and estrogen receptors are steroid-dependent. Neuroendocrinology *56*, 74–84.

Galant, R., Walsh, C.M. and Carroll, S.B. (2002). Hox repression of a target gene: extradenticle-independent, additive action through multiple monomer binding sites. Development *129*, 3115–3126.

Gilbert, J.J. (2002). Endogenous regulation of environmentally induced sexuality in a rotifer: a multigenerational parental effect induced by fertilization. Freshwat. Biol. *47*, 1633–1641.

Gilbert, J.J. (2003). Environmental and endogenous control of sexuality in a rotifer life cycle: developmental and population biology. Evol. Dev. *5*, 19–24.

Gilbert, J.J. and Schröder, T. (2004). Rotifers from diapausing, fertilized eggs: unique features and emergence. Limnol. Oceanogr. *49*, 1341–1354.

Gillis, J.A., Dahn, R.D. and Shubin, N.H. (2009). Shared developmental mechanisms pattern the vertebrate gill arch and paired fin skeletons. Proc. Natl. Acad. Sci. U.S.A. *106*, 5720–5724.

Ginsburg, S. and Jablonka, E. (2010). The evolution of associative learning: a factor in the Cambrian explosion. J. Theor. Biol. *266*, 11–20.

Gorbman, A. and Davey, K. (1991). Endocrines. In: *Neural and Integrative Animal Physiology*, C.L. Prosser (ed.), fourth edition. Wiley-Liss, pp. 693–754.

Gotoh, A., Sameshima, S., Tsuji, K., Matsumoto, T. and Miura, T. (2005). Apoptotic wing degeneration and formation of an altruism-regulating glandular appendage (gemma) in the ponerine ant *Diacamma* sp. from Japan (Hymenoptera, Formicidae, Ponerinae). Dev. Genes Evol. *215*, 69–77.

Granger, N.A., Sturgis, S.L., Ebersohl, R., Geng, C. and Sparks, T.C. (1996). Dopaminergic control of corpora allata activity in the larval tobacco hornworm, *Manduca secta*. Arch. Insect Biochem. *32*, 449–466.

Guarino, F.M., Paulesu, L., Cardone, A., Bellini, L., Ghiara, G. and Angelini, F. (2002). Endocrine activity of the corpus luteum and placenta during pregnancy in *Chalcides chalcides* (Reptilia, Squamata). Gen. Comp. Endocrinol. *111*, 261–270.

Guilette, L.J., Jr. (1993). The evolution of viviparity in lizards. BioScience *43*, 742–751.

Guillette, L.J. (2005). Morphology of the reproductive tract in a lizard exhibiting incipient viviparity (*Sphenomorphus fragilis*) and its implications for the evolution of the reptilian placenta. J. Morphol. *212*, 163–173.

Haase, E. (1993). Zur Wirkung von Androgenen auf die Induktion des Schlichtkleids bei kastrierten Stockerpeln (*Anas platyrhynchos*). J. Ornithol. *134*, 191–195.

Haase, E., Ito, S. and Wakamatsu, K. (1995). Influences of sex, castration, and androgens on the eumelanin and pheomelanin contents of different feathers in wild mallards. Pigment Cell Res. *8*, 164–170.

Hagelin, J.C. (2001). Castration in gambel's and scaled quail: ornate plumage and dominance persist, but courtship and threat behaviors do not. Horm. Behav. *39*, 1–10.

Haines, L. and Currie, P.D. (2001). Morphogenesis and evolution of vertebrate appendicular muscle. J. Anat. *199*, 205–209.

Hall, M., Lindholm, A.K. and Brooks, R. (2004). Direct selection on male attractiveness and female preference fails to produce a response. BMC Evol. Biol. *4*, p. 1.

Hammel, H.T., Jackson, D.C., Stolwijk, J.A.J., Hardy, J.D. and Stroeme, S.B. (1963). Temperature regulation by hypothalamic proportional control with an adjustable set point. J. Appl. Physiol. *18*, 1146–1154.

Hastings, I.M. (1994). Manifestations of sexual selection may depend on the genetic basis of sex determination. Proc. Biol. Sci. *258*, 83–87.

Hegstrom, C.D., Riddiford, L.M. and Truman, J.W. (1998). Steroid and neuronal regulation of ecdysone receptor expression during metamorphosis of muscle in the moth, *Manduca sexta*. J. Neurosci. *18*, 1786–1794.

Heil, P., Bronchti, G., Wollberg, Z. and Scheich, H. (1991). Invasion of visual cortex by the auditory system in the naturally blind mole rat. Neuroreport *2*, 735–738.

Hennig, R.M. and Weber, T. (1997). Filtering of temporal parameters of the calling song by cricket females of two closely related species: a behavioral analysis. J. Comp. Physiol. A *180*, 621–630.

Heyers, D., Zapka, M., Hoffmeister, M., Wild, J.M. and Mouritsen, H. (2010). Magnetic field changes activate the trigeminal brainstem complex in a migratory bird. Proc. Natl. Acad. Sci. U.S.A. *107*, 9394–9399.

Highnam, K.C. and Mordue, A.J. (1974). Induced changes in neurosecretory activity of adult female *Schistocerca gregaria* in relation to feeding. Gen. Comp. Endocrinol. *22*, 519–525.

Hildebrand, C., Fried, K., Tuisku, F. and Johansson, C.S. (1995). Teeth and tooth nerves. Prog. Neurobiol. *45*, 165–222.

Hinchliffe, J.R. (2002). Developmental basis of limb evolution. Int. J. Dev. Biol. *46*, 835–845.

Hogan, K.A., Ambler, C.A., Chapman, D.L. and Bautch, V.L. (2004). The neural tube patterns vessels developmentally using the VEGF signaling pathway. Development *131*, 1503–1513.

Holliday, R. (1987). The inheritance of epigenetic defects. Science *238*, 163–170.

Holliday, R. and Pugh, J.E. (1975). DNA modification mechanisms and gene activity during development. Science *187*, 226–232.

Houde, A.E. (1992). Sex-linked heritability of a sexually selected character in a natural population of *Poecilia reticulata* (Pisces: Poeciliidae) (guppies). Heredity *69*, 229–235.

Houle, M., Sylvestre, J.R. and Lohnes, D. (2003). Retinoic acid regulates a subset of Cdx1 function *in vivo*. Development *130*, 6555–6567.

Hudspeth, A.J. (1997). How hearing happens. Neuron *19*, 947–950.

Ikeya, T., Galic, M., Belawat, P., Nairz, K. and Hafen, E. (2002). Nutrient-dependent expression of insulin-like peptides from neuroendocrine cells in the CNS contributes to growth regulation in *Drosophila*. Curr. Biol. *12*, 1293–1300.

Jablonka, E. (1994). Inheritance systems and the evolution of new levels of individuality. J. Theor. Biol. *170*, 301–309.

Jablonka, E. and Lamb, M.J. (1989). The inheritance of acquired epigenetic variations. J. Theor. Biol. *139*, 69–83.

Jaenisch, R. and Bird, A. (2003). Epigenetic regulation of gene expression: how the genome integrates intrinsic and environmental signals. Nat. Genet. *33*, 245–254.

Jernvall, J., Keraenen, S.V.E. and Thesleff, I. (2000). Evolutionary modification of development in mammalian teeth: quantifying gene expression patterns and topography. Proc. Natl. Acad. Sci. U.S.A. *97*, 14444–14448.

Jokusch, E.L. and Ober, K.A. (2004). Hypothesis testing in evolutionary developmental biology: a case study from insect wings. J. Hered. *95*, 382–396.

Kalb, J., Egelhaaf, M. and Kurtz, R. (2006). Robust integration of motion information in the fly visual system revealed by single cell photoablation. J. Neurosci. *26*, 7898–7906.

Kassai, Y., Munne, P., Hotta, Y., Penttilä, E., Kavanagh, K., Ohbayashi, N., et al. (2005). Regulation of mammalian tooth cusp patterning by ectodin. Science *309*, 2067–2070.

Kawakami, Y., Capdevila, J., Buscher, D., Itoh, T., Esteban, C.R. and Belmonte, J.C.I. (2001). WNT signals control FGF-dependent limb initiation and AER induction in the chick embryo. Cell *104*, 891–900.

Keefe, D.L. (2002). Sex hormones and neural mechanisms. Arch. Sex Behav. *31*, 401–403.

Kettunen, P., Løes, S., Furmanek, T., Fjeld, K., Kvinnsland, I.H., Behar, O., et al. (2005). Coordination of trigeminal axon navigation and patterning with tooth organ formation: epithelial-mesenchymal interactions, and epithelial Wnt4 and Tgfß1 regulate semaphorin 3a expression in the dental mesenchyme. Development *132*, 323–334.

Kimball, R.T. and Ligon, J.D. (1999). Evolution of avian plumage dichromatism from a proximate perspective. Am. Nat. *154*, 182–193.

Kirkpatrick, M. and Hall, D.W. (2004). Sexual selection and sex linkage. Evolution *58*, 683–691.

Kirwin, S.M., Bhandari, V., Dimatteo, D., Barone, C., Johnson, L., Paul, S., et al. (2006). Leptin enhances lung maturity in the fetal rat. Pediatr. Res. *60*, 200–204.

Knosp, W.M., Scott, V., Bächinger, H.P. and Stadler, H.S. (2004). HOXA13 regulates the expression of bone morphogenetic proteins 2 and 7 to control distal limb morphogenesis. Development *131*, 4581–4592.

Koch, P.B., Merk, R., Brakefield, P.M., and French-Constant, R.H. (2002), Ecdysone receptors link pattern and process in butterfly eyespot development. *XV International Ecdysone Workshop*. 30 June to 6 July 2002. Kolymbari Crete, Greece (abstract).

Koch, P.B., Merk, R., Reinhardt, R. and Weber, P. (2003). Localization of ecdysone receptor protein during colour pattern formation in wings of the butterfly *Precis coenia* (Lepidoptera: Nymphalidae) and co-expression with distal-less protein. Dev. Genes Evol. *212*, 571–584.

Kondo, B., Baker, J.M. and Omland, K.E. (2004). Recent speciation between the Baltimore oriole and the black-backed oriole. The Condor *106*, 674–680.

Kremen, C. and Nijhout, H.F. (1989). Juvenile hormone controls the onset of pupal commitment in the imaginal disks and epidermis of *Precis coenia* (Lepidoptera: Nymphalidae). J. Insect Physiol. *35*, 603–612.

Kristjansson, B. (2005). Rapid morphological changes in threespine stickleback, *Gasterosteus aculeatus*, in freshwater. Environ. Biol. Fishes *74*, 357–363.

Kronnie, G.T. and Reggiani, C. (2002). Skeletal muscle fibre type specification during embryonic development. J. Muscle Res. Cell Motil. *23*, 65–69.

Kunert, G. and Weisser, W.W. (2005). The importance of antennae for pea aphid wing induction in the presence of natural enemies. B. Entomol. Res. *95*, 125–131.

Kunert, G., Otto, S., Roese, U.S.R., Gershenzon, J. and Weisser, W.W. (2005). Alarm pheromone mediates production of winged dispersal morphs in aphids. Ecol. Lett. *8*, 596–603.

Laurikkala, J., Kassai, Y., Pakkasjärvi, L., Thesleff, I. and Itoh, N. (2003). Identification of a secreted BMP antagonist, ectodin, integrating BMP, FGF, and SHH signals from the tooth enamel knot. Dev. Biol. *264*, 91–105.

Laurin, M., Girondot, M. and de Ricqles, A. (2000). Early tetrapod evolution. Trends Ecol. Evol. *15*, 118–123.

Leevers, S.J. (2001). Growth control: invertebrate insulin surprises! Curr. Biol. *11*, R209–R212.

Lewis, F.P. and Fullard, J.H. (1996). Neurometamorphosis of the ear in the gypsy moth, *Lymnatria dispar*, and its homologue in the earless forest tent caterpillar moth, *Malacosoma disstria*. J. Neurobiol. *31*, 245–262.

Li, T.-R. and White, K.P. (2003). Tissue-specific gene expression and ecdysone-regulated genomic networks in *Drosophila*. Dev. Cell *5*, 59–72.

Liu, L. and Laufer, H. (1996). Isolation and characterization of sinus gland neuropeptides with both mandibular organ inhibiting and hyperglycemic effects from the spider crab *Libinia emarginata*. Arch. Insect Biochem. *32*, 375–385.

Lillie, F.R. (1942). On the development of feathers. Biol. Rev. *17*, 247–266.
Lloyd, G.E.R. (1999). *Aristotle: The Growth and Structure of His Thought.* Cambridge University Press, p. 72.
Løes, S., Kettunen, P., Kvinnsland, H. and Luukko, K. (2002). Mouse rudimentary diastema tooth primordia are devoid of peripheral nerve fibers. Anat. Embryol. *205*, 187–191.
Long, J.A. and Gordon, M.S. (2004). The greatest step in vertebrate history: a paleobiological review of the fish-tetrapod transition. Physiol. Biochem. Zool. *77*, 700–719.
Lorenzon, S., Edomi, P., Giulianini, P.G., Mettulio, R. and Ferrero, E.A. (2005). Role of biogenic amines and cHH in the crustacean hyperglycemic stress response. J. Exp. Biol. *208*, 3341–3347.
Lu, H.-C., Revelli, J.-P., Goering, L., Thaller, C. and Eichele, G. (1997). Retinoid signalling is required for the establishment of a ZPA and for the expression of Hoxb-8, a mediator of ZPA formation. Development *124*, 1643–1651.
Lumsden, A.G. (1988). Spatial organization of the epithelium and the role of neural crest cells in the initiation of the mammalian tooth germ. Development *103*, 155–169.
Luukko, K. (1997). Immunohistochemical localization of nerve fibres during development of embryonic rat molar using peripherin and protein gene product 9.5 antibodies. Arch. Oral Biol. *42*, 189–195.
Luukko, K. (1998). Neuronal cells and neurotrophins in odontogenesis. Eur. J. Oral Sci. *106*(Suppl. 1), 80–93.
Luukko, K., Sariola, H., Saarma, M. and Thesleff, I. (1997). Localization of nerve cells in the developing rat tooth. J. Dent. Res. *76*, 1350–1356.
Lyons, K. and Ezaki, M. (2009). Molecular regulation of limb growth. J. Bone Joint Surg. Am. *91*(Suppl. 4), 47–52.
Ma, L.-H., Gilland, E., Bass, A.H. and Baker, R. (2010). Ancestry of motor innervation to pectoral fin and forelimb. Nat. Commun. *1*, 1–8.
Maden, M., Sonneveld, E., van der Saag, P.T. and Gale, E. (1998). The distribution of endogenous retinoic acid in the chick embryo: implications for developmental mechanisms. Development *125*, 4133–4144.
Makhijani, K., Kalyani, C., Srividya, T. and Shashidhara, L.S. (2007). Modulation of Decapentaplegic gradient during haltere specification in *Drosophila*. Dev. Biol. *302*, 243–255.
Maklad, A. and Fritzsch, B. (2003). Development of vestibular afferent projections into the hindbrain and their central targets. Brain Res. Bull. *60*, 497–510.
Mank, J.E., Hall, D.W., Kirkpatrick, M. and Avise, J.C. (2006). Sex chromosomes and male ornaments: a comparative evaluation in ray-finned fishes. Proc. Biol. Sci. *273*, 233–236.
Marlétaz, F., Holland, L.Z., Laudet, V. and Schubert, M. (2006). Retinoic acid signaling and the evolution of chordates. Int. J. Biol. Sci. 2, 38–47.
Marquardt, T., Ashery-Padan, R., Andrejewski, N., Scardigli, R., Guillemot, F. and Gruss, P. (2001). Pax6 is required or the multipotent state of retinal progenitor cells. Cell *105*, 43–55.
Masumura, M., Satake, S., Saegusa, H. and Mizoguchi, A. (2000). Glucose stimulates the release of bombyxin, an insulin-related peptide of the silkworm *Bombyx mori*. Gen. Comp. Endocrinol. *118*, 393–399.
Mathies, T. and Andrews, R.M. (1995). Thermal and reproductive biology of high and low elevation populations of the lizard *Sceloporus scalaris*: implications for the evolution of viviparity. Oecologia *104*, 101–111.
Mathies, T. and Andrews, R.M. (2000). Does reduction of the eggshell occur concurrently with or subsequent to the evolution of viviparity in phrynosomatid lizards? Biol. J. Linn. Soc *71*, 719–736.

Matsuda, R. (1987). *Animal Evolution in Changing Environments, with Special Reference to Abnormal Metamorphosis*. Wiley-Interscience, New York, NY.

Mayr, E. (1942). *Systematics and the Origin of Species*. Columbia University Press, New York, NY, p. 120.

Mayr, E. (1961). Cause and effect in biology. Science *134*, 1501–1506.

McBrayer, Z., Ono, H., Shimell, M.-J., Parvy, J.-P., Beckstead, R.B., Warren, J.T., et al. (2007). Prothoracicotropic hormone regulates developmental timing and body size in *Drosophila*. Dev. Cell *13*, 857–871.

McCaffery, P. and Dräger, U.C. (1994). Hot spots of retinoic acid synthesis in the developing spinal cord. Proc. Natl. Acad. Sci. U.S.A. *91*, 7194–7197.

McKinnell, I.W., Turmaine, M. and Patel, K. (2004). Sonic Hedgehog functions by localizing the region of proliferation in early developing feather buds. Dev. Biol. *272*, 76–88.

McNaught, M.K. and Owens, I.P.F. (2002). Interspecific variation in plumage colour among birds: species recognition or light environment. J. Evol. Biol. *15*, 505–514.

Mercader, R., Leonardo, E., Piedra, M.E., Martinez, A.C., Ros, M.A. and Torres, M. (2000). Opposing RA and FGF signals control proximodistal vertebrate limb development through regulation of *Meis* genes. Development *127*, 3961–3970.

Mic, F.A. and Duester, G. (2003). Patterning of forelimb bud myogenic precursor cells requires retinoic acid signaling initiated by RALDH-2. Dev. Biol. *264*, 191–201.

Mic, F.A., Sirbu, I.O. and Duester, D. (2004). Retinoic acid synthesis controlled by Raldh-2 is required early for limb bud initiation and then later as a proximodistal signal during apical ectodermal ridge formation. J. Biol. Chem. *279*, 26698–26706.

Milan, M. and Cohen, S.M. (2003). A re-evaluation of the contributions of apterous and notch to the dorsoventral lineage restriction boundary in the *Drosophila* wing. Development *130*, 553–562.

Miner, A.L., Rosenberg, A.J. and Nijhout, H.F. (2000). Control of growth and differentiation of the wing imaginal disk of *Precis coenia* (Lepidoptera: Nymphalidae). J. Insect Physiol. *46*, 251–258.

Mitsiadis, T.A. and Graf, D. (2009). Cell fate determination during tooth development and regeneration. Birth Defects Res. C *87*, 199–211.

Mitsiadis, T.A., Angeli, I., James, C., Lendahl, U. and Sharpe, P.T. (2003). Role of Islet1 in the patterning of murine dentition. Development *130*, 4451–4460.

Miura, T. (2005). Developmental regulation of caste-specific characters in social-insect polyphenism. Evol. Dev. *7*, 122–129.

Moczek, A.P. and Nagy, L.M. (2005). Diverse developmental mechanisms contribute to different levels of diversity in horned beetles. Evol. Dev. *7*, 175–185.

Moczek, A.P. and Nijhout, H.F. (2003). Rapid evolution of a polyphonic threshold. Evol. Dev. *5*, 259–268.

Monier, B., Astier, M., Sémériva, M. and Perrin, L. (2005). Steroid-dependent modification of Hox function drives myocyte reprogramming in the *Drosophila* heart. Development *132*, 5283–5293.

Moodie, G.E.E. and Reimchen, T.E. (1976). Phenetic variation and habitat differences in *Gasterosteus* populations of the Queen Charlotte Islands. System. Biol. *25*, 49–61.

Mora, C.V., Davison, M., Wild, J.M. and Walker, M.M. (2004). Magnetoreception and its trigeminal mediation in the homing pigeon. Nature *432*, 508–511.

Mouritsen, H. and Ritz, T. (2005). Magnetoreception and its use in bird navigation. Curr. Opin. Neurobiol. *15*, 406–414.

Mouritsen, H., Janssen-Bienhold, U., Liedvogel, M., Feenders, G., Stalleicken, J., Dirks, P., et al. (2004). Cryptochromes and neuronal-activity markers colocalize in the retina of migratory birds during magnetic orientation. Proc. Natl. Acad. Sci. U.S.A. *101*, 14294–14299.

Mouritsen, H., Feenders, G., Liedvogel, M., Wada, K. and Jarvis, E.D. (2005). Night-vision brain are in the migratory songbirds. Proc. Natl. Acad. Sci. U.S.A. *102*, 8339–8344.

Muheim, R., Baeckman, J. and Akesson, S. (2002). Magnetic compass orientation in European robins is dependent on both wavelength and intensity of light. J. Exp. Biol. *205*, 3845–3856.

Mukoyama, Y., Gerber, H.-P., Ferrara, N., Gu, C. and Anderson, D.J. (2005). Peripheral nerve-derived VEGF promotes arterial differentiation via neuropilin 1-mediated positive feedback. Development *132*, 941–952.

Munyiri, F.N., Shintani, Y. and Ishikawa, Y. (2004). Evidence for the presence of a threshold weight for entering diapause in the yellow-spotted longicorn beetle, *Psacothea hilaris*. J. Insect Physiol. *50*, 295–301.

Nagaraju, G.P.C., Prasad, G.L.V. and Reddy, P.S. (2005). Isolation and characterization of mandibular organ-inhibiting hormone from the eyestalks of freshwater crab, *Oziotelphusa senex senex*. Int. J. Appl. Sci. Eng *3*, 61–68.

Namba, R., Pazdera, T.M., Cerrone, R.L. and Minden, J.S. (1997). *Drosophila* embryonic pattern repair: how embryos respond to bicoid dosage alteration. Development *124*, 1393–1403.

Necker, R., Rehkamper, G. and Nevo, E. (1992). Electrophysiological mapping of body representation in the cortex of the blind mole rat. Neuroreport *3*, 505–508.

Neubüser, A., Peters, H. and Balling, R. (1997). Antagonistic interactions between FGF and BMP signaling pathway: a mechanism for positioning the sites of tooth formation. Cell *99*, 247–254.

Newman, S.A. and Müller, G.B. (2000). Epigenetic mechanisms of character origination. J. Exp. Zool. *288*, 304–317.

Niemiller, M.L., Fitzpatrick, B.M. and Miller, B.T. (2008). Recent divergence with gene flow in Tennessee cave salamanders (Plethodontidae: *Gyrinophilus*) inferred from gene genealogies. Mol. Ecol. *17*, 2258–2275.

Nijhout, H.F. (1996). Focus on butterfly eyespot development. Nature *384*, 209–210.

Nijhout, H.F. (1999). Polyphenic development in insects. Bioscience *49*, p. 181192.

Nijhout, H.F. (2003). Development and evolution of adaptive polyphenisms. Evol. Dev. *5*, 9–18.

Nijhout, H.F. and Grunert, L.W. (2002). Bombyxin is a growth factor for wing imaginal disks in *Lepidoptera*. Proc. Natl. Acad. Sci. U.S.A. *99*, 15446–15450.

Nilsson, D.-E. (1996). Eye ancestry: old genes for new eyes. Curr. Biol. *6*, 39–42.

Nilsson, D.-E. (2004). Eye evolution: a question of genetic promiscuity. Curr. Opin. Neurobiol. *14*, 407–414.

Nishihara, A., Fujii, M., Sampath, T.K., Miyazono, K. and Reddi, A.H. (2003). Bone morphogenetic protein signaling in articular chondrocyte differentiation. Biochem. Biophys. Res. Commun. *301*, 617–622.

Niswander, L. (2002). Interplay between the molecular signals that control vertebrate limb development. Int. J. Dev. Biol. *46*, 877–881.

Niwa, N., Akimoto-Kato, A., Niimi, T., Tojo, K., Machida, R. and Hayashi, S. (2010). Evolutionary origin of the insect wing via integration of two developmental modules. Evol. Dev. *12*, 168–176.

Oda, S., Tatarazako, N., Watanabe, H., Morita, M. and Iguchi, T. (2005). Production of male neonates in *Daphnia magna* (Cladocera, Crustacea) exposed to juvenile hormones and their analogs. Chemosphere *61*, 1168–1174.

Olivera-Martinez, I., Thélu, J., Teillet, M.-A. and Dhouailly, D. (2001). Dorsal dermis development depends on a signal from the dorsal neural tube, which can be substituted by Wnt-1. Mech. Dev. *100*, 233–244.

Olmstead, A.W. and LeBlanc, G.A. (2002). Juvenoid hormone methyl farnesoate is a sex determinant in the crustacean *Daphnia magna*. J. Exp. Zool. *293*, 736–739.

Omland, K.E. (1997). Examining two standard assumptions of ancestral reconstructions: repeated loss of dichromatism in dabbling ducks (Anatini). Evolution *51*, 1636–1646.

Oostra, V., de Jong, M.A., Invergo, B.M., Kesbeke, F., Wende, F., Brakefield, P.M., et al. (2011). Translating environmental gradients into discontinuous reaction norms via hormone signalling in a polyphenic butterfly. Proc. Biol. Sci. *278*, 789–797.

Owens, I.P.F. and Short, R.V. (1995). Hormonal basis of sexual dimorphism in birds: implications for new theories of sexual selection. Tends Ecol. Evol. *10*, 44–47.

Painter, D., Jennings, D.H. and Moore, M.C. (2002). Placental buffering of maternal steroid hormone effects on fetal and yolk hormone levels: a comparative study of a viviparous lizard, *Sceloporus jarrovi*, and an oviparous lizard, *Sceloporus graciosus*. Gen. Comp. Endocrinol. *127*, 105–116.

Parker, J. and Johnston, L.A. (2006). The proximate determinants of insect size. J. Biol. *5*, 15.

Paulesu, L. (1997). Cytokines in mammalian reproduction and speculation about their possible involvement in nonmammalian viviparity. Microsc. Res. Tech. *38*, 188–194.

Paulesu, L., Bigliardi, E., Paccagnini, E., Ietta, F., Cateni, C., Guillaume, C.P., et al. (2005). Cytokines in the oviparity/viviparity transition: evidence of the interleukin-1 system in a species with reproductive bimodality, the lizard *Lacerta vivipara*. Evol. Dev. *7*, 282–288.

Peters, H. and Balling, R. (1999). Teeth: where and how to make them. Trends Genet. *15*, 59–65.

Peterson, R.S., Yarram, L., Schlinger, B.A. and Saldanha, C.J. (2005). Aromatase is pre-synaptic and sexually dimorphic in the adult zebra finch brain. Proc. Biol. Sci. *272*, 2089–2096.

Phillips, B.L. and Shine, R. (2004). Adapting to an invasive species: toxic cane toads induce morphological change in Australian snakes. Proc. Natl. Acad. Sci. U.S.A. *101*, 17150–17155.

Phillips, B.L. and Shine, R. (2005). The morphology, and hence impact, of an invasive species (the cane toad, *Bufo marinus*): changes with time since colonization. Anim. Conserv. *8*, 407–413.

Phillips, B.L. and Shine, R. (2006). An invasive species induces rapid adaptive change in a native predator: cane toads and black snakes in Australia. Proc. Biol. Sci. *273*, 1545–1550.

Pigliucci, M. (2001). *Phenotypic Plasticity: Beyond Nature and Nurture*. John Hopkins University Press, Baltimore, MD.

Plikus, M., Zeichner-David, M., Mayer, J.A., Reyna, J., Bringas, P., Thewissen, J.G., et al. (2005). Morphoregulation of teeth: modulating the number, size, shape and differentiation by tuning Bmp activity. Evol. Dev. *7*, 440–457.

Prum, R.O. and Brush, A.H. (2002). The evolutionary origin and diversity of feathers. Q. Rev. Biol. *77*, 261–295.

Qualls, C.P. (1996). Influence of the evolution of viviparity on eggshell morphology in the lizard *Lerista bougainvillii*. J. Morphol. *228*, 119–125.

Raff, R.A. and Kaufman, T.C. (1983). *Embryos, Genes, and Evolution*. Macmillan, New York, NY, pp. 67–78.

Reeve, H.K. and Pfennig, D.W. (2003). Genetic biases for showy males: are some genetic systems especially conducive to sexual selection?. Proc. Natl. Acad. Sci. U.S.A. *100*, 1089–1094.

Reif, W.-E. (1982). Evolution of dermal skeleton and dentition in vertebrates: the odontode-regulation theory. Evol. Biol. *15*, 287–368.
Reznick, D., Meredith, R. and Collette, B.B. (2007). Independent evolution of complex life history adaptations in two families of fishes, live-bearing halfbeaks (Zenarchopteridae, Beloniformes) and Poeciliidae (Cyprinodontiformes). Evolution *61*, 2570–2583.
Riccomagno, M.M., Martinu, L., Mulheisen, M., Wu, D.K. and Epstein, D.J. (2002). Specification of the mammalian cochlea is dependent on Sonic hedgehog. Genes Dev. *16*, 2365–2378.
Riccomagno, M.M., Takada, S. and Epstein, D.J. (2005). Wnt-dependent regulation of inner ear morphogenesis is balanced by the opposing and supporting roles of Shh. Genes Dev. *19*, 1612–1623.
Richardson, M.K., Jeffery, J.E. and Tabin, C.J. (2004). Proximodistal patterning of the limb: insights from evolutionary morphology. Evol. Dev. *6*, 1–5.
Riddle, R.D., Johnson, R.L., Laufer, E. and Tabin, C. (1993). Sonic hedgehog mediates the polarizing activity of the ZPA. Cell *75*, 1401–1416.
Rider, C.V., Gorr, T.A., Olmstead, A.W., Wasilak, B.A. and LeBlanc, G.A. (2005). Stress signaling: coregulation of hemoglobin and male sex determination through a terpenoid signaling pathway in a crustacean. J. Exp. Biol. *208*, 15–23.
Riggs, A.D. (1975). X inactivation, differentiation and DNA methylation. Cytogenet. Cell Genet. *14*, 9–25.
Ritz, T., Adem, S. and Schulten, K. (2000). A model for photoreceptor-based magnetoreception in birds. Biophys. J. *78*, 707–718.
Ritz, T., Thalau, P., Phillips, J.B., Wiltschko, R. and Wiltschko, W. (2004). Resonance effects indicate a radical-pair mechanism for avian magnetic compass. Nature *429*, 177–180.
Robertson, R.M., Pearson, K.G. and Reichert, H. (1982). Flight interneurons in the locust and the origin of the insect wings. Science *217*, 177–179.
Robinson, G.E. and Vargo, E.L. (1997). Juvenile hormone in adult eusocial Hymenoptera: gonadotropin and behavioral pacemaker. Arch. Insect Biochem. *35*, 559–583.
Roper, A. and Zann, R. (2006). The onset of song learning and song tutor selection in fledgling zebra finches. Ethology *112*, 458–470.
Ross, S.A., McCaffery, P.J., Drager, U.C. and De Luca, L.M. (2000). Retinoids in embryonal development. Physiol. Rev. *80*, 1021–1054.
Rulifson, E.J., Kim, S.K. and Nusse, R. (2002). Ablation of insulin-producing neurons in flies: growth and diabetic phenotypes. Science *296*, 1118–1120.
Saitoh, Y., Saitoh, H., Ohtomo, H and Mizuno, S. (1991). Occupancy of the majority of DNA in the chicken W chromosome by bent-repetitive sequences. Chromosoma *101*, 32–40.
Salazar-Ciudad, I. and Jernvall, J. (2002). A gene network model accounting for development and evolution of mammalian teeth. Proc. Natl. Acad. Sci. U.S.A. *99*, 8116–8120.
Sameshima, S.-Y., Miura, T. and Matsumoto, T. (2004). Wing disc development during caste differentiation in the ant *Pheidole megacephala* (Hymenoptera: Formicidae). Evol. Dev. *6*, 336–341.
Sanyal, S., Jansen, H.G., de Grip, W.J., Nevo, E. and de Jong, W.W. (1990). The eye of the blind mole rat, *Spalax ehrenbergi*. Rudiment with hidden function? Invest. Ophthalmol. Vis. Sci *31*, 1398–1404.
Sarkar, C., Chakroborty, D., Mitra, R.B., Banerjee, S., Dasgupta, P.S. and Basu, S. (2004). Dopamine *in vivo* inhibits VEGF-induced phosphorylation of VEGFR-2, MAPK, and focal adhesion kinase in endothelial cells. Am. J. Physiol. Heart Circ. Physiol. *287*, H1554–H1560.

Sasaki, R., Nakasuji, F. and Fujisaki, K. (2002). Environmental factors determining wing form in lygaeid bug, *Dimorphopterus japonicus* (Heteroptera: Lygaeidae). Appl. Entomol. Zool. *37*, 329–333.

Sasaki, R., Nakasuji, F. and Fujisaki, K. (2003). Seasonal changes in wing dimorphism of the lygaeid bug *Dimorphopterus japonicus* (Heteroptera: Lygaeidae) in relation to environmental factors. Entomol. Sci. *6*, 63–70.

Schlichting, C.D. and Pigliucci, M. (1998). *Phenotypic Evolution: A Reaction Norm Perspective*. Sinauer Associates, Inc. Publishers, Sunderland, MA.

Schlinger, B.A. and Arnold, A.P. (1993). Estrogen synthesis *in vivo* in the adult zebra finch: additional evidence that circulating estrogens can originate in brain. Endocrinology *133*, 2610–2616.

Schlinger, B.A., Fivizzani, A.J. and Callard, G.V. (1989). Aromatase, 5 alpha- and 5 beta-reductase in brain, pituitary and skin of the sex-role reversed Wilson's phalarope. J. Endocrinol. *122*, 573–581.

Schneider, R.A. and Helms, J.A. (2003). The cellular and molecular origins of beak morphology. Science *299*, 565–568.

Sears, K.E., Behringer, R.R., Rasweiler, J.J., IV and Niswander, L.A. (2006). Development of bat flight: morphologic and molecular evolution of bat wing digits. Proc. Natl. Acad. Sci. U.S.A. *103*, 6581–6586.

Sgrò, C.M. and Partridge, L. (2000). Evolutionary responses of the life history of wild-caught *Drosophila melanogaster* to two standard methods of laboratory culture. Am. Nat. *156*, 341–353.

Shaw, S.R. (1994). Detection of airborne sound by a cockroach 'vibration detector': a possible missing link in insect auditory evolution. J. Exp. Biol. *193*, 13–47.

Sherman, A. and Dickinson, M.H. (2004). Summation of visual and mechanosensory feedback in *Drosophila* flight control. J. Exp. Biol. *207*, 133–142.

Shine, R. (1983). Reptilian viviparity in cold climates: testing the assumptions of an evolutionary hypothesis. Oecologia *57*, 397–405.

Shubin, N., Tabin, C. and Carroll, S. (1997). Fossils, genes and the evolution of animal limbs. Nature *388*, 639–648.

Silverin, B., Baillien, M., Foidart, A. and Balthazart, J. (2000). Distribution of aromatase activity in the brain and peripheral tissues of passerine and nonpasserine avian species. Gen. Comp. Endocrinol. *117*, 34–53.

Sim, C. and Denlinger, D.L. (2009). A shut-down in expression of an insulin-like peptide, ILP-1, halts ovarian maturation during the overwintering diapause of the mosquito *Culex pipiens*. Insect Mol. Biol. *18*, 325–332.

Simionato, E., Ledent, V., Richards, G., Thomas-Chollier, M., Kerner, P., Coornaert, D., et al. (2007). Origin and diversification of the basic helix-loop-helix gene family in metazoans: insights from comparative genomics. BMC Evol. Biol. *7*, p. 33.

Simionato, E., Kerner, P., Dray, N., Le Gouar, M., Ledent, V., Arendt, D., et al. (2008). Atonal- and achaete-scute-related genes in the annelid *Platynereis dumerilii*: insights into the evolution of neural basic-helix-loop-helix genes. BMC Evol. Biol. *8*, p. 1.

Simmons, N.B. (2005). An eocene big bang for bats. Science *307*, 527–528.

Smith, S.A. and Shine, R. (1997). Intraspecific variation in reproductive mode within the scincid lizard *Saiphos equalis*. Aust. J. Zool. *45*, 435–445.

Snell, T.W., Kubanek, J., Carter, W., Payne, A.B., Kim, J., Hicks, M.K., et al. (2006). A protein signal triggers sexual reproduction in *Brachionus plicatilis* (Rotifera). Mar. Biol. *149*, 763–773.

Sordino, P., van der Hoeven, F. and Duboule, D. (1995). *Hox* gene expression in teleost fins and the origin of vertebrate digits. Nature *375*, 678–681.

Soukup, V., Epperlein, H.-H., Horácek, I. and Cerny, R. (2008). Dual epithelial origin of vertebrate oral teeth. Nature *455*, 795–798.

Steigenga, M.J., Zwaan, B.J., Brakefield, P.M. and Fischer, K. (2005). The evolutionary genetics of egg size plasticity in a butterfly. J. Evol. Biol. *18*, 281–289.

Stelzer, C.P. and Snell, T.W. (2003). Induction of sexual reproduction in *Brachionus plicatilis* (Monogononta, Rotifera) by a density-dependent chemical cue. Limnol. Oceanogr. *48*, 939–943.

Stenmark, K.R. and Gebb, S.A. (2003). Lung vascular development: breathing new life into an old problem. Am. J. Respir. Cell Mol. Biol. *28*, 133–137.

Stewart, J.R. and Thompson, M.B. (2000). Evolution of placentation among squamate reptiles: recent research and future directions. Comp. Biochem. Physiol. A Mol. Integr. Physiol. *127*, 411–431.

Strahl, B.D. and Allis, C.D. (2000). The language of covalent histone modifications. Nature *403*, 41–45.

Stratford, T., Horton, C. and Maden, M. (1996). Retinoic acid is required for the initiation of outgrowth in the chick limb bud. Curr. Biol. *6*, 1124–1133.

Stratford, T., Logan, C., Zile, M. and Maden, M. (1999). Abnormal anteroposterior and dorsoventral patterning of the limb bud in the absence of retinoids. Mech. Dev. *81*, 115–125.

Streelman, J.T., Webb, J.F., Albertson, R.C. and Kocher, T.D. (2003). The cusp of evolution and development: a model of cichlid tooth shape diversity. Evol. Dev. *5*, 600–608.

Surget-Groba, Y., Heulin, B., Guillaume, C.P., Thorpe, R.S., Kupriyanova, L., Vogrin, N., et al. (2001). Intraspecific phylogeography of *Lacerta vivipara* and evolution of viviparity. Mol. Phylogenet. Evol. *18*, 449–459.

Tanaka, M., Tamura, K. and Ide, H. (1996). Citral, an inhibitor of retinoic acid synthesis, modifies chick limb development. Dev. Biol. *175*, 239–247.

Tatarazako, N., Oda, S., Watanabe, H., Morita, M. and Iguchi, T. (2003). Juvenile hormone agonists affect the occurrence of male *Daphnia*. Chemosphere *53*, 827–833.

Teotónio, H. and Rose, M.R. (2000). Variation in the reversibility of evolution. Nature *408*, 463–466.

Teotónio, H., Matos, M. and Rose, M.R. (2002). Reverse evolution of fitness in *Drosophila melanogaster*. J. Evol. Biol. *15*, 608–617.

Thesleff, I. and Sharpe, P. (1997). Signalling networks regulating dental development. Mech. Dev. *67*, 111–123.

Thompson, M.B. and Speake, B.K. (2006). A review of the evolution of viviparity in lizards: structure, function and physiology of the placenta. J. Comp. Physiol. B *176*, 179–189.

Tollet, J., Everett, A.W. and Sparrow, M.P. (2001). Spatial and temporal distribution of nerves, ganglia, and smooth muscle during the early pseudoglandular stage of fetal mouse lung development. Dev. Dyn. *221*, 48–60.

Tomoyasu, Y., Wheeler, S.R. and Denell, R.E. (2005). *Ultrabithorax* is required for membranous wing identity in the beetle *Tribolium castaneum*. Nature *433*, 643–647.

Tone, M., Nakano, N., Takao, E., Narisawa, S. and Mizuno, S. (1982). Demonstration of W chromosome-specific repetitive DNA sequences in the domestic fowl, *Gallus g. domesticus*. Chromosoma *86*, 551–569.

Torday, J.S. and Rehan, V.K. (2002). Stretch-stimulated surfactant synthesis is coordinated by the paracrine actions of PTHrP and leptin. Am. J. Physiol. Lung. Cell. Mol. Physiol. *283*, L130–L135.

Torday, J.S., Ihida-Stansbury, K. and Rehan, V.K. (2009). Leptin stimulates *Xenopus* lung development: evolution in a dish. Evol. Dev. *11*, 219–224.

Trainor, P.A., Sobieszczuk, D., Wilkinson, D. and Krumlauf, R. (2002). Signalling between the hindbrain and paraxial tissues dictates neural crest migration pathways. Development *129*, 433–442.

Tucker, A.S. and Lumsden, A. (2004). Neural crest cells provide species-specific patterning information in the developing branchial skeleton. Evol. Dev. *6*, 32–40.

Tucker, A.S., Matthews, K.L. and Sharpe, P.T. (1998). Transformation of tooth type induced by inhibition of BMP signaling. Science *282*, 1136–1138.

Tuisku, F. and Hildebrand, C. (1994). Evidence for a neural influence on tooth germ generation in a polyphyodont species. Dev. Biol. *165*, 1–9.

van Staaden, M.J. and Römer, H. (1998). Evolutionary transition from stretch to hearing organs in ancient grasshoppers. Nature *394*, 773–776.

Vargo, E.L. (1998). Primer pheromones in ants. In: *Pheromone Communication in Social Insects. Ants, Wasps, Bees, and Termites*, R.K. Vander Meer, M.D. Breed, K.E. Espelie and M.L. Winston (eds.). Westview Press, Boulder, CO, pp. 293–313.

von der Malsburg, C. (2002). Self-organization in the brain. In: *The Handbook of Brain Theory and Neural Networks*, M. Arbib (ed.), second edition. The MIT Press, Cambridge, MA, p. 6.

Wagner, G.P. and Chiu, C.H. (2001). The tetrapod limb: a hypothesis on its origin. J. Exp. Zool. *291*, 226–240.

Waters, D.A. (2003). Bats and moths: what is there left to learn? Physiol. Entomol. *28*, 237–250.

Weatherbee, S.D., Halder, G., Kim, J., Hudson, A. and Carroll, S. (1998). Ultrabithorax regulates genes at several levels of the wing-patterning hierarchy to shape the development of the *Drosophila* haltere. Genes Dev. *12*, 1474–1482.

Weatherbee, S.D., Nijhout, H.F., Grunert, L.W., Halder, G., Galant, R., Selegue, J., et al. (1999). *Ultrabithorax* function in butterfly wings and the evolution of insect wing patterns. Curr. Biol. *9*, 109–115.

Weatherbee, S.D., Behringer, R.R., Rasweiler, J.J. and Niswander, L.A. (2006). Interdigital webbing retention in bat wings illustrates genetic changes underlying amniote limb diversification. Proc. Natl. Acad. Sci. U.S.A. *103*, 15103–15107.

Weisser, W.W., Braendle, C. and Minoretti, N. (1999). Predator-induced morphological shift in the pea aphid. Proc. Biol. Sci. *266*, 1175–1181.

Wellik, D.M. and Capecchi, M.R. (2003). Hox10 and Hox11 genes are required to globally pattern the mammalian skeleton. Science *301*, 363–367.

West-Eberhard, M.J. (1986). Alternative adaptations, speciation, and phylogeny. Proc. Natl. Acad. Sci. U.S.A. *83*, 1388–1392.

West-Eberhard, M.J. (1989). Phenotypic plasticity and the origins of diversity. Annu. Rev. Ecol. Syst. *20*, 249–278.

West-Eberhard, M.J. (2003). *Developmental Plasticity and Evolution*. Oxford University Press, Oxford, p. 35.

Wheeler, D. and Nijhout, H.F. (1983). Soldier determination in *Pheidole bicarinata*: effect of methoprene on caste and size within caste. J. Insect Physiol. *29*, 847–854.

Wheeler, D.E., Buck, N. and Evans, J.D. (2006). Expression of insulin pathway genes during the period of caste determination in the honey bee, *Apis mellifera*. Insect Mol. Biol. *15*, 597–602.

Widelitz, R.B. (2008). Wnt signaling in skin organogenesis. Organogenesis *4*, 123–133.

Wigglesworth, V.B. (1973). Evolution of insect wings and flight. Nature *246*, 127–129.

Williford, A., Stay, B. and Bhattacharya, D. (2004). Evolution of a novel function: nutritive milk in the viviparous cockroach, *Diploptera punctata.* Evol. Dev. *6*, 67–77.

Wiltschko, W. and Wiltschko, R. (2002). Magnetic compass orientation in birds and its physiological basis. Naturwissenschaften *89*, 445–452.

Wiltschko, W. and Wiltschko, R. (2007). Magnetoreception in birds: two receptors for two different tasks. J. Ornithol. *148*, S61–S76.

Wiltschko, W., Möller, A., Noll, C. and Wiltschko, R. (2004). Light-dependent magnetoreception in birds: analysis of the behaviour under red light after pre-exposure to red light. J. Exp. Biol. *207*, 1193–1202.

Wiltschko, W., Munro, U., Ford, H. and Wiltschko, R. (1993). Red light disrupts magnetic orientation of migratory birds. Nature *364*, 525–527.

Wittbrodt, J. (2002). Brain patterning and eye development. EMBL Research Reports 2002, p. 93.

Wolpert, L., Beddington, R., Lawrence, P. and Jessell, T.M. (1998). *Principles of Development.* Current Biology Ltd. and Oxford University Press, Oxford.

Wu, Q. and Brown, M.R. (2006). Signaling and function of insulin-like peptides in insects. Annu. Rev. Entomol. *51*, 1–24.

Xu, X., Zhou, Z.-H. and Prum, R.O. (2001). Branched integumental structures in *Sinornithosaurus* and the origin of feathers. Nature *410*, 200–204.

Xu, Y., He, J., Wang, X., Lim, T.M. and Gong, Z. (2000). Asynchronous activation of 10 muscle-specific protein (MSP) genes during zebrafish somitogenesis. Dev. Dyn. *219*, 201–215.

Yager, D.D. (1999). Structure, development, and evolution of insect auditory systems. Microsc. Res. Tech. *47*, 380–400.

Yoshida, M.-A., Shigeno, S., Tsuneki, K. and Furuya, H. (2010). Squid vascular endothelial growth factor receptor: a shared molecular signature in the convergent evolution of closed circulatory systems. Evol. Dev. *12*, 25–33.

Zakany, J. and Duboule, D. (2007). The role of Hox genes during vertebrate limb development. Curr. Opin. Genet. Dev. *17*, 359–366.

Zapka, M., Heyers, D., Hein, C.M., Engels, S., Schneider, N.L., Hans, J., et al. (2009). Visual but not trigeminal mediation of magnetic compass information in a migratory bird. Nature *461*, 1274–1277.

Zentilin, L., Puligadda, U., Lionetti, V., Zacchigna, S., Collesi, C., Pattarini, L., et al. (2010). Cardiomyocyte VEGFR-1 activation by VEGF-B induces compensatory hypertrophy and preserves cardiac function after myocardial infarction. FASEB J. *24*, 1467–1478.

Zera, A.J. and Denno, R.F. (1997). Physiology and ecology of dispersal polymorphism in insects. Annu. Rev. Entomol. *42*, 207–232.

Zera, A.J. and Tiebel, K.C. (1988). Brachypterizing effect of group rearing, juvenile hormone III and methoprene in the wing-dimorphic cricket, *Gryllus rubens.* J. Insect Physiol. *34*, 489–498.

Zou, H.S., Juan, C.C., Chen, S.-C., Wang, H.-Y. and Lee, C.-Y. (2003). Dopaminergic regulation of crustacean hyperglycemic hormone and glucose levels in the hemolymph of the crayfish *Procambarus clarkii.* J. Exp. Zool. A Comp. Exp. Biol. *289 A*, 44–52.

Zuarte-Luis, V. and Hurle, J.M. (2002). Programmed cell death in the developing limb. Int. J. Dev. Biol. *46*, 871–876.

14 Evolution by Loss

I think there can be little doubt that use in our domestic animals strengthens and enlarges certain parts, and disuse diminishes them; and that such modifications are inherited.

Darwin (1859)

Vestigialization of Metazoan Structures

Any organ, part, or other animal structure is necessary for performing specific function(s) but changes in environment or conditions of life may, sometimes, make some functions unnecessary and the organs used for performing these functions will be used less or not used at all, thus, becoming an evolutionary encumbrance. The loss of a structure after it becomes unnecessary is related to the evolutionary disadvantage that the cost of maintaining that structure represents. Obviously, an evolutionary pressure will arise for losing it.

In the "Origin," Darwin pointed out:

> *I believe that disuse has been the main agency; that it has led in successive generations to the gradual reduction of various organs, until they have become rudimentary,—as in the case of the eyes of animals inhabiting dark caverns, and of the wings of birds inhabiting oceanic islands, which have seldom been forced to take flight, and have ultimately lost the power of flying.*
>
> Darwin (*1859, p. 454*)

Darwin reemphasized this in the "Recapitulation":

> *Disuse, aided sometimes by natural selection, will often tend to reduce an organ, when it has become useless by changed habits or under changed conditions of life.*
>
> Darwin (*1859, p. 479*)

There is a consensus that loss of function precedes the loss of structure used to perform that function (loss of, for example, eyes or sight in cave-dwelling animals, limbs in snakes, legs in aquatic mammals, or back fins in amphibians). The fact that adaptive changes in behaviors precede changes in morphology, physiology, and life histories is neither surprising nor unpredictable: behavior is the most plastic component of animal phenotype.

As pointed out in Chapter 8, animals respond immediately to environmental changes by changing their behavior. The changed behavior may contribute to their survival under changed conditions and to buying time until adaptive morphological, physiological, and life history characters arise. So, the presumptive aquatic mammals had to switch from a walking to a swimming behavior before losing their limbs and

Epigenetic Principles of Evolution. DOI: 10.1016/B978-0-12-415831-3.00014-8

acquiring general fish morphology; snakes' ancestors adopted a burrowing lifestyle before losing their limbs; and birds in predator-free islands had to adopt a walking way of locomotion before evolving reduced wings and losing the ability to fly.

Loss of structures as a result of disuse under changed conditions of life is a widespread phenomenon in the animal kingdom. J.B.S. Haldane believed that regressive evolution in animals might have been as much as 10 times more frequent than the progressive evolution (Fong et al., 1995).

Commonly, evolutionary loss is not an "all-or-none" process but the end result of an orderly process of vestigialization of the part or organ.

Charles Darwin believed that natural selection was not involved in vestigialization of organs in metazoans, since useless organs would not be selected for or against:

> *Rudimentary organs, from being useless, are not regulated by natural selection, and hence are variable.*
>
> *Darwin (1872)*

To a similar conclusion, but from a different perspective, came Gans:

> *Natural selection acts upon the totality of the organism, hence there should be no a priori reason for assuming that the organ that is vestigialized is indeed the primary target of selection.*
>
> *Gans (1975)*

Vestigialization of Limbs in Squamates

Loss of limbs and digits in squamata reptiles implies preceding states of vestigialization that persisted for periods of million years. Vestigialization of limbs occurred frequently in the course of evolution of the group. Limb-reduced reptile squamates have snake-like body form and may be grouped in two ecomorphs: long-tailed surface dwellers and short-tailed burrowers. The group consists of about 3,000 known reptile squamates (snakes and lizards) on earth, which have repeatedly and independently experienced limb reduction in every major continental region (Wiens et al., 2006).

Limb vestigialization often appears as a transitional state to limblessness. Squamate reptiles display various degrees of limb vestigialization and a reduced number of digits. The periods of persistence of the vestigialized state average between 9 and 63 million years (Brandley et al., 2008). Estimates from other authors suggest greater rates of loss of limbs in the *Lerista* species of Australia, where the estimated period of time for the full loss of limbs is less than 3.6 million years, but "limb loss could have occurred over only part of this interval" (Skinner et al., 2008). Limb reduction proceeds by a distal-to-proximal sequence of limb bones, as is observed in aquatic mammals and birds (Lande, 1978).

There is no evidence to correlate vestigialization of limbs with any changes in genes or DNA, and the relatively frequent evolutionary reversions (see Chapter 15) of limbs suggests that all important genes and gene regulatory networks (GRNs) involved in limb development are conserved and functionally intact in limbless forms.

Simplification of the Brain and Morphology in Plethodontid Salamanders

Most biologists used to believe, and still do, in the existence of an evolutionary trend that ascending the evolutionary tree, the morphology and the nervous system of metazoans become more complex. However, this rule does not seem to apply, at least, to amphibians. Plethodontid salamanders of the tribe Bolitoglossini, comprising 180 species, display the highest degree of secondary simplification of the nervous system and of the body in general (Roth et al., 1993). Secondary simplification of the brain is also observed in some other amphibians and lungfish (Roth et al., 1997).

In many respects, their brain is less differentiated than the brains of all salamander species and the brains of most fish, including lampreys and hagfish, a fact that is at variance with the position of the group in the evolutionary tree. They have the lowest number of neurons per volume unit and the lowest level of differentiation and migration in the nervous system; they have also lost lungs, most of the larval stages in the egg, as well as an aquatic larval stage (Roth et al., 1992).

Both the brain simplification and morphological simplification in Bolitoglossini are secondary rather than plesiomorphic. This fact poses another extremely difficult problem to the neo-Darwinian paradigm, for the neo-Darwinian prediction that morphological evolution is related to evolution of genes and the number of genes is obviously contradicted by the "paradox" that the secondary simplification of the morphology in amphibians is associated with a vast expansion of their genome. It is questioned:

> *Why should these evolutionarily successful vertebrates have reduced the complexity of their brains and sense organs, when the trend has been toward increased complexity in other lineages?*
>
> *Roth et al. (1992)*

Without elaborating, attempts are made to relate this paradox to paedomorphosis (Roth et al., 1992). While it is clear that a correlation among the simplification of the nervous system, the morphology, and paedomorphosis exists, there is no evidence that the last is the cause of the simplification of the nervous system in the Bolitoglossini group of plethodontid salamanders. The reverse may equally be true. Let us remember that, as a stage of metamorphosis, paedomorphosis is under control of the the central nervous system (CNS), not the other way around.

Loss of Animal Structures in Nature

One of the major chapters of the evolutionary loss of structures in animals, the so-called regressive evolution, is related to their switching to parasitic forms. Transition to parasitic mode of living often involved loss of whole organs or even systems of organs, as is the case, for example, with many parasitic worms that have lost their limbs, eyes, digestive tracts, and respiratory organs. However, the evolutionary loss of phenotypic traits in parasites is out of the scope of the present work.

Loss of Wings in Insects

Wings in insects are a monophyletic adaptation, and loss of wings is secondary to their evolution. Despite the clear evolutionary advantages of wings, loss of wings has occurred thousands of times in insects, and wingless insects represent about 5% of the extant insect species (Whiting et al., 2003). The fact that GRNs for wing development are conserved in wingless insects for more than 300 million years suggests that the loss of wings is related to an epigenetic inactivation of these GRNs in wingless insects rather than to any changes in genes that these networks consist of. We know for a fact that, most commonly, activation/inactivation of GRNs in insects is under hormonal control. And we also know that the basic hormones involved in the development and loss of wings in insects, ecdysteroids and juvenile hormone, are under strict cerebral control via neuropeptides, prothoracicotropic hormone (PTTH), and allatostatins/allatotropins, respectively.

Loss of Wings in Phasmids

One impressive case of loss of wings in insects is that of phasmids (order Phasmatodea). Although ancestral conditions of the order had been wingless, later in their evolution they independently developed wings in as many as four cases. Now, about 60% of the 3,000 species of this group of three families and ~500 genera have reduced wings or are wingless (Whiting et al., 2003).

Evolution of wings in phasmids is not a *de novo* event but a reversion to the lost ancestral phasmid wings (Whiting et al., 2003). Having lost wings early in their phylogeny, later phasmids gained wings and again became wingless a number of times. Now, most of them are wingless species that have lost fully (both fore- and hind wings) or partially (hind wings) their wings:

> *Entomologists have long assumed that re-evolution of wings in apterous lineages was impossible, because functional wings require complex interactions among multiple structures, and the associated genes would be free to accumulate mutations in wingless lineages, effectively blocking the path for any future wing reacquisition.*
>
> *Whiting et al. (2003)*

There is no evidence that evolution and loss of flight in insects is related to changes affecting the function of any of the genes involved in wing development. Moreover, it has been repeatedly observed that even after the loss of the wings, not only GRNs and genes relevant to wing development, but even the flight muscles and neural circuits determining the flight behavior, are conserved in wingless insects:

> *Studies of flight motor patterns in flying and non-flying phasmids indicate that the non-flying phasmids have retained the neural structures and basic functional circuitry required for flight .... Our results support the hypothesis that the developmental pathway for wing formation evolved only once in insect diversification, but that wings evolved many times by silencing and re-expressing this pathway in different lineages during insect evolution.*
>
> *Whiting et al. (2003)*

That the loss or reversal of wings in insects requires no changes in genes or DNA is also corroborated by the polyphenism observed in some insects, such as *Lopaphus*, which exhibit both partially winged and wingless condition in individuals of the same genotype (Whiting et al., 2003).

Loss of the Gasbladder in Fish

Most of the fish families that have no gasbladder live in the bottom of the water or in the deep sea, where buoyancy is not needed (McCune and Carlson, 2004). The function of the gasbladder in fish, however, is not restricted to buoyancy but is also used for hearing and sound production (McCune and Carlson, 2004).

Gasbladder loss occurred in 9 of the 14 extant teleost superorders. In 79 of the 425 extant families of teleost fish, the gasbladder is absent in at least one species. In 25 families, there were multiple species in at least two genera that lacked the gasbladder. Most taxa (60 families) that lack a gasbladder are either benthic (live on or at the bottom) or deep-sea fish. The loss of gasbladder in fish seems to be not related to gene mutations:

> *All the bladderless mutations in this study are lethals. Thus, the mutations we have identified are not the actual mutations that have led to loss of the gas bladder in living teleosts.*
>
> *McCune and Carlson (2004)*

There are teleost fish species, such as tuna, which, while sharing a common genotype, *sometimes*, have reduced gasbladder or even lack it completely (McCune and Carlson, 2004). This adds to the empirical evidence that reduction of gasbladder, and even its absence, are not related to changes in genes or genetic information.

From a neo-Darwinian view, unexplainable is not only the reduction or lack of gasbladder in a proportion of individuals that share the same genotype with the rest of population that have a gasbladder. Unexplainable from this view is also the repeated independent emergence and loss of this organ in the course of evolution.

Loss of Teeth in Birds

The loss of dentition in birds is one of the most enigmatic and one of the major losses of organs (affecting the whole class Aves) in vertebrate evolution. The loss is thought to have occurred about 60–80 mya (Chen et al., 2000; Mitsiadis et al., 2003a,b). Now it is empirically demonstrated that even after such a long time since loss of dentition, birds have conserved the odontogenesis developmental program and GRN.

In 1980, Kollar and Fisher combined chick presumptive embryonic epithelium with mouse molar mesenchyme of neural crest origin and cultivated the recombinant *in vitro* in the anterior chamber of the mouse eye. They observed that structures similar to the mouse tooth developed, and concluded that the loss of teeth in birds is not related to any changes in genes or loss of genetic information (Kollar and Fisher, 1980).

This was corroborated by results of neural tube homotopic transspecific transplantations *in vivo*. The mouse cranial neural crest was grafted in place of chick. Chicks grafted with a mouse neural crest of mesencephalon and rhombencephalon (series B and C) origin developed tooth structures, while those grafted with a neural crest of prosencephalon origin did not, suggesting that only the first is involved in odontogenesis.

Neo-Darwinian Explanation

Theoretically, it is incomprehensible as to what would be the selective advantages of losing teeth or why the natural selection would eliminate toothed individuals. The argument that weighty dentition would be disadvantageous for flying animals is not convincing, for dentition has not been disadvantageous to other flying animals, such as bats.

The standard neo-Darwinian interpretation of the mechanism of loss of teeth in birds would be that it is a result of accumulation of relevant mutations in odontogenic genes that, under the action of natural selection, led to the inactivation of the GRN for odontogenesis. However, no evidence has been presented that would suggest that such mutations occurred in genes involved in GRNs for tooth formation in birds or that GRNs for tooth development are absent or nonfunctional in Aves. Moreover, contrary to the neo-Darwinian prediction on gene mutations being the cause of the loss of teeth, solid experimental evidence shows that currently, *ca.* 80 million years after the loss of dentition in birds, embryonic bird epithelium is capable of forming teeth when supplied with neural crest cells from the mouse midbrain. This fact unequivocally shows that even during such an evolutionarily long period of time, despite unavoidable mutations that might have been accumulated, odontogenic genes and GRNs are fully functional in these toothless animals.

Epigenetic Explanation

Tooth development results from interactions between oral epithelium and underlying ectomesenchyme cells of cranial neural crest origin. It has been observed that, although birds have lost dentition, during ontogeny they go through initial stages of odontogenesis, similar to those observed during mammal tooth development, suggesting that they have retained the ancestral odontogenetic signaling pathway. Experimental evidence shows that they do not form teeth because somehow they are prevented from expressing Bmp4, and hence genes *Msx1* and *Msx2* (Chen et al., 2000). Moreover, genes that are expressed during odontogenic activity of neural crest cells (e.g., *Pax9*, *Msx1*, *Barx1*, *MK*) in mammals such as mice are functionally intact in Aves; only the ability of avian neural crest cells to express these genes is lost.

In a classic experiment of homotopic transplantation of the murine neural tube from the midbrain into chick embryos, it was observed that migration of the donor (mouse) neural crest cells to the mandibular and maxillar processes of the developing chick embryos leads to formation of tooth-like germ structures in the latter. This clearly suggests not only that the murine neural crest cells possess inducers of tooth formation in chick epithelial cells but also that the latter possess functionally unchanged odontogenic genes. Indeed, expression of *Msx1*, *Barx1*, and *MK* genes by the transplanted murine neural crest cells induces expression of BMP4, Shh, and FGF8 and odontogenesis in chick epithelial cells of mandibular and maxillar processes (Mitsiadis et al., 2003a,b; Figure 13.37 in the previous chapter).

All of this suggests that the loss of dentition in Aves is result of a nongenetic, epigenetic-regulatory loss of ability of their neural crest cells to secrete signaling molecules necessary for chick epithelium to initiate odontogenesis.

What is the cause of the loss of the ability of the chick neural crest cells to induce tooth formation? At a theoretical level, one might argue that the fact that it is the neural tube/CNS that provides neural crest cells with information on "what to do" in the sites of their migration suggests that the chick neural tube/CNS ceased to provide odontogenic information to these cells. Since no changes in the function of the key odontogenic genes are involved, it may be safely inferred that the change is determined by an epigenetic change in the chick neural tube/CNS.

Loss of Tetrapod Limbs

Loss of limbs has occurred in three of four tetrapod classes (amphibians, reptiles, and mammals). It represents one of the most extreme morphological changes in the history of tetrapods (Lande, 1978) and has been associated especially with elongation of the body and increase in the number of vertebrae.

It is believed that the loss of limbs occurred in response to new ways of locomotion as a result of a change in the lifestyle of tetrapods. This seems to have been the case with transition of reptiles to a burrowing lifestyle and reptant locomotion, which made their limbs useless. The loss and reduction in size of limbs in tetrapods was thought to have been a gradual process of sequential loss of limb components in the reverse order (distal-to-proximal) of their formation during the individual development (proximal-to-distal). Later studies, however, have shown that often evolutionary processes of body elongation, reduction of limb size, and reduction of digits occurred almost simultaneously (Wiens and Slingluff, 2001).

Loss of limbs has occurred repeatedly and independently in a large number of reptile species. Generally, forelimbs and pectoral girdle are lost before the hind limbs and pelvic girdle. Loss of limbs in reptiles is associated with (Lande, 1978; Cohn and Tickle, 1999), and preceded by (Lande, 1978), body elongation. In turn, body elongation results from two different mechanisms: trunk elongation, related to subterranean dwelling, and tail elongation related to surface dwelling (Wiens and Slingluff, 2001). There is no consensus on the rates of evolution of limblessness in snakes. Two contrasting hypotheses have been proposed: one positing sudden loss of limbs (Cohn and Tickle, 1999), and the other stating that the loss has been gradual (Wiens and Slingluff, 2001).

Loss of Limbs in Amphibians and Reptiles

Total loss of both pairs of limbs occurred several times in amphibians and reptiles. About 150 amphibian species of Caeciilidae family of the monophyletic order Gymnophiona (Apoda) are limbless. They populate tropical forests. The loss of limbs in this group is believed to have resulted from transition to subterranean mode of living (Summers and O'Reilly, 1997), and a number of caecilians presently are fossorial rather than aquatic species.

Among reptiles, only the superorder Squamata has limbless species. Snakes are always functionally limbless. Four lizard families consist mainly of species with vestigialized limbs, and three of the four families of reptiles of the suborder Amphisbaenia have extremely reduced limbs (Gans, 1975).

There is no evidence that the loss or reduction of limbs in amphibians and reptiles is related to any changes in any relevant genes or DNA.

Loss of Limbs in Snakes

Between 2,700 (Coates and Ruta, 2000) and 3,000 (Wiens et al., 2006) extant snake species are currently known. In a study on 261 species of squamate reptiles, it was observed that snake-like body forms (short-tailed burrowers and long-tailed surface dwellers) evolved independently 25 times (Wiens et al., 2006).

Snakes evolved from limbed terrestrial ancestors (Greene and Cundall, 2000; Tchernov et al., 2000). The possibility of a reverse, aquatic-to-terrestrial, origin of snakes evolving from marine voracious reptiles has also been suggested (Coates and Ruta, 2000), but a terrestrial-to-marine transition is more likely as a common theme of tetrapods switching to the aquatic mode of living (Greene and Cundall, 2000).

Most biologists consider the loss of limbs in snakes to be a result of adaptation to a burrowing or surface-dwelling lifestyle. Phylogenetic conclusions contradict the widely held "subterranean" theory of snake origins and instead imply that burrowing snakes (scolecophidians and anilioids) acquired their fossorial adaptations after the evolution of the snake-like body form and jaw apparatus in a large aquatic or (surface-active) terrestrial ancestor (Scanlon and Lee, 2000). As pointed out by Gans (1975), the potential adaptive value of the transition to snake-like morphology has not been well established, but it is generally assumed that snake-like body shape facilitates locomotion underground and in dense grass (Wiens et al., 2006).

Gans believes that a switch to an undulatory locomotion might have given rise to both body elongation and loss of limbs observed in so many species of amphibians, reptiles, and mammals.

The fact that tail loss, to various extents (from two-thirds of its length to total loss), was observed in 58% of individuals of a large population of tiger snakes in western Australia (Aubret et al., 2005) suggests that the species is in the process of losing the tail. Whether you call this a form of developmental polymorphism or even penetrance is of little importance. What scientifically matters in this case is the fact that certain proportions of individuals of the same genotype, under the same environmental conditions, display different phenotypes. This clearly contradicts the basic tenet of the neo-Darwinian paradigm that evolution of limblessness, as any other evolutionary change, requires accumulation of favorable mutations in relevant genes. Logically, this suggests that a nongenetic mechanism induces the loss of limbs in this snake species.

Loss of Forelimbs in Pythons

Pythons have no forelimbs, but they develop reduced hind limbs. Anatomical transformations in python limbs have been sudden, rather than gradual, and are related to the progressive expansion of *Hox* gene expression patterns (Cohn and Tickle, 1999).

The loss of forelimbs in pythons is believed to be related to an anterior expansion of expression pattern of *Hox* genes. Hind limb buds are initiated in pythons, but the zone of polarizing activity (ZPA) does not develop, and the ectoderm does not form an apical ectodermal ridge (AER) in the region where the limb bud emerges in tetrapods, even though all of the signaling genes responsible for their development are present. This is believed to be caused by changes in mesodermal *Hox* gene expression (Cohn and Tickle, 1999).

In contrast to the forelimbs, pythons develop hind limb buds and rudimentary hind limbs with truncated pelvic girdle and femur. However, they cannot express sonic hedgehog (Shh) because they have no AER, and they do not express in their ectoderm AER-related genes, *Dlx* (*Distal-less*), *Fgf2*, and *Msx* (Cohn and Tickle, 1999). This does not mean that these genes are not functional, for they are expressed in other organs of the python embryo.

The python hind limb mesenchyme can be experimentally induced to form an AER and express Shh by application of fibroblast growth factor (FGF). The fact that the python mesenchyme from the hind limb is functional when grafted to a chick embryo wing (Cohn and Tickle, 1999) proves beyond doubt that the python hind limb bud possesses all genes involved in the initial development of tetrapod limbs, and the loss of hind limbs in pythons is not related to any change in the function of limb-inducing genes.

A glimpse at the embryonic expression domain of the *HoxC-6* and *HoxC-8* genes in chicks and python embryos shows that whereas in chicks, expression of these genes takes place along the trunk, with interruptions at the levels of the forelimbs and hind limbs, in python embryos, these genes are expressed uninterruptedly anteriorly but are not expressed at the level of hind limbs (Figure 14.1).

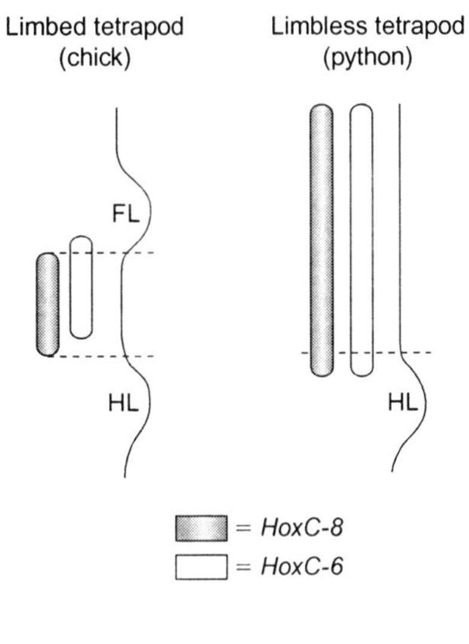

Figure 14.1 The distribution of *HoxC-8* and *HoxC-6* in a limbed tetrapod (embryos of the common fowl) and a snake (embryos of the python). The expression boundaries are extended slightly more posteriorly, and much more anteriorly, in python than in chick embryos.
Abbreviations: FL, forelimb buds; HL, hind limb buds.
Source: From Bejder and Hall (2002).

Expression of *HoxC-8* is under control of RA pathway, and in the case of limb bud development, the pattern and sites of *HoxC-8* expression are negatively controlled by RA secreted by brachial spinal nerves that innervate the limb bud. This suggests an antagonistic relationship between the *Hox* gene expression along the trunk and limb development.

It has been argued that small changes in the regulatory sequences of *HoxC-8* gene enhancers between the mice and chicks may be the cause of differences in the region of the development of limb buds in the embryos of two species (Belting et al., 1998), but such changes in sequences will unavoidably accumulate over the time if they do not lead to the loss of gene function. Investigators have not concluded whether the changes in sequences of *HoxC-8* gene enhancers in mice and chicks are the cause of the pattern of expression or are a normal result of the long divergent evolution of these species that did not affect the expression. This is why investigators themselves cautioned: "Additional experiments will be required to determine the specificity of nucleotide changes in the regulation of *HoxC-8* expression pattern and correlated modifications of the body plan" (Belting et al., 1998).

Besides, and predictably, differences in the enhancer are also observed between *HoxC-8* enhancers in mice and whales (a 4 bp deletion), but as two of the same group of investigators admit, they have found no correlation between the sequences of the baleen whale *HoxC-8* enhancer and any specific morphological trait that evolved in this species (Shashikant et al., 1998).

With changes in genes and enhancers excluded as the cause of vestigialization and loss of limbs in pythons, the remaining alternative explanation is an epigenetic-regulatory mechanism. Having shown that *Hox* gene expression domains along the body axis determine the absence of forelimbs and vestigialization of hind limbs in pythons, now we must remember that patterns of expression of *Hox* genes in vertebrates determined by the patterns of expression of RA along the body axis, in which the neural tube and motor neurons, as was shown earlier, play a crucial role. (See for further information in Section Role of the Nervous System in Limb Development, Chapter 13.)

Loss/Reduction of Limbs in Aquatic Mammals

Paleontological evidence shows that the ancestral forms of modern cetaceans, such as *Pakicetus inachus* of Early Eocene (*ca.* 58–48 mya) in Pakistan, may have been land tetrapods exhibiting all of the typical features of terrestrial mammals (Figure 14.2). A later stage (*ca.* 47 mya) in the evolution of cetaceans in the fossil evidence is exemplified by *Ambulocetus natans*, which shows signs of transition to aquatic morphology characterized by reduction of forelimbs but still retains well-developed hind limbs with webbed feet, reminiscent of hind limbs of the sea otter. It probably swam by vertical axial undulations of the spine, while using hind limbs like a fluke. The next stage (*ca.* 40 mya) in the evolution of cetaceans is the elongation of the body and increase in the number of vertebrae as well as marked vestigialization of hind limbs (*Basilosaurus*) indicating adaptation to a fully aquatic life (Thewissen and Bajpai, 2001). At a final stage of evolution of cetaceans, flukes evolved, and the swimming by axial undulation was complemented by tail oscillations.

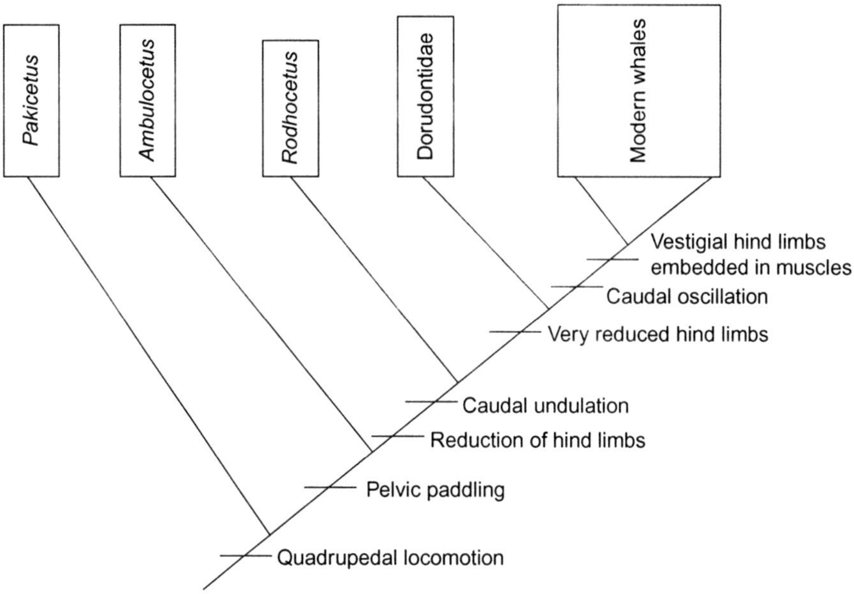

Figure 14.2 Evolution of the changes in swimming mode during cetacean evolution. Modern whales comprise baleen and toothed whales.
Source: From Bejder and Hall (2002).

Evolution of terrestrial mammals into marine swimmers followed, and/or was correlated with, changes in locomotory and other behaviors that the aquatic life imposed. Adaptive changes in the locomotory behavior and accompanying changes in morphology in the course of evolution of aquatic mammals can be illustrated with eclectic examples of modern animals that presumably are in the process of the evolutionary adaptation to the aquatic life (Figures 14.3–14.7).

In short, minks (*Mustela vison*) paddle quadrupedally (Figure 14.4), and freshwater otters (*Lontra canadensis*) (Figure 14.5) swim mainly with their hind limbs (pelvic paddling), although they derive some additional lift from the tail (pelvic undulation). The sea otter, *Enhydra lutris* (Figure 14.6), uses its highly asymmetrical feet as the propelling surfaces, but most of the power for the movements comes from undulations of the vertebral column (pelvic undulation) rather than from the muscles of the hind limbs. The giant South American freshwater otter, *Pteronura brasiliensis*, uses caudal undulations: sinusoidal motions of the vertebral column, like a wave moving through the entire spine, power a long and narrow tail that is dorsoventrally flat (Figure 14.7). No otter swims like a modern cetacean, but the swimming mode of *Pteronura* approximates whale swimming. Modern cetaceans differ from *Pteronura* in having a rigid body with most of the movement concentrated at one point: undulation, thus, became oscillation. In addition, modern cetaceans evolved a fluke (Thewissen and Bajpai, 2001).

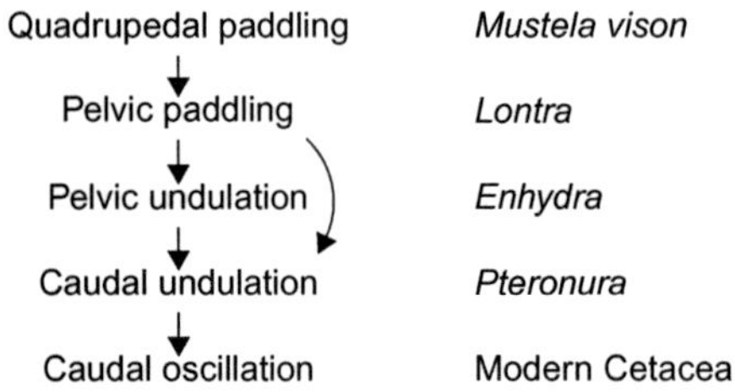

Figure 14.3 Hypothesis for the evolution of the caudal oscillation swimming mode of modern Cetacea, based on Thewissen and Fish (1997). Different swimming modes are listed in the left column, and arrows indicate transitions that can be predicted on the basis of efficiency considerations. Modern mustelids swim using various modes, and cetaceans probably went through these modes sequentially during their evolutionary history. Morphological study indicates that *Ambulocetus* was probably a pelvic paddler or caudal undulator and that *Kutchicetus* was mainly a caudal undulator.
Source: From Thewissen and Bajpai (2001).

Figure 14.4 The American mink, *Mustela vison.*
Source: From American Mink Mustela vison Biopix; http://www.biopix.com/photo.asp?photoid = 50933&photo = mustela-vison.

Given that the loss of limbs is a process of adaptation to new conditions of living (aquatic, fossorial, or dense grass environment), animals first had to learn new modes of limbless locomotion (e.g., lateral undulation, swimming undulation, concertina). The process of learning may have been facilitated by the fact that ancestral motor patterns and fixed motor patterns (FAPs) during evolution are conserved, and motor circuits generating these FAPs could be activated under stressful habitat

Figure 14.5 The northern river otter, *Lontra canadensis*.
Source: From Wikipedia: The Free Encyclopedia; http://en.wikipedia.org/wiki/File:RiverOtter SwimmingOregonZoo.jpg.

Figure 14.6 The sea otter, *Enhydra lutris*.
Source: From BBC Nature Wildlife: http://www.bbc.co.uk/nature/life/Sea_otter#p009c4qv.

conditions. Indeed, all of the basic modes of locomotion, swimming (undulatory waves passing down the body), crawling, and lateral undulation are functions of a single motor pattern circuit that evolved in invertebrate ancestors. The motor pattern for swimming was not lost in tetrapods adapted to terrestrial life, and most tetrapods are still capable of learning to swim.

Concertina locomotion (Figure 14.8A) implies that some part of the body is fixed on the ground in order to push the rest of the body forward. While this form of locomotion is widespread among burrowing snakes, another form, the so-called internal concertina, has been adopted by many caecilians, limbless snake-like amphibians (Figure 14.8B). This mode of locomotion consists in undulatory movements performed by vertebral column only (not the body as a whole).

Figure 14.7 The giant South American freshwater otter, *Pteronura brasiliensis*.
Source: From http://www.thefullwiki.org/Pinniped

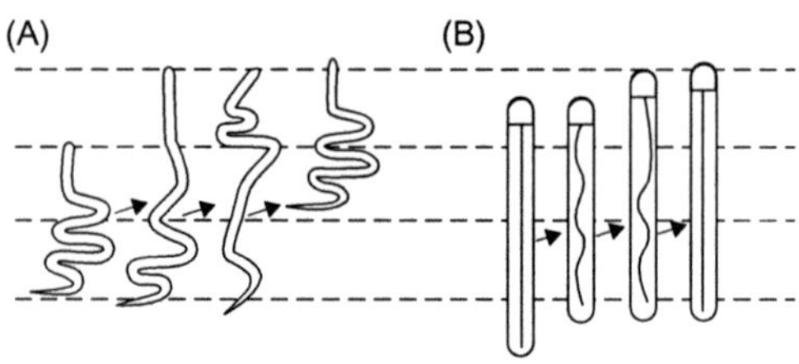

Figure 14.8 A comparison of normal concertina (A) and internal concertina (B). Concertina is shown in a snake traversing a smooth surface. Internal concertina is shown in a caecilian moving in a burrow. The vertebral column and skull are superimposed on the outline of the caecilian.
Source: From Summers and O'Reilly (1997).

Changes in the locomotory behavior preceded and facilitated vestigialization and loss of hind limbs in the evolution of these aquatic mammals from terrestrial tetrapods (Figure 14.9). Two crucial steps in this process were a reduction of the time of expression of Shh in the hind limb bud and later a loss of ZPA in the hind limbs (Thewissen et al., 2006).

It is noteworthy that during the ontogeny, cetaceans develop hind limb buds showing all the initial steps of terrestrial mammal limb bud development, including cell differentiation, formation of both signaling centers, the AER and ZPA, innervation, and secretion of FGF8, before entering the regression stage, as a result of Shh suppression. It is believed that the evolutionary reduction of the expression of Shh in the limb bud of aquatic mammals (and the corresponding limb reduction) started

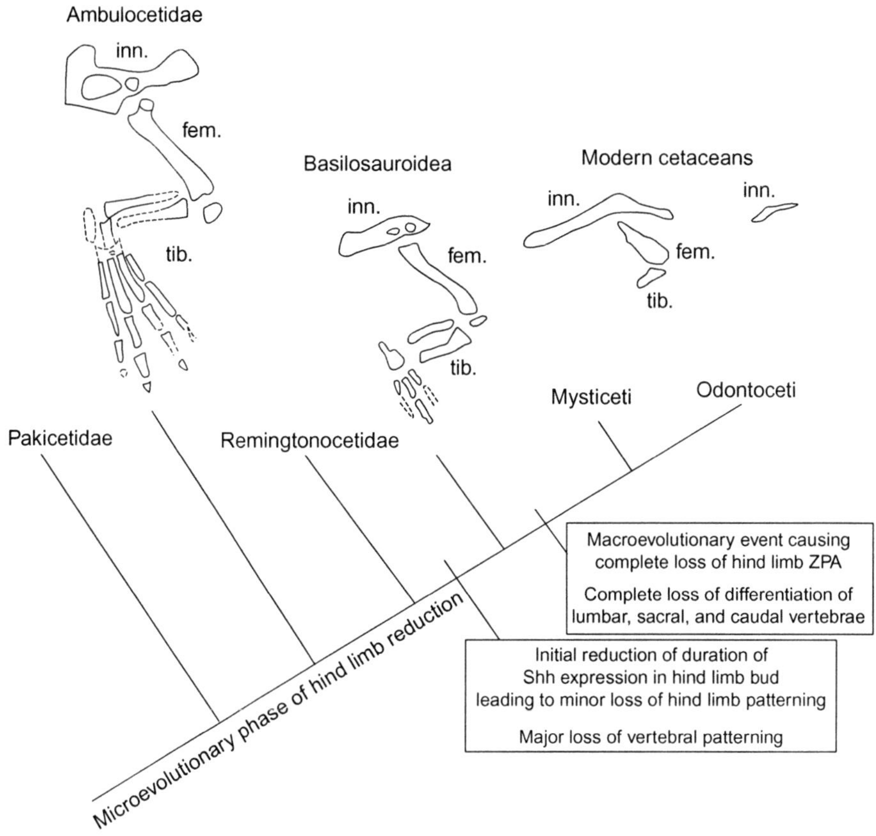

Figure 14.9 Simplified phylogeny of cetaceans discussed here with evolutionary events indicated. Hind limbs represent fossil ambulocetid *Ambulocetus*, fossil basilosauroid *Basilosaurus*, and two modern mysticetes (Bowhead whale and Sei whale, respectively). In most odontocetes, the only hind limb element preserved is the innominate, as in the Sei whale. Labeled bony elements of the hind limb are innominate (inn.), femur (fem.), and tibia (tib.). *Source*: From Thewissen et al. (2006).

ca. 41 mya, whereas the total loss of Shh expression (and resulting loss of hind limbs) occurred *ca.* 34 mya (Thewissen et al., 2006).

Embryos of the limbless river mammal *Stenella attenuata* also develop limb buds. The limb bud grows to reach a length of 10–30cm before starting regressive processes that lead to their reduction and total disappearance. Moreover, individuals with vestigial hind limbs are observed, at a low frequency, among the adult populations of the spotted dolphin (Sedmera et al., 1997). The sperm whale (*Physeter catadon*) is the only toothed whale with a hind limb skeleton (15cm long), although its expression is variable (Hall, 1995). Reduction of limb development in whales is extreme, but rudiments of the tetrapod bones are present and 37% of individuals of the Antarctic population of minke whale have ossified femoral rudiment. Sometimes the following are also observed:

> *Atavistic skeletal elements can be surprisingly complete; 79 cm long bones in 125 cm long left and right "hindlimbs" in a female humpback whale.*
>
> *Bejder and Hall (2002)*

As for evolution of flippers from terrestrial mammalian limbs in the spotted dolphin, like other cetaceans, these organs, apparently adaptations for aquatic life, have retained the inner mammalian limb structure, except for a marked increase in the number of phalanges, which is clearly an adaptation for the aquatic life (Sedmera et al., 1997).

The brief review of the normal development of limbs in tetrapods in Chapter 13 (Section Role of the Nervous System in Limb Development) may be illuminating on the possible mechanisms of the evolutionary loss of limbs in limbless vertebrate groups.

Neo-Darwinian Explanation of Loss of Limbs

No changes in the function of the "limb-determining" or other relevant genes, other key limb-inducing genes or regulatory regions are involved in the loss of limbs in vertebrates, as is proven, among other things, by the fact that most of the limbless species initially activate the genes, form AER and ZPA, and even develop limbs to advanced stages before arresting their development or starting the programmed cell death of limb tissues.

From the neo-Darwinian view, the occurrence of such radical morphological differences as the presence and absence of limbs, e.g., exteriorization and posteriorization of limb buds, between species that have functionally unchanged all of the limb-determining genes (including *Hox* genes) is unexplainable at best.

Epigenetic Explanation

A look at expression patterns of *HoxC-8* gene shows that in both chicks and mice embryos it is expressed in the midthoracic mesoderm and in the brachial region of the neural tube. However, the anterior boundary of expression extends less anteriorly in chicks than in mice, thus determining the longer cervical region, more posterior appearance of limb bud as well as the smaller number of thoracic segments in chickens (Figure 14.10). It is noteworthy that the anterior boundary of *HoxC-8* expression in both species coincides with the site of origin of the brachial nerves that innervate limbs (Bejder and Hall, 2002). Also remember, expression of *Hox* genes in general, and *HoxC-8* in particular, are regulated by RA, which downregulates expression of posterior *Hox* genes along the embryonic anterior–posterior and causes respective truncation of the embryo (Kessel, 1992).

Genes for RA synthesis enzymes in vertebrates have not changed. What has changed is the spatiotemporal pattern of expression of RA in limbed and limbless tetrapods as well as in chickens and mice, as shown in Figure 14.10. This change is

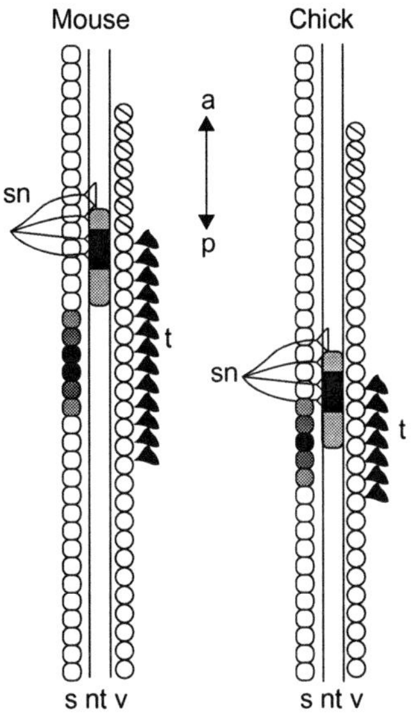

Figure 14.10 Schematic comparison of *HoxC-8* expression in chicken and mouse in relation to morphological landmarks. Cervical, thoracic, and lumbar regions of the vertebral column and the brachial region of the neural tube are indicated. Brachial spinal nerves C6, C7, C8, and T1 in mouse and C13, C14, C15, and T1 in chicken are shown. Shaded region in somites and neural tube represent *HoxC-8* expression. Regions of highest expression are indicated in dark shades. The double-headed arrow indicates the anteroposterior orientation of the body axis. *Abbreviations*: a, anterior; nt, neural tube; p, posterior; s, somites; sn, spinal nerves; t, thoracic vertebrae; v, vertebrae.
Source: From Belting et al. (1998).

clearly nongenetic (all of the limb-inducing genes are present and functional in both limbed and limbless species).

Where may be the source of the epigenetic information that is used for these adaptive changes in expression patterns of *Hox* and other genes involved in the development of limbs or leading to limblessness in tetrapods?

The evidence presented in this chapter as well as in Chapter 13 (Section Role of the Nervous System in Limb Development) on the evolution of limbs in vertebrates shows that RA signals from the neural tube and local innervation are essential for the development of limbs in tetrapods.

In the process of vertebrate limb loss and reduction are also involved mechanisms of programmed cell death, which are epigenetically regulated as well. The process of apoptosis that leads to regression of the limb bud is known to be related to the fact that the AER does not secrete FGF, especially FGF8 and FGF4 (Boulet et al., 2004).

As shown earlier (Sections Apoptosis in Invertebrates and Neural Control of Apoptosis in Chapter 5), the programmed cell death during the individual development is epigenetically determined via signal cascades that ultimately originate in the nervous system.

Loss of Lungs in Salamanders

Loss of lungs in aquatic salamanders is an illustration of the old (also Darwinian) idea on the role of "disuse" as a cause of loss of organs in animals. Ancestors of modern lungless plethodontid salamanders were lunged ambystomatid-like forms. Lungless plethodontid salamanders have evolved independently at least five times from lunged ancestors in the Mesozoic (251–65 mya) in rapidly flowing upland

Appalachian streams. The loss of lungs in these salamanders seems to have been an adaptation to the oxygen-rich swift streams for decreasing the risk of downstream drift, where lungs were maladaptive because of the buoyancy. This hypothesis has found further support by the recent discovery of a lungless frog in Borneo, which also inhabits rapid cold oxygen-rich streams (Bickford et al., 2008). The loss of lungs was favored under the circumstances of a parallel evolution of cutaneous respiration in these species (Fong et al., 1995).

There is no evidence to relate the loss of lungs in plethodontid salamanders to any changes in genes, and no neo-Darwinian explanation is known to the author. We must admit that there are no sufficient empirical data for reconstructing a developmental mechanism of the evolution of lunglessness in salamanders. However, in general theoretical terms, it may be argued that signals for inactivating the developmental pathways that induce lung development in salamanders might have been of neural origin, as is suggested by the fact that the whole processes of organogenesis, including lung development, in metamorphosizing salamanders are neurally regulated (see Section Neural Control of Metamorphosis in Amphibians in Chapter 6); what controls the development of a morphology must be responsible for its evolutionary change.

Loss of Eyes in the Mole Rat Spalax ehrenbergi

The fossorial rodent mole rat, *S. ehrenbergi*, has very rudimentary eyes covered by skin. It does not respond (Sanyal et al., 1990), or shows only little sensitivity, to light stimuli, but, in the usual sense, it is blind (Bronchti et al., 1991; Necker et al., 1992). Reduction of the eye and the optic nerve in this species is correlated with a shift in the function of the visual dorsal lateral geniculate nucleus (LGBd) and in a part of the visual cortices, which in this blind species, in the absence of visual input (Heil et al., 1991), compensatorily receive auditory and somatosensory input (Necker ct al., 1992; Bronchti ct al. 2002; Figurc 14.11).

There is evidence, however, that the rudimentary eyes of the mole rat have acquired another function. Removal of the rudimentary eyes disturbs photoperiodic perception. This suggests that the harderian gland that has replaced the eye structure, in the process of its evolutionary loss, may have been reorganized into a functionally pineal-like organ for photoperiodic regulation and is included in the endocrine pathways mediating photoperiodicity (Sanyal et al., 1990; Cernuda-Cernuda et al., 2002).

There are no indications that gene mutations, changes in allele frequencies, or genetic recombination might have been involved in the loss of structure and function of eyes and in the modification of the structure and function of the respective brain centers.

Loss of Characters in Cave-Dwelling Animals

Life in dark caves usually leads to an evolutionary pressure for losing certain characters and acquiring troglomorphic (from ancient Greek *trogle*, cave) characters (Table 14.1).

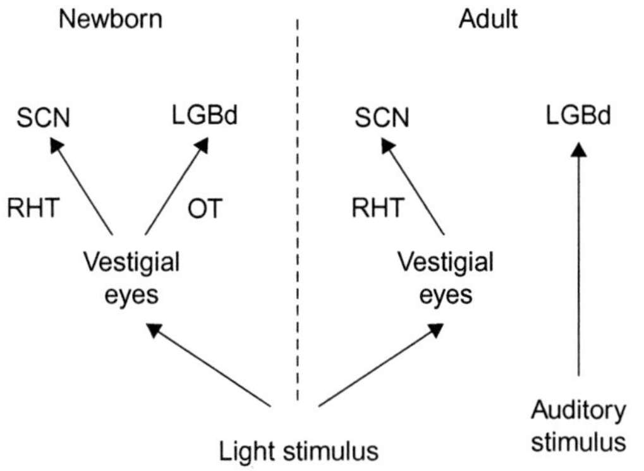

Figure 14.11 A schematic diagram of the circadian (SCN, suprachiasmatic nucleus) region and part of the visual (LGBd, dorsal lateral geniculate nucleus) system in newborn and adult mole rats. In the adult mole rat, the connection between the vestigial eyes to the LGBd degenerates, and instead this region receives auditory stimuli. *Abbreviations*: OT, optic tract; RHT, retinohypothalamic tract.
Source: From Kimchi and Terkel (2002).

Table 14.1 Catalog of "Troglomorphic" Features

Morphological	Physiological	Behavioral
Reduced, diminished, or lost		
Eyes, ocelli	Metabolism	Photoresponse
Visual brain centers	Circadian rhythms	Aggregation
Pigmentation	Fecundity	Response to alarm substances
Pineal organ		Aggression
Body size		
Cuticles (terrestrial arthropods)		
Scales (fishes)		
Swimbladder (fishes)		
Enlarged, enhanced, or exaggerated		
Chemo- and mechanoreceptors	Life span	
Appendages	Lipid storage	
Body size	Metabolism	
	Egg volume	

Source: From Romero and Green (2005).
These are the characters that frequently differ from those in closely related epigean organisms. Troglomorphic organisms may display only a few, some, or all of these characters. Some characters may differ in either direction (e.g., some troglomorphic fish display reduced metabolism, while other species exhibit an increase).

One of the most widespread evolutionary phenomena is loss of eyes in animal species living in dark caves, where normal photoreceptive eyes are of little use, if any.

Most of the 50,000–100,000 obligate cave-dwelling species (arachnids, insects, crustaceans, fish, and salamanders) have lost their eyes (Fong et al., 1995).

Out of about 28,000 fish species known in the world, 299 are reported to be cave dwelling (Romero et al., 2009).

Loss of Eyes in Astyanax fasciatus (mexicanus)*: Epigenetics of an Evolutionary Event*

The Mexican teleost fish species, *Astyanax mexicanus*, exists in two forms: an eyed surface-dwelling form (epigean) and an eyeless cave-dwelling form (hypogean). Both morphs are interfertile, although usually in nature they are spatially isolated. Over the last 10,000 years, at least four times, cavefish populations of *A. mexicanus* independently evolved various degrees of loss of eyes (Dowling et al., 2002; Jeffery, 2005). Correlated with the loss of the eye structure and function and with reduction of the size of the optic tecta (the visual processing center in the brains of fish, amphibians, and reptiles), the cave-dwelling morph has also evolved new behaviors, various degrees of body depigmentation as well as several constructive characters in jaws, teeth, and taste buds, mechanosensory system of cranial neuromasts, compensating for the lack of eyes (Teyke, 1990). Changes also occurred in the number of rib-bearing thoracic vertebrae in the axial skeleton (Dowling et al., 2002).

There are no remarkable differences in the early development of the eye anlagen between the blind hypogean and eyed epigean embryos besides the eye size and proportions. The divergence becomes apparent during the growth stage of the eye, when the embryos of the blind morph fail to enter that stage, and the vestigial eye is covered by the growing regional skin (Figure 14.12). Simultaneously, regressive processes start with the programmed cell death (apoptosis) taking place in the lens and later in the retina.

The development of the eye in invertebrates and vertebrates in general depends on a "conserved Pax6 dependent mechanism" (Quiring et al., 1994) that is operative at early stages of development (Tomarev et al., 1997). *Pax6* gene is expressed in both sides of the midline of the anterior part of the neural plate. Anteriorly, the *Pax6* expression domains fuse to form the forebrain and optic anlagen. Secretion of Shh by midline tissues is also essential for the development of ventral eye structures (Zhang and Yang, 2001), and the initial signals for the development of eye anlagen originate in the neural plate/neural tube.

In cavefish, investigators found that all of oculogenic genes are functional, and all of them are expressed normally:

> *It appears that eye gene cascades are completely operational in cavefish embryos prior to the general transcriptional shutdown that occurs after the beginning of apoptosis.*
>
> *Jeffery (2005)*

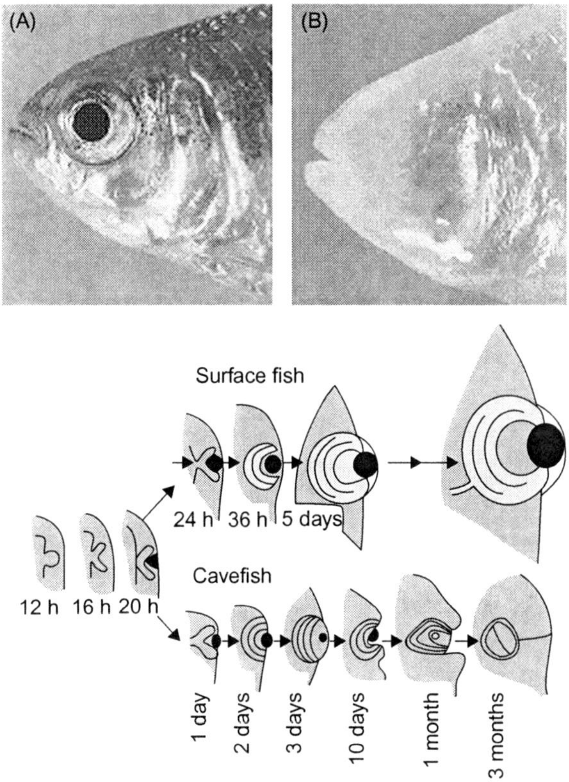

Figure 14.12 Eye development and degeneration in *Astyanax mexicanus*. Surface fish (A) and cavefish (B) adults. Diagram showing the timing of eye growth and development in surface fish (top) and eye degeneration in cavefish (bottom).
Source: From Jeffery (2005).

Role of the nervous system in lens development. Lens formation in vertebrates requires the presence of the optic vesicle (Furuta and Hogan, 1998), which is an extension of the forebrain neuroepithelium (Behesti et al., 2009), the precursor of the retina. As first observed by Spemann by the beginning of the twentieth century, the optic vesicle is necessary for the development of the lens. At the site of the physical contact with the anterior side of the forebrain (optic vesicle), the head ectoderm is induced to form the lens placode. The lens GRN is neurally activated by signals from the optic vesicle/retina and pigment cells (Weaver and Hogan, 2001; Reza and Yasuda, 2004a; Adler and Canto-Soler, 2007; Figure 14.13).

Unilateral ablation of prospective retinal region of the neural plate prevents formation of the optic cup (and expression of these genes) and lens formation in the operated side of the lateral head ectoderm (Kamachi et al., 1998). As mentioned earlier, the optic vesicle is an extension of the brain neuroepithelium, hence the fact that no lens develops if the presumptive lens ectoderm does not come into contact with

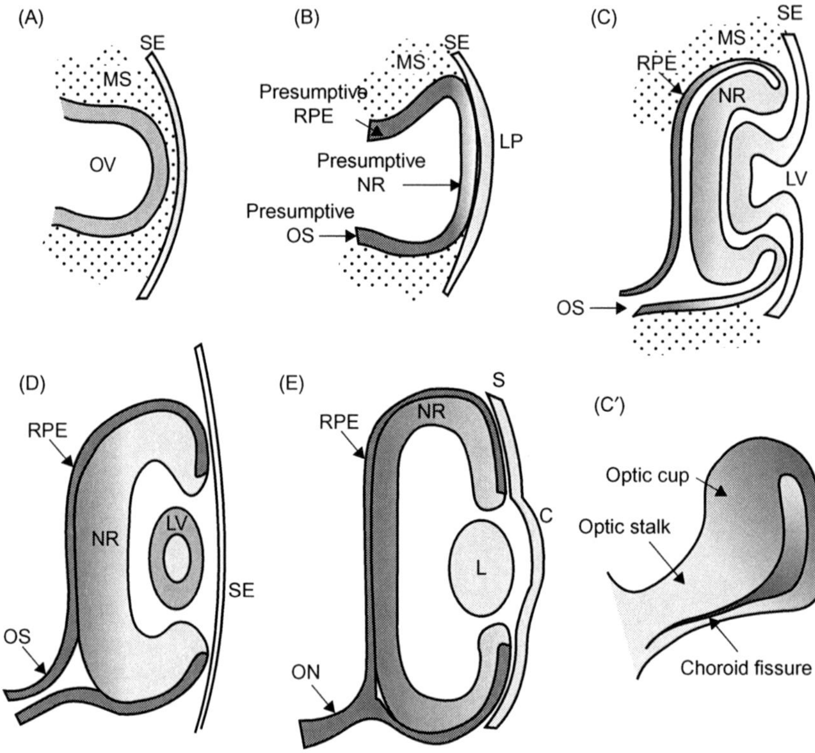

Figure 14.13 Schematic representation of vertebrate eye development. (A) The optic vesicle forms as an evagination from the diencephalon. (B) Upon contact with the surface ectoderm, the optic vesicle becomes patterned into presumptive RPE, neural retina, and optic stalk; the surface ectoderm in turn forms the lens placode. (C) The optic vesicle and lens placode invaginate, giving rise to the optic cup and the lens vesicle, respectively. (C′) The ventral region of the invaginating optic vesicle forms the choroid fissure. (D–E) Transition from early to mature optic cup. The lens vesicle loses its cavity and becomes a solid structure; the neural retina and the pigment epithelium become apposed, reducing ependymal cavity to a virtual space; the optic stalk gives rise to the optic nerve, and the surface ectoderm adjacent to the lens gives rise to the corneal epithelium. *Abbreviations*: C: cornea; L: lens; LP: lens placode; LV: lens vesicle; MS: mesenchyme; NR: neural retina; ON: optic nerve; OS: optic stalk; OV: optic vesicle; RPE: retinal pigment epithelium; S: sclera; SE: surface ectoderm.
Source: From Adler and Canto-Soler (2007).

the optic vesicle suggests that signals from the forebrain trigger the development of the lens in the process of the formation of the eyecup.

The retina is a source of lens-inducing factors (Arresta et al., 2005), and both the optic vesicle and retina are demonstrated to be capable of inducing synthesis of crystalline lens (Cannata et al., 2008). Essential for the lens fiber development is Pax6 released from the neural retina (Reza and Yasuda, 2004b) and BMP4 (Furuta and Hogan, 1998).

Experimental transplantation of the lens vesicle of epigean eyed fish to the embryos of blind cave morph induces the development of eye structures. However, the offspring of eyeless fish, experimentally transformed into eyed fish, are functionally blind (Romero et al., 2003) due to the loss of the function and changes in the structure of the optic tecta, the main visual centers. Commonly, the optic nerve in eyeless fish is still connected with the brain, but the fact that cases of the loss of optic nerve have also been observed (Wilkens, 1970) shows that variation in developmental pathways or gene expression, not changes in genes *per se*, are responsible for the loss of the optic nerve.

The arrest of the eye development in embryos of the blind form of cavefish is related to the lateral expansion of the expression domains of the *shh* and *tiggy-winkle hedgehog* (*twhh*) genes.

Significant progress has been made recently in understanding the mechanics of the apoptotic processes leading to the evolutionary loss of eyes in the cavefish. This progress is partly related to recognition of the role of chaperones in cavefish apoptosis. The two hsp90 isoforms (α and β) have different expression patterns in the eyed and eyeless morphs of *A. mexicanus*. Expression of hsp90α in eyeless morphs reaches its highest level just prior to the fragmentation of nuclei of the dying cells in the lens, and the lens apoptosis is blocked by administration of hsp90α inhibitors. Both of these experimental facts suggest that hsp90α has an important role in inducing lens apoptosis in the cavefish.

It has been hypothesized that hspα performs its apoptotic function by interfering with the activity of an antiapoptotic factor (Hooven et al., 2004). In other experiments, it has been demonstrated that nicotine induces hspα, and the latter is the mediator of the nicotine-induced apoptosis in human cells (Wu et al., 2002).

Neo-Darwinian Explanation

From the neo-Darwinian view, the evolutionary loss of eyes in the hypogean form of *A. mexicanus* requires, as a *sine qua non*, new genetic information to be transmitted from eyeless parents to the offspring. But more than 50 years of the studies of the loss have provided no evidence on any changes in eye genes or DNA (Jeffery, 2005); all of these genes are functionally normal in both the blind cavefish and its epigean eyed form.

> *The developmental evidence does not support an evolutionary model that proposes loss of function of the genes involved in early eye development and/or eradication of the embryonic eye to conserve energy.*
>
> *Jeffery (2005)*

A prediction of the neo-Darwinian paradigm would be that the loss of eyes during the embryonic development would be associated with a downregulation of expression of oculogenic genes. Contrary to this prediction, many of these genes are upregulated in the cavefish rather than in the surface fish (Jeffery, 2005).

The *neutral mutations hypothesis* is equally unfit for explaining loss of eyes in *A. mexicanus*. According to that hypothesis, under conditions of darkness in caves, where sight is not useful, mutations in genes that are involved in eye formation, but do not affect the development of other structures, might accumulate through the genetic drift. The latter would make it possible for neutral mutant alleles to be fixed in cavefish populations. But even theoretically, the genetic drift would need evolutionarily long periods of time "to fix eyeless alleles," whereas the loss of eyes in cavefish, which occurred four times in *A. mexicanus*, took only a "moment" (*ca.* 10,000 years) by evolutionary standards.

Even if, for the sake of argument, one were to accept that theoretically it would be possible for neutral mutations for eyelessness to occur and accumulate, there is no evidence to suggest that alleles for eye loss are accumulated in the eyeless morph. On the contrary, as pointed out above, all of the genes involved in the eye formation of the cavefish have remained functional, as functional as in the epigean form. Hence,

> *experiments provide evidence against the neutral mutation hypothesis as an evolutionary mechanism for eye degeneration.*
>
> *Jeffery (2005)*

The hypothesis of energy conservancy is a hypothesis of indirect selection. It proposes that loss of eyes under conditions of darkness would offer a selective advantage by setting free energy for the development of sensory organs and other "constructive" traits that evolved in cavefish. That loss of eyes in cavefish is advantageous is a mere truism that does not contribute to our understanding of the causes of the loss. It is argued that

> *several lines of evidence argue against the possibility that cavefish eye development is blocked to conserve energy. First, cavefish males and females show the same degree of eye reduction, although the high cost of egg production might be expected to dictate a greater degree of eye reduction in females, as has been reported in cave-adapted beetles. Second, cave fish populations inhabiting pools under bat colonies do not appear to be food-limited, yet they show significant eye regression. Third, the manner of eye degeneration in* Astyanax *cavefish does not appear to be economical. Instead of undergoing eye loss at a very early stage, the cavefish eye develops to a relatively mature stage prior to the beginning of degeneration, presumably at high energetic cost.*
>
> *Jeffery (2005)*

Finally, the hypothesis that sees the evolutionary eye loss as a byproduct of the need for better feeding apparatus in caves (Jeffery, 2005) explains the benefits of the evolutionary change but does not address the most essential fact of the evolutionary change, i.e., whether the new information for the eye loss is mutational (which the author of the hypothesis denies) or epigenetic.

In summary, it may be said that all of the neo-Darwinian hypotheses presented above fail to account for the exceptionally rapid and repeated loss of eyes in *A. mexicanus*.

Epigenetic Explanation

Before considering the possible developmental mechanisms of evolution of eyelessness in cavefish, let us glimpse the recent evidence suggesting that eyelessness in fish is an evolutionarily plastic trait. Some populations of *A. mexicanus*, and other cavefish as well, show a remarkable polyphenism; within the same population, blind, eyed, and intermediate eye morphologies exist (Romero and Green, 2005).

Exposure of the larvae of the eyed, eyeless, and hybrid forms of *A. mexicanus* to light or darkness for 1 month leads to dramatic phenotypic changes, such as development of eyes in the eyeless form and enlargement of eyes in the eyed form, suggesting that the photic stimulus influences the developmental pathways of eye formation (Figure 14.14). Remember, the only known way that light may influence developmental pathways is the neural way.

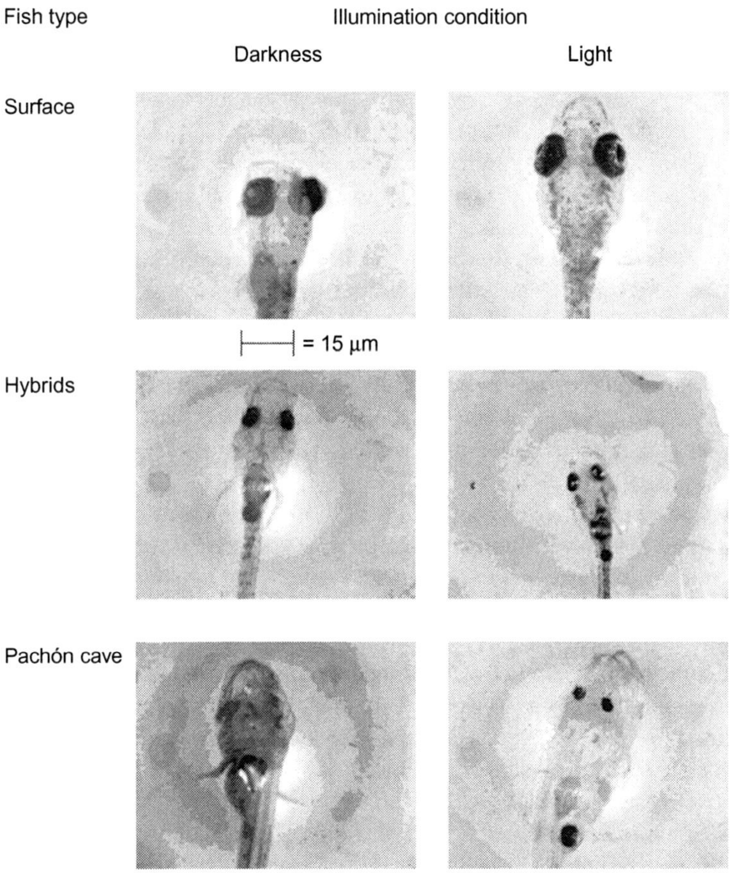

Figure 14.14 Variation in developmental responses to light exposure of larval surface, cave, and hybrid *Astyanax fasciatus*. Larvae were reared in continuous darkness or continuous light for 30 days, beginning when they were 24 h old. All three forms reveal an effect of light in the development of their eye tissues and the number of melanophores. The difference is particularly dramatic in the cavefish larvae.
Source: From Romero and Green (2005).

The ability of troglomorphic individuals to regain some of the lost eye tissue and pigmentation when experimentally exposed to light illustrates the retention of a substantial capability for phenotypic plasticity, even if under natural conditions they seem to represent an ecotype (Romero and Green, 2005). The fact that the initial stages of the development of the eye anlage, including development of the crystalline lens, take place normally in the eyeless morphs suggests that all of the basic genes involved in eye formation (*Pax6*, *Shh*, *Sox2*, and *Sox3*) are normal and functionally unchanged (Jeffery, 2005).

The other fact that in some other vertebrates, lens formation occurs only in the presence of the retina (Goss, 1969; Furuta and Hogan, 1998; Reza and Yasuda, 2004a,b; Arresta et al., 2005), and the fact that the best substitute for the eyecup in lens regeneration experiments is the adjacent brain (Goss, 1969) implies that neural signals are necessary for lens development. It is likely that a failure of the retina to send that neural signal may be the proximate cause of the arrest of development of lens in the hypogean form. In turn, suppression of the development of the lens in the cavefish is the proximate cause of the arrest, or even lack, of development of the iris, cornea, and retinal pigment epithelium, as is indicated by the fact that implantation of the epigean embryo lens in the optic cup of hypogean embryos induces formation of those optic structures in the presumptive eyeless cavefish.

Unlike the eyed epigean morph, in the hypogean embryos, the *Pax6* expression domains on both sides of the anterior midline of the neural plate are reduced in size and, consequently, so are their optic anlagen (Strickler et al., 2001). This reduction results from an expansion of the Shh expression domain in the midline of the neural tube (Jeffery, 2005). Not only no differences exist in the function of Shh gene, or its product, between the surface and cave forms of the *Astyanax* species, but the genomic structures of hedgehog (Hh) are also conserved from invertebrates to vertebrates (Wang et al., 2007).

Expansion of the Shh expression domain in the neural plate leads to the arrest of the crystalline lens development, to the programmed cell death of the lens vesicle and the overlaying presumptive cornea, and finally to the sinking of these optic structures into the orbits (Figure 14.15). Thus, the difference in the pattern of expression of *Pax6* in the incipient nervous system is the *earliest relevant difference* observed in the developmental pathways of sighted and blind morphs of *A. mexicanus*.

Nevertheless, this is not to say that the diverging point is the ultimate cause of the evolutionary loss of eyes. The expansion of the *Shh* expression domain along the midline of the neural plate is considered to be necessary for inducing degenerative processes that lead to regression of the developing eye, starting with the lens apoptosis. According to Yamamoto et al. (2004), expansion of hh signaling results in hyperactivation of downstream genes, lens apoptosis, and arrested eye growth and development. This is corroborated by the fact that these features can be mimicked in the surface fish by overexpressing *twhh* and/or *shh*, supporting the role of *hh* signaling in the evolution of cavefish eye regression. It is noteworthy that *twhh* gene is exclusively expressed in the neural tube, in distinction from *shh* that is expressed both in the neural tube and notochord (Ekker et al., 1995). The *hh* expression domain

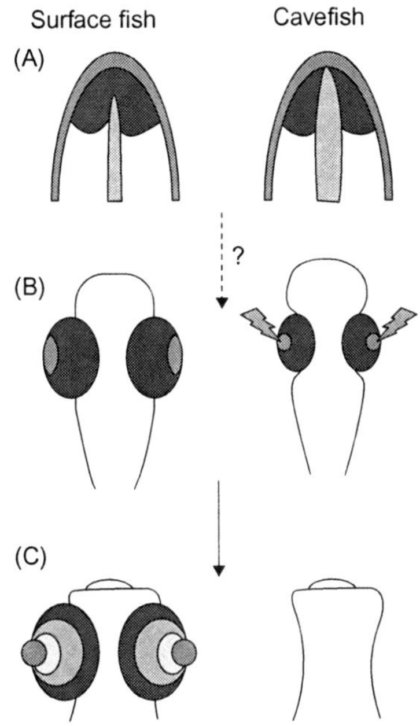

Figure 14.15 A possible mechanism for eye degeneration in cavefish. (A–C) Successive events of eye development and degeneration. (A) Neural plate stage (~10–12 h postfertilization (hpf)) when Shh midline signaling (green) is enlarged in cavefish, and the subsequent eyefield delineated by *Pax6* expression (purple) is smaller in cavefish. The lens will differentiate from the placodal field (nonneural ectoderm; pink) at the border of the neural plate. (B) Lens apoptosis in cavefish, starting at ~24 hpf, while the neural retina develops in a relatively normal fashion, with an attempt to generate retinal layers (concentric dotted lines).
(C) Adult stage, when the cavefish eyes are regressed, have sunk into the orbits and are covered by the skin.
Source: From Rétaux et al. (2008).

in the embryonic midline is almost twice as wide in blind than eyed fish, and the hh overexpression can phenocopy cavefish eye degeneration.

What, then, could induce expansion of the Shh expression domain in the midline of the neural plate of the hypogean form of *Astyanax*?

Bear in mind that the formation of the neural plate at the end of the gastrula stage takes place before the beginning of the large-scale expression of zygotis genes, implying that it is epigenetically determined by parental cytoplasmic factors deposited in gametes. Hence, the ultimate cause of the loss of eyes in *A. mexicanus* is epigenetic information provided parentally to gamete(s) in the form of cytoplasmic factors.

In our case, two likely scenarios of transmission of the epigenetic information for eyelessness in the offspring of cavefish may be imagined.

The first mechanism would relate this with transmission via gamete(s) of parental Shh, and the second would posit that the embryonic CNS is epigenetically programmed to produce increased amounts of Shh (it is known that the neural tube is a major producer of Shh; Hamade et al., 2006).

According to the first hypothesis, expansion of the Shh expression in the midline of the neural plate in cavefish results from parental Shh (or changes in the quantity/spatial distribution of the parentally provided Shh) in gamete(s). Provision of gametes with Shh is known to occur in metazoans, and in fish particularly. The zebrafish is known to deposit *Shh* mRNAs in the eggs (Chen et al., 2001), and so is the common carp (*Cyprinus carpio*) (Wang et al., 2007). If this were the case for the cavefish then, given the fact that Shh induces transcription of the *Shh* gene, it is tempting to believe that the parentally provided Shh protein might determine expansion of the Shh expression domain in the neural plate of the eyeless offspring.

According to the second hypothesis, the reduced domains of the *Pax6* expression in the neural plate result from increased secretion of Shh proteins by the midline of the prechordal plate (Jeffery, 2005) or by the neural floor plate and the notochord, as occurs in mice embryos (Thibert et al., 2003), under control of different regulators (Jeong and Epstein, 2003). Constitutive expression of hh in the ventral midline of the neural floor plate is crucial for dorsoventral patterning of the zebrafish brain (Ekker, 1995). It is noteworthy that the process of neural induction starts during the blastula stage, i.e., much earlier than thought so far, before the gastrula stage and formation of the mesoderm (Wessely et al., 2001; Kuroda et al., 2004). Ever-increasing evidence shows that the neural plate is not mesodermally induced by the Nieuwkoop center via Spemann organizer, but maternally determined by maternal factors deposited in the animal pole, as suggested by the fact that *Xenopus* embryos lacking mesoderm can still develop the CNS.

As pointed out earlier, formation of eye anlagen initially proceeds normally in cavefish embryos, and critical for the evolutionary loss of eyes in *A. mexicanus* is the programmed cell death of eye structures at the beginning of the stage of eye growth. It is demonstrated that the "process of eye degeneration is controlled by signals emanating from outside the eye itself" (Jeffery, 2005), that is, from the neural plate in the form of Shh (and twhh). It has been shown that Shh may act both as an inducer (Sanz-Ezquerro and Tickle, 2000; Yamamoto et al., 2004) and an inhibitor (Podlasek et al., 2007; Bond et al., 2008) of apoptosis. The treatment of the optic cup with *Shh* mRNA is demonstrated to induce programmed cell death in the eye structures of the eyed epigean form of *A. mexicanus*.

What may determine the inherited change of Shh expression pattern in the neural plate/neural tube?

There are at least two ways of regulation of Shh synthesis in the nervous tissue: secretion of RA (retinoic acid) and direct neural activity. Synthesis of neuronal RA is negatively regulated by neural activity and increases after the blockade of the neural activity (Aoto et al., 2008). Experimental evidence shows that Shh synthesis is negatively regulated by the neural activity, and neural activity blockade leads to increased Shh expression (Bond et al., 2008). In peripheral nerves, this results from the activity of another protein, hedgehog-interacting protein, that is transported by peripheral nerves to particular tissues, where it acts as a negative regulator of Shh (Angeloni et al., 2009).

Thus, the adaptive change in expression domain of Shh may be determined by the neural activity in the neural plate/neural tube during the cavefish embryogenesis.

It may be said that in both scenarios, the loss of eyes results from some specific changes in the epigenetic information:

- parental cytoplasmic factors) in gamete(s) of the hypogean form or
- neural modification of expression patterns of the Shh in the midline of the neural plate that leads to expansion of the expression of Shh in the neural plate.

In both scenarios, the eyelessness, as an evolutionarily new trait, is transmitted to the offspring by an as-of-yet unidentified, epigenetic change in the gamete(s).

The epigenetic transmission of evolutionary loss of eyes from parents to the offspring in *Astyanax*, via maternal cytoplasmic factors, is anything but surprising. Given the fact that the process of the deposition of maternal factors in the egg cell in a number of described cases in invertebrates (Handler and Postlethwait, 1977; Raikhel and Lea, 1985; Webb and Denlinger, 1998) and vertebrates (Lipar and Ketterson, 2000; Sockman et al., 2001; Gil et al., 2004; Hayward and Wingfield, 2004) is demonstrated to be neurally regulated, and neural regulation may be a general mechanism of deposition of those factors in metazoan egg cells (Cabej, 2004), it may be syllogistically concluded that the evolution of eye regression in the hypogean form of *A. mexicanus* is ultimately determined by parental neural mechanisms.

Based on the above facts and arguments, let us try to reconstruct tentatively the signal cascades and flow of the epigenetic information for eye loss in the hypogean form of *A. mexicanus*:

- The parental CNS determines a specific change in the spatial patterning and/or quantity of deposited maternal *Shh mRNA* and other dorsal axis–related *mRNAs* in the egg/sperm cell of the hypogean form. That such phenomena have occurred in the evolution of fish is empirically demonstrated by the recent evidence that the common carp (*C. carpio*) deposits hh transcripts in its eggs (Wang et al., 2007).
- Translation of the maternal *Shh* mRNA in early blastomeres of the animal hemisphere (Hainski and Moody, 1992; Pandur et al., 2002) and in the ectoderm of the presumptive neural plate leads to expanded expression of *Shh* gene in the neural plate.
- The expanded expression domain of the *Shh* in the anterior midline of the neural plate shrinks *Pax6* domain, causing underdevelopment of the eye anlage, the optic vesicle, and the optic cup (Strickler et al., 2001; Yamamoto et al., 2004).
- Neural signals from the optic vesicle (neural retina) induce the initial development of the lens vesicle and, consequently, lens-related structures (cornea, iris, and retinal pigment epithelium), from the ectoderm.
- Structural reorganizations in the embryonic brain, including midbrain and hindbrain, are involved in the process of eye loss, as is suggested by the fact that implantation of the lens from epigean embryos into the optic cup of hypogean embryos also leads to reorganization of those parts of the CNS (Soares, 2004).
- Increased secretion of Shh/twhh from the neural plate/CNS midline and neural retina induces apoptosis and degeneration of the lens (in the blind cave-dwelling fish, *Phreatichthys andruzzi*, eye degeneration, after initial rapid development, starts with a reduction in the rate of proliferation of neuroblasts in the retinal anlage; Berti et al., 2001), preventing the development of lens-dependent eye structures, thus leading to sinking of the eyes into the eye orbits.

The above tentative reconstruction of events leading to eye loss suggests that the new information necessary for the loss of this organ in the cavefish is neural by origin and, hence, epigenetic by nature. It results from a neurally determined change in the spatial organization of neuralizing/dorsalizing maternal factors in the eggs of the hypogean form.

Loss of Pigmentation in the Cavefish A. mexicanus

As a consequence of living in darkness, the hypogean morph of the teleost fish *A. mexicanus* lost not only its eyes but its pigmentation as well.

The body pigmentation in this fish depends on the presence of pigment cells, melanophores, in the skin. These pigment cells, as well as two other *Astyanax* types of pigment cells, iridophores and xanthophores, originate from neural crest cells that form in the neural keel, a structure that forms by the infolding of the neural plate (Papan and Campos-Ortega, 1994) under the influence of Hh signaling that affects the medial and lateral neurogenesis (Takamiya and Campos-Ortega, 2006).

Morphologically, melanoblasts in cavefish resemble melanophores and even are capable of producing melanin when provided with L-dopa (McCauley et al., 2004). Various cavefish populations differ widely from each other in the degree of depigmentation and in the proportion of melanophores to the total number of melanoblasts. While the pigmented epigean form of *A. mexicanus* has a 1:2 ratio of melanoblasts to melanophores, the Curva cavefish has a 8:1 ratio, and the Pachón cavefish has no melanophores at all and has the smallest number of melanoblasts of any cavefish (McCauley et al., 2004).

From a neo-Darwinian view, as early as 1957, Sadoglu hypothesized that depigmentation in *Astyanax* was related to a mutation in an unidentified gene. Later, it was hypothesized that mutations in two unidentified genes might be involved in the evolutionary depigmentation of the cavefish. Both hypotheses are incompatible with the fact that depigmentation in *Astyanax* is not an "all-or-none" process (melanophores are still produced at a low proportion), as would be expected when one or two unfunctional genes would be involved, but it is an ongoing epigenetic process of gradual loss of ability to differentiate melanoblasts into melanophores, as indicated by the wide range of variation of the melanophore-to-melanoblast ratio.

In an attempt to overcome such difficulties, it was later proposed that evolutionary depigmentation of *Astyanax* is a result of accumulation of neutral mutations, especially at a late step of the metabolic pathway of melanin synthesis. The fact that hypogean fish give birth to offspring that produce a proportion of melanophores rejects the hypothesis that neutral gene mutations may be involved in depigmentation of cavefish. Furthermore, melanoblasts in various populations of cave-dwelling *Astyanax*, including the one that produces no melanophores at all, synthesize melanin when provided with L-dopa, indicating that all of them have conserved the tyrosinase, which catalyzes various steps of melanin biosynthesis from tyrosine.

The facts that almost all of the depigmented *Astyanax* cavefish are capable of forming melanoblasts, and that melanophores and melanoblasts are capable of synthesizing melanin when provided with the amino acid neurotransmitter L-dopa, clearly show that there is no change in any gene that causes the evolutionary depigmentation and regression of pigment cells in these fish.

For these reasons, the new tendency is to see the depigmentation of the cavefish as an epigenetically determined evolutionary change. It is suggested that in cave morphs of *A. mexicanus*, the melanogenesis cascade is not blocked "because of a missing genetic component" but because of a nongenetic cause:

> *A permanent block in tyrosine accessibility seems to have occurred during cavefish evolution.*
>
> *McCauley et al. (2004)*

Loss of the Male Conspicuous Coloration in Lizards

Sexual dichromatisms, differences in body color according to the sex, have evolved in many phrynosomatid lizards, in which males have conspicuously blue-colored throat and belly as a courtship signal or as a warning to predators. Repeated losses, and less frequent gains, of sexually dichromatic coloration were found in a study on 130 lizard species. Loss of sexual dichromatism was found to be related to ground dwelling, due to increased predation in such habitats.

The loss of male conspicuous coloration seems paradoxical, given that the sexual selection would favor evolution and maintenance of male conspicuous coloration. It is argued that in this case, as well as in many other described cases of the loss of conspicuous coloration of plumage in birds, the cause of the loss of the male conspicuous coloration is not genetic but is a consequence of changes (reduction or loss) in female mating preferences (Wiens, 1999).

In turn, these preferences are determined by the mate recognition system, comprising sensory organs and their pathways to the CNS. As will be explained in some detail later (see Section Mate Recognition System in Chapter 19), changes in mate preferences result from inherited changes in the activity of specific neural circuits rather, not related to any genetic changes.

Loss of Sexual Dichromatism in Birds

In the northern hemisphere, some bird species are dichromatic, and some are monochromatic. Monochromatism is believed to be a derived character. Phylogenetical evidence shows that loss of dichromatism has occurred repeatedly in ducks, with a stronger tendency for losing rather than gaining it. The same is true for passerine birds, where dichromatism is lost three times more often than gained, and, according to Peterson (1996), in birds in general, dichromatism is lost five times more often than gained (Omland, 1997).

This seems to contradict the general neo-Darwinian belief that dichromatism is related to sexual selection for colorful conspicuous plumage. For this reason, some evolutionists, like Mayr (1942), resorted to gene (allelic) drift as a possible agent of the evolutionary loss of dichromatism. But the hypothesis is not substantiated. Observational evidence, for example, shows that ducks and mallard species that lose dichromatism retain the gene for the pigment in a functional state, as may be inferred from the persistence of the yellow-green color of the bill after the loss of plumage dichromatism. Besides,

> *monochromatic species in five of the six major clades of Anas (all except the green-wing clade) show evidence of vestigial features of the bright dichromatic plumage of their Northern relatives.*
>
> *Omland (1997)*

One plausible mechanism of the loss of dichromatism in birds would be an epigenetic mechanism involving changes in female sensory biases, followed by the action of natural selection. Not only has this hypothesis found greater empirical support, but it seems to account rationally for the frequent loss of female preferences (Ryan, 1998; Wiens, 2001), based on the relative evolutionary plasticity of the neural circuits determining animal behavior (see on the evolution of female preferences in Chapter 19).

Loss of Stages in Complex Life Cycles in Insects

Many aphids, including many species of Pemphiginae subfamily, show dispersal polymorphism.

Pemphigus betae is an aphid with a complex life history. It has a spring gall-forming phase on the narrowleaf cottonwood, *Populus angustifolia*, and a summer root phase on the secondary host plants of the genus *Rumex*. In autumn, with the drop in temperature as the only known cue, winged insects from root colonies fly to deposit their sexual generation on *P. angustifolia*. However, in response to crowding, *P. betae* may skip a phase (the first host, cottonwood tree) of its life cycle by producing a wingless parthenogenetic generation that feeds on roots of *Rumex* and goosefoot plants of the genus *Chenopodium*. This is the phenomenon of anholocycly. The non-migrating root colonies reproduce in the spring in the roots of the same *Rumex* plant.

Populations of this species in the Weber canyon, Utah, also show a clear tendency to switch to the reduced, one-host life cycle in the upper elevations of the canyon (Moran and Whitham, 1988; Moran et al., 1993). It is observed that even clones with identical histories and genotype show very different natural tendencies for producing winged migrants (Moran et al., 1993).

The mechanism of this radical change in the life history and in the morphology (winged/wingless individuals) is not known. What is certainly known is that no changes in genes are involved in producing it and that the development/suppression of wings in insects is *ultimately* neurally, i.e., epigenetically, determined. (See on the wing polyphenisms and experimental polyphenisms in insects in Chapter 10, and on the evolution of wings in insects in Chapter 13.)

Loss of Adult Stage of Development—Paedomorphosis in Insects

Many insects exhibit paedogenesis (neoteny), i.e., they reach sexual maturity during the larval stage and do not metamorphose into the adult form. Facultative paedogenesis in insects arose at least six times (four times in Diptera alone), twice independently in gall midges, *Heteropeza pygmaea* and *Mycophila speyeri* of the Cecidomyiidae (Diptera) family (Hodin and Riddiford, 2000). The only detectable difference between the paedomorphic and metamorphic species is larval expression of the functional ecdysone receptors, EcRs, and ultraspiracle (USP) in paedomorphic species.

From the neo-Darwinian point of view, changes in genes responsible for metamorphosis would be necessary for the parallel evolution of paedomorphosis in these midge species. The fact that the ecdysone pathway responsible for entering metamorphosis is conserved not only in species of the family Cecidomyiidae but across the insect taxa refutes that neo-Darwinian explanation.

An epigenetic explanation, based on the present knowledge of the neurohormonal mechanisms of metamorphosis in insects, seems to be plausible. The functional receptor (EcR + USP) responsible for metamorphosis in insects is activated by ecdysone secretion by the prothoracic gland (and by the activity of nerve endings), which in turn is cerebrally regulated by secretion of the neurohormone PTTH.

Loss of Terrestrial Mature Stage in Amphibians—Paedomorphosis

Loss of terrestrial stage by reaching reproductive maturity while still in a larval stage has occurred both in urodeles (salamanders and newts) and anurans (frogs). Salamanders of the genera *Necturus* and *Siren*, in North America, and *Proteus* (subterranean cave salamanders), in Europe, have completely lost the ability to metamorphose, hence they are known to be obligatory paedomorphic. In distinction from them, most salamander species of the genus *Ambystoma* are facultatively paedomorphic, i.e., under certain environmental or laboratory conditions they can switch from paedomorphosis to full metamorphic development.

The *Ambystoma tigrinum* complex consists of species of salamanders that, during the last few million years, have independently evolved several times obligate and facultative paedomorphosis from the ancestral metamorphic state (Shaffer and Voss, 1996; Figure 14.16).

The mechanism of paedomorphosis can be understood only in the context of the general mechanism of metamorphosis. Metamorphosis in salamanders is stimulated by a surge in the level of the hormone thyroxine determined by a signal cascade that starts in the salamander's brain (Figure 14.17). The timing of the activation of the cascade is determined by the

> *hypothalamic maturation comprising neurons of several regulatory centers and culminating at the time of the secretory surge.*
>
> *Rosenkilde and Ussing (1996)*

Paedomorphic salamanders fail to generate the characteristic burst of hypothalamic stimulation for activating the thyroid axis. This seems to be the main mechanism behind the axolotl (*Ambystoma mexicanum*) paedomorphosis (Rosenkilde and Ussing, 1996). The hypothalamus regulates reproductive morphology and physiology while evading its role as regulator of metamorphosis (Figure 14.18).

The regulatory role of the brain in the process of metamorphosis in salamanders is not limited to the activation of the hypothalamic–pituitary–thyroid axis (thyrotropin-releasing hormone, TRH → thyroid-stimulating hormone, TSH, → thyroid hormones, T3 and T4). There is evidence suggesting that, via the

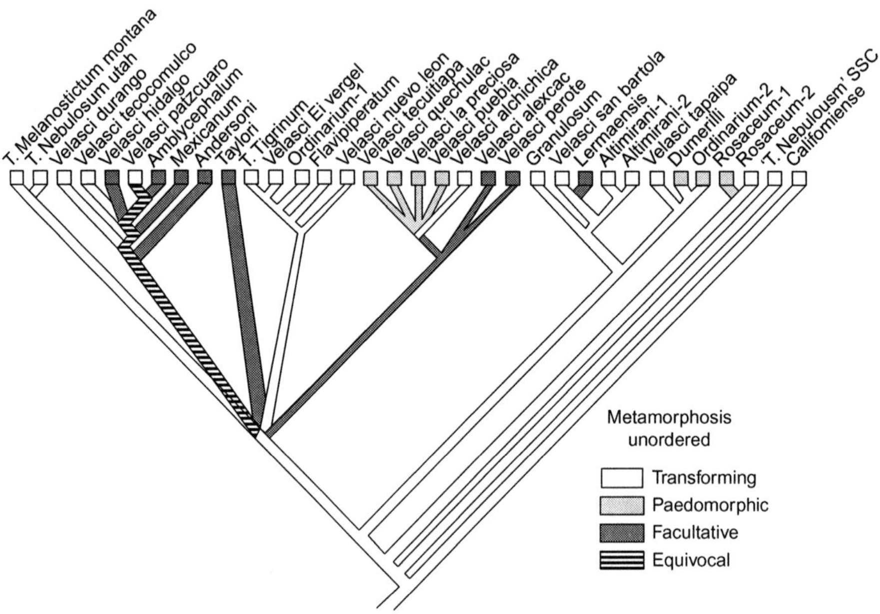

Figure 14.16 A reconstruction of the evolution of life history mode in the tiger salamander complex. Metamorphosis is treated as an unordered character with three states: transforming, facultative (both conditions found in a single population), and paedomorphic. Taxon names are the species or subspecies of *Ambystoma*, followed by the general locality of the sample. *Source*: From Shaffer and Voss (1996).

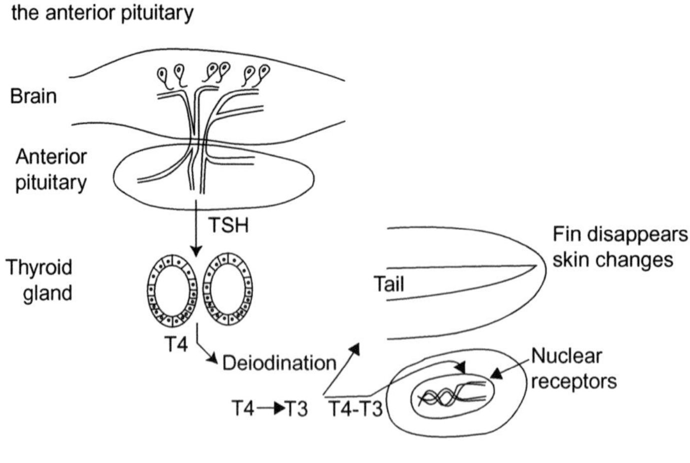

Figure 14.17 Neurohormonal mechanism of metamorphosis in salamanders. *Source*: From Rosenkilde and Ussing (1996).

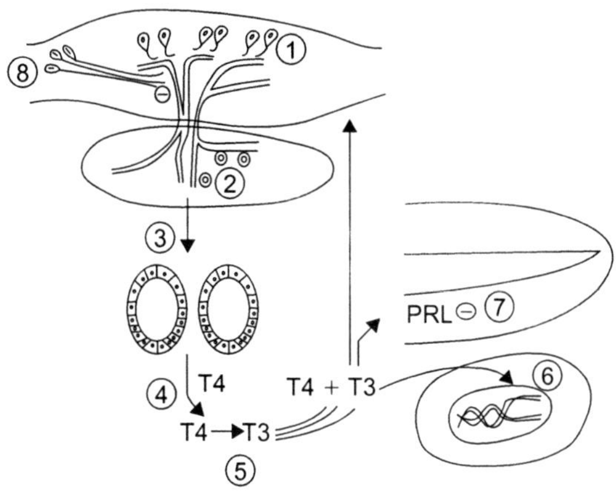

Figure 14.18 Diagrammatic representation of a mechanism of paedomorphosis in salamanders. The T4 surge occurring at the stage when toes differentiate shows (1) that TSH-stimulating neurons have matured and are able to secrete; (2) that the TSH neurons are able to secrete and stimulate the thyroid; and (3) that this gland is sensitive to TSH; and (4) able to secrete thyroxine to a high plasma level. The immersion experiments show that both young and older larvae are (5) able to respond to T3 with metamorphosis, but (6) the ability to activate thyroid hormone by deiodination of T4 to T3 is delayed, compared to metamorphosing species. Finally, two possible inhibitors are suggested by some experiments: Inhibition by prolactin, most probably (7) at the tissue level, or (8) a cerebral inhibitor, acting at the pituitary stimulating neurons. *Abbreviations*: PRL, prolactin; T3, the active deiodinated form of thyroxine; T4, thyroid prohormone—thyroxine; TSH, pituitary thyroid-stimulating hormone.
Source: From Rosenkilde and Ussing (1996).

hypothalamic–pituitary axis, the brain controls the antagonist effects of prolactin on metamorphosis and, via the hypothalamic–pituitary–adrenal axis, controls the agonist effect of corticoids (by increasing the number of T3 receptors) (Rosenkilde and Ussing, 1996).

Metamorphosis has been experimentally induced in paedomorphic salamanders by administration of thyroid hormones, but it can also be induced by manipulations at every level of the neurohormonal cascade. Thyroid hormone (T4) implanted in the brain is 10 times more active in inducing metamorphosis than when administered intravenously.

In addition to neurohormonal manipulations, experimental metamorphosis in paedomorphic salamanders is induced by stressful conditions (capture stress and conditions of captivity) that cause general disturbance in the CNS or by increasing the environmental temperature (Rosenkilde and Ussing, 1996).

Cases of spontaneous metamorphosis in paedomorphic salamanders have also been reported, corroborating the idea that no changes in genes are necessary for transition from metamorphosis to paedomorphosis and vice versa.

Neo-Darwinian Explanation

The fact that paedomorphic salamanders spontaneously or under stressful conditions can revert to the ancestral state of metamorphosis unequivocally proves that they possess the functionally intact ancestral "metamorphosis genes" and developmental mechanisms of metamorphosis. Hence, a neo-Darwinian explanation (e.g., gene mutations, recombination, or changes in allele frequencies) seems impossible.

Epigenetic Explanation

The essential question on evolution of paedomorphosis is: Where is the signal cascade that determines the metamorphosis disrupted in paedomorphic salamanders?

The fact that all of the hormones of the signal cascade for metamorphosis are normal and functional in these species suggests that the disturbance may be at the initial neural signals that activate the cascade. The theoretical inference that the disruption has occurred at a cerebral level is corroborated by empirical evidence:

1. Neurobiological disturbances in the brain, related to stressful conditions (capture and captivity), induce paedomorphic individuals to perform metamorphosis.
2. Brain implants are 10 times more efficient in comparison to systemic administration of hormones in inducing metamorphosis.

The hypothalamic neurons respond to the surge in thyroid hormone (Figure 14.19) by removing an inhibitor, thus enabling them to secrete TRH.

Why do these neurons not respond to the production of thyroxine in paedomorphic salamanders? Hypothalamic neurons self-activate and secrete TRH in response to low premetamorphic levels of thyroxine. The fact that they do not respond that way in paedomorphic salamanders suggests that the hypothalamus may have

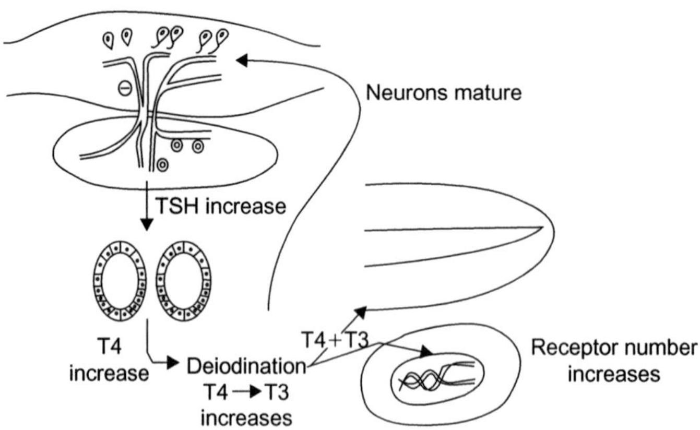

Figure 14.19 Focal points of the activity increase in the thyroid axis leading up to metamorphosis.
Source: From Rosenkilde and Ussing (1996).

adaptively heightened the set point for responding to the hormone. The changes in set points are a well-known epigenetic function of the hypothalamus in vertebrates.

Loss of Physiological Characters in *Drosophila melanogaster*

Loss of Resistance to Environmental Stressors in D. melanogaster

Under laboratory conditions, *D. melanogaster* lost rapidly (within 3 years) the resistance to environmental stressors, starvation and desiccation. For starvation, the mean time to 50% mortality declined from 50.1 to 35.9 h, and for desiccation it shifted from 14.3 to 9.8 h.

> *The rapidity of the response suggested that mutation accumulation could not account for it.*
>
> *Hoffmann et al. (2001)*

Selection for early reproduction as well leads to the loss of these traits, although that character is inversely related to the resistance to environmental stressors (Hoffmann et al., 2001), indicating that selection does not act on genes or genetic material.

The extraordinarily short time of the evolution of the above characters, under laboratory conditions, clearly suggests that a nongenetic mechanism is responsible for evolution of the above traits.

Loss of Behaviors

Loss of the Acoustic Startle Response in Moths Endemic to Bat-Free Habitats

It might be predicted that moths that are no longer under bat-predation threat, over time, will lose the ability to hear bat echolocation calls. The hypothesis is validated in a study on seven species of day-flying moths (Notodontidae: Dioptinae) that have evolved from species sensitive to hearing echolocation calls in Venezuela. These diurnal species are currently in different stages of the reduction or loss of hearing: two of them have normal ears, two have reduced hearing at bat-specific frequencies, and the remaining three exhibit advanced or complete loss of high-frequency hearing (Fullard et al., 1997).

Studies for testing the hypothesis that in moths of bat-free areas, gradual decrease of sensitivity of the auditory system and, over time, deafness will evolve have been conducted on noctuid moths in the Pacific islands of French Polynesia, Tahiti and Moorea, where no gene flow from populations of bat-inhabited areas has occurred. These islands have been bat-free since they emerged 0.25–1.75 mya (Fullard et al., 2007).

While moths that have recently immigrated to these bat-free islands have normal auditory sensitivity and flight behavior, moths that have anciently migrated to these same bat-free islands, although still in possession of ears, have lost the auditory sensitivity, exhibit partial deafness, and have lost the acoustic startle response (ASR). It is believed that the initial step in the process of the decline of the auditory sensitivity and loss of the flight interruption behavior has been "the decoupling of the sensory input from the neural pathways that evoke behaviour" (Fullard et al., 2004).

Neuroanatomical examinations of vestigial networks in other insects suggest that cellular events underlying this decoupling involve the sensory neurons (e.g., reduction in receptor cell terminal arborizations (Arbas, 1983a; Riede et al., 1990) and/or in the interneurons that process these inputs (Arbas, 1983b)). It was proposed that the anti-bat flight defenses of noctuid moths are bimodal with

> *the most sensitive auditory cell (A1) evoking controlled flight away from an approaching bat and the less sensitive cell (A2) activating the sudden erratic flight which constitutes the ASR. It is therefore possible that the extinction of ASR in Tahitian moths may be the result of a single regressive event at the level of A2 cell.*
>
> *Fullard et al. (2004)*

The ASR-evoking A2 neuron is not lost but is still present in the endemic Tahitian moths, and the only difference of this neuron with A2 neurons of moths with normal hearing at bat-specific frequencies is that in Tahitian moths, the A2 neuron has increased the auditory sensitivity set point (threshold) to ultrasounds from 25 to 30 kHz so that it responds with reduced firing to the bat echolocation call stimulus, thus failing to perform the ancestral ASR behavior. Moths that arrived earlier in Tahitian islands are in more advanced stages of the process of the loss of ears (Fullard et al., 2007).

The decoupling of the sensory input from the neural pathways evoking the ASR, a phenomenon that is also observed in cases of the loss of flight in insects, suggests that inactivation of circuits determining specific behaviors is the first step in the process of evolutionary loss of morphological characters in metazoans, and this is in line with the prediction of the epigenetic paradigm that evolution of a phenotype usually starts with changes in the behavior(s) related to that phenotype. In our particular case of moths in bat-free Tahiti islands, the loss of ASR behavior may be a prelude to an ongoing process of simplification or vestigialization of the morphology of the moth auditory system.

The fact that in numerous known and described cases, evolution of new behaviors, as products of evolution of new or modified circuits, is the first step in species recognition (see Section Mate Recognition System in Chapter 19) and evolutionary diversification, suggests that the process of the decline of the auditory sensitivity and loss of flight interruption behavior in Tahiti and Moorea islands, French Polynesia, may have started with the degeneration of neural circuits. Neural circuits are the most malleable components of the auditory system. As Fullard points out:

> *It could also be that the most "expensive" components of a functional auditory system exist within the CNS circuits to which it connects. These circuits, and the behaviors they control, might be lost or inhibited at an earlier evolutionary stage than the cheaper peripheral sensory structures that activate them.*
>
> *Fullard (1994)*

The hearing loss in moths of the bat-free Pacific islands may be an illustration of the normal process of the regression that precedes the evolutionary loss of structures. It starts with the evolutionary loss of behaviors regulated by specific neural circuits, i.e., involves epigenetic changes in neural circuits rather than changes in genetic information.

References

Adler, R. and Canto-Soler, M.V. (2007). Molecular mechanisms of optic vesicle development: complexities, ambiguities and controversies. Dev. Biol. *305*, 1–13.

Angeloni, N.L., Bond, C.W., Monsivais, D., Tang, Y. and Podlasek, C.A. (2009). The role of hedgehog-interacting protein in maintaining cavernous nerve integrity and adult penile morphology. J. Sex. Med. *6*, 2480–2493.

Aoto, J., Nam, C.I., Poon, M.M., Ting, P. and Chen, L. (2008). Synaptic signaling by all-*trans* retinoic acid in homeostatic synaptic plasticity. Neuron *60*, 308–320.

Arbas, E.A. (1983a). Neural correlates of flight loss in a Mexican grasshopper, *Barytettix psolus*. I. Motor and sensory cells. J. Comp. Neurol. *216*, 369–380.

Arbas, E.A. (1983b). Neural correlates of flight loss in a Mexican grasshopper, *Barytettix psolus*. II. DCMD and TCG interneurons. J. Comp. Neurol. *216*, 381–389.

Arresta, E., Bernardini, S., Gargioli, C., Filoni, S. and Cannata, S.M. (2005). Lens-forming competence in the epidermis of *Xenopus laevis* during development. J. Exp. Zool. A Comp. Exp. Biol. *303*, 1–12.

Aubret, F., Bonnet, X. and Maumelat, S. (2005). Tail loss, body condition and swimming performances in tiger snakes, *Notechis ater occidentalis*. J. Exp. Zool. A Comp. Exp. Biol. *303*, 894–903.

Behesti, H., Papaioannou, V.E. and Sowden, J.C. (2009). Loss of Tbx2 delays optic vesicle invagination leading to small optic cups. Dev. Biol. *333*, 360–372.

Bejder, L. and Hall, B.K. (2002). Limbs in whales and limblessness in other vertebrates: mechanisms of evolutionary and developmental transformation and loss. Evol. Dev. *4*, 445–458.

Belting, H.-G., Shashikant, C.S. and Ruddle, F.H. (1998). Modification of expression and *cis*-regulation of Hoxc8 in the evolution of diverged axial morphology. Proc. Natl. Acad. Sci. U.S.A. *95*, 2355–2360.

Berti, R., Durand, J.P., Becchi, S., Brizzi, R., Keller, N. and Ruffat, G. (2001). Eye degeneration in the blind cave-dwelling fish *Phreatichthys andruzzi*. Can. J. Zool. *79*, 1278–1285.

Bickford, D., Iskandar, D. and Barlian, A. (2008). A lungless frog discovered on Borneo. Curr. Biol. *18*, R374–R375.

Bond, C., Tang, Y. and Podlasek, C.A. (2008). Neural influences on sonic hedgehog and apoptosis in the rat penis. Biol. Reprod. *78*, 947–956.

Boulet, A.M., Moon, A.M., Arenkiel, B.R. and Capecchi, M.R. (2004). The roles of Fgf4 and Fgf8 in limb bud initiation and outgrowth. Dev. Biol. *273*, 361–372.

Brandley, M.C., Huelsenbeck, J.P. and Wiens, J.J. (2008). Rates and patterns in the evolution of snake-like body form in squamate reptiles: evidence for repeated re-evolution of lost digits and long-term persistence of intermediate body forms. Evolution *62*, 2042–2064.

Bronchti, G., Rado, R., Terkel, J. and Wollberg, Z. (1991). Retinal projections in the blind mole rat: a WGA-HRP tracing study of a natural degeneration. Dev Brain Res. *58*, 159–170.

Bronchti, G., Heil, P., Sadka, R., Hess, A., Scheich, H. and Wollberg, Z. (2002). Auditory activation of "visual" cortical areas in the blind mole rat (*Spalax ehrenbergi*). Eur. J. Neurosci. *16*, 311–329.

Cabej, N.R. (2004). *Neural Control of Development*. Albanet, Dumont, NJ, pp. 87–100.

Cannata, S.M., Bernardini, S., Filoni, S. and Gargioli, C. (2008). The optic vesicle promotes cornea to lens transdifferentiation in larval *Xenopus laevis*. J. Anat. *212*(5), 621–626.

Cernuda-Cernuda, R., DeGrip, W.J., Cooper, H.M., Nevo, E. and Garcia-Fernandez, J.M. (2002). The retina of *Spalax ehrenbergi*: novel histologic features supportive of a modified photosensory role. Invest. Ophthalmol. Vis. Sci. *43*, 2374–2383.

Chen, Y., Zhang, Y., Jiang, T.-X., Barlow, A.J., St., Amand, T.R., Hu, Y., et al. (2000). Conservation of early odontogenic signaling pathways in Aves. Proc. Natl. Acad. Sci. U.S.A. *97*, 10044–10049.

Chen, W., Burgess, S. and Hopkins, N. (2001). Analysis of the zebra fish smoothened mutant reveals conserved and divergent functions of hedgehog activity. Development *128*, 2385–2396.

Coates, M. and Ruta, M. (2000). Nice snake, shame about the legs. Trends Ecol. Evol. *15*, 503–507.

Cohn, M.J. and Tickle, C. (1999). Developmental basis of limblessness and axial patterning in snakes. Nature *399*, 474–479.

Darwin, C.R. (1859). *The Origin of Species by Means of Natural Selection or The Preservation of Favored Races in the Struggle for Life*, first edition. John Murray, London, first issue, p. 454.

Darwin, C.R. (1859). Ibid. p. 134.

Darwin, C.R. (1872). *The Origin of Species by Means of Natural Selection or The Preservation of Favored Races in the Struggle for Life*, sixth edition. John Murray, London (additions and corrections, p. 131).

Dowling, T.E., Martasian, D.P. and Jeffery, W.R. (2002). Evidence for multiple genetic forms with similar eyeless phenotypes in the blind cavefish, *Astyanax mexicanus*. Mol. Biol. Evol. *19*, 446–455.

Ekker, S.C., Ungar, A.R., Greenstein, P., von Kessler, D.P., Porter, J.A., Moon, R.T., et al. (1995). Patterning activities of vertebrate hedgehog proteins in the developing eye and brain. Curr. Biol. *5*, 944–955.

Fong, D.W., Kane, T.C. and Culver, D.C. (1995). Vestigialization and loss of nonfunctional characters. Annu. Rev. Ecol. Syst. *26*, 249–268.

Fullard, J.H. (1994). Auditory changes in noctuid moths endemic to a bat-free habitat. J. Evol. Biol. *7*, 435–445.

Fullard, J.H., Dawson, J.W., Daniel Otero, L. and Surlykke, A. (1997). Bat-deafness in day-flying moths (Lepidoptera, Notodontidae, Dioptinae). J. Comp. Physiol. A *181*, 477–483.

Fullard, J.H., Ratcliffe, J.M. and Soutar, A.R. (2004). Extinction of the acoustic startle response in moths endemic to a bat-free habitat. J. Evol. Biol. *17*, 856–861.

Fullard, J.H., Ratcliffe, J.M. and ter Hofstede, H. (2007). Neural evolution in the bat-free habitat of Tahiti: partial regression in an anti-predator auditory system. Biol. Lett. *3*, 26–28.

Furuta, Y. and Hogan, B.L.M. (1998). BMP4 is essential for lens induction in the mouse embryo. Genes Dev. *12*, 3764–3775.

Gans, C. (1975). Tetrapod limblessness: evolution and function corollaries. Am. Zool. *15*, 455–467.

Gil, D., Leboucher, G., Lacroix, A., Cue, R. and Kreutzer, M. (2004). Female canaries produce eggs with greater amounts of testosterone when exposed to preferred male song. Horm. Behav. *45*, 64–70.

Goss, R.J. (1969). *Principles of Regeneration*. Academic Press, New York, NY.
Greene, H.W. and Cundall, D. (2000). Limbless tetrapods and snakes with legs. Science *287*, 1939–1941.
Hainski, A.M. and Moody, S.A. (1992). *Xenopus* maternal mRNAs from a dorsal animal blastomere induce a secondary axis in host embryos. Development *116*, 347–355.
Hall, B.K. (1995). Atavisms and atavistic mutations. Nat. Genet. *10*, 126–127.
Hamade, A., Deries, M., Begemann, G., Bally-Cuif, L., Genêt, C., Sabatier, F., et al. (2006). Retinoic acid activates myogenesis *in vivo* through Fgf8 signalling. Dev. Biol. *289*, 127–140.
Handler, A.M. and Postlethwait, J.H. (1977). Endocrine control of vitellogenesis in *Drosophila melanogaster*: effects of the brain and corpus allatum. J. Exp. Zool. *202*, 389–402.
Hayward, L.S. and Wingfield, J.C. (2004). Maternal corticosterone is transferred to avian yolk and may alter growth and adult phenotype. Gen. Comp. Endocrinol. *135*, 365–371.
Heil, P., Bronchti, G., Wollberg, Z. and Scheich, H. (1991). Invasion of visual cortex by the auditory system in the naturally blind mole rat. Neuroreport *2*, 735–738.
Hodin, J. and Riddiford, L.M. (2000). Parallel alterations in the timing of ovarian ecdysone receptor and ultraspiracle expression characterize the independent evolution of larval reproduction in two species of gall midges (Diptera: Cecidomyiidae). Dev. Genes Evol. *210*, 358–372.
Hoffmann, A.A., Hallas, R., Sinclair, C. and Partridge, L. (2001). Rapid loss of stress resistance in *Drosophila melanogaster* under adaptation to laboratory culture. Evolution *55*, 436–438.
Hooven, T.A., Yamamoto, Y. and Jeffery, W.R. (2004). Blind cavefish and heat shock protein chaperones: a novel role for hsp90α in lens apoptosis. Int. J. Dev. Biol. *48*, 731–738.
Jeffery, W.R. (2005). Adaptive evolution of eye degeneration in the Mexican blind cavefish. J. Hered. *96*, 185–196.
Jeong, Y. and Epstein, D.J. (2003). Distinct regulators of Shh transcription in the floor plate and notochord indicate separate origins for these tissues in the mouse node. Development *130*, 3891–3902.
Kamachi, Y., Uchikawa, M., Collignon, J., Lovell-Badge, R. and Kondoh, H. (1998). Involvement of Sox1, 2 and 3 in the early and subsequent molecular events of lens induction. Development *125*, 2521–2532.
Kessel, M. (1992). Respecification of vertebral identities by retinoic acid. Development *115*, 487–501.
Kimchi, T. and Terkel, J. (2002). Seeing and not seeing. Curr. Opin. Neurobiol. *12*, 728–734.
Kollar, E.J. and Fisher, C.C. (1980). Tooth induction in chick epithelium: expression of quiescent genes of enamel synthesis. Science *207*, 93–95.
Kuroda, H., Wesely, O. and De Robertis, E.M. (2004). Neural induction in *Xenopus*: requirement for ectodermal and endomesodermal signals via chordin, noggin, beta-catenin, and cerberus. PloS Biol. *2*, p. e92.
Lande, R. (1978). Evolutionary mechanisms of limb loss in tetrapods. Evolution *32*, 73–92.
Lipar, J.L. and Ketterson, E.D. (2000). Maternally derived yolk testosterone enhances the development of the hatching muscle in the red-winged blackbird *Agelaius phoeniceus*. Proc. Biol. Sci. *267*, 2005–2010.
Mayr, E. (1942). *Systematics and the Origin of Species*. Columbia University Press, New York, p. 120.
McCauley, D.W., Hixon, E. and Jeffery, W.R. (2004). Evolution of pigment cell regression in the cavefish *Astyanax*: a late step in melanogenesis. Evol. Dev. *6*, 209–218.
McCune, A.R. and Carlson, R. (2004). Twenty ways to lose your bladder: common natural mutants in zebrafish and widespread convergence of swim bladder loss among teleost fishes. Evol. Dev. *6*, 246–259.

Mitsiadis, T.A., Angeli, I., James, C., Lendahl, U. and Sharpe, P.T. (2003). Role of Islet1 in the patterning of murine dentition. Development *130*, 4451–4460.

Mitsiadis, T.A., Chéraud, Y., Sharpe, P. and Fontaine-Pérus, J. (2003). Development of teeth in chick embryos after mouse neural crest transplantations. Proc. Natl. Acad. Sci. U.S.A. *100*, 6541–6545.

Moran, N.A. and Whitham, T.G. (1988). Evolutionary reduction of complex life cycles: loss of host-alternation in *Pemphigus* (Homoptera: Aphidae). Evolution *42*, 717–728.

Moran, N., Seminoff, J. and Johnstone, L. (1993). Genotypic variation in propensity for host alternation within a population of *Pemphigus betae* (Homoptera: Aphididae). J. Evol. Biol. *6*, 691–705.

Necker, R., Rehkamper, G. and Nevo, E. (1992). Electrophysiological mapping of body representation in the cortex of the blind mole rat. Neuroreport *3*, 505–508.

Omland, K.E. (1997). Examining two standard assumptions of ancestral reconstructions: repeated loss of dichromatism in dabbling ducks (Anatini). Evolution *51*, 1636–1646.

Pandur, P.D., Sullivan, S.A. and Moody, S.A. (2002). Multiple maternal influences on dorsoventral fate of *Xenopus* animal blastomeres. Dev. Dyn. *225*, 581–587.

Papan, C. and Campos-Ortega, J.A. (1994). On the formation of the neural keel and neural tube in the zebrafish *Danio (Brachydanio) rerio*. Roux Arch. Dev. Biol. *203*, 178–186.

Podlasek, C.A., Meroz, C.L., Tang, Y., McKenna, K.E. and McVary, K.T. (2007). Regulation of cavernous nerve injury-induced apoptosis by sonic hedgehog. Biol. Reprod. *76*, 19–28.

Quiring, R., Walldorf, U., Kloter, U. and Gehring, W.J. (1994). Homology of the eyeless gene of *Drosophila* to the small eye gene in mice and aniridia in humans. Science *265*, 785–789.

Raikhel, A.S. and Lea, A.O. (1985). Hormone-mediated formation of the endocentric complex in mosquito oocytes. Gen. Comp. Endocrinol. *57*, 422–433.

Rétaux, S., Pottin, K. and Alunni, A. (2008). Shh and forebrain evolution in the blind cavefish *Astyanax mexicanus*. Biol. Cell *100*, 139–147.

Reza, H.M. and Yasuda, K. (2004). Lens differentiation and crystallin regulation: a chick model. Int. J. Dev. Biol. *48*, 805–817.

Reza, H.M. and Yasuda, K. (2004). The involvement of neural retina Pax6 in lens fiber differentiation. Dev. Neurosci. *26*, 318–327.

Riede, K., Kaemper, G. and Hoefler, I. (1990). Tympana, auditory thresholds, and projection areas of tympanal nerves in singing and silent grasshoppers (Insecta, Acridoidea). Zoomorphology *109*, 223–230.

Romero, A. and Green, S.M. (2005). The end of regressive evolution: examining and interpreting the evidence from cave fishes. J. Fish Biol. *67*, 3–32.

Romero, A., Green, S.M., Romero, A., Lelonek, M.M. and Stropnicky, K.C. (2003). One eye but no vision: cave fish with induced eyes do not respond to light. J. Exp. Zool. B. Mol. Dev. Evol. *300*, 72–79.

Romero, A., Zhao, Y. and Chen, X. (2009). The hypogean fishes of China. Environ. Biol. Fish *86*, 211–278.

Rosenkilde, P. and Ussing, A.P. (1996). What mechanisms control neoteny and regulate induced metamorphosis in urodeles? Int. J. Dev. Biol. *40*, 665–673.

Roth, G., Dicke, U. and Nishikawa, K. (1992). How do ontogeny, morphology and physiology of sensory systems constrain and direct the evolution of amphibians. Am. Nat. *139*, S105–S124.

Roth, G., Nishikawa, K.C., Naujoks-Manteuffel, C., Schmidt, A. and Wake, D.B. (1993). Paedomorphosis and simplification in the nervous system of salamanders. Brain Behav. Evol. *42*, 137–170.

Roth, G., Nishikawa, K.C. and Wake, D.B. (1997). Genome size, secondary simplification, and the evolution of the brain in salamanders. Brain Behav. Evol. *50*, 50–59.

Ryan, M.J. (1998). Sexual selection, receiver biases, and the evolution of sex differences. Science *281*, 1999–2003.

Sanyal, S., Jansen, H.G., de Grip, W.J., Nevo, E. and de Jong, W.W. (1990). The eye of the blind mole rat, *Spalax ehrenbergi*. Rudiment with hidden function? Invest. Ophthalmol. Vis. Sci. *31*, 1398–1404.

Sanz-Ezquerro, J.J. and Tickle, C. (2000). Autoregulation of Shh expression and Shh induction of cell death suggest a mechanism for modulating polarising activity during chick limb development. Development *127*, 4811–4823.

Scanlon, J.D. and Lee, M.S.Y. (2000). The Pleistocene serpent *Wonambi* and the early evolution of snakes. Nature *403*, 416–420.

Sedmera, D., Misek, I. and Klima, M. (1997a). On the development of Cetacean extremities: I. Hind limb rudimentation in the Spotted dolphin (*Stenella attenuata*). Eur. J. Morphol. *35*, 25–30.

Sedmera, D., Misek, I. and Klima, M. (1997b). On the development of Cetacean extremities: II. Morphogenesis and histogenesis of the flippers in the spotted dolphin (*Stenella attenuata*). Eur. J. Morphol. *35*, 117–123.

Shaffer, H.B. and Voss, S.R. (1996). Phylogenetic and mechanistic analysis of a developmentally integrated character complex: alternate life history modes in ambystomatid salamanders. Am. Zool. *36*, 24–35.

Shashikant, C.S., Kim, C.B., Borbély, M.A., Wang, W.C.H. and Ruddle, F.H. (1998). Comparative studies on mammalian *Hoxc8* early enhancer sequence reveal a baleen whale-specific deletion of a *cis*-acting element. Proc. Natl. Acad. Sci. U.S.A. *95*, 15446–15451.

Skinner, A., Lee, M.S.Y. and Hutchinson, M.N. (2008). Rapid and repeated limb loss in a clade of scincid lizards. BMC Evol. Biol. *8*, p. 310.

Soares, D. (2004). Source localization of environmental cues in crocodilians and teleost. In: *The 7th International Congress of Neuroethology*, August 8–13, 2004, Nyborg, Denmark, PL6.

Sockman, K.W., Schwabl, H. and Sharp, P.J. (2001). Regulation of yolk-androgen concentrations by plasma prolactin in the American kestrel. Horm. Behav. *40*, 462–471.

Strickler, A.G., Yamamoto, Y. and Jeffery, W.R. (2001). Early and late changes in *Pax6* expression accompany eye degeneration during cavefish development. Dev. Genes Evol. *211*, 138–144.

Summers, A.P. and O'Reilly, J.C. (1997). A comparative study of locomotion in the caecilians *Dermophis mexicanus* and *Typhlonectes natans* (Amphibia: Gymnophiona). Zool. J. Linn. Soc. Lond. *121*, 65–76.

Takamiya, M. and Campos-Ortega, J.A. (2006). Hedgehog signalling controls zebrafish neural keel morphogenesis via its level-dependent effects on neurogenesis. Dev. Dyn. *235*, 978–997.

Tchernov, E., Rieppel, O., Zaher, H., Polcyn, M.J. and Jacobs, L.L. (2000). A fossil snake with limbs. Science *287*, 2010–2012.

Teyke, T. (1990). Morphological differences in neuromasts of the blind cave fish *Astyanax hubbsi* and the sighted fish *Astyanax mexicanus*. Brain Behav. Evol. *35*, 23–30.

Thewissen, J.G.M. and Bajpai, S. (2001). Whale origins as a poster child for macroevolution. Bioscience *51*, 1037–1049.

Thewissen, J.G.M. and Fish, F.E. (1997). Locomotor evolution in the earliest cetaceans: functional model, modern analogues, and paleontological evidence. Paleobiology *23*, 482–490.

Thewissen, J.G.M., Cohn, M.J., Stevens, L.S., Bajpai, S., Heyning, J. and Horton, W.E., Jr. (2006). Developmental basis for hind-limb loss in dolphins and origin of the cetacean bodyplan. Proc. Natl. Acad. Sci. U.S.A. *103*, 8414–8418.

Thibert, C., Teillet, M.A., Lapointe, F., Mazelin, L., Le Douarin, N.M. and Mehlen, P. (2003). Inhibition of intraepithelial patched-induced apoptosis by Sonic hedgehog. Science *301*, 843–846.

Tomarev, S.I., Callaerts, P., Kos, L., Zinovieva, R., Halder, G., Gehring, W. and Piatigorsky, J. (1997). Squid Pax-6 and eye development. Proc. Natl. Acad. Sci. U.S.A. *94*, 2421–2426.

Wang, N., Sun, Y-H., Liu, J., Wang, Y-P. and Zhu, Z.Y. (2007). Molecular characterization of common carp (*Cyprinus carpio*) Sonic Hedgehog and discovery of its maternal expression. Dev. Genes Evol. *217*, 299–305.

Weaver, M. and Hogan, B. (2001). Powerful ideas driven by simple tools: lessons from experimental embryology. Nat. Cell Biol. *3*, E165–E167.

Webb, M.-L.Z. and Denlinger, D.L. (1998). GABA and picrotoxin alter expression of a maternal effect that influences pupal diapause in the flesh fly, *Sarcophaga bullata*. Physiol. Entomol. *23*, 184–191.

Wessely, O., Agius, E., Oelgeschlaeger, M., Pera, E.M. and De Robertis, E.M. (2001). Neural induction in the absence of mesoderm: beta-catenin-dependent expression of secreted BMP antagonists at the blastula stage in *Xenopus*. Dev. Biol. *234*, 161–173.

Whiting, M.F., Bradler, S. and Maxwell, T. (2003). Loss and recovery of wings in stick insects. Nature *421*, 264–267.

Wiens, J.J. (1999). Phylogenetic evidence for multiple losses of a sexually selected character in phrynosomatid lizards. Proc. Biol. Sci. *266*, 1529–1536.

Wiens, J.J. (2001). Widespread loss of sexually selected traits: how the peacock lost its spots. Trends Ecol. Evol. *16*, 517–523.

Wiens, J.J. and Slingluff, J.L. (2001). How lizards turn into snakes: a phylogenetic analysis of body-form evolution in anguid lizards. Evolution *55*, 2303–2318.

Wiens, J.J., Brandley, M.C. and Reeder, T.W. (2006). Why does a trait evolve multiple times within a clade? Repeated evolution of snakelike body form in squamate reptiles. Evolution *60*, 123–141.

Wilkens, H. (1970). Der bau des auges cavernicoler sippen von *Astyanax fasciatus* (Characidae, Pisces). Beitrag zur Problematik degenerativer Evolutionsprozesse. Dev. Genes Evol. *166*, 54–75. (abstract).

Wu, Y.-p., Kita, K. and Suzuki, N. (2002). Involvement of human heat shock protein 90α in nicotine-induced apoptosis. Int. J. Cancer *100*, 37–42.

Yamamoto, Y., Stock, D.W. and Jeffery, W.R. (2004). Hedgehog signaling controls eye degeneration in blind cavefish. Nature *431*, 844–847.

Zhang, Z.M. and Yang, X.J. (2001). Temporal and spatial effects of sonic hedgehog signaling in chicken eye morphogenesis. Dev. Biol. *233*, 271–290.

15 Evolution by Reverting to Ancestral Characters

The most probable hypothesis to account for the reappearance of very ancient characters is that there is a tendency in the young of each successive generation to produce the long-lost character, and that this tendency, for unknown causes sometimes prevails.

Darwin (1859, p. 166)

Evolutionary Reversions: The Course of Evolution Is Not Unidirectional

Idea of reversion to ancestral states as a mode of metazoan evolution was embraced by Darwin (1859), who even spoke of a Law of reversion. He considered the occurrence of atavisms as a proof of the possibility and reality of reversion of ancestral characters (Darwin, 1859, p. 166).

After Darwin, in 1883, a Belgian biologist, Louis Antoine Marie Joseph Dollo (1857–1931), formulated the law of irreversibility of evolutionarily lost traits, known as Dollo's law in his honor. As defined by one of its coauthors, the Law says: "What in the course of ages has phylogenetically disappeared cannot again recur" (Hall, 1998). This law explicitly denies the possibility of evolutionary reversions.

At that time, under circumstances of wide gaps in the paleontological record, Dollo's law raised no doubts about its validity, and later it was incorporated into the general neo-Darwinian scheme of evolution. It was argued that a lost trait cannot reappear after long periods of time, because, in the absence of selection on genes responsible for the trait, genes will mutate to such an extent that would become non-functional. It was estimated that the silenced genes might retain their function for no longer than 6 million years (Marshall et al., 1994) because

> *degradation of genetic information is sufficiently fast that genes or developmental pathways released from selective pressure will rapidly become non-functional.*
>
> *Marshall et al. (1994)*

One should keep in mind that the estimation is made for a single gene, whereas in most cases more than one gene are involved in the development of morphological traits. Empirical evidence, however, has shown that genes may remain functional for periods of time longer than 6 million years. (Odontogenic genes in birds, for example, are still functional presently, ~80 million years after the loss of dentition in this vertebrate class.)

Epigenetic Principles of Evolution. DOI: 10.1016/B978-0-12-415831-3.00015-X

Now, more than one century after the formulation of Dollo's law, adequate solid evidence on reversion of ancestral structures is accumulated, and hardly any biologists would reasonably question the occurrence of evolutionary reversions, which now seem to have been a leitmotif in the evolution of the Animal Kingdom.

In the meantime, rare attempts to validate Dollo's law have been made. Bull and Charnov present what they call "7 *possible* examples" of irreversible evolution: all-female parthenogenesis (thelytoky), polyploidy, selfing in hermaphroditic populations, dioecy evolved from hermaphroditism, heteromorphic sex chromosomes, Muller's ratchet, and haplodiploidy.

It is noteworthy that their list does not include the irreversibility of *morphological* characters, the most visible aspect of evolution and diversity of the animal world. They believe that the irreversibility of the above non-morphological characters can be deduced from the uniqueness and irreversibility of the history of living organisms (Bull and Charnov, 1985). By inferring *possible* examples (!) of irreversible evolution "from the uniqueness and irreversibility of the history of living organisms," the authors only make a circular reasoning for the irreversibility of evolution is nothing but the irreversibility of the history of living organisms.

Theoretical arguments against the Bull and Charnov's examples aside, recent empirical evidence shows that one of the assumed impossible reversions, the re-evolution of sexuality from parthenogenesis, occurred in mites of the Crotoniidae family. These sexually reproducing mites evolved from the parthenogenetically reproducing ancestors of the family Camisiidae, and investigators believe that this case defies Dollo's law of irreversibility of evolution and proves that "parthenogenesis is not necessarily an evolutionary dead end" (Domes et al., 2007).

Bernard Rensch (1900–1990) considered two striking exceptions from the principle of the "phylogenetical irreversibility" as questionable (Rensch, 1960, p. 125). First, the transformation of the heterodont teeth (teeth of different morphology) of the primeval whales (*Archaeoceti*) into isodont teeth of modern whales, which represents a return to the reptilian dentition from which mammal heterodentition originated, and second, the reappearance of the undifferentiated type of vertebral column in snakes and slow worms (legless lizards of Anguidae and Amphisbaenidae families). However, he argued, "such reversibilities seem to be extremely rare in the major steps of transspecific evolution." Rensch's admission that these evolutionary reversions occurred contradicts Dollo's law. The difficulties in explaining their origin compelled him to "downgrade" Dollo's law into a "rule."

Rensch (1960) attempted to relate such phenomena of reversible evolution to the reversibility of mutations, but this would raise a serious theoretical objection. Such "macroreversions" are difficult to conceive as products of adaptive mutations (all of the experimentally produced mutations are deleterious or have no adaptive value) of a single gene, for such transformations require multiple adaptive mutations. Furthermore, there is no evidence of mutations in genes related to transition from heterodonty to isodonty and from differentiated to undifferentiated type of vertebral column. On the contrary, as shown earlier, the gene regulatory networks (GRNs) and genes involved in other transitions are conserved across the vertebrate taxa.

There is some confusion over the meaning of the "evolutionary reversion" of a species to its ancestral phenotype. Rensch (1960, p. 125) believed that what reappears during reversions is only the general appearance of the character, not the identical structure.

Probably every biologist would agree that "identical" return to ancestral features is inherently impossible. However, by recognizing the occurrence of evolutionary reversions, biologists do not imply any "identical" return to the ancestral state. When we speak of a trait, such as a fin, head, tail, or tooth, we use such words to describe structures that, being distinct in different animal species, are related to each other by the origin, function, and patterning. No principle of identicalness could apply in determining whether a structure is of the same kind, because no identical structures (*sensu stricto*) could ever evolve.

The concept of "identicalness" in our biological context is irrelevant and cannot be a defining criterion of evolutionary reversions. So, for example, huge as they are, differences between the head of a mammal and that of a fish, they are not essential enough to force us to look for new terms for describing the same organ in the two classes. When we say that phasmid insects have lost and regained their wings, this is a clear statement of evolutionary reversion of a structure with a specific function that was lost somewhere in phylogeny no matter whether, or how much, it differs from the ancestral structure. In this regard, evolutionary reversions are as real as losses of structures. It would not make much sense to look at the modern metazoans for structures that are "identical" to the ones their ancestors had. If the modern biology does not apply any criterion of "identicalness" to the study of homologous organs or parts, why should we expect evolutionary reversions to be identical to ancestral structures?

The impossibility of returning to a structure that would be identical to its ancestral state has its specific causal basis. It is related to the unavoidable differences that evolve over time in their developmental and genetic contexts. Remember that not only in different species and different individuals, but even in the different parts of the same organism, different developmental contexts may determine different patterns of gene expression and phenotypic outcomes. So, for example, the *pdm* and *apterous* genes show distinctive patterns of expression in wings and legs in *Drosophila* (Cohen et al., 1992; Ng et al., 1995; Averof and Cohen, 1997).

Hence, what is to be expected in the cases of evolutionary reversals is not "identicalness" to ancestral structures but rather recurrence of ancestral "design," with the last word used, in Webster's meaning, to describe "instructions for making something which leave the details to be worked out." It is namely a "common design" related to a common developmental pathway, executed under different developmental and genetic contexts, that makes us viscerally think of the locomotor appendages of reptiles, birds, and mammals as limbs, despite the obvious differences in their structure and morphology.

In the light of the modern knowledge on the relationship between species' genome and its morphology, it is not the genes or groups of genes involved in the formation of a biological structure that count, but the patterns of their expression. Evolutionary reversions would, thus, necessarily differ somehow from the ancestral original because:

1. The evolution of the genome implies quantitative and qualitative changes in genes and overall organization of the genome. It creates a new and *different genetic context*, which may affect relevant GRNs and the result of their activation.
2. The biochemical and cytological environments in which the products of genes will act and interact also may change in the course of evolution. A *different developmental context* will arise, which may also influence the phenotypic outcome.

Thus, the changed genetic developmental background would lead to unavoidable differences in the phenotypic results, and evolutionary reversions will not be identical to the ancestral structures. What reverts is the ancestral morphological design (in the above Webster's meaning) rather than an identical structure.

Hence, evolutionary reversion consists in reappearance of a phenotype that is similar, rather than identical, to the lost ancestral phenotype. It arises as a result of activation of a suppressed ancestral developmental pathway under conditions of the changed developmental and genetic context.

This definition allows us to predict that:

1. Evolutionary reversions can occur whenever an inactivated ancestral developmental pathway is be reactivated.
2. Evolutionary reversions may be reproducible and can be experimentally induced.

By the early 1970s, biologists came to realize that evolutionary reversions were induced by other than random mechanisms (Regal, 1977).

From present-day knowledge, it can be argued that extant species have conserved, in a functionally unaffected state, genes involved in the development of ancestral structures.

Confronted with numerous experimental and observational examples of evolutionary reversions in the animal world, some biologists have argued that reactivation should be restricted to genes that remain mutationally not inactivated for relatively short periods of time of up to 6 million years (Raff, 1996).

Evidence accumulated in the meantime is at variance with this prediction. The North American wildcat, *Lynx*, re-evolved a lost third cusp in the carnassial teeth of the lower jaw after 20 million years, and birds have conserved in a functionally intact state genes involved in teeth development for 70–80 million years after having lost dentition (Mitsiadis et al., 2003).

Most developmental pathways are remarkably conserved in metazoans. Hence, it would be evolutionarily advantageous to activate those conserved pathways than to re-evolve them. This is what some of the experiments mentioned by Raff seem to suggest. In one of those experiments, it has been demonstrated that a relatively simple treatment with testosterone in an all-female species of fish induces a number of complex processes: "reactivation" of "complex morphogenetic pathways" and reversion of the "lost" complex morphologies of "complex insemination apparatus" and male body proportions and pigmentation as well as male sexual behavior and spermatogenesis.

Another example: whales lost limbs 40–50 mya, but cases of atavistic reappearance of hind limbs, although in a reduced form, in whales show that genes, GRNs, and developmental pathways determining hind limb development are still present and functional in these marine mammals.

It is clear that, over time, genes unavoidably evolve via mutations, but rapidly accumulating evidence indicates that evolutionary changes of the phenotype result from specific changes in developmental pathways and in patterns of gene expression rather than on changes in genes.

Loss and reversion of a structure implies not silencing or loss of genes; most genes are conserved in the course of phylogeny because they are necessary for the development of other structures in the body. Hundreds and thousands of genes are involved in the development of each structure in the animal body (e.g., ~2,500 genes are involved in the development of eye), and most of those genes are involved in the development of most of the rest of animal structures.

Reviewing a number of experimental studies on the loss and induction of organs in amphibians, Hall came to the following conclusion:

> *An organ may be lost without loss of the entire developmental system for producing that organ ... Loss of organs is often mediated through modification (not loss) of inductive reactions.*
>
> *Hall (1998)*

Metazoan organisms develop (or prevent the ectopic development of) different structures by specifically activating different developmental pathways in different parts of the body, although the same genes are present all over the body. They succeed in doing this because they are capable to selectively switch off/on different developmental pathways in different regions of the body.

Atavisms: Ancestral Developmental Pathways May Be Conserved and Reactivated

Atavisms are sudden reversions to ancestral morphological features in very small proportions of individuals of a population. According to de Beer, the fundamental criterion of an atavistic structure is morphological resemblance to that of an ancestor, regardless of its genetic basis (Lande, 1978).

No "hypothesis of reverse mutations" could explain their origin; firstly, because no reverse mutation is known to occur systematically at frequencies some atavisms occur, and secondly, emergence of atavistic structures requires reactivation and occurrence of "useful" mutations simultaneously in more than one gene, and thirdly, there is no hint, let alone evidence, of any "atavistic" reverse mutations.

The suddenness of the appearance of lost ancestral structures taking place during atavisms proves that:

1. While losing structures, metazoans can still conserve developmental pathways for the lost ancestral structures and re-evolve them.
2. Reversion to ancestral structures does not require new or changed genes or genetic information.

Regarding the atavistic appearance of hind limbs in marine mammals, Raff (1996, p. 393) estimated that "relatively weak selection could lead to limb reduction and virtual loss in as little as one million years." He also estimated that atavistic appearance of hind limbs in marine mammals is retained for as long as 10^6 to 10^7 generations. Lande (1978) notes that the process of vestigialization of hind limbs in whales may have taken a few million years until they were lost ~40 mya, but atavistic recovery of hind limbs still occurs after this very long period of time.

Among atavisms recognized by Rensch in his *Evolution Above the Species Level* are the formation of a fourth toe (which is normally reduced) in guinea pigs, appearance of rudimentary hind limbs in whales and dolphins, formation of supernumerary nipples in mammals (Rensch, 1960, p. 125), secondary lack of shells in snails, and secondary development of a cap-shaped shell in snails such as *Ancylus* (Rensch, 1960, p. 126).

Rare cases of atavistic development of hind limbs have been reported to occur in the humpback whale, *Megaptera nodosa*, and the sperm whale, *Physeter catadon*, with an estimated frequency of 0.02% of the general population (Lande, 1978). In another study, 37% of a population of 72 individuals of minke whale, *Balaenoptera acutorostrata*, developed a bony femoral rudiment. *Balaena mysticetus* even develops a vestigial femur and tibia. Skeletal elements, distinct from rudimentary pelvic girdle, appear in humpback whales at a frequency of 1:5,000, and completely developed hind limbs have been observed in another whale (Bejder and Hall, 2002).

The number of digits is five in amphibians (some forms have four), but in mammals it varies from five (humans) to one (horse). It is well known that with a certain frequency, horses develop two additional toe bones, one on each side.

One in several hundred pintail ducks in the Kerguelen islands of the southern hemisphere shows many markings of the northern pintail, *Anas acuta* (Omland, 1997).

Musculus iliofemoralis externus (IFE) atavistically regularly reappears in individuals of many bird species in Hawaii, Australia, and New Zealand, and atavisms of *musculus caudiliofemoralis pars iliofemoralis* are recorded in birds in the USA and in the Tuamotu Archipelago in the Pacific Ocean (Hall, 1998, p. 290).

Evidence is presented on the anomalous reappearance of ancestral muscles in individuals of species of other birds and mammals that currently lack these muscles. Such is the case with *musculus caudiliofemoralis* observed on the left side of an individual of the bird *Artamus leucorhynchus*; *musculus abductor cruris caudalis* in the hind limb of the rodent jerboa (*Jaculus jaculus*); *musculus latissimus dorsi pars caudalis* in bird wings has been found on both sides and/or on one side in individuals of the passerine bird, *Thraupis palmarum*; two cases of re-establishment of *musculus IFE* are observed in birds of the family of sturnids (Raikow et al., 1979).

An atavism in humans is the sudden appearance of the "werewolf syndrome" (*congenital generalized hypertrichosis*, characterized by a very intense hair growth all over the human body). It is assumed that the developmental pathway for hair coverage was silenced after humans diverged from our primate ancestors, but occasionally it is reactivated to produce the atavism.

Neo-Darwinian Explanation of Atavisms

There is no serious *neo-Darwinian interpretation* of atavistic reversions, and none of the known neo-Darwinian mechanisms of evolutionary change (gene mutations, genetic recombinations, neutral mutations, gene drift, and the implied natural selection) is applicable as explanations for the occurrence of atavisms. Even if the highly speculative idea that genes have been silenced during the phylogeny may be reactivated to produce atavisms were to be proven to be true, it will not fit into the neo-Darwinian paradigm.

Epigenetic Explanation of Atavisms

One of the basic tenets of the epigenetic theory of evolution presented in this work is that loss of various phenotypic (behavioral, morphological, physiological, and life history) characters is not necessarily associated with loss or changes in relevant genes or developmental pathways. As is extensively shown in Chapters 10 and 11, switching of developmental pathways, involving no changes in genes, for producing alternative (in some cases inherited) phenotypes, is a widespread phenomenon in metazoans. Such switches to alternative developmental pathways, or even to ancestral developmental pathways in some cases, has been possible to induce experimentally (see on experimentally induced reversions later in this chapter), and these cases represent nothing less than experimental atavisms or reversions.

What takes place in the cases of the appearance of atavisms in nature is that a "forbidden" developmental pathway is unpredictably activated. From this perspective, atavisms can be considered to result from accidental activation of developmental pathways that have been switched off in the course of the species phylogeny.

Evolutionary Reversions in Nature

Reversion of Shell Coiling in Gastropods

Although evolution of shell coiling in gastropods has been associated with several adaptive advantages, it has been repeatedly lost and the taxa that lost shell coiling have been considered to be unable to revert to the coiled shell because of the developmental and constructional constraints that evolved after the loss of shell coiling, which presumably restricted the number of possible morphologies and prevented the reversion to regular coiling (Gould and Robinson, 1994).

However, reliable evidence shows that evolutionary uncoiling of shells in gastropods does not represent an evolutionary dead end. Paleontologists have described a great number of evolutionary convergences in the shell form of molluscs. As a result of the volcanic character of the Steinheim basin in southwestern Germany and of periodic appearance of warm springs, the snail *Gyraulus multiformis* has been subject to environment temperatures that varied widely. Accordingly (as a consequence of the appearance and disappearance of the warm springs), the snail changed its

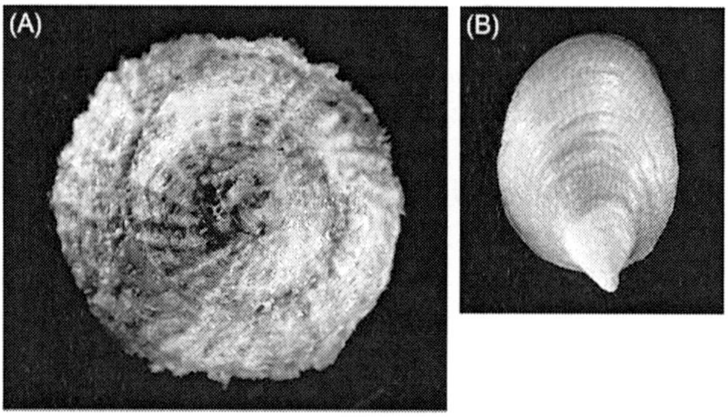

Figure 15.1 Coiled *Trochita calyptraeformis* (A) and a typically uncoiled, bilaterally symmetrical *Crepidula* species *Crepidula norrisiarum* (B).
Source: From Collin and Cipriani (2003).

shell form from that of a flat helix to a high helix and back to the original flat helix. Paleontological studies of Hennig and Fejervary have shown that the return to ancestral forms of shells is a widespread phenomenon in ammonites (Beurlen, 1937).

Reversions of coiled shells have been identified in two populations of *Trochita calyptraeformis* and possibly in three other species of *Zegalerus* and *Sigapatella* (*Zegalerus tenuis*, *Sigapatella terranovae*, and *S. novazealandica*). *Trochita* appeared first during Miocene, 23.8–20.5 mya, and paleontological evidence shows that it lost shell coiling during the evolution, whereas the modern *Trochita calyptraeformis* has re-evolved regular shell coiling (Collin and Cipriani, 2003; Figure 15.1).

Investigators believe that the mechanism of shell coiling is retained in the larval stage of these molluscs, and a heterochronic mechanism may have been operating in these cases. They also believe that

> *the genetic and developmental pathways for shell coiling have been retained in the larval stages of the uncoiled.*
>
> *Collin and Cipriani (2003)*

Reversion of Cartilaginous Skeleton in Fish

Embolomers and earliest stegocephals had fully ossified endocranium while later, in stegocephals and amphibians, endocranium turned cartilaginous like that of their earlier ancestors (Beurlen, 1937). Ostracoderms, the oldest class of fish, lived in freshwater for nearly 500 million years. They had no jaws and no paired fins, but they had bony plates and scales. Their survivors, the living cyclostomes, have no bones on the skin or any other place. It was thought that the "enzymatic complex necessary for the deposit of bone is no longer present" in cyclostomes (Kent, 1973). Cartilaginous fish derive from bony fish (Carter, 1967).

The cartilaginous skeleton of sharks also is a secondary development, and the bone seen in the skeleton of ostracoderms, placoderms, and the first bony fish is truly primitive (Colbert and Morales, 1991). The cartilaginous skeleton of sharks, thus, is a reversion to the cartilaginous skeleton of primitive chordates.

There is no scientific evidence to relate any changes in genes/DNA to these evolutionary reversions. The conservation of developmental programs of the lost characters and their reactivation suggests that epigenetic mechanisms are responsible for these evolutionary reversions.

Reversion of the Hydrodynamic Body Shape in Marine Mammals

The most impressive and well-known example of the re-evolution of ancestral body shapes is the adaptation of terrestrial mammals to the aquatic life. In the process of adaptation to aquatic habitat, they evolved fish-like shapes by shortening of the neck and head, addition of vertebrae, reduction of the pelvic girdle, loss of hind limbs, isolation of ear-bones, and other developments.

Biologists believe that the evolution of the first whale forerunners from their ungulate terrestrial mammal ancestors occurred about 60 mya. This is the estimated age of fossils of small primeval whales, which still possessed carnivorous heterodont dentition and articulate mobile cervical vertebrae. Fifty-two million years ago, a whale ancestor, *Ambulocetus natans*, existed that still used its hind- and forelimbs for locomotion. It was still able to walk on land, probably similar to modern sea lions, and to swim by undulating its spine and propelling itself by feet movements.

All cetaceans (e.g., whales, dolphins, and porpoises) are recognized to have evolved from terrestrial mammals. They have developed a fish-like body shape as well as fins and fluke. Within no more than 10 million years, these terrestrial mammals evolved into a whole order of marine mammals (Eldredge, 1989). Given that smaller morphological changes necessary for evolution of *Hyracotherium* to a horse took 50–60 million years (Wesson, 1991), it is reasonable to think that there should be some factor that facilitated the evolutionary reversion from terrestrial ungulates to modern aquatic mammals. As has been repeatedly pointed out, this may be related to conservation in an inactive form of the developmental pathways of their marine ancestors.

Such a major evolutionary event as this rapid appearance of a whole group of aquatic mammals would be unthinkable from the point of view of the neo-Darwinian gradualism that implies the occurrence and accumulation of numerous useful point mutations under the action of natural selection. Some calculations made by Wesson (1991) give a general idea of the improbability of evolution of the whale from its terrestrial ancestor via gene mutations:

> *By Mayr's calculation, in a rapidly evolving line an organ may enlarge about 1 to 10 percent per million years, but the organs of the whale-in-becoming must have grown about ten times more rapidly over 10 million years. Perhaps 300 generations are required for a gene substitution. Moreover, mutations need to occur many times, even with considerable selective advantage, in order to have a good chance of becoming fixed. Considering the length of whale generations, the rarity with which the needed mutations are likely to appear, and the multitude of mutations needed to convert a land animal into a whale, it is easy to conclude that gradualist natural selection of random variations cannot account for this animal.*
>
> *Wesson (1991)*

If the changes that occurred in the morphology, physiology, and behavior of marine mammals belong to the category of changes involving more than one gene, any evolutionist would rightly ask, "How could such an almost improbable event occur more than once, i.e., in whales, dolphins, porpoises, and seals?"

Miracles ruled out, no convincing answer could be given to the above question based on the tenets of the neo-Darwinian paradigm.

Reversion of Wings in Stick Insects

Reversion of ancestral wings in wingless insects has been considered impossible because of the complexity of the events that would enable their re-evolution:

> *Entomologists have long assumed that re-evolution of wings in apterous lineages was impossible, because functional wings require complex interactions among multiple structures, and the associated genes would be free to accumulate mutations in wingless lineages, effectively blocking the path for any future wing reacquisition.*
>
> *Whiting et al. (2003)*

The order of stick insects, phasmids, consists of 3,000 described species belonging to three families with about 500 genera. Sixty percent of the extant phasmid species are fully or partially wingless. Although their ancestral condition is wingless, phasmids have independently evolved wings at least 4 times (Whiting et al., 2003). Wings in phasmids did not evolve *de novo* but are the result of activation of an ancestral wing-patterning pathway. This was possible because the wingless insects have conserved the neural structures and basic functional circuitry required for flight (Whiting et al., 2003).

However, the study has been challenged by Trueman et al. (2004), who believe that, while representing an important advance in our understanding of insect and gene evolution, it is at variance with the long-held view that wings evolved only once in insects and have been repeatedly lost (Trueman et al., 2004). In turn, Whiting et al. responded to the critique by arguing that their analyses, as well as those of Trueman et al., suggest that both parsimony and likelihood methods support the notion that the ancestral state in stick insects has been wingless, and wings have independently evolved many times in this group, and wing reversions represent an evolutionary phenomenon that is more widespread than generally assumed (Whiting and Whiting, 2004).

Other biologists also believe that

> *the studies of these insects illustrate that the basic blueprints for complex developmental structures can remain largely intact even over large evolutionary spans (i.e., radiations of higher level taxonomic groups).*
>
> *Porter and Crandall (2003)*

Reversion of wings in phasmids involved no changes in genes. Despite the loss of wings, insects have conserved the wing-patterning pathway and GRNs and, as pointed out earlier, the neural circuitry necessary for flight. The reactivation of the wing-developmental pathway, in the absence of genetic changes, is epigenetically determined, as most regulatory processes of gene activation are.

Reversion of Mandible in a Collembolan Insect

Insect mandibles are complex characters that have been considered to be stable characters. However, insects of the Brachystomellidae family, order Collembola, have lost mandibles several times, independently and adaptively, as a result of switching to a sucking feeding behavior for ingesting food particles in liquid suspension. Recently, it has been reported that this complex character has also re-evolved at least once, in the case of *Probrachystomellides nicolaii* (Najt et al., 2005).

Reversion of Eyes in Ostracods

Two podocopid ostracodes (bivalved crustaceans), *Dutoitella lesleyae* Dingle and *Poseidonamicus whatley* Dingle, currently live in shallow water around Marion Island, between South Africa and Antarctica, at a depth of 113–474 and 240–355 m, respectively. Scientific evidence suggests that they have colonized this habitat by migrating from sea regions deeper than 600–900 m, where ostracodes are functionally blind. These sighted ostracodes evolved from blind ostracodes of Eocene (56–34 mya) in response to the lighter aquatic habitat they moved in (Dingle, 2003). The reversion of eyes in these crustaceans implies that the complex ancestral eye-patterning developmental pathways have been conserved for long evolutionary periods in deep sea–dwelling eyeless ostracodes, despite the occurrence of unavoidable changes in genes.

Reversion of Limbs in Snakes

An interesting instance of reversion of limbs in snakes is inferred from the study of a 95-million-year-old fossil snake from the Middle East. It represents the most extreme hind limb development seen so far in snakes. The limb consists of tibia, fibula, tarsals, metatarsals, and phalanges. The snake is nested with basal snakes, macrostomatans, which retained rudimentary hind limbs, and represents a reversion to the ancestral limbed state (Tchernov et al., 2000).

Reappearance of Musculus IFE in the Bowerbird Loria loriae *and the New Zealand Thrush,* Turnagra capensis

Musculus IFE is lost in most Passeriform birds. It occurred in Loria's Bird-of-Paradise (*Loria loriae*) and the New Zealand thrush (*Turnagra capensis*; Turnagridae). Based on extensive supporting evidence, Raikow et al. (1979) concluded that the loss of the muscle in these birds "was limited to its expression in the phenotype," and they believe that

> *the loss of a structure in the phenotype does not necessarily mean that the genetic information controlling its development has also been eliminated from the genome.*
>
> *Raikow et al. (1979)*

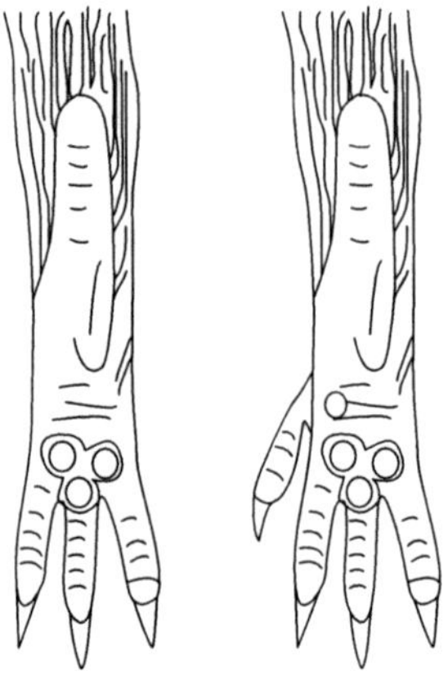

Figure 15.2 Sketches of left hind feet of guinea pigs. The normal three-toed condition is shown at the left. At the right is a foot with well-developed little toe and a corresponding plantar tubercle.
Source: From Wright (1934).

Reversion of an Ancestral Digit in Guinea Pigs

All of the species of the family of Caviidae, including the guinea pig, have lost one digit (digit I) in the front feet and two digits (I and V) in the hind feet. The fact that the same is observed in species of the related family of Hydrochoeridae suggests that the loss may have occurred in their common ancestor, i.e., about 20 mya. Nevertheless, it is not uncommon that individuals with an extra digit are born from normal guinea pigs. By selection and inbreeding of such individuals, by the beginning of the twentieth century, Castle succeeded in creating a strain of guinea pigs that constantly produced offspring with the same fully developed extra digit (Figure 15.2). The breeding results could be accounted for by neither recessive nor dominant inheritance (Wright, 1934).

Evolutionary Reversion of Life Histories

Shaffer studied a number of metamorphosing, facultative, and permanently paedomorphic (neotenic) ambystomatid salamander species from the Mexican highlands that diverged from their common metamorphosing ancestor some 0.5–1 mya. Paedomorphic species occasionally have been shown to revert to metamorphosing species (Schaffer, 1984).

Species of an amphibian group, *Stereospondylii*, showed the same disgust for the terrestrial life when, after 150 million years of amphibian status, they headed back to the sea to live a wholly aquatic life (Taylor, 1983).

No change in genes involved in the signal cascade determining metamorphosis has occurred in the paedomorphic and facultative paedomorphic salamanders.

Evolutionary Reversions in Experiments

Experimental Reversion of Teeth in Birds

Loss of teeth in birds occurred about 70–80 mya (Mitsiadis et al., 2003). Bird embryos develop only transient rudimentary thickenings of epithelium, reminiscent of dental thickening in other vertebrates. The avian mandibular neural crest-derived non-odontogenic mesenchyme (but not, for example, limb bud mesenchyme) can be experimentally transformed into dental mesenchyme, form ectopic tooth buds and express the same vertebrate odontogenic genes, *Msx-1*, *Msx-2*, and *Bmp-4*, after heterospecific recombination with early mouse embryo odontogenic epithelium (Wang et al., 1998).

Three decades ago, Kollar and Fisher (1980) grafted chick epithelium with mouse molar mesenchyme. They found ten aberrant structures consisting of dentin and odontoblasts in molar-like configurations. Complete teeth structure developed in four grafts. They came to the conclusion that the loss of teeth in Aves resulted not from a loss of genetic coding for enamel synthesis in the oral epithelium but from an alteration in the tissue interactions required for ontogenesis.

Later biologists succeeded to revert dentition in chicks by homotopic transplantation in chick embryos of specific segments of mouse neural tube. It was observed that, 1–2 days after transplantation, mouse neural crest cells migrate to the mandibular and maxillar regions of the chick embryo. Investigators believe that the main factor that led to the loss of dentition in Aves is the failure of their oral epithelium to express BMP4 (Chen et al., 2000). This suggests that the main functions of the mouse neural crest mesenchyme may be the activation of BMP4 expression in the chick oral epithelium:

> *Neural crest cells may play a role in the activation of BMP4 and Shh expression in tooth-forming sites of the murine oral epithelium.*
>
> *Mitsiadis et al. (2003)*

Chick–mouse chimera formed teeth that morphologically were of the mouse type (Mitsiadis et al., 2003). The fact that embryonic chick oral epithelium is able to properly interact with the GRN of the mouse showed that the GRN of the oral chick epithelium is conserved despite the very long time of the loss of dentition in Aves.

What is the relative role of oral epithelium and neural crest in developing dentition and shaping tooth morphology? Which is the sender of instructions, and which is the receiver of these instructions? Experimental evidence has shown that both the oral epithelium and the neural crest–derived mesenchyme are indispensable for tooth formation. Whereas the oral epidermis forms initial teeth primordia, the neural crest–derived mesenchyme provides the information necessary for their development into teeth of specific structure and morphology. The fact that from the same oral epithelium in the mandibular and maxillar Anlagen develop different teeth made investigators propose that teeth type differences are determined by

> *different origins of the neural crest cells that populate these components of the first branchial arch. Fate mapping in avian and mouse embryos shows that the mandible is mainly composed of CNC (cranial neural crest—N.C.) cells that migrate from the midbrain with some contribution from rhombomeres 1 and 2. The maxillary ectomesenchymal cells are derived from CNC cells migrating from both the midbrain and the forebrain. Such a difference in axial origin might explain the different responses of these cells to epithelial signals.*
>
> *Ferguson et al. (2000)*

Neo-Darwinian Explanation

From a neo-Darwinian view, the loss of dentition in Aves would result from gradual accumulation of mutations in genes that determine odontogenesis.

Even for traits that are determined by a single gene in metazoans, the maximal estimated time after which a silenced gene could be reactivated for producing the lost ancestral trait is 6 million years (Raff, 1996). Since a relatively large number of genes are involved in the development of most phenotypic traits, including teeth, reversion of dentition in birds would be possible only for periods of time shorter than 6 million years. Consequently, both the natural reversion and the experimental induction of teeth development in species of this vertebrate class would be impossible.

The experimental induction of teeth, about 60–80 million years after they are lost in birds, refutes the Dollo's law and the neo-Darwinian prediction that reversion of teeth in birds is impossible.

Contrary to the neo-Darwinian prediction, empirical evidence shows that even after many million years, these genes are present and functional in birds although mutations that do not affect the function of proteins they code for, unavoidably have occurred in the course of the long evolution.

Epigenetic Explanation

The fact that in experiments, the chick epithelium produced teeth of the mouse type shows that the mouse neural crest–derived mesenchyme provided the epigenetic information for odontogenesis and that the bird oral epithelium has conserved genes and gene networks determining odontogenesis. With changes in genes or genetic information excluded as possible cause of the loss of dentition in birds, the only logical alternative for explaining the loss of teeth in birds is a regulatory change in the properties of the neural crest cells migrating from the bird midbrain and forebrain to the mandibular and maxillar regions of the chick. Long ago, in birds, these neural crest cells ceased providing the epigenetic information necessary for inducing expression of Bmp4 and Shh in tooth-forming sites of the chick oral epithelium.

Reversion of Ancestral Genetic Systems in Insects

Genetic systems in insects are different and interchangeable. Along with the basal state of diplodiploidy (populations consist of individuals of two sexes, with each sex in possession of the diploid set of chromosomes), many insects exhibit thelytoky (Gr. *thèlys*—female and *tókos*—offspring) where workers or queens produce eggs that when unfertilized produce diploid females. In haplodiploidy, the offspring consist of two sexes, one of them diploid (females) and the other haploid (males). Haploid males are produced by unfertilized eggs (arrhenotoky) or by elimination of paternal chromosomes during spermatogenesis or after fertilization (pseudoarrhenotoky). This second form of haplodiploidy, the pseudoarrhenotoky, is still enigmatic: biologists have argued on selective advantages of haplodiploidy by elimination of paternal chromosomes but have been unable to learn anything about the mechanism of selective elimination of paternal chromosomes. An attempt has been made to explain elimination of paternal chromosomes with the activity of maternally transmitted bacteria by preventing chromosome decondensation in male-determining sperm nuclei of male zygotes (Normark, 2004).

Evolutionary transitions and reversions to haplodiploidy are known, but there is no evidence of the involvement of changes in genes or genetic mechanisms in their evolution. Thelytokous insect species are of recent evolutionary origin. Thelytoky was believed to have evolved from spontaneous mutations occurring in natural populations, but the discovery that Wolbachia infection can produce thelytoky in normal diplodiploid populations has invalidated that neo-Darwinian explanation.

Transitions from one genetic system to another and reversions to ancestral systems have occurred relatively often with a clear trend toward transitions to obligate all-female systems in insects (Normark, 2003).

Reversion of Life History Characters

Reversion of Ancestral Modes of Development in Gastropods

Species of calyptrate gastropods with feeding larvae may lose that stage and transform into direct-developing species. Direct-developing species with nurse eggs have the potential to transition to an alternative mode of development (Collin, 2004).

Although the loss of complex morphological characters has been considered irreversible, Collin (2004) came to the conclusion that three gastropod clades (*C. aculeata*, *C. onyx*, and *C. dilatata* groups) have evolved planktotrophy rapidly from the direct development, probably from ancestral groups of direct development with nurse eggs, as opposed to those with direct development from large eggs. *Crepipatella fecunda* also seem to have re-evolved biphasic development (Collin et al., 2007).

Recently, a case has been reported of reversion of complex development (egg → tadpole → adult) from direct development in *Gastrotheca* marsupial frogs (Wiens et al., 2007).

Reversion of Direct and Biphasic Development in Plethodontid Salamanders

In most amphibians, the life cycle comprises the aquatic larval phase and the terrestrial reproductive phase. However, in the course of evolution, many amphibian species switched to the direct development, depositing their eggs in land. So, for example, all of the more than 500 species of the genus *Eleutherodactylus* are direct-developing frogs, and the plethodontid salamanders have independently reversed to the direct development from the biphasic (aquatic and terrestrial) life cycle. Other amphibians have repeatedly reversed from metamorphosis to direct development. This has occurred in frogs and toads, gymnophyonans (caecilians), and caudates (salamanders).

Due to the complexity of the process, the evolutionary reversion from direct development to the biphasic life cycle with an aquatic free-living larval phase and a reproductive terrestrial phase has been considered "unlikely." Contrary to this belief, evidence suggests that at least 20 species of the desmognathine genus have recently reversed to the biphasic life cycle. Reversion to the ancestral biphasic life cycle in desmognathines occurred only in northeastern North America. It is speculated that the increased competition with terrestrial plethodontides exerted the evolutionary pressure necessary for the transition of desmognathines to the aquatic phase of the life cycle, which is estimated to have occurred ~10 mya.

It is suggested that the reinvasion of the aquatic habitat enabled the observed rapid diversification of desmognathines (Chippindale et al., 2004; Figure 15.3; Chippindale and Wiens, 2005).

Neo-Darwinian Explanation

No changes in genes or genetic information have ever been identified in relation to transition from the direct development to the biphasic life cycle in desmognathines. Hence, no neo-Darwinian mechanism of evolution (gene mutations, gene drift, and recombination) can offer an explanans for the repeated occurrence of evolutionary reversions from direct development to the biphasic life cycle with free-swimming larvae.

Epigenetic Explanation

Embryological studies have shown that the direct-developing desmognathines, and their plethodontine ancestors, retain in the egg the larval hyobranchial apparatus, which is essential for respiration and feeding in the water. In contrast, the closely related species of bolitoglossine plethodontids, which develop only very vestigialized hyobranchial apparatus, and species of the genus *Eleutherodactylus*, which do not develop that larval morphology in the egg, have never returned to the biphasic development (Chippindale et al., 2004).

Thus, conservation of hyobranchial apparatus in direct-developing desmognathines seems to have been the crucial factor enabling them to reverse to the ancestral biphasic (aquatic + terrestrial) life cycle. Ecological stress, resulting from the intense competition in the terrestrial habitat, is believed to have been the major external factor of this evolutionary reversion in the life history of salamanders (Chippindale et al., 2004).

With the genetic factors and genetic mechanisms excluded from the involvement in the evolutionary transition to biphasic life cycle, the remaining explanation is an epigenetic reactivation of the specific ancestral metamorphosis developmental pathway. Recall, metamorphosis and its developmental pathways in amphibians are under strict cerebral control, especially via the hypothalamus–pituitary–thyroid axis (see Section Neural Control of Metamorphosis in Amphibians in Chapter 5).

Reversion of Ancestral Reproductive Modes in Vertebrates

Amphibians have switched back from viviparity to oviparity (their eggs hatch in the environment), but an amphibian species of the Bufonidae family (order Anura),

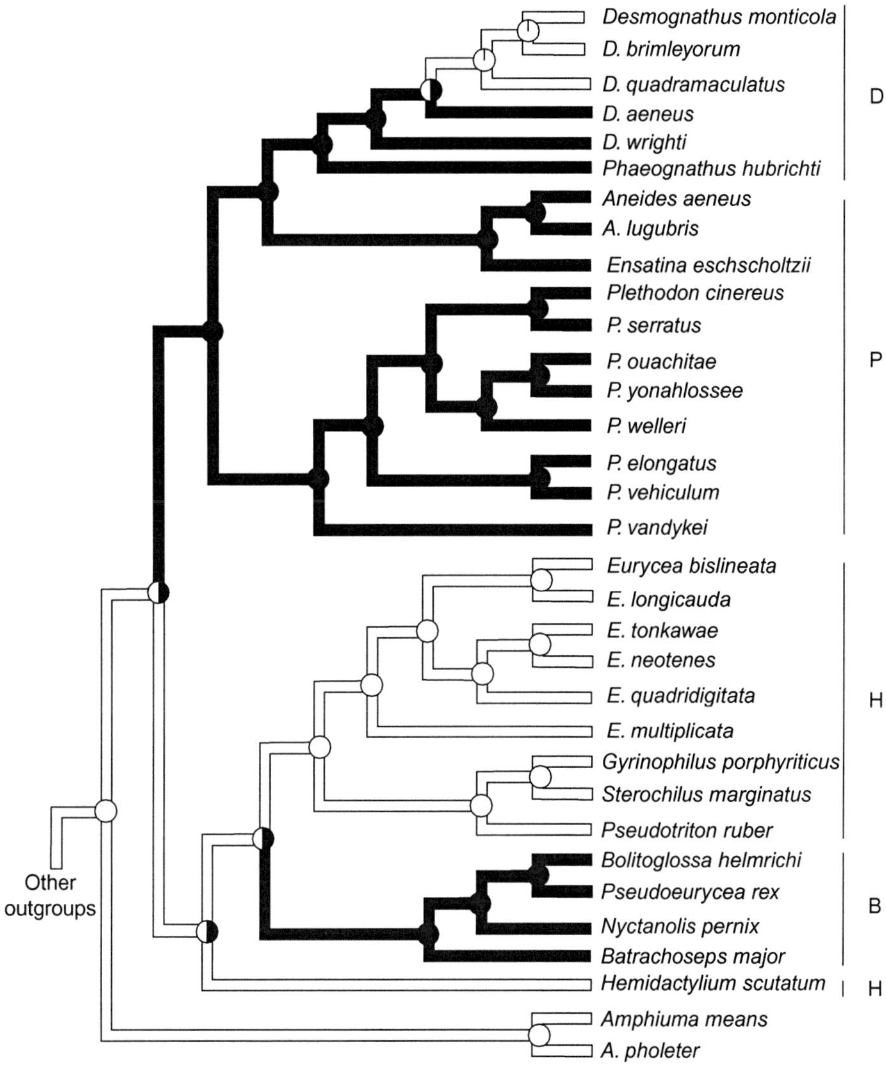

Figure 15.3 Phylogeny of plethodontid salamanders, showing parsimony and maximum likelihood-based reconstructions of ancestral developmental modes. Topology is that of the single most parsimonious tree based on 123 non-molecular and 2,998 mitochondrial and nuclear sequence characters. Branch shading reflects the single most parsimonious reconstruction for ancestral developmental mode with amphiumids coded as biphasic; light branches represent free-living aquatic larvae, and dark branches represent direct development. Pie charts at nodes indicate likelihood-based probability of biphasic life cycle (white) versus direct development (black). *Abbreviations*: B, Bolitoglossini; D, Desmognathinae; H, Hemidactyliini; and P, Plethodontini.
Source: From Chippindale et al. (2004).

Nectophrynoides viviparus, and populations of at least two salamander species (*Salamandra salamandra* and *S. algira*) are viviparous, with larvae remaining in the uteri, and young launched onto the land fully metamorphosed (Kent, 1973, p. 38). *Salamandra atra* secretes nutritive substances and produces eggs on which its viviparous young feed during the prenatal life.

As already mentioned, in 98 occasions reptiles (especially snakes) non-genetically switched back from oviparity to viviparity.

When ichthyosauri started aquatic life and could not make use of the sun warmth to hatch their eggs, they also switched back to viviparity, i.e., their eggs hatched inside mother's body. However, in some mammals, such as monotremes, reproduction remains oviparous. (They lay eggs, which hatch in the environment.) And still higher, placental mammals are remarkably adapted to a perfect viviparous development.

The strange loss and re-evolution of oviparity, ovoviviparity, and viviparity in vertebrate classes seem neither to have always been influenced by any evolutionary pressure nor to have gradually arisen. The repeated pattern of switching to alternative modes of reproduction suggests that vertebrates may have conserved ancestral developmental pathways responsible for ancestral modes of reproduction.

Reversion to Ancestral Oviparity in Sharks and Rays

Oviparity, a complex life history trait, is the ancestral reproductive mode in these fish groups, but currently most of their species are viviparous. Transition to viviparity in sharks and rays occurred independently in 12–15 cases. Two cases of reversion from viviparity to oviparity have been identified among these fish, in skates of the family Rajidae (25% of all species) and in the zebra shark, *Stegostoma fasciata* (Dulvy and Reynolds, 1997).

Reversion of Viviparity in Reptiles

It is commonly assumed that reptiles evolved viviparity from oviparity, but the reverse has not occurred. This reversion has been considered to be unlikely because it would entail the evolution of complex structural and physiological adaptations necessary for nutrition, oxygen supply, and special maternal hormonal mechanisms.

An analysis of the reproductive mode in reptiles has shown that it has changed a minimum of 49 times in squamates ("lizards" and snakes), with 35 forward transitions, 5 reversions, and 9 undetermined transitions (Lee and Shine, 1998).

Based on strong phylogenetic evidence, and the evolution of parity, it has been concluded that populations of the bimodal reproducing (viviparously and oviparously) European lizard species, *Zootoca vivipara*, are not monophyletic and that although viviparity evolved only once (with ancestral reproduction mode being oviparous), a number of reversions to oviparity occurred in various populations of this species (Figure 15.4). It is believed that repeated transitions to parity modes of this lizard are related to climatic changes (warmer climate favoring transition to oviparity, and colder climates, viviparity) that occurred during Pleistocene in the continent (Surget-Groba et al., 2006).

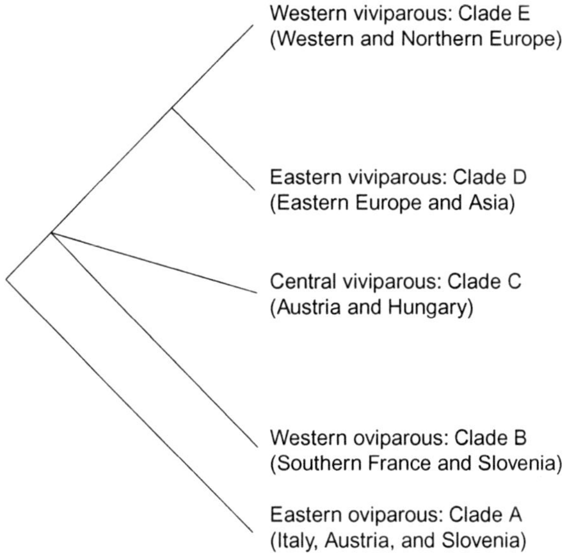

Figure 15.4 Phylogenetic relationships between the oviparous and viviparous strains of *Zootoca vivipara*.
Source: From Surget-Groba et al. (2006).

Reversion of oviparity in populations of a single species makes unlikely involvement of changes in relevant genes in the reversion and, consequently, makes neo-Darwinian interpretation inapplicable.

Re-Evolution of Oviparity in Bovides

Boas have evolved viviparity from an ancestral viviparity. Not long ago, biologists believed that the 41 species of the Old and New World boas are viviparous, but in the last decade, at least the Arabian sand boa, *Eryx jayakari*, and the Saharan sand boa, *E. muelleri*, are described to be viviparous. *Eryx jayakari* evolved oviparity about 60 million years after transition of the group to viviparity (Lynch and Wagner, 2010).

Reversion of Arboreal Carabides to the Ground-Dwelling Habitat

Carabides (ground beetles) represent a large family of terrestrial predators with more than 40,000 species. Arboreal carabides, which evolved from ground-dwelling forms, evolved some morphological characteristics (e.g., large hemispheric eyes, elongated prothorax; long elytra, long legs) as adaptations to conditions of living under bark or on leaves. It was generally believed that reversion from arboreality to the ground-dwelling life was impossible because of the impossibility of reversion to ground-dwelling morphology.

Recently, Ober (2003) found that reversion from arboreality to ground dwelling has occurred in all the phylogenies she studied, and she concluded:

> *Reversal may be a common evolutionary process, and evolution of new ecological interactions or evolution into new habitats may not inhibit further evolution or reversals.*
>
> *Ober (2003)*

Among other insects, weevils, a very large group of 60,000 species, being ancestrally (>200 mya) gymnosperm feeders, have shifted to angiosperm hosts and then back to gymnosperms (Marvaldi et al., 2002).

Experimental Reversion of Ancestral Characters

Experimental Reversion of Ancestral Characters in Drosophila

Strains of *Drosophila melanogaster* kept under laboratory conditions for decades (hundreds of generations) have diverged from the wild strain of origin in several biochemical, physiological, and life history characters. When these populations were experimentally returned to the ancestral environmental conditions, they reverted, to various degrees, to most of the lost ancestral characters (e.g., starvation resistance, reproduction time, developmental time, dry body weight, lipid content) within 20 generations (Teotónio and Rose, 2000; Teotónio et al., 2002).

Three different patterns of reversion to ancestral states for different characters were observed in these experiments. Flies reverted to ancestral type very rapidly (in several to 20 generations) for some characters, whereas the reversion for other characters required up to 50 generations, and for some characters the reversion was incomplete. The incomplete reversions were related neither to epistasis nor linkage disequilibrium nor to the absence or insufficiency of genetic variation, as is indicated by the fact that experimental hybrids did not exhibit higher reversibility (Teotónio and Rose, 2000).

Selection for late-life reproduction was associated with increased longevity and stress resistance. Reversion to the ancestral state of early-life reproduction occurred within 20 generations (Service et al., 1988), and by the hundredth generation, reversion to the ancestral state was observed for all of the studied characters (Graves et al., 1992).

Doubts have been expressed as to whether these convergences approximate the primitive state (Porter and Crandall, 2003), but, as argued earlier, no evolutionary reversion could produce phenotypes that are identical to the ancestral phenotype; morphological resemblance and functional similarity, rather than morphological identicalness, is the determining criterion of evolutionary reversions.

Neo-Darwinian Explanation

Elementary knowledge from genetics and evolution theory suggests that, in the above cases, the information necessary for "approximating the primitive state" is impossible to be acquired in an "evolutionary instant" of 20–50 generations. No gene mutations, gene recombinations, or changes in allele frequencies have been related to the rapid experimental reversions to ancestral characters in *D. melanogaster* under laboratory conditions. However, attempts to find a neo-Darwinian explanation without invoking the above neo-Darwinian mechanisms have been made. So, for example, as for the pattern of rapid reversion to ancestral state of starvation resistance, investigators write:

> *The rapid reversion of starvation resistance and early fecundity indicates that they were under influence of pleiotropic alleles generating a negative genetic correlation between these two characters, a correlation demonstrated in a sibling analysis by Service and Rose (1985). As a result of this correlation, these characters rapidly moved toward their ancestral values during reverse evolution, because selection focused on early fertility.*
>
> *Teotónio and Rose (2001)*

The authors do not elaborate on what this hypothetical "negative genetic correlation" consists in, nor do they illustrate it with any examples of how it might work. With no changes in genes and in genetic information, it is difficult to imagine why these "negative genetic correlations" arise in reverting populations but not the control populations. A tentative hypothesis for explaining the reversion (an unknown) with undemonstrated existence of pleiotropic alleles (another unknown) is devoid of explanatory power.

As for the slow incomplete reversion observed for some characters, we are told:

> *The slow response of these characters and their eventual convergence on ancestral values, after more than 100 generations in the ancestral environment, may have been a result of mutation accumulation, because this process is expected to affect evolution noticeably after a considerable time.*
>
> *Teotónio and Rose (2001)*

The fact that such mutations are not demonstrated to occur, that gene mutations have an extremely low frequency of occurrence, and that the frequency of "useful" mutations is still several orders smaller, all tell us that systematic reversion of these ancestral states in laboratory populations of *D. melanogaster* did not involve mutational changes. This is definitely felt by the investigators themselves, for immediately thereafter they admit:

> *There is no data as to the effect (additive or epistatic) of particular novel mutations.*
>
> *Teotónio and Rose (2001)*

Epigenetic Explanation

Discussing the causes of slower tempo of some evolutionary reversions, investigators inquired whether this may be related to the lack of genetic variability and epistasis. Based on their hybridization experiments, they conclude:

> *If lack of genetic variability was restricting reverse evolution, randomly mating hybrid populations should be freed of this constraint, because accumulation of identical genetic changes in populations of different evolutionary history is highly*

> *unlikely. Also, if epistasis led the derived populations to converge on strong evolutionary attractors, producing stasis under reverse evolution, the large perturbation to gene frequencies caused by hybridization should allow some stalled populations to escape from these attractor states. But the results showed no difference between uncrossed and hybrid populations.*
>
> *Teotónio and Rose (2001)*

The exclusion of genetic factors (gene mutation, genetic variability, genetic recombination, and epistatic interactions) as possible causative agents of reversion to ancestral states of investigated characters leaves open the possibility of the involvement of epigenetic factors in these evolutionary reversion experiments. However, lack of studies on the mechanisms of the examined characters makes it impossible to reconstruct the epigenetic mechanism of evolutionary reversion of these characters in *Drosophila*.

Several facts would justify focusing our attention on epigenetic factors as possible causal agents of evolutionary reversions induced by the return to ancestral conditions in laboratory strains of *Drosophila*.

First, the return to the ancestral environment of *Drosophila* populations under laboratory conditions implies that these populations are subject to an environmental stress that, according to this epigenetic theory of evolution, is a universal trigger of evolutionary changes in metazoans. As shown earlier, stress conditions sometimes lead to developmental instability and to behavioral changes, which generally are the first step in the process of morphological evolution. Needless to say, the only known way that environmental stressors influence the development and evolution of characters in metazoans is via the nervous system (see Chapter 7, Evolution and Stress Responses to Changes in Environment).

Second, the experiments on the evolutionary reversion of various biochemical, physiological, and life history characters in *Drosophila* spp. unambiguously show that these evolutionary events take place rapidly (reversion to ancestral states in some individuals occurs within a few generations), contrary to what is predicted by neo-Darwinian hypotheses. Empirical evidence in other species (e.g., phenotypic plasticity, predator-induced defenses, experimental reversion of "hip glands" in voles (see below)) shows that the only way of inducing such "sudden" inherited changes in phenotypic characters is by activating inactive ancestral pathways via neurohormonal cascades starting in the CNS. Hence, it is plausible that only conservation in inactive state of ancestral developmental pathways makes the extraordinary rapid reversals to ancestral characters in the *Drosophila* laboratory strains possible.

Experimental Reversion of "Hip Glands" in Voles

The presence of specialized sebaceous glands in the skin is characteristic of a number of mammal species, including small mammal voles. But two species of voles, *Microtus pennsylvanicus* and *M. longicaudus*, lack specialized posterolateral skin glands (hip glands).

By losing the glands, these species have conserved both genes and developmental pathways involved in their development. Three weeks after subcutaneous administration (by injection or implantation) of appropriate doses of the hormone testosterone, animals (seven of eight males of *M. pennsylvanicus*, and six of eight males and all of the five females of *M. longicaudus*) developed "hip glands" in species-specific regions: *M. pennsylvanicus*, toward the tail, in a region that is characteristic for *M. montanus*, and individuals of *M. longicaudus* developed "hip glands" in both, more anterior (toward the flanks) and more posterior (toward the tail) positions (Jannett, 1975).

This spectacular example of induction of the reversion of an ancestral morphological character demonstrates that developmental pathways for developing "hip glands" in these vole species are conserved, although the glands have been lost somewhere in the course of their phylogeny.

Neo-Darwinian Explanation

This experiment refutes any imaginable neo-Darwinian explanation; the emergence of "hip glands" took place within the lifetime of animals, thus excluding the involvement of neo-Darwinian mechanisms (gene mutations, gene drift, and genetic recombination) and is reproducible.

Epigenetic Explanation

The fact that a hormonal treatment induces formation in voles of a number of glands that the species has lost in the course of its phylogeny proves that a certain level of the testosterone is both necessary and sufficient for the development of these glands. Given the position of the hormones of the peripheral endocrine glands as downstream elements of signal cascades along the hypothalamic–pituitary axes, the reconstructed developmental cascade responsible for the formation of the "hip glands" in other vole species may look as follows:

neural signals from medial cortex → hypothalamic GnRH neurons
→ pituitary FSH → testosterone.

The signal cascade suggests that the absence of "hip glands" in *M. pennsylvanicus* and *M. longicaudus* is caused by a local block of the action of testosterone.

Reversion of Sexuality in Parthenogenetic Lizards

Treatment of developing embryos of a parthenogenetic all-female lizard species, *Cnemidophorus uniparens*, with fadrozole (CGS16949A) a potent inhibitor of aromatase activity in mammals, induces production of male offspring. Experimentally produced males develop normal male genital tract and are fertile. Production of *C. uniparens* male lizards indicates that the genes required for male sexual differentiation have not been lost or functionally changed in this parthenogenetic lizard (Wibbels and Crews, 1994). In other words, parthenogenetic females can produce male individuals and female individuals from the same genotype, i.e., no "male genes" are necessary for developing the male phenotype.

Treatment with the fadrozole on day 20 of incubation of *C. uniparens* embryos produced all-female offspring, whereas the same treatment on day 5 produced only males (Wennstrom and Crews, 1995).

Sudden experimental transformation of a whole all-female parthenogenetic population into sexually reproducing population (comprising male and female individuals) cannot be accounted for from a neo-Darwinian view: no changes in genes, no gene drift, no genetic recombination, and no selection are involved in this radical populational transformation.

The mechanism of reversion is clearly non-genetic, involving epigenetic changes in the sex-determining neurohormonal hypothalamus–pituitary–gonadal axis, including regulation of aromatase level by the brain. In this respect, it is somewhat illuminating that *Cnemidophorus uniparens* is an evolutionary descendant of the sexually reproducing species, *C. inornatus* and that some of the most relevant changes between them are discovered in the hypothalamic brain areas (Woolley et al., 2004).

Modified Ancestral Structures Reappear Stepwise

Evolutionary reversions, as opposed to the cases of transgenerational developmental plasticity, require a relatively large number of generations. As shown earlier in this chapter, experimentally induced evolutionary reversions of ancestral biochemical and life history characters (of very recent origin) in *Drosophila* take from several to 50 generations to occur. Moreover, not all of the individuals of a population may succeed in reverting to the ancestral character, implying that natural selection may have a critical role in the process of evolutionary reversions.

Given that the characters studied in the cases of experimental reversions in *Drosophila* are relatively simple and very recently acquired characters, it may be imagined that the time and the number of generations necessary for the reversion of complex traits is supposed to be greater, more so with reversions of Baupläne. The available paleontological evidence on evolutionary reversions (also the analysis of the available evidence from studies on animals assumed to be in the process of such evolutionary transformation) suggests that evolutionary reversion is a process, often a stepwise process, rather than an event.

It is generally assumed that metazoan Baupläne arose stepwise and that the observed stepwise reversion to ancestral Baupläne might reflect the way these ancestral Baupläne have evolved.

For illustrating the stepwise character of reappearance of ancestral morphologies, let us return to the well-known example of cetaceans (whales, porpoises, and dolphins). Paleontological record substantiates a number of intermediate forms that appeared in the course of their transformation into modern aquatic Ceataceans (see also Section Loss/Reduction of Limbs in Aquatic Mammals, Chapter 14).

Sequential adoption of five different modes of swimming behavior preceded five corresponding basic stages of morphological transformation of terrestrial quadrupeds into marine mammals with streamlined body (Thewissen and Fish, 1997).

First, the quadrupedal paddling of their terrestrial ancestors. It stimulated elongation of the body characterized by some modifications of vertebrae as reflected in the paleontological record by *Ambulocetus* and as it may be seen in present-day freshwater otters (Figure 14.5). The second mode of swimming was facilitated by the elongation of the body that enabled the pelvic paddling, which was followed by further elongation of the body but also by reduction of hind limbs as it is seen in the fossils of *Rodhocetus* or, *in vivo*, in sea otters (flattened head and tail, palmated feet, and webbed toes) (Figure 14.6). This made possible the transition to the third mode of swimming by pelvic undulation (undulation of the vertebral column) as the main propelling force during swimming, and made hind limb paddling less important, thus leading to further reduction of hind limbs and increased size of the tail. This morphology facilitated another mode of swimming by caudal undulation and undulatory movements of the entire vertebral column and the dorso-ventrally flattened tail. This was characteristic not only for *Rodhocetus* but also for *Dorudontidae*, the extinct family of whale ancestors. In modern mammals, this stage is exemplified by the giant South American freshwater otter, *Pteronura brasiliensis* (Figure 14.7).

By the Upper Eocene, an elongation of snout, comparable to that of the extinct fish-eating reptiles, occurred. The nostrils migrated dorsally toward the top of the skull while the number and form of teeth (except for front teeth) remained essentially the same as in the primitive placental skull. By the end of Eocene occurred the reduction and gradual loss of hind limbs, remodeling of the front limbs into short steering flippers, further vestigialization of hind limbs in many species, the re-emergence of the dorsal fin, transformation of the flattened tail into a horizontal fluke, adaptive modification of the hearing apparatus with transformation of the auditory ossicles into bulla, and even, in some species, the reversion of the ancestral pisciform dorsal fin.

The transition of terrestrial mammals to an aquatic fish-eating life began some time during lower Eocene in a long and stepwise process that was completed by the Upper Oligocene.

At the present time, we are probably witnessing a similar processes of gradual adaptation to a frilly aquatic life of some carnivorous (fish-eating) species such as seals (aquatic carnivorous mammals of the families Phocidae and Otariidae) and otters (several species of the genus *Lutra*) of which many species already have evolved aquatic adaptations, such as a relatively streamlined body, flattened head, short palmated feet, webbed toes, and horizontally flattened tail that already facilitates swimming.

With no changes in genes/GRNs, no genetic recombination or gene drift events related to the loss and reversion of limbs being ever suggested, any attempt to explain from a neo-Darwinian perspective evolutionary reversions observed in the morphology of aquatic mammals would be delusional.

References

Averof, M. and Cohen, S.M. (1997). Evolutionary origin of insect wings from ancestral gills. Nature *385*, 627–630.

Bejder, L. and Hall, B.K. (2002). Limbs in whales and limblessness in other vertebrates: mechanisms of evolutionary and developmental transformation and loss. Evol. Dev. *4*, 445–458.

Beurlen, K. (1937). *Die stammesgeschichtlichen Grundlagen der Abstammungslehre.* Gustav Fischer Verlag, Jena, pp. 32–34.

Bull, J.J. and Charnov, E.L. (1985). On irreversible evolution. Evolution *39*, 1149–1155.

Carter, G.S. (1967). *Structure and Habit in Vertebrate Evolution.* University of Washington Press, Seattle, p. 68.

Chen, Y-P., Zhang, Y., Jiang, T-X., Barlow, A.J., St. Amand, T.R. and Hu, Y., et al. (2000). Conservation of early odontogenic signaling pathways in Aves. Proc. Natl. Acad. Sci. USA *97*, 10044–10049.

Chippindale, P.T. and Wiens, J.J. (2005). Re-evolution of the larval stage in the plethodontid salamander genus *Desmognathus*. Herpetol. Rev. *36*, 113–117.

Chippindale, P.T., Bonnet, R.M., Baldwin, A.S. and Wiens, J.J. (2004). Phylogenetic evidence for a major reversal of life-history evolution in plethodontid salamanders. Evolution *58*, 2809–2822.

Cohen, B., McGuffin, M.E., Pfeifle, C., Segal, D. and Cohen, S.M. (1992). *apterous*, a gene required for imaginal disc development in *Drosophila* encodes a member of the LIM family of developmental regulatory proteins. Genes Dev. *6*, 715–729.

Colbert, E.H. and Morales, M. (1991). *Evolution of Vertebrates.* Wiley–Liss, p. 44.

Collin, R. (2004). Phylogenetic effects, the loss of complex characters, and the evolution of development in calyptraeid gastropods. Evolution *58*, 1488–1502.

Collin, R. and Cipriani, R. (2003). Dollo's Law and the re-evolution of shell coiling. Proc. Biol. Sci. *270*, 2551–2555.

Collin, R., Chaparro, O.R., Winkler, F. and Véliz, D. (2007). Molecular phylogenetic and embryological evidence that feeding larvae have been reacquired in a marine gastropod. Biol. Bull. *212*, 83–92.

Darwin, C. (1859). *The Origin of Species by Means of Natural Selection or the Preservation of Favored Races in the Struggle for Life*, first edition. John Murray, London, p. 196, first issue.

Darwin, C. (1859). Ibid. p. 166.

Dingle, R.V. (2003). Some palaeontological implications of putative, long-term, gene reactivation. J. Geol. Soc. London *160*, 815–818.

Domes, K., Norton, R.A., Maraun, M. and Scheu, S. (2007). Reevolution of sexuality breaks Dollo's law. Proc. Natl. Acad. Sci. U.S.A. *104*, 7139–7144.

Dulvy, N.K. and Reynolds, J.D. (1997). Evolutionary transitions among egg-laying, live-bearing and maternal inputs in sharks and rays. Proc. Biol. Sci. *264*, 1309–1315.

Eldredge, N. (1989). *Macroevolutionary Dynamics: Species, Niches, and Adaptive Peaks.* McGraw Hill, New York, NY, p. 23.

Ferguson, C.A., Yucker, A.S. and Sharpe, P.T. (2000). Temporospatial cell interactions regulating mandibular and maxillary arc patterning. Development *127*, 403–412.

Gould, S.J. and Robinson, B.A. (1994). The promotion and prevention of recoiling in a maximally snaillike vermetid gastropod: a case study for the centenary of Dollo's law. Palaeobiology *20*, 368–390.

Graves, J.L., Tolson, E.C., Jeong, C., Vu, L.N. and Rose, M.R. (1992). Desiccation, flight, glycogen, and postponed senescence in *Drosophila melanogaster*. Physiol. Zool. *65*, 268–286. (abstract).

Hall, B.K. (1998). *Evolutionary Developmental Biology*, second edition. Chapman & Hall, London, p. 357.

Jannett, F.J., Jr. (1975). "Hip glands" of *Microtus pennsylvanicus* and *M. longicaudus* (Rodentia: Muridae), voles "without" hip glands. Syst. Zool. *24*, 171–175.

Kent, G.C. (1973). *Comparative Anatomy of the Vertebrates.* The C.V. Mosby Co., Saint Louis, p. 30.

Kollar, E.J. and Fisher, C.C. (1980). Tooth induction in chick epithelium: expression of quiescent genes for enamel synthesis. Science *207*, 93–95.

Lande, R. (1978). Evolutionary mechanisms of limb loss in tetrapods. Evolution *32*, 73–92.

Lee, S. and Shine, R. (1998). Reptilian viviparity and Dollo's law. Evolution *52*, 1441–1450.

Lynch, V.J. and Wagner, G.P. (2010). Did egg-laying boas break Dollo's law? Phylogenetic evidence for reversal to oviparity in sand boas (*Eryx*: Boidae). Evolution *64*, 207–216.

Marshall, C.R., Raff, E.C. and Raff, R.A. (1994). Dollo's law and the death and resurrection of genes. Proc. Natl. Acad. Sci. U.S.A. *91*, 12283–12287.

Marvaldi, A.E., Sequeira, A.S., O'Brien, C.W. and Farrell, B. (2002). Molecular and morphological phylogenetics of weevils (Coleoptera, Curculionoidea): do niche shifts accompany diversification?. Syst. Biol. *51*, 761–785.

Mitsiadis, T.A., Chéraud, Y., Sharpe, P. and Fontaine-Pérus, J. (2003). Development of teeth in chick embryos after mouse neural crest transplantations. Proc. Natl. Acad. Sci. U.S.A. *100*, 6541–6545.

Najt, J., Weiner, W.M. and Grandcolas, P. (2005). Phylogeny of the Rachystomellidae (Collembola)—were mandibles ancestrally absent and did they reappear in this family? Zool. Scr. *34*, 305–312.

Ng, M., Diaz-Benjumea, F.J. and Cohen, S.M. (1995). Nubbin encodes a POU-domain protein required for proximal-distal patterning in the *Drosophila* wing. Development *121*, 589–599.

Normark, B.B. (2003). The evolution of alternative genetic systems. Annu. Rev. Entomol. *48*, 397–423.

Normark, B.B. (2004). Haplodiploidy as an outcome of coevolution between male-killing cytoplasmic elements and their hosts. Evolution *58*, 790–798.

Ober, K.A. (2003). Arboreality and morphological evolution in ground beetles (Carabidae: Harpalinae): testing the taxon pulse model. Evolution *57*, 1343–1358.

Omland, K.E. (1997). Examining two standard assumptions of ancestral reconstructions: repeated loss of dichromatism in dabbling ducks (Anatini). Evolution *51*, 1636–1646.

Porter, M.L. and Crandall, K.A. (2003). Lost along the way: the significance of evolution in reverse. Trends Ecol. Evol. *18*, 541–547.

Raff, R.A. (1996). *The Shape of Life—Genes, Development, and Evolution of Animal Form.* The University of Chicago Press, Chicago and London, p. 390.

Raikow, R.J., Borecky, S.R. and Berman, S.L. (1979). The evolutionary re-establishment of a lost ancestral muscle in the bowerbird assemblage. Condor *81*, 203–206.

Regal, P.J. (1977). Evolutionary loss of useless features: is it molecular noise suppression? Am. Nat. *111*, 129–133.

Rensch, B. (1960). *Evolution Above the Species Level.* Columbia University Press, New York, p. 125.

Schaffer, H.B. (1984). Evolution in a paedomorphic lineage. Evolution *38*, 1194–1206.

Service, P.M. and Rose, M.R. (1985). Genetic covariation among life history components: the effect of novel environments. Evolution *39*, 943–945.

Service, P.M., Hutchinson, E.W. and Rose, M.R. (1988). Multiple genetic mechanisms for the evolution of senescence. Evolution *42*, 708–716.

Surget-Groba, Y., Heulin, B., Guillaume, C.P., Puky, M., Semenov, D., Orlova, V., et al. (2006). Multiple origins of viviparity, or reversal from viviparity to oviparity? The European common lizard (*Zootoca vivipara*, Lacertidae) and the evolution of parity. Biol. J. Linn. Soc. *87*, 1–11.

Taylor, G.R. (1983). *The Great Evolution Mystery*. Harper & Row Publishers, New York and London, p. 61.

Tchernov, E., Rieppel, O., Zaher, H., Polcyn, M.J. and Jacobs, L.L. (2000). A fossil snake with limbs. Science *287*, 2010–2012.

Teotónio, H. and Rose, M.R. (2000). Variation in the reversibility of evolution. Nature *408*, 463–466.

Teotónio, H. and Rose, M.R. (2001). Perspective: reverse evolution. Evolution *55*, 653–660.

Teotónio, H., Matos, M. and Rose, M.R. (2002). Reverse evolution of fitness in *Drosophila melanogaster*. J. Evolution. Biol. *15*, 608–617.

Thewissen, J.G.M. and Fish, F.E. (1997). Locomotor evolution in the earliest cetaceans: functional model, modern analogues, and paleontological evidence. Paleobiology *23*, 482–490.

Trueman, J.W.H., Pfeil, B.E., Kelchner, S.A. and Yeates, D.K. (2004). Did stick insects really regain their wings? Syst. Entomol. *29*, 138–139.

Wang, Y-H., Upholt, W.B., Sharpe, P.T., Kollar, E.J. and Mina, M. (1998). Odontogenic epithelium induces similar molecular responses in chick and mouse mandibular mesenchyme. Dev. Dyn. *213*, 386–397.

Wennstrom, K.L. and Crews, D. (1995). Making males from females: the effects of aromatase inhibitors on a parthenogenetic species of whiptail lizard. Gen. Comp. Endocr. *99*, 316–322.

Wesson, R. (1991). *Beyond Natural Selection*. The MIT Press, Cambridge MA, London, p. 52.

Whiting, M.F. and Whiting, A.S. (2004). Is wing recurrence really impossible? A reply to Trueman et al. Syst. Entomol. *29*, 140–141.

Whiting, M.F., Bradler, S. and Maxwell, T. (2003). Loss and recovery of wings in stick insects. Nature *421*, 264–267.

Wibbels, T. and Crews, D. (1994). Putative aromatase inhibitor induces male sex determination in a female unisexual lizard and in a turtle with temperature-dependent sex determination. J. Endocrinol. *141*, 295–299.

Wiens, J.J., Kuczynski, C.A., Duellman, W.E. and Reeder, T.W. (2007). Loss and re-evolution of complex life cycles in marsupial frogs: does ancestral trait reconstruction mislead? Evolution *61*, 1886–1911.

Woolley, S.C., Sakata, J.T. and Crews, D. (2004). Tracing the evolution of brain and behavior using two related species of whiptail lizards: *Cnemidophorus uniparens* and *Cnemidophorus inornatus*. ILAR J. *45*, 46–53.

Wright, S. (1934). An analysis of variability in number of digits in an inbred strain of Guinea pigs. Genetics *19*, 506–536.

16 Neural Crest–Determined Evolutionary Novelties

The proper program of events governing the migration of crest may need first to be established in the hindbrain, to allow migratory crest cells to interpret and respond to environmental signals set up through a series of tissue interactions.

Trainor, Sobieszczuk, Wilkinson, Krumlauf (2002)

The neural crest is believed to have evolved from an invertebrate neural structure. It is critically involved in the development of numerous, often *de novo*, organs in vertebrates. To a large extent, the neural crest is responsible for the unprecedented rapid rates of vertebrate evolution.

As for its embryonic origin, neural crest cells and nerve cells derive from a common stock of precursor cells and differentiate almost simultaneously in the neural tube. Neural crest cells then delaminate from the neural tube to start an ordered migration throughout the animal body to their target sites where they direct formation of organs and other structures.

Empirical evidence shows that, before leaving the neural tube, neural crest cells are provided with information not only for finding their way through the maze leading to the target sites, but also on what they must transform themselves into, as well as for regulating differentiation and proliferation of cells in the target site. Evidently, the information that neural crest cells are provided with is epigenetic information (for they share the same genetic information with all of the rest of the cells throughout the animal body).

Neural crest cells provided the developmental repertory of vertebrates with a new "do-it-yourself" mechanism in addition to the "instructionist" mode of control of development accomplished by communicating developmental information via the brain–hypothalamic–pituitary–peripheral glands axes and local innervation. The neural crest cell–derived structures are among the most malleable structures in vertebrates.

Neural crest cells are essentially involved in the development of almost all vertebrate organs. The omnipresence of neural crest cells in the process of individual development, the accelerated evolution, and the malleability of the neural crest–induced organs indicate the important role that they played in the evolution of vertebrates.

Neural Crest–Determined Evolution in Vertebrates

The rise of vertebrates marks a new stage in metazoan evolution, characterized by accelerated evolutionary rates and increased complexity of structure and function with the morphological diversity as its most visible and amazing manifestation.

Epigenetic Principles of Evolution. DOI: 10.1016/B978-0-12-415831-3.00016-1

Transition from invertebrates to vertebrates is characterized not only by accelerated tempo of evolution but also by an unprecedented evolution of *de novo* structures.

These changes in the tempo and trends of evolution of vertebrates are inextricably related to the advent of the neural crest. All of the extant vertebrate species possess a neural crest, and the gene regulatory networks inducing formation of neural crest are conserved across the vertebrate taxa (Meulemans and Bronner-Fraser, 2004). The neural crest, as a specialized neural tube population of cells, is involved in the individual development both as a contributor of cells to various organs and parts throughout the animal body and as a source of inductive signals for morphogenesis and organogenesis at the target sites.

Origin of the Neural Crest

The neural crest develops from/on the neural tube at the area of its contact with ectoderm. Some authors believe that neural crest cells detach from the dorsal portion of the neural tube (Sohal et al., 1998). One strong argument in favor of this idea comes from the experimental evidence that, in response to ablation of the cranial neural crest (CNC), the neural tube regenerates neural crest cells with all of the capacities of the original, species-specific properties. The regeneration of the neural crest from the dorsal neural tube occurs only at the axial level of the ablated neural crest (Scherson et al., 1993).

Not only developmental evidence but phylogenetic evidence as well suggests that the neural crest is a derivative of the nervous system. According to Gans and Northcutt (1983), the evolutionary precursor of both the neural crest and neurogenic epidermal placodes is the epidermal nerve plexus of "protochordates." In ascidians, for example, nerve plexus neurons are identified, which, like the vertebrate olfactory placode neurons, migrate to the brain (Stone and Hall, 2004).

It seems plausible that the reason the neural crest took over the functions of mesoderm in the facial cranium has been the failure of the mesoderm to respond to the evolutionary pressure for morphological transformations of the head and jaws that the new predatory style of life of vertebrates required. Vertebrates responded to that pressure by evolving a neural structure, probably an epidermal nerve plexus, which evolved into the neural crest. Hall also believes that the neural crest could

> *have existed initially as an epidermal nerve plexus or net controlling ciliary function during movement and filter-feeding. With increasing muscle-based locomotion, the dorsal nerve cord took over innervative control of locomotion, freeing the epidermal nerve cells for other functions.*
>
> *Hall (1999)*

The neural crest appeared ~450 million years ago, when the earliest vertebrates evolved, i.e., around the time of quadruplication of the *Hox* gene cluster that occurred in this group and even earlier. Structures homologous to the neural tube are found in protochordates (Corbo et al., 1997), ascidian urochordates (Baker and Bronner-Fraser,

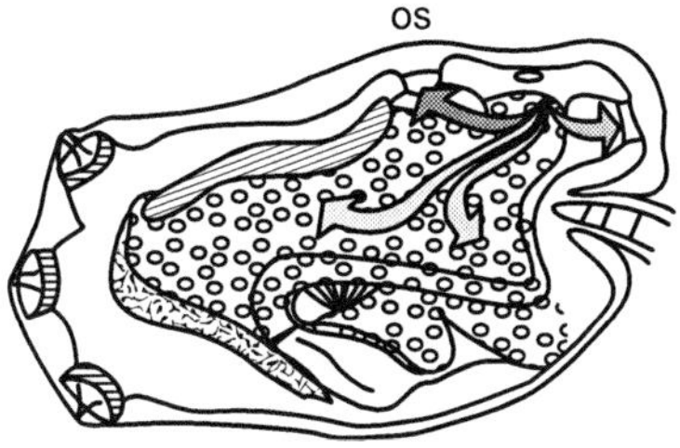

Figure 16.1 Diagram showing the three pathways of cell migration in *E. turbinata*: anterior to oral siphon primordium (red), posterior to anal siphon primordium (blue), and ventral to the larger part of the head and future adult body (yellow). (For interpretation of the references to color in this figure legend, the reader is referred to the web version of this book.)
Source: From Jeffery (2007).

1997; Jeffery et al., 2004), and amphioxus (Baker and Bronner-Fraser, 1997), all of them presumed to be evolutionary precursors of the vertebrate neural crest. Migratory neural crest–like cells (NCLCs) in the ascidian urochordate *Ecteinascidia turbinata* emerging from the neural tube/central nervous system (CNS) in an ordered manner migrate into the body wall below the epidermis, where they, like vertebrate neural crest cells, differentiate into pigment cells (Jeffery et al., 2004). NCLCs migrate from near the dorsal neural tube in three directions (Figure 16.1).

There are several major facts indicating that the neural crest played a crucial role in the evolution of the morphology of vertebrates.

First, the appearance of the neural crest as a new neural formation coincides with the unprecedented burst of morphological *innovations* and increased structural complexity characterizing the vertebrate evolution.

Second, neural crest cells migrate to highly specific parts of the animal body to participate in molding local structures.

Third, morphological traits determined and molded by neural crest cells are among the most malleable vertebrate structures.

All of the above justify the idea that the neural crest has been "at the centre stage of the vertebrate evolutionary play" (Hall, 1998).

The "do-it-yourself" mode of fashioning new morphologies adopted by the neural crest was a developmental "invention" with revolutionary consequences for evolution of vertebrates. It enabled the extraordinary evolutionary malleability of organs or parts in whose development the neural crest is involved. Under appropriate conditions, these structures can dramatically change, sometimes within a small number of generations.

The neural crest has been essential for the evolution of almost all of the new vertebrate morphological features: jaws, pharyngeal jaws in cichlid fish, mammal middle ear ossicles, shell in turtles, feathers in birds, constantly growing incisors in rodents, placenta and viviparity in mammals and reptiles, respectively, lungs in tetrapods, etc. (Hall, 1999, pp. 216–217).

Development of the Neural Crest

Generally, it is believed that the neural crest develops between the neural tube and the ectoderm after the neural tube breaks off the ectoderm. Its cells delaminate from the neural tube and migrate to specific sites in animal body to form or contribute to formation of the most different cells, organs, and tissues.

By the middle of the twentieth century, it was thought that formation of the neural crest requires inducing signals from the mesoderm. However, Moury and Jacobson demonstrated that ectopic transplantation of ectoderm to the axolotl (*Ambystoma mexicanum*) neural tube induces formation of the neural crest (Moury and Jacobson, 1989), and both epidermal and neural plate cells are differentiated into neural crest cells (Moury and Jacobson, 1990).

The classical view that the neural crest precursors are a distinct population between the epithelium and epidermis also is at variance with its evolutionarily neural origin, and recent analyses have demonstrated that neural crest cells and the neural tissue derive from the same cytological precursors. Individual precursor cells within the neural folds can give rise to epidermal-, neural crest–, and neural tube derivatives.

Neural induction determines division of the ectoderm into neural ectoderm (neural plate), the remaining ectoderm, which develops into epidermis, and a border region between them that becomes the neural crest (Selleck and Bronner-Fraser, 1995; Gammill and Bronner-Fraser, 2003; Figure 16.2).

By tracing back the developmental origins of the signals for neural crest formation, one finally arrives at the maternal factors of neural induction, proneural genes, which stimulate expression of Delta and its receptor, Notch (Itoh et al., 2003), in the neural plate. *Notch signals from the neural plate, via induction of the Hairy drive expression of Bmp, Wnt, and Fgf genes, in the underlying mesoderm and adjacent nonneural ectoderm, thus determining lateral inhibition of neurogenesis.* These secreted factors, as well as signals from the neural plate (Dlx5), induce expression of "neural plate border specifiers" (*Zic* factors, *Pax3/7*, *Dlx5*, and *Msx1/2*) (Meulemans and Bronner-Fraser, 2004).

Next, the neural plate border specifiers stimulate expression of "neural crest specifiers" (*Slug/Snail*, *AP-2*, *FoxD3*, *Sox10*, *Sox9*, and *c-Myc*), which form a dense network of interacting elements that makes possible expression of the "neural crest effector genes" (Muelemans and Bronner-Fraser, 2004), which in turn determine formation of mature neural crest cells just before delaminating and starting their migration to strictly determined regions all over the animal body.

Empirical evidence shows that the midbrain–hindbrain junction and the brain-inductive signaling center influence the fate of adjacent neural crest (Trainor et al., 2002a). Neural crest cells are transformed from epithelial into mesenchyme cells, which is typical for many neuronal cells (Baker and Bronner-Fraser, 1997).

After delaminating from the neural tube/CNS, neural crest cells start migrating to particular regions of the embryo, where they direct formation of numerous structures and organs by differentiating themselves into a variety of cell types and by inducing differentiation and proliferation of local cells.

As pointed out, both neurons of the neural tube and neural crest cells derive from common precursor cells. The distinction between neuroepithelial cells of the neural

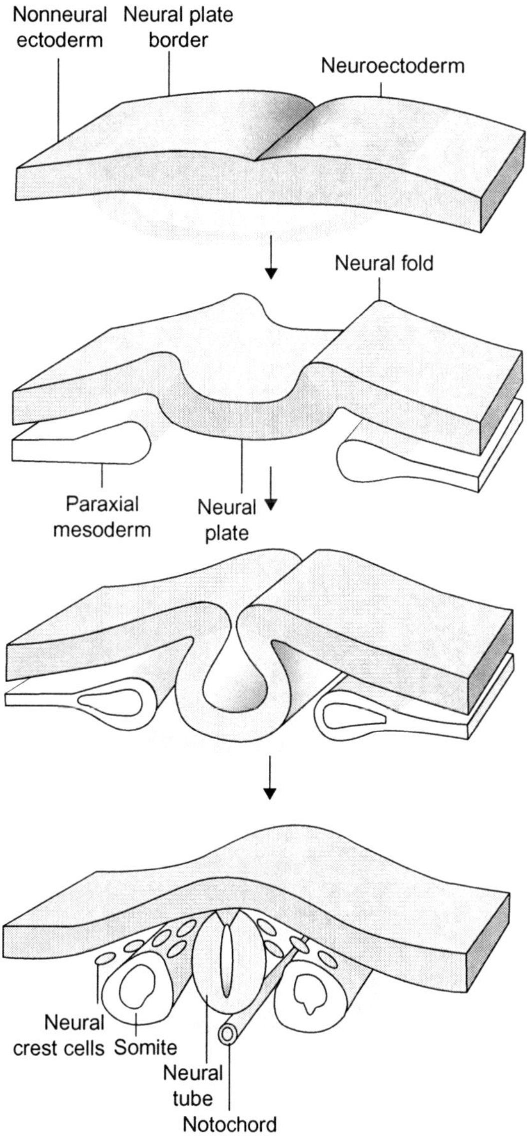

Figure 16.2 Border induction and neurulation. The neural plate border (green) is induced by signaling between the neuroectoderm (purple) and the nonneural ectoderm (blue) and from the underlying paraxial mesoderm (yellow). During neurulation, the neural plate borders (neural folds) elevate, causing the neural plate to roll into a neural tube. Neural crest cells (green) delaminate from the neural folds or the dorsal neural tube (shown), depending on the species and axial level. (For interpretation of the references to color in this figure legend, the reader is referred to the web version of this book.)
Source: From Gammill and Bronner-Fraser (2003).

tube and neural crest cells might not be as well defined as is generally believed. After ablation of a part of neural crest in chick embryos, the adjacent neural tube produces a migratory population of cells that gives rise to neural crest cells:

> *The neural tube cells ventral to the ablation, which normally would not form neural crest cells, regulate to reform the missing regions of the neural tube and the neural crest after ablation of their dorsal neighbors.*
>
> *Scherson et al. (1993)*

This view is also supported by the observation that neural crest cells are produced by some spinal cord neurons (Sohal et al., 1998). These experimental facts, and others to be presented later, suggest that the neural tube may be the producer of neural crest cells and the provider of the epigenetic information to the neural crest cells before they delaminate from the neural tube.

Migration of Neural Crest Cells to Target Sites

The unique morphogenetic role of migrating neural crest cells in different regions of the body, where they form some of the most complex vertebrate structures, suggests that:

1. Neural crest cells are in possession of positional information about the sites where they must go.
2. They "know" what they must do in their target sites, as is proven by the fact that they transform themselves and the cytological elements in the region into specific cell types and often do so even when directed to other regions of the body or even when transplanted to embryos of other species.

How are neural crest cells oriented while navigating through meandering pathways to their destination? A general idea on the mechanism has been obtained from studies on axon conductance.

The prevailing opinion is that it is the cellular and biochemical environment that directs the growth of an axon to its target cells. While it is true that the environment provides biochemical cues for their journey, there is a tendency to ignore the essential fact that only neurons (no other cells) can "recognize" these cues and follow them.

It is believed that axons find their way to the target cells through their *growth cone* with their finger-like appendices, which keep extending and retracting until they find a *specific* cell membrane receptor molecule and adhere to the cell or form a synapse with it if it is another neuron. Netrin-1 is an attractive cue for axons, but its action is mediated by Ca^{2+} influx from both the extracellular environment and internal stores (Hong et al., 2000). Netrin-1 does not serve as an attractant for axons when the plasma membrane Ca^{2+} channels are blocked. The nerve growth cone uses Ca^{2+} (calcium ions) as directional cues for attraction or repulsion. Attractant molecules, such as netrin-1, increase concentration of extracellular Ca^{2+}. The rise of the cAMP concentration in the growth cone decreases the concentration of Ca^{2+}. By integrating global and local Ca^{2+} signals, the growth cone determines by itself the direction of growth (Zheng, 2000).

It is important, however, to bear in mind that axons of different neurons at the same location may follow different itineraries, use different cues, and select different target cells, suggesting that different axons may possess different information or "instructions" on where to go (Lodish et al., 1996).

On the way to their destination, axons actively seek guidance factors (Delcomyn, 1998). Contrary to the conventional wisdom, the ability of axons to identify these factors speaks in favor of the hypothesis that they are in possession of information for recognizing them, and against any instructive role of the cellular environment.

The Source of Epigenetic Information of Neural Crest Cells

Migration of neural crest cells to the target sites throughout the vertebrate body is also a directed process. Similarly to the axon conductance, neural crest cells use local cues as landmarks to thread their way through the long intercellular labyrinths. The neural crest cells "recognize" and use those cues as a Theseus thread to their destinations.

As mentioned earlier, the fact that, among other migrating cell types, only they can identify and use these cues implies that only neural crest cells have the information to recognize and make use of them. Only they are capable of deciphering, reading, or understanding the "content" of those cues. Hence, the information for their migration is not in the cellular environment they navigate through, not in the "cues" themselves, which *per se* are meaningless, but in the neural crest cells that can recognize them. In a crude analogy, regular switching of colors in a traffic light is a cue rather than information. It is our brain that generates the roadway information by giving meaning to these color-switching patterns as compared to squirrels, which sometimes have to pay the highest price for lacking that information. Similarly, but at a different organismal level, migrating birds use natural data (Earth magnetism) to find their way to migration places. The variations of the magnetic field are data not information showing the way to migration sites. Migrating birds use the magnetic field data as cues, because they have information or a "map," which we humans lack, on the variation of the magnetic field in the course of migration.

At the proper time, during the embryonic development, cells from specific regions of the neural crest migrate orderly to specific regions of the body. Upon arrival at the specific site, they use the *preexisting* information, their "intrinsic biases," in Hall's expression, to differentiate into cell types characteristic of the structures they form. There is evidence that the "migration program" of neural crest cells to the target sites (branchial arches) is coded in the hindbrain:

> *Signals influencing the generation and migration of neural crest are established by interactions between the hindbrain and these tissues ... neural crest survival, emigration and migration into the arches may result from the sequential interactions initiated in the hindbrain during neural plate stages. Hence, in a manner similar to what we have found in studying cranial neural crest plasticity in A–P patterning, the proper program of events governing the migration of crest may need first to be established in the hindbrain (my emphasis—N.C.), to allow migratory crest cells to interpret and respond to environmental signals set up through a series of tissue interactions.*
>
> *Trainor et al. (2002b)*

Neural crest cells possess information not only about the itinerary to their destination but, before leaving the neural tube, they are provided with information for inducing specific cell differentiation and morphogenesis at the sites of migration. In a series of elegant experiments, Schneider and Helms (2003) have shown that by transplanting quail neural crest cells to ducks, and the reverse, it is possible to produce offspring that inherit the beak morphology of the graft donor (quail beaks in ducks, and duck beaks in quails). In their interpretation:

> *Neural crest cells provide patterning information for beak morphology. Not only do neural crest cells direct their own morphogenesis, they also pattern non-neural crest beak tissues in a manner characteristic of the donor species.*
>
> *Schneider and Helms (2003)*

They have also demonstrated that the grafted donor cells bring to, and impose on, the host their developmental program. They alter in a donor-specific manner the patterns of gene expression of the host tissues (Schneider and Helms, 2003).

Additional evidence on the role of the neural tube/CNS as provider of epigenetic information of the neural crest cells comes from experimental transplantation of neural crest cells to heterotopic regions (rhombomeres) of the brain. When neural crest cell progenitors from the midbrain–hindbrain region are transplanted caudally, they continue to induce structures they used to induce in their characteristic sites of migration, thus leading, for example, to formation of ectopic bones of the donor type (Figure 16.3).

Now, to summarize: the neural tube/hindbrain provides neural crest cells with epigenetic information for the itinerary they must follow and for the developmental processes they must induce in the target sites.

Neural Crest and Vertebrate Morphogenesis

From the neural crest that forms on the neural tube starts a mass migration of neural crest cells. They are at the origin of many cell types and contribute to the formation of a great number of tissues and organs. A group of them from the developing brain and spinal cord go in direction of ectoderm to be differentiated into pigment cells (melanocytes). The rest of them follow an internal path between the neural tube and somites to reach down to the deepest layers and form sensory and sympathetic ganglia (Figure 16.4). The pluripotent neural crest cells differentiate into nerve cells, glial cells, adrenergic cells, pigment cells (leukophores, erythrophores, melanophores, iridiophores, and xanthophores), and chondrocytes.

Once the hindbrain neural crest cells exit the neural tube, they are maintained in segregated streams (Noden and Trainor, 2005). In mice, the hindbrain-derived neural crest cells migrate in three segregated streams adjacent to the even-numbered rhombomeres, compartments of developing hindbrain (Trainor et al., 2002b), but odd and even rhombomeres produce equal number of neural crest cells. The neural crest cell free zones are at the level of rhombomeres 3 and 5 (physically restricted by the invaginating otic placode). Creation of neural crest cell free zones is necessary for maintaining the separation of neural crest cells of different origin and function (Trainor et al., 2002a). The pluripotent neural crest cells contribute to the development and morphology of regions where they migrate not only by transforming themselves into different types of cells, but also by transforming the local cells in appropriate types of cells. So, for example, initiation of chondrogenesis in the pharyngeal endoderm requires direct contact with the neural crest cells (Hall, 1999, p. 137).

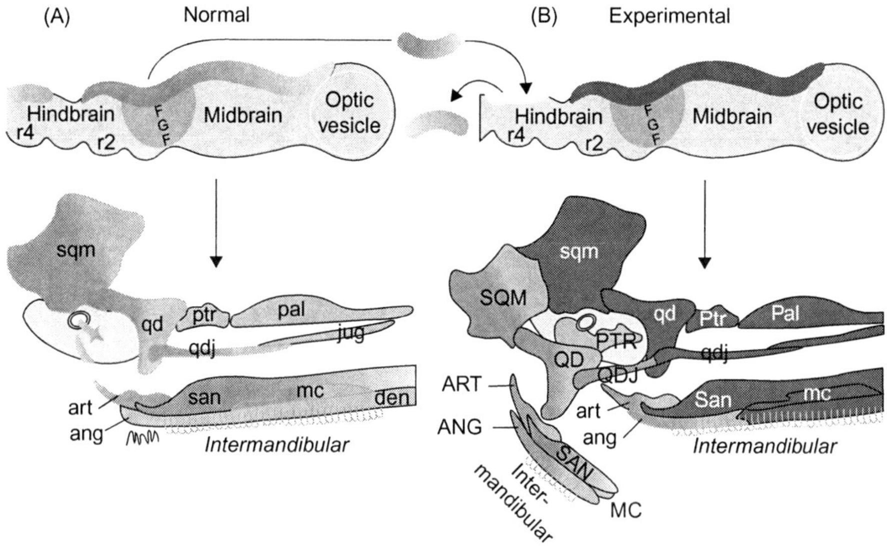

Figure 16.3 The morphogenesis of branchial neural crest tissues results from both autonomous, prespecified and dependent, acquired properties. (A) The contributions of neural crest cells from the mesencephalic, metencephalic, and rhombomere 4 levels to the first branchial arch skeleton. Note that the proximal (caudal) parts of the articular and angular bones are derived from second arch (rhombomere 4) crest cells that became contiguous with first arch (rhombomere 2)–derived crest cells dorsal to the first pharyngeal pouch. (B) Presumptive proximal first arch neural crest precursors were grafted in place of second arch crest precursors. Most formed ectopic first arch skeletal structures in the second arch location (labeled in uppercase letters), and directed myogenic cells entering the second arch to form first arch-specific muscles, for example, the INTERMANDIBULAR. At interfaces with neighboring, untransplanted crest cells, however, grafted crest cells relinquished their prespecified biases and cooperatively formed structures anatomically correct for their new location, for example, the retroarticular cartilage and proximal angular bone in the first branchial arch. *Abbreviations*: Ang, angular; Art, articular; Den, dentary; Mc, mandibular cartilage; Pal, palatine; Ptr, pterygoid; Qd, quadrate; Qju, quadratojugal; San, surangular; Sqm, squamosal.
Source: From Noden and Trainor (2005).

Migrating neural crest cells themselves serve as precursors of neuroblasts, from which neurons of the spinal ganglia, nerve cells of the autonomic vegetative system, peripheral ganglia neurons, and ganglia cells that accompany nervous processes differentiate, such as Schwann cells, but also nonneural cells of some of the faster evolving/recurring parts of the vertebrate body, such as the visceral skeleton, which previously was believed to derive from mesoderm.

Of neural crest origin are also cartilaginous elements of the jaws and pharyngeal arches and their sound-transmitting derivatives of the inner ear: incus, malleus, and stapes, as well as the laryngeal and tracheal skeleton, the dental papilla, producing bone-like material to the tooth germs. Of similar neural crest origin are most

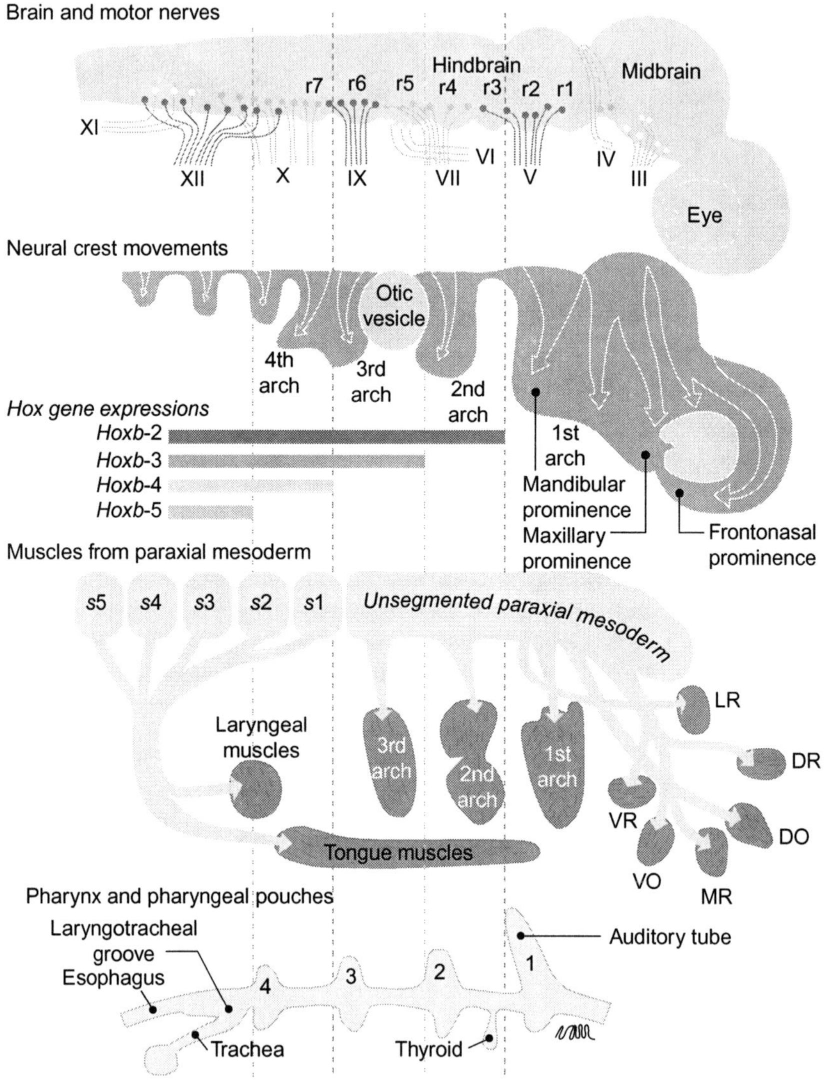

Figure 16.4 Staggered, lateral views of all internal tissue layers in an early avian embryo. These illustrate the changes in locations of each population and the spatial relations among them. Neural crest progenitors, cranial nerves, and myogenic primordia for each branchial arch all arise at the same axial level and maintain this close registration throughout their dorsoventral movements. For example, crest cells that will populate the second branchial arch arise from the same axial location (rhombomere 4) as the seventh cranial nerve and the second arch muscles it will innervate. By contrast, the periocular neural crest, the extraocular muscles, and the motor nerves that innervate them all arise at separate axial locations and do not establish stable relations until all have reached their sites of terminal differentiation. *Source*: From Noden and Trainor (2005).

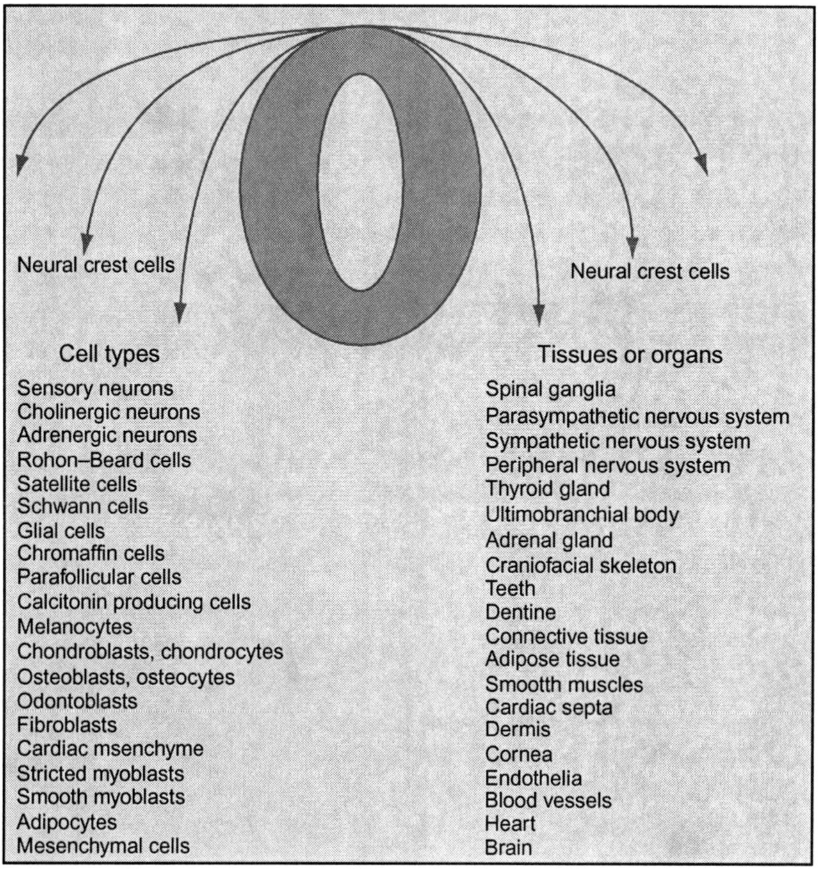

Figure 16.5 Neural crest cell derivatives. Derived from the neural tube, neural crest cells are a pluripotent, mesenchymal population that migrates extensively and gives rise to a vast array of cell types, tissues, and organs. Given the wide variety of differentiative fates and a limited capacity for self-renewal, neural crest cells are often considered to be a stem cell–like population.
Source: From Trainor et al. (2003).

bones of the face and of the brain case, pigment cells of the skin, feathers, and hairs (Figure 16.5).

Neural crest cells also produce the dorsal fins of fish (Müller, 1996b) and, in all likelihood, the reemerged dorsal fins in aquatic mammals and ichthyosaurus as well as characteristic crests developing in numerous reptiles, including dinosaurs. The neural crest also provides neurons, mesenchymal cells, cells of the aortic pulmonary conotruncal septa, valves, and major vessels to the developing heart in chick (Hall, 1999). They also determine formation of neurohormonal cells, such as cells of the adrenal medulla and neuroendocrine cells of the digestive tract. The unique role of migrating neural crest cells in different regions of the animal body to form there some of the most typical vertebrate features clearly suggests that:

1. Neural crest cells possess positional–structural information for the organs or parts, whose formation they are involved in.
2. Neural crest cells induce differentiation and proliferation of neighboring cells to participate in formation of those parts or tissues.
3. Evolution of neural crest–induced structures is based on the intrinsic properties of neural crest cells, which are provided with the necessary epigenetic information before leaving the neural tube/CNS.

Now let us illustrate, with some representative examples, the rapid evolutionary changes in vertebrate morphology, in response to changes in conditions of living, in which the neural crest was essentially involved.

Neural Crest–Determined Evolutionary Novelties in Vertebrates

The Vertebrate Head

Underscoring the innovative character of the vertebrate head, Gans and Northcutt (1983) considered it a "new head" and put forward the hypothesis that the more anterior (or rostral) part of the head, including sense organs, prosencephalon and mesencephalon, together with the corresponding region of the skull, is derived from the neuroectoderm. They point out:

> *It is more difficult to understand why the neural crest should function like mesoderm in the anterior head. This is clearly the most striking change coincident with vertebrate origins. To make a crude analogy: the vertebrate head may be conceived as an addition to the existing body of protochordates.*
>
> *Gans and Northcutt (1983)*

The whole "prechordal" skull is virtually of neural crest origin, and the contribution of the mesoderm to the "new head" is reduced essentially to muscles and a limited posterior (or caudal) part of the skull (Figure 16.6). As a vertebrate neural crest–derived innovation, the *cranium* is probably the greatest evolutionary expression of the potential of the exploratory neural crest (Gerhart and Kirschner, 1997, p. 553).

Neural crest cells originating from rhombomeres (transient segments of the hindbrain) go to the head, where they differentiate into the cytological elements of bones and cartilages of the head. On the way, they do not mingle.

The pathway of transformation of the CNCs into skeletogenic components in successive embryonic stages of development of chick and mouse is presented in Figure 16.7.

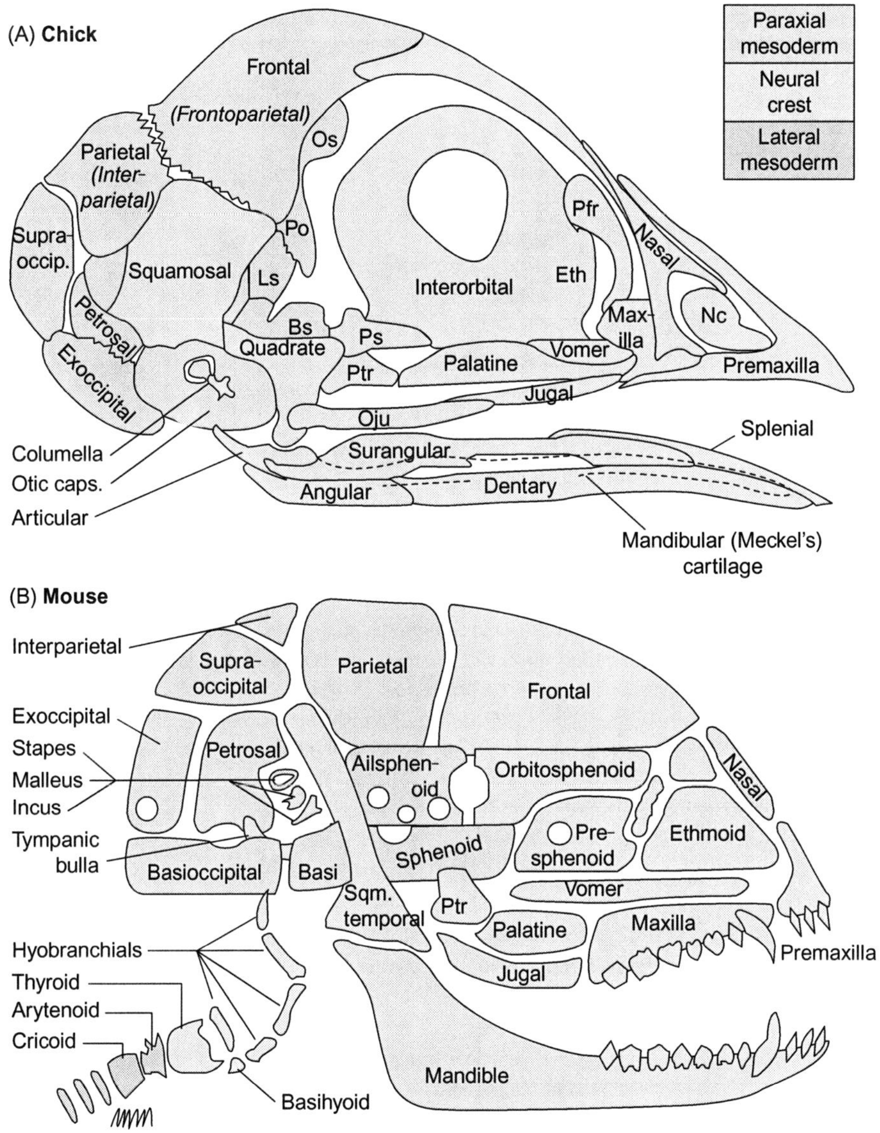

Figure 16.6 Schematic chick and mouse skulls showing the contributions of neural crest, paraxial and lateral mesoderms to the cranial skeleton. The avian map is based on transplantation and retroviral lineage tracings in the chick embryo; hyobranchial structures, all of which are derived from neural crest cells, are not shown. The mouse map is based largely on the location of neural crest cells, as identified by expression of *LacZ* driven by a *Wnt1* promoter in *cre-lox* transgenic embryos (Jiang et al., 2002). Origins of mouse laryngeal cartilages are, by extrapolation, from avian data, with the caveat that birds do not have a thyroid cartilage. *Abbreviations*: Ang, angular; Art, articular; Bs, basisphenoid; Den, dentary; Eth, ethmoid; Lac, lacrimal; Ls, laterosphenoid*; Mc, mandibular cartilage; Nc, nasal capsule; Os, orbitosphenoid*; Pal, palatine; Pfr, prefrontal; Po, postorbital; Ps, presphenoid; Ptr, pterygoid; Qd, quadrate; Qju, quadratojugal; San, surangular; Sqm, squamosal; *regions of the pleurosphenoid. (For interpretation of the references to color in this figure legend, the reader is referred to the web version of this book.)
Source: From Noden and Trainor (2005).

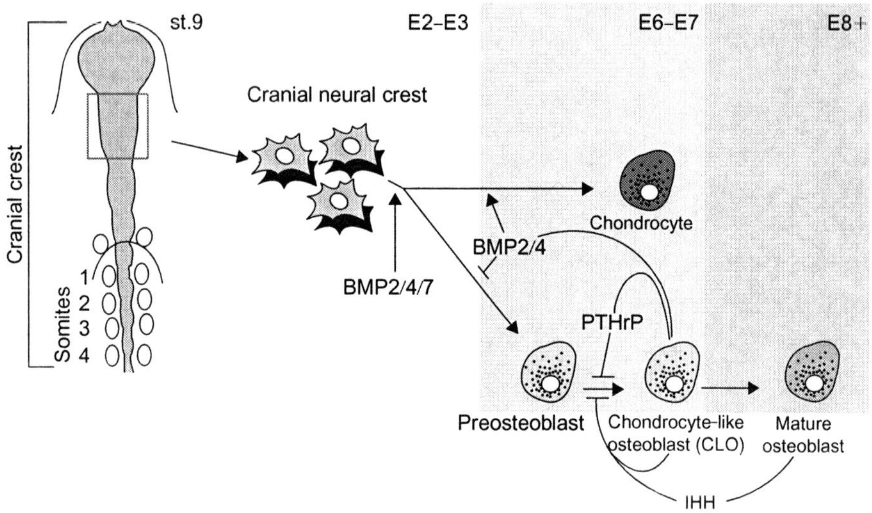

Figure 16.7 During craniofacial development, mesencephalic CNC cells migrate to populate mesenchyme of the future face and skull. Cells of the early cranial skeletogenic condensations depend on BMP2/4/7 activities to form preosteoblastic progenitors, whereas high levels of BMP2 and/or BMP4 alone induced a chondrogenic fate. Differentiation into the chondrocyte-like osteoblasts is regulated by both IHH (Indian hedgehog) and PTHrP (parathyroid hormone-related protein) activities.
Source: From Abzhanov et al. (2007).

CNCs are also responsible for the migration, patterning, and differentiation of muscle precursors, and blockade of the CNC causes anomalies in the development of head skeletal muscles.

> *CNC cells control craniofacial development by regulating positional interactions with mesoderm-derived muscle progenitors that together shape the cranial musculo-skeletal architecture in vertebrate embryos.*
>
> *Rinon et al. (2007)*

Evolution of the Vertebrate Jaw

The evolution of ear ossicles (malleus, incus, stapes, and the tympanic bone), a key morphological innovation that took place in the class of mammals, is one of the most enigmatic transformations of vertebrates (Figure 16.8). By increasing predation capabilities, the advent of the jaw played a determining role in the evolutionary success of vertebrates.

The vertebrate jaw derives from a deep modification of the elements of the first pair of branchial arches *of Agnatha* that took place some 500 million years ago. Its evolution coincided with the appearance of the neural crest.

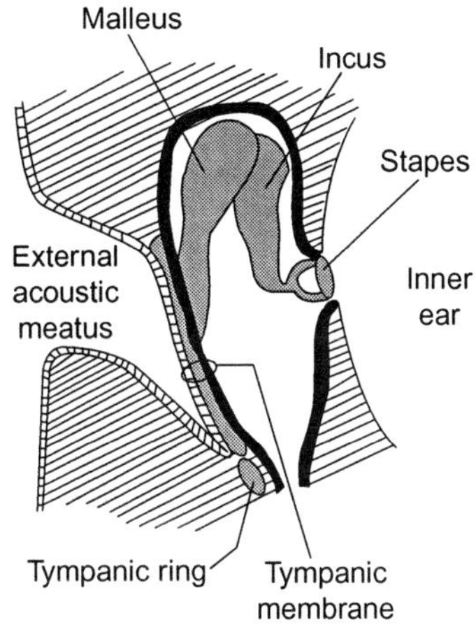

Figure 16.8 Diagrammatic representation of the otic region of a typical mammal. Note the trilaminar structure of the tympanic membrane (eardrum), which is formed by the epithelial layers of the external acoustic meatus and the tubotympanic recess, along with an intervening layer of cells derived from the first and second pharyngeal arches. The tympanic membrane is anchored by the tympanic ring. The three auditory ossicles (malleus, incus, and stapes) conduct vibrations from the tympanic membrane across the middle ear cavity and transmit these vibrations into the inner ear.
Source: From Mallo and Gridley (1996).

Attempts are made to speculatively relate its evolution with the observed duplication of *Wnt-3*, *Wnt-5*, *Wnt-7*, and *Wnt-10* genes in the lineage leading to craniates (Sidow, 1992). However, as Hall (1998, p. 258) points out, the expression pattern of those genes cannot be a basis for evolution of jaws.

The mandible is formed by the dentary and Meckel's cartilage. The dentary consists of four parts, each formed by a separate population of neural crest cells, which form *condensations* of cells that must reach a critical mass before the morphogenesis of dentary takes place.

As pointed out earlier, structures in whose formation neural crest cells play a dominant role are characterized by a relatively high phenotypic and evolutionary plasticity. The dentary does not represent an exception from the rule. The mandible still shows a great variation in morphology and is strongly influenced by environmental factors. In a very interesting case on the Isle of May in Scotland, it was observed that hybrids between the house and feral mice born 18 months after the introduction of the house mouse on the island evolved modified mandibular morphology, and investigators estimated that only 57% of the variation was genetically determined (Scriven and Bauchau, 1992).

Evolution of the middle ear ossicles is a result of an evolutionary pressure for hearing under conditions of the new sound-transmission medium (from water to air) related to transition of vertebrates to terrestrial life.

The evolution of mammals from reptiles was characterized by a continuous growth of one of the lower jaw bones, the dentary, at the expense of all six other bones, of which their lower jaw consists. The increase in size of the dentary brought it into contact with the skull and formed the modern squamoso-mandibular joints, thus freeing quadrate and articular bones of their interconnecting function and moving them posteriorly. The evolutionary trend toward freeing reptile lower jaw bones of their articulating function and their posterior displacement to form middle ear ossicles coincided with a heterochronic change in the arrival of neural crest cells in the region where the Meckel's cartilage forms.

Paleontological Evidence of the Evolution of the Middle Ear Ossicles

Paleontological evidence shows that middle ear ossicles evolved not later than 195 million years ago (Early Jurassic), which is the age of *Hadrocodium*, the earliest-known taxon that lost the mandibular attachment to the middle ear ossicles (Luo, 2001).

The mandibular precursors of the middle ear ossicles were used for mechanosensation in premammalian groups, such as cynodonts and *Morganucodon*, which continued to conserve these "reptilian" jaw bones, but separated by a cartilage mass. An intermediate reduced bone detached from the dentary is found in some mammal fossils of the Early Cretaceous in China (Wang et al., 2001). Schematized evolution of mandibles and middle ear ossicles from corresponding reptile jaw bones is shown in Figure 16.9.

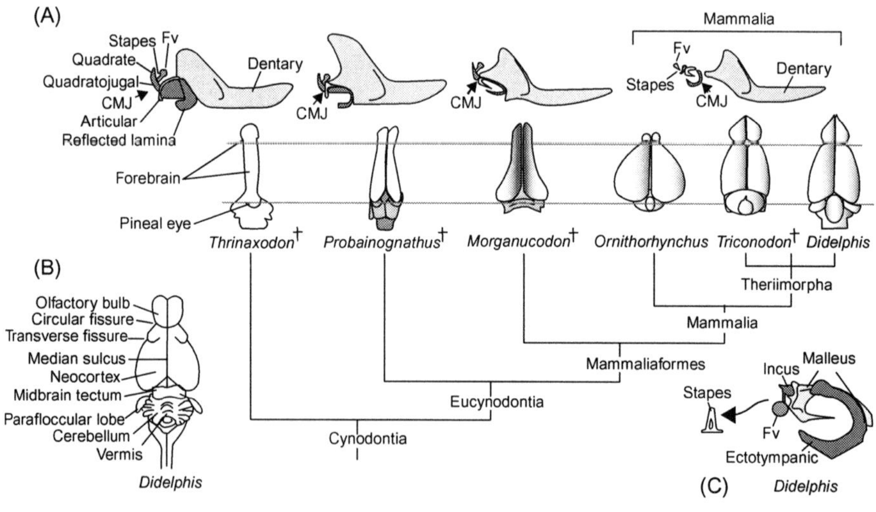

Figure 16.9 (A) Evolution of the mammalian mandible and middle ear (right lateral view) plotted on a phylogeny of selected mammals and their closest extinct relatives. (B) Right lateral view of auditory chain of *Didelphis*; the stapes is rotated and offset from between the incus and fenestra vestibulae of the inner ear. Crosses signify extinct species. *Abbreviations*: FV, fenestra vestibuli; CMJ, craniomandibular joint.
Source: From Rowe (1996).

Whether the evolutionary transformation of the lower jaw bones into middle ear ossicles in mammals has occurred only once, in their common ancestor (Rowe, 1996), or more than once in different groups, after their divergence from the common ancestor, has been a controversial issue in modern biology. Given the complexity of the structure of the middle ear ossicles and of the process of the transformation of mandibular bones into middle ear ossicles, most biologists believed that this evolution occurred only once to the common ancestor of all the mammals. This hypothesis of monophyletic origin of middle ear ossicles seems to have been refuted by recent paleontological evidence.

Rich et al. (2005) in Australia found a 115 million-year-old fossil jaw of an Early Cretaceous monotreme, *Teinolophos trusleri*, considered to be an extinct relative of the modern Australian monotremes, platypus and echidna (Martin and Luo, 2005). The fossil has a trough in which postdentary reduced bones, homologous to the mammalian ear ossicles, were housed, implying that although the reduced bones might have been used for hearing, they were still an integral part of the jaw, in the mandibular trough. They represent, thus, an important link between the mandibular bones and fully transformed mammalian middle ear ossicles in monotremes (Rich et al., 2005). In view of the facts that monotremes, to which *Teinolophos trusleri* belongs, split off as a separate mammalian group more than 150 million years ago and that modern monotremes have middle ear ossicles, the presence of the bones homologous to the middle ear ossicles still attached to the mandible in *T. trusleri* proves that middle ear ossicles in monotremes evolved independently from other groups of mammals.

The freeing of the mammalian ear bones from the lower jaw may have occurred more often than can be conclusively documented at present (Rich et al., 2005), and Martin and Luo (2005) believe that middle ear ossicles have evolved independently three times in mammals: in marsupials, in placentals, and in monotremes.

Ontogeny of the Middle Ear Ossicles

The evolutionary trend for freeing reptile lower jaws from articulating functions in the process of their evolution into mammalian middle ear ossicles is supported by evidence on the development of ear ossicles during ontogeny (Figure 16.10).

Middle ear ossicles are of CNC origin. The hindbrain (rhombomeres 1 and 2) and caudal midbrain supply neural crest cells for the malleus and incus, whereas the rhombomere 4 is at the origin of CNC cells for the stapes (Köntges and Lumsden, 1996; Mallo, 2003). Migration of these CNC cells to the target sites marks the start of developmental processes leading to formation of middle ear ossicles.

> *In mammalian development the auditory chain arises connected to the mandible but later detaches, recapitulating the phylogenetic transformation. In modern didelphid development, the auditory chain reaches mature size by the third week after birth and is then separated from the jaw and displaced caudally as the neocortex grows for another 9 weeks.*
>
> *Rowe (1996)*

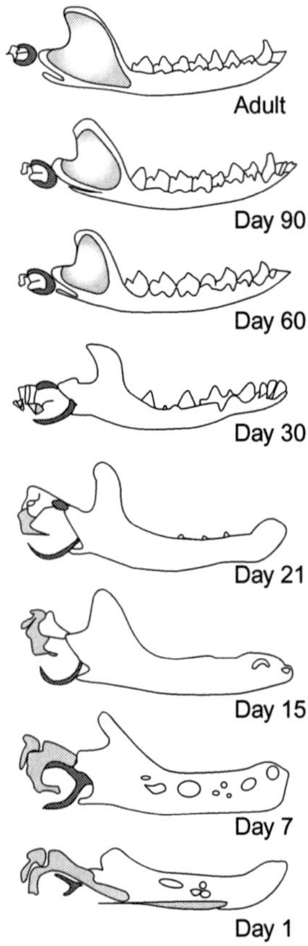

Figure 16.10 Development and relative growth of the didelphid mandibular arch of *Monodeplhis domestica*. Embryonic cartilage is in blue, the tympanic ring or ectotympanic bone is in red, and uncolored regions have become mineralized. (For interpretation of the references to color in this figure legend, the reader is referred to the web version of this book.)
Source: From Rowe (1996).

Meckel's cartilage arises as a result of achondrogenic epithelial–mesenchymal interaction posterior to the dentary. The reduced posterior part of the Meckel's cartilage develops into the malleus anlage, while its middle portion degenerates later in the ontogeny. Only in mammals, where the transformation of lower jaw elements into middle ear ossicles occurs, the epithelial–mesenchymal interaction begins after neural crest cells reach the region (Smith and van Nievelt, 1997) of the middle ear.

During the mammal embryogenesis, the aboral part of Meckel's cartilage reaches the squamosa bone, and the middle part of the cartilage atrophies, whereas ossification of the aboral part leads to formation of the middle ear ossicles: incus, malleus, tympanum, and stapes.

Neo-Darwinian Explanation

With no changes in genes involved in the process of evolution of the mammal middle ear ossicles from reptile jaw bones, it is not surprising that no neo-Darwinian mechanism on the evolution of the ossicles has been presented.

Epigenetic Explanation

Two epigenetic explanations of the process of transformation of the jaw bones into mammal middle ear ossicles have been noted that

1. It resulted from expansion of the brain, which, by widening the gap between the middle ear and the reduced jaw bones, pulled these bones apart from the mandible in mammals (Luo, 2001). This epigenetic hypothesis seems to have been refuted by paleontological evidence that the brain sometimes is less expanded in fossils that have detached middle ear ossicles than in mammal fossils with ossicles still attached to the dentary (Wang et al., 2001).
2. It is related to separation of the reduced postdentary ossicles:

> *Reduction of the PDU increasingly weakened its tie to the dentary until a critical point was reached where the dentary, while erecting to a more vertical position during ontogeny, no longer seized the PDU, which was moored at the basicranium by connective tissue. This hypothesis is similar to the detaching mechanism of the ear ossicles in marsupials, without requiring brain expansion as the initial trigger.*
>
> *Wang et al. (2001)*

None of the above hypotheses addresses the fundamental cause of the reduction in size of postdentary ossicles. Convergent evolution of such complex structures as middle ear ossicles (they are believed to have independently evolved three times), their simultaneous coordinated reduction in size, and their adaptive posterior displacement in mammals suggest that evolution of ossicles might not have been as contingent as is conventionally imagined. If one were to admit that formation of the middle ear ossicles is not the only possible solution to the problem of hearing in mammals, then the question arises: What could this strong bias to the same solution to hearing problems in mammals be related to?

Even more essential is the question: Where did the epigenetic information (=signals) for the simultaneous reduction of the size of three reptile bones come from?

Biased changes in the size and in migration sites of the postdentary bones cannot occur randomly but need information, a kind of information that obviously is different from the genetic information for the primary structure of proteins. The crucial role of the neural crest cells in the process of the formation of the postdentary ossicles suggests that in searching for the source of that information, one should focus on the function of these cells.

Neural crest cells represent the basic building blocks of the middle ear ossicles. These cells come from the midbrain and hindbrain (rhombomeres 1, 2, and 4) (Köntges and Lumsden, 1996; Mallo, 2003). It is a well-known fact that before leaving the neural tube, neural crest cells are provided with information not only for migration (Trainor et al., 2002b) to the site of ossicle formation but also for determining the shape and size of bones (Schneider and Helms, 2003; Tucker and Lumsden, 2004):

> *With regards to shape of the resulting cartilage elements, the patterning cues reside in the neural crest before migration.*
>
> *Tucker and Lumsden (2004)*

All the above suggests that the evolutionary changes in morphology, morphometry, and the developmental behavior of the postdentary bones that evolved into middle ear ossicles may have been determined by changes in the epigenetic information, which is provided to the neural crest cells before they leave the neural tube.

Neural Crest in Evolution of Dentition in Vertebrates

Neural crest played a crucial role in the evolution of vertebrate dentition. The crucial element in the development of teeth in this group are neural crest cells which, after migrating to the presumptive dentition regions, form the odontogenic ectomesenchyme, which activates the overlying oral epithelium and interacts with it.

Critical for the evolutionary change in morphology, and in animal phenotype in general, is the source of information for the change. (The change is not random but adaptive.) Adequate empirical evidence shows that this information is provided by the odontogenic neural crest cells (Lumsden, 1988) originating from the posterior midbrain (Imai et al., 1996) and by the local innervation (Tuisku and Hildebrand, 1994; Hildebrand et al., 1995; Figures 13.39 and 13.40) of the presumptive teeth region.

With genes and genetic processes not known to be involved in evolution of dentition in vertebrates (of all the odontogenic genes and gene regulatory networks are conserved across vertebrate taxa) and the neural crest being pivotal in evolution of teeth, an epigenetic change in the information that the neural crest cells are provided with before leaving the neural crest is a plausible explanation of the evolution of dentition in vertebrates.

An Evolutionary Event in Darwin's Finches

Darwin's finches of the Galapagos Islands are a group of birds that exhibit unusual variability in the beak form and size, despite the fact that all of them are closely related and recently diverged from a common ancestor.

One of these finch species, *Geospiza conirostris*, on Isla Genovesa, Galapagos, is currently in the process of evolving into two incipient species. The finch population of the island is segregated in two phenotypically distinct groups. One group, designated as group A, consists of individuals with longer and narrower beaks, and group B, of individuals with shorter but wider beaks. Male individuals of the A type, with longer and narrower beaks, "feed from cactus fruits by drilling a hole with the beak, removing seeds and eating the surrounding arils," whereas the B type males "were observed ripping open cactus pads with their beaks to feed on the exposed insect larvae and pupae." Additionally, both groups differ in the type of song they sing, and this difference is the most important cause of the reproductive isolation in sympatry that characterizes two *G. conirostris* populations of the island that seem to be in the process of splitting in two incipient species.

The rapid evolution of beaks in Darwin's finches represents a rare opportunity for the study of the process of the evolutionary change. One of the first things that comes to mind when considering the exceptional evolutionary plasticity of the beak in Darwin's finches is the critical role of the neural crest cells in development of this organ. Hence, a brief review of the present knowledge on the cytological and molecular factors involved in the beak development may provide some clues about the mechanisms of evolutionary changes in Darwin's finches' beak morphology.

The Role of the Neural Crest in Beak Development

After delaminating from the forebrain, the beak-inducing neural crest cells start an orderly migration to the beak region, where they differentiate into specific beak cells and induce molecular and cytological transformation of cells in the region of the presumptive beak.

The central regulatory role that neural crest cells play in determining beak morphology is demonstrated in numerous experimental studies, especially in studies on interspecific neural crest transplantation at particular embryonic stages in birds (Figure 16.11).

Homotopic transplantation of the neural tube at specific times of embryogenesis between quails (*Coturnix coturnix japonica*) and ducks (*Anas platyrhynchos*) leads to development of structures (Schneider and Helms, 2003) and plumage color patterns (Kinutani and Le Douarin, 1985) typical of the donor species rather than of the host.

Tucker and Lumsden (2004) focused on the bilateral and unilateral embryonic transplantation of distinct neural crest populations from Japanese quail and duck. They observed that homotopic grafting with quail midbrain and rhombomeres 1+2 (which supply the neural crest cells for the lower beak) into the duck embryo neural tube led to production of chimerae with smaller (quail type) lower beaks (Figure 16.12). This is because the lower beak is determined by the midbrain neural crest, which is of quail origin, whereas the upper beak is determined by more anterior parts of the brain, which come from the duck. They also observed that not only the size but also the shape of the entoglossum resembled that of the quail. Unilateral grafts led to the development of the donor shape only on the grafted side.

Unilateral homotopic grafting of the quail neural crest from rhombomeres 1 to 4 in duck embryos produced retroarticular process (RAP) of the quail type (because RAP is derived from the rhombomere 1 neural crest) (Tucker and Lumsden, 2004). Investigators conclude:

> *Contrary to previous reports claiming that facial skeletal patterning information resides solely in the endoderm and that the neural crest is a passive player in the patterning process (Couly et al. 2002), our results show that neural crest cells do indeed contribute to the final pattern, albeit in a less extensive way that originally proposed by Noden (1983).*
>
> *Tucker and Lumsden (2004)*

Experimental heterotopic transplantation of the first arch neural crest cells from chicks and quails into the excised second and third arch neural crest of the chick gave rise to ectopic development of beak-like structures on the ventrolateral side of the neck.

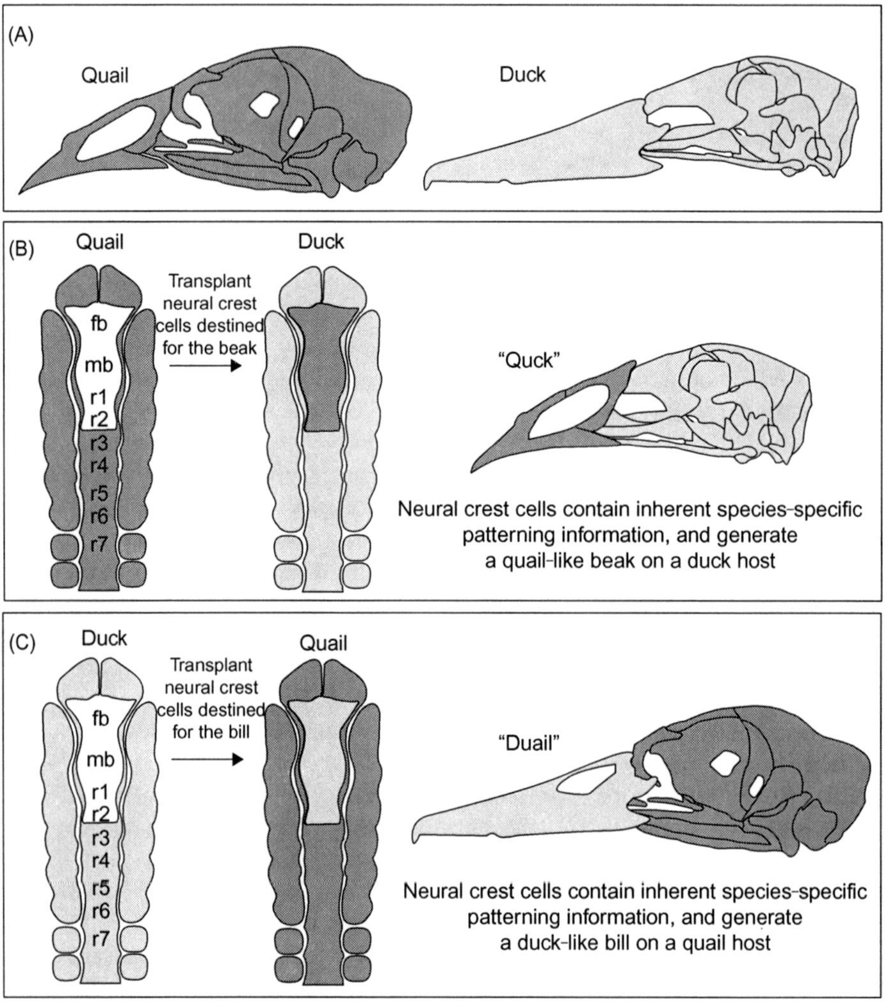

Figure 16.11 Transplantation experiments provide evidence that the neural crest inherently contains species-specific patterning information. (A) Quail and duck embryos exhibit distinct anatomical features. For example, quails exhibit a shorter, narrower beak compared with the longer, broader duck bill. (B) When quail neural crest cells from forebrain (fb), midbrain (mb), and rhombomeres 1 and 2 (r1, r2) are transplanted into a duck host, a quail-like beak develops in lieu of a duck's bill. (C) When duck neural crest cells are transplanted into a quail host, the quail develops a duck-like bill.
Source: Figure reproduced courtesy of *Nature*; from Tapadia et al. (2005).

The fact that duck crest cells respond to the chick endodermal signals in a duck-like manner and vice versa has been interpreted as a proof of "the existence of a species-specific prepattern in the neural crest cells" (Tucker and Lumsden, 2004).

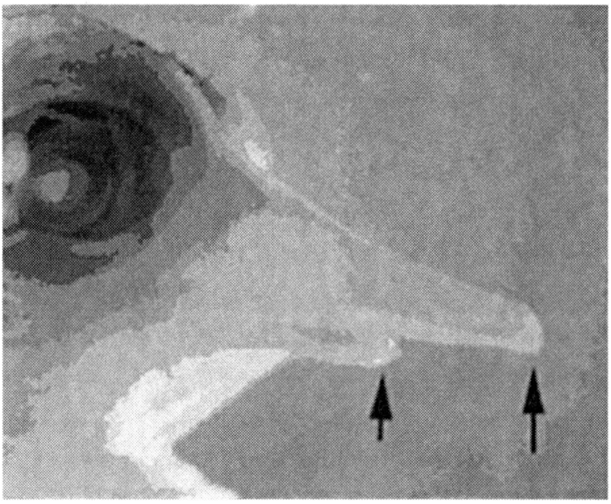

Figure 16.12 Duck head at E13 grafted with quail midbrain (1 R1/2) neural crest showing reduced size of lower beak when compared with upper beak. This is because the lower beak is determined.
Source: From Tucker and Lumsden (2004).

Neural crest cells provide the epigenetic information for the size and the shape of the beak in these bird species. In compliance with previous experimental evidence, they believe that the CNC is

> *morphogenetically prespecified in respect of its branchial skeletal derivatives, that is, that information for the number, size, shape, and position of its individual elements is already determined in these cells when they are still in the neural folds. This positional information would somehow be preserved after they delaminate from the neural tube and migrate into the branchial arches, before being read out as spatial pattern of chondrogenesis and osteogenesis .... The ability to form species-specific patterns of craniofacial skeletal tissue thus appears to be an inherent property of the neural crest, expressed as species-specific responses to endodermal signals.*
> *Tucker and Lumsden (2004)*

Tucker and Lumsden (2004) are not the only investigators who, based on experiments of orthotopic neural crest transplantation from anterior midbrain rhombomeres of quail donors into ducks, have shown that the form of the beak in chimeras is determined by the donor of the neural crest rather than by the host. Earlier, Hu et al. seem to have implied the instructive role of the CNS on neural crest cells:

> *Cranial neural crest destined for the FNP (frontonasal process—N.C.) migrate over the forebrain rather than past the pharyngeal endoderm, making it unlikely that signals from this tissue impact neural crest skeletal precursors en route to the FNP.*
> *Hu et al. (2003)*

However, no consensus seems to exist yet on the instructive role of the neural crest and on the brain origin of information for their migration. Based on experiments on the patterning of the facial skeleton, earlier Couly et al., while acknowledging the instructive role of neural crest cells, have questioned the idea that neural crest cells are in possession of all of the information necessary for the development of the facial skeleton:

> *The neural crest cells that form the bones and the cartilages do not themselves possess all the information necessary for patterning the facial skeleton.*
> *Couly et al. (2002)*

Recent studies on the role of neural crest cells in shaping beak morphology in Darwin's finches have shed new light on the mechanism of evolution of beak morphology (Abzhanov et al., 2004). Most relevant factors examined, including various forms of BMP (bone morphogenetic protein), Shh, and Fgf8, are functionally identical in different species of finches. The only distinctive factor that investigators have been able to identify is an epigenetic change, an increased and earlier expression of the BMP4 in the mesenchyme of the embryos of finch species with deeper and wider beak. (Bmp2 and Bmp7 expression was correlated with the size but not with the shape of the beak.) By stage 26, embryos of *Geospiza magnirostris*, which are characterized by deeper and wider beak, showed higher levels of BMP4 than embryos of the rest of finch species (Figure 16.13).

By stage 29, embryos of all three ground finch species, *G. fuliginosa*, *G. scandens*, and *G. magnirostris*, showed increased expression of BMP4, with *G. magnirostris* still maintaining the higher levels. At that stage, BMP4 was found to have been expressed in the mesenchyme of the developing beak of other species that have shallow and narrow beaks (*G. conirostris*, *G. scandens*, and *G. conirostris*) (Abzhanov et al., 2004).

In comparative interspecific studies, Wu et al. (2004) found that increased expression of Bmp4 occurs in two sites of the presumptive beak in chick embryos but in only one in duck embryos, suggesting that differences in the shape and size of the beak between these species result from the differences in patterns of Bmp4 expression. By using Noggin for increasing or decreasing BMP4 expression, they were able to manipulate the size of the beak (Wu et al., 2004).

Another group of investigators found that by manipulating expression of BMP4, similar changes in the beak morphology, in both form and size, are also produced in chick embryos (Helms et al., 2005).

Another correlation was observed between the elevated level of calmodulin and elongated beak morphology. Abzhanov et al. (2006) found that calmodulin expression was higher in the distal–ventral mesenchyme of the frontonasal processes of long-beaked finches than in processes of the short-beaked finches. Thus, they have identified the effect of the combined action of Bmp4 and calmodulin expression in determining the shape and size of the beaks in Darwin's finches (Figure 16.14).

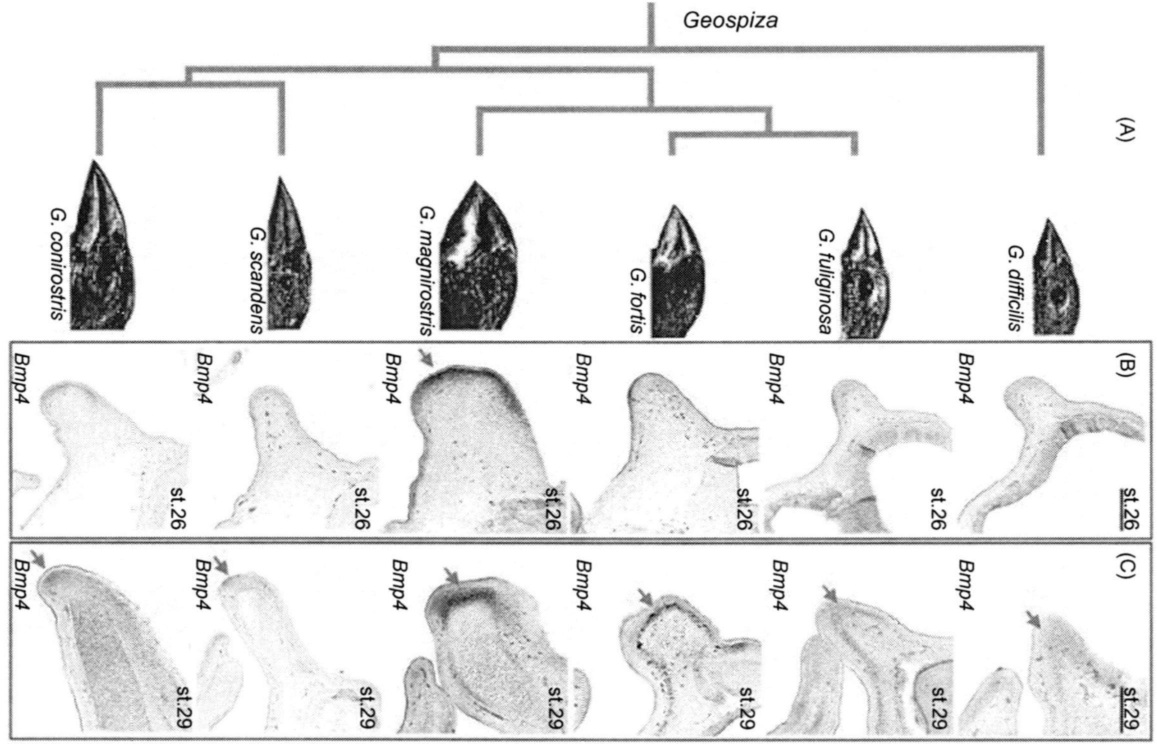

Figure 16.13 (A) *G. difficilis* is the most basal species of the genus *Geospiza*, and the rest of the species form two groups: ground and cactus finches, with distinct beak morphologies. (B) At stage 26, *Bmp4* is strongly expressed in a broad distal–dorsal domain in the mesenchyme of the upper beak prominence of *G. magnirostris* and at significantly lower levels in *G. fortis* and *G. conirostris*. No *Bmp4* was detected in the mesenchyme of *G. difficilis*, *G. fuliginosa*, and *G. scandens*. (C) At stage 29, *Bmp4* continues to be expressed at high levels in the distal beak mesenchyme of *G. magnirostris*. Broad domains of *Bmp4* expression are detectable around prenasal cartilages of *G. fuliginosa* and *G. fortis*. A small domain of strong *Bmp4* expression is also found in the most distal mesenchyme of *G. conirostris*, and weaker expression is seen in *G. scandens* and *G. fortis* (arrows). Scale bars: 1 mm in (B) and 2 mm in (C).
Source: From Abzhanov et al. (2004).

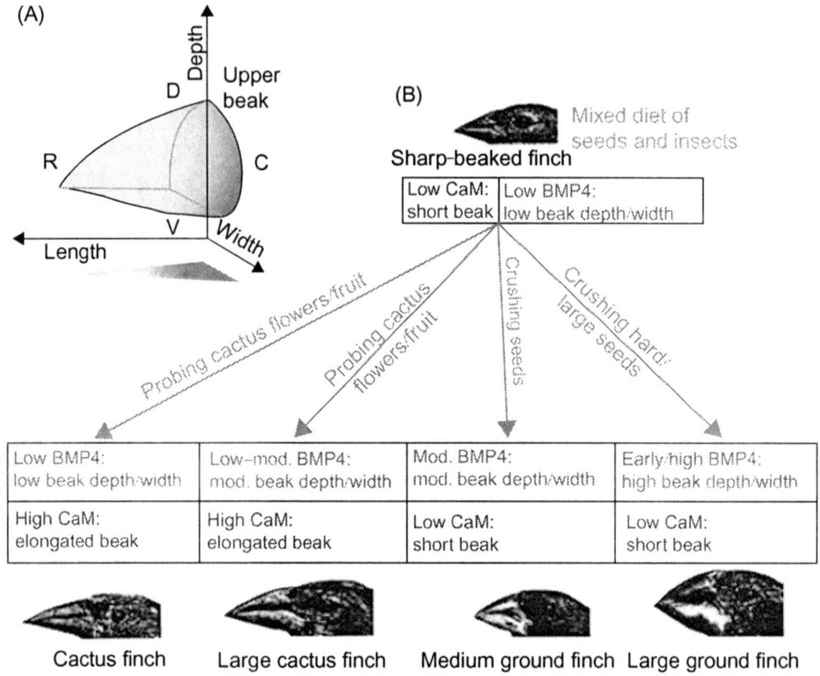

Figure 16.14 BMP- and CaM-dependent signaling regulates growth along different axes, facilitating the evolution of distinct beak morphologies in Darwin's finches. (A) Developing avian beak is a three-dimensional structure that can change along any of the growth axes. (B) A beak of the sharp-beaked finch reflects a basal morphology for *Geospiza*. The model for BMP4 and CaM involvement explains development of both elongated and deep/wide beaks of the more derived species. *Abbreviations*: C, caudal; CaM, calmodulin; D, dorsal; R, rostral; V, ventral.
Source: From Abzhanov et al. (2006).

Neo-Darwinian Explanation of Beak Evolution in Darwin's Finches

The sudden establishment of the inherited correlation between the beak morphology/morphometry and the singing patterns observed in two sympatric populations of *G. conirostris*, on Isla Genovesa, Galapagos, is enigmatic from the neo-Darwinian point of view. No separation of sympatric populations could be sustained under conditions of the random mating of individuals from two populations.

In all likelihood, the reproductive isolation between the narrow-long-beaked and the wider-shorter-beaked populations is maintained by the differences these populations have evolved in the types of songs they sing.

Based not only on general considerations that changes in behavior are not induced by changes in genes (see especially Section Animal Behavior Is Not Determined by Genes, Chapter 8) but also on the very short time and the unavoidable random mating in sympatry, accumulation of gene mutations can be excluded as a possible cause of reproductive isolation between the two populations. Leading investigators deny the likelihood that changes in genes may be responsible for the evolution of the reproductive isolation of the two morphologically indistinguishable sympatric populations of *G. conirostris* on the island (Grant and Grant, 1997). They found that hybrids between species of Darwin's species are viable and fertile to the same degree the offspring of nonhybrid matings are (Grant and Grant, 1997). They firmly conclude that speciation of Darwin's finches

> *continues in sympatry .... Differences in beak size between closely related species tend to be greater in sympatry than in allopatry .... Because the morphological traits are highly heritable, natural selection in one generation led to an evolutionary response in the next. The magnitude of the response was such as to probably dwarf any genetic changes occurring over the short term by drift and immigration.*
>
> *Grant and Grant (1997)*

There is no evidence of any genetic incompatibility between the two groups, which potentially can hybridize and produce viable and fertile offspring. There is no evidence that genetic recombination or changes in allele frequencies between the two groups have played any role. There is no evidence that the inherited variability in these traits is genetically determined. In short, there is an insuperable problem for the neo-Darwinian paradigm: the evolution and the maintenance of the two song types in two sympatric bird populations of a small island.

Above all, no neo-Darwinian mechanism would allow for evolution of two incipient species within the same small island, i.e., in the conditions of sympatry and gene flow between populations.

Epigenetic Explanation of Evolution of Beak in Darwin's Finches

Podos (2001) made an attempt to mechanistically explain changes in singing patterns of various species of Darwin's finches on the Galapagos Islands as a result of the respective changes in beak morphology. While there is no question that the beak morphology and morphometry imposes constraints on bird's singing, we should not forget that the birds' singing patterns and song learning are functions of the song circuit with song control centers in the brain as the central mechanism of song production. While the size and the shape of the beak can influence the quality of the song, the type of the song is determined by the song control centers and the song circuit, including the vocal apparatus. With the same beak, a bird can generate a very wide range of song types and lengths. The patterns of the movements of the singing apparatus that produce any type of song are determined in the bird's brain rather than in the beak.

The study of Grant and Grant (1997) has not been designed to prove whether the resulting change in the beak size might have been an appropriate morphological response of birds to the changed conditions of living. It has been taken for granted that it resulted from the action of natural selection on existing genetic variability in the beak shape and size. However interesting, the value of their study lies not only in what it proved but even more in what it did not:

- It did not prove that the inherited change in the beak size was a result of accumulation of gradual changes, and
- It did not exclude the possibility that the recorded change in the beak size was a sudden adaptive response to the changed condition in the environment (availability of big or small seeds). There is no convincing evidence in their studies that the hereditary variability on which the natural selection acted preexisted, hence the possibility exists that the phenotypic variability rose in response to the changed conditions of living.

It would not take a great deal of imagination to envision, in line with the theory, a hypothetical mechanism by which birds could rapidly adapt their beak morphology to two alternative (e.g., dry and wet) conditions of living. Under wet conditions, the absence of big seeds and abundance of small seeds would have a two-fold effect:

1. A high death rate of big-beaked individuals, and
2. A change in the behavior of the surviving small-beaked individuals.

The stress condition imposed by the drastic change in the weather conditions is associated with, and determined by, neurohormonal mechanisms. There is evidence that neurohormonal mechanisms of stress may induce inherited changes in developmental pathways and produce evolutionary changes without changes in genes (see Section Environmental Stress Induces Evolutionary Changes Without Changes in Genes, Chapter 7). The stress condition may trigger the neurohormonal system to increase the release of a signal, which, through neural and circulation pathways, can reach the sexual cells. This signal may contain instructions for neural crest cells to induce the development in the offspring of a beak that is different in size and shape.

How, under conditions of sympatry, such a high degree of segregation between the two groups of the same spatially uninterrupted population on a small island might have evolved? It is about four phenotypic characters (size and shape of beaks, feeding behavior, and song type) that are expressed in two different forms in two groups of the same population on a small island of the Galapagos archipelago. Let us remember that song production in birds is a function of song circuits that are under control of the brain song centers (see Section The Song Circuit in the Brain of Birds, Chapter 19).

We should bear in mind that not only within the group of Darwin's finches but even between bird species that are not closely related, such as chickens and ducks, gene regulatory networks (GRNs) and developmental pathways for beak development are well conserved. With genes and developmental pathways determining the development of beaks in Darwin's finches being, functionally at least, unchanged, with genetic recombination and changes in allele frequencies excluded as possible agents in the evolution, one is left with no alternative explanation but accepting that the sudden split of the original population of *G. conirostris* into two populations distinct in morphology and morphometry is determined by changes in the behavior of the neural crest cells. At a deeper molecular level, these changes are epigenetically determined not by mutational changes but by epigenetic changes in the spatiotemporal patterns of expression of BMPs (BMP4 and, to a lesser extent, by BMP2 and

BMP7) and calmodulin. This is definitely demonstrated and, as pointed out earlier, investigators have succeeded in manipulating the size and shape of the beak in Darwin's finches and produce beak shapes and forms resembling those of species in nature, by simply modifying expression of these two inducers (Abzhanov et al., 2006).

The size and shape of the beak is determined by neural crest cells, which before starting their migration from the neural tube to the target sites (in our case, in the presumptive beak) are provided with epigenetic information on where to go and what to do (Trainor et al., 2002; Schneider and Helms, 2003; Tucker and Lumsden, 2004). Both the upper and lower beak (as well as cranium) in birds develop from the CNC cells from the forebrain, the midbrain, and the brain rhombomeres 1 and 2.

The most important aspect of the formation of two sympatric populations of *G. conirostris* is the evolution of two different song types on which mate choice and reproductive isolation of these populations is based. How this divergence in the singing of males of two populations arose is not known. What is known is that no changes in genes and genetic mechanisms are involved in this process.

The sudden evolution of a new type of song among males of *G. conirostris* on the island created conditions for a split of the original population in two. Females that happen to prefer the new type of song will tend to mate with males that happen to sing that new song type. This led to formation of two separate fertilization systems and marked the beginning of a speciation process in sympatry. Thus, based on female mate preferences for the song types, two reproductively isolated populations emerged within the original population of finches on the island starting a process of speciation under conditions of sympatry, without spatial separation or geographical isolation of populations.

Based on a considerable number of described cases of sudden evolution of reproductive isolation without changes in genes between populations in sympatry (see Chapter 19, Epigenetics of Sympatric Speciation—Speciation as a Mechanism of Evolution), it may be inductively assumed that an epigenetic change in the properties of song circuits and song control centers in one of the sympatric populations led to the reproductive isolation of the two populations of *G. conirostris* on Isla Genovesa.

Evolution of the Adult Pigment Patterns in Danio Fish

Larval and adult patterns of melanophore stripes in the lineage of danio fish are different. The change in stripe patterns in most cases is related to the fact that, in distinction from the larval patterns that result from spatial arrangement of embryonic crest cells migrating to the skin (which transform into melanocytes and other pigment cells) from the neural crest, in the development of adult patterns postembryonic neural crest–derived stem cells are also involved (Quigley et al., 2004). In *Danio rerio*, the larval stripe pattern appears as early as 3 days postfertilization and remains for about 2 weeks. Then neural crest–derived latent stem cell precursors are transformed into melanophores and migrate to the flank region, thus patterning the adult stripes in the danio lineage. Another danio species, *Danio nigrofasciatus*, develops its adult stripe pattern differently. The adult stripe pattern is almost entirely determined by the larval neural crest–derived melanophores, which persist to adulthood, with the metamorphic melanophores only minimally involved in the adult stripe patterning. It is believed that the ancestral state is the one observed in *D. rerio* (Quigley et al., 2004).

Obviously, no changes in genes and no genetic variability or genetic processes whatsoever are involved in the development of the pigment patterning of these fish and in the evolution of the new form of patterning in *D. nigrofasciatus* from the ancestral *D. rerio*-like patterning. Moreover, there are no differences even in the melanophores, cytological elements determining stripe patterning and colors. Both species produce melanophores.

The evolutionary change in stripe patterning is purely epigenetic, i.e., related to changes in the behavior and patterns of organization of melanophores in these species. Remember, it is experimentally determined that the epigenetic information for migration of neural crest cells to the target sites in the skin, for their transformation into melanophores and for skin patterning, is provided to the neural crest cells before they leave the neural crest/CNS.

Evolution of the Cardiac Tract Outflow

The population of neural crest cells involved in formation of the complex morphology of the cardiac outflow tract originates from the (cardiac) neural crest. They populate the aortic pulmonary septum and conotruncal cushions prior to and during overt septation of the outflow tract. In both mammals and birds, the neural crest–derived cells contribute to the development of the arch of aorta, the carotid arteries (Figure 16.15).

Formation of smooth muscles of the coronary arteries in avian embryos, which previously was attributed to epicardium, is also determined by neural crest cells (Jiang et al., 2000).

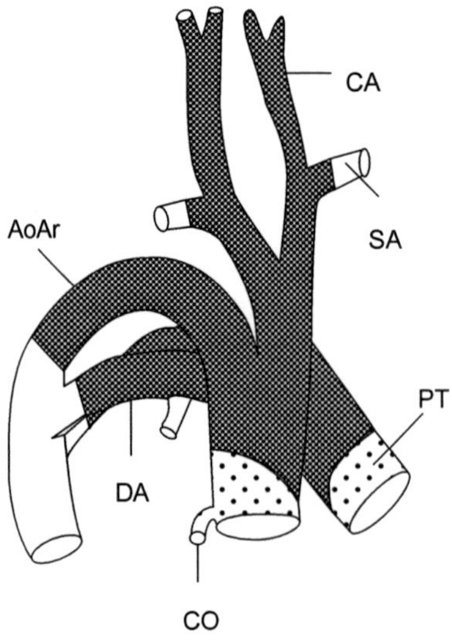

Figure 16.15 Schematic representation of neural crest cell contribution to the developing arterial system in the stage 40 chick embryos. *Abbreviations*: AoAr, aortic arch; AoS, aortic sac; AsAo, ascending aorta; BA, brachiocephalic artery; CA, carotid artery; CO, coronary artery; DA, ductus arteriosus; DAo, dorsal aorta; DsAo, descending aorta; PA, pulmonary artery; PT, pulmonary trunk; SA, subclavian artery; and III, IV, and VI, pharyngeal arch arteries. *Source*: From Bergwerff et al. (1998).

Neo-Darwinian Explanation

No neo-Darwinian explanation is possible as long as there is no evidence or indication that changes in existing genes or evolution of new genes are related to the evolution of the cardiac tract outflow.

Epigenetic Explanation

The evolution of the cardiac outflow tract is the result of changes in behavior of the neural crest cells coming from the cardiac neural crest. The plausibility of the epigenetic explanation derives from the fact that the migration of neural crest cells involved in the patterning of the cardiac tract outflow and the differentiation of them into cytological elements of the tract is predetermined by the *epigenetic* information that the cardiac neural crest cells are provided with, before leaving the neural tube/CNS (see Section Migration of Neural Crest Cells to Target Sites earlier in this chapter).

Evolution of the Forebrain Vasculature

The forebrain represents a morphological innovation of the CNS of higher vertebrates. Its evolution was associated with an expansion of the brain vasculature. While components of the dorsal posterior brain vasculature derive from the paraxial cephalic mesoderm, the later evolution of the ventral–anterior vasculature is determined by migratory neural crest cells from a specific brain region (Figure 16.16).

It appears plausible that there was a causal relationship between the construction of neural crest cells–dependent vascular domain and the continued expansion of the anterior brain and head over the course of evolution (Etchevers et al., 2001).

Neo-Darwinian Explanation

As is to be expected, no neo-Darwinian hypothesis has been presented for explaining the patterning of the new vasculature without changes in genes.

Epigenetic Explanation

Two epigenetic scenarios of the evolution of head vasculature may be considered:

1. Neural crest cells are programmed to direct the development of the vasculature before leaving the CNS, as it usually occurs in neural crest–derived structures, and
2. An experimentally identified neurogenic mechanism of the vasculature patterning may also be operational in the development of the forebrain vasculature: in the process of its development, the head innervation secretes vascular endothelial growth factors, which serves as a template for the parallel development of the vasculature (see Section Vasculogenesis and Angiogenesis, Chapter 5).

Neural Crest–Determined Development of Feathers in Bird's Head

Feathers in birds develop from feather buds in sequential rows that form pterylae ("capital tracts" in the craniofacial region) (see also Section Molecular Mechanism of Feather Development, Chapter 1).

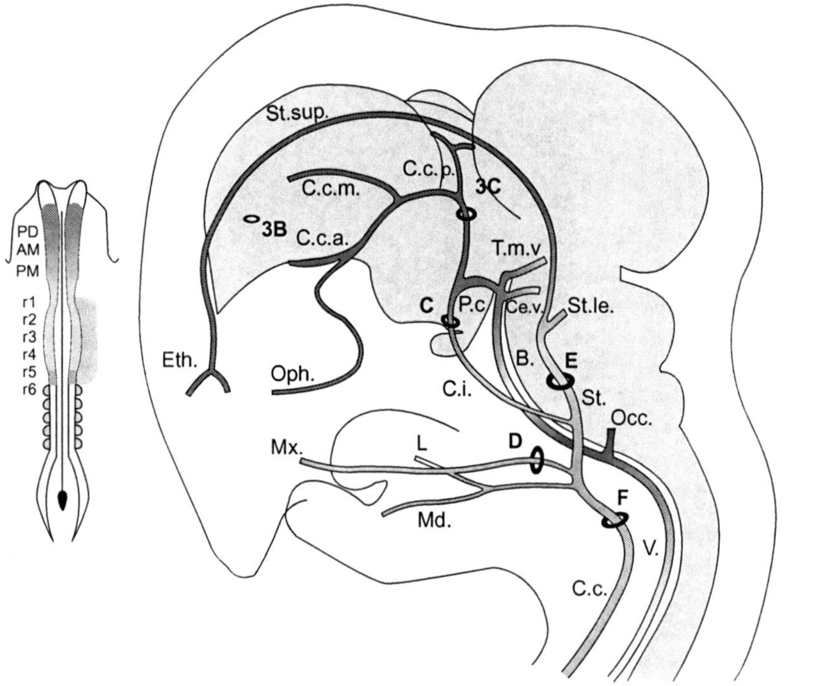

Figure 16.16 Cephalic neural crest cells and mesoderm (blank) contribute to the musculoconnective wall of separate arterial trees in the head. Darker shading corresponds to cells derived from posterior diencephalic (PD), anterior and posterior mesencephalic (AM, PM) neural folds; lighter shading corresponds to rhombomeres r1, r2, r4, and r5; and black to r6 (in the vascular media of a schematic E7.5 chicken head). Boundaries overlap between domains ensured by neural crest cells (NCCs) of given origins in vessel walls. Blank denotes cells of mesodermal origin, with a sharp boundary between the two at the diencephalon/ mesencephalon junction.
Source: From Etchevers et al. (2001).

The dermis in the craniofacial region of birds arises from neural crest mesenchyme, whereas the epidermis originates from the embryonic ectoderm. Initially, the neural crest mesenchyme of the craniofacial region forms a dermis layer underneath the ectoderm-derived epithelium, stimulating formation of an epidermal placode from the adjacent epithelium.

Homotopic transplantation of the premigratory neural crest cells from the midbrain and rostral hindbrain at the same embryonic stage, HH9.5, between quail and duck embryos, induces appearance of donor-specific spatial pattern and timing of the development of feathers on the head of these bird species (Figure 16.17). These experiments demonstrate the overriding regulatory capabilities of the neural crest–derived mesenchyme and the developmental plasticity of the head epithelium (Eames and Schneider, 2005).

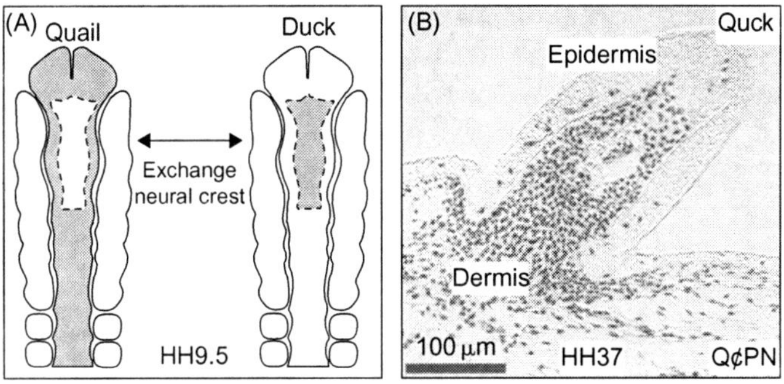

Figure 16.17 (A) Neural crest cells were cut either bilaterally (as shown) or unilaterally from the rostral neural tube and exchanged between quail and duck embryos stage-matched at HH9.5. Among other derivatives, these cells are destined to form much of the craniofacial dermis. (B) Chimeric "quck" feather follicles contain duck host epidermis and quail-derived donor dermis stained black with an anti-quail antibody (Q¢PN). Individual quail-derived melanocytes associated with the duck host epidermis are present. Scale bar: 100 μm. *Source*: From Eames and Schneider (2005).

The initial signal released by the neural crest–derived dermis for starting feather morphogenesis is BMP (bone morphogenetic factor), but it is possible that other genes important in feather development may be regulated by the donor-derived dermis, too (Eames and Schneider, 2005).

In both species, the quail and duck, expression of genetic markers is stage-matched, but in chimeric quck (neural crest from midbrain of quail transplanted to duck midbrain), the timing of gene expression and histogenesis in the dermis and epidermis was delayed by three stages. In duail chimerae (duck neural crest transplanted to quail midbrain) as well, the dermis and epidermis were transformed according to the identity of the donor species. Investigators came to the following conclusion:

> *Transplant experiments validate the instructive role of the neural crest-derived dermis as a primary source of spatiotemporal patterning information in the capital tracts, and suggest that the epidermis has only a permissive role .... Neural crest cells function as the dominant source of spatial and temporal patterning information via the regulation of genes essential to cranial feather morphogenesis.*
>
> *Eames and Schneider (2005)*

The evolution of colors and patterns of feathers and plumage in birds is not related to any relevant changes in genes or genetic variability. The same genes seem to have been used for producing the most diverse feathers and plumages. The neural crest cells involved in feather and plumage formation in birds provide the epigenetic information that enables the diversity of feathers and plumages that arose in the course of bird evolution. In turn, neural crest cells are provided with that information before leaving the neural tube/CNS.

References

Abzhanov, A., Protas, M., Grant, B.R., Grant, P.R. and Tabin, C.J. (2004). Bmp4 and morphological variation of beaks in Darwin's finches. Science *305*, 1462–1465.

Abzhanov, A., Kuo, W.P., Hartmann, C., Grant, B.R., Grant, P.R. and Tabin, C.J. (2006). The calmodulin pathway and evolution of elongated beak morphology in Darwin's finches. Nature *442*, 563–567.

Abzhanov, A., Rodda, S.J., McMahon, A.P. and Tabin, C.J. (2007). Regulation of skeletogenic differentiation in cranial dermal bone. Development *134*, 3133–3144.

Baker, V.H. and Bronner-Fraser, M. (1997). The origins of the neural crest. Part II: an evolutionary perspective. Mech. Dev. *69*, 13–29.

Bergwerff, M., Verberne, M.E., DeRuiter, M.C., Poelmann, R.E. and Gittenberger-de-Groot, A.C. (1998). Neural crest cell contribution to the developing circulatory system implications for vascular morphology? Circ. Res. *82*, 221–231.

Corbo, J.C., Erives, A., Di Gregorio, A., Chang, A. and Levine, M. (1997). Dorsoventral patterning of the vertebrate neural tube is conserved in a protochordate. Development *124*, 2335–2344.

Couly, G., Creuzet, S., Bennaceur, S., Vincent, C. and Le Douarin, N.M. (2002). Interactions between Hox-negative cephalic neural crest cells and the foregut endoderm in patterning the facial skeleton in the vertebrate head. Development *129*, 1061–1073.

Delcomyn, F. (1998). *Foundations of Neurobiology*. W.H. Freeman and Co., New York, NY, p. 251.

Eames, B.F. and Schneider, R.A. (2005). Quail–duck chimeras reveal spatiotemporal plasticity in molecular and histogenic programs of cranial feather development. Development *132*, 1499–1509.

Etchevers, H.C., Vincent, C., Le Douarin, N.M. and Couly, G.F. (2001). The cephalic neural crest provides pericytes and smooth muscle cells to all blood vessels of the face and forebrain. Development *128*, 1059–1068.

Gammill, L.S. and Bronner-Fraser, M. (2003). Neural crest specification: migrating into genomics. Nat. Rev. Neurosci. *4*, 795–805.

Gans, C. and Northcutt, R.G. (1983). Neural crest and the origin of vertebrates: a new head. Science *220*, 268–274.

Gerhart, J. and Kirschner, M. (1997). *Cells, Embryos, and Evolution*. Blackwell Science, p. 544.

Gilbert, S.F. (2000). *Developmental Biology*, sixth edition. Sinauer Associates, Inc. Publishers, Sunderland, MA.

Grant, P.R. and Grant, B.R. (1997). Genetics and the origin of bird species. Proc. Natl. Acad. Sci. U.S.A. *94*, 7768–7775.

Hall, B.K. (1998). *Evolutionary Developmental Biology*, second edition. Chapman & Hall, London , p. 166.

Hall, B.K. (1999). *The Neural Crest in Development and Evolution*. Springer-Verlag, New York, NY, p. 42.

Helms, J.A., Cordero, D. and Tapadia, M.D. (2005). New insights into craniofacial morphogenesis. Development *132*, 851–861.

Hildebrand, C., Fried, K., Tuisku, F. and Johansson, C.S. (1995). Teeth and tooth nerves. Prog. Neurobiol. *45*, 165–222.

Hong, K., Nishiyama, M., Henley, J., Tessier-Lavigne, M. and Poo, M. (2000). Calcium signalling in the guidance of nerve growth by netrin-1. Nature *403*, 93–98.

Hu, D., Marcucio, R.S. and Helms, J.A. (2003). A zone of frontonasal ectoderm regulates patterning and growth in the face. Development *130*, 1749–1758.

Imai, H., Osumi-Yamashita, N., Ninomiya, Y. and Eto, K. (1996). Contribution of early-emigrating midbrain crest cells to the dental mesenchyme of mandibular molar teeth in rat embryos. Dev. Biol. *176*, 151–165.

Itoh, M., Kim, C.H., Palardy, G., Oda, T., Jiang, Y.J. and Maust, D., et al. (2003). Mind bomb is a ubiquitin ligase that is essential for efficient activation of Notch signaling by Delta. Dev. Cell. *4*, 67–82.

Jeffery, W.R. (2007). Chordate ancestry of the neural crest: new insights from ascidians. Semin. Cell Dev. Biol. *18*, 481–491.

Jeffery, W.R., Strickler, A.G. and Yamamoto, Y. (2004). Migratory neural crest-like cells form body pigmentation in a urochordate embryo. Nature *431*, 696–699.

Jiang, X., Rowitch, D.H., Soriano, P., McMahon, A.P. and Sucov, H.M. (2000). Fate of the mammalian cardiac neural crest. Development *127*, 1607–1616.

Kinutani, M. and Le Douarin, N.M. (1985). Avian spinal cord chimeras. I. Hatching ability and posthatching survival in homo- and heterospecific chimeras. Dev. Biol. *111*, 243–255.

Köntges, G. and Lumsden, A. (1996). Rhombencephalic neural crest segmentation is preserved throughout craniofacial ontogeny. Development *122*, 3229–3242.

Lodish, H., Baltimore, D., Berk, A., Zipursky, S.L., Matsudaira, P. and Darnell, J. (1996). *Molecular Cell Biology*, third edition. Scientific American Books, New York, NY, p. 1177.

Lumsden, A.G. (1988). Spatial organization of the epithelium and the role of neural crest cells in the initiation of the mammalian tooth germ. Development *103*(Suppl.), 155–169.

Luo, Z.-X. (2001). A new mammaliaform from the early Jurassic and evolution of mammalian characteristics. Science *292*, 1535–1541.

Mallo, M. (2003). Formation of the outer and middle ear, molecular mechanisms. Curr. Top. Dev. Biol. *57*, 85–113.

Mallo, M. and Gridley, T. (1996). Development of the mammalian ear: coordinate regulation of formation of the tympanic ring and the external acoustic meatus. Development *122*, 173–179.

Martin, T. and Luo, Z.-X. (2005). Homoplasy in the mammalian ear. Science *307*, 861–862.

Meulemans, D. and Bronner-Fraser, M. (2004). Gene-regulatory interactions in neural crest evolution and development. Dev. Cell *7*, 291–299.

Moury, J.D. and Jacobson, A.G. (1989). Neural fold formation at newly created boundaries between neural plate and epidermis in the axolotl. Dev. Biol. *133*, 44–57.

Moury, J.D. and Jacobson, A.G. (1990). The origins of neural crest cells in the axolotl. Dev. Biol. *141*, 243–253.

Müller, W.A. (1996). *Developmental Biology*. Springer, New York, pp. 237–238.

Noden, D.M. and Trainor, P.A. (2005). Relations and interactions between cranial mesoderm and neural crest populations. J. Anat. *207*, 575–601.

Podos, J. (2001). Correlated evolution of morphology and vocal signal structure in Darwin's finches. Nature *409*, 185–188.

Quigley, I.K., Turner, J.M., Nuckels, R.J., Manuel, J.L., Budi, E.H., MacDonald, E.L., et al. (2004). Pigment pattern evolution by differential deployment of neural crest and post-embryonic melanophore lineages in *Danio* fishes. Development *131*, 6053–6059.

Rich, T.H., Hopson, J.A., Musser, A.M., Flannery, T.F. and Vickers-Rich, P. (2005). Independent origins of middle ear bones in monotremes and therians. Science *307*, 910–914.

Rinon, A., Lazar, S., Marshall, H., Büchmann-Møller, S., Neufeld, A., Elhanany-Tamir, H., et al. (2007). Cranial neural crest cells regulate head muscle patterning and differentiation during vertebrate embryogenesis. Development *134*, 3065–3075.

Rowe, T. (1996). Coevolution of the mammal middle ear and neocortex. Science *273*, 651–654.

Scherson, T., Serbedzija, G. and Bronner-Fraser, M. (1993). Regulative capacity of the cranial neural tube to form neural crest. Development *118*, 1049–1061.

Schneider, R.A. and Helms, J.A. (2003). The cellular and molecular origins of beak morphology. Science *299*, 565–568.

Scriven, P.N. and Bauchau, V. (1992). The effect of hybridization on mandible morphology in an island population of the house mouse. J. Zool. *226*, 573–583.

Selleck, M.A. and Bronner-Fraser, M. (1995). Origins of the avian neural crest: the role of neural plate–epidermal interactions. Development *121*, 525–538.

Sidow, A. (1992). Diversification of the *Wnt* gene family on the ancestral lineage of vertebrates. Proc. Natl. Acad. Sci. U.S.A. *89*, 5098–5102.

Smith, K.K. and van Nievelt, A.F.H. (1997). Comparative rates of development in monodelphis and didelphis. Science *275*, 683–684.

Sohal, G.S., Ali, M.M., Galileo, D.S. and Ali, A.A. (1998). Emigration of neuroepithelial cells from the hindbrain neural tube in the chick embryo. Int. J. Dev. Neurosci. *16*, 477–481.

Stone, J.R. and Hall, B.K. (2004). Latent homologues for the neural crest as an evolutionary novelty. Evol. Dev. *6*, 123–129.

Tapadia, M.D., Cordero, D.R. and Helms, J.A. (2005). It's all in your head: new insights into craniofacial development and deformation. J. Anat. *207*, 461–477.

Trainor, P.A., Ariza-McNaughton, L. and Krumlauf, R. (2002a). Role of the isthmus and FGFs in resolving the paradox of neural crest plasticity and prepatterning. Science *295*, 1288–1291.

Trainor, P.A., Melton, K.R. and Manzanares, M. (2003). Origins and plasticity of neural crest cells and their roles in jaw and craniofacial evolution. Int. J. Dev. Biol. *47*, 541–553.

Trainor, P.A., Sobieszczuk, D., Wilkinson, D. and Krumlauf, R. (2002b). Signalling between the hindbrain and paraxial tissues dictates neural crest migration pathways. Development *129*, 433–442.

Tucker, A.S. and Lumsden, A. (2004). Neural crest cells provide species-specific patterning information in the developing branchial skeleton. Evol. Dev. *6*, 32–40.

Tuisku, F. and Hildebrand, C. (1994). Evidence for a neural influence on tooth germ generation in a polyphyodont species. Dev. Biol. *165*, 1–9.

Wang, Y., Hu, Y., Meng, J. and Li, C. (2001). An ossified Meckel's cartilage in two cretaceous mammals and origin of the mammalian middle ear. Science *294*, 357–362.

Wu, P., Jiang, T.-X., Suksaweang, S., Widelitz, R.B. and Chuong, C.-M. (2004). Molecular shaping of the beak. Science *305*, 1465–1466.

Zheng, J.Q. (2000). Turning of nerve growth cones induced by localized increases in intracellular calcium ions. Nature *403*, 89–93.

17 Evolutionary Convergences: The Trend Toward Sameness in Metazoan Evolution

There can be little doubt that the tendency to vary in the same manner has often been so strong that all the individuals of the same species have been similarly modified without the aid of any form of selection.

Darwin (1872)

Predictability in Metazoan Evolution: Evolutionary Sameness

Biological evolution, as a process of adaptation of living organisms to different or changing conditions in environment, leads to diversification of their form and function. Controversy, however, arises in relation to the nature of phenotypic diversification; is it contingent or laws and rules are involved in determining the evolutionary diversification and the order observed in the living world. The prevailing view is the neo-Darwinian tenet that evolution of the living world is a product of spontaneous, randomly occurring mutations in genes, the raw material on which natural selection acts for shaping evolutionary changes that adapt living organisms to their environment. This neo-Darwinian view is commonly identified with Darwin's concept on the role of natural selection in evolution. This claim, however, does not faithfully represent Darwin's concept of organic evolution.

> *Darwin did not reduce the causal basis of organic evolution to natural selection but he believed in a tendency of individual of the same species "to vary in the same manner ... without the aid of any form of selection"*
>
> *Darwin (1872)*

The neo-Darwinian concept that randomly occurring favorable changes will be positively selected and transmitted to future generations has found its expression in the popular idea that rerunning the tape of life would lead to a completely different organic world. It views organic evolution on Earth to be accidental and unpredictable. Hence, divergence is the prevailing trend in the evolution of living things.

The biological diversification is not the exclusive trend of evolution in metazoans. Along it, for a long time, biologists have noticed an opposing tendency toward common patterns of phenotypic diversification. Metazoans often share phenotypic traits that, by origin, may be homologous, derived from the last common ancestor, or homoplasious. Different as they are by origin, homology and homoplasy are similar as far as the phenotypic result is concerned, for both imply the sameness or resemblance in

Epigenetic Principles of Evolution. DOI: 10.1016/B978-0-12-415831-3.00017-3

morphology or structure (Meyer, 1998; Wake, 1998) and, most importantly, both often may result from activation of similar developmental pathways.

Distinction between the homology and homoplasy (convergence, parallel evolution, evolutionary reversions, and vestigializations) is mainly based on whether the "same" structure existed or not in the last common ancestor rather than on genes, or gene regulatory networks and developmental pathways involved in their appearance, which often are similar. In fact, comparative studies have shown "an unexpected degree of conservation of genes and genetic programs," which made some students consider the distinction to be "somewhat artificial" (Meyer, 1998). While homology implies the evolutionary continuity and conservation of an ancestral trait, homoplasy is derived phenotypic sameness to which two or more taxa reach independently.

The recently introduced umbrella term of *sameness* is used here synonymously with *homoplasy*. Difficulties arise frequently in discriminating between the parallel evolution and evolutionary convergences, due to the uncertainty of identifying the ancestral forms of present species. For this reason, and because evolutionary vestigializations and losses of phenotypic characters are considered in Chapter 14, evolutionary convergences alone are considered here.

From an evolutionary point of view, sameness may result from the fact that solutions to evolutionary pressures may be limited, or, in the process of evolutionary transformation, different species may end up with the same solution.

Evolutionary Convergences in the Animal Kingdom

Evolutionary convergences imply the independent development of the same character in two or more lineages not linked by common descent. The concept of the morphological convergence implies a general visually perceived phenotypic similarity rather than identicalness of convergent structures. Convergences are evolutionary innovations that commonly result from activation of different developmental pathways.

Evolutionary convergence seems to be a more widespread phenomenon than is generally realized (Hodin, 2000), and the ubiquity of the phenomenon in the kingdom Animalia makes its study crucial for understanding mechanisms of evolutionary change. From a theoretical standpoint, the ubiquity of evolutionary convergences suggests that evolution of living forms might not be simply a contingent process of exclusively unpredictable outcomes.

Convenient and updated information on the ubiquity of evolutionary convergence and the pervasiveness of convergent evolutionary phenomena may be found in a review by Barlow (2003). Among the most eloquent examples of convergent evolution included in that review are:

- Evolution of wings from forelimbs in birds and mammals (bats);
- Evolution of flightlessness in insects and birds after migrating to islands;
- Independent evolution of flightless vegetarian birds in Africa (ostriches), South America (rheas), Australia (emus), and Madagascar ("elephant birds");
- Repeated independent evolution of gas bladders in fish and female octopuses;

- Evolution of venomous bite in snakes and at least two small Caribbean mammals;
- Evolution of bioluminescence in numerous deep-sea fish and insects;
- Echolocation (ultrasonic hearing) in insects, birds (several owl species), and mammals (whales and bats);
- Independent evolution (10 times) of venomous sting in taxa ranging from lower vertebrates, such as coelenterates (jellyfish) to arthropods (molluscs, spiders, and insects), vertebrates such as fish, reptilians (snakes), and even mammals (male duckbill platypus);
- Use of magnetically charged particles of magnetite for orientation during migration in butterflies, fish, and birds;
- Evolution of external organs for producing auditory signals and songs in insects and American tropical birds (manakins);
- Evolution of eusociality, i.e., living in colonies, implying division of labor and castes of distinct morphological, behavioral, life history, and other traits.

Convergent Evolution of Eyes

From the neo-Darwinian view, evolution of complex structures, such as eyes through accumulation of small gradual changes under the action of natural selection, could not have happened more than once. Nevertheless, the eye is an often-quoted example of convergence thought to have independently evolved about 40–70 times in the course of metazoan evolution (Salvini-Plawen and Mayr, 1977).

We know that the development of eyes in species as different as humans and *Drosophila* involves, and is under control of the same basic homologous genes *Pax-6*, which, despite mutations, to which it has been subject, has amazingly remained functionally unchanged after more than 500 million years of separate evolution of these taxa (Quiring et al., 1994). With the same basic control gene and with the same highly conserved gene, regulatory network organisms belonging to very remotely related taxa, such as mammals and cephalopods (octopuses and squids), have evolved similar camera eye structures (Figure 17.1).

While insects and other invertebrates have five times independently evolved compound eyes, with numerous ommatidia as the basic unit of visual reception, the camera eye evolved independently at least seven times in both vertebrates and invertebrates.

Convergence of Electrical Organs and Electroreception in Fish

Striking examples of convergence are observed in electric fish of distantly related groups, such as Mormyriformes of Africa and Gymnotiformes of South America, which share no common electrogenic or electrosensory ancestor (Figure 17.2). Special structures and physiological mechanisms for emitting and receiving electrical signals have independently evolved several times in these two groups. Both groups have also independently evolved sinusoidal wave-type electrical organ discharges (EODs) of constant rates and pulse-type, separated by long intervals;

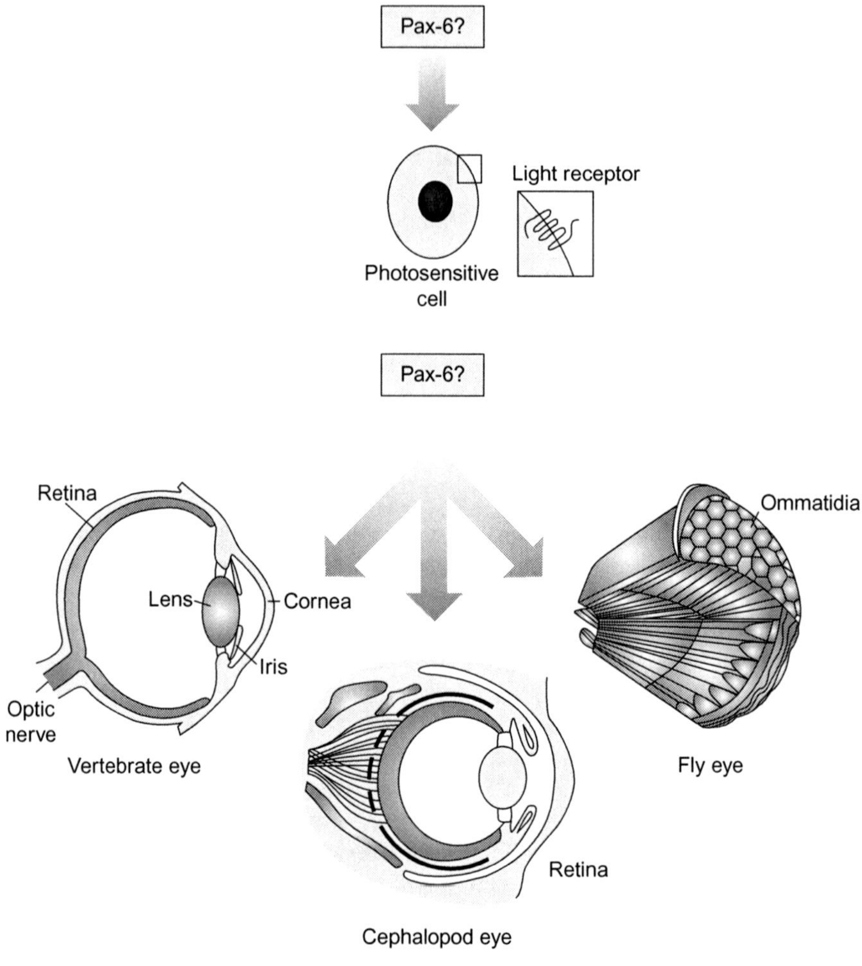

Figure 17.1 Controller of all eyes? Possible sites of action of *Pax-6* in the development of three very different types of eyes—human (vertebrate), octopus (cephalopod), and the compound eye of *Drosophila*.
Source: From Zucker (1994).

furthermore, both groups have evolved three types of electroreceptors with distinct functions.

Differences in the neural basis of differential phase comparisons are identified even within the group of mormyrids [in *Gymnarchus*, comparisons are made in the hindbrain, whereas in *Brienomyrus*, they are made in the midbrain (see also section "The Structure of the System of Electroreception in Fish" in Chapter 19)].

Finally, and more surprisingly, both distant groups have independently evolved similar neural networks in the brain for changing their own EOD frequencies in cases of jamming from conspecific EODs (Nishikawa, 2002).

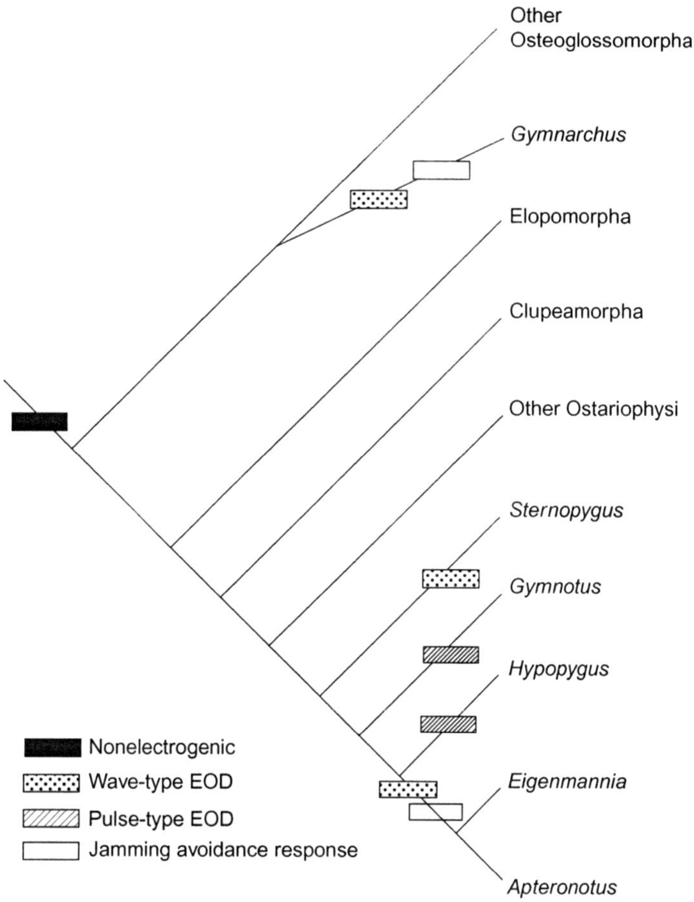

Figure 17.2 Cladogram illustrating convergent evolution of wave-type electric organ discharge (EOD) and jamming avoidance responses in African mormyrid and South American gymnotid fishes (based on Alves-Gomes et al., 1995). The common ancestor of mormyrids and gymnotids was nonelectrogenic. Wave-type EODs and jamming avoidance responses evolved independently in *Gymnarchus* and in the common ancestor of *Eigenmannia* and *Apteronotus*. *Sternopygus* possesses a wave-type EOD but lacks a jamming avoidance response.
Source: From Nishikawa (2002).

Evolutionary Convergences of the Nervous System

These convergences are believed to have arisen from similar intrinsic pressures for increased computational capabilities (Carr and Soares, 2002; Eisthen and Nishikawa, 2002; Nishikawa, 2002). Striking similarities are observed among brain stem circuits encoding auditory signals in birds and mammals (Carr and Soares, 2002). As for the widespread convergences in the structure and physiology of the central nervous systems, it is thought that they result from the properties of neural circuits rather than any changes in genes. (Changes in properties of neural circuits are related to changes in organization of neural circuits, which involve no changes in genes.)

> *One reason that convergence is so common in biological world may be that the evolutionary appearance of novel functions is associated with constraints, for example in the algorithms used for a given neural computation. Convergence in functional organization may thus reveal basic design features of neural circuits in species that possess unique evolutionary histories but use similar algorithms to solve basic computational problems.*
>
> *Nishikawa (2002)*

Endbulbs (arborized synaptic endings of the ascending auditory nerve fibers in the brain) are believed to have "emerged as an adaptation for accurate transmission of phase information for frequencies above 500 Hz, perhaps associated with the development of hearing in land vertebrates" (Carr and Soares, 2002). Studies on the structure of endbulbs have shown that they have evolved convergently in reptiles, birds, and mammals (Figure 17.3).

> *[M]orphological similarities might not support homology, but rather similar computational demands, and we can argue that the neurons of nucleus laminaris and MSO (mammalian medial superior olive—N.C.) might have converged upon their similar form. In another example, it appears that large somatic terminals on NM (nucleus magnocellularis—N.C.) or bushy cells are an ancestral feature of amniote auditory nerve.*
>
> *Carr and Soares (2002)*

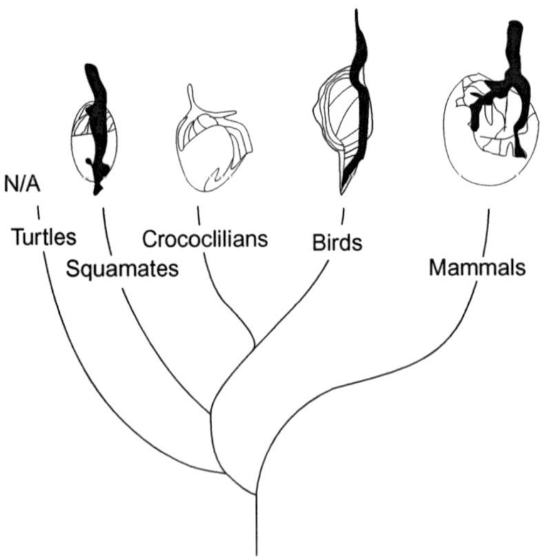

Figure 17.3 Large somatic terminals appear in NM (nucleus magnocellularis—N.C.) of birds, crocodilians, and the alligator lizard, as well as the mammalian ventral cochlear nucleus. These terminals are similarly elaborate in birds and mammals, and fractal analysis reveals similar morphological complexity (Ryugo et al., 1996; Carr et al., 1997). This complexity appears to have developed in parallel in both birds and mammals.
Source: From Carr and Soares (2002).

Convergence of Cerebellum-Like Structures

Among cerebellum-like structures that have independently evolved in vertebrates are the dorsal octavolateral nucleus (DON), medial octavolateral nucleus (MON), the marginal layer of the optic tectum, the electrosensory lobe (ELL), the rostrolateral nucleus (RLN) of the thalamus, the dorsal cochlear nucleus (DCN), and the cerebellum proper that evolved in almost all craniates (Bell, 2002). Similarities among these cerebellum-like structures can be best explained by convergence, rather than by homology, whereas evolution of ELL is considered to have resulted from the evolutionary reversal of the electroceptive DON of more basal vertebrates (Bell, 2002). It has not been claimed or alluded that changes in existing genes or the evolution of new genes have been involved in the evolution of the cerebellum-like structures.

Convergence of Ear Asymmetry in Owls

The circuitry for sound localization in birds and mammals has evolved convergently rather than inherited homologously from their common ancestor, as is conventionally assumed (Carr and Soares, 2002). Some owl species (e.g., the barn owl, *Tyto alba*) are the only vertebrates that have evolved asymmetrical ears, i.e., ears that are asymmetrically positioned, one lower and the other higher on each side of the head. This causes the sound of the prey to reach the owl's ears at different times, enabling the bird, by calculating this interaural (Lat. *auris*, ear) time difference (ITD) in circuits of the auditory system, to determine, in darkness, with a high degree of precision, the position of the prey.

Thus, asymmetrical-eared owls have the unique advantage of precision-preying under conditions of deep darkness, in the absence of visual and other cues, relying exclusively on sounds generated by the prey, whereas owls with symmetrical ears, under such conditions, cannot even fly. The neuroanatomy of the auditory pathways is similar in both the symmetrical- and asymmetrical-eared owls. The relevant difference is the fact that asymmetrical-eared owls have in the nucleus laminaris 10 times more (10,000) neurons than chickens. Asymmetrical ears evolved independently four to seven times among owls (Nishikawa, 2002; Figure 17.4).

In general, brain nuclei involved in the processing of binaural auditory cues are also similar between owls with symmetrical and asymmetrical ears, except that owls with asymmetrical ears have greater acuity of sound localization. In this system, as in others, small changes in existing neural pathways appear to underlie the emergence of novel abilities, such as the ability to catch prey in total darkness using auditory cues alone (Nishikawa, 2002). It has never been suggested that new or changed genes are involved in the night hunting ability of the barn owl.

Convergence of the Ballistic Tongue

The ballistic tongue is an adaptation of the tongue for capturing prey in distance. The ballistic tongue and the respective behavior of its protraction and retraction back to

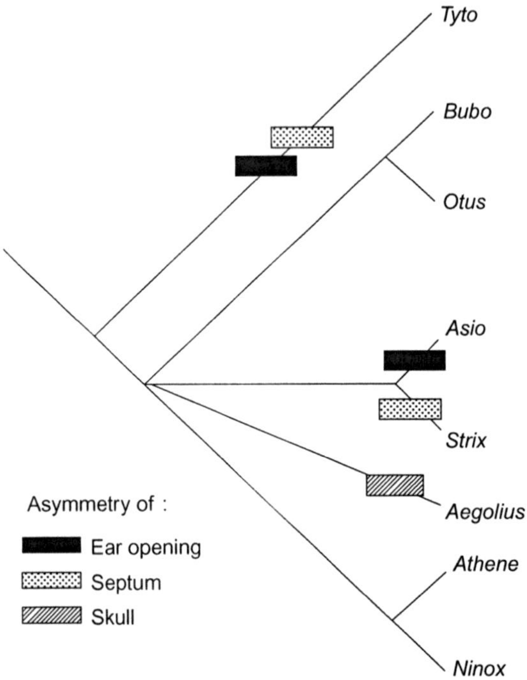

Figure 17.4 Cladogram illustrating convergent evolution of ear asymmetry among owls. *Tyto* and *Asio* possess asymmetries of the soft tissues of the ear opening, and *Tyto* and *Strix* possess asymmetries of the septum. Only *Aegolius* possesses asymmetries of the cranium. Intrageneric variation (not shown) is present in *Bubo* and *Strix*. Given this level of variation in ear anatomy, ear asymmetry may have evolved as many as seven times independently among owls.
Source: From Nishikawa (2002).

the mouth have independently evolved many times among anuran amphibians and salamanders. (Only among anuran amphibians it has evolved at least eight times.) (Deban et al., 1997; Nishikawa, 2002). The chameleon can project its tongue as far as 1.5 times its body length (de Groot and van Leeuwen, 2004).

The motor that generates the energy for protraction and retraction of the ballistic tongue are paired subarcualis rectus muscles (Deban et al., 2007), innervated by sensory neurons. These neurons that innervate tongue epithelial mechanoreceptors have independently evolved four to five times among anurans with ballistic tongue projection (Nishikawa, 2002; Figure 17.5).

Different mechanisms of neuromuscular control of tongue protraction have led to the evolution of three different mechanisms of tongue pulling (mechanical pulling, inertial elongation, and hydrostatic elongation) (Nishikawa, 2002).

Convergent transition to ballistic tongue projection in anuran amphibians was associated with convergent evolution of hypoglossal afferents, and the latter is based on minor changes in connections of sensory neurons that control muscle activity (Nishikawa, 2002).

Among salamanders, members of the genus *Hydromantes* of the Plethodontidae family are also known to have and use ballistic tongue to shoot rapidly and with the greatest accuracy (Deban et al., 1997; Figure 17.6). There is no evidence or hypothesis that evolution of the ballistic tongue may be related to any change in genes or DNA.

Figure 17.5 Cladogram illustrating convergent evolution of ballistic tongue projection (via inertial elongation) and hypoglossal afferents among anurans. The common ancestor of anurans used mechanical pulling to protract its tongue and lacked hypoglossal afferents. Inertial elongation and hypoglossal afferents that modulate jaw muscle coactivation evolved up to five times independently among anurans.
Source: From Nishikawa (2002).

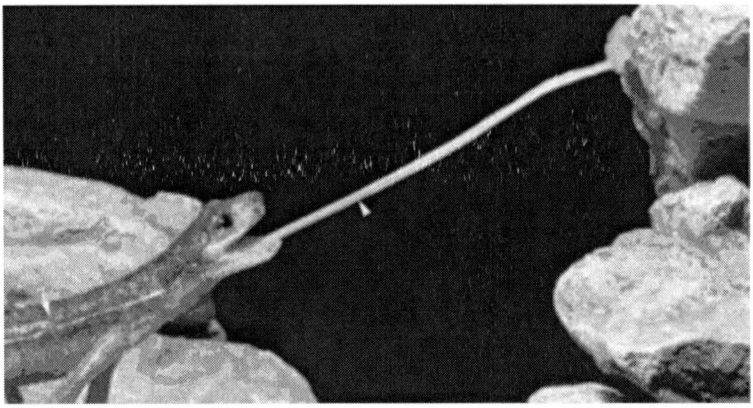

Figure 17.6 Live *Hydromantes supramontis* firing its tongue at a housefly. The tongue pad almost engulfs the fly yet is not fully extended. Arrow indicates the posterior extent of the tongue protractor muscle, which is visible as a bulge beneath the skin. Arrowhead points to the posterior ends of the tongue skeleton, which extends forward to the tongue pad. It has left the mouth entirely and is anchored by the retractor muscles.
Source: From Deban et al. (1997).

Convergence of Plumage Patterns in Orioles

The study of plumage in orioles has revealed that this trait has homoplastically evolved often and very rapidly in direction of the reversion to ancestral states (Omland and Lanyon, 2000). Results of an earlier study by the middle of the twentieth century showing convergence of plumage patterns in orioles were ignored because of apparent incompatibility with the neo-Darwinian hypotheses on the evolution of signaling traits and the irreversibility of evolution.

While the runaway hypothesis and good gene hypothesis of sexual selection would predict divergence in the evolution of mate signaling, the reconstruction of the plumage evolution in orioles, to the contrary, has shown that it is the convergence and reversal to ancestral plumage patterns that dominates the rapid evolution of the trait in these birds.

The widespread convergence of plumage patterns in sympatry in orioles is problematic because it could not be explained by sexual selection and selection for crypsis in the same or similar habitats. Sympatric evolution of orioles shows *overall* convergence without convergence in overall appearance, which is different from what would be expected from selection under similar conditions. Furthermore, several allopatric oriole species of North America living in regions as different as desert and forest regions also have converged to similar overall patterns despite the widely different environmental conditions, clearly suggesting that something different from selection is responsible for their convergence to the same Baltimore-type pattern.

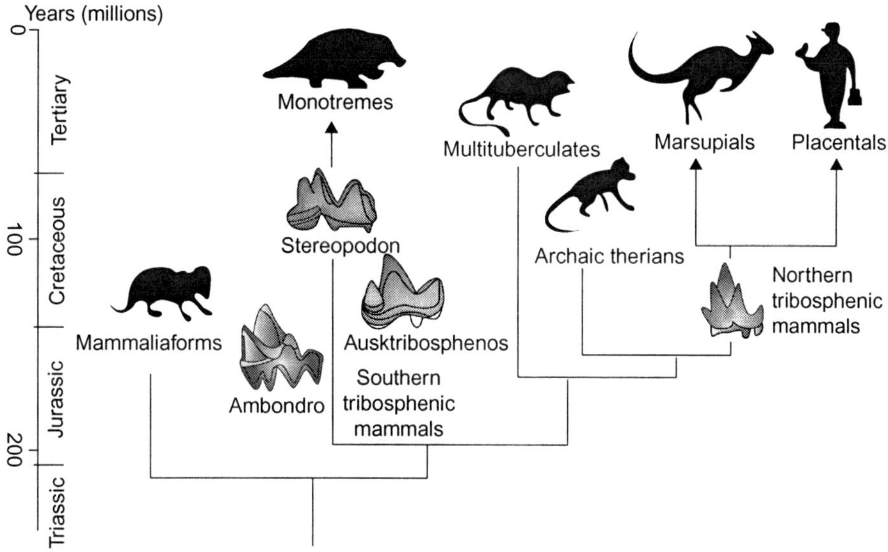

Figure 17.7 "Unique" mammalian molars actually may have evolved twice.
Source: From Stokstad (2001).

> *Overall convergence in pattern without convergence in overall appearance is less consistent with any of the possible selective explanations based on sympatry or similar environments ... The convergence in overall plumage types among distantly related orioles results from a combination of convergence, reversal, and retained ancestral states in individual characters.*
>
> *Omland and Lanyon (2000)*

With the divergence of signaling traits predicted by the runaway and good gene hypotheses being empirically rejected, an alternative epigenetic explanation is warranted. The epigenetic paradigm would predict that the phenotypic convergence of plumage patterns in orioles is a result of the convergence of mating preferences determined by neural circuits specifying these preferences and mating behaviors, which require no changes in genes or genetic information (see sections "Evolution of Mating Preferences" and "Evolution of Receiver Biases" in Chapter 19).

Convergence of the Tribosphenic Molar Teeth

Tribosphenic (three-cusped) molar teeth evolved independently twice in mammals, once in ambondros (middle Jurassic, Madagascar) and ausktribosphenos (early Cretaceous, Australia) as can be seen in modern monotremes (insectivorous monotremes and platypuses, with the latter having no teeth as adults) and later in placentals and marsupials in early Cretaceous in the northern hemisphere (Zhe-Xi et al., 2001; Figure 17.7).

Multifaceted Convergence in Tunas and Lamnid Sharks

Tunas (*Thunnus* genus) and sharks of the Lamnidae family are large-sized, highly specialized predator fish that diverged from their common ancestor ~400 Mya. Ever since, they have convergently evolved regional endothermy (ability to warm particular regions of the body via capillary nets formed of veins and arteries running in parallel where veins transfer to arteries metabolic heat for warming up the region), similar biochemical features, modifications of blood vessels, and similar musculotendinous designs related to the high-speed swimming, which other fish taxa lack. In these groups, red muscles are disconnected from white muscle sheets and moved from subcutaneous position toward the center of the torso, and all the force of their contraction is transferred exclusively to the tail via specialized long tendons (Shadwick, 2005; Figure 17.8).

Convergence to a Common Mode of Reproduction in Frogs

Another impressive example of evolutionary convergence in metazoans is the unusual mode of reproduction that evolved independently in seven amphibian genera, which comprise about 60 species of frogs in rain forests and mountains of Central America and tropical South America. Their eggs develop in a special pouch on their mother's back. In some of these species, the adult male individuals transport eggs from the female's cloaca into the back pouch by fertilizing them with their sperm. The embryonic development in the pouch resembles mammal pregnancy in its two-stage hormonal regulation of pregnancy and in the respective parturition behavior. The structure of the inner pouch is also reminiscent of placenta. Amino acid comparison data show that these frogs evolved between 40 and 80 million years ago in the areas where they now live (del Pino, 1989). There is no indication that the evolutionary convergence of frogs to this reproduction mode may be related to any change in genes.

Convergence of Overall Body Form in Aquatic Mammals and the Reptilian *Ichthyosaurus*

The return to the ancestral aquatic habitat in species of two different classes such as reptiles (the extinct ichthyosaurus) and mammals induced reversion of the ancestral fish-like body (Figures 17.9 and 17.10).

Their anterior appendages were shortened and flattened and ultimately evolved into paddles and in both of them a remarkable increase in the number of phalanges (about two dozen per digit) occurred (Kent, 1973). The characteristic fish-like body of ichthyosaurus, porpoise (a small cetacean), and shark, respectively representing classes of reptiles, mammals, and fish, have always been textbook examples of convergent evolution.

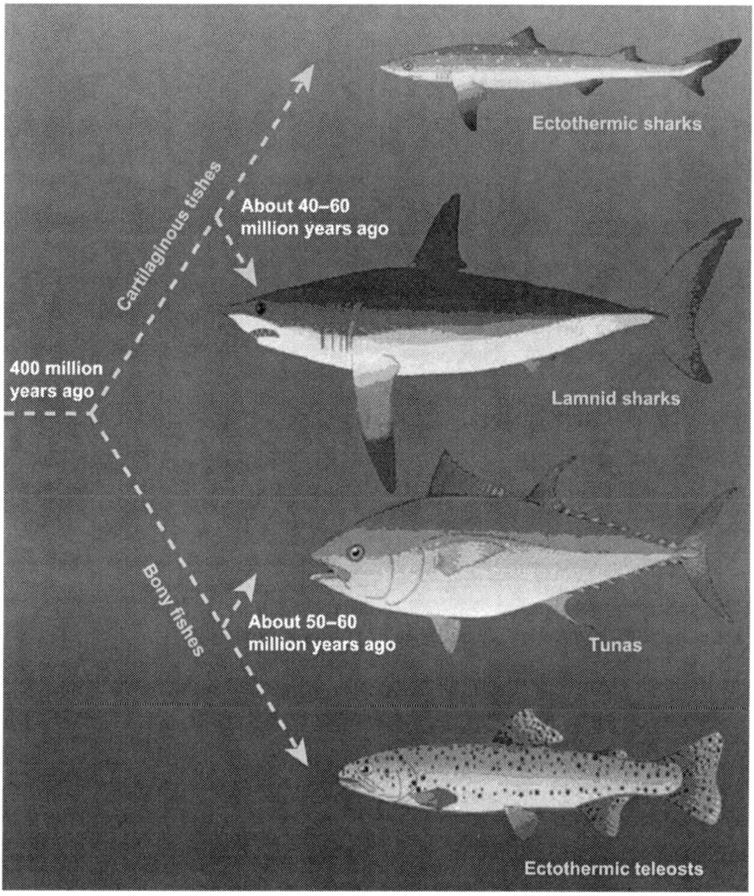

Figure 17.8 Cartilaginous and bony fish diverged about 400 million years ago, making sharks and tunas distant relatives. Most fish are ectothermic, but tunas and lamnid sharks both have regional endothermy and other traits that aid their speed. These near-identical traits arose independently, in a dramatic case of convergent evolution, probably owing to similar selection pressures that began to affect both groups of fish about 40–60 million years ago.
Source: From Shadwick (2005).

Ichthyosaurus' immediate ancestor was a terrestrial reptile. Transition to the marine habitat brought about shortening of their limbs and limb bones, disappearance of joints between them, replacement of phalanges by numerous fin rays, and evolution of fish-like fins. Transformation of tetrapod limbs into fish-like fins, transformation of their tail into a vertically positioned tail fin, and emergence of a back fin in some species were basic steps in the process of evolution of the tetrapod reptile into a fish-like reptile.

Looking at the shark likeness of certain mammals and shark-like adaptations developed by these mammals that shifted from terrestrial to marine habitats (e.g.,

Figure 17.9 Historical restoration of *Ichthyosaurus communis*.
Source: From http://paleontology.ac/atelier/Ichthyosaurus.html

Figure 17.10 Dolphins.
Source: From http://dsc.discovery.com/news/2008/06/10/dolphins-comeback.html

whales, dolphins, and porpoises) and then at the reptile ichthyosaurus, which has undergone similar shifts, one marvels at the reappearance of similar shark morphology in such very remotely related taxa. No evidence has been provided that reversion of these animals to ancestral fish-like shape and adaptations is related to specific changes in genes or regulatory sequences.

Convergence of Placental Mammals and Australian Marsupials

More than 100 million years ago, the Pangea, the supercontinent, began splitting into present continents. Australia broke off the Old World about 135 million years ago, and South America separated from Africa ~65 Mya, i.e., before the beginning of the mammal dominance and the evolution of placental mammals. Surprisingly, the evolution of mammals in the already separated continents very often followed similar patterns and led to emergence in Australia of numerous species that are morphologically very similar to those that evolved in the Old World. One often-quoted example is the Tasmanian marsupial wolf, which is considered to be a "carbon copy" of the European timber wolf. Both of them have a common ancestor (Milton, 1997). One group of South American carnivores, the *Borhyaenidae*, also is anatomically similar to the Tasmanian wolf (Wills, 1989). Unpredictable similarities are also observed between many other mammals in Australia and the Old World, such as koalas and bears, sugar gliders, and flying squirrels (Figure 17.11).

In view of the widely diverse environmental conditions in the Old World, Australia, and South America, as well as of the long geological time since the split of these continents from Pangea, convergence in the morphology of placentals and marsupials would not have been predicted from the neo-Darwinian view.

Paleontological Evidence for Convergence

It is very interesting to look back at the neglected view of paleontologists of the 1920 and 1930s on evolutionary convergences in the paleontological record. It clearly suggests that early paleontologists believed that the process of evolutionary convergence was not related to convergent evolution of genotypes. Evolution of ichthyosaurus and cetaceans to a shark-like shape was considered to be a "latent homoplasy" (Beurlen, 1937) of living forms

> *linked through specific laws of development of the form that are common for the whole group of species ... all the conservative and extremely iterative forms from the beginning to the end are the same, only their phenotypic expression of the genotype is different.*
>
> *(1937, p. 66)*

Convergence, thus, was considered to be a form of expression of common structure and function involving no changes in genes.

The whole *Ornithis* group of dinosaurs acquired early the bipedal posture, which led to a modification of the reptile pelvis analogous to that of birds, but their descendants *Ceratops* (late Cretaceous dinosaur) and *Stegosaurus* returned to the quadrupedal locomotion. The requirements of that originally reptile posture were met by those species not by undoing previously evolved bipedal pubis bones (those bones still were present). Instead, both *Ceratops* and *Stegosaurus* evolved an additional bone

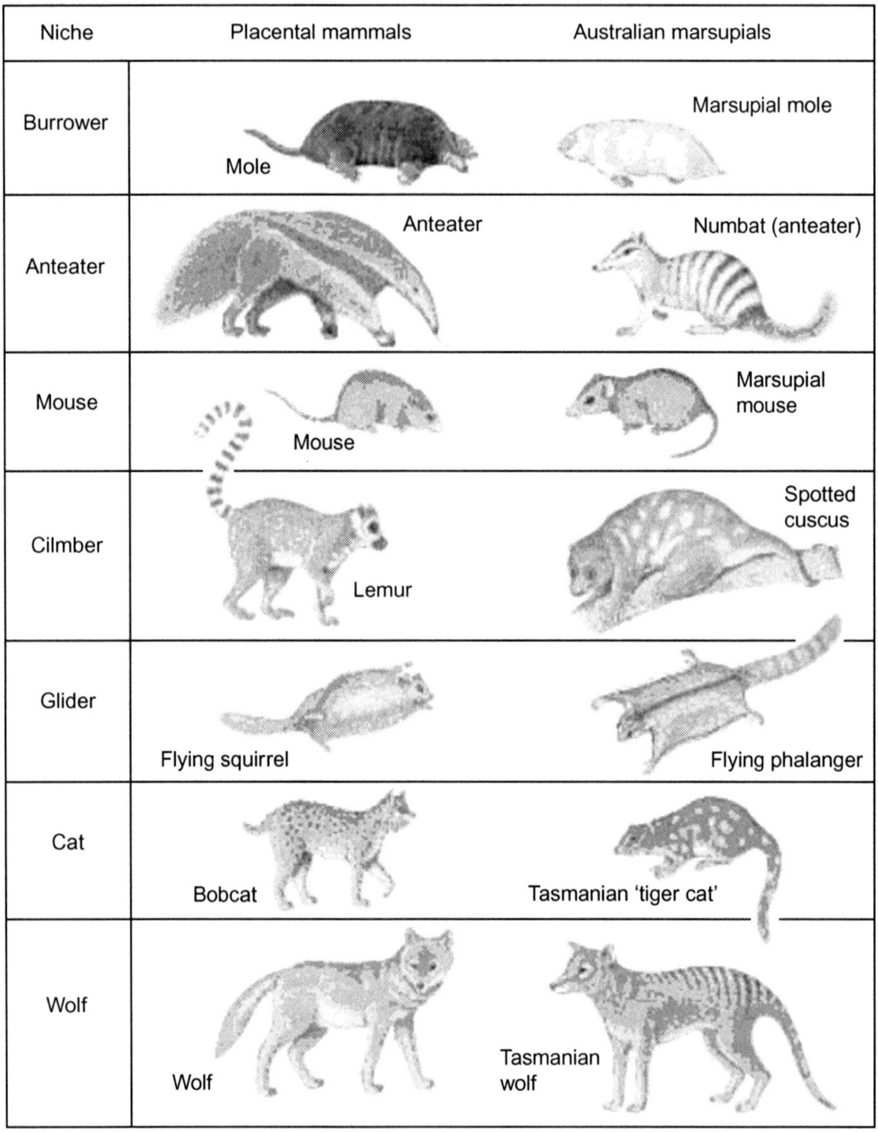

Figure 17.11 Convergent evolution of morphology in placental and marsupial species in the Old World and Australia.
Source: From Dr. George Johnson's Backgrounders (Internet).

to carry out the functions of the original pubis bones, the so-called *praepubis*. What happened with these extinct species is not uniquely restricted to them. This led to conclusion that crocodiles descended from *sphenosuchids*, which have been bipedals. And this is why modern crocodiles inherit a fully developed praepubis (Beurlen, 1937, p. 28). The same occurred with the ancestors of pterosaurs. In both groups, bipedalism has been only a transitional episode (Beurlen, 1937, p. 57). In both of them, a bird-like backward positioning of pelvis has occurred.

Neo-Darwinian Explanation of Convergent Evolution

According to the neo-Darwinian view, species living in similar environmental conditions experience similar evolutionary pressures, and natural selection, acting on the random changes or on the existing genetic variability, leads to similar phenotypic solutions.

The idea that similar conditions of living would systematically lead to evolution of similar characters in different species not linked by common descent is confronted with some objections:

> First, the extremely low frequency of gene mutations, when multiplied by several lower orders of the occurrence of "useful" mutations, makes the possibility of the same useful mutations in genes occurring in two metazoan species highly unlikely.
> Second, and equally importantly, it has never been possible to demonstrate that a particular change in a gene or in a number of genes led to an evolutionary convergence in nature.
> Third, numerous convergences have evolved in species inhabiting regions of the earth that are ecologically very different and, consequently, have been subjected to different "evolutionary pressures."

The triple convergence in morphology of placentals and marsupials that occurred on three widely different continents (Europe, Australia, and South America) obviously cannot be explained reasonably with the similarity of conditions of living or evolutionary pressures to which these animals were subject to.

Epigenetic Explanation of Evolutionary Convergences

As pointed out earlier, Darwin himself believed in the possibility of convergent evolution based on directed change "without the aid of any form of selection" rather than on random variability (gene mutations in modern terminology):

> *If the varying individual did not actually transmit to its offspring its newly acquired character, it would undoubtedly transmit to them, as long as the existing conditions remained the same, a still stronger tendency to vary in the same manner.*
>
> *Darwin (1872)*

Evolutionary phenotypic changes in metazoan morphology may imply modification of existing developmental pathways, reactivation of ancestral pathways, or evolution of new pathways. It is important to bear in mind that neither modification of existing pathways nor evolution of new developmental pathways implies obliteration or irreversibility of previous or ancestral developmental pathways; gene products involved in these pathways generally are still present and functionally unchanged. Only the spatiotemporal pattern of their expression is epigenetically changed and regulated.

The epigenetic paradigm would relate the occurrence of evolutionary convergences with the similarity of solutions to problems arising from the adverse effects of environmental conditions, with the limited number of developmental algorithms (signal cascades) and constraints on modification of these algorithms. Since these signal cascades start with neural signals, one basic prediction from the view of the epigenetic paradigm would be that the frequency of evolutionary convergences would dramatically increase with the evolution of the nervous system, coinciding with the Cambrian explosion.

This prediction seems to have been validated in a recent study by Vermeij showing that before the Cambrian explosion, evolutionary convergences have been rare events, but later evolutionary innovations appear repeatedly. Vermeij estimated that only 23.3% of first 56 convergences occurred during the first 2.5 billion years of life on Earth, and 76.7% during the last 0.5 billion years since the Cambrian explosion. Only 4% of the singular nonconvergent innovations have occurred during the last ~250 million years, indicating a clear tendency of increased frequency of evolutionary convergences in the course of metazoan evolution (Vermeij, 2006):

> *If these inferences are correct, they would imply that history during its early phases was substantially more contingent, that is, more dependent on singular circumstances, than are more recent historical episodes. In other words, unique "frozen accidents" were more common in the very distant past than in more recent times.*
>
> *Vermeij (2006)*

No relation seems to exist between frequencies of gene mutations and the increased frequency of evolutionary convergences in metazoans.

What might have determined the abrupt change from the singularity to repeatability of evolutionary innovations after the Cambrian explosion? The increasing trend toward evolutionary convergence in metazoans may be related to the great informational revolution that characterized and enabled their evolution, the advent of the epigenetic information, and the evolution of the nervous system. In contrast to the randomness of the production of the new genetic information via gene mutations, the epigenetic information is not random but is generated in neural circuits. Hence, it is replicable and reemergent. The nonrandomness of epigenetic information may be the causal basis of the ubiquity of the phenotypic convergence observed in the metazoan world.

References

Alves-Gomes, J.A., Ortí, G., Haygood, M., Heiligenberg, W. and Meyer, A. (1995). Phylogenetic analysis of the South American electric fishes (Order Gymnotiformes) and the evolution of their electrogenic system: a synthesis based on morphology, electrophysiology, and mitochondrial sequence data. Mol. Biol. Evol. *12*, 298–318.

Barlow, C. (2003). Let There Be Sight! A Celebration of Convergent Evolution. Available online: www.thegreatstory.org (revised March 2007).

Bell, C.C. (2002). Evolution of cerebellum-like structures. Brain Behav. Evol. *59*, 312–326.

Beurlen, K. (1937). *Die Stammesgeschichtlichen Grundlagen der Abstammungslehre.* Gustav Fischer Verlag, Jena, p. 65.

Carr, C.E. and Soares, D. (2002). Evolutionary convergence and shared computational principles in the auditory system. Brain Behav. Evol. *59*, 294–311.

Carr, C.E., Kubke, M.F., Massoglia, D.P., Cheng, S.M., Rigby, L.L. and Moiseff, A. (1997). Development of temporal coding circuits in the barn owl. In: *Psychophysical and Physiological Advances in Hearing*, A.R. Palmer, A. Rees, Q. Summerfield and R. Meddis (eds.). Whurr Publishers, London, pp. 344–351.

Darwin, C.R. (1872). *The Origin of Species by Means of Natural Selection or The Preservation of Favored Races in the Struggle for Life*, sixth edition. John Murray, London, p. 72.

Deban, S.M., Wake, D.B. and Roth, G. (1997). Salamander with a ballistic tongue. Nature *389*, 27–28.

Deban, S.M., O'Reilly, J.C., Dicke, U. and van Leeuwen, J.L. (2007). Extremely high-power tongue projection in plethodontid salamanders. J. Exp. Biol. *210*, 655–667.

de Groot, J.H. and van Leeuwen, J.L. (2004). Evidence for an elastic projection mechanism in the chameleon tongue. Proc. Biol. Sci. *271*, 761–770.

del Pino, E.M. (1989). Marsupial frogs. Sci. Am. *260*, 110–118.

Eisthen, H.L. and Nishikawa, K.C. (2002). Convergence: Obstacle or opportunity? Brain Behav. Evol. *59*, 235–239.

Hodin, J. (2000). Plasticity and constraints in development and evolution. J. Exp. Zool. *288*, 1–20.

Kent, G.C. (1973). *Comparative Anatomy of the Vertebrates*. C.V. Mosby Co., Saint Louis, p. 180.

Meyer, A. (1998). Homology and homoplasy: the retention of genetic programmes. *Symposium on Homology, held at the Novartis Foundation*, London, July 21–23, 1998 (abstract).

Milton, R. (1997). *Shattering the Myths of Darwinism*. Park Street Press, Rochester, Vermont, p. 192.

Nishikawa, K.C. (2002). Evolutionary convergence in nervous systems: insights from comparative phylogenetic studies. Brain Behav. Evol. *59*, 240–249.

Omland, K.E. and Lanyon, S.M. (2000). Reconstructing plumage evolution in orioles (*Icterus*): repeated convergence and reversal in patterns. Evolution *54*, 2119–2133.

Quiring, R., Walldorf, U., Kloter, U. and Gehring, W.J. (1994). Homology of the eyeless gene of *Drosophila* to the Small eye gene in mice and Aniridia in humans. Science *265*, 785–789.

Ryugo, D.K., Wu, M.M. and Pongstaporn, T. (1996). Activity-related features of synapse morphology: a study of endbulbs of Held. J. Comp. Neurol. *365*, 141–158.

Salvini-Plawen, L.v. and Mayr, E. (1977). On the evolution of photoreceptors and eyes. Evol. Biol. *10*, 207–263.

Shadwick, R.E. (2005). How tunas and lamnid sharks swim: an evolutionary convergence. Am. Sci. *93*, 524–531.

Stokstad, E. (2001). Tooth theory revises history of mammals. Science *291*, p. 26.

Vermeij, G.J. (2006). Historical contingency and the purported uniqueness of evolutionary innovations. Proc. Natl. Acad. Sci. USA *103*, 1804–1809.

Wake, D.B. (1998). Homoplasy, homology and the problem of "sameness" in biology. *Symposium on Homology, held at the Novartis Foundation*, London, July 21–23, 1998 (abstract).

Wills, C. (1989). *The Wisdom of the Genes*. Basic Books, Inc., New York, p. 40.

Zhe-Xi, L, Cifelli, R.L. and Kielan-Jaworowska, Z. (2001). Dual origin of tribosphenic mammals. Nature *409*, 53–57.

Zucker, C.S. (1994). On the evolution of eyes: Would you like it simple or compound?. Science *265*, 742–743.

18 Species and Allopatric Speciation

Species are merely those strongly marked races or local forms which, when in contact, do not intermix.

Wallace (1865)

The Concept of Species

Darwin considered species to be qualitatively not different from varieties, and he used the term *species* for "the sake of convenience" rather than for describing a taxonomic unit:

> *I look at the term species as one arbitrarily given, for the sake of convenience, to a set of individuals closely resembling each other, and that it does not essentially differ from the term variety, which is given to less distinct and more fluctuating forms. The term variety, again, in comparison with mere individual differences, is also applied arbitrarily, for convenience sake.*
>
> *Darwin (1859a)*

The post-Darwinian biology has generally recognized the real existence of species as an evolutionarily discrete group of individuals, but spatial and temporal criteria that deny the real existence of species are still in currency (Mallet, 2007). However, Darwin's concept contains an important element of truth in the meaning that species, however, discrete, are evolutionarily transient states of supraindividual organization and evolution of multicellular organisms.

In the neo-Darwinian era, the species was universally recognized as the fundamental taxonomic unit of evolution. One of the most widely accepted concepts of species is the biological species concept (BSC), introduced by Dobzhansky, who considered species to be:

> *That stage in the evolutionary process at which the once actually or potentially interbreeding array of forms becomes segregated in two or more separate arrays which are physiologically incapable of interbreeding.*
>
> *Dobzhansky (1935)*

The concept was later echoed and developed by Mayr:

> *Species are groups of actually or potentially interbreeding natural populations which are reproductively isolated from other such groups.*
>
> *Mayr (1942)*

Epigenetic Principles of Evolution. DOI: 10.1016/B978-0-12-415831-3.00018-5

At the core of the BSC is the notion of reproductive isolation. BSC implies and proclaims that allopatry is the necessary and sufficient condition of speciation.

The BSC, like about two dozens other concepts of species, is far from universal as far as its applicability to various forms of life is concerned. It is not applicable to most unicellular prokaryotes, or to a great number of asexual, parthenogenetically reproducing invertebrates, self-pollinating plants, and many interspecifically hybridizing plants.

Dobzhansky's and Mayr's definitions of species are inseparable from the neo-Darwinian assumption that geographic isolation is the necessary and sufficient condition for the establishment of reproductive isolation and speciation. This implies that reproductive isolation of one species from another results from accumulation of genetic changes between populations that have been geographically isolated (allopatric speciation) for long periods of time. Needless to say, Darwin did not relate speciation to allopatry.

According to the BSC, allopatrically formed species are postzygotically isolated, i.e., even when they secondarily come in contact and can interbreed, they are incapable of producing fertile hybrids. This prediction is not always validated by observations in nature. Ever-increasing evidence is showing that many well-established vertebrate and invertebrate species are reproductively *not isolated*, i.e., they are "potentially interbreeding" groups and, in many cases, their offspring gives birth to viable and fertile offspring. This shows that the reproductive incompatibility and postzygotic isolation as criteria for biological species definition are not universally true.

Later, due to inherent flaws of the definition and problems arising from recognition of the BSC, numerous attempts have been made to redefine the concept of species. (24 named species concepts were known until in 1997 (de Queiroz, 2005).) It is not within the scope of this discussion to review different species definitions still in currency in the biological community.

By emphasizing the role of reproductive isolation in the process of speciation, the BSC underestimates (while implying) the most important cohesive force of the species, the mate recognition system, which represents the deeper causal mechanism responsible for establishing reproductive isolation between populations of the original species. Reproductive isolation is determined by neurocognitive mating decisions on the basis of mate preferences.

Observations on a number of morphologically and genetically sibling (cryptic) species lend some weight to Paterson's (1985) recognition species concept (RSC), which, in distinction from the BSC, views the *reproductive compatibility within the group* as a basic criterion of species definition. Paterson's RSC maintains that species is the most inclusive population of biparental organisms sharing a common fertilization system, and it *is based on mate preferences* and the specific mate recognition system (matching of copulatory organs, courtship behavior, and neuroendocrine communication, as well as temporal sequence of associating reproductive events). The concept emphasizes the reproductive unity of the group rather than its differences with other groups. The reproductive isolation results from a change in the mate recognition system, which creates "a common fertilization system" in a group of individuals of a species.

The species recognition concept is also not applicable to asexual organisms. Both RSC and BSC would deny the status of species to many organisms that, under natural conditions, do not hybridize but can hybridize and produce fertile hybrids under special conditions.

In an attempt to overcome difficulties arising from almost all of the definitions that used to emphasize particular properties of species for species definition, recently de Queiroz proposed another concept of species, according to which the evolutionary divergence between lineages leading to speciation first leads to quantitative divergences between lineages, which may produce distinctive character states over time. de Queiroz envisages species formation as a process of continuous divergence of populations that, over time, leads to sequential, but temporally not orderly, acquisition of specific properties (e.g., phenetic distinction, ecological distinction, changes in mate recognition system, prezygotic isolation, postzygotic isolation, and differences between diverging lineages) (de Queiroz, 2005). While attempting to reconcile widely different definitions of *species*, this concept, à la Darwin, blurs the difference between varieties and species.

The Neo-Darwinian Theory and the Problem of Speciation

In developing his theory of evolution via natural selection, C. Darwin, for obvious reasons, did not attempt to elaborate on the process of speciation, the crux of the process of metazoan evolution. He believed that speciation in plants and animals occurs in sympatry, implying that no geographic isolation was necessary for the process to take place.

After de Vries' saltationist theory of species formation, known as the *mutational theory*, by the 1930s, the field of evolutionary theory was dominated by the neo-Darwinian thought. In the neo-Darwinian definition, evolution was, essentially, a statistical process of changes in allele frequencies, under the action of natural selection. Constrained by the fact that genetic changes that could occur in natural populations would be diluted under conditions of panmixis, neo-Darwinians had to postulate that geographical isolation of evolving populations was a necessary condition of speciation in nature, although Darwin envisaged speciation as a process occurring in sympatry. Allopatric populations, under the action of natural selection, accumulate changes in the genotype, leading to their genetic divergence, postzygotic reproductive isolation and, finally, to the formation of new species. Accordingly, geographic isolation, the absence of physical contact between populations, is an indispensable condition of speciation.

However, discrete, and obviously different from each other, species were seen as demonstration of gradualism in the great scheme of evolution of the living world. This gradualist view has only rarely been challenged, despite the fact that it has never been seriously attempted to elaborate theoretically on the mechanism of speciation or to trace back the details of the process of accumulation of gradual changes, under the action of natural selection, presumably leading to reproductive isolation of populations.

Nevertheless, the unmistakable links and signs of transition between species and other higher taxa in nature do not, by themselves, prove the belief that accumulation of gradual hereditary changes is a mechanism, let alone the only mechanism, of speciation. Contrary to the mainstream of the biological thought, during the last 140 years, a number of biologists, especially from the field of paleontology, have continually challenged the gradualist concept and presented evidence that seemed to contradict the view of gradual speciation. But amid the unprecedented enthusiasm engendered by the triumph of the classical and molecular genetics and the advent of the neo-Darwinian paradigm, for most of the twentieth century, their claims remained "voices in the desert."

From the neo-Darwinian perspective, formation of new species and higher taxa, the macroevolution in general, was seen as nothing more than an extension, or an unavoidable finalization, of the process of gradual accumulation of microevolutionary (mutational) changes in genes. Accordingly, the difference between two processes lies not in the mechanisms used, or in the causal basis, but in results alone. And the difference in the results is a function of time. Any ongoing microevolutionary process, at some point in time, sooner or later, would produce macroevolutionary events. In a well-known aphorism, microevolution plus sufficient time equals macroevolution.

The neo-Darwinian gradualistic concept of evolution took a serious blow in early 1970s, when Eldredge and Gould, based on paleontological evidence, proposed their theory of *punctuated equilibria* that evolution is characterized by long periods of stasis (Eldredge, 1971) with no observable changes in morphology and sudden, relatively short, periods of evolutionary changes (Eldredge and Gould, 1972). They proposed that the lack of transitional forms in the paleontological record was "an accurate reflection of evolution" rather than a result of the imperfect record (Gould, 1991). They argued that Darwin had to emphasize the incremental, gradual character of evolutionary changes in his confrontation with the prevailing concept of the immutability and permanence of species (Eldredge, 1985).

It is true that Darwin related the evolutionary gradualism with the action of natural selection, but the fact that Darwin himself repeatedly and explicitly stated that natural selection is not the only mechanism of evolution (Darwin, 1859b, p. 6; 1872) indicates that he accepted the possibility of nongradual evolution.

Beginning in the early 1980s, studies of the Cambrian fauna showed that ~540 Mya, within a short period of ~10 million years, all the known extant and many more extinct phyla evolved with extraordinary rapidity. This "Cambrian explosion" of metazoan life was a major blow to, and an empirical refutation of, the neo-Darwinian principle of gradualism as the only way of evolution and speciation. While the role of mutations in molecular evolution, i.e., in changing properties of proteins, is sufficiently and satisfactorily substantiated, it has never been demonstrated that particular mutations in a gene or in a number of genes have been responsible for an adaptive morphological change in metazoans.

Adequate experimental and observational evidence has also invalidated the neo-Darwinian tenet that geographic isolation, i.e., the spatial separation of populations, under the action of natural selection, is a necessary and sufficient condition of speciation. Allopatric populations of a species are commonly capable of hybridizing and giving birth to fertile offspring, which refutes the neo-Darwinian hypothesis of the evolution postzygotic isolation in allopatry as a condition of speciation. Not only allopatric populations of a species but also, in some cases, named species belonging to the same genus, and even to different genera, which have been geographically separated for evolutionarily long periods of time since they diverged from their common ancestor, are still capable of hybridizing and producing fertile offspring in F1.

None of the neo-Darwinian agents of evolution (gene mutations, genetic recombination, and changes in allele frequencies) has been demonstrated to affect speciation events in sympatry or allopatry, and often sibling species are both morphogenetically and genetically almost indistinguishable (Purves and Orians, 1987a). Sometimes, sibling species that are reproductively isolated show no genetic differences, and often, genetic differences between individuals and populations of the same *Drosophila* species are greater than those between sibling species:

> *Gel electrophoretic techniques reveal that sibling species of Drosophila share nearly all their alleles. The prevalent allele at most loci is nearly always found at a frequency of at least 0.01 in other closely related species. The vast majority of differences among the species is based upon variability already present within the individuals of a species. All of several hundred species of this genus that have evolved in Hawaii during the past 5–6 million years are very similar morphologically and genetically.*
>
> *Purves and Orians (1987a)*

This is also the case with other invertebrate taxa:

> *The genetic variation among the 200 Lake Victoria species is less than that within a single species of horseshoe crab; and yet the morphological variation among them is extensive.*
>
> *Levin (1999)*

Empirical evidence also shows that speciation does not necessarily require long periods of time (for accumulation of "favorable mutations") to occur. Many species are known to have evolved in "evolutionary instants" of a few thousand years. For example, formation of an estimated 500 extant and extinct species of cichlid fish, of which 200 are endemic (not found elsewhere), in Lake Victoria is estimated to have occurred sometime between now and 750,000 years ago (Seehausen et al., 1997) (cichlid fish in East African lakes), and it took one to two centuries for new *Rhagoletis* species to evolve (Linn et al., 2003).

These facts suggest that genetic changes might not be related to, or be necessary for, prezygotic reproductive isolation. And the question naturally arises whether the minimal amount of genetic and morphologic difference between those species (which often is smaller than the variability already present within the species) is *prespeciational* or *postspeciational* by origin.

From a morphological point of view, it is within everyone's realm of experience that various races of domestic animals, created through artificial selection by man during the last few millennia, are often more different between them than from their wild conspecifics. Very often, even under natural conditions, various races of a species in sympatry differ so much phenotypically from each other that for long periods of time they have been viewed as distinct species. This has been the case, for example, with the deer mouse (*Peromyscus maniculatus*), the most widely distributed small animal in North America, which exhibits extraordinary regional variations in coat color, which matches the background color of soils, and in tail and foot length, which determine how readily the deer mouse can climb (Purves and Orians, 1987b, p. 1051).

Speciation Modes According to the Spatial Occurrence

Darwin adopted a sympatric approach to the problem of speciation, but post-Darwinian evolutionists distinguished between two main forms of speciation: allopatric speciation, taking place under conditions of spatial isolation of populations, and sympatric speciation, occurring within populations sharing the same habitat. However, in line with the basic neo-Darwinian tenet of accumulation of changes in genes and alleles as a condition of speciation, their approach required spatial isolation, for under conditions of sympatry, the panmixis, random mating would not allow accumulation of genetic differences between two populations. Hence, from a neo-Darwinian view, sympatric speciation is not more than a theoretical possibility.

The classification of modes of speciation according to spatial occurrence is a phenomenological classification, for the same causes of speciation may be involved in both allopatry and sympatry.

Allopatric Speciation

Darwin believed that species form by gradual transformation of local varieties as a result of accumulation, under action of natural selection, of indefinite changes occurring in animal populations. Hence, according to Darwin, no geographical isolation was necessary for speciation to take place (Mayr, 1992).

For neo-Darwinians, the panmictic system of mating of sympatry was an almost insuperable barrier to the process of speciation. The very low frequency of mutations in living organisms, multiplied by the very low proportion of favorable mutations to the total number of mutations, reduce by several orders of magnitude the likelihood of each individual mating with an individual of the opposite sex with the same mutation. Accordingly, panmixis will tend to level off or "dilute" the very rare "favorable mutations" that could arise. The solution to the dilemma came in the form of the idea of geographic isolation as a *necessary* condition for genetic and reproductive isolation. Thus, the emerging allopatric model of speciation, almost by consensus, was recognized as the predominant, if not the exclusive, way of speciation.

Allopatric speciation has been considered to be the most important form of speciation by most evolutionary biologists (Bush, 1975) and "the exclusive mode of speciation among birds and mammals" (Mayr, 2001).

Now, most biologists believe that allopatric speciation has played a bigger role in the speciation and formation of higher taxa than sympatric speciation (Turelli et al., 2001), with the latter only recently having attracted serious attention and stimulated intensive research in evolutionary biology.

Allopatric speciation implies formation of new species via reproductive isolation that results from spatial separation (e.g., separation of lakes by newly emerging stripes of land, geological depressions that may separate a continuous territory into islands, migration into new islands, formation of glaciers, or course change of rivers) of the population of a single species in two or more populations, thus leading to divergent changes in their gene pools and rise of postzygotic reproductive barriers. In Ernst Mayr's view, allopatry or geographic isolation is a critical condition for the speciation process to begin (Mayr, 1942).

The spatial separation of populations, by preventing gene flow between them, leads to accumulation of divergent changes in genes and allele frequencies in those populations. These genetic differences can progress to such an extent that developmental incompatibilities leading to postzygotic isolation and reproductive isolation of geographically isolated populations may arise. When such populations come into contact again, only inviable hybrids, if any, will be produced. The loss of fertility of hybrids results in the reproductive isolation of the two populations that, according to the biological concept of species, from this stage on, represent two incipient species. This form of allopatric speciation determined by geographic isolation is considered by most neo-Darwinians as the predominant form of speciation in nature.

Despite the plausibility of the geographic isolation as a means of allopatric speciation, the scientific evidence in support of geographic species is scarce and inadequate for drawing the conclusion that it is the most important, let alone "the exclusive mode of speciation" of any metazoan group.

Among the few examples of demonstrated allopatric speciation is a recently reported case of mice in the Madeira island, off the northwestern coast of Africa, where six chromosomally distinct mice (*Mus musculus domesticus*) races have evolved during the last five centuries. (The island was populated by mice for the first time during the fifteenth century (Britton-Davidian et al., 2000).) Such a chromosomal evolution is an extraordinary phenomenon in metazoans, and its evolutionary advantage and its role in their evolution are not yet determined (Capanna and Castiglia, 2004). Authors of the report do not explicitly state whether these races represent different mouse species.

Another example of geographical speciation is the example of different species of Darwin's finches that have evolved in the Galapagos Islands as a result of geographic isolation, but again with a serious problem from the neo-Darwinian viewpoint: despite the geographical isolation, these species still are not isolated reproductively, they are not authentic species for they do not fit the neo-Darwinian criterion of reproductive isolation as a defining property of species.

While it is theoretically possible that under conditions of geographic isolation, two populations of a species could diverge from the original type and from each other to such an extent that they become reproductively isolated, the supporting scientific evidence is scarce.

Controversy also arises as to whether natural barriers in the real world are so widespread or numerous enough to account for many millions of extant and extinct species and higher taxa that evolved on the Earth. The fact that land and water bodies, where the bulk of speciation processes took place represent mainly geographic continua rather than separated chunks of the Earth's lithosphere and hydrosphere, does not lend support to the hypothesis that geographic isolation might have been the main source of speciation.

The idea that the postzygotic reproductive isolation is a necessary first step in speciation, as a fundamental tenet of the neo-Darwinian view of speciation, is refuted by empirical evidence that species that were separated long ago still can hybridize and produce fertile offspring. Hybridization does not necessarily lead to inviability and sterility, for we know of numerous examples when populations, races, and even species (e.g., various sibling species, cichlid fish, Darwin's finches) constantly produce viable and fertile offspring, without interfering in the integrity of their groupings. A great number of species are known that, although reproductively isolated under natural conditions, have never evolved postzygotic isolation (genetic incompatibility). Fertile hybrids are produced not only between species of the same genus but also even between species belonging to different genera and, sometimes, even between species of different families. As much as 10% of bird named species produce fertile hybrids with other bird species (Weiner, 1994).

Verification of Allopatric Models of Speciation

The Basic Allopatric Model

Voluminous work has been conducted for half a century in order to verify the basic neo-Darwinian thesis that evolving genetic differences between the isolated populations are the cause of reproductive (pre- or postzygotic) isolation. The results of this extensive work have been reviewed by Rice and Hostert (1993), on which this discussion is mainly based.

One basic assumption of allopatric models is that speciation is facilitated when the number of individuals in the isolated population is reduced. This idea derives from Ernst Mayr's claim that evolutionary changes occur more rapidly in small founder populations living in the periphery of the habitat and under entirely different biotic and abiotic conditions, which creates opportunities for new niches (Mayr, 1992).

In *Drosophila*, an assessment of the genetic drift (significant deviations from the normal frequency of an allele that may occur in small groups of individuals) as a potential factor of reproductive isolation could be made by examining whether inbred lines in *Drosophila* species show any degree of reproductive isolation. Results of the experimental work have been equivocal, both in support of and against the role of the sampling drift in reproductive isolation:

> *Overall studies of mating among inbred strains suggest that sampling drift can both contribute to or detract from isolation among populations.*
>
> *Rice and Hostert (1993)*

A pleiotropy/genetic hitchhiking hypothesis posits that divergent selection can lead to prezygotic reproductive isolation of allopatric populations. Accordingly, reproductive isolation may be a correlational result of pleiotropy or genetic hitchhiking, when alleles selected for divergent characters happen to be correlated to alleles for assortative mating. Several difficulties arise if pleiotropy/genetic hitchhiking would be accepted. Let's start with cases of prezygotic isolation.

First, theoretically, it might be argued that this correlation in the case of genetic hitchhiking requires a combination of a mutator gene and, close to it, a gene which under the influence of the mutator gene undergoes a "useful" mutation and, hence, gives the species an advantage over other individuals. (Natural selection would favor both of these genes since they are linked.) This result is strictly conditioned: it will happen only *if* the above genes exist and are linked. In the absence of any evidence that it is the case even in a single example, the hypothesis is far from validated at best.

Second, experimental data that indirectly support this model of reproductive isolation in *Drosophila*, *Musca domestica*, and others are ambiguous and are opposed by a considerable number of other experiments.

Third, even in the cases when these experiments have resulted in a considerable increase of homotypic mating relative to random mating (which represents prezygotic reproductive isolation) of individuals from divergently selected allopatric populations (Kilias et al., 1980), mate preferences for individuals of the same origin are ultimately an expression of a mating bias, which are determined by neural cognitive systems, which are demonstrated to change without any change in genes or genetic information (see section "Neural-cognitive Mechanisms of Sympatric Reproductive Isolation," Chapter 19).

Numerous studies have focused on divergent selection as a potential factor of reproductive isolation between populations of a species in allopatry. This form of natural selection results from differences in the habitat of two populations, which generate different evolutionary pressures. Divergent selection can also lead to postzygotic reproductive isolation, which may be related to changes in the fitness and fertility of hybrids in different environments, or it may be an effect of genetic and developmental incompatibilities arising in diverging populations over time. However,

> *Postmating effects, especially as observed in hybrid sterility and "hybrid breakdown" are incidental to the speciation process. They do not appear to serve to actively reinforce reproductive isolation.*
>
> *Carson (1985)*

There is scarce evidence that the postzygotic reproductive isolation may be related mainly to the presumed inviability/low viability and infertility of hybrids (inbreeding depression) between populations in the state of incipient species. However, one should be reminded that hybrid inviability and sterility are not universal phenomena and hybridization often has positive effects (heterosis effect) on the viability of the hybrid offspring, and hundreds of cases are known of viable and fertile hybrids arising from closely related species in fish and insects without affecting the integrity of species, even for evolutionarily long periods of time.

Somewhat more convincing sounds the hypothesis put forward by Turelli et al. (2001) that sexual selection may be the cause of reproductive isolation between allopatric populations because during their geographical isolation, mate preferences in both populations may diverge from their original form:

> *However, natural selection need not be involved in speciation via sexual selection: examples are changes in male genitalia or postmating, prezygotic reproductive isolation (e.g., sperm-egg incompatibility), both of which can be driven by male-male competition or by female behavioral or biochemical preference.*
>
> *Turelli et al. (2001)*

They believe that the sensory drive, a mechanism of sympatric speciation, can naturally lead to premating isolation of allopatric populations (Turelli et al., 2001).

The Reinforcement Model

According to this model, populations in allopatry diverge to such an extent that after the secondary contact they show various degrees of prezygotic reproductive isolation because the interbreeding produces hybrid offspring of low viability and fertility. Thus, genetic divergence that evolves in allopatry is the cause of reproductive isolation observed in the secondary contact between two populations. The premating isolation and reproductive character displacement (greater expression of mating signals and mating biases in sympatric populations than allopatric populations) observed in these populations in sympatry are only by-products of the genetic divergence previously evolved in allopatry. Although during the secondary contact individuals of two divergent populations continue to mate, prezygotic isolation of two populations is reinforced by the low fitness of hybrids.

One counterargument to the model is that even if a genetic correlation would be established between the hybrid inviability and prezygotic isolation, which is not always the case, recombination of respective genes in hybrids would break that correlation down (Sanderson et al., 1992).

Experiments on reinforcement have provided mixed results. Reinforcement is reported to have been observed in a number of experiments. However, reports on laboratory experiments with *Drosophila* that have led to prezygotic reproductive isolation are of doubtful relevance, for in most of them a strong artificial selection was exerted against the hybrids (the so-called "destroy-the-hybrids" experiments), which *artificially* reduces the experimental populations into postzygotically isolated populations.

An interesting case of the reinforcement model has been described by Saetre et al. (2003) in their study on the role of the evolution of sex chromosomes in the postzygotic and prezygotic isolation in two bird species, the flycatcher *Ficedula hypoleuca* and *Ficedula albicollis*. They note that these species, after the last glacial period, came, and still are, into secondary contact with each other. They observed that reinforcement does occur in the hybrid zone, and they believe that genes for hybrid inviability and "genes for species recognition" in this particular case are linked to the sexual Z chromosome.

> *Linkage between genes affecting hybrid fertility and traits used in species recognition can explain why reinforcement operates despite rather extensive introgression and recombination of autosomal genes.*
>
> *Saetre et al. (2003)*

However, experiments with the tetiigonid bushcricket *Ephippiger ephippiger* have failed

> *to support any particular role for maternally derived sex-linked genes ... Reinforcing selection is unlikely to be involved in the absence of postmating isolation.*
>
> *Ritchie (2000)*

Even if other linkage-related cases of reinforcement model will be found, it is unlikely that it may be a widespread model of speciation. The coincidence or linkage of "genes for" hybrid inviability and mate recognition cannot be a frequent phenomenon. Moreover, no gene for mate recognition has ever been demonstrated to exist, and, as will be shown later, species recognition, which determines the prezygotic isolation, is not a function of any particular genes. Species recognition is the behavioral output of the processing of sensory input in specific mating behavior circuits (see section "Mate Recognition System" in Chapter 19).

Finally, the abundant evidence on the viability of hybrids and against reinforcement clearly outweighs the evidence on hybrid inviability/sterility and on reinforcement.

Most of the experiments for inducing reproductive isolation by disruptive selection in the laboratory have failed (Thoday and Gibson, 1970). However, working on large populations of *Drosophila melanogaster*, in the absence of genetic correlation between the selected character (sternopleural bristle number) and the tendency for reproductive isolation, Barker and Karlsson (1974) concluded that *intense* disruptive selection "may lead to 'assortative mating'" (Barker and Karlsson, 1974). However, in their review, Rice and Hostert reported that the available laboratory evidence "provides no support for the reinforcement model of speciation" (Rice and Hostert, 1993).

Empirical support for the role of reinforcement in speciation is very limited (Butlin and Ritchie, 1991). Even the supporters of the reinforcement model consider it to be *theoretically possible* but conditionally. It can occur only "if substantial premating and/or postmating divergence have previously evolved" (Turelli et al., 2001).

Studies on American field crickets, *Gryllus texensis* and *Gryllus rubens*, have shown that these cryptic sister species exhibit complete prezygotic isolation while lacking postzygotic isolation. The prezygotic isolation is determined by the fact that females mate their conspecific males because they recognize and prefer their species-specific pulse rates. This is proven by the fact that, under natural conditions in sympatry, these species do not hybridize. Although in the laboratory they can produce fertile hybrids of intermediate male song and female preference, no character displacement in male signals or female responses has been observed (Gray and Cade, 2000).

Even if one were to take it for granted that the reinforcement model does play a role in reproductive isolation and speciation in allopatry, as seems to have been shown in a few cases, it does this via assortative mating (individuals of each divergent population are biased to mate individuals of the opposite sex of their own population), which arises more rapidly (Gray and Cade, 2000) than the reinforcement. Besides, hybrids between them, in most cases, are viable and fertile.

Divergence-with-Gene-Flow Model

Surprisingly, the idea of the necessity of prevention of gene flow for speciation in allopatry (Mayr, 1963, and others) seems to have been abandoned, and all models of divergence-with-gene-flow require a certain level of gene flow as a positive feedback for further divergence and adaptation of diverging populations to local conditions. It is believed that an initial low level of reproductive isolation as a result of the appearance of assortative mating would reduce gene flow between populations. At later stages of the process of divergence/speciation, that gene flow would facilitate stabilizing selection, which then automatically leads to further reduction and prevention of gene flow between diverging reproductively isolated populations. Based on experimental evidence, it has been concluded that

> *any environmental context generating restricted gene flow and producing strong, discontinuous, and multifarous divergent selection should also frequently lead to speciation via pleiotropy/hitchhiking.*
>
> *Rice and Hostert (1993)*

While the idea that such conditions lead to experimental prezygotic reproductive isolation seems plausible, it has been impossible in these experiments to demonstrate a concrete correlation between the genes for characters that are selected, and genes that presumably lead to prezygotic isolation. The idea, itself, of a correlation is speculative, inferred by the highly doubtful, indeed false, concept that assortative mating, mate recognition, and mate preferences are genetically determined (see later in Chapter 19 on mate recognition system and its evolution).

Experiments for verifying this model of speciation show that, ultimately, various degrees of reproductive isolation in animals are determined by assortative mating, i.e., by preferential mating of individuals with individuals of the opposite sex that are perceived to be like themselves.

Assortative mating, as opposed to random mating, is based on sensory identification of those individuals and is a function of the mate recognition system, which depends not on genes and whose evolution is dissociated from evolution of genes. It depends on mating preferences and mating behaviors, which are epigenetically determined as a result of processing of reproductive and mating behavior stimuli in respective neural circuits.

Peripatric Speciation

Peripatric speciation would result from physical isolation of a founder peripheral population that does not exchange genes with the population of origin (!). The small number of individuals of the founder may differ in allele frequency (genetic drift) from the overall population and under new conditions of spatial separation from the parental population, over time, may accumulate additional genetic changes leading to reproductive isolation from original population. There is no reliable evidence of speciation under conditions of peripatry.

This form of speciation is also highly conjectural, and it has not been possible to substantiate with empirical evidence.

The Founder Effect Model

When the population that is geographically isolated is small, chances are that the founders will diverge faster (if they will not undergo extinction, to which they are exposed because of the small numbers and the loss of alleles as a result of inbreeding) from the population of origin because of the genetic drift, leading to peripatric speciation. This hypothesis is testable, but no direct empirical evidence from experiments and nature has been presented to prove the speciation potential of the founder effect.

Experimental evidence shows that the sampling drift (sampling error) or genetic drift may not have that important role in the process of allopatric speciation that biologists have proclaimed for a long time. The sampling drift hypothesis states that very small populations (or, as a minimum, a single inseminated female) that are separated from the original population tend to rapidly diverge genetically from the original population ("founder effect"). Theoretically, this would be the case, for example, when a very small group of individuals of a species somehow succeed in colonizing a distant island. The accelerated speciation in such cases is attributed to the sampling drift, which may dislocate the population to a new equilibrium. If this assumption on the "founder effect" were true, it would be easy to prove in laboratory. Attempts to do this have not been spared, but experimental evidence seems to be inconclusive at best (Rice and Hostert, 1993).

In a ten-year experiment and 100 generations with 59 populations of *Drosophila pseudoobscura*, designed to prove the possibility of inducing reproductive isolation via the founding effect, investigators admit to having failed the attempt:

> *We conclude that our experiment provides no empirical support for the proposition that genetic and demographic processes associated with the founding of a population by very few individuals are very likely to cause speciation.*
>
> *Moya et al. (1995)*

Results of experimental tests on the effect of populational bottlenecks (drastic reduction of the number of individuals that diminishes the genetic variability) in laboratory indicate that founder effects have no causal role in speciation (Charlesworth, 1995).

Another genetic study on evolution of 13 extant species of Darwin's finches in Galapagos islands suggests that the founding population of finches, when they presumably colonized Galapagos islands, was not as small as predicted by the founder effect hypotheses; the most conservative estimation ("in all probability an underestimate") shows that the founding group comprised at least 30 individuals (Vincek et al., 1997).

The Bottleneck Model

This model essentially posits that drastic reductions in the number of individuals in a population, the so-called bottlenecks, may decrease the genetic variability, but the ensuing homozigosity enables a new increase in the genetic variability, creating conditions for potential "genetic revolutions" via reorganization of the gene pool:

> *This "genetic revolution," released by the isolation of the founder population, may well have the character of a chain reaction. Changes in any locus will in turn affect the selective values at many other loci, until finally the system has reached a new state of equilibrium.*
>
> *Mayr (1954)*

But the concept of "genetic revolution" has not been rationalized, and Mayr has never explained what the "reorganization of the gene pool" consists of, if anything else beyond changes in allele frequencies. It would be expected that the bottleneck mechanism would have been operational in cases of rapid speciation, but studies on the classical examples of rapid speciation of cichlid fish in great East African lakes, for example,

> *Give no support to the notion that the species have recently passed through genetic bottlenecks.*
>
> *van Oppen et al. (1998)*

Attempts to reproduce bottleneck conditions in laboratory by selecting mated females for creating isofemale lines and reproducing them for a number of generations have been unsuccessful (Rice and Hostert, 1993). The empirical evidence for substantiating allopatric models is inadequate at best. Hence,

> *It might be time for a re-evaluation of the geographical classification of speciation modes in favor of one based primarily on evolutionary mechanisms.*
>
> *Via (2001)*

Ecological Speciation

Proponents of the ecological speciation believe that this form of speciation may occur in both allopatry and sympatry (Schluter, 2001; Rundle and Nosil, 2005). Ecological speciation implies reproductive isolation of populations inhabiting separate niches as a result of divergent selection of ecologically influenced characters. It may lead to mainly premating isolation, mainly postmating isolation, or a combination of the two (Schluter, 2001). It may result from natural selection, from sexual selection and from direct selection on premating isolation (reinforcement) (Figure 18.1). According to the

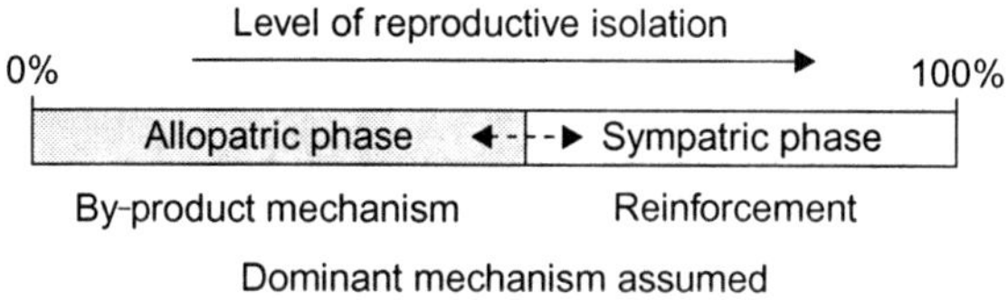

Figure 18.1 The classic scenario of an ecological speciation event, from beginning to end. Reproductive isolation builds in allopatry (dark shaded) as an incidental by-product of adaptation to alternative environments (by-product mechanism). Reinforcement of premating isolation, driven by reduced hybrid fitness, completes the speciation process during the sympatric phase (light shaded). The timing of secondary contact is flexible (indicated by arrows at the boundary between the allopatric and sympatric phases).
Source: From Schluter (2001).

model, being under different evolutionary pressures, populations in different regions accumulate genetic differences, which may lead to their mutual reproductive isolation. Hence, the pre- and postmating reproductive isolation is a by-product of divergent selection.

Even the first step of the model of ecological speciation, the divergent selection alone has not been demonstrated to occur. Pleiotropy and linkage disequilibrium as genetic mechanisms believed to be involved in the divergent selection during ecological speciation and reproductive isolation, despite decade-long attempts to identify them, also remain speculative and elusive (Rundle and Nosil, 2005).

However, this form of speciation requires prevention of gene flow between populations, which is difficult to imagine, and more so to occur under conditions of sympatry.

The problem with the ecological speciation is that it initially requires an evolutionary starter, the geographic isolation, attainment of a considerable degree of reproductive isolation, and reduced hybrid fitness, which, in complex, are difficult to be traced in nature.

Recall that from a *scientific* point of view, the concepts of "genes for divergent ecology" and "genes for mate choice" are purely speculative, for such genes have never been demonstrated to exist.

References

Barker, J.S.F. and Karlsson, L.J.E. (1974). Effects of population size and selection intensity on responses to disruptive selection in *Drosophila melanogaster*. Genetics *78*, 715–735.

Britton-Davidian, J., Catalan, J., da Graça Ramalhinho, M., Ganem, G., Auffray, J.C., Capela, R., et al. (2000). Rapid chromosomal evolution in island mice. Nature *403*, p. 158.

Bush, G.L. (1975). Modes of animal speciation. Ann. Rev. Ecol. Syst. *6*, 339–364.

Butlin, R.K. and Ritchie, M.G. (1991). Variation in female mate preference across grasshopper hybrid zone. J. Evol. Biol. *4*, 227–240.

Capanna, E. and Castiglia, R. (2004). Chromosomes and speciation in *Mus musculus domesticus*. Cytogenet. Genome Res. *105*, 375–384.
Carson, H.L. (1985). Unification of speciation theory in plants and animals. Syst. Bot. *10*, 380–390.
Charlesworth, B. (1995). Speciation: down the bottleneck? Curr. Biol. *5*, 995–996.
Darwin, C.R. (1859a). *The Origin of Species by Means of Natural Selection or the Preservation of Favored Races in the Struggle for Life*, first edition. John Murray, London, 1st issue, p. 52.
Darwin, C.R. (1859b). Ibid., p. 6.
Darwin, C.R. (1872). *The Origin of Species by Means of Natural Selection or the Preservation of Favored Races in the Struggle for Life*, sixth edition. John Murray, London, with additions and corrections, p. 72.
Dobzhansky, Th. (1935). A critique of the species concept in biology. Philos. Sci. *2*, 344–355. (In: Ayala, F.J. and Fitch, W.M. (1997). Genetics and the origin of species: an introduction. P. Natl. Acad. Sci. U.S.A. *94*, 7691–7697.)
de Queiroz, K. (2005). Ernst Mayr and the modern concept of species. Proc. Natl. Acad. Sci. U.S.A. *102*, 6600–6607.
Eldredge, N. (1971). The allopatric model and phylogeny in paleozoic invertebrates. Evolution *25*, 156–167.
Eldredge, N. (1985). *Time Frames: The Rethinking of Darwinian Evolution and the Theory of Punctuated Equilibria*. Simon & Schuster, New York, NY, p. 189.
Eldredge, N. and Gould, S.J. (1972). Punctuated equilibria: an alternative to phyletic gradualism. In: *Models in Paleobiology*, T.J.M. Schopf (ed.). Freeman, Cooper & Co., San Francisco, CA, pp. 82–115.
Gould, S.J. (1991). Opus 200. Nat. Hist. *100*, 12–18.
Gray, D.A. and Cade, W.H. (2000). Sexual selection and speciation in field crickets. Proc. Natl. Acad. Sci. U.S.A. *97*, 14449–14454.
Kilias, G., Alahiotis, S.N. and Pelecanos, M. (1980). A multifactorial genetic investigation of speciation theory using *Drosophila melanogaster*. Evolution *34*, 730–737.
Levin, R. (1999). *Patterns of Evolution*. Scientific American Library, p. 83.
Linn, C., Jr., Feder, J.L., Nojima, S., Dambroski, H.R., Berlocher, S.H. and Roelofs, W. (2003). Fruit odor discrimination and sympatric host race formation in *Rhagoletis*. Proc. Natl. Acad. Sci. U.S.A. *100*, 11490–11493.
Mallet, J. (2007). Species, concepts of. p. 1–15. Available online: http://www.ucl.ac.uk/taxome/jim/Sp/species.pdf.
Mayr, E. (1942). *Systematics and the Origin of Species*. Columbia University Press, New York, NY, p. 120.
Mayr, E. (1954). Change of genetic environment and evolution. In: *Evolution as a Process*, J Huxley, A.C. Hardy and E.B. Ford (eds.). Allen & Unwin, London, pp. 157–180.
Mayr, E. (1963). *Animal Species and Evolution*. Belknap Press of Harvard University Press, Cambridge, MA, p. 604.
Mayr, E. (1992). Speciational evolution or punctuated equilibria. In: *The Dynamics of Evolution*, A. Somit and S. Peterson (eds.). Cornell University Press, New York, NY, pp. 21–48.
Mayr, E. (2001). *What Evolution Is*. Basic Books, New York, NY, p. 176.
Moya, A., Galiana, A. and Ayala, F.J. (1995). Founder-effect speciation theory: failure of experimental corroboration. Proc. Natl. Acad. Sci. U.S.A. *92*, 3983–3986.

Paterson, H.E.H. (1985). The recognition concept of species. In: *Species and Speciation, Transvaal Museum Monograph*, E.S. (1985). Vrba (ed.). vol. 4., pp. 21–29.

Purves, W.K. and Orians, G.H. (1987a). *Life—The Science of Biology*, second edition. Sinauer Associates Inc., Sunderland, MA, pp. 1049–1050.

Purves, W.K. and Orians, G.H. (1987b). *Life—The Science of Biology*, second edition. Sinauer Associates Inc., Sunderland, MA, p. 1051.

Rice, W.R. and Hostert, E.E. (1993). Laboratory experiments on speciation: what have we learned in 40 years? Evolution *47*, 1637–1653.

Ritchie, M.G. (2000). The inheritance of female preference functions in a mate recognition system. Proc. Biol. Sci. *267*, 327–332.

Rundle, H.D. and Nosil, P. (2005). Ecological speciation. Ecol. Lett. *8*, 336–352.

Sanderson, N., Szymura, J.M. and Barton, N.H. (1992). Variation in mating call across the hybrid zone between the fire-bellied toads *Bombina bombina* and *B. variegata*. Evolution *46*, 595–607.

Saetre, G.-P., Borge, T., Lindroos, K., Haavie, J., Sheldon, B.C., Primmer, C. and Syvänen, A.-C. (2003). Sex chromosome evolution and speciation in *Ficedula* flycatchers. Proc. Biol. Sci. *270*, 53–59.

Schluter, D. (2001). Ecology and the origin of species. Trends Ecol. Evol. *16*, 372–380.

Seehausen, O., van Alphen, J.J.M. and Witte, F. (1997). Cichlid fish diversity threatened by eutrophication that curbs sexual selection. Science *277*, 1808–1811.

Thoday, J.M. and Gibson, J.B. (1970). The probability of isolation by disruptive selection. Am. Nat. *104*, 219–230.

Turelli, M., Barton, N.H. and Coyne, J.A. (2001). Theory and speciation. Trends Ecol. Evol. *16*, 330–343.

van Oppen, M.J.H., Turner, G.F., Rico, C., Robinson, R.L., Deutsch, J.C., Genner, M.J., et al. (1998). Assortative mating among rock-dwelling cichlid fishes supports high estimates of species richness from Lake Malawi. Mol. Ecol. (UCLA Sym. Bi.) *7*, 991–1001.

Via, S. (2001). Sympatric speciation in animals: the ugly duckling grows up. Trends Ecol. Evol. *16*, 381–390.

Vincek, V., O'Huigin, C., Satta, Y., Takahata, Y., Boag, P.T., Grant, P.R., et al. (1997). How large was the founding population of Darwin's finches? Proc. Biol. Sci. *264*, 111–118.

Wallace, A.R. (1865). On the phenomena of variation and geographical distribution as illustrated by the Papilionidae of the Malayan region. Trans. Linn. Soc. London *25*(1–71), p. 12.

Weiner, J. (1994). *The Beak of the Finch: A Story of Evolution in Our Time*. Knopf, New York, NY, pp. 198–199.

19 Epigenetics of Sympatric Speciation—Speciation as a Mechanism of Evolution

Speciation is a major cause of the diversification of living things.

West-Eberhard (2003)

Recognition of the Reality of Sympatric Speciation

The prevailing neo-Darwinian concept is that speciation occurs under conditions of geographic isolation of populations, via gradual accumulation of changes in allele frequencies as a result of gene mutations, genetic recombination, or gene drift, under the action of natural selection. In the process, geographically separated populations become reproductively isolated, so that later, even on secondary contact, populations evolve as separate species.

It is important to bear in mind, however, that this concept of the primacy of allopatric speciation was deduced from theoretical considerations (that speciation needs prevention of gene flow between populations and that under conditions of sympatry gene flow between populations will unavoidably occur) rather than from empirical evidence. Under this geographic orthodoxy, two basic conditions were sanctioned as necessary for speciation: physical separation of populations and the derived reinforcement upon secondary contact of incipient species.

For a long time since the 1940s, evolutionary studies have been focusing only on the geographical separation of populations, the "60-year-old blind alley" (Mallet, 2001), as almost the exclusive means of reproductive isolation that is necessary for speciation. Sympatric speciation has been rejected by evolutionists (Felsenstein, 1981) or has been considered only as a theoretical possibility of speciation with little, if any, role in the evolution of metazoans. While concluding that there is little evidence for sympatric speciation in island birds, Coyne and Price (2000) doubt whether it may represent an important mechanism of speciation.

Beginning in the late 1980s, however, despite the theoretical restrictions imposed by evolutionary genetics studies, investigators are increasingly accepting the possibility that sympatric speciation may have played a greater role in evolution and speciation than was generally assumed. Now it seems that the pendulum is swinging the other way, and the gravity center of evolutionary studies is definitely shifting toward sympatric mechanisms of speciation (Via, 2001).

Epigenetic Principles of Evolution. DOI: 10.1016/B978-0-12-415831-3.00019-7

Experimental studies on the process of speciation under laboratory conditions have led to the conclusion that reproductive isolation, as a condition of speciation, may occur with or without allopatry.

Rice and Hostert (1993)

Most theoretical studies are still guided by the assumption that disruptive selection within the population is necessary for the process of speciation via reproductive isolation to occur in sympatry. Seehausen et al. (1999), for example, think that disruptive selection acting on the existing color polymorphism might have been the cause of the rapid speciation in many cichlid fish species in East African lakes with relatively good visual conditions, but they also believe that the cause of the disruptive selection is an intrinsic property, which ultimately is related to neurocognitive functions of neural circuits determining mate preferences and mate choices. However, all of the experimental work for inducing reproductive isolation in laboratory populations via disruptive selection has failed, with probably a single exception (Thoday and Gibson, 1970; Fry, 2003).

Recently, models are presented of sympatric speciation without disruptive selection, in the absence of physical barriers or predators (Higashi et al., 1999). Most importantly, empirical evidence is accumulating, indicating that sympatric speciation occurred and is still occurring in nature (Schliewen et al., 1994). Studies in many insects provide strong evidence that a number of sensory (visual, olfactory, auditory) signal traits are used for creating separate mating systems within original populations, leading to their reproductive isolation in the process of incipient sympatric speciation. Now, sympatric speciation is demonstrated in an adequate number of cases and is set on a firmer theoretical ground (Tregenza and Butlin, 1999).

Sympatric Speciation: No Changes in Genes Are Involved

The bone of contention on sympatric speciation is the issue of reproductive isolation, which from the neo-Darwinian view cannot arise between two sympatric populations, i.e., in the absence of spatial separation and the resulting prevention of gene flow between them. From an orthodox neo-Darwinian perspective, and from the view of the biological species concept (BSC), the sympatric speciation is impossible, for the gene flow between populations in sympatry would prevent their divergent evolution (Felsenstein, 1981).

However, it may be argued, first, that, to a certain extent, gene flow occurs (Bush and Smith, 1998 and references therein: Feder et al., 1994; Taylor et al., 1997) not only between populations of a species, but even between species in nature, and there is no criterion on how much gene flow would be admissible for two populations to speciate or be considered true species. Second, we know of numerous examples of changes in phenotypic characters that depend not on changes in genes but on changed patterns of gene expression alone.

Analysis of the mtDNA from six named subspecies of a wide-ranging species of cactus wren in southern California and the Baja California Peninsula revealed only two, instead of six, mtDNA groups. This phenomenon is common in birds in general. A survey of 41 named bird species revealed that only 3% of avian subspecies represent independent evolutionary units. It seems rational, after Crandall, to believe that subspecies represent inherited adaptive phenotypic variations of species that are not related to changes at the genetic level. Indeed, if these bird subspecies had evolved as independent evolutionary units, it would be expected that they would have evolved in more than one character, which generally is not the case (Zink, 2004). Studies on continental European bird subspecies show that the independence of morphological evolution from changes in genes is, by far, a more widespread phenomenon than generally recognized:

> *97% of these species, distinct evolutionary units, failed the test of congruence, i.e., show no differences in their mitochondrial DNA.*
>
> *Zink (2004)*

In general, there is no reliable evidence on existence of a relationship between the degree of morphological and physiological changes and speciation: sibling species both phenotypically and genetically are almost identical to each other and still represent separate species on their own, while different races of dogs created by artificial selection are so distinctive from each other and still belong to a single species.

On theoretical grounds, it would be assumed that sympatric speciation requires a special mechanism of dividing a continuous panmictic population in two. The only such mechanism that has been empirically substantiated is that of sudden changes in the mating behavior of a group of individuals, which leads to a separate mating system and reproductive isolation of the group from the rest of the sympatric population. That a behavioral mechanism takes the lead in the process of incipient speciation is not surprising if one bears in mind that behavior is the most plastic component of the animal phenotype.

The change in animal mating behavior may arise in the form of a shift in mating preferences. Sudden shifts in mating preferences, affecting whole populations, are observed not only in the course of evolution but within the lifetime of an animal. Such a shift in mating preferences might automatically lead to formation of two separate mating systems and two reproductively isolated populations within the range of the original panmictic population.

Certainly, this transition is impossible to occur from the neo-Darwinian view, according to which behaviors, as all other phenotypic traits, are determined by specific genes. Accordingly, under sympatric conditions, i.e., under conditions of panmixis, because of the gene flow between populations, changes in genes cannot lead to formation of two separate mating systems.

Many biologists are embarrassed by the still-prevailing idea of the existence of particular genes for mate choice. The reasoning goes: genes for mate preference will be mixed up by genetic recombination during the earlier stages of the speciation process, when mating between individuals of different populations may still occur. To overcome this difficulty, another assumption is made: genes for mate signaling and for mate preference are so close to each other in chromosomes that the probability of getting recombined is negligible (McCune and Lovejoy, 1998).

No evidence has ever been presented to substantiate the idea of genes for mate preference or mate signaling.

However, granted that such genes really exist, it is highly improbable that nature would have arranged genes for both mate choice and mate signaling in the same chromosome and, more so, even so close to each other as to prevent their recombination for the sake of the theory, to facilitate resolving our recombination problem. Certainly, even highly improbable events may happen once or twice, but rapid sympatric speciation has been far from a rare event: in East African cichlid fish alone, it has occurred repeatedly and independently in hundreds of cases.

Adequate observational and experimental evidence shows that sympatric speciation did occur and is still occurring. Biologists should understand that it is not the nature that must throw in the towel. Their difficulty stems from the faulty premise of the existence of the illusory "genes for mate choice" and "genes for mate signaling."

Sympatric speciation as a fact of nature invalidates all theoretical restrictions on its occurrence. Although unambiguous cases of sympatric speciation were described about 40 years ago (Bush, 1969), because of the conceptual constraints imposed by the neo-Darwinian paradigm, only recently has the sympatric speciation been recognized by most biologists as a real and widespread mechanism of speciation. This is what should be expected when theoretical concepts, built on insufficient (if any) empirical evidence, take priority over scientifically established facts.

Confronted with solid evidence on the occurrence of sympatric speciation, especially from studies on the explosive speciation of hundreds of cichlid species, with some of them having evolved several thousand years ago in East African lakes, many biologists are theoretically reconsidering the likelihood of sympatric speciation.

Speciation in sympatry is not a theoretical possibility but a demonstrated mechanism of evolution. Recent studies on two sibling species of cichlid fish of Lake Victoria, Africa, have shown that female mating preferences in these species are highly heritable, thus confirming the idea that female mate choice has not only been necessary but also sufficient for enabling reproductive isolation and the explosive speciation process that the cichlid fish have experienced very recently in East African lakes (Haesler and Seehausen, 2005).

Two incipient species of *Drosophila melanogaster* live in two slopes, the north-facing slope and the south-facing slope, of the "Evolution Canyon," Israel. The canyon is only 100 m wide at the bottom and 400 m at the top, i.e., within the *Drosophila* flight range of several kilometers a day. Populations in two opposing slopes of the canyon, however, have diverged in mate preference, body size, oviposition, and thermal preference (Michalak et al., 2001). There is no evidence on genetic differences between the two incipient species.

Now, at the beginning of the twenty-first century, the situation has changed dramatically:

> *It has become clear that the traditional geographical classification of speciation modes is no longer appropriate to capture the essential complexity of many speciation processes (e.g., Doebeli and Dieckmann 2003; Mizera and Meszéna 2003). By emphasizing adaptive processes, rather than restricting attention to biogeographical*

> *patterns of diversification, theoretical and experimental speciation research have taken off again to new shores.*
>
> *Doebeli et al. (2005)*

Natural Selection and Sexual Selection in Sympatric Speciation

What is natural selection: a result (reproductive success) or a process? Is it an effect or a cause? This question needs an unequivocal answer, for the imprecise and ambiguous use of the concept of natural selection unavoidably influences the scientific inferences on its role in evolution.

In Darwin's original meaning, natural selection is the visible *result* of the "struggle for life," or the competition between organisms for limited resources. Two main factors determine that result: the number of offspring produced by an individual/population and the offspring's survival rate until reproductive age:

> *As many more individuals of each species are born than can possibly survive; and as, consequently, there is a frequently recurring struggle for existence, it follows that any being, if it vary however slightly in any manner profitable to itself, under the complex and sometimes varying conditions of life, will have a better chance of surviving, and thus be naturally selected.*
>
> *Darwin (1859, p. 5)*

The genius uses here the concept of natural selection as a result of the struggle for life rather than a cause or agent of evolution.

In the interaction organism–environment, the organism is the active, rather than the passive, element, as it is manifested in its inherent tendency to avoid harmful agents of environment and generate phenotypic (heritable or not) changes in order to adapt to its environment. It is the organism that, in Darwin's vivid expression, is committed to a "struggle for life," whose result, survival, or the death of the organism will depend not simply on the environment but also, to a great extent, on the ability of the organism to maintain its homeostasis under conditions of that particular environment or generate phenotypic changes that will adapt it to the inhospitable environment.

The result of the "struggle for life" will depend not simply on the conditions of living. The simple fact that in a group of individuals living in the same environment some individuals survive while others perish suggests that there is something intrinsic to these organisms that determines their death or survival to reproduction. Natural selection as a result of the Darwinian "struggle for life" is ultimately determined by factors intrinsic to the "struggling" organisms living in the same habitat.

From another perspective, and accidental events aside, survivors in the "struggle for life" will be individuals that happen to be in possession of better morphological, physiological, and behavioral adaptations. As it is well known, the rise of adaptations in the process of evolution is determined intrinsically, by the organism itself, not extrinsically, by the environment. The success in the struggle for life is earned by the better-adapted organism rather than awarded by the environment.

Darwin also coined the term "sexual selection," for designating another important factor of evolution. During the sexual selection, it is the choosing sex that really selects its mating partner among a number of individuals of the opposite sex.

Under sexual selection, Darwin understood the process of evolution of secondary sexual traits in animals that make them preferred mates, thus leaving more offspring than less-preferred individuals (Darwin, 1859, pp. 88–90, 156–157).

> *Sexual selection depends on the success of certain individuals over others of the same sex, in relation to the propagation of the species; whilst natural selection depends on the success of both sexes, at all ages, in relation to the general conditions of life.*
>
> *Darwin (1871)*

Darwin did not relate sexual selection to the process of speciation, and a number of authors have expressed doubts about the possibility that sexual selection alone can lead to sympatric speciation (Panhuis et al., 2001; Turelli et al., 2001; Arnegard and Kondrashov, 2004).

Sometimes doubts have been raised about the possibility of speciation by sexual selection because it may evolutionarily make unnecessary other modes of speciation:

> *Many morphologically and genetically similar species differ markedly in mating signals and preferences. It might be inferred from this that sexual selection has driven speciation, but to do so requires that other modes of divergence be excluded.*
>
> *Panhuis et al. (2001)*

In recent years, the action of the Darwinian sexual selection in evolution of sexual traits has been extended to include a role of sexual selection in metazoan speciation.

In 1983, West-Eberhard proposed a broader concept, according to which sexual selection represents a subset of social competition for resources, which leads to rapid divergence of sexual traits, which can evolve without ecological differences between isolated populations. The rapid divergence of sexual traits may accelerate the process of speciation, what is believed to have occurred in *Anolis* lizards where interspecific variation in dewlaps and displays is mainly determined by the social selection (West-Eberhard, 1983).

It has been argued that sexual selection is ultimately reduced to "social selection." This hypothesis posits that sexual selection is a result of the reproductive social behavior which, ultimately, is determined by the intrinsic trend for increasing the number of the offspring, a process in which unique social systems arise (Roughgarden, 2005; Roughgarden et al., 2006).

The reproductive success of males depends on the number of matings, while the success of the female depends on the number of eggs she lays or the embryos she carries. Obviously, the reproductive success of the competing males and females is not determined by environmental factors, which are generally similar to competing individuals. It depends instead on the intrinsically determined factors: mate preferences for females and mate visual, acoustic, olfactory, and other sensory signals for males. It can lead to reproductive isolation of populations.

> *Sexual selection has the potential to lead to rapid divergence between populations, it can be independent of environmental differences, and it is predisposed to generate reproductive isolation because of its direct effect on traits involved in mate recognition.*
>
> *Panhuis et al. (2001)*

As is known, Darwin introduced the term "sexual selection" to express his idea on a form of selection that leads to evolution of secondary sexual characters, but he did not suggest that it is related to speciation. Sexual selection, as Darwin conceived it, results "from differential mating success among individuals within a population" (Panhuis et al., 2001).

The question now is whether sexual selection is a factor of reproductive isolation. If yes, what determines the sexual selection?

What Darwin understood as sexual selection is behavioral expression of an underlying cause, of the mate choice, which in turn is a function of the activity of the mate recognition system comprising both male signaling and female responses to it (Butlin and Ritchie, 1991). Hence, sexual selection is not a cause on its own but a result and manifestation of mating preferences and mate choices (Panhuis et al., 2001; Ryan and Rand, 2003), hence, as a concept, it lacks explanatory power of its own. Thus, sexual selection is ultimately determined by neurocognitive mechanisms. This is why sexual selection is an exclusive metazoan phenomenon, unknown in plantae or unicellulars, which have no known cognitive properties.

In searching for a causal explanation of evolution, reproductive isolation, and speciation one should not exclusively focus on, or be distracted by, their consequences. We should not lose sight of the causes behind their effects. This is why in the following I will deal in some length with the nature and origin of the mate choice.

Neurocognitive Populational Breakup

All of the direct observational evidence on processes of sympatric speciation points to the fact that it starts with changes in sexual behavior, which make a group of individuals exhibit mating biases toward members of the group and discriminate against the rest of the individuals in the overall population. Such shifts in mating biases precede changes in genes, in morphology and physiology taking place in the process of sympatric speciation.

Beginning from the 1960s, evidence on the occurrence of sympatric evolution has shown that neither geographical isolation nor "genetic revolutions" are necessary for speciation (Bush, 1975) in metazoans.

Sympatric speciation, unlike allopatric speciation, takes place in spatially continuous populations. Separate mating systems in sympatric, spatially continuous, populations arise as a result of specific changes in behavior of a group of individuals, whose preference for mating with individuals of the opposite sex displaying certain phenotypic characters leads to their neurosensory isolation from the rest of the original population. Theoretically, formation of separate mating groups within the range of a species may result from:

- Sensory-determined preferences for particular individuals of the group, displaying specific distinctive phenotypic trait, or
- Sensory-determined preferences for specific niches of particular groups of individuals.

In all likelihood, the above neurally determined changes in mating preferences and/or neurally determined preferences for host plants represent the basic mechanisms of sympatric speciation.

No changes in genes are involved in evolution of mate preferences (see Section Evolution of Receiver Biases later in this chapter).

Reproductive Isolation via Mate Choice

On the basis of sexual selection is the mate choice, i.e., the bias that the choosing sex (usually females) exhibits for mating with individuals of the opposing sex displaying particular phenotypic characters. A strictly preference-based, discriminatory mating occurring within a separate group of individuals of a population can, theoretically at least, lead to splitting of an original population in two reproductively isolated populations. Thus, it is possible that a speciation process may start in sympatry, with a change in mate preferences for certain sexual characters of a group of individuals. Indeed, empirical evidence suggests that sudden shifts in mating preferences of a group of individuals may be the most frequently used mechanism of reproductive isolation of a group of individuals within the original population.

Mate choice is a neurobiologically determined decision. It results from processing in mating behavior circuits of various sensory (e.g., visual, olfactory, acoustic, tactile) information received from prospective mates. An intrinsic drive to influence these decisions stimulates the prospective mates of the opposite sex to display all of their sexual characters and reproductive behaviors, but we are still far from a real understanding of the nature of this decision-making information processing.

Changes in properties of neural circuits may lead to changes in mate preferences for certain phenotypic characters. This suggests that mating behavior circuits in the central nervous system (CNS) may be potential initiators of sexual isolation, reproductive isolation, and speciation. Mating decisions and their heritable changes may have important evolutionary effects (Phelps et al., 2006).

Mate choice implies mate recognition, which in turn is a function of the mate recognition system. There is no consensus among biologists on whether animals use the same traits for mate choice and for discriminating between conspecific and heterospecific individuals. Some observations suggest that *Drosophila heteroneura* uses different traits for discriminating conspecifics from heterospecific individuals and for mate choice (Boake et al., 1997) but most of the evidence shows that mate recognition and reproductive isolation from heterospecific individuals may be based on the same traits (Ryan and Rand, 1993).

For instance, sympatric white butterflies, *Pieris protodice* and *Pieris occidentalis*, do not hybridize under natural sympatric conditions, although no postmating isolating mechanism exists for preventing their hybridization. *P. protodice* has a wing-melanin pattern different from that of *P. occidentalis*, and they use a wing-melanization pattern for both interspecific recognition and mate recognition (Wiernasz and Kingsolver, 1992).

Experiments with two *Drosophila* species (*D. montana* and *D. lummei*) have shown that the male courtship song is used by females for both recognizing conspecifics and for mate choice (Saarikettu et al., 2005). Another example is that of male individuals of the *repleta* group of *Drosophila* flies, which likewise use their song both for recognizing conspecifics and for stimulating mating behavior in females (Ewing and Miyan, 1986).

Olfactory mating signals in *Drosophila* spp. are cuticular hydrocarbons. From studies on hybrids of two *Drosophila* species (*D. serrata* and *D. birchii*), it was concluded that the mechanism of mate recognition by cuticular hydrocarbons is responsible for both mate choice and evolution of reproductive isolation in *Drosophila* (Blows and Allan, 1998). Similarly, in frogs the advertisement call is used for both species recognition and mate choice and has functioned as a mechanism of the premating reproductive isolation of the group (Ryan and Rand, 2003).

This invalidates the neo-Darwinian prediction that traits (signals) used for mate choice must be different from traits, whose evolution leads to reproductive isolation of diverging populations. Another neo-Darwinian prediction that the traits used for the reproductive isolation of the group would evolve gradually as a result of accumulation of genetic changes in populations is also refuted by the evidence that reproductive isolation between two populations often arises suddenly as a result of changes in reproductive behavior without changes in genes.

Evolution of Mating Preferences

Early evolutionary biologists doubted whether sexual selection through female choice occurred at all, because they did not think that female choice could evolve. In the past 20 years, however, evolution of female choice has been demonstrated to occur in a large number of species, including various insects, fish, birds, and reptiles.

Female (or male less often) choice is the behavioral expression of an intrinsic mating preference/bias for species-specific and individual (visual, auditory, olfactory) signals that the female receives via the sensory organs and perceives *in its brain*. Three main hypotheses for the evolution of female choice have been proposed: Fisher's runaway hypothesis, the "good gene" hypothesis, and the sensory exploitation hypothesis. All of them focus on the preservation and spread of mating preferences by natural selection, paying very little attention to, if not disregarding, the cause and mechanism of the change of mating preference, which is the key to the process of reproductive isolation and speciation.

Fisher's Runaway Hypothesis

The hypothesis predicts that females that display mating preference for, and mate with, males with particular trait(s) provide the male carrier of that trait with a selective advantage by producing more offspring with the same signal trait and similar mating preference. In turn, this leads to positive selection of the mating preference (hence, the designation "runaway") for exaggerated signaling traits up to a point where the cost of the development of the trait overweighs the advantage, i.e., "until checked by severe counterselection" or "disadvantageous consequences" (Lande, 1981; West-Eberhard, 1983; Kokko et al., 2003). Thus, the female preferences and male signal traits coevolve. Female preference may exert direct selection on male traits, especially by favoring female biases for male traits that may be related to fertility and parental care (Kokko et al., 2003). Evolution of female preferences is an indirect result of direct selection and evolution of male signaling traits.

It is not known whether "runaway" could occur in natural populations (Kirkpatrick and Ryan, 1991). Despite the large amount of empirical work on indirect benefits of mate choice, the fundamental prediction of the hypothesis that mating with attractive males increases the net offspring fitness has not been empirically tested (Kokko et al., 2003).

Fisher's runaway hypothesis of evolution of mating preferences is still waiting to be validated.

The "Good Gene" Hypothesis

The "good gene" hypothesis holds that male attractiveness is an indicator of the presence of "good genes" for higher viability in males. Hence, female mating bias is genetically correlated with male mating signals and is indirectly selected as a result of the direct selection of genes for increased fitness in males (good genes).

The good gene and runaway hypotheses posit that female preferences may be influenced by natural selection if they lead to reproductive advantage, as would occur when female preference may be correlated with signals of males of higher fitness qualities.

Although it is possible that in some cases, genes for hybrid inviability and mate recognition are linked to the same chromosome, it is hard to believe that this genetic correlation may be a widespread phenomenon as to be speciationally relevant.

Studies on the female preference for the swordtail in poeciliid fish, *Xiphophorus helleri* and *Priapella olmecae*, belonging to two sister genera, have shown that the female preference for swordtail is stronger in the species that has not evolved swordtail, what is the opposite of what would be expected from both Fisher's runaway and the good gene model (Basolo, 1998). Additional evidence on differential deposition of maternal hormones and other substances in eggs of birds mated with preferred males also contradicts the good gene model and has led investigators to the conclusion that results of these experiments show that the father has no effect on the condition and fitness of the offspring (Balzer and Williams, 1998; Cunningham and Russell, 2000).

Some cases of possible relevance in support of the good gene hypothesis come from comparative studies showing that the fluctuating asymmetry in elaborate feather ornaments in swallows (*Hirundo rustica*) is negatively related to the size of the ornament. In those studies, it was observed that males with larger symmetric tails mated earlier and were reproductively more successful so that the female preference for males with larger and symmetric ornaments is believed to be related to higher quality of these males (Møller, 1992).

The hypothesis of good genes selection predicts that evolution of receiver biases is a by-product of natural selection because signals that are preferred are genetically correlated with other traits of higher evolutionary fitness. This prediction is refuted by experiments showing that

> *trait evolution and preference evolution are often decoupled in sexual selection, that they need not evolve through genetic correlation, nor are the response properties of the receiver tightly matched to the properties of the signal, as a lock and key would be matched. Analogies between animal communication systems and human-engineered systems often stress the necessity of tightly matched signals and receivers. Studies of receiver biases suggest that such analogies might not be broadly applicable. The receiver's past history might bias neural processing strategies toward those that are merely sufficient to enhance the receiver's evolutionary fitness but are not optimal engineering solutions. Furthermore, tightly matched signal-receiver systems might have a selective disadvantage if they constrain the receiver's ability to accommodate meaningful population variation.*
>
> *Ryan (1998)*

The runaway and good gene hypotheses would predict that signals from the sender are necessary for maintaining female preferences. This prediction also is not substantiated, while several examples invalidating it have been presented. For example, the all-female species of the poeciliid fish, *Poecilia formosa*, has the same preference for body size that females of its ancestral dioecious species have, and it uses sperm from other species to fertilize its eggs, but the heterospecific male genome is not incorporated in the genome of the offspring. This indicates that *P. formosa* has been able to retain its ancestrally inherited mate preference, in the absence of male signaling, for evolutionarily long periods of time (Ryan, 1998).

Female guppies of the species *Poecilia reticulata* prefer to mate with males with larger orange spots. Selection experiments for male attractiveness and female preferences with *P. reticulata* for three generations also failed to show results that would be expected according to the above hypotheses (Hall et al., 2004). There is evidence that, contrary to the previous belief, the preference for males with large orange spots is not related to any higher fitness of these males but it is related to a general sensory preference of females of this species for orange-colored objects, including food (Rodd et al., 2002).

Physalaemus pustulosus is a monophyletic group of frogs in which females exhibit preferences for four call traits, suggesting that no correlation between the evolution of genes and the evolution of mating signals has taken place.

Generally, it may be said that there is no adequate evidence in support of the idea that preferred or "sexy" males possess better genes and lead to an increase of fitness in the offspring.

Sensory Exploitation Hypothesis

According to the sensory exploitation (sensory bias) hypothesis of evolution of mate preferences, female preferences antecede male signaling traits and males that evolve their signaling traits in order to match the female preferences. Hence, according to the hypothesis, female preferences evolve independently of male signaling and are not under direct or indirect action of natural selection.

The sensory exploitation hypothesis is a *noncorrelational hypothesis* that seems to have found considerable empirical support. It holds that males evolve signaling traits for exploiting preexisting mating biases of the females, and no genetic correlation is necessary for evolution of female mating preferences and male mating signals. The prediction of the hypothesis that receiver's (most commonly females) biases precede evolution of respective signaling traits of the mate is substantiated in a number of cases and is in line with the present knowledge of the evolution of the mate recognition system.

It is generally believed that no genetic correlation is necessary for male signaling traits to evolve in response to specific female preferences.

> *The sensory exploitation hypothesis suggests that, contrary to coevolution through genetic correlation, a trait and a preference in sexual selection—or, more generally, a signal and a receiver in animal communication—can evolve out of concert, with the evolution of one component lagging behind that of the other.*
>
> *Ryan (1998)*

The receiver bias is a product of the neurosensory system (the system of perception of the external world by senses including neural pathways of transmission of information and perception produced in respective regions of the CNS), which evolves independently of evolution of genes. The hypothesis would predict that the sender of the signals evolves its signals for exploiting preexisting mate preferences of the receiver.

Evidence that female mate preferences evolve for females' own benefit, generally for increasing females' fitness, is scarce. One example of this evolution may be females of several species that prefer mating with males that provide nest sites or care for the young (Kirkpatrick and Ryan, 1991). While this hypothesis is more attractive to many biologists, an explanation has not been provided on how the male can evolve mating signals to satisfy female mating biases.

The visual, auditive, and other signals, emanated from the sender (most of times a male), are received by sensory organs of the receiver, which converts them into patterns of electrical signals and transmits them for processing in neural circuits, where the input is compared with the standard neurocognitive mating preference before making any mate-choice decision. This gives us some hints on the surprising, and still unexplained, phenomenon of the sudden appearance and changes in mating preferences, in particular on the divergence of mate preferences without changes in genes. Such sudden changes in mating preferences could be predicted in view of the neurally determined high plasticity of the function of the "mate recognition system."

> *Something needs to happen to make females choose and thereby make it worthwhile for males to display.*
>
> *Kokko et al. (2003)*

While we are not aware of *what* exactly happens, certainly we know *where* it happens: in the neural circuits for mate choice, and there is where the "nudge" comes from. Both male display and female choice are mating behaviors that, like any other behaviors, have a nongenetic, neural substrate and neurocognitive basis.

All three models (Fisher's runaway, the "good gene," and the sensory exploitation) would predict that, over time, a correlation between the female preferences and male signals will evolve.

Models have shown that all three hypotheses for the evolution of preference are internally valid; i.e., they could work. Testing the external validity of these hypotheses (i.e., the probability for generalization), however, has proven troublesome (Ryan and Rand, 1993).

All three hypotheses are still waiting to be validated.

Neurocognitive Mechanisms of Sympatric Reproductive Isolation

The reproductive isolation, as the first stage in the process of speciation, offers no selective advantage, and natural selection can explain neither why speciation happens nor *why* and *how* mating biases change and evolve. An understanding of sympatric reproductive isolation and speciation requires a comprehensive view on the sensory bias as inducers of reproductive isolation and sympatric speciation rather than their influence on evolution of secondary sexual traits that is still dominating studies on "sexual selection."

Evolution of Receiver Biases

Mating choices are expression of sensory biases. As behavioral characters, sensory biases represent behavioral output of the processing of mate sensory input in neural circuits. While we know that these biases are inherited and evolve as species-specific characters, we do not know how they are fashioned in the nervous system. Mating biases or preferences, as neurobiological products of the activity of the nervous system, underlie the mating behavior and behavioral characters (see also Sections Neural Basis of Animal Behavior and Animal Behavior Is Not Determined by Genes in Chapter 8).

Based on general considerations on receiver's biases as manifestations of neurobiological processes, one would agree with Enquist and Arak (1993, 1994) that these biases are a necessary outcome of sensory processing in neural circuits. The intrinsic properties of these circuits might determine the curious phenomena of preferences for exaggerated sender's signaling traits and the origin of attractivity, including preference for symmetry.

Female preferences are neurally determined mating biases resulting from integration and processing of sensory (e.g., visual, auditory, olfactory, tactile) stimuli communicated by the sender (usually the male) to the receiver (usually the female). Male stimuli may be perceived by the female as attractive, unattractive, or even repelling. We have no real idea of the brain mechanisms that assign certain traits—the attributes of being attractive or "sexy." We do not know how these sensory stimuli are manipulated in the brains of these species to produce the attractiveness, mating biases, and preferences on which mating decisions are made.

In general terms, it may be predicted that males with "attractive" characters—characters that induce the female's pleasure—will have a greater chance of mating and leaving offspring and consequently will increase their representation over generations and lead to maintenance or evolution of the male "attractive" characters, such as singing, color, and color patterning of the body.

Focusing almost entirely on male signals, sexual selection theory, in general, did not consider the origin of female biases and preferences. Understanding signal evolution under sexual selection is not the whole issue. There is no doubt that female choice can drive in males evolution of traits that are more attractive to females. The difficult, and quite controversial, issue is why females have evolved receiver properties that make one trait more attractive than another (Ryan and Rand, 1993).

What, essentially, is the attractiveness of individuals of one sex, which stimulates sexual preference of the other?

From our human experience, we know that anything that is attractive to our sight or hearing is beautiful, but we have only vague ideas on those *subjective* rules or criteria on forms, colors, and patterns that produce the sense or feeling of beauty and attractiveness.

A human endeavor for perfection has found its expression in the development of fine arts, where exaggeration of real proportions and patterns has always been, and still is, an essential part of artistic trends, which are reminiscent of the exaggeration of the signaling traits in the animal world.

Symmetry in signaling and other traits is an essential component of the sender's attractiveness. It has been described for females, but evidence that males prefer symmetry has also been presented (Hansen et al., 1999).

In regard to the origin of the preference for symmetry, Enquist and Arak reason:

> *One problem faced by animals is the need to recognize objects in different positions and orientations in the visual field. An object viewed from a particular location is focused on the retina as an "image," which is a geometrical transformation of the object itself. An intriguing idea is that the need to generalize many such transformations of the same object may lead to preferences for symmetry and the evolution of symmetrical signals.*
>
> *Enquist and Arak (1994)*

From results of their experiments with artificial neural networks, Enquist and Arak (1994) concluded that preference for symmetry evolved from an evolutionary pressure for recognizing signals, regardless of their position in the visual field. Commenting on the results of their experiments on female biases in artificial neural networks, and their relation to the female preference in nature, they pointed out:

> *It is an interesting thought that all nervous systems built for recognition may share certain general biases which result from hidden properties of the recognition system. Indeed, many elaborate signals that occur in nature are often as impressive to human observers as they appear to be to the intended recipient. Darwin's idea that a "sense of the beautiful" is an inherent, aesthetic property of animal nervous systems may be not far from the truth. In Darwin's own words, "When we behold a male bird elaborately displaying his graceful plumes or splendid colours ... it is impossible to doubt that [the female] admires the beauty of her male partner."*
>
> *Enquist and Arak (1994)*

According to another hypothesis, preferences for symmetry evolve as a by-product of cognitive processes where the sum of fluctuating asymmetries is averaged to zero asymmetry, which is symmetry (Enquist and Arak, 1994). The hypothesis seems to have found empirical support in a few experiments. Untrained starlings exhibit no preference for symmetry (Swaddle, 1999), but when these birds were trained to detect a set of asymmetric images, they developed preference for symmetry. Such examples suggest that preference for symmetry may develop on the basis of learning mechanisms, without any correlation between this trait and the fitness (Swaddle et al., 2004). So, for example, no relation to male fitness or quality has been found in experiments on symmetry and attractiveness of the human face (Rhodes et al., 1999).

A generalization probably could be made for birds. Many studies have shown that they exhibit a clear preference for symmetry (Enquist and Arak, 1994; Swaddle and Cuthill, 1994) and for what humanly are considered to be ornaments in their morphological characters. It has been hypothesized, and limited evidence is presented, that both of these preferences are used as indicators of male quality. Since developmental asymmetry is considered to be a negative indicator of fitness, and since in experiments an inverse correlation has been found to exist between the size of the ornament and the asymmetry, it has been concluded that female preference for exaggerated mating signals may be an indicator of male good quality (Møller, 1992). This sensory bias for symmetry "may account for the observed convergence on symmetrical forms in nature and decorative art" (Enquist and Arak, 1994).

The idea has been expressed that mate preference for symmetry is evolutionarily selected and that the prevailing explanation of the evolution of preferences for symmetry in bilateral traits is that it results from selection against fluctuating asymmetry, which in turn is a consequence of developmental disturbances and developmental stress. However, experiments with artificial neural networks suggest that these preferences may evolve in the absence of any correlation between the symmetry of male signaling traits and male fitness, and that the female preference for symmetry evolved as a result of selection for mate recognition (Johnstone, 1994).

Attempts have been made to relate general preference for symmetry with the concept of *Gestalt* (German for *shape*) by relating the preference for symmetry to the *Gestalt* law of simplicity. According to the *Gestalt* concept, percepts precede sensory experience, with the latter acting as a stimulus for activating preexisting percepts. These percepts may be related with the specificity of organization of neural structures in the brain. Some evidence in support of this explanation comes from studies on infants, which seem to validate Helmholtz's prediction that it is the organization of the inner ear that makes humans find pleasure in harmony and dislike discordance in music (Ryan, 1998).

Let us return, now, to the general problem of the attractiveness. In a model by Phelps et al. (2006), animals convert the input of visual, auditory, and olfactory signals (morphological and courtship signals) from possible mates into neural equivalents suitable for comparison before deciding to choose the most attractive mate that exceeds some minimal criterion. In their model, mate choice and species recognition are fundamentally similar, in the sense that both rely on the same mechanisms (Phelps et al., 2006). The authors wonder

> *whether shifts in preference acuity or threshold reflect changes in the number or nature of neurons that process sensory information or assign it affective value.*
>
> *Phelps et al. (2006)*

The answer to the question is "Yes," based not only on theoretical consideration of the behavior as a product of the activity of neural circuits but also on empirical evidence. For example, the marine plainfin midshipman fish, *Porichthys notatus*, has two male morphs displaying differences in body size, gonad/body weight index, reproductive tactics, and vocal motor traits. The type I males reach reproductive maturity at smaller body size than the type II males and females. Immunocytochemical examination revealed that the size and number of GnRH neurons in the hypothalamic preoptic area (POA) was 50–100% greater in adults than in juveniles, and changes in the POA phenotype are correlated with the development of type I and type II males in *P. notatus*. The temporal pattern of such changes in POA may represent the proximate mechanism for the development of alternative male reproductive morphs (Grober et al., 1994).

These results suggest that historical processes "shape the design of communication systems" and determine receiver biases. Based on analogies between the behavior of neural networks and female responses, investigators believe that circuits are represented by "reciprocal connections of the torus semicircularis and auditory thalamus." (Phelps and Ryan, 1998, 2000).

In most cases, the "choosy" sex is the female, and male signaling traits are selected under the influence of female preferences for certain male traits. However, "exceptions to the rule" are known. There are cases in insect, fish, amphibian, and bird species, when males are also attracted to, or show preference for, females displaying specific traits. For example, both males and females of the rock sparrow, *Petronia petronia*, have a yellow breast patch. In females, the size of the patch is correlated with the fecundity (the annual number of broods), and males prefer mating with females having larger patches, implying that the female trait is sexually selected.

Experimental reduction of breast patches in female rock sparrows leads to declining sexual interest of males for these females (Griggio et al., 2005).

Controversy exists over the possibility of coevolution of male signaling traits and female preferences. In a comparison between two sister species of the swordtail fish, *Xiphophorus multilineatus* and *Xiphophorus nigrensis*, Morris and Ryan (1996) observed that females of *X. nigrensis* responded similarly to males with vertical bars of *X. multilineatus* and to conspecific males without bars. In contrast, females of *X. nigrensis* found males with bars of the sister species to be more attractive than those without bars of their own species, what speaks against coevolution of male signaling and female preferences, that male traits and female preferences "do not coevolve via genetic correlations" (Ryan, 1998).

However, examples of correlation and coevolution of male traits and female preferences have also been presented. For example, the stalk-eyed fly, *Cyrtodiopsis dalmanni*, exhibits strong sexual eye dimorphism. Artificial selection for long and short eye span in males for 13 generations led to a shift of the female preference for long-eye-span males in females of the long-eye-span line and unselected line, while females of the short-eye-span line continued to prefer short-eye-span males (Wilkinson and Reillo, 1994). Experiments with artificial neural networks also suggest that female biases and male signals may coevolve.

Enquist and Arak (1993) have proposed a hypothesis on the evolution of the preference for exaggerated male secondary sexual traits, based on experiments with artificial neural networks. They found that artificial neural networks may be trained to recognize certain images. Although no recognition system can be trained to identify all of the possible variations of those images (there are too many to identify specifically), they acquire the ability to "generalize," that is, to classify new variants of images into groups of images they have recognized before and to respond to them. When the network trained to recognize, a bird image was presented with images of exaggerated long-tailed (and long-winged) conspecific males, they gave a weaker response to short-tailed, heterospecific males and did not respond at all to tailless bird images. This implies that biases in neural networks arise unavoidably. Investigators also attempt to explain the conclusions of a number of studies that female biases precede the evolution of specific male signals (Enquist and Arak, 1993).

It is assumed that sexual preferences of receivers for specific sexual traits of the sender arise and evolve via selection, but the substantiating evidence is scarce and equivocal. To the contrary, evidence that evolution of receiver sexual preferences is not related to selection has also been accumulated, and studies with neural networks have shown that these biases in the animal's brain might arise as an emergent property of neural processing of external signals in the absence of any training selective context. In many respects, artificial neural networks of neuronoid units have often been found to behave like real nervous systems. For example, in experiments with túngara frog calls, these neural networks have been able to respond to the call, and, based on such experiments, it is suggested that the call may have evolved as a by-product of a sensory system for species recognition. In the context of the role of mating calls in the reproductive isolation, this implies that higher speciation rates may be related to the rapid evolution of mating signals (Phelps and Ryan, 1998).

Most investigators believe that mate preference precedes the evolution of mating signals, and it is the latter that evolves to match the existing mate preference. Besides the example of evolution of preference for swordtail in *P. olmecae*, mentioned earlier, in favor of this hypothesis, a number of other examples testify to the occurrence of female preferences in species with males that lack the preferred trait. So, for example, females of common grackles, *Quiscalus quiscula*, have a preference for repertoires of four song types, although the conspecific males sing only one song type. It is believed that female preference antedates the evolution of male song repertoires (Searcy, 1992). Natural selection or sexual selection cannot explain the evolution of such examples of female biases.

Female biases for mating signals may change over evolutionary time, and they may be related to the diversity of male mating signaling. One example of evolution of female biases comes from a study on females of two sister genera, *Xiphophorus* and *Priapella*. *Xiphophorus* males with swordtail attract females of both *X. helleri* and *P. olmecae*, although swords in *Xiphophorus* evolved after divergence of the two genera. Thus, *Priapella* females may have evolved the preference for swordtail, although its conspecific males have not evolved swordtail.

Receiver biases can even change during the lifetime of a single individual, as is the case with females of the satin bowerbirds (*Ptilonorhynchus violaceus*), which display different mating preferences during different stages of life (Coleman et al., 2004).

Female preferences are innate traits, but they can be modified by experience. Exposure of female fish of the green swordtail to predation, for instance, makes them switch mate preference to swordless fish from the original state of long-sword preference (Johnson and Basolo, 2003). Early life experiences in guppies modify female mate preference to orange male coloration.

Cases of reversal of the lost ancestral mate preferences have also been described. Visual exposure of the female guppy, *Poecilia reticulata*, to its cichlid predator, *Cichlasoma biocellatum*, induces her to revert to the initial preference for brighter males or to become unreceptive (Gong and Gibson, 1996).

Mating preferences can be experimentally changed by administration of hormones. For example, administration of human chorion gonadotropin, a ligand of the pituitary luteinizing hormone, in neotropical túngara frogs increases the female receptivity (measured by the number of responses and the time lapse between call and response) and permissiveness (the likelihood of responding to the less-attractive calls) (Lynch et al., 2006).

From a neo-Darwinian standpoint, i.e., from the view that mate preferences are determined by genes, such sudden changes are inexplicable. But such changes become quite understandable from the epigenetic view that mate preferences are a function of neural circuits, which display a relatively high plasticity at both the developmental and evolutionary levels.

Evolution of Sender's Signaling

In most cases, females are receivers of male signals, i.e., they are the choosy sex, whereas males are senders of signals and objects of female mating preferences.

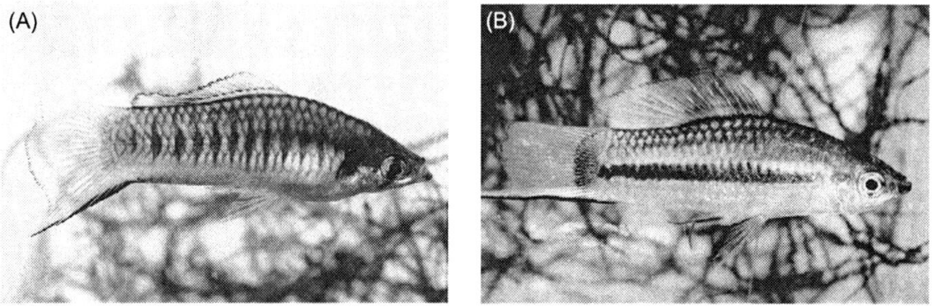

Figure 19.1 (A) Vertical bars on a large *X. multilineatus* male. (B) *X. nigrensis* males do not have vertical bars.
Source: From Morris and Ryan (1996).

A number of phenotypic (visual, acoustic, olfactory, behavioral) traits serve as sexual signals.

Evolution of male signaling may be a direct (sensory exploitation hypothesis) or indirect ("good gene" or runaway hypotheses) result of selection on male sexual phenotypic traits, which show a high level of plasticity. In many described cases, male signals evolve independently of female biases.

Mating signals can also evolve independently of the female biases. One example of evolution of mating signals decoupled from female biases is observed in experiments with two sibling species of the northern swordtail fish of the genus *Xiphophorus*—*X. multilineatus* and *X. nigrensis* (Figure 19.1). These species have a pattern of dark vertical bars that functions as a signal for attracting females and for deterring rival males. Vertical bars are present in the species *X. multilineatus* but absent in *X. nigrensis*. Nevertheless, females of the latter respond to this signal and show a preference for it, although it is absent in males of their own species. Given that both species come from a common ancestor, the presence or absence of bars in them may have resulted from the loss of bars in *X. nigrensis* or the gain of bars in *X. multilineatus*.

Male courtship in *Drosophila* is an elaborate innate behavior, a fixed action pattern, released by pheromonal volatile and nonvolatile sensory inputs from receptive female flies. The courtship consists of a series of standard sequential movements and actions, such as movement in the direction of the female, such as tapping the female with forelegs, singing the species-specific song, while extending and vibrating one of the wings, licking male genitalia, and curling the abdomen for copulation (Baker et al., 2001; Demir and Dickson, 2005).

Recent identification of the neural circuit that is responsible for male courtship behavior in *Drosophila* shows that the behavior is based on expression of the *fru* gene in ~20 groups of neurons (Baker et al., 2001), not only in the CNS but also in the olfactory, gustatory, and auditory systems of male flies. Male courtship behavior is associated with synaptic transmission in all neurons expressing *fru*. Experimental silencing of the *fru* olfactory receptor neurons (ORNs) reduces the courtship behavior, implying that ORNs respond to female volatile pheromones. Stockinger et al. (2005) believe that most, or probably all, of the *fru* neurons in the nervous system, at the sensory, central, and motor levels, might be directly involved in expression of the male courtship behavior:

> *These neurons may be directly interconnected in a circuit that extends from sensory input through to motor output.*
>
> *Stockinger et al. (2005)*

All three genes involved in the courtship behavior (*fru*, *dsx*, and *tra*) are spliced in sex-specific mode, that is, they produce different neuropeptides in the nervous system of males and females. By inducing male splicing of *fru* alone, it is possible to generate normal male courtship in *Drosophila* females (Demir and Dickson, 2005).

Male display traits are often lost in reptiles and in mammals. It has been proposed that the loss may result from the loss of female biases for those traits or from increased predation risk that the presence of these traits may represent. So, for example, in sexually dimorphic phrynosomatid lizards, male signaling traits (colored patches in belly) are repeatedly lost (Wiens, 1999).

Loss of male signaling traits, while the female preference for the traits is still present, also is a very widespread phenomenon in nature, a fact that rejects the possibility of existence in these cases of a genetic correlation between male signaling and receiver preferences.

Wiens has found that the widespread loss of male traits and reductions of female preferences are inconsistent with the Fisher's runaway model, but lends only some support for the "good gene" hypothesis and greater support for the sensory bias hypothesis (Wiens, 2001).

How is female mating preference related to male mate signaling? Given that male advertising and female preference are behaviors, the relationship between two processes is determined in the brain. This fact, however, does not tell us anything on the crucial question of how new male mating signals arise (not how they are selected, which, theoretically at least, presents no problem to understand). There is adequate evidence that visual (morphological in general, including color and color patterning of the body), acoustic, olfactory, tactile, or courtship-ritualistic signals are epigenetically controlled, generated, and communicated via neural pathways.

While female preferences seem to influence the evolution of male "ornaments," less attention has been paid to the influence of male preferences on female ornamentation, a phenomenon that is common in fish. For example, males of the three-spined stickleback, *Gasterosteus aculeatus*, and of the brook stickleback, *Culaea inconstans*, prefer females with nuptial coloration. The coloration of the belly of two-spotted female gobies, *Gobiusculus flavescens*, to a certain degree, evolved as a response to male preferences for that nuptial coloration, which correlates with the female fecundity (Figure 19.2). It is suggested that the female nuptial coloration in this species evolved before male ornamentation (Amundsen and Forsgren, 2001).

Nuptial coloration, orange belly coloration, and transparency of the belly that makes colorful ovaries more visible to the mate are neurohormonally regulated in

Figure 19.2 Mutual courtship display of two-spotted gobies *Gobiusculus flavescens*. The female (upper) displays her colorful belly by bending the body toward the male.
Source: From Amundsen and Forsgren (2001).

female gobies by the neurotransmitter noradrenaline, as well as by melanin, prolactin, and α-melanocyte stimulating hormone (Sköld et al., 2008). Thus, evolution of mate signaling in these species involved no changes in genes but is related to evolution of neurohormonal mechanisms.

Mate Recognition System

Mate recognition system is the system of communication of signals between individuals of opposing sexes in dioecious species for making mating decisions and mating, based on the receiver's or mutual preferences (Butlin and Ritchie, 1991; Ritchie, 2000). One of the most impressive manifestations of the function of the mate recognition system is the lek ritual practiced in many bird species; during the season of the reproductive activity, males congregate at specific sites, known as *leks*, to display their physical qualities, especially their mating signals, to females, which are privileged to choose a male.

The basic lekking behavior is conserved among many bird species. Color patterns and display behaviors in lekking birds have repeatedly evolved to maximize male conspicuousness under the local conditions during lekking, and to minimize it under the rest of the circumstances.

In fish, the lekking ritual has been observed among guppies (Endler and Thery, 1996). Mate recognition system often shows considerable plasticity. Most females have no fixed threshold or criterion to evaluate males. Although receiver's biases are innate, a cultural factor, the imitation of mating choice of other females, may play a role in determining the female's choice, as has been demonstrated in experiments with female guppies, *Poecilia reticulata*. Having an innate preference for orange color, female guppies change their mating choice by imitating mating choice of other females. So, for example, when, after viewing a model female preferring the less-colorful male, female guppies were exposed to males that differ by 12%, 24%, or 40% in orange coloration of the body, in contrast with their innate preference for orange coloration, they chose less-colorful males. However, they did not imitate the model females that chose less-colorful males, when males they had to choose from differed by more than 40% in orange body coloration, thus suggesting that cultural factors (imitation of other females) did supersede the innate preference only in cases when the difference in orange body coloration of males is less than 40% (Dugatkin, 1996).

Mate recognition system displays some degree of habituation as a "subjective" component in mate choice. Females of the field cricket, *Gryllus lineaticeps*, which normally prefer male songs of higher chirp rates, when exposed to more attractive songs respond less than normally to less-attractive songs (Wagner et al., 2001). Other times, differences or the "subjectivity" in mate choice may result from differences in the information that each individual may obtain on various possible mates. This is the case, for example, with female marine iguanas:

> *While for one female a particular male may reveal the highest display score, the same male may only be the second or third best for another female. We consider this distinction between complete, objective information and incomplete, subjective information highly important. We suggest that variance in mate choice in many animal mating systems can be partly or largely attributed to such incomplete information. Thus, even if each female makes the "perfect" choice in her eyes, the cost of female choice will prevent females from gathering complete information to make the objectively "perfect" choice.*
>
> *Wikelski et al. (2001)*

Interestingly, to further enhance that conspicuousness, some birds have also evolved behavioral traits that make the conspicuousness more visible (Marchetti, 1993). So, for example, during courtship, male golden manakins (tropical birds in America), *Manacus vitellinus*, in order to make more visible the golden patches of their plumage, build and decorate bowers, clean leaf litter and vegetation in the background, and display courtship behavior (Uy and Endler, 2004). To the female satin bowerbird (*Ptilonorhynchus violaceus*), the bower quality is believed to convey information for assessing the quality of the potential mate (body size, ectoparasite load), and the ultraviolet plumage coloration carries messages about the growth rate and invasion by blood parasites (Doucet and Montgomerie, 2003; Figure 19.3).

An experimental study on the relative role of olfactory and visual cues in recognition of conspecific males has been conducted on three endemic species of pupfish (*Cyprinodon beltrani*, *Cyprinodon labiosus*, and *Cyprinodon maya*) in Lake Chichancanab (Mexico). These morphospecies have evolved very recently, and their diversification started after the lake desiccated approximately 8,000 (4,000–12,000) years ago. Despite marked differences in morphology (especially conspicuous are morphological changes in their feeding apparatus), these species show "very little genetic change," which is within the limits of the normal intraspecific genetic

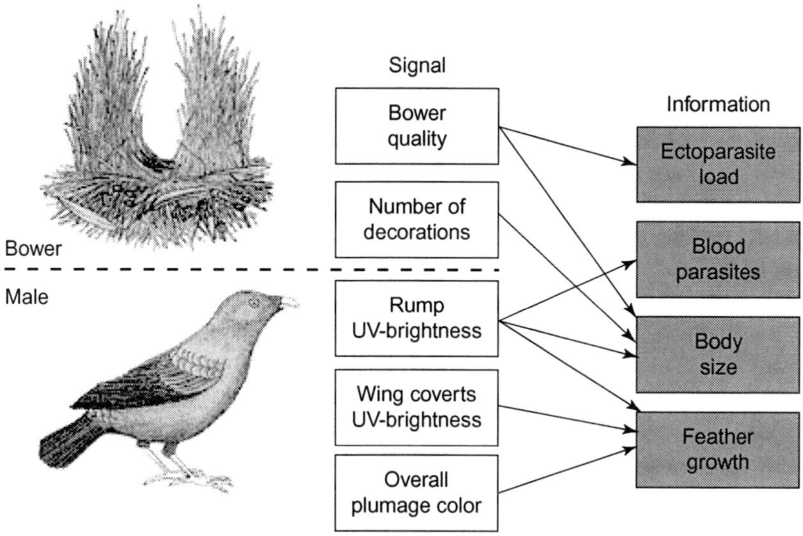

Figure 19.3 Associations between five male attributes (signals) potentially used by females in mate choice and four indicators of male quality (information) in the satin bowerbird. Arrows show variables that significantly ($p \leq 05$ in multiple regression models) predicted each target quality indicator (Doucet and Montgomerie, 2003).

variation (Strecker and Kodric-Brown, 1999), suggesting that no genetic changes are responsible for the evolution of morphological differences and mate recognition systems between them. These species are obviously in a divergent process of evolving distinct visual and olfactory cues for recognition of conspecific from heterospecific individuals. However, they still show a preference for their own males when exposed to olfactory cues from both their males and *C. beltrani* males, suggesting that both morphospecies have diverged more in the olfactory part (production and perception of chemical signals) of the recognition system than in the visual part. (The nuptial color of *C. beltrani* and *C. labiosus* males are still very similar.)

According to Strecker et al. (1996), *these morphospecies show no genetic divergence*. Even though, reproductively, they are still incompletely isolated and interbreed (Strecker and Kodric-Brown, 1999), they display differences in morphology and in olfactory signaling as well as an incipient preference for mating with their own males. The divergence in signaling characters and their perception seems to have advanced more in the third morphospecies, *C. maya*, which has a mate recognition system based on both visual and olfactory cues, displays complete mate preference for conspecifics, and has evolved reproductive isolation.

Sympatric sibling cichlid species *Pundamilia nyereri* and *P. pundamilia* are morphologically similar and populate the same habitat in Lake Victoria, East Africa. However, they are reproductively isolated based on differences in body coloration. Results of experiments with laboratory hybrids between two species led investigators to the conclusion that female preferences are highly heritable, and this heritability is determined by more than one, but no more than five, separate loci (Haesler and Seehausen, 2005).

They infer this conclusion from results in F1 and F2 hybrids, which show that differences in female mating preferences are heritable, thus excluding any possible involvement in female mating preferences of heritable epigenetic factors which, as we show later, are determinants of mating preferences. Thus, the existence of these genes is not demonstrated but assumed by exclusion, without presenting the rationale behind the exclusion of any possible role of epigenetic factors in the results of hybrid mating preferences. This is more surprising in view of the fact that the authors themselves acknowledge the role of "up- or down-regulation of opsin gene expression" and "information processing" in the results of mating preferences displayed by hybrids:

> *Colour vision is important in cichlid communication .... It is perceivable that a switch in visual colour preference can be achieved with modifications to the visual system, for example through up- or down-regulation of opsin gene expression. However, mate preferences could also be determined at higher levels of information processing.*
>
> *Haesler and Seehausen (2005)*

But both "up- or down-regulation of opsin gene expression" and "higher levels of information processing" are not genetic but epigenetic phenomena that imply no changes in genes or genetic information. Furthermore, the conclusion that mating preferences in these species are related to changes in 1–4 unidentified genes contradicts the authors' assertion that these species evolved in sympatry (the continued gene flow and the evolutionarily short period of time would obviously not allow for "favorable changes" in up to four "genes of mate preferences" to occur). Besides, it cannot be imagined *why the selection would favor the reproductive isolation and speciation* in this particular case and in general. The difficulty becomes insuperable when one bears in mind that it is not about an isolated case but about similar formation of hundreds of species of cichlids in Lake Victoria, *in sympatry*.

There are more probable events that might lead to change in preferences; as argued earlier, mate preferences and switches between different preferences take place without changes in genes, but by changes in the properties of neural circuits that result from changes in the synaptic morphology of neural circuits.

Earlier, it was mentioned that the genus *Xiphophorus* of the poeciliid fish consists of swordtailed (with a colored extension of the caudal fin) and unsworded species. A receiver's bias for swordtails has existed in the common ancestor of both genera *Xiphophorus* and *Priapella*, before diverging from their common ancestor, probably before the evolution of the swordtail. This bias is not sex limited, i.e., not only the females, as is generally the case, but males, as well, show the same preference for sworded females, suggesting that both sexes may share receiver's mechanisms for mating responses. Females of five species in both genera (the green swordtail, *X. helleri*, as well as *X. variatus*, and *X. maculatus* in the first genus and *P. olmecae* in the second) display preference for conspecific sworded males.

Not all males in these species are attracted by sworded females: while *P. olmecae* males are attracted by sworded females, *X. helleri* males respond negatively to sworded conspecific females (Basolo, 2002), and this discrimination of males against sworded females makes it clear why female green swordtails (*X. helleri*) do not currently have swords.

From the evolutionary viewpoint, it is important the fact that female preferences may be influenced by the environment. As already mentioned, experiments for determining female preferences after visual exposure of the guppy, *Poecilia reticulata*, to its cichlid predator have shown that the female guppies changed the preference from dull males, which they normally prefer, to bright males (Gong and Gibson, 1996).

By influencing the ability of the perceptual system to receive and identify mating signals, the environment may also play a role in mate recognition and in discriminating between conspecific and heterospecific individuals. Evolution of visual traits that contrast with their background is helpful in this respect. So, it is observed that birds living in darker habitats have evolved more conspicuous bright coloration and patterns.

There is no evidence that the plasticity and evolution of the mate recognition system in animals is related or correlated with changes in genes, allele frequencies, or other genetic mechanisms.

As a rule, animals prefer conspecific traits, but females of some species prefer traits of males of related species rather than traits of their own males. Such is the above-mentioned case of two closely related swordtail species of tropical American freshwater fish. *X. nigrensis* has swordtail and consists of individuals of three size classes, and males show a preference for mating with conspecifics of large size.

A closely related, but allopatric species, *X. pygmaeus*, consists of individuals of small size only, which lack the swordtail and display a different mating courtship. When *X. pygmaeus* females have a chance to choose between males of their own species and heterospecific *X. nigrensis* males, they prefer heterospecific females as mating partners. Investigators believe that the preference of *X. pygmaeus* females for the heterospecific courtship of *X. nigrensis* males indicates that the full courtship of the latter was shared by the common ancestor of both species, and that *X. pygmaeus* males lost the character, its females retained the preference for the lost ancestral trait (Ryan and Wagner, 1987; Ryan, 1998). Their observations suggest:

> *Trait evolution and preference are often decoupled in sexual selection, that they need not evolve through genetic correlation, nor are the response properties of the receiver tightly matched to the properties of the signal, as a lock and key would be matched.*
>
> *Ryan (1998)*

The absence of genetic correlation between the evolution of a trait and the preference for it is also observed in other cases. Such, for example, is the above-mentioned case of two fish species, *X. nigrensis* and *X. multilineatus*. Males of the latter species have bars, but *X. nigrensis* lack such bars. However, females of both species were responsive to bars of *X. multilineatus*, although males of *X. nigrensis* lack bars. Contrary to the incongruence observed in the evolution of signal and female response in both species, males of both species have evolved more congruently: *X. nigrensis* males that lack bars are nonresponsive to bars, whereas males of *X. multilineatus*, which have bars, are responsive to bars (Morris and Ryan, 1996).

In 1984, West-Eberhard proposed that receiver biases precede evolution of mating signals in the opposite sex (West-Eberhard, 1984). One typical example of receiver biases preceding evolution of mate signaling is the case of two poeciliid species, *X. helleri* and *P. olmecae*, belonging to two sister genera, where the preference for sword is stronger in the species that has evolved no sword (*P. olmecae*), than in the species that evolved it. This fact may suggest that:

1. The bias for sword has been present in their common ancestor, or
2. The bias for sword has evolved (becoming stronger in *P. olmecae*).

The second possibility may have theoretical implications for the evolution of both female mating preferences and male mating signals.

Studies on a sexual/asexual system of poeciliid fish composed of two bisexual species, Atlantic molly (*Poecilia mexicana*) and sailfin molly (*Poecilia latipinna*), which are closely related, on the one hand, and an asexual all-female (gynogenetic) fish, the Amazon molly (*Poecilia formosa*), on the other, have led to some observations with important implications for validation of the most important hypotheses on evolution and maintenance of female preferences. The gynogenetic Amazon molly (*P. formosa*) is believed to have evolved between 10,000 and 100,000 years ago or ~30,000 and 300,000 generations, as a result of hybridization of the aforementioned molly species. This species has no males. The female uses sperm of males from parental species to fertilize and activate her eggs, but normally males do not contribute genetically to the offspring of *P. formosa*.

It has been observed that *P. formosa* shows a bias for mating with large-body males of the parental species. The fact that males neither contribute genes nor resources nor nesting nor parental care to the offspring suggests that the cause of the evolution of this preference cannot be of any possible advantage related to "good genes" of the large-body males. It also suggests that Fisher's runaway hypothesis that female preference somehow is genetically correlated with male signaling is not validated in this case, for no genetic material or benefit is provided by males to the offspring of gynogenetic Amazon mollies. The fact that no advantages or increased fitness are provided to the offspring by mating with the large-body males indicates that no selection for female preference has been involved in the evolution of the mating behavior.

As far as the hypothesis of direct selection via sensory exploitation is concerned, there is no evidence that by mating large-body males, *P. formosa* produces a larger number of or more viable offspring. Investigators believe that the preference for large-body males in *P. formosa* is inherited from its parental species and not determined by selection for fitness (Marler and Ryan, 1997).

Investigators offer two alternative explanations for the failure of the species to alter mate preference for large males. First, it may be related to a pleiotropic effect (the female preference is a by-product of a selection for larger body size as a protection against predators), or second, the cost of the mating is so small that a longer time would be necessary for selecting for smaller body males (Marler and Ryan, 1997). However, earlier experimental work had suggested that the beneficiaries of the *P. formosa* preference for large-body males of sailfin mollies (*Poecilia latipinna*) are the latter because females of this species increase their own preference for large-body conspecific males by copying mate preference of gynogenetic females (Schlupp et al., 1994).

Neurocognitive Mechanism of Reproductive Isolation

Visual-Cognitive Mechanism of Reproductive Isolation

Neural Reception of Visual Signaling

Visual perception of signaling traits is one of the most important means for discriminating between conspecific and heterospecific individuals as well as for mate choice. Visual stimuli are converted into electrical signals in the retina and from there are transmitted for processing to the lateral geniculate nucleus and still higher to the primary and secondary visual cortex. In the process, the light stimuli are filtered and further transformed in little-known ways, in order to produce the visual perception, which is a neurobiological interpretation of the light information. Different animal species are receptive to different wavelengths of light reflected by the Umwelt.

Fish use sensory recognition and sensory preferences for identifying and choosing conspecific individuals of the opposite sex as a mechanism of reproductive isolation under sympatric conditions. The visual system of fish is adapted to the environment. In cichlid fish, the system has evolved to adapt to the spectral transmission in the aquatic environment in general but is also adapted to the degree of the transparency of the water, in order to better identify mating color signaling and body patterning.

There is evidence that visual cues, mainly male coloration, are used for mate recognition in cichlid fish (Kornfield and Smith, 2003). Four closely related species of the cichlid fish *Pseudotropheus zebra* group are morphologically indistinguishable and show no differences in their courtship behavior, suggesting that the color pattern may be the only distinctive component of their mate recognition system (Couldridge and Alexander, 2002).

Body color and patterning are key visual cues used by female cichlid fish of the great East African lakes for discriminating between conspecific males and males of closely related species that are morphologically very similar. Females also use those cues for mate choice (Carleton et al., 2005).

Females of the mbuna fish of Lake Malawi, as well as rock-dwelling cichlids in Lake Victoria, East Africa, prefer the larger and more active males, but despite the role of body size, the male body color is the most important of the visual cues these fish species use for mate recognition and is believed to have played a leading role in the extraordinary rapid evolution of these East African lake species (Danley and Kocher, 2001).

Males of the cichlid fish *Pseudotropheus callainos* court only conspecific females. It was assumed that males of this species during the courtship emit sounds, which are detectable by females, but the use of auditory cues in mate recognition in this species seems to be unlikely. The fact that males of this species cannot discriminate between conspecific and heterospecific females that are similar in body color indicates that no acoustic or chemical cues are used for mate recognition by East African cichlid fish (Knight and Turner, 1999).

Males of two closely related fish species of the *Haplochromis nyererei* complex have distinct body colors. Males of one species are blue, and the other's are red. Each of them mates with conspecific females only. When color distinction between species was experimentally masked by monochromatic light, they mated nonassortatively, indicating that body coloration is the only cue used for mate recognition. However, females of both species mated more frequently with blue males, which are larger and have higher display rates, suggesting that in the absence of the color cue, females use body size and display rates as mating criteria (Seehausen and van Alphen, 1998). Two morphospecies of the Caribbean fish of the genus *Hypoplectrus* (*H. unicolor* and *H. gemma*) show only minimal genetic difference but display distinct color patterns. Living in sympatry, they mate exclusively like with like (Barreto and McCartney, 2008).

The dragon lizard, *Ctenophorus decresii*, consists of a monophyletic group of sibling species (*C. decresii*, *C. fionni*, *C. rufescens*, *C. tjantjalka*, and *C. vadnappa*). These species use complex displays for social and sexual communication behavior. While these displays involve the ventral side of the body, throat, chest, and other areas, which are conspicuously colored, the colors and patterns of the dorsal part of their body, head included, have evolved to match the habitat background for reducing the avian predation risk (Stuart-Fox et al., 2004). This is considered to be an example of combined effects of the natural and sexual selection acting, respectively, on the dorsal and ventral parts of the body for body color and patterning.

In birds, especially, visual cues are the most important element of mate-choice decisions, and they are believed to have played an important role in evolution of the class Aves.

Neural Reception and Processing of Bioluminescent Signals

Bioluminescence is a widespread phenomenon in Animalia, in both invertebrates (insects, worms, and arachnids) and vertebrates (fish). It evolved as an adaptation to the dark environment in deep-sea species and for intraspecific communication in insects. Some insect species use bioluminescent flash signals for sexual communication. Fireflies also use bioluminescent flash for luring and deceiving their prey.

In the living world, the ability to generate "cold light" has evolved independently about 30 times, with insects alone having "invented" it at least three times.

Fireflies of the Lampyridae family (order Coleoptera) are nocturnal flies that emit bioluminescent flash signals as components of their courtship behavior. Firefly flash signals are of short duration (a few milliseconds), and they are very diverse and species specific. In the firefly, *Photinus greeni*, both males and females have lanterns (light organs) surrounded by tracheoles consisting of a layer of photocytes

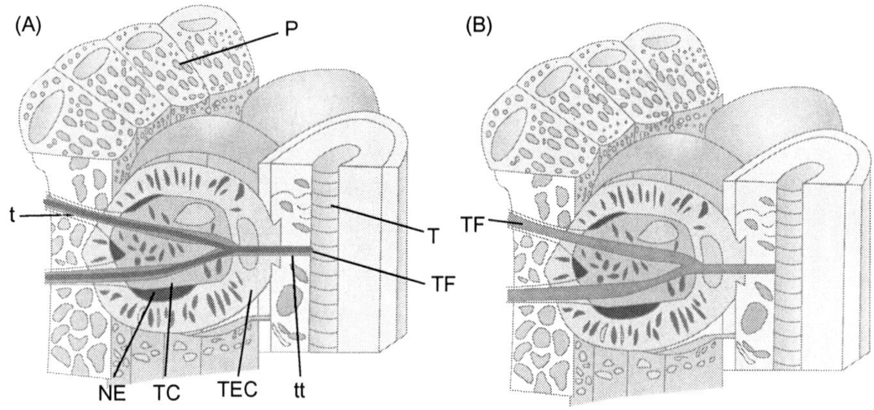

Figure 19.4 Diagrams depicting changes in tracheolar fluid length responsible for controlling oxygen access to the photocytes. (A) Increased fluid length during no light emission. (B) Decreased fluid length during light emission. Neural stimulation leads to a transient increase in the osmotic potential of the tracheolar cell, resulting in decreased tracheolar fluid levels. The resultant decreased diffusional barrier allows greater oxygen supply to the photocytes, relieving intracellular anoxia and enabling light emission. (Molecular oxygen is required.) As tracheolar cell osmotic potential returns to the resting state, tracheolar fluid levels increase, oxygen diffusion to the photocytes is decreased, intracellular anoxia occurs in the photocytes, and light emission is inhibited. *Abbreviations*: NE, nerve ending; P, photocytes; t, tracheole; T, trachea; TC, tracheolar cell; TEC, tracheal end cell; TF, tracheolar fluid. *Source*: From Timmins et al. (2001).

containing reactants that, in the presence of oxygen, emit bioluminescent flash signals (Figure 19.4).

A neural signal in the form of the neurotransmitter octopamine, released by nerve endings of four motoneurons of the terminal abdominal ganglion (A8) that innervates both light organs (Christensen et al., 1983), induces production in tracheole cells of nitric oxide synthase, which synthesizes nitric oxide (NO) (Figure 19.5). The latter enters photocytes (P), where it removes O_2 from photocyte mitochondria. By gating oxygen into peroxysomes, it triggers there a chain of reactions: under the action of luciferase, luciferin is transformed into an intermediate luciferin-adenylil that, by reacting with O_2, produces oxyluciferin, which produces light by emitting photons:

$$\begin{aligned}&\text{Luciferase} - \text{luciferin} - \text{AMP} + O_2\\&\qquad\rightarrow \text{Luciferase} + \text{oxyluciferin} + CO_2 + \text{AMP} + \text{light}\end{aligned}$$

Thus, by regulating pulses of the neurotransmitter octopamine release, the firefly CNS determines the insect's flash patterns (Greenfield, 2001; Trimmer et al., 2001). Flashing patterns are species specific and also sex specific. They are the most important visual element of mating courtship in the *Photinus* group. Females choose their mating partner based on the assessment of the flash signals emitted by males.

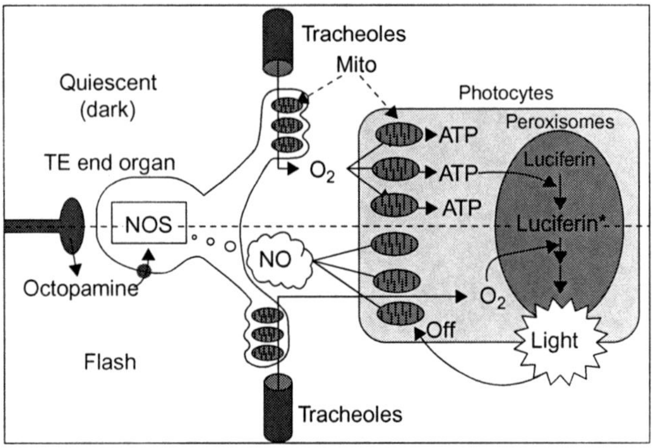

Figure 19.5 Illustration of the scheme for NO interaction with mitochondria in the mechanism for oxygen gating for on/off switching of firefly flashing.
Source: From Aprille et al. (2004).

Although there are no morphological differences between males competing for females, as a rule, females prefer males that have higher flash pattern rates (Demary et al., 2006). Flash patterns are determined by a central pattern generator, and individual flashes are neurally triggered by the release of the neurotransmitter octopamine.

The decision for choosing the mate is made by the female firefly after receiving, analyzing, and assessing mating signals from a number of competing males. The male to be chosen is the one whose mating signals match better with the preferred model of the flash pattern encoded in the insect's brain.

Behavioral studies on *Photinus ignites* have shown that females of this species display preference for longer and brighter male flash signals, and these traits are positively correlated with the size of spermatophore at the beginning of the mating season. However, the preference declines sharply when the duration of the flash exceeds the range of species-specific norm (Cratsley and Lewis, 2003), thus preventing the possibility of mating with heterospecific males.

Insects of the Lampyridae family have species-specific models of encoded flash patterning in the CNS, but they can imitate heterospecific flashing patterns for luring individuals of prey species of the *Photinus* group. This is the case with the predator female fireflies, *Photuris versicolor*, which can modulate their flash signals to mimic female flash responses of their prey, the firefly *Photinus tanytoxus* (Trimmer et al., 2001).

From an evolutionary point of view, the central control of the flashing patterns in these insects represents a potential mechanism for reproductive isolation between sympatric populations found under different evolutionary pressures and for ensuing evolutionary and speciation processes.

Since changes in the bioluminescent flashing patterns involve no changes in genes, it is logical to assume that evolutionary changes in bioluminescent flashing patterns leading to reproductive isolation and incipient speciation result from epigenetic changes in the function of neural circuits determining generation of these signals.

Olfactory-Cognitive Mechanism of Reproductive Isolation

Neural Reception and Processing of Olfactory Signals

Olfactory systems provide animals with a key sense, not only for detecting predators, preys, and food, but also for discriminating between conspecific and heterospecific individuals. In many aquatic and terrestrial species, mate choice is determined, partially or totally, by olfactory cues.

Olfactory signals may have been important factors in the reproductive isolation, speciation, and evolution of many species, such as *Drosophila* spp., and probably in the rapid formation of hundreds of species of cichlid fish in East African lakes. A good idea about the importance of the system in the life and evolution of animals can be obtained from the fact that as many as 4% of genes in higher eukaryotes code for olfactory proteins (Firestein, 2001).

Olfactory systems emerged early in the evolution of metazoans as a part of the nervous system and in its main features are conserved across metazoans, from insects to mammals. Axons of the ORNs that express the same receptor molecules, in *Drosophila* converge to the same glomeruli in the antennal lobe (AL), thus forming a spatial odor map in the fly brain. Odorants bind multiple olfactory receptors; hence, it is possible that the representation of olfactory stimuli is combinatorial (Jefferis et al., 2004).

In *Drosophila*, ORN axons connect with dendrites of the second-order projection neurons (PNs) in the AL in a strictly determined pattern (Figure 19.6). PNs, in turn, send axons to higher brain centers. Contrary to what was generally believed, it is not the axons of the ORNs that select the dendrites or determine the synaptic morphology there; PNs specify their dendrites and create a prototype of the adult glomerular map before their presynaptic partners, axons of the ORNs, reach these neurons. It may be possible, however, that axons of the ORNs help to refine the specificity of the preformed connections (Jefferis et al., 2004).

Prespecification of dendrites in PNs determines the hardwiring of the olfactory system and the resulting behavioral responses of flies to odorants (Jefferis et al., 2001). How the ORN axons find their corresponding dendrites among the numerous PN dendrites is unknown. According to one hypothesis,

> *ORN axons and PN dendrites have substantial autonomous patterning ability ... the two proto-maps interact during development to generate the final mature glomerular organization.*
>
> *Jefferis et al. (2004)*

In a study of the anatomy of the AL of more than 30 endemic Hawaiian species of the family Drosophilidae, it was found that two (out of 51 identifiable) glomeruli were enlarged in the AL of males only. This trait evolved independently in 37 Drosophilidae species of two genera (*Drosophila* and *Scaptomyza*, derived from the first *Drosophila* species that migrated to Hawaii between 1 and 2 mya). This sexual dimorphism of *Drosophila* brains is estimated to have arisen sometime between 0.4 and 1.9 mya (Kondoh et al., 2003).

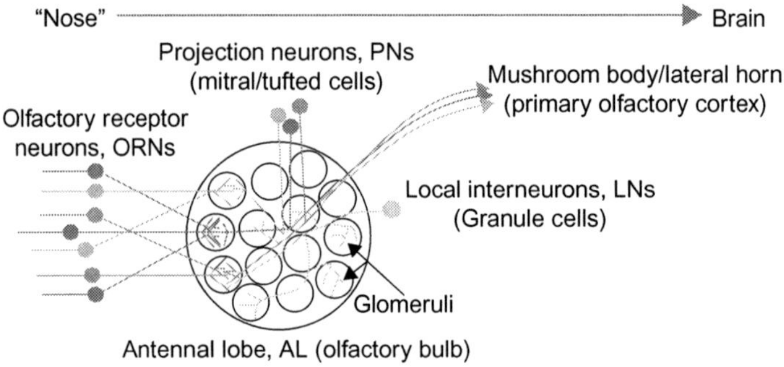

Figure 19.6 Organization of the mature *Drosophila* antennal lobe (AL) with vertebrate counterparts in parentheses. Each differently shaded circle represents olfactory receptor neurons (ORNs) expressing a particular seven-transmembrane span receptor or their postsynaptic projection neuron (PN) partners.
Source: From Jefferis et al. (2004).

The idea that the brain sexual dimorphism is related to male-specific genes (Kondoh et al., 2003) is rejected by earlier experiments of Schneiderman et al. (1986) and Rossler et al. (1999). Based on the results of those experiments, it is concluded that not the differences in any sex-specific genes but male and female ORC (olfactory receptor neuron) axons are involved in determining the position, anatomical features, and innervation of sexually dimorphic glomeruli (Schneiderman et al., 1986).

The olfactory system is an important component of the mate recognition system with crucial functions for the mate choice. It detects and recognizes species-specific olfactory signals, above all pheromones that are used for coordinating the reproductive behavior and activity of mating partners in invertebrates and vertebrates.

In *Drosophila melanogaster*, olfactory signals (pheromones) released by males are received by sensory nerves and transmitted to the brain (AL) for processing in the olfactory neural circuit. The olfactory circuit in *Drosophila* not only is responsible for identification of odors but it is also necessary for the courtship behavior of the fly that is determined by another circuit, which, in male flies, comprises a number of neurons of a larger group of neurons that have been identified to express two specific genes (Stockinger et al., 2005).

Secretion of pheromones in insects is neurally regulated by cerebral neuropeptides, pheromonotropic hormones, of which the most extensively studied is the pheromone biosynthesis activating neuropeptide (PBAN) family of neuropeptides secreted in the insect brain (Altstein et al., 1993).

Another pheromonotropic hormone, isolated from the moth *Helicoverpa zea*, genus *Heliothis*, is pheromonotropic melanizing peptide (Hez-PMP), which induces pheromone release and melanization of the insect in a dose-dependent mode. Besides ganglia, the neuropeptide is also released by the esophageal nerve (Raina et al., 2003).

The biosynthesis and secretion of the sex pheromone in the hemolymph of the noctuid insects, *Spodoptera littoralis* and *Mamestra brassicae*, also may involve PBAN or PBAN-like neuropeptides. The control and regulation of the synthesis and secretion of these neuropeptides is determined by a "neural input from the ventral nerve cord" on the pheromone gland (Iglesias et al., 1998).

The fly has ~1,300 ORNs in the antennae and maxillary palps. Each of these neurons expresses only one of the 43 types of olfactory receptor molecules. Each ORN projects an axon to one of the specific glomeruli in the AL, the equivalent of the vertebrate olfactory bulb. From the AL, 150–200 PNs transmit olfactory signals to the mushroom body (MB) calyx and to the lateral horn. When a group of antennal ORNs bind a specific molecule, it activates a corresponding group of PNs in the AL glomeruli, which is then inhibited by inhibitor circuits that form combinatorially by local interneurons (Ng et al., 2002).

Key elements in the olfactory pathway are PNs, which connect ALs with the lateral protocerebrum (LPR), not only directly but indirectly as well. In order for a male insect to initiate and perform male courtship behavior, it is necessary to receive the specific olfactory and chemosensory cues or visual cues of a conspecific virgin female. The initiation of the male courtship behavior is related not to the presence or activity of the MB (that behavior is also initiated in the MB-ablated males) but to a cluster of some excitatory and some inhibitory neurons in the LPR. These neurons seem to integrate the sensory information that triggers the male courtship behavior (Broughton et al., 2004). Thus, the chemosensory and visual information for activating the male courtship behavior in *Drosophila* reaches the LPR in two ways: the indirect pathway via the MB calyx (responsible for the experience-dependent olfactory processing) and a direct pathway (responsible for experience-independent olfactory processing). Ablation of the MB calyx, i.e., the block of the indirect pathway, does not impair the experience-independent functions of odor detection and male courtship behavior but affects the experience-dependent functions of odor- and courtship conditioning. Unlike the ablation of MB calyx, the block of synaptic transmission by tetanus toxin light chain (TeTxLC) impairs both odor detection and courtship behavior.

The basic organization of the olfactory system in metazoans, from insects to humans, is well conserved. Mammals also use olfactory cues for mate choice. For example, both humans and mice prefer the odor of individuals that possess dissimilar antigens (cell surface glycoproteins) of the major histocompatibility complex (MHC), a group of extremely polymorphic genes involved in the immunological mechanism of recognition of "self from non-self." Mice can recognize individuals that differ only in their MHC system but otherwise are genetically identical to them. Recognition of their own individual MHC phenotype in other individuals allows them to avoid inbreeding (Penn and Potts, 1998, 1999).

Olfactorily Determined Reproductive Isolation in Sympatry

Chemosensory signals and the receptive olfactory system in insects play an important role in initiating the process of reproductive isolation between sympatric populations. Experiments on two recently diverged sympatric cichlid fish species (*Pseudotropheus emmiltos* and *P. fainzilberi*) from Lake Malawi have shown that

P. emmiltos females use olfactory signals and preferences, to a much larger extent than the male body color, for distinguishing their conspecific mating partners from *P. fainzilberi* males.

Based on the results of these experiments and other relevant evidence on the role of olfactory and auditory cues in the Mexican pupfish (Strecker and Kodric-Brown, 1999) and *Drosophila* (Ortiz-Barrientos et al., 2004), it has been concluded that changes in olfactory signals and olfactory preferences have played a great role in the rapid speciation of these fish (Plenderleith and Ryan, 2005) and *Drosophila*.

Other observations suggest that olfactory cues have diverged in the course of the evolution of poeciliid swordtail species *Xiphophorus cortezi* and the closely related species, *X. nigrensis* (McLennan and Ryan, 1997).

Male plethodontid salamanders, during the courtship, before releasing sperm, deliver to the females pheromones secreted by the mental gland. These pheromones (sodefrin and sodefrin-like peptides) increase female sexual receptivity. Most salamander species use a 50–100 million-year-old ancestral "scratching" behavior to deliver pheromones to females. They scratch the female's back with the protruding premaxillary teeth (PPT) and then rub the mental gland on the scratched region. About 19 mya, species of the eastern *Plethodon* clade entered a period of divergent evolution (Figure 19.7). Males of *P. welleri* and *P. wehrlei* lost the PPT and deliver pheromones by application of the mental gland to the nares of the female or by head-rubbing behavior. Males of another plethodontic salamander species, *P. glutinosus*, also have lost the premaxillary teeth, but they deliver pheromones by slapping their posteriorly displaced mental glands on the nares of the female salamanders (Palmer et al., 2007).

Male salamanders of the Plethodontidae family secrete a protein pheromone that contains the plethodontid receptivity factor (PRF) (Rollmann et al., 1999), which increases female receptivity during courtship interactions. The pheromone also contains another protein, known as plethodon modulatory factor (PMF). These two proteins represent 85% of the total protein content of the pheromone.

While PRF has a stimulating effect on female receptivity (expressed in the form of shorter courtship and mating time), PMF has the opposite effect of prolonging the courtship and mating time. Each of the two main components of the pheromone, PRF and PMF, binds to receptors in separate vomeronasal neurons, which transmit their separate information for processing in the brain, thus regulating mating behavior (Wirsig-Wiechmann et al., 2006).

Another essential pheromone component in plethodontides is the protein sodefrin precursor-like factor (SPF), which has been recruited for a pheromonal function in this group about 50–100 mya, that is much earlier than PRF, which existed in other plethodontid salamanders. This fact supports the notion that evolution of genes in metazoans is decoupled from the evolution of morphology and behavior (Palmer et al., 2007; Cabej, 2011).

The evolutionary transition from the use of SPF to the use of PRF in salamanders is correlated with a change in the properties of the olfactory circuits, which evolved for receiving and perceiving the PRF, instead of SPF, as a pheromone.

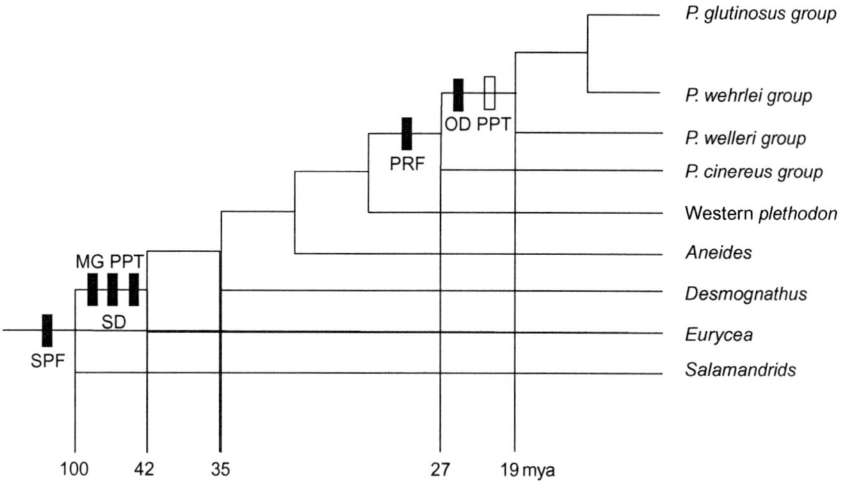

Figure 19.7 Cladogram showing the relationships of various clades of plethodontid salamanders and the evolution of characters involved in courtship pheromone delivery. Approximate divergence times are shown at bottom. Small rectangular boxes show the point of origin (solid) or loss (open) of various characters. *Abbreviations*: MG, mental gland; OD, olfactory delivery of courtship pheromones; PPT, protruding premaxillary teeth; PRF, plethodontid receptivity factor; SD, scratching delivery of courtship pheromones; SPF, sodefrin precursor-like factor.
Source: From Palmer et al. (2007).

Auditory-Cognitive Reproductive Isolation

Neural Reception and Processing of Acoustic Signals

Male singing in crickets is regulated by a brain center located in the command neurons of the anterior protocerebrum. During the processing of song patterns, changes in the cytosolic Ca^{2+} occur in parallel with the chirp rhythm in the auditory interneurons. Acoustic signals of the male song are used as mating signals by female crickets (Hedwig, 2005). There is empirical evidence on the role of courtship auditory cues in mate choice and discrimination of conspecific from heterospecific individuals in insects.

In amphibians, the receiver processes auditory signals (mating calls) in a part of the midbrain, *torus semicircularis*, which is considered to be homologous to the mammalian *inferior colliculus*. Coding of acoustic communication signals is also believed to take place in this part of the brain.

The response of the auditory midbrain of female túngara frogs, *Physalaemus pustulosus*, to mating calls of conspecific, heterospecific, and irrelevant calls has been evaluated by the expression of the gene *egr-1* in various regions of the *torus semicircularis*. The pattern of expression in response to those stimuli is different and specific for the laminar, midline, and principal nuclei. Within these nuclei, a difference is observed in the patterns of *egr-1* expression in the midline and principal nuclei, on the one hand and the laminar nucleus on the other (Hoke et al. 2004). Another part of the amphibian brain that is crucially involved in the signal receiver response to auditory mating calls is the hypothalamus, which is anatomically and functionally connected to *torus semicircularis*. Several hypothalamic nuclei are involved in neurohormonal regulation of reproductive behavior (Figure 19.8).

A correlation exists between the levels of *egr-1* expression in the auditory midbrain and in the hypothalamus. Differences in *egr-1* expression, in response to auditory signals, are also observed between the different regions of the hypothalamus that are involved in the behavioral responses to auditory inputs (Hoke et al., 2005).

Specific responses of individual hypothalamic regions are derivatives of the responses of the auditory nuclei to which they are connected, whereas the functional connectedness within the hypothalamus is an emergent property modulated by the relevance of the social context (Hoke et al., 2005; Figure 19.8).

The principles of parallel processing and distributed functional networks in the frog hypothalamus are remarkably similar to those processes posited in cognitive neuroscience, including perception, memory, and decision making. Acoustic signals in frogs are received by two peripheral auditory organs, the amphibian papilla for low-frequency sounds and basilar papilla for higher frequencies. The auditory input is processed in the hypothalamic auditory nuclei, the auditory nuclei of the midbrain, and the thalamus.

In the majority of birds, it is the male that performs singing, and it is observed that males have two main song nuclei in the forebrain, the high vocal center (HVC) and the nucleus robustus archistriatalis. In the course of evolution, with the increase

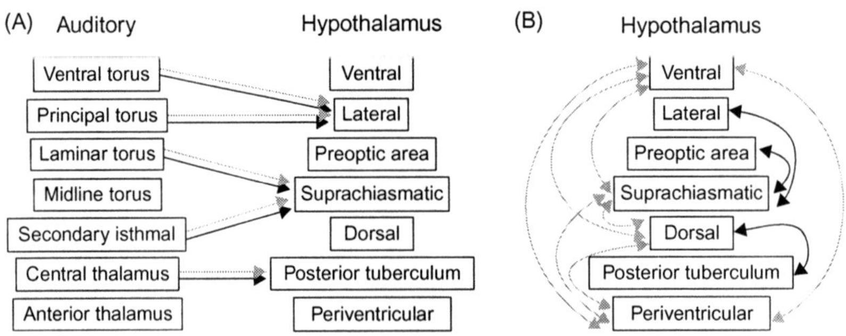

Figure 19.8 Functional connectivity of the hypothalamus. Gray arrows show significant relationships in frogs that heard irrelevant acoustic stimuli (*P. enesefae* whine and chuck only), and black arrows indicate relationships in frogs exposed to behaviorally relevant stimuli (conspecific whine and whine–chuck). (A) The egr-1 levels in midbrain and thalamic nuclei implicated in auditory processing are significant predictors of hypothalamic expression patterns. Relationships between auditory and hypothalamic regions do not vary with relevance of stimulus. (B) The egr-1 correlations between hypothalamic regions differ based on behavioral relevance of acoustic stimulus.
Source: From Hoke et al. (2005).

of the song repertoire, a corresponding increase in the volume of HVC occurred. For example, in male sedge warblers, HVC volume is seven times greater than in females. The sexual dimorphism of the brain arises experience-independently, but brain differences are also observed between birds reared in isolation and those exposed to songs (Leitner et al., 2002). In all likelihood, these remarkable changes in the brain structure between males and females are epigenetically determined and inherited, for no genes in sex chromosomes have been identified to be involved in the process of development of brain sexual dimorphism. Not only the brain centers but also the muscles regulating singing in males are larger than in females.

In singing birds in general, brain nuclei responsible for singing increase their volume during the breeding season.

The Song Circuit in the Brain of Birds

The song production is function of the "song system," a group of functionally related brain nuclei and respective connecting pathways, which are similar in all groups of birds that have evolved learned song (e.g., songbirds, parrots, and hummingbirds). Two distinct song circuits or impulse pathways that start from the HVC have evolved within the song system in birds: the posterior descending pathway (PDP), or song production circuitry (HVC → RA → nXIIts → muscles of syrinx, the song organ), which is responsible for both the acquisition and production of learned song, and the anterior forebrain pathway or song learning circuitry (HVC → Area X → DLM → LMAN → RA), which is necessary for song acquisition only (Nottebohm, 2005; Figure 19.9).

The neural circuitry in the brain song control system is determined by brain-intrinsic mechanisms, and recent evidence shows that the song repertoire in birds is an automatic result of the spontaneous activity of neural circuits in the brain song nuclei. This conclusion has been drawn from the fact that even when kept in complete acoustic isolation, or when birds are experimentally deafened, they can still develop the species-specific song repertoire in an experience-independent way (Leitner et al., 2002). However, this may not be universally true; other experiments with fledgling zebra finches have shown that formation of the motor phase of the song neural circuitry is determined within 10 days after birth, when it learns tutor's song, and becomes fully functional by 35 days of age (Roper and Zann, 2006).

Evidence on factors involved in regulation of the song control system in birds is also contradictory. While some reports suggest that animals respond to the gonadal testosterone by modifying the structure and function of the song control system (Brenowitz, 2004), other studies show that the development of the neural song system in zebra finch males is regulated by brain factors rather than influenced by circulating hormones (Gahr, 2004; Wade and Arnold, 2004). Besides the neurosteroid pathway in the brain, which regulates the production of sex hormones, it is important to remember that even changes in gonadal testosterone levels, ultimately, represent downstream results of neuroendocrine cascades starting in the brain, along the hypothalamic–pituitary–gonadal axis.

Neural song circuits are determined before hatching, and their establishment does not depend on experience. How are these neural song circuits in male birds

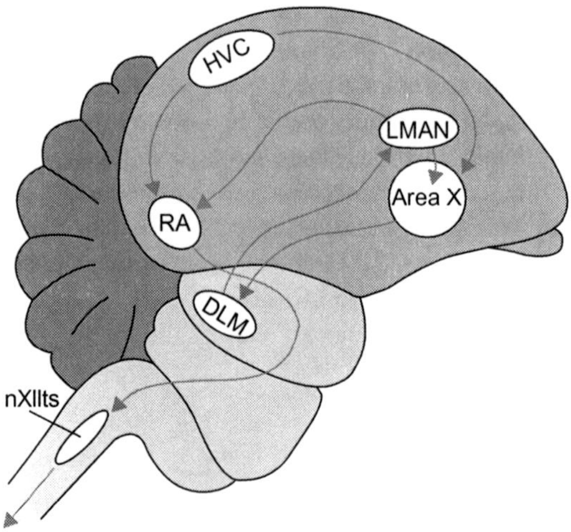

Figure 19.9 The song system of songbirds. Nucleus HVC feeds information into two pathways that ultimately lead to the neurons in the tracheosyringeal half of the hypoglossal nucleus (nXIIts) that project to vocal muscles. HVC projects to nucleus RA directly (PDP) and indirectly via Area X, the dorsolateral anterior thalamic nucleus (DLM), and LMAN (AFP) in a manner that shares similarities with the mammalian pathway cortex → basal ganglia → thalamus → cortex. *Abbreviations:* AFP, anterior forebrain pathway; DLM, dorsolateral anterior thalamic nucleus; LMAN, lateral magnocellular nucleus of the nidopallium; nXIIts, the tracheosyringeal motor nucleus in the brain stem; PDP, posterior descending pathway; RA, robust nucleus of the arcopallium.
Source: From Nottebohm (2005).

determined and established? There is no reason to believe that they are established differently from the ways in which other neural circuits in the brain form: they are designed during the individual development and fine-tuned during the later life by learning and experience. Their establishment during the individual development seems to take place experience-independently, "according to the brain's best guess."

Neural song circuits in the brains of birds show remarkable plasticity, in both the size of song control nuclei (which become larger during the breeding season) and the synaptic connections, plasticity that manifests itself in changed patterns and duration of songs during the breeding and nonbreeding seasons. Changes in neural song circuits are causes of the observed seasonal variations in songs (Brenowitz, 2004).

Acoustically Determined Reproductive Isolation in Sympatry

One of the first forms of interindividual communication in metazoans has probably been based on the production and perception of vibrations and sounds. A colerrhynchan insect, *Hackeriella veitchi*, in Northeastern Australia, believed to be of a Gondwanan relict insect lineage, emits vibrational signals for interindividual communication. This suggests that vibrational signaling by a simple tymbal (sound-producing organ) observed in this insect, and probably other mechanisms of signal production (percussion and stridulation), evolved in four groups of Hemiptera as early as 230 mya (Hoch et al., 2006).

Some insects have evolved special organs for producing ultrasounds of above 20,000 vibrations per minute and use them as cues in mate choice. For example, males of the lesser wax moth emit such ultrasonic signals in 100–120 pairs of pulses/second, which have a phonotactic effect on females. The female brain has evolved such a high acoustic resolution power that it can distinguish even acoustic signals that last for as little as 150 μs. Female preferences for pulse amplitude, rate, length, and length intervals are greater than the average values of male populations. However, when the pulse rate increased to 142 pairs/second, the female preference leveled off and later decreased because such sounds last less than the length of an action potential, and auditory neurons of *A. grisella* females cannot generate 142 action potentials for per second (Jang and Greenfield, 1996). Male crickets produce their song by rhythmic wing movements under control and regulation of the command neurons descending from a particular center in the anterior protocerebrum (Hedwig, 2005).

Hennig (2003) observed that females of two sibling cricket species, *Teleogryllus oceanicus* and *T. commodus*, have evolved two different temporal filters, the first species a period filter, and the second a pulse-duration filter. The rapid evolution of homologous circuits in two sibling species of such different properties of pattern analysis and temporal filters, after diverging from the common ancestor, is believed to have resulted from changes in the properties of neural circuits responsible for conspecific call recognition. There is no evidence of any changes in genes being involved in this rapid evolutionary divergence.

Gray and Cade (2000) studied the acoustically determined speciation process in the case of the North American field cricket species, *Gryllus texensis* (formerly *Gryllus integer*) and *Gryllus rubens*, two cryptic sister species that are morphologically indistinguishable, living both in sympatry and allopatry. They are prezygotically completely isolated from each other while showing no postzygotic isolation. They do not hybridize under natural conditions, although they can produce viable hybrids with equal fertility to that of parental species. The prezygotic reproductive isolation and the evolution of these sister species seems to have been based exclusively on the evolution of divergent male songs and corresponding female preferences in two sympatric populations of the common ancestral species. These cryptic species differ between them in both male and female preferences. Based on the above, it is concluded that evolution of these two species from their common ancestor is the result of sexual selection (Gray and Cade, 2000).

Males of insect sibling species *Neoconocephalus robustus* and *Neoconocephalus bivocatus* generate exceptionally fast calls with pulse rates of ~200 pulses/s and ~175 pulses/s, respectively. Females of both species are highly selective about conspecific male calls, but recognition mechanisms are "strikingly different" despite their close relationship as sibling species. Females of *N. robustus* respond to a continuous, very fast conspecific call of 200 pulses/s with very short intervals, which is recognized as such but may be impossible to be faithfully encoded by their sensory system.

N. bivocatus females, on the other hand, are attracted to their own conspecific calls with a pulse rate of 175 pulses/s, but the sensory system of the insect merges every two consecutive pulses into one pulse pair, by ignoring the interval between each pair, thus creating 87 pairs of pulses. By halving that fast and difficult-to-recognize pulse rate, the insect is able to recognize its conspecific call (Deily and Schul, 2004). Obviously, no changes in genes can be related to the neurobiologically determined ability of insects to "halve" the pulse rate.

Two subspecies of the grasshopper, *Chorthippus parallelus*, live on both sides of the Pyrenees (southwestern France and the Iberian Peninsula). Their mate recognition system uses acoustic signals and possibly olfactory signals. A study of the female preference of the grasshoppers on both sides of the hybrid zone has shown that the female homogamic preference differs abruptly over a distance of 1 km, suggesting that selection for female preference operates in the hybrid zone (Butlin and Ritchie, 1991).

At least three sympatric cichlid species (*Pseudotropheus zebra*, *P. callainos*, and *P.* "zebra gold") in Lake Malawi (formed ~1 mya) generate different species-specific acoustic signals that are distinct with regard to the pulse rate and frequency and are used as cues for mate choice, although there is no evidence that these courtship acoustic signals have led to their sympatric speciation (Amorim et al., 2004).

Males of the túngara frog, *Physalaemus pustulosus*, add a chuck to the basic whine component of their mating call. Adding of this suffix is characteristic of bigger males, so that by preferring calls with chucks, indirectly, females choose larger male mates. The whine component is processed in the amphibian papilla, the auditory structure of the inner ear, and chuck component of higher frequency is processed in the basilar papilla of the female frogs. The amphibian papilla may function for species recognition and basilar papilla may be specifically involved in determining mate preferences. The tuning of the basilar papilla for sensing chucks represents an ancestral state, so the preference for chuck evolved before the evolution of chucks in male calls and, most importantly, its evolution involved no changes in genes.

> *The reproductive advantage gained by preference for whines with lower frequency chucks is an evolutionary effect due to the way frog brains work and how they have evolved.*
>
> *Autumn et al. (2002)*

Earlier, it was suggested that the trend for becoming more complex and the variation of the amphibian inner ear contributed to the divergence of mating calls and, consequently, to the speciation in frogs (Ryan, 1986).

In a series of studies, it is proven that evolution of courtship song and reproductive isolation in the group *Drosophila willistoni*, consisting of six sibling species (Gleason and Ritchie, 1998), and evolution of mating calls in túngara frogs are not related to genetic evolution (Pröhl et al., 2006).

Birds may be hearing-only birds or hearing-and-vocalizing birds (e.g., hummingbirds, parrots, and songbirds). The latter are capable of vocal learning by imitating other birds or even nonavian species in the case of parrots. This capability is related to the evolution of specialized forebrain structures in these birds (Figure 19.10).

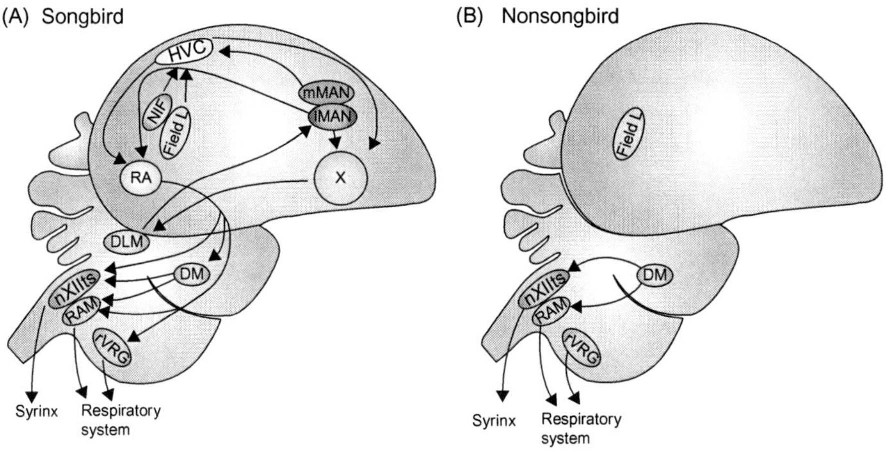

Figure 19.10 The brains of songbirds and nonsongbirds differ. These schematic diagrams of parasagittal views of the brains of a songbird (A) and a nonsongbird (B) illustrate the dramatic differences between them. Songbirds have an elaborate network of interconnected forebrain nuclei that form an interface between auditory input (which converges on field L, the primary auditory projection region in the avian forebrain) and vocal output, which is produced in the syrinx, the avian vocal organ. Nonsongbirds also have field L, and they can produce vocalizations in the syrinx, but they do not have the network of forebrain nuclei that songbirds have. *Abbreviations*: DLM, nucleus dorsolateralis anterior, pars medialis; DM, dorsomedial nucleus of the midbrain nucleus intercollicularis; HVC, high vocal center; lMAN, lateral magnocellular nucleus of the anterior nidopallium; mMAN, medial magnocellular nucleus of the anterior nidopallium; NIF, nucleus interface of the nidopallium; nXIIts, tracheosyringeal portion of the nucleus hypoglossus; RA, robust nucleus of the arcopallium; RAM, nucleus retroambigualis; rVRG, rostroventral respiratory group; X, Area X.
Source: Adapted from Bolhuis and Gahr (2006).

Seven brain centers are related to vocalization, of which five centers in telencephalon (caudomedial neostriatum, caudomedial hyperstriatum, dorsocaudal neostriatum, intermediate archistriatum, and caudal paleostriatum) one in thalamus (dorsointermediate nucleus of the posterior thalamus), and one in mesencephalon (dorsal part of the lateral mesencephalic nucleus) (Jarvis et al., 2000).

In vocal-learning birds, such as hummingbirds, songbirds, and parrots, these structures have evolved independently, are well conserved among species, and comprise seven vocal-control nuclei in seven different regions of the forebrain. (In hummingbirds, there is an additional, eighth, nucleus in the mesencephalon.) Not surprisingly, these structures are absent in hearing-only, vocal nonlearner, avian species. Given the fact that a group of *seven structures* has independently and similarly evolved in three of 23 avian orders, it is suggested that

> *the evolution of these structures is under strong epigenetic constraints; in which case, similar structures may have also evolved in vocal learning mammals (humans, cetaeans and bats).*
>
> *Jarvis et al. (2000)*

Evidence presented in this subsection shows that neural mechanisms of releasing, receiving, and processing of auditory signals determine reproductive isolation among a number of closely related species of insects. That evidence also suggests that these neural mechanisms change and determine reproductive isolation of natural populations without changes in genes.

Electrocognitive Mechanism of Reproductive Isolation

Electrogenesis in Fish

In 1951, Lissmann suggested that the ability of a fish occurring in the Nile River, *Gymnarchus niloticus*, to avoid obstacles when swimming backward, i.e., outside its visual field, may be related to the presence of an electric organ, which "may enable the animal to detect objects in the vicinity of its body" (Lissmann, 1951). Ever since, the study of electric signals in fish has been the object of a special interest.

Elasmobranch fish (rays, skates, and sharks), a group of ~800 species, use favorable aquatic media for electrical prey detection and intraspecific communication. Electrogenesis, production and emission of electrical signals, is a defining feature of tropical mormyriform fish (the genus *Mormyridae* with about 200 electric species and the genus *Gymnarchidae*, with only one species of electric fish) in Africa and gymnotiformes with 115 nominal species in seven fish families of the South and Central America, as well as a number of catfish (Hopkins, 1999).

These teleost fish have specialized electrical organs, consisting of muscle-derived cells or electrocytes, which in mormyrids are represented by multinucleated structures of several centimeters in diameter. The electric organs produce weak electrical pulses and form an electromotor system for transmitting electrical signals for electrical guidance, reproductive behavior, and intraspecific communication.

Electrical signals used for interindividual communication in mormyrid fish consist of electric organ discharge (EOD) pulses, with a constant waveform that is believed to serve species recognition, while the sequence of pulse intervals is variable (100–300 ms) and expresses sender's identity and motivation (Carlson and Hopkins, 2004).

The electrical organ is innervated by electromotor neurons (EMNs) of the spinal cord, which receive signals from medullary neurons that, in turn, are induced by signals from the fish brain. Ultimately, EODs are determined by specific central nuclei in the CNS. A medullary command nucleus (CN) receives inputs from the precommand nucleus (PCN) and the dorsal posterior nucleus (Carlson, 2002).

A simple neural circuit that starts with the CN, via the medullary relay nucleus, activates spinal EMNs, triggering production of EODs (Figure 19.11).

Regulation of the normal baseline rhythm of 100–300 ms EOD intervals is a function of the PCN and DP, which, in turn, receive input from the dorsal subdivision of the ventroposterior nucleus of the torus semicircularis (Carlson and Hopkins, 2004).

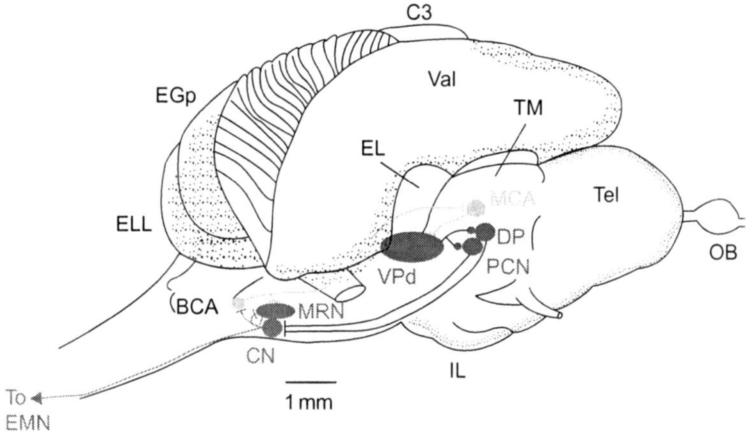

Figure 19.11 Sagittal schematic showing the functional neuroanatomy of the mormyrid electromotor system. Excitatory terminals are identified by flat lines, inhibitory terminals by solid circles. Black denotes medullary electromotor nuclei, dark gray denotes mesencephalic and diencephalic electromotor nuclei, and light gray denotes corollary discharge nuclei. *Abbreviations*: BCA, bulbar command-associated nucleus; C3, third cerebellar lobule; CN, command nucleus; DP, dorsal posterior nucleus of the thalamus; EGp, *eminentia granularis pars posterior*; EL, exterolateral nucleus of the torus semicircularis; ELL, electrosensory lateral line lobe; EMN, electromotor neurons; IL, inferior lobe of the hypothalamus; MCA, mesencephalic command-associated nucleus; MRN, medullary relay nucleus; OB, olfactory bulb; PCN, precommand nucleus; Tel, telencephalon; TM, *tectum mesencephali*; Val, valvula of the cerebellum; VPd, dorsal subdivision of the ventroposterior nucleus of the torus semicircularis.
Source: From Carlson and Hopkins (2004).

The Structure of the System of Electroreception in Fish

Mormyriform fish have cutaneous electroreceptors of three types: ampullary electroreceptors, used for electrolocation of preys and predators, mormyromast electroreceptors for electrolocation within close range, and *Knollenorgans* (German for *tuberous organ*), electroreceptors for reception of EODs from other fish, which are used for interindividual communication (Hopkins, 1999).

The first anatomic descriptions of ampullary organs in elasmobranch fish (sharks, skates, and rays) appeared by the middle of the seventh century, but it took three centuries until, by the early 1960s, biologists discovered that the ampullary organ has electrosensory functions (Tricas, 2001).

Knollenorgan electroreceptors are broadly tuned to the species-specific EOD spectrum. Damages of the Knollenorgan pathway to the brain prevent fish from performing various communication behaviors. When electric fish emit EODs, the Knollenorgan pathway in the CNS is blocked so that its own EODs are not transmitted for processing in the brain (Arnegard et al., 2005).

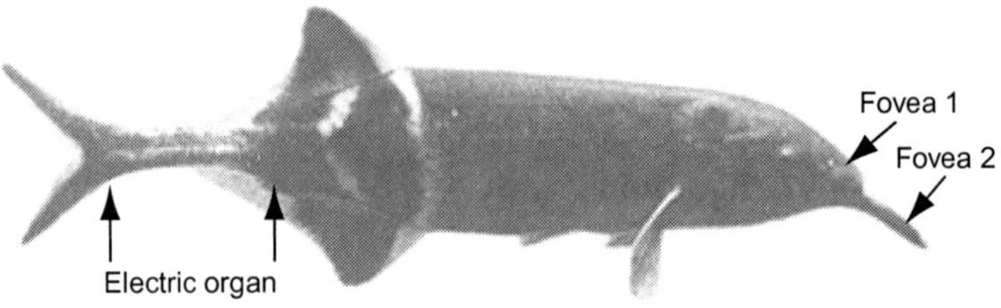

Figure 19.12 *Gnathonemus petersii* seen from the side. The locations of the electric organ and of the two proposed electric foveae at the nasal region and the Schnauzenorgan are marked. *Source*: From von der Emde (2006).

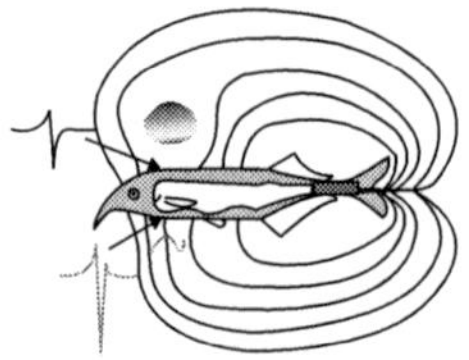

Figure 19.13 Schematic drawing of a vertical section of the electrical current lines during an EOD around a *G. petersii*. The electric organ in the caudal peduncle is drawn in black; the electroreceptive skin areas are shown in gray. Field distortions caused by two types of objects are indicated. Dorsally, a nonconducting stone decreases field line density, while a ventrally placed worm, a better conductor than the water, increases it. Examples of the locally occurring EODs are shown. The stone decreases local EOD amplitude and does not cause any waveform distortions. In contrast, the worm increases amplitude and distorts the local EOD.
Source: From von der Emde (2006).

The electromotor system has coevolved with the sensory electroreceptor system. Besides, the mormyrid *Gnathonemus petersii*, as well as some other South American and African species, are capable of emitting electrosignals and perceiving their reflection in the skin electroreceptors. These fish have two small electroceptive pits or foveae (Latin plural of *fovea*, small pit) resembling the visual fovea in the retina in the Schnauzenorgan (German *Schnauze*, muzzle), which help them to form an "electrical image" of objects with strong contextual effects, by determining the distance and the three-dimensional shape of the object (Caputi and Budelli, 2006; von der Emde, 2006; Figures 19.12 and 19.13).

Electrosensory units of these fish are ampullae of Lorenzini (Figure 19.14). Each of these units consists of a group of subdermal alveoli that are connected to the environment via a 1 mm long canal (Tricas, 2001). The lining of the ampulla consists of a layer of sensory cells, with their kinocilia projecting to the lumen of the ampulla, and the supporting cells. The receptor cells are innervated by primary afferent neurons, which encode the amplitude and frequency of electrical signals and transmit them to the brain (Tricas, 2001), where a perception of the source of electrical stimuli is generated.

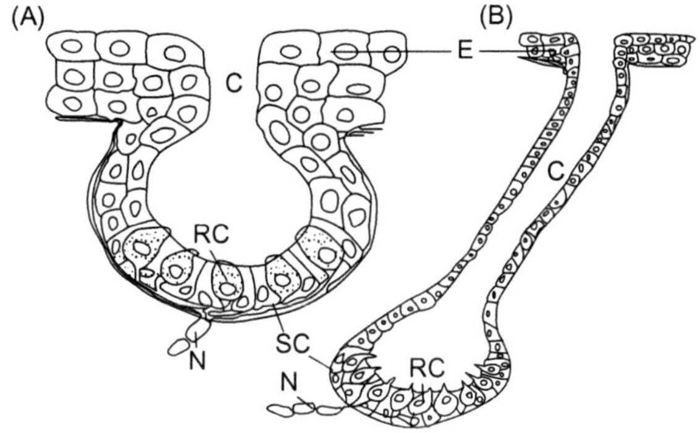

Figure 19.14 Schematic diagram of sensory organs utilized in electroreception in fishes. (A) Representative form of an ampulla common to freshwater teleosts. (B) Generic example of an ampulla of Lorenzini from a freshwater stingray. *Abbreviations*: C, canal; E, epidermis; N, nerve; RC, receptor cells; SC, supportive cells (Collin and Whitehead, 2004).

It is believed that electric signaling, based on the ampullary electrosensory systems such as receptors of ampullae of Lorenzini, appeared very early and repeatedly (in fish alone, it evolved independently at least six times) in vertebrate evolution (according to Bullock et al., 1982, in Bodznick et al., 2003).

The electrosensory system allows fish and amphibians to recognize prey, predators, conspecifics, and mates as well as to regulate social behavior. The primary afferent neurons encode the data on electrical amplitude and frequency and transmit these data to the brain (Tricas, 2001), in the dorsal octavolateral nucleus (DON) on the dorsolateral wall of the hindbrain (Bodznick et al., 2003).

Reception of electrical signals emitted by other conspecific and heterospecific fish must be first filtered from the noise before it is transmitted to higher centers for processing in the midbrain, where that information is mapped into the lateral mesencephalic nucleus (LMN) and the anterior mesencephalic nucleus (AMN).

The processing of the electrosensory information is done in the electrosensory lateral line lobe (ELL) and higher brain centers (Figure 19.15).

The object's image initially may form in the ELL, then the distance from the object may be calculated in torus semicircularis and tuned in optic tectum (Assad et al., 1999).

Electrosensory Communication in Social and Reproductive Behavior

EOD diversity in electric fish stems from differences in the shape and duration of the wave but also from the pulse polarity. The form of the wave depends on the type of

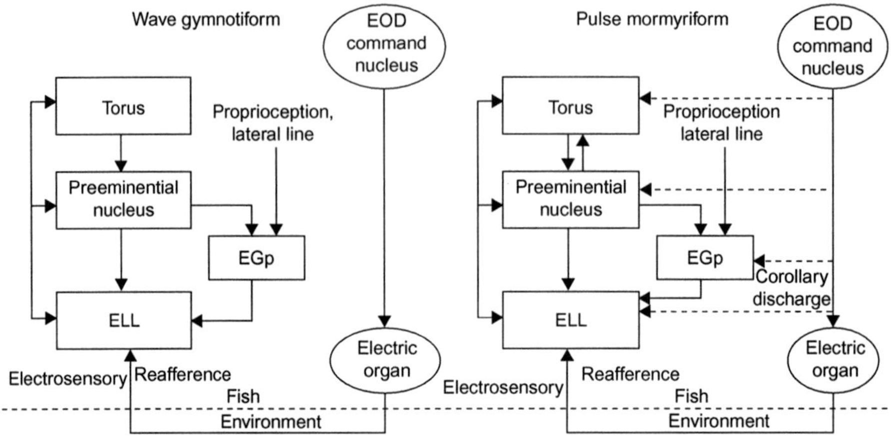

Figure 19.15 Central electrosensory pathways of wave gymnotiform and pulse mormyriform fish. In both gymnotiform and mormyrid fish, information from electrosensory afferents is relayed first to the electrosensory lateral line lobe (ELL) and then to higher stages of electrosensory processing in the preeminential nucleus and the torus. Cells in the preeminential nucleus give rise to two feedback pathways, one directly to ELL and the other via granule cells of the eminentia granularis posterior (EGp). EGp also receives inputs from other sensory modalities and motor systems. In mormyrid fish, electric organ corollary discharge (EOCD) inputs linked to the occurrence of the EOD are relayed to multiple stages of electrosensory processing. Mormyrids also have a projection from the preeminential nucleus to the torus that is not present in gymnotiforms. Ascending electrosensory and motor command pathways are shown as black lines, EOCD pathways are shown as dashed lines, and descending pathways are shown as gray lines.
Source: From Sawtell et al. (2005).

cells in the electric organ. The duration of the wave varies between 100 μs and 10 ms (Hopkins, 1999).

In the electric fish that show sexual dimorphism for EOD, differences are also observed in the morphology of electrocytes (Hopkins and Bass, 1981), but there are fish species that show the same EOD sexual dimorphism, although the morphology of electrocytes in the electric organs seems similar in males and females (Mills et al., 1992).

The existence of differences between male and female adults in the EOD waveform produced by electrocytes (Bass et al., 1986) suggests that these fish also use electrical signaling for mate recognition (Bodznick et al., 2003; Figure 19.16).

Both males and females of electric fish can modulate their EODs according to specific social and physiological situations. A correlation between the reproductive behavior (courtship and spawning) and the electrical displays of males and females is observed in the mormyrid fish, *Pollimyrus isidori*. Females start the courtship behavior and spawning in the nest region of the male, with the male abandoning the usual aggressive behavior. At that time, the female switches from the resting and hiding pattern of its sex-specific EODs to a new EOD pattern for the whole period of the courtship and spawning. A similar switch performs the male, which stops his

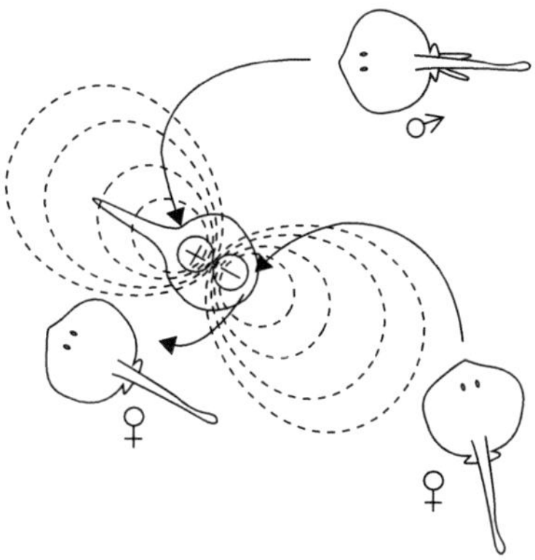

Figure 19.16 Mating stingrays, *Urolophus halleri*, in shallow waters of the Sea of Cortez orient to a buried plastic model during playback of the low-frequency bioelectric field recorded from a female. Males approach, explore, and sometimes dig up buried models (as they do with actual females) in an attempt to mate. Females also locate and approach the model and often bury next to it.
Source: Adapted from Bodznick et al. (2003); From Collin and Whitehead (2004).

nocturnal sound production only after the end of spawning, when it starts singing (Bratton and Kramer, 1989). Females make mate-choice decisions based on EOD waveform characteristics (Feulner et al., 2009).

In the bulldog fish, *Marcusenius macrolepidotus*, no courtship takes place, and the female starts spawning immediately after being allowed to enter the male's territory, but during spawning bouts, both fish coordinate intervals between EODs (Werneyer and Kramer, 2005).

Duration of EODs reflects the social status of male fish (high-ranking males have longer EODs, and lower-ranking ones have shorter EODs), as assessed by males themselves, and is used in the communication between them. Changes in the social status as well as changes in hormonal levels are correlated with respective changes in the EOD duration, making this trait very plastic (Carlson et al., 2000).

Studies on the form and duration of EOD waveforms in fish led to the identification of new species within what previously, based on morphological criteria, have been considered to be single species (Figure 19.17).

Changes in the electrosensory system of fish also occur during their individual development. The sensitivity of the electrosensory system in the Atlantic stingray, *Dasyatis sabina*, changes with age, which enables the developing fish to avoid predators when still young, and to detect prey and choice mates in adulthood (Sisneros and Tricas, 2002). Obviously, these changes in electrosensitivity are epigenetically determined, i.e., they involve no changes in genetic information.

Stingrays produce a standing direct current (dc) bioelectric field, which they use for detecting and locating conspecifics or mates. The highest frequency sensitivity of the primary afferents in the clearnose skate (*Raja eglanteria*) and the little skate (*Raja erinacea*) is similar to the pulse rate of their respective EODs, indicating the importance of electrocommunication in their social and reproductive behaviors (Tricas and Sisneros, 2004). The round stingray, *Urolophus halleri*, does not have a special electric organ, but its bioelectric field arises from standing ionic potentials at various sites on the skin and buccal epithelium, and this bioelectric field during mating season is used by males as the primary cue for localizing their female mates buried under the sandy bottom of shallow shoreline (Tricas et al., 1995).

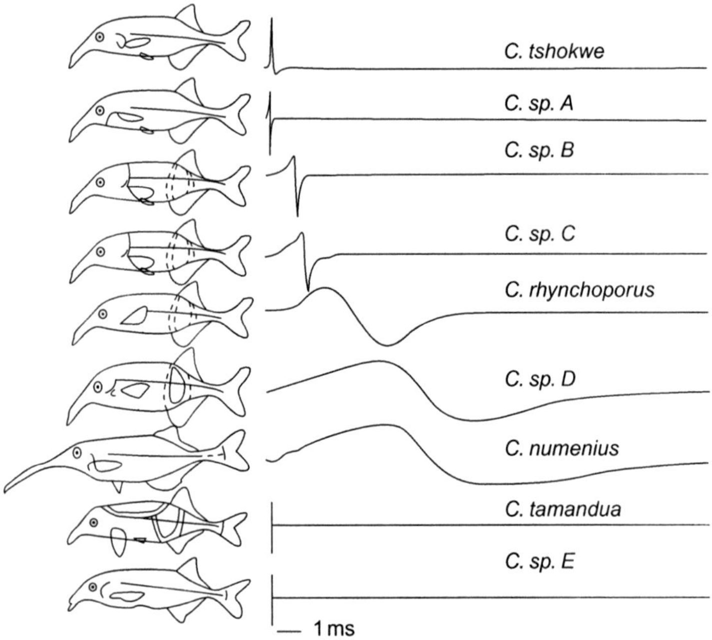

Figure 19.17 Systematists disagree on the number of species in the genus *Campylomormyrus*, but recent recordings of EOD waveforms suggest there may be even more species than the latest revision, which recognized 18 species. Only four of the nine species represented here can be unambiguously identified by reference to the type specimens, most of which are housed in the Musée Royal de l'Afrique Central in Tervuren, Belgium. The remainder represents forms whose identity is uncertain.
Source: From Hopkins (1999).

In males of the Atlantic stingrays, *Dasyatis sabina*, a correlation exists between the seasonal changes in the androgen levels in body fluids and changes in the dental sexual dimorphism, sexual behavior, and aggressive behavior, as well as the neurophysiological changes in the electrosensitivity. These electrosensory changes enhance the ability of juvenile fish to avoid predators, improve the ability of adults to locate prey, coordinate social behavior in both sexes, and help adult males to detect and identify mates (Sisneros and Tricas, 2000).

Arnegard et al. (2005) examined pairs of three sympatric morphs (types I, II, and III) of the *Brienomyrus* flock of species of mormyrid fish that have radiated recently in East Africa. These morphs are genetically and morphologically indistinguishable but are clearly distinct in their EOD patterns (rate and waveform). Moreover, they use the same microhabitat—what obviously excludes the possibility of separate gene pools—and action of different evolutionary pressures:

> *Type I and type II/III are conspecific signal types comprising undifferentiated gene pools in each regionally defined and phenotypically polymorphic population.*
>
> Arnegard et al. (2005)

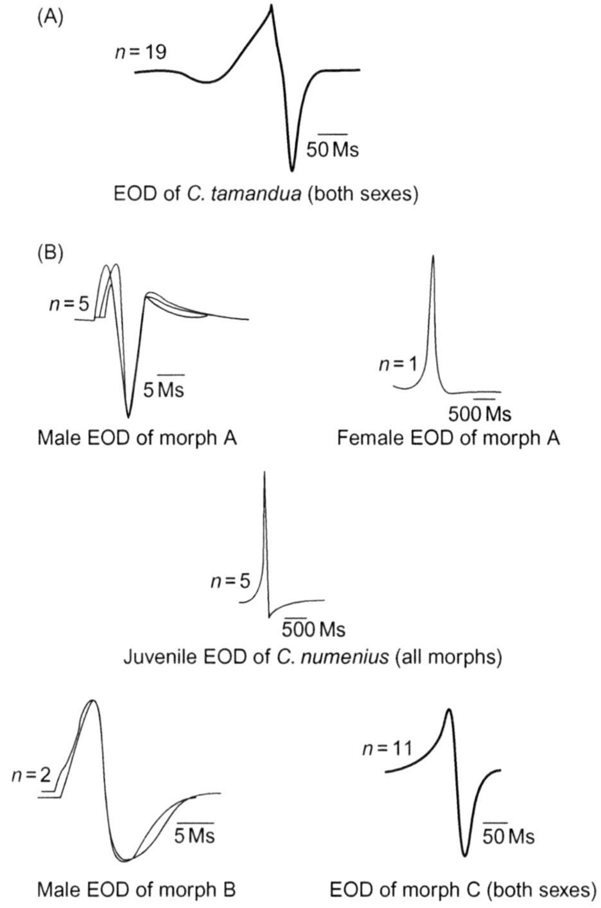

Figure 19.18 Overlays of amplitude-normalized EODs (*n*, number of individuals per overlay). (A) Common EOD type of *C. tamandua.* (B) Common juvenile and different adult EOD types of three different morphs (A, B, and C) of *C. numenius*. Note the different time scales. *Source*: From Feulner et al. (2006).

Reproductive isolation in sympatry of fish of the *Campylomormyrus* genus of Central African river basin is correlated with differences that these species show in waveform types of their EODs. *Campylomormyrus tamandua* is characterized by common EOD in male and female individuals (Figure 19.18). *C. numenius* displays a common EOD waveform type in juveniles, but adult individuals generate three different EOD types. It is observed that these distinct waveform EOD types are each correlated with a different fish morphotype (morphotypes A, B, and C). These sympatric morphotypes are reproductively isolated. Based on the study of microsatellite loci, investigators have concluded that these "morphotypes" of *C. numenius* in fact represent three cryptic species (Feulner et al., 2006).

Evolution of Electrosignals, Electrocognitive Isolation of Populations, and Sympatric Speciation in Fish

Behavioral studies have shown that the ability of electric fish to distinguish between species-specific and -nonspecific EODs as well as between female and male EOD waveforms (Hopkins and Bass, 1981; Arnegard et al., 2005) may have played a role in evolution of electric fish; it may have been used not only to recognize conspecifics and for social interactions between them but also as a mechanism of reproductive isolation between groups of individuals or populations in the process of speciation. This seems to be supported by observations on more than 20 sympatric species of the *Brienomyrus* fish flock in the Gabon River, with each of them producing a species-specific EOD waveform.

Nested within the Gabon *Brienomyrus* species flock of the Gabon River is the *magnostipes* complex, consisting of sympatric morphologically indistinguishable morphs (type I and types II and III) of the same size. Reproductively, morphotypes of this complex are not isolated but express distinct EODs, although they are genetically indistinguishable at several nuclear loci. Each morph of the magnostipes complex generates its specific EOD (Figure 19.19).

However, males of the type II, when exposed to playback EODs of both types I and II, respond preferably to type II, whereas type I males show no preference (Arnegard et al., 2005, 2006). In view of the striking morphological and genetic similarities between sympatric morphs, while they are still interbreeding, the dramatic differences in the EOD waveform and EOD-mediated species recognition during courtship within the *Brienomyrus* stock suggest that they are in the process of sympatric speciation (Arnegard et al., 2005; Figure 19.20).

> *Two lines of evidence and a well-studied central nervous system mechanism suggest a general role for EODs in species isolation in the Gabon* Brienomyrus *flock. Firstly, each of the two dozen or more morphologically distinct species that have already been discovered produces a different, species-typical EOD waveform. Secondly,* in situ *electrical playback experiments have demonstrated EOD-mediated species recognition in the context of courtship in* Brienomyrus *sp. vad. Lastly, the EODs of type I and type II/III appear to differ sufficiently for waveform discrimination via a neural pathway described by Xu-Friedman and Hopkins (1999; and references therein). Therefore, instances of dramatic signal difference between sympatric morphs appear somewhat paradoxical in this group of fish when genetic evidence for reproductive isolation between them is lacking.*
>
> *Arnegard et al. (2005)*

Evolution of electrogenic and electrosensory organs in gymnotiform fish of South America and in freshwater mormyrid fish of Africa represent striking examples of evolutionary convergence. That evolution has occurred independently in both groups, and yet they exhibit extraordinary similarities. In both groups, the production of EODs is controlled by a ventral midline nucleus, which is known as *pacemaker nucleus* in gymnotiforms and *medullary* CN in mormyrids. In both groups, neurons

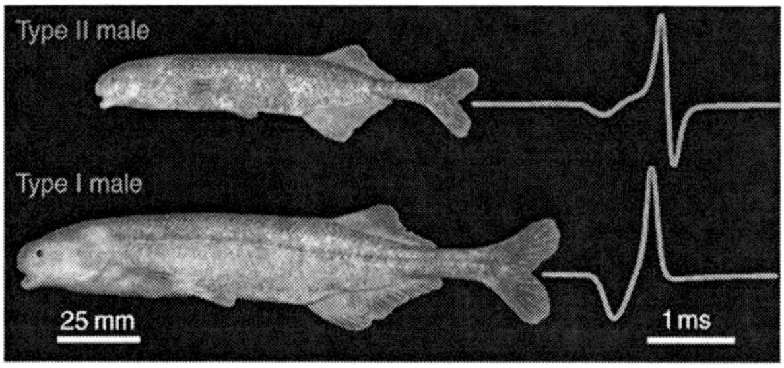

Figure 19.19 Adult sizes of sympatric *magnostipes* complex morphs from the Makokou region of the Ivindo River. Photographs of a type II male (above: specimen 5945; SL = 105 mm) and a type I male (below: specimen 5944; SL = 147 mm) collected from Loa-Loa Rapids, showing their elongated EODs (Arnegard et al., 2006).

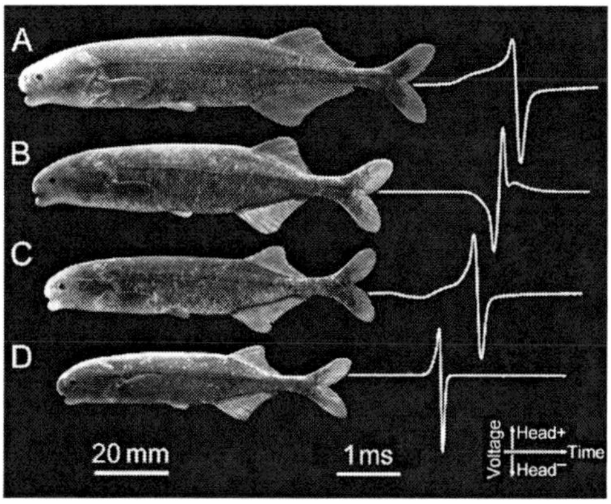

Figure 19.20 Examples of sympatric assemblage of morphologically similar mormyrids from the *Brienomyrus* species flock of Gabon. Photographs are shown next to voltage traces of electric organ discharges (EODs) recorded from the same individuals: (A) adult male; (B) type I morph, specimen 3025, adult female or nonbreeding male, CU80231; (C) type II morph, specimen 3014, adult female or nonbreeding male, CU80358; and (D) *sn3*, specimen 3027, adult female or nonbreeding male, CU80356. Scale bars of 20 mm and 1 ms are indicated. *Source*: From Arnegard et al. (2005).

of this nucleus project to adjacent relay neurons, whose axons innervate EMNs in the spinal cord (Carlson and Hopkins, 2004).

Studies conducted on these fish seem to exclude genetic factors from a possible involvement in the process of the evolutionary divergence of these morphs in sympatry:

> *Instances of dramatic signal difference between sympatric morphs appear somewhat paradoxical in this group of fish when genetic evidence for reproductive isolation between them is lacking ... Striking genetic similarity between coexisting morphs provides the signature of a fully sympatric process, whether it involves incipient speciation or the maintenance of phenotypic dimorphism by some other mechanism.*
>
> *Arnegard et al. (2005)*

Neo-Darwinian Explanation of Sympatric Reproductive Isolation and Speciation of Brienomyrus *spp.*

From the paradigmatic neo-Darwinian view, it would be predicted that evolution of *Brienomyrus* spp. in sympatry is an improbable, if not impossible, event because:

1. Inherited phenotypic changes, including behavioral changes in electrical communication, require changes in genes or allele frequencies, and such changes are not there.
2. The unavoidable gene flow between populations that use the same microhabitat will prevent genetic isolation as a condition of reproductive isolation and speciation.

Epigenetic Explanation of Sympatric Reproductive Isolation and Speciation of Brienomyrus *spp.*

Repeated evolution of electrolocation in fish was facilitated by the suitability of aquatic environment for electric communication. But why did the electrolocation evolve first in fish and in other aquatic vertebrates (aquatic amphibians and even in a semiaquatic mammal) but not in aquatic invertebrates? One plausible explanation would be that electrocognition requires a complex algorithm that probably is beyond the computational capabilities of the invertebrate CNS/neural net. In this context, let us remember that electrogenic fish devote a considerable part of the brain to analysis and interpretation of electric signals that invertebrates probably cannot afford to. This seems to be supported by a recent study on diversification of electric fish (Carlson et al., 2011).

Fish use electric signals for both species and mate recognition. Evolution of electric signals and preference for the changed signals in groups of individuals may lead to reproductive isolation of these groups, their divergence, and speciation in sympatry. The processing of electric signal information is made in a region of the fish brain, the exterolateral nucleus of the torus semicircularis (EL). A recent study showed that species of clade A (subfamily Mormyrinae) EL evolved a larger and more complex structure, EL anterior/EL posterior (ELa/ELp), while it remained anatomically constant in species of the family Petrocephalinae. Evolution of ELa/ELp in the clade A enhanced signal discrimination, accelerated rates of evolution of electric signals, and species diversification. Additionally, clade A, but not Petrocephalinae species, established developmentally flexible electrocyte stalks, which increased their ability to generate new signal variants (Carlson et al., 2011). Investigators observed that the rate of diversification in clade A was at least three to five times higher than in the other lineage. They attribute the "explosive diversification" of species of clade A to the evolution of ELa/ELp and conclude that

> *neural innovations can drive the diversification of signals and promote speciation … brain evolution directly promotes diversification.*
>
> *Carlson et al. (2011)*

Brain, Behavior, and Evolution

Animal behavior is a product of the computational and motor activity of specific circuits and is closely related to cognitive functions (e.g., learning, memory, and decision making) of the brain. The idea that a change in behavior is the first stage in the process of speciation is neither new nor consistently rejected (West-Eberhard, 1989; Price et al., 2003). As early as 1963, Ernst Mayr admitted the role of behavior in initiating the process of speciation:

> *A shift into a new niche or adaptive zone is, almost without exception, initiated by a change in behavior. The other adaptations to the new niche, particularly the structural ones, are acquired secondarily. With habitat and food selection—behavioral phenomena—playing a major role in the shift into new adaptive zones, the importance of behavior in initiating new evolutionary events is self-evident.*
>
> *Mayr (1963)*

By moving to another area or by expanding their range, animals may be subject to the speciational effects of geographic isolation and gene drift.

Being the most plastic of phenotypic characters in metazoans, as a rule, animal behavior precedes the morphological and physiological adaptation:

> *The impression that behavior takes the lead in evolution is commonplace .... Adaptive behavioral plasticity is expected to evolve more readily than does adaptive morphological plasticity .... The evolution of an adaptive plastic morphological response in animals, by contrast, requires a cue operating early enough in ontogeny to trigger the development of the appropriate morphology .... Behavior during development can extensively influence morphology.*
>
> *West-Eberhard (1989)*

In line with this idea, the evolutionary change starts with neural processing of the external/internal stimuli from which the new adaptive behavior arises. However, from a conventional view, it is difficult to see how this new behavior and the neural mechanism that produces it may be related to the future change in morphology.

While changes in behavior allow animals to adapt rapidly to the environment, such changes may also enable them to enter new niches or adaptive zones and expand their geographical range. In doing so, animals are subject to new selection pressures that facilitate divergent evolution and speciation processes (Wcislo, 1989; West-Eberhard, 1989). Indeed, experiments on *Drosophila* kept under total darkness for 800 generations led to a number of inherited changes in behavioral (phototaxis, olfaction, daily rhythms) and morphological traits (Wcislo, 1989). However, behavioral plasticity and innovativeness, by adapting metazoans to the changed conditions in the environment, not only can enhance, but also can inhibit, the rate of evolutionary change (Price et al., 2003; Paenke et al., 2007).

One logical consequence of the role of animal behavior in evolution would be that the brain, the determinant of animal behavior, might have been crucially involved in metazoan evolution. Further, it would be predicted that a correlation should exist between the evolution of the brain and metazoan evolution in general. This prediction is at the core of the hypothesis of the behavioral drive.

Behavioral Drive or Brain Size–Environmental Change Hypothesis

Darwin may have been the first to observe a positive correlation between the position of the species in the tree of life and the rate of evolutionary change:

> *The productions of the land seem to change at a quicker rate than those of the sea, of which a striking instance has lately been observed in Switzerland. There is some reason to believe that organisms, considered high in the scale of nature, change more quickly than those that are low: though there are exceptions to this rule.*
>
> *Darwin (1859, p. 313)*

After looking upon the anatomical evolution of birds in the context of the low accumulation of point mutations in class Aves and the time elapsing since their first appearance, Wyles et al. (1983) came to the conclusion that the morphological evolution of birds has been much faster than other vertebrate classes, except for mammals. Table 19.1 shows a correlation between the relative size of the brain and the rates of evolutionary change in selected vertebrates.

To explain their observation on the unusually rapid evolution of bird morphology, Wyles et al. (1983) proposed the hypothesis of "behavioral drive," according to which the relatively higher proportions of brains in relation to body weight in birds and mammals, in comparison with fish, reptiles, and amphibians, make these classes behaviorally more innovative. Behavioral drive hypothesis predicts that large-brained animals, being behaviorally more flexible and innovative, have comparatively higher evolutionary rates and will have more clades. Accordingly, acquisition of new behaviors allows them to extend the range of their habitat and adopt new habitats, finding thus themselves under new evolutionary pressures.

Table 19.1 Brain Size in Relation to Rate of Anatomical Evolution

Taxonomic Group	Relative Brain Size	Anatomical Rate
Homo	114	>10
Homonoids	26	2.5
Songbirds	23	1.6
Other mammals	12	0.7
Other birds	4.3	0.7
Lizards	1.2	0.25
Frogs	0.9	0.23
Salamanders	0.8	0.26

Source: According to Wyles et al. (1983).

Evidence in favor of the behavioral drive hypothesis is rapidly accumulating. Reader and Laland (2002) have shown that large-brained primates, which have greater learning capabilities and use tools, have had comparatively higher rates of morphological evolution (Reader and Laland, 2002). Still echoing pioneering work of Wyles et al. (1983), two decades later, Nikolakakis et al. (2003) found that in birds, the brain size is positively correlated with the number of species per clade. In a similar comparative study, Sol et al. (2005b) found that Holarctic (living in northern areas of the earth) passerine birds with relatively larger brains have more subspecies than small-brained species, adding further empirical support to the behavioral drive hypothesis. Having excluded the possibility that behavioral flexibility might lead to accelerated evolution, investigators conclude:

> *Avian lineages that have larger brains and exhibit a higher propensity for innovative behaviors tend to contain more species than less flexible lineages.*
>
> *Sol et al. (2005b)*

In another study, it was found that in birds, larger brains were positively related to success in new environments, and the innovative behavior was positively related to the relative brain size (Sol et al., 2005a, 2008).

A number of other studies have shown that innovative behavior, related to the brain cognitive functions, enables animals to enter new adaptive zones. Lefebvre et al. (2004) have shown that innovative behavior, both in birds and primates, is positively correlated with the size of areas of the brain involved in sensory integration and learning, the hyperstriatum ventrale and neostriatum in birds, and the isocortex and striatum in primates. Furthermore, they have found a remarkable convergent evolution in brain-cognition organization of both classes during more than 300 million years since their divergence began (Lefebvre et al., 2004).

All of the above evidence suggests that behavioral innovation and flexibility, which seem to be positively related to the evolution of the brain and nervous system in general, seem to be the driving forces of evolution.

Neo-Darwinian Explanation of the Correlation Between the Brain Size and Evolutionary Rates

From the neo-Darwinian point of view, the correlation between brain size and the rates of evolution could be an *indirect* result of the increase in brain size and behavioral innovativeness: species with larger brains and higher behavioral plasticity are capable of extending their range, thus being subject to more diverse selection pressures, which, consequently, lead to more diverse phenotypes, new species, and higher taxa.

However correct this prediction might be at first sight, a serious countervailing argument has been repeatedly presented. It has been argued theoretically that adaptive changes in behavior on entering into a new niche increase the fitness, remove or minimize evolutionary pressures, and hide the existing variability from natural selection, thus preventing, rather than stimulating, evolution in animals (Huey et al., 2003; Dukas, 2004). This Bogert hypothesis of inhibition of evolutionary change by behavioral plasticity has been validated by a null model (Huey et al., 2003). It is argued that the behavioral flexibility and learning, by widening species range and by enabling individuals of a species to enter new niches and adaptive zones, enhance their survival chances (Dukas, 2004), thus decreasing the evolutionary pressure for morphological evolution.

From the neo-Darwinian view, this is a paradoxical situation: empirical evidence shows that large-brained animals, which are behaviorally more flexible, speciate more rapidly when they should not. But paradoxes are mental, rather than real, situations arising from apparent discrepancy between observational facts and the existing explanatory paradigm. If the evidence about the correlation between brain size and behavioral plasticity, on the one hand, and the rates of evolution on the other, is real (which, by all accounts, it seems to be), there is no alternative solution to the paradox: rejection of the neo-Darwinian hypothesis on the cause of the correlation between the brain size and the rates of evolution and speciation.

Epigenetic Explanation of the Correlation Between the Brain Size and Evolutionary Rates

Studies on the positive correlation between the brain size and behavioral innovation, on the one hand, and the rate of evolutionary change and speciation, on the other, have shown that the effect of the increase in body size and new adaptive behaviors is not mediated by expansion of the species range.

> *The independence of behavior and geographical range size is an unexpected result, given the long-held idea that flexible behaviors favor a species' establishment in new regions (Mayr, 1965) and the support for this idea in recent comparative studies in birds.*
>
> *Sol et al. (2005b)*

In examples of rapid speciation in birds with relatively bigger brains, it may be safely said that the speciation occurred in sympatry because there are no real geographical barriers for birds in continents where speciation occurred. As shown earlier, cases of sympatric speciation in which named bird species show no genetic differences can only be explained on the assumption that evolutionary phenotypic divergence and genetic divergence are separate processes.

If neither genetic differences nor increased range size nor greater behavioral flexibility are proximate causes of the observed correlation between brain size and the accelerated rates of evolutionary change, one is left with no alternative but admit:

- the epigenetic nature of the factor(s) determining the correlation, and
- the possible role of the progress in neurocognitive mechanisms, related to brain size in the process of speciation.

Indeed, the available evidence shows that phenotypic changes in the process of incipient sympatric speciation start with inherited behavioral changes in mating preferences and the mating behavior of the diverging population. Both mating preferences and sexual behavior are neurobiological products resulting from the activity of behavioral neural circuits, rather than products of gene activity. Morphological and genetic changes start accumulating only after the establishment of behaviorally determined reproductive isolation.

Neurocognitive Sympatric Speciation

Although allopatric speciation is theoretically possible and seems to have occurred, contrary to the conventional wisdom, most of the *scientifically* proven cases of speciation are those related to the reproductive isolation occurring in sympatry. Premating isolation in sympatry seems to be the most frequent of demonstrated factors in speciation in nature. In most of the described cases, the sympatric premating isolation is behavior dependent: incipient species avoid the interbreeding, while being capable of interbreeding and producing fertile hybrid offspring under natural conditions. They use specific (e.g., visual, olfactory, and auditory) cues to discriminate against breeding non-self-like individuals. This sexual avoidance behavior is determined by the activity of neural circuits involved in the discrimination of self-like from non-self-like individuals.

Thus, premating reproductive isolation, as the first step in the process of speciation, is a new and exclusive property of eumetazoans, related to the evolution of the nervous system in the kingdom Animalia. Changes in mating behavior and mating sensory signals represent the most frequently documented mechanism of reproductive isolation in sympatry.

Both the basic allopatric and reinforcement models (suggesting incipient speciation occurring in allopatry and, later, after secondary contact, reduced hybrid viability) of speciation posit that reproductive isolation of populations in the process of speciation occurs under conditions of geographic separation and in the absence of gene flow or partial gene flow between populations, which is necessary for accumulation of genetic changes that, over time, lead to postzygotic reproductive isolation.

Neo-Darwinians have considered sympatric speciation to be unlikely or to have been skeptical of whether it can ever occur. However, recently, two theoretical models have been developed (Dieckmann and Doebeli, 1999; Kondrashov and Kondrashov, 1999) in order to eliminate the discrepancy between the neo-Darwinian theoretical prediction of the impossibility of sympatric speciation under conditions of gene flow, on the one hand, and the facts of occurrence of sympatric speciation, on the other. Both models are based on conditions that are not demonstrated to exist in nature. The models are criticized for requiring "unlikely conditions" (Drès and Mallet, 2002), but construction of models based on assumptions that are not likely to exist in nature seems to be heuristically of little use.

Vast observational and experimental evidence suggests that neither of the neo-Darwinian requirements of speciation (physical separation prevention of gene flow or preliminary accumulation of genetic differences between populations) are necessary for reproductive isolation to take place. Populations of a species that are genetically similar (sharing a common gene pool) may create, in sympatry, a separate fertilization system and enter the process of speciation by rapidly evolving nongenetic changes in mate preferences or courtship behavior.

Neurocognitive Basis of Sympatric Speciation

In 1979, West-Eberhard related the Darwinian idea of sexual selection to the evolution of preferences for mating with individuals of the opposite sex that express certain morphological characters that are perceived to be of their own kind (West-Eberhard, 1979). This would enable the speciation process to take place even under conditions of sympatry.

Neurocognitive segregation of within-populations groups and the incipient speciation implies not competition (West-Eberhard, 1983; Ryan, 1998) and rivalry but establishment of a neurobiologically determined preference for certain phenotypes in individuals that leads to creation of a separate fertilization system and reproductive isolation of a group of individuals from the rest of the population.

Thus, changes in mate preferences in a group of individuals within a population may lead to sympatric reproductive isolation of the group from the rest of the population.

In a partial list of *possible* examples of speciation based on mate signaling, Panhuis et al. (2001) include the sex comb row number in the Hawaiian *Drosophila silvestris*; pulse rate of male song and female preference in North American crickets *Gryllus texensis* and *G. rubens*; male color pattern and female preference in the guppy fish *Poecilia reticulata*; sexually dimorphic coloration and corresponding female preference in haplochromine cichlid fish in Lake Victoria; male strut displays, body size, and feather morphology in the bird *Centrocerus urophasianus*; and male bower display and female preference for it in the Vogelkop bowerbird *Amblyornis inornatus*, among other examples (Panhuis et al., 2001).

Changes in mate preferences, however, might not be the only means of sympatric speciation. Other factors, such as changes in host preferences, can also lead to sympatric speciation as is demonstrated in some cases in insects. Different as they might appear at first, both processes of changes in mate preferences and changes in host preferences have a common causal basis, that is, both processes are determined by changes in the properties of behavioral neural circuits, which are responsible for mate preferences and host preferences, respectively.

Following the Darwinian concept of sexual selection as a process that was involved in the evolution of sexual traits, but not in speciation, here I interchangeably use terms *sensory-driven sympatric speciation* or *neurocognitive sympatric speciation* to describe the process of speciation that starts in sympatric populations with changes in mate choices, mate preferences, host preferences, and sensory signals.

Inherited changes in mating signals and mate preferences, *per se* (i.e., not involving any form of Darwinian competition) may serve as a behavioral mechanism of sympatric speciation. Such sudden changes in mating signals (e.g., visual, olfactory, acoustic, electric) and corresponding mate preferences have been shown to lead to evolution of incipient species and arguably represent a driving force of sympatric speciation.

In view of the fact that behavior is the most plastic of all phenotypic (morphological, physiological, and life history) characters, it is not surprising that the process of speciation starts with the appearance of differences in mate choice and sexual behavior. Differences in genes and in visible sexual characters are products of the separate evolution of reproductive isolation of populations in sympatry. This is also the reason why incipient species and sibling species show no differences in relevant genes in the early stages of reproductive isolation. Genetic differentiation between populations in sympatry is an effect, rather than a cause, of reproductive isolation.

Reproductive isolation of a population from the rest of the populations of a species, while still within the species range, implies selective mating of individuals of the group with self-like individuals only or preferably. The mechanism of recognition of "alikeness" is a perceptual, neural, and epigenetic, not genetic, mechanism. It enables them to avoid interbreeding with individuals of the parental population that are not perceived to be self-like, leading to the reproductive isolation of the group as the first step in the process of speciation.

Female preference is a behavioral trait that is related to, and exists for the sake of, some male morphological or behavioral trait, sometimes known as *mating traits* (Boughman et al., 2005). It results from properties of neural circuits in which sensory stimuli (e.g., visual, auditory, olfactory, tactile) are integrated and processed. The female preference is at the basis of the female mate choice.

With females being considered to be the choosy sex, and males being less discriminating in mate choice, investigation of male mating preferences has been almost neglected. Recent evidence, however, shows that in a number of cases, males are the choosy sex and that male mate choice is not less important than female mate choice (Wong et al., 2005). Studies on swordtail fish, as well, have shown that males of the fish *Xiphophorus birchmanni* show higher sensitivity to olfactory stimuli released from conspecific and heterospecific, but closely related, species, compared to females. Males of *X. birchmanni* are attracted to odors of conspecific females, show only limited preference for females of the closely related species *X. malinche* (with which they interbreed in sympatry), and ignore females of another closely related species, *X. variatus*, as possible mates (Wong et al., 2005).

Neurocognitive Sympatric Speciation in Nature

Neurocognitive Sympatric Speciation in Insects

Neurocognitive sympatric speciation in insects may be determined by changes in sexual behaviors (*sexual* neurocognitive sympatric speciation) or by changes in nonsexual behaviors (*nonsexual* neurocognitive sympatric speciation) of groups of individuals within a population.

Sexual Neurocognitive Sympatric Speciation

Among insects, the genus *Drosophila* offers some of the most spectacular examples of sympatric speciation. Out of the total of 1,500 *Drosophila* species described throughout the world, almost 500 have sympatrically evolved in the Hawaiian islands within about one to several million years since these islands emerged. This explosive speciation took place in sympatry as a result of changes in song patterns and courtship patterns in *Drosophila* populations. Let us illustrate this with a few examples.

Zimbabwe (Z) lines of *Drosophila melanogaster* in East Africa represent one of the most authentic cases of incipient speciation in nature. Females of these lines, under natural conditions, do not mate with males from cosmopolitan (other continents) M (*melanogaster*) lines. In contrast, females of the cosmopolitan races mate with males of their own and males of Zimbabwe populations (Takahashi and Ting, 2004). Their reciprocal crosses display almost no reproductive isolation and produce viable fertile hybrids (Hollocher et al., 1997). The viability and fertility of their hybrids in F_1 and F_2 and the polymorphism of sexual behavior in males and females of Zimbabwe populations suggest that they may be in an initial stage of speciation determined by changes in the mating preferences of female African flies (Wu et al., 1995).

The mechanism of the changed mate preference of the Z line is still unknown. A group of investigators reported that, in experiments of substitutions of *ds2* (desaturase-2) locus with $ds2^Z$ allele of Z flies, a decline in cold resistance and an increase of starvation resistance occurs in M (cosmopolitan line) flies and that the substitution may be responsible for the sexual isolation between them (Greenberg et al., 2003). Investigators believe that the reproductive isolation is a pleiotropic effect of the ds2 on the cuticular hydrocarbon profile of the flies. After repeating these experiments, another group of researchers failed to confirm any effect of ds2 on cold resistance and starvation resistance or any influence of cuticular carbons on the sexual behavior of M and Z populations (Takahashi and Ting, 2004; Coyne and Elwyn, 2006a,b).

Despite the controversial evidence on the cause of reproductive isolation between *Drosophila* M and Z lines, by consensus it is admitted that these lines represent two reproductively isolated populations in the stage of incipient species occurring in sympatry.

North American field crickets *Gryllus texensis* and *Gryllus rubens* are two cryptic sister species living in sympatry and allopatry in an area extending from the south central to the southeastern part of the North American continent, the first being the dominant species in the western part of the area, and the second on the eastern part. Males of both species are morphologically indistinguishable, with the pulse rate of calling song being the only known distinctive character on which their reproductive isolation is based. Although viable and fertile hybrids between two species may be produced in laboratory, hybridization in nature does not take place or occurs very rarely and cannot succeed because the intermediate variant of the hybrid song does not attract males of either species. Even if the reproductive isolation is not absolute, and a low degree of gene flow between populations of the two species does occur, their reproductive isolation in nature is based on prezygotic mechanisms. The neo-Darwinian prediction of hybrid infertility or inviability as well as character displacement (predicted by the reinforcement model) have not been shown to exist in nature (Gray and Cade, 2000).

Nonsexual Neurocognitive (Host Plant Shifting) Sympatric Speciation

Changes in nonsexual behaviors in insects may determine a tendency for shifting to new hosts for mating and living. Such shifts isolate reproductively particular groups from the rest of the population and lead to rapid formation of new races and species. Due to the fact that speciation in such cases takes place under conditions of spatial isolation (different tree species), this often is considered to be an ecological mechanism of reproductive isolation. Others consider such cases of host shifting to be a form of reproductive isolation in allopatry. However, the deepest causal basis for the switch of insect populations from the ancestral host plant to a new host plant is not a spatial separation or geographic isolation but a new behavior that is neurobiologically determined by processing of sensory (e.g., visual or olfactory) cues in neural circuits.

1. *Rhagoletis pomonella* is an apple fruit fly. It evolved from *Rhagoletis* hawthorn feeding flies, very recently, in less than two centuries after the introduction of apple trees in America. The shift of *R. pomonella* from its original hawthorn host to apples was first described, more than one century ago, by Walsh in 1867, as a probable case of sympatric speciation, but it was Bush who argued that apple and hawthorn flies and four or more other groups represent a species complex that has sympatrically radiated via host plant shifts (Linn et al., 2003).

   The reproductive isolation of *R. pomonella* from the original hawthorn feeding race is based not on genetic changes but on a chemosensorially determined increased preference for apple fruit volatile compounds and decreased preference for hawthorn volatiles. This epigenetic switch of preferences from hawthorn to apple volatile compounds made possible the evolution of a new sibling species from the original stock of hawthorn flies within 150 years (Linn et al., 2004).

   Ripening apple fruits emanate volatile compounds (with butyl hexanoate being the key olfactory signal), which serve *Rhagoletis* apple flies as a long-range cue for detecting apple trees, whereas visual cues become dominant at shorter distances for localizing apple fruits. In a process of specialization for using different host plants, by evolving new host preferences, one (or more) ancestral population of *R. pomonella* entered a stage of reproductive isolation from the rest of the populations. Over time, the isolated population succeeded in adapting its life history to the phenology of the host plants, thus leading to present-day status of sibling species.

   Within the extremely short time of existence as a separate species, *R. pomonella* has evolved two distinct morphs: one with longer ovipositors when reared on hawthorn, and one with shorter ovipositors when reared on apples (Bush, 1969). The apple race of *R. pomonella* exhibits a high host fidelity that strongly restricts interbreeding with the hawthorn race to as little as 4–6% per generation (Feder et al., 1994; Feder, 1998).

   Now *R. pomonella* is a group of six sibling species believed to have evolved in sympatry, very recently in the evolutionary time scale. Each of these species feeds on a particular nonoverlapping species of plants, using volatile compounds as olfactory cues for recognizing its respective host plant.

2. *Rhagoletis suavis* also comprises a group of six species, each of which is specialized on a distinct host walnut species of the genus *Juglans*, without host shifts beyond *Juglans* of North America.

   Attempts have been made to explain the different rates of speciation in these two groups of fruit fly species by assuming that different mechanisms of speciation have been in action in each of these species complexes. It is hypothesized that adaptation to walnut husks, which contain juglone, a very potent phenolic toxin, may have required specializations that have prevented any host shift beyond the *Juglans* genus, a phenomenon that also occurred in *Drosophila pacea*, which, for normal larval development, requires a sterol found in its senita cactus host plant (Bush and Smith, 1998 and references therein). Coevolution of

R. suavis species and their respective walnut hosts is excluded as a possibility because all of the modern species of the *Juglans* genus arose ~40 mya, much earlier than the estimated 2–5 million years since the appearance of *suavis* species (Bush and Smith, 1998).

In contrast to the evolution of *suavis* species, evolution of six sibling species of *R. pomonella* took place within the last two centuries.

3. The larch budmoth, *Zeiraphera diniana* (Lepidoptera: Tortricidae), has two host races feeding, respectively, on the European larch (*Larix decidua*) and Cembran pine (*Pinus cembra*) in the mountains of Europe. Both races interbreed in captivity and may also hybridize in nature. These races differ in a number of traits, but the most important for mate choice and reproductive isolation between them are differences they show in mating signals, pheromones released by females and in the male response to them. The host plant of the larch budmoth, *Zeiraphera diniana*, serves as an indirect cue for mate finding and assortative mating. Males respond to the female pheromone, but the response is stronger when the call comes from a tree of their own or from a neighborhood consisting of such trees than when coming from nonspecific host plants (Emelianov et al., 2003).

4. A stem-galling tephritid fly, *Eurosta solidaginis*, is a species with two host races living on two goldenrod hosts of the genus *Solidago* (*S. altissima* and *S. gigantea*) growing in the same habitat. Both races are electrophoretically distinct only at the level of host races. Their hybrid offspring are viable and fertile. Females inject an egg into an unexpanded leaf of the host plant.

Both species differ in the time of emergence, but difference in phenology of emergence cannot sustain the observed reproductive isolation. Populations associated with one host emerge 10–14 days earlier than the populations of the other host, thus increasing chances of maintaining the reproductive isolation of two host races. They also show strong mating and oviposition preferences, and these seem to be the main factors for their reproductive isolation. The two races are in an intermediate stage of the sympatric speciation (Craig et al., 1993).

5. The European corn borer, *Ostrinia nubilalis* (Lepidoptera: Crambidae), introduced to the American continent almost five centuries ago. Now this species consists of two sympatric species/host races, one feeding on maize (*Zea mays*), and the other on mugwort (*Artemisia vulgaris*). No genetic differences exist between their mitochondrial DNAs, and their pheromone binding proteins are identical. Both host races show diverging male mating preferences: the eastern population with males attracted to females with *trans*-11-tetradecenyl acetate as prevailing pheromone and western population males attracted to females with the isomere *cis*-11-tetradecenyl acetate as main component of the pheromone mix. The two sibling species do not interbreed (Bush, 1975), although a very low level of gene flow and hybridization between the two races seems to take place in nature. Both the difference in the time of emergence (the maize race emerges ten days later than the mugwort race) and different pheromones released by females of both races allow males to recognize females of their own race, thus promoting their reproductive isolation (Thomas et al., 2003).

6. Two closely related Australian fruit fly species, members of the family Drosophilidae, *Scaptodrosophila hibisci* and *S. aclinata*, court and mate on flowers of different *Hibiscus* species (rosemallow, flowering plants of the family Malvaceae) hosts. They only show genetic differences of the same magnitude with allopatric populations of *S. hibisci*, suggesting that these sympatric populations are in the process of incipient speciation (Barker, 2005).

7. A monophyletic group of forest-dwelling *Laupala* crickets in Hawaii experienced a speciation rate of 4.17 species per million years, which is 26 times faster than the average speciation rate of speciation of arthropods and has second fastest rate described after the one observed in evolution of cichlid fish of the East African lakes. The explosive speciation is still going on. It is based on evolution of female preference to mate conspecific male crickets whose courtship song's pulse rate is characteristic for each species. Courtship songs, thus, enable crickets to avoid interbreeding with heterospecific crickets of the *Laupala* group (Mendelson and Shaw, 2005).

Neo-Darwinian Explanation of Sympatric Speciation by Host Plant Shifting

According to the neo-Darwinian paradigm, the first step in the process of speciation is accumulation of differences in the gene pool of two populations, as a result of the prevention/reduction of the gene flow between two populations that are geographically isolated from each other. Over time, the accumulation of genetic changes eventually leads to genetic incompatibility and postzygotic reproductive isolation of populations. The majority of biologists dealing with the problem of speciation still regard the gradual accumulation of favorable genetic variation, hence speciation is not a sudden event (Maynard Smith, 1989). Accordingly, the speciation process requires

1. geographic isolation of populations, which would prevent gene flow and
2. long periods of time for accumulation of the extremely rare "useful mutations" in relevant genes.

Evolution of *Rhagoletis pomonella* complex from the original hawthorn *Rhagoletis* spp. within an evolutionary instant of about two centuries, along with the fact that this evolution took place in sympatry, refute both neo-Darwinian predictions. Attempts continue to be made for explaining the contradiction between the neo-Darwinian paradigm and the facts on sympatric speciation. Feder et al. (2005) have developed the "reticulate" model of speciation of *R. pomonella* sibling species complex. Contrary to the previous evidence, they speculate in length that the case of *Rhagoletis* speciation *may* involve allopatry, which *may* have produced the necessary genetic variation for later host shifting (Feder et al., 2005).

There is no evidence that this long sequence of hypothetic events might have happened, and the time required for the presumed sequence of events required by the reticulate model is incompatible with the extremely short period of time within which the *R. pomonella* species complex evolved. Since none of the events in the assumed sequence of events in this reticulate scenario has been substantiated in the case of speciation of the *Rhagoletis* group, it is hard to understand why one should resort to such a highly speculative allopatric model when experimental evidence clearly suggests that the insect has shifted the host under conditions of sympatry.

While the theoretical possibility of the reticulate scenario cannot be rejected, it is the burden of supporters of the scenario to prove that it is not a pure conjecture. Until this is done, Ockham's razor suggests that speciation occurred in sympatry by host shifting.

Bush and Smith (1998) see the evolution of the *Rhagoletis* species complex as a case of ecological speciation in sympatry. They believe that the ecological speciation resulted from a sudden shift of the insect's host preference. But they do not deal with the origin of the new behavior for host shifting, which implies two neurocognitive phenomena, the preference for the new host and recognition of the host. Essentially, that shift, as already pointed out, is a behavioral shift that results from changes in properties of the neural circuits determining that change in the insect's behavior.

They believe that

> *genes at relatively few key loci affecting mate recognition, habitat choice, and fitness in alternative habitats can be sufficient to initiate the process of ecological race formation and speciation.*
>
> *Bush and Smith (1998)*

This assertion raises serious objections. First, neither population genetics knowledge nor empirical evidence would support the idea that several sibling species of *R. pomonella* and necessary changes in genes could evolve from a common ancestral hawthorn-infesting species within an extraordinarily short period of time of ~150 years/generations.

Second, no changes have been identified in any of the "relatively few key loci," and electrophoretic analysis has only identified several differences in not-so-key allozyme loci (Berlocher, 1999).

Third, but not less importantly, host preferences and, especially, mate preferences are determined not by genes or their products but are epigenetically determined by neurocognitive processes taking place in the CNS.

However, Bush and Smith come very close to accepting the role of the neural circuits in the insect brain in determining the host shift, reproductive isolation, and speciation of the *Rhagoletis* species complex:

> *Host selection is mediated by both visual and olfactory cues. Ultimately selection of a host for mating and oviposition is determined by specific chemical cues emanating from the host plant and their fruits. Changes in host perception can have a direct effect on mate choice.*
>
> *Bush and Smith (1998)*

Needless to say, "changes in host perception" are not related to any changes in genes but are neurally, epigenetically determined in neural circuits.

Epigenetic Explanation of Sympatric Speciation by Host Plant Shifting

The key element for initiating host shift–related sympatric speciation is the emergence of the neurally determined preference for a new host, which leads to adoption of the new plant as a natural host.

The shift to a new host plant of a group of individuals or a population may lead to prezygotic reproductive isolation from the rest of the original population. Once this occurs, the phenotypic changes of different kinds will unavoidably take place in the process of the evolutionary adaptation to the new host. That this adaptation may occur rapidly is suggested by the fact that the larvae of parasitic insects from several genera can develop in laboratory on unnatural plant hosts (Dres and Mallet, 2002).

The apple race *R. pomonella* evolved its preference for a blend of apple fruit volatiles within 150 years (Linn et al., 2003). Hybrid flies between various *Rhagoletis* sibling species lose the ability to respond to the volatile substances emanated by the host tree in doses that elicit that response in parents and, consequently, cannot recognize or distinguish them from volatiles of other plants. Investigators have argued that this is not a result of any developmental imbalance related to hybridization because

> *if the altered behavior of F1 is a byproduct of a general developmental imbalance, then hybrids would be expected to exhibit various other phenotypic abnormalities with detrimental fitness consequences, in addition to their reduced olfactory response.*
>
> *Linn et al. (2004)*

They believe that this may result from a rise in the threshold of the behavioral response by olfactory circuits in hybrid flies (Linn et al., 2004). In view of the fact that flies recognize their host trees based on processing of olfactory and visual cues as well as on learning, it is logical to conclude that in F_1 something has changed in the properties of the neural circuits where the processing of these signals takes place. Indeed, there is experimental evidence substantiating this prediction. ORN responses of F_1 hybrids of *Rhagoletis* races show ORN response profiles that are different from those of each of the parents, a fact that may contribute to the reproductive isolation by decreasing mating chances of hybrids (Olsson et al., 2006a).

Given the absence of evidence on changes in relevant genes in these cases of incipient speciation in populations of *R. pomonella*, it has been hypothesized that changes in host preferences may be related to changes in the number or types of neurons that receive olfactory signals. Empirical evidence clearly shows that, contrary to the hypothesis, differences in host preference among *Rhagoletis* populations are not related to any alterations in the number or class of receptor neurons responding to host volatiles (Olsson et al., 2006b). There is evidence that these receptor neurons are generally resistant to evolutionary change. Studies in nine sibling species of the *Drosophila melanogaster* subgroup have shown that ligand affinity and action potential amplitude have been conserved to a remarkable degree over millions of years and across changing environments (Stensmyr et al., 2003).

Empirical evidence also shows that the only observable difference in the case of apple, hawthorn, and flowering dogwood-origin populations of *R. pomonella* is only an epigenetic change in the *sensitivity* of ORNs and in the firing patterns of these neurons, suggesting that these changes may be the cause of the divergent host preferences and host plant shifts of these populations (Olsson et al., 2006c). Hence, there is reason to believe that host shifts in *Rhagoletis* spp. are consequences of nongenetic changes in the structure (and properties related to the structure) of the neural circuits that process the host plant volatile substances. We have already seen that neurocognitive processes on which insect preferences are based are plastic and that plasticity may be a source of the evolutionary rapid host plant shifts:

> *Each fly taxon has a similar capacity for detecting all volatiles, regardless of host species. Thus, prior to host shifts, fly populations appear to have already possessed the ability to detect novel host volatiles and did not evolve new odor receptors for these volatiles. Instead, switched preferences for novel fruit volatiles and avoidance*

> *of ancestral hosts may have been established through alterations in the central processing centers of the brain (mainly the antennal lobe and protocerebrum) or modifications in glomerular innervation and connectivity to higher processing centers. Thus, many insects may have an innate chemosensory potential for rapidly changing their host affiliation that, when coupled with appropriate variation in host-related performance (survivorship), could help trigger rapid race formation and speciation.*
>
> *Dambroski et al. (2005)*

It may be safely said that host plant shifts are related not to any changes in genes but only to a changed behavior and preference for the new host. This changed preference is an epigenetic affect of "alterations in the central processing centers of the brain."

Neurocognitive Sympatric Speciation in Fish

Cichlid fish probably represent the largest catalog of examples of speciation and reproductive isolation without geographical or ecological separation. More than 3,000 species of cichlid fish are known to exist worldwide, making this group the largest family of species in vertebrates. Almost 2,000 cichlid species of fish have evolved in the lakes of East Africa (Victoria, Malawi, and Tanganyika) during "the very recent evolutionary past" (Kocher, 2004).

Lake Victoria, the largest lake in Africa, harbors more than 500 endemic species of haplochromine cichlid fish generally believed to have originated from a single common ancestor (Meyer et al., 1990) in less than 12,000 years since the lake dried up (Johnson et al., 1996; Nagl et al., 2000). Some authors believe that the lake did not entirely dry up and that the speciation from the common ancestor started earlier, about 100,000 years ago (Verheyen et al., 2003; Fryer, 2004). Nevertheless, both estimations indicate that the rate of speciation of cichlid fish in this tropical lake has been exceptional. No recent genetic bottlenecks did occur, and experimental interbreeding of these endemic forms still produces viable and fertile hybrids (van Oppen et al., 1998).

The explosive speciation of cichlid fish in Lake Victoria is unpredictable and unexplainable from any neo-Darwinian view of speciation as a process of accumulation and spread of changes in genes or allele frequencies under the action of natural selection. No changes in alleles or relevant gene mutations have been reported, and no postzygotic reproductive isolation between present species has evolved since the time of divergence from their common ancestor. They still intercross and produce hybrid offspring that are viable and fertile (Seehausen et al., 1997b).

In Lake Malawi, more than 500 endemic cichlid fish species are identified, all of which evolved from a common ancestor not earlier than 700,000 years ago. For a long time, allopatric speciation has been considered to be the most important, if not the only, mechanism of speciation in Lake Malawi, since sympatric speciation did not fit to the neo-Darwinian models of speciation (Shaw et al., 2000). It was believed that the large number of endemic fish species arose via "intralacustrine microallopatric speciation" as a result of differential preferences in diet or microhabitat specialization. Most recent observations have not confirmed the hypothesis, suggesting that sexual selection may be responsible for the explosive cichlid speciation in the lake. Laboratory experiments have shown that reproductive isolation of three morphologically similar species of rock-dwelling cichlid fish, living sympatrically in the lake, relies not on postzygotic but on prezygotic (sexual behavior) barriers between them (Knight et al., 1998).

It is estimated that *Tropheus gracilor*, a rock-dwelling cichlid fish in Lake Malawi, evolved ~1,000–2,300 years ago, whereas another cichlid fish, *T. tropheops*, some 17,200 years ago (Won et al., 2005). Radiations of cichlid fish species in Lake Malawi and Victoria

> *have occurred in such a short period of time (and in spite of significant levels of gene flow) that classic models of speciation do not easily explain cichlid evolution.*
>
> *Kocher (2004)*

Based on the evidence about the genetic population structure, the lack of regional variation in morphology, and male breeding colors, investigators suggest that

> *populations of the pelagic cichlids are potentially single panmictic units within the lake, with little, if any, opportunity for or indication of allopatric isolation and genetic divergence.*
>
> *Shaw et al. (2000)*

Out of twelve endemic cichlid fish species inhabiting Lake Barombi Mbo, only three miles in diameter, laying on the crater of a volcano in Cameroon, seven species belong to four endemic genera, and the volcano is only a few thousand years old (Schliewen et al., 2001). Nine endemic fish species have sympatrically evolved in the Lake Bermin, Cameroon ($0.5\,km^2$ area). Lake Ejagham, a small 5,000-year-old nonvolcanic lake in Cameroon, harbors five phenotypically similar, but distinct, cichlid fish of genus *Tilapia* (Schliewen et al., 2001).

Another example of sympatric speciation has been described for cichlid fish in a less-than-23,000-year-old crater lake in Nicaragua. Lake Apoyo, 5 km in diameter, contains the widespread cichlid species, *Amphilophus citrinellus*, and an endemic species, *A. zaliosus*. Genetic analyses have shown that the lake was colonized only once by an ancestor of *A. citrinellus*, and they have excluded the possibility of secondary colonization of the crater lake. The small size of the lake and the homogenous habitat shared totally by both species rule out the possibility of allopatric divergence; only sympatric speciation would explain formation of the new species within $<$10,000 years in this small crater lake (Barluenga et al., 2006).

In British Columbia, each of the five glacial lakes formed 12,000 − 15,000 years ago harbors two different stickleback fish species.

All of the known examples of sympatric speciation in fish suggest that formation of new species is not related to any gradual accumulation of changes in genes as cause of reproductive isolation. There is no evidence that neo-Darwinian founder effects have played any role in the process of rapid speciation of cichlid fish in the great lakes of East Africa (Kornfield and Smith, 2000). There are no postmating reproductive barriers between these species as would be predicted by the neo-Darwinian view of speciation. The only mechanism that maintains and perpetuates existence of these species is mate choice (Seehausen et al., 1997a), based on the recognition of, and preference for, species-specific mate signals—visual (Seehausen et al., 1997a; Seehausen and van Alphen, 1998), olfactory (Plenderleith et al., 2005), or even auditory (Amorim et al., 2004).

The "ecological selection" propounded by Kocher also requires conditions that are not really "likely" in these lakes, because the gene flow in the barrierless East African lakes occurs, and neither the presumed "strong linkage of genes for an ecological trait and mate preference" nor any genes for ecological traits or for mate preferences have ever been scientifically demonstrated to exist.

"Ecological selection" is observed in different stream-resident (smaller body size) and anadromous (larger body size) populations of the three-spined stickleback, *Gasterosteus aculeatus*. In laboratory experiments, it is demonstrated that females of different populations of this species prefer to mate individuals of their own ecotype (originating from similar habitats, no matter how distant). But, let us remember that selective mating is based on the perception of the body size rather than on male body coloration (McKinnon et al., 2004) and that perception and mate preferences are cognitive phenomena, functions of neural circuits, not related to the presence or absence of specific genes.

Kocher (2004) himself casts serious doubts on mathematical models of sympatric speciation when he admits that these models are founded on the premise of the existence and selection of genes that have not been demonstrated to exist (Kocher, 2004).

We should always bear in mind that changes in mate preferences or biases result from epigenetic factors, i.e., from changes in computational properties of neural circuits, which imply no changes in genes or allele frequencies, Moreover, experimental evidence from captive cichlid fish from Lake Malawi has shown that mate choice alone, without involvement of ecological cues, such as habitat choice and seasonality, can determine assortative mating and reproductive isolation (Knight et al., 1998).

But if, in fact, we know of *no speciation genes* as of yet, if we still have not established the necessary links between the genetic models and empirical observations of speciation, and if already-sophisticated genomic technologies have not helped us fill that gap, the question arises: Is this situation the result of imperfection or inappropriateness of the modern gene technology or of the methodological approach to the problem? In a Kuhnian mood, one would ask: Are the numerous cases of genetically inexplicable speciation phenomena related to the still-low level of our understanding, or do they represent counterinstances requiring a new concept on causal bases of speciation?

Neurocognitive Sympatric Speciation in Salamanders

A rapid burst of speciation that took place ~5 mya in eastern North America, especially in the Appalachian Mountains, led to differentiation of 35 species of Plethodon salamanders. It coincided with a shift in the mode of delivery of male courtship pheromones, released from the mental gland, to the female nares instead of the ancient "scratching" mode when the pheromone was delivered to the female's blood by abrading her skin with the premaxillary teeth.

Species of the glutinosus group (now consisting of 22 species) use olfactory cues for mate recognition and for species recognition, although they also hybridize (Wiens et al., 2006). Contrary to predictions of the "hybrid swarm" hypothesis, investigators found that hybridization is a result, rather than the cause, of rapid speciation (Wiens et al., 2006).

The only relevant change coinciding with the rapid speciation of the glutinosus group of salamanders is the sudden epigenetic change in sexual behavior. There is no evidence of changes in genes or regulatory sequences being involved in this speciational explosion of plethodontid salamanders.

Neurocognitive Sympatric Speciation in Birds

Two populations of the Vogelkop bowerbirds (*Amblyornis inornatus*) build bowers and decorate them in strikingly different ways. Females from each of the two populations prefer bowers and decorations of males from their own population, thus establishing a behaviorally conditioned reproductive isolation in this initial stage of the speciation process that is actually taking place in two populations of the same species (Uy and Borgia, 2000). Individuals of both populations are morphologically identical and show only little genetic difference of recent origin, but there is no evidence of any differences in genes related to the bower-building behavior.

Despeciation or Fusion of Species

Rare cases of the "reverse" process of two metazoan species merging into one are observed in nature. This occurs exclusively between sympatric species. Kraak et al. (2001) reported an increase in the proportion of hybrids between a pair of "benthic" (bottom-dwelling) and "limnetic" (top-dwelling) species of the three-spined sticklebacks, *Gasterosteus aculeatus*, in Lake Enos, southeastern Vancouver Island, Canada, from 1% to 12–17% within 7–15 years and concluded that the benthic and limnetic forms of the fish were collapsing and that a "new" species was evolving from their hybridization (Kraak et al., 2001). Their conclusion was corroborated later by another group (Taylor et al., 2006). The species' breakdown seems to be asymmetrical in the sense that the limnetic form is introgressed into the benthic form (Gow et al., 2006).

A neo-Darwinian explanation would hardly be applicable in the cases of species diffusion for the sterility, and low viability of hybrids predicted by the paradigm would not lead to fusion but to gradual extinction of both species.

There are indications that some extant vertebrate species, such as the red wolf (*Canis rufus*), evolved from introgressive hybridization between the gray wolf and coyote (Roy et al., 1996), reminding us once again of the unreliability of the reproductive isolation as a basic criterion of biological definition of species.

References

Amorim, M.C.P., Knight, M.E., Stratoudakis, Y. and Turner, G.F. (2004). Differences in sounds made by courting males of three closely related Lake Malawi cichlid species. J. Fish Biol. *65*, 1358–1371.

Amundsen, T. and Forsgren, E. (2001). Male mate choice selects for female coloration in a fish. Proc. Natl. Acad. Sci. U.S.A. *98*, 13155–13160.

Aprille, J.R., Lagace, C.J., Modica-Napolitano, J. and Trimmer, B.A. (2004). Role of nitric oxide and mitochondria in control of firefly flash. Integr. Comp. Biol. *44*, 213–219.

Arnegard, M.E. and Kondrashov, A.S. (2004). Sympatric speciation by sexual selection is unlikely. Evolution *58*, 222–237.

Arnegard, M.E., Bogdanowicz, S.M. and Hopkins, C.D. (2005). Multiple cases of striking genetic similarity between alternative electric fish signal morphs in sympatry. Evolution *59*, 324–343.

Arnegard, M.E., Jackson, B.S. and Hopkins, C.D. (2006). Time-domain signal divergence and discrimination without receptor modification in sympatric morphs of electric fishes. J. Exp. Biol. *209*, 2182–2198.

Assad, C., Rasnow, B. and Stoddard, P.K. (1999). Electric organ discharges and electric images during electrolocation. J. Exp. Biol. *202*, 1185–1193.

Autumn, K., Ryan, M.J. and Wake, D.B. (2002). Integrating historical and mechanistic biology enhances the study of adaptations. Q. Rev. Biol. *77*, 383–408.

Baker, B.S., Taylor, B.J. and Hall, J.C. (2001). Are complex behaviors specified by dedicated regulatory genes? Reasoning from *Drosophila*. Cell *105*, 13–24.

Balzer, A.L. and Williams, T.D. (1998). Do female zebra finches vary primary reproductive effort in relation to male attractiveness? Behaviour *135*, 297–309.

Barker, J.S.F. (2005). Population structure and host-plant specialization in two *Scaptodrosophila* flower-breeding species. Heredity *94*, 129–138.

Barluenga, M., Stoelting, K.N., Salzburger, W., Muschick, M. and Meyer, A. (2006). Sympatric speciation in Nicaraguan crater lake cichlid fish. Nature *439*, 719–722.

Barreto, F.S. and McCartney, M.A. (2008). Extraordinary AFLP fingerprint similarity despite strong assortative mating between reef fish color morphospecies. Evolution *62*, 226–233.

Basolo, A.L. (1998). Evolutionary change in a receiver bias: a comparison of female preference functions. Proc. Biol. Sci. *265*, 2223–2228.

Basolo, A.L. (2002). Congruence between the sexes in preexisting receiver responses. Behav. Ecol. *13*, 832–837.

Bass, A.H., Segil, N. and Kelley, D.B. (1986). Androgen binding in the brain and electric organ of a mormyrid fish. J. Comp. Physiol. A. *159*, 535–544.

Berlocher, S.H. (1999). Host race or species? Allozyme characterization of the "flowering dogwood fly" a member of the *Rhagoletis pomonella* complex. Heredity *83*, 652–662.

Blows, M.W. and Allan, R.A. (1998). Levels of mate recognition within and between two *Drosophila* species and their hybrids. Am. Nat. *152*, 834–837.

Boake, C.R.B., DeAngelis, M.P. and Andreadis, D.K. (1997). Is sexual selection and species recognition a continuum? Mating behavior of the stalk-eyed fly *Drosophila heteroneura*. Proc. Natl. Acad. Sci. U.S.A. *94*, 12442–12445.

Bodznick, D., Montgomery, J. and Tricas, T.C. (2003). Electroreception: extracting behaviorally important signals from noise. In: *Sensory Processing in Aquatic Environments*, S.P. Collin and N.J. Marshall (eds.). Springer-Verlag, New York, NY, pp. 389–403.

Bolhuis, J.J. and Gahr, M. (2006). Neural mechanisms of birdsong memory. Nat. Rev. Neurosci. *7*, 347–357.

Boughman, J.W., Rundle, H.D. and Schluter, D. (2005). Parallel evolution of sexual isolation in sticklebacks. Evolution *59*, 361–373.

Bratton, B.O. and Kramer, B. (1989). Patterns of the electric organ discharge during courtship and spawning in the mormyrid fish, *Pollimyrus isidori*. Behav. Ecol. Sociobiol. *24*, 349–368.

Brenowitz, E.A. (2004). Plasticity of the adult avian song control system. Ann. N.Y. Acad. Sci. *1016*, 560–585.

Broughton, S.J., Kitamoto, T. and Greenspan, R.J. (2004). Excitatory and inhibitory switches for courtship in the brain of *Drosophila melanogaster*. Curr. Biol. *14*, 538–547.

Bullock, T.H., Northcutt, R.G. and Bodznick, D.A. (1982). Evolution of electroreception. Trends Neurosci. *5*, 50–53.

Bush, G.L. (1969). Sympatric host race formation and speciation in frugivorous flies of the genus *Rhagoletis* (Diptera: Tephritidae). Evolution *23*, 237–251.

Bush, G.L. (1975). Modes of animal speciation. Annu. Rev. Ecol. Syst. *6*, 339–364.

Bush, G.L. and Smith, J.J. (1998). The genetics and ecology of sympatric speciation: a case study. Res. Popul. Ecol. *40*, 175–187.

Butlin, R.K. and Ritchie, M.G. (1991). Variation in female mate preference across grasshopper hybrid zone. J. Evol. Biol *4*, 227–240.

Cabej, N.R. (2011). Neural control of gene recruitment in metazoans. Dev. Dyn. *240*, 1–8.

Caputi, A.A. and Budelli, R. (2006). Peripheral electrosensory imaging by weakly electric fish. J. Comp. Physiol. A Neuroethol. Sens. Neural Behav. Physiol. *192*, 587–600. (abstract).

Carleton, K.L., Parry, J.W.L., Bowmaker, J.K., Hunt, D.M. and Seehausen, O. (2005). Colour vision and speciation in Lake Victoria cichlids of the genus *Pundamilia*. Mol. Ecol. *14*, 4341–4353.

Carlson, B.A. (2002). Neuroanatomy of the mormyrid electromotor control system. J. Comp. Neurol. *454*, 440–455.

Carlson, B.A. and Hopkins, C.D. (2004). Central control of electric signaling behavior in the mormyrid *Brienomyrus brachyistius*: segregation of behavior-specific inputs and the role of modifiable recurrent inhibition. J. Exp. Biol. *207*, 1073–1084.

Carlson, B.A., Hopkins, C.D. and Thomas, P. (2000). Androgen correlates of socially induced changes in the electric organ discharge waveform of a mormyrid fish. Horm. Behav. *38*, 177–186.

Carlson, B.A., Hasan, S.M., Hollmann, M., Miller, D.B., Harmon, L.J. and Arnegard, M.E. (2011). Brain evolution triggers increased diversification of electric fishes. Science *332*, 583–586.

Christensen, T.A., Sherman, T.G., McCaman, R.E. and Carlson, A.D. (1983). Presence of octopamine in firefly photomotor neurons. Neuroscience *9*, 183–189.

Coleman, S.W., Patricelli, G.L. and Borgia, G. (2004). Variable female preferences drive complex male displays. Nature *428*, 742–745.

Collin, S.P. and Whitehead, D. (2004). The functional roles of passive electroreception in non-electric fishes. Anim. Biol. *54*, 1–25.

Couldridge, V.C.K. and Alexander, G.J. (2002). Color patterns and species recognition in four closely related species of Lake Malawi cichlid. Behav. Ecol. *13*, 59–64.

Coyne, J.A. and Elwyn, S. (2006a). Does the *desaturase-2* locus in *Drosophila melanogaster* cause adaptation and sexual isolation? Evolution *60*, 279–291.

Coyne, J.A. and Elwyn, S. (2006b). Desaturase-2, environmental adaptation, and sexual isolation in Drosophila melanogaster. Evolution *60*, 626–627.

Coyne, J.A. and Price, T.D. (2000). Little evidence for sympatric speciation in island birds. Evolution *54*, 2166–2171.

Craig, T.P., Itami, J.K., Abrahamson, W.G. and Horner, J.D. (1993). Behavioral evidence for host-race formation in *Eurosta solidaginis*. Evolution *47*, 1696–1710.
Cratsley, C.K. and Lewis, S.M. (2003). Female preference for male courtship flashes in *Photinus ignitus* fireflies. Behav. Ecol. *14*, 135–140.
Cunningham, E.J. and Russell, A.F. (2000). Egg investment is influenced by male attractiveness in the mallard. Nature *404*, 74–77.
Dambroski, H.R., Linn, C., Jr., Berlocher, S.H., Forbes, A.A., Roelofs, W. and Feder, J.L. (2005). The genetic basis for fruit odor discrimination in *Rhagoletis* flies and its significance for sympatric host shifts. Evolution *59*, 1953–1964.
Danley, P.D. and Kocher, T.D. (2001). Speciation in rapidly diverging systems: lessons from Lake Malawi. Mol. Ecol. *10*, 1075–1086.
Darwin, C.R. (1859). *The Origin of Species by Means of Natural Selection or The Preservation of Favored Races in the Struggle for Life*, first edition. John Murray, London (first issue).
Darwin, C.R. (1871). *The Descent of Man, and Selection in Relation to Sex*, first edition, vol. 2. John Murray, London, p. 398.
Deily, J.A. and Schul, J. (2004). Recognition of calls with exceptionally fast pulse rates: female phonotaxis in the genus *Neoconocephalus* (Orthoptera: Tettigoniidae). J. Exp. Biol. *207*, 3523–3529.
Demary, K., Michaelidis, C.I. and Lewis, S.M. (2006). Firefly courtship: behavioral and morphological predictors of male mating success in *Photinus greeni*. Ethology *112*, 485–492.
Demir, E. and Dickson, B.J. (2005). *fruitless* splicing specifies male courtship behavior in *Drosophila*. Cell *121*, 785–794.
Dieckmann, U. and Doebeli, M. (1999). On the origin of species by sympatric speciation. Nature *400*, 354–357.
Doebeli, M., Dieckmann, U., Metz, J.A.J. and Tautz, D. (2005). What we have also learned: adaptive speciation is theoretically plausible. Evolution *59*, 691–695.
Doucet, S.M. and Montgomerie, R. (2003). Multiple sexual ornaments in satin bowerbirds: ultraviolet plumage and bowers signal different aspects of male quality. Behav. Ecol. *14*, 503–509.
Drès, M. and Mallet, J. (2002). Host races in plant-feeding insects and their importance in sympatric speciation. Philos. Trans. R. Soc. B Biol. Sci. *357*, 471–492.
Dugatkin, L.A. (1996). Interface between culturally based preferences and genetic preferences: female mate choice in *Poecilia reticulata*. Proc. Natl. Acad. Sci. U.S.A. *93*, 2770–2773.
Dukas, R. (2004). Evolutionary biology of animal cognition. Annu. Rev. Ecol. Evol. S *35*, 347–374.
Emelianov, I., Simpson, F., Narang, P. and Mallet, J. (2003). Host choice promotes reproductive isolation between host races of the larch budmoth *Zeiraphera diniana*. J. Evol. Biol. *16*, 208–218.
Endler, J.A. and Thery, M. (1996). Interacting effects of lek placement, display behavior, ambient light, and color patterns in three neotropical forest-dwelling birds. Am. Nat. *148*, 421–452.
Enquist, M. and Arak, A. (1993). Selection of exaggerated male traits by female aesthetic senses. Nature *361*, 446–448.
Enquist, M. and Arak, A. (1994). Symmetry, beauty and evolution. Nature *372*, 169–172.
Ewing, A.W. and Miyan, J.A. (1986). Sexual selection, sexual isolation and the evolution of song in the *Drosophila repleta* group of species. Anim. Behav. *34*, 421–429.
Feder, J. (1998). The applemaggot fly, *Rhagoletis pomonella*: Flies in the face of conventional wisdom about speciation? In: *Endless Forms: Species and Speciation*, D.J. Howard and S.H. Berlocher (eds.). Oxford University Press, New York, pp. 130–144.
Feder, J.L., Opp, S.B., Wlazlo, B., Reynolds, K., Go, W. and Spisak, S. (1994). Host fidelity is an effective barrier between sympatric races of the apple maggot fly. Proc. Natl. Acad. Sci. U.S.A. *91*, 7990–7994.

Feder, J.L., Xie, X., Rull, J., Velez, S., Forbes, A. and Leung, B., et al. (2005). Mayr, Dobzhansky, and Bush and the complexities of sympatric speciation. Proc. Natl. Acad. Sci. U.S.A. *102*, 6573–6580.

Felsenstein, J. (1981). Skepticism towards Santa Rosalia, or why are there so few kinds of animals? Evolution *35*, 124–238.

Feulner, P.G.D., Kirschbaum, F., Schugardt, C., Ketmaier, V. and Tiedemann, R. (2006). Electrophysiological and molecular genetic evidence for sympatrically occurring cryptic species in African weakly electric fishes (Teleostei: Mormyridae: *Campylomormyrus*). Mol. Phylogenet. Evol. *39*, 198–208.

Feulner, P.G., Plath, M., Engelmann, J., Kirschbaum, F. and Tiedemann, R. (2009). Electrifying love: electric fish use species-specific discharge for mate recognition. Biol. Lett. *5*, 225–228.

Firestein, S. (2001). How the olfactory system makes sense of scents. Nature *413*, 211–218.

Fry, J.D. (2003). Multilocus models of sympatric speciation: Bush versus Rice versus Felsenstein. Evolution *57*, 1735–1746.

Fryer, G. (2004). Speciation rates in lakes and the enigma of Lake Victoria. Hydrobiologia *519*, 167–183.

Gahr, M. (2004). Hormone-dependent neural plasticity in the juvenile and adult song system: what makes a successful male? Ann. N.Y. Acad. Sci. *1016*, 684–703.

Gleason, J.M. and Ritchie, M.G. (1998). Evolution of courtship song and reproductive isolation in the *Drosophila willistoni* species complex: do signals diverge the most quickly? Evolution *52*, 1493–1500.

Gong, A. and Gibson, R.M. (1996). Reversal of a female preference after visual exposure to a predator in the guppy, *Poecilia reticulate*. Anim. Behav. *52*, 1007–1015.

Gow, J.L., Peichel, C.L. and Taylor, E.B. (2006). Contrasting hybridization rates between sympatric three-spined sticklebacks highlight the fragility of reproductive barriers between evolutionarily young species. Mol. Ecol. *15*, 739–752.

Gray, D.A. and Cade, W.H. (2000). Sexual selection and speciation in field crickets. Proc. Natl. Acad. Sci. U.S.A. *97*, 14449–14454.

Greenberg, A.J., Moran, J.R., Coyne, J.A. and Wu, C-I. (2003). Ecological adaptation during incipient speciation revealed by precise gene replacement. Science *302*, 1754–1757.

Greenfield, M.D. (2001). Missing link in firefly bioluminescence revealed: no regulation of photocyte respiration. Bioessays *23*, 992–995.

Griggio, M., Valera, F., Casas, A. and Pilastro, A. (2005). Males prefer ornamented females: a field experiment of male choice in the rock sparrow. Anim. Behav. *69*, 1243–1250.

Grober, M.S., Fox, S.H., Laughlin, C. and Bass, A.H. (1994). GnRH cell size and number in a teleost fish with two male reproductive morphs: sexual maturation, final sexual status and body size allometry. Brain Behav. Evol. *43*, 61–78.

Haesler, M.P. and Seehausen, O. (2005). Inheritance of female mating preference in a sympatric sibling species pair of Lake Victoria cichlids and implications for speciation. Proc. Biol. Sci. *272*, 237–245.

Hall, M., Lindholm, A.K. and Brooks, R. (2004). Direct selection on male attractiveness and female preference fails to produce a response. BMC Evol. Biol. *4*, p. 1.

Hansen, L.T.T., Amundsen, T. and Forsgren, E. (1999). Symmetry: attractive not only to females. Proc. Biol. Sci. *266*, 1235–1240.

Hedwig, B. (2005). Pulses, patterns and paths: neurobiology of acoustic behavior in crickets. J. Comp. Physiol. A Neuroethol. Sens. Neural. Behav. Physiol. *192*, 677–689.

Hennig, R.M. (2003). Acoustic feature extraction by cross-correlation in crickets? J. Comp. Physiol. A Neuroethol. Sens. Neural. Behav. Physiol. *189*, 589–598.

Higashi, M., Takimoto, G. and Yamamura, N. (1999). Sympatric speciation by sexual selection. Nature *402*, 523–526.

Hoch, H., Deckert, J. and Wessel, A. (2006). Vibrational signalling in a Gondwanan relict insect (Hemiptera: Coleorrhyncha: Peloridiidae). Biol. Lett. *2*, 222–224.

Hoke, K.L., Burmeister, S.S., Fernald, R.D., Rand, A.S., Ryan, M.J. and Wilczynski, W. (2004). Functional mapping of the auditory midbrain during mate call reception. J. Neurosci. *24*, 11264–11272.

Hoke, K.L., Ryan, M.J. and Wilczynski, W. (2005). Acoustic social cues shift functional connectivity in the hypothalamus. Proc. Natl. Acad. Sci. U.S.A. *102*, 10712–10717.

Hollocher, H., Ting, C.T., Wu, M.L. and Wu, C.I. (1997). Incipient speciation by sexual isolation in *Drosophila melanogaster*: extensive genetic divergence without reinforcement. Genetics *147*, 1191–1202.

Hopkins, C.D. (1999). Signal evolution in electric communication. In: *The Design of Animal Communication*, M.D. Hauser and M. Konishi (eds.). MIT Press, Cambridge, MA, pp. 461–491.

Hopkins, C.D. and Bass, A.H. (1981). Temporal coding of species recognition signals in an electric fish. Science *212*, 85–87.

Huey, R.B., Hertz, P.E. and Sinervo, B. (2003). Behavioral drive versus behavioral inertia in evolution: a null model approach. Am. Nat. *161*, 357–366.

Iglesias, F., Pilar-Marco, M., Jacquin-Joly, E., Camps, F. and Fabrias, G. (1998). Regulation of sex pheromone biosynthesis in two noctuid species, *S. littoralis* and *M. brassicae*, may involve both PBAN and the ventral nerve cord. Arch. Insect Biochem. *37*, 295–304.

Jang, Y. and Greenfield, M.D. (1996). Ultrasonic communication and sexual selection in wax moths: female choice based on energy and asynchrony of male signals. Anim. Behav. *51*, 1095–1106.

Jarvis, E.D., Ribeiro, S., da Silva, M.L., Ventura, D., Vielliard, J. and Mello, C.V. (2000). Behaviourally driven gene expression reveals song nuclei in hummingbird brain. Nature *406*, 628–632.

Jefferis, G.S., Vyas, R.M., Berdnik, D., Ramaekers, A., Stocker, R.F., Tanaka, N.K., et al. (2004). Developmental origin of wiring specificity in the olfactory system of *Drosophila*. Development *131*, 117–130.

Jefferis, G.S.X.E., Marin, E.C., Stocker, R.F. and Luo, L. (2001). Target neuron prespecification in the olfactory map of *Drosophila*. Nature *414*, 204–208.

Johnson, J.B. and Basolo, A.L. (2003). Predator exposure alters female mate choice in the green swordtail. Behav. Ecol. *14*, 619–625.

Johnson, T.C., Scholz, C.A., Talbot, M.R., Kelts, K., Ricketts, R.D. and Ngobi, G., et al. (1996). Late Pleistocene desiccation of Lake Victoria and rapid evolution of cichlid fishes. Science *273*, 1091–1093.

Johnstone, R.A. (1994). Female preference for symmetrical males as a by-product of selection for mate recognition. Nature *372*, 172–175.

Kirkpatrick, M. and Ryan, M.J. (1991). The evolution of mating preferences and the paradox of the lek. Nature *350*, 33–38.

Knight, M.E. and Turner, G.F. (1999). Reproductive isolation among closely related Lake Malawi cichlids: can males recognize conspecific females by visual cues? Anim. Behav. *58*, 761–768.

Knight, M.E., Turner, G.F., Rico, C., van Oppen, M.J.H. and Hewitt, G.M. (1998). Microsatellite paternity analysis on captive Lake Malawi cichlids supports reproductive isolation by direct mate choice. Mol. Ecol. *7*, 1605–1610.

Kocher, T.D. (2004). Adaptive evolution and explosive speciation: the cichlid fish model. Nat. Rev. Genet. *5*, 288–298.

Kokko, H., Brooks, R., Jennions, M.D. and Morley, J. (2003). The evolution of mate choice and mating biases. Proc. Biol. Sci. *270*, 653–664.

Kondoh, Y., Kaneshiro, K.Y., Kimura, K. and Yamamoto, D. (2003). Evolution of sexual dimorphism in the olfactory brain of Hawaiian *Drosophila*. Proc. Biol. Sci. *270*, 1005–1013.

Kondrashov, A.S. and Kondrashov, F.A. (1999). Interactions among quantitative traits in the course of sympatric speciation. Nature *400*, 351–354.

Kornfield, I. and Smith, P.F. (2000). African cichlid fishes: model systems for evolutionary biology. Ann. Rev. Ecol. Syst *31*, 163–196.

Kraak, S.B.M. (2001). Increased number of hybrids between benthic and limnetic three-spined sticklebacks in Enos Lake, Canada; the collapse of a species pair? J. Fish Biol. *58*, 1458–1464.

Lande, R. (1981). Models of speciation by sexual selection on polygenic traits. Proc. Natl. Acad. Sci. U.S.A. *78*, 3721–3725.

Lefebvre, L., Reader, S.M. and Sol, D. (2004). Brains, innovations and evolution in birds and primates. Brain Behav. Evol. *63*, 233–246.

Leitner, S., Nicholson, J., Leisler, B., DeVoogd, T.J. and Catchpole, C. (2002). Song and the song control pathway in the brain can develop independently of exposure to song in the sedge warbler. Proc. Biol. Sci. *269*, 2519–2524.

Linn, C., Feder, J.L., Nojima, S., Dambroski, H.R., Berlocher, S.H. and Roelofs, W. (2003). Fruit odor discrimination and sympatric host race formation in *Rhagoletis*. Proc. Natl. Acad. Sci. U.S.A. *100*, 11490–11493.

Linn, C.E., Dambroski, H.R., Feder, J.L., Berlocher, S.H., Nojima, S. and Roelofs, W.L. (2004). Postzygotic isolating factor in sympatric speciation in *Rhagoletis* flies: reduced response of hybrids to parental host-fruit odors. Proc. Natl. Acad. Sci. U.S.A. *101*, 17753–17758.

Lissmann, H.W. (1951). Continuous electrical signals from the tail of a fish, *Gymnarchus niloticus* Cuv. Nature *167*, 201–202.

Lynch, K.S., Crews, D., Ryan, M.J. and Wilczynski, W. (2006). Hormonal state influences aspects of female mate choice in the Túngara Frog (*Physalaemus pustulosus*). Horm. Behav. *49*, 450–457.

Mallet, J. (2001). The speciation revolution. J. Evol. Biol. *14*, 887–888.

Marchetti, K. (1993). Dark habitats and bright birds illustrate the role of the environment in species divergence. Nature *362*, 149–152.

Marler, C.A. and Ryan, M.J. (1997). Origin and maintenance of a female mating preference. Evolution *51*, 1244–1248.

Maynard Smith, J. (1989). *Evolutionary Genetics*. Oxford University Press, Oxford, p. 276.

Mayr, E. (1963). *Animal Species and Evolution*. Harvard University Press, Cambridge, MA, p. 604.

McCune, A.R. and Lovejoy, N.R. (1998). The relative rate of sympatric and allopatric speciation in fishes: tests using DNA sequence divergence between sister species and among clades. In: *Endless Forms: Species and Speciation*, D.J. Howard and S.H. Berlocher (eds.). Oxford University Press, New York, NY, pp. 291–308.

McKinnon, J.S., Mori, S., Blackman, B.K., David, L., Kingsley, D.M. and Jamieson, L., et al. (2004). Evidence for ecology's role in speciation. Nature *429*, 294–298.

McLennan, D.A. and Ryan, M.J. (1997). Responses to conspecific and heterospecific olfactory cues in the swordtail *Xiphophorus cortezi*. Anim. Behav. *54*, 1077–1088.

Mendelson, T.C. and Shaw, K.L. (2005). Rapid speciation in an arthropod. Nature *433*, 375–376.

Meyer, A., Kocher, T.D., Basasibwaki, P. and Wilson, A.C. (1990). Monophyletic origin of Lake Victoria cichlid fishes suggested by mitochondrial DNA sequences. Nature *347*, 550–553.

Michalak, P., Minkov, I., Helin, A., Lerman, D.N., Bettencourt, B.R. and Feder, M.E., et al. (2001). Genetic evidence for adaptation-driven incipient speciation of *Drosophila melanogaster* along a microclimatic contrast in "Evolution Canyon," Israel. Proc. Natl. Acad. Sci. U.S.A. *98*, 13195–13200.

Mills, A., Zakon, H.H., Marchaterre, M.A. and Bass, A.H. (1992). Electric organ morphology of *Sternopygus macrurus*, a wave-type, weakly electric fish with a sexually dimorphic EOD. J. Neurobiol. *23*, 920–932.

Møller, A.P. (1992). Female swallow preference for symmetrical male sexual ornaments. Nature *357*, 238–240.

Morris, M.R. and Ryan, M.J. (1996). Sexual difference in signal-receiver coevolution. Anim. Behav. *52*, 1017–1024.

Nagl, S., Tichy, H., Mayer, W.E., Takezaki, N. and Klein, J. (2000). The origin and age of haplochromine fishes in Lake Victoria, East Africa. Proc. Biol. Sci. *267*, 1049–1061.

Ng, M., Roorda, R.D., Lima, S.Q., Zemelman, B.V., Morcillo, P. and Miesenböck, G. (2002). Transmission of olfactory information between three populations of neurons in the antennal lobe of the fly. Neuron *36*, 463–474.

Nikolakakis, N., Sol, D. and Lefebvre, L. (2003). Behavioural flexibility predicts species richness in birds, but not extinction risk. Anim. Behav. *65*, 445–452.

Nottebohm, F. (2005). The neural basis of birdsong. PLoS Biol. *3*(5), e164.

Ortiz-Barrientos, D., Counterman, B.A. and Noor, M.A. (2004). The genetics of speciation by reinforcement. PLoS Biol. *2*, e416.

Olsson, S.B., Linn, C.E., Jr., et al., Michel, A., Dambroski, H.R., Berlocher, S.H. and Feder, J.L., et al. (2006a). Receptor expression and sympatric speciation: unique olfactory receptor neuron responses in F1 hybrid *Rhagoletis* populations. J. Exp. Biol. *209*, 3729–3741.

Olsson, S.B., Linn, C.E., Jr. and Roelofs, W.L. (2006b). The chemosensory basis for behavioral divergence involved in sympatric host shifts. I: characterizing olfactory receptor neuron classes responding to key host volatiles. J. Comp. Physiol. A Neuroethol. Sens. Neural Behav. Physiol. *192*, 279–288.

Olsson, S.B., Linn, C.E., Jr. and Roelofs, W.L. (2006c). The chemosensory basis for behavioral divergence involved in sympatric host shifts II: olfactory receptor neuron sensitivity and temporal firing pattern to individual key host volatiles. J. Comp. Physiol. A Neuroethol. Sens. Neural Behav. Physiol. *192*, 289–300.

Paenke, I., Sendhoff, B. and Kawecki, T.J. (2007). Influence of plasticity and learning on evolution under directional selection. Am. Nat. *170*, E47–E58.

Palmer, C.A., Watts, R.A., Houck, L.D., Picard, A.L. and Arnold, S.J. (2007). Evolutionary replacement of components in a salamander pheromone signaling complex: more evidence for phenotypic-molecular decoupling. Evolution *61*, 202–215.

Panhuis, T.M., Butlin, R., Zuk, M. and Tregenza, M. (2001). Sexual selection and speciation. Trends Ecol. Evol. *16*, 364–372.

Penn, D. and Potts, W. (1998). MHC-disassortative mating preferences reversed by cross-fostering. Proc. Biol. Sci. *266*, 1299–1306.

Penn, D.J. and Potts, W.K. (1999). The evolution of mating preferences and major histocompatibility complex genes. Am. Nat. *153*, 145–164.

Phelps, S.M. and Ryan, M.J. (1998). Neural networks predict response biases of female túngara frogs. Proc. Biol. Sci. *265*, 279–285.

Phelps, S.M. and Ryan, M.J. (2000). History influences signal recognition: neural network models of túngara frogs. Proc. Biol. Sci. *267*, 1633–1639.

Phelps, S.M., Rand, A.S. and Ryan, M.J. (2006). A cognitive framework for mate choice and species recognition. Am. Nat. *167*, 28–42.

Plenderleith, M., van Oosterhout, C., Robinson, R.L. and Turner, G.F. (2005). Female preference for conspecific males based on olfactory cues in a Lake Malawi cichlid fish. Biol. Lett. *1*, 411–414.

Price, T.D., Qvarnström, A. and Irwin, D.E. (2003). The role of phenotypic plasticity in driving genetic evolution. Proc. Biol. Sci. *270*, 1433–1440.
Pröhl, H., Koshy, R., Mueller, U., Rand, A.S. and Ryan, M.J. (2006). Geographic variation and behavioral traits in túngara frogs in a zone of secondary contact. Evolution *60*, 1669–1679.
Raina, A.K., Kingan, T.G. and Kochansky, J.P. (2003). A pheromonotropic peptide of *Helicoverpa zea*, with melanizing activity, interaction with PBAN, and distribution of immunoreactivity. Arch. Insect Biochem. *53*, 147–157.
Reader, S.M. and Laland, K.N. (2002). From the cover: social intelligence, innovation, and enhanced brain size in primates. Proc. Natl. Acad. Sci. U.S.A. *99*, 4436–4441.
Rhodes, G., Roberts, J. and Simmons, L.W. (1999). Reflections on symmetry and attractiveness. Psych. Evol. Gen. *1*, 279–296.
Rice, W.R. and Hostert, E.E. (1993). Laboratory experiments on speciation: what have we learned in 40 years? Evolution *47*, 1637–1653.
Ritchie, M.G. (2000). The inheritance of female preference functions in a mate recognition system. Proc. Biol. Sci. *267*, 327–332.
Rodd, F.H., Hughes, K.A., Grether, G.F. and Baril, C.T. (2002). A possible non-sexual origin of mate preference: are male guppies mimicking fruit? Proc. Biol. Sci. *269*, 475–481.
Rollman, S.M., Houck, L.D. and Feldhoff, R.C. (1999). Proteinaceous pheromone affecting female receptivity in a terrestrial salamander. Science *285*, 1907–1909.
Roper, A. and Zann, R. (2006). The onset of song learning and song tutor selection in fledgling zebra finches. Ethology *112*, 458–470.
Rossler, W., Randolph, P.W., Tolbert, L.P. and Hildebrand, J.G. (1999). Axons of olfactory receptor cells of transsexually grafted antennae induce development of sexually dimorphic glomeruli in *Manduca sexta*. J. Neurobiol. *38*, 521–541.
Roughgarden, J. (2005). The myth of sexual selection. California Wild—The Magazine of the California Academy of Sciences *58*(3).
Roughgarden, J., Oishi, M. and Akcay, E. (2006). Reproductive social behaviour: cooperative games to replace sexual selection. Science *311*, 965–969. http://researcharchive.calacademy.org/calwild/2005summer/stories/sexual.html
Roy, M.S., Geffen, E., Smith, D. and Wayne, R.K. (1996). Molecular genetics of pre-1940 red wolves. Conserv. Biol. *10*, 1413–1424.
Ryan, M.J. (1986). Neuroanatomy influences speciation rates among anurans. Proc. Natl. Acad. Sci. U.S.A. *83*, 1379–1382.
Ryan, M.J. (1998). Sexual selection, receiver biases, and the evolution of sex differences. Science *281*, 1999–2003.
Ryan, M.J. and Rand, A.S. (1993). Sexual selection and signal evolution: the ghost of biases past. Phil. Trans. R. Soc. Lond. B *340*, 187–195.
Ryan, M.J. and Rand, A.S. (2003). Mate recognition in túngara frogs: a review of some studies of brain, behavior, and evolution. Acta Zool. Sinica *49*, 713–726.
Ryan, M.J. and Wagner, W.E., Jr. (1987). Asymmetries in mating preferences between species: female swordtails prefer heterospecific males. Science *236*, 595–597.
Saarikettu, M., Liimatainen, J.O. and Hoikkala, A. (2005). The role of male courtship song in species recognition in *Drosophila montana*. Behav. Genet. *35*, 257–263.
Sawtell, N.B., Williams, A. and Bell, C.C. (2005). From sparks to spikes: information processing in the electrosensory systems of fish. Curr. Opin. Neurobiol. *15*, 437–443.
Schliewen, U.K., Tautz, D. and Paabo, S. (1994). Sympatric speciation suggested by monophyly of crater lake cichlids. Nature *368*, 629–632.
Schliewen, U., Rassmann, K., Markmann, M., Markert, J., Kocher, T. and Tautz, D. (2001). Genetic and ecological divergence of a monophyletic cichlid species pair under fully sympatric conditions in Lake Ejagham, Cameroon. Mol. Ecol. *10*, 1471–1488.

Schlupp, I., Marler, C. and Ryan, M.J. (1994). Benefit to male sailfin mollies of mating with heterospecific females. Science *263*, 373–374.

Schneiderman, A.M., Hildebrand, J.G., Brennan, M.M. and Tumlinson, J.H. (1986). Transsexually grafted antennae alter pheromone-directed behaviour in a moth. Nature *323*, 801–803.

Searcy, W.A. (1992). Song repertoire and mate choice in birds. Am. Zool. *32*, 71–80.

Seehausen, O. and van Alphen, J.J.M. (1998). The effect of male coloration on female mate choice in closely related Lake Victoria cichlids (*Haplochromis nyererei* complex). Behav. Ecol. Sociobiol. *42*, 1–8.

Seehausen, O., van Alphen, J.J.M. and Witte, F. (1997a). Cichlid fish diversity threatened by eutrophication that curbs sexual selection. Science *277*, 1808–1811.

Seehausen, O., van Alphen, J.J.M. and Witte, F. (1999). Can ancient colour polymorphisms explain why some cichlid lineages speciate rapidly under disruptive sexual selection? Belg. J. Zool. *129*, 43–60.

Seehausen, O., Witte, F., Katunzi, E.F.B., Smits, J. and Bouton, N. (1997b). Patterns of the remnant cichlid fauna in southern Lake Victoria. Conserv. Biol. *11*, 890–905.

Shaw, P.W., Turner, G.F., Idid, M.R., Robinson, R.L. and Carvalho, G.R. (2000). Genetic population structure indicates sympatric speciation of Lake Malawi pelagic cichlids. Proc. Biol. Sci. *267*, 2273–2280.

Sisneros, J.A. and Tricas, T.C. (2000). Androgen-induced changes in the response dynamics of ampullary electrosensory primary afferent neurons. J. Neurosci. *20*, 8586–8595.

Sisneros, J.A. and Tricas, T.C. (2002). Ontogenetic changes in the response properties of the peripheral electrosensory system in the Atlantic stingray (*Dasyatis sabina*). Brain Behav. Evol. *59*, 130–140.

Sköld, H.N., Amundsen, T., Svensson, P.A., Mayer, I., Bjelvenmark, J. and Forsgren, E. (2008). Hormonal regulation of female nuptial coloration in a fish. Horm. Behav. *54*, 549–556.

Sol, D., Duncan, R.P., Blackburn, T.M., Cassey, P. and Lefebvre, L. (2005a). Big brains, enhanced cognition, and response of birds to novel environments. Proc. Natl. Acad. Sci. U.S.A. *102*, 5460–5465.

Sol, D., Stirling, D.G. and Lefebvre, L. (2005b). Behavioral drive or behavioral inhibition in evolution: subspecific diversification in holarctic passerines. Evolution *59*, 2669–2677.

Sol, D., Bacher, S., Reader, S.M. and Lefebvre, L. (2008). Brain size predicts the success of mammal species introduced into novel environments. Am. Nat. *172*(Suppl. 1), S63–S71.

Stensmyr, M.C., Dekker, T. and Hansson, B.S. (2003). Evolution of the olfactory code in the *Drosophila melanogaster* subgroup. Proc. Biol. Sci. *270*, 2333–2340.

Stockinger, P., Kvitsiani, D., Rotkopf, S., Tirián, L. and Dickson, B.J. (2005). Neural circuitry that governs *Drosophila* male courtship behavior. Cell *121*, 795–807.

Strecker, U. and Kodric-Brown, A. (1999). Mate recognition systems in a species flock of Mexican pupfish. J. Evol. Biol. *12*, 927–935.

Strecker, U., Meyer, C.G., Sturmbauer, C. and Wilkens, H. (1996). Genetic divergence and speciation in an extremely young species flock in Mexico formed by the genus *Cyprinodon* (Cyprinodontidae, Teleostei). Mol. Phylogenet. Evol. *6*, 143–149.

Stuart-Fox, D.M., Moussalli, A., Johnston, G.R. and Owens, I.P.F. (2004). Evolution of color variation in dragon lizards: quantitative tests of the role of crypsis and local adaptation. Evolution *58*, 1549–1559.

Swaddle, J.P. (1999). Limits to length asymmetry detection in starlings: implications for biological signalling. Proc. Biol. Sci. *266*, 1299–1303.

Swaddle, J.P. and Cuthill, I.C. (1994). Preference for symmetric males by female zebra finches. Nature *367*, 165–166.

Swaddle, J.P., Che, J.P.K. and Clelland, R.E. (2004). Symmetry preference as a cognitive by-product in starlings. Behaviour *141*, 469–478.

Takahashi, A. and Ting, C.T. (2004). Genetic basis of sexual isolation in *Drosophila melanogaster*. Genetica *120*, 273–284.

Taylor, E.B., McPhail, J.D. and Schluter, D. (1997). History of ecological selection in sticklebacks: uniting experimental and phylogenetic approaches. In: *Molecular evolution and adaptive radiation*, T.J. Givnish and K.J. Systsma (eds.). Cambridge University Press, Cambridge, pp. 511–534.

Taylor, E.B., Boughman, J.W., Groenenboom, M., Sniatynski, M., Schluter, D. and Gow, J.L. (2006). Speciation in reverse: morphological and genetic evidence of the collapse of a three-spined stickleback (*Gasterosteus aculeatus*) species pair. Mol. Ecol. *15*, 343–355.

Thoday, J.M. and Gibson, J.B. (1970). The probability of isolation by disruptive selection. Am. Nat. *104*, 219–230.

Thomas, Y., Bethenod, M.T., Pelozuelo, L., Frérot, B. and Bourguet, D. (2003). Genetic isolation between two sympatric host-plant races of the European corn borer, *Ostrinia nubilalis* Hübner. I. Sex pheromone, moth emergence timing, and parasitism. Evolution *57*, 261–273.

Timmins, G.S., Robb, F.J., Wilmot, C.M., Jackson, S.K. and Swartz, H.M. (2001). Firefly flashing is controlled by gating oxygen to light-emitting cells. J. Exp. Biol. *204*, 2795–2801.

Tregenza, T. and Butlin, R.K. (1999). Speciation without isolation. Nature *400*, 311–312.

Tricas, T.C. (2001). The neuroecology of the elasmobranch electrosensory world: why peripheral morphology shapes behavior. Environ. Biol. Fish. *60*, 77–92.

Tricas, T.C. and Sisneros, J.A. (2004). Ecological functions and adaptations of the elasmobranch electrosense. In: *The Senses of Fish: Adaptations for the Reception of Natural Stimuli*, G. von der Emde, J. Mogdans and B.G. Kapoor (eds.). Narosa Publishing House, New Delhi, pp. 308–329.

Tricas, T.C., Michael, S.W. and Sisneros, J.A. (1995). Electrosensory optimization to conspecific phasic signals for mating. Neurosci. Lett. *202*, 129–132.

Trimmer, B.A., Aprille, J.R., Dudzinski, D.M., Lagace, C.J., Lewis, S.M. and Michel, T., et al. (2001). Nitric oxide and the control of firefly flashing. Science *292*, 2486–2488.

Turelli, M., Barton, N.H. and Coyne, J.A. (2001). Theory and speciation. Trends Ecol. Evol. *16*, 330–343.

Uy, J.A.C. and Borgia, G. (2000). Sexual selection drives rapid divergence in bowerbird display traits. Evolution *54*, 273–278.

Uy, J.A.C. and Endler, J.A. (2004). Modification of the visual background increases conspicuousness of golden collared manakin displays. Behav. Ecol. *15*, 1003–1010.

van Oppen, M.J.H., Turner, G.F., Rico, C, Robinson, R.L, Deutsch, J.C. and Genner, M.J., et al. (1998). Assortative mating among rock-dwelling cichlid fishes supports high estimates of species richness from Lake Malawi. Mol. Ecol. *7*, 991–1001.

Verheyen, E., Salzburger, W., Snoeks, J. and Meyer, A. (2003). Origin of the superflock of cichlid fishes from Lake Victoria, East Africa. Science *300*, 325–329.

Via, S. (2001). Sympatric speciation in animals: the ugly duckling grows up. Trends Ecol. Evol. *16*, 381–390.

von der Emde, G. (2006). Non-visual environmental imaging and object detection through active electrolocation in weakly electric fish. J. Comp. Physiol. A *192*, 601–612.

Wade, J. and Arnold, A.P. (2004). Sexual differentiation of the zebra finch song system. Ann. N.Y. Acad. Sci. *1016*, 540–549.

Wagner, W.E., Jr., Smeds, M.R. and Wiegman, D.D. (2001). Experience affects female responses to male song in variable field cricket *Gryllus lineaticeps* (Orthoptera, Gryllidae). Ethology *107*, 769–776.

Wcislo, W.T. (1989). Behavioral environments and evolutionary change. Annu. Rev. Ecol. Syst. *20*, 137–169.

Werneyer, M. and Kramer, B. (2005). Electric signaling and reproductive behavior in a mormyrid fish, the bulldog *Marcusenius macrolepidotus* (South African form). J. Ethol. *23*, 113–125.

West-Eberhard, M.J. (1979). Sexual selection, social competition, and evolution. Proc. Am. Philos. Soc. *123*, 222–234.

West-Eberhard, M.J. (1983). Sexual selection, social competition, and speciation. Q. Rev. Biol. *58*, 155–183.

West-Eberhard, M.J. (1984). Sexual selection, competitive communication and species-specific signals in insects. In: *Insect Communication*, T. Lewis (ed.). Academic Press, Toronto, pp. 283–324.

West-Eberhard, M.J. (1989). Phenotypic plasticity and the origins of diversity. Annu. Rev. Ecol. Syst. *20*, 249–278.

West-Eberhard, M.J. (2003). *Developmental Plasticity and Evolution*. Oxford University Press, Oxford, U.K, p. 526.

Wiens, J.J. (1999). Phylogenetic evidence for multiple losses of a sexually selected character in phrynosomatid lizards. Proc. Biol. Sci. *266*, 1529–1536.

Wiens, J.J. (2001). Widespread loss of sexually selected traits: how the peacock lost its spots. Trends Ecol. Evol. *16*, 517–523.

Wiens, J.J., Engstrom, T.N. and Chippindale, P.T. (2006). Rapid diversification, incomplete isolation, and the “speciation clock” in North American salamanders (genus *Plethodon*): testing the hybrid swarm hypothesis of rapid radiation. Evolution *60*, 2585–2603.

Wiernasz, D.C. and Kingsolver, J.G. (1992). Wing melanin pattern mediates species recognition in *Pieris occidentalis*. Anim. Behav. *43*, 89–94.

Wikelski, M., Carbone, C., Bednekoff, P.A., Choudhury, S. and Tebbich, S. (2001). Why is female choice not unanimous? Insights from costly mate sampling in marine iguanas. Ethology *107*, 623–638.

Wilkinson, G.S. and Reillo, P.R. (1994). Female choice response to artificial selection on an exaggerated male trait in a stalk-eyed fly. Proc. R. Soc. Lond. B *255*, 1–6.

Wirsig-Wiechmann, C.R., Houck, L.D., Wood, J.M., Feldhoff, P.W. and Feldhoff, R.C. (2006). Male pheromone protein components activate female vomeronasal neurons in the salamander *Plethodon shermani*. BMC Neurosci. *7*, p. 26.

Won, Y.-J., Sivasundar, A., Wang, Y. and Hey, J. (2005). On the origin of Lake Malawi cichlid species: a population genetic analysis of divergence. Proc. Natl. Acad. Sci. U.S.A. *102*, 6581–6586.

Wong, B.B.M., Fisher, H.S. and Rosenthal, G.G. (2005). Species recognition by male swordtails via chemical cues. Behav. Ecol. *16*, 818–822.

Wu, C., Hollocher, H., Begun, D.J., Aquadro, C.F., Xu, Y. and Wu, M. (1995). Sexual isolation in *Drosophila melanogaster*: a possible case of incipient speciation. Proc. Natl. Acad. Sci. U.S.A. *92*, 2519–2523.

Wyles, J.S., Kunkel, J.G. and Wilson, A.C. (1983). Birds, behavior and anatomical evolution. Proc. Natl. Acad. Sci. U.S.A. *80*, 4394–4397.

Zink, R.M. (2004). The role of subspecies in obscuring avian biological diversity and misleading conservation policy. Proc. Biol. Sci. *271*, 561–564.

Ad/73.

CPSIA information can be obtained at www.ICGtesting.com
Printed in the USA
BVOW070431211211

278824BV00003B/28/P

9 780124 158313